Automotive Mechanics
Fundamentals

HOW and WHY of the Design, Construction, and Operation of Modern Automotive Systems and Units

GREGORY'S AUTOMOTIVE PUBLICATIONS

SYDNEY

AUTOMOTIVE MECHANICS
Fundamentals

Published by Gregory's Automotive Publications
a division of Universal Press Pty Ltd
1 Waterloo Road, Macquarie Park, New South Wales 2113 Australia

Printed by Paramount Printing Pte Ltd

First Edition 1988
Reprinted 1989
Reprinted 1991
Reprinted 1992
Reprinted 1993
Reprinted 1994
Reprinted 1995
Reprinted 1996
Reprinted 1997
Reprinted 1998

ISBN 0 85566 626 9

National Library of Australia Cataloguing-in-Publication Data
    Automotive Mechanics Fundamentals
    1. Automobiles - Design and construction
    No. 629.2'3

Original text and illustrations courtesy the Goodheart-Willcox Company Inc. U.S.A.
Authors of original material James E. Duffy, Martin T. Stockel and Martin W. Stockel.
Text and illustrations in this edition compiled and metricated by Gregory's Automotive
Publications, Australia.

INTRODUCTION

GREGORY'S AUTOMOTIVE MECHANICS FUNDAMENTALS provides you with a thorough understanding of the Fundamentals of Design, Construction, and Operation of Modern Automotive Systems and Units. Later books will explain in more detail how to service and repair these systems and units safely and correctly.

Each system or unit is approached by starting with basic theory, then parts are added until the system or unit is complete. By following this procedure, the function of each system or unit assembly is explained and its relationship to the complete car is made clear.

Learning objectives are provided at the beginning of each chapter so you will know what you are expected to learn by studying the chapter. At the end of each chapter, you will find "Review Questions" covering the content of the chapter.

Hundreds of up-to-date informative illustrations have been used. Important areas are featured in these drawings, and many are exaggerated to place emphasis on parts being discussed.

Non-related, unimportant features are minimised or completely removed. Spot colour is also used and is carefully co-ordinated with the specific needs of each illustration. This adds a positive visual impact that enhances both reader interest and learning.

Safety Precautions and Safe Work Practices are stressed in red where jobs discussed involve the risk of serious accident. In this way safety instruction becomes meaningful. You will better understand and appreciate the need and value of safety.

This book is a valuable guide to anyone interested in cars. Those who are interested in preparing for a career in the automotive trade will find the text a "must". Experienced mechanics can use this text as a refresher course on the latest technology. Car owners, who need a general guide to mechanics, will also find this book to be outstanding.

CONTENTS

ACKNOWLEDGEMENT

Gregory's Scientific Publications would like to thank the following staff from the NSW Department of TAFE who have given special support and assistance with areas of text presentation for this book:

Rick Cleary (Head of School, Automotive and Aircraft Engineering)
Allan Kinghorne (Head, Division of Automotive Engineering)
Alan Gibson (Head of Division, Automotive Engineering)
Alan Ireland (Head of Branch)
Tom Clapham (Head Teacher, Plant Mechanics)
Roger Beaton (Teacher of Automotive Engineering).

Chapter 1

THE MOTOR CAR

After studying this chapter, you will be able to:
□ Identify and locate the most important parts of a car.
□ Describe the purpose for fundamental automotive systems.
□ Explain the interaction of automotive systems.
□ Describe major motor car design variations.
□ Comprehend the following text chapters with a minimum amount of difficulty.

Over the years, the motor car has become a very complicated machine. It is no longer the simple "horseless carriage" of the past. In fact, the modern car has about 15,000 different parts. With all of these parts and millions of cars on the road today, properly educated mechanics are always in demand.

This chapter begins your study of automotive mechanics by introducing the major parts of a car. By learning about main components, you will be better prepared to study the rest of this book which explains everything in more detail. Many seemingly unrelated parts can affect each other. This makes a quick summary of the motor car helpful to your full understanding of automotive mechanics.

AUTOMOTIVE SYSTEMS

An automotive system is a group of related parts that perform a specific function. For example, a car's steering wheel is part of the steering system. This system of parts allows the driver to turn the front wheels. The brake pedal is one part in the braking system. This system allows the driver to slow or stop the car. Fig. 1-1 illustrates several major systems.

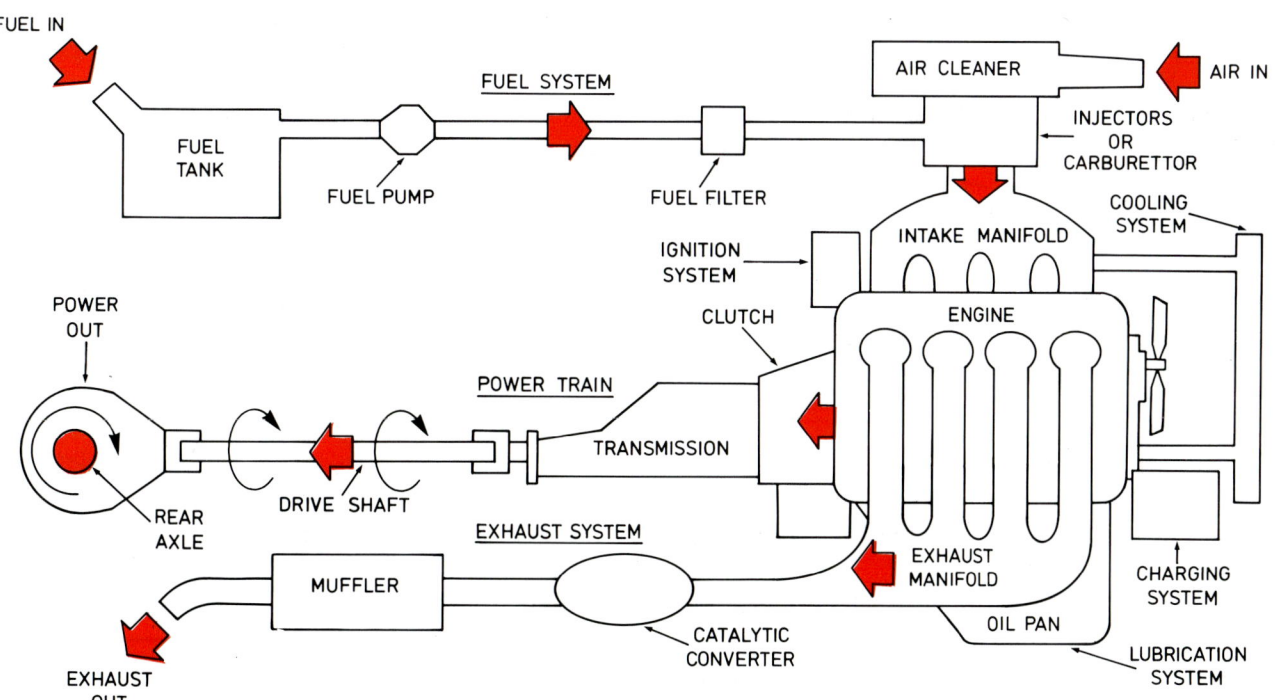

Fig. 1-1. Note location of parts. Study flow of fuel, air, exhaust, and power.

The systems of a motor car can be categorised into six major divisions:

1. BODY AND CHASSIS SYSTEMS (body, frame, suspension, steering, braking, and other systems) to support, stop, and enclose the parts of the car.
2. ENGINE SYSTEMS (engine mechanical, fuel, cooling, lubrication systems) to provide power for the car.
3. ELECTRICAL SYSTEMS (ignition, charging, starting, lighting systems) that operate the electrical devices in the car.
4. POWER TRAIN SYSTEMS (clutch, transmission, drive shaft and rear axle assembly or transaxle and axle shafts) that use engine power to propel the car.
5. EMISSION CONTROL SYSTEMS (crankcase ventilation, catalytic converter, fuel vapour storage, air injection, and other systems) for reducing air pollution produced by the car.
6. ACCESSORY SYSTEMS (air conditioning and other optional systems) for increasing passenger comfort and convenience.

BODY AND CHASSIS

The body and chassis are the two largest sections of a car, as illustrated in Fig. 1-2. The car body serves as an attractive covering for the chassis and also forms the passenger compartment. The chassis generally includes everything but the car body.

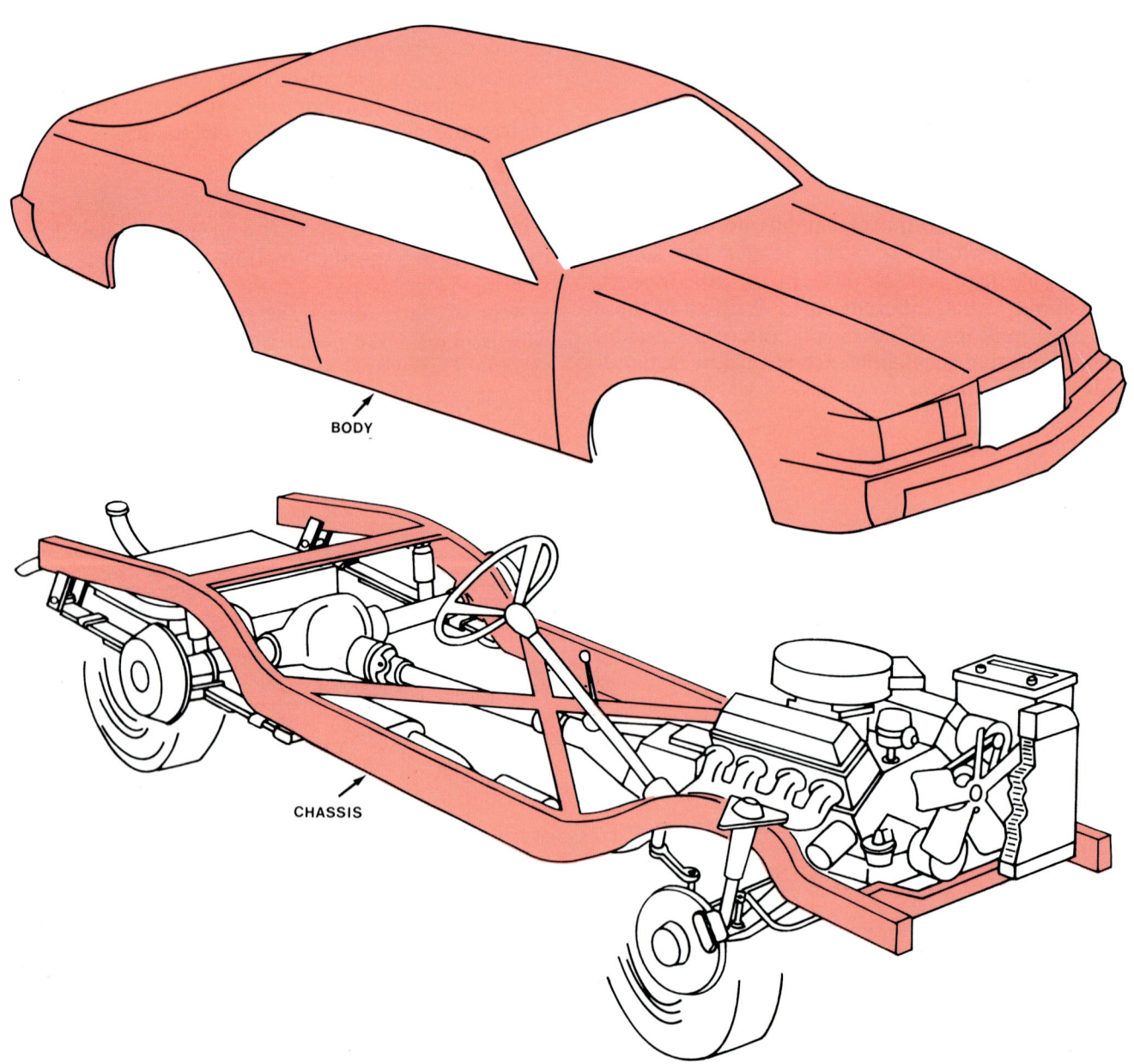

Fig. 1-2. Body is sheet metal, fibreglass, or plastic covering over chassis. Chassis includes framework and other major components.

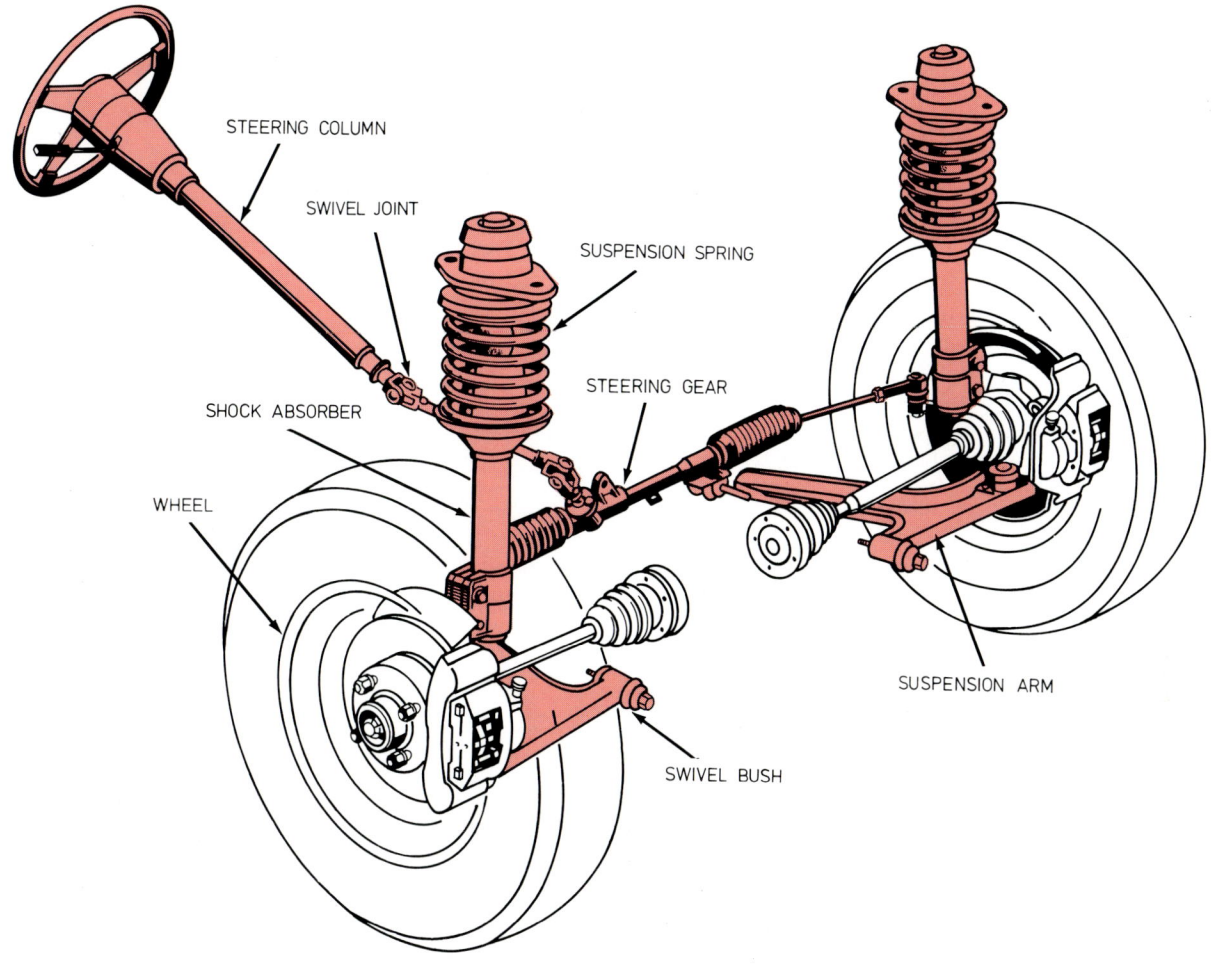

STEERING COLUMN

SWIVEL JOINT

SUSPENSION SPRING

SHOCK ABSORBER

STEERING GEAR

WHEEL

SUSPENSION ARM

SWIVEL BUSH

Fig. 1-3. Suspension and steering systems mount on frame. Study part names. (VW)

Frame

The term frame refers to a very strong, steel structure that supports the other components of the car. Some cars have a frame separate from the body. Many cars use the internal body structure as a frame. This is called unitised construction, monocoque construction, or unibody. Some inner body sections are strengthened so that they can support the engine, suspension, and other major parts.

Suspension system

The suspension system allows the car's wheels to move up and down with little or no body movement. This helps the car ride smoothly and safely. The suspension system must also prevent the car from leaning excessively in turns. As you can see in Fig. 1-3, various springs, swivel bushes, and arms make up the suspension system.

Steering system

The steering system allows the driver to control

vehicle direction by turning the front wheels right or left. It uses a series of gears, swivel joints, and rods. Study the names of the parts in Fig. 1-3.

Brake system

The brake system produces friction to reduce speed or stop the car. Fig. 1-4 shows the fundamental parts of a brake system. When the driver presses the brake pedal, fluid pressure expands the brake mechanism on each wheel. The brake mechanisms then produce friction that resists wheel rotation.

ENGINE

The engine provides the energy to propel (move) the car and operate the other systems, Fig. 1-5. Most car engines consume petrol or diesel fuel. The fuel burns in the engine to produce heat. The heat causes expansion and pressure. The pressure can then be used to move the parts of the engine and produce power.

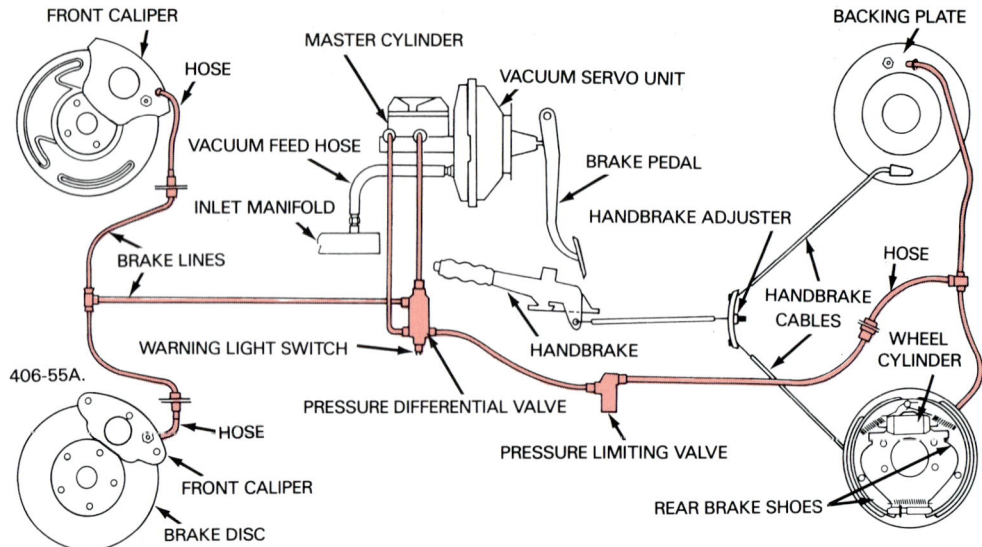

FRONT CALIPER
HOSE
MASTER CYLINDER
VACUUM SERVO UNIT
BACKING PLATE
VACUUM FEED HOSE
BRAKE PEDAL
INLET MANIFOLD
HANDBRAKE ADJUSTER
BRAKE LINES
HANDBRAKE CABLES
HOSE
WHEEL CYLINDER
WARNING LIGHT SWITCH
406-55A.
HOSE
HANDBRAKE
FRONT CALIPER
PRESSURE DIFFERENTIAL VALVE
BRAKE DISC
PRESSURE LIMITING VALVE
REAR BRAKE SHOES

Fig. 1-4. When brake pedal is pressed, pressure is placed on a confined fluid. The fluid pressure operates on each wheel. Emergency brake is mechanical system for applying rear wheel brakes.

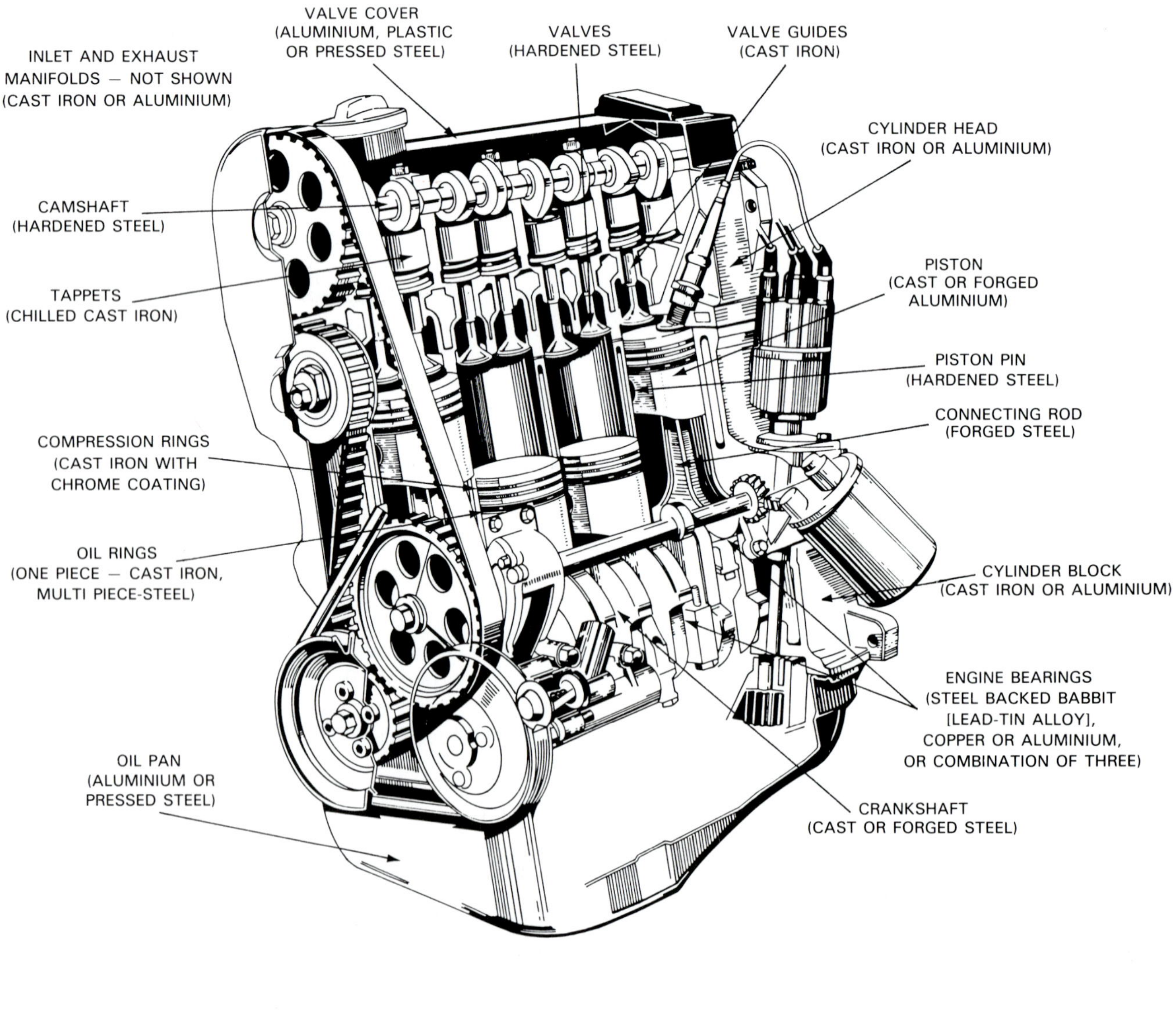

VALVE COVER
(ALUMINIUM, PLASTIC
OR PRESSED STEEL)
VALVES
(HARDENED STEEL)
VALVE GUIDES
(CAST IRON)
INLET AND EXHAUST
MANIFOLDS — NOT SHOWN
(CAST IRON OR ALUMINIUM)
CYLINDER HEAD
(CAST IRON OR ALUMINIUM)
CAMSHAFT
(HARDENED STEEL)
PISTON
(CAST OR FORGED
ALUMINIUM)
TAPPETS
(CHILLED CAST IRON)
PISTON PIN
(HARDENED STEEL)
CONNECTING ROD
(FORGED STEEL)
COMPRESSION RINGS
(CAST IRON WITH
CHROME COATING)
OIL RINGS
(ONE PIECE — CAST IRON,
MULTI PIECE-STEEL)
CYLINDER BLOCK
(CAST IRON OR ALUMINIUM)
ENGINE BEARINGS
(STEEL BACKED BABBIT
[LEAD-TIN ALLOY],
COPPER OR ALUMINIUM,
OR COMBINATION OF THREE)
OIL PAN
(ALUMINIUM OR
PRESSED STEEL)
CRANKSHAFT
(CAST OR FORGED STEEL)

Fig. 1-5. An engine commonly burns petrol or diesel fuel to produce power for car. Note part names and materials from which parts are commonly made.

An engine is usually located in the front of the chassis. With the heavy engine in the front, the car is safer in a head-on collision. A few cars have the engine mounted in the rear. Refer to Fig. 1-6.

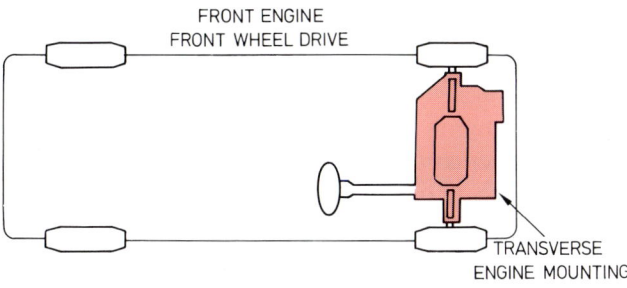

FRONT ENGINE
FRONT WHEEL DRIVE

TRANSVERSE
ENGINE MOUNTING

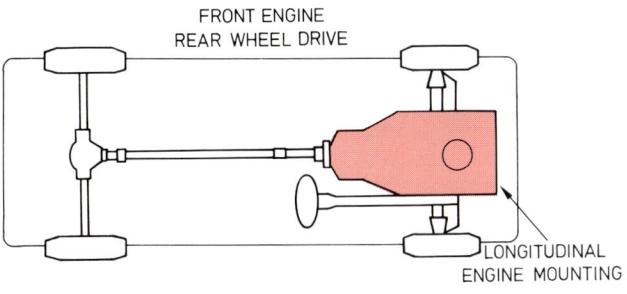

FRONT ENGINE
REAR WHEEL DRIVE

LONGITUDINAL
ENGINE MOUNTING

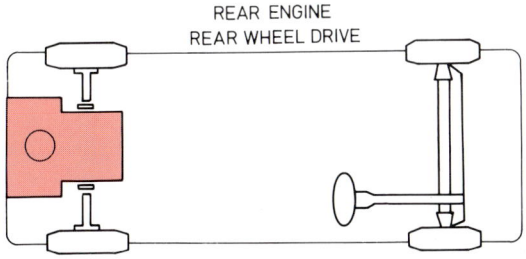

REAR ENGINE
REAR WHEEL DRIVE

Fig. 1-6. Engine can be located in front or rear of car.

Basic engine parts

The basic parts of a one-cylinder engine are shown in Fig. 1-7. Refer to this illustration as each part is introduced.
1. The block holds all of the other engine parts.
2. The cylinder is a round hole bored (machined) in the block. It guides piston movement.
3. The piston transfers the energy of combustion (burning of air-fuel mixture) to the connecting rod.
4. The rings seal the small gap around the sides of the piston. They keep combustion pressure and oil from leaking between the piston and cylinder wall (cylinder surface).
5. The connecting rod links the piston to the crankshaft.
6. The crankshaft changes the reciprocating (up and down) motion of the piston and connecting rod into useful rotary (spinning) motion.

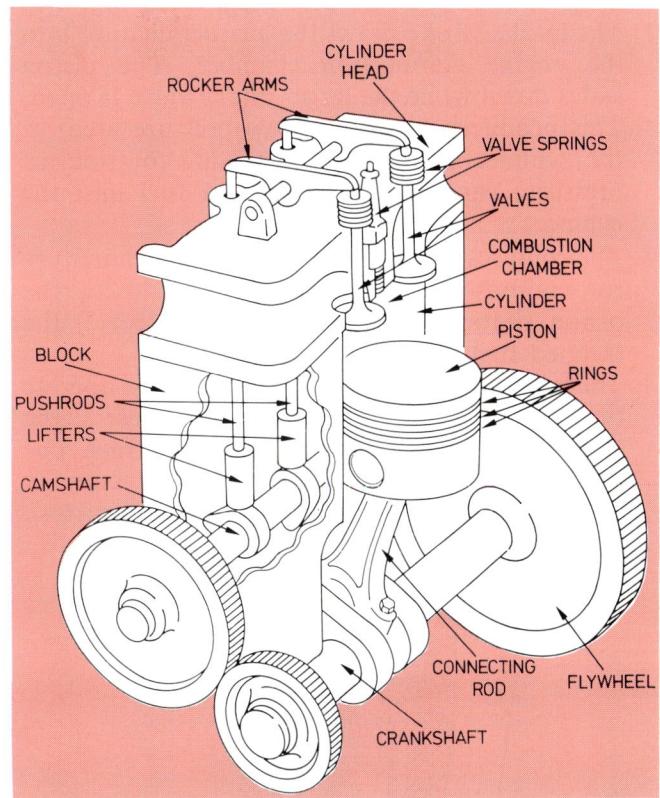

Fig. 1-7. Memorise basic parts of this one cylinder engine.

7. The cylinder head covers and seals the top of the cylinder. It also holds the valves, rocker arms, and sometimes the camshaft.
8. The combustion chamber is a small cavity (enclosed area) between the top of the piston and the bottom of the cylinder head. The burning of the air-fuel mixture occurs in the combustion chamber of the engine.
9. The valves open and close to control the flow of fuel mixture into and exhaust out of the combustion chamber.
10. The camshaft controls the opening of the valves.
11. The valve springs keep the valves closed when they do not need to be open.
12. The rocker arms transfer camshaft action to the valves.
13. The lifters ride on the camshaft and transfer motion via the push rods to the other parts of the valve train.
14. The flywheel helps keep the crankshaft turning smoothly.

Four-stroke cycle

Motor cars normally use four-stroke cycle engines. Four separate piston strokes (up or down movements) are needed to produce one cycle (complete series of events). The piston must slide up, down, up and down again to make one power producing event.

As the four strokes are described, study Fig. 1-8.

1. The intake stroke draws the air-fuel mixture into the engine combustion chamber. The piston slides down while the larger intake valve is open. This produces a vacuum (low pressure area) in the cylinder. Atmospheric pressure (outside air pressure) can then force air and fuel into the engine.

2. The compression stroke prepares the fuel mixture for combustion. With both valves closed, the piston slides up and compresses (squeezes) the trapped fuel mixture.

3. The power stroke produces the energy to operate the engine. With both valves still closed, the spark plug arcs (sparks) and ignites the fuel. The burning fuel expands and develops pressure in the combustion chamber and top of the piston.

This pushes the piston down with enough force to keep the crankshaft spinning until the next power stroke.

4. The exhaust stroke must remove the burned gases from the engine. The piston slides up while the exhaust valve is open. Since the intake valve is closed, the burned fuel mixture is pushed out of the engine.

During engine operation, these four strokes are repeated over and over. With the help of the heavy flywheel, this action produces smooth power output at the engine crankshaft.

Obviously, other devices are needed to lubricate the engine parts, operate the spark plug, cool the engine, and provide the correct fuel mixture. These topics will be discussed shortly.

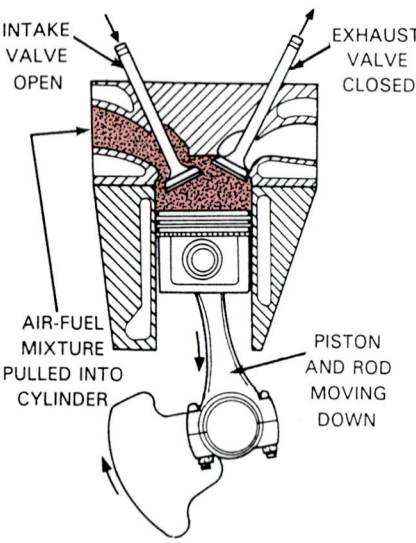

Intake stroke. Intake valve open. Exhaust valve closed. Piston slides down forming vacuum in cylinder. Atmospheric pressure pushes air and fuel into combustion chamber.

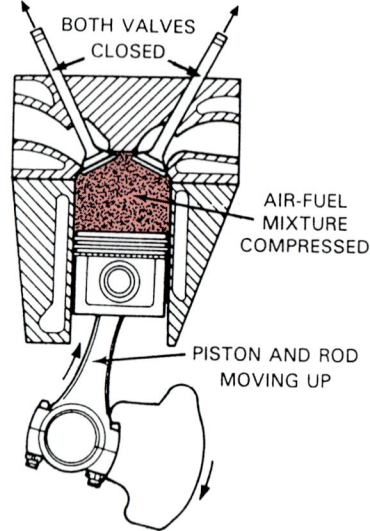

Compression stroke. Both valves are closed. Piston slides up and pressurises air-fuel mixture. This readies mixture for combustion.

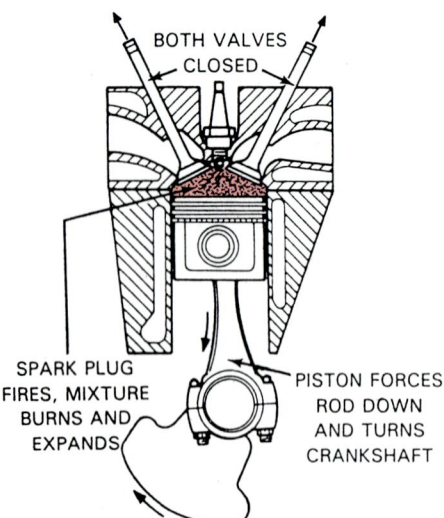

Power stroke. Spark plug sparks. Air-fuel mixture burns. High pressure forces piston down with tremendous force. Crankshaft rotates under power.

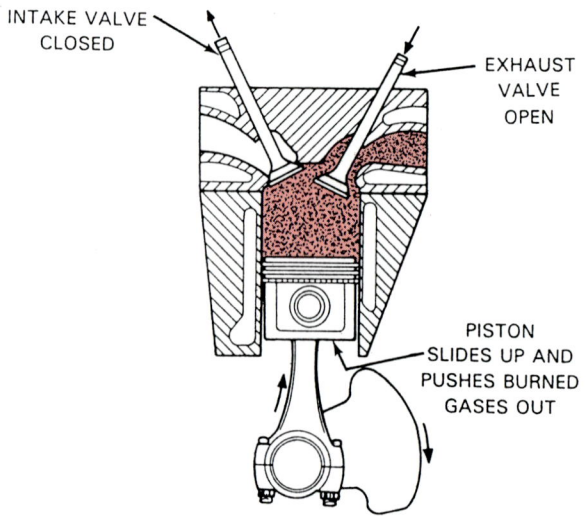

Exhaust stroke. Exhaust valve opens. Intake valve remains closed. Piston slides up, pushing burned gases out of cylinder. This prepares combustion chamber for another intake stroke.

Fig. 1-8. Petrol engine four-stroke cycle. Study series of events.

Automotive engines

Unlike the one-cylinder engine just discussed, auto engines are multiple-cylinder engines. They have more than one piston and cylinder. Cars commonly have 4, 6, or 8-cylinder engines. The additional cylinders smooth engine operation because there is less time between each power stroke. This also increases power output from the engine.

Fig. 1-9 pictures actual automotive engines. Study the shape, location, and relationship of the major parts.

A

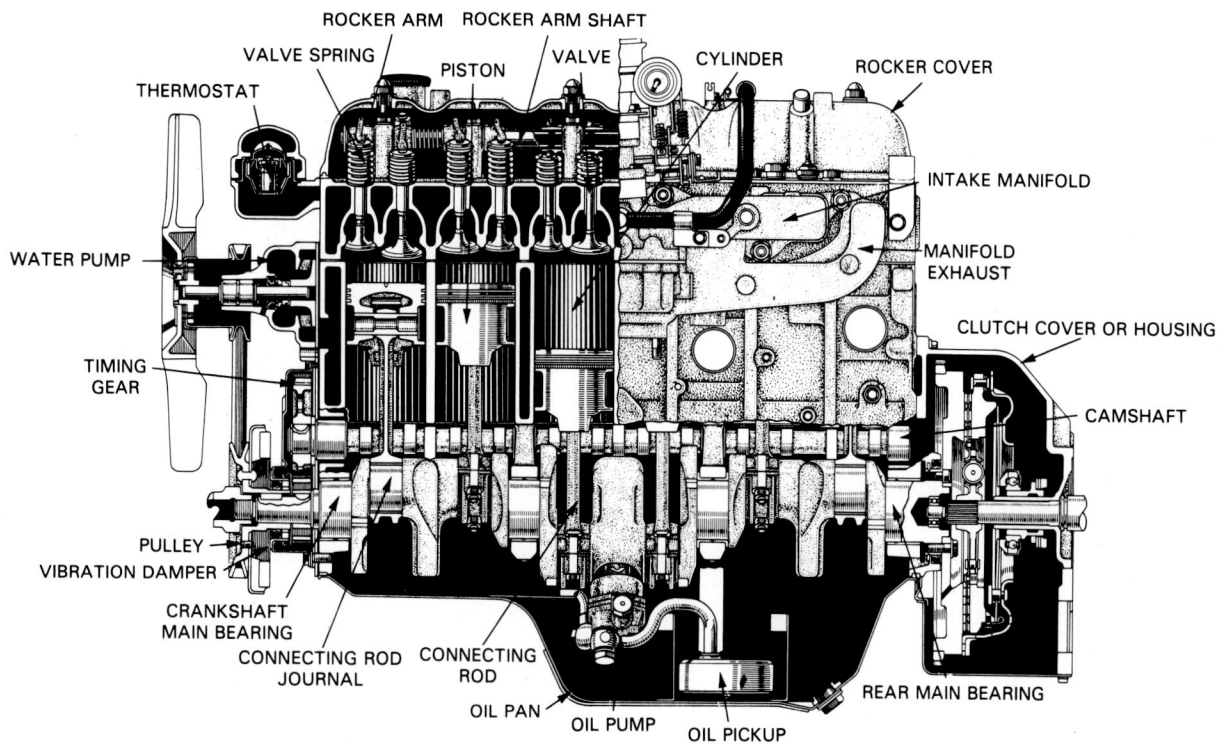

B

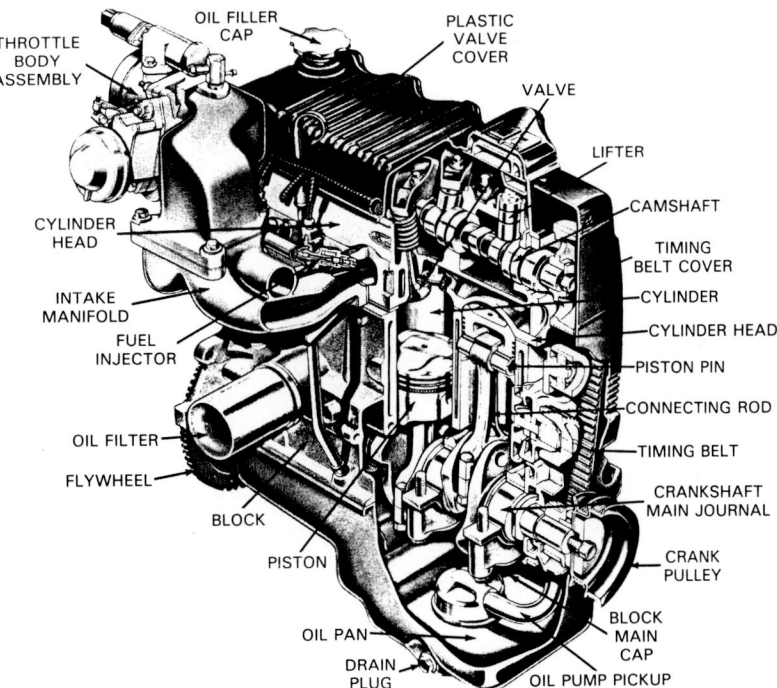

Fig. 1-9. Automotive engines are multi-cylinder engines. Locate major parts and visualise their operation. A — Cross section of typical six cylinder, overhead valve Toyota engine. B — Cutaway view of four cylinder overhead camshaft Ford engine.

FUEL SYSTEM

The fuel system must provide the correct mixture of air and fuel for efficient engine operation. It must add just the right amount of fuel to the air entering or in the cylinder. This assures that a very volatile (burnable) mixture enters the combustion chambers.

The fuel system must also alter the air-fuel ratio (percentage of air and fuel) with changes in operating conditions (engine temperature, speed, load).

There are three basic types of automotive fuel systems: carburettor, petrol injection, and diesel injection. The petrol types (carburettor and petrol injection) are the most common. Look at Fig. 1-10.

Carburettor fuel system

The carburettor fuel system uses engine vacuum (suction) to draw fuel into the engine. The amount of airflow through the carburettor controls how much fuel is used. This automatically maintains the correct air-fuel ratio, Fig. 1-10A.

The fuel pump draws fuel out of the tank and delivers it to the carburettor. The engine's intake strokes form a vacuum inside the intake manifold and carburettor. This causes petrol to be drawn from the carburettor and into the engine.

The carburettor throttle valve (air valve) is connected to the driver's accelerator pedal. When the pedal is pressed, the throttle valve opens. This allows more air to flow through the carburettor, pulling more fuel into the air.

The throttle can be opened or closed to control engine speed and power output.

Petrol injection

Modern petrol injection systems use a computer, engine sensors, and electrically operated injectors (fuel valves) to meter fuel into the engine. See Fig. 1-10B.

An electrical fuel pump keeps a constant fuel pressure at the injectors. The computer, depending upon electrical data from the sensors, opens the injectors for the correct amount of time. Fuel sprays into and mixes with the air entering the combustion chambers.

Like a carburettor, a throttle valve is used to control airflow, engine speed, and engine power. When the throttle is open, the computer holds the injectors open longer, allowing more fuel to spray out. When the throttle is closed, the computer opens the injectors for only a short period of time.

Diesel injection

A diesel fuel system is a mechanical system that forces diesel oil (not petrol) directly into the combustion chambers. A diesel does NOT use spark plugs like a petrol engine. Instead, it uses extremely high compression stroke pressure to heat the air in the combustion chamber. The air is squeezed until hot enough to ignite the fuel. Refer to Fig. 1-10C.

When the mechanical pump sprays the diesel fuel into a combustion chamber, the hot air causes the fuel to begin to burn. The burning fuel expands and forces the piston down on the power stroke.

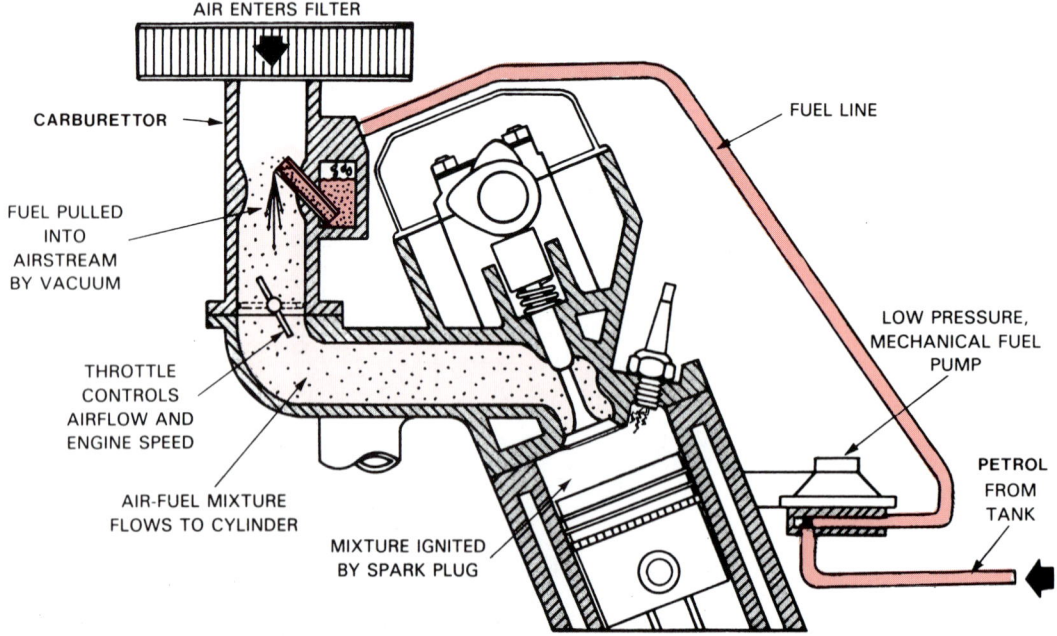

A — Carburettor fuel system. Fuel pump fills carburettor with fuel. When air flows through carburettor, fuel is pulled into engine in correct proportions. Throttle valve controls airflow and engine power output.

Fig. 1-10. Three basic types of fuel systems. Compare differences.

The Motor Car

PETROL INJECTION SYSTEM

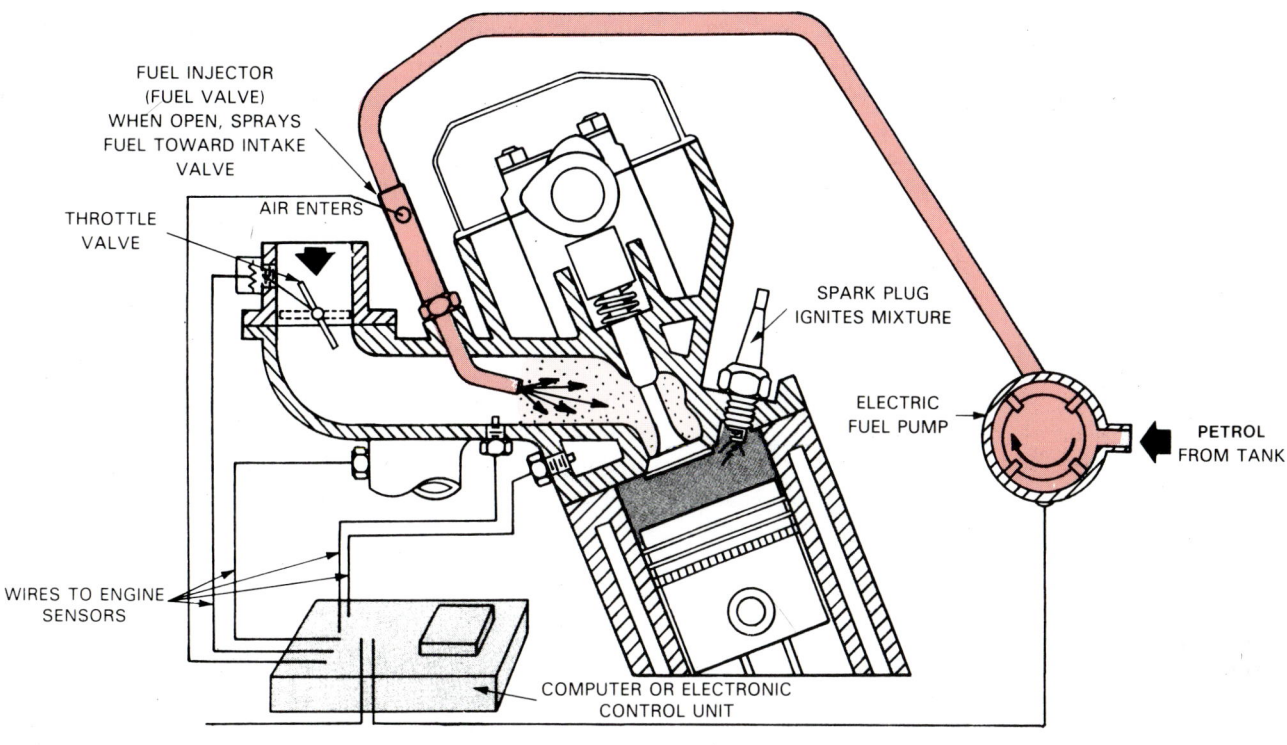

FUEL INJECTOR (FUEL VALVE) WHEN OPEN, SPRAYS FUEL TOWARD INTAKE VALVE

THROTTLE VALVE

AIR ENTERS

SPARK PLUG IGNITES MIXTURE

ELECTRIC FUEL PUMP

PETROL FROM TANK

WIRES TO ENGINE SENSORS

COMPUTER OR ELECTRONIC CONTROL UNIT

B —Petrol injection system. Engine sensors feed information (electrical signals) to computer about engine conditions. Computer can then open injector right amount of time. This maintains correct air-fuel ratio. Spark plug ignites fuel.

DIESEL INJECTION SYSTEM

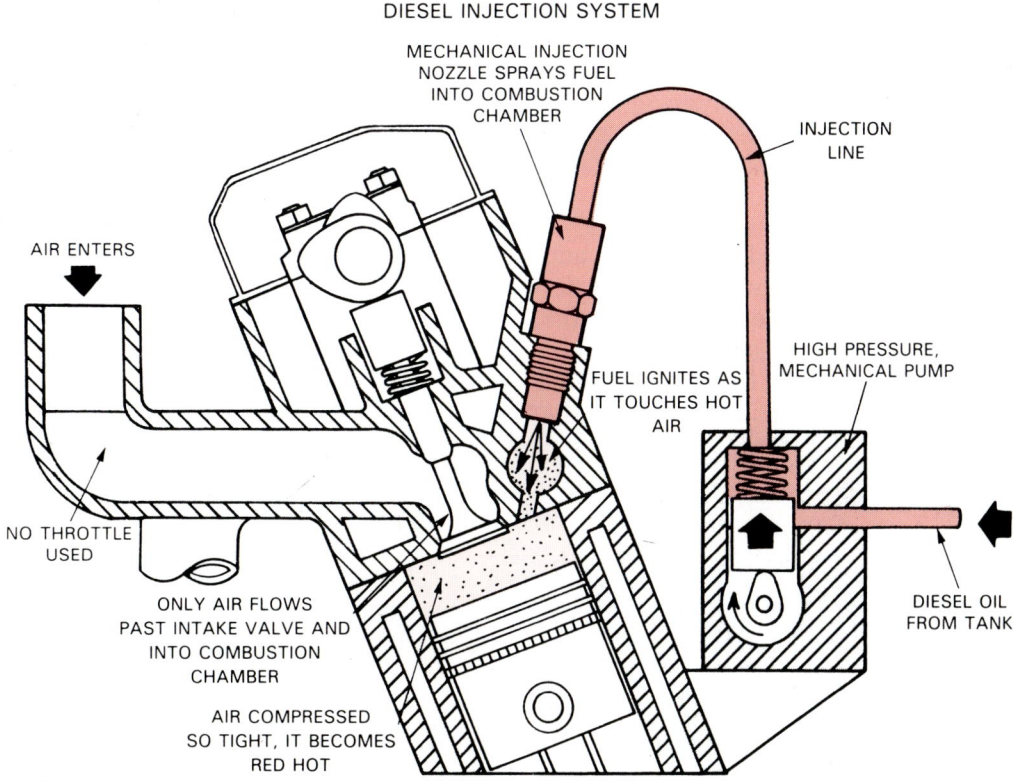

MECHANICAL INJECTION NOZZLE SPRAYS FUEL INTO COMBUSTION CHAMBER

INJECTION LINE

AIR ENTERS

FUEL IGNITES AS IT TOUCHES HOT AIR

HIGH PRESSURE, MECHANICAL PUMP

NO THROTTLE USED

DIESEL OIL FROM TANK

ONLY AIR FLOWS PAST INTAKE VALVE AND INTO COMBUSTION CHAMBER

AIR COMPRESSED SO TIGHT, IT BECOMES RED HOT

C — Diesel injection system. High pressure mechanical pump sprays fuel directly into combustion chamber. Piston squeezes and heats air enough to ignite diesel fuel. Fuel begins to burn as soon as it touches heated air. Note that no throttle valve or spark plug is used. Amount of fuel injected into chamber controls engine power and speed.

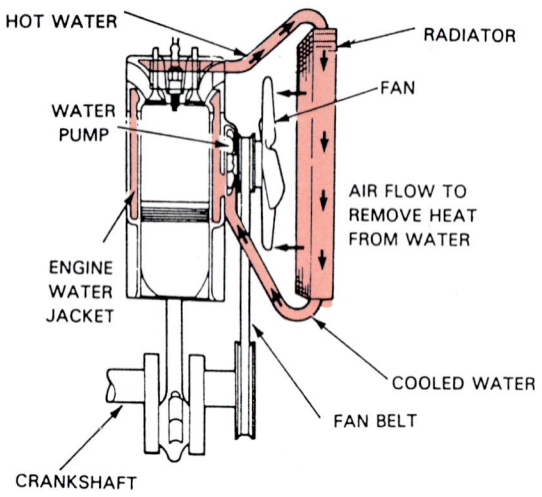

HOT WATER
RADIATOR
FAN
WATER PUMP
AIR FLOW TO REMOVE HEAT FROM WATER
ENGINE WATER JACKET
COOLED WATER
FAN BELT
CRANKSHAFT

Fig. 1-11. Cooling system must protect engine from heat of combustion. Combustion heat could melt and ruin engine parts. System must also maintain constant operating temperature and speed warmup. Study part names.

COOLING SYSTEM

An engine cooling system maintains a constant engine operating temperature. It removes excess heat to prevent engine damage and also speeds engine warm-up. Look at Fig. 1-11.

The water pump forces coolant (water and antifreeze solution) through the inside of the engine. The coolant collects heat from the hot engine parts and carries it back to the radiator. The radiator allows the coolant heat to transfer into the outside air. The thermostat controls coolant flow and engine temperature.

LUBRICATION SYSTEM

The engine lubrication system reduces friction and wear between internal engine parts. It circulates filtered motor oil to high friction points in the engine. In Fig. 1-12, study the parts and operation of a lubrication system. Note how the oil pump pulls oil out of the pan and pushes it to the parts of the engine.

ELECTRICAL SYSTEM

The car's electrical system consists of several sub-systems: ignition system, starting system, charging system, lighting system, and other systems.

Ignition system

An ignition system is needed on petrol engines to ignite the air-fuel mixture. It produces extremely high voltage that operates the spark plugs. A very hot electric arc jumps across the tip of the spark plug at the correct time. This causes the engine's air-fuel mixture to burn and produce power. Study Fig. 1-13.

With the ignition switch ON and the engine running, the distributor produces tiny electrical signals for the amplifier or electronic control unit (electronic circuit). One signal is produced for each power stroke. The electronic control unit amplifies (increases) these pulses into on/off current signals for the ignition coil.

By turning the coil current on and off, the coil can produce a high voltage output to "fire" the spark plugs. When the ignition key is turned off, the coil stops functioning and the engine stops running.

Starting system

The starting system has an electric motor that rotates the engine crankshaft until the engine starts and runs on its own power. See Fig. 1-14A.

The battery provides the electricity for the starting system. When the key is turned to start, current flows to the parts of the starting system. The electric starting motor gear engages a gear on the engine flywheel. This spins the crankshaft. As soon as the engine starts, the starting system is shut off.

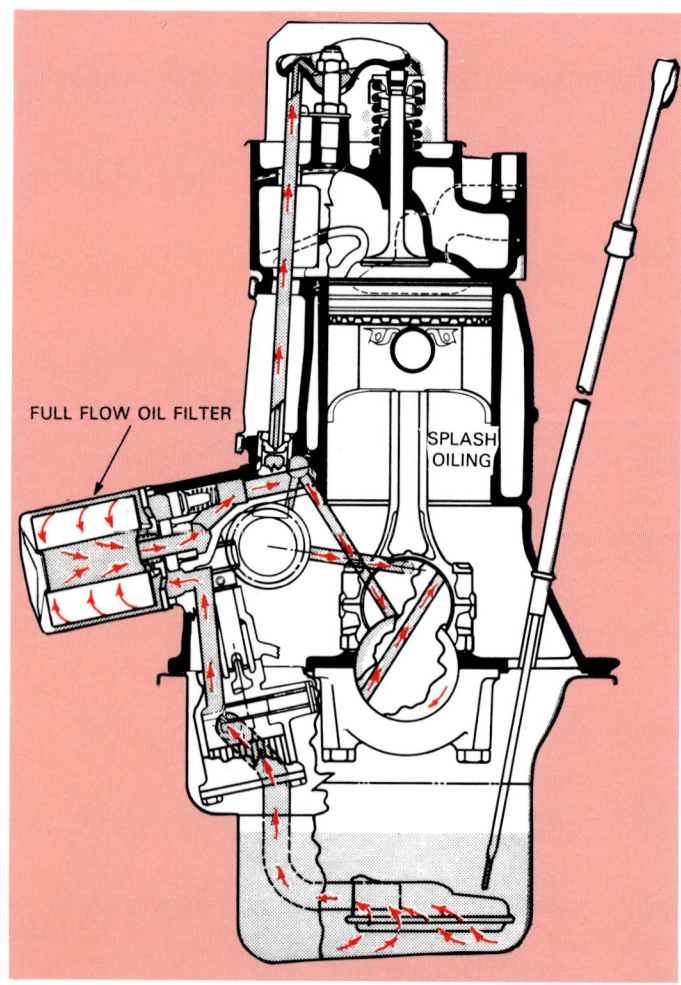

FULL FLOW OIL FILTER
SPLASH OILING

Fig. 1-12. Lubrication system uses oil to reduce friction. Pump forces oil to high friction points.

Charging system

The charging system is needed to replace electrical energy drawn from the battery during starting system operation. To re-energise the battery, the charging system forces electric current back into the battery.

The fundamental parts of this system are shown in Fig 1-14B. Study them!

When the engine is running, a fan belt spins the alternator pulley. The alternator (generator) can then produce electricity for the battery and other electrical needs of the car. A voltage regulator is used to control the output of the alternator.

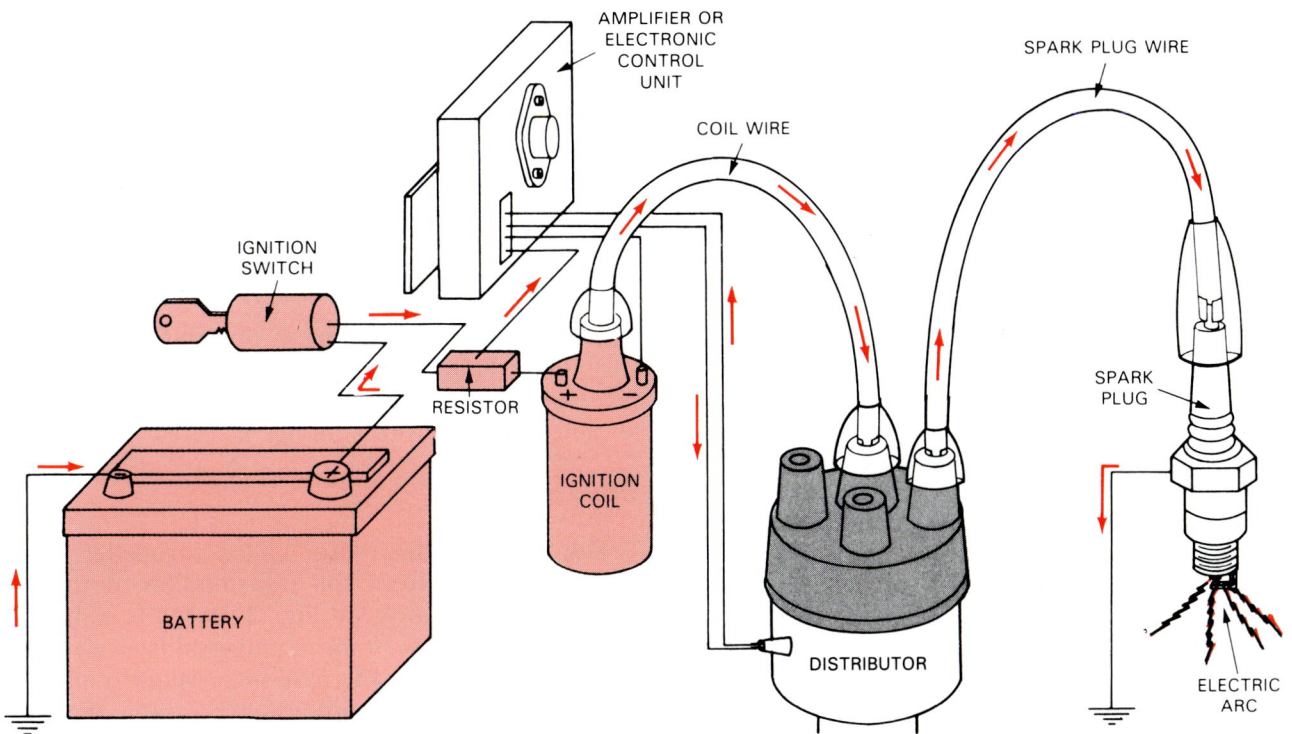

Fig. 1-13. Ignition system is used on petrol engines to start combustion. Spark plug must fire at exactly the correct time during power stroke. Distributor operates amplifier. Amplifier operates ignition coil. Coil produces high voltage for spark plugs.

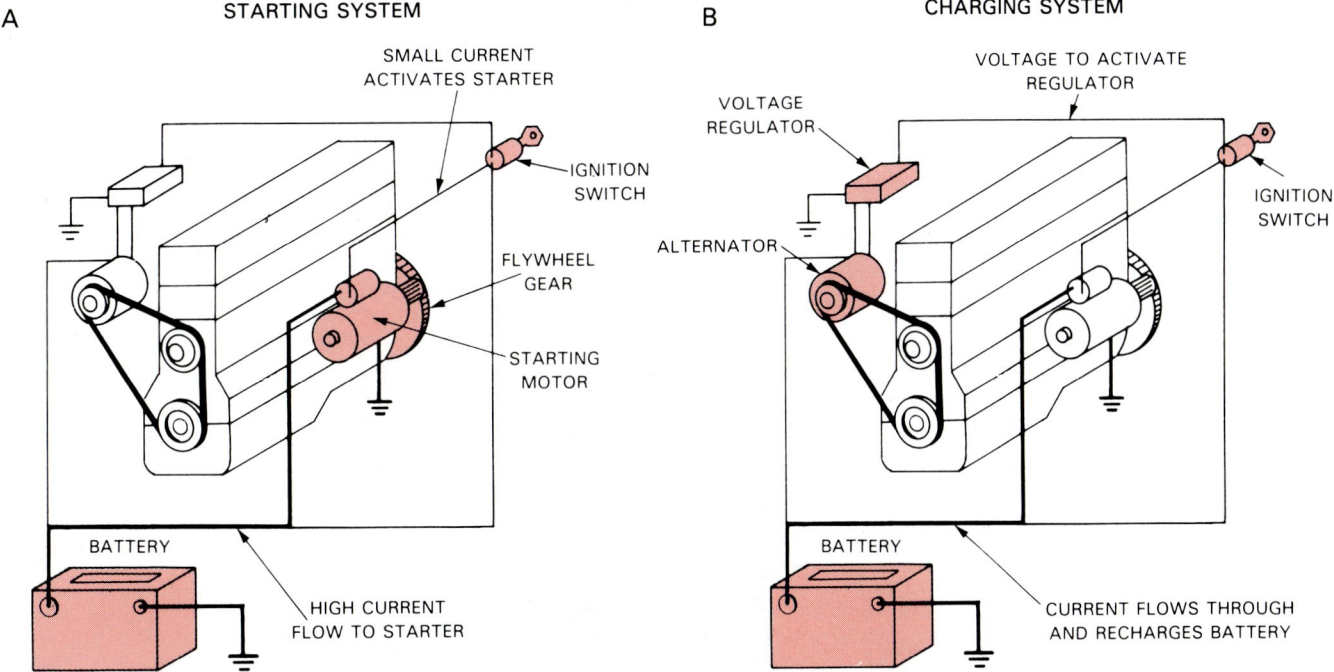

Fig. 1-14. Note basic actions and components of starting and charging systems.

EXHAUST SYSTEM

The exhaust system quietens engine operation and routes exhaust gases to the rear of the car. Fig. 1-15 illustrates the basic parts of an exhaust system. Trace the flow of exhaust gases through the system. Learn the names of the parts.

EMISSION CONTROL SYSTEMS

Various emission control systems are used to reduce the amount of toxic (poisonous) substances that enter the atmosphere (air surrounding earth). Some systems prevent fuel vapour from entering the outside air. Other emission systems remove toxic chemicals from the engine exhaust.

POWER TRAIN

The power train transfers turning force from the engine crankshaft to the car's drive wheels. Depending upon whether the car has rear-wheel or front-wheel drive or a manual or automatic transmission, power train designs vary.

Fig. 1-16 shows power trains for both front and rear-wheel drive vehicles. Compare the two.

Clutch

A clutch allows the driver to engage or disengage the engine and transmission. It is used with a manual (hand-operated) transmission or transaxle.

When the driver presses the clutch pedal, the clutch releases and the engine crankshaft no longer turns the transmission input shaft.

When the clutch pedal is released, the clutch locks the flywheel and transmission input shaft together. This causes engine power to rotate the parts of the drive line and propel the car.

Manual transmission

A manual transmission allows the driver to change gear ratios and engine torque going to the drive wheels, Fig. 1-17. It allows the car to accelerate quickly in lower transmission gears. It also provides good fuel economy in higher gears.

Automatic transmission

An automatic transmission does NOT have to be shifted by hand. It uses an internal hydraulic (oil pressure) system to shift gears. Elementary parts of an automatic transmission are pictured in Fig. 1-18.

Drive shaft

The drive shaft, also called propeller shaft, transfers power from the transmission to the rear axle assembly. Look at Fig. 1-19. It is a hollow, metal tube with two or more universal (swivel) joints. The universals allow the rear suspension to move up and down without bending or breaking the drive shaft.

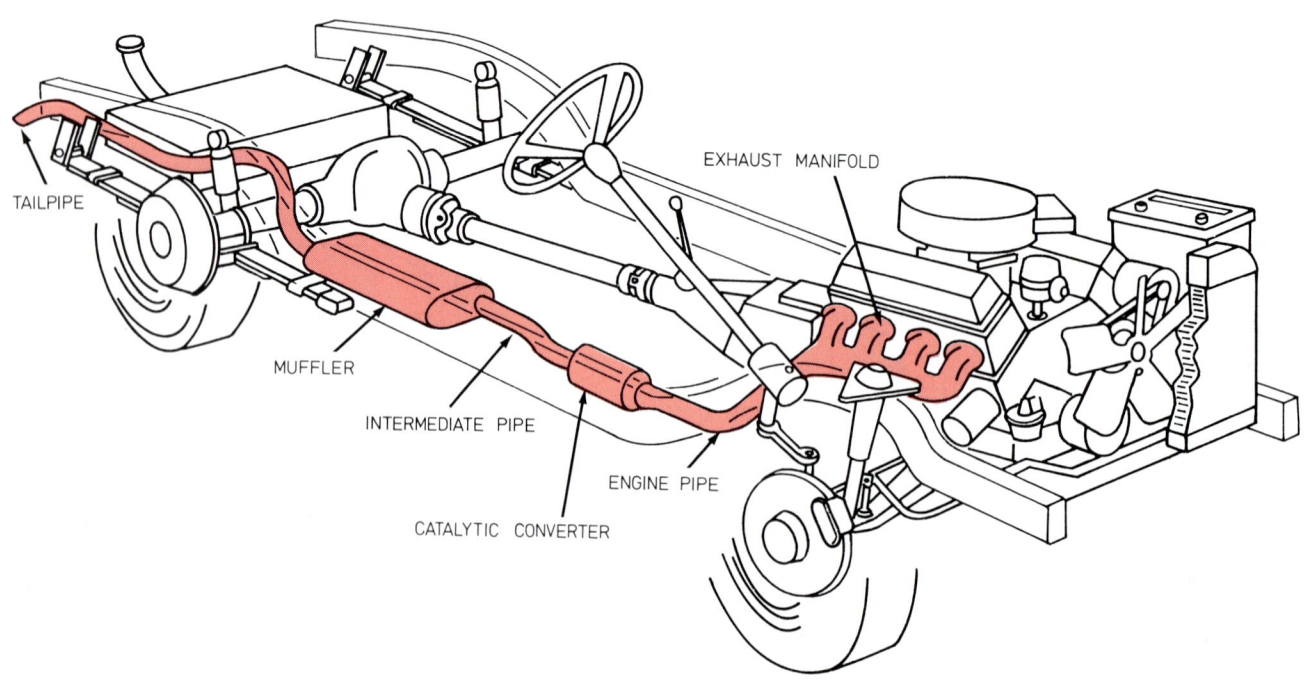

TAILPIPE

EXHAUST MANIFOLD

MUFFLER

INTERMEDIATE PIPE

ENGINE PIPE

CATALYTIC CONVERTER

Fig. 1-15. Exhaust system carries burned gases to rear of car. It also reduces noise of engine.

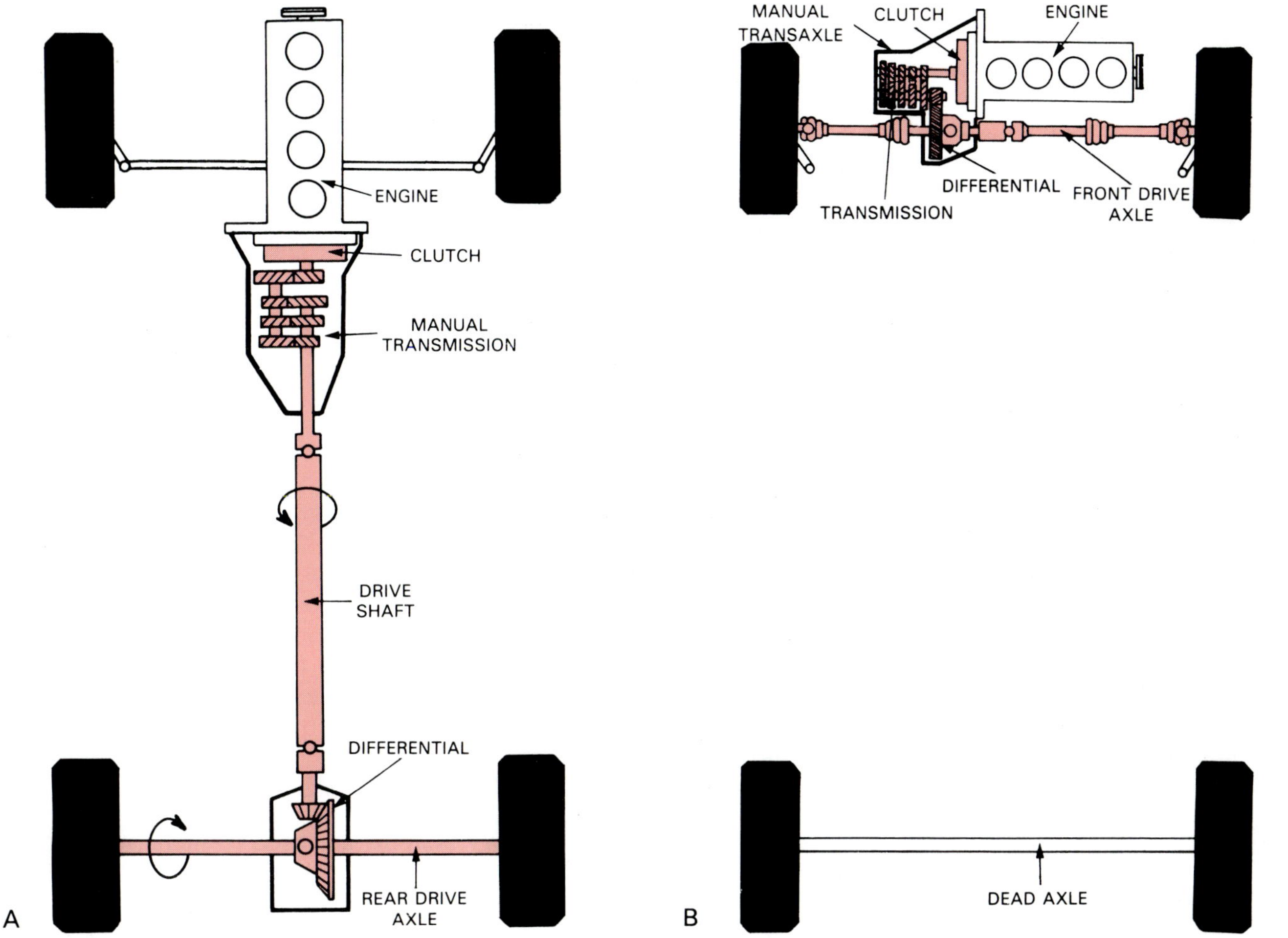

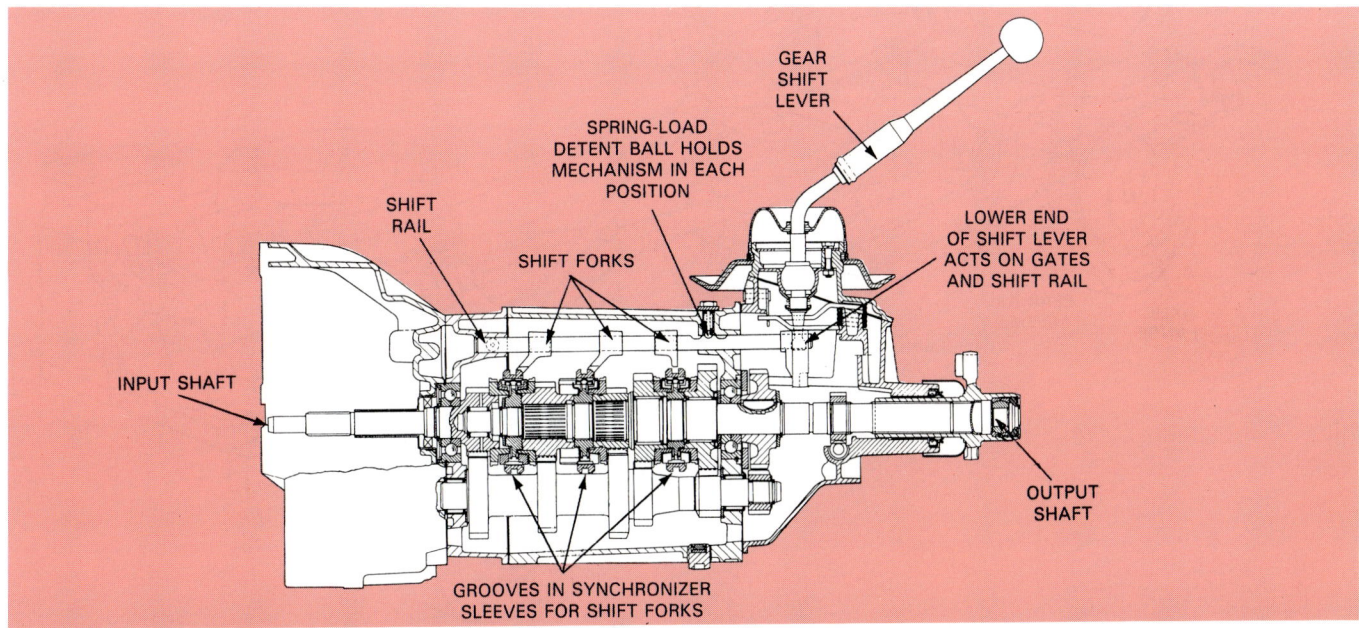

MANUAL TRANSAXLE
CLUTCH
ENGINE
TRANSMISSION
DIFFERENTIAL
FRONT DRIVE AXLE

ENGINE
CLUTCH
MANUAL TRANSMISSION
DRIVE SHAFT
DIFFERENTIAL
REAR DRIVE AXLE

A

DEAD AXLE

B

Fig. 1-16. Power train transfers engine power to drive wheels. Study differences between the common types of systems. A — Front engine, rear-wheel drive. B — Front engine, front-wheel drive.

GEAR SHIFT LEVER

SPRING-LOAD DETENT BALL HOLDS MECHANISM IN EACH POSITION

SHIFT RAIL

SHIFT FORKS

LOWER END OF SHIFT LEVER ACTS ON GATES AND SHIFT RAIL

INPUT SHAFT

OUTPUT SHAFT

GROOVES IN SYNCHRONIZER SLEEVES FOR SHIFT FORKS

Fig. 1-17. Manual transmission uses gears and shafts to allow car to accelerate quickly. Speed of output shaft compared to speed of input shaft varies in each gear position. This allows driver to change amount of torque going to wheels.

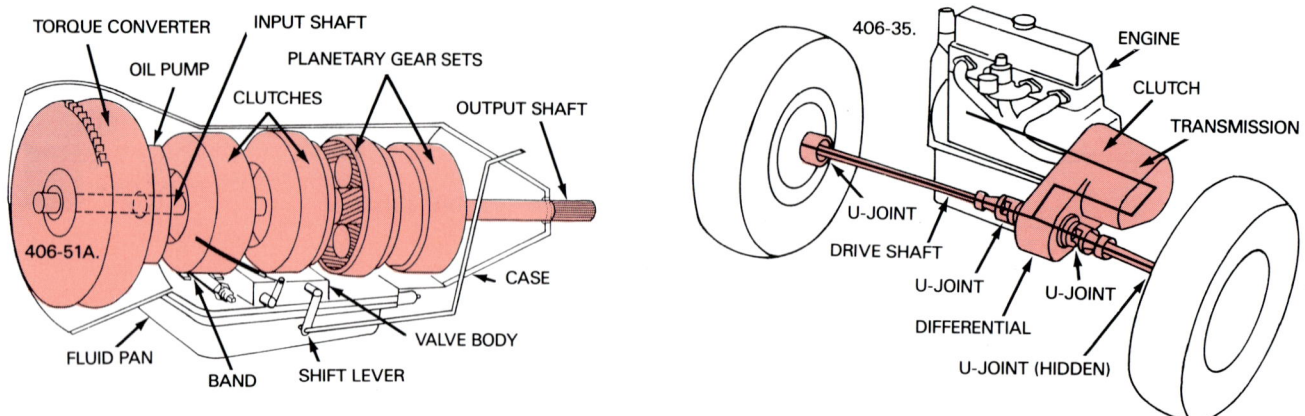

Fig. 1-18. Automatic transmission serves same function as manual transmission. However, it uses an electronic or oil pressure system to change gears.

Fig. 1-20. Front-wheel drive car does not have a drive shaft and rear drive axle assembly. Complete drive line is in front. (Typical)

Rear axle assembly

The rear axle assembly contains a differential and two axles. The differential is a set of gears and shafts that transmit power from the drive shaft to the axles. The axles are steel shafts that connect the differential and drive wheels, Fig. 1-19.

Transaxle

A transaxle contains a transmission and a differential in one case. It is commonly used with front-wheel drive vehicles. Fig. 1-20. Automatic and manual transaxles are available. They operate on the same basic principles as transmissions.

Fig. 1-21 illustrates the internal parts of a modern transaxle assembly.

ACCESSORY SYSTEMS

Accessory systems are used to increase driver and passenger comfort and convenience. Common accessory systems are the air conditioner, power seats and power windows.

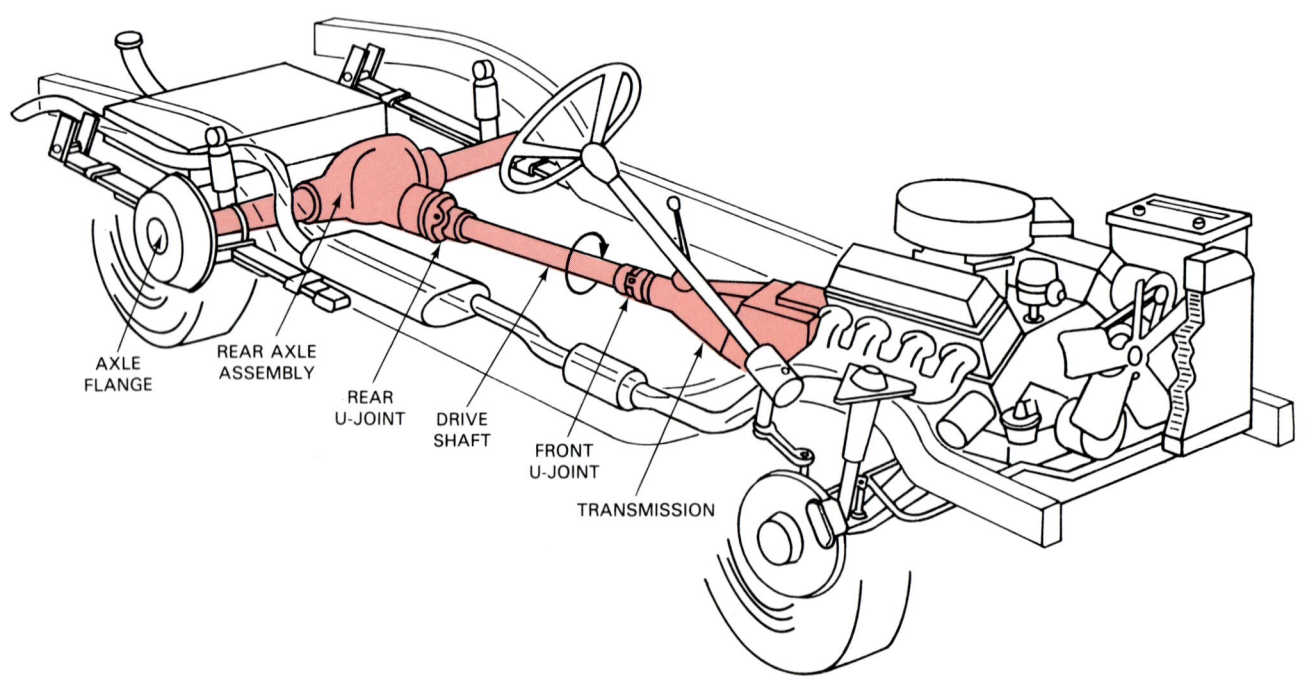

Fig. 1-19. Drive shaft sends power to rear axle assembly. Rear axle assembly contains differential and two axles that turn rear drive wheels.

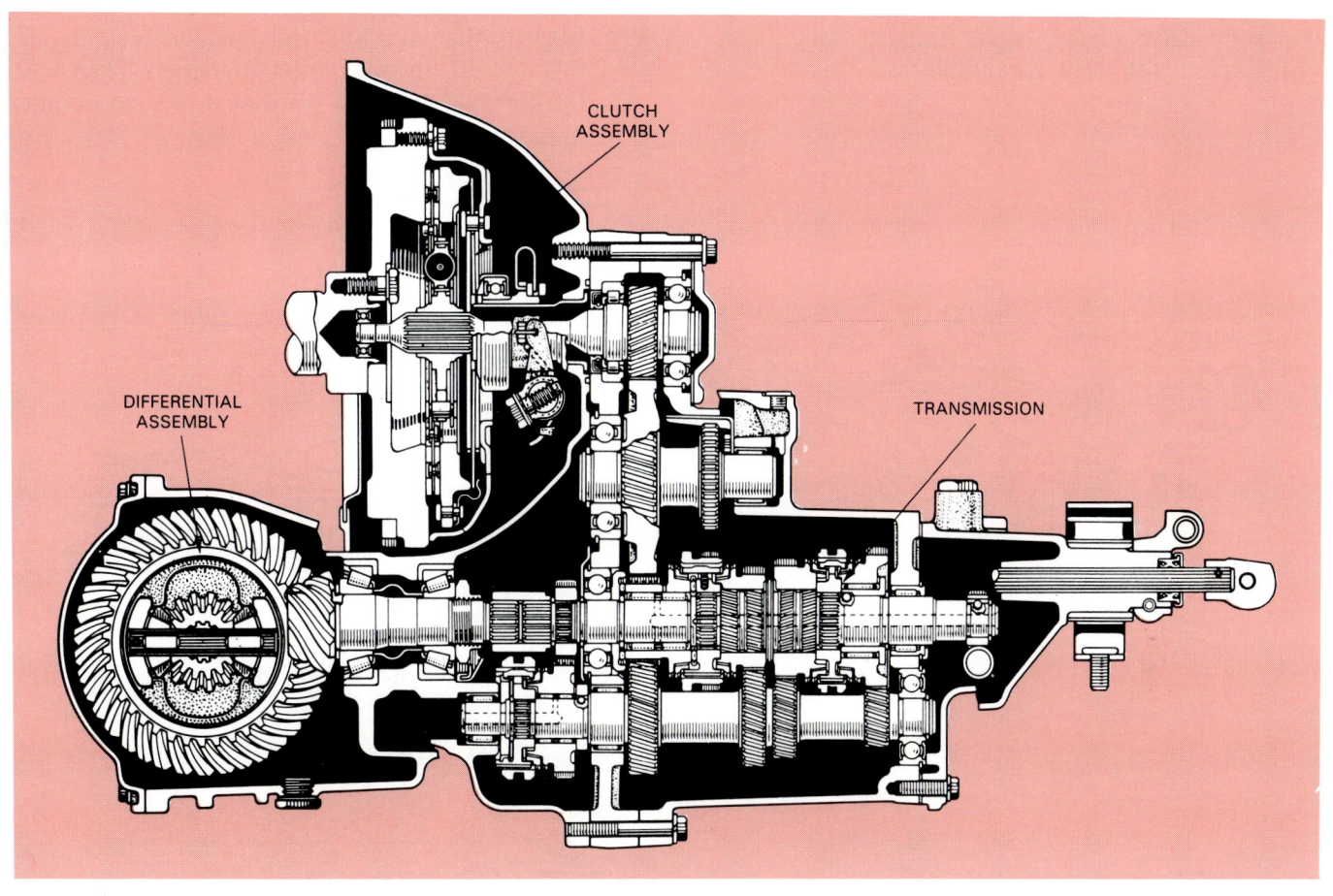

Fig. 1-21. Transaxle contains a transmission and a differential.

KNOW THESE TERMS

System, Chassis, Frame, Unibody, Suspension system, Steering system, Brake system, Engine, Four-stroke cycle, Multi-cylinder engine, Carburettor fuel system, Petrol injection, Diesel injection, Cooling system, Lubrication system, Ignition system, Starting system, Charging system, Exhaust system, Emission control system, Power train, Clutch, Manual transmission, Automatic transmission, Drive shaft, Rear axle assembly, Transaxle, Accessory systems.

REVIEW QUESTIONS

1. What is an automotive system?
2. List the six categories of automotive systems.
3. The suspension system mounts the car's wheels solid on the frame. True or False?
4. Which of the following is NOT part of an engine?
   a. Block.
   b. Piston.
   c. Muffler.
   d. Crankshaft.
5. Explain the engine's four-stroke cycle.
6. Most car engines are multiple-cylinder engines. True or False?
7. List and describe the three common types of fuel systems.
8. A diesel engine does NOT use spark plugs. True or False?
9. The car's electrical system consists of:
   a. Ignition, starting, lubrication, lighting and other systems.
   b. Ignition, charging, lighting, hydraulic, and other systems.
   c. Lighting, charging, starting, ignition, and other systems.
   d. None of the above are correct.
10. The _____ _____ system reduces the amount of toxic substances entering the atmosphere.
11. What is the difference between a manual and automatic transmission?
12. A single, one-piece drive shaft rotates the drive wheels on most front-wheel drive cars. True or False?

13. A rear _____ _____ assembly contains a set of solid drive _____ and _____.
14. Explain the term "transaxle".

15. List three accessory systems.
16. Identify the parts illustrated below. Write from 1 through 10 on your sheet of paper. Then write the correct letter and word next to each number.

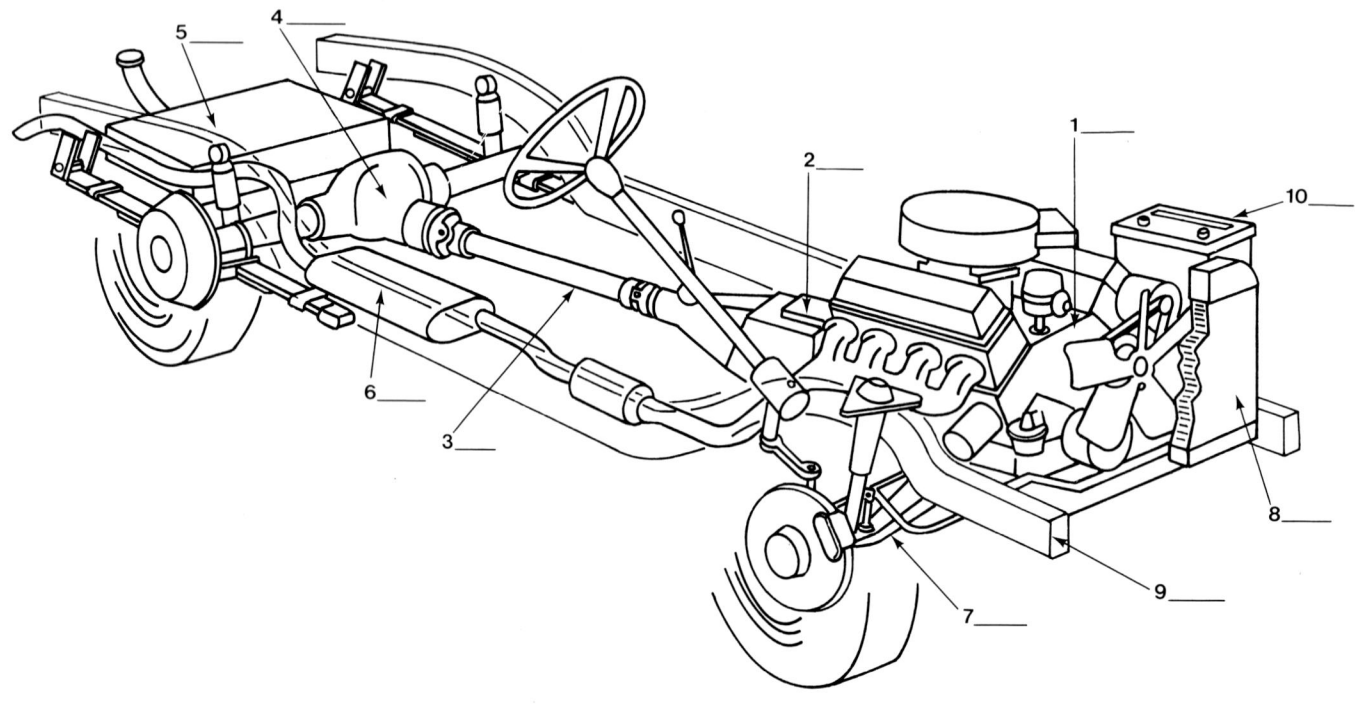

Fig. 1-22. Can you identify these parts?

A. TRANSAXLE
B. FRAME
C. DRIVE LINE
D. BATTERY
E. MUFFLER

F. VAPOUR SEPARATOR
G. ENGINE
H. REAR AXLE ASSEMBLY
I. TRANSMISSION

J. FUEL TANK
K. WATER PUMP
L. RADIATOR
M. SUSPENSION ARM

Chapter 2

USING AUTOMOTIVE TECHNICAL PUBLICATIONS

After studying this chapter, you will be able to:
☐ Describe the different types of service manuals.
☐ Find and use the service manual index and contents section.
☐ Explain the different kinds of information and illustrations used in a service manual.
☐ Describe the three basic types of trouble-shooting charts found in service manuals.
☐ Summarise the other kinds of service publications found in an auto shop.

Modern cars contain thousands of parts. Many of these parts are assembled to close tolerances (fits) and require precise assembly and adjustments. Sometimes a mechanic needs technical information to properly repair a car. In these cases, the mechanic refers to a service manual.

SERVICE MANUALS (WORKSHOP MANUALS)

Service manuals, also called workshop manuals, are books with detailed information on how to repair a car. They have step-by-step procedures, specifications, diagrams, part illustrations, and other data for each car model. Every automotive workshop will normally have a set of service manuals. They help mechanics with difficult repairs. Refer to Fig. 2-1.

Manufacturers workshop manuals are written in very concise, technical language. They are designed to be used by well-trained mechanics. After completing your studies in this text, you should be well prepared to understand service manuals. You will then find a service manual to be one of your MOST IMPORTANT TOOLS.

Service manual types

There are two types of service manuals: Manufacturer's manuals and non-manufacturer's manuals.
Manufacturer's manuals are published by the various car makers (Ford, General Motors Holden, Mitsubishi, Toyota, Nissan, Honda and Subaru etc).

When the manufacturer produces a particular model vehicle usually a workshop manual is also produced to cover that particular model. The manual may be produced in various forms, namely:
1. A single manual or volume to cover all systems on the vehicle.
2. A manual in two or more volumes (which can be any number of volumes up to six) to cover all systems of the car.
3. A manual in supplement form which can be in one or more volumes. With supplement type manuals only procedures which are different to the previous model vehicle are normally covered.

When working with supplements it also means that the manual or manuals for the previous model or models will also be needed.

Fig. 2-1. Service manuals will answer almost any repair question. They are essential reference tools of mechanic.

Non-manufacturer's manuals are sold by companies other than the major car makers (Gregory's Scientific Publications and Haynes etc). These volumes are like the manufacturer's manuals but are not quite as comprehensive.

In the non-manufacturers manuals service operations such as automatic transmission overhaul and differential overhaul are usually not covered.

Service manual sections

A service manual is divided into sections such as: general information, engine, transmission, and electrical. See Fig. 2-2. You need to understand these sections.

The general information section of a manufac-turer's manual helps you with vehicle identification.

An important topic in this section is the vehicle identification (ID) number which provides data about the car. It is commonly used when ordering parts. The ID NUMBER on the plate contains a code. The manual will explain what each part of the number code means.

Look at Fig. 2-3. The ID number tells you engine type, transmission type, and other useful information.

The general information section will also provide important information regarding the vehicle compliance plate, data plate and engine number identification.

The repair sections of a service manual cover the major systems of a car. These sections explain how

INTRODUCTION

How To Use This Manual

This manual is divided into 16 sections. The first page of each section is marked with a black tab that lines up with one of the thumb index tabs on the front and back covers. You can quickly find the first page of each section without looking through a full table of contents. The symbols printed at the top corner of each page can also be used as a quick reference system.

Each section includes:

1. A table of contents, or an exploded view index showing:
   - Parts disassembly sequence.
   - Bolt torques and thread sizes.
   - Page references to descriptions in text.
2. Disassembly/assembly procedures and tools.
3. Inspection.
4. Testing/troubleshooting.
5. Repair.
6. Adjustments.

Special Information

WARNING   Indicates a strong possibility of severe personal injury or loss of life instructions are not followed.

CAUTION: Indicates a possibility of personal injury or equipment damage if instructions are not followed.

NOTE: Gives helpful information to make the job easier.

CAUTION: Detailed descriptions of *standard* workshops procedures, safety principles and service operations are not included. Please note that this manual does contain warnings and cautions against some specific service methods which could cause PERSONAL INJURY, or could damage a vehicle or make it unsafe. Please understand that these warnings cannot cover all conceivable ways in

General Info	**i**
Special Tools	**tools**
Specifications	**specs**
Maintenance	
Engine	
Engine Electrical	
Cooling	
Fuel	
Emission Controls	
Transaxle	

Fig. 2-2. Service manual is divided into several repair sections. This is shown by portion of a table of contents page. Read introduction and special information. It is typical of most manuals. (Honda)

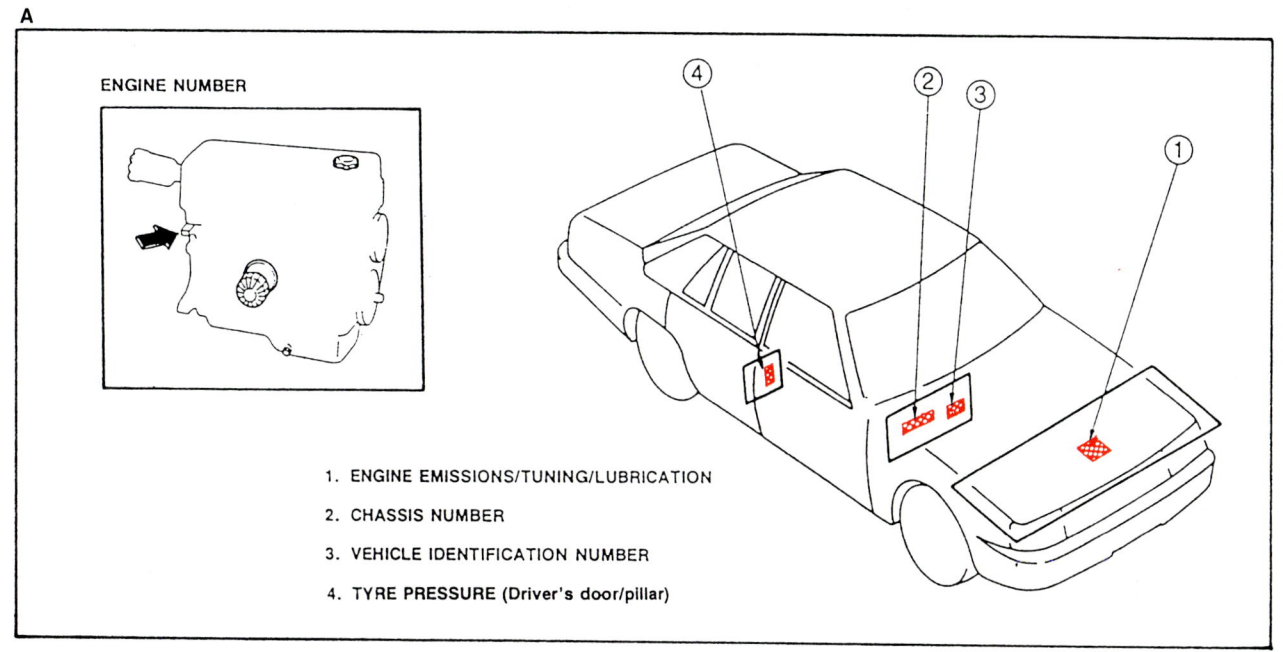

1. ENGINE EMISSIONS/TUNING/LUBRICATION

2. CHASSIS NUMBER

3. VEHICLE IDENTIFICATION NUMBER

4. TYRE PRESSURE (Driver's door/pillar)

VEHICLE IDENTIFICATION

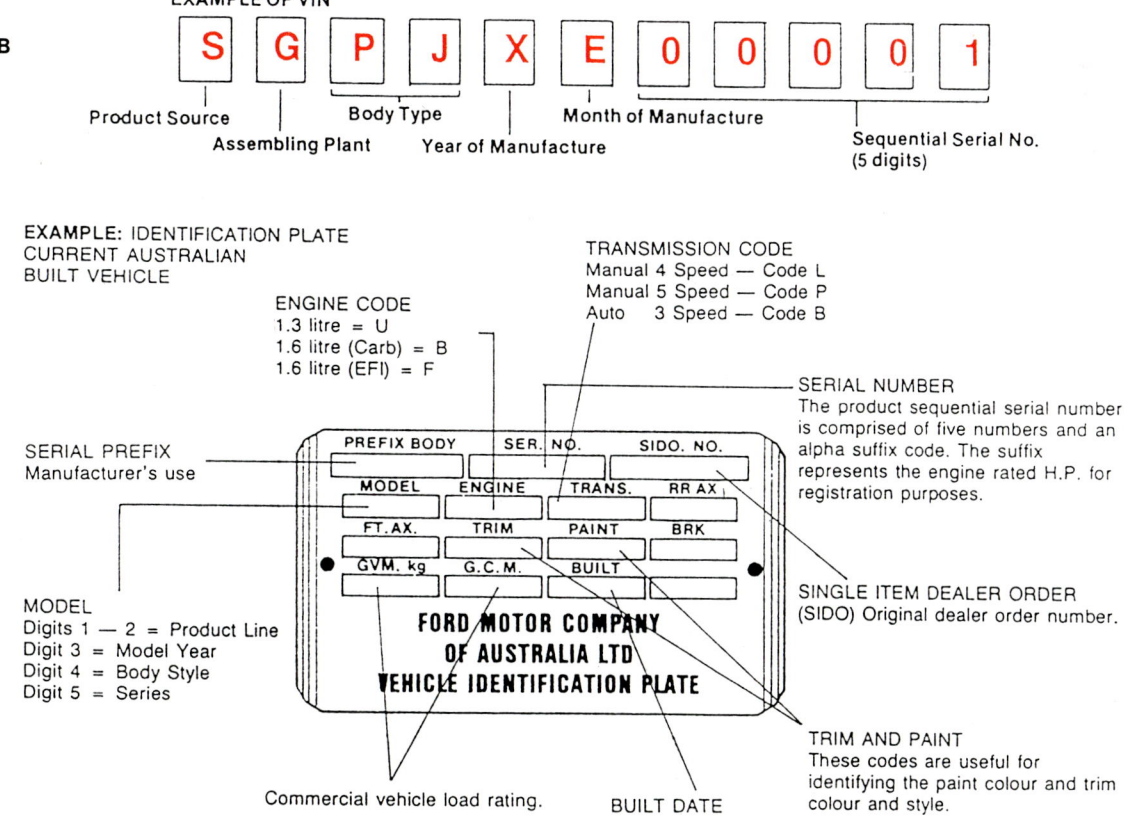

Fig. 2-3. A — Vehicle identification tag is normally located in the engine compartment on the bulkhead. B — Vehicle ID number on tag is a code. Service manual will explain code as shown. (Ford.)

CLUTCH and TRANSMISSION SPECIAL TOOLS

TOOL NUMBER & DESCRIPTION	ILLUSTRATION
49-0813-310 CENTERING TOOL, CLUTCH DISC	
49-0500-330 INSTALLER, TRANSMISSION BEARING	
49-0259-440 TURNING HOLDER, MAINSHAFT	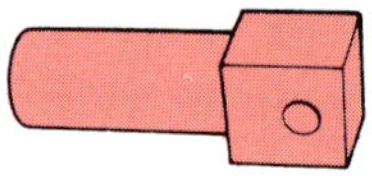
49-0862-350 GUIDE, SHIFT FORK ASSEMBLY	

Fig. 2-4. Service manual explains special tool numbers. Note these special tools for clutch and transmission repairs. (Mazda)

to diagnose (recognise) problems, inspect, test, and repair each system. One page may describe how to remove the engine. Another page might say how to disassemble the engine.

Specifications (bolt tightening limits, capacities, clearances, operating temperatures) are given in the repair sections. They are commonly used during service and repair operations.

The repair sections also refer to special tools (tools for a limited number of repair tasks). These tools will normally be listed in the repair instructions and may be pictured at the end of the manual section. Refer to Fig. 2-4 for an example.

Service manual illustrations

Various types of illustrations are used to supplement (go along with) the written information in a service manual. Some show how to measure part wear or install a part. Others show an exploded (disassembled) view of parts. Fig. 2-5 shows the most common types of service manual illustrations.

When using a service manual, you will find the pictures essential for full understanding of the procedures and specifications. They show you what the parts look like, how they fit together, where leaks might occur, or how a part works.

Service manual diagrams

Diagrams are drawings used when working with electrical circuits, vacuum hoses, and hydraulic circuits. They represent how wires, hoses, passages, and parts connect together.

Wiring diagrams show how the car's wiring connects to the electrical components. See Fig. 2-5C. This subject is covered later in the text.

Vacuum diagrams, like wiring diagrams, help the mechanic trace and determine how vacuum hoses connect to the engine and vacuum-operated devices. Fig. 2-5D shows a vacuum hose illustration.

Hydraulic diagrams show how fluid (usually oil) flows in a circuit or part. They are helpful in understanding how a component operates or how to troubleshoot problems. They are commonly given for automatic transmissions and power steering systems.

Service manual abbreviations

Abbreviations are letters that stand for words. They are often used in service manuals. Sometimes, abbreviations are explained as soon as they are used. They may also be explained at the front or rear of the manual in a chart.

Since abbreviations vary from one service manual to another, this textbook only uses universally accepted abbreviations. It does NOT use those that only apply to one car manufacturer. Fig. 2-6 gives some abbreviations used by one manufacturer.

Troubleshooting (diagnosis) charts

Troubleshooting or diagnosis charts give steps (inspections, tests, measurements, and repairs) for finding and correcting problems in an automobile. If the source of the problem is hard to find, a troubleshooting chart should be used. It will guide you to the most common causes for specific problems.

There are three basic types of troubleshooting charts: Tree chart, Block chart, and Illustrated type chart.

A tree diagnosis chart, Fig. 2-7, provides a logical sequence for what should be inspected or tested when trying to solve a repair problem.

For instance, if a horn will not work, the top of the tree chart may tell you to check the horn's fuse. Then, if the fuse is good, it may have you measure the voltage going to the horn. You can work your way down the "tree" until the problem is fixed.

A block diagnosis chart, Fig. 2-8, lists conditions (problem symptoms), causes (problem sources) and corrections (repairs needed) in columns.

One example, if an engine overheats (runs hot), the most common cause would be the top listing (loss of coolant in this example). You would check the coolant level and, if needed, perform the listed tasks (fill the radiator and check for leaks). If the coolant level was OK, you would go to the next listing.

An illustrated diagnosis chart uses pictures, symbols, and words to guide the mechanic through a

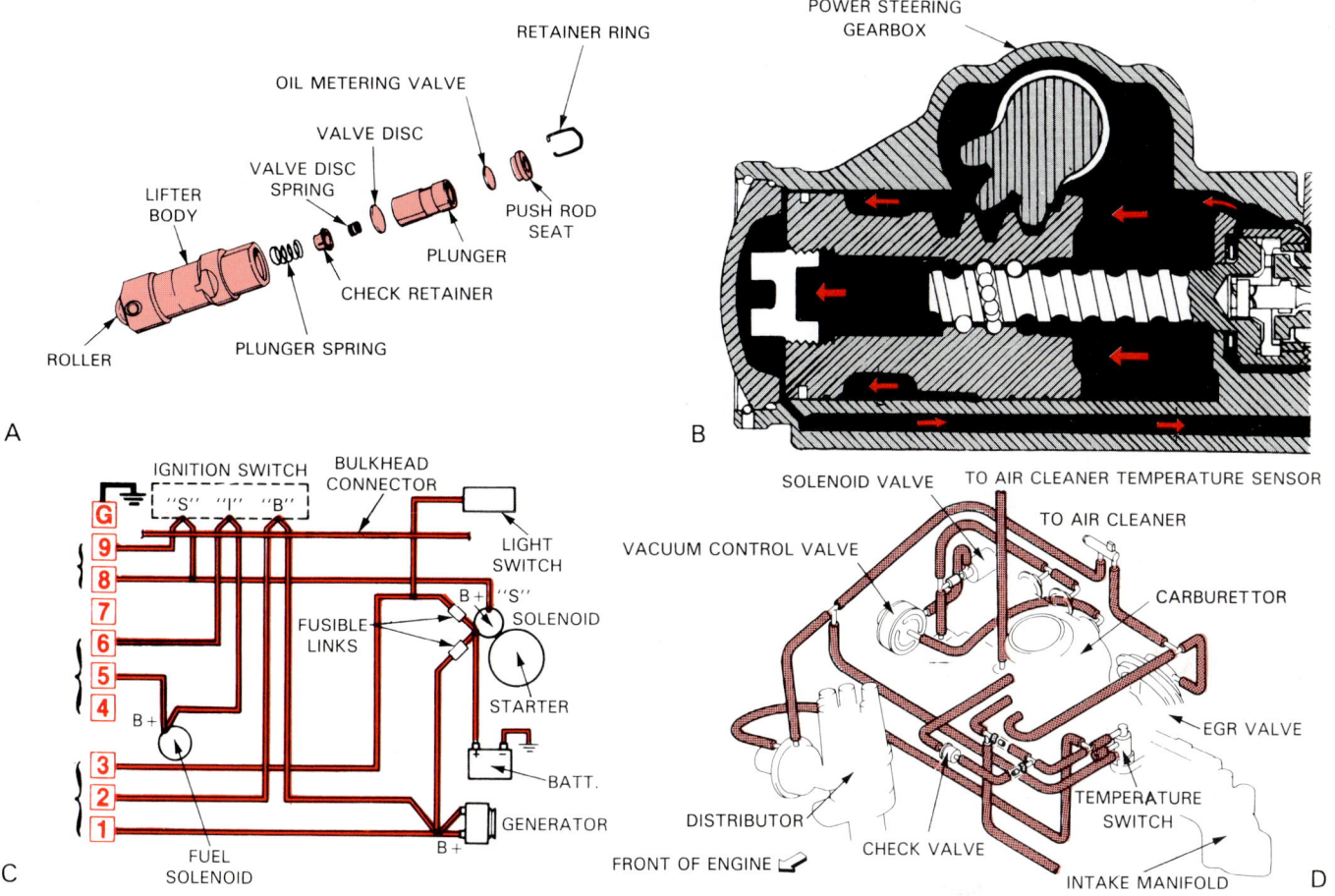

Fig. 2-5. Typical service manual illustrations. A — Exploded view shows how parts fit together. B — Operational illustration shows how part functions. C — Wiring diagram shows how wires connect to components. D — Vacuum diagram shows how hoses connect to components.

ABBREVIATIONS USED IN THIS MANUAL	
A/C	Air Conditioner
AI	Air Injection
A/T	Automatic Transmission
BTDC	Before Top Dead Centre
EGR	Exhaust Gas Recirculation
EVAP	Evaporative (Emission Control)
EX	Exhaust (manifold, valve)
Ex.	Except
IN	Intake (manifold, valve), Inch
IG	Ignition
MC	Mixture Control
MP	Multipurpose
M/T	Manual Transmission
O/S	Oversized
PCV	Positive Crankcase Ventilation
P/S	Power Steering
SC	Spark Control
SST	Special Service Tool
STD	Standard
S/W	Switch
TDC	Top Dead Centre
TP	Throttle Positioner
U/S	Undersized
W/O	Without

Fig. 2-6. These are samples of abbreviations used by one car maker. (Toyota)

sequence of tests. This type of troubleshooting chart is illustrated in Fig. 2-9.

If an engine oil pressure gauge shows low oil pressure for example, the chart shows you exactly what to do step by step, until the problem is corrected. This type of diagnosis chart not only tells you what to do; it shows you HOW TO DO IT.

USING A SERVICE MANUAL

To use a service manual follow these three basic steps:

1. Locate the right manual for the particular vehicle model being worked on. Some manuals come in sets or volumes. Others have all the information required in the one manual. If you are working on the engine find the volume or section that gives the information on the engine you are working on. Take care here as sometimes there may be two or more different engines which can be fitted to the vehicle.

2. Turn to the table of contents or the index. This will help you quickly find the needed information. NEVER thumb through a manual looking for a subject.

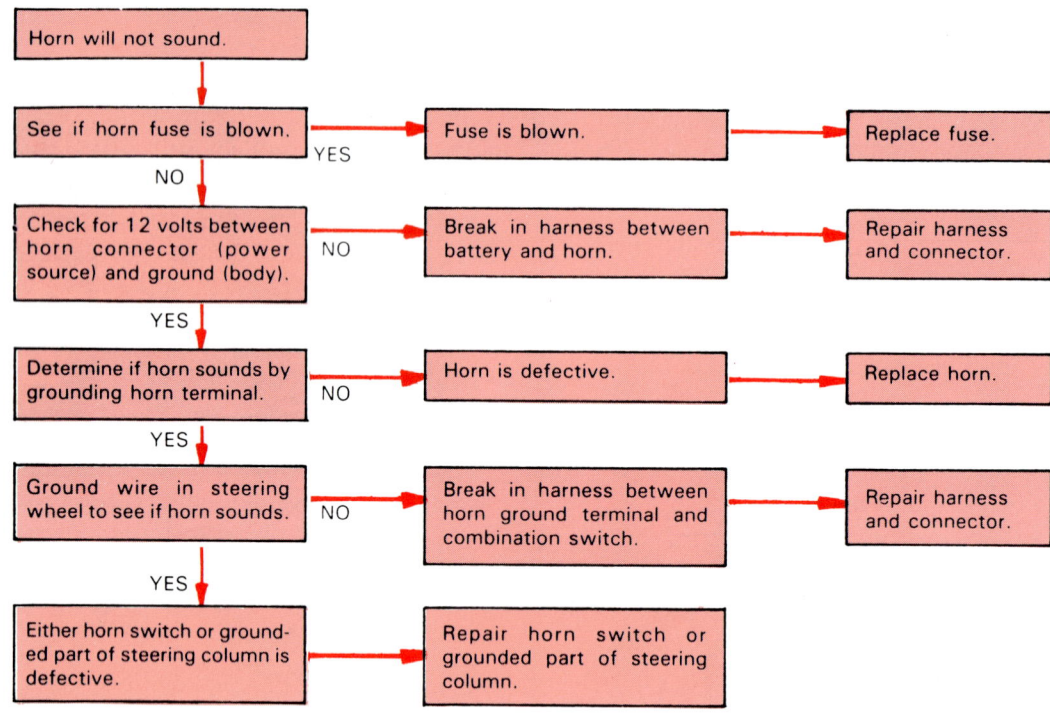

Fig. 2-7. Tree diagnosis chart starts at top and guides you through repair operations.

Condition	Possible Cause	Correction
• Loss of coolant	• Pressure cap and gasket	• Inspect, wash gasket, and test. Replace only if cap will hot hold pressure test specifications.
	• Exhaust leakage	• Pressure test system.
		• Inspect hose, hose connections, radiator, edges of cooling system gaskets, core plugs, drain plugs, transmission oil cooler lines, water pump, heater system components. Repair or replace as required.
	• Internal leakage	• Check for obvious restrictions.
		• Check torque of head bolts. Retorque if necessary.
		• Disassemble engine as necessary — check for: cracked intake manifold, blown head gaskets, warped head or block gaskets surfaces, cracked cylinder head, or engine block.
• Engine overheats	• Low coolant level	• Fill as required. Check for coolant loss.
	• Loose fan belt	• Adjust
	• Pressure cap	• Test. Replace if necessary.
	• Radiator or A/C condenser obstruction	• Remove bugs and leaves.
	• Closed thermostat	• Test, replace if necessary.
	• Fan drive clutch	• Test, replace if necessary.
	• Ignition	• Check timing and advance. Adjust as required.
	• Temperature gauge or cold light	• Check electrical circuits and repair as required.
	• Engine	• Check water pump and block for blockage.
	• Exhaust system	• Check for restrictions.
• Engine fails to reach normal operating temperature	• Thermostat stuck open.	• Test, replace if necessary.
	• Temperature gauge or cold light inoperative	• Check electrical circuits and repair as required. Refer to electrical section.

Fig. 2-8. Block diagnosis chart lists conditions, causes, and corrections in columns. Read to right to match causes and corrections with condition. (Ford)

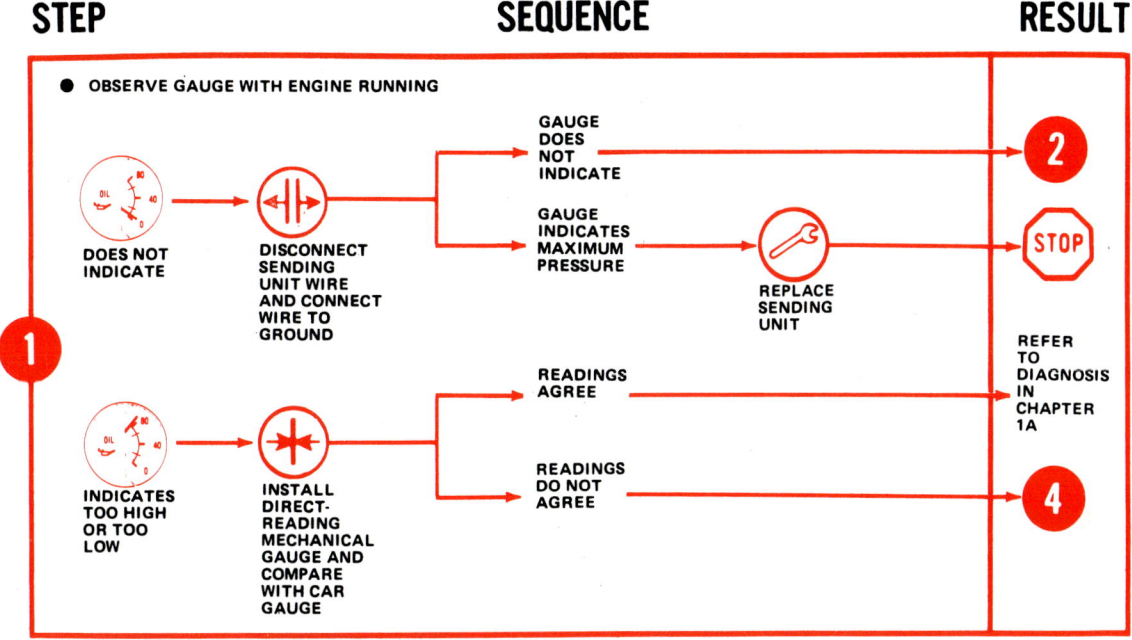

PROBLEM: OIL PRESSURE GAUGE DOES NOT INDICATE CORRECTLY

STEP   SEQUENCE   RESULT

● OBSERVE GAUGE WITH ENGINE RUNNING

DOES NOT INDICATE → DISCONNECT SENDING UNIT WIRE AND CONNECT WIRE TO GROUND

GAUGE DOES NOT INDICATE → **2**

GAUGE INDICATES MAXIMUM PRESSURE → REPLACE SENDING UNIT → **STOP**

1

INDICATES TOO HIGH OR TOO LOW → INSTALL DIRECT-READING MECHANICAL GAUGE AND COMPARE WITH CAR GAUGE

READINGS AGREE →

READINGS DO NOT AGREE → **4**

REFER TO DIAGNOSIS IN CHAPTER 1A

Fig. 2-9. Illustrated diagnosis chart uses small illustrations and symbols to show how to find and correct problems.

3. Use the page listings given at the beginning of each repair section. Most manuals have small contents tables at the beginning of each section. This will help you find a topic quickly.
4. Read the procedures carefully. A service manual will give highly detailed instructions. You must NOT overlook any step or the repair may fail.
5. Study the manual illustrations closely. The pictures in a service manual contain essential information. They cover special tools, procedures, torque values, and other data essential to the repair.

SERVICE PUBLICATIONS

A service manual is just one kind of book which contains technical information for a vehicle. Other types, called service publications, include: owner's manuals, technical bulletins, flat rate manuals (labour estimating books) and parts books.

Owner's manual

An owner's manual is a small booklet given to the purchaser of a new car. It contains basic information on starting the engine, maintaining the car, and operating vehicle accessories.

Flat rate manual

A flat rate manual is needed to calculate how much labour to charge the customer for a repair. It contains an estimate of HOW MUCH TIME a specific repair should take. This time can then be multiplied by the shop's hourly labour rate to find the labour charge in dollars. Using the flat rate manual, you will be able to give the customer an estimate of cost before the actual repair.

Technical bulletins

Technical bulletins help the mechanic stay up-to-date with recent technical changes, repair problems, and other service related information. Usually, only a few pages long, they are mailed to the service manager who passes them to the mechanics. Technical bulletins are published by auto manufacturers and equipment suppliers.

KNOW THESE TERMS

Service manual, Manufacturer's manual, Non manufacturer's manual, Vehicle identification number, Special tools, Diagrams, Abbreviations, Troubleshooting chart, Service publications.

REVIEW QUESTIONS

1. What is a service manual?
2. Which of the following is NOT a service manual containing information on car repairs?
   a. Manufacturer's manual.
   b. Owner's manual.
   c. Non manufacturer's manual.
   d. All of the above are correct.
3. The general information section of most manufacturer's service manuals contains an explanation about the vehicle identification number. True or False?
4. Specifications are given in all service manuals. True or False?
5. Explain the purpose of the following.
   a. Wiring diagrams.
   b. Vacuum diagrams.
   c. Hydraulic diagrams.
6. List ten common abbreviations and explain them.
7. The following is NOT a common type of troubleshooting chart.
   a. Track diagnosis chart.
   b. Tree diagnosis chart.
   c. Block diagnosis chart.
   d. Illustrated diagnosis chart.
8. Write the five basic steps for using a service manual.
9. A _____ _____ manual is needed to calculate how much labour to charge for a repair.
10. _____ _____ help the mechanic stay up-to-date with recent technical changes, repair problems, and other service related information.

Chapter 3

DICTIONARY OF AUTOMOTIVE TERMS

A

ABSOLUTE ZERO: A state in which no heat is present. Believed to be –273.16°C.

AC: Alternating current.

ACCELERATOR: Floor pedal used to control, through linkage, throttle valve in carburettor.

ACCELERATOR PUMP: Small pump, located in carburettor, that sprays additional petrol into air stream during acceleration.

ACCUMULATOR (Air Conditioning): Receiver-dehydrator combination.

ACCUMULATOR PISTON (Automatic Transmission): Unit designed to assist the servo to apply brake band quickly, yet smoothly.

ACETYLENE: Gas commonly used in welding or cutting operations.

ACKERMAN PRINCIPLE: Bending outer ends of steering arms slightly inward so that when car is making a turn, inside wheel will turn more sharply than outer wheel. This principle produces toe-out on turns.

ADDITIVE: Solution, powder, etc., added to petrol, oil, grease, etc., in an endeavour to improve characteristics of original product.

ADVANCE (Ignition timing): To set ignition timing so a spark occurs earlier or more degrees before tdc.

AIR BLEED: An orifice or small passageway designed to allow a specific amount of air to enter a moving column of liquid (such as fuel).

AIR CLEANER: Device used to remove dust, abrasive, etc., from air being drawn into an engine, compressor, power brake, etc.

AIR CONDITIONING: Controlling the temperature, movement, cleanliness, and humidity of air.

AIR COOLED: An object cooled by passing a stream of air over its surface.

AIR DAM: A device placed beneath front bumper to reduce amount of air turbulence and drag under car.

AIR FILTER: A device through which air is drawn to remove dust, dirt, etc.

AIR FOIL: Device, similar to a stubby wing, mounted onto a racing car or dragster to provide high speed stability. The air foil is mounted in a horizontal position.

AIR-FUEL RATIO: Ratio (by weight or by volume) between air and petrol that makes up engine fuel mixture.

AIR GAP (Regulator): Distance between contact armature and iron core that when magnetized, draws armature down.

AIR GAP (Spark Plugs): Distance between centre and side electrodes.

AIR HORN (Accessory): Warning horn operated by compressed air.

AIR HORN (Carburettor): Top portion of air passageway through carburettor.

AIR INJECTION REACTION: An emission control system used to lower levels of carbon monoxide and hydrocarbon emissions. Accomplished by injecting a stream of air into exhaust stream near exhaust valve.

AIR POLLUTION: Contamination of earth's atmosphere by various natural and manufactured pollutants such as smoke, gases, dust, etc.

AIR SPRING: Container and plunger separated by air under pressure. When container and plunger attempt to squeeze together, air compresses and produces a spring effect. Air spring has been used on some suspension systems.

ALIGN: To bring various parts of unit into correct positions in respect to each other or to a predetermined location.

ALLOY: Mixture of two or more materials.

ALNICO MAGNET: Magnet using (A1) aluminium, (Ni) nickel, and (Co) cobalt in its construction.

ALPHA NUMERIC (Tyre): Tyre size designation using letters to indicate load carrying capacity. Example: GR 78-15. The letter G = load/size relationship. R = radial construction. 78 = height to width ratio. 15 = rim diameter in inches.

ALTERNATING CURRENT (AC): Electric current that first flows one way in circuit and then other. Type used in homes.

ALTERNATOR: Device similar to generator but which produces AC current. The AC must be rectified before reaching the car's electrical system.

ALTERNATOR-REGULATOR: Voltage regulator, control box.

AMBIENT TEMPERATURE: Temperature of air surrounding an object.

AMMETER: Instrument used to measure rate of current flow in amperes.

AMPERE: Unit of measurement used in expressing rate of current flow in a circuit.

AMPERE HOUR CAPACITY: Measurement of storage battery ability to deliver specified current over specified length of time.

ANEROID: A device, such as a barometer bellows, that neither contains nor uses a liquid.

ANEROID BAROMETER: A device to measure atmospheric pressure. It utilises an aneroid capsule (bellows) that expands and contracts with differences in pressure.

ANGLE OF APPROACH: The steepest angle that a vehicle can drive up on from a level surface without dragging the front end of vehicle.

ANGLE OF DEPARTURE: The steepest angle that a vehicle can descend and drive off onto a level surface without dragging rear of vehicle.

ANNEAL: To remove hardness from metal. Heat steel to a red colour, then allow it to cool slowly. Unlike steel, copper is annealed by heating, and then plunging it immediately into cold water.

ANODE: In an electrical circuit—the positive pole.

ANTIBACKFIRE VALVE: Valve used in air injection reaction (exhaust emission control) system to prevent backfiring during period immediately following sudden deceleration.

ANTIFREEZE: Chemical added to cooling system to prevent coolant from freezing in cold weather.

ANTIFRICTION BEARING: Bearing containing rollers or balls plus an inner and outer race. Bearing is designed to roll, thus minimising friction.

ANTIKNOCK (Fuel): Indicates various substances, such as tetraethyl lead, that can be added to petrol to improve its resistance to knocking (spark knock, pre-ignition, or detonation).

ANTIPERCOLATOR: Device for venting vapours from main discharge tube, or well, of a carburettor.

ARC OR ELECTRIC WELDING: Welding by using electric current to melt both metal to be welded and welding rod or electrode that is being added.

ARCING: Electricity leaping the gap between two electrodes.

ARMATURE (Relay, regulator, horn, etc.): The movable part of the unit.

ARMATURE (Starter or generator): The portion that revolves between the pole shoes, made up of wire windings on an iron core.

ASBESTOS: Heat resistant and nonburning fibrous mineral widely used for brake shoes, clutch linings, etc. Do not inhale asbestos fibres or dust. Believed to cause cancer.

ASPECT RATIO (Tyre): Ratio between height and width of a tyre.

ASYMMETRIC TYRE TREAD: Tyre tread with a non-symmetrical (uneven) pattern. Designed to reduce tyre vibration and noise.

ATMOSPHERE: Layer of gases (referred to as air) surrounding earth.

ATMOSPHERIC PRESSURE: Pressure exerted by atmosphere on all things exposed to it. Around 101 kPa at sea level.

ATOM: Tiny particle of matter made up of electrons, protons, and neutrons. Atoms or combinations of atoms make up molecules. The electrons orbit around the centre or nucleus made up of the protons and neutrons.

ATOMISE: To break a liquid into tiny droplets.

AUTOMATIC CHOKE: A carburettor choke device that automatically positions itself in accordance with carburettor needs.

AUTOMATIC LEVELLING CONTROL: Pressure system which maintains proper body height with changes in load.

AUTOMATIC TRANSMISSION: A transmission that shifts itself. Shift range and points are determined by road speed, quadrant position, engine loading, altitude, etc.

AWL (Tyre): Sharp pointed steel tool used to probe cuts, nail holes, etc., in tyres.

AXIAL: Direction parallel to shaft or bearing hole centre line.

AXLE (Full-floating): Axle used to drive rear wheels. It does not hold them on nor support them.

AXLE (Semi or one-quarter floating): Axle used to drive wheels, hold them on, and support them.

AXLE (Three-quarter floating): Axle used to drive rear wheels as well as hold them on. It does not support them.

AXLE END GEARS: Two gears, one per axle, that are splined to the inner ends of drive axles. They mesh with and are driven by "spider" gears.

AXLE HOUSING: Enclosure for axles and differential. It is partially filled with differential fluid (oil).

AXLE RATIO: Relationship or ratio between the number of times the propeller shaft must revolve to turn the axle drive shafts one turn.

AXLE SHAFT: Drive shaft, half shaft.

AXLE SHAFT BEARING: Rear wheel bearing, rear axle race, rear hub bearing.

B

BACKFIRE (Exhaust system): Passage of unburned fuel mixture into exhaust system where it is ignited and causes an explosion (backfire).

BACKFIRE (Intake system): Burning of fuel mixture in intake manifold. May be caused by faulty timing, crossed plug wires, leaky intake valve, etc.

BACKING PLATE: Brake plate, back plate.

BACKLASH: Amount of "play" between two parts. In case of gears, it refers to how much one gear can be moved back and forth without moving gear into which it is meshed.

BACK PRESSURE: Refers to resistance to flow of exhaust gases through exhaust system.

BACKUP LIGHTS: White lights attached to the rear of a vehicle. Lights are turned on whenever transmission is placed in reverse. Provides illumination for backing at night.

BAFFLE: Obstruction used to slow down or divert the flow of gases, liquids, sound, etc.

BALANCE (Tyre): See Static Balance and Dynamic Balance.

BALK RING: Ring attached to balk gear hub in overdrive unit. It lets gear rotate while ring is stationary.

BALL AND TRUNION: Type of universal joint using needle bearing mounted balls which swivel inside sockets.

BALL BEARING (Antifriction): Bearing consisting of an inner and outer hardened steel race separated by a series of hardened steel balls.

BALL JOINT: Flexible joint utilising ball and socket type of construction. Used in steering linkage setups, steering knuckle pivot supports, etc.

BALL JOINT ROCKER ARMS: Rocker arms that instead of being mounted on shaft, are mounted upon ball-shaped device on end of stud.

BALL JOINT STEERING KNUCKLE: Steering knuckle that pivots on ball joints instead of on a kingpin.

BALLAST RESISTOR: Resistor constructed of special type wire, properties of which tend to increase or decrease voltage in direct proportion to heat of wire.

BAROMETRIC PRESSURE: Atmospheric pressure as determined by a barometer.

BASE CIRCLE: As applied to camshaft—lowest spot on cam. Area of cam directly opposite lobe.

BATTERY: Electrochemical device for producing electricity.

BATTERY CAPACITY: Rating of current output of battery. Determined by plate size, number of plates, and amount of acid in electrolyte.

BATTERY CHARGING: Process of renewing battery by passing electric current through battery in reverse direction.

BATTERY — MAINTENANCE-FREE: A battery so designed as to not need any additional water during its normal service life.

BATTERY RATING: Standardised measurement of battery's ability to deliver electrical energy under specified conditions.

BATTERY RESERVE CAPACITY: The time in minutes that a new fully charged battery will supply a constant load of 25 Amps without the voltage falling below 10.5 volts for a 12 volt battery and 5.25 for a 6 volt battery.

BATTERY SHEDDING: Grids losing active materials that fall off and settle in bottom of battery case.

BATTERY STATE OF CHARGE: Electrical energy available from a given battery in relation to that which would be normally available if battery were fully charged.

BATTERY VOLTAGE: Determined by the number of cells. Each cell has 2.1V. Three cells will produce a 6 V battery and six cells a 12 V battery.

BEAD (Tyre): Steel wire reinforced portion of tyre that engages the wheel rim.

BEARING: Area of unit in which contacting surface of a revolving part rests.

BEARING CLEARANCE: Amount of space left between shaft and bearing surface. This space is for lubricating oil to enter.

BELL HOUSING (Clutch housing): Metal covering around flywheel and clutch, or torque converter assembly.

BELTED BIAS: Tyre plies crisscrossed; belts beneath tread area.

BENDIX TYPE STARTER DRIVE: A self-engaging starter drive gear. Gear moves into engagement when starter starts spinning and automatically disengages when starter stops.

BEVEL GEAR: Gear in which teeth are cut in a cone shape, as found in axle end gears.

BEVEL SPUR GEAR: Gear in which teeth are cut in a cone shape. Teeth are aligned with cone centreline, as found in some differential gears.

BEZEL: Crimped edge of metal that secures glass face to an instrument.

BIAS PLY: Tyre plies crisscross; belts not used under tread area.

BIG END BEARINGS: Connecting rod bearings, con-rod bearings.

BI-METAL SPRING: Thermostat spring.

BLEEDING: Removing air, pressure, fluid, etc., from a closed system, as in air conditioning.

BLEEDING THE BRAKES: Refers to removal of air from hydraulic system. Bleeder screws are loosened at each wheel cylinder, (one at a time) and brake fluid is forced from master cylinder through lines until all air is expelled.

BLOCK: Part of engine containing cylinders.

BLOW-BY: Refers to escape of exhaust gases past piston rings.

BLOWER: Supercharger.

BLUEPRINTING (Engine): Dismantling engine and reassembling it to EXACT specifications.

BODY PUTTY: Material designed to smooth on dented body areas. Upon hardening, putty is dressed down and area painted.

BOILING POINT: Exact temperature at which a liquid begins to boil.

BONDED BRAKE LINING: Brake lining that is attached to brake shoe by adhesive.

BONNET: Hood, engine compartment lid.

BOOSTER: Device incorporated in car system (such as brakes and steering), to increase pressure output or decrease amount of effort required to operate or both.

BOOT: Luggage compartment in sedan car. Also term for flexible cover over drive shaft joints and gearshift lever etc.

BORE: May refer to cylinder itself or to diameter of the cylinder.

BORE DIAMETER: Diameter of cylinders.

BORING: Renewing cylinders by cutting them out to a specified size. Boring bar is used to make cut.

BORING BAR (Cylinder): Machine used to cut engine cylinders to specific size. As used in garages, to cut worn cylinders to a new diameter.

BOSS: A rib or enlarged area designed to strengthen a certain portion or area of an object.

BOTTLED GAS: LPG (Liquified Petroleum Gas) gas compressed into strong metal tanks. When confined in tank under pressure, gas is in liquid form.

BOUND ELECTRONS: Electrons in inner orbits around nucleus of atom. They are difficult to move out of orbit.

BOURDON TUBE: Circular, hollow piece of metal used in some instruments. Pressure on hollow section causes it to attempt to straighten. Free end then moves needle on gauge face.

BOXED FRAME: A frame in which two channel shaped rails are welded together (open side to open side) forming a box shape providing great strength.

BOXED ROD: Connecting rod in which I-beam section has been stiffened by welding plates on each side of the rod.

BRAKE ANCHOR: Steel stud upon which one end of brake shoes is either attached to or rests against. Anchor is firmly affixed to backing plate.

BRAKE ANTI-ROLL DEVICE: Unit installed in brake system to hold brake line pressure when car is stopped on hill. When car is stopped on hill and brake pedal released, anti-roll device will keep brakes applied until either clutch is released or, as on some models, accelerator is depressed.

BRAKE BACKING PLATE: Rigid steel plate upon which brake shoes are attached. Braking force applied to shoes is absorbed by backing plate.

BRAKE BAND: Band, faced with brake lining, that encircles a brake drum. Used on several parking brake installations.

BRAKE BLEEDING: See Bleeding the Brakes.

BRAKE CYLINDER: See Wheel Cylinder.

BRAKE — DISC TYPE: Braking system that instead of using conventional brake drum with internal brake shoes, uses steel disc with caliper type lining application. When brakes are applied, section of lining on each side of spinning disc is forced against disc thus imparting braking force. This type of brake is very resistant to brake fade.

BRAKE DRUM: Cast iron or aluminium housing, bolted to wheel, that rotates around brake shoes. When shoes are expanded, they rub against machined inner surface of brake drum and exert braking effect upon wheel.

BRAKE DRUM LATHE: Machine to refinish inside of a brake drum.

BRAKE FADE: Reduction in braking force due to loss of friction between brake shoes and drum. Caused by heat buildup.

BRAKE FEEL: Discernible, to driver, relationship between the amount of brake pedal pressure and the actual braking force being exerted. Special device is incorporated in power brake installations to give driver this feel.

BRAKE FLUID: Special fluid used in hydraulic brake systems. Never use anything else in place of regular fluid.

BRAKE FLUSHING: Cleaning brake system by flushing with alcohol or brake fluid. Done to remove water, dirt, or any other contaminant. Flushing fluid is placed in master cylinder and forced through lines and wheel cylinders where it exits at cylinder bleed screws.

BRAKE HORSEPOWER: Measurement of actual usable horsepower delivered at crankshaft. Commonly computed using an engine on a dynamometer.

BRAKE HOSE: Flexible hose which connects brake lines to wheel cylinders. It allows suspension movement without damage.

BRAKE LINE: Steel tubing which carries brake fluid from master cylinder to wheel cylinders. Also called Bundy tube.

BRAKE LINING: Friction material fastened to brake shoes. Brake lining is pressed against rotating brake drum thus stopping car.

BRAKE — PARKING or EMERGENCY: Brake used to hold car in position while parked. One type applies rear brake shoes by mechanical means and other type applies brake band to brake drum installed in drive train.

BRAKE — POWER: Conventional hydraulic brake system that utilises either engine vacuum or hydraulic pressure to operate a power piston. Power piston applies pressure directly to master cylinder piston. This reduces amount of pedal pressure driver must exert to stop car.

BRAKE SHOE GRINDER: Grinder used to grind brake shoe lining so it will be square to and concentric with brake drum.

BRAKE SHOE HEEL: End of brake shoe adjacent to anchor bolt or pin.

BRAKE SHOE TOE: Free end of shoe, not attached to or resting against an anchor pin.

BRAKE SHOES: Part of brake system, located at wheels, upon which brake lining is attached. When wheel cylinders are actuated by hydraulic pressure, they force brake shoes apart and bring lining into contact with drum.

BRAKES — POWER BOOSTER (Hydraulic): Brake booster employing hydraulic pressure from the power steering system to apply force to master cylinder.

BRAZE: To join two pieces of metal together by heating edges to be joined and then melting drops of brass or bronze on area. Unlike welding, this operation is similar to soldering, only a higher melting point material is used.

BREAK-IN: Period of operation between installation of new or rebuilt parts and time in which parts are worn to the correct fit. Driving at reduced and varying speed for a specified distance to permit parts to wear to the correct fit.

BREAK-OVER ANGLE: Included angle that a vehicle can cross over without dragging underneath centre section. Break-over angle is determined by vehicle ground clearance and wheelbase.

BREAKER (Tyre): Rubber or fabric (or both) strip placed under tread to provide additional protection for main tyre carcass.

BREAKER ARM: Movable arm upon which one of breaker points is affixed.

BREAKER POINTS (Ignition): Pair of movable points that are opened and closed to break and make the primary circuit.

BREATHER PIPE: Pipe opening into interior of engine. Used to assist ventilation. Pipe usually extends downward to a point just below engine so passing air stream will form a partial vacuum thus assisting in venting engine.

BROACH: Bringing metal surface to desired shape by forcing multiple-edged cutting tool across surface.

BRONZE: An alloy basically consisting of tin and copper.

BRUSH: Pieces of carbon, or copper, that rub against the commutator in generator and starter motor.

BULB: Lamp, globe.

BULKHEAD: Scuttle, firewall.

BUMPERS: Bumper bar, fender.

BUMP RUBBER: Bumper, bump stop, rebound rubber.

BURNISH: To bring a surface to a high shine by rubbing with hard, smooth object.

BUSHING: Bearing for shaft, spring shackle, piston pin, etc., of one piece construction which may be removed from part.

BUTANE: Petroleum gas that is liquid, when under pressure. Often used as engine fuel in trucks.

BUTTERFLY VALVE: Valve in carburettor that is so named due to its resemblance to insect of same name.

BY-PASS: To move around or detour regular route or circuit taken by air, fluid, electricity, etc.

BY-PASS FILTER: Oil filter that constantly filters PORTION of oil flowing through engine.

BY-PASS VALVE: Valve that can open and allow fluid to pass through in other than its normal channel.

C

CALIBRATE: As applied to test instruments— adjusting dial needle to correct zero or load setting.

CALIPER (Brake): Disc brake component which forms cylinder and houses piston and brake pads (linings). It produces clamping action on rotating disc to stop car.

CALIPERS (Inside and Outside): Adjustable measuring tool placed around or within an object and adjusted until it just contacts. It is then withdrawn and distance measured between contacting points.

CALORIE (Gram): A unit of heat. Amount of heat required to raise the temperature of one gram of water 1 deg. centigrade.

CALORIFIC VALUE: Measurement of the heating value of fuel.

CALORIMETER: Measuring instrument used to determine amount of heat produced when a substance is burned; also friction and chemical change heat production.

CAM: Offset portion of shaft that will, when shaft turns, impart motion to another part such as valve lifters.

CAM ANGLE or DWELL (Ignition): Number of degrees breaker cam rotates from time breaker points close until they open again.

CAM GROUND: Piston ground slightly egg-shaped. When heated, it becomes round.

CAMBER: Tipping top of the wheel centreline outward produces positive camber. Tipping wheel centreline inward at top produces negative camber. When camber is positive, tops of tyres are further apart than bottoms.

CAMSHAFT: Shaft with cam lobes (bumps) used to operate valves.

CAMSHAFT GEAR: Gear that is used to drive camshaft.

CANDELA: Measurement of light producing ability of light bulb.

CAPACITANCE: Property of condenser that permits it to receive and retain an electrical charge.

CAPACITOR: See Condenser.

CARBON: Used to describe hard, or soft, black deposits found in combustion chamber, on plugs, under rings, on and under valve heads, etc.

CARBONISE: Building up of carbon on objects such as spark plugs, pistons, heads, etc.

CARBON MONOXIDE: Deadly, colourless, odourless, and tasteless gas found in engine exhaust. Formed by incomplete burning of hydrocarbons.

CARBON PILE: Refers to amperage or voltage regulator utilising a stack of carbon discs in its construction.

CARBURETTOR: Device used to mix petrol and air in correct proportions.

CARBURETTOR ADAPTER: Adapter used to fit or place one type of carburettor on an intake manifold that may not be originally designed for it. Also used to adapt four-barrel carbs to two-barrel manifolds.

CARBURETTOR CIRCUITS: Series of passageways and units designed to perform a specific function—idle circuit, full power circuit, etc.

CARBURETTOR ICING: Formation of ice on throttle plate or valve. As fuel nozzles feed fuel into air horn, it turns to a vapour. This robs heat from air and when weather conditions are just right (fairly cold and quite humid), ice may form.

CARBURISING FLAME: Welding torch flame in which there is an excess of acetylene.

CARDAN JOINT: Type of universal joint.

CARRIER BEARINGS: Bearings upon which differential case is mounted.

CASE-HARDENED: Piece of steel that has had outer surface hardened while inner portion remains relatively soft.

CASTER: Tipping top of kingpin either forward or toward the rear of car. When tipped forward, it is termed negative caster. When tipped toward rear, it is termed positive caster.

CASTING: Pouring molten metal into a mould to form an object.

CASTLE or CASTELLATED NUT: Nut having series of slots cut into one end, into which cotter pin may be passed to secure nut.

CATALYTIC CONVERTER: Device used in exhaust system to reduce harmful emissions. Catalyst in converter may be coated with palladium, platinum, and rhodium. Catalyst may be of oxidising and/or reducing design.

CATHODE: In electric circuit—the negative pole.

CELL (Battery): Individual (separate) compartments in battery which contain positive and negative plates suspended in electrolyte. Six-volt battery has three cells, twelve-volt battery six cells.

CELL CONNECTOR: Lead strap or connection between battery cell groups.

CENTIGRADE: Thermometer on which boiling point of water is 100 deg. and freezing point is 0 deg.

CENTRELINE: Imaginary line drawn lengthwise through centre of an object.

CENTRE LINK: Also called relay rod or connecting link, it transfers motion from pitman arm to tie rods.

CENTRE OF GRAVITY: Point in object, if through which an imaginary pivot line were drawn, would leave object in balance. In car, the closer the weight to the ground, the lower the centre of gravity.

CENTRE STEERING LINKAGE: Steering system utilising two tie rods connected to steering arms and to central idler arm. Idler arm is operated by drag link that connects idler arm to pitman arm.

CENTRIFUGAL ADVANCE (Distributor): Unit designed to advance and retard ignition timing through action of centrifugal force.

CENTRIFUGAL CLUTCH: Clutch that utilises centrifugal force to expand a friction device on driving shaft until it is locked to a drum on driven shaft.

CENTRIFUGAL FORCE: Force which tends to keep moving objects travelling in straight line. When moving car is forced to make a turn, centrifugal force attempts to keep it moving in straight line. If car is turning at too high a speed, centrifugal force will be greater than frictional force between tyres and road and the car will slide off the road.

CERAMIC FILTER: Filtering device utilising porous ceramic as filtering agent.

CETANE NUMBER: Measurement of diesel fuel performance characteristics.

CFM: Cubic feet per minute. A measure of air flow.

CHAMFER: To bevel (or a bevel on) edge of an object.

CHANGE OF STATE: Condition in which substance changes from a solid to a liquid, a liquid to a gas, a liquid to a solid, or a gas to a liquid.

CHARCOAL CANISTER: Emission control device containing activated charcoal granules. Used to store petrol vapours from tank and carburettor. When engine is started, stored vapours are drawn into cylinders and burned.

CHARGE (Air Conditioning): A given or specified (by weight) amount of refrigerant.

CHARGE (Battery): Passing electric current through battery to restore it to active (charged) state.

CHARGE RATE: Electrical rate of flow, in amperes, passing through the battery during charging.

CHARGING (Air Conditioning): Inserting the specified charge (amount) of refrigerant into the air conditioning system.

CHASE: To repair damaged threads.

CHASSIS: Generally, chassis refers to frame, engine, front and rear axles, springs, steering system and petrol tank. In short, everything but body and fenders.

CHASSIS DYNAMOMETER: See Dynamometer.

CHECK VALVE: Valve that opens to permit passage of fluid or air in one direction and closes to prevent passage in opposite direction.

CHILLED IRON: Cast iron possessing hardened outer skin.

CHOKE: Butterfly valve located in carburettor used to enrichen mixture for starting engine when cold.

CHOKE STOVE: Heating compartment in or on exhaust manifold from which hot air is drawn to automatic choke device.

CID: Cubic Inch Displacement.

CIRCUIT (Electrical): Source of electricity (battery), resistance unit (headlight, etc.) and wires that form path for flow of electricity from source through unit and back to source.

CIRCUIT BREAKER (Lighting System): Protective device that will make and break flow of current when current draw becomes excessive. Unlike fuse, it does not blow out but vibrates on and off thus giving driver some light to stop by.

CLEARANCE: Given amount of space between two parts—between piston and cylinder, bearing and journal, etc.

CLOCKWISE: Rotation to right as that of clock hands.

CLOSED COOLING SYSTEM: Type of system which uses an overflow tank.

CLOSED LOOP FUEL SYSTEM: A fuel system in which air-fuel ratio is constantly adjusted in relationship to hydrocarbon content of exhaust. An oxygen sensor in exhaust, working through an electronic control unit, alters either carburettor jet size or fuel injection system pulse width.

CLUSTER, COUNTER GEAR, LAYGEAR: Cluster of gears that are all cut on one long gear blank. Cluster gears ride in bottom of transmission. Cluster provides a connection between transmission input shaft and output shaft.

CLUTCH: Device used to connect or disconnect flow of power from one unit to another.

CLUTCH DIAPHRAGM SPRING: Round dish-shaped piece of flat spring steel. Used to force pressure plate against clutch disc in some clutches.

CLUTCH DISC: Part of clutch assembly splined to transmission clutch or input shaft. Faced with friction material. When clutch is engaged, disc is squeezed between flywheel and clutch pressure plate.

CLUTCH EXPLOSION: Clutches have literally flown apart (exploded) when subjected to high rpm. Scatter shield is used on competition cars to protect driver and spectators from flying parts if clutch explodes.

CLUTCH HOUSING or BELL HOUSING: Cast iron or aluminium housing that surrounds flywheel and clutch mechanism.

CLUTCH LINKAGE: Mechanism which transfers movement from clutch pedal to throw-out fork.

CLUTCH PEDAL FREE TRAVEL: Specified distance clutch pedal may be depressed before throw-out bearing actually contacts clutch release fingers.

CLUTCH PILOT BEARING: Small bronze bushing, or in some cases ball bearing, placed in end of crankshaft or in centre of flywheel depending on car, used to support outboard end of transmission input shaft.

CLUTCH PRESSURE PLATE: Part of a clutch assembly that through spring pressure, squeezes clutch disc against flywheel thereby transmitting driving force through the assembly. To disengage clutch, pressure plate is drawn away from flywheel via linkage.

CLUTCH SEMI-CENTRIFUGAL RELEASE FINGERS: Clutch release fingers that have a weight attached to them so that at high rpm release fingers place additional pressure on clutch pressure plate.

CLUTCH THROW-OUT FORK: Device or fork that straddles throw-out bearing and used to force throw-out bearing against clutch release fingers.

COIL (Ignition): Unit used to step up battery voltage to point necessary to fire spark plugs.

COIL CORE: Multilayered mass of iron around which coil windings are wrapped. Some designs place core outside of windings.

COIL SPRING: Section of spring steel rod wound in spiral pattern or shape. Widely used in both front and rear suspension systems.

COLD: Little or no perceptible heat.

COLD START INJECTOR: Fuel injector which sprays additional fuel for cold engine starting.

COLLAPSED (Piston): Piston whose skirt diameter has been reduced due to heat and forces imposed upon it during service in engine.

COMBUSTION: Process involved during burning.

COMBUSTION CHAMBER: Area above piston with piston on tdc. Head of piston, cylinder, and head form the chamber.

COMBUSTION CHAMBER VOLUME: Volume of combustion chamber (space above piston with piston on tdc) measured in cc (cubic centimetres).

COMMUTATOR: Series of copper bars connected to armature windings. Bars are insulated from each other and from armature. Brushes (as in generator or starter) rub against rotating commutator.

COMPENSATING JET: Secondary jet.

COMPENSATING PORT: Small hole in brake master cylinder to permit fluid to return to reservoir.

COMPENSATOR VALVE (Automatic Transmission): Valve designed to increase pressure on brake band during heavy acceleration.

COMPOUND: Two or more ingredients mixed together.

COMPRESSION: Applying pressure to a spring, or any springy substance, thus causing it to reduce its length in direction of compressing force. Applying pressure to gas, thus causing reduction in volume.

COMPRESSION CHECK: Testing compression in all cylinders at cranking speed. All plugs are removed, compression gauge placed in one plug hole, throttle cracked wide open and engine cranked until gauge no longer climbs. Compression check is a fine way in which to determine condition of valves, rings and cylinders.

COMPRESSION GAUGE: Gauge used to test compression in engine cylinders.

COMPRESSION RATIO: Relationship between cylinder volume (clearance volume) when piston is on tdc and cylinder volume when piston is on bdc.

COMPRESSION RINGS: Top piston rings, generally two, designed to seal between piston and cylinder to prevent escape of gas from combustion chamber.

COMPRESSION STROKE: Portion of piston's movement devoted to compressing the fuel mixture trapped in engine's cylinder.

COMPRESSOR (Air Conditioning): Device using a series of pistons to raise pressure of refrigerant in system. Also causes refrigerant to flow through system.

COMPRESSOR PROTECTION SWITCH (Air Conditioning): A heat and/or pressure sensitive switch to protect compressor from damage caused by loss of oil and over-heating.

COMPUTERISED IGNITION: Ignition system using sensors which feed electrical information to computer. Computer then controls ignition system and sometimes other functions (carburettor, fuel injection, transmission) for maximum efficiency.

CONCENTRIC: Two or more circles so placed as to share common centre.

CONDENSATION: Moisture, from air, deposited on a cool surface.

CONDENSE: Turning vapour back into liquid.

CONDENSER (Ignition): Unit installed between breaker points and coil to prevent arcing at breaker points. Condenser has ability to absorb and retain surges of electricity.

CONDENSER (Refrigeration): Unit in air conditioning system that cools hot compressed refrigerant and turns it from vapour into liquid.

CONDUCTION: Transfer of heat from one object to another by having objects in physical contact.

CONDUCTOR: Material forming path for flow of current.

CONE CLUTCH: Clutch utilising cone-shaped member that is forced into a cone-shaped depression in flywheel, or other driving unit, thus locking two together. Although no longer used on cars, cone clutch finds some applications in small riding tractors, heavy power mowers, etc.

CONNECTING ROD: Connecting link between piston and crankshaft.

CONNECTING ROD BEARINGS: Inserts which fit into connecting rod and run on crankshaft journals.

CONNECTING ROD CAP: Lower removable part of rod which holds lower bearing insert.

CONSTANT DRIVE OVERDRIVE: An overdrive unit that, when engaged, is constantly connected to drive wheels and will not permit freewheeling (coasting).

CONSTANT MESH GEARS: Gears that are always in mesh with each other.

CONSTANT VELOCITY UNIVERSAL JOINT: Universal joint so designed as to effect smooth transfer of torque from driven shaft to driving shaft without any fluctuations in speed of driven shaft.

CONTACT POINTS also called BREAKER POINTS: Two removable points or areas that when pressed together, complete circuit. These points are usually made of tungsten, platinum or silver.

CONTINUOUS FUEL INJECTION: A fuel injection system in which injectors are always open and as such, feed fuel constantly. Amount of fuel delivered can be determined by an airflow sensor.

CONTRACTION (Thermal): Reduction in size of object when cooled.

CONTROL ARM: Wishbone, suspension arm, link, track control arm.

CONTROL ARM BALL JOINT: Upper ball joint, lower ball joint.

CONTROL ARM SHAFT: Fulcrum pin, fulcrum shaft, pivot pin.

CONTROL RACK: Toothed rod inside mechanical injection pump which rotates pump plunger to control quantity of injected fuel.

CONVECTION: Transfer of heat from one object to another when hotter object heats surrounding air and air in turn heats other object.

COOLANT: Liquid in cooling system.

COOLING SYSTEM: System, air or water, designed to remove excess heat from engine.

CORE: When referring to casting—sand unit placed inside mould so that when metal is poured, core will leave a hollow shape.

CORONA (Electrical): Luminous discharge of electricity visible near surface of an electrical conductor under high voltage.

CORRODE: Removal of surface material from object by chemical action.

COUNTERBALANCE: Weight attached to some moving part so that the part will be in balance.

COUNTERBORE: Enlarging hole to certain depth.

COUNTERCLOCKWISE: Rotation to the left as opposed to that of clock hands.

COUNTERSHAFT: Intermediate shaft that receives motion from one shaft and transfers it to another. It may be fixed (gears turn on it) or it may be free to revolve.

COUNTERSINK: To make a counterbore so that head of a screw may set flush, or below the surface.

COUPLING: Connecting device used between two objects so motion of one will be imparted to other.

COUPLING POINT: This refers to point at which both pump and turbine in torque converter are travelling at same speed. The drive is almost direct at this point.

COURTESY LAMP: Interior light, roof light.

CRANKCASE: Part of engine that surrounds crankshaft. Not to be confused with the pan which is a thin steel cover that is bolted to crankcase.

CRANKCASE DILUTION: Accumulation of unburned petrol in crankcase. Excessively rich fuel mixture or poor combustion will allow some of the petrol to pass down between pistons and cylinder walls.

CRANKCASE VENTILATION: Process of drawing clean air through interior of engine to remove blow-by gases and other fumes.

CRANKING MOTOR: Starter Motor. Device to revolve engine crankshaft to start engine. Works through a gear engaging another gear on flywheel.

CRANKSHAFT: Shaft running length of engine. Portions of shaft are offset to form throws to which connecting rods are attached. Crankshaft is supported by main bearings.

CRANKSHAFT GEAR: Gear mounted on front of crankshaft. Used to drive camshaft gear.

CRANKSHAFT THROW: Offset part of crankshaft where connecting rods fasten.

CROSS AND ROLLER: Type of universal joint using a centre cross (spider) mounted in needle bearings.

CROSS SHAFT (Steering): Shaft in steering gearbox that engages steering shaft worm. Cross shaft is splined to pitman arm.

CROWNWHEEL: Ring gear, final drive gear, spiral drive gear.

CRUDE OIL: Petroleum in its raw or unrefined state. It forms the basis of petrol, engine oil, diesel oil, kerosene, etc.

CU. IN. (C.I.): Cubic inch.

CUNO FILTER: Filter made up of a series of fine discs or plates pressed together in a manner that leaves very minute space between discs. Liquid is forced through these openings to produce straining action.

CURRENT: Movement of free electrons through conductor.

CUTOUT (Regulator): Device to connect or disconnect generator from battery circuit. When generator is charging, cutout makes circuit. When generator stops, cutout breaks circuit. Also referred to as cutout relay, and circuit breaker.

CYCLE: Recurring period during which series of events take place in definite order.

CYLINDER: Hole, or holes, in cylinder block that contain pistons.

CYLINDER BLOCK: Crankcase, engine block. See Block.

CYLINDER BORE: See Bore.

CYLINDER HEAD: Metal section bolted on top of block. Used to cover tops of cylinders. In many cases cylinder head contains the valves. Also forms part of combustion chamber.

CYLINDER HONE: Tool that uses an abrasive to smooth out and bring to exact measurements components such as engine cylinders, wheel cylinders, bushings, etc.

CYLINDER LINER: See Cylinder Sleeve.

CYLINDER SLEEVE: Replaceable cylinder. It is made of a pipe-like section that is either pressed or shrunk into the block.

CYLINDER STROKE: See Stroke.

D

DAMPER: A unit or device designed to remove or reduce vibration, oscillation, etc., of a moving part, fluid, air, etc.

DASHBOARD: Part of body containing driving instruments, switches, etc.

DASHPOT: Unit utilising cylinder and piston, or cylinder and diaphragm, with small vent hole, to retard or slow down movement of some part.

DC (Electrical): Direct Current.

DEAD AXLE: Axle that does not rotate but merely forms base upon which to attach wheels.

DECELERATION: The process of slowing down in rotational speed, forward speed, etc.

DECELERATION VALVE: It feeds air into intake manifold to prevent backfiring during deceleration.

DECIBEL: A unit of measurement used to indicate a sound level or to indicate the difference in specific sound levels.

DEFLECTION RATE (Springs): Measurement of force, in Newton's, required to compress leaf spring a distance of one millimetre.

DEGLAZER: Abrasive tool used to remove glaze from cylinder walls so that a new set of rings will seat.

DEGREE (Circle): 1/360 part of a circle.

DEGREE WHEEL: Wheel-like unit attached to engine crankshaft. Used to time valves to a high degree of accuracy.

DEHYDRATE: To dry out. Remove moisture.

DEMAGNETISE: Removing residual magnetism from an object.

DEPOLARISE: Removal of residual magnetism thereby destroying or removing the magnetic poles.

DESICCANT: Material, such as silica-gel, placed within a container to absorb and retain moisture.

DETENT BALL AND SPRING: Spring loaded ball that snaps into a groove or notch to hold some sliding object in position.

DETERGENT: Chemical added to engine oil to improve its characteristics (sludge control, nonfoaming, etc.).

DETONATION: Fuel charge firing or burning too violently, almost exploding.

DIAGNOSIS: Process of analysing certain symptoms, readings, etc., in order to determine underlying reason for trouble at hand.

DIAL GAUGE OR INDICATOR: Often used precision micrometer type instrument that indicates exact reading via needle moving across dial face.

DIAPHRAGM: Flexible cloth-rubber sheet stretched across an area thereby separating two different compartments.

DICHLORODIFLUOROMETHANE: Refrigerant-12 used in air conditioning system.

DIE (Forming): One of a matched pair of hardened steel blocks that are used to form metal into a desired shape.

DIE (Thread): Tool for cutting threads.

DIE CASTING: Formation of an object by forcing molten metal, plastic, etc., into a die.

DIESEL ENGINE: Internal combustion engine that uses diesel oil for fuel. True diesel does not use an ignition system but injects diesel oil into cylinders. Piston compresses air so tightly that air is hot enough to ignite diesel fuel without spark.

DIESEL INJECTION PUMP: Mechanically operated fuel pump which develops high pressure to force fuel out of injectors and into combustion chambers.

DIESELING: Condition in which engine continues to run after ignition key is turned off. Also called "running on".

DIFFERENTIAL: Unit that drives both rear axles at same time but allows them to be driven at different speeds when negotiating turns.

DIFFERENTIAL CASE (Carrier): Steel unit to which the crown wheel gear is attached. Case drives spider gears and forms an inner bearing surface for axle end gears.

DIFFERENTIAL PINION: Spider pinion, pinion gear, planet wheel.

DIFFERENTIAL SIDE GEAR: Axle shaft gear, spider gear, sun wheel.

DIFFERENTIAL WINDING: A secondary winding in an electrical device that is wound in a reverse manner as related to the primary (main) windings.

DIMMER SWITCH OR DIPPER SWITCH: Foot or hand operated switch that operates headlight low and high beams. Also called dipper switch.

DIODE: Unit having ability to pass electric current readily in one direction but resisting current flow in the other.

DIPSTICK: Metal rod that passes into oil sump.

Used to determine quantity of oil in engine.

DIRECT CURRENT (DC): Electric current that flows steadily in one direction only.

DIRECT DRIVE: Such as high gear when crankshaft and drive shaft revolve at same speed.

DIRECT FUEL INJECTION: Fuel is sprayed directly into combustion chamber.

DIRECTIONAL STABILITY (Steering): Ability of car to move forward in straight line with minimum of driver control.

DISCHARGE (Battery): Drawing electric current from battery.

DISCHARGE PRESSURE (Air Conditioning): Pressure of refrigerant as it leaves compressor.

DISCHARGE SIDE (Air Conditioning): The high pressure section of the air conditioning system extending from the compressor to the expansion valve.

DISC WHEEL: Wheel constructed of stamped steel.

DISPLACEMENT: Total volume of air displaced by piston travelling from bdc to tdc.

DISTILLATION: Heating a liquid and then catching and condensing the vapours given off by heating process.

DISTRIBUTION TUBES (Cooling System): Tubes used in engine cooling area to guide and direct flow of coolant to vital areas.

DISTRIBUTOR (Ignition): Unit designed to make and break the ignition primary circuit and to distribute resultant high voltage to proper cylinder at correct time.

DISTRIBUTOR CAP (Ignition): Insulated cap containing central terminal with series (one per cylinder) of terminals that are evenly spaced in circular pattern around central terminal. Secondary voltage travels to central terminal where it is then channelled to one of outer terminals by the rotor.

DIVERTER VALVE: A device used in the air injection system to divert air away from the injection nozzles during periods of deceleration. Prevents "backfiring".

DOHC: Refers to an engine with double (two) overhead camshaft.

DOOR GLASS: Door window, window glass.

DOUBLE FLARE: End of tubing, especially brake tubing, has a flare so made that flare area utilises two wall thicknesses. This makes a much stronger joint from safety standpoint.

DOWEL PIN: Steel pin, passed through or partly through, two parts to provide proper alignment.

DOWNDRAFT CARBURETTOR: A carburettor in which air passes downward through carburettor into intake manifold.

DOWNSHIFT: Shifting to lower gear.

DRAG LINK: A steel rod connecting pitman arm to one of steering knuckles. On some installations drag link connects pitman arm to a centre idler arm.

DRAW (Electrical): Amount of electrical current

required to operate electrical device.

DRAW (Forming): To form (such as wire) by pulling wire stock through series of hardened dies.

DRAW (Temper): Process of removing hardness from a piece of metal.

DRAW-FILING: Filing by passing file, at right angles, up and down the length of work.

DRIER (Receiver-Drier): Tank, containing desiccant, inserted in air conditioning system to absorb and retain moisture.

DRILL: Tool used to bore holes.

DRILL PRESS: Non-portable machine used for drilling.

DRIVE-FIT: Fit between two parts when they must be literally driven together.

DRIVELINE: Propeller shaft, universal joints, etc., connecting transmission output shaft to axle pinion gear shaft.

DRIVEN PLATE: Clutch plate, driven disc, clutch disc.

DRIVE PINION: Final drive pinion, bevel pinion, spiral drive pinion.

DRIVE OR PROPELLER SHAFT SAFETY STRAP: A metal strap or straps, surrounding drive shaft to prevent shaft from falling to ground in event of a universal joint or shaft failure.

DRIVEN PLATE FACINGS: Driven plate linings, clutch linings, friction facings.

DRIVE SHAFT: Shaft connecting transmission output shaft to differential pinion shaft.

DRIVE TRAIN: All parts that generate power (engine) and transmit it to road wheels (transmission, clutch, drive shaft, differential, drive axles).

DRIVING LIGHTS: Auxiliary headlights, often very bright, that can be used to increase amount of illumination provided by regular headlights.

DROP CENTRE RIM: Centre section of rim being lower than two outer edges. This allows bead of tyre to be pushed into low area on one side while the other side is pulled over and off the flange.

DROP FORGED: Part that has been formed by heating steel blank red hot and pounding it into shape with a powerful drop hammer.

DROPPED AXLE: Front axle altered so as to lower the frame of car. Consists of bending axle at outer ends. (Solid front axle.)

DRY CELL OR DRY BATTERY: Battery (like flashlight battery) that uses no liquid electrolyte.

DRY CHARGED BATTERY: Battery with plates charged but lacking electrolyte. When ready to be placed in service, electrolyte is added.

DRY FRICTION: Resistance to movement between two unlubricated surfaces.

DRY SLEEVE: Cylinder sleeve application in which sleeve is supported in block metal over its entire length. Coolant does not touch sleeve itself.

DRY SUMP: Instead of letting oil throw-off drain into a regular oil pan sump, system collects and pumps this oil to a remote (separate) container or sump.

DRY WEIGHT: Weight of vehicle without fluid (oil, fuel, water) in various units.

DUAL BRAKES: Tandem or dual master cylinder to provide separate brake system for both front and rear of car.

DUAL BREAKER POINTS (Ignition): Distributor using two sets of breaker points to increase cam angle so that at high engine speeds, sufficient spark will be produced to fire plugs.

DUALS: Two sets of exhaust pipes and mufflers — one for each bank of cylinders.

DWELL: See Cam Angle.

DYNAMIC BALANCE: When centreline of weight mass of a revolving object is in same plane as centreline of object, that object would be in dynamic balance. For example, weight mass of the tyre must be in the same plane as centreline of wheel.

DYNAMO: Another word for generator.

DYNAMOMETER: Machine used to measure engine power output. Engine dynamometer measures power at crankshaft and chassis dynamometer measures power output at wheels.

E

EARTH (Electrical): Term for ground.

EARTH WIRE: Term for ground wire.

ECCENTRIC (Off Centre): Two circles, one within the other, neither sharing the same centre. A protrusion on a shaft that rubs against or is connected to another part.

ECONOMISER VALVE: Fuel flow control device within carburettor.

ELECTRIC ASSIST CHOKE: A choke utilising an electric heating unit to speed up its opening time.

ELECTRIC FUEL PUMP: Fuel pump operated by electricity and electric motor. Normally mounted in or near fuel tank.

ELECTROCHEMICAL: Chemical (battery) production of electricity.

ELECTRODE (Spark Plug): Centre rod passing through insulator forms one electrode. The rod welded to shell forms another. They are referred to as centre and side electrodes.

ELECTRODE (Welding): Metal rod used in arc welding.

ELECTROLYTE: Sulphuric acid and water solution in battery.

ELECTROMAGNET: Magnet produced by placing coil of wire around steel or iron bar. When current flows through coil, bar becomes magnetised and will remain so as long as current continues to flow.

ELECTROMAGNETIC: Magnetic (generator) production of electricity.

ELECTRON: Negatively charged particle that makes up part of the atom.

ELECTRON THEORY: Accepted theory that electricity is flow of electrons from one area to another.

ELECTRONIC: Refers to electrical circuits or units employing transistors, magnetic amplifiers, computers, etc.

ELECTRONIC FUEL INJECTION: An electric solenoid type injector, engine sensors, computer, etc., used to control fuel spray into engine.

ELECTRONIC IGNITION: Ignition system with no conventional breaker points. Primary circuit is broken by magnetic pickup and electronic control unit.

ELECTROPLATE: Process of depositing gold, silver, chrome, nickel, etc., upon an object in special solution and then passing an electric current through solution. Object forms one terminal, special electrode the other. Direct current is used.

ELEMENT (Battery): Group of plates. Three elements for a six volt and six elements for the twelve volt battery. The elements are connected in series.

ELLIOT TYPE AXLE: Solid bar front axle on which ends span or straddle steering knuckle.

EMISSIONS: By-products of automotive engine combustion that are discharged into atmosphere. Major pollutants are oxides of nitrogen, carbon monoxide, hydrocarbons, and various particulates. Term also includes vapour (hydrocarbon) loss from fuel tank and carburettor.

END PLAY: Amount of axial (lengthwise) movement between two parts, end float.

ENERGY (Physics): Capacity for doing work.

ENGINE ADAPTER: Unit that allows a different engine to be installed in a car — and still bolt up to original transmission.

ENGINE (Auto): Device that converts heat energy into useful mechanical motion.

ENGINE DISPLACEMENT (Size): Volume of space through which head of piston moves in full length of its stroke — multiplied by number of cylinders in engine. Result is given in cubic centimetres, or cubic inches.

ENGINE MOUNTS: Pads made of metal and rubber which hold engine to frame.

ENGINE STEADY: Stabiliser, support.

EP LUBRICANT (Extreme Pressure): Lubricant compounded to withstand very heavy loads imposed on gear teeth.

ESC: Electronic Spark Control.

ETHYL PETROL: Petrol to which Ethyl fluid has been added to improve petrol's resistance to knocking. Slows down burning rate thereby creating a smooth pressure curve that will allow the petrol to be used in high compression engines.

ETHYLENE GLYCOL: Chemical solution added to cooling system to protect against freezing.

EVAPORATION: Process of a liquid turning into a vapour.

EVAPORATION CONTROL SYSTEM: Emission control system designed to prevent petrol vapours from escaping into atmosphere from tank and carburettor.

EVAPORATOR: Unit in air conditioning system used to transform refrigerant from a liquid to a gas. It is at this point that cooling takes place.

EXCITE: To pass an electric current through a unit such as field coils in generator.

EXHAUST CUTOUT: Y-shaped device placed in exhaust pipe ahead of muffler. Driver may channel exhaust through muffler or out other leg of the Y where exhaust passes out without going through the muffler.

EXHAUST GAS ANALYSER: Instrument used to check exhaust gases to determine combustion efficiency.

EXHAUST GAS RECIRCULATION: Admitting a controlled amount of exhaust gas into intake manifold during certain periods of engine operation. This lowers combustion flame temperature, thus reducing level of nitrogen oxides emission.

EXHAUST MANIFOLD: Connecting pipes between exhaust ports and exhaust pipe.

EXHAUST PIPE: Pipe connecting exhaust manifold to muffler.

EXHAUST SENSOR: A device placed in exhaust stream to measure oxygen content. It can, through an electronic control unit, be used to alter air-fuel ratios, engine timing, etc.

EXHAUST STROKE: Portion of piston's movement devoted to expelling burned gases from cylinder.

EXHAUST SYSTEM: Parts which carry engine exhaust to rear of car — exhaust manifold, pipes, muffler, and catalytic converter.

EXHAUST VALVE (Engine): Valve through which burned fuel charge passes on its way from cylinder to exhaust manifold.

EXPANSION TANK (Cooling System): A tank, connected to cooling system, into which water can enter or leave as needed during coolant heating (expansion) or cooling (contraction).

EXPANSION VALVE (Air Conditioning): Device used to reduce pressure and meter flow of refrigerant into evaporator.

F

FAHRENHEIT: Thermometer on which boiling point of water is 212 deg. and freezing point is 32 deg. above zero.

FAN: A device designed to create a moving stream of air. Generally employed for cooling purposes.

FARAD: Unit of capacitance; capacitance of condenser retaining one coulomb of charge with one volt difference of potential.

FEELER GAUGE: Thin strip of hardened steel, ground to an exact thickness, used to check clearances between parts.

FENDER: Mudguard, wing.

FENDER SKIRT: Plate designed to cover portion of rear fender wheel opening.

FERROUS METAL: Metal containing iron or steel.

F-HEAD ENGINE: Engine having one valve in the head and the other in the block.

FIBREGLASS: Mixture of glass fibres and resin that when cured (hardened) produces a very light and strong material. Used to build boats, car bodies, repair damaged areas, etc.

FIELD: Area covered or filled with a magnetic force.

FIELD COIL: Insulated wire wrapped around an iron or steel core. When current flows through wire, strong magnetic force field is built up.

FIELD FRAME: That portion of an electrical generator or motor upon which field coil is wound.

FILAMENT: Fine wire inside light bulb that heats to incandescence when current passes through it. The filament produces the light.

FILLET: Rounding joint between two parts connected at an angle.

FILTER: Device designed to remove foreign substances from air, oil, gasoline, water, etc.

FINAL DRIVE RATIO: Overall gear reduction (includes transmission, overdrive, auxiliary transmission, etc., gear ratio as well as axle ratio) at front or rear wheels.

FINISHING STONE (Hone): Fine stone used for final finishing during honing.

FIRE WALL: Metal partition between driver's compartment and engine compartment, also called bulkhead.

FIRING ORDER: Order in which cylinders must be fired — 1, 5, 3, 6, 2, 4, etc.

FIRST SPEED: Low gear, first gear, bottom gear.

FIT: Contact area between two parts.

FLARING TOOL: Tool used to form flare connections on tubing.

FLASH POINT: The point in the temperature range at which a given oil will ignite and flash into flame.

FLAT ENGINE: Engine in which cylinders are on a horizontal plane. This reduces overall height and enables them to be used in locations where vertical height is restricted. Also called horizontally opposed cylinder engine.

FLAT HEAD: Engine with all the valves in block.

FLAT SPOT: Refers to a spot experienced during an acceleration period where the engine seems to "fall on its face" for a second or so and will then begin to pull again.

FLEXIBLE BRAKE PIPE: Brake hose, flexible connector.

FLOAT: Unit in carburettor bowl that floats on top of fuel. It controls inlet needle valve to produce proper fuel level in bowl, can be of hollow metal, plastic, or cork.

FLOAT BOWL: The part of the carburettor that acts as a reservoir for petrol and in which the float is placed.

FLOAT CIRCUIT: That portion of the carburettor devoted to maintaining a constant level of fuel in carburettor. Consists of float bowl, float, inlet valve, etc.

FLOAT LEVEL: Height of fuel in carburettor float bowl. Also refers to specific float setting that will produce correct fuel level.

FLOODING: Condition where fuel mixture is overly rich or an excessive amount has reached cylinders. Starting will be difficult and sometimes impossible until condition is corrected.

FLOW METER: Sensing device which measures flow of air or liquid.

FLUID COUPLING: Unit that transfers engine torque to transmission input shaft through use of two vaned units (called a torus) operating very close together in a bath of oil.

FLUTE: Groove in cutting tool that forms a passageway for exit of chips removed during the cutting process.

FLUX (Magnetic): Lines of magnetic force moving through magnetic field.

FLUX (Soldering, Brazing): Ingredient placed on metal being soldered or brazed, to remove and prevent formation of surface oxidation which would make soldering or brazing difficult.

FLYWHEEL: Relatively large wheel that is attached to crankshaft to smooth out firing impulses. It provides inertia to keep crankshaft turning smoothly during periods when no power is being applied. It also forms a base for starter ring gear and in many instances, for clutch assembly.

FLYWHEEL RING GEAR: Gear on outer circumference of flywheel. Starter drive gear engages ring gear and cranks engine.

FOG LIGHTS: Amber or clear lamps specially designed to provide better visibility in fog. Are usually mounted as close to road as is feasible.

FOOT POUND: Measurement of work involved in lifting one pound one foot.

FOOT POUND (Torque): One pound pull one foot from centre of an object.

FORCE: Pressure (pull, push, etc.) acting upon body that tends to change state of motion, or rest, of the body.

FORCE-FIT: Same as drive-fit.

FORGE: To force piece of hot metal into desired shape by hammering.

FOUR BANGER, SIX BANGER, ETC.: Four cylinder, six cylinder engine, etc.

FOUR-ON-THE-FLOOR: Four-speed manual transmission with floor mounted shift.

FOUR-STROKE CYCLE ENGINE: Engine requiring two complete revolutions of crankshaft to fire each piston once.

FOUR-WHEEL DRIVE: Vehicle, such as Jeep, in which front wheels, as well as rear, may be driven.

FRAME: Portion of automobile upon which body rests and to which engine and springs are attached. Generally constructed of steel channels, also called chassis.

FRAME (Conventional): Strong, steel members run from front to rear of body.

FRAME (Integral): Car body serves as portion or all of frame.

FRAME RAILS: Structural sections of the car frame. Often specifically used to refer to two outside longitudinal sections.

FREE ELECTRONS: Electrons in outer orbits around nucleus of atom. They can be moved out of orbit comparatively easily.

FREEWHEEL: Usually refers to action of car on downgrade when overdrive over-running clutch is slipping with resultant loss of engine braking. This condition will only occur after overdrive unit is engaged but before balk ring has activated planetary gearset.

FREEZING: When two parts that are rubbing together heat up and force lubricant out of area, they will gall and finally freeze or stick together.

FREON-12: Gas used as cooling medium in air conditioning and refrigeration systems.

FREQUENCY: The rate of change in direction, oscillation, cycles, etc., in a given time span.

FRICTION: Resistance to movement between any two objects when placed in contact with each other. Friction is not constant but depends on type of surface, and normal pressure holding two objects together, etc.

FRICTION BEARING: Bearing made of babbitt, bronze, etc. There are no moving parts and shaft that rests in bearing merely rubs against friction material in bearing.

FUEL: Combustible substance that is burned within (internal) or without (external) an engine so as to impart motion to pistons, vanes, etc.

FUEL ACCUMULATOR: Spring loaded diaphragm device which dampens fuel pressure pulsations, muffles noise, and helps maintain residual pressure with engine off.

FUEL BOWL: Storage area in carburettor for extra fuel.

FUEL DISTRIBUTOR: Device which meters fuel to injectors at correct rate of flow for engine conditions.

FUEL FILTER: A device that removes dirt, rust particles, and in some cases, water from fuel before it moves into carburettor or fuel injection system.

FUEL GAUGE: A device to indicate the approximate amount of fuel in tank.

FUEL INJECTION: Fuel system that uses no carburettor but sprays fuel either directly into cylinders or into intake manifold just outside of the cylinders.

FUEL LINE: That portion of the fuel system, consisting of tubing and hose, that carries fuel from tank to carburettor or injection system.

FUEL MIXTURE: Mixture of petrol and air. An average mixture, by weight, would contain 15 parts of air to one part of petrol.

FUEL PULSATION: Fuel pressure variations due to fuel pump action.

FUEL PUMP: Vacuum device, operated either mechanically or electrically, that is used to draw petrol from tank and force it into carburettor.

FUEL TANK: A large tank of steel or plastic, used to store a supply of fuel aboard vehicle.

FULCRUM: Support on which a lever pivots in raising an object.

FULL-FLOATING AXLE: Rear drive axle that does not hold wheel on nor does it hold wheel in line or support any weight. It merely drives wheel. Used primarily on trucks.

FULL-FLOW OIL FILTER: Oil filter that filters ALL of oil passing through engine — before it reaches the bearings.

FULL PRESSURE SYSTEM: Type of oiling or lubrication system using an oil pump to draw oil out of a sump and force it through passages in engine.

FULL-TIME FOUR-WHEEL DRIVE: Setup in which all four wheels are driven — all the time — off road or on. Addition of a third differential, located at transfer case, permits front and rear wheels to operate at different speeds.

FULL-TIME TRANSFER CASE: Four-wheel drive transfer case that drives all four wheels all the time. Two-wheel drive is not possible. Such systems permit four-wheel drive on dry, hard surfaced roads by incorporating a differential in transfer case unit.

FUSE: Protective device that will break flow of current when current draw exceeds capacity of fuse.

FUSIBLE LINK: A special wire inserted into a circuit to provide protection in event of overloading, shorts, etc. Overloads will melt wire and break circuit. Unlike a regular fuse, fusible link will permit overloading for a short time before melting.

FUSION: Two metals reaching the melting point and flowing or welding themselves together.

G

GALVANOMETER: Instrument used to measure pressure, amount of, and direction of an electric current.

GAS: A non-solid material. It can be compressed. When heated, it will expand and when cooled, it will contract. (Such as air.)

GAS BURNER or GASSER: Competition car with engine set up to operate on standard pump petrol instead of an alcohol, nitro, etc., mixture.

GASKET: Material placed between two parts to ensure proper sealing.

GASSING: Small hydrogen bubbles rising to top of battery electrolyte during battery charging.

GAWR: See Gross Axle Weight Rating.

GEAR: Circular object, usually flat edged or cone-shaped, upon which a series of teeth have been cut. These are meshed with teeth of another gear and when one turns, it also drives the other.

GEAR RATIO: Relationship between number of turns made by driving gear to complete one full turn of driven gear. If driving gear turns four times to turn driven gear once, gear ratio would be 4 to 1.

GEAR SHIFT LEVER: Change speed lever, gear selector lever, gear change lever.

GENERATOR or DYNAMO: Electromagnetic device for producing DC electricity.

GENERATOR REGULATOR: See Regulator.

GLASS: Term used for the material "fibre-glass."

GLASS LIFT CHANNEL: Window lift channel.

GLASS PACK MUFFLER: Straight through (no baffles) muffler utilising fibreglass packing around perforated pipe to deaden exhaust sound.

GLASS REGULATOR: Window lift regulator, window winder.

GLASS RUN CHANNEL: Glass channel glass runner, bailey channel.

GLAZE: Highly smooth, glassy finish on cylinder walls.

GLAZE BREAKER or DEGLAZER: Abrasive tool used to remove glaze from cylinder walls prior to installation of new piston rings.

GLOW PLUG: A heating device placed in a diesel engine precombustion chamber to facilitate cold engine starting. When engine is cold, an electric current is passed through plug causing it to glow red hot. This helps ignite compressed fuel.

GOVERNOR: Device designed to automatically control speed or position of some part.

GPM: Gallons Per Minute.

GRADIENT: Angle of hill. A 20 percent gradient would be a hill that would rise two feet for every ten forward feet of travel. This is determined by rise as a percentage of forward travel.

GRID: Lead screen or plate to which battery plate active material is affixed.

GRIND: To remove metal from an object by means of revolving abrasive wheel, disc or belt.

GROSS AXLE WEIGHT RATING: Total load carrying capacity of a given axle (front or back) setup. Weight rating can be expressed as rating at springs (total load on springs) or at ground (total load measured where tyre meets ground). Weight at ground rating includes weight of tyres, wheels, axle, and springs.

GROUND (Battery): Terminal of battery connected to metal framework of car. In this country, NEGATIVE terminal is grounded.

GROWLER: Instrument used in testing starter and generator armature.

GUDGEON PIN: Term for piston pin, wrist pin.

GUM (Fuel system): Oxidised portions of fuel that form deposits in fuel system or engine parts.

GUSSET: A metal piece, usually of flat plate, used to strengthen a joint between two parts such as frame rails. Usually welded or riveted into place.

GVW: Gross Vehicle Weight. Total weight of vehicle including vehicle passengers, load, etc. Used as indicator of how heavy vehicle can be loaded (GVW minus vehicle curb weight = payload).

GVWR: Gross Vehicle Weight Rating. See GVW.

H

HALF-MOON KEY: Driving key serving same purpose as regular key but it is shaped somewhat like a half circle.

HALOGEN BULB: Light bulb in which tungsten filament is surrounded by a halogen gas such as iodine, bromine, etc. Bulb glass is quartz to withstand intense heat.

HAND BRAKE: Hand operated brake which prevents vehicle movement while parked by applying rear wheel brakes or transmission brake.

HARMONIC BALANCER: See Vibration Damper.

HARNESS: Loom.

HAZARD WARNING SYSTEM: Emergency flashers.

HEADLIGHTS OR HEADLAMPS: Main driving lights used on front of vehicle.

HEADLINING: Roof lining, head cloth.

HEAD PRESSURE: See Discharge Pressure.

HEAT CROSSOVER (V-Type Engine): Passage from one exhaust manifold up, over, and under carburettor and on to other manifold. Crossover provides heat to carburettor during engine warmup.

HEAT ENGINE: Engine operated by heat energy released from burning fuel.

HEAT EXCHANGER: Device, such as radiator, either used to cool or heat by transferring heat from one object to another.

HEAT RANGE (Spark Plugs): Refers to operating temperature of given style plug. Plugs are made to operate at different temperatures depending upon thickness and length of porcelain insulator as measured from sealing ring down to tip.

HEAT RISER: Area, surrounding portion of the intake manifold, through which exhaust gases can pass to heat fuel mixture during warmup.

HEAT SINK: Device used to prevent overheating of

electrical device by absorbing heat and transferring it to atmosphere.

HEAT STOVE: Sheet metal housing around a portion of exhaust manifold. An intake pipe from housing provides hot air to carburettor air cleaner when needed. Can also mean a small shrouded depression in exhaust manifold from which hot air may be drawn to automatic choke housing.

HEAT TREATMENT (Metal): Application of controlled heat to metal object in order to alter its characteristics (toughness, hardness, etc.).

HEEL (Brake): End of brake shoe which rests against anchor pin.

HEEL (Gear Tooth): Wide end of tapered gear tooth such as found in differential gears.

HELICAL: Spiraling shape such as that made by a coil spring.

HELICAL GEAR: Gear that has teeth cut at an angle to centre line of gear.

HEMI: Engine using hemispherical-shaped (half of globe) combustion chambers.

HEMISPHERICAL COMBUSTION CHAMBER: A round, dome-shaped combustion chamber that is considered by many to be one of the finest shapes ever developed. Hemispherical-shape lends itself to use of large valves for improved breathing and suffers somewhat less heat loss than other shapes.

HERRINGBONE GEARS: Two helical gears operating together and so placed that angle of the teeth form a "V" shape.

Hg: Abbreviation for the word MERCURY.

HIGH COMPRESSION HEADS: Cylinder head with smaller combustion chamber area thereby raising compression. Head can be custom built or can be a stock head milled (cut) down.

HIGH-RISE MANIFOLD: Intake manifold designed to mount carburettor or carburettors, considerably higher above engine than is done in standard manifold, done to improve angle at which fuel is delivered.

HIGH TENSION: High voltage from ignition coil. May also indicate secondary wire from the coil to distributor and wires from distributor to plugs.

HONE: To remove metal with fine grit abrasive stone to precise tolerances.

HORIZONTAL-OPPOSED ENGINE: Engine possessing two banks of cylinders that are placed flat or 180 deg. apart.

HORN PUSH: Horn switch, horn button.

HORSEPOWER (Brake): See Brake Horsepower.

HOSE CLAMPS: Devices used to secure hoses to their fittings.

HOSES: Flexible rubber tubes for carrying water, oil, air, and other fluids.

HOTCHKISS DRIVE: Method of connecting transmission output shaft to differential pinion by using open drive shafts. Driving force of rear wheels is transmitted to frame through rear springs or through link arms connecting rear axle housing to frame.

HOT SPOT: Localised area in which temperature is considerably higher than surrounding area.

HOT WIRE: Wiring around ignition switch so as to start car without key.

Wire connected to battery or to some part of electrical system in which a direct connection to battery is present. A current-carrying wire.

HUB (Wheel): Unit to which wheel is bolted.

HYATT ROLLER BEARING (Antifriction): Similar to conventional roller bearing except that rollers are hollow and are split in a spiral fashion from end to end.

HYDRAULIC: Refers to fluids in motion. Hydraulics is science of fluid in motion.

HYDRAULIC BRAKES: Brakes operated by hydraulic pressure. Master cylinder provides operating pressure transmitted via steel tubing to wheel cylinders or pistons that in turn apply brake shoes to brake drums and/or discs.

HYDRAULIC LIFTER: Valve lifter that utilises hydraulic pressure from engine's oiling system to keep it in constant contact with both camshaft and valve stem. They automatically adjust to any variation in valve stem length.

HYDRAULICS: The science of liquid in motion.

HYDROCARBON: A mixture of hydrogen and carbon.

HYDROCARBON — UNBURNED: Hydrocarbons that were not burned during the normal engine combustion process. Unburned hydrocarbons make up about 0.1 percent of engine exhaust emission.

HYDROCARBONS: Combination of hydrogen and carbon atoms. All petroleum based fuels (petrol, kerosene, etc.) consist of hydrocarbons.

HYDROMETER: Float device for determining specific gravity of electrolyte in a battery. This will determine the state of charge.

HYDROPNEUMATIC SUSPENSION: Suspension system using both a liquid (oil) and compressed air for springing.

HYGROSCOPIC: Ability to absorb moisture from air.

HYPOID GEARING: System of gearing wherein pinion gear meshes with ring gear below centre line of ring gear. This allows a somewhat lower drive line thus reducing hump in the floor of car. For this reason hypoid gearing is used in differential on many cars.

I

ICING: Formation of ice (under certain atmospheric conditions) on throttle plate, air horn walls, etc., caused by lowering of fuel mixture temperature as it passes through air horn.

ID: Inside diameter.

IDLE: Indicates engine operating at its normal slow speed with throttle closed.

IDLE JET: Slow running jet, low speed jet.

IDLER ARM: Steering system part that supports one end of centre link.

IDLER GEAR: A gear, between two other gears, that is driven by one and drives other. This permits both driving and driven gear to rotate in same direction with no change in gear ratio.

IDLING MIXTURE SCREW: Volume control screw, mixture control screw, mixture timing screw.

IGNITION: Lighting or igniting fuel charge by means of a spark (petrol engine) or by heat of compression (diesel engine).

IGNITION SWITCH: Key operated switch in driver compartment for connecting and disconnecting power to ignition and electrical system.

IGNITION SYSTEM: Portion of car electrical system, designed to produce a spark within cylinders to ignite fuel charge. Consists basically of battery, key switch, resistor, coil, distributor, points or electronic switching, condenser, spark plugs and necessary wiring.

IGNITION TIMING: Refers to relationship between exact time a plug is fired and position of piston in degrees of crankshaft rotation.

I-HEAD ENGINE: Engine having both valves in the head.

IMPACT WRENCH: An air, or electrical driven wrench that tightens or loosens nuts, cap screws, etc., with series of sharp, rapid blows.

IMPELLER: Wheel-like device upon which fins are attached. It is whirled to pump water, move and slightly compress air, etc.

INCLUDED ANGLE (Steering): Angle formed by centre lines drawn through steering axis (kingpin inclination) and centre of wheel (camber angle) as viewed from front of car. Combines both steering axis and camber angles.

INDEPENDENT SUSPENSION: A suspension system that allows each wheel to move up and down without undue influence on other wheels.

INDICATED POWER (ip): Measure of power developed by burning fuel within cylinders.

INDIRECT FUEL INJECTION: Fuel is sprayed into intake manifold.

INDUCTION: Imparting of electricity into one object, not connected, to another by the influence of magnetic fields.

INERTIA: Force which tends to keep stationary object from being moved, and tends to keep moving objects in motion.

INERTIA SWITCH: An electrical switch designed to be operated by a sudden movement, such as that caused by a collision.

INHIBITOR: Substance added to oil, water, gas, etc., to prevent action such as foaming, rusting, etc.

INJECTOR: Refers to valve mechanism (used in fuel injection system) that squirts or injects measured amount of petrol into intake manifold in vicinity of intake valve, In diesel engine, fuel is injected directly into combustion chamber.

INJECTOR TIMING (Diesel): Relationship between instant of fuel injection in any one cylinder, to position of piston in degrees of crankshaft rotation.

IN-LINE ENGINE: Engine in which all cylinders are arranged in straight row.

INPUT SHAFT: Clutch shaft, first motion shaft, main drive shaft, primary shaft, main drive gear, spigot shaft.

INSERT BEARING: Removable, precision made bearing which ensures specified clearance between bearing and shaft.

INSTRUMENT LAMP: Panel light.

INSULATION: Material used to reduce transfer of noise (sound insulation), heat (heat insulation), electricity (electrical insulation).

INSULATOR (Electrical): Material that will not (readily) conduct electricity.

INTAKE MANIFOLD: Connecting tubes between base of carburettor and port openings to intake valves.

INTAKE STROKE: Portion of piston's movement devoted to drawing fuel mixture into engine cylinder.

INTAKE VALVE (Engine): Valve through which fuel mixture is admitted to cylinder.

INTEGRAL: Part of. (The cam lobe is an integral part of camshaft.)

INTERMEDIATE GEAR: Any gear in auto transmission between 1st and high.

INTERMITTENT: Not constant but occurring at intervals.

INTERMITTENT FUEL INJECTION: Fuel system in which injectors are only open for short periods and remain closed rest of the time. Amount of fuel delivered is controlled by how long injector is held open.

INTERNAL COMBUSTION ENGINE: Engine that burns fuel within itself as means of developing power.

INTERNAL GEAR: A gear with teeth cut on an inward facing surface. Example: Outer gear in a planetary gearset. Teeth face inward towards centre.

ION: Electrically charged atom or molecule produced by electrical field, high temperature, etc.

IONISE (Air): To convert wholly or partly into ions. This causes air to become a conductor of electricity.

J

JACKSHAFT: A shaft used between two other shafts.

JET: Small hole or orifice used to control flow of

petrol in various parts of carburettor.

JOULE: Metre-kilogram-second (mks) unit of energy or work equal to a force of 1 newton applied through a distance of 1 metre.

JOURNAL: Part of shaft prepared to accept a bearing. (Connecting rod, main bearing.)

K

KEY: Parallel-sided piece inserted into groove cut part way into each of two parts, which prevents slippage between two parts.

KEYWAY: Slot cut in shaft, pulley hub, wheel hub, etc. Square key is placed in slot and engages a similar keyway in mating piece. Key prevents slippage between two parts.

KICKDOWN SWITCH: Electrical switch that will cause transmission, or overdrive unit, to shift down to lower gear. Often used to secure fast acceleration.

KILL SWITCH: Special switch designed to shut off ignition in case of emergency.

KILOMETRE: Metric measurement equivalent to 1,000 metres.

KINGPIN: Hardened steel pin that is passed through the steering knuckle and axle end. The steering knuckle pivots about the kingpin.

KINGPIN or STEERING AXIS INCLINATION: Tipping the tops of the kingpins inward towards each other. This places the centreline of steering axis nearer centreline of tyre-road contact area.

KNOCKING (Bearing): Noise created by part movement in a loose or worn bearing.

KNOCKING (Fuel): Condition, accompanied by audible noise, that occurs when petrol in cylinders burns too quickly. Also referred to as detonation.

KNUCKLE: A part utilising a hinge pin (kingpin, swivel pin) that allows one part to swivel around another part. An example is a steering knuckle.

KNURL: To roughen surface of piece of metal by pressing series of cross-hatched lines into the surface and thereby raising area between these lines.

L

LACQUER (Paint): Fast drying automotive body paint.

LAMINATED: Something made up of many layers.

LAMP: Light.

LAND: Metal separating a series of grooves.

LANDS (Ring): Piston metal between ring grooves.

LAP: One complete trip around race track or route laid out for racing.

LAP or LAPPING: To fit two surfaces together by coating them with abrasive and then rubbing them together.

LATENT HEAT: Amount of heat (joules/kg) beyond boiling or melting point, required to change

liquid to a gas, or a solid to a liquid.

LATENT HEAT OF EVAPORATION: Amount of heat (joules/kg) required to change a liquid to a vapour state without elevating vapour temperature above that of the liquid.

LAYGEAR: Cluster gear, second motion gear, intermediate gear.

LAYSHAFT: Cluster gear shaft, second motion shaft, intermediate shaft, counter shaft.

LEAD BURNING: Connecting two pieces of lead by melting edges together.

LEADED PETROL: Petrol containing tetraethyl lead, an antiknock additive. It must not be used in vehicles with catalytic converters.

LEAF SPRING: Suspension spring made up of several pieces of flat spring steel. Varying numbers of leaves (individual pieces) are used depending on intended use. Some vehicles use single leaf in each rear spring.

LEAN MIXTURE (Fuel): A fuel mixture with an excessive amount of air in relation to fuel.

LENS: Glass, crystal.

LETTER DRILLS: Series of drills in which each drill size is designated by letter of alphabet — A, B, C, etc.

LEVER: A rigid bar or shaft pivoting about a fixed fulcrum (shaft, pin, etc.). It is used to increase force or to transmit or change motion.

LEVERAGE: Increasing force by utilising one or more levers.

L-HEAD ENGINE or SIDE VALVE: Engine having both valves in block and on same side of cylinder.

LIGHTENED VALVES: Valves in which all possible metal has been ground away to reduce weight. This will allow higher rpm without valve float.

LIMITED-SLIP DIFFERENTIAL: Differential unit designed to provide superior traction by transferring driving torque, when one wheel is spinning, to wheel that is not slipping.

LININGS: Shoe facings, friction facings.

LINKAGE: Movable bars or links connecting one unit to another.

LIQUID LINE (Air Conditioning): High-pressure liquid refrigerant line between receiver-dehydrator and expansion valve.

LIQUID TRACTION: Special liquid applied to tyres of drag racers to provide superior traction.

LIQUID-VAPOUR SEPARATOR: Tank which prevents liquid fuel from entering vapour line.

LIQUID WITHDRAWAL (LPG): Drawing LPG from bottom of tank to ensure delivery of liquid LPG. Withdrawal from top of tank will deliver LPG in the gaseous state.

LITRE: Metric measurement of capacity — equivalent to 1,000 millilitres.

LIVE AXLE: Axle upon which wheels are firmly affixed. Axle drives the wheels.

LIVE WIRE: See Hot Wire.

LOAD RANGE (Tyre): Letter system (A, B, C, etc.) used to indicate specific tyre load and inflation limit.

LONG and SHORT ARM SUSPENSION: Suspension system utilising upper and lower control arm. Upper arm is shorter than lower. This is done so as to allow wheel to deflect in a vertical direction with a minimum change in camber.

LONGITUDINAL LEAF SPRING: Leaf spring mounted so it is parallel to length of car.

LOUVRE: Ventilation slots such as sometimes found in hood of automobile.

LOW BRAKE PEDAL: Condition where brake pedal approaches too close to floorboard before actuating the brakes.

LOW LEAD FUEL: Petrol containing not much more than 0.13 grams of tetraethyl lead per litre.

LOW PIVOT SWING AXLE: Rear axle setup that attaches differential housing to frame via a pivot mount. Conventional type of housing and axle extend from differential to one wheel. The other side of differential is connected to other driving wheel by a housing and axle that is pivoted at a point in line with differential to frame pivot point.

LOW PRESSURE LINE (Air Conditioning): Low pressure refrigerant line between evaporator outlet and compressor.

LPG: Liquefied petroleum gas.

LUBRICANT: Any material, usually of a petroleum nature such as grease, oil, etc., that is placed between two moving parts in an effort to reduce friction.

LUBRICATION: Reducing friction between two parts by coating them with oil, grease, etc.

LUG (Engine): To cause engine to labour by failing to shift to a lower gear when necessary.

LUGGAGE COMPARTMENT: Boot.

M

MACPHERSON STRUT: Front end suspension system in which wheel assembly is attached to a long, telescopic strut. Strut permits wheels to pivot and move up and down. Strut also acts as a shock absorber.

MAG: Magneto.

MAGNAFLUX: Special chemical process used to check parts for cracks.

MAGNET (Permanent): Piece of magnetised steel that will attract all ferrous material. Permanent magnet does not need electricity to function and will retain its magnetism over a period of years.

MAGNETIC CLUTCH (Air Conditioning): Electromagnetic clutch that engages or disengages air conditioning compressor pulley.

MAGNETIC FIELD: Area encompassed by magnetic lines of force surrounding either a bar magnet or electromagnet.

MAGNETO: Engine driven unit that generates high voltage to fire spark plugs. It needs no outside source of power such as battery.

MAGS or MAG WHEEL: Lightweight, sporty wheels made of magnesium. Term mag is often applied to aluminium, and aluminium and steel combination wheels.

MAIN BEARINGS: (Engine): Bearings supporting crankshaft in cylinder block.

MAIN BEARING SUPPORTS: Steel plate installed over main bearing caps to increase their strength for racing purposes.

MAIN DISCHARGE TUBE: Carburettor fuel passage from bowl to air horn.

MAIN JET: Primary jet.

MAINSHAFT: Third motion shaft, output shaft, secondary shaft.

MANDREL: Round shaft used to mount stone, cutter, saw, etc.

MANIFOLD: Pipe or number of pipes connecting series of holes or outlets to common opening. See Exhaust and Intake Manifold.

MANIFOLD HEAT CONTROL VALVE: Valve placed in exhaust manifold, or in exhaust pipe, that deflects certain amount of hot gas around base of carburettor to aid in warmup.

MANOMETER: Instrument to measure pressure usually (vacuum).

MANUAL CONTROL VALVE (Transmission): Hand (linkage) operated valve which controls oil flow and transmission gear selection.

MASTER CYLINDER: Part of hydraulic brake system in which pressure is generated.

MASTER ROD (Radial Engine): Primary or main connecting rod to which other connecting rods are attached.

MATTER: Substance making up physical things occupying space, having weight, and perceptible to the senses.

MECHANICAL BRAKES: Service brakes that are actuated by mechanical linkage connecting brakes to brake pedal.

MECHANICAL EFFICIENCY: Engine's rating as to how much potential power is wasted through friction within moving parts of engine.

MECHANICAL FUEL INJECTION: A mechanically driven pump forces fuel into engine.

MECHANICAL FUEL PUMP: Engine mounted pump operated by eccentric device.

MEDALLION: Name plate, cover, motif, emblem.

MEGOHM: 1,000,000 ohms.

MESH: To engage teeth of one gear with those of another.

METAL FATIGUE: Crystallising of metal due to vibration, twisting, bending, etc. Unit will eventually break. Bending a piece of wire back and forth to break it is a good example of metal fatigue.

METERING ROD: Movable rod used to vary

opening area through carburettor jet.

METRIC SIZE: Units made to metric system measurements.

METRIC SYSTEM: A decimal system of measurement based on metre (area, length, volume), litre (capacity), and gram (weight and mass).

MICRO: When the word "micro" precedes measurement units, such as watt, ampere, etc., it means one-millionth of that unit.

MICROFARAD: 1/1,000,000 farad.

MICROMETER (Inside and Outside): Precision measuring tool that will give readings in hundredths of a millimetre.

MIKE: Either refers to micrometer or to using micrometer to measure an object.

MILL: To remove metal through use of rotating toothed cutter.

MILLIMETRE: Metric measurement equivalent to 1/1,000 of a metre.

MILLING MACHINE: Machine that uses variety of rotating cutter wheels to cut splines, gears, keyways, etc.

MISFIRE: Fuel charge in one or more engine cylinder which fails to fire or ignite at proper time.

MODULATOR (Transmission): Pressure control or adjusting valve used in hydraulic system of automatic transmission.

MOLECULE: Smallest portion that matter may be divided into and still retain all properties of original matter.

MONOBLOCK: All cylinders cast as one unit.

MOTOR: Electrically driven power unit (electric motor). Term is often incorrectly applied to internal combustion engine.

MOTOR (Generator): Attaching generator to battery in such a way it revolves like an electric motor.

MOULD: Hollow unit into which molten metal is poured to form a casting.

MOULDING: Chrome strip, finishing strip.

MPH: Miles per hour.

MUFFLER (Air Conditioning): Device which reduces pumping noise and vibration in system.

MUFFLER (Exhaust): Unit through which exhaust gases are passed to quiet sounds of running engine.

MULTIPLE DISC CLUTCH: Clutch utilising several clutch discs in its construction.

MULTI-VISCOSITY OILS: Oils meeting SAE requirements for both low temperature requirements of light oil and high temperature requirements of heavy oil. Example: (SAE 10W-30).

MUTUAL INDUCTION: Creating voltage in one coil by altering current in another coil nearby.

O

NC THREADS: National Coarse thread sizes.

NEEDLE BEARING (Antifriction): Roller type bearing in which rollers have very narrow diameter in relation to their length.

NEEDLE VALVE: Valve with long, thin, tapered point that operates in small hole or jet. Hole size is changed by moving needle valve in or out.

NEGATIVE TERMINAL: Terminal (such as on battery) from which current flows on its path to positive terminal.

NEUTRON: Neutral charge particle forming part of an atom.

NEWTON'S THIRD LAW: For every action there is an equal, and opposite reaction.

NF THREADS: National Fine thread sizes.

NITROGEN OXIDES (NOx): In combustion process, nitrogen from air combines with oxygen to form nitrogen oxides.

NONFERROUS METALS: All metals containing no iron — except in very minute quantities.

NON-SERVICEABLE: A part or device whose design and construction does not permit rebuilding. Can also be used to indicate a part of the device that is no longer fit for use.

NORTH POLE (Magnet): Magnetic pole from which lines of force emanate; travel is from north to south pole.

NOZZLE: Opening through which fuel mixture is directed into carburettor air stream.

NUMBER DRILLS: Series of drills in which each size is designated by number (0-80).

NUMBER PLATE LAMP: Licence plate light.

N

OCTANE RATING: Rating that indicates a specific petrol's ability to resist detonation.

OD: Outside diameter.

ODOMETER: Device used to measure and register number of kilometres or miles travelled by car.

OFF-ROAD VEHICLE: Vehicle designed to operate in rough country (hills, sand, mud, etc.) without benefit of regular roads.

OHM: Unit of measurement used to indicate amount of resistance to flow of electricity in a given circuit.

OHMMETER: Instrument used to measure amount of resistance in given unit or circuit. (In ohms.)

OHM'S LAW: Formula for calculating electrical values in a circuit.

OIL BATH AIR CLEANER: Air cleaner that utilises a pool of oil to ensure removal of impurities from air entering carburettor.

OIL BURNER: Engine that consumes an excessive quantity of oil.

OIL — COMBINATION SPLASH and PRESSURE SYSTEM: Engine oiling system that uses both pressure and splash oiling to accomplish proper lubrication.

OIL CONTROL RING (Piston): A piston ring designed to prevent excessive oil consumption by scraping excess oil from cylinder and returning it to oil pan sump. Usually lower (furthest down in cylinder) ring or rings.

OIL COOLER: Device used to remove excess heat from engine and/or transmission oil. Can be air or water cooled design.

OIL DIPSTICK: See Dipstick.

OIL FILTER: Device used to strain oil in engine thus removing abrasive particles.

OIL — FULL PRESSURE SYSTEM: Engine oiling system that forces oil, under pressure, to moving parts of engine.

OIL GALLERY: Pipe or drilled passageway in engine used to carry engine oil from one area to another.

OIL GAUGE: A dash mounted device that indicates engine oil pressure in kilopascals (kPa) or pounds per square inch (psi). Can be of either electrical or Bourdon tube design.

OIL PAN: See Pan.

OIL PICKUP: Connects to oil pump and extends into bottom of oil pan. Oil is drawn through pickup into pump.

OIL PUMP: Device used to force oil, under pressure, to various parts of the engine. It is driven by gear on camshaft.

OIL PUMP (Rotor type): Inner rotor, outer rotor.

OIL PUMP (Gear type): Drive gear, driven gear.

OIL PUMPING: Condition wherein an excessive quantity of oil passes piston rings and is consumed in combustion chamber.

OIL RING: Normally bottom piston ring which scrapes excess oil off cylinder wall.

O-RING: Neoprene O-ring, oil seal.

OIL SEAL: Device used to prevent oil leakage past certain area.

OIL SLINGER: Device attached to revolving shaft so that all oil passing that point will be thrown outward where it will return to point of origin.

OIL — SPLASH SYSTEM: Engine oiling system that depends on connecting rods to dip into oil troughs and splash oil to all moving parts.

OIL STRAINER (Engine): A fine wire mesh screen through which oil entering oil pump is drawn. It will remove larger particles of dirt or other abrasives.

OIL SUMP: That portion of oil pan that holds supply of engine oil.

ONE-WAY CLUTCH: It locks shaft in one direction and allows rotation in other direction.

OPEN CIRCUIT: Circuit in which a wire is broken or disconnected.

OPEN CIRCUIT VOLTAGE (Battery): Cell voltage when battery has not completed circuit across posts and is not receiving or delivering energy.

ORIFICE: A small hole or restricted opening used to control flow of petrol, air, oil, etc.

OSCILLATING ACTION: Swinging action such as

that in pendulum of a clock.

OSCILLOSCOPE: Testing unit which projects visual reproduction of the ignition system spark action onto screen of cathode-ray tube.

OTTO CYCLE: Four-stroke cycle consisting of intake, compression, firing, and exhaust strokes.

OUTPUT SHAFT: Shaft delivering power from within mechanism. Shaft leaving transmission, attached to propeller shaft, is transmission output shaft.

OVERDRIVE: Unit utilising planetary gearset so actuated as to turn drive shaft about one-third faster than transmission output shaft.

OVERDRIVE STAGES: Locked-out; direct, free-wheeling; and overdrive.

OVERHEAD CAMSHAFT: Camshaft mounted above the head, driven by long timing chain or belt.

OVERHEAD VALVES: Valves located in head.

OVERRUNNING CLUTCH: Clutch mechanism that will drive in one direction only. If driving torque is removed or reversed, clutch slips.

OVERRUNNING CLUTCH STARTER DRIVE: Starter drive that is mechanically engaged. When engine starts, overrunning clutch operates until drive is mechanically disengaged.

OVERSQUARE ENGINE: Engine in which bore diameter is larger than length of stroke.

OVERSTEER: Tendency for car, when negotiating a corner, to turn more sharply than driver intends.

OXIDES OF NITROGEN (NOx): Undesirable exhaust emission, especially prevalent when combustion chamber flame temperatures are high.

OXIDISE (Metal): Action where surface of object is combined with oxygen in air to produce rust, scale, etc.

OXIDISING FLAME: Welding torch flame in which an excess of oxygen exists. Free or unburned oxygen tends to burn molten metal.

OXYGEN: Gas, used in welding, made up of colourless, tasteless, odourless, gaseous element oxygen found in atmosphere.

P

PAN: Thin stamped cover bolted to the bottom of crankcase. It forms a sump for engine oil and keeps dirt, etc., from entering engine.

PANEL: Skin.

PAPER AIR CLEANER: Air cleaner that makes use of special paper through which air to carburettor is drawn.

PARABOLIC REFLECTOR: A light reflector (concave mirror) that emits parallel light rays. Bulb filament must be located at focal point of parabola.

PARALLEL CIRCUIT: Electrical circuit with two or more resistance units so wired as to permit current to flow through both units at same time. Unlike series circuit, current in parallel circuit does not have to pass through one unit to reach the other.

PARALLELOGRAM STEERING LINKAGE: Steering system utilising two short tie rods connected to steering arms and to a long centre link. The link is supported on one end on an idler arm and the other end is attached directly to pitman arm. Arrangement forms a parallelogram shape.

PARKING BRAKE: Hand operated brake which prevents vehicle movement while parked by locking rear wheels, or transmission output shaft.

PARKING LIGHTS: Small lights on both front and rear of vehicle. Usually red colour at rear and white at front. Used so that vehicle will be more visible during dark hours.

PARTICULATES (Lead): Tiny particles of lead found in engine exhaust emissions when leaded fuel is used.

PART-TIME TRANSFER CASE: Four-wheel drive transfer case that permits either four-wheel or two-wheel drive.

PASCAL'S LAW: "When pressure is exerted on confined liquid, it is transmitted undiminished."

PAWL: Stud or pin that can be moved or pivoted into engagement with teeth cut on another part — such as parking pawl on automatic transmission that can be slid into contact with teeth on another part to lock rear wheels.

PAYLOAD: Amount of weight that may be carried by vehicle. Computed by subtracting vehicle curb weight from GVW.

PEEN: To flatten out end of a rivet, etc., by pounding with round end of a hammer.

PENETRATING OIL: Special oil used to free rusted parts so they can be removed.

PERIPHERY: Outside edge or circumference.

PERMANENT MAGNET: Magnet capable of retaining its magnetic properties over very long period of time.

PETROL: Hydrocarbon fuel used in the internal combustion engine.

PETROLEUM: Raw material from which petrol, kerosene, lube oils, etc., are made. Consists of hydrogen and carbon.

PHILLIPS HEAD SCREW: Screw having a fairly deep cross slot instead of single slot as used in conventional screws.

PHOSPHOR-BRONZE: Bearing material composed of tin, lead, and copper.

PHOTOCHEMICAL: Relates to branch of chemistry where radiant energy (sunlight) produces various chemical changes.

PHOTOCHEMICAL SMOG: Fog-like condition produced by sunlight acting upon hydrocarbon and carbon monoxide exhaust emissions in atmosphere.

PICKUP COIL: Device in electronic type distributor which senses engine speed (distributor rotation) and sends electrical pulses to control unit.

PIEZOELECTRIC IGNITION: System of ignition that employs use of small section of ceramic-like material. When this material is compressed, even a very tiny amount, it emits a high voltage that will fire plugs. This system does not need a coil, points, or condenser.

PILOT SHAFT: Dummy shaft that is placed in a mechanism as a means of aligning parts. It is then removed and regular shaft installed.

PINGING: Metallic rattling sound produced by the engine during heavy acceleration when ignition timing is too far advanced for grade of fuel being burned.

PINION (Gear): Small gear either driven by or driving a larger gear.

PINION CARRIER: Part of rear axle assembly that supports and contains pinion gear shaft.

PINION FLANGE: Drive coupling.

PIPES: Exhaust system pipes.

PISTON: Round plug, open at one end, that slides up and down in cylinder. It is attached to connecting rod and when fuel charge is fired, will transfer force of explosion to connecting rod then to crankshaft.

PISTON BOSS: Built-up area around piston pin hole.

PISTON COLLAPSE: Reduction in diameter of piston skirt caused by heat and constant impact stresses.

PISTON DISPLACEMENT: Amount (volume) of air displaced by piston when moved through full length of its stroke.

PISTON EXPANSION: Increase in diameter of piston due to normal piston heating.

PISTON HEAD: Portion of piston above top ring.

PISTON LANDS: Portion of piston between ring grooves.

PISTON PIN or GUDGEON PIN or WRIST PIN: Steel pin that is passed through piston. Used as base upon which to fasten upper end of connecting rod. It is round and is usually hollow.

PISTON RING: Split ring installed in a groove in piston. Ring contacts sides of ring groove and also rubs against cylinder wall thus sealing space between piston and wall.

PISTON RING (Compression): Ring designed to seal burning fuel charge above piston. Generally there are two compression rings per piston and they are located in two top ring grooves.

PISTON RING (Oil Control): Piston ring designed to scrape oil from cylinder wall. Ring is of such design as to allow oil to pass through ring and then through holes or slots in groove. In this way oil is returned to pan. There are many shapes and special designs used on oil control rings.

PISTON RING END GAP: Distance left between ends of the ring when installed in cylinder.

PISTON RING EXPANDER: See Ring Expander.

PISTON RING GROOVE: Slots or grooves cut in piston head to receive piston rings.

PISTON RING SIDE CLEARANCE: Space between sides of ring and ring lands.

PISTON SKIRT: Portion of piston below rings. (Some engines have an oil ring in skirt area.)

PISTON SKIRT EXPANDER: Spring device placed inside piston skirt to produce an outward pressure which increases diameter of skirt.

PISTON SKIRT EXPANDING: Enlarging diameter of piston skirt by inserting an expander, by knurling outer skirt surface, or by peening inside of piston.

PITMAN ARM: Short lever arm splined to steering gear cross shaft. Pitman arm transmits steering force from cross shaft to steering linkage system.

PITS: Area at a race track for fuelling, tyre changing, making mechanical repairs, etc.

PIT STOP: A stop at the pits by racer, for fuel, tyres, repairs, etc.

PIVOT: Pin or shaft about which a part moves.

PLANET CARRIER: Part of a planetary gearset upon which planet gears are affixed. Planet gears are free to turn on hardened pins set into carrier.

PLANET GEARS: Gears in planetary gearset that are in mesh with both ring and sun gear. Referred to as planet gears in that they orbit or move around central or sun gear.

PLANETARY GEARSET: Gearing unit consisting of ring gear with internal teeth, sun or central pinion gear with external teeth, and series of planet gears that are meshed with both the ring and the sun gear.

PLATES (Battery): Thin section of lead peroxide or porous lead. There are two kinds of plates — positive and negative. The plates are arranged in groups, in an alternate fashion, called elements. They are completely submerged in the electrolyte.

PLATINUM: Precious metal sometimes used in the construction of breaker points. It conducts well and is highly resistant to burning.

PLAY: Movement between two parts.

PLEXIGLAS: Trade name for an acrylic plastic, made by the Rhom and Haas Co.

PLIES (Tyre): Layers of rubber impregnated fabric that make up carcass or body of tyre.

PLUG GAPPING: Adjusting side electrode on spark plug to provide proper air gap between it and the centre electrode.

PLY RATING (Tyres): Indication of tyre strength (load carrying capacity). Does not necessarily indicate actual number of plies. Two-ply four-ply rating tyre would have load capacity of a four-ply tyre of same size but would have only two actual plies.

P-METRIC (Tyre): Tyre size designation based on international standards. Example: P 155/80R13. P = passenger car use. 155 = section width in millimetres. 80 = height to width ratio. R = radial construction. 13 = wheel rim diameter in inches.

POLARISING (Generator): Process of sending quick surge of current through field windings of generator in direction that will cause pole shoes to assume correct polarity. This will ensure that the generator will cause current to flow in same direction as normal.

POLARITY (Battery Terminals): Indicates if the battery terminal (either one) is positive or negative (plus or minus) (+ or -).

POLARITY (Generator): Indicates if pole shoes are so magnetised as to make current flow in a direction compatible with direction of flow as set by battery.

POLARITY (Magnet): Indicates if end of a magnet is north or south pole (N or S).

POLE (Magnet): One end, either north or south, of a magnet.

POLE SHOES: Metal pieces about which field coil windings are placed. When current passes through windings, pole shoes become powerful magnets. Example: pole shoes in a generator or starter motor.

PONY CAR: Small, sporty car along the lines of the Mustang, Firebird, Camaro, etc.

POPPET: A spring-loaded ball engaging one or more indentations. Used to hold one object in position in relation to another.

POPPET VALVE: Valve used to open and close valve port entrances to engine cylinders.

PORCELAIN (Spark Plug): Material used to insulate centre electrode of spark plug. It is hard and resistant to damage by heat.

POROSITY: Small air or gas pockets, or voids, in metal.

PORT: Openings in engine cylinder head for intake and exhaust flow during engine operation; opened and closed by valves.

To smooth out, align, and somewhat enlarge intake passageway to the valves.

POSITIVE CRANKCASE VENTILATION: System which prevents crankcase vapours from being discharged directly into atmosphere.

POSITIVE TERMINAL: Terminal (such as on battery), to which current flows.

POST (Battery): Round, tapered lead posts protruding above top of battery to which battery cables are attached.

POTENTIAL: An indication of amount of available energy.

POTENTIOMETER: Variable resistor with three connections. One connection (called wiper) slides along resistive unit. Can be used as a voltage divider.

POUR POINT: Lowest temperature at which fluid will flow under specified conditions.

POWER: Time rate at which energy is converted into work.

POWER (Frictional): Amount of power lost to engine friction.

POWER (Gross): Maximum power developed by engine without a fan, air cleaner, alternator, exhaust system, etc.

POWER (Net): Maximum power developed by engine equipped with fan, air conditioning, air cleaner, exhaust system, and all other systems and items normally present when engine is installed in car.

POWER BOOSTER (Brakes): Engine vacuum or power steering fluid operated device on bulkhead which increases brake pedal force on master cylinder during stops.

POWER TAKE OFF: A spot or place on transmission or transfer case from which an operating shaft from another unit (such as a winch) can be driven. Usually consists of a removable plate that exposes a drive gear.

POWER TO WEIGHT FACTOR: Relationship between total weight of car and power available. By dividing weight by power, number of kilogrammes to be moved by one kilowatt is determined. (kg/kW). This factor has a great effect on acceleration, petrol consumption and all around performance.

POWER UNIT (Auto): The vehicle engine and transmission.

POWER STEERING: Steering system utilising hydraulic pressure to increase the driver's turning effort. Pressure is utilised either in gearbox itself or in hydraulic cylinder attached to steering linkage.

POWER STEERING PUMP: Belt driven pump which produces pressure for power steering system.

POWER or FIRING STROKE: Portion of piston's movement devoted to transmitting power of burning fuel mixture to crankshaft.

PPM (Parts-Per-Million): Term used in determining extent of pollution existing in given sample of air.

PRACTICAL EFFICIENCY: Amount of power delivered to drive wheels.

PRECISION INSERT BEARING: Very accurately made replaceable type of bearing. It consists of an upper and lower shell. The shells are made of steel to which a friction type bearing material has been bonded. Connecting rod and main bearings are generally of precision insert type.

PREHEATING: Application of some heat prior to later application of more heat. Cast iron is preheated to avoid cracking when welding process is started. A coil (ignition) is preheated prior to testing.

PREHEATING (Metal): Process of raising temperature of metal to specific level before starting subsequent operations such as welding, brazing, etc.

PREIGNITION: Fuel charge being ignited before proper time.

PRELOADING: Adjusting antifriction bearing so it

is under mild pressure. This prevents bearing looseness under a driving stress.

PRESS-FIT: Condition of fit (contact) between two parts that requires pressure to force parts together. Also referred to as drive or force fit.

PRESSURE BLEEDER: Device that forces brake fluid, under pressure, into master cylinder so that by opening bleeder screws at wheel cylinders, all air will be removed from brake system.

PRESSURE CAP: Special cap for radiator. It holds a predetermined amount of pressure on water in cooling system. This enables water to run hotter without boiling.

PRESSURE DIFFERENTIAL SWITCH: Hydraulic switch in brake system that operates brake warning light in dashboard.

PRESSURE LIMITING VALVE: Brake pressure regulator, pressure reduction valve, pressure conscious reducing valve.

PRESSURE PLATE: Clutch drive plate.

PRESSURE PLATE COVER: Clutch housing, clutch cover.

PRESSURE RELIEF VALVE: Valve designed to open at specific pressure. This will prevent pressures in system from exceeding certain limits.

PRIMARY CIRCUIT (Ignition System): Low voltage (6 or 12 volt) part of ignition system.

PRIMARY CUP: Main cup, main rubber seal.

PRIMARY, FORWARD, or LEADING BRAKE SHOE: Brake shoe installed facing front of car. It will be a self-energising shoe.

PRIMARY WINDING (Coil): Low voltage (6 or 12 volt) winding in ignition coil. The primary winding is heavy wire; secondary winding uses fine wire.

PRIMARY WIRES: Wiring which serves low voltage part of ignition system. Wiring from battery to switch, resistor, coil, distributor points.

PRINTED CIRCUIT: Electrical circuit made by connecting units with electrically conductive lines printed on a panel. This eliminates actual wire and task of connecting it.

PROGRESSIVE LINKAGE: Carburettor linkage designed to open throttle valves of multiple carburettors. It opens one to start and when certain opening point is reached, it will start to open others.

PRONY BRAKE: Device utilising a friction brake to measure power output of engine.

PROPANE (LPG): Petroleum product, similar to and often mixed with butane, useful as engine fuel. May be referred to as LP-Gas.

PROPELLER SHAFT: Shaft connecting transmission output shaft to differential pinion shaft.

PROPORTIONING VALVE (Brakes): Valve in brake line which keeps rear wheels from locking up during rapid stops.

PROTON: Positive charge particle, part of atom.

PULL IT DOWN (Engine): Term often used in reference to dismantling and overhauling an engine.

PULSATION DAMPER: Device to smooth out fuel pulsations or surges from pump to carburettor.

PULSE AIR INJECTION: Emission control system which feeds air into exhaust gases by using pressure pulsations of exhaust system.

PULSE WIDTH: Often used to describe length of time a fuel injector is held open. Pulse being electric current applied to the injector winding and width being length of time current is allowed to flow. The wider the pulse, the more fuel delivered.

PUMP: A device designed to cause movement of water, fuel, air, etc., from one area to another.

PUMPING THE ACCELERATOR PEDAL: Forcing accelerator up and down in an endeavour to provide extra petrol to cylinders. This is often cause of flooding.

PURGE: Removing impurities from system. See Bleeding.

PUSH ROD: Rod that connects valve lifter to rocker arm. Used on overhead valve installations.

Q

QUADRANT (Gearshift): A gear position indicator often using a shift lever actuated pointer. Can be marked PRND21 (3-speed), PRND321 (4-speed), etc.

QUADRA-TRAC: See Full Time Four-Wheel Drive.

QUENCHED (Flame): Flame front in combustion chamber being extinguished as it contacts colder cylinder walls. This sharply elevates hydrocarbon emissions.

QUENCHING: Dipping heated object into water, oil, or other substance, to quickly reduce temperature.

QUICKSILVER: Metal mercury. Often used in thermometers.

R

RACE (Bearing): Inner or outer ring that provides a contact surface for balls or rollers in bearing.

RACE CAMSHAFT: Camshaft, other than standard, designed to improve performance by altering cam profile. Provides increased lift, faster opening and closing, earlier opening and later closing, etc.

RACING SLICK: Type of tyre used in ''drag racing'' as well as some ''stock car'' applications. Tread surface of tyre is completely smooth, for maximum rubber contact with track surface.

RACK AND PINION GEARBOX (Steering): Steering gear utilising pinion gear on end of steering shaft. Pinion engages long rack (bar with teeth along one edge). Rack is connected to steering arms via rods.

RADIAL (Direction): Line at right angles (perpendicular) to shaft, cylinder, bearing, etc., centre line.

RADIAL COMPRESSOR: A small air conditioning compressor using reciprocating pistons working at right angles to shaft and spaced around shaft in radial fashion.

RADIAL ENGINE: Engine possessing various numbers of cylinders so arranged that they form circle around crankshaft centreline.

RADIAL TYRE: Plies parallel and at right angle to tread, belts under tread area.

RADIATION: Transfer of heat from one object to another when hotter object sends out invisible rays or waves that upon striking colder object, cause it to vibrate and thus heat.

RADIATOR (Engine Cooling): A device used to remove heat from engine coolant. It consists of a series of finned passageways. As coolant moves through passages, heat is conducted to fins where it transfers to a stream of air forced through fins.

RADIATOR CAP: Pressure cap that fits on radiator neck. It keeps coolant from boiling.

RADIUS: Distance (in a straight line) from centre of a circle or circular motion, to a point on edge (circumference).

RADIUS RODS: Rods attached to axle and pivoted on frame. Used to keep axle at right angles to frame and yet permit an up and down motion.

RAIL: Dragster built around a relatively long pipe frame. Often the only body panels used are around the driver's cockpit area.

RAM AIR: Air ''scooped'' up by an opening due to vehicle forward motion.

RAM INDUCTION: Using forward momentum of car to scoop air and force it into carburettor via a suitable passageway.

RAM INTAKE MANIFOLD: Intake manifold that has very long passageways that at certain speeds aid entrance of fuel mixture into cylinders.

RATED POWER (Engine): Indication of power load that may safely be placed upon engine for prolonged periods of time. This would be somewhat less than the engine maximum power.

RATIO: Fixed relationship between things in number, quantity, or degree. For example, if fuel mixture contains one part of petrol for fifteen parts of air, ratio would be 15 to 1.

REACTOR: See Stator.

REAM: To enlarge or smooth hole by using round cutting tool with fluted edges.

REAR AXLE (Banjo Type): Rear axle housing from which differential unit may be removed while housing remains in place on car.

REAR AXLE HOUSING (Split Type): Rear axle housing made up of several pieces and bolted together. Housing must be split apart to remove differential.

REAR GLASS: Back light, rear screen, back window.

RECEIVER-DRIER: See Drier.

RECIPROCATING ACTION: Back-and-forth movement such as action of pistons.

RECIRCULATING BALL WORM AND NUT: Very popular type of steering gear. It utilises series of ball bearings that feed through and around and back through grooves in worm and nut.

RECTIFIER: Device used to change AC (alternating current) into DC (direct current).

RED LINE: Top recommended engine rpm. If a tachometer is used, it will have a mark (Red line) indicating maximum rpm.

REDUCING FLAME: Welding flame in which there is an excess of acetylene.

REDUCTION: (Gear): A gear that increases torque by reducing rpm of a driven shaft in relation to that of driving shaft.

REFRIGERANT: Liquid used in refrigeration systems to remove heat from evaporator coils and carry it to condenser.

REFRIGERANT-12: Name applied to refrigerant generally used in automotive air conditioning systems.

REFRIGERANT OIL: Special oil which lubricates air conditioning compressor.

REGULATOR (Electrical): Device used to control generator voltage and current output.

REGULATOR (Gas or Liquid): Device to reduce and control pressure.

RELATIVE HUMIDITY: Actual amount of moisture in a given sample of air compared to total amount that sample could hold (at same temperature).

RELAY: Magnetically operated switch used to make and break flow of current in circuit. Also called "cutout, and circuit breaker".

RELEASE BEARING: Throw-out bearing, thrust bearing, withdrawal bearing.

RELEASE BEARING PLATE: Throw-out bearing plate, thrust plate.

RELEASE LEVER: Clutch fork, release fork, throw-out lever, withdrawal lever.

RELIEF VALVE: Release valve.

RELIEVE: Removing, by grinding, small lip of metal between valve seat area and cylinder — and removing any other metal deemed necessary to improve flow of fuel mixture into cylinder. Porting is generally done at same time.

RELUCTOR: A component in electronic ignition system distributor. It is affixed to the distributor shaft and triggers magnetic pickup. This, in turn, triggers control unit which breaks coil primary circuit causing coil to "fire".

RESIDUAL MAGNETISM: Magnetism remaining in an object after removal of any magnetic field influence.

RESISTANCE (Electrical): Measure of conductor's ability to retard flow of electricity.

RESISTOR: Device placed in circuit to lower voltage. It will also decrease flow of current.

RESISTOR SPARK PLUG: Spark plug containing resistor designed to shorten both capacitive and inductive phases of spark. This will suppress radio interference and lengthen electrode life.

RESONATOR: Small muffler-like device that is placed into exhaust system near end of tail pipe. Used to provide additional silencing of exhaust.

RETAINER or VALVE COLLET: Small unit that snaps into a groove in end of valve stem. It is designed to secure valve spring, valve spring retaining washer, and valve stem together. Some are of a split design, some of a horseshoe shape, etc.

RETARD (Ignition Timing): To set the ignition timing so that spark occurs later or less degrees before TDC.

RETURN SPRING: Pull-off spring, retractor spring.

REVERSE-ELLIOT TYPE AXLE: Solid bar front axle on which steering knuckles span or straddle axle ends.

REVERSE FLUSH: Cleaning cooling system by forcing clean water through system in a direction opposite to that of normal flow.

REVERSE IDLER GEAR: Gear used in transmission to produce a reverse rotation of transmission output shaft.

RHEOSTAT: A variable type resistor used to control current flow.

RICARDO PRINCIPLE: Arrangement in which portion of combustion chamber came in very close contact with piston head. Other portion, off to one side, contained more space. As the piston neared TDC on compression stroke, fuel mixture was squeezed tightly between piston and head thus causing mixture to squirt outward into larger area in very turbulent manner. This produced a superior mixture and allowed compression ratios to be raised without detonation.

RICH MIXTURE (Fuel): A fuel mixture with an excessive amount of fuel in relation to air.

RIDING THE CLUTCH: Riding the clutch refers to driver resting a foot on clutch pedal while car is being driven.

RIM (Tyre): The outer portion of a wheel upon which tyre is mounted.

RING (Chrome): Piston ring on which the outer edge has a thin layer of chrome plate.

RING (Pinned): Steel pin, set into piston, is placed in space between ends of ring. Ring is thus kept from moving around in groove, usually two stroke engines.

RING EXPANDER: Spring device placed under rings to hold them snugly against cylinder wall.

RING GAP: Distance between ends of piston ring when installed in cylinder.

RING GEAR: Large gear attached to differential carrier or to outer gear in planetary gear setup.

RING GROOVES: Grooves cut into piston to accept rings.

RING JOB: Reconditioning cylinders and installing new rings.

RING RIDGE: Portion of cylinder above top limit of ring travel. In a worn cylinder, this area is of smaller diameter than remainder of cylinder and will leave ledge or ridge that must be removed.

RIVET: Metal pin used to hold two objects together. One end of the pin has head and other end must be set or peened over.

ROAD FEEL: Feeling imparted to steering wheel by wheels of car in motion. This feeling can be very important in sensing and predetermining vehicle steering response.

ROCKER ARM: Arm used to direct upward motion of push rod into a downward or opening motion of valve stem. Used in overhead valve installations.

ROCKER ARM SHAFT: Shaft upon which rocker arms are mounted.

ROCKER COVER: Valve cover, tappet cover.

ROCKER PANEL: Section of car body between front and rear fenders and beneath doors.

ROCKWELL HARDNESS: Measurement of the degree of hardness of given substance.

ROD: Refers to an engine connecting rod.

ROD CAP: Lower removable half of connecting rod big end.

RODDING THE RADIATOR: Top and sometimes the bottom tank of the radiator is removed. The core is then cleaned by passing a cleaning rod down through tubes. This is done when radiators are quite clogged with rust, scale, and various mineral deposits.

ROLL BAR: Heavy steel bar that goes from one side of frame, up and around behind the driver, and back down to the other side of frame. It is used to protect driver in the event that the car rolls over.

ROLLER BEARING: Bearing utilising a series of straight, cupped, or tapered rollers engaging an inner and outer ring or race.

ROLLER CLUTCH: Clutch, utilising series of rollers placed in ramps, that will provide drive power in one direction but will slip or freewheel in the other direction.

ROLLER TAPPETS or LIFTERS: Valve lifters that have roller placed on end contacting camshaft. This is done to reduce friction between lobe and lifter. They are generally used when special camshafts and high tension valve springs have been installed.

ROLLING RADIUS: Distance from road surface to centre of wheel with vehicle moving under normal load. Rolling radius is dependent on tyre size.

ROLLOVER VALVE: Valve in fuel delivery line to prevent escape of raw fuel during an accident in which car is upside down.

ROOM TEMPERATURE: An enclosed space air temperature of around 20–22°C (68–72°F).

ROTARY ENGINE: Piston engine in which the crankshaft is fixed (stationary) and in which cylinders rotate around crankshaft.

ROTARY ENGINE (Wankel): Internal combustion engine which is not of a reciprocating (piston) engine design. Central rotor turns in one direction only and yet effectively produces required intake, compression, firing, and exhaust strokes.

ROTARY FLOW (Torque Converter): Movement of oil as it is carried around by pump and turbine. Rotary motion is not caused by oil passing through pump, to turbine, to stator, etc., as is case with vortex flow. Rotary flow is at right angles to centre line of converter whereas vortex flow is parallel (more or less depending on ratio between speeds of pump and turbine).

ROTARY MOTION: Continual motion in circular direction such as performed by crankshaft.

ROTOR ARM or ROTOR BUTTON (Distributor): Cap-like unit placed on end of distributor shaft. It is in constant contact with distributor cap central terminal and as it turns, it will conduct secondary voltage to each of the outer terminals in turn.

ROUGHING STONE (Hone): Coarse stone used for quick removal of material during honing.

RUNNING-FIT: Fit in which sufficient clearance has been provided to enable parts to turn freely and to receive lubrication.

RUNNING ON: See Dieseling.

RUNOUT: Refers to a rotating object, surface of which is not revolving in a true circle or plane. Runout can be measured in a radial (at right angles to axial centreline of object) direction or in a lateral (lengthwise to centreline) direction.

S

SAE: Society of Automotive Engineers.

S.A.E. CRANKING CURRENT: Internationally recognised S.A.E. Cranking Performance Test: The discharge load in Amperes which a new fully charged battery at 18°C can deliver for 30 seconds and maintain a voltage of 1.2 volts per cell or higher.

SAFETY FACTOR: Providing strength beyond that needed, as an extra margin of insurance against part failure.

SAFETY HUBS: Device installed on the rear axle to prevent wheels from leaving car in event of a broken axle.

SAFETY RIM: Rim having two safety ridges, one on each lip, to prevent tyre beads from entering drop centre area in event of a blowout. This feature keeps tyre on rim.

SAFETY VALVE: Valve designed to open and relieve pressure within a container when container pressure exceeds predetermined level.

SAND BLAST: Cleaning by the use of sand propelled at high speeds in an air blast.

SAYBOLT VISCOMETER: Instrument used to determine fluidity or viscosity (resistance to flow) of an oil.

SCALE (Cooling System): Accumulation of rust and minerals within cooling system.

SCATTER SHIELD: Steel or nylon guard placed around bell or clutch housing to protect driver and spectator from flying parts in event of part failure at high rpm. Such a shield is often placed around transmissions and differential units.

SCAVENGING: Referring to a cleaning or blowing out action in reference to the exhaust gas.

SCHRADER VALVE: Valve, similar to spring loaded valve used in tyre stem, used in car air conditioning system service valves.

SCORE: Scratch or groove on finished surface.

SCREW EXTRACTOR: Device used to remove broken bolts, screws, etc., from holes.

SEAL: Device which prevents oil leakage around moving part.

SEALED BEAM HEADLIGHT: Headlight lamp in which lens, reflector, and filament are fused together to form single unit.

SEALED BEARING: Bearing that has been lubricated at factory and then sealed. It cannot be lubricated during service.

SEAT: Surface upon which another part rests or seats. Example: Valve seat is matched surface upon which valve face rests.

SEAT (Rings): Minor wearing of piston ring surface during initial use. Rings then fit or seat properly against the cylinder wall.

SECONDARY CIRCUIT: (Ignition System): High voltage part of ignition system.

SECONDARY CUP: Piston seal.

SECONDARY JET: Compensating jet.

SECONDARY, REVERSE, or TRAILING BRAKE SHOE: Brake shoe that is installed facing rear of car.

SECONDARY WIRES: High voltage wire from coil to distributor central tower and from outer towers to spark plugs.

SECTION MODULUS: Relative structural strength measurement of member (such as frame rail) that is determined by cross-sectional area and member shape.

SECTION WIDTH: (Tyre): Overall width minus height of any lettering or pattern extending outward from sidewalls.

SECTOR SHAFT: Roller shaft, pitman arm shaft, drop arm shaft.

SEDIMENT: Accumulation of matter which settles to bottom of a liquid.

SEIZE: See Freezing.

SELECTOR SHAFT: Selector rod, selector rail, shift rail, shift rod, shifter shaft.

SELF-ENERGISING: Brake shoe (sometimes both shoes) that when applied develops wedging action that actually assists or boosts braking force applied by wheel cylinder.

SELF-INDUCTION (Electromagnetic): Creation of voltage in a circuit by varying current in circuit.

SEMI-ELLIPTICAL SPRING: Spring, such as commonly used on truck rear axles, consisting of one main leaf and number of progressively shorter leaf springs.

SEMI-FLOATING AXLE: Type of axle commonly used in modern car. Outer end turns wheel and supports weight of car; inner end which is splined, "floats" in differential gear.

SEPARATORS (Battery): Wood, rubber, or plastic sheets inserted between positive and negative plates to prevent contact.

SERIES CIRCUIT: Circuit with two or more resistance units so wired that current must pass through one unit before reaching other.

SERIES-PARALLEL CIRCUIT: Circuit of three or more resistance units in which a series and a parallel circuit are combined.

SERVICEABLE: A part or unit whose design and construction permit disassembly for purposes of rebuilding. Can also be used to indicate a part or unit whose condition is such that it can still be used.

SERVICE MANAGER: Person in charge of overall shop or garage operation.

SERVICE VALVES: Valves in air conditioning system which allow system to be charged (filled), evacuated (emptied), and tested with pressure gauges.

SERVO (Transmission): Oil operated device used to push or pull another part — such as tightening the transmission brake bands.

SERVO ACTION: Brakes so constructed as to have one end of primary shoe bearing against end of secondary shoe. When brakes are applied, primary shoe attempts to move in the direction of the rotating drum and in so doing applies force to the secondary shoe. This action, called servo action, makes less brake pedal pressure necessary and is widely used in brake construction.

SHACKLE: Device used to attach ends of a leaf spring to frame.

SHAVE (Engine): Removal of metal from contact surface of cylinder head or block.

SHIFT FORKS: Devices that straddle grooves cut in sliding gears. Fork is used to move gear back and forth on shaft.

SHIFT MECHANISM: Device for changing transmission gears range.

SHIFT POINT: Point, either in engine rpm or road speed, at which transmission should be shifted to next gear.

SHIFT RAILS: Sliding rods upon which shift forks

are attached. Used for shifting the transmission (manual).

SHIFT RANGE (4-wheel drive): Used to refer to two-speed transfer case gear position. Case can be shifted into HIGH RANGE (no gear reduction) or LOW RANGE (around two to one reduction).

SHIM: A thin piece of brass, steel, etc., inserted between two parts so as to adjust distance between them. Sometimes used to adjust bearing clearance.

SHIMMY: Front wheels shaking from side to side.

SHOCK ABSORBER: Oil filled device used to control spring oscillation in suspension system.

SHORT BLOCK: Engine block complete with crankshaft and piston assemblies.

SHORT or SHORT CIRCUIT: Refers to some "hot" portion of the electrical system that has become grounded. (Wire touching a ground and providing a completed circuit to the battery.)

SHRINK-FIT: Fit between two parts which is so tight, outer or encircling piece must be expanded by heating so it will fit over inner piece. In cooling, outer part shrinks and grasps inner part securely.

SHROUD: Metal enclosure around fan, engine, etc., to guide and facilitate flow of air.

SHUNT: An alternate or by-pass portion of an electrical circuit, parallel connection.

SHUNT WINDING: Wire coil forming an alternate or bypass circuit through which current may flow.

SIDE-DRAFT CARBURETTOR: Carburettor in which air passes through carburettor into intake manifold in a horizontal plane.

SIDEWALL: Part of tyre between tread and bead, usually has size and rating information.

SIGHT GLASS: Clear glass window in air conditioning line which lets mechanic check refrigerant for air bubbles (low refrigerant) and moisture (pink colour).

SILENCER: Muffler, expansion box.

SILVER SOLDER: Similar to brazing except that special silver solder metal is used.

SINGLE-BARREL, DOUBLE-BARREL, and FOUR-BARREL CARBURETTORS: Number of throttle openings or barrels from the carburettor to the intake manifold.

SINTERED BRONZE: Tiny particles of bronze pressed tightly together so that they form a solid piece. The piece is highly porous and is often used for filtering purposes.

SIPE: Small slits in tyre tread designed to increase traction. Also called kerfs.

SKID PLATE: Stout metal plate or plates attached to underside of vehicle to protect oil pan, transmission, fuel tank, etc., from damage caused by "grounding out" on rocks, curbs and road surface.

SKIVING: Cutting away a portion of tyre tread to correct out-of-round problem.

SLANT ENGINE: In-line engine in which cylinder block has been tilted from vertical plane.

SLAVE CYLINDER: Clutch cylinder, operating cylinder, actuating cylinder.

SLEEVE (Cylinder): See Cylinder Sleeve.

SLIDING-FIT: See Running-Fit.

SLIDING GEAR: Transmission gear splined to the shaft. It may be moved back and forth for shifting purposes.

SLIP ANGLE: Difference in actual path taken by a car making a turn and path it would have taken if it had followed exactly as wheels were pointed.

SLIP JOINT: Joint that will transfer driving torque from one shaft to another while allowing longitudinal movement between two shafts.

SLUDGE: Black, mushy deposits throughout interior of the engine. Caused from mixture of dust, oil, and water being whipped together by moving parts.

SMOG: Fog made darker and heavier by chemical fumes and smoke.

SNAP RING: Split ring snapped into a groove in a shaft or in a groove in a hole. It is used to hold bearings, thrust washers, gears, etc., in place.

SNUBBER: Device used to limit travel of some part.

SODIUM VALVE: Valve in which stem has been partially filled with metallic sodium to speed up transfer of heat from valve head, to stem and then to guide and block.

SOHC: Engine with single overhead camshaft.

SOLDERING: Joining two pieces of metal together with lead-tin mixture. Both pieces of metal must be heated to ensure proper adhesion of melted solder.

SOLENOID: Electrically operated magnetic device used to operate some unit. Movable iron core is placed inside of coil. When current flows through coil, core will attempt to centre itself in coil. In so doing, core will exert considerable force on anything it is connected to.

SOLID AXLE: Single beam runs between both wheels. May be used on either front or rear of car.

SOLID STATE: An electrical device, such as a regulator, that has no moving parts. Such units use transistors, diodes, resistors, etc., to perform all electrical functions.

SOLVENT: Liquid used to dissolve or thin other material. Examples: Alcohol thins shellac; petrol dissolves grease.

SPARK: Bridging or jumping of a gap between two electrodes by current of electricity.

SPARK ADVANCE: Causing spark plug to fire earlier by altering position of distributor breaker points in relaton to distributor shaft.

SPARK ARRESTOR: Device used to prevent sparks (burning particles of carbon) from being discharged from exhaust pipe. Usually used on off-road equipment to prevent forest fires.

SPARK GAP: Space between centre and side electrode tips on a spark plug.

SPARK KNOCK: See Preignition.

SPARK PLUG: Device containing two electrodes across which electricity jumps to produce a spark to fire fuel charge.

SPECIALISATION: When a mechanic concentrates in one particular area of auto repair — brakes, tune-up, transmission, engines.

SPECIFIC GRAVITY: Term for the ratio of density of a substance to that of water, also called Relative Density. A fully charged automotive battery should have a HYDROMETER reading of 1.250 to 1.290 meaning the ELECTROLYTE is 1.250 and 1.290 times as dense as water. If half-charged the reading is about 1.190; when discharged about 1.1. These readings vary with the temperature.

SPEED: Time rate of motion without regard to direction. Forward speed (mph or km/h) of a vehicle, rotational speed (rpm) of an engine, etc.

SPEEDOMETER: Instrument used to determine forward speed of an auto in kilometres per hour or miles per hour.

SPIDER GEARS: Small gears mounted on shaft pinned to differential case. They mesh with, and drive, the axle end gears.

SPIGOT BEARING: Pilot bearing, support bearing.

SPINDLE (Wheel): Machined shaft upon which inside races of front wheel bearings rest. Spindle is an integral part of steering knuckle.

SPIRAL BEVEL GEAR: Ring and pinion setup widely used in automobile diffentials. Teeth of both ring and pinion are tapered and are cut on a spiral so that they are at an angle to centre line of pinion shaft.

SPLINE: Metal (land) remaining between two grooves. Used to connect parts.

SPLINED JOINT: Joint between two parts in which each part has a series of splines cut along contact area. The splines on each part slide into grooves between splines on other part.

SPLIT MANIFOLD: Exhaust manifold that has a baffle placed near its centre. An exhaust pipe leads out of each half.

SPONGY PEDAL: When there is air in brake lines, or shoes that are not properly centred in drums, brake pedal will have a springy or spongy feeling when brakes are applied. Pedal normally will feel hard when applied.

SPOOL BALANCE VALVE (Automatic Transmission): Hydraulic valve that balances incoming oil pressure against spring control pressure to produce a steady pressure to some control unit.

SPOOL VALVE: Hydraulic control valve shaped somewhat like spool upon which thread is wound.

SPORTS CAR: Term commonly used to describe a relatively small, low slung, car with a high performance engine.

SPOT WELD: Fastening parts together by fusing, at various spots. Heavy surge of electricity is passed through the parts held in firm contact by electrodes.

SPRAG CLUTCH: Clutch that will allow rotation in one direction but that will lock up and prevent any movement in the other direction.

SPRING (Main Leaf): Long leaf on which ends are turned to form an "eye" to receive shackle.

SPRING BOOSTER: Device used to "beef" up sagged springs or to increase the load capacity of standard springs.

SPRING CAPACITY AT GROUND: Total vehicle weight (sprung and unsprung) that will be carried by spring bent or deflected to its maximum normal loaded position.

SPRING CAPACITY AT PAD: Total vehicle sprung weight that will be carried by spring bent or deflected to its normal fully loaded position.

SPRING LOADED: Device held in place, or under pressure from a spring or springs.

SPRING STEEL: Heat treated steel having the ability to stand a great amount of deflection and yet return to its original shape or position.

SPRING WINDUP: Curved shape assumed by rear leaf springs during acceleration or braking.

SPROCKET: Toothed wheel used to drive chain or cogged belt.

SPRUNG WEIGHT: Weight of all parts of car that are supported by suspension system.

SPUR GEAR: Gear on which teeth are cut parallel to shaft.

SPURT OR SQUIRT HOLE: Small hole in connecting rod big end that indexes (aligns) with oil hole in crank journal. When holes index, oil spurts out to lubricate cylinder walls.

SQ. IN.: Square Inch.

SQUARE ENGINE: Engine in which bore diameter and stroke are of equal dimensions.

STABILISER BAR: Transverse mounted spring steel bar that controls and minimises body lean or tipping on corners.

STALL: To stop rotation or operation.

STAMPING: Sheet metal part formed by pressing between metal dies.

STARTER (Engine): Electric motor which uses a geardrive to crank (rotate) engine for starting.

STARTER PINION GEAR: Small gear on end of starter shaft that engages and turns large flywheel ring gear.

STARTER SOLENOID: Large electric relay that makes and breaks the electrical connection between the battery and starting motor.

STARTING SYSTEM: Parts (starter motor, gear drive, switch, solenoid, wires, battery, etc.) involved in system used to crank car for starting.

STATIC BALANCE: When a tyre, flywheel, crankshaft, etc., has an absolutely even distribution of weight mass around axis of rotation, it will be in static balance. For example, if front wheel is jacked up and tyre, regardless of where it is placed, always slowly turns and stops with the same spot down, it would not be in static balance. If, however, wheel remains in any position in which it is placed, it would be in static balance. (Bearings must be free, no brake drag, etc.)

STATIC ELECTRICITY: Electricity generated by friction between two objects. It will remain in one object until discharged.

STATIC PRESSURE (Brakes): Certain amount of pressure that always exists in brake lines — even with brake pedal released. Static pressure is maintained by a check valve.

STATIC RADIUS: Distance from road surface to centre of wheel with vehicle normally loaded, at rest.

STATIC SUPPRESSION: Removal or minimising of unwanted electromagnetic waves that cause radio static interference (hissing, crackling, etc.).

STATOR: Small hub, upon which series of vanes are affixed in radial position, that is so placed that oil leaving torque converter turbine strikes stator vanes and is redirected into pump at an angle conducive to high efficiency. Stator makes torque multiplication possible. Torque multiplication is highest at stall when the engine speed is at its highest and the turbine is standing still.

STEERING ARMS: Arms, either bolted to, or forged as an integral part of steering knuckles. They transmit steering force from tie rods to knuckles, thus causing wheels to pivot.

STEERING AXIS INCLINATION: See Kingpin Inclination.

STEERING CONNECTING ROD: Relay rod, drag link.

STEERING GEAR: Gears, mounted on lower end of steering column, used to multiply driver turning force.

STEERING GEOMETRY: Term sometimes used to describe various angles assumed by components making up front wheel turning arrangement, camber, caster, toe-in, etc.

Also used to describe related angles assumed by front wheels when car is negotiating a curve.

STEERING KNUCKLE: Inner portion of spindle affixed to and pivoting on either a kingpin or on upper and lower ball joints.

STEERING KNUCKLE ANGLE: Angle formed between steering axis and centreline of spindle. This angle is sometimes referred to as Included Angle.

STEERING LINKAGE: Various arms, rods, etc., connecting steering gear to front wheels.

STEERING SHAFT: Steering column.

STEERING SYSTEM: All parts (steering wheel, shaft, gears, linkage, etc.) used in transferring motion of steering wheel to front wheels.

STETHOSCOPE: Device (such as used by doctors) to detect and locate abnormal engine noises. Very handy tool for troubleshooter.

STICK SHIFT: Transmission that is shifted manu-

ally through use of various forms of linkage. Often refers to upright gearshift stick that protrudes through floor.

Either floor or steering column mounted manual shift device for transmission.

STOCK CAR: Car as built by factory.

STOICHIOMETRIC FUEL MIXTURE: A fuel mixture in which proportions of air and fuel are such as to permit complete burning. The ideal mixture for any given engine and set of conditions.

STOPLIGHT or STOPLAMP: Warning lights, red in colour, attached to rear of vehicle. Stoplights come on whenever brake pedal is depressed.

STORAGE BATTERY: Another term for car battery. See Battery.

STRESS: To apply force to an object. Force or pressure and object is subjected to.

STRIPING TOOL: Tool used to apply paint in long narrow lines.

STROBOSCOPE: See Timing Light.

STROKE: Distance piston moves when travelling from TDC to BDC.

STROKED CRANKSHAFT: Crankshaft, either special new one or stock crank reworked, that has connecting rod throws offset so that length of stroke is increased.

STROKER: Engine using crankshaft that has been stroked.

STUB AXLE: Swivel axle.

STUB AXLE SUPPORT: Steering knuckle support, swivel link, control arm link.

STUD: Metal rod with threads on both ends.

STUD PULLER: Tool used to install or remove studs.

SUCTION: See Vacuum.

SUCTION LINE: See Low Pressure Line.

SUCTION THROTTLING VALVE: Valve placed between air conditioning evaporator and compressor which controls evaporator pressure to provide maximum cooling without icing evaporator core.

SULPHATION: Formation of lead sulphate on battery plates.

SUMP: Part of oil pan that contains oil.

SUN GEAR: Centre gear around which planet gears revolve.

SUPER CAR: Car with high power engine that will provide fast acceleration and high speed.

SUPERCHARGER: Unit designed to force air, under pressure, into cylinders. Can be mounted between carburettor and cylinders or between carburettor and atmosphere.

SUPERHEAT SWITCH: See Compressor Protection Switch.

SURE-GRIP or LIMITED SLIP DIFFERENTIAL: High traction differential which causes both axles to rotate under power.

SUSPENSION STRUT: Suspension leg, MacPherson Strut.

SWAY BAR: See Stabiliser Bar.

SWEATING: Joining two pieces of metal together by placing solder between them and then clamping them tightly together while heat, sufficient to melt the solder, is applied.

SWING AXLE: Independent rear suspension system in which each driving wheel can move up or down independently of other. Differential unit is bolted to frame and various forms of linkage are used upon which to mount wheels. Drive axles, utilising one or more universal joints, connect differential to drive wheels.

SWITCH (Electric): A device to make (complete) or break (interrupt) flow of current through a circuit.

SYNCHRO HUB: Clutch hub.

SYNCHROMESH TRANSMISSION: Transmission using device (synchromesh) that synchronises speeds of gears that are being shifted together. This prevents "gear grinding". Some transmissions use synchromesh on all shifts, while others synchronise second and high gearshifts.

SYNCHRONISE: To bring about a timing that will cause two or more events to occure simultaneously; plug firing when the piston is in correct position, speed of two shafts being the same, valve opening when piston is in correct position, etc.

SYNCHRO PLATE: Synchro bar, synchro key.

SYNCHRO RING: Baulk ring, synchro cone.

SYNCHRO SLEEVE: Clutch sleeve.

SYNCHRO SPRING: Clutch spring, energising spring.

T

TACHOMETER: Device used to indicate speed of engine in rpm.

TAIL LIGHT or TAIL LAMP: Lights, red colour, attached to rear of vehicle. Lights operate in conjunction with headlights and park lights.

TAIL PIPE: Exhaust piping running from muffler to rear of car.

TANK GAUGE SENDER UNIT: Variable resistor device in fuel tank. It operates fuel gauge in dashboard.

TAP: To cut threads in a hole, or can be used to indicate fluted tool used to cut threads.

TAP AND DIE SET: Set of taps and dies for internal and external threading — usually covers a range of the most popular sizes.

TAPERED ROLLER BEARING (Antifriction): Bearing utilising series of tapered, hardened steel rollers operating between an outer and inner hardened steel race.

TAPPET (Hydraulic or solid): Valve lifter, cam follower.

TAPPET NOISE: Noise caused by lash or clearance between valve stem and rocker arm or between valve stem and valve lifter.

TEFLON: Plastic with excellent self-lubricating (slippery) bearing properties.

TEMPER: To effect a change in physical structure of piece of steel through use of heat and cold.

TEMPERATURE GAUGE: Dash mounted instrument to indicate temperature of engine coolant. May be operated electrically or by Bourdon tube.

TENSION: Pulling or stretching stress applied to an object.

TERMINAL: Connecting point in electric circuit. When referring to battery, it would indicate two battery posts.

T-HEAD ENGINE: Engine having intake valve on one side of cylinder and exhaust on other.

THERMAL EFFICIENCY: Percentage of heat developed in burning fuel charge that is actually used to develop power determines thermal efficiency. Efficiency will vary according to engine design, use, etc. If an engine utilises great deal of heat to produce power, its thermal efficiency would be high.

THERMISTOR: Resistor that changes its resistance in relation to temperature fluctuations.

THERMOSTAT: Temperature sensitive device used in cooling system to control flow of coolant in relation to temperature.

THERMOSTATICALLY CONTROLLED AIR CLEANER (TAC): An emission control device used to control temperature of air entering air cleaner. Cleaner receives heated air during engine warmup.

THERMOSTATIC SWITCH: A switch that is actuated by temperature changes.

THIRD BRUSH (Generator): Generator in which a third, movable brush is used to control current output.

THREE-WAY CATALYTIC CONVERTER: Converter, sometimes called dual converter, that combines both an oxidising and reducing catalyst. Controls NOx, CO, and HC emissions.

THROTTLE RETURN DASHPOT: Carburettor device which slows throttle closing and prevents stalling.

THROTTLE STOP SCREW: Idling speed screw.

THROTTLE VALVE: Butterfly in carburettor. It is used to control amount of fuel mixture that reaches cylinders.

THROTTLE VALVE SENSOR AND THROTTLE POSITION SENSOR: Sensor which measures amount of throttle valve opening and provides information for fuel injection computer.

THROTTLE VALVE (Transmission): Valve in automatic transmission which controls oil flow to regulator plug.

THROW: Offset portion of crankshaft designed to accept connecting rod.

THROWING A ROD: When an engine has thrown a connecting rod from crankshaft, major damage is usually incurred.

THRUST: A pushing force exerted against one body by another.

THRUST BEARING: Bearing designed so as to resist side pressure.

THRUST SURFACE: That portion (surface) of a part that either receives or transmits a force from or to another part.

THRUST WASHER: Bronze or hardened steel washer placed between two moving parts. The washer prevents longitudinal movement and provides a bearing surface for thrust surfaces of parts.

TIE ROD: Rod, or rods, connecting steering arms together. When tie rod is moved, wheels pivot.

TIE ROD BALL JOINT: Tie rod end, track rod end.

TIG: Gas tungsten arc welding (Tungsten Inert Gas).

TIMED FUEL INJECTION: Fuel injection is timed to occur when intake valve opens.

TIMING: The act of co-ordinating two or more separate events or actions in relation to each other. Example: Timing the firing of plug to piston position on compression stroke.

TIMING BELT: A flexible, toothed belt used to rotate camshaft.

TIMING CHAIN: Drive chain that operates camshaft by engaging sprockets on camshaft and crankshaft.

TIMING CHAIN TENSIONER: Chain tightener.

TIMING COVER: Cover over timing chain, belt, or gear mechanism. May contain front crankshaft oil seal.

TIMING GEARS: Both the gear attached to the camshaft and the gear on the crankshaft. They provide a means of driving the camshaft.

TIMING LIGHT: Stroboscopic unit that is connected to secondary circuit to produce flashes of light in unison with firing of specific spark plug. By directing these flashes of light on whirling timing marks, marks appear to stand still. By adjusting distributor, timing marks may be properly aligned, thus setting timing.

TIMING MARKS (Ignition): Marks, usually located on vibration damper, used to synchronise ignition system so plugs will fire at precise time.

TIMING MARKS (Valves): One tooth on either the camshaft or crankshaft gear will be marked with an indentation or some other mark. Another mark will be found on other gear between two of teeth. Two gears must be meshed so that marked tooth meshes with marked spot on other gear.

TIMING SPROCKETS: Chain type sprockets on crankshaft and camshaft.

TINNING: Coating piece of metal with a very thin layer of solder.

TOE-IN: Having front of wheels closer together than the back (front wheels). Difference in measurement across front of wheels and the back will give amount of toe-in.

TOE-OUT: Having front of wheels further apart than the back.

TOE-OUT ON TURNS: When car negotiates a curve, inner wheel turns more sharply and while wheels remain in this position, a condition of toe-out exists.

TOGGLE SWITCH: Switch actuated by flipping a small lever either up and down or from side to side.

TOLERANCE: Amount of variation permitted from an exact size or measurement. Actual amount from smallest acceptable dimension to largest acceptable dimension.

TOOTH HEEL (Differential Ring Gear): Wider outside end of tooth.

TOOTH TOE (Differential Ring Gear): Narrower inside end of tooth.

TOP GEAR: High gear, high speed.

TOP OFF: Fill a container to full capacity.

TORQUE: Turning or twisting force such as force imparted on drive line by engine.

TORQUE BAR or ROD: An articulated bar between frame and drive axle housing designed to relieve leaf springs of axle torque (twisting) strain. Prevents axle windup and/or hop during heavy acceleration or braking. Also called Traction Bar.

TORQUE CONVERTER: Unit, quite similar to fluid coupling, that transfers engine torque to transmission input shaft. Unlike fluid coupling, torque converter can multiply engine torque. This is accomplished by installing one or more stators between torus members. In torque converter, driving torus is referred to as "pump" and driven torus as "turbine".

TORQUE (Gross): Maximum engine torque developed by engine without fan, air cleaner, alternator, exhaust system, etc.

TORQUE (Net): Maximum torque developed by engine equipped with fan, air cleaner, exhaust system, and all other systems or units normally present when engine is installed in the vehicle.

TORQUE MULTIPLICATION (Automatic Transmission): Increasing engine torque through the use of a torque converter.

TORQUE TUBE DRIVE: Method of connecting transmission output shaft to differential pinion shaft by using an enclosed drive shaft. Drive shaft is enclosed in torque tube that is bolted to rear axle housing on one end and is pivoted through a ball joint to rear of transmission on other. Driving force of rear wheels is transferred to frame through torque tube.

TORQUE WRENCH: Wrench used to draw nuts, cap screws, etc., up to specified tension by measuring torque (turning force) being applied.

TORSIONAL VIBRATION: Twisting and untwisting action developed in shaft. It is caused either by intermittent applications of power or load.

TORSION BAR: Long spring steel rod attached in such a way that one end is anchored while other is free to twist. If an arm is attached at right angles to free end, any movement of arm will cause rod or bar to twist. Bar's resistance to twisting provides a spring action. Torsion bar replaces both coil and leaf springs in some suspension systems.

TORSION BAR SUSPENSION: Suspension system that makes use of torsion bars in place of leaf or coil spring.

TORUS: Fluid coupling rotating member. There are two — driving and driven torus.

TRACK: Distance between front wheels or distance between rear wheels. They are not always the same.

TRACTION BAR: Articulated bar or link attached to both frame and rear axle housing to prevent spring windup (with resultant wheel hop) during heavy acceleration or braking.

TRACTION DIFFERENTIAL: See Limited-Slip Differential.

TRACTION (Tyre): Frictional force generated between tyre and road. Necessary for braking, steering, and driving.

TRAMP: Hopping motion of front wheels.

TRANSAXLE: Drive setup in which transmission and differential are combined into a single unit.

TRANSDUCER: Vacuum regulator actuated or controlled electrically. A device that converts an input signal (electrical) into an output signal (diaphragm movement) of a different form.

TRANSFER CASE: Gearbox, driven by transmission, that will provide driving force to both front and rear propeller shafts on four-wheel drive vehicle.

TRANSFORMER: Electrical device used to increase or decrease voltage. Car ignition coil transforms voltage from 12 volts to upward of 30,000 volts.

TRANSISTOR: Electrical device made of semi-conducting material and using at least three electrical connections. Often used as a switching device.

TRANSISTOR IGNITION: Form of ignition system utilising transistors and a special coil. Conventional distributor and point setup is used. With transistor unit, voltage remains constant, thus permitting high engine rpm without resultant engine "miss". Point life is greatly extended as transistor system passes a very small amount of current through points.

TRANSMISSION: Device that uses gearing or torque conversion to effect a change in ratio between engine rpm and driving wheel rpm. When engine rpm goes up in relation to wheel rpm, more torque but less speed is produced. Reduction in engine rpm in relation to wheel rpm produces a higher road speed but delivers less torque to driving wheels.

TRANSMISSION ADAPTER: A unit that allows a different make or year transmission to be bolted up to original engine.

TRANSMISSION (Automatic): Transmission that automatically effects gear changes to meet varying road and load conditions. Gear changing is done through series of oil operated clutches and bands.

TRANSMISSION (Standard or Conventional): Transmission that must be shifted manually to effect a change in gearing.

TRANSMISSION BRAKE: A brake of either drum or disc design, working at transmission. Stops rotaton of propeller shaft.

TRANSVERSE LEAF SPRING: Leaf spring mounted so it is at right angles to length of car.

TREAD (Tyre): Portion of tyre which contacts roadway.

TREAD WIDTH (Tyre): Distance between outside edges of tread as measured across tread surface.

TRIP ODOMETER: Auxiliary odometer that may be reset to zero at option of driver. Used for keeping track of distance covered on trips up to one thousand kilometres.

TROUBLESHOOTING: Diagnosing engine, transmission, etc., problems by various tests and observations.

TRUNNION: One of two pivots, bearings, etc., placed opposite to each other so as to permit a swivelling or tilting action of some part. Example: Universal joint trunnion (yoke bearings) that allow cross to swivel.

TUBE CUTTER: Tool used to cut tubing by passing a sharp wheel around and around tube.

TUBELESS (Tyre): Tyre constructed for use without inner tube, valve stem snaps into and seals in wheel rim.

TUNE-UP: Process of checking, repairing, and adjusting carburettor, spark plugs, points, belts, timing, etc., in order to obtain maximum performance from engine.

TURBINE: Wheel upon which series of angled vanes are affixed so moving column of air or liquid will impart a turning motion to wheel.

TURBINE ENGINE: Engine that utilises burning gases to spin a turbine, or series of turbines, as a means of propelling the car.

TURBOCHARGER: Exhaust powered supercharger.

TURBULENCE: Violent, broken movement or agitation of a fluid or gas.

TURNING RADIUS: Diameter of circle transcribed by outer front wheel when making a full turn.

TURN SIGNAL LAMP: Direction indicator light, flasher light, trafficator light.

TURN SIGNAL RELAY: Flasher unit.

TURN SIGNAL SWITCH: Direction indicator switch, flasher switch, trafficator switch, indicator stalk.

TV ROD: Throttle valve rod that extends from foot

throttle linkage to throttle valve in automatic transmission.

TWIST DRILL: Metal cutting drill with spiral flutes (grooves) to permit exit of chips while cutting.

TWO-STROKE CYCLE ENGINE: Engine requiring one complete revolution of crankshaft to fire each piston once.

TYRE BALANCE: In that tyres turn at relatively high speeds, they must be carefully balanced both for static and dynamic balance.

TYRE BEAD: Portion of tyre that bears against rim flange. Bead has a number of turns of steel wire in it to provide great strength.

TYRE CASING: Main body of tyre exclusive of tread.

TYRE PLIES: Layers of nylon, rayon, etc., cloth used to form casing. Many car tyres are two ply with a four ply rating. Two ply indicates two layers of cloth or plies.

TYRE ROTATION: Moving front tyres to rear and rear to front to equalise any wear irregularities.

TYRE SIDEWALL: Portion of tyre between tread and bead.

TYRE SIZE: Given on tyre sidewall as coded letter-number designation of size, section height, and diameter across bead.

TYRE TREAD: Part of tyre that contacts road.

U

UNDERCOATING: Soft deadening material sprayed on underside of car, under hood, boot lid, etc.

UNDER-SQUARE ENGINE: Engine in which bore diameter is smaller than length of stroke.

UNDERSTEER: Tendency for car, when negotiating a corner, to turn less sharply than driver intends.

UNITISED BODY: Car body in which body itself acts as frame.

UNIVERSAL JOINT: Flexible joint that will permit changes in driving angle between driving and driven shaft.

UNIVERSAL JOINT CROSS: Joint trunnion.

UNIVERSAL JOINT YOKE: Universal joint fork.

UNLEADED PETROL: Petrol not containing tetraethyl lead. Must be used with vehicles equipped with a catalytic converter.

UNSPRUNG WEIGHT: All parts of car not supported by suspension system; wheels, tyres, etc.

UPDRAFT CARBURETTOR: Carburettor in which the air passes upward through the carburettor into the intake manifold.

UPSET: Widening of diameter through pounding.

UPSHIFT: Shifting to a higher gear.

V

VACUUM: Enclosed volume in which air pressure is below that of surrounding atmospheric pressure.

VACUUM ADVANCE UNIT (Distributor): Unit designed to advance and retard ignition timing through action of engine vacuum working on a diaphragm.

VACUUM BOOSTER: Small diaphragm vacuum pump, generally in combination with fuel pump, that is used to bolster engine vacuum during acceleration so vacuum operated devices will continue to operate.

VACUUM GAUGE: Gauge used to determine amount of vacuum existing in a chamber.

VACUUM MODULATOR: Device which uses engine vacuum to control throttle valve in automatic transmission.

VACUUM MOTOR: A device, utilising a vacuum operated diaphragm, which causes movement of some other unit.

VACUUM PUMP: Diaphragm type of pump used to produce vacuum.

VACUUM RUNOUT POINT: Point reached when vacuum brake power piston has built up all the braking force it is capable of with vacuum available.

VACUUM SERVO UNIT: Servo, power brake unit, brake booster unit.

VACUUM SWITCH (Electric): An electrical switch that is operated by vacuum.

VACUUM TANK: Tank in which a vacuum exists. Generally used to provide vacuum to power brake installation in event engine vacuum cannot be obtained. Tank will supply several brake applications before vacuum is exhausted.

VALVE: Device used to either open or close an opening. There are many different types.

VALVE CLEARANCE (Engine): Space between end of valve stem and actuating mechanism (rocker arm, lifter, etc).

VALVE DURATION: Length of time, measured in degrees of engine crankshaft rotation, that valve remains open.

VALVE FACE: Outer lower edge of valve head. The face contacts the valve seat when the valve is closed.

VALVE FLOAT or BOUNCE: Condition where valves in engine are forced back open before they have had a chance to seat. Brought about (usually) by extremely high rpm.

VALVE GRINDING: Renewing valve face area by grinding on special grinding machine.

VALVE GUIDE: Hole through which stem of poppet valve passes. It is designed to keep valve in proper alignment. Some guides are pressed into place and others are merely drilled in block or in head metal.

VALVE HEAD (Engine): Portion of valve above stem.

VALVE-IN-HEAD ENGINE: Engine in which both intake and exhaust valves are mounted in the cylinder head and are driven by push rods or by an overhead camshaft.

VALVE KEEPER or VALVE KEY or VALVE RETAINER or VALVE COLLET: Small unit that snaps into a groove in end of valve stem. It is designed to secure valve spring, valve spring retaining washer, and valve stem together. Some are of a split design, some of a horseshoe shape, etc.

VALVE LASH: Valve tappet clearance or total clearance in the valve operating train with cam follower on camshaft base circle.

VALVE LIFT: Distance a valve moves from full closed to full open position.

VALVE LIFTER or CAM FOLLOWER: Unit that contacts end of valve stem and camshaft. Follower rides on camshaft and when cam lobes move it upward, it opens valve.

VALVE MARGIN: Width of edge of valve head between top of valve and edge of face. Too narrow a margin results in pre-ignition and valve damage through overheating.

VALVE OIL SEAL: Neoprene rubber ring placed in groove in valve stem to prevent excess oil entering area between stem and guide. There are other types of these seals.

VALVE OVERLAP: Certain period in which both intake and exhaust valve are partially open. (Intake is starting to open while exhaust is not yet closed.)

VALVE PORT: Opening, through head or block, from intake or exhaust manifold to valve seat.

VALVE ROTATOR: Unit that is placed on end of valve stem so that when valve is opened and closed, the valve will rotate a small amount with each opening and closing. This gives longer valve life.

VALVE SEAT: Area onto which face of poppet seats when closed. Two common angles for this seat are forty-five and thirty degrees.

VALVE SEAT GRINDING: Renewing valve seat area by grinding with a stone mounted upon a special mandrel.

VALVE SEAT INSERT: Hardened steel valve seat that may be removed and replaced.

VALVE SPRING: Coil spring used to keep valves closed.

VALVE STEM (Engine): Portion of valve below head. The stem rides in the guide.

VALVE TAPPET: Adjusting screw to obtain specified clearance at end of valve stem (tappet clearance). Screw may be in top of lifter, in rocker arm, or in the case of ball joint rocker arm, nut on mounting stud acts in place of a tappet screw.

VALVE TIMING: Adjusting position of camshaft to crankshaft so that valves will open and close at the proper time.

VALVE TRAIN: Various parts making up valve and its operating mechanism.

VALVE UMBRELLA: Washer-like unit that is placed over end of the valve stem to prevent the entry of excess oil between the stem and the guide. Used in valve-in-head installations.

VANE: Thin plate affixed to rotatable unit to either throw off air or liquid, or to receive thrust imparted by moving air or liquid striking the vane. In the first case it would be acting as a pump and in the second case as a turbine.

VAPOUR: Gaseous state of a subtance usually a liquid or solid. Example: Steam.

VAPOURISATION: Breaking petrol into fine particles and mixing it with incoming air.

VAPOUR LOCK: Boiling or vapourising of the fuel in the lines from excess heat. Boiling will interfere with movement of the fuel and will in some cases, completely stop the flow.

VARIABLE PITCH STATOR: Stator that has vanes that may be adjusted to various angles depending on load conditions. Vane adjustment will increase or decrease efficiency of stator.

VARIABLE VENTURI: A carburettor venturi whose opening size can be varied to meet changing engine speed and loading needs.

VARNISH: Deposit on interior of engine caused by engine oil breaking down under prolonged heat and use. Certain portions of oil deposit themselves in hard coatings of varnish.

V-BELT: V shaped belt commonly used to drive alternator, water pump, power steering pump, and air conditioning compressor.

VELOCITY: Time rate of motion. Speed with which an object moves as measured in metres per second, kilometres per hour, etc.

VENTILATOR GLASS: Quarter light, no draught ventilator, vent glass, quarter glass, flipper window.

VENTURI: That part of a tube, channel, pipe, etc., so tapered as to form a smaller or constricted area. Liquid, or a gas, moving through this constricted area will speed up and as it passes narrowest point, a partial vacuum will be formed. Taper facing flow of air is much steeper than taper facing away from flow of air. Venturi principle is used in carburettor.

VIBRATION DAMPER: Round weighted device attached to front of crankshaft to minimise torsional vibration.

VISCOSIMETER or VISCOMETER: Device used to determine viscosity of a given sample of oil. Oil is heated to specific temperature and then allowed to flow through set orifice. Length of time required for certain amount to flow determines oil's viscosity.

VISCOSITY: Measure of oil's ability to pour. (Thick, thin.)

VISCOSITY INDEX: Measure of oil's ability to resist changes in viscosity when heated.

VOLATILE: A substance that evaporates (turns to vapour) easily. Example: Petrol.

VOLATILITY: Property of petrol, alcohol, etc., to evaporate quickly and at relatively low temperatures.

VOLT: Unit of electrical pressure or force that will move a current of one ampere through a resistance of one ohm.

VOLTAGE: Difference in electrical potential between one end of a circuit and the other. Also called EMF (electromotive force). Voltage causes current to flow.

VOLTAGE DROP: Lowering of voltage due to excess length of wire, undersize wire, etc.

VOLTAGE REGULATOR: See Regulator — Voltage.

VOLTMETER: Instrument used to measure voltage in given circuit. (In volts.)

VOLUME: Measurement, in cubic centimetres, cubic metres, etc., of amount of space within a certain object or area.

VOLUMETRIC EFFICIENCY: Comparison between actual volume of fuel mixture drawn in on intake stroke and what would be drawn in if cylinder were to be completely filled.

VORTEX: Mass of whirling liquid or gas.

VORTEX FLOW (Torque Converter): Whirling motion of oil as it moves around and around from pump, through turbine, through stator and back into pump and so on.

V-TYPE ENGINE: An engine in which cylinders are arranged in two separate banks (rows) and set at an angle (V-shape) to each other when viewed from front.

VULCANISATION: Process of heating compounded rubber to alter its characteristics — making it tough, resilient, etc.

W

WANDERING (Steering): Condition in which front wheels tend to steer one way and then another.

WANKEL ENGINE: Rotary combustion engine that utilises one or more three-sided rotors mounted on drive shaft operating in specially shaped chambers. Rotor turns constantly in one direction yet produces an intake, compression, firing, and exhaust stroke.

WARNING LAMP: Indicator light.

WATER DETECTOR: Sensor in diesel fuel system which warns driver of water contamination of fuel.

WATER JACKET: Area around cylinders and valves that is left hollow so that water may be admitted for cooling.

WATER PUMP: The pump, usually a centrifugal type, used to circulate coolant throughout cooling system.

WATER PUMP IMPELLER: Rotor.

WEATHERSTRIP: Weathershield, glazing channel, door belt weatherstrip.

WEDGE: Engine using wedge-shaped combustion chamber.

WEDGE COMBUSTION CHAMBER: Combustion chamber utilising wedge shape. It is quite efficient and lends itself to mass production and as a result is widely used.

WEIGHT (Curb): Weight of vehicle (no passengers) with all systems (fuel, cooling, lubrication) filled.

WEIGHT (Shipping): Basic vehicle weight including all standard items but without fuel or coolant.

WEIGHT (Sprung): See Sprung Weight.

WEIGHT DISTRIBUTION: Percentage of total vehicle weight as carried by each axle (front and rear).

WELCH PLUG: Core plug, expansion plug.

WELD: To join two pieces of metal together by raising area to be joined to point hot enough for two sections to melt and flow together. Additional metal is usually added by melting small drops from end of metal rod while welding is in progress.

WET FRICTION: Resistance to movement between two lubricated surfaces.

WET SLEEVE: Cylinder sleeve application in which water in cooling system contacts a major portion of sleeve itself.

WHEEL ALIGNER: Device used to check camber, caster, toe-in, etc.

WHEEL ALIGNMENT: Refers to checking or adjusting various angles involved in proper placement or alignment of both front and rear wheels.

WHEEL BALANCER: Machine used to check wheel and tyre assembly for static and dynamic balance.

WHEELBASE: Distance between centre of front wheels and centre of rear wheels.

WHEEL BEARING: Ball or roller bearings on which wheel hub rotates.

WHEEL BRAKE: A brake operating at wheel, either drum or disc design.

WHEEL CYLINDER: Part of hydraulic brake system that receives pressure from master cylinder and in turn applies brake shoes to drums.

WHEEL CYLINDER CUP: Wheel cylinder rubber, wheel cylinder seal.

WHEEL NUT or BOLT: Nuts or bolts used to fasten wheel to hub.

WHEEL TRAMP: Hopping action of rear wheels during heavy acceleration.

WIDE TREADS, WIDE OVAL, etc: Wide tyres. Tyre height (bead to tread surface) is about 70 percent of tyre width across outside of carcass.

WINCH: An electrically or mechanically driven drum that will wind in a length of cable. Used to remove vehicles from mud, ascend very steep slopes, pull logs, etc.

WINDSCREEN: Alternative term for windshield.

WINDSCREEN WIPER: Windshield wiper.

WIRING DIAGRAM: Drawing showing various electrical units and wiring arrangement necessary for them to function properly.

WISHBONE: Control arm, suspension arm.

WITNESS MARKS: Punch marks used to position or locate some part in its proper spot.

WOBBLE PLATE: A round, flat plate with a shaft passing through its centre. Plate is affixed to shaft at an angle to shaft centreline. When shaft turns, plate will rotate but will also wobble from side to side. Used on axial air conditioning compressor to operate pistons. Also called SWASH PLATE.

WORK: A force applied to a body causing body to move.

WORM GEAR: Coarse, spiral shaped gear cut on shaft. Used to engage with and drive another gear or portion of a gear. As used in steering gearbox, it often engages cross shaft via a roller or by a tapered pin.

WORM AND ROLLER: Type of steering gear utilising a worm gear on steering shaft. A roller on one end of cross shaft engages worm.

WORM AND SECTOR: Type of steering gear utilising worm gear engaging sector (a portion of a gear) on cross shaft.

WORM AND TAPER PIN: Type of steering gear utilising worm gear on steering shaft. End of cross shaft engages worm via taper pin.

WRIST PIN: See Piston Pin.

Y

YIELD STRENGTH (Elastic Limit): Maximum force (in N, KN, MN, etc.,) that can be sustained by given member and have that member return to its original position, length, shape, etc., when force is removed.

YOKE: Slotted or split end of an object that straddles and is fastened to another.

Z

ZENER DIODE: A silicone diode that serves as a voltage controlled rectifier (allows electrical current flow in ONE direction only) until voltage being applied to diode attains a certain level (called Zener voltage) at which point diode becomes conductive.

COMMON ABBREVIATIONS

abdc: After bottom dead centre.
ABS: Antilock braking system.
AC: Alternating current.
AFC: Air flow controlled.
APS: Air preheat system.
atdc: After top dead centre.
bbdc: Before bottom dead centre.
bdc: Bottom dead centre.
bhp: Brake horsepower.
bp: Brake power.
btdc: Before top dead centre.
cc: Cubic centimetre.
CIS: Continuous injection system.
CO: Carbon monoxide.
CSI: Cold start injection.
CU: Control unit.
CVJ: Constant velocity joint.
Cyl: Cylinder.
DC: Direct current.
Deg: Degrees.
DFI: Digital fuel injection.
ECU: Electronic control unit.
EEC: Evaporative emission control.
EFI: Electronic fuel injection.
EGR: Exhaust gas recirculation.
emf: Electromotive force (Voltage).
EP: Extreme pressure.
ESC: Electronic spark control.
fhp: Frictional horsepower.
fp: Frictional power.
FWD: Front wheel drive.
HC: Hydrocarbons.
HC: High compression.
HD: Heavy duty.
HT: High tension.
ihp: Indicated horsepower.
imep: Indicated mean effective pressure.
ip: Indicated power.
LC: Low compression.
LH: Left hand.
LT: Low tension.
MS: Manufacturers' specification.
NOx: Oxides of nitrogen.
OHC: Overhead camshaft.
OHV: Overhead valves.
PCV: Positive crankcase ventilation.
psi: Pounds per square inch.
PTO: Power take off.
RH: Right hand.
rpm: Revolutions per minute.
SC: Single carburettor.
SCS: Speed control switch.
TAC: Thermostatically controlled air cleaner.
TC: Twin carburettor.
TCS: Transmission controlled spark.
tdc: Top dead centre.
TRS: Transmission regulated spark.
TVS: Thermal vacuum switch.
UJ: Universal joint.
VAC: Vacuum advance control.
wp: Wheel power.

Chapter 4

FASTENERS, TORQUE WRENCHES

After studying this chapter, you will be able to:
- ☐ Identify automotive fasteners.
- ☐ Properly select fasteners.
- ☐ Torque fasteners to specifications when needed.
- ☐ Repair damaged or broken fasteners.

INCREASING IMPORTANCE

In the modern car, various components are subjected to heavy loads, high frequency vibration, heat, and severe stress. As a result, fastener (nuts, bolts, screws) design, material, and torque settings have assumed a position of major importance.

It is important that the mechanic becomes familiar with the various types, uses, and proper installation of fasteners.

READ CAREFULLY

Be sure to read this chapter carefully, study the various fasteners, their markings, and uses until you can recognise them immediately. Pay particular attention to the section on torque wrenches.

MACHINE SCREWS

Machine screws are used without nuts. They are passed through one part and threaded into another. When drawn up, the two parts are then held in firm contact. Fig. 4-1 illustrates the use of a cap screw (machine screw with a hexagonal head).

There are many different types of machine screws and screw heads. Fig. 4-2 shows a number of those in common use.

SELF-TAPPING OR SHEET METAL SCREWS

Sheet metal screws are used to fasten thin metal parts together and for attaching various items to

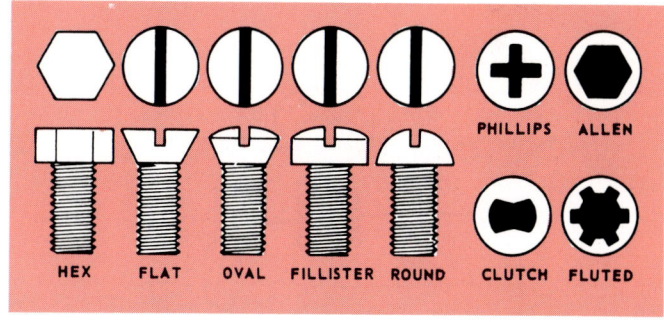

Fig. 4-2. Typical machine screws. Four heads at right illustrate various openings for turning tools.

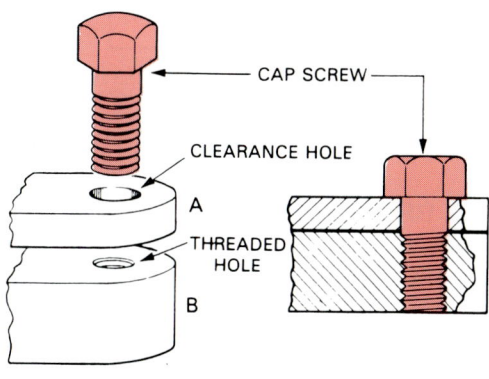

Fig. 4-1. Cap screw. Cap screw is passed through clearance hole in part A and threaded into part B.

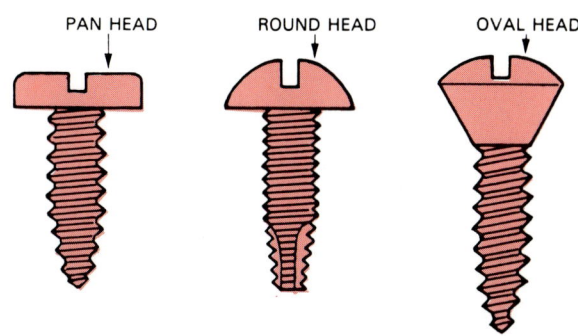

Fig. 4-3. Typical sheet metal screws.

sheet metal. They are much faster and less expensive than bolts, Fig. 4-3.

To use a sheet metal screw, punch a hole in the sheet metal. The hole should be slightly smaller than the screw minor diameter (diameter of screw if threads were ground off). A punched hole is better than a drilled hole because the punched hole attempts to close as the screw is tightened. This provides added gripping power, Fig. 4-4.

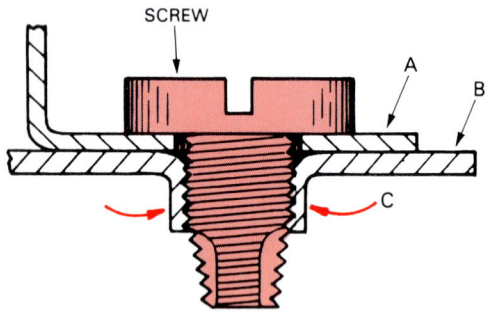

Fig. 4-4. Screw passes freely through A and cuts threads in punched hole in B. When screw tightens, punched metal draws up and in, providing a secure grip, C.

BOLT

A bolt is a metal rod that has a head at one end and a screw thread for a nut at the other. The bolt is passed through the parts to be joined. Then, the nut is installed and drawn up. This holds the parts together, Fig. 4-5.

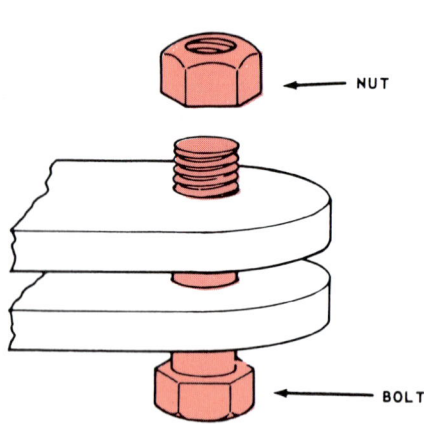

Fig. 4-5. Using a bolt to hold two parts together.

STUDS

A stud is a metal rod threaded on both ends. The stud is turned into a threaded hole in a part. The other part is slipped over the stud. A nut is turned down on the stud to secure the part. Studs are available in many lengths and diameters. Some have

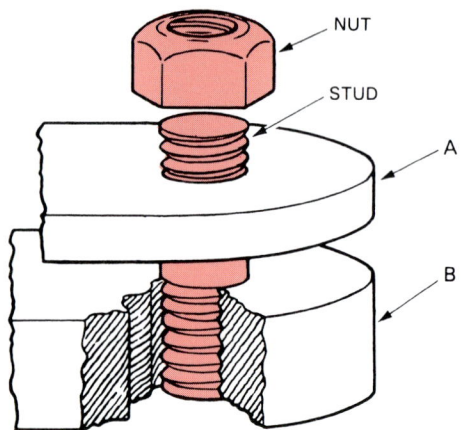

Fig. 4-6. Stud threaded into B. Part A slipped over stud. Nut is placed on stud and tightened.

a coarse thread on one end, a fine thread on the other. Others have the same thread on both ends. In some cases, this thread may run the full length of the stud, Fig. 4-6.

A stud extractor should be used to install or remove studs. Be careful not to damage the threads. If no stud extractor is available, place two nuts on the stud and "jam" them together (turn the top one clockwise, the bottom counter-clockwise until they come together). Place a spanner on the lower nut to remove the stud and on the upper nut to install. See Fig. 4-7.

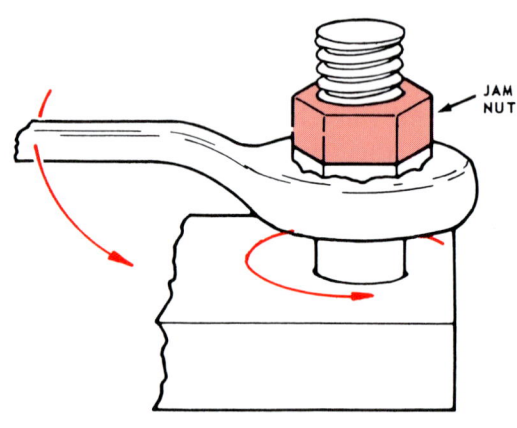

Fig. 4-7. Using jam nuts and spanner to remove stud.

REMOVING BROKEN STUDS OR SCREWS

There are several methods for removing broken fasteners. If a portion of the fastener projects above the work, it may be gripped with vise-grip pliers, or a small pipe wrench and backed out.

Where the portion protruding is not sufficient to grasp with pliers or a pipe wrench, flat surfaces may be filed to take a spanner. Also, a slot may be cut to allow the use of a screwdriver, Fig. 4-8,A.

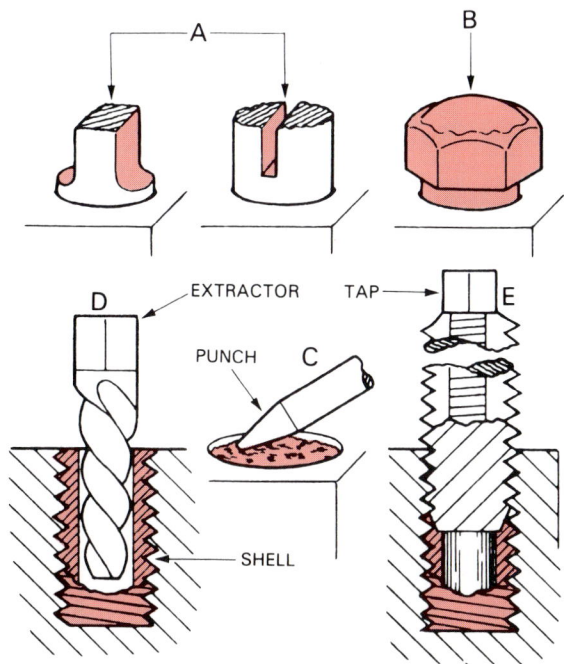

Fig. 4-8. Methods used in removing broken stud. A — Stud slotted or filed flat. B — Nut welded on. C — Punch used to unscrew broken piece. D — Screw extractor. E — Using a tap to remove shell.

Another method is to drill a hole in a section of flat steel, place it over the broken stud and weld the strip to the stud. A nut large enough to fit over the stud can also be welded on. The arc welder does the job quickly and with a minimal amount of heating, Fig, 4-8, B.

WHEN WELDING, BE CAREFUL OF FIRE AND DAMAGE TO PARTS.

When the stud is broken off flush or slightly below the surface, you may use a thin, sharp pointed punch. Try driving the broken section in a counter-clockwise direction. Sometimes the stud will turn out easily. If you are not getting results, stop and try another method, Fig. 4-8, C.

A screw extractor can often be used with good results. Centre-punch in the EXACT centre of the stud. Drill through the stud with a small diameter drill. Then, run a drill through that is slightly smaller than the stud minor diameter. Lightly tap the extractor into the shell that remains and back it out with a spanner. The sharp edges on the flutes will grip the shell. Do not exert enough turning force to break the extractor. It could present a real problem since it is hardened. See Fig. 4-8, D.

In the event the methods previously described fail, select the proper tap size drill and after running it through the stud shell, carefully tap out the hole. If done properly, the tap will remove the shell threads leaving the original threads in the hole undamaged, Fig. 4-8, E.

When drilling, drill through the stud only. Do not drill beyond as you may damage the part. If working on a setup where metal chips may fall into a housing, coat the drill and tap with a heavy coat of sticky grease. The chips will adhere to the grease and tools.

REMOVING DAMAGED NUTS

Occasionally, nuts will be difficult to remove due to rust, dirt, and corrosion. When this happens, there are several methods you may use to assist removal: HEAT (a fire hazard, be careful!), NUT SPLITTER (CRACKER), or HACKSAW. See Fig. 4-9.

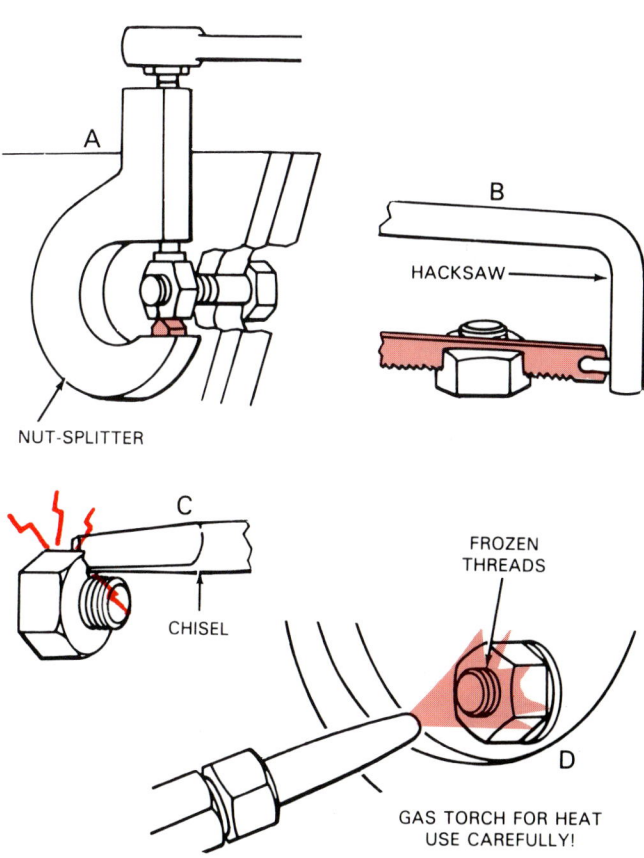

Fig. 4-9. Four common methods of removing stubborn nuts. A — Nut splitter. B — Hacksaw. C — Chisel. D — Gas torch.

USE PENETRATING OIL

Regardless of the method of removal, it is a good idea to apply penetrating oil (special light oil used to free rusty and dirty parts) to the area and give it a few minutes to work in. If heat is not injurious to the part, an application of heat will also help.

Use caution not to overheat. If in doubt as to the effects — do not apply heat.

NEVER USE A TORCH NEAR A FUEL TANK, BATTERY, OR OTHER FLAMMABLE MATERIALS.

REPAIRING THREADS

Occasionally threads, both external and internal, are only partially stripped. In such cases they can be readily cleaned up through the use of a die nut or a tap. See Fig. 4-10.

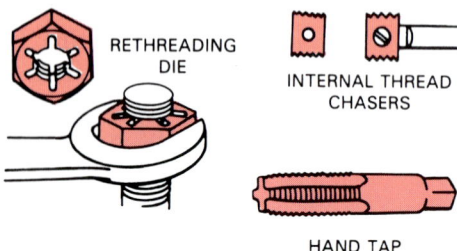

Fig. 4-10. Some thread restoring tools.

When threads in holes are damaged beyond repair, one of four things can be done:

1. The hole may be drilled and tapped to the next suitable oversize. Then, a larger diameter cap screw or stud can be installed. Use a chart to determine the proper size (tap size) to use. A clearance drill (drill the size of the bolts major diameter) must be passed through the attaching part to allow an oversize cap screw to be used, Fig. 4-11.

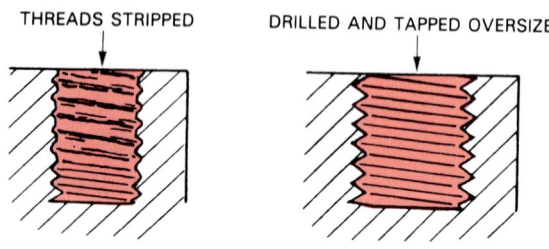

Fig. 4-11. Repairing stripped thread by drilling and tapping to next oversize.

2. The hole may be drilled and tapped to accept a threaded plug. The plug should also be drilled and tapped to the original screw size. A special selftapping plug already threaded to the original size may be used. You merely drill a hole to the specified size. Run the threaded plug into a hole using a cap screw and jam nut. When fully seated, the jam nut is loosened and the cap screw removed, Fig. 4-12.

3. Another method makes use of a patented coil wire insert called a Heli-Coil. The hole is drilled then tapped with a special tap. A Heli-Coil is then inserted. This brings the hole back to its original diameter and thread, Fig. 4-13.

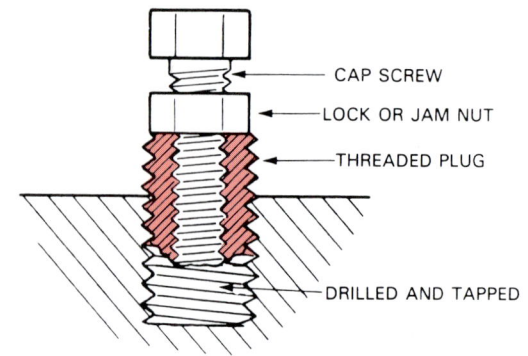

Fig. 4-12. Inserting threaded plug to repair stripped threads.

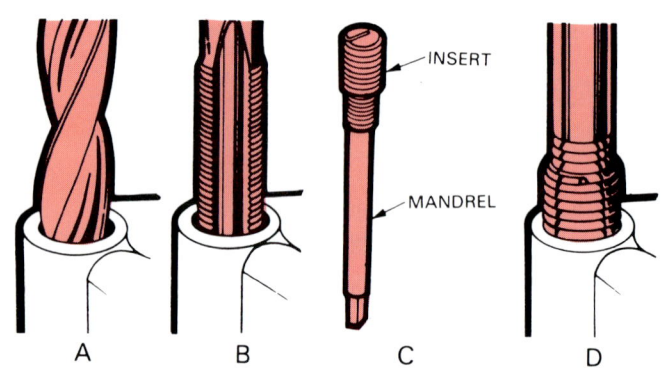

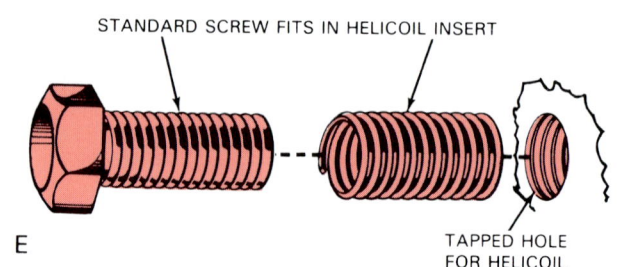

Fig. 4-13. Using a helicoil to repair stripped threads. A — Drill hole oversize. B — Tap hole oversize. C — Mount helicoil on mandrel. D — Thread helicoil into hole. E — Helicoil allows use of original size bolt.

REMEMBER

When removing a broken screw or repairing stripped threads, proceed carefully. A frantic or careless attempt at repair can often cause serious and costly trouble.

NUTS

Nuts are manufactured in a variety of sizes and styles. Nuts for automotive use are generally hexagonal in shape (six sided). They are used on bolts, studs, and obviously must be of the correct diameter and thread pitch. The most common ones are pictured in Fig. 4-14. Study their names closely.

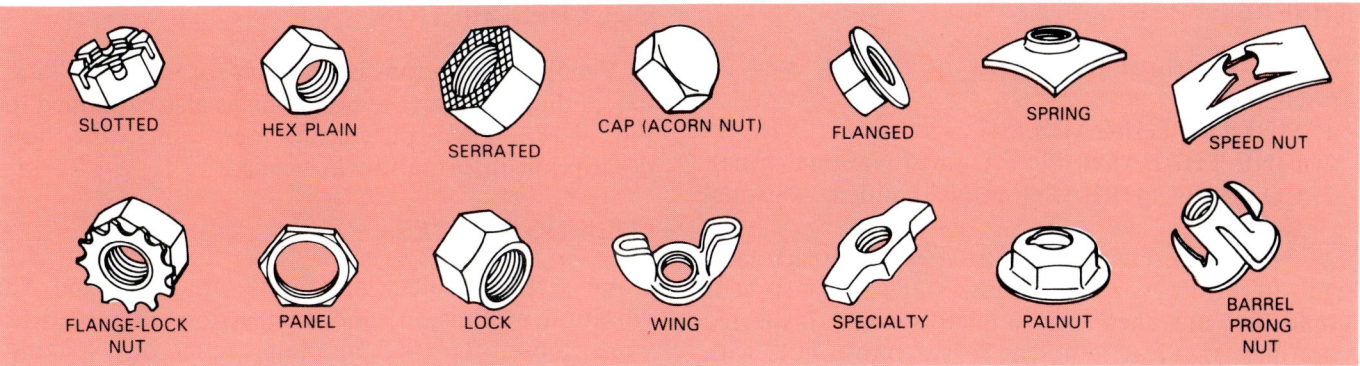

Fig. 4-14. All of these nut types are used in autos. Study them!

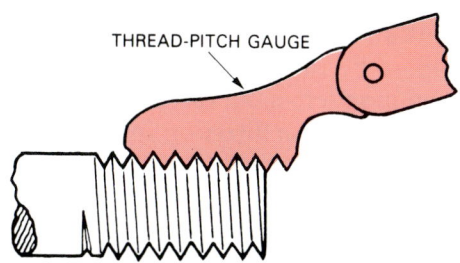

Fig. 4-16. Using a thread-pitch gauge to determine the type of thread and pitch.

BOLTS AND SCREWS TERMINOLOGY

Bolts and screws come in various sizes, grades (strengths), and thread types. It is important to be familiar with these differences.

The most important bolt dimensions are:

1. Bolt size is a measurement of the outside diameter of the bolt threads. See Fig. 4-15.
2. Bolt head size is the distance across the flats or outer sides of the bolt head. It is the same as the wrench size.
3. Bolt length is measured from the bottom of the bolt head to the threaded end of the bolt.
4. Thread pitch is the same as thread coarseness. With metric fasteners it is the distance between each thread in millimetres. With unified fasteners, it is the number of threads per inch. Refer again to Fig. 4-15. The pitch can be determined by using a thread pitch gauge as shown in Fig. 4-16 and Fig 4-17.

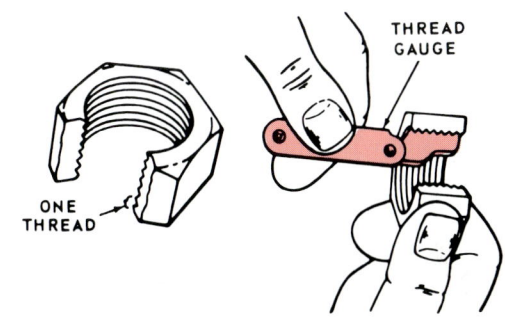

Fig. 4-17. Thread-pitch gauge being used to check nut for type of thread and pitch.

METRIC SYSTEM (Bolt M12-1.75x25)

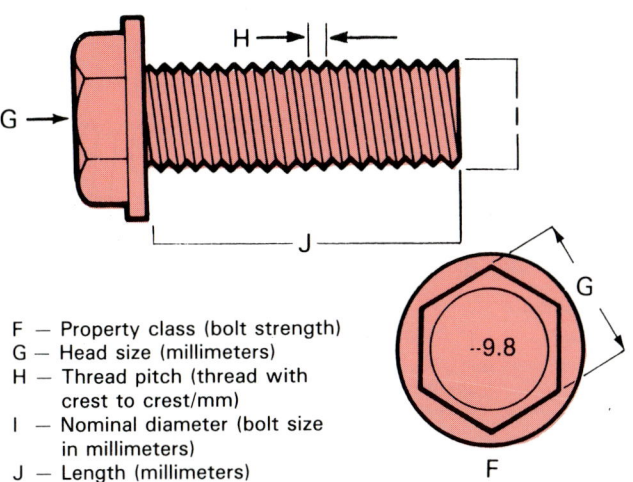

F — Property class (bolt strength)
G — Head size (millimeters)
H — Thread pitch (thread with crest to crest/mm)
I — Nominal diameter (bolt size in millimeters)
J — Length (millimeters)

UNIFIED SYSTEM (Bolt, 1/2-13x1)

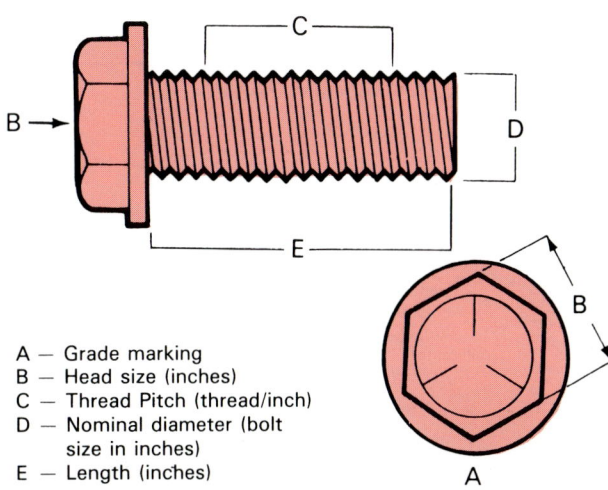

A — Grade marking
B — Head size (inches)
C — Thread Pitch (thread/inch)
D — Nominal diameter (bolt size in inches)
E — Length (inches)

Fig. 4-15. Bolt terminology. Compare terms used with metric and unified bolts.

THREAD TYPES

There are three basic types of threads used on fasteners:
1. METRIC THREADS (SI).
2. FINE THREADS (UNF-Unified National Fine).
3. COARSE THREADS (UNC-Unified National Coarse).

Never interchange thread types or thread damage will result. As shown in Fig. 4-18, metric threads could be mistaken for unified if not inspected carefully. If a metric bolt is forced into a hole with fine threads, either the bolt or part threads will be ruined.

Bolts and nuts also come in right and left-hand threads. With common right-hand threads, the fastener must be turned clockwise to tighten. With the less common left-hand threads, turn the fastener in a counter-clockwise direction to tighten. The letter "L" may be stamped on fasteners with left-hand threads.

BOLT GRADE (HEAD MARKINGS)

Tensile strength or grade refers to the amount of pull or stretch a fastener can withstand before breaking. Tensile strengths can vary. Bolts are made of different metals, some stronger than others.

Bolt head markings, also called grade markings, specify the tensile strength of the bolt. Conventional bolts are marked with lines or slash marks. The more lines, the stronger the bolt. A metric bolt is marked with a numbering system. The larger the number, the stronger the bolt. Look at Fig. 4-18.

DANGER! Never replace a high grade bolt with a lower grade bolt. The weaker bolt could easily snap, possibly causing part failure and a dangerous situation.

BOLT DESCRIPTION

A bolt description is a series of numbers and letters that describe the bolt. This is also explained in Fig. 4-18. When purchasing new bolts, the bolt description information in needed.

LOCKING DEVICES

As screws, bolts, and nuts are subjected to vibration, expansion, and contraction, they tend to work loose. To prevent this, numerous locking devices have been developed. These may be an integral part of the screw or nut, or may be a part placed under, through, and around the screw or nut. Epoxy cement is sometimes used.

SELF-LOCKING NUTS

Some nuts are designed to be self-locking. This is accomplished in various ways but all share the same principle. They create friction between the threads of the bolt or stud and the nut, Fig. 4-19.

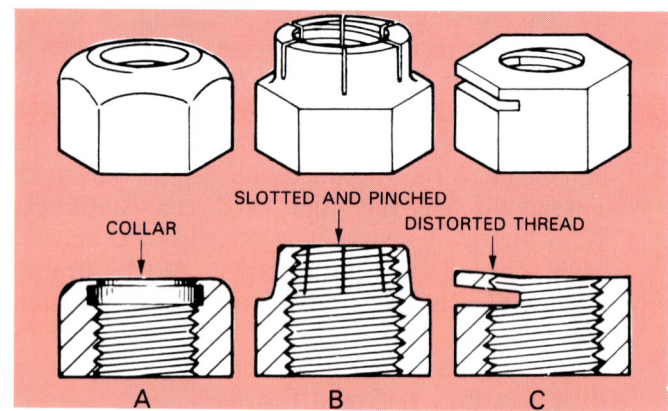

Fig. 4-19. Self-locking nuts. A — Soft collar type. B — Top section slotted and pinched together. C — Slot to distort upper thread area.

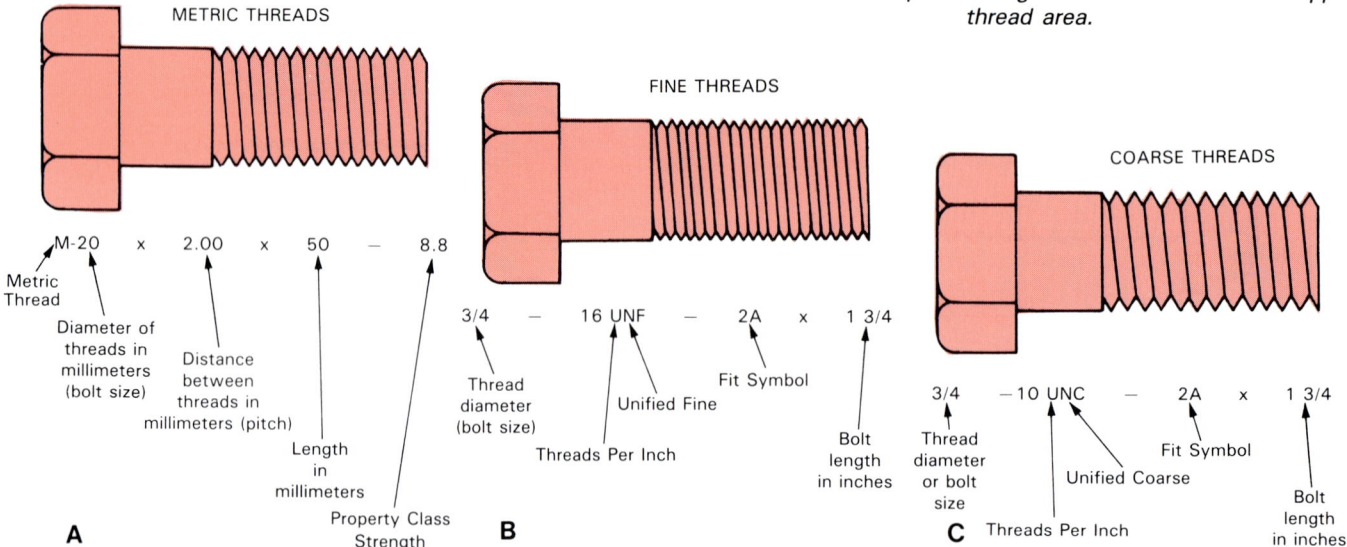

Fig. 4-18. Bolt designation number gives information about bolt. Number is commonly used when purchasing new bolts.

In Fig. 4-19, nut A utilises a collar of soft metal, fibre, or plastic. As the bolt threads pass up through the nut, they must force their way through the collar. This jams the collar material tightly into threads, locking the nut in place.

In B, the nut upper section is slotted and the segments are forced together. When the bolt passes through the nut, it spreads the segments apart thus producing a locking action.

Detail C shows a single slot in the side of the nut. The slot may be forced open or closed during manufacture, thus distorting the upper thread. This will create a jamming effect when bolt threads pull nut threads back into alignment. A crimped nut is shown in Fig. 4-53.

SELF-LOCKING SCREWS

Some cap screws have heads that are designed to spring when tightened to produce a self-locking effect. Occasionally, the threaded end of a cap screw will be split and the halves slightly bent outward. When threaded into a hole, the halves are forced together. This creates friction between the threads.

LOCK WASHERS

A lock washer is used under the nut and grips both the nut and the part surface. The three basic designs are the internal, external, and the plain.

When using lock washers with die cast or aluminium parts, a plain steel non-locking washer is frequently used under the lock washer. It prevents damage to the part, Fig. 4-20.

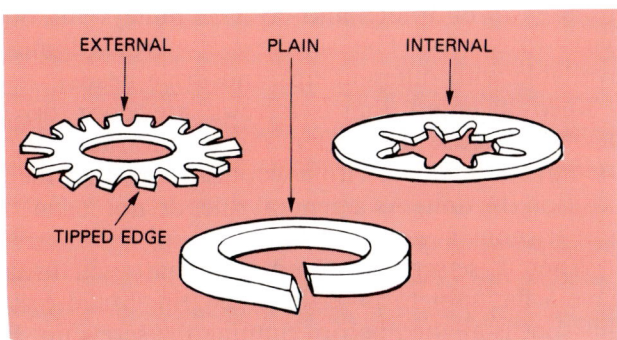

Fig. 4-20. Typical lockwashers. Not illustrated is another type that uses both internal and external fingers. Tipped edges provide gripping power in the "off" direction.

PALNUT

The palnut locking device is constructed of thin, stamped steel. It is designed to bind against the threads of the bolt when installed. In use, the palnut is spun down into contact with the regular nut (open side of palnut away from regular nut) with the fingers. Once firmly in contact with the nut, it is

given one-half turn. Do not tighten beyond ONE-HALF TURN as the effectiveness of the palnut will be destroyed. The one-half turn draws the steel fingers towards the nut causing them to jam into the threads, Fig. 4-21.

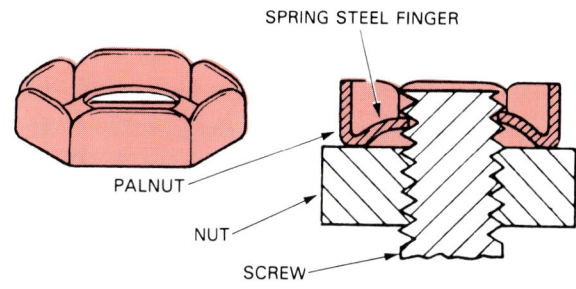

Fig. 4-21. Palnut. Half-turn jams steel fingers against threads.

COTTER PIN OR SPLIT PIN

Cotter pins are used both with slotted and castle nuts as well as on clevis pins and linkage ends. Use as thick a cotter pin as possible. Cut off the surplus length and bend the ends as shown in Fig. 4-22. If necessary, they may be bent around the sides of the nut. Make certain that the bent ends will not interfere with some part.

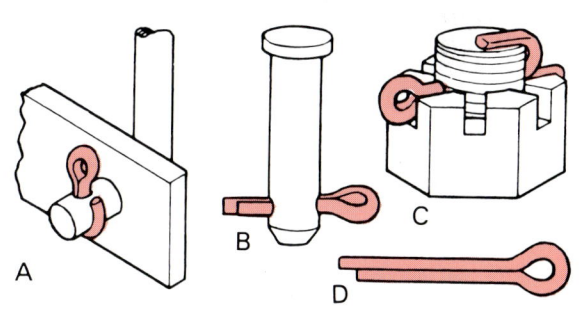

Fig. 4-22. Uses of cotter pin. A - Linkage. B — Clevis pin. C — Slotted hex nut. D — Typical cotter pin.

KEYS, SPLINES, AND PINS

Keys, splines, and pins are used to attach gears, sprockets, and pulleys to shafts so that they rotate as a unit. When a key or pin is used, the unit being attached to the shaft is generally fixed for no end to end movement. Splines will allow, when desired, longitudinal movement while still causing the parts to rotate together. In some cases pins are used to fix shafts in housings to prevent end movement and rotation. Fig. 4-23.

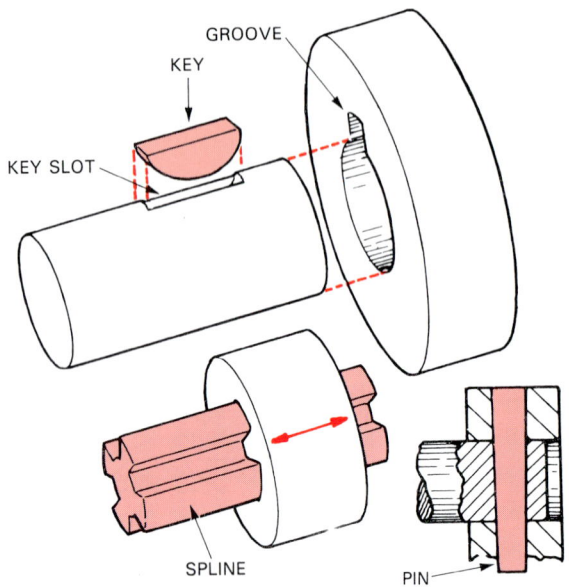

Fig. 4-23. Key, spline, and pin. Note that the spline allows end movement. The pin fixes the shaft to the housing, allowing no movement. The key is commonly referred to as a woodruff key or a half-moon key.

LOCKING PLATES AND SAFETY PINS

Locking plates are made of thin sheet metal. The plate is generally arranged so that two or more screws pass through it. The metal edge or tab is then bent up snugly against the bolt. Various patterns are used.

Occasionally screws will be locked with safety wire (soft or ductile wire). The wire is passed from screw to screw in such a manner as to exert a clockwise pull.

Never reuse safety wire and always dispose of locking plates on which the tabs are fatigued (ready to crack), Fig. 4-24.

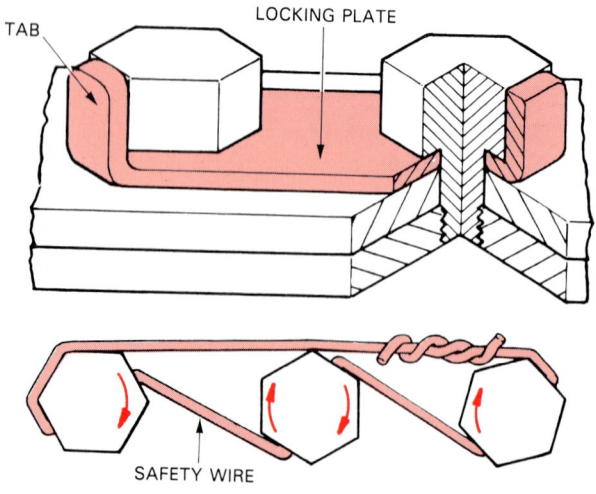

Fig. 4-24. Locking plate and safety wire. Tabs must be bent firmly against cap screw flat to prevent rotation.

SNAP RINGS

Snap rings are used to position shafts, bearings, gears, and other similar parts. There are both internal and external snap rings of numerous sizes and shapes.

The snap ring is made of spring steel. Depending on the type, it must be expanded or contracted to be removed or installed. Special snap ring pliers are used.

Be careful when installing or removing snap rings because overexpansion or contraction will distort and ruin them. If a snap ring is sprung out of shape — throw it away. NEVER attempt to pound one back into shape. Never compress or expand snap rings any more than necessary. Above all, do not pry one end free of the groove and slide it along the shaft. This may ruin the ring, Fig. 4-25.

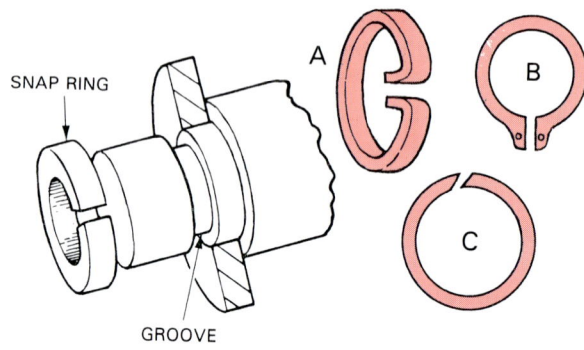

Fig. 4-25. Snap rings. A — Flat internal type. B — External. C — Round external. There are many shapes and sizes of rings.

SETSCREWS (GRUB SCREWS)

Setscrews, which are also known as grub screws, are used to both lock and position pulleys and other parts to shafts. The setscrew is hardened and is available with different tips and drive heads.

Keep in mind that setscrews are poor driving devices because they often slip on the shaft. When used in conjunction with a woodruff key, they merely position the unit. As a general rule, do not install any unit without a woodruff key.

When a setscrew is used, the shaft will usually have a flat spot to take the screw tip. Make certain this spot is aligned before running the screw up, Fig. 4-26.

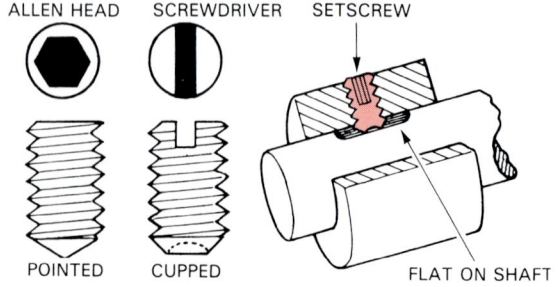

Fig. 4-26. Typical setscrews. Setscrews are hardened and they should be run up very tightly.

RIVETS

Rivets are made of various metals, including brass, aluminium, and soft steel. They find many applications on an automobile. They are installed cold so that there is no contraction that would allow side movement between parts. Fig. 4-27 shows several types of rivets.

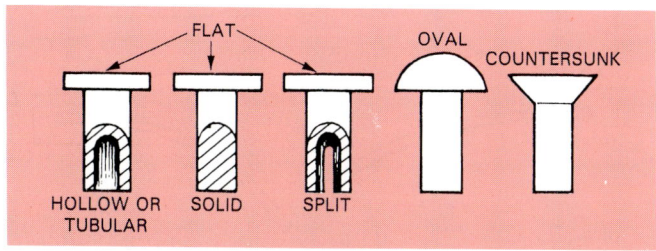

Fig. 4-27. Several types of rivets.

When using rivets, there are several important considerations. The two parts to be joined must be held tightly together before and during riveting. The rivet should fit the hole snugly. The rivet material must be in keeping with the job to be done. The rivet must be of the correct type (flat head, oval). The rivet should be set with a tool (rivet set) designed for the purpose. Fig. 4-28 illustrates the settings of a solid and a tubular rivet.

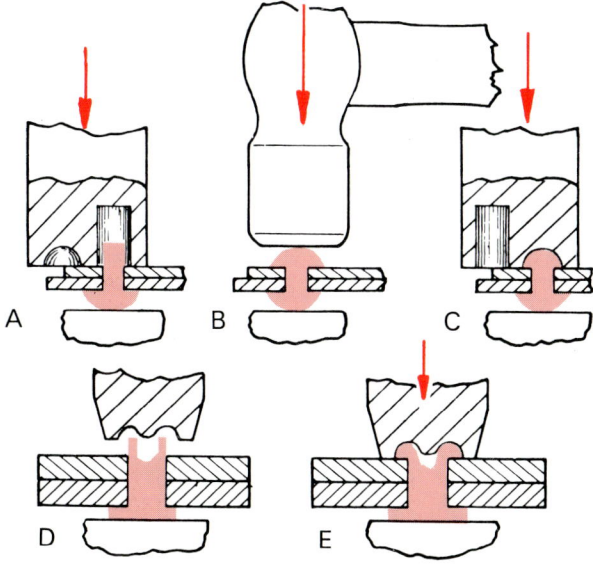

Fig. 4-28. Setting rivets. A — Pieces brought together and rivet seated. B — Rivet bulged. C — Rivet crowned and set. D — Set used for tubular rivet. E — Set forced down, crowning rivet as shown.

POP RIVETS

When one side of the work to be riveted is inaccessible, pop rivets may be used. They can be set from the outside and make blind rivets practical. Fig.4-29 illustrates the use of one form of pop rivet.

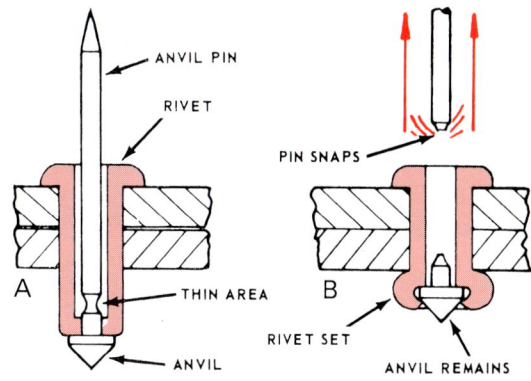

Fig. 4-29. Installing a pop rivet. A — Pop rivet in place. B. — Rivet tool has pulled anvil pin outward, pulling parts together, setting rivet and snapping off pin.

The pop rivet is inserted through the parts to be joined, a hand-operated setting tool, Fig.4-30, is placed over the rivet anvil pin. When the handles are closed, the anvil pin is pulled outward. As the anvil is drawn outward, the rivet head is forced against the work and the hollow stem is set. The setting process draws the two parts tightly together. Further pressure on the tool handles causes the anvil pin to snap off just ahead of the anvil. The anvil remains in the set area.

Fig. 4-30 shows a pop rivet tool being used to attach seat back trim.

Fig. 4-30. Pop rivet tool in use.

OTHER FASTENERS

In addition to fasteners already discussed, there are numerous other specialised type fasteners such as hose clamps, C-washers, clevis pins, and spring lock pins. Many types are pictured in Fig. 4-53.

FASTENERS SHOULD BE TORQUED

To better understand the reason for, and the proper application of, controlled torque, the mechanic should be familiar with several important terms. Read the definitions which follow carefully. These terms will be used a great deal in this section.

TORQUE: Torque is a turning or twisting force exerted upon an object — in this case, the fastener. It is measured in newton-metres. See Fig. 4-31.

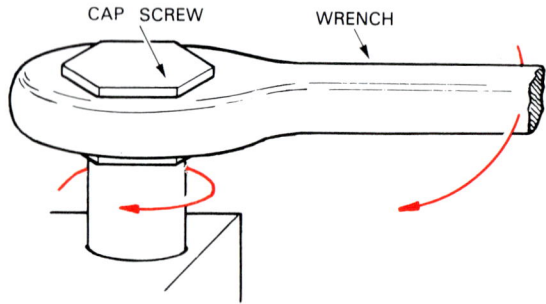

Fig. 4-31. Torque. Torque or a twisting force being applied to a cap screw with a box end wrench.

TENSION: Tension is a pulling force. When a cap screw is tightened, it actually stretches about 0.025 mm per 13,500 kg of tension, due to the tension being applied, Fig. 4-32.

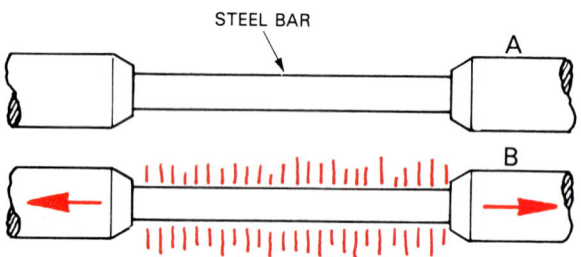

Fig. 4-32. Tension. A — Steel bar placed in jaws of a test machine. B — Jaws moving apart, creating a pull or tension on bar.

ELASTIC LIMIT: The amount or distance an object can be distorted (compressed, bent, stretched) and still return to the same dimension when the force is removed. Fig. 4-33.

DISTORTION: The normal shape or configuration of an object being changed or altered due to the application of some force or forces, Fig. 4-34.

TENSILE STRENGTH: the amount of pull an object will withstand before breaking, Fig. 4-35.

RESIDUAL TENSION: The stress remaining in an elastic object that has been distorted and not allowed to return to its original dimension, Fig. 4-36.

ELASTICITY: The ability of an object to return, after distortion, to its original shape and dimensions once the distortive force has been removed, Fig. 4-37.

COMPRESSION: A force tending to compress or squeeze an object, Fig. 4-38.

COLD FLOW: This refers to the tendency of an object under compression to expand outward thus reducing its thickness in the direction of compression, Fig. 4-39.

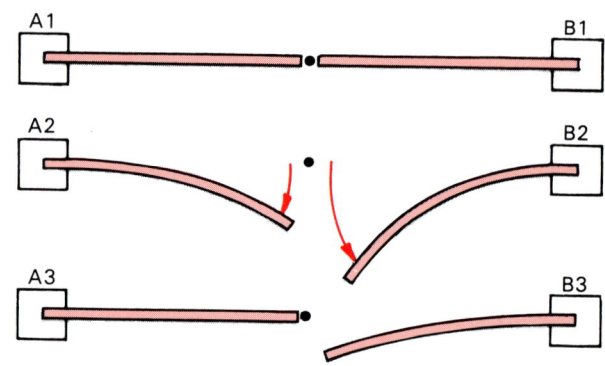

Fig. 4-33. Elastic limit. Bars in A1 and B1 at rest. Note that they are aligned with the black dot. In A2 the bar is bent within elastic limit and when pressure is removed it springs back to its normal (A3) position. Bar in B2 is bent beyond its elastic limit and when pressure is removed, the bar springs only part way back as in B3.

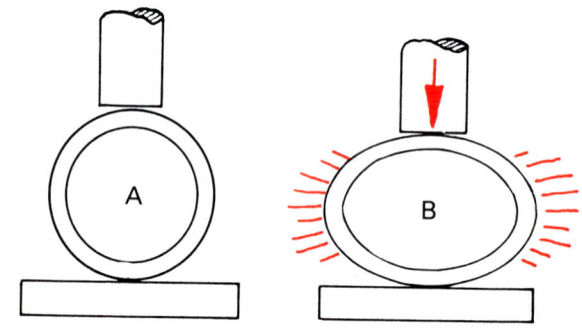

Fig. 4-34. Distortion. A — Hydraulic ram about to engage round steel ring. B — Pressure from ram bends or distorts ring.

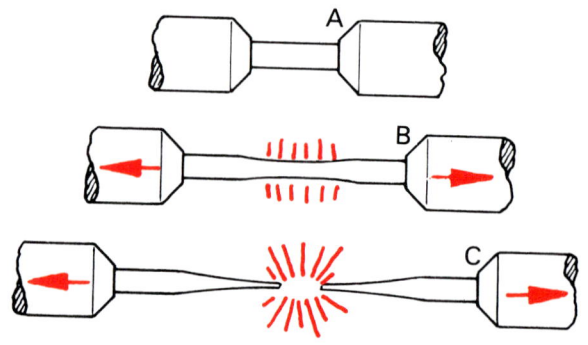

Fig. 4-35. Tensile strength. A — Bar of steel in a test machine. B — Heavy tension applied exceeding elastic limit, causing bar to stretch. C — Increased pull finally snaps bar as tension exceeds tensile strength.

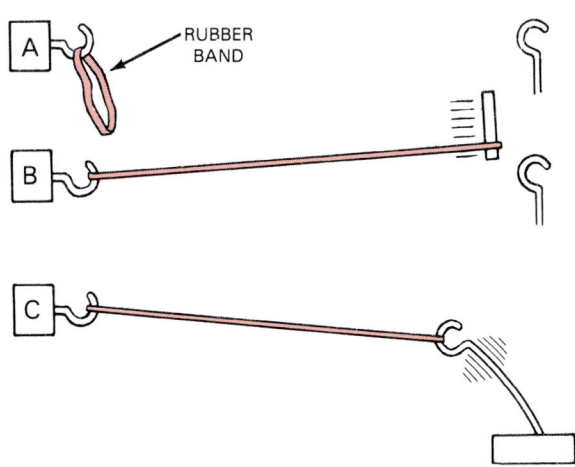

Fig. 4-36. Residual tension. A — Rubber band at rest; no residual tension. B — Band being pulled (distorted) out to engage spring steel hook. C — Band attempts to return to original dimensions, creating a pull (residual tension) and bending hook. Within its elastic limit, steel is more elastic than rubber.

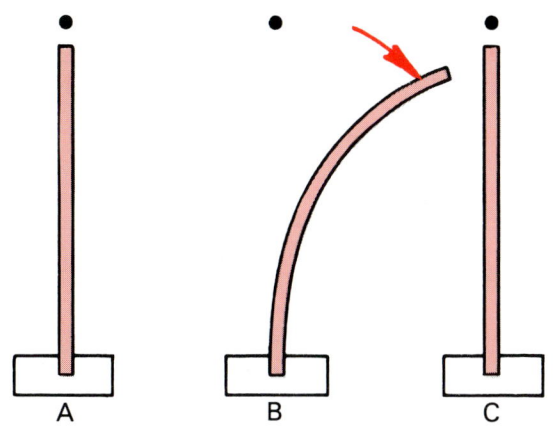

Fig. 4-37. Elasticity. A — Original position of bar. B — deflected by pressure. C — No pressure, and bar returns to original position.

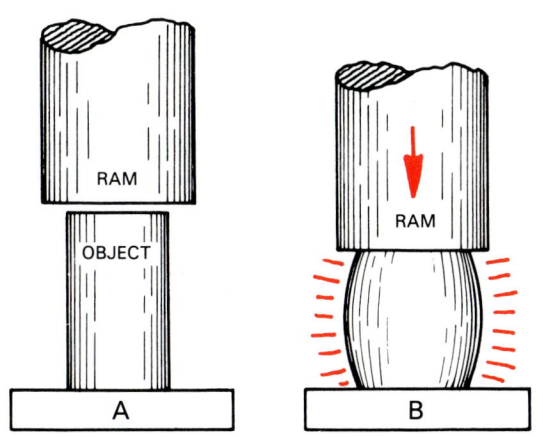

Fig. 4-38. Compression. A — Object at rest. B — Object under compression as ram builds up pressure.

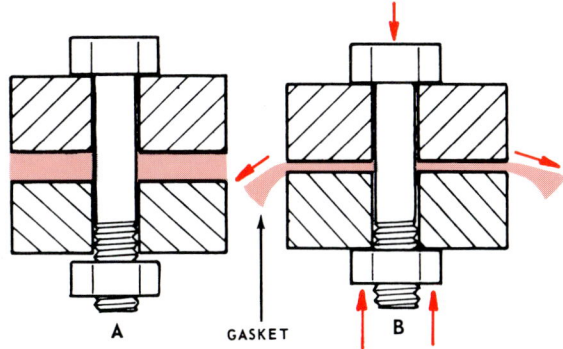

Fig. 4-39. Cold flow. A — The nut is not tight and there is no compressive force on gasket. B — Nut is tightened, compressing gasket and causing it to flow outward as thickness decreases.

HOOKE'S LAW: This law states that the amount of distortion (lengthening, shortening, bending, twisting), as long as it is kept within the elastic limits of the material, will be directly proportional to the applied force. This forms the basis for spring scales and torque wrenches, Fig. 4-40.

HIGH PRESSURE LUBRICANT: A lubricant that continues to reduce friction between two objects even when they are forced together under heavy pressure.

TORQUE FASTENERS

To understand the VITAL NECESSITY of torquing, we should first establish what we want to accomplish by tightening fasteners. Once this is clear, the reason for the use of a torque wrench becomes obvious.

We tighten fasteners to hold parts together. On the surface, this seems like a simple statement, but there is more here than meets the eye. When we say to hold parts together, we are, in effect, saying that once

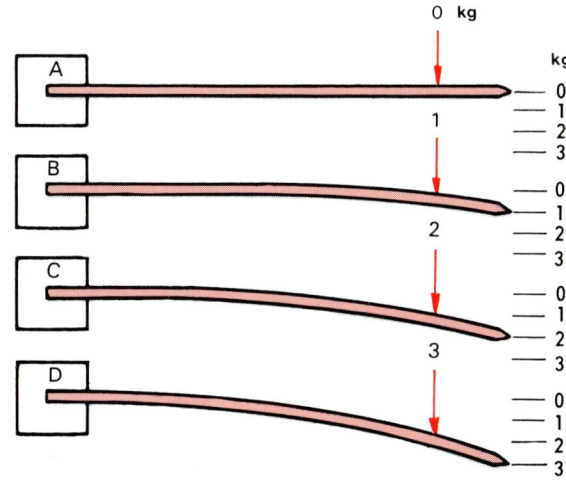

Fig. 4-40. Hooke's Law. Note that as weight on spring bar is increased, there is a proportionate movement on scale. This would continue until bar was deflected past its elastic limit.

together, the parts remain that way. When drawn together, the parts should not be distorted. The fasteners should not be overtightened to the point where they will fail in service. They must be tightened enough to prevent them from working loose, and perhaps being sheared or pounded apart. Oil, fuel, and water leaks should not occur.

Let's assume that a "shoddy" mechanic has just completely assembled an engine with a "guess" method of tightening. Here is what COULD HAPPEN to the engine:

1. Cylinders out-of-round.
2. Connecting rod and main bearings egg shaped.
3. Cylinder head warped.
4. Valve guides forced out of alignment.
5. Camshaft bearing centreline out.
6. Crankshaft centreline out.
7. All engine components affected to some extent.

In addition, blown head gaskets, oil, water and air leaks, and broken connecting rods can plague the job.

Obviously, the amount of distortion will vary depending on the stresses set up within the assembly. Even at best, ring, piston, valve, and bearing wear will be accelerated. The job will fail in service long before it should.

PROPER FASTENER TENSION

The first thing to keep in mind is that all car manufacturers publish torque specifications. They should be followed. Each company has spent a great deal of time and money determining the fastener torque for their products that will give the best results. When using torque charts, make sure they pertain to the job at hand.

It has been found that for the vast majority of applications, a fastener should be tightened until it has built up a tension within itself that is around 50 to 60 percent of its elastic limit.

When the fastener has been drawn up to this point, it will not be twisted off. It will retain enough residual tension to continue to exert pressure on the parts and will resist loosening. Steel bolts and cap screws will stretch about .025 mm for each 14,000 kg of tension. Like a rubber band, the tendency to return to their normal length provides continuous clamping effect.

FASTENER MATERIAL

As previously mentioned, most bolts and screws have radial lines or numbers on the head that indicates tensile strength. When replacing a fastener, use a quality at least equal to that originally used. You will find that the more critical the application (main bearing, connecting rod) the better the quality.

HOW FASTENER TORQUE IS MEASURED

To secure recommended torque, a measuring tool called a TORQUE WRENCH is a "must". The torque wrench will measure the torque (twisting force) applied to the fastener. Single round beam, double round beam, and single tapered beam type of torque wrenches are shown in Fig. 4-41.

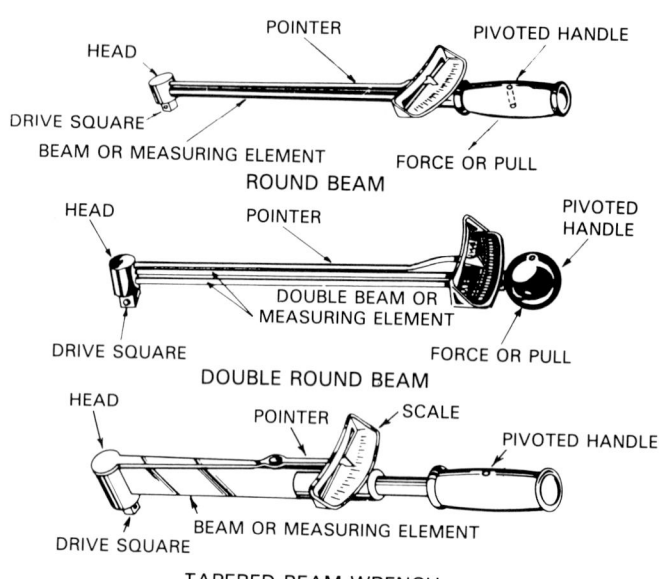

Fig. 4-41. Torque wrenches. These are all beam type wrenches, all widely used, durable, and accurate.

HOW A TORQUE WRENCH WORKS

The torque wrench uses Hooke's Law in its construction. By deflecting (bending), a steel beam (in some cases a coil spring) the relationship between the pull on the handle (torque) and the amount of beam deflection is readily established.

When the head is attached to the fastener and the handle is pulled, the flexible beam is bent. The pointer rod, being attached to the solid wrench head, is not bent. Since the scale is attached to the handle element, it follows the flexible beam, thus moving the scale under the pointer end. The scale is calibrated so that the operator can see how much torque is being applied.

The torque is expressed in newton metre (Nm) on the scale. One newton metre of torque is equal to the torque produced by 1 newton of force acting at a perpendicular distance of 1 metre from an axis of rotation. See Fig. 4-42.

Torque wrenches are available with a sensing device in addition to the scale. This warns the user that a preset torque has been reached without reading the scale. Various types of sensing devices such as a light or audible click are employed. When a torque wrench must be used in a position that makes reading the scale difficult or impossible, the sensing device is handy.

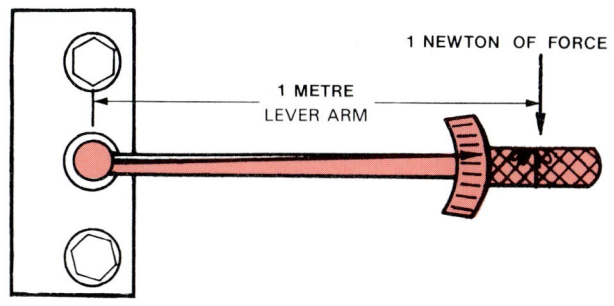

Fig. 4-42. One newton metre (1 Nm) of torque equals one newton of force acting at a distance of one metre.

TORQUE WRENCH RANGE

Torque wrenches are made in different sizes or ranges. Ideally the mechanic should have a 1-20 Nm, 10-200 Nm and 60-400 Nm torque wrench.

A torque wrench will produce BEST results if used somewhere near the middle half of its range. For example a 10-200 Nm torque wrench would give the most accurate readings from around 50 to 150 Nm. By having several ranges of torque wrenches, the mechanic will also find that this will offer several lengths. The shorter ones can be useful in restricted areas.

RANGE CAN BE ALTERED BY ADAPTER

Say you have a 1-200 Nm torque wrench available and the torque recommendation is 250 Nm. This is obviously beyond the range of the wrench. It can still be used, however, through the use of an adapter to lengthen the effective range.

If the lever length (distance from the centre of the wrench head to the pivot point on the handle) is 0.50 metre and you used an adapter bar of equal length, the torque being applied would be double that shown on the scale. If the adaptor was 0.250 metre or half as long as the lever length, the torque would be one and a half times that shown on the scale. A handy formula to determine applied torque when using an adapter or extension is:

$$\frac{\text{Dial reading} \times (L + A)}{L}$$

$$= \text{Torque applied to fastener.}$$

(L) is length in metres from centre of handle pivot to centre of wrench head.

(A) is length in metres from centre of wrench head to end of adapter. (A must be measured parallel to centreline of wrench).

Fig. 4-43 shows three adapter setups. Notice that the effective length (L + A) is always measured parallel to the centreline of the wrench. REMEMBER: when using adapters or extensions, be certain of their exact length. Do not forget that length and torque are directly related, Fig. 4-43.

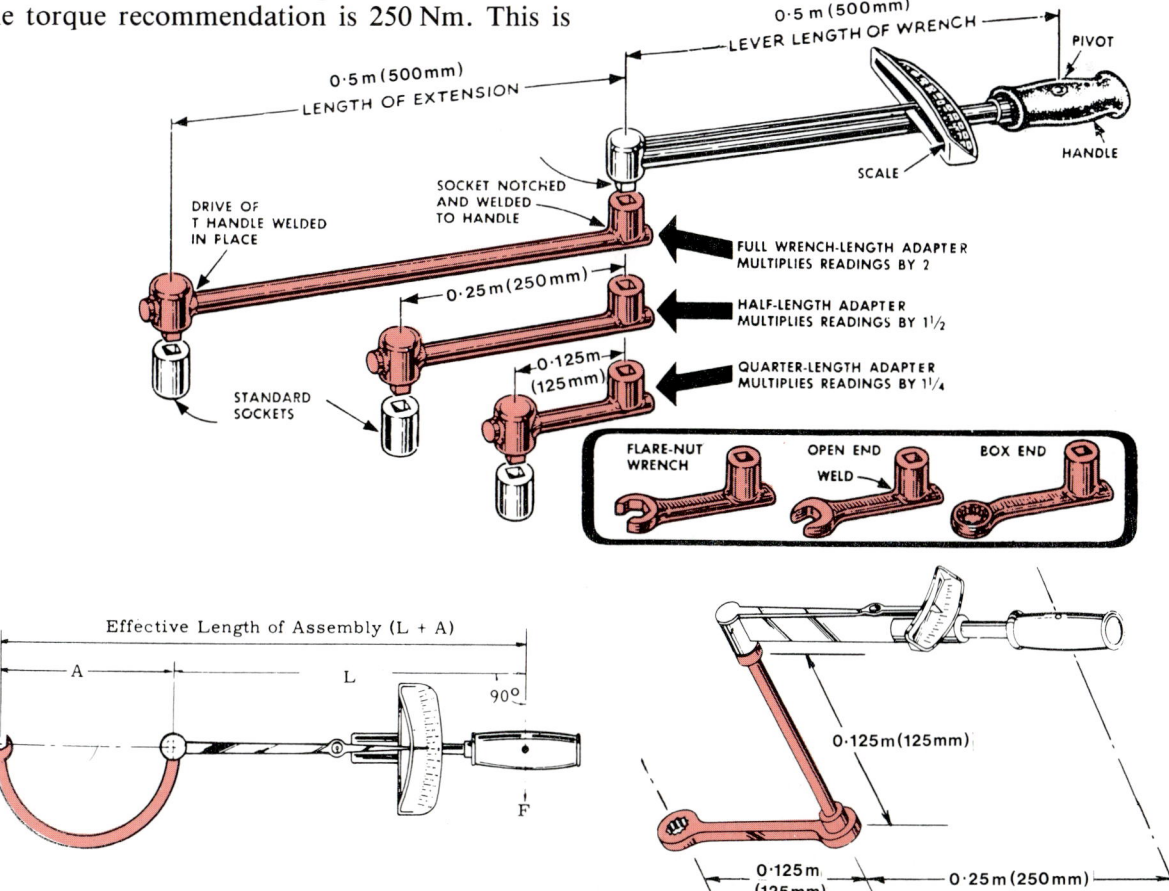

Fig. 4-43. Torque wrench adapters.

USING TORQUE WRENCH

After determining the proper torque and selecting a suitable range torque wrench, you are ready to proceed. Be sure to observe the following:

THREADS MUST BE CLEAN. The threads on the bolt or screw as well as those in the nut or hole, must be absolutely clean. Rust, carbon, and dirt will cause galling and improper tension. An accurate torque reading with dirty threads is impossible.

USE HIGH TEMPERATURE LUBRICANT. Unless the use of a lubricant is specifically forbidden (due to possibility of area contamination or need of a special sealant) always apply a high pressure lubricant to the threads and to the area where the nut or cap screw head contacts the part.

Refer to manufacturer recommendations to find out which high strength lubricant is suitable.

The use of lubricant will prevent or reduce the possibility of galling, seizing (sticking) or stripping. It will assure that the fastener torque has created the proper tension. It should be mentioned that the lubricant, while making the fasteners easier to remove at some future date, will not (if torqued properly) cause them to loosen in service. To the contrary, the increased tensioning for the same torque reading will actually cause the fastener to remain more secure.

USE PROPER LOCKING DEVICE

Unless a self-locking nut or cap screw is being used, make certain the recommended lock washer is in place. When running a fastener up against the softer metals, the use of a plain, flat washer between the lock washer and the part is often specified. This prevents the part from being "chewed" up and allows proper torquing without crushing the part.

CHECK FASTENERS

Be careful to check fasteners for correct diameter, pitch, and length. When installing cap screws, make certain they will not bottom (strike bottom of a threaded hole), in a blind hole (hole not drilled clear through part).

Also, they must not protrude into a housing and damage a part of the unit.

REMEMBER: Stripped threads, broken screws, loose parts, and damaged units can result. Be careful!

In A, Fig. 4-44, the screw has bottomed leaving the part loose. Continued torquing could twist off the screw. In B, the screw protruded into case and damaged gear. In C, coarse thread screw, jammed into hole with fine threads, cracked part, Fig. 4-44.

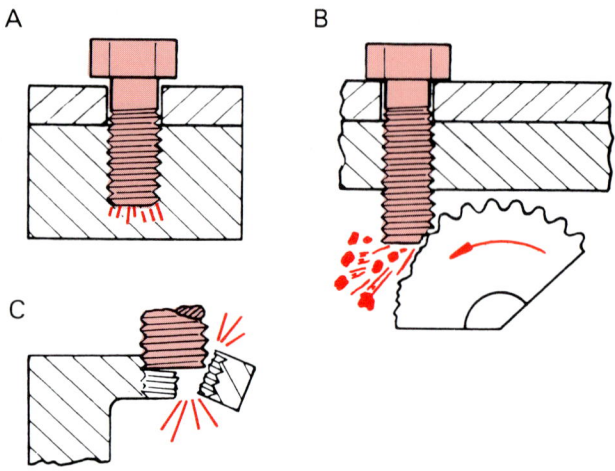

Fig. 4-44. Check fasteners! Make certain that fasteners are of correct diameter, length, and with sufficient thread of correct pitch.

Some fasteners serve an additional purpose. For instance, a head bolt or cap screw may be drilled for passage of oil, or a cap screw may have a threaded hole in the head to which another assembly is attached. Be careful to insert these fasteners in the correct place.

FOLLOW RECOMMENDED SEQUENCE

Where a number of fasteners are used to secure a part (such as a cylinder head), the proper sequence (order) of tightening should be followed. Fig. 4-45 illustrates the head bolt tightening sequence for one model engine. Always follow the manufacturer's specifications. See Figs. 4-45 and 4-46.

If no sequence chart can be obtained, it is usually advisable to start in the centre and work out to the ends. The chart in Fig. 4-47 illustrates this technique.

On some assemblies, it is advisable to use a crisscross sequence. Always avoid starting in one spot and tightening one after another in a row.

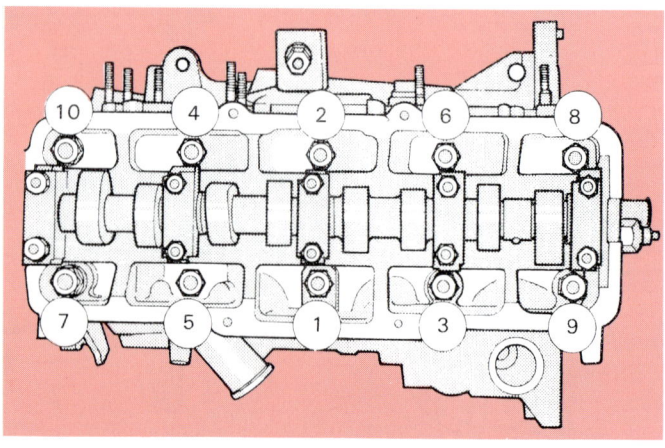

Fig. 4-45. Cylinder head bolt tightening sequence for one specific engine.

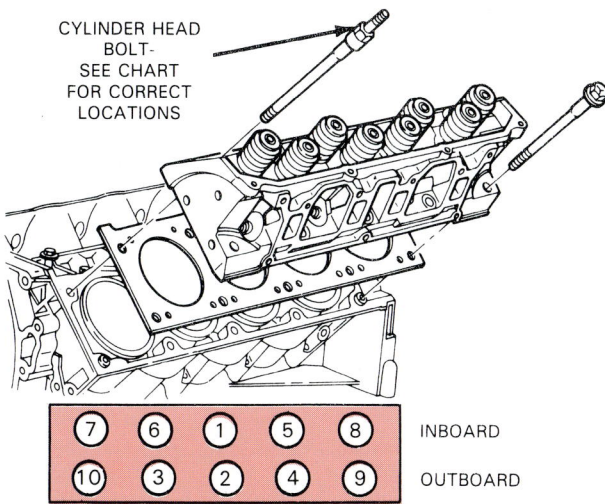

Fig. 4-46. Another cylinder head bolt tightening sequence.

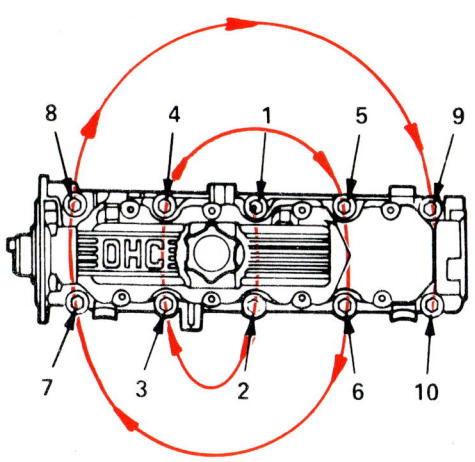

Fig. 4-47. Head bolt tightening sequence when no special recommendation is available.

Remember that the object is to tighten the parts so that an even stress is set up. At the same time, a crisscross pattern will allow the parts to be drawn together so that their mating surfaces will contact evenly, Fig. 4-48.

Would a good fit be acquired if you followed the sequence shown in Fig. 4-49?

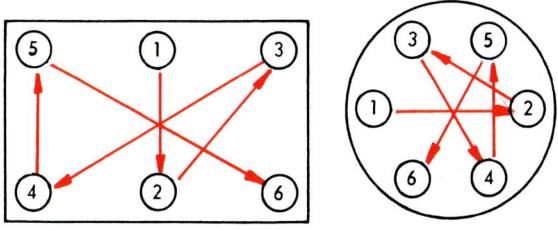

Fig. 4-48. Tightening bolts in crisscross sequence.

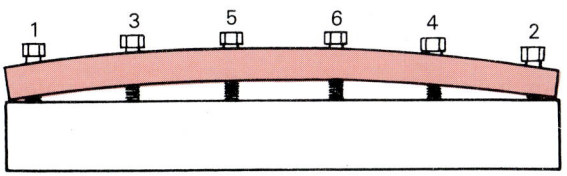

Fig. 4-49. Wrong sequence in tightening fasteners. This sequence would produce a very poor fit!

Quite obviously, if this sequence is followed, the two ends would be clamped down first. When the centre bolts were tightened, the part could not flatten out. In order to flatten, it must spread outward. In order to do this, the ends must be free.

TORQUE IN FOUR STEPS

Always run the fasteners up snug (do not overtighten) with a regular spanner. Then, observe the following four steps:

1. Run each fastener, in the proper sequence, up to one-third of the recommended torque setting.
2. Repeat the process running up to two-thirds of the setting.
3. Repeat, running every fastener up to full torque.
4. This is a very important and frequently overlooked step — often to the embarrassment of the mechanic when the unit fails. REPEAT STEP THREE TO BE POSITIVE YOU HAVE NOT MISSED A FASTENER!

HOLDING THE TORQUE WRENCH

Where possible (it saves skinned knuckles), PULL on the wrench. Keep your hand on the handle. If using a pivoted handle, keep the handle from tipping in against the wrench. This is important as the pivot is where the pull should be for exact readings. Items A and B, in Fig, 4-50, show the correct hand position. In C, the mechanic has placed the right hand on one end of the handle tipping it and causing interference with wrench action. D shows an extension in place on the handle. This should never be done.

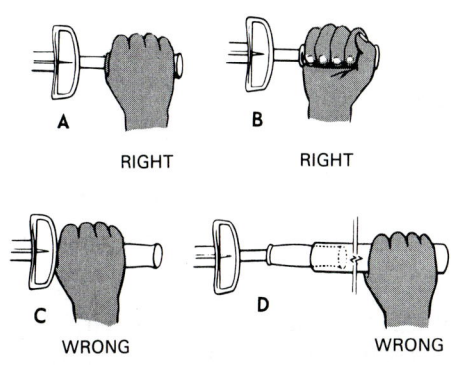

Fig. 4-50. Grasp the torque wrench properly.

PULLING THE WRENCH

When using a beam-type torque wrench, especially the single round beam, be careful to pull so that the beam is bent only in the direction of travel. If the wrench is bent up or down while pulling, the indicator point can drag on the scale, thus impairing the reading.

Place the palm of the left hand on the head of the wrench to counterbalance the pull on the handle. Allow your palm to turn with the wrench.

Fig. 4-51 illustrates the use of the left hand for balance. In this case, both an adapter and extension are being used.

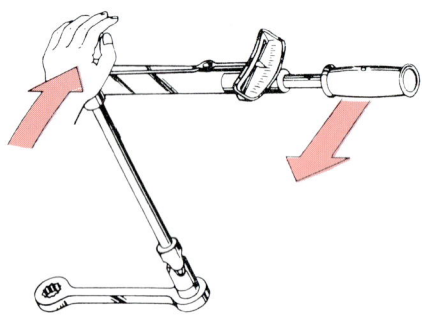

Fig. 4-51. Use the palm of the hand on the head of the wrench to balance the pull on the handle.

STICKING

Quite often, when nearing full torque value, you will hear a popping sound. The fastener will seem to stick and stop turning. If you increase pressure on the wrench, it may run up to full torque without moving the fastener.

You will find that when a fastener has stuck, the torque required to start it moving (break-away torque), is much higher than that required to keep it moving. This indicates that break-away torque is not a true fastener torque.

When sticking occurs, run the fastener in an off-direction (about one-half turn) until it breaks free. Then, with a smooth and steady pull, sweep the wrench handle around in a tightening direction. STOP when the required torque is reached.

RUN-DOWN TORQUE

Self-locking nuts, slightly damaged threads, or foreign material will cause the fastener to turn with some degree of resistance before it begins drawing parts together. This is called run-down torque.

If at all noticeable, add this run-down torque to the recommended torque. Determine run-down torque only during the last one or two turns of the fastener. When a fastener is first started, it may show considerable resistance. However, by the time it reaches bottom, this may have lessened or disappeared.

CAUTION! Whenever a fastener shows undue resistance — remove it and make sure it is the right length, diameter, and has the proper thread pitch.

WHEN TORQUE RECOMMENDATIONS ARE NOT AVAILABLE

The mechanic should try to secure the car manufacturer's recommended torque for the specific job. If, however, it is not available, consult a chart such as the one in Fig. 4-52. You will note that by using the head markings and diameter, an approximate torque setting may be determined.

TIGHTENING TORQUE OF STANDARD BOLT

Grade	Bolt or nut size	Bolt or nut nominal diameter mm	Pitch mm	Tightening torque N.m
4T	M6	6.0	1.0	3 - 4
	M8	8.0	1.25	8 - 11
			1.0	8 - 11
	M10	10.0	1.5	16 - 22
			1.25	16 - 22
	M12	12.0	1.75	26 - 36
			1.25	30 - 40
	M14	14.0	1.5	46 - 62
7T	M6	6.0	1.0	6 - 7
	M8	8.0	1.25	14 - 18
			1.0	14 - 18
	M10	10.0	1.5	25 - 35
			1.25	26 - 36
	M12	12.0	1.75	45 - 61
			1.25	50 - 68
	M14	14.0	1.5	76 - 103
9T	M6	6.0	1.0	8 - 11
	M8	8.0	1.25	19 - 25
			1.0	20 - 27
	M10	10.0	1.5	36 - 50
			1.25	39 - 51
	M12	12.0	1.75	65 - 88
			1.25	72 - 97
	M14	14.0	1.5	109 - 147

Fig. 4-52. Typical chart showing tightening torque of standard metric bolt.

Keep in mind that if the fastener is threaded into aluminium, brass, or thin metal, the torque figures may have to be reduced to prevent stripping.

RETORQUING

On some assemblies, such as cylinder heads and manifolds, the car manufacturer may recommend that they be retorqued after a certain period of operation. A service manual for the particular vehicle will show the proper interval and procedure.

SUMMARY

The expert mechanic is vitally concerned with fastener design, application, and torque. He or she realises that, to a great extent, the success or failure of the work depends upon the proper use of fasteners.

There are many types of fasteners: screws that thread into a part, bolts that pass through the parts and require nuts, studs that thread into the part, and also, nuts and sheet metal screws that cut their own threads. Fig. 4-53.

Threaded fasteners are identified by material, thread pitch, diameter, length of thread, type, etc. Steel bolts may use markings on the head to indicate material and tensile strength.

The removal of broken fasteners can cause difficulty unless done properly. Various methods are used.

When threads in a hole are damaged beyond repair, the hole may be drilled and tapped:

1. To the next suitable oversize and a larger cap screw installed.
2. To accept a threaded plug.
3. To accept a patented coil wire insert.

See Fig. 4-54 which shows tap drill sizes for Metric (SI), Unified National Coarse and Unified National Fine threads.

Snap rings, rivets, clevis pins, keys, and splines are nonthreaded fasteners.

Fasteners tend to loosen in service. Self-locking nuts, various lock washers, safety wire, locking plates, and cotter pins are some of the methods of keeping fasteners tight.

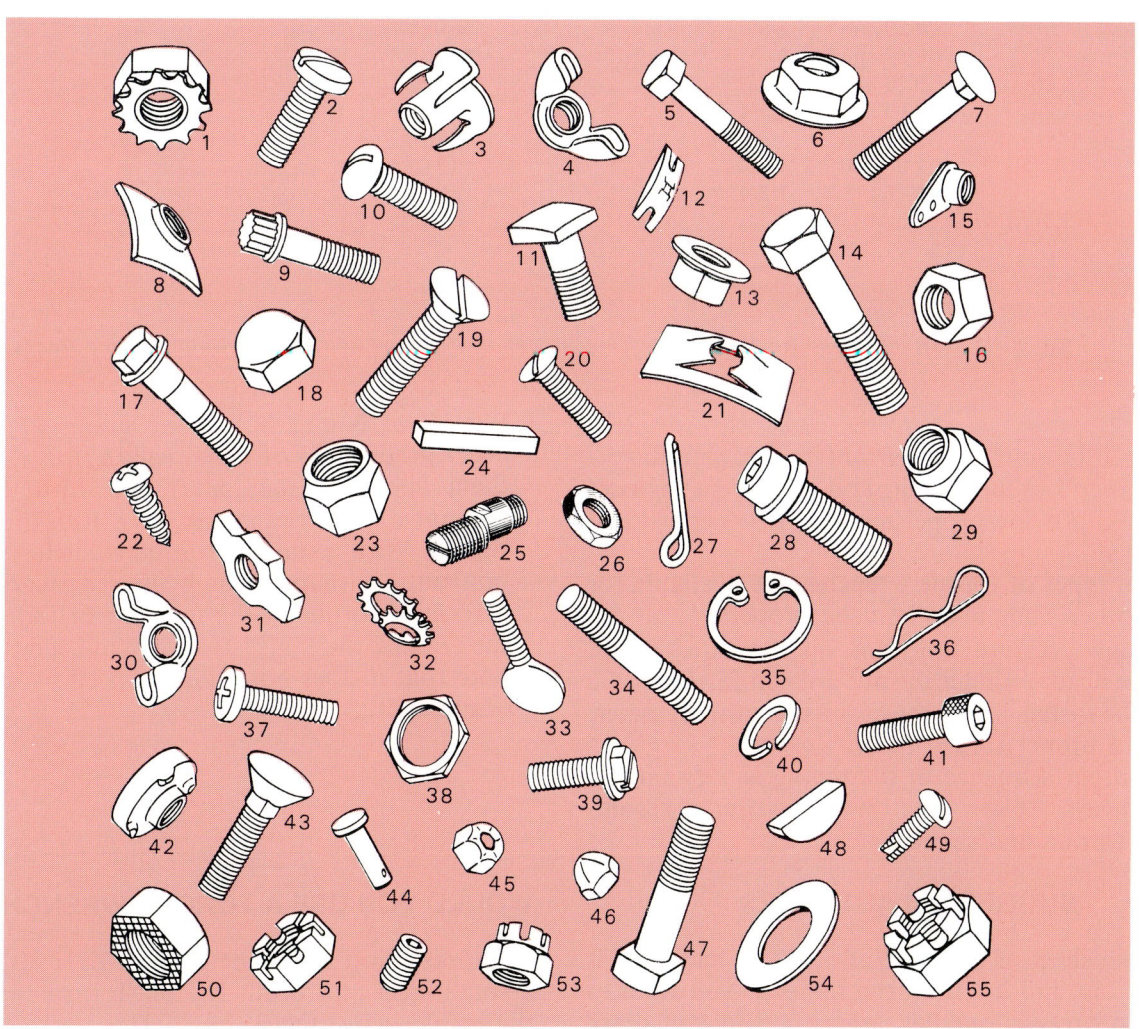

Fig. 4-53. An assortment of fasteners. Although terminology can vary somewhat, these are commonly used names: 1 — Flange-lock nut. 2 — Fillister head machine screw. 3 — Barrel prong nut. 4 — Wing nut. 5 — Cap screw. 6 — Palnut. 7 — Carriage bolt. 8 — Spring nut. 9 — 12-point head bolt. 10 — Round head machine screw. 11 — Askew-head bolt. 12 — Single thread nut. 13 — Flanged nut. 14 — Cap screw. 15 — Anchor nut. 16 — Plain hex nut. 17 — Hex flange screw. 18 — Acorn (cap) nut. 19 — Flat head screw. 20 — Small flat head screw. 21 — Speed nut. 22 — Sheet metal screw. 23 — Locking nut. 24 — Key. 25 — Offset (eccentric) stud. 26 — Thin nut. 27 — Cotter pin. 28 — Socket head bolt. 29 — Locking nut. 30 — Wing nut. 31 — Specialty nut. 32 — Toothed lock washers. 33 — Thumbscrew. 34 — Stud. 35 — Snap ring. 36 — Spring lock pin. 37 — Cross head machine screw. 38 — Panel nut. 39 — Flanged hex slotted head screw. 40 — Split lock washer. 41 — Hex socket head bolt. 42 — Welded nut. 43 — Plow bolt. 44 — Clevis pin. 45 — Open top acorn nut. 46 — Closed top acorn nut. 47 — Square head cap screw. 48 — Woodruff key. 49 — Self-tapping screw. 50 — Serrated nut. 51 — Slotted nut. 52 — Set screw. 53 — Castle nut. 54 — Flat washer. 55 — Castle nut.

I.S.O. METRIC

Nominal Diameter	Pitch	Tapping Drill
2.0	0.4	1.60 mm
2.5	0.45	2.05 mm
3.0	0.5	2.50 mm
3.5	0.6	2.90 mm
4.0	0.7	3.30 mm
4.5	0.75	3.70 mm
5.0	0.8	4.20 mm
6	1.0	5.00 mm
7	1.0	6.00 mm
8	1.0	7.00 mm
8	1.25	6.80 mm
9	1.25	7.80 mm
10	1.0	9.00 mm
10	1.25	8.80 mm
10	1.5	8.50 mm
11	1.5	9.50 mm
12	1.25	10.80 mm
12	1.75	10.20 mm
14	1.25	12.80 mm
14	1.5	12.50 mm
14	2.0	12.00 mm
16	1.5	37/64
16	2.0	35/64

U.N.C.

Nominal Diameter	T.P.I.	Tapping Drill
3 (.099)	48	2.00 mm
4 (.112)	40	2.25 mm
5 (.125)	40	2.60 mm
6 (.138)	32	2.75 mm
8 (.164)	32	3.40 mm
10 (.190)	24	3.80 mm
12 (.216)	24	4.40 mm
1/4	20	5.10 mm
5/16	18	6.60 mm
3/8	16	5/16
7/16	14	9.40 mm
1/2	13	27/64
9/16	12	31/64
5/8	11	17/32
3/4	10	21/32
7/8	9	49/64
1	8	7/8
1.1/8	7	63/64
1.1/4	7	1.7/64
1.3/8	6	1.7/32
1.1/2	6	1.21/64
1.3/4	5	1.35/64
1	4.1/2	1.25/32

U.N.F.

Nominal Diameter	T.P.I.	Tapping Drill
3 (.099)	56	2.10 mm
4 (.112)	48	2.35 mm
5 (.125)	44	2.65 mm
6 (.138)	40	2.90 mm
8 (.164)	36	3.50 mm
10 (.190)	32	4.10 mm
12 (.216)	28	4.60 mm
3/16	32 UNS	4.00 mm
1/4	28	5.50 mm
5/16	24	6.90 mm
3/8	24	8.50 mm
7/16	20	25/64
1/2	20	29/64
9/16	18	33/64
5/8	18	37/64
3/4	16	11/16
7/8	14	13/16
1	14 N.S.	15/16
1	12	59/64
1.1/8	12	1.3/64
1.1/4	12	1.11/64
1.3/8	12	1.19/64
1.1/2	12	1.27/64

Fig. 4-54. Tap drill sizes for Metric (SI), Unified National Coarse and Unified National Fine threads.

Fastener tension is important to prevent distortion, to keep fasteners tight, and to prevent fastener failure. To provide proper tension, fasteners should be torqued.

Several types of torque wrenches are available for this purpose. They must be used properly.

Use high pressure lubricant on the threads and under the head or under the nut area on fasteners. Be certain the fastener is of the correct length, diameter, and has a proper thread pitch.

The proper sequence of tightening is very important. Always follow the manufacturer's recommended torque and sequence.

SUGGESTED ACTIVITIES

1. Take a sheet of paper. Wad it into a ball. Pull it back out and lay it on the table. If you were to try to press it out flat, where would you place your hands (fastener) first? In what direction (sequence) would you move them? Try it. How does this compare to tightening sequence?
2. Using a regular spanner, turn up several 10 mm screws to what you would guess to be (20 Nm) of torque. Take a torque wrench and break them loose. Watch the scale carefully to determine the torque required to start them. Even though this will be different than true torque, how even were they? Was it close to 20 Nm?
3. Place two 6 mm bolts (one grade 4T and the other grade 10T) of equal length in a vice. Keep them about 50 mm apart and with the same amount of material in the jaws. Run the vice up tightly. With a suitable torque wrench, turn each bolt until it snaps. Watch the scale carefully to determine torque at the moment of failure. Was the reading the same? If not, Why? You will also note that it does not take much effort to snap a 6 mm bolt.

WOULD YOU USE A TORQUE WRENCH?

Suppose you are to be carried aloft 20 stories on a small steel platform. The platform is attached to the cable with ONE bolt. This bolt MUST be torqued to 200 Nm. At 210 Nm it will break in mid-air and at 140 Nm it will slip. Would you use a torque wrench? I think you would.

REMEMBER: YOUR REPUTATION AS A MECHANIC MAY WELL BE JUDGED ON THE FAILURE OF ONE SUCH BOLT. KEEP IT SAFE. ALWAYS FOLLOW RECOMMENDED TORQUE AND USE A TORQUE WRENCH!

KNOW THESE TERMS

Machine screw, Sheet metal screw, Jam nut, Screw extractor, Penetrating oil, Heli-Coil, Hex, Slotted, Pitch, Tensile strength, Head marking, Thread series, Lockwasher, Cotter pin, Snap ring, Setscrews, Torque, Hooke's Law, Torque wrench, Newton-metres, Tightening sequence, Run-down torque, Retorque.

REVIEW QUESTIONS

1. Screws require the use of nuts. True or False?
2. Sheet metal screws should be threaded into a hole about the size of their major diameter. True or False?
3. Drilling is considered superior to punching holes in which sheet metal screws are to be inserted. True or False?
4. A stud has _____ on _____ ends.
5. Studs are best installed with pliers. True of False?
6. Name three methods that may be used to remove broken screws or studs.
7. How can a stripped hole be repaired? Name three methods.
8. The number 7 stamped on the head of a bolt indicates that it has greater tensile strength than a bolt with the number 10. True or False?
9. How can the thread pitch on a screw be determined?
10. Name two popular thread series.
11. Describe two kinds of self-locking nuts.
12. Name the three basic types of lock washers.
13. To use a palnut, run it down to the nut, open side away, and then give it _____ _____.
14. All fasteners have threads. True or False?
15. A spline and a woodruff key both act as driving mechanism or device. True or False?
16. What is a lock plate?
17. Snap rings should NEVER be reused. True or False?
18. A setscrew usually has a hexagonal head. True or False?
19. When a rivet is used, the rivet should be _____ in the hole. The parts must be _____ together and a _____ _____ should be used.
20. Torque and tension are one and the same. True or False?
21. Define the following: 1. Elastic Limit. 2. Distortion. 3. Tensile Strength. 4. Torque. 5. Tension. 6. Residual Tension. 7. Compression. 8. Elasticity. 9. Hooke's Law. 10. High Pressure Lubricant.
22. List three reasons for proper fastener tension.
23. Proper tension is best achieved by using a _____ _____ to tighten fasteners.
24. Why use lubricant on fastener threads?
25. Torquing should be in three initial steps. Fasteners drawn up to _____ of recommended torque then to _____ _____ and finally to _____ torque.
26. What is the important fourth step in torquing?
27. Which of the following torque wrenches would you use to tighten a bolt to 100 Nm?
    a. 1-20 Nm.
    b. 10-20 Nm.
    c. 60-400 Nm.
28. What effect will an adapter have on a torque wrench reading?
29. Describe how sticking during the final torquing should be handled.
30. To allow the user to torque fasteners when the position makes seeing the scale impossible, a _____ device is used.
31. Always PUSH a torque wrench. True or False?
32. Once fasteners have been properly torqued, they will never need to be torqued again. True or False?
33. What is a torque chart?
34. Torque, for automotive use, is measured in _____ _____.

Chapter 5

PRECISION MEASURING TOOLS, MEASUREMENT CHARTS

After studying this chapter, you will be able to:
☐ Identify common measuring tools.
☐ Select the appropriate measuring tool for the job.
☐ Use precision measuring tools.
☐ Properly maintain precision measuring tools.
☐ Use measurement conversion charts.

The auto mechanic must be thoroughly familiar with the precision measuring tools used in the trade. Many jobs involve checking sizes, clearances, and alignments.

A careless or inaccurate measurement can be costly, both in money and customer relation — to say nothing, damaging the mechanic's reputation.

QUALITY TOOLS

When selecting measuring tools that will be used for a period of years, it pays to buy top quality tools. The initial cost will obviously be higher. However, considering the importance of accuracy, the longer life span of superior tools easily justifies the extra cost.

STORAGE

It is advisable to keep your measuring tools in a protective case. Also, keep them in an area that will not be subjected to excessive moisture or heavy usage. See Fig. 5-1.

After each use, wipe the tool down with a lightly oiled, lint-free, clean cloth. Never dip a precision measuring tool in solvent (unless it is being completely dismantled). Do not use an air hose for cleaning precision measuring tools.

HANDLING

When using a measuring tool, place it in a clean area where it will not fall or be struck by other tools. Never pry, hammer, or force the tools. REMEMBER: they are PRECISION tools — keep them that way!

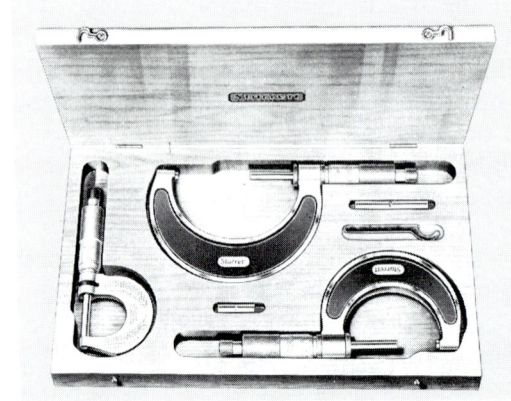

Fig. 5-1. This micrometer case provides excellent protection for tools. (Starrett)

CHECK FOR ACCURACY

It is good practice to occasionally check precision tools for accuracy. They may be checked against a tool of known accuracy or by using special gauges.

If a tool is accidentally dropped or struck by some object, immediately check it for accuracy. Adjustments for wear or very minor damage are provided on many measuring tools. Follow the manufacturer's instructions.

OUTSIDE MICROMETER

The outside micrometer (mike) is used to check the diameter of pistons, pins, crankshafts, and other machined parts. The most commonly used metric micrometer reads in one hundredths of a millimetre. With this micrometer it is easy to estimate as close as one-half of a hundredth of a millimetre.

Some metric micrometers can measure to within two thousandths of a millimetre. This type uses a vernier scale.

A cutaway view of a typical outside micrometer is shown in Fig. 5-2. Learn the names of the parts and their relationship to the operation of the tool.

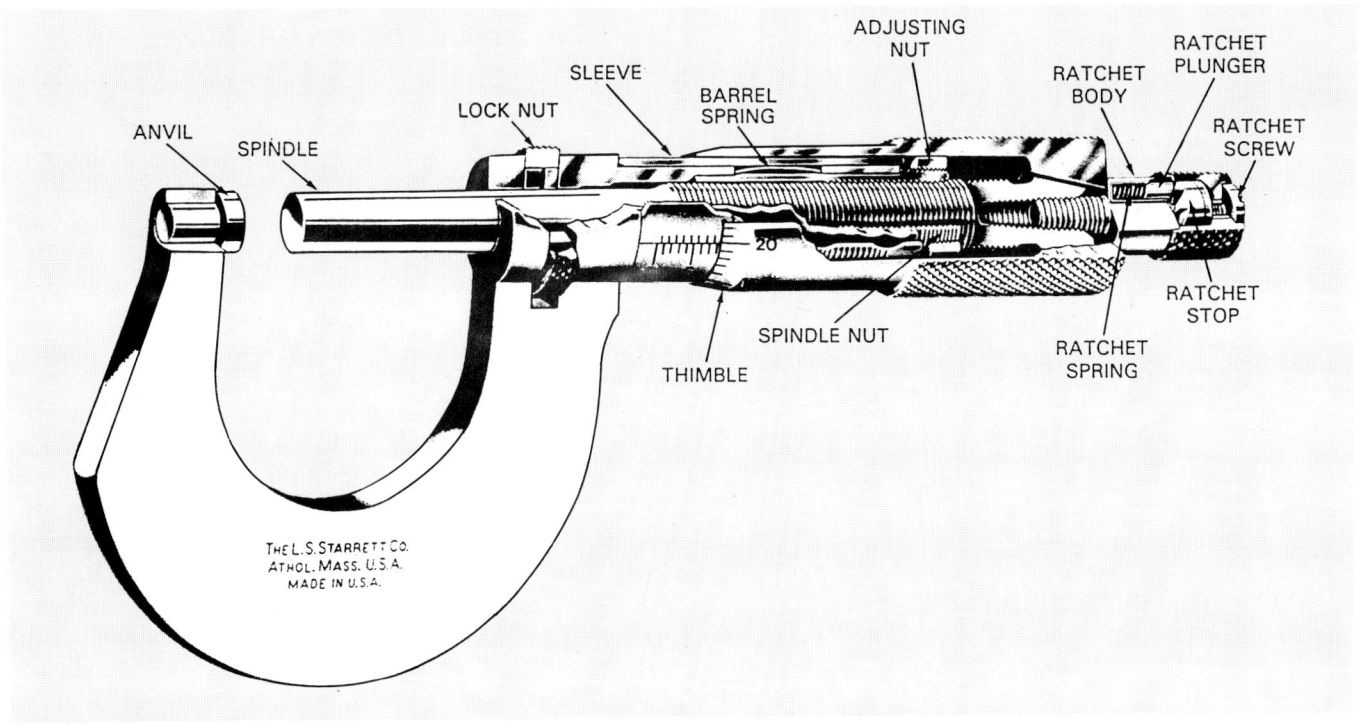

Fig. 5-2. Cutaway view of an outside micrometer. Learn the names of the various parts.

MICROMETER RANGE

Usually each micrometer is designed to produce readings over a range of 25 millimetres. Ideally, the auto mechanic should obtain a set of six micrometers covering sizes 0-25 mm, 25-50 mm, 50-75 mm, 75-100 mm, 100-125 mm, 125-150 mm. Fig. 5-3 shows a set of twelve micrometers covering 0-300 mm.

It would be less expensive to purchase only two micrometers, a 0-100 mm and 50-150 mm, both with interchangeable anvils. However the multirange micrometer is more bulky and is less convenient to use. See Fig. 5-4.

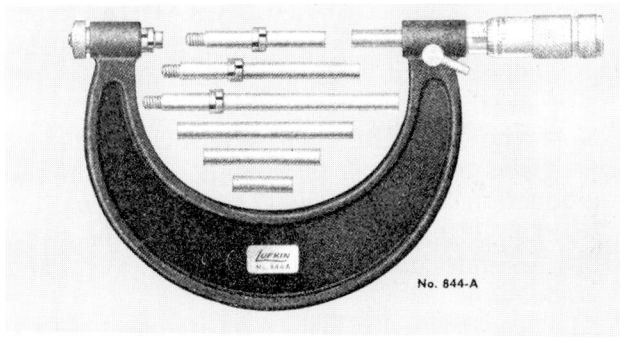

Fig. 5-4. A multiple range micrometer. By using proper anvil, this micrometer covers a range of from 0 mm to 100 mm. (Lufkin)

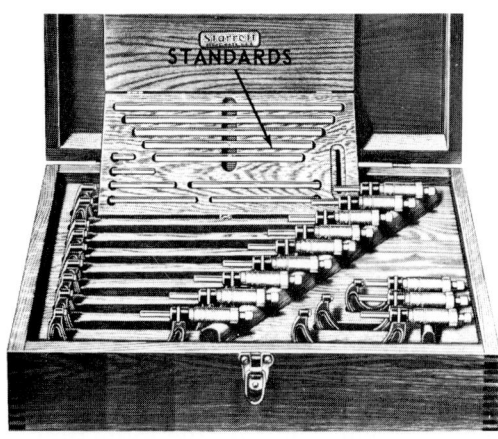

Fig. 5-3. Cased set of twelve outside micrometers. Note box of standards for checking accuracy of each "mike".

READING THE METRIC MICROMETER

To read a metric micrometer, follow the four steps given below:
1. Read the SLEEVE NUMBER. Each sleeve number equals 1.00 mm (2 sleeve numbers = 2.00, 3 = 3.00). See Fig. 5-5.
2. Count and record the SLEEVE GRADU-ATIONS visible to the right of the sleeve number. Each sleeve line equals 0.50 mm (2 sleeve lines = 1.00, 3 = 1.50).
3. Read the THIMBLE GRADUATION lined up with the horizontal sleeve line. Each thimble graduation equals 0.01 mm (2 thimble gradu-ations = 0.02, 3 = 0.03).
4. ADD the values obtained in the previous 3 steps. This will give you the metric micrometer reading, as in Fig. 5-5.

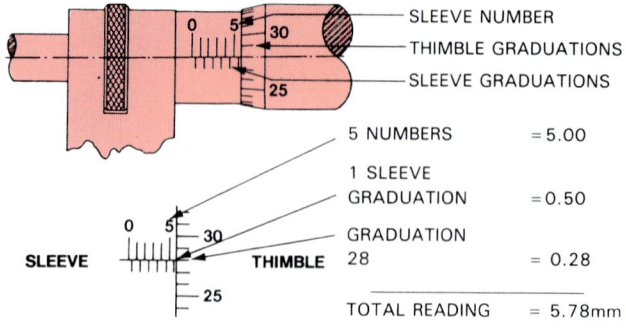

5 NUMBERS	= 5.00
1 SLEEVE GRADUATION	= 0.50
GRADUATION 28	= 0.28
TOTAL READING	= 5.78mm

Fig. 5-5. Study basic steps for reading a micrometer graduated in hundredths of a millimetre. Read number, then sleeve graduations then thimble. Add these three values to obtain reading.

READING THE IMPERIAL MICROMETER

To read an imperial micrometer, follow the four steps listed below:
1. Note the LARGEST NUMBER visible on the micrometer sleeve (barrel). Each number equals .100 in. (2 = .200, 3 = .300, 4 = .400). This is illustrated in Fig. 5-6.
2. Count the number of FULL GRADUATIONS, to the right of the sleeve number. Each full sleeve graduation equals .025 in. (2 full lines = .050, 3 = .075).
3. Note the THIMBLE GRADUATION aligned with the horizontal sleeve line. Each thimble graduation equals .001 in. (2 thimble graduations = .002, 3 = .003). Round off when the sleeve line in not directly aligned with a thimble graduation.
4. ADD the decimal values from steps 1, 2, and 3. Also, add any full inches. This will give you the micrometer reading, Fig. 5-6.

READING A MICROMETER GRADUATED IN TWO THOUSANDTHS OF A MILLIMETRE

This type of micrometer is called a vernier micrometer and can measure to within 0.002mm. That is two thousandth of a millimetre. In order to achieve this an additional VERNIER scale is marked on the sleeve of the micrometer.

The same reading technique is used to read this type of micrometer. Instead of estimating fractions of a hundredth between thimble marks the vernier scale can be used.

The vernier consists of thin lines scribed parallel to the sleeve long line. They are marked 0-10 in increments of 2. When the thimble marks do not fall in line with the long sleeve line, indicating a fraction of a hundredth of a millimetre, carefully examine the vernier lines. One of the vernier lines will be aligned with one of the thimble marks. When you have discovered the specific vernier line that is aligned, the number of that vernier line indicates the number of two thousandths to be added to your initial thimble reading. Look at Fig. 5-7.

Where there is no coincidence of lines, then the intermediate thousandths can be estimated, ie. if the reading lies between four and six, then the additional thousandth reading would be 0.005 mm.

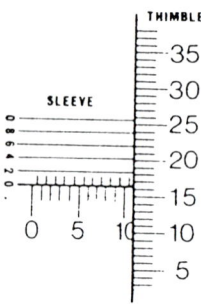

Fig. 5-7. In this reading the 6 line is aligned with the thimble mark. This means that 0.006 or six thousandths has to be added to the initial reading.

WHEN USING ANY MEASURING TOOL

Always thoroughly clean the work to be measured. This assures accurate readings and reduces wear on the working tips of the tool.

USING OUTSIDE MICROMETER

When measuring small objects, grasp the micrometer in the right hand. At the same time, insert the object to be measured between the anvil and spindle. While holding the work against the anvil, turn the thimble with the thumb and forefinger until

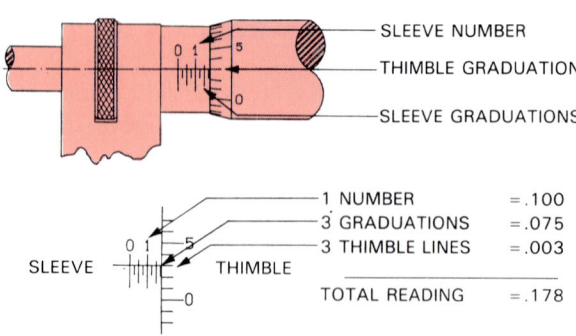

1 NUMBER	= .100
3 GRADUATIONS	= .075
3 THIMBLE LINES	= .003
TOTAL READING	= .178

Fig. 5-6. Study basic steps for reading a micrometer graduated in thousandths of an inch. Read number, then sleeve graduations, then thimble. Add these three values to obtain reading. (Starrett)

the spindle engages the object. Do not clamp the micrometer tight. Use only enough pressure on the thimble to cause the work to JUST FIT between the anvil and spindle. Slip the object in and out of the micrometer while giving the thimble a final adjustment. The work must slip through the micrometer with a VERY LIGHT FORCE.

When satisfied that your adjustment is correct, read the micrometer setting. BE CAREFUL THAT YOU DO NOT MOVE THE ADJUSTMENT, Fig. 5-8.

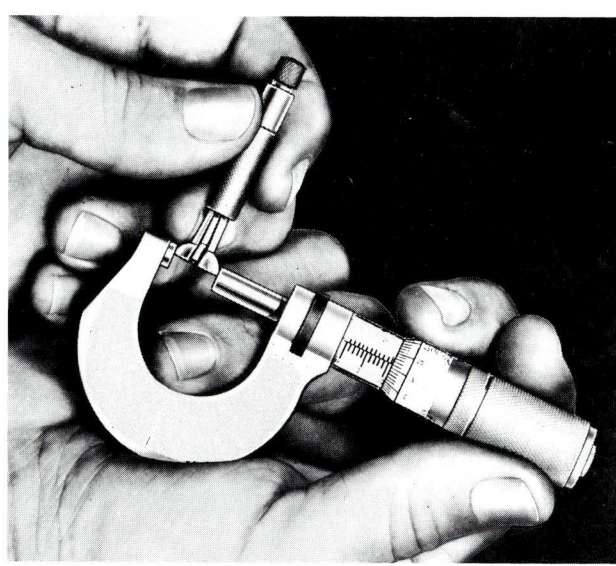

Fig. 5-8. Miking a small hole gauge. Heel of hand supports micrometer frame while thumb and forefinger turn thimble. (Starrett)

To measure larger objects, grasp the frame of the micrometer and slip the micrometer over the work while adjusting the thimble. Slip the mike back and forth until very light resistance is felt, Fig. 5-9.

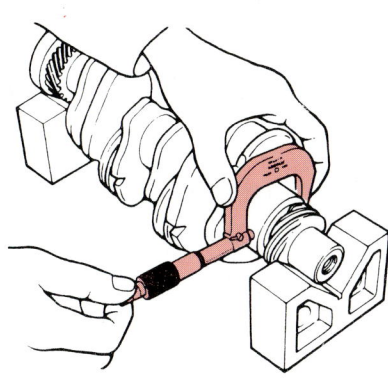

Fig. 5-9. Miking a crankshaft. Notice how mike is held.

Some micrometers have a ratchet clutch knob on the end of the thimble. It allows the user to bring the spindle down against the work with the same amount of tension each time.

As the micrometer is slipped back and forth over the work, it should be rocked from side to side slightly. This will make certain the spindle cannot be closed an additional amount, Fig. 5-10.

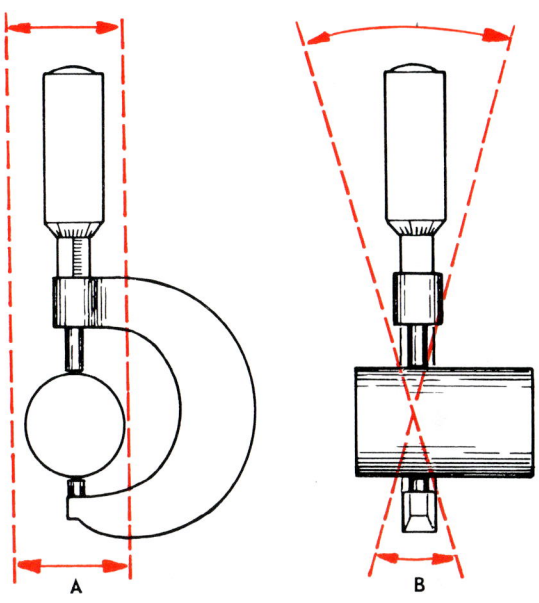

Fig. 5-10. — A Micrometer is slipped back and forth over object. B — Micrometer is rocked from side to side to make certain smallest diameter is found. Rocking is actually very slight.

PRACTICE IS NECESSARY

Measure objects of a known diameter until you have mastered the feel of using a micrometer. Keep practicing until you are completely confident of your readings. REMEMBER — A MECHANIC MUST BE ABLE TO MAKE ACCURATE MICROMETER READINGS.

HANDLE THE MICROMETER WITH CARE. NEVER STORE A MIKE WITH THE ANVIL AND SPINDLE TIP TOUCHING (this encourages rusting between tips). CLEAN YOUR WORK BEFORE MEASURING.

INSIDE MICROMETER

The inside micrometer is used for making measurements in cylinder bores, brake drums, large bushings, and similar parts, Fig. 5-11.

An inside mike is read in the same manner as the outside micrometer. The same feel is required. When measuring, rock the inside mike from side to side. At the same time, keep the anvil firmly against one side of the bore. While the free end is being rocked, it

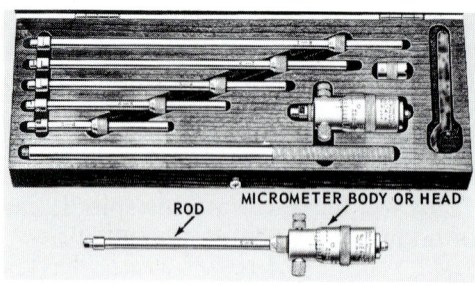

Fig. 5-11. Inside micrometer. By changing rods, this set will measure from 50 to 200 mm. (Starrett)

must also be tipped in and out. The rocking allows you to locate the widest part of the bore. The tipping assures that the micrometer is at right angles to the bore, Fig. 5-12.

An extension handle permits the use of an inside micrometer in a bore too small to hand hold the tool.

MICROMETER DEPTH GAUGE

This is a handy tool for reading the depth of slots, splines, counterbores, and holes. See Fig. 5-13.

To use this tool, the base is pressed against the work (after cleaning) and the spindle is run down into the hole to be measured.

It is read like an outside micrometer. The only difference is the sleeve marks run in a reverse direction, Fig. 5-14.

DIAL GAUGE OR INDICATOR

The dial indicator is a precision tool designed to measure movements in hundredths of a millimetre.

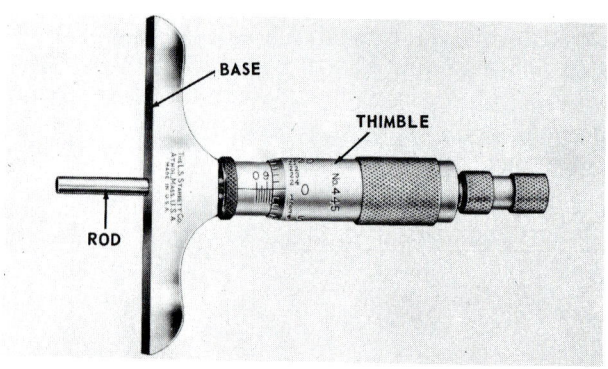

Fig. 5-13. Micrometer depth gauge. The range can be increased by using longer rods.

Some common uses are checking end play in shafts, backlash between gears, valve lift, shaft run-out, and taper in cylinders.

Use care in the handling of this tool as it is sensitive and easily damaged. When not in use, keep it in a protective case.

Dial indicator faces are calibrated either in hundredths of a millimetre or thousandths of an inch.

Various type dial markings are available. Ranges (distance over which indicator can be used) vary, depending upon the instrument, Fig. 5-15.

Various mounting arms, swivels, and adapters are provided so that the indicator can be used on various set ups.

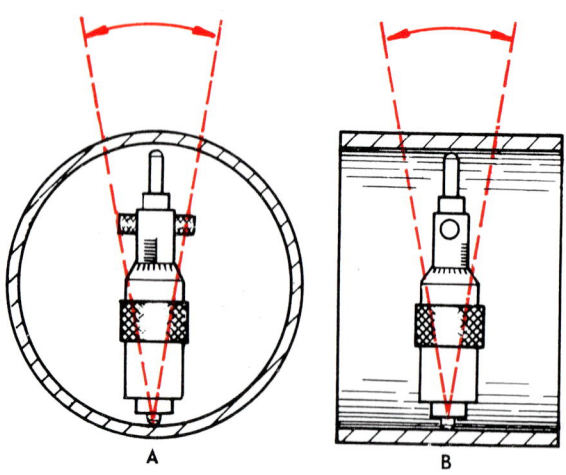

Fig. 5-12. A — Micrometer must be rocked from side to side. B — While at the same time, it must be tipped. Both movements are relatively slight.

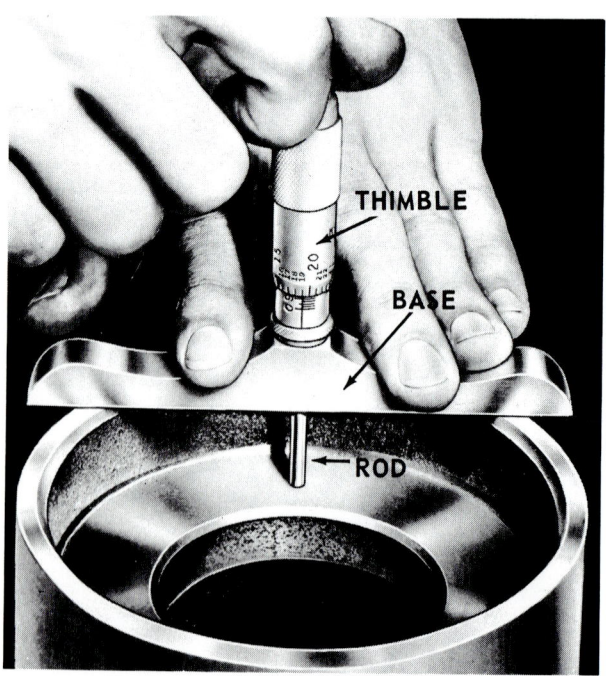

Fig. 5-14. Using the micrometer depth gauge. The base is held firmly against work and thimble turned until rod contacts shoulder.

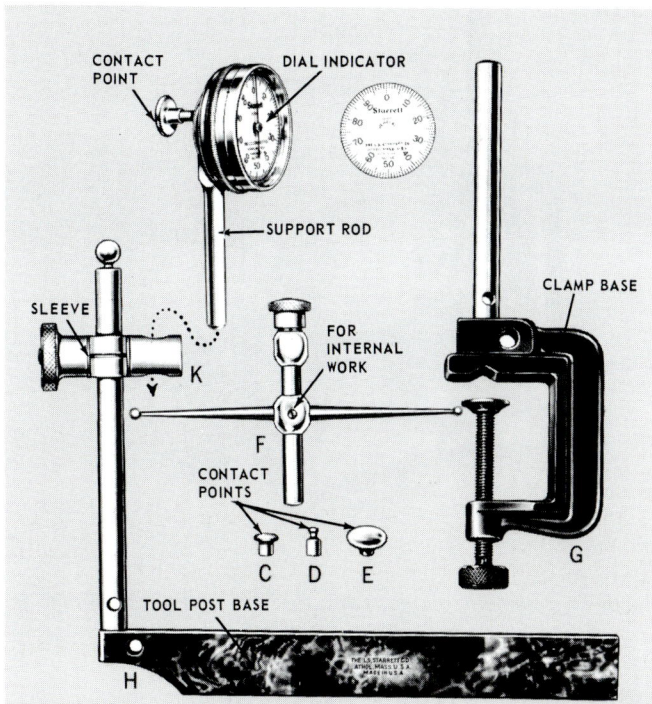

Fig. 5-15. Dial indicator and holding attachments. (Starrett)

When using a dial indicator, be certain that it is firmly mounted and that the standard (actuating rod) is parallel to the plane (direction) of movement to be measured, Fig. 5-16.

Place the rod end against the work to be measured. Force the indicator toward the work. Make the indicator needle travel far enough around the dial so that movement in either direction can be read. The dial face can then be turned to line the 0 mark with the indicator needle. Be sure that the indicator range (limit of travel) will cover the movement anticipated. Ranges usually run from

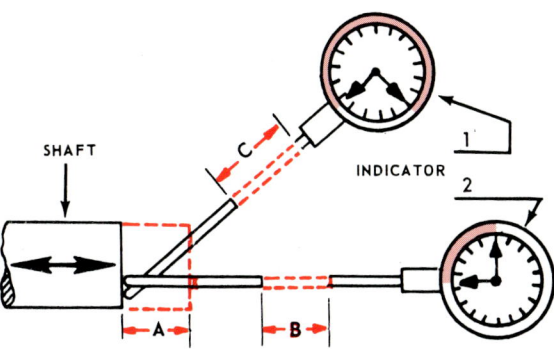

Fig. 5-16. Indicator 1 set up is NOT parallel to movement of shaft. When shaft moves distance A, indicator rod moves distance C, giving a false reading for shaft end play. Indicator 2 is parallel and shaft movement A causes indicator rod to move distance B, producing an accurate reading.

around 5 millimetre to 20 millimetre depending on the instrument.

Fig. 5-17, 5-18, and 5-19 illustrate typical dial indicator set ups.

OTHER DIAL INDICATOR TOOLS

Two other valuable measuring tools utilise a dial indicator as part of their construction. They are the

Fig. 5-17. Checking timing gear backlash with a dial indicator. Indicator rod is angled to place it in line with gear rotation.

Fig. 5-18. Using a dial indicator to determine piston top dead centre.

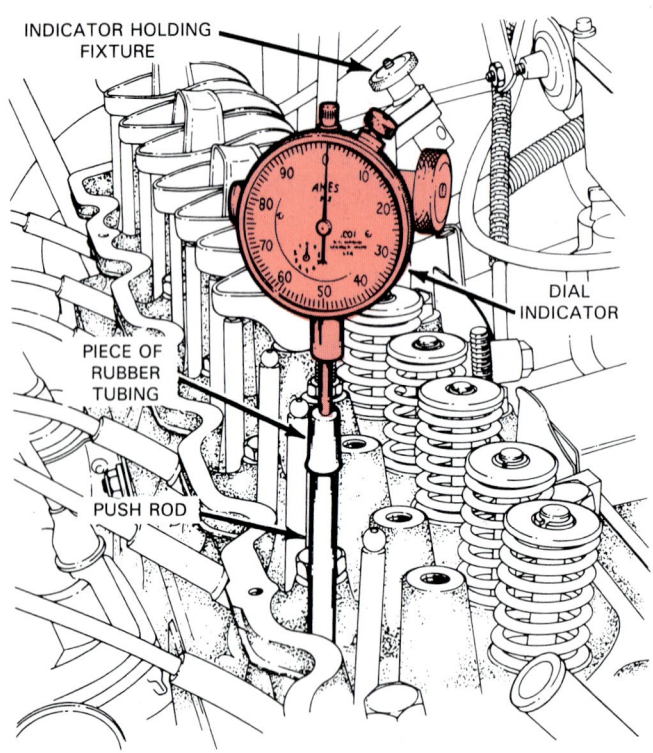

Fig. 5-19. Using a dial indicator to measure camshaft lobe lift.

out-of-roundness gauge and the cylinder gauge. The out-of-roundness gauge is used to check connecting rod big end bores. This can be done with an inside mike but this special gauge makes the job easier and faster, Fig. 5-20.

The cylinder gauge makes the checking of cylinder bore size, taper, and out-of-roundness quick and accurate, Fig. 5-21.

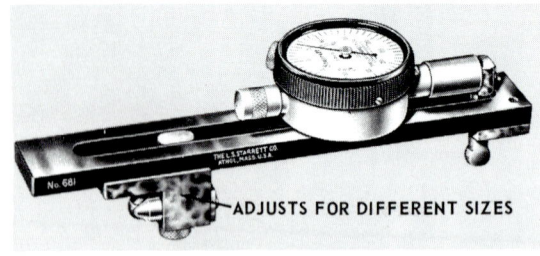

Fig. 5-20. Out-of-roundness gauge.

OTHER USEFUL MEASURING TOOLS

In addition to the precision tools already discussed, there are a number of other tools that a mechanic should own. Keep in mind that, in your work as an auto mechanic, a number of measurements, varying from a few thousandths to several metres will be required.

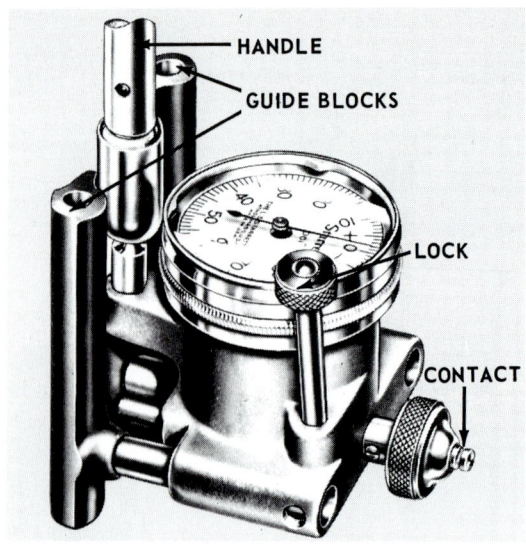

Fig. 5-21. Cylinder gauge. Only a short section of handle is shown. (Starrett)

INSIDE AND OUTSIDE CALIPERS

Calipers are useful tools for measurements when accuracy is not critical. Fig. 5-22 illustrates a pair of outside calipers.

Fig. 5-23 shows inside caliper. The inside caliper is used to measure the diameter of holes. To determine the reading, hold the calipers on an accurate steel rule. Careful measuring across the points (very light touch) with an outside micrometer will give a more accurate reading.

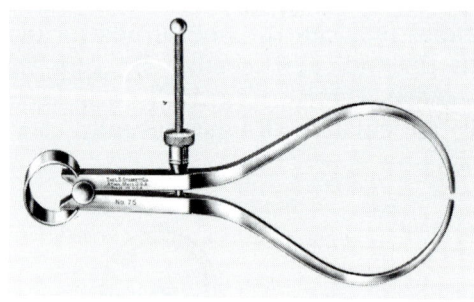

Fig. 5-22. Outside caliper.

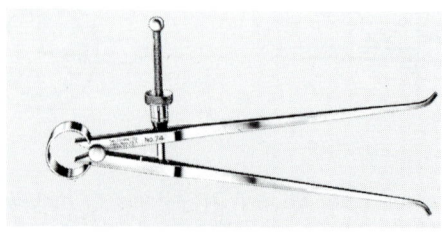

Fig. 5-23. Inside caliper.

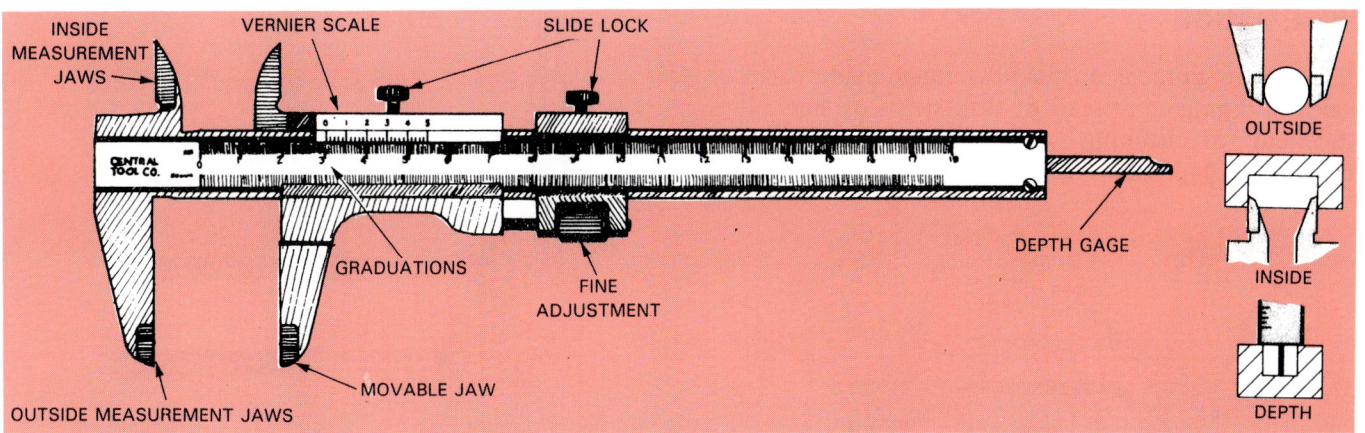

Fig. 5-24. Vernier caliper is handy measuring tool. It will quickly check inside, outside and depth measurements.

VERNIER CALIPERS

The vernier caliper is a very useful precision measuring instrument capable of obtaining inside, outside, and depth readings. Because these calipers are highly accurate, they should be handled and used with great care. Always store them in a protective case when not in use.

The vernier caliper shown in Fig. 5-24 will measure objects up to 150 mm. It is a popular size for mechanics.

The main scale of the caliper is graduated in millimetres and numbered each centimetre. The vernier scale is divided into 50 divisions over a distance of 49 mm, each division equalling 49/50th of a millimetre, the difference between a division on the main scale and the vernier scale being 0.02 mm (1/50th mm). See sample reading below.

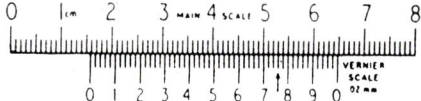

MAIN SCALE	15.00 mm
VERNIER DECIMAL NUMERAL	.70 mm
COINCIDENT LINE	
(3 DIV'S OF .02 mm)	.06 mm
READING	15.76 mm

DIVIDERS

Dividers are somewhat like calipers but have straight shanks and pointed ends. They are handy for making circles and taking surface measurements. Fig. 5-25 illustrates a pair of dividers.

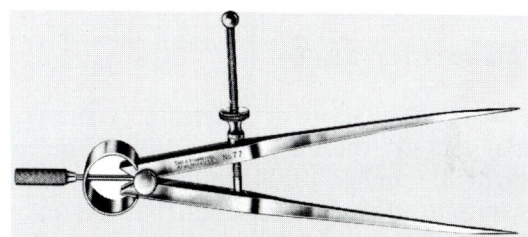

Fig. 5-25. Dividers. Points must be sharp.

FEELER GAUGES

Feeler or thickness gauges are thin strips of specially hardened and ground steel with the thickness marked in hundredths of a millimetre or thousandths of an inch. They are used to check clearances between two parts (valve gap, piston ring side and end gap clearance for example). They come in sets as shown in Fig. 5-26. There are also nonferrous versions.

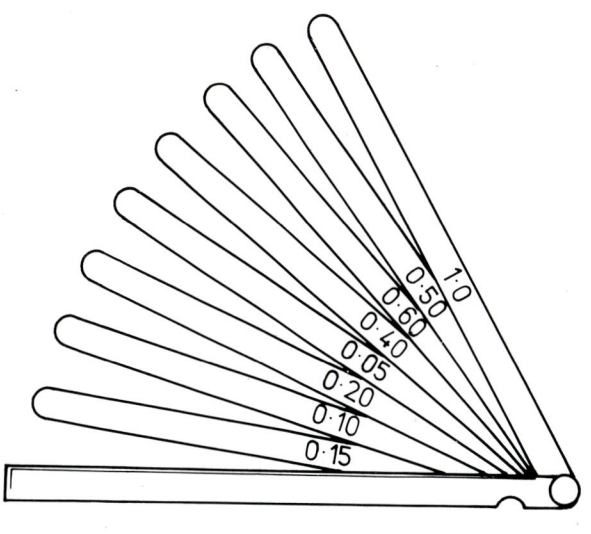

Fig. 5-26. Feeler gauge set.

WIRE GAUGE

The wire gauge is a thickness gauge using wires of varying diameter instead of thin strips of steel. It is excellent for checking spark plug gaps and distributor point gap. See Fig. 5-27.

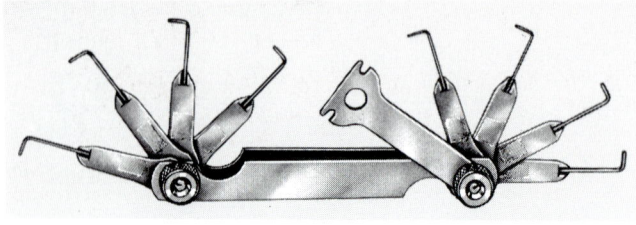

Fig. 5-27. Wire gauge set for checking spark plug gap.

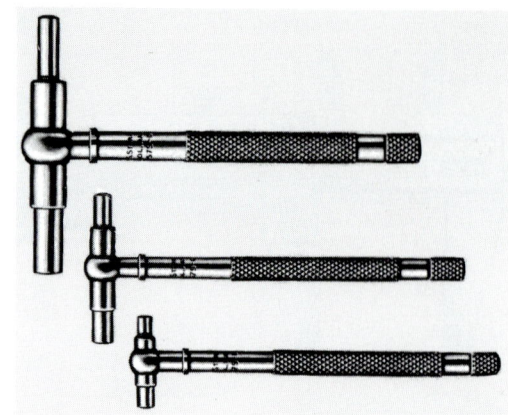

Fig. 5-29. Telescoping gauges.

THREAD PITCH GAUGE

This is a handy tool for determining the type of particular screw threads. Fig. 5-28. The gauge consists of a number of blades which have indentations on one edge. The user matches the unknown thread with different blades until the one with the correct profile is found.

On metric thread pitch gauges the number on the gauge blade is the thread pitch in mm.

STEEL RULES

Other measuring tools that can be used to good advantage include:
1. Steel rules. Blade lengths most preferred by mechanics are 150 millimetres (pocket size) and 300 millimetre (general purpose) and marked in ½ millimetre graduations.
2. A three metre pocket tape rule.
3. A combination square, which if desired, can include protractor head and centre head.
4. A thin steel hook rule with a sliding steel head. See Figs. 5-30, 5-31 and 5-32.

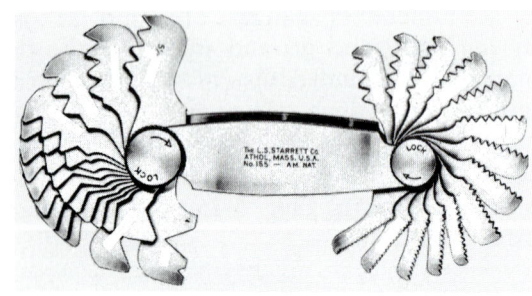

Fig. 5-28. Thread gauge.

TELESCOPING GAUGE

The telescoping gauge is an accurate tool for measuring inside bores of engine connecting rods and main bearings. To use this tool, the plungers are compressed and locked by turning the knurled screw on the handle. The gauge is placed inside the bore. The plungers are then released until they contact the bore walls. Next, they are locked and the tool is removed. An outside micrometer is used to measure across the plungers for an accurate checking of bore size. Telescoping gauges have different ranges and may be purchased in sets. The proper feel for using this tool will be the same as the inside micrometer, Fig. 5-29.

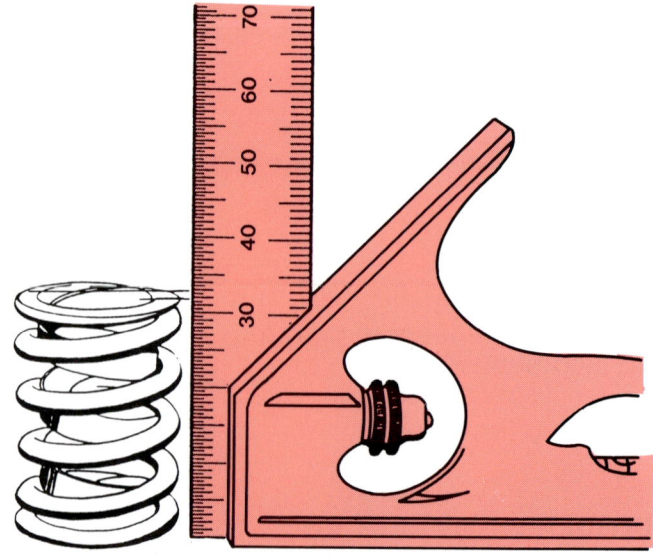

Fig. 5-30. Combination square is needed when scale must be parallel to part.

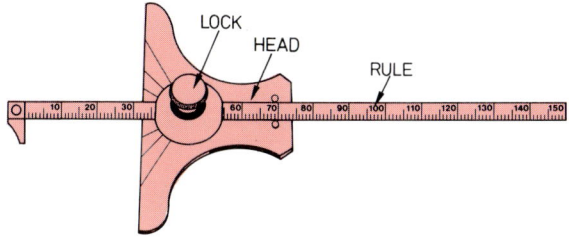

Fig. 5-31. Hook rule with sliding depth and angle head.

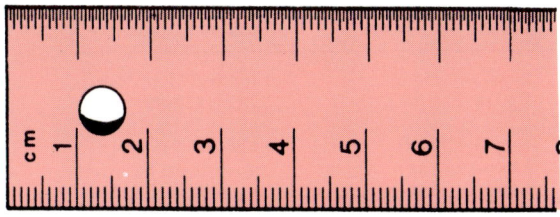

Fig. 5-32. Steel rule.

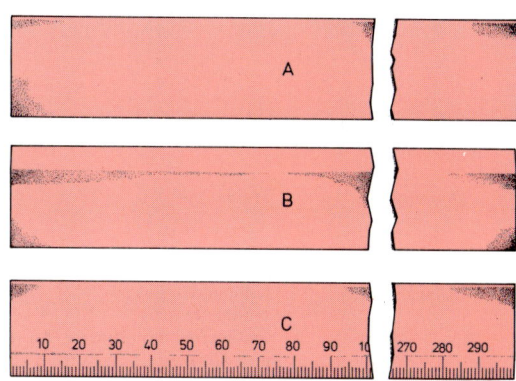

Fig. 5-34. Steel straightedge. A — Square edge. B — Bevel edge. C — Bevel and ruled edge. These are available in different lengths.

SPRING SCALE

Two spring scales, which read in newtons (N), are a 'must'. These are needed to determine force such as contact point spring tension and pull on feeler strips when fitting pistons. Look at Fig. 5-33.

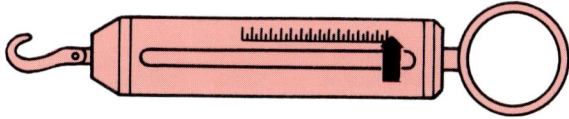

Fig. 5-33. Spring scale. A must in every tool kit.

STEEL STRAIGHTEDGE

An accurate steel straightedge is used to check part warpage. It should be long enough to span the length of an engine block or cylinder head for checking these parts for warpage. Be careful when handling and storing a straightedge. It must not be damaged, Fig. 5-34.

TEMPERATURE IS IMPORTANT

Many specifications for measurements will state room temperature, an exact temperature, or engine at normal running temperature. Remember that all metals contract and expand in direct proportion to their temperature. This makes it imperative that temperature specifications be followed when making precision measurements and settings. Your measuring tools themselves can be affected by extremes of heat and cold. If your tools must be used when very cold or very hot, check them for accuracy before using.

AUTOMOTIVE MEASURING SYSTEMS

The two automotive measuring systems are the SI Metric System and the Imperial or English system. Australia and almost all other countries in the world use the SI Metric System. The U.S.A. is one of the last countries to go metric but it also is now slowly replacing its system with the metric system.

The (SI) Metric measuring system uses a power of 10 for all basic units. It is a far simpler and more logical system than the Imperial system. Computation often requires nothing more than adding zeros or moving a decimal point. For instance, one metre equals 10 decimetres, 100 centimetres or 1000 millimetres.

However as there are still quite a number of vehicles running on Australian roads that were produced 'in imperial' a mechanic must still be conversant with the old system, at least to a point where he or she is capable of using measuring system conversion charts.

A measuring system conversion chart is needed when changing from one measuring system to another: inches to millimetres, gallons to litres and so on. See Fig. 5-35.

A decimal conversion chart is also commonly used by a mechanic to interchange and find equal values for fractions, decimals and millimetres. See Fig. 5-36.

A temperature conversion chart is also handy to convert degrees fahrenheit to degrees celsius or vice versa. A temperature conversion chart is also included in Fig. 5-36.

MEASURING SYSTEM CONVERSION CHART

Quantity	Metric Unit	Imperial Unit	Conversion Factors (Approximate)	
			Metric to Imperial Units	Imperial to Metric Units
LENGTH	millimetre (mm) or centimetre (cm)	inch(in)	1 cm = 0.394 in	1 in = 25.4 mm
	centimetre (cm) or metre (m)	foot (ft)	1 m = 3.28 ft	1 ft = 30.5 cm
	metre (m)	yard (yd)	1 m = 1.09 yd	1 yd = 0.914 m
	kilometre (km)	mile	1 km = 0.621 mile	1 mile = 1.61 km
MASS	gram (g)	ounce (oz)	1 g = 0.0353 oz	1 oz = 28.3g
	gram (g) or kilogram (kg)	pound (lb)	1 kg = 2.20 lb	1 lb = 454g
	tonne (t)	ton	1 tonne = 0.984 ton	1 ton = 1.02 tonne
AREA	square centimetre (cm²)	square inch (in²)	1 cm² = 0.155 in²	1 in² = 6.45 cm²
	square centimetre (cm²) or square metre (m²)	square foot (ft²)	1 m² = 10.8 ft²	1 ft² = 929 cm²
	square metre (m²)	square yard (yd²)	1 m² = 1.20 yd²	1 yd² = 0.836 m²
	hectare (ha)	acre (ac)	1 ha = 2.47 ac	1 ac = 0.405 ha
	square kilometre (km²)	square mile (sq. mile)	1 km² = 0.386 sq. mile	1 sq. mile = 2.59 km²
VOLUME	cubic centimetre (cm³)	cubic inch (in³)	1 cm³ = 0.0610 in³	1 in³ = 16.4 cm³
	cubic decimetre (dm³) or cubic metre (m³)	cubic foot (ft³)	1 m³ = 35.3 ft³	1 ft³ = 28.3 dm³
	cubic metre (m³)	cubic yard (yd³)	1 m³ = 1.31 yd³	1 yd³ = 0.765 m³
	cubic metre (m³)	bushel (bus)	1 m³ = 27.5 bus	1 bus = 0.0364 m³
VOLUME (fluids)	millilitre (mL)	fluid ounce (fl oz)	1 mL = 0.0352 fl oz	1 fl oz = 28.4 mL
	millilitre (mL) or litre (L)	pint (pt)	1 litre = 1.76 pint	1 pint = 568 mL
	litre (L) or cubic metre (m³)	gallon (gal)	1 m³ = 220 gallons	1 gal = 4.55 litre
FORCE	newton (N)	pound-force (lbf)	1 N = 0.225 lbf	1 lbf = 4.45 N
PRESSURE	kilopascal (kPa)	pound per square inch (psi)	1 kPa = 0.145 psi	1 psi = 6.89 kPa
(for meteorology)	millibar (mb)	inch of mercury (inHg)	1 mb = 0.0295 inHg	1 inHg = 33.9 mb
ANGULAR VELOCITY	radian per second (rad/s)	revolution per minute (r/min, rpm)	1 rad/s = 9.55 r/min	1 r/min = 0.105 rad/s
VELOCITY	kilometres per hour (km/h)	miles per hour (mph)	1 km/h = 0.621 mph	1 mph = 1.61 km/h
TEMPERATURE	Celsius temp (°C)	Fahrenheit temp (°F)	$°F = (\frac{9}{5} × °C) + 32$	$°C = \frac{5}{9} × (°F - 32)$
ENERGY	kilojoule (kJ)	British thermal unit (Btu)	1 kJ = 0.948 Btu	1 Btu = 1.06 kJ
	megajoule (MJ)	therm	$1 MJ = 9.48 × 10^{-3} therm$	1 therm = 106 MJ
POWER	kiloWatt (kW)	horsepower (hp)	1 kW = 1.34 hp	1 hp = 0.746 kW
TORQUE	newton metres (Nm)	pounds-feet (lbs-ft)	1 Nm = 0.738 lbs-ft	1 lbs-ft = 1.35 Nm
FUEL ECONOMY	litres per 100 km (L/100 km)	miles per gallon (mpg)	$L/100 km = \frac{282.48}{mpg}$	$mpg = \frac{282.48}{L/100 km}$
ENGINE CAPACITY	litres (L)	cubic inches (ci)	1L = 61 ci	1 ci = 0.16L
VACUUM	kilopascal (kPa)	inches of mercury (in Hg)	1 kPa = 0.295 in Hg	1 in Hg = 3.386 kPa

Fig. 5-35. Measuring system conversion chart. To convert from one system to another multiply known value by conversion factor. This will give an approximate equal value.

DECIMAL AND TEMPERATURE CONVERSION CHART

INCHES			DECIMALS	MILLI-METRES	INCHES TO MILLIMETRES		MILLIMETRES TO INCHES		FAHRENHEIT & CENTIGRADE			
					Inches	m.ms.	m.ms.	Inches	°F	°C	°C	°F
		1/64	.015625	.3969	.0001	.00254	0.001	.000039	-20	-28.9	-30	-22
	1/32		.03125	.7937	.0002	.00508	0.002	.000079	-15	-26.1	-28	-18.4
		3/64	.046875	1.1906	.0003	.00762	0.003	.000118	-10	-23.3	-26	-14.8
1/16			.0625	1.5875	.0004	.01016	0.004	.000157	-5	-20.6	-24	-11.2
		5/64	.078125	1.9844	.0005	.01270	0.005	.000197	0	-17.8	-22	-7.6
	3/32		.09375	2.3812	.0006	.01524	0.006	.000236	1	-17.2	-20	-4
		7/64	.109375	2.7781	.0007	.01778	0.007	.000276	2	-16.7	-18	-0.4
1/8			.125	3.1750	.0008	.02032	0.008	.000315	3	-16.1	-16	3.2
		9/64	.140625	3.5719	.0009	.02286	0.009	.000354	4	-15.6	-14	6.8
	5/32		.15625	3.9687	.001	.0254	0.01	.00039	5	-15.0	-12	10.4
		11/64	.171875	4.3656	.002	.0508	0.02	.00079	10	-12.2	-10	14
3/16			.1875	4.7625	.003	.0762	0.03	.00118	15	-9.4	-8	17.6
		13/64	.203125	5.1594	.004	.1016	0.04	.00157	20	-6.7	-6	21.2
	7/32		.21875	5.5562	.005	.1270	0.05	.00197	25	-3.9	-4	24.8
		15/64	.234375	5.9531	.006	.1524	0.06	.00236	30	-1.1	-2	28.4
1/4			.25	6.3500	.007	.1778	0.07	.00276	35	1.7	0	32
		17/64	.265625	6.7469	.008	.2032	0.08	.00315	40	4.4	2	35.6
	9/32		.28125	7.1437	.009	.2286	0.09	.00354	45	7.2	4	39.2
		19/64	.296875	7.5406	.01	.254	0.1	.00394	50	10.0	6	42.8
5/16			.3125	7.9375	.02	.508	0.2	.00787	55	12.8	8	46.4
		21/64	.328125	8.3344	.03	.762	0.3	.01181	60	15.6	10	50
	11/32		.34375	8.7312	.04	1.016	0.4	.01575	65	18.3	12	53.6
		23/64	.359375	9.1281	.05	1.270	0.5	.01969	70	21.1	14	57.2
3/8			.375	9.5250	.06	1.524	0.6	.02362	75	23.9	16	60.8
		25/64	.390625	9.9219	.07	1.778	0.7	.02756	80	26.7	18	64.4
	13/32		.40625	10.3187	.08	2.032	0.8	.03150	85	29.4	20	68
		27/64	.421875	10.7156	.09	2.286	0.9	.03543	90	32.2	22	71.6
7/16			.4375	11.1125	.1	2.54	1	.03937	95	35.0	24	75.2
		29/64	.453125	11.5094	.2	5.08	2	.07874	100	37.8	26	78.8
	15/32		.46875	11.9062	.3	7.62	3	.11811	105	40.6	28	82.4
		31/64	.484375	12.3031	.4	10.16	4	.15748	110	43.3	30	86
1/2			.5	12.7000	.5	12.70	5	.19685	115	46.1	32	89.6
		33/64	.515625	13.0969	.6	15.24	6	.23622	120	48.9	34	93.2
	17/32		.53125	13.4937	.7	17.78	7	.27559	125	51.7	36	96.8
		35/64	.546875	13.8906	.8	20.32	8	.31496	130	54.4	38	100.4
9/16			.5625	14.2875	.9	22.86	9	.35433	135	57.2	40	104
		37/64	.578125	14.6844	1	25.4	10	.39370	140	60.0	42	107.6
	19/32		.59375	15.0812	2	50.8	11	.43307	145	62.8	44	112.2
		39/64	.609375	15.4781	3	76.2	12	.47244	150	65.6	46	114.8
5/8			.625	15.8750	4	101.6	13	.51181	155	68.3	48	118.4
		41/64	.640625	16.2719	5	127.0	14	.55118	160	71.1	50	122
	21/32		.65625	16.6687	6	152.4	15	.59055	165	73.9	52	125.6
		43/64	.671875	17.0656	7	177.8	16	.62992	170	76.7	54	129.2
11/16			.6875	17.4625	8	203.2	17	.66929	175	79.4	56	132.8
		45/64	.703125	17.8594	9	228.6	18	.70866	180	82.2	58	136.4
	23/32		.71875	18.2562	10	254.0	19	.74803	185	85.0	60	140
		47/64	.734375	18.6531	11	279.4	20	.78740	190	87.8	62	143.6
3/4			.75	19.0500	12	304.8	21	.82677	195	90.6	64	147.2
		49/64	.765625	19.4469	13	330.2	22	.86614	200	93.3	66	150.8
	25/32		.78125	19.8437	14	355.6	23	.90551	205	96.1	68	154.4
		51/64	.796875	20.2406	15	381.0	24	.94488	210	98.9	70	158
13/16			.8125	20.6375	16	406.4	25	.98425	212	100.0	75	167
		53/64	.828125	21.0344	17	431.8	26	1.02362	215	101.7	80	176
	27/32		.84375	21.4312	18	457.2	27	1.06299	220	104.4	85	185
		55/64	.859375	21.8281	19	482.6	28	1.10236	225	107.2	90	194
7/8			.875	22.2250	20	508.0	29	1.14173	230	110.0	95	203
		57/64	.890625	22.6219	21	533.4	30	1.18110	235	112.8	100	212
	29/32		.90625	23.0187	22	558.8	31	1.22047	240	115.6	105	221
		59/64	.921875	23.4156	23	584.2	32	1.25984	245	118.3	110	230
15/16			.9375	23.8125	24	609.6	33	1.29921	250	121.1	115	239
		61/64	.953125	24.2094	25	635.0	34	1.33858	255	123.9	120	248
	31/32		.96875	24.6062	26	660.4	35	1.37795	260	126.6	125	257
		63/64	.984375	25.0031	27	690.6	36	1.41732	265	129.4	130	266

Fig. 5-36. Decimal conversion chart is commonly used in workshops. It allows you to interchange fractions, decimals and millimetres. A temperature conversion chart is included in right hand columns. It allows you to convert degrees Fahrenheit to Centigrade and vice versa.

SUMMARY

It is important for you to select and correctly use measuring tools. This will help you take highly accurate measurements, and is a MUST for all auto mechanics.

Precision tools require cleanliness, careful handling, and proper storage.

The mechanic should own, or have available, outside and inside micrometers, micrometer depth gauge, dial indicator set up, inside and outside calipers, vernier calipers, dividers, feeler gauges, wire gauges, screw thread gauge, telescoping gauge, steel rules, a straightedge, and spring scales.

Other specialised measuring tools may be acquired as the need dictates.

SUGGESTED PRACTICE JOBS

A. Practice reading both a metric and imperial micrometer until you can make a correct reading every time.
B. Use an outside micrometer to measure several objects of known size.
C. Measure the inside diameter of a cylinder of a known size with your inside micrometer.
D. Using a depth gauge, measure the distance from the surface of a cylinder head to the top of a valve guide. (Valve-in-head engine.)
E. Check the run out on a camshaft by using a dial indicator.
F. Measure the inside diameter of a gudgeon pin bore using a telescoping gauge and an outside micrometer.
G. Check the accuracy of an outside micrometer by using a STANDARD (measuring rod of exact length) furnished for this purpose.
H. Check the accuracy of an inside mike by using the outside micrometer you have just checked with the standard.
I. Check the gap between spark plug electrodes by using a wire gauge.
J. Determine the type and size of a bolt thread by using a screw thread gauge.
K. Using a spring scale, determine the amount of tension required to move a feeler strip while fitting a piston to its cylinder bore.
L. With a straightedge, check the surface of a cylinder block for warpage.

SUGGESTED ACTIVITIES

1. Place a gudgeon pin in the freezer compartment of a refrigerator. When thoroughly cold, remove, wipe and quickly measure both the diameter and length using an outside micrometer (Hold the gudgeon pin with a cloth.) Write down your readings.

Now place the gudgeon pin in boiling water. When hot, remove, dry, and quickly recheck diameter and length. Was there a difference? If so, how much? What does this indicate?
2. Explain how to read a micrometer to a friend that does not know how. Have your friend try a reading and continue to help until your friend does it correctly. By doing this, you will reinforce your own knowledge.

KNOW THESE TERMS

Micrometer, Depth gauge, Small hole gauge, Dial indicator, Magnetic stand, Out-of-roundness gauge, Cylinder gauge, Caliper, Dividers, Feeler gauge, Telescoping gauge, Spring scale, Steel straightedge.

REVIEW QUESTIONS

1. When using a micrometer, make sure that the tool is clamped around the work tightly. True or False?
2. Measuring tools are rustproof. True or False?
3. A micrometer should be checked _____ _____ if accidentally dropped.
4. An inside micrometer is read in the same fashion as the outside micrometer. True or False?
5. To measure an object 88.90 mm in diameter, you would use a micrometer with a range of _____ to _____.
6. Name the best tool to handle each of the following measurements:
   a. Diameter of a gudgeon pin.
   b. Diameter of a cylinder bore.
   c. Distance from face of cylinder head to valve guide top.
   d. End play in crankshaft.
   e. Diameter of gudgeon pin bore in a piston.
   f. Connecting rod big end bore diameter.
   g. Lash (free movement or play) between two gears.
   h. The type of thread on a bolt.
   i. Clearance between the valve stem and rocker. (Valve-in-head engine.)
   j. Diameter of an exhaust pipe.
   k. Spark plug gap.
   l. Disc brake rotor runout.
   m. Length of a muffler.
   n. Distance between the fan blades and radiator.
   o. Engine block surface for warpage.
7. The two measuring systems are the _____ and _____ systems.
8. What is a measuring system conversion chart used for?
9. A decimal system conversion chart is used to _____ and find equal values for _____ and _____.

WHAT IS YOUR OPINION?

A person has just applied for a job as a mechanic. The garage has a reputation for excellent work. The owner is interested; there is an opening; the pay is good. So the owner introduces the applicant to you, who, as shop supervisor, will be expected to evaluate this person's worth as a mechanic.

You walk to a nearby work bench, open your tool chest and lay out a selection of measuring tools. You indicate a specific cylinder bore you would like miked, and inform the applicant to choose the tools and make the measurement.

The applicant picks up an inside caliper and a 150 mm steel rule, adjusts the caliper in the bore, then places the caliper on the face of the steel rule. After some squinting, the applicant informs you that the bore diameter is "just a whisker over 100 millimetres".

The actual bore diameter is 101.600 mm. What do you think of the applicant's ability? Will you recommend hiring this person? If not, why?

REMEMBER:

No one can be termed a top-notch mechanic who is not familiar with and competent in the use of measuring tools. You can be proud of your ability to make precision measurements, it is the mark of a fine mechanic!

Chapter 6

TOOL INDENTIFICATION AND USE

After studying this chapter, you will be able to:
☐ Identify the most common automotive tools.
☐ Describe the proper use of automotive tools.
☐ List the safety rules to follow when using certain tools.
☐ Select the correct tools for the job.

Today's mechanics must be familiar with and understand the use of a large number of tools. Proper tool selection will improve both the quality and speed of any repair operation. Many repair jobs, in fact, would be exceedingly difficult to perform without the right tools for the job.

Good mechanics own a wide selection of quality tools. They constantly strive to add to their collection and take great pride in keeping their tools clean and orderly.

WHAT TO LOOK FOR IN TOOLS

Some mechanics prefer one brand of tools; some, another. You will find, however, that mechanics agree on several important features that will be found in quality tools.

TOOL MATERIAL

The better tools are made of high strength alloy steel and, as such, can be made without a great deal of bulk. These tools are light and easy to use in tight quarters. Heavy, "fat" tools are useless on many jobs.

The quality alloy also gives the tool great strength. Its working areas will stand abuse, and useful life will be greatly extended. Quality material makes it possible for the manufacturer to give a good guarantee on the tool.

TOOL CONSTRUCTION

Quality tools receive superior heat treatment. Their openings and working surfaces are held to closer tolerances. Sharp edges are removed and the tools are carefully polished. This imparts a finish that is easy to wipe clean and a tool that is more comfortable to use.

Tools worthy of a good mechanic will be slim, efficient and properly designed for the job at hand. They will be strong, easy to clean and a pleasure to use. A good guarantee plus efficient service and parts replacement will be offered. This can be an important feature; otherwise usable tools often have to be discarded due to the loss or failure of some minor part that is no longer available.

REMEMBER: In buying tools, you usually get just what you pay for. Fine tools are a good investment.

USE OF TOOLS

Use tools designed for the job at hand. Keep your tools orderly and clean in a good, roll-type cabinet, tool chest and "tote" tray (a small tray that may be carried to the job to keep a few selected tools close at hand). This will assist you in keeping tools in good shape and readily available.

Any tool subjected to rust should be cleaned and lightly oiled. Cutting tools, such as files and chisels, should be separated to preserve the cutting edges. Delicate measuring tools should be placed in protective containers. Keep heavy tools by themselves, and arrange tools so those most often used are handy.

Repeat — KEEP YOUR TOOLS CLEAN. You cannot hope to successfully assemble a fine piece of machinery with grubby tools. The slightest bit of dirt or abrasive that finds its way into moving parts can create havoc when the unit is placed in operation. Tools are a good indication of a mechanic's worth. Dirty, beat up and jumbled tools reveal that the owner has a lot to learn. The quality of this mechanic's finished work is likely to be just as shabby as the tool used on the job.

TYPES OF TOOLS

This chapter does not attempt to cover all the tools used in the automotive trade. Large garages often utilise hundreds of special tools designed for specific jobs, models and units. Basic tools commonly used will be discussed.

HAMMERS

Mechanics find great use for ball pein, plastic-tipped and brass-tipped hammers. The ball pein is used for general striking work and is available in weights ranging from 100 or 120 grams to over 1 kg. Brass and plastic-tipped hammers are used where there is danger of the steel ball pein marring the surface. Fig. 6-1.

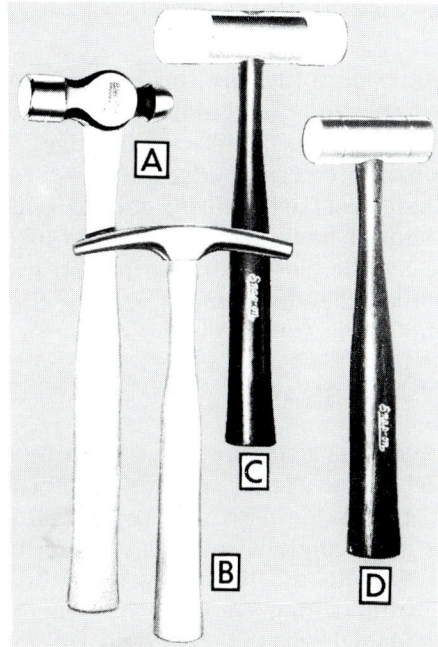

Fig. 6-1. Several types of hammers. A - Ball pein. B — Upholstery. C — Plastic-tipped. D — Brass-tipped. (Snap-on Tools)

CAUTION!

BE CAREFUL WHEN USING A HAMMER. DO NOT SWING THE HAMMER IN A DIRECTION THAT WOULD ALLOW IT TO STRIKE SOMEONE IF IT SLIPS FROM YOUR GRASP. KEEP THE HAMMER HANDLE TIGHT IN THE HAMMER HEAD, AND KEEP THE HANDLE CLEAN AND DRY.

CHISELS

Several sizes and types of chisels are essential for cutting bolts, rivets, etc., Fig. 6-2, A. When chiselling grasp the chisel firmly and strike it squarely. Attempt to keep the fingertips around the chisel body to prevent it from flying from your grasp. On the other hand, avoid grasping it too tightly because a poorly aimed blow may cause serious injury to the hand. You may want to use a chisel holder in some situations. The holder will allow the use of larger hammers and greater striking force without the chance of injuring your fingers or hand. One type of chisel holder is shown in Fig. 6-2, B.

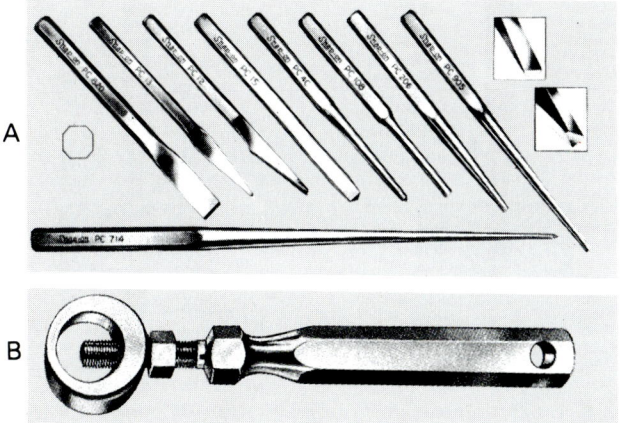

Fig. 6-2. A — Note chisel and punch assortment. B — Chisel holder, by holding punch or chisel, protects hands when striking with hammer. (Snap-on Tools)

Keep the cutting edge of the chisel sharp, and the striking surface chamfered (edges properly tapered) to reduce the danger of mushroom (smashed end) particles flying about. Fig. 6-3. Wear goggles when using a chisel.

The flat cold chisel, illustrated in Fig. 6-3, is used for general cutting. Special chisels such as the cape, half-round and diamond point are used when their shape fits a definite need.

PUNCHES

A starting punch is a punch that tapers to a flat tip. It is used in starting to punch out pins, and to drive out rivets after the heads have been cut off.

Once the pin has been started, the starting punch (because of its taper) can no longer be used. A drift punch, which has the same diameter for most of its length, is used to complete the job.

A pin punch is similar to a drift punch but is smaller in diameter. Fig. 6-4.

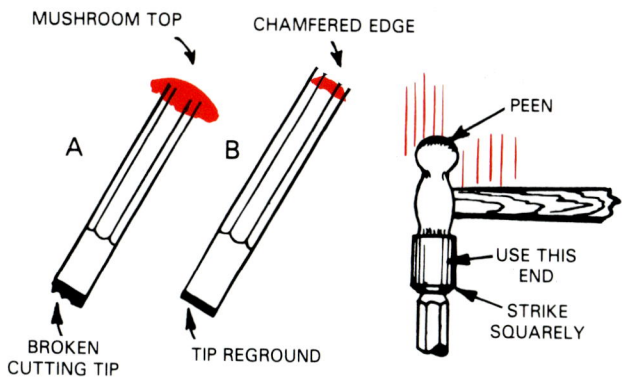

Fig. 6-3. Keep chisels in good condition. Chisel in A is in a dangerous condition. B is same chisel after grinding.

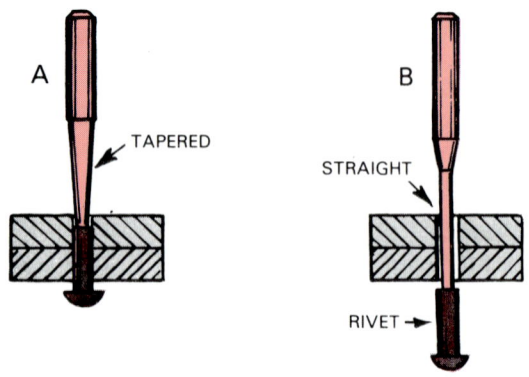

Fig. 6-4. Starting and drift punches. Starting punch A is used to start rivet from hole. Drift punch B is used to drive rivet from hole.

An aligning punch, which has a long, gradual taper, is useful to shift parts and bring corresponding holes into alignment. Fig. 6-5.

A centre punch is used to mark the material before drilling. It leaves a small, V-shaped hole that assists in aligning the drill bit. A centre punch may also be used to mark parts so they will be assembled in the correct positions. Fig. 6-6.

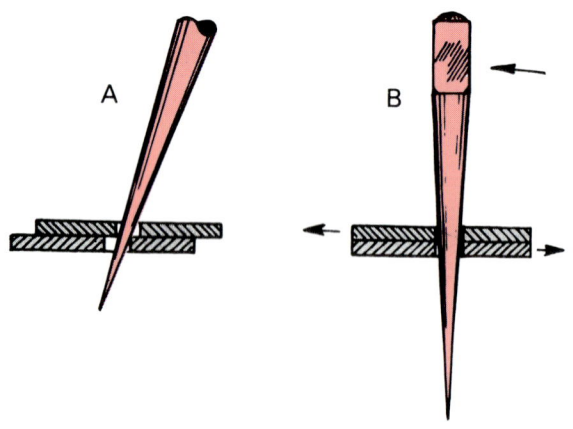

Fig. 6-5. Aligning punch. This punch is pushed through holes in A, then pulled upright and pushed deeper into holes in B. This lines up the two holes.

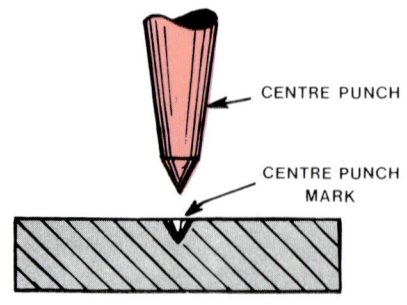

Fig. 6-6. Centre punch is used for marking parts and starting drills.

Be careful when sharpening chisels and punches. Keep the edges at the proper angle and avoid turning them blue (overheating) on the grinder. Overheating will draw (remove) the temper and render the tip soft and useless. Grind slowly and keep quenching (dipping) the tool in water.

CAUTION!

WEAR GOGGLES WHEN GRINDING OR STRIKING A CHISEL OR PUNCH.

FILES

Commonly used files are the flat mill, half-round, round, square, and triangular. All come in different sizes and with fine-to-coarse cutting edges. Some files have one or more cutting edges. Some files have one or more safe (surface with no cutting edges) sides.

A file should have a handle firmly affixed to the tang. The handle gives a firm grip, and it eliminates the danger of the sharp tang piercing the hand. A typical mill file is shown in Fig. 6-7.

FILE CUT AND SHAPE

When a file has a single series of cutting edges that are parallel to each other, the file is referred to as a single-cut file. A file with two sets of cutting edges that cross at an angle is called a double-cut file. See Fig. 6-8.

A file, from rough to smooth, depends on the number and size of the cuts. Fig. 6-9 illustrates several classifications of cut for both single and double-cut files.

Common file shapes are illustrated in Fig. 6-10. The thin, flat, point file also shown in Fig. 6-10 is used to file distributor and regulator points.

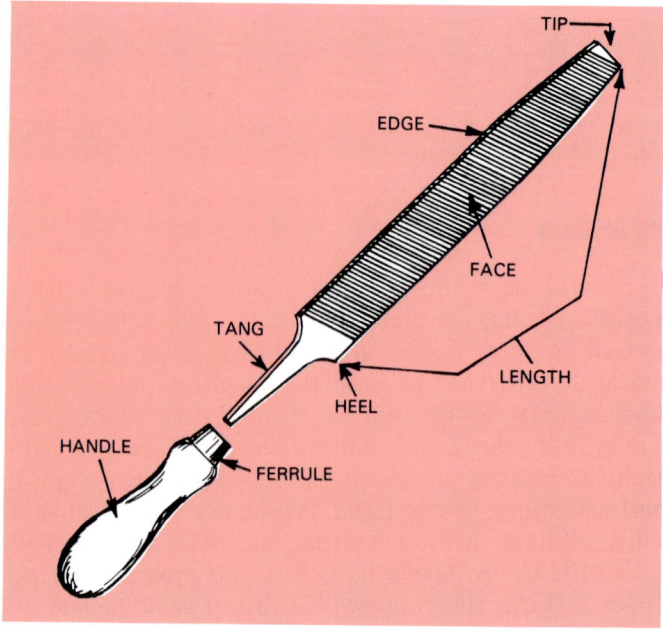

Fig. 6-7. Typical mill file. Single-cut.

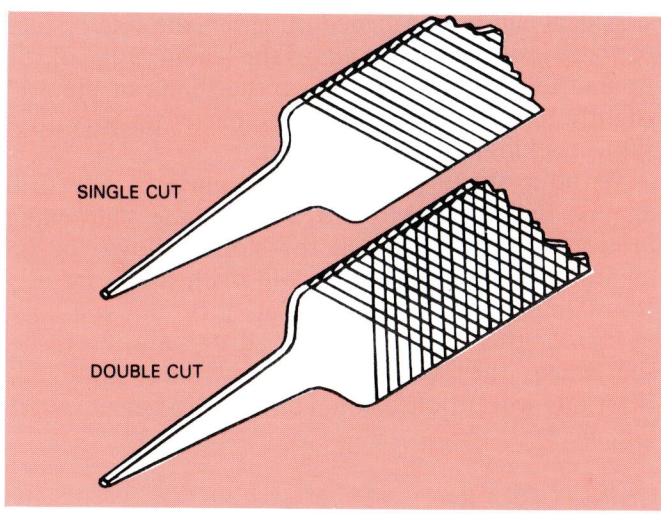

Fig. 6-8. Single and double-cut files.

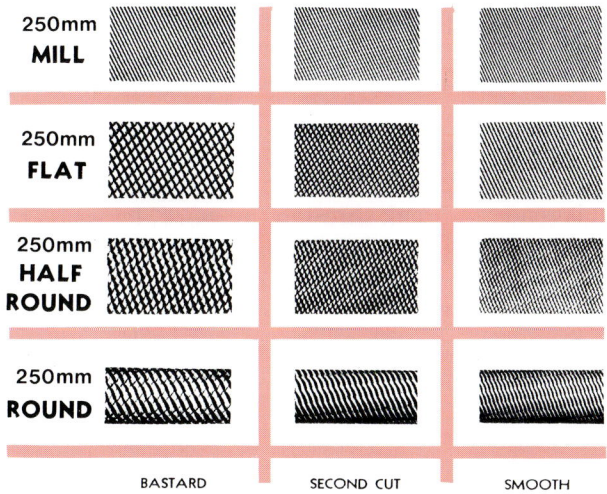

Fig. 6-9. File cuts. Three different file cuts: bastard, second-cut and smooth.

USING A FILE

When filing, hold the file with both hands. On the forward cut, bear down with pressure sufficient to produce good cutting. On the return stroke, raise the file to avoid battering the cutting edges. One hand is on the handle, while the other hand grasps the tip of the file.

Control the file to prevent it from rocking. Practice is essential. Remember that a file is NOT a crude tool. In the hands of an expert, a file becomes a precision tool.

Choose a file that is correct for the job at hand. Use a coarse cut file on soft material to prevent clogging. A slight drag on the return stroke, when filing soft material, will help keep the cutting teeth clear of filings.

Chalking (rubbing file with chalk) will help remove oil from the file and keep it from clogging. A file card (short bristle wire hand brush) is helpful in

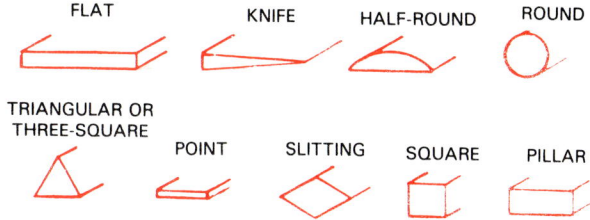

Fig. 6-10. Nine common file shapes, handy for the mechanic.

cleaning the file. An occasional tap of the handle will also help.

Keep file clean and free of oil and grease. Always use a handle! Install by striking handle of file in place, on a solid object. Do not hammer on file, it may break. Some handles screw on.

When you have finished filing, place the file in a dry spot, protected from contact with other files and tools. A good file is a fine tool. Treat it with respect, and you will do more and better work.

There are many files for special purposes, such as those used in body and fender repair. Swiss needle files are miniature files that come in almost all shapes and are very handy for delicate operations.

ROTARY FILES

A selection of rotary files is often found in the mechanic's toolbox. These files are designed to be chucked in an electric hand drill. They are useful in working in "blind" holes, etc., where a normal file is useless. Fig. 6-11.

DRILLS

A twist drill is used by mechanics for drilling holes. Twist drills are available in part millimetre sizes, fractional inch sizes, letter sizes (A to Z) and number sizes (wire gauge, numbers 1 to 80).

For general automotive shop use a set of metric drills 1.0 to 13 mm in 0.5 mm rises (totalling 25 drills), plus a set of fractional inch sizes 1/16 to 1/2 in 64th rises (totalling 29 drills), plus a set of number drills 1 to 60 (totalling 60) will suffice.

A number 1 drill has a diameter of 0.228 in. The number 60 drill has a diameter 0.040 in.

Twist drills with a straight shank are used in hand drills and portable electric drills. For heavy duty power drills, the taper shank drill is often used. See Fig. 6-12.

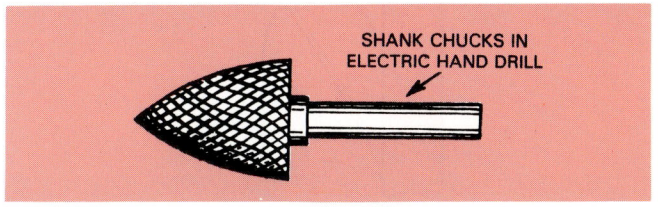

Fig. 6-11. Rotary file.

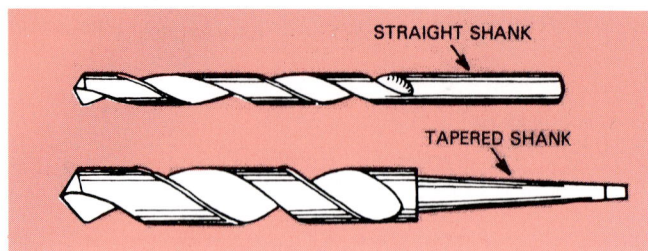

Fig. 6-12. Drill shanks.

Drills are commonly furnished in carbon steel and high speed steel. Carbon drills will require more frequent sharpening, and they will not last nearly as long as high speed steel drills.

SHARPENING DRILLS

Different cutting angles are used to make drills more efficient in various metals. Lip clearance angles also vary.

When sharpening a drill, keep in mind that both cutting lips must be the same length and angle. Avoid overheating the drill. If a carbon drill turns blue from overheating, it will be worthless until the blue area of the drill is ground off.

The general purpose drill should have the lips cut at 59 to 60 deg. to the centreline of the drill. Lip clearance should be around 12 deg. Fig. 6-13.

Proper drill grinding must be learned by practice. Use an old drill to practice on. A simple drill grinding gauge is pictured in Fig. 6-14. This gauge will help you maintain proper lip length and angle.

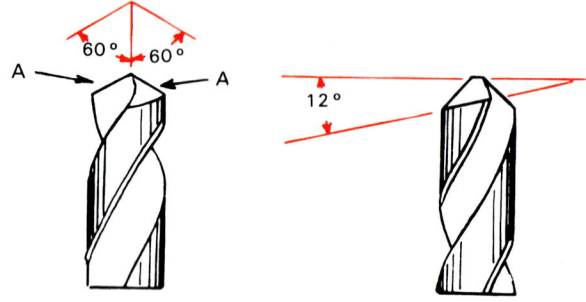

Fig. 6-13. Drill lip angles. Lengths at A must be equal. Angles should be the same on both sides.

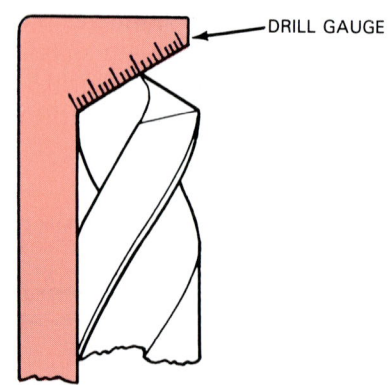

Fig. 6-14. Using a drill gauge. Gauge checks angle and length of tip. (Starrett)

A drill can be sharpened on either the side or face of the grinder. Start the cut at the leading edge of the lip and finish at the heel. Keep the shank of the drill slightly lower than the tip. The cut is taken with a slight rocking, pivoting motion.

Without starting the grinder, place a new drill against the stone. Move it through the sharpening motion and see how well the stone follows the lip angles. Try your skill on an old drill, then carefully compare it with a new one. Keep at it until you have mastered this skill. If your drills are correctly sharpened, they will cut readily without grabbing. Both lips will produce equal amounts of metal swarf Fig. 6-15.

USING DRILLS

Always securely fasten the piece to be drilled. Chuck the drill, Fig. 6-16, tightly and centre punch the spot to be drilled. After the drill has started, apply a small quantity of cutting oil. Oil will make cutting easier, faster and will lengthen drill life. When drilling thin metal or cast iron, oil is not required.

Hold the drill at the proper angle and apply enough pressure to achieve good cutting. When the drill is ready to break through the work, let up on the pressure to avoid grabbing. On thin metal, hold the work down, as well as keep it from turning. Thin work attempts to "climb" up the drill.

SAFETY HINTS

Large portable electric drills develop considerable torque, Fig. 6-17. Make certain you have a firm grip before engaging the drill.

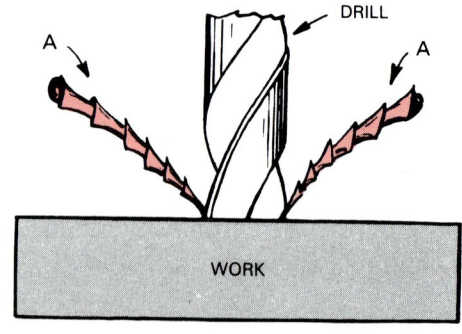

Fig. 6-15. Drill cutting properly. Swarf, A, of equal size indicates both lips are cutting.

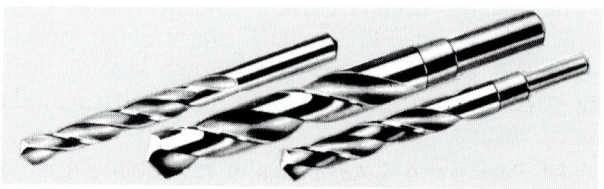

Fig. 6-16. Twist drills.

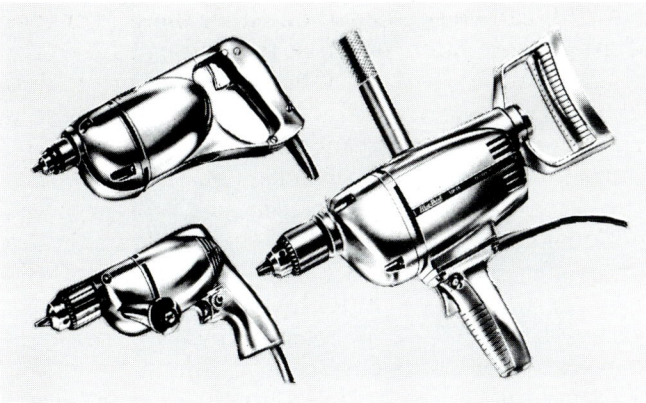

Fig. 6-17. Portable electric drills. Note different handle arrangements.

Never install or remove a drill in a chuck without unplugging the cord. If it is accidentally started when you are grasping the drill or the chuck key, you could be injured.

Keep loose clothing (sleeves, pants legs, ties, etc.) away from the drill.

Always secure the work firmly. If the drill grabs and the work is loose, the work can swing around with a vicious cutting force.

Wear goggles when grinding drills.

Never use a power tool that is not properly grounded, and do not use power tools when standing on a wet surface.

REAMERS

Reamers are used to enlarge or shape holes. They produce holes that are more accurate in size and smoother than those produced by drills. The reamer should not be used to make deep cuts in metal. A cut a few hundredths of a millimetre at a time is all that should be attempted.

Some reamers are adjustable in size, while others are of a fixed size. Tapered reamers are handy for removing burrs, assisting in tap starting, etc.

Use cutting oil when reaming. Always turn a reamer clockwise, both when entering and when leaving the hole. Several different types of reamers are shown in Figs. 6-18 and 6-19.

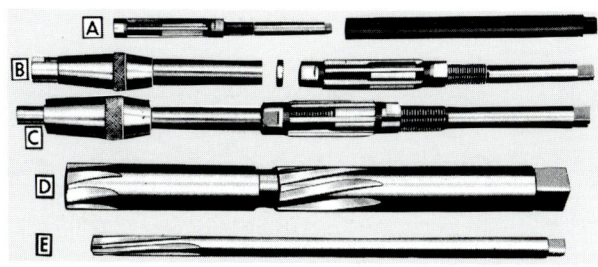

Fig. 6-18. Reamers. A — Adjustable. B, C — Adjustable with a pilot to help align reamer. D — Kingpin reamer. E — Valve guide reamer. (Snap-on Tools)

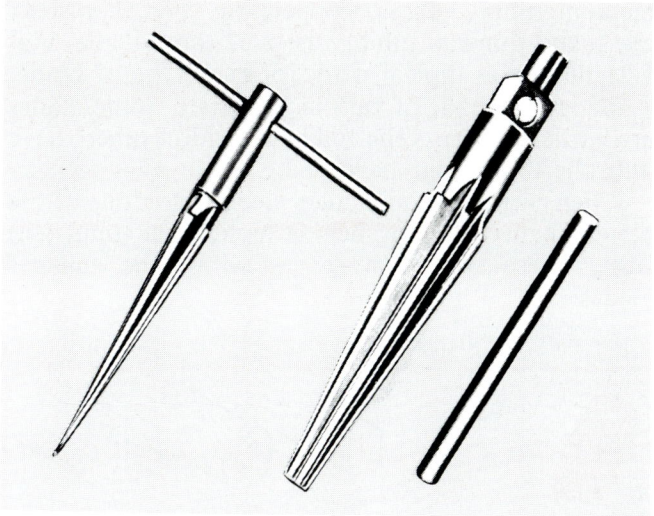

Fig. 6-19. Tapered reamers. (Snap-on Tools)

IMPACT WRENCH

A power-operated (electric or air) impact wrench is a must for fast work. Parts can be removed and replaced in a fraction of the time required with hand wrenches.

Although the impact wrench in no way replaces hand-operated wrenches, it can be used often and should be included in the tool collection. Various sizes are available. Fig. 6-20 pictures an air-operated impact wrench.

HACKSAWS

The hand hacksaw is a much used tool. It is excellent for cutting bolts, tubing, etc.

Hacksaw blades are generally furnished with 14, 18, 24, or 32 teeth per inch. The 14-tooth blade is handy for cutting fairly thick articles. The 18-tooth blade handles work of medium thickness. The 24-tooth blade is best used on heavy sheet metal,

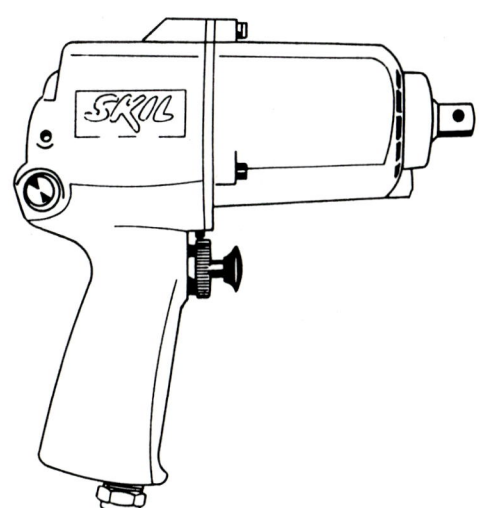

Fig. 6-20. An impact (power) wrench is a must for every mechanic. This wrench is air-operated. (Skil)

medium tubing, brass, copper, etc. For thin sheet metal and thinwall tubing, use a 32-tooth blade. Fig. 6-21 illustrates typical work for each type of blade.

Blades are made of various materials. Some blades are hardened across the full blade width; others have only the tooth edge hardened.

When cutting through alloy steels, select one of the better quality blades, such as high speed (tungsten) steel. These will outwear several of the cheaper blades.

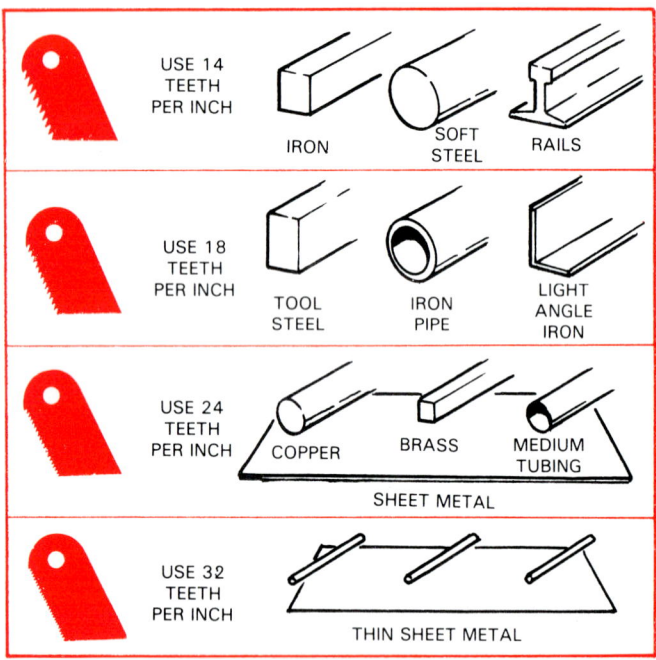

Fig. 6-21. The type of work to be cut determines the best blade tooth per inch selection. These are typical.

USING THE HACKSAW

When hacksawing, keep the blade tight in the saw frame. Make certain the teeth cutting edges face away from the handle. Fig. 6-22. See that the work is held securely to prevent slippage that would result in a broken blade.

Grasp the hacksaw frame firmly, one hand on the grip and the other on the front. Apply enough pressure on the forward stroke to ensure cutting, then, raise the blade slightly on the return stroke. Avoid starting the cut on a sharp edge; this may chip the saw teeth.

Try to saw at a speed that will produce 40 to 50 strokes per minute (forward strokes). A little oil will help.

SPECIAL HACKSAWS

The auto mechanic will sometimes find use for a special hacksaw that allows cutting in restricted quarters. A hole saw, operated by an electric drill is

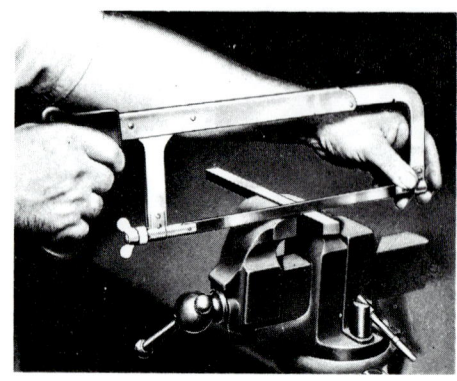

Fig. 6-22. When using a hacksaw, blade must be tight, teeth must face away from handle. Hold work securely and cut at 40 — 50 strokes per minute. (Starrett)

also handy for drilling large holes in thin sheet metal Fig. 6-23.

A conventional hacksaw, as well as two special, tight-quarter saws, are pictured in Fig. 6-24.

VICE

A vice is a work-holding device. A typical bench vice is shown in Fig. 6-25. If the work is easily marred, place special copper covers over the serrated steel jaws.

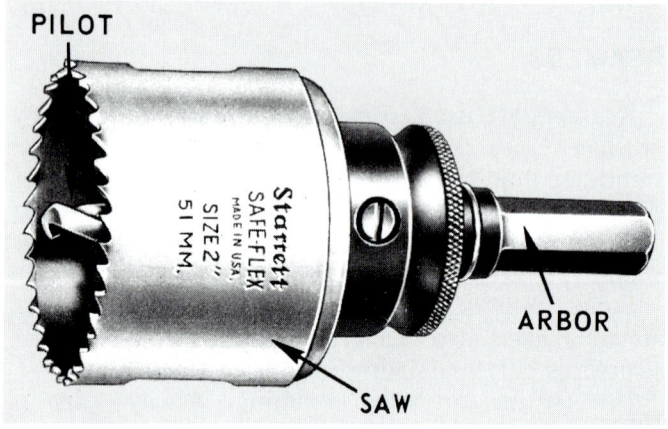

Fig. 6-23. Hole saws are handy for cutting openings in thin metal, as well as for making large diameter holes.

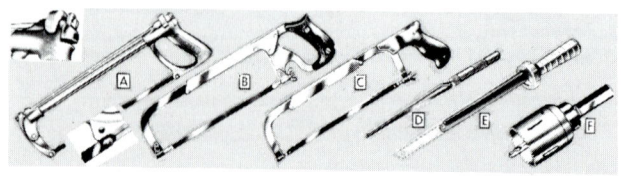

Fig. 6-24. Hacksaw types. A, B, C — Standard type hacksaw frames. D — Thin junior type for close quarters. E — Jab type will hold broken blade sections. It is handy in restricted areas. F — Hole saw used to drill holes in thin sheet metal.

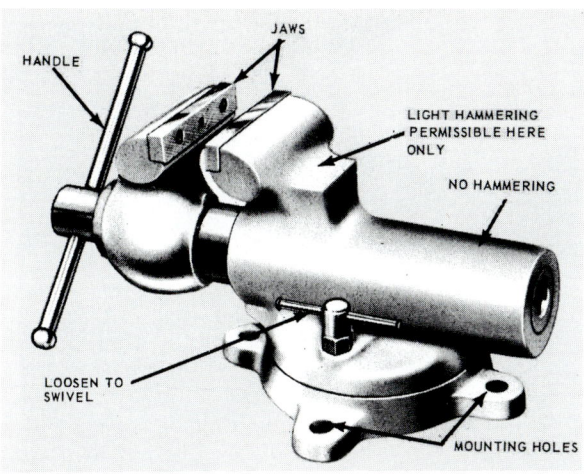

Fig. 6-25. Typical bench vice.

Keep your vice clean. Oil all working parts. Do not use a vice for an anvil, and do not hammer the handle to tighten the jaws on the work.

TAPS

Taps are used for cutting internal threads. Fig. 6-26. For general garage use a set of taper, intermediate and bottoming taps should be provided. Each type of tap will have to be provided in I.S.O. Metric, National Fine and National Coarse threads, as well as the various sizes, 3.5 mm, 4.0 mm etc.

The taper tap is used to tap completely through the hole. Notice that it has a long gradual taper that allows the tap to start easily.

The intermediate tap is used to tap threads partway through.

The bottoming tap is used to cut threads all the way to the bottom of a blind hole. The intermediate tap should precede the bottoming tap, since the bottoming tap will not start well.

The intermediate tap is the most widely used. It will work satisfactorily, except in the case of running threads completely to the bottom of a blind hole. Fig. 6-26.

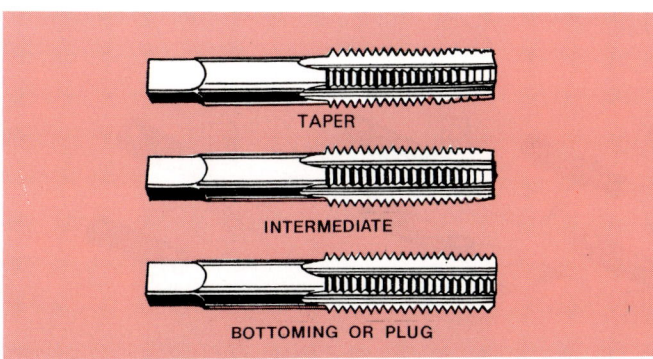

Fig. 6-26. Three kinds of taps.

USING THE TAP

After determining the diameter and type of thread of the screw or stud that will enter the tapped hole, use a tap drill chart to find what size hole to drill.

For example, suppose you find that you want a threaded hole for a 10 mm I.S.O. metric stud with a thread pitch of 1.25 mm. Referring to the I.S.O. Metric chart in Fig. 6-27, you will find the 10 mm nominal diameter in the left hand or first column, the thread pitch in the second column and the tapping drill size in the last column. This means that for a 10 mm stud with a 1.25 mm pitch thread you must drill a 8.80 mm diameter hole.

If the hole is to be tapped partway through, use a 10 mm x 1.25 intermediate tap. Place the tap in a tap handle.

Carefully start the tap in the hole. Place some tap lubricant on the tap. After threading the tap in one or two turns, back the tap up about a quarter to one-half turn to break the chip. Repeat this process as your tapping continues.

Be careful the hole does not clog with chips. It may be necessary to withdraw the tap and remove the chips. Taps are quite brittle. Use them with care, and make certain you use the proper size tap drill.

What size hole would you drill to tap threads for a 8 mm 1.0 pitch bolt? (Answer 7.0 mm) Figs. 6-27 and 6-28.

TAPPING DRILL SIZE CHART
I.S.O. METRIC

Nominal Diameter	Pitch	Tapping Drill
2.0	0.4	1.60 mm
2.5	0.45	2.05 mm
3.0	0.5	2.50 mm
3.5	0.6	2.90 mm
4.0	0.7	3.30 mm
4.5	0.75	3.70 mm
5.0	0.8	4.20 mm
6	1.0	5.00 mm
7	1.0	6.00 mm
8	1.0	7.00 mm
8	1.25	6.80 mm
9	1.25	7.80 mm
10	1.0	9.00 mm
10	1.25	8.80 mm
10	1.5	8.50 mm
11	1.5	9.50 mm
12	1.25	10.80 mm
12	1.75	10.20 mm
14	1.25	12.80 mm
14	1.5	12.50 mm
14	2.0	12.00 mm

Fig. 6-27. I.S.O. metric tapping drill chart.

DIES

Dies are used to cut external threads. A die of the correct size is placed in a diestock (handle) and turned. Use lubricant, back up every one or two turns, and keep free of chips.

Often dies are adjustable in size, so you can enlarge or reduce (slightly) the outside diameter of a threaded area.

Both taps and dies should be cleaned, lightly oiled and placed in a protective box for storage.

SPECIAL TAPS AND DIES

The mechanic will also find use for a few special taps and dies as illustrated in Fig. 6-29. One form of rethreading tool is shown in A. It is placed on the thread and turned. B shows an internal thread chaser used to clean up dirty or damaged internal threads. A thread restorer, C, is handy for quickly reconditioning external threads. The axle rethreader, D, is placed around the good thread area, clamped shut, and is then turned back over the damaged area.

Nut or rethreading dies, E and F, can be turned on a damaged thread. An ordinary ring spanner can be used to turn them. G and H show spark plug hole taps. These are very handy to clean up damaged or carboned plug hole threads. A combination tap and die set for tube flare fittings is illustrated in J. A combination tap and die set is pictured in K.

Fig. 6-28. Tapping operation must be performed carefully. Keep tap square with hole; use lubrication when required and keep backing tap to break chip. Use proper size tap drill.

CLEANING TOOLS

Various cleaning tools are pictured in Fig. 6-30. These different brushes and scrapers provide valuable assistance in the thorough cleaning of parts.

Carbon cleaning brushes that may be operated in a small electric drill are shown in Fig. 6-31. A mandrel, D, is also shown. It is used to hold a small wire wheel.

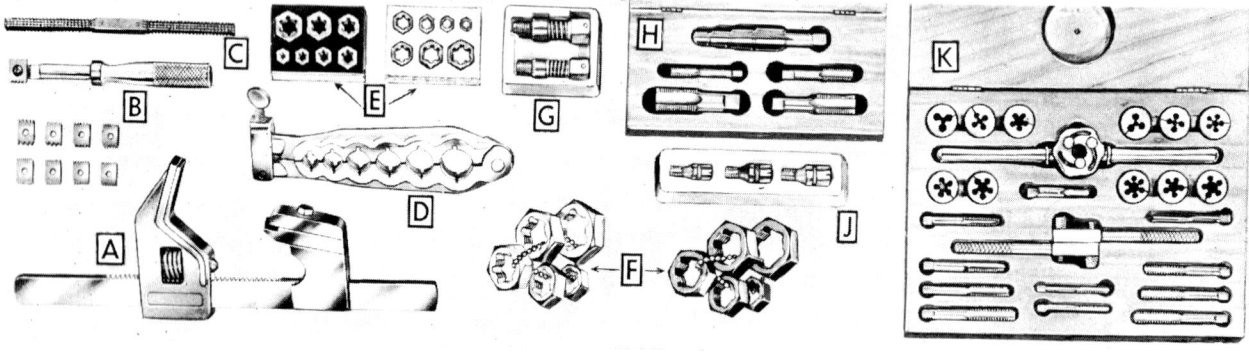

Fig. 6-29. Taps and dies.

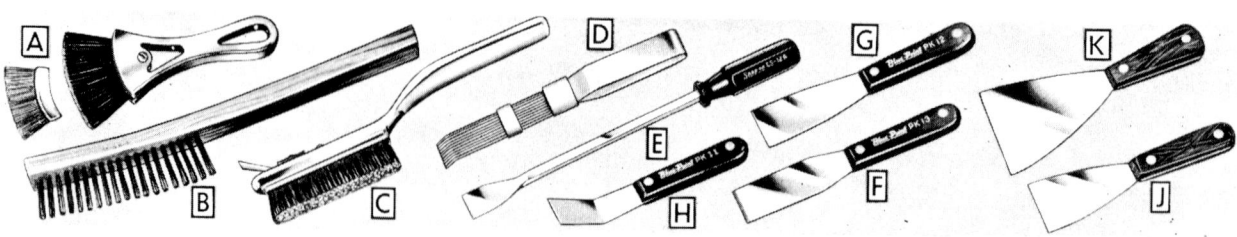

Fig. 6-30. Cleaning tools. A — Sparkproof bristle brush. B — Wire scratch brush. C — Wire brush and scraper combined. D — Flexible fingers carbon scraper. E — Rigid carbon scraper. F, G, H — Scraper blades. J, K — Putty knives.

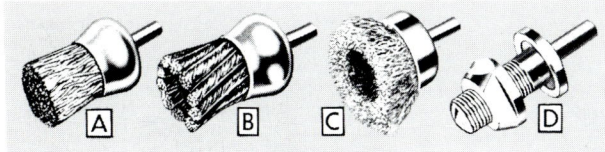

Fig. 6-31. Power carbon cleaning brushes. D is a mandrel that may be chucked in the electric hand drill to operate a wire wheel.

A wire wheel that may be used on a portable electric drill or mounted on a grinder mandrel is shown in Fig. 6-32.

CAUTION!

ALWAYS WEAR GOGGLES WHEN USING POWER GRINDERS OR WIRE WHEELS.

PLIERS

A mechanic needs a good variety of pliers. Study the pliers shown in Figs. 6-33 through 6-36. Learn the names and uses of each pliers.

Never use pliers in place of a wrench. Use pliers designed for the job. Avoid cutting hardened parts

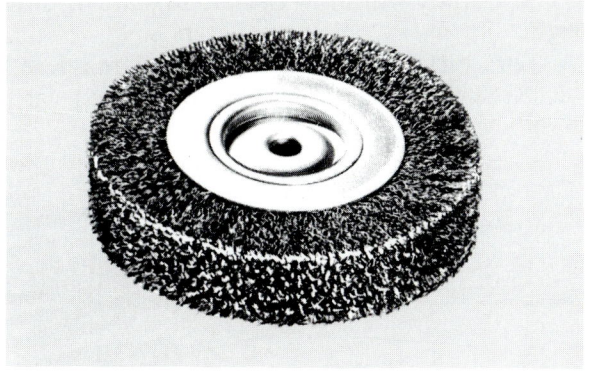

Fig. 6-32. Power wire wheel.

with your combination, diagonal or other cutting pliers. Needle nose pliers are delicate; use them carefully.

SCREWDRIVERS

Screwdrivers of various lengths and types are required. The standard, Phillips, Reed and Prince,

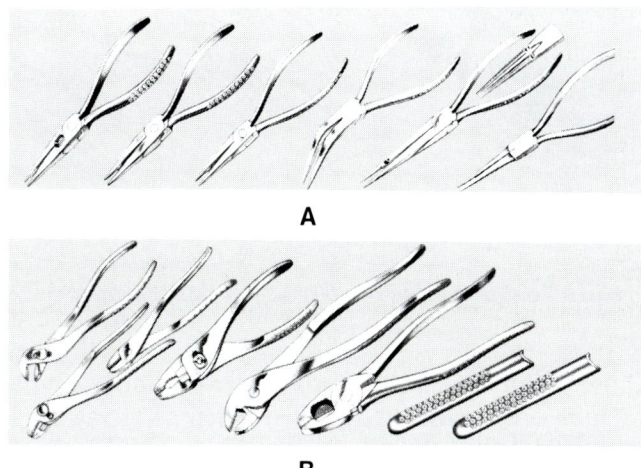

Fig. 6-33. A — Needle nose pliers. B — Assorted pliers.

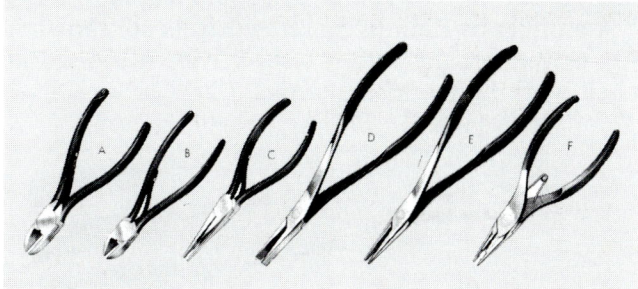

Fig. 6-34. Assorted pliers. A, B — Long nose diagonal cutters. C — Needle nose. D — Duck bill. E — Short needle nose. F — Cutter with long reach cutting jaws.

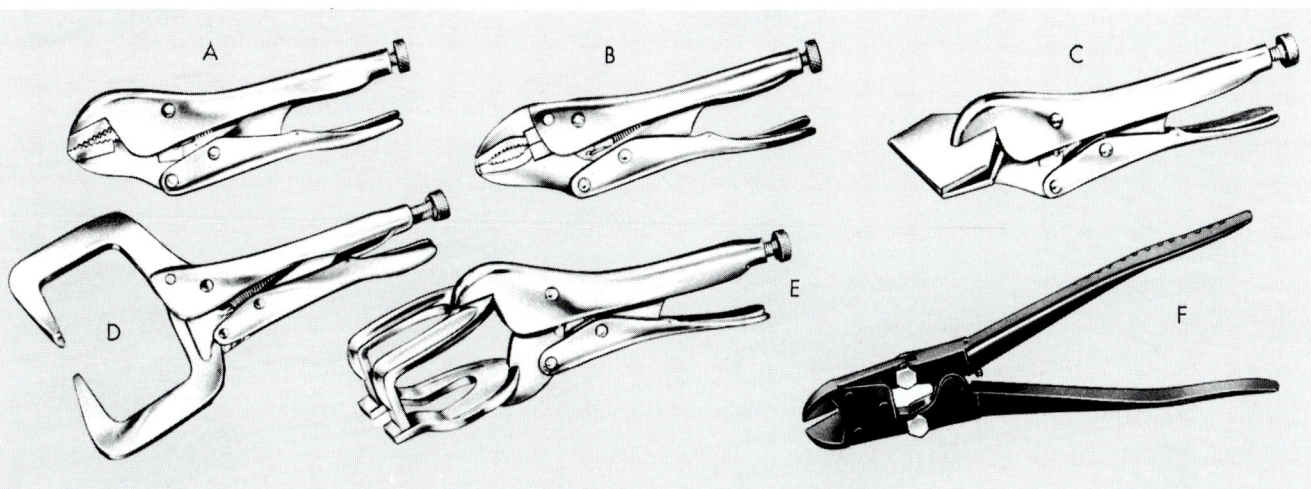

Fig. 6-35. Vice-grip pliers, A, B — Standard vice-grip. C — Vice-grip bending tool. D — vice-grip C — Clamp. E — Vice-grip welding pliers. F — High leverage cutter for wire and small bolts.

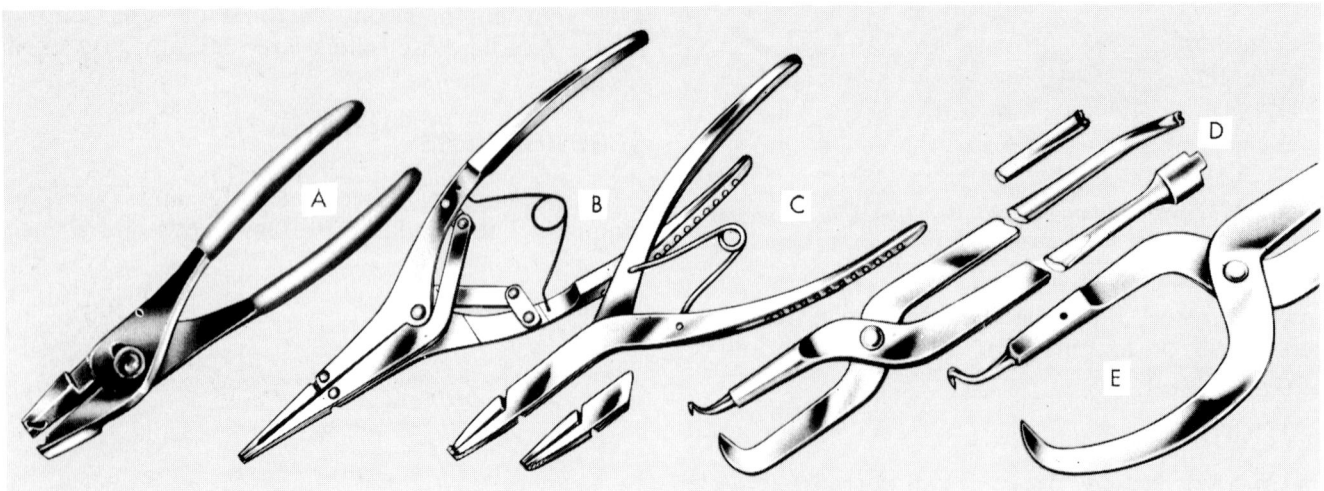

Fig. 6-36. Special purpose pliers. A — Hose clamp pliers. B — Parallel jaw lock ring pliers. C — Lock ring and snap ring pliers. D and E — Brake spring pliers.

Clutch, and Torx-bit tips will cover the various types of jobs ordinarily encountered when servicing cars.

It is wise to have in your tool kit several large and heavy types that will stand some prying and hammering.

The standard tip screwdriver is pictured in Fig. 6-37. It can have a round or square shank.

A number of Phillips tip screwdrivers are shown in Fig. 6-38. Note the stubby screwdriver. It is essential when working in close quarters.

The clutch tip or, as it is sometimes called, the "butterfly", is illustrated in Fig. 6-39.

A set of Reed and Prince tip screwdrivers is shown in Fig. 6-40. At first glance, it looks like a Phillips tip. Close study will show the two tips are of different design.

A Torx-bit type socket set is illustrated in Fig. 6-41. Note the shape of a Torx-bit.

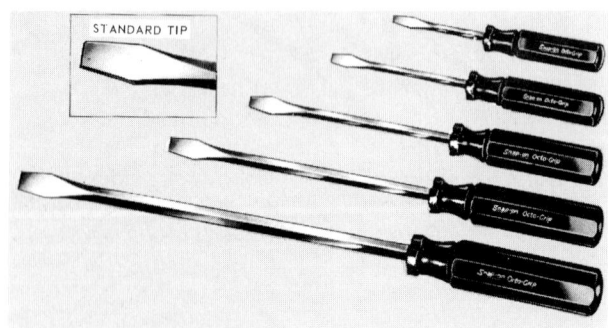

Fig. 6-37. Standard tip screwdrivers.

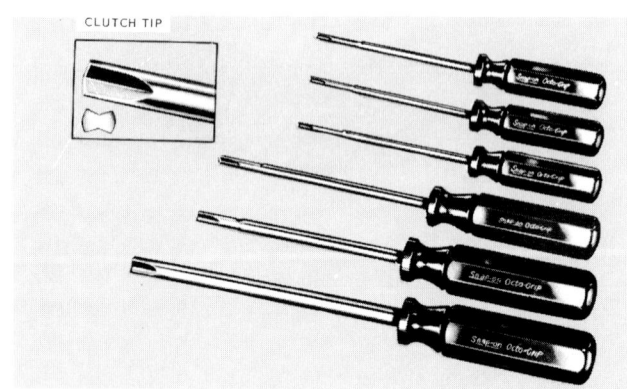

Fig. 6-39. Clutch tip screwdrivers.

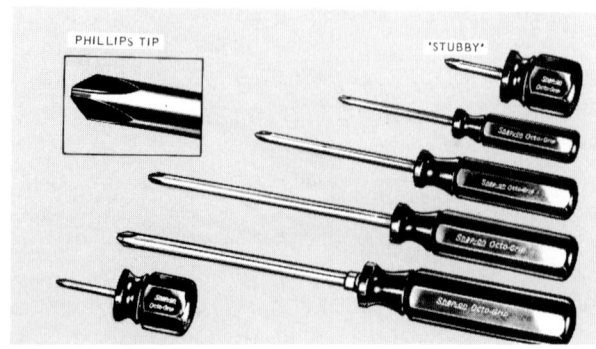

Fig. 6-38. Phillips tip screwdrivers.

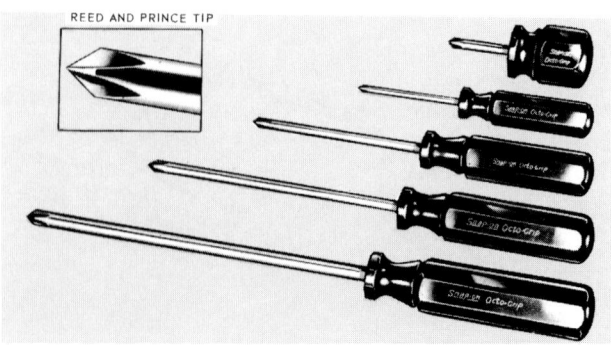

Fig. 6-40. Reed and Prince tip screwdrivers.

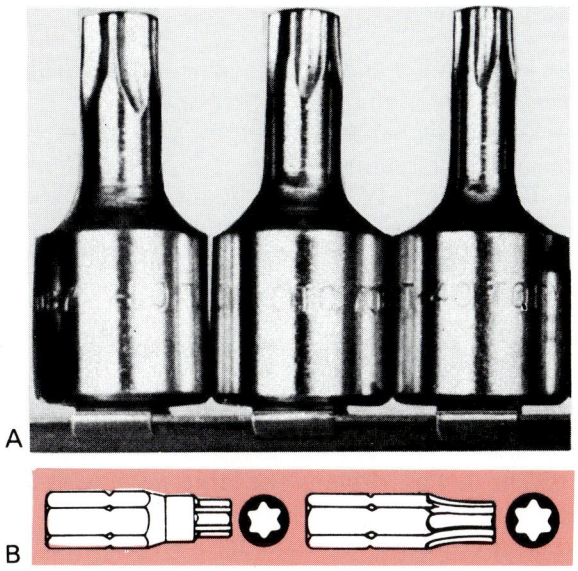

Fig. 6-41. A — Torx-bit drivers which can be used with ratchets, speed handles, etc. They provide a very positive grip. B — Torx replacement bits.

OTHER TYPES

An assortment of offset screwdrivers make it easy to remove and drive screws in different areas. Look at Fig. 6-42.

The long, thin shank, electrical-type screwdriver is handy for working on small intricate assemblies. Another essential screwdriver is the type designed to hold screws while they are being started.

USING SCREWDRIVERS

As with all tools, use a screwdriver that is designed for the job. Unless the screwdriver is designed for the purpose, do not use a hammer or pry with it.

If the screwdriver tip becomes worn or damaged, grind it slowly to avoid overheating the metal. Attempt to retain the original shape. On the standard tip, do not grind a sharp taper. This will cause the screwdriver to ride up out of the slot.

CAUTION!

WHEN HOLDING SMALL UNITS IN THE HAND, DO NOT PUSH DOWN ON THE SCREWDRIVER: IT MAY SLIP AND PIERCE YOUR HAND. IF WORKING ON ELECTRICAL EQUIPMENT, DISCONNECT THE BATTERY. WHERE IT IS NOT PRACTICABLE TO DISCONNECT THE BATTERY USE AN INSULATED (FULL LENGTH) SCREWDRIVER.

RING SPANNERS

The ring spanner is an excellent tool because it grips the nut on all sides. This reduces the chances of

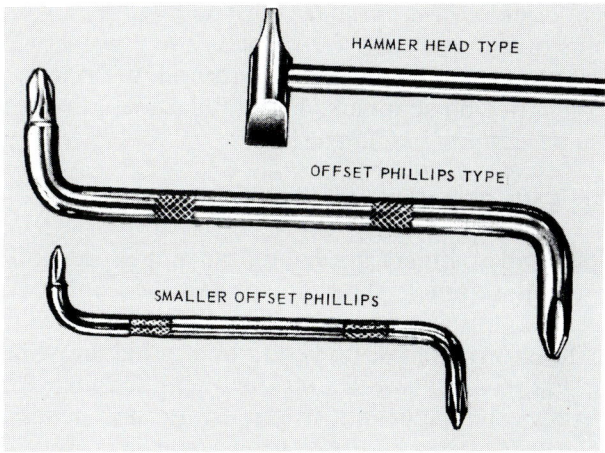

Fig. 6-42. Offset screwdrivers.

slipping with resultant damage to the nut and possibly the hand.

A ring spanner is designed with either a 6 or 12-point opening. For general use, the 12-point works well. It allows the spanner to be removed and replaced without moving the handle over such a long swing. For stubborn jobs, damaged nuts, or when there is danger of collapsing the nut, the 6-point will do a better job. It grips the nut across each flat, reducing slippage to a minimum.

The ring spanner is available in either a double offset or a 15 deg. angle offset. Sets in standard size openings, as well as standard lengths, will handle most jobs. Sets are available in midget length for small work in cramped areas. Each end on the ring spanner is a different size.

CAREFUL!

IN USING ALL SPANNERS, PULL, DO NOT PUSH. IF YOU MUST PUSH ON THE WRENCH, PUSH WITH THE PALM OF THE HAND WITH THE FINGERS OUTSPREAD TO AVOID SMASHING THEM IF THE SPANNER SLIPS.

Fig. 6-43 shows several ring spanners. Detail A has an extra deep offset for additional clearance. B is a

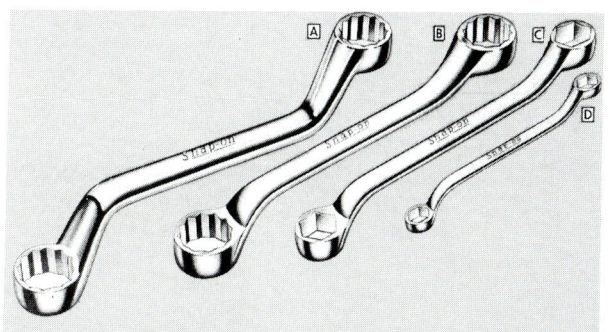

Fig. 6-43. Ring spanners. A — Double offset, 12-point. B — Standard offset, 12-point. C and D — Standard offset, 6-point.

typical double offset. C and D are double offset with 6-point openings. Can you see why the 6-point will grasp the nut or cap screw more securely? These are short length ring spanners. The standard length is of the same design, but longer.

OPEN END SPANNERS

Open end spanners are handy but not as dependable as ring spanners. They grasp the nut on only two of its flats, and are subject to slipping under a heavy pull. There are places, however, where they must be used. Fig. 6-44 illustrates an open end spanner for adjusting valve tappets; it is long with a thin profile. The standard open end spanner is somewhat shorter and of a huskier design.

The open end spanner has the head set at an angle. When the spanner has travelled as far as allowable, it may be flipped over and placed on the nut in an arc of 30 deg. Each end has a different size opening. Open ends are also available with the heads offset more than 15 deg.

COMBINATION RING AND OPEN END SPANNER

The combination spanner has a ring head on one end and an open end on the other. Generally, the open end is offset 15 deg. from the spanner centreline. Both ends are of the same size.

The combination spanner makes a very convenient tool since the ring end can be used both for breaking loose and final tightening. The open end is for faster fastener removal or installation. Fig. 6-45.

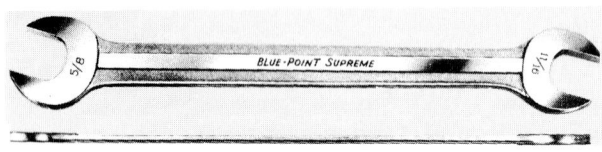

Fig. 6-44. Open end spanner.

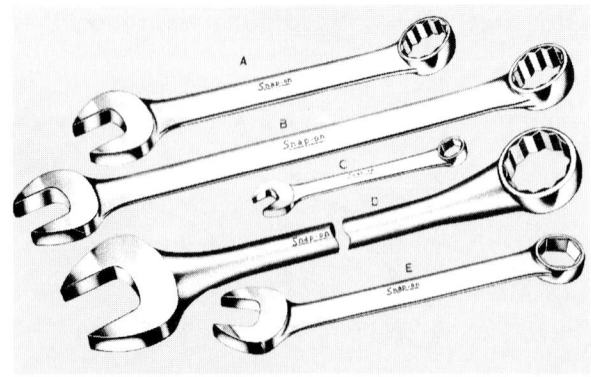

Fig. 6-45. Combination ring and open end spanners.
A — Short length. B — Long length. C — Midget length.
D — Large nut size. E — Double offset, both ends set off at
15 deg. angle from centreline to spanner. (Snap-on Tools)

SOCKET SPANNERS

A socket spanner is very convenient and, in most instances, is faster than the other spanners.

Sockets are available with 6-point, 12-point, and double-square openings. The 1/4 in. (6.35 mm), 3/8 in. (9.52 mm), 1/2 in. (12.70 mm), and 3/4 in. (19.05 mm), drives will cover a wide range. The drive size refers to the size of the square hole into which the socket handle fits. The larger the drive, the heavier and bulkier the socket.

The 1/4 in. drive is for small work in difficult areas. The 3/8 in. drive will handle a lot of general work where the torque (tightness) requirements are not too high. The 1/2 in. drive is for all-around service. The 3/4 in. drive is for large work with high torque settings.

Sockets are furnished in standard length; also deep sockets for work such as spark plugs requiring a longer than ordinary reach. Swivel sockets are good for angle work. Fig. 6-46.

Sockets should be kept clean, including the inside, and stored according to size.

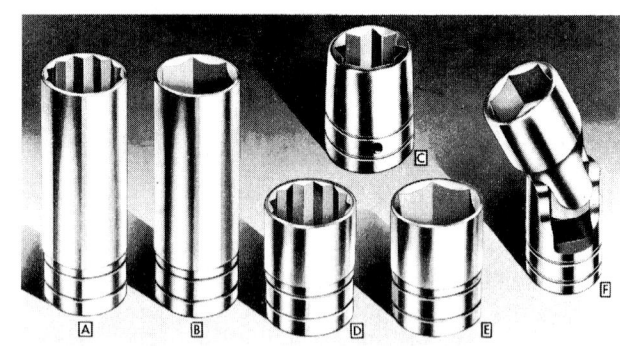

Fig. 6-46. Sockets. A — Deep, 12-point. B — Deep, 6-point.
C — Double-square. D — Standard, 12-point. E — Standard,
6-point. F — Swivel.

SOCKET HANDLES

Various handles are available for sockets. The speed handle is used for fast operation because it can be turned rapidly.

Flex handles in varying lengths allow the socket to be turned with great force, and at odd angles.

Extension bars of different lengths allow the mechanic to lengthen the socket setup to reach difficult areas

The sliding T-handle varies the handle length. See Fig. 6-47.

The ratchet handle is most versatile. It allows the user to either tighten or remove a nut without removing the socket. On the backstroke, the handle "ratchets". Fig. 6-48.

A ratchet that may be used with the speed handle, flex handle, etc., as well as an adapter that allows the

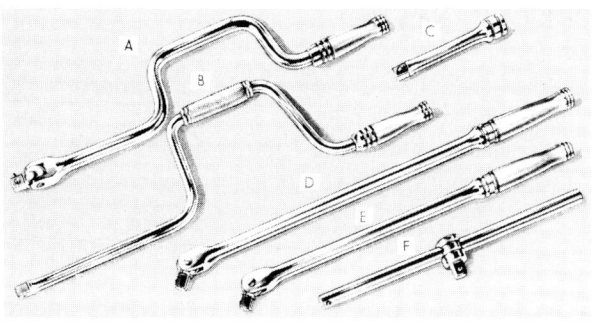

Fig. 6-47. Socket handles. A — Swivel head speed handle. B — Standard speed handle. C — Extension bar. D — Long flex handle. E — Short flex handle. F — Sliding T-handle.

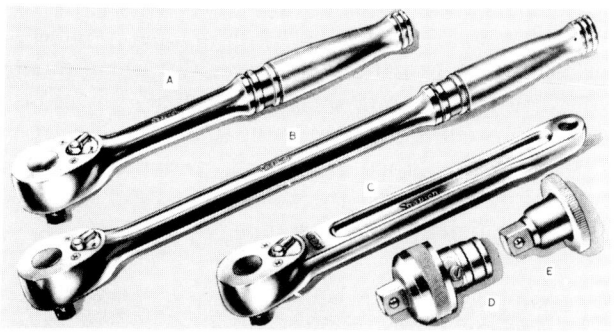

Fig. 6-48. Socket ratchet handles. A — Short ratchet. B — Long ratchet, C — Ratchet. D — Ratchet adapter. E — Ratchet spinner.

mechanic to spin the socket with fingers until too tight, is also shown in Fig. 6-48.

SOCKET ATTACHMENTS

Many socket attachments are available. Such items as screwdriver heads, drag link sockets, pan screw sockets, crowfoot attachments, Allen wrench heads, etc., all combine to make a socket set a fine tool. A few attachments are shown in Fig. 6-49.

PIPE WRENCHES

Pipe wrenches, both inside and outside, are useful for grasping large or irregular objects. They are strong and exert a powerful holding force. A and B, Fig. 6-50.

CRESCENT OR ADJUSTABLE WRENCH

The Crescent wrench is handy in that it can be adjusted for size. However, it tends to slip, so it is a poor wrench to use for most jobs when other tools are available. When used, adjust the jaws firmly.

Also, make certain the wrench is placed so that the pull on the handle is toward the bottom side. This relieves heavy pressure on the adjustable jaw. See C, Fig. 6-50.

OTHER USEFUL SPANNERS

The flexible head socket spanner, A in Fig. 6-51, as well as the ratcheting ring spanner, B, are useful additions to the mechanic's toolbox.

Cylinder head crowfoot spanners are used for the removal of cylinder head nuts and cap screws. Note the various shapes. Fig. 6-52.

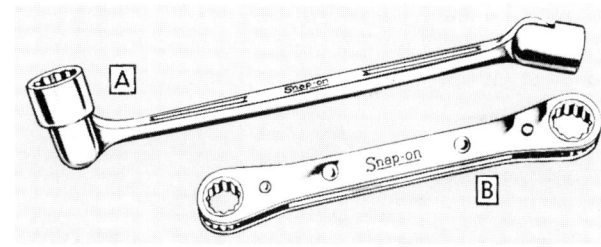

Fig. 6-51. Flex head socket and ratchet ring spanners. A — Flex head socket. B — Ratchet ring.

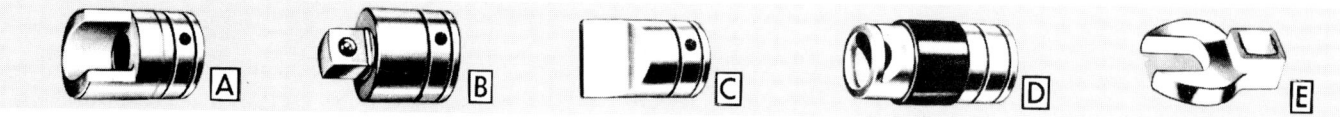

Fig. 6-49. Socket attachments. A — Adapter. B — Handle size adapter. C — Drag link socket. D — Pan screwdriver. E — Crowfoot.

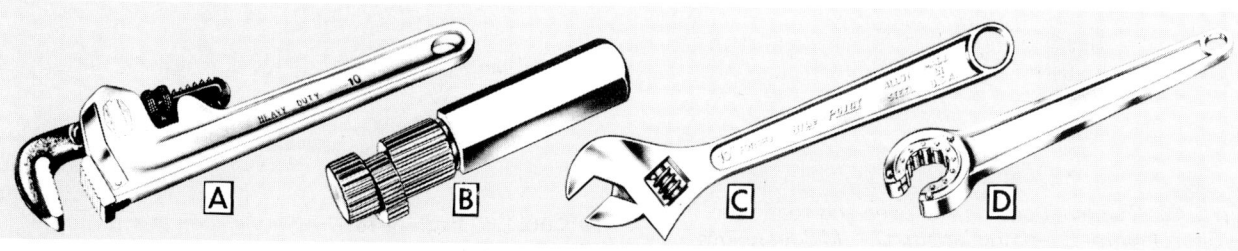

Fig. 6-50. Wrenches. A — Pipe wrench. B — Inside pipe wrench. C — Adjustable wrench. D — Cam-lock wrench.

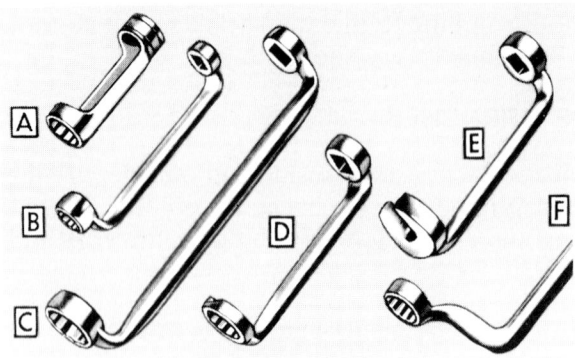

Fig. 6-52. Cylinder head crowfoot spanners.

A stud remover, obviously, is used to remove various studs. Two types are shown in Fig. 6-53.

Allan and fluted wrenches are often used in tightening or removing setscrews, cap screws, etc. They are strong and provide a good grip. Fig. 6-54.

TORQUE WRENCHES

The torque wrench is a MUST for all mechanics. It will enable you to tighten bolts, nuts, etc., to exact torque (tightness) specifications supplied by the manufacturer.

It is of utmost importance that bolts be pulled up to proper specifications. Improper and varying torque on units or assemblies will cause distortion. When certain torque is specified (heads, main, and

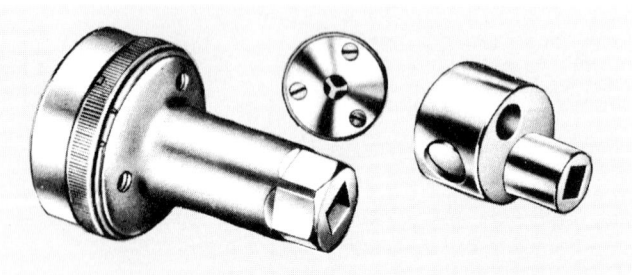

Fig. 6-53. Stud removers. D — Three jaw stud puller. E — Wedge type stud puller. (Snap-on Tools)

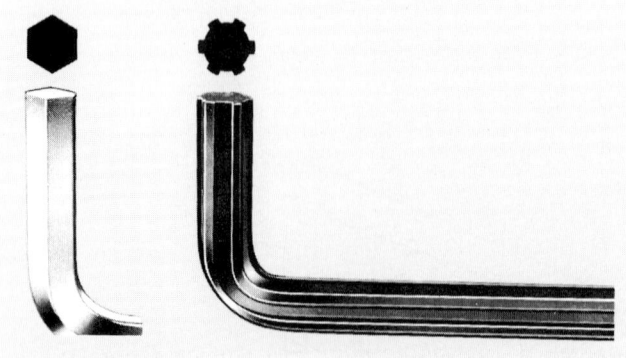

Fig. 6-54. Allen and fluted wrenches. Hex Allen wrench (left) is more popular type.

rod bearings, etc.), use an accurate torque wrench. Fig. 6-55.

Torque wrenches are available in numerous sizes with either a single scale for reading only in Metric (Nm) or with a dual scale for reading in both metric (Nm) and Imperial (lb-ft. or lb-in. depending on the size).

For general automotive work the mechanic will require three or four torque wrenches which will cover a torque range from 1Nm to approximately 400 Nm. There are also special torque wrenches. One such wrench is pictured in Fig. 6-56.

PULLERS

Auto mechanics will find use for a wide variety of pullers. They are used for pulling gears, bearings hubs, etc. An assortment of pullers is shown in Fig. 6-57. Note the various sizes.

Another excellent puller is the slide hammer type. The puller jaws grasp the work, and the weighted slide is banged against the stop. This type of puller is fast and efficient on many types of jobs. Fig. 6-58.

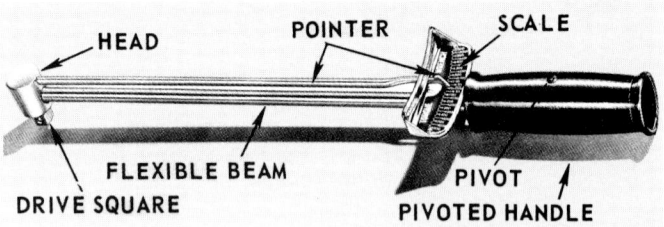

Fig. 6-55. One popular design torque wrench. Torque wrenches ensure correct fastener tension.

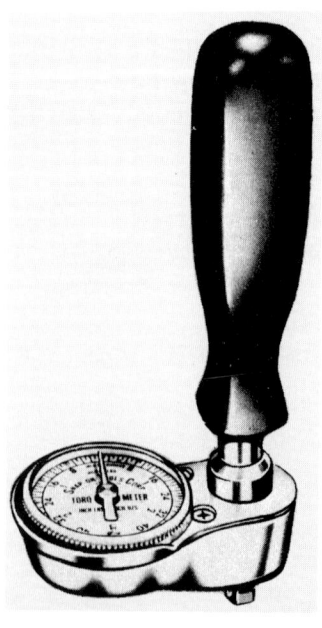

Fig. 6-56. Small, hand operated, dial indicator torque wrench. This wrench is very handy for small jobs such as torquing screws, bolts, etc., used on carburettors, automatic transmission valve bodies, etc.

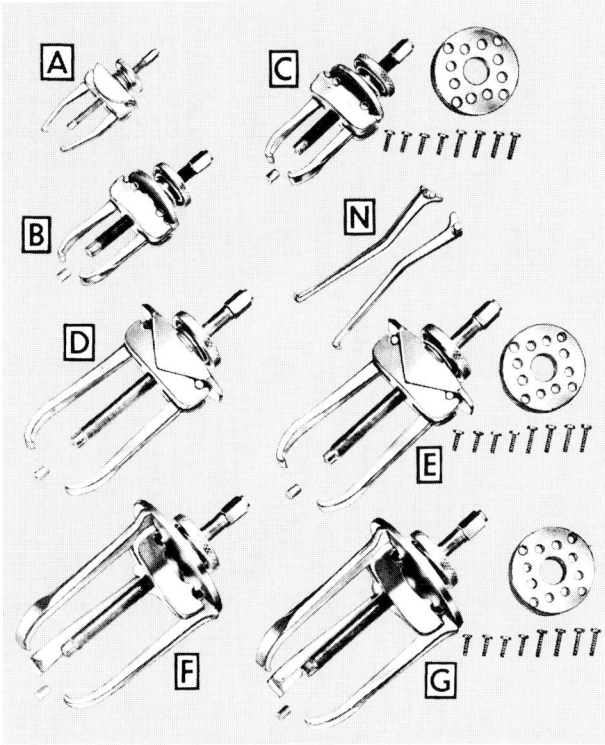

Fig. 6-57. Pullers. Pullers are available in various styles and sizes.

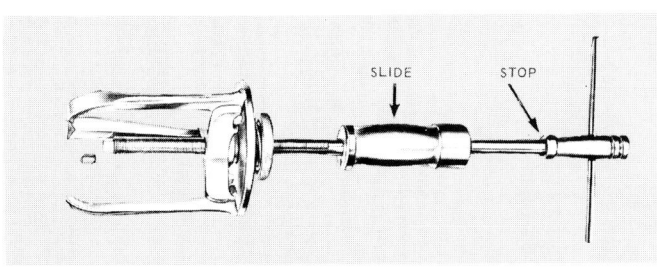

Fig. 6-58. Slide hammer puller. Slide is hammered on stop to develop powerful pulling force.

SPECIAL PURPOSE TOOLS

Other tools shown in accompanying illustrations are considered special purpose tools. Fig. 6-59 pictures brake service tools. Fig. 6-60 features engine service tools. Some tools for regulator and distributor work are shown in Fig. 6-61.

Useful tyre and wheel tools are illustrated in Fig. 6-62. Body tools are shown in Fig. 6-63. Some very handy tools that enable the mechanic to see in difficult areas, as well as remove loose parts where the hand will not reach, are pictured in Fig. 6-64.

A tubing cutter and a tube flaring tool are shown in Fig. 6-65. All tube fittings should be removed and replaced with special tubing flare nut wrenches. They are designed to provide a maximum amount of contact. Slide the open point over the tube and then down on the nut. Fig. 6-66. Miscellaneous hand tools are illustrated in Fig. 6-67.

SOLDERING EQUIPMENT

For soldering all automotive wiring, use RESIN CORE solder only. Acid core solder leaves a residue that will cause corrosion in electrical units, and should never be used for wiring. Acid core is satisfactory for radiator work.

All parts to be soldered should be clean and bright. Heat the parts with the soldering iron and add solder. Make certain the solder bonds properly. Cool and wipe clean. Tape soldered wiring with plastic tape. Irons are shown in Fig. 6-68.

MEASURING TOOLS

Mechanics are often called on to make PRECISION measurements. To do this, they must have, and understand the use of: an outside micrometer,

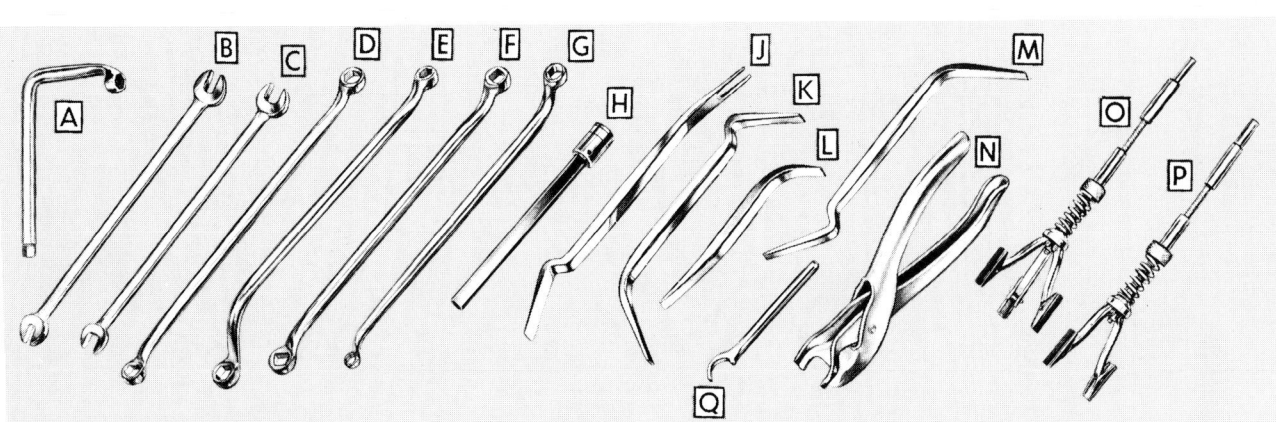

Fig. 6-59. Brake service specialty tools. A — Right angle bleeder wrench. B — Bendix eccentric cam wrench. C — Ford eccentric cam wrench. D, E — Offset-head bleeder wrenches. F — Taunus brake wrench. G — Compact car bleeder wrench. H, I, J, K, L, & M Various brake shoe adjusting tools. N — C - washer pliers. O — Three arm hone for brake cylinders. P — Two arm hone for brake cylinders. Q — Parking brake tool.

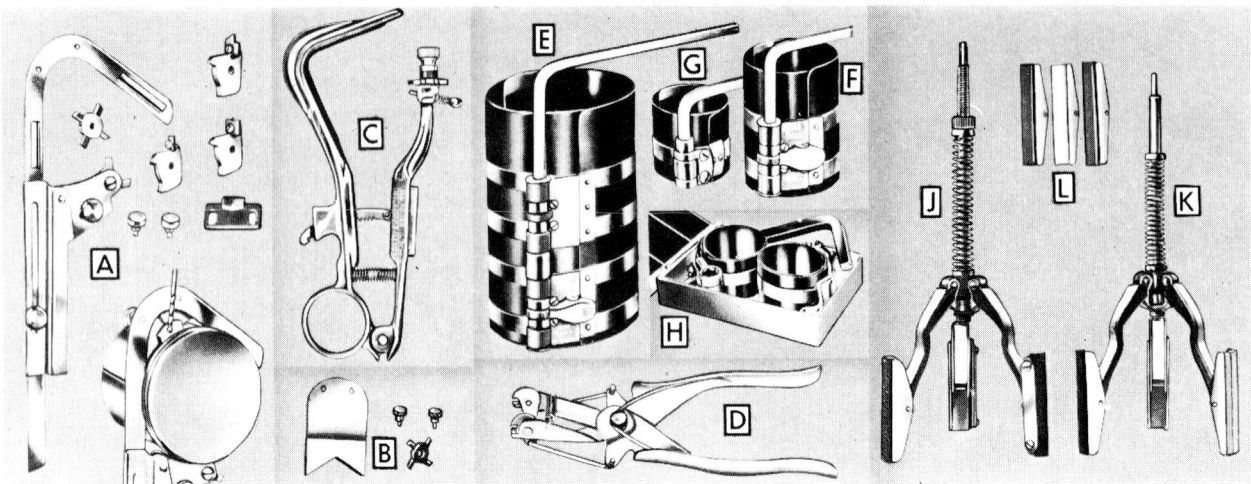

Fig. 6-60. Piston and cylinder service tools. A — Ring groove cleaner and cutter. B — Ring groove cleaner attachments. C — Ring groove cleaner. D — Piston ring spreader. E, F, G, H — Ring compressors. J, K — Cylinder wall deglazers. L — Deglazer stones. (Snap-on Tools)

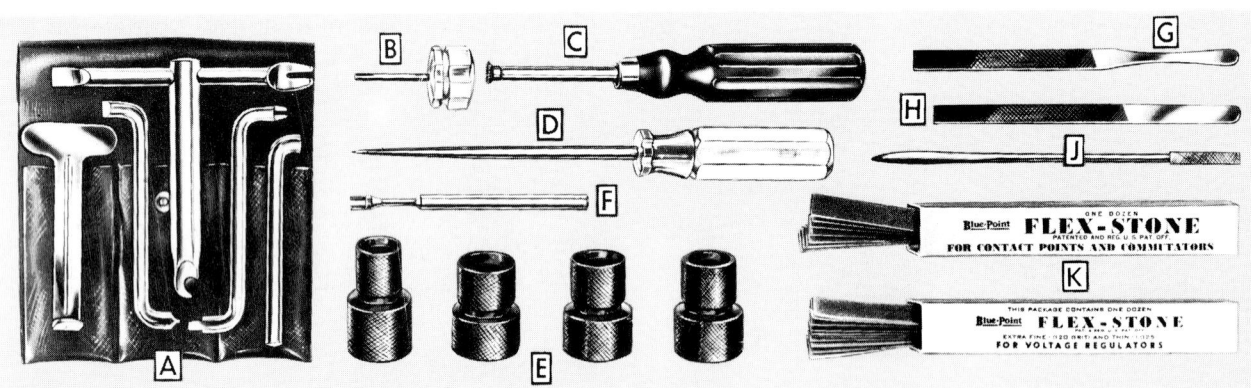

Fig. 6-61. Regulator and distributor tools. A — Voltage regulator tools. B — Hex tool for distributor point adjustment. C — Distributor cleaning brush. D — Neon spark tester. E — Distributor point rubbing block surfacers. F — Distributor spring tension tool. G — Distributor point file. H — Regulator point file. J — Point riffler file. K — Flex stones for cleaning points. (Snap-on Tools)

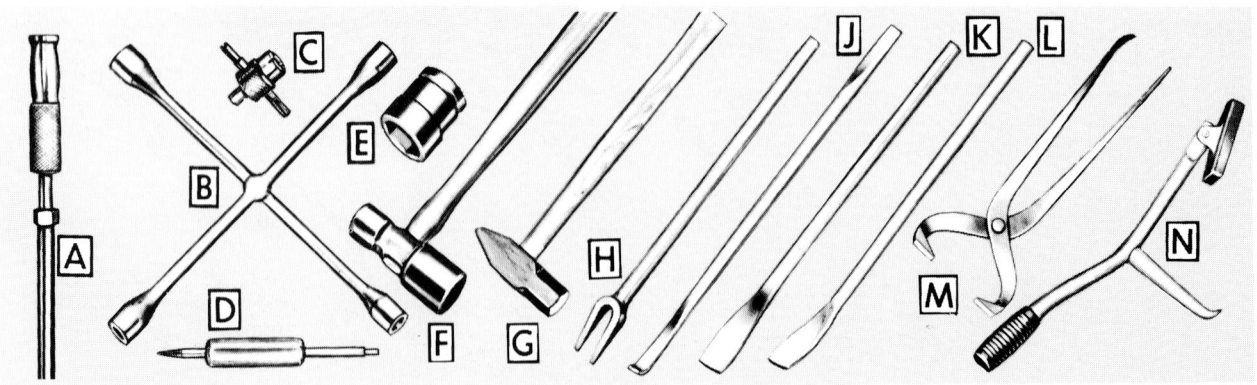

Fig. 6-62. Tyre and wheel tool. A — Tyre bead remover. B — Wheel nut wrench. C, D — Valve core tools. E — Impact socket. F — Rubber-tipped steel hammer. G — Cross pein hammer. H — Tie rod separator. J — Split rim tool. K, L — Tyre removing tools. M — Grease cap tool. N — Hub cap tool.

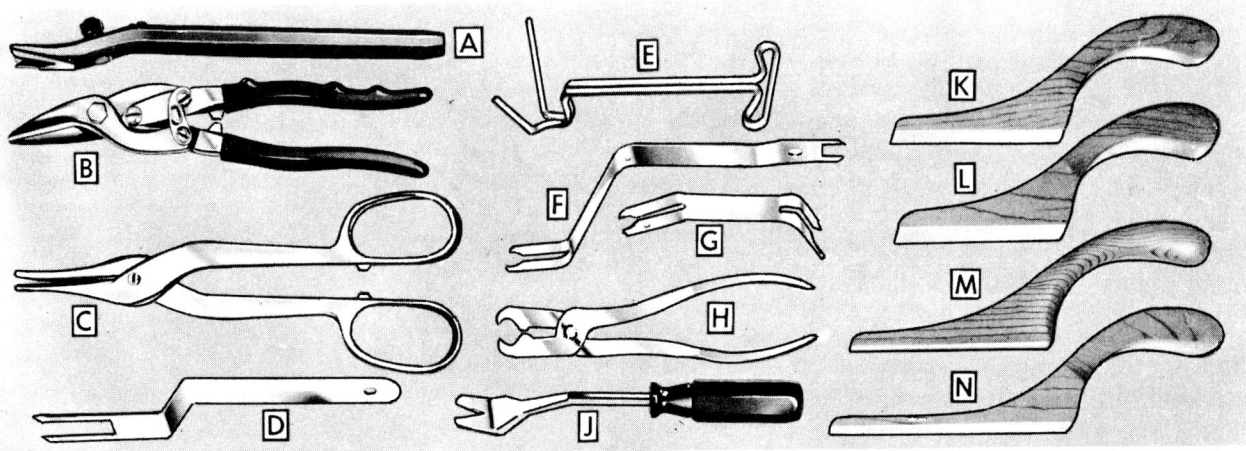

Fig. 6-63. Body tools. A — Body panel cutter. B — Metal cutting shears. C — Tin snips. D, E, F, G — Door handle tools. H — Door handle pliers. J — Door panel remover. K, L, M, N — Body solder paddles.

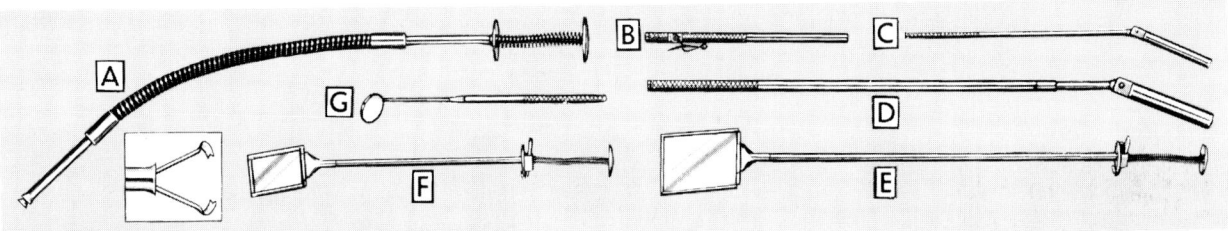

Fig. 6-64. Probing tools. Various mirrors and pickup tools for probing into areas that are otherwise inaccessible.

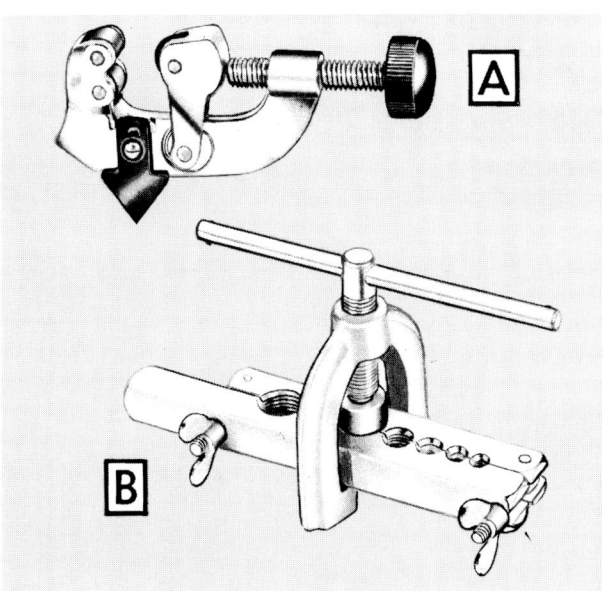

Fig. 6-65. Tubing tools. A — Tubing cutter. B — Tubing flaring tool.

and inside micrometer, dial gauges, calipers, depth gauges, combination square. See chapter 5 for detailed information on precision measuring tools.

SPECIALISED TOOLS

Space limitations will not permit covering the many specialised tools such as those shown in Fig. 6-69. The student would be wise to procure

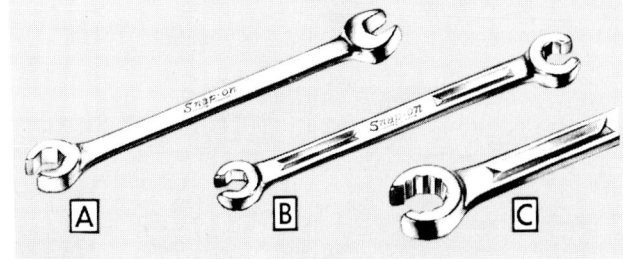

Fig. 6-66. Tubing wrenches. A, B — Six-point. C — Twelve-point.

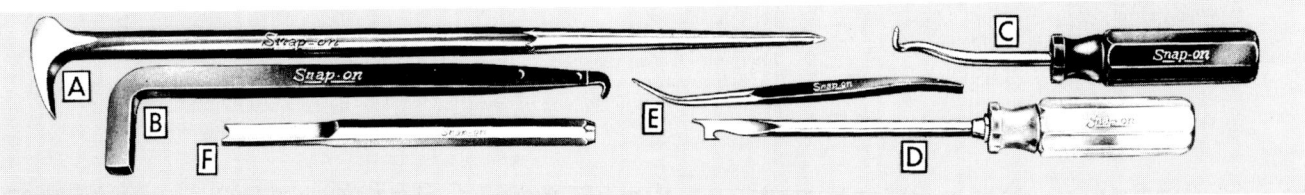

Fig. 6-67. Other mechanics hand tools. A — Pry bar, B — Cotter pin remover. C, D, E — Cotter pin removers. F — Bushing cutter.

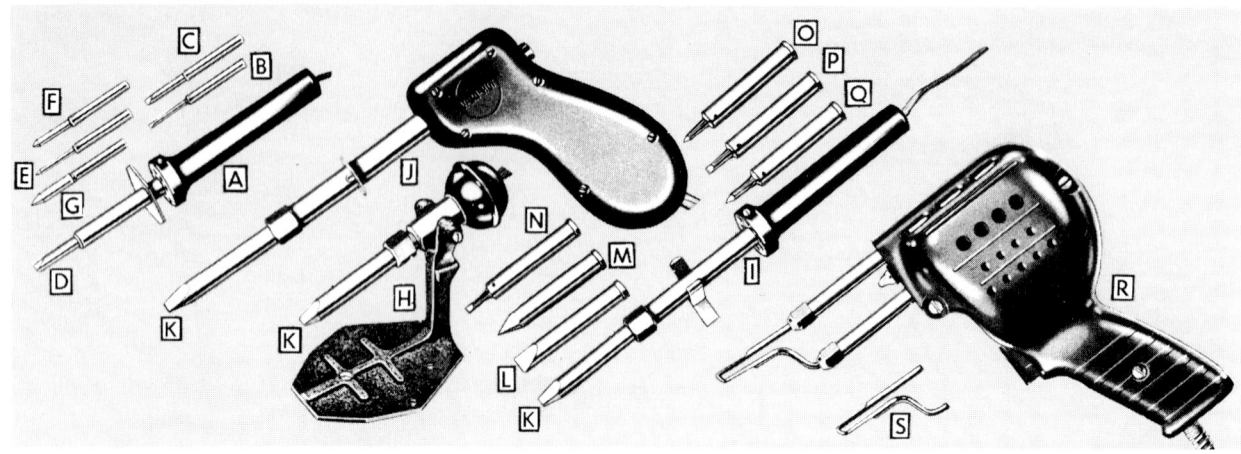

Fig. 6-68. Soldering irons. Several types of irons and tips, all electric.

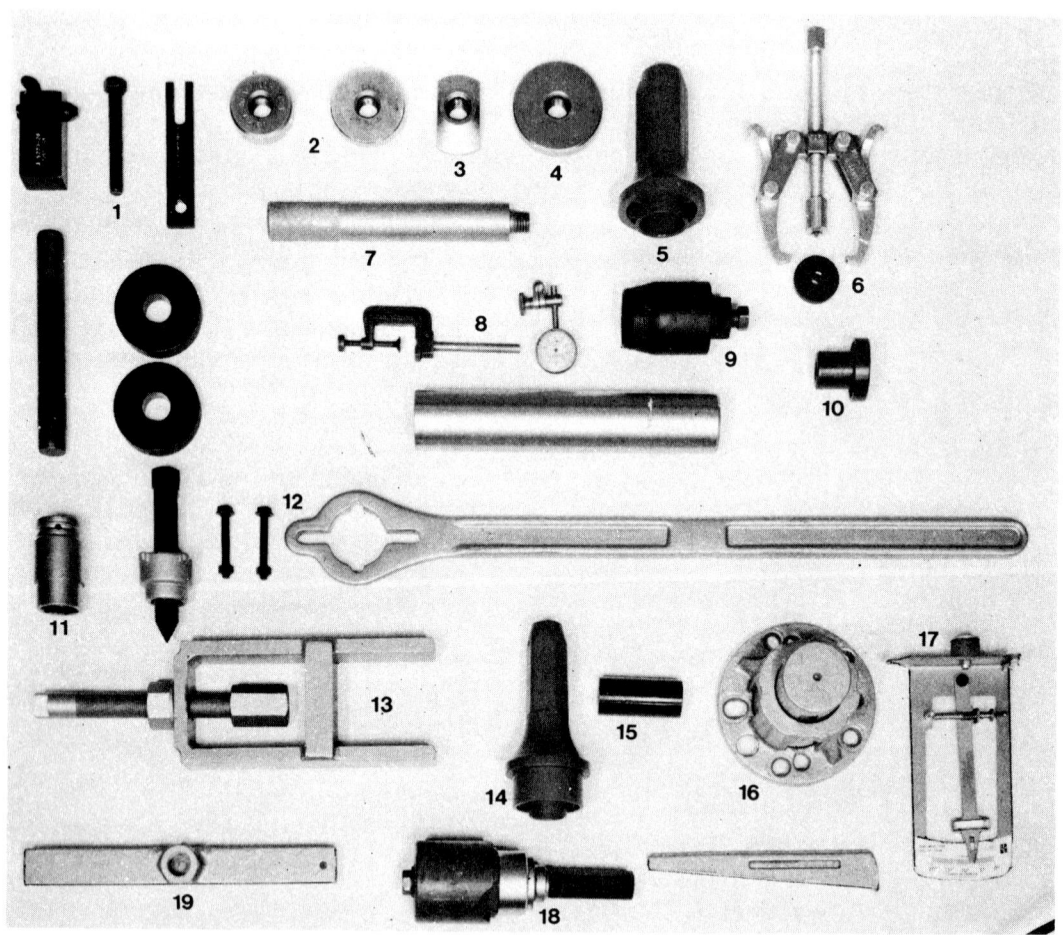

1 — Pinion setting gauge.	8 — Dial indicator set.	14 — Pinion oil seal installer.
2 — Pinion setting gauge adaptors.	9 — Pinion oil seal remover.	15 — Rear hub puller spacer.
3 — Rear pinion bearing remover.	10 — Differential bearing installer.	16 — Rear hub puller and wedge.
4 — Rear pinion bearing cup installer.	11 — Companion nut socket.	17 — Pinion angle gauge (Inclinometer).
5 — Pinion oil seal installer.	12 — Companion flange holder and remover.	18 — Bushing service set.
6 — Differential side bearing puller.	13 — Axle shaft remover.	19 — End play checking tool.
7 — Driver handle.		

Fig. 6-69. Specialised tools can make repair operations much easier, faster, and better. This group of tools is designed for drive axle work.

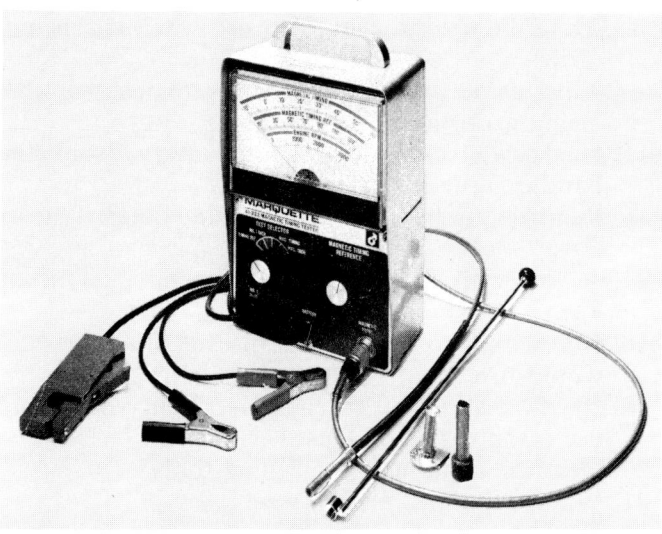

Fig. 6-70. Engine ignition magnetic timing tester.

catalogues from tool manufacturers. Study the catalogues to learn tool identification and practice.

ELECTRICAL TEST INSTRUMENTS

The mechanic must be familiar with and skilled in the use of many electrical testing instruments. The VOLTMETER, AMMETER, OHMMETER, OS-CILLOSCOPE, TIMING LIGHT, TACH-OMETER, etc., are all very important to proper diagnosis, service, and repair.

Some instruments, such as the magnetic ignition timing device, shown in Fig. 6-70, are designed as individual "hand-held" units. Others like the engine analyser, pictured in Fig. 6-71, combine a number of instruments into a single stand.

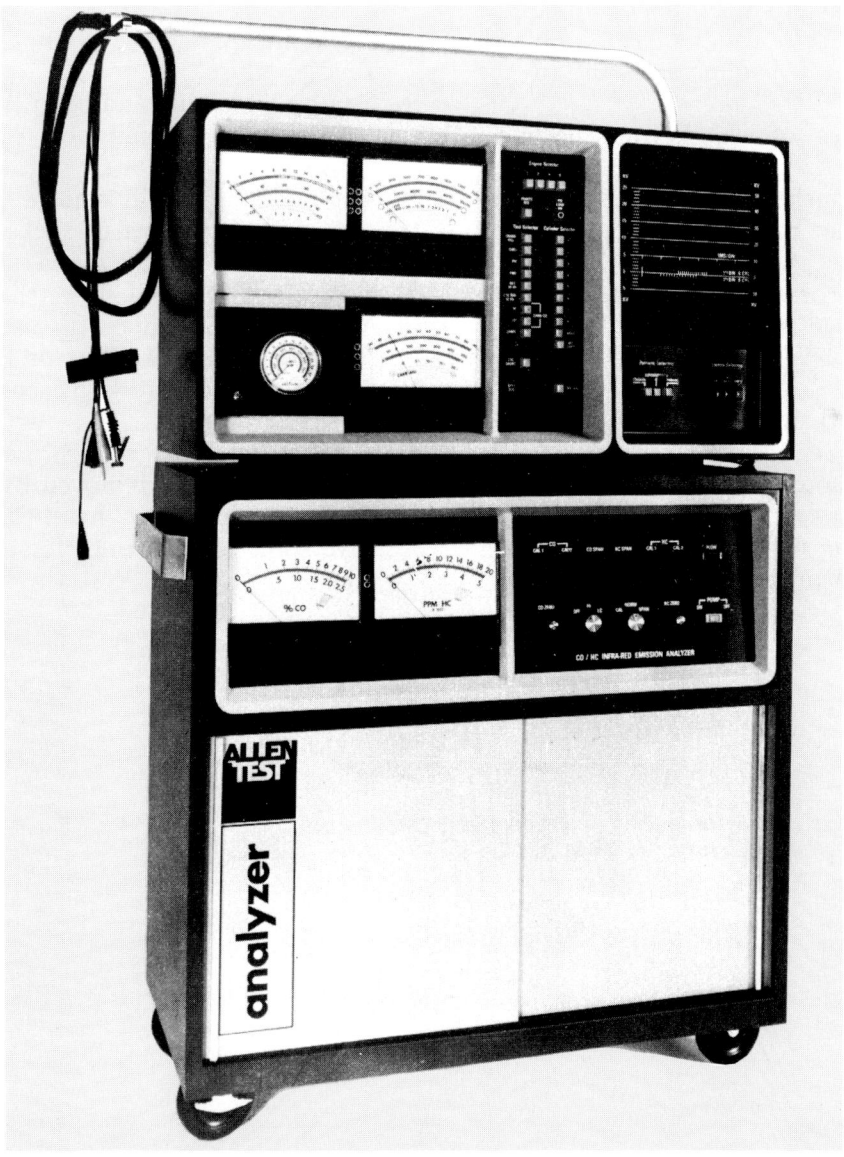

Fig. 6-71. This engine analyser incorporates a number of different instruments that help the mechanic make many rapid, accurate tests and measurements. (Allen)

REMEMBER:

GOOD TOOLS make your work easier, faster and more efficient. Use them properly and give them good care. If treated with intelligence they will give you many years of satisfying use.

KNOW THESE TERMS

Hammer, Chisel, Punch, File, Drill, Reamer, Impact wrench, Hacksaw, Vice, Tap, Die, Pliers, Screwdriver, Ring spanner, Open end spanner, Combination wrench, Socket spanner, Extension, Ratchet, Torque wrench, Allen wrench, Puller.

REVIEW QUESTIONS

1. What material is commonly used in construction of quality tools?
2. Name three things to look for when buying tools.
3. Give a few general rules for the use of tools.
4. Name three types of hammers useful to the mechanic.
5. Name several types of chisels the mechanic should have.
6. Of what use are punches?
7. What is the difference between starting, drift and aligning punches?
8. A file with two sets of cutting edges at angles to each other is referred to as a _____ file.
9. Give some safety rules for the use of hammers, chisels, and files.
10. Give two methods that are helpful in keeping a file clean.
11. Drills made of carbon steel will last longer than those made of high speed steel. True or False?
12. For general drilling in steel, what lip angles would you recommend?
13. Describe the procedure to use when sharpening drills.
14. Give some safety precautions to observe when using a portable hand drill.
15. Hacksaw teeth should point away from the handle. True or False?
16. Reamers are used for drilling large holes. True or False?
17. A vice will also be useful as an anvil. True or False?
18. If you wish to cut internal threads in a hole, you would use a tap. True or False?
19. If you wish to cut threads on a bolt, you would use a tap. True or False?
20. Describe some useful cleaning tools.
21. When grinding, always wear _____.
22. Name five types of pliers.
23. List six types of screwdrivers.
24. A ring spanner is one of the best to use. True or False?
25. Open end spanners grip a nut on all but one flat. True or False?
26. Combination ring and open end spanners have a different size opening on each end. True or False?
27. A spark plug would require _____ socket.
28. A 6.35 mm drive set of sockets would be handy for removing head bolts. True or False?
29. Name four types of socket handles.
30. A pipe wrench may be used for gripping a crankshaft journal. True or False?
31. For tightening head bolts, connecting rod bolts, etc., always use a _____ wrench.
32. Describe an Allen wrench.
33. Gears are usually removed by using a _____.
34. See how many other hand tools you can name.
35. Give some general rules for the use and care of tools.

Chapter 7

POWER TOOLS AND EQUIPMENT

After studying this chapter, you will be able to:
☐ List the most commonly used power tools and equipment.
☐ Describe the uses for power tools and equipment.
☐ Compare the advantages of one type of tool over another.
☐ Explain safety rules that pertain to power tools and equipment.

To be a productive mechanic in today's auto workshop, you must know when and how to use power tools and equipment. They increase the ease and speed of many repair operations.

Power tools are tools using electricity, compressed air, or hydraulics (liquid confined under pressure). Larger workshop tools such as: floor jacks, parts cleaning tanks, and steam cleaners, are classified as workshop equipment.

This chapter stresses the importance of properly selecting and using power tools and equipment. They can be very dangerous if misused. Always follow the operating instructions for the particular tool or piece of equipment. If in doubt, ask your instructor for a demonstration.

AIR COMPRESSOR

An air compressor is the source of compressed (pressurised) air for the automotive workshop. Look at Fig. 7-1. An air compressor normally has an electric motor that spins an air pump. The air pump forces air into a large, metal storage tank. Metal air lines feed out from the tank to several locations in the workshop. The mechanic can then connect flexible air hoses to the metal lines.

An air compressor turns ON and OFF automatically to maintain a preset pressure in the system. DANGER! Workshop air pressure is usually around 700 to 1000 kPa. This is enough pressure to severely injure or kill. Respect workshop air pressure!

Air Hoses

High pressure air hoses are connected to the metal lines from the air compressor. Since they are flexible, they allow the mechanic to take a source of air pressure to the car being repaired. Quick-disconnect type couplings are used on air hoses. To connect or disconnect an air hose, slide back the outer fitting sleeve and push or pull on the hose.

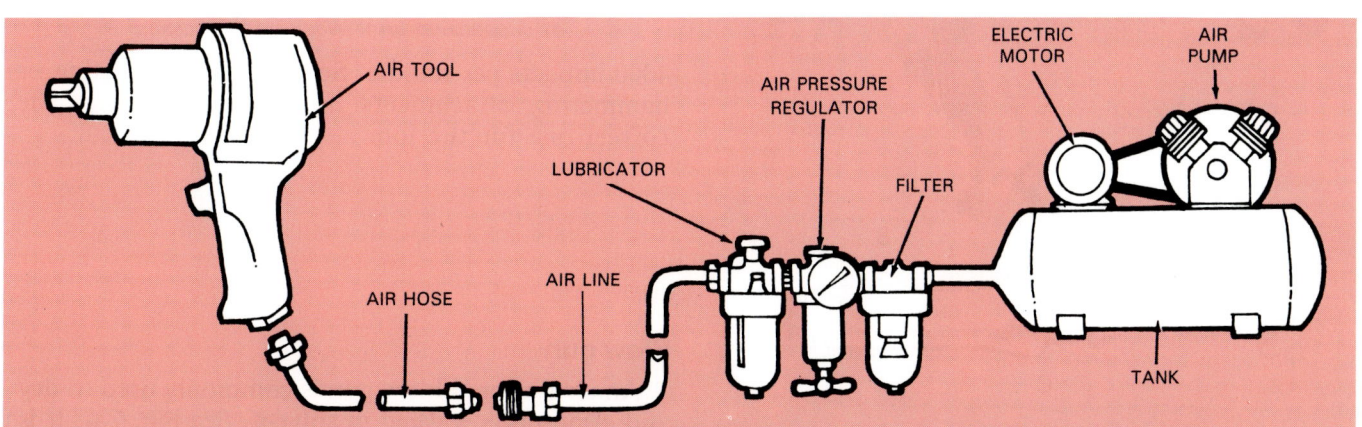

Fig. 7-1. Basic parts of a typical workshop air pressure system. Air compressor develops air pressure. Filter removes moisture. Regulator allows mechanic to control system pressure. Metal line and flexible hose carry air to tool.

AIR TOOLS

Air tools, also called pneumatic tools, use air pressure for operation. They are labour saving tools, well worth their cost.

Always lubricate an air tool before use. Squirt a few drops of air tool oil (light oil) into the air inlet fitting. This protects the internal parts of the tool, increasing service life and tool power.

Air wrenches (impact wrenches)

Air wrenches or impact wrenches provide a very fast means of installing or removing threaded fasteners. Look at Fig. 7-2A and B. An impact wrench uses compressed air to rotate a driving head. The driving head holds a socket which fits on the fastener head.

A button or switch on the air wrench controls the direction of rotation. In one position, the impact tightens the fastener. The impact loosens the fastener in the other direction.

Impact wrenches come in 3/8, 1/2, 3/4 and 1 in. drive sizes. A 3/8 drive is ideal for smaller bolts. The 1/2 in. drive is general purpose for medium to large fasteners. The 3/4 and 1 in. drive impact is for extremely large fasteners and is NOT commonly used in an automotive workshop.

CAUTION! Until you become familiar with the operation of an air wrench, be careful not to over-tighten bolts and nuts or leave them too loose. It is easy to strip or break fasteners with an air tool.

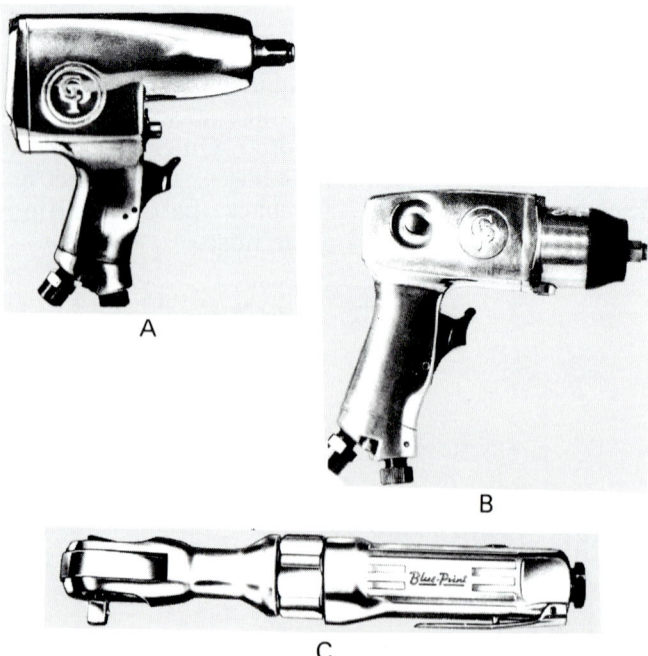

Fig. 7-2. Air wrenches. A — 1/2 in. drive impact wrench. B — 3/8 in. drive impact wrench. C — 3/8 in. drive air ratchet.

Air ratchet

An air ratchet is a special impact type wrench designed for working in tight quarters. Look at Fig. 7-2C. It is very slim and will fit into small areas. For instance, an air ratchet is commonly used when removing water pumps. It will fit between the radiator and engine easily.

An air ratchet normally has a 3/8 in. drive. It does not have very much turning power. Final tightening and initial loosening must be done by hand.

Impact sockets and extensions

Special impact sockets and extensions must be used with air wrenches. They are thicker and much stronger than conventional sockets and extensions. A conventional socket can be ruined or broken by the hammering blows of an impact wrench.

Special impact sockets and extensions are easily identified. They are usually FLAT BLACK, not chrome.

Air hammer (Chisel)

An air hammer or chisel is useful during various driving and cutting operations. Look at Fig. 7-3. The air hammer is capable of producing about 1000 to

Fig. 7-3. Air hammer is being used to quickly drive bushing out of suspension arm. Wear safety glasses.

4000 impacts per minute. Several different cutting or hammering attachments are available. Select the correct one for the job.

CAUTION! Never turn an air hammer ON unless the tool is pressed tight against the work piece. If not, the tool head can fly out of the hammer with great force — as if shot from a gun!

Blow gun

An air powered blow gun is commonly used to dry and clean parts washed in solvent. See Fig. 7-4. It is also used to blow dust and loose dirt off a part before disassembly.

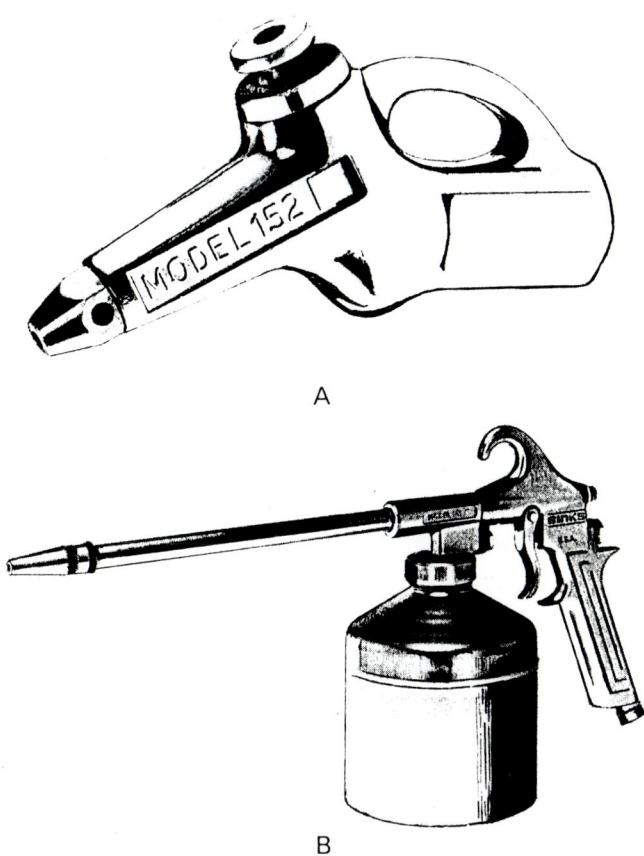

A

B

Fig. 7-4. A — Blow gun, commonly used to blow parts clean and dry after washing in solvent. B — Solvent gun can be used to wash parts.

When using a blow gun, wear eye protection. Direct the blast of air away from yourself and others. Do not blow brake and clutch parts clean. These parts contain asbestos, a cancer causing substance.

Air drill

An air drill is excellent for many repairs because of its power output and speed adjustment capabilities. Its power and rotating speed can be set to match the job at hand. Look at Fig. 7-5. With the right attachment, air drills can drill holes, grind, polish, and clean parts.

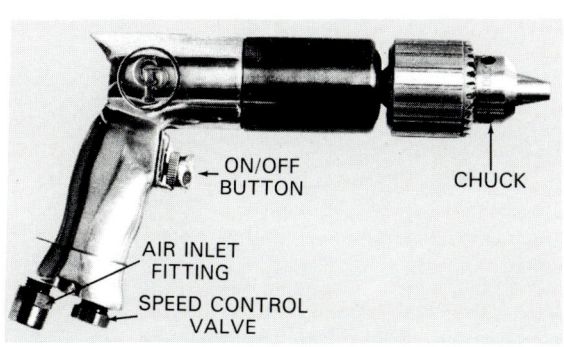

Fig. 7-5. Air drill speed can be adjusted. It is capable of very high turning force.

A rotary brush, Fig. 7-6, is used in an air or electric drill for rapid cleaning of parts. It will quickly rub off old gasket material, carbon deposits on engine parts, and rust, with a minimum amount of effort.

CAUTION! Only use a high speed type rotary brush in an air drill. A brush designed for an electric drill may fly apart. To be safe, always adjust an air drill to the SLOWEST ACCEPTABLE SPEED.

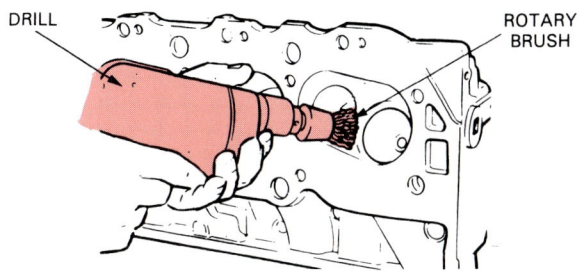

Fig. 7-6. Rotary brush is commonly used in a drill for cleaning off old gaskets or carbon. Wear eye protection!

A rotary file or stone can be used in either an air drill, electric drill, or an air (die) grinder, Fig. 7-7. It is handy for removing metal burrs and nicks.

Make sure the stone is not turned too fast by the air tool. Normally the speed specifications (maximum allowable rotating speed) will be printed on the file or stone container.

Fig. 7-7. Die grinder with a high speed stone installed. This tool is used for removing burrs and for other smoothing operations. (Robert Bosch)

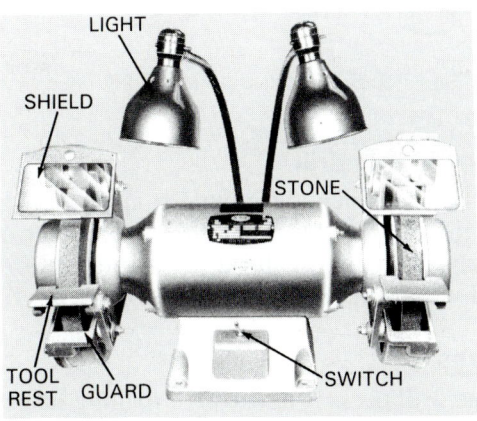

Fig. 7-8. Bench grinder stone is used to sharpen tools. Brush can be used to clean and polish small parts. Keep shields, tool rests, and guards in place.

BENCH GRINDER

A bench grinder, Fig. 7-8, can be used for grinding, cleaning, or polishing operations. The hard grinding wheel is used for sharpening or deburring. The soft wire wheel is for cleaning and polishing.

A few BENCH GRINDER RULES to follow are:

1. Wear eye protection and keep your hands away from the stone and brush.
2. Keep the tool rest adjusted close to the stone and brush. If NOT up close, the part can catch in the grinder.
3. Do NOT use the wire wheel to clean soft metal parts (aluminium pistons or brass bushings, for example). The rubbing, abrasive action of the wheel can remove metal, scuff, and ruin the part. Use a solvent and a dull hand scraper on soft metal parts that could be damaged.
4. Make sure the grinder shields are in place.

DRILLS

A twist drill or drill bit is used to drill holes in metal and plastic parts. They fit into either an electric or air powered drill, Fig. 7-9. Drill bits are commonly made of either carbon steel or high speed steel. High speed steel is better because of its resistance to heat. It will not lose its hardness when slightly overheated.

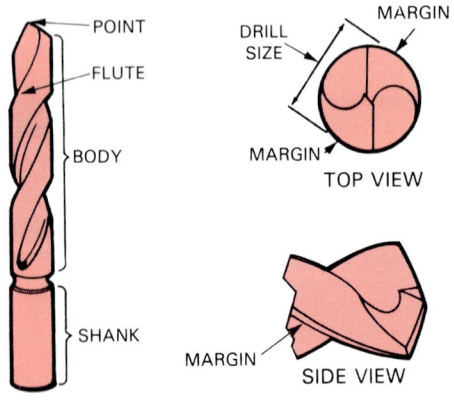

Fig. 7-9. Study basic parts of drill bit.

Portable electric drill

A drill bit is chucked (mounted) and rotated by an electric drill, Fig. 7-10. A special key must be used to tighten the drill bit in the drill. A portable electric drill will work fine on most small drilling operations.

Drill press

A large drill press is needed for drilling large holes, deep holes, or a great number of holes in several parts, Fig. 7-11. The drill press handle allows the bit to be forced into the work with increased force. Also, the drill chuck will accept very large bits.

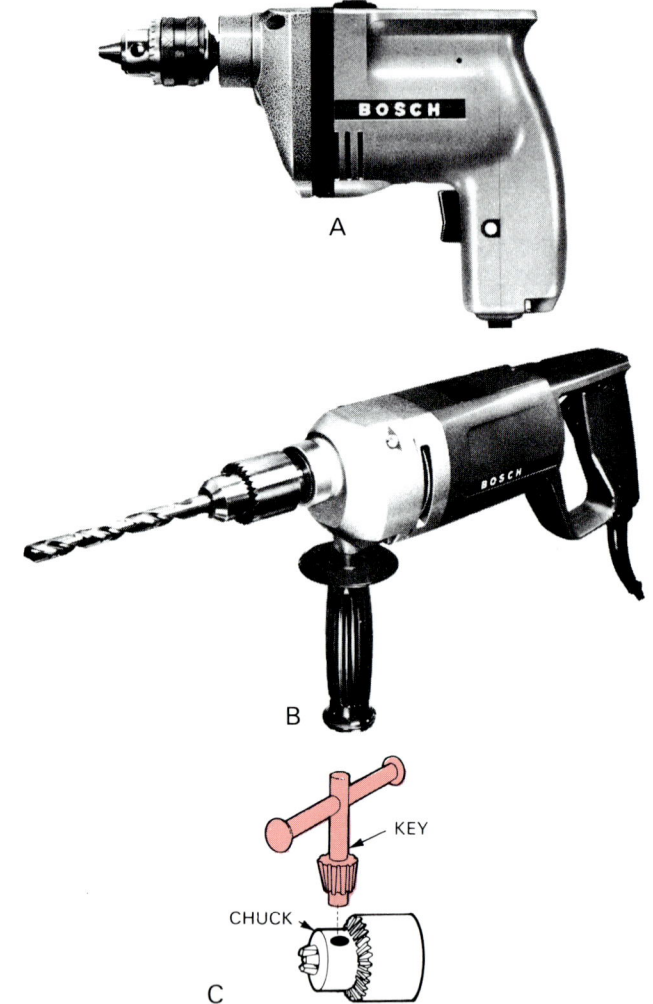

Fig. 7-10. Portable electric drills. A — Small 10 mm drill. B — Larger, 13 mm drill, C — Key is used to tighten bit in chuck. (Robert Bosch)

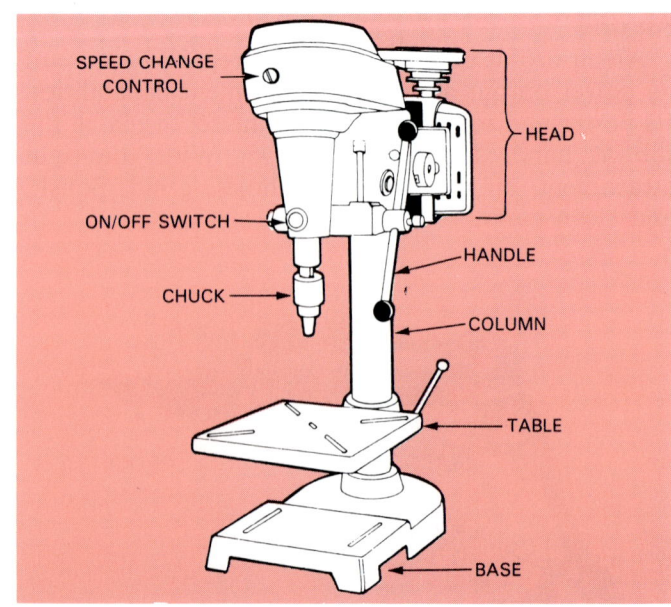

Fig. 7-11. Note parts of drill press. It is for drilling deep or large holes when part will fit on table.

A few DRILL PRESS RULES to follow include:

1. Secure the part to be drilled in a vice or with G-clamps.
2. Use a centre punch to indent the part and start the hole.
3. Remove the key before turning on the drill.
4. To prevent possible injury, release drilling pressure right before the bit breaks through the bottom of the part. A drill bit tends to catch when breaking through. This can cause the drill or part to rotate dangerously.
5. Oil the bit as needed.

TYRE CHANGER

A tyre changer is a common piece of equipment used to remove and replace tyres on wheels. Some are hand-operated and others use air pressure. Do not attempt to operate a tyre changer without proper supervision. Follow the directions provided with the changer.

HYDRAULIC BOTTLE JACKS

Hydraulic bottle jacks, which are available in various lift capacities, are a very useful piece of workshop equipment. They are handy, not only for raising a vehicle, but also for supporting 'on vehicle' components when dismantling or assembling. Fig. 7-12A.

FLOOR JACKS

A floor jack is also used to raise either the front, sides, or rear of a car. Look at Fig. 7-12B.

To avoid vehicle damage, place the jack head under a solid part of the car (frame, suspension arm, rear axle). If you are NOT careful, it is very easy to smash an oil pan, muffler, floor pan, or other sheet metal part.

The car should be free to roll while being raised. After raising, place the car on jack stands and block wheels.

Normally, to raise the car, you must turn the jack handle or knob clockwise and pump the handle. To lower, turn the pressure relief valve counterclockwise slowly.

JACK STANDS

Jack stands support a car during repair. After raising the car with a jack, place stands under the vehicle, Fig. 7-12C. It is NOT SAFE to work under a car held by a floor jack.

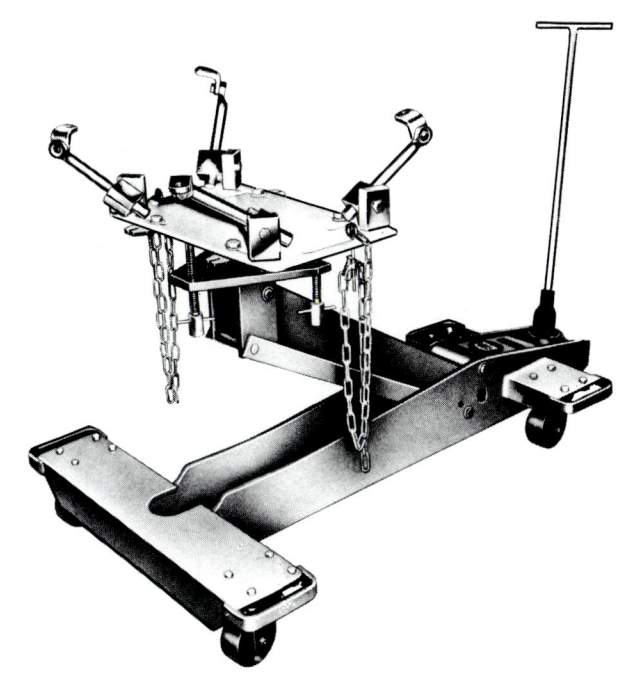

Fig. 7-13. Transmission jack is a floor jack modified to accept and hold a transmission or transaxle. Special hardware is used to secure unit to jack head.

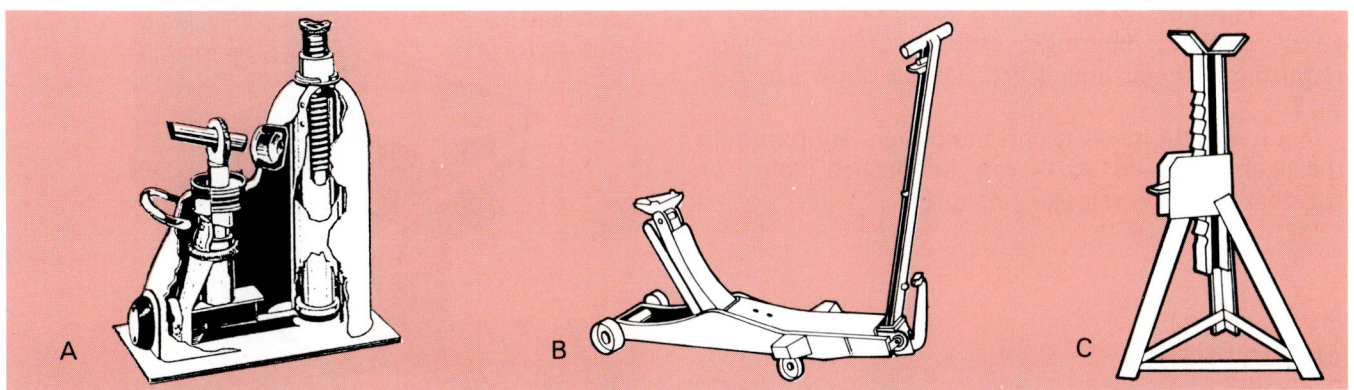

Fig. 7-12. Vehicle lifting equipment. A — Bottle jack. B — Floor jack is for raising car only. C — Jack stands are needed before working under car.

TRANSMISSION JACK

Special transmission jacks are designed for removing and installing transmissions. One type is similar to a floor jack. However, the head is enlarged to fit the bottom of a transmission, Fig. 7-13.

Another type of transmission jack is used when the car is raised on a vehicle hoist. It has a long post which can reach high into the air to support the transmission.

ENGINE CRANE

A portable engine crane is used to remove and install car engines, Fig. 7-14. It has a hydraulic hand jack for raising and a pressure release valve for lowering. An engine crane is also handy for lifting heavy engine parts (intake manifolds, cylinder heads), transmissions, and transaxles.

Fig. 7-14. Hydraulic engine crane can be used to lift heavy objects such as engines, transmissions, transaxles, rear axle assemblies.

HYDRAULIC PRESS

An hydraulic press is used to install or remove gears, pulleys, bearings, seals, and other parts requiring high pushing force. One is shown in Fig. 7-15.

An hydraulic press uses a hand jack. By pumping the jack, press-fit parts can be pushed apart or together, a valve releases pressure.

NOTE! An hydraulic press can exert TONNES OF FORCE. Wear face protection and use recommended procedures.

ARBOR PRESS

An arbor press performs the same function as an hydraulic press but at lower pressures. It is a hand-operated, mechanical press for smaller jobs.

ENGINE STAND

An engine stand is used to hold a car's engine while being overhauled (rebuilt) or repaired. The engine bolts to the stand. For convenience, the engine can usually be rotated and held in different positions.

COLD SOLVENT TANK

A cold solvent tank, Fig. 7-16, removes grease and oil from parts. After removing all old gaskets and scraping off excess grease, you can scrub the parts

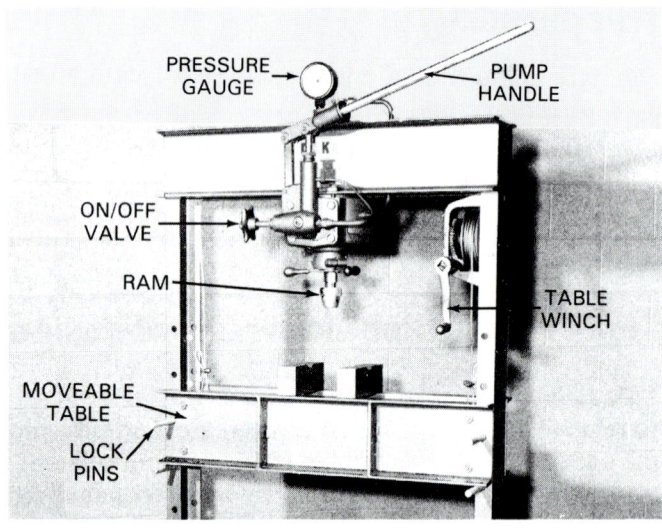

Fig. 7-15. Hydraulic press is needed for numerous pressing operations. It is commonly used to remove and install bearings, bushings, seals, and other pressed-on parts. Note! Double-check lock pins before using press. If not installed, cables in unit could snap.

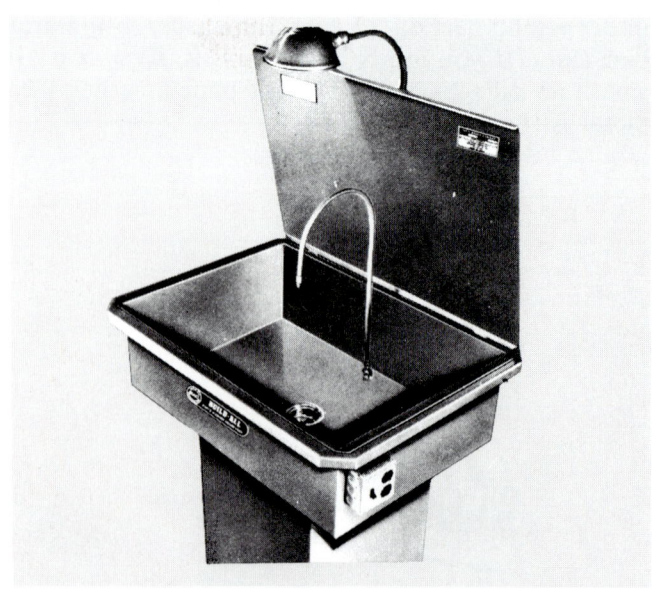

Fig. 7-16. Cold solvent tank is used for removing oil and light grease from parts. Unit sprays filtered solvent on parts. Rub parts with brush for rapid cleaning.

Fig. 7-17. High pressure washer will remove grease buildup on outside of assemblies before disassembly.

clean in the solvent. A blow gun is normally used to remove the solvent.

STEAM CLEANER AND HIGH PRESSURE WASHER

A steam cleaner or high pressure washer is used to remove heavy deposits of dirt, grease, and oil from the outside of large assemblies (engines, transmissions, transaxles). They provide an easy and rapid method of cleaning large units BEFORE DISASSEMBLY. Look at Fig. 7-17.

DANGER! A steam cleaner operates at relatively high pressures and temperatures. Follow your safety rules and specific operating instructions.

OXYACETYLENE TORCH

An oxyacetylene torch outfit can be used to cut, bend, and weld or braze (join) metal parts, Fig. 7-18A. Its rapid cutting action is extremely beneficial. For example, a cutting torch is often used to remove old, deteriorated exhaust systems. Tremendous heat is produced by burning acetylene gas and oxygen.

ARC WELDER

An arc welder is also used to weld metal parts together, Fig. 7-18B. It uses high electric current and the resulting arc to produce welding heat.

If at all possible, you should take a welding course. DO NOT attempt to weld or cut until properly trained. A good motor mechanic should also be proficient at welding.

SOLDERING GUN

A soldering gun or iron is normally used to solder (join) wires, Fig. 7-19. An electric current heats the tip of the gun, then, the hot gun tip can be used to heat the wires and melt solder. When the solder solidifies (hardens), a strong, solid connection is produced.

BATTERY CHARGER

A battery charger is used to recharge (energise) a discharged (de-energised) car battery. It forces current back through the battery. Normally, the red charger leads connects to the battery positive (+) terminal. The black charger leads connects the negative (-) battery terminal.

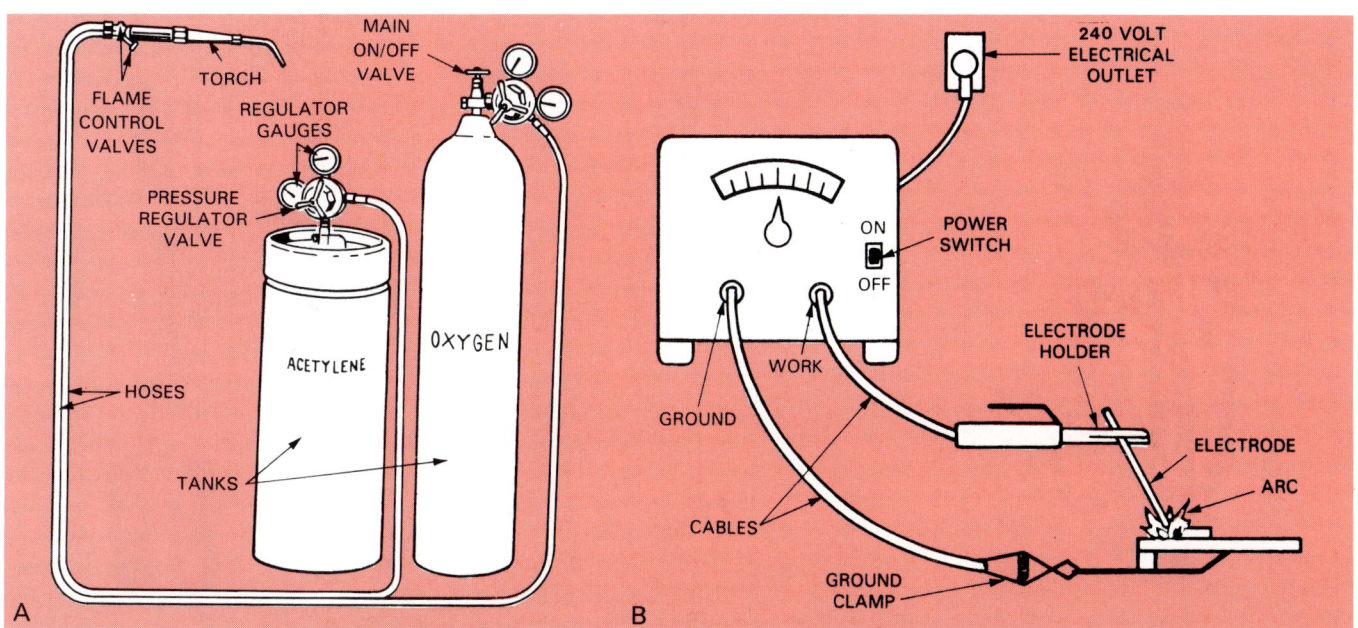

Fig. 7-18. A — Oxyacetylene outfit can be used for cutting or welding metal. B — Study basic parts of arc welder.

WARNING! Always connect the battery charger leads to the battery BEFORE turning the charger ON. This will prevent sparks that could ignite any battery gas. The gases around the top of a battery can EXPLODE violently.

LEAD LIGHT

A lead light, Fig. 7-20, provides a portable source of illumination (light). The light can be taken to the repair area under the car, hood, engine, or anywhere it is dark.

PULLERS

Pullers are needed to remove seals, gears, pulleys, steering wheels, axles, and other pressed-on parts. A few puller types are pictured in Fig. 7-21.

DANGER! Pullers can exert ENORMOUS FORCE. They must be used properly to prevent injury or part damage. Wear eye protection!

JUMPER CABLES

Jumper cables are used to start a car with a dead (discharged) battery. The cables can be connected between the dead battery and another car's battery. This will let you crank and start the car. See Fig. 7-22.

When connecting jumper cables, connect positive to positive and negative to negative. Also, keep sparks away from the dead battery. Connect the negative cable to the vehicle frame so that any sparks will not occur near the battery.

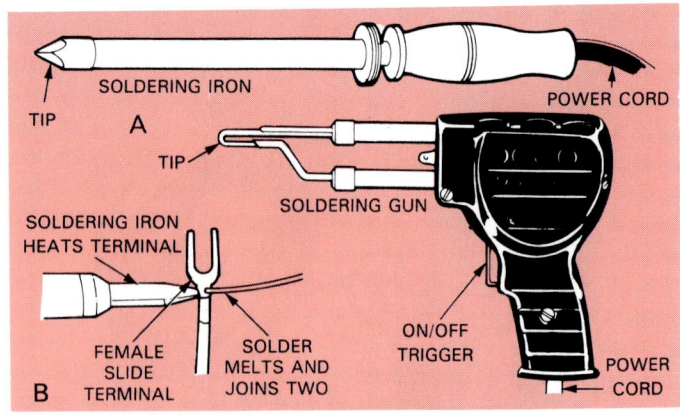

Fig. 7-19. A — Soldering iron and soldering gun. B — Soldering iron or gun produces enough heat to melt solder for joining wires and small metal terminals.

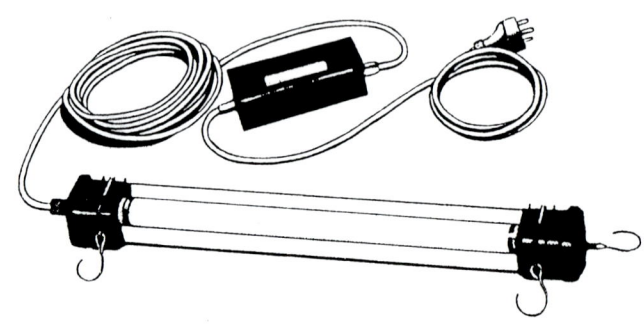

Fig. 7-20. Typical workshop fluorescent lead light. The one illustrated is a roll proof type. For safety reasons it is good policy to use a transformer at the mains outlet so low voltage lighting can be used.

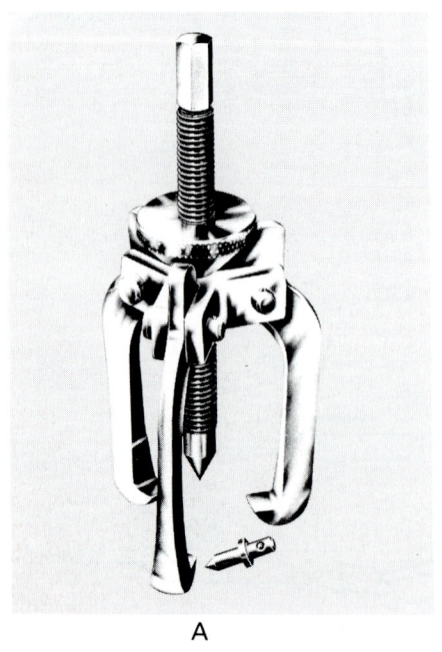

A

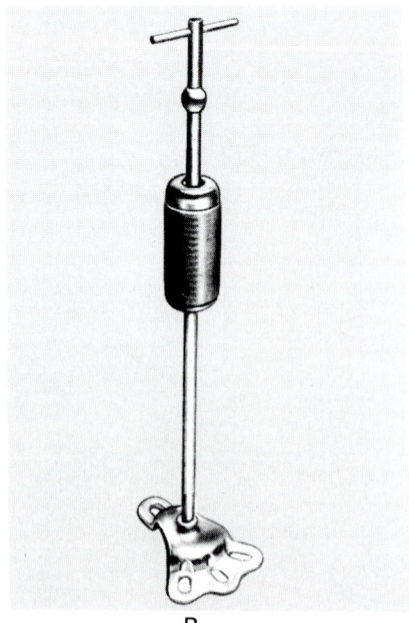

B

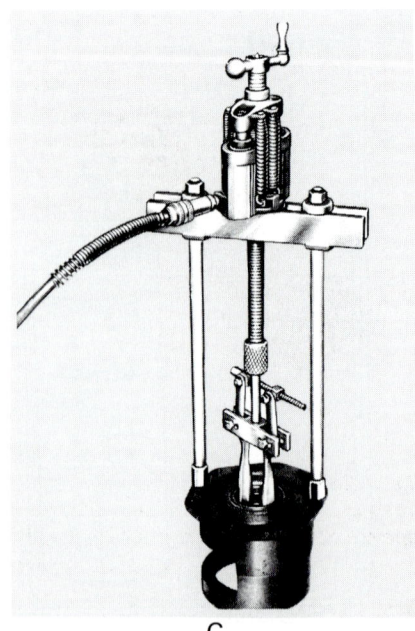

C

Fig. 7-21. A — Three-jaw puller. B — Slide hammer puller. C — Power puller.

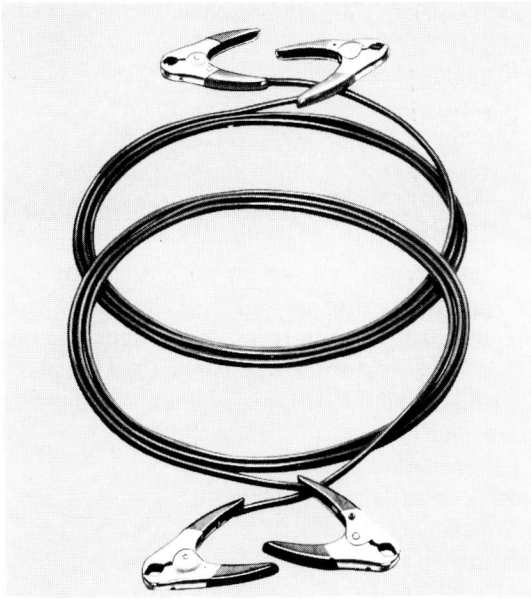

Fig. 7-22. Jumper cables are for emergency starting of cars. Connect red lead to positive terminal of both batteries. Black is for negative and ground.

CREEPER

A creeper is useful when working under a car supported on jack stands, Fig. 7-23A. It lets the mechanic easily roll under the car without getting dirty.

Stool creeper

A stool creeper allows the mechanic to sit while working on parts low to the ground. See Fig 7-23B. For example, a stool creeper is often used on brake repairs. The brake parts and tools can be placed on the creeper. The mechanic can sit and still be eye level with the brake assembly.

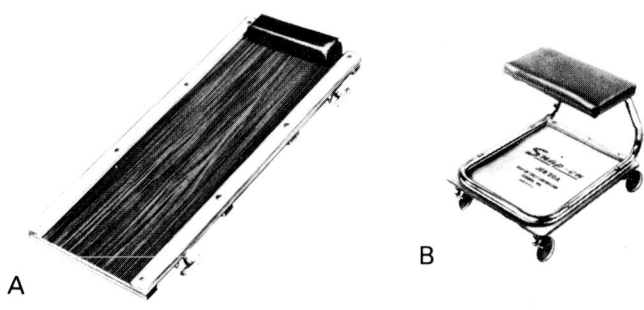

Fig. 7-23. A — Creeper is for working under car. B — Stool creeper is commonly used during brake and suspension repairs. You can sit on the stool and store tools on bottom. (Snap-On Tools)

ROLL-AROUND CART

A large roll-around cart or table is handy for taking a number of tools to the job. One is pictured

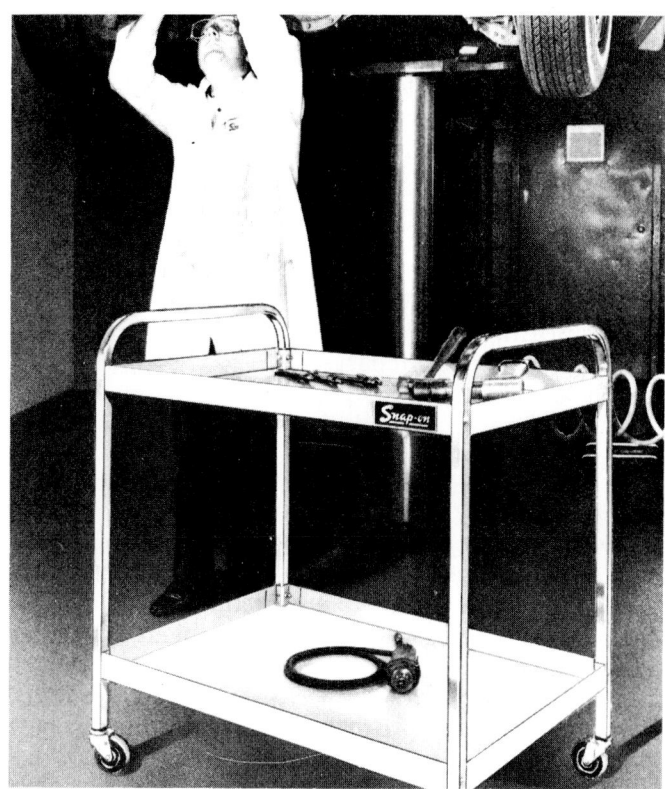

Fig. 7-24. Roll-around cart allows you to take several tools to car. This saves several trips to tool box.

in Fig. 7-24. Since the mechanic is working on a car raised on a hoist, the cart positions all of the needed tools within "hands reach". This saves time and effort.

FENDER COVERS

Fender covers are placed over a car's fenders, upper grille, or other body section for protection. They protect the car's paint from nicks and scratches. See Fig. 7-25A. Never lay your tools on a car's painted surface.

Seat Covers

Seat covers are placed over car seats to protect them from dirt, oil, and grease that might be on your

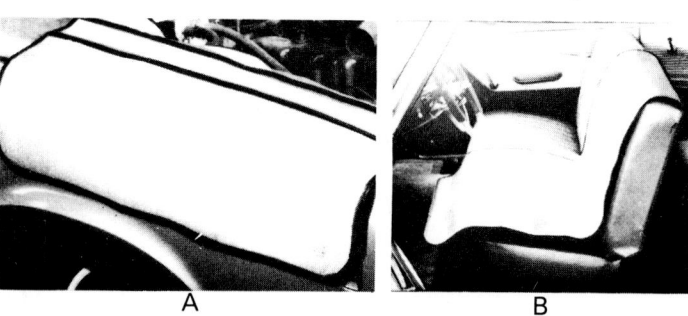

Fig. 7-25. Always take good care of a customer's car. A — Fender covers protect paint from nicks and dents. B — Seat cover protects upholstery from dirty work clothes. (Snap-On Tools)

work clothes. The covers are normally used while driving the car in and out of the workshop or while working in the passenger compartment. Look at Fig. 7-25B.

KNOW THESE TERMS

Air compressor, Air tool, Impact socket, Blow gun, Rotary brush, Engine crane, Hydraulic press, Solvent tank, Battery charger, Lead light, Puller, Jumper cables, Creeper, Fender cover.

REVIEW QUESTIONS

1. Power tools use _____, _____, _____, or _____ as sources of energy.
2. Shop air pressure is only about 170 kPa and cannot cause injuries. True or False?
3. Which of the following is NOT a commonly used air tool?
   a. Impact wrench.
   b. Air ratchet.
   c. Air chisel.
   d. Air saw.
4. A _____ _____ is used to blow dirt off parts and to dry parts after cleaning.
5. A rotary brush is used in an electric or air drill for rapid cleaning of parts. True or False?
6. List four important rules for a bench grinder.
7. List five important rules for a drill press.
8. Use this tool to support the car while working under the car.
   a. Floor jack.
   b. Jack stands.
   c. Transmission jack.
   d. Bumper jack.
9. Explain the use of a solvent tank.
10. What are pullers used for?

Chapter 8

WORKSHOP SAFETY

After studying this chapter you will be able to:
☐ Describe the typical layout and areas of an automotive workshop.
☐ List the types of accidents that can occur in an automotive workshop.
☐ Explain how to prevent automotive workshop accidents.
☐ Describe general safety rules for the automotive workshop.

An automotive workshop can be a very safe and enjoyable place to work. However, if basic safety rules are NOT followed, an automotive workshop can be very dangerous.

Every year, a number of mechanics are injured or killed on the job. Most of these accidents resulted from a broken safety rule. The injured mechanics learned to respect safety rules the hard way, by experiencing a painful injury. You must learn to respect safety rules the easy way, by studying and following the safety rules given in this book.

NOTE: Specific safety rules on hand tools, power tools, equipment, and special operations are given elsewhere in the book. It is much easier to understand and remember these rules when they are covered fully.

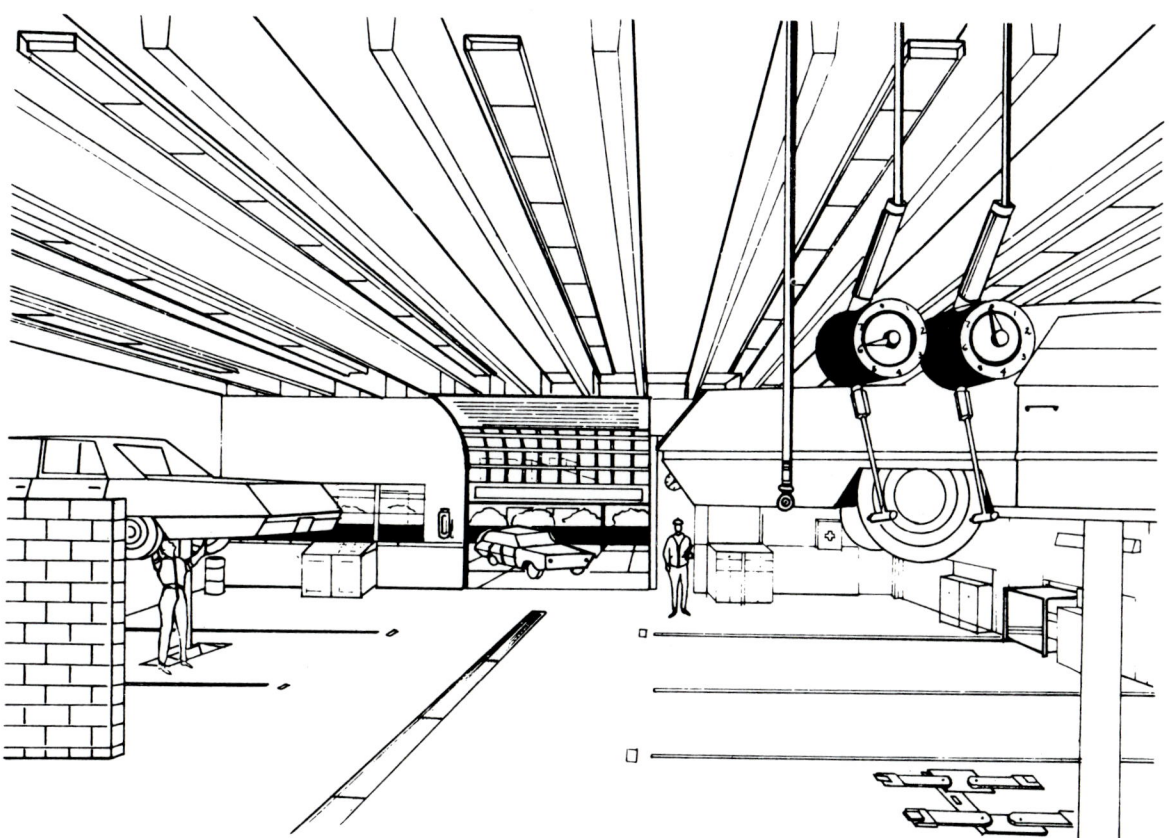

Fig. 8-1. An automotive workshop is a highly organised area. It will help you work more efficiently, if you are familiar with everything in your workshop.

AUTOMOTIVE WORKSHOP AREAS

There are several different areas in an automotive workshop. You must know their names and the rules that apply to each.

The automotive workshop can be divided into the following work areas:

1. WORKSHOP REPAIR AREA.
2. WORK BAYS.
3. LUBE BAY (HOIST AREA).
4. FRONT END BAY (WHEEL ALIGNMENT BAY).
5. TOOL ROOM.
6. OUTSIDE WORK AREA.
7. WORKSHOP CLASSROOM.
8. LOCKER ROOM (DRESSING ROOM).

Workshop repair area

The workshop repair area includes any location in the workshop where repair operations are performed. See Fig. 8-1. It normally includes every area EXCEPT the classroom, locker room, and tool room. It is important that you learn your workshop layout and organisation to improve work efficiency and safety.

Work bays

A work bay is a small area where a car can be brought into the workshop for repairs. Look at Fig. 8-1. Each work bay is sometimes numbered and marked off with lines painted on the floor.

Lube bay (hoist)

The lube bay contains a hoist for raising a car into the air. Refer to Fig. 8-2. It is handy for working under the car (draining oil, greasing front end parts, repairing exhaust systems etc).

Remember these HOIST SAFETY RULES:
1. Obtain a demonstration from the workshop foreperson and get permission before using the hoist.
2. Centre the car on the hoist, as described in a service manual, Fig. 8-3. Raise the car slowly!
3. Check ceiling clearance before raising trucks and campers. Make sure the vehicle roof does not hit overhead pipes, lights or the ceiling.
4. Make sure the safety catch is engaged. Do not walk under the hoist without the catch locked into position, Fig. 8-4.

Front end bay (wheel alignment bay)

The front end bay or wheel alignment bay is another specialised bay used to work on a car's steering and suspension systems. One is shown in

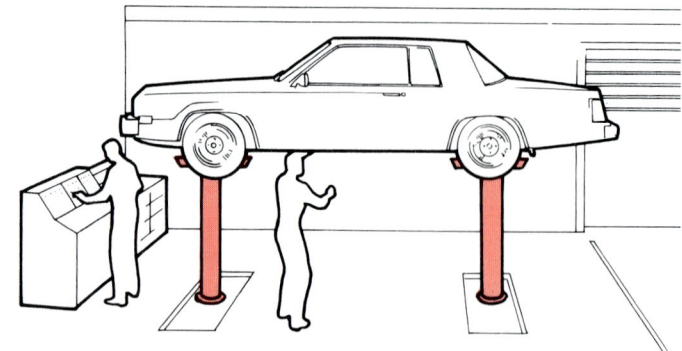

Fig. 8-2. Lube bay which usually contains hoist is handy for many repairs on parts under car. It is commonly used when changing oil, greasing car, exhaust system repairs.

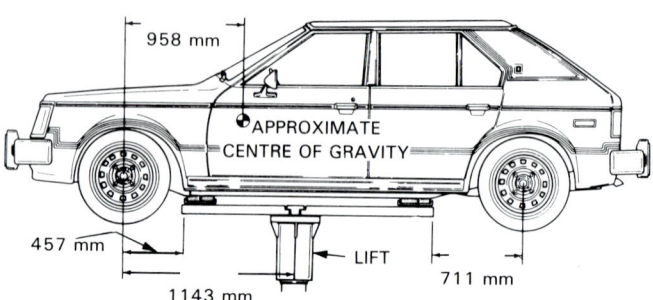

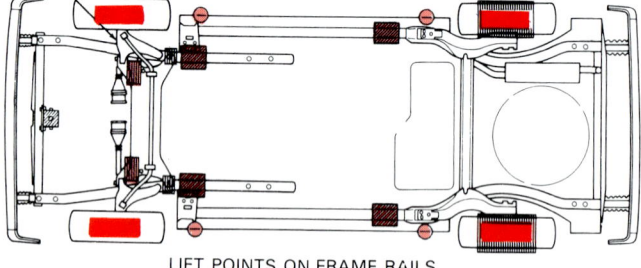

LIFT POINTS ON FRAME RAILS
■ TWIN POST LIFT POINTS
▨ FRAME CONTACT OR FLOOR JACK
■ DRIVE ON HOIST
● SCISSORS JACK (EMERGENCY) LOCATIONS

Fig. 8-3. Follow service manual directions when raising car on hoist. Note specific lifting instructions for this car.

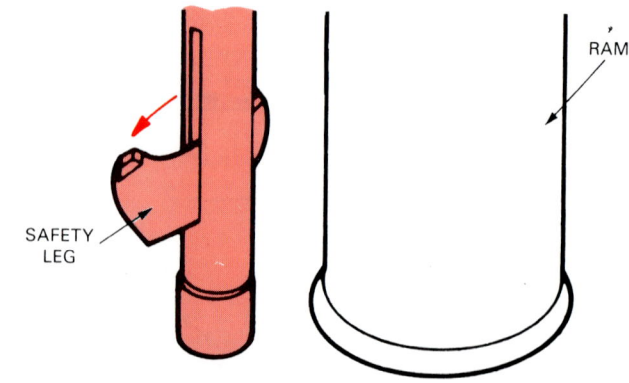

Fig. 8-4. Most hoists have a safety catch. It must be engaged before working under car.

Fig. 8-5. It may contain a special tool board and equipment used when replacing worn suspension parts, steering parts, and for adjusting wheel alignment.

When using wheel alignment equipment, the car should be brought onto the equipment slowly and carefully. Someone should GUIDE THE DRIVER and help keep the tyres centred on the equipment. As with other complicated and potentially dangerous equipment, obtain a full demonstration from the person in charge before working.

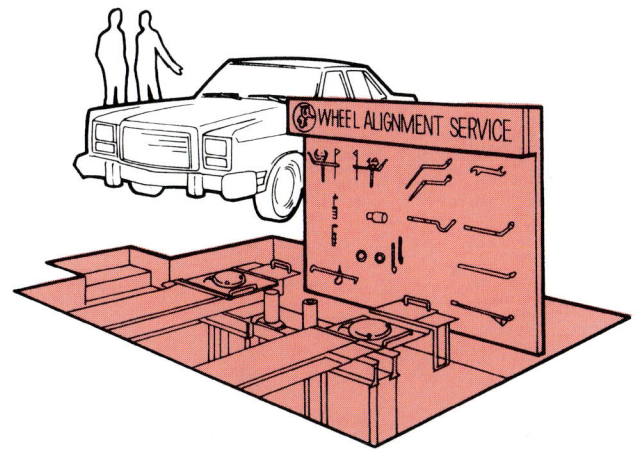

Fig. 8-5. A front end bay or wheel alignment bay is used in most larger workshops. It is needed for steering and front end repairs.

Tool room

The tool room is normally a separate area adjacent to (next to) the main workshop. It is used to store workshop tools, small equipment, and supplies (nuts, bolts, oil).

If the worshop is large enough to have a store person looking after the tool room this person will be responsible for keeping track of all tools and equipment. Every tool checked out of the tool room must be recorded and called in before the end of the day or week, as the case may be.

Normally, the tools will hang on the walls of the tool room for easy access. Each tool will have a painted silhouette (shadow) which indicates where each tool is kept, Fig. 8-6.

The workshop foreperson or person in charge of the tool room will detail specific tool room policies and procedures.

Outside work area

Some workshops have an outside work area adjacent to the garage overhead doors. In good weather this area can be used for automotive repairs.

Always raise the workshop doors all the way and bring cars through the doors very slowly. Check the height of trucks and campers to make sure that they will clear the doors.

Locker room

The locker room or dressing room is usually located adjacent to the main workshop. It provides an area for changing into your work clothes. It may even have shower facilities. Always do your part to keep the locker room CLEAN and ORDERLY.

TYPES OF ACCIDENTS

Basically, you should be aware of and try to prevent six kinds of accidents.
1. FIRES
2. EXPLOSIONS
3. ASPHYXIATION (airborne poisons)
4. CHEMICAL BURNS
5. ELECTRIC SHOCK
6. PHYSICAL INJURIES

ATTENTION! If an accident or injury ever occurs in the workshop, notify the person in charge immediately. Use common sense on deciding to get a fire extinguisher or take other action.

Fires

Fires are terrible accidents capable of causing instant and permanent scar tissue. There are numerous combustible substances (petrol, oily rags, paints, thinners) found in an automotive workshop. Any of these flammables are capable of producing a fire.

Petrol is, by far, the most dangerous and often underestimated flammable liquid in an automotive workshop. Petrol has astonishing potential for causing a tremendous fire. Just a cup full of petrol can instantly engulf a car in flames.

A few PETROL SAFETY RULES include:

1. Store petrol and other flammable liquids in approved, sealed containers.
2. When disconnecting a car's fuel line or hose, wrap a shop rag around the fitting to keep fuel from squirting or leaking.
3. Disconnect the vehicle battery before working on a fuel system.
4. Wipe up petrol spills immediately. Do not place sand, or sawdust or other absorbent on petrol because the absorbent will become highly flammable.
5. Keep any source of heat away from the parts of a fuel system.
6. Never use petrol as a cleaning solvent.

Oily rags can also start fires. Soiled rags should be stored in an approved safety receptacle.

Paints, thinners, and other combustible materials should be stored in a fire cabinet. Also, never allow flammables near a source of sparks (grinder), flame (welder or water heater), or heat (furnace for example).

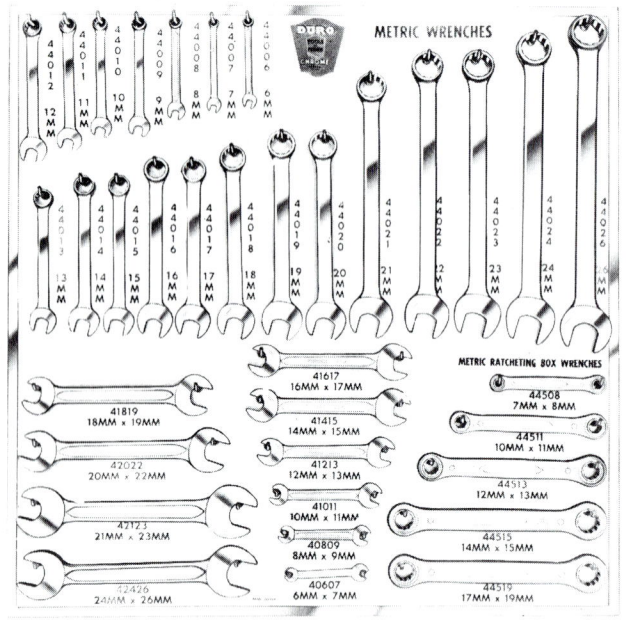

Fig. 8-6. Always keep all workshop tools clean and organised. Make sure you replace every tool in its correct storage location. (Duro Tools)

Note the location of all fire extinguishers in your workshop. A few seconds of time can be a "life time" during a fire!

Electrical fires can result when a "hot wire" (wire carrying current to a component) touches ground (vehicle frame or body). The wire can begin to heat up, melt the insulation, and burn. Then, other wires can do the same. Dozens of wires could burn up in a matter of seconds.

To prevent electrical fires, always disconnect the battery when told to do so in a service manual.

Explosions

Several types of explosions are possible in an automotive workshop. You should be aware of these sources of sudden death and injury.

Car batteries can explode! Hydrogen gas can surround the top of car batteries being charged or discharged (used). This gas is highly explosive. The slightest spark or flame can ignite and cause the battery to explode. Chunks of battery case and acid can blow into your eyes and face. Blindness, facial cuts, acid burns, and scars can result. Fig. 8-7.

Fuel tanks can explode, even seemingly empty ones! A drained fuel tank can still contain fuel gum and varnish. When this gum is heated and melts, it can emit vapours which can ignite.

Keep sparks and heat away from fuel tanks. When a fuel tank explodes, one side will usually blow out. Then, the tank will shoot across the workshop as if shot out of a cannon. You or other workers could be killed or seriously injured.

Various other sources can cause shop explosions. For example, special sodium-filled engine valves, oxy/acetylene bottles, LPG filled bottles can all explode if mishandled.

Asphyxiation

Asphyxiation is caused by breathing toxic or poisonous substances in the air. Mild cases of asphyxiation will cause dizziness, headaches, and vomiting. Severe asphyxiation can cause death.

The most dangerous source of asphyxiation in an

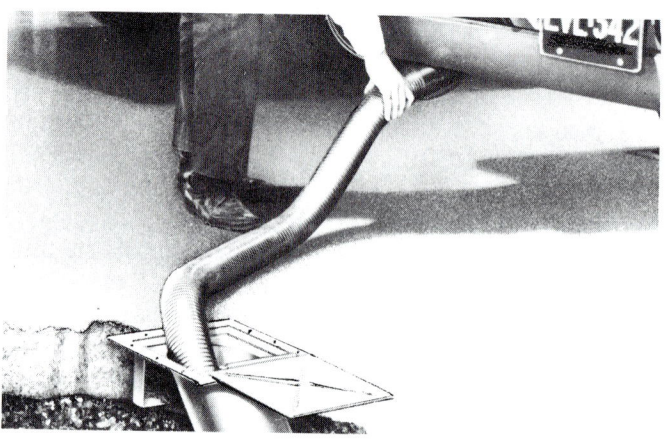

Fig. 8-8. Place an exhaust hose over tailpipe of any car running in enclosed workshop. This will prevent shop from filling with poisonous fumes. (Kent-Moore)

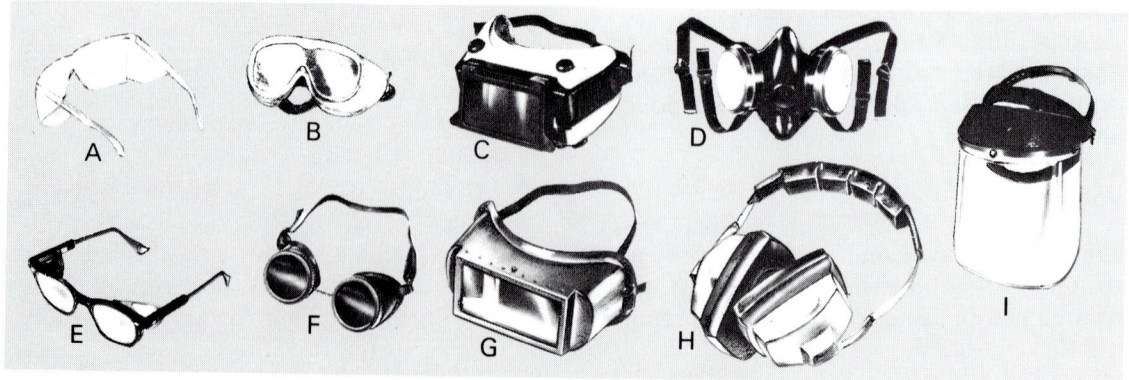

Fig. 8-7. Wear approved eye and face protection when needed. A — Safety glasses. B — Safety glasses with side shields. C — Goggles. D — Respirator. E, F, G — Welding goggles . H — Noise mufflers. I — Face shield. (Snap-On Tools)

automotive workshop is an automotive engine. An engine's EXHAUST GASES ARE DEADLY POISON. As shown in Fig. 8-8, connect a shop vacuum or suction hose to the tailpipe of any car being operated in the workshop. Also, make sure the exhaust evacuation system is turned ON.

Other workshop substances which are harmful include asbestos (brake lining dust, clutch disc dust) and paint spray. Respirators (filter masks) should be worn when working around any kind of airborne impurities. Refer to Fig. 8-7G.

Chemical burns

Various solvents (part cleaners), battery acid, and other substances which are commonly used in a workshop can cause chemical burns to the skin and eyes. Always read the directions on chemicals.

Carburettor decarbonising cleaner for example, is super powerful and can severely burn your hands in a matter of seconds. Wear rubber gloves and suitable eye protection when using carburettor cleaner. If a skin or eye burn occurs, seek first aid immediately and follow label directions.

Electric shock

Electric shock can occur when using improperly grounded electric power tools. Never use an electric tool unless it has a functional EARTH PRONG (third prong on the plug socket). This prevents current from accidentally passing through your body. Also, never use an electric tool on a wet workshop floor.

Physical injury

Physical injuries (cuts, broken bones, strained backs) can result from hundreds of different accidents. As a mechanic, you must constantly think and evaluate every repair technique. Decide whether a particular operation is safe or dangerous and take action as required.

For instance, if you are pulling on a hand wrench as hard as you can and the bolt will not turn, STOP! Find another tool that is larger, has more leverage, and is safer. This mental attitude will help prevent injuries and improve your mechanical abilities as well.

GENERAL SAFETY RULES

Listed are several general safety rules that should be remembered and followed at all times.

1. WEAR EYE PROTECTION during any operation that could endanger your eyes. This would include operating power tools, working around a running car engine, carrying batteries, and so on.

2. A "CLOWN CAN KILL!" In other words, avoid anyone who does not take workshop work seriously. Remember, a joker is "an accident just waiting to happen".

3. KEEP YOUR WORKSHOP ORGANISED. Return all tools and equipment to their proper storage areas. Never lay tools, creepers, or parts on the floor.

4. DRESS LIKE A MECHANIC, not like a jewellery model. Remove rings, bracelets, necklaces, watches, and other jewellery. They can get caught in the engine fans, belts, driveshafts — tearing off flesh, fingers, chunks of hair, and ears. Also roll up long sleeves and secure long hair, they too can get caught in spinning parts.

5. NEVER CARRY SHARP TOOLS or parts in your pockets. They can puncture the skin.

6. WEAR FULL FACE PROTECTION when grinding, welding, and during other operations where severe hazards are present.

7. WORK LIKE A PROFESSIONAL, not like a "crazy monkey". When learning automotive mechanics, it is easy to get excited about your work. However avoid working too fast. You could overlook a repair procedure or safety rule.

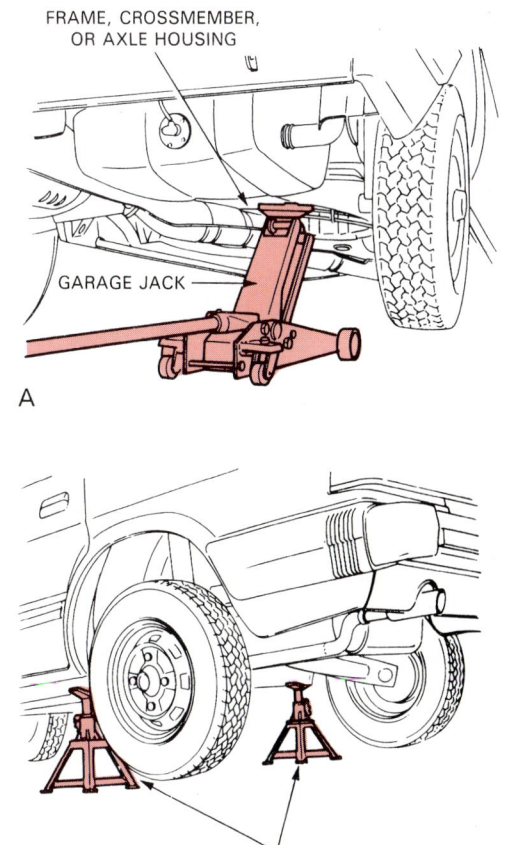

FRAME, CROSSMEMBER, OR AXLE HOUSING

GARAGE JACK

A

JACK STANDS

B

Fig. 8-9. Never work under a car only supported by a floor jack. A — Jack is only used for initial lifting. B — Jack stands are for securing car before working. Place them under recommended lift points. (Subaru)

8. USE THE RIGHT TOOL FOR THE JOB! There is usually a "best tool" for each repair task. Always be thinking about whether a different tool will work better than another, especially when you run into difficulty.

9. KEEP GUARDS OR SHIELDS IN PLACE. If a power tool has a safety guard, use it.

10. LIFT WITH YOUR LEGS, not your back. There are many assemblies that are very heavy. When lifting, bend at your knees while keeping your back straight. On extremely heavy assemblies (transmissions, engine blocks, rear axles, transaxles) use a portable crane.

11. USE ADEQUATE LIGHTING. A portable workshop light not only increases working safety, it increases working speed and precision.

12. VENTILATE WHEN NEEDED. Turn on the workshop ventilation fan any time fumes are present in the workshop.

13. NEVER STIR UP ASBESTOS DUST! Asbestos dust (particles found in brake and clutch assemblies) are powerful CANCER-CAUSING AGENTS. Do NOT use compressed air to blow the dust off these parts.

14. JACK UP A CAR SLOWLY AND SAFELY. A car can weigh between one and two tonne. Never work under a car unless it is supported by jack stands. It is NOT safe to work under a car held only by a floor jack. See Fig. 8-9.

15. DRIVE SLOWLY WHEN IN THE WORK-SHOP AREA. With all of the other mechanics and cars in the workshop, it is very easy to have an accident.

16. REPORT UNSAFE CONDITIONS TO THE PERSON IN CHARGE. If you notice any type of hazard, let the person in charge know about it.

17. STAY AWAY FROM ENGINE FANS! The fan on a car engine is like a SPINNING KNIFE. It can inflict serious injuries. Also, if a part or tool is dropped into the fan, it can fly out and hit someone.

18. RESPECT RUNNING ENGINES. When a car engine is running, make sure that the transmission is in 'park', the parking brake is 'on' and the wheels are 'chocked'. If the car were to be accidentally knocked into gear, it could run over you or a friend.

19. NO SMOKING! No one should smoke in an automotive workshop. Smoking is a serious fire hazard, considering fuel lines, cleaning solvents, paints, and other flammables may be exposed.

20. OBTAIN PERMISSION before using any new or unfamiliar power tool, hoist, or other workshop equipment. Also, have the person in charge give you a demonstration before you use the equipment.

KNOW THESE TERMS

Work bay, Lube bay, Hoist, Front end bay, Electrical fire, Asphyxiation, Chemical burns, Asbestos.

REVIEW QUESTIONS

1. List four safety rules to follow when using a vehicle hoist.

2. A _____, _____, _____, or _____, _____, _____ is used when working on a car's steering and suspension systems. It has special equipment for aligning the wheels of a car.

3. Describe the most common and dangerous flammable found in the auto workshop?

4. What is an electrical fire?

5. Car batteries can explode. True or False?

6. Asbestos, found in brakes and clutches, is harmful and can cause _____.

7. Which of the following cannot cause electrical shock?
   a. Missing earth prong on cord.
   b. Using electric drill on wet floor.
   c. Using electric tools with an earth prong.

8. Explain what must be done to prevent physical injuries.

9. If you are pulling on a wrench as hard as you can and the wrench does NOT turn, what should you do to prevent injury?

10. List 20 general safety rules.

Chapter 9

GASKETS, SEALANTS, SEALS, ADHESIVES

After studying this chapter, you will be able to:
☐ Describe gasket construction, materials, and application.
☐ Explain the selection and use of sealants and adhesives.
☐ Describe the construction and installation of seals.

IMPORTANT AND WIDELY USED

Gaskets and seals are used throughout the car. They confine petrol, oil, water, and other fluids, in addition to air and vacuum, to specific units or areas. They keep dust, dirt, water, and other foreign materials out of various parts. They play an important part in the proper functioning and service life of all components.

Unfortunately, the importance of the proper selection, preparation, and installation of gaskets and seals is not always clearly understood. In addition to these duties, they effect torque and tension, part alignment and clearance, temperature, compression ratios, and lubrication. REMEMBER: THE FAILURE OF GASKETS, SEALANTS, SEALS, OR ADHESIVES CAN CAUSE EXTENSIVE DAMAGE AND EXPENSE. STUDY THE MATERIAL IN THIS CHAPTER CAREFULLY AND APPLY THE INFORMATION TO YOUR WORK!

GASKET

A gasket is a flexible piece of material, or in some cases, a soft sealant, placed between two or more parts. When the parts are drawn together, any irregularities (warped spots, scratches, dents) will be filled by the gasket material to produce a leakproof joint. See Fig. 9-1.

GASKET MATERIALS

Many materials are used in gasket construction. Steel, aluminium, copper, asbestos, cork, rubber

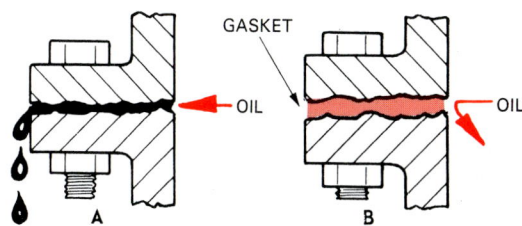

Fig. 9-1. Gasket stops leaks. A — Assembly has no gasket. Irregularities on part mating surfaces allow leakage. B — Same assembly is shown but with a gasket. Irregularities are filled and leak is stopped.

(synthetic), paper, felt, and liquid silicone. The materials can be used singly or in combination.

Gasket material compressibility (how easily it flattens under pressure) varies widely. The gasket must compress to some extent to effect a seal. However, excessive compressibility will cause the gasket to extrude (cold flow outward and reduce thickness in direction of compression) or reduce its thickness beyond a specified point.

The gasket material selected will depend on the specific application, temperature, type of fluid to be confined, smoothness of mating parts, fastener

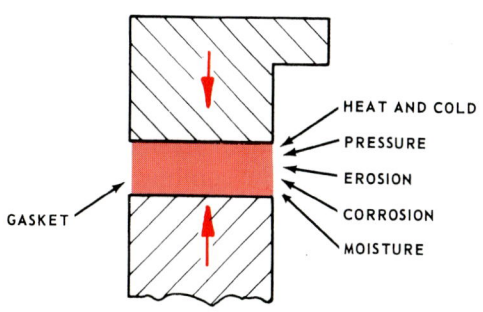

Fig. 9-2. Gasket must withstand many forces. Destructive forces shown, in addition to others not illustrated, are constantly attempting to destroy gasket.

tension, pressure of confined fluid, material used in construction of mating parts, and part clearance relationship. All of these affect the choice of gasket material and design.

When constructing or selecting gaskets, give careful thought to these factors and choose wisely. Fig. 9-2 illustrates some of the destructive forces that the gasket must resist to function properly.

GASKET CONSTRUCTION

Some gaskets are of very simple construction. The engine top water outlet, for example, can use a medium thickness, chemically treated, fibrous paper gasket. Unit loading (pressure between mating parts) is light, temperature medium, coolant pressure low, and the coolant presents only mild problems, Fig. 9-3.

Fig. 9-3. Simple paper gasket. The paper is soft, tough, and water resistant.

As the sealing task becomes more difficult, gasket construction becomes more involved. The exhaust manifold to exhaust pipe gasket, where used, is more complex. Unit loading pressure is higher with corrosive flames, gases, and high temperatures attempting to destroy the gasket. This gasket, in two basic types, uses asbestos and steel in its construction, Fig. 9-4.

Perhaps the most complicated gasket, in terms of materials used and construction techniques, is the cylinder head gasket. Unit pressure is tremendous, combustion temperatures and pressures are very high. The head gasket must also seal against coolant, oil, and corrosive gases.

There are several basic designs in common use. Asbestos, steel, copper, and rubber may be used in their construction.

One type of multiple-layer gasket is shown in A, Fig. 9-5. A steel centre core, perforated to produce tiny gripping hooks, is placed between two sheets of specially prepared asbestos. Steel or copper grommets are placed around the combustion chamber and coolant openings to assist in sealing. The entire gasket is then formed in a one-piece unit.

In B, Fig. 9-5, an asbestos centre core is placed between two sheets of steel or copper. Note that the edges are rolled to produce a grommet.

The single layer beaded or corrugated type of gasket shown in C, Fig. 9-5, is popular on high compression engines. A single sheet, around

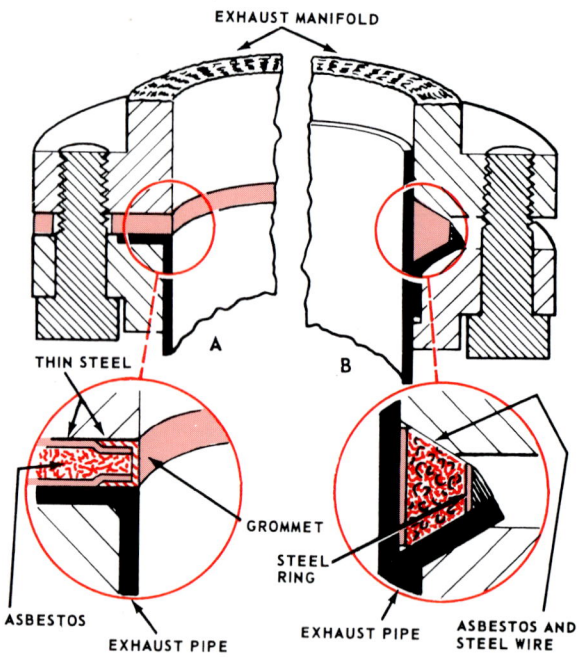

Fig. 9-4. Exhaust manifold gaskets. A — Gasket has an asbestos centre with a thin steel outer layer. Note how inner edge is protected with a steel grommet. B — Gasket is made up of asbestos and steel wire. A thin steel outer ring can also be used for additional strength.

0.50 mm thick, is stamped to produce a beaded edge around combustion chamber and fluid openings. This particular one is given an aluminium coating, about 0.025 mm thick, on both sides to assist in sealing and to prevent corrosion. This type of gasket requires accurate and smooth block-to-head surfaces. The aluminium coated steel gasket will withstand high temperatures and pressures quite successfully. In addition, it will not produce torque loss (gasket becoming thinner under continued fastener tension thereby reducing bolt tension and torque).

LOCALISED UNIT LOADING

To produce higher unit loading around the combustion chambers or any other opening, a copper wire can be inserted between the top and bottom layers, near the edge. The remainder of the gasket tends to compress more readily. This creates the desired pressure around the opening, D, Fig. 9-5.

Another technique used to produce localised unit pressure or loading is shown in E, Fig. 9-5. This type uses a copper or soft iron grommet around the rolled edges.

Coolant and oil openings are sometimes sealed by placing special rubber or neoprene grommets in the gasket openings. These are highly resilient and maintain constant pressure around the openings, F, Fig. 9-5.

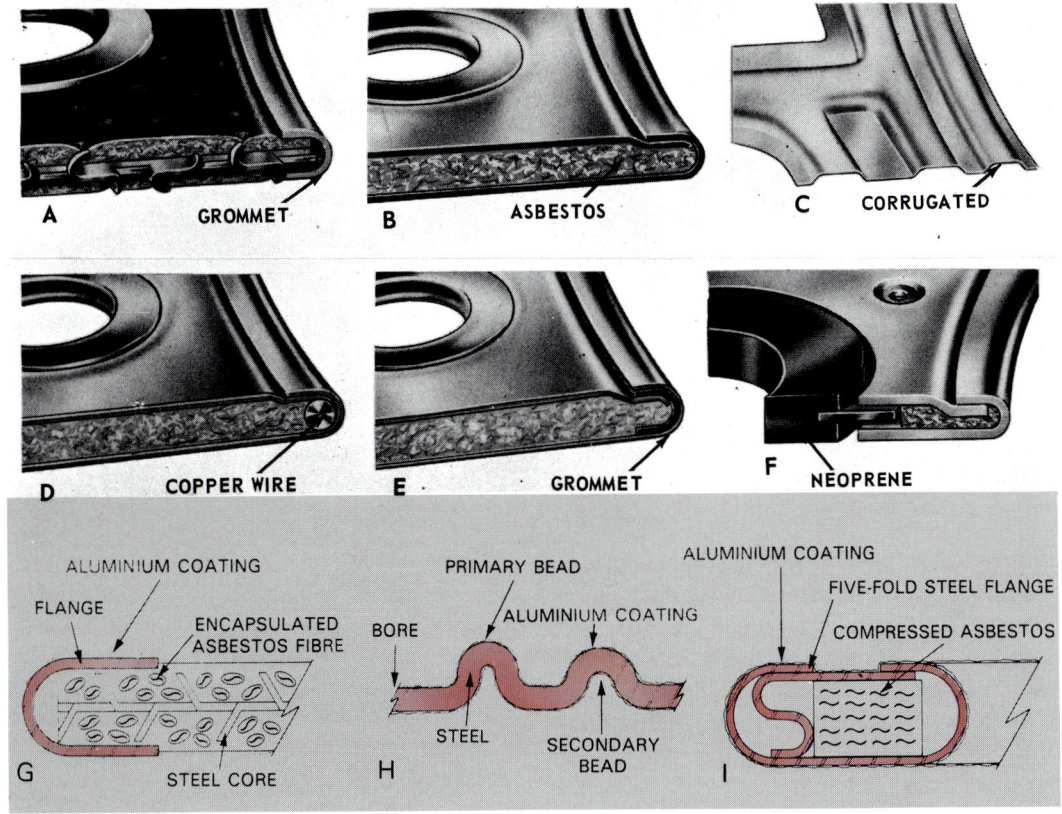

Fig. 9-5. Some of the different methods employed in head gasket construction.

Several other techniques employed in head gasket design and construction are shown in G, H, and I, Fig. 9-5. A soft-seal surface composition gasket is shown in G. Note the use of encapsulated asbestos fibre and an aluminium coating over the outside surface for better sealing. The use of a sealer is recommended with this gasket. An embossed steel type (similar to C), also called "shim" gasket, is pictured in H. It is produced from sheet steel approximately 0.50 mm in thickness. It provides good strength in the bead area and fine sealing capabilities when properly installed. The use of a sealer is advised. This type of gasket is very popular today on original equipment engines.

Fig. 9-5, I, is a metal clad, sandwich gasket. It is constructed of sheet steel wrapped around a compressed asbestos centre. The steel is coated with aluminium on the outside to aid in sealing. Use of a sealer is also advisable.

Fig. 9-6 shows sheets used to make a cylinder head gasket.

GASKETS OFTEN COME IN SETS

Gaskets are often ordered in sets. For engine work gaskets are available in a VALVE REGRIND SET (which includes all gaskets necessary to do a valve grind job). CONVERSION SET (which includes gaskets for the lower half of the engine, or gaskets which are needed where the crankshaft is removed),

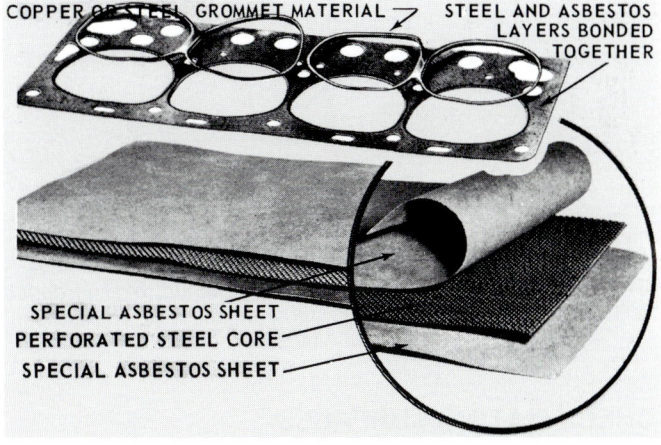

Fig. 9-6. One type of head gasket construction.

and FULL ENGINE SET (which includes all gaskets necessary to do a complete engine overhaul). See Fig. 9-7. Sets for transmission, carburettor, and differential are available separately. Single gaskets for some specific parts are also available. Gasket sets also usually include necessary oil seal replacements.

GASKET INSTALLATION TECHNIQUES

After selecting a gasket material and construction, there are a few important installation considerations. Regardless of the suitability of the gasket, if not properly installed, it will fail.

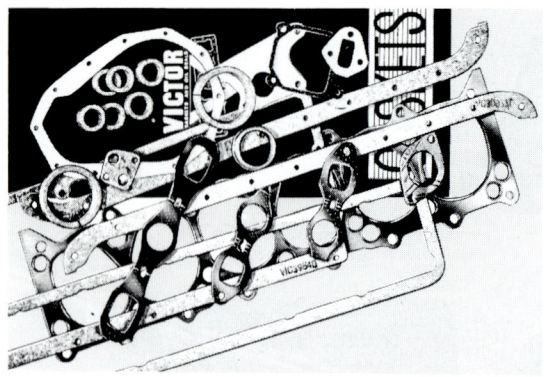

Fig. 9-7. Engine overhaul gasket set. This set is for a six cylinder engine.

NEVER REUSE A GASKET

Once a gasket has been in service it will lose a great deal of its resiliency. When removed, it will not return to its original thickness. If reused, it will fail to compress and seal properly. Gasket costs, as related to part and labour costs, are small. The professional mechanic does not even consider using old gaskets. Fig. 9-8 demonstrates how the use of old gaskets will produce leaks.

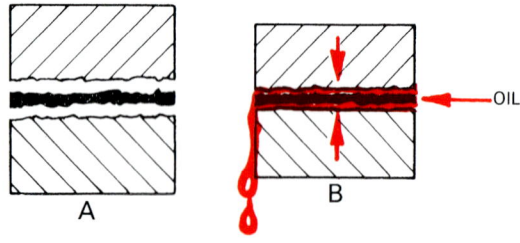

Fig. 9-8. Used gaskets will not work! A — Used gasket is positioned. B — When the parts are tightened, old, hardened gasket cannot compress and fill irregularities. The results: LEAKS!

CHECK MATING SURFACES

After thorough cleaning, inspect both part mating surfaces to detect any nicks, dents, pieces of old gasket or sealer, burrs, dirt, or warpage that may make proper sealing impossible. Refer to Fig. 9-9.

CHECK GASKET FOR PROPER FIT

Place the gasket on the part to determine if it fits properly. On the more complicated setups, such as cylinder head gaskets, make certain the gasket is right side up, proper end forward, and that bolt, coolant, and other openings are clear and in proper alignment. Occasionally you may notice that the gasket coolant openings may be slightly larger or smaller than the ports in the block or head. This

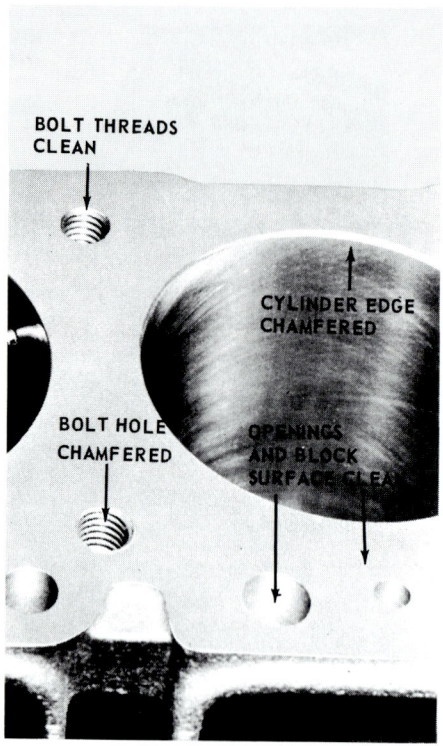

Fig. 9-9. Check mating surface. Notice that cylinder block surface is clean, smooth, and that all openings are clean.

gasket may be designed to fit several models or to restrict or improve coolant circulation. Check out these situations carefully.

Head gaskets for the left and right bank on some V-8 engines are interchangeable. Others are not. Many head gaskets have the word TOP and occasionally the word FRONT stamped on them, Fig. 9-10.

SOME GASKETS TEND TO SHRINK OR EXPAND

Paper and cork type gaskets that have been stored for some time tend to either lose or pick up moisture depending on storage conditions. Loss of moisture can cause them to shrink. Excess moisture can expand them. In either case, when checking for proper fit, they will show signs of misalignment.

This condition can be corrected by soaking shrunken gaskets in water for a few minutes or by placing expanded gaskets in a warm (not over 65-95°C) spot. Check them occasionally to prevent over-doing the treatment, Fig. 9-11.

CHAMFERING SCREW HOLES MAY BE NECESSARY

When installing head gaskets, examine the screw holes in the block. If the threads run right up to the very top, it is a good idea to chamfer them lightly. Then, run the proper size tap in and out of the holes.

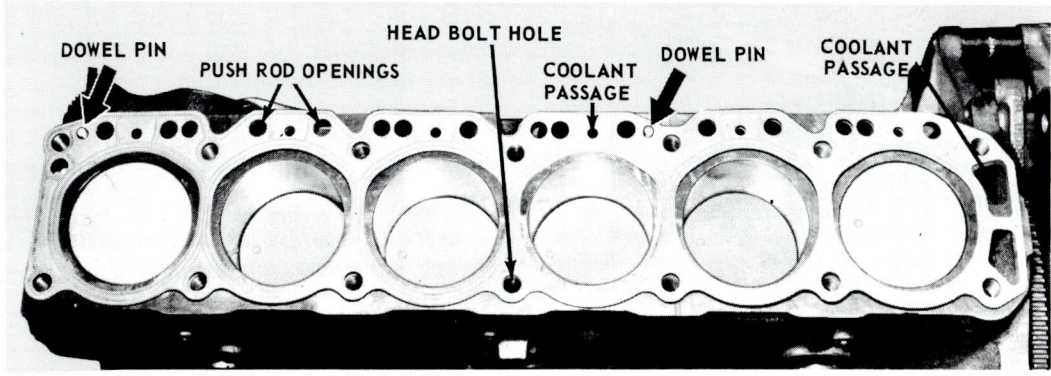

Fig. 9-10. Checking head gasket for proper fit. Dowel pins hold head gasket in place and align cylinder head to block. All openings are in proper alignment. This is a single layer, beaded steel gasket.

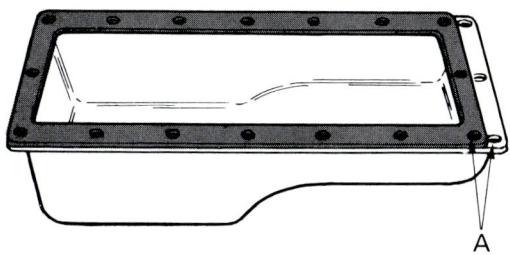

Fig. 9-11. Pan gasket has shrunk. Gasket has dried out, producing shrinkage. Note in A how screw holes fail to match. Soaking will salvage this gasket.

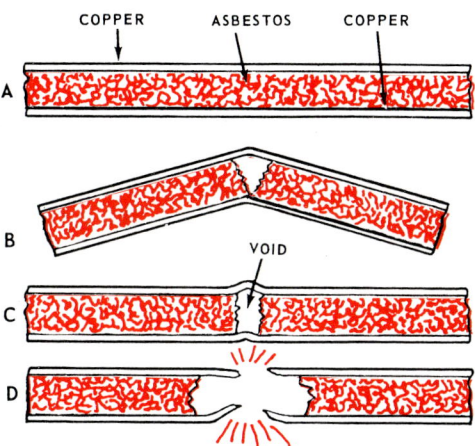

Fig. 9-12. Creased gasket. A — Multiple-layer head gasket. B — Gasket has been creased and the centre packing pulled apart. C — Gasket straightened producing void. D — Gasket has "blown" in service.

The chamfer prevents the top thread from being pulled above the block surface. Blow out the holes with compressed air.

WHEN USING AN AIR HOSE FOR CLEANING, ALWAYS WEAR GOGGLES. SMALL PARTICLES CAN BE THROWN WITH GREAT FORCE — BE CAREFUL!

EACH GASKET SHOULD BE CHECKED

Carefully inspect the gasket itself for dents, dirt, cracks, and folds. A minor crease in a cork or paper gasket usually does not render it useless. However, when checking head gaskets, BEWARE of ALL creases. If bent sharply, do not attempt to straighten it. The inner layer may be separated and cause failure. A gentle bend will not ruin the gasket — sharp kinks and creases will. Fig. 9-12 illustrates what happens when a multiple-layer head gasket is creased and then straightened.

MAKING A GASKET

A simple paper or combination cork and rubber gasket can be made. First, trace the pattern. Then, cut with scissors or lay the gasket material on the part and gently tap along the edges with a brass hammer. Screw holes can also be tapped lightly with the pein end of the ball pein hammer. Do not tap hard enough to damage the threads. Gasket punches can also be used to make neat screw holes. It will help hold the material in place if you tap out the corner holes and start these screws before tapping around the edges, Fig. 9-13.

HANDLE GASKETS WITH CARE

Gaskets should be stored flat, in their containers, and in an area where they will not be bent or struck. Storage space should not be subjected to extremes of temperature or humidity. Handle gaskets carefully. Do not attempt to force them to fit. If a gasket is accidentally cracked or torn — throw it away.

USE OF SEALANTS

A new, properly installed gasket will usually produce a leakproof joint. However, mating surfaces

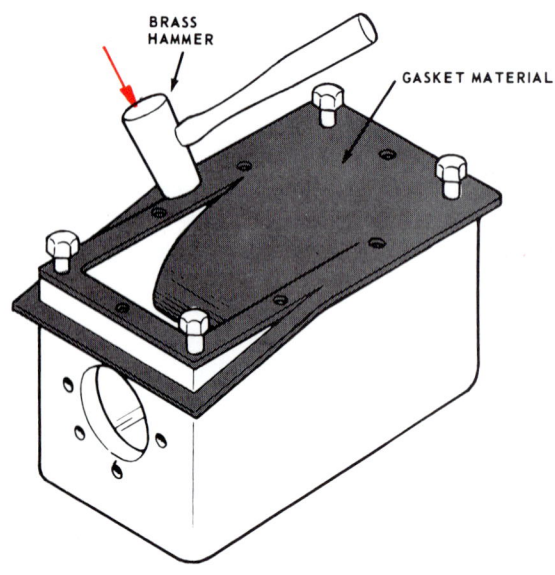

Fig. 9-13. Making a gasket. Four corner screws hold gasket material in place while tapping. A ball pein hammer is used for holes.

are not always true. Corners can present problems. Torque loss can reduce pressure on the gasket surface. Gaskets may shrink slightly and minute part shifting can break the seal. For these reasons, it is generally considered good practice to use a sealant on MOST gaskets.

The addition of a sealant helps hold the gaskets in place during assembly. Also, small cracks, indentations, and corner voids are sealed. In short, the use of a good sealant provides additional assurance that the joints will be leakproof.

REMEMBER: A small amount of oil seepage will, due to engine heat, spread over a large area. This produces a messy looking job and is certain to deposit oil dribbles on the customer's garage floor — hardly a good advertisement for any workshop.

SEALANT

Gasket sealer or sealant is a liquid or semiliquid material that is sprayed, brushed, or spread on the gasket surface. Various types, having different properties, are available. Some set hard and others remain pliable. Most, but not all, are highly resistant to oil, water, fuel, grease, antifreeze, mild acid, and salt solutions. Resistance to heat and cold vary, but in general, most sealers are adequate in this respect for all uses other than exhaust applications.

The mechanic should be thoroughly familiar with sealers and their properties and uses. The chart in Fig. 9-14 lists various sealants, properties, and recommended uses for various products. Sealant manufacturers will be happy to provide the mechanic with specific recommendations for using their products.

The use of too much sealer is generally worse than using none at all. Excess sealer is squeezed out of the joint and can clog water, fuel, and oil passages. A

THIN coat is ample. On some oil pan gaskets with corners difficult to seal, a small dab where the gaskets meet is permissible.

In general, a nonhardening, flexible sealer will produce the desired results.

Some parts with extremely small holes or ports, such as carburettors and automatic transmission valve bodies, can be rendered useless if ANY sealant is squeezed into the openings. In cases such as this, do not use a sealant.

In any specific application, be sure to follow the manufacturer's recommendations.

USING RUBBER GASKETS

Rubber gaskets are highly resilient and will usually do a good job of sealing without the addition of a sealer. In fact, rubber gaskets tend to extrude (squeeze out) under pressure when a sealer is used. Unless a sealant is specifically recommended, a rubber gasket should be installed WITHOUT SEALER.

HOLDING GASKET DURING ASSEMBLY

Where a sealant is used, the gasket will usually stay in place during assembly.

If sealant is not being used and the gasket tends to slip, the gasket can be held in place with a thin coat of grease. On rubber gaskets use grease or sealant only at a few small spots.

Some parts, such as oil pans, can be difficult to assemble without disturbing gasket position. In some cases, in addition to using a sealant, it is advisable to tie the gasket with thin soft string. The parts may be tightened with the string in place. Patented gasket holders are also available and work well.

In other instances, such as cylinder head installation, guide pins are used to hold the gasket in alignment.

Make certain the gasket is correctly installed and that it remains in alignment during assembly. See Fig. 9-15.

USE PROPER SEQUENCE AND TORQUE WRENCH

After running all fasteners up snug, tighten them in the proper sequence as recommended in the chapter on fasteners. First, tighten to one-third torque, second to two-thirds torque, third to full torque.

Improper sequence and torque, in addition to snapping fasteners and parts, producing distortion, will very likely cause the gasket to fail to seal. Excessive torque can place the gasket under too much pressure. This can cause it to extrude badly, Fig. 9-16 shows the results of improper tightening procedures as related to gasket sealing.

PRODUCT	USES	RESISTS	TEMPERATURE PRESSURE/STRENGTH RANGE	SETS
LOCTITE 510 (GASKET ELIMINATOR)	Eliminator for a wide variety of preformed, pre-cut gaskets.	Water, engine oil, gear oil, transmission fluid, petrol, diesel, kerosene and Freon 12.	-57°C to 204°C	Flexible
LCCTITE 515 (MASTER GASKET SEALANT)	Eliminator for gaskets in flange type joints.	Water, engine oil, alcohol, unleaded petrol, toluene, 1,1,1, trichlorethane.	-50°C to 150°C	Flexible
PERMATEX No. 1 (FORM-A-GASKET)	Permanent assemblies, repair gaskets, fills uneven surfaces, seals thread connections, repairs cracked batteries.	Water, steam, kerosene, petrol, oil, grease, mild acid, alkali and salt solutions, aliphatic hydrocarbons, antifreeze mixtures.	-18°C to 204°C	Hard
PERMATEX No. 2 (FORM-A-GASKET)	Semi-permanent assembly work.	"	-18°C to 204°C 34, 450 kPa	Flexible
PERMATEX No. 3 AVIATION (FORM-A-GASKET)	Sealing of close fitting parts. It is easy to apply on irregular surfaces.	"	-18°C to 204°C 34, 450 kPa	Flexible
DOW CORNING (SILASTIC GASKET)	Eliminator for a wide variety of preformed, pre-cut gaskets.	Water, engine oil, grease, antifreeze.	-65°C to 235°C	Flexible
LOCTITE 767 (ANTI-SEIZE LUBRICANT)	Lubricates gears, chains, cables, sprockets, levers, pivots, rollers, valve stems.	Galling and corrosion, wear in heavy pressure, applications.	-54°C to +760°C	—
LOCTITE (QUICK METAL)	Reducing clearances between worn components, stops bearing spinout.	Oils, cutting fluids and chlorinated solvents.	0°C to 150°C Compressive strength: 860 to 1,550 MPa	Hard
LOCTITE 222 (SUPER SCREW LOCK)	To lock screws, nuts and studs, with long thread engagements.	Water, oils, diesel, petrol, organic solvents and refrigerants.	0°C to 150°C Shear strength (max): 6,890 kPa	Hard
LOCTITE 242 (SUPER NUT LOCK)	To lock screws, nuts, and studs, with medium length thread engagements.	"	0°C to 150°C Shear strength: 6,890 to 13,790 kPa	Hard
LOCTITE 262 (SUPER STUD LOCK)	To lock screws, nuts, and studs with short length thread engagements.	"	0°C to 150°C Shear strength: 10,300 to 15,500 kPa	Hard
LOCTITE 601 (SUPER RETAINING COMPOUND)	To retain ball races, seals, pulleys, gears, sprockets, bushes, to reduce key and spline clearances.	"	0°C to 150°C Shear strength: 13,780 to 20,670 kPa	Hard
LOCTITE 495 (INSTANT ADHESIVE NON-METALS)	Joining acrylic, ABS, polyamide, rigid vinyl and rubbers together.	Atmospheric agents.	-55°C to 85°C	Hard
LOCTITE 496 (INSTANT ADHESIVE METALS)	Joining aluminium to PVC, steel to steel.	Fatigue and vibration.	-55°C to 85°C	Hard
DOW CORNING (SILASTIC AUTO SEALANT)	Sealing, bonding and repair to trim, weather strips and vinyl roofing.	Water, dirt and dust ingress.	-65°C to 235°C	Flexible
DOW CORNING (SILASTIC WIND-SCREEN SEALANT)	Repair of leaks in front, rear, and fixed side screens, sunroofs and weather strips.	Shrinking, cracking and perishing.	-65°C to 235°C	Flexible

Fig. 9-14. Sealant and adhesive chart.

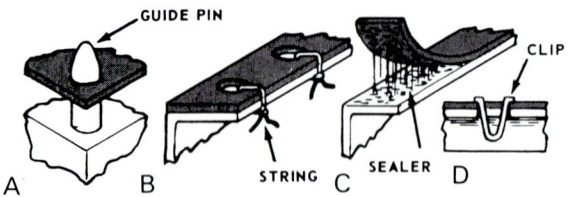

Fig. 9-15. Holding gasket in place. It is important that gaskets be held in alignment during assembly.

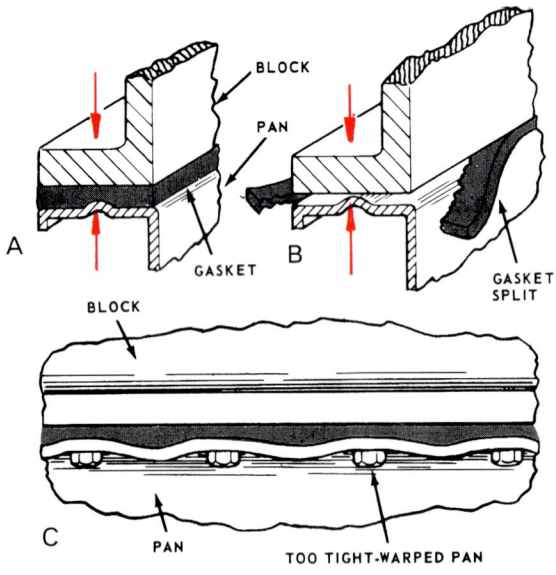

Fig. 9-16. Overtightening will cause damage. A — Proper fastener tension. B — Excessive tightening has split cork pan gasket. C — Excessive tension has warped oil pan flange.

STAMPED PARTS REQUIRE EXTRA CARE

Relatively thin stamped parts such as rocker arm covers, oil pans, and some timing covers, if bent along the engaging edge, must be straightened before installation. Place the part edge on a smooth, solid metal surface. Gently tap to straighten the bent sections. When installing, do not overtighten as the parts will be bent again, Fig. 9-17.

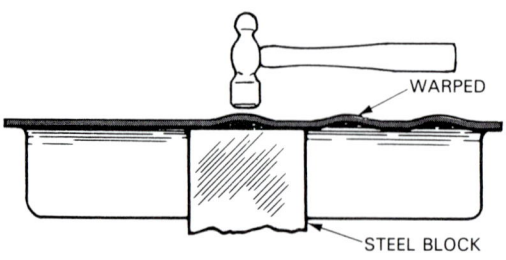

Fig. 9-17. Straighten warped flange. Warped edges cause leaks. Straighten them before installation.

REMEMBER THESE STEPS IN PROPER GASKET INSTALLATION

1. Clean parts, fasteners, and threaded holes.
2. Remove all burrs, bent edges, and excessive warpage and check for dents and scratches.
3. Select a new gasket of the correct size and type.
4. Check the gasket for fit.
5. Where sealant is used, apply a THIN coat of the correct sealant on one side of the gasket. Place the gasket with the coated side against the part. Spread a THIN coat on the uncoated side. Do not slop sealant into parts. Wipe off excess.
6. If alignment difficulty is anticipated during assembly, secure the gasket by additional means.
7. Carefully place mating part in place.
8. Coat threads of fasteners with antiseize (unless prohibited). Install in their PROPER location and run up snug.
9. Torque the fasteners in proper sequence.
10. If necessary, retorque after a specified length of time.

ANALYSE GASKET FAILURE

When a gasket fails in service, there has to be a reason for the failure. If you do not detect the reason, your own installation might fail also. The following simple steps will help you find the underlying cause of the failure:

1. Ask the owner about any unusual conditions. Try to determine if the gasket failed suddenly or over a period of time.
2. Before dismantling, check fastener torque with a torque wrench. You can loosen each one and notice the reading at break-away. This will be somewhat less than true torque. Another method is to carefully mark the position of the head of the screw or nut in relationship to the part (use a sharp scribe). Back the nut off about one-quarter turn. Carefully retighten until the scribed lines are exactly in alignment. If done properly, this will give you a fair indication of torque at the time of failure.

   If the torque is significantly below that specified, this could well be the cause of failure. If torque varies from fastener to fastener, this too could be the cause. **ALWAYS ALLOW AN ENGINE TO COOL BEFORE REMOVAL OF A CYLINDER HEAD.** A cylinder head can be warped by removing it when too hot.
3. Following dismantling, carefully blot off any grease, oil, dirt, and carbon from the gasket. Do not rub or wash the gasket immediately, as this may remove tell-tale signs.

   Inspect the gasket for signs of uneven pressure, burning, corrosion, cracks, or voids. Check to determine if the gasket is of the correct material and type for the job.

4. Inspect the mating parts for warpage and burrs. ALWAYS TRY TO FIND THE CAUSE OF GASKET FAILURE SO YOU MAY EFFECT A CORRECTION WHEN INSTALLING A NEW GASKET.

RETORQUE

Constant fastener tension and the expansion and contraction of parts will tend to further compress a gasket. This will leave the fasteners below proper torque. In a critical application, such as a head gasket, it can cause gasket failure unless the fasteners are retorqued after a period of time.

A service manual for the particular vehicle will usually indicate retorquing procedure if it is recommended.

OIL SEALS

An oil seal can be used to confine fluids, prevent the entry of foreign materials, and separate two different fluids.

An oil seal is secured to one part while the sealing lip allows the other part to rotate or reciprocate (move).

Oil seals are used throughout the mechanical parts of the car. Engine, transmission, drive line, differential, wheels, steering, brakes, and accessories all use seals in their construction.

OIL SEAL CONSTRUCTION AND MATERIALS

Seals are made up of three basic parts. A metal container or case, the sealing element, and a small spiral spring called the GARTER spring.

Sealing elements are usually made of synthetic rubber or leather. Synthetic rubber seals are displacing leather in most applications. The rubber seal can be made to close tolerances and can be given special configurations (shapes). Also, specific wear and heat resistant properties can be imparted.

In the rubber oil seal, the sealing element is bonded to the case. The element rubs against the shaft. The case holds it in place and in alignment. The garter spring forces the seal lip to conform to minor shaft runout (wobble) while at the same time maintaining constant and controlled pressure on the lip. Fig. 9-18 illustrates typical oil seal construction.

VARIOUS DESIGNS ARE USED

Many different element and lip shapes are provided. Each is designed to provide the best seal for a specific task. Fig. 9-19 shows several designs. Notice that more than one lip can be used. The outside diameter, or one edge, may be coated with rubber to provide better OD (outside diameter) sealings.

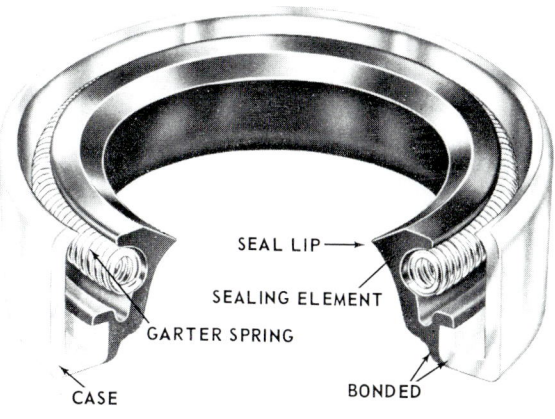

Fig. 9-18. Typical oil seal construction. (Victor)

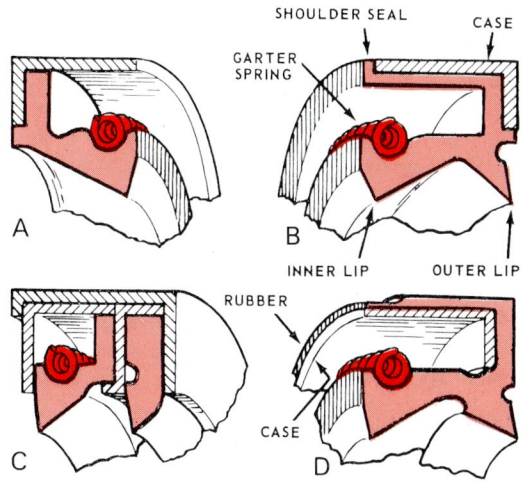

Fig. 9-19. Oil seal designs. A — Single lip. B — Double lip with rubber shoulder seal. Inner lip controls oil and outer lip keeps out dust and water. C — Double lip. Both lips control oil. D — Double lip with rubber outer coat to assist outside diameter sealing.

OTHER TYPES OF OIL AND GREASE SEALS

Engine rear main bearing oil seals are usually constructed in two halves. They may be made of graphite impregnated asbestos wicking or synthetic rubber. Some grease (not oil) seals use a felt sealing element. Occasionally, a combination will use an inner rubber seal and a felt outer seal, Fig. 9-20.

OIL SEAL REMOVAL

Seals may be removed by prying, driving or pulling, depending on the location.

Before removal, notice the depth to which the seal was installed. As with a gasket, inspect the seal after removal for any signs of unusual wear or hardening.

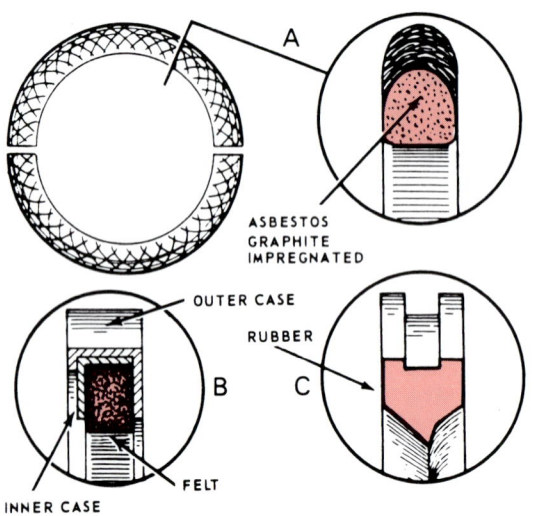

Fig. 9-20. Other seal types. A — Main bearing (rear) seal made of asbestos wicking. Both upper and lower halves fit into grooves in block and cap. B — Typical grease seal using a felt sealing ring. C — Synthetic rubber main bearing oil seal. Rubber O-rings (not shown) are used in several areas. They are simple round rubber rings.

DO NOT REUSE SEALS. WHEN UNITS ARE DOWN FOR SERVICE, REPLACE THE SEALS. Use care to avoid damage to seal bore during seal removal. Such damage can cause leaks, difficult installation, and damage to new seals. See Fig. 9-21.

SEAL INSTALLATION

After removing the old seal, carefully clean the seal recess or counterbore. Inspect for nicks and burrs. Compare the old seal with the new one to make certain you have the proper replacement. The OD must be the same. The ID (inside diameter) may be slightly smaller in the new seal as it has not been spread and worn. The width can vary a little.

COAT WITH NONHARDENING SEALER

Coat the inside of the seal housing with a THIN coat of nonhardening sealer. If there is too much sealer, the seal may scrape it off as it enters, causing the surplus to drip down on the shaft and sealing lip. This can cause seal failure, Fig. 9-22.

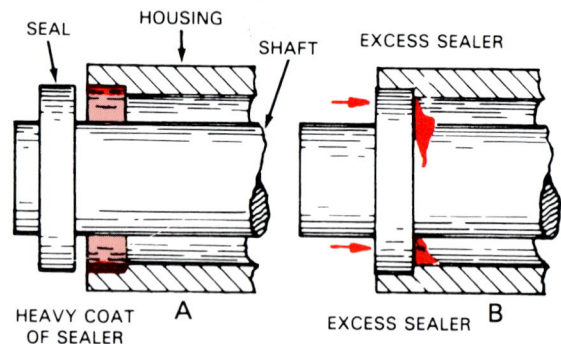

Fig. 9-22. Apply sealer sparingly! A — Seal counterbore has been given a heavy coat of sealer. B — When the seal is driven into the counterbore, excess sealer will be forced out onto shaft and seal lips. In addition to ruining seal, this could clog some opening in mechanism.

DRIVING THE SEAL WITH NO SHAFT PRESENT

After preparing the seal housing place the seal squarely against the opening WITH THE SEAL LIP FACING INWARD OR TOWARD THE AREA IN WHICH THE FLUID IS BEING CONFINED. If the lip faces the other way, it will probably leak, Fig. 9-23.

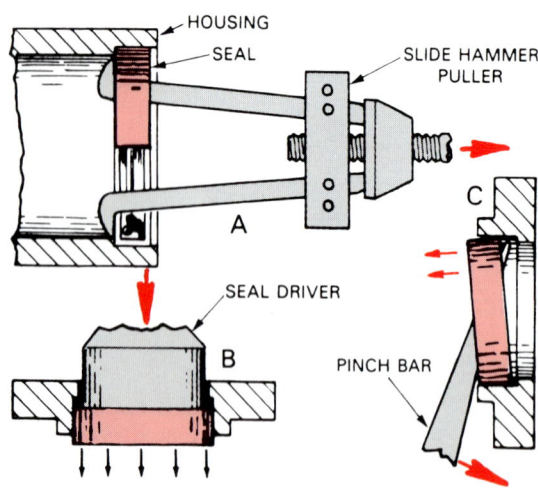

Fig. 9-21. Seal removal. A — Slide hammer puller jaws are pushed through seal and then expanded. Operating slide hammer will pull seal out. B — A seal driver can often be used. C — Many seals can be "popped out" with a small pinch bar. When a seal must be removed, while a shaft is present, a hollow threaded cone is threaded into seal. The cone, attached to a slide hammer, will withdraw seal.

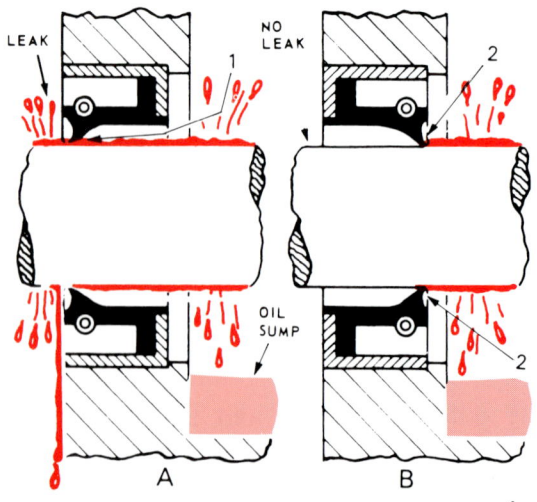

Fig. 9-23. Seal lip must face fluid! A — Seal has been installed backwards. Lip faces away from fluid causing fluid 1 to force seal lip from shaft, causing leakage. B — Seal is correctly installed with lip facing fluid. Pressure at 2 forces seal against shaft, preventing a leak.

USE SUITABLE DRIVER

The driver should be just a little smaller (about 0.50 mm) than the seal OD when the seal will be driven below the surface. If the seal is to be driven flush (even with surface), the driver can be somewhat wider. In any case, the driver should contact the seal near the outer edge only. NEVER STRIKE THE INNER PORTION OF A SEAL. This might bend the flange inward and distort the sealing element, Fig. 9-24.

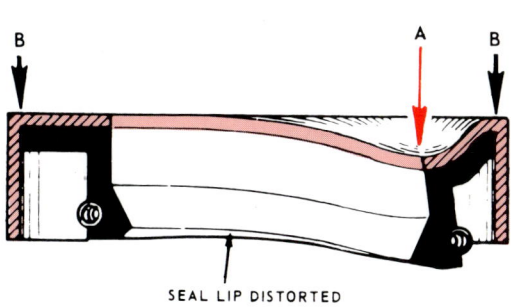

Fig. 9-24. Damaged seal. Seal case badly distorted by careless installation. Punch struck case at A. All driving force should be applied at B. This seal would leak badly.

If a seal driving set is not available, a section of pipe of the correct diameter can be used. Make sure the ends are square. If a hammer is used to start a seal, follow it up with a drift punch. Be careful to strike at different spots (near the outer edge) each time. If the seal begins to tip, strike the high side.

REMEMBER: A SEAL IS EASILY DAMAGED THROUGH IMPROPER INSTALLATION — BE CAREFUL!

DRIVE SEAL TO PROPER DEPTH

If a locating shoulder is used, drive the seal snugly against it. This is especially important if the seal inner edge has a rubber sealing compound designed to flatten against the shoulder. See B, Fig. 9-19.

When no shoulder is used, keep the seal square and stop at the specified depth. If you drive it in too far, you may ruin it while attempting to pull it back.

WHEN SEAL LIP MUST SLIDE OVER SHAFT DURING INSTALLATION

When driving a seal that must slip over a shaft, use care to see that the sealing lip is not nicked or abraded.

If a plain shaft (no keyway, splines, or holes) is involved, check the shaft carefully for burrs and nicks.

If any are found, remove them by polishing (shoe shine motion) with CROCUS cloth (a very fine abrasive). Examine the shaft surface where the sealing lips will operate. It must be smooth at this point.

If the end of the shaft is chamfered (bevelled), polish the chamfered area. If there is no chamfer then use a mounting bullet or thimble. See Fig. 9-25.

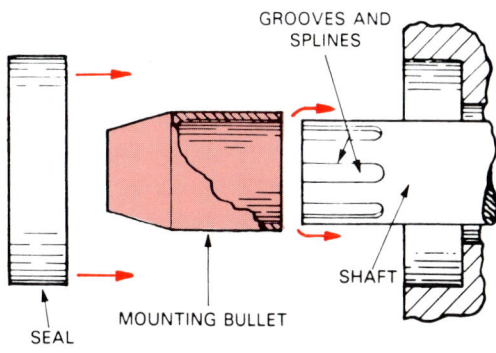

Fig. 9-25. Installing seal using mounting bullet. Bullet or sleeve is placed over shaft and seal can then be installed without lip damage by spline edges.

Once the shaft is free of scratches, wipe it CLEAN and apply a film of oil to the full length. Place a small amount of oil or soft grease on the seal lip and inner face. With the seal lip facing toward the fluid to be confined (coat counterbore with a thin coat of sealer), carefully slip the sealing lips over the chamfer onto the shaft. Slide the seal along the shaft until it engages the counterbore. Using a suitable driver, seat the seal, Fig. 9-26.

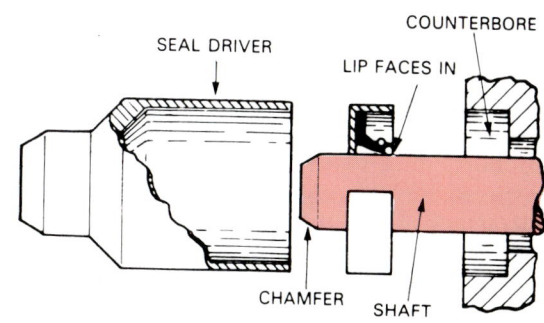

Fig. 9-26. Installing a seal over a plain shaft. Seal will start over chamfered shaft end without damage. Shaft must be smooth, clean, and oiled.

MOUNTING SLEEVES AND BULLETS

When driving a seal that must first slide over a keyway, drilled hole, splines or shaft square end, a mounting sleeve or bullet should ALWAYS be used. This will prevent damage to the seal lip. Fig. 9-25 illustrates the proper setup. The OD of the mounting sleeve should not be much over (1.00 mm) larger than the shaft or the seal lips will be spread excessively.

In the event no mounting tools are available, one may be quickly made by using shim stock (thin brass sheets in various thicknesses).

Wrap the stock tightly around the shaft (one wrap with a small lap) and trim off. Tin the lap with a soldering iron. File the lapped edge after soldering. Then, smooth with abrasive cloth. Bend the leading edge inward and it is ready to use, Fig. 9-27.

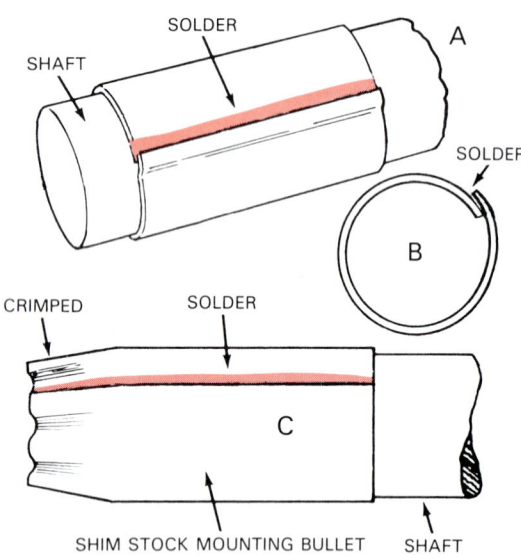

Fig. 9-27. Shim stock mounting sleeve. A — Sleeve formed and soldered. B — Edge sanded smooth. C — Sleeve installed and leading edge crimped. All edges must be smooth.

REMEMBER THESE STEPS IN SEAL INSTALLATION

1. Clean seal counterbore, remove nicks and burrs and coat with a VERY THIN layer of nonhardening sealer.
2. Inspect shaft, polish burrs and scratches with CROCUS cloth. Pay particular attention to the area where the seal lip will operate.
3. Check the new seal for correct size and type.
4. Lubricate the sealing element and shaft.
5. If needed, install mounting tool on shaft.
6. Push seal, LIP EDGE TOWARD FLUID, up to counterbore.
7. Using a suitable driver, seat the seal. Make certain it is in to the proper depth and is square with the bore.

IMPORTANT!

The seal must be a drive or press fit in the counterbore. A seal that slides in easily will leak.

When the housing has air vents to relieve pressure build up, make sure they are open. If clogged, pressure within the housing will force the lubricant past the best of seals.

If the shaft is installed after the seal, observe the same precautions against seal damage.

Cleanliness here, as in all automotive service, is important.

If a new seal is improperly installed and must be removed, throw it away. Use another new seal.

O-RING SEALS — CONSTRUCTION

O-ring seals are solid, round (doughnut shaped), and are usually made from an elastomer (synthetic rubber or plastic). They are used to create a seal between two parts, close off passageways, prevent the loss or transfer of fluids, and help retard the entry of dust and water. See Fig. 9-28.

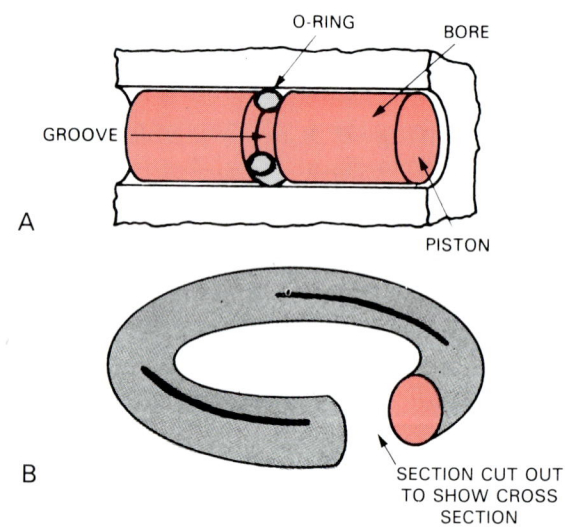

Fig. 9-28. O-rings. A — Note how O-ring is fitted into groove in piston. This allows piston movement while maintaining a seal. B — Typical O-ring construction.

O-RING OPERATION

Because the O-ring is composed of a somewhat soft, pliable material, it seals when slightly squeezed between two surfaces. If the O-ring is also sealing under pressure, the pressure itself will aid in deformation (causing it to deform) of the ring, further making a final seal. They can be used to seal both static (nonmoving) and dynamic (moving) parts. Fig. 9-29, A, shows an O-ring correctly installed in a static application. Fig. 9-29, B, illustrates an O-ring sealing between moving parts.

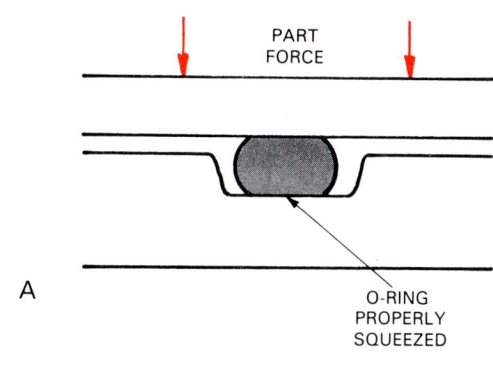

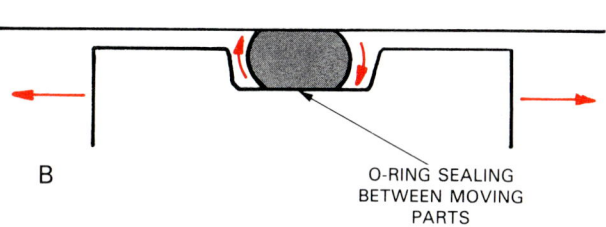

Fig. 9-29. When an O-ring is installed, it is slightly squeezed. Note oblong shape. The O-ring attempts to return to its round shape and thus maintains constant pressure to form a leakproof seal. A — Static seal. B — Dynamic seal.

O-RING INSTALLATION STEPS

1. Make sure the new O-ring is the correct size and that it is compatible with the fluid being sealed.
2. Clean the area where the O-ring is to be installed thoroughly.
3. Inspect the O-ring grooves or notches for burrs and nicks which could damage the new ring. Dress any sharp areas with a fine abrasive stone. Again, thoroughly clean the area to remove all metal and stone particles.
4. Check the shaft or spool (if used) for sharp edges and nicks. Remove all damaged spots with a fine abrasive stone or cloth. Reclean the area thoroughly.
5. Before installation, lubricate the O-ring with the same type of fluid used in the part or system.
6. Install the O-ring. Protect it from sharp edges and other parts. Do not stretch it more than necessary.
7. Be sure the parts are correctly aligned before mating to avoid damage to the O-ring.
8. Make a final check after the O-ring is installed to be sure there are no leaks and that the parts move correctly.

O-RING FAILURE DIAGNOSIS

Improper handling, installation, and applications will reduce O-ring service life. Fig. 9-30 illustrates some common O-ring failures. Be sure to follow manufacturer's recommendations when replacing or working with O-rings. When replacing O-rings that have failed in service, try to diagnose the reason behind the failure. Fig. 9-30 shows several common causes.

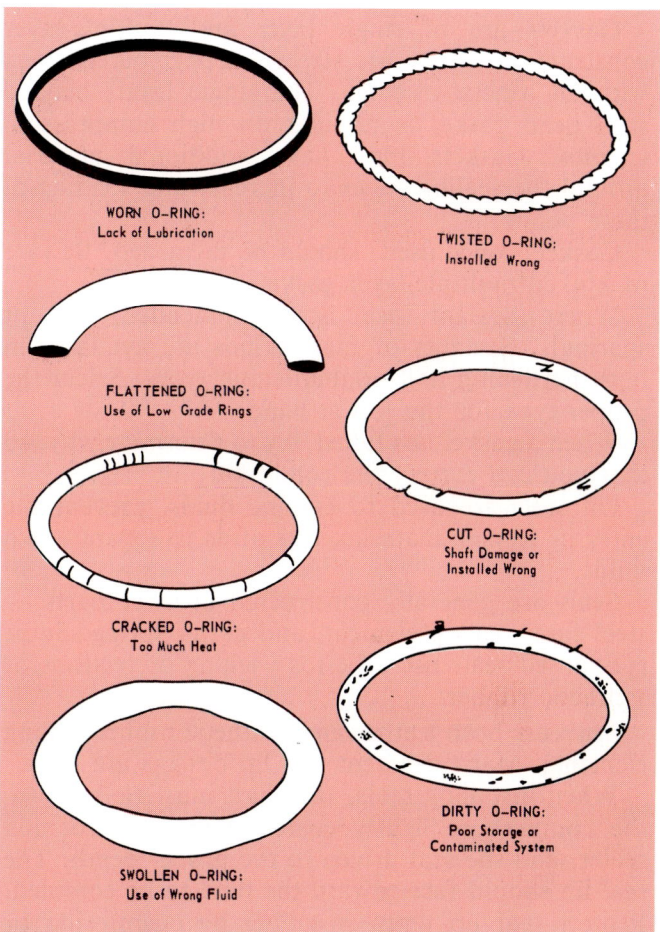

Fig. 9-30. Some typical causes of O-ring failure.

ADHESIVES

The modern automobile uses many types of adhesives to secure a varied array of parts. The adhesive material is generally a liquid or semiliquid substance. It can be spread on (with its own dispenser or other suitable tool), brushed on, or sprayed on.

Most adhesives, when dry, form a hard bond. Others remain somewhat pliable (rubbery). They can be removed with special removers or thinners.

Adhesive uses include: securing weather-stripping, underhood fibreglass pads, body side moulding, and inside car rear view mirror bases (glued to the windshield).

Follow manufacturer's recommendations when using a specific adhesive.

SUMMARY

Gaskets and seals are used throughout the car. Their selection, preparation, and installation is of critical importance.

Gaskets provide leakproof joints. They are made of paper, cork, rubber, asbestos, steel, and copper. Different materials or combinations of materials are needed for specific applications.

Gaskets are of single layer and multiple-layer construction. Many use steel or copper outer layers with an asbestos centre. The single layer, beaded steel head gasket is popular on high compression engines. Gaskets may have additional material around the sealing edges to increase unit loading at these points.

Gaskets, once used, should be discarded. Beware of kinked multiple-layer gaskets.

Where sealant use is recommended, use it sparingly. Sealants of many kinds are available in both hardening and nonhardening types. Select the proper type for the job at hand.

When a gasket has failed, try to determine why, so that you can correct the condition.

Oil seals are used to confine fluids, prevent the entry of foreign material, and often to separate two fluids.

Seals are generally constructed in three parts — steel case, sealing element, and garter spring. Some specialised seals use asbestos wicking or sections of synthetic rubber.

Seals use both leather and synthetic rubber sealing elements. Many different seal lip designs are used.

When installing seals, the shaft must be smooth, the counterbore lightly coated with nonhardening sealer, and the seal driven to the proper depth. The seal lip should face toward the fluid to be confined. Protect seal lip when installing by chamfering or using special mounting tools. Always use a suitable driver. Lubricate seal and shaft before installing the seal. Cleanliness must be observed at all times.

O-rings are solid, pliable round rings. They are used to seal between parts and stop the entry of dust and water. They seal best when slightly compressed. Be sure to carefully follow installation instructions so the O-ring will not be damaged. Use care when handling. Always follow manufacturer's recommendations when working with and replacing an O-ring. Never substitute or reuse O-rings.

The modern vehicle makes use of many different adhesives to secure parts. The adhesive material is usually a liquid or semi-liquid.

Adhesives, when dry, can form a hard or pliable bond, depending upon the chemicals used. Follow manufacturer's recommendations when using or removing adhesives.

SUGGESTED ACTIVITIES

1. Determine how many separate gaskets are used on a 6 cylinder engine. List the materials used in their construction.
2. Make a gasket by placing the gasket material over the part and tapping around the edges and holes.
3. Obtain a head gasket that has BLOWN (failed). Examine it carefully and see if you can determine the cause. List some of the possible causes of head gasket failure.
4. With a torque wrench and following specifications, go over the fasteners on an engine that has been in service for some time. Were they torqued to specifications? If not, what had happened during service? What part could the gaskets have played in this torque change?
5. Check this same car for oil, fuel, and water leaks. Do not overlook the transmission, rear axle, and brake lines. Is the car free of leaks? If leaks are present, what could be the major cause?
6. Inspect some used oil seals and O-ring seals that have failed in service. What shape are they in? What had happened to them? Discount damage that may have been incurred during removal.

SO WHAT'S A LITTLE LEAK?

You might ask why a chapter is devoted to such "trifles" as gaskets and seals. It might seem that they are so simple that a passing mention would be enough. Surely they are not that important and if some part leaks well — so what's a little leak!

The facts are that proper gasket and seal selection and installations are actually VERY IMPORTANT. Every repair job is made up of series of steps or operations — some large and some small. All operations, including the little things, are very important.

Leaks are not only messy and create poor customer relations, they cause part failure and expensive comebacks, plus real damage to the reputation of both garage and mechanic. In fact, even a minor leak may cost someone's life!

Let's take the case of mechanic "X" (unfortunately, there are too many mechanics of this type). Assigned to a brake job, this mechanic had replaced the master cylinder and rear wheel cylinders, repaired the front calipers, machined the rear drums and front discs, installed new brake shoes and pads, and replaced front wheel seals and rear axle seals.

Upon completion, the mechanic bled and adjusted the brakes, checked for fluid leaks, and after road testing declared the job complete. The customer, a sales representative, took delivery.

Several weeks later, the representative was returning home. The mountain road was dark. Its wet surface shimmered in the glare of the headlights. Rounding a curve a rock slide loomed out of the night. The representative did not panic. The car's speed was not excessive, and though it would be touchy, there was time to stop.

Considering the slippery conditions, the driver pressed hard on the brake pedal, but not too hard. The car began to slow, and then it happened. The left rear wheel grabbed, locked up tight, lost traction, and sent the car into a violent slide.

The driver released the brake pedal, spun the wheel, stopped the skid, and reapplied the brakes. Another lockup, another terrifying skid, but now it was too late. The car struck the corner of the rock slide with a sickening thud, bounced high in an arcing skid, and plunged off the highway.

The driver was lucky and lived through the crash. Subsequent study of the accident disclosed that machanic "X" had driven the left rear axle seal in so that it was cocked to one side. The rear axle lubricant had worked through and fouled the brake lining.

Being a mechanic takes intelligence, training, technical knowledge, and attention to details. If you ever hear someone say, "So what's a little leak". — YOU TELL THEM!

KNOW THESE TERMS

Gasket, Mating surface, Chamfering, Sealant, Anti-seize, VRS, Oil seals.

REVIEW QUESTIONS

1. Define the word GASKET.
2. Give two important reasons for installing gaskets.
3. List seven materials that are used in gasket construction.
4. Name four factors that influence the service life of a gasket.
5. Gaskets are of either _____ layer or _____ layer construction.
6. A gasket that must resist great heat will often use _____ in its construction.
7. Define the term UNIT LOADING.
8. The beaded steel head gasket is used on _____ _____ engines.
9. What features in gasket construction provide higher localised unit loading?
10. If you plan a complete engine repair job you would order an _____ set.
11. Old gaskets generally can be reused with success. True or False?

12. Always clean and check both _____ surfaces before installing a gasket.
13. A gasket that has shrunk can often be brought back to size by _____ in _____.
14. A sharp crease in a multiple-layer gasket, if it is straightened out, will not harm the gasket. True or False?
15. Of what value is a gasket sealer?
16. Sealers are of the _____ or _____ type.
17. When applying sealer, always use a liberal amount. True or False?
18. Sealer should ALWAYS BE USED. True or False?
19. Name three ways of holding a gasket in place during part assembly.
20. What effect will incorrect torque and sequence have on the gasket sealing properties?
21. Bent mating surfaces on steel stampings should be _____ before _____.
22. List seven important steps in proper gasket installation.
23. Why should the mechanic try to determine the reason for gasket failure?
24. The typical oil seal is made in _____ parts.
25. These parts are the _____ _____, the _____ _____, and the _____ _____.
26. Leather sealing elements are more widely used than synthetic rubber. True or False?
27. Draw a cross section of a single lip oil seal.
28. All oil seals are of one piece construction. True or False?
29. Describe three methods of removing an oil seal.
30. Place a small quantity of nonhardening sealer on the lips of each seal before installing. True or False?
31. Oil seal lip should face the fluid to be confined. True or False?
32. Describe a suitable oil seal driver.
33. Nicks and scratches on a shaft should be removed by polishing with _____ _____.
34. How are seal lips protected when the seal must slide over a splined, keyed, or drilled shaft?
35. Give seven important steps in proper seal installation.
36. Once a part has been torqued, the pressure will always remain constant. True or False?
37. O-rings are usually made of an _____.
38. O-rings can be used on parts that move or on parts that do not move. True or False?
39. List eight steps that should be followed when installing O-rings.
40. Some adhesives when dry, form a _____ bond, while others remain somewhat _____.
41. List three parts that can be attached to the vehicle with adhesives.

Chapter 10

ANTIFRICTION BEARINGS

After studying this chapter, you will be able to:

☐ List the different kinds of antifriction bearings.
☐ Explain the advantages of each type of antifriction bearing.
☐ Describe service procedures for antifriction bearings.

CONSTRUCTION

The antifriction type bearing utilises rolling elements (ball or rollers) to reduce friction through rolling contact. In most applications, the rollers or balls are placed between inner and outer rings. The rolling elements are separated by a cage or separator, generally made of stamped steel. The cage prevents the elements from bunching and sliding against each other. In the case of separable (can be taken apart) bearings, the cage prevents the loss of the elements.

The balls or rollers, as well as the inner and outer rings, are hardened and ground to assure proper contact and clearance.

Needle bearings (long, thin rollers) often use only an outer shell. In some needle roller applications, the bore and shaft are hardened then ground and placed in direct contact with the rollers.

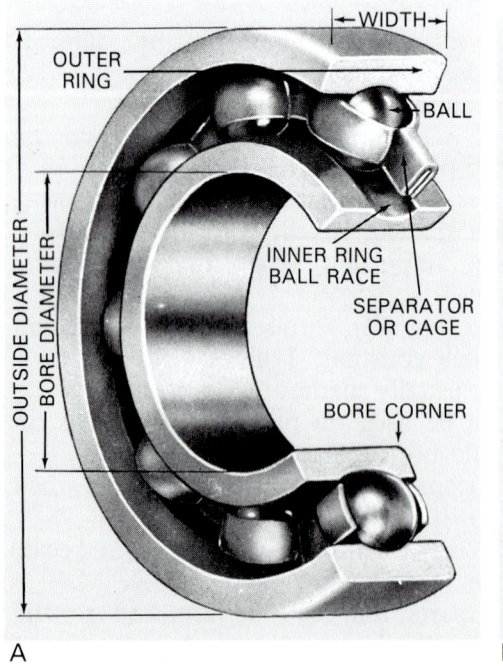
A

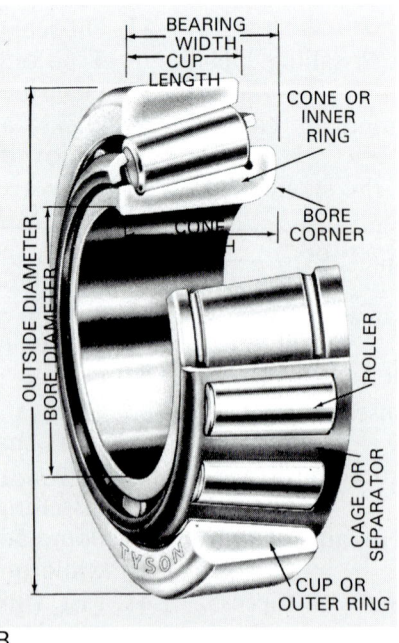
B

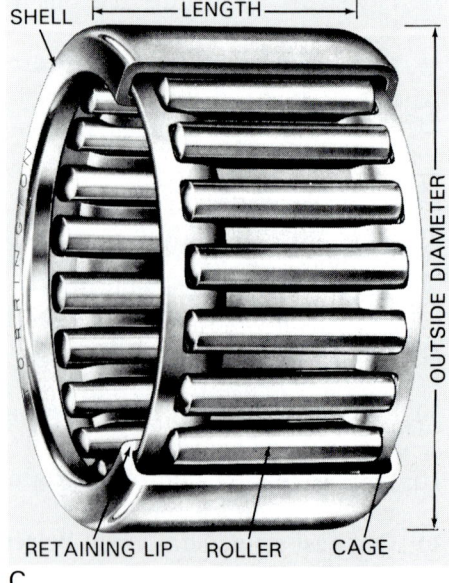
C

Fig. 10-1. Three types of antifriction bearings. A — Typical ball bearing construction. Note how cage keeps balls evenly spaced. (Nice) B — Roller bearing. This bearing uses tapered roller design. Outer ring is separate. (SKF) C — Caged needle bearing. Rollers in this bearing operate against outer shell and in direct contact with hardened, ground shaft surface. (Torrington)

THREE BASIC TYPES

Bearings are commonly divided into three types: BALL, ROLLER, and NEEDLE. Each type has certain applications it serves best. The ball bearing produces the least amount of friction, but for a given size, does not have the load carrying ability of the roller. All three types are used in automotive construction. Fig. 10-1 illustrates the three types. Learn the names of the parts.

LOADING DESIGN

Bearings are designed to handle RADIAL, THRUST, or a combination of both radial and thrust loads. Radial designs handle loads at right angles to the axis of the bearing. Thrust designs handle loads parallel to the axis. Combination designs handle loads from any direction. Fig. 10-2 shows the loading designs.

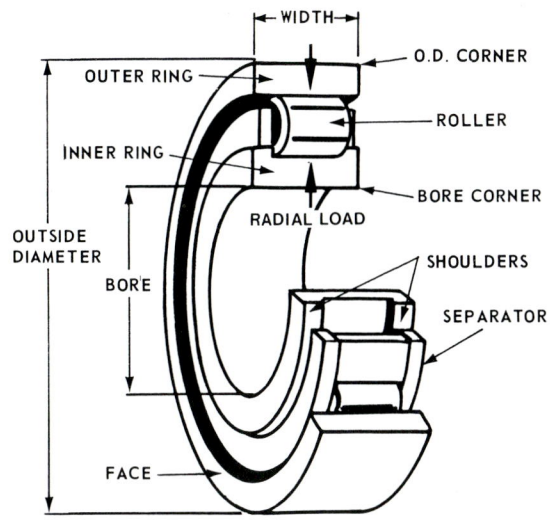

Fig. 10-3. Straight roller bearing designed for radial load only. (AFBMA)

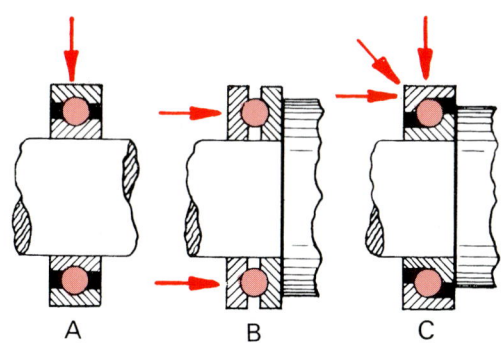

Fig. 10-2. Loading designs. A — Radial. B — Thrust. C — Combination radial and thrust. Arrows in colour indicate direction of load.

VARIATIONS

There are many variations of the three basic types. Each different design attempts to meet a specific demand. The installation may call for light or heavy loads, high or low speeds, radial, thrust, or a combination loading. By understanding the problems involved and the type of bearing best suited, the mechanic will be greatly aided in all bearing work.

Some of the more common variations are the straight roller, spherical roller, tapered roller, deep groove ball, angular contact ball, multiple row, and self-aligning.

STRAIGHT ROLLER

The straight roller is designed to handle heavy RADIAL loads. In most designs, it will handle little or no thrust, Fig. 10-3.

SPHERICAL ROLLER

The rollers in this bearing are of curved or spherical shape. It will handle HEAVY radial loads and MODERATE thrust loads. It is self-aligning to a degree, Fig. 10-4.

Fig. 10-4. Spherical roller bearing. Note "barrel" shape of rollers. (SKF)

TAPERED ROLLER

The tapered roller is the most widely used of the rolling bearings as it will carry both HEAVY thrust and radial loads. The apex of the angles formed by both the rollers and raceways, if extended, would meet on a common axis. This allows the roller to follow the tapered raceways with no bind or skidding. Common practice is to secure the rollers to the cone with a steel cage. The cone raceway is indented to form a lip that keeps the rollers centred. The cup is then separable. Look at Figs. 10-1, B, and 10-5.

DEEP GROOVE BALL

The deep groove ball bearing will handle HEAVY radial and MODERATE thrust loads. Neither the inner or outer ring is separable, Fig. 10-6.

ANGULAR CONTACT BALL

This ball bearing will handle both HEAVY thrust and radial loads. The balls are contained within a cage. Both inner and outer rings are separable, Fig. 10-7.

MULTIPLE ROW BEARING

Bearings can employ two or more rows of balls or rollers so that heavier loads, both radial and thrust, can be carried. They can also be designed to provide for thrust loads in BOTH directions, Fig. 10-8.

THRUST DIRECTION

You will note that several of the bearings shown will sustain thrust in ONE direction only. Thrust in the opposite direction would force the rings apart. By using two or more bearings, facing in opposite directions, thrust in either direction can be handled. See Fig. 10-9.

THRUST BEARING

The bearings shown in Fig. 10-10 are designed to handle THRUST forces only.

SELF-ALIGNING BEARINGS

When there is a possibility, or in some instances, a desirability, of permitting either housing or shaft misalignment during operation, a self-aligning bearing is used. This bearing will allow a degree of tilt without distorting the bearing elements. Both internal and external self-aligning bearings are shown in Fig. 10-11.

BEARING IDENTIFICATION

Most bearings are marked with a part number for ease of replacement. The number is usually on the face of the rings. If necessary, replacement bearing size can be checked by careful measurement.

BEARING SEALS

Bearings can be open on both sides or sealed on one or both sides. Sealing on one side is often used to help confine lubricant and to prevent the entry of dirt. When both sides are sealed, the bearing is lubricated during assembly and no lubricant can be added in the field. See Fig. 10-12.

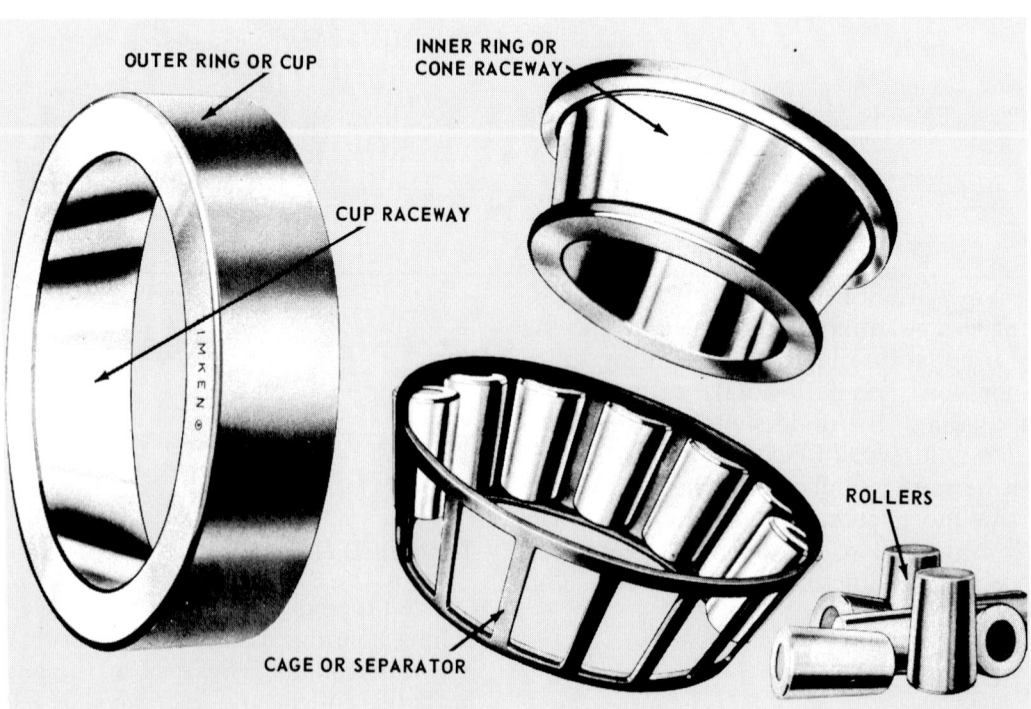

Fig. 10-5. Tapered roller bearing parts. Once assembled, this particular bearing will have a separable outer ring but rollers, cage, and inner ring will be one unit.

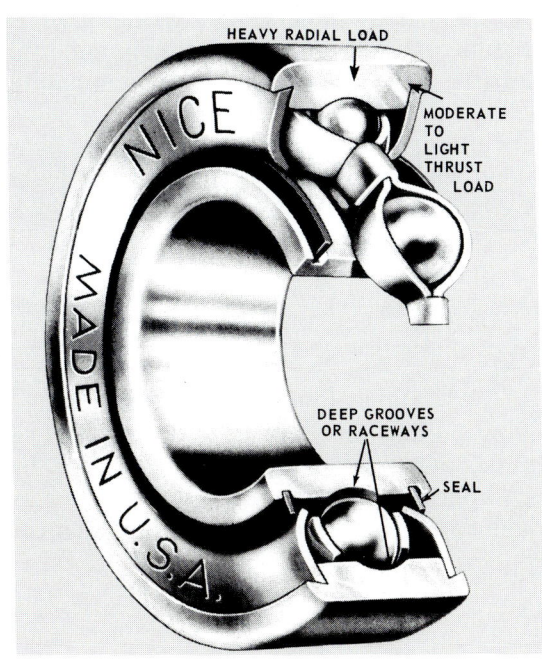

Fig. 10-6. Deep groove ball bearing. Note the use of seals on both sides.

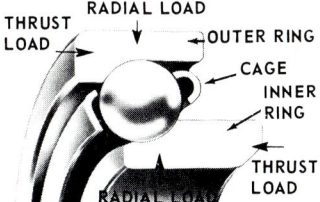

Fig. 10-7. Angular contact ball bearing. This type is often used as a car front wheel bearings.

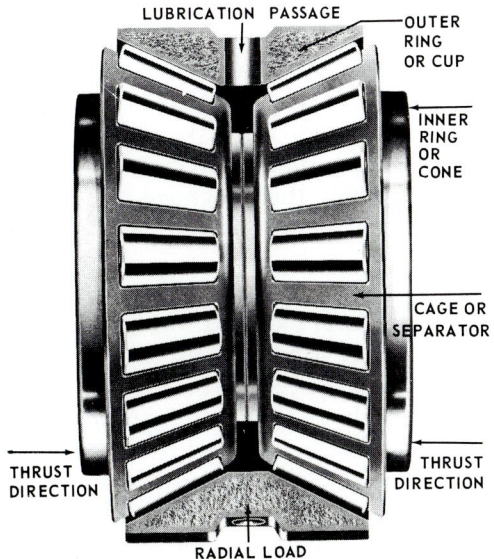

Fig. 10-8. Double row, tapered roller bearing. The outer ring is one piece, the inner rings are separate. (Timken)

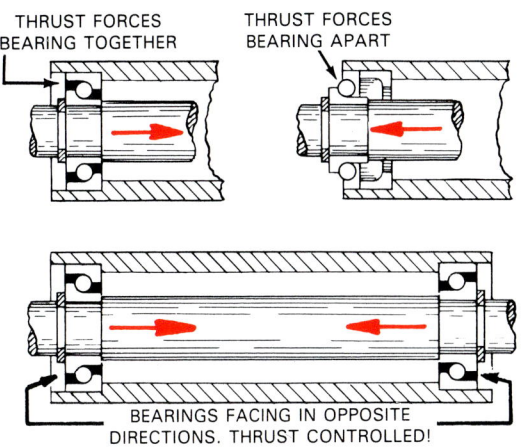

Fig. 10-9. By using two bearings, thrust in either direction is controlled. Arrows indicate thrust direction.

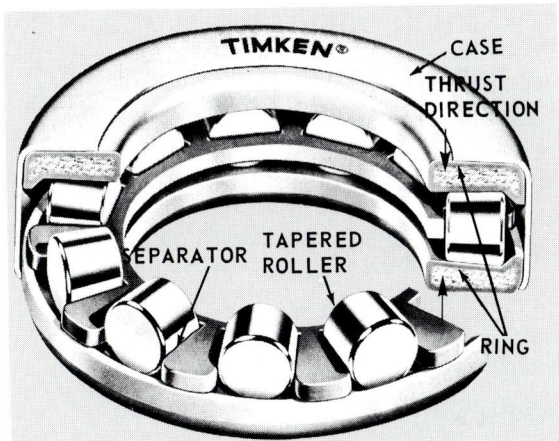

A

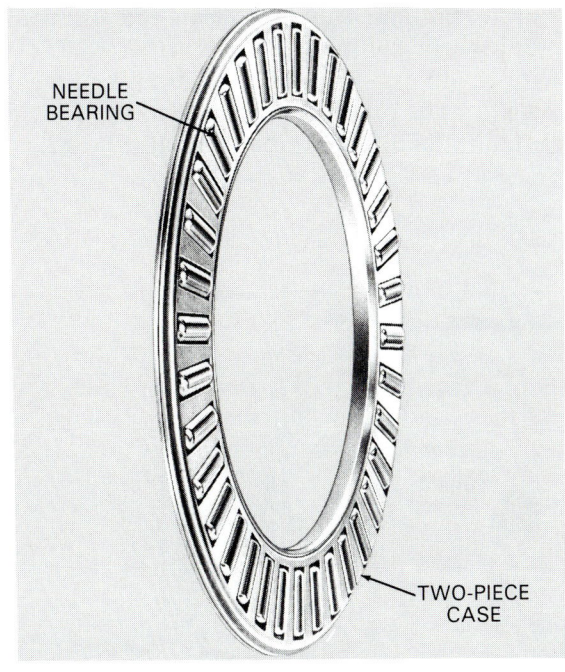

B

Fig. 10-10. A — Tapered roller thrust bearing. B — Needle thrust bearing. Note how two-piece case also acts as needle separator. (Timken, Torrington)

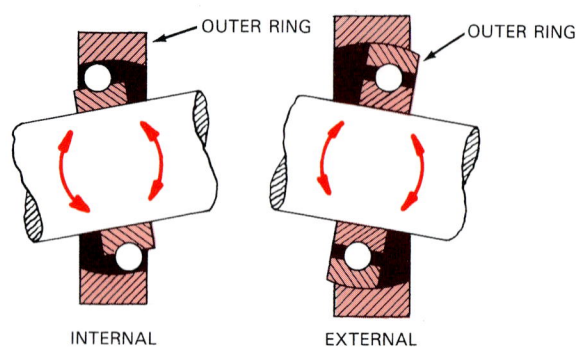

Fig. 10-11. Internal and external self-aligning bearings. Note how shaft is free to tip. External design will handle heavier loads as ball has a wider contact area with outer ring.

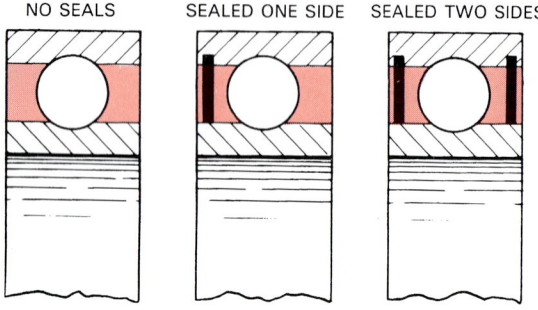

Fig. 10-12. Bearing seal construction.

REMOVING BEARINGS

Prior to pulling bearings, clean the surrounding area to prevent contamination.

Bearings are generally best removed with mechanical or hydraulic pushing or pulling tools. They exert a heavy and STEADY force, Fig. 10-13.

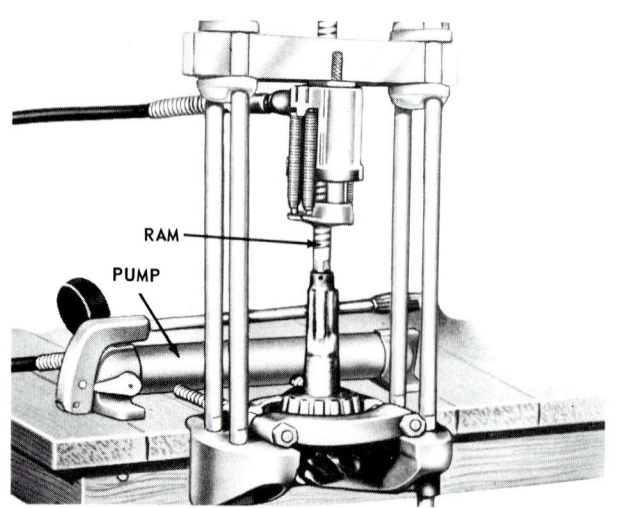

Fig. 10-13. Removing differential pinion shaft bearing with hydraulic puller.

Without such pullers, or where their use is impossible or undesired, a suitable hammer and soft steel drifts, sleeves, and cup drivers will handle many jobs.

Any attempt to pull or install a bearing by exerting force on the free (not tight) ring is apt to chip the balls or rollers. The ring itself could crack and fly apart. There are some instances, as you will see later, that require force on either the free ring or rolling elements. However, WHENEVER POSSIBLE, EXERT THE FORCE ON THE TIGHT RING ONLY.

Fig. 10-14 shows both the right and wrong way of applying pulling force. Note that in A, the supporting puller plate rests on the free outer ring. In B, the plate supports the inner ring only, thus avoiding damage to the outer ring and rolling elements.

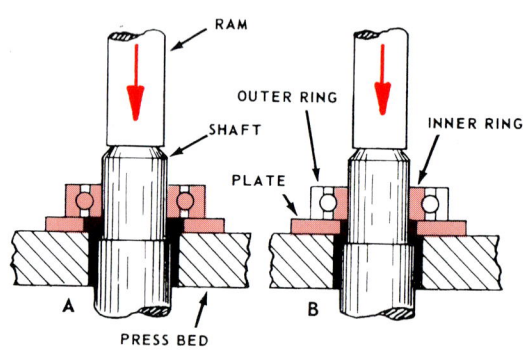

Fig. 10-14. Pulling setups. A — Wrong as force is applied through free outer ring and rolling elements. B — Correct. Force is through tight ring only.

WHEN INNER RING CANNOT BE GRASPED

Occasionally, the bearing inner ring is pressed against a shoulder that is as wide or wider than the ring. In the case of the tapered roller bearing, a special segmented (made in parts) adapter ring can be used. It applies the pulling force to the ends of the rollers while forcing them against the cone. This allows the bearing to be removed without damage, Fig. 10-15.

Another type of puller, especially adapted for axle shaft bearing work, is pictured in Fig. 10-16. A split sleeve, with pulling rings, is used. The axle shaft passes up through a section of tubing. The puller sleeve grasps both bearing and tubing. The top section of the tubing is fastened to a heavy plate on the bed of the press. As pressure is applied to the shaft end, it is forced through the tube to pull the bearing. Note that the entire bearing is shrouded or shielded to protect the operator from flying parts if the bearing should explode. This puller will remove both tapered roller and ball bearings, Fig. 10-16.

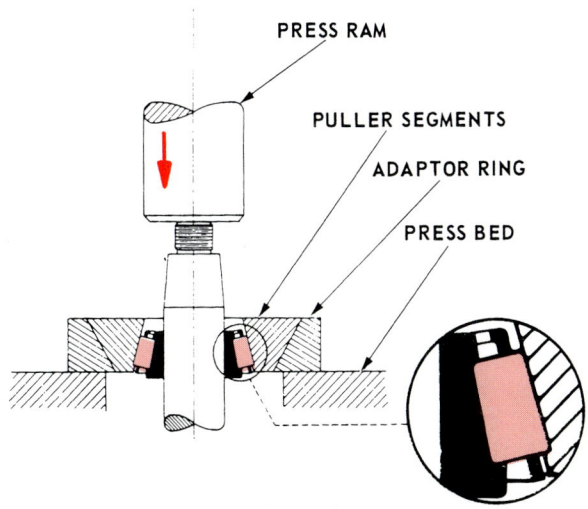

Fig. 10-15. Pulling bearing by applying pressure through rollers. Magnified portion at lower right shows how end of roller is grasped by puller segments. (Timken)

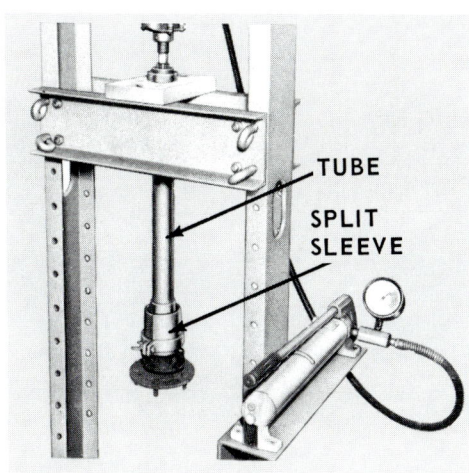

Fig. 10-16. Removing axle shaft bearing with special puller.

WHEN BEARING CANNOT BE GRASPED

There are instances in which a retaining plate or dust shield is so close to, or surrounding, the bearing that it is impossible to grasp it. In these cases, you must grind away a portion of the inner ring (protect shaft with a metal sleeve). Cut out the cage and remove the elements. The outer ring can then be removed, exposing the inner ring for grasping.

Unhardened retaining rings are sometimes used to hold bearings in place. They are best removed by notching with a sharp chisel. This will loosen them enough for easy removal. See Fig. 10-17.

Inner bearing rings can also be removed by partial grinding or by cutting part way through with an acetylene cutting torch. WRAP THE SHAFT, ON

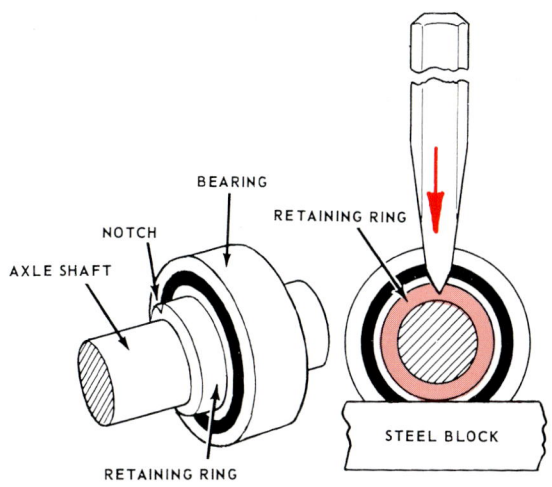

Fig. 10-17. Removing bearing retaining ring by notching with a chisel.

BOTH SIDES OF THE BEARING, WITH WET CLOTHS TO PREVENT HEATING. CUT ONLY PARTWAY THROUGH. The ring is then squeezed tightly in a vice and struck with a hammer where indicated by the arrow in Fig. 10-18. This will crack the ring and allow it to be pulled. WEAR SAFETY GOGGLES WHEN STRIKING BEARING PARTS.

Always pull bearings whenever possible. AVOID GRINDING AND CUTTING WITH A TORCH, UNLESS ABSOLUTELY NECESSARY.

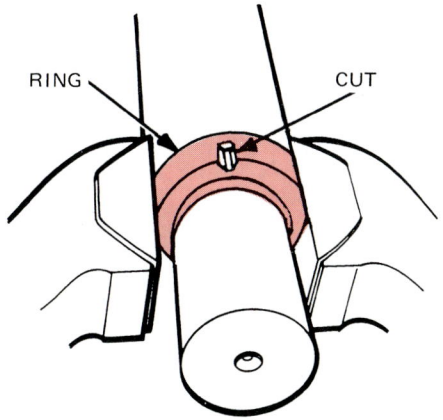

Fig. 10-18. Bearing inner ring partially cut and then squeezed in a vice. Strike with a hammer where indicated by arrow.

KEEP BEARING PARTS TOGETHER

When a separable bearing is removed, keep the parts together. Under no circumstances should bearing elements be mixed.

GENERAL RULES FOR BEARING REMOVAL

1. Exert force, where possible, on the tight ring.
2. Use pullers of the correct size and shape.

3. Mount puller to exert force in a line parallel to the bearing axis.
4. Use unhardened, mild steel drifts and sleeves.
5. Never strike the outer or free ring.
6. Use care to avoid damage to the shaft or housing.
7. If necessary to hammer a shaft, use a brass, lead, or plastic hammer.
8. Keep all parts of one bearing together.

WATCH OUT!

PULLING BEARINGS, BOTH WITH PRESSURE OR STRIKING TOOLS, CAN BE A DANGEROUS OPERATION. BEARINGS UNDER SUCH PRESSURE CAN SHATTER AND SEND PIECES FLYING OUTWARD WITH LETHAL FORCE. WHENEVER POSSIBLE, SHIELD THE BEARING. WEAR SAFETY GOGGLES. KEEP OTHER PERSONNEL AWAY FROM WORK AREA.

CLEANING BEARINGS

When the bearing is removed, wipe off all surplus grease or oil. Soak in kerosene or solvent. A regular cleaning tank with tray and solvent hose is ideal. If none is available, a clean bucket will suffice, Fig. 10-19.

Fig. 10-19. Tray full of bearings being placed in kerosene.

CAUTION!

NEVER USE PETROL OR OTHER VOLATILE FLUIDS FOR CLEANING. THEY ARE ROUGH ON HANDS AND WILL IGNITE READILY. DO NOT USE CARBON TETRACHLORIDE AS IT PRODUCES POISONOUS FUMES.

While some bearings are soaking, brush each in turn with a nylon bristle brush and blow out the worst of the grease. Continue soaking and brushing until the bearing looks clean. Blow the bearing out again. If any sign of grease is visible, soak, brush, and blow out once more.

DO NOT SPIN!

NEVER SPIN A BEARING WITH AIR PRESSURE. NOT ONLY WILL IT DAMAGE THE BEARINGS, IT CAN ALSO BE DANGEROUS. WHEN THE OUTER RING OF A SEPARABLE BEARING IS REMOVED THE ROLLING ELEMENTS ARE HELD TO THE CENTRE RING WITH THE CAGE. IF THE CAGE AND ROLLERS ARE SPUN, THE TREMENDOUS CENTRIFUGAL FORCE GENERATED CAN CAUSE ONE OR MORE ELEMENTS TO FLY OUTWARD WITH VIOLENT FORCE.

When certain the bearing is CLEAN, rinse in a container of CLEAN kerosene and blow dry. Look at Fig. 10-20.

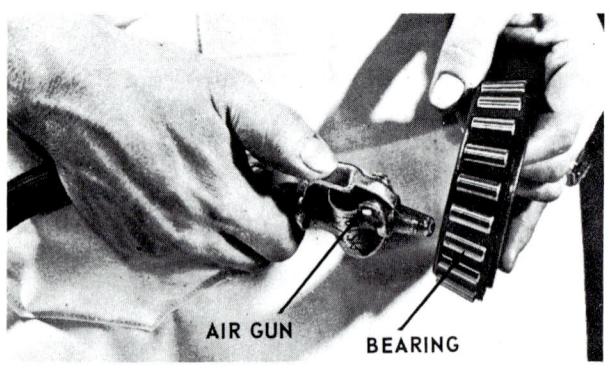

Fig. 10-20. Using clean, dry air, blow bearing dry. Do not allow bearing to spin. (Timken)

USE CLEAN, DRY AIR

Most air compressor systems are equipped with a filter and moisture trap. Service them often. Directing a stream of air into a white cloth will show if dirt or oil is present.

DO NOT WASH SEALED BEARINGS

When a bearing is factory packed and completely sealed on both sides, it must not be washed. Wipe off the outside with a clean, dry cloth. Washing will dilute the lubricant and lead to early failure.

CLEAN WORK AREA IS A MUST

Once the bearings are cleaned and dried, take them to a CLEAN work area. It is a good idea to reserve

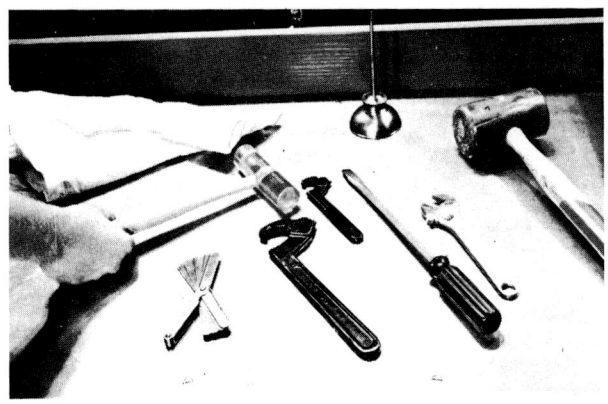

Fig. 10-21. Ideal bearing work area. (SKF)

a section where this assembly area will be free of dusty air, grinding machines, and steam cleaning moisture. Fig. 10-21 pictures an ideal work section.

BEARING DEFECTS

Prior to discussing checking procedures, it is wise to familiarise yourself with some of the most common bearing defects. Refer to Fig. 10-39.

As is the case with friction bearings, DIRT is the number one enemy of ball and roller bearings. It will cause scratching, pitting, and rapid wear. Other common defects include spalling, brinelling, overheating, cracked rings, broken cages, damaged seals, and corroded areas.

SPALLING

Foreign particles, overloading, and normal wear over an extended period can lead to spalling. Spalling starts when tiny areas fracture and flake off. These small flakes are carried around in the bearing causing more flaking. Advanced flaking or spalling will produce large craters, Fig. 10-22.

Fig. 10-22. Badly spalled inner ring. (AFBMA)

BRINELLING

Brinelling is the term used to describe a series of dents or grooves worn in one or both rings. The grooves run across the raceway and are usually spaced at regular intervals. Once brinelling starts (often from inadequate lubrication) a fine reddish iron oxide powder is formed. As the powder is carried around, it increases the wear rate. Fig. 10-23 shows a badly brinelled outer shell.

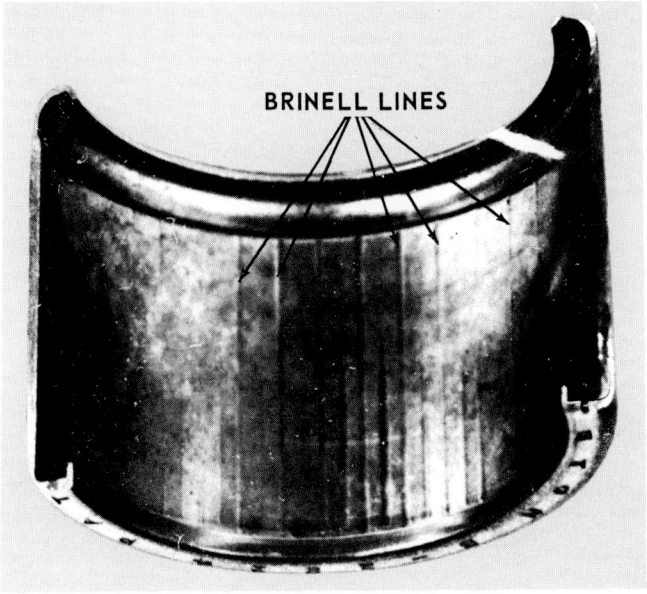

Fig. 10-23. Brinelled needle bearing shell.

OVERHEATING

Overheating will break down the physical properties of the bearing and cause rapid failure. Lack of lubrication, improper lubrication, and poor adjustment are the principal causes. The bearing rings and rolling elements which have been overheated will have a blue or brownish-blue discolouration. See Fig. 10-24.

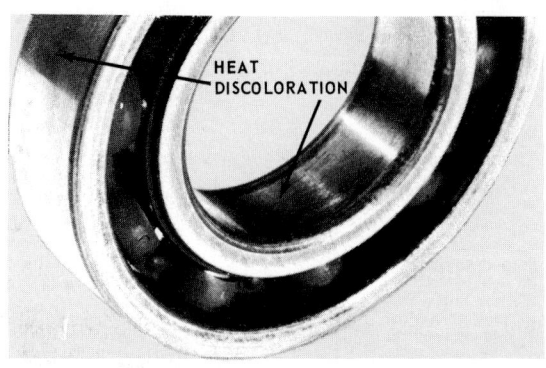

Fig. 10-24. Overheated bearing. Note discolouration.

Fig. 10-25. Cracked inner ring.

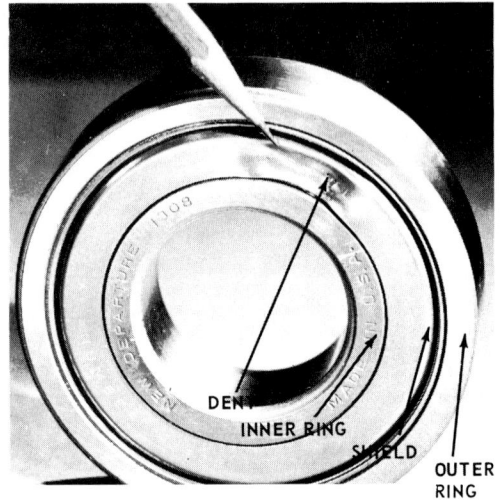

Fig. 10-27. Badly dented bearing shield or seal.

CRACKED RINGS

One or both rings may be cracked. Improper removal or assembly techniques and wrong bore or shaft size are common causes. Look at Fig. 10-25.

BROKEN OR DENTED CAGE

Improper removal and assembly procedures will often result in a dented or broken cage. Pieces of dirt and metal chips will also cause cage breakage. Refer to Fig. 10-26.

Fig. 10-26. Broken cage.

DENTED SHIELDS

As with a broken cage, careless assembly often produces dented shields. This could also damage the cage as well as cause binding and lubricant loss. See Fig. 10-27.

CORROSION

The entry of moisture (often from air hose), wrong or contaminated lubricant, or storage near corrosive vapours can produce corrosion in the bearing. A bearing remaining static (not being rotated) for an extended time, often corrodes, Fig. 10-28.

DIRT WEAR

If the dirt is very fine, it will have a lapping effect (removal of surface metal through fine abrasive

Fig. 10-28. Corroded bearing.

action) that will leave the rolling elements and raceways with a dull, matte (nonreflecting) finish. Larger dirt particles will produce scratches and pits.

ELECTRICAL PITTING

Electric motor or generator bearings are sometimes pitted by the passage of current (from an internal short or from static electricity) through the bearing. The minute arcing produces numerous tiny pits. Fig. 10-29 illustrates the effect of electrical pitting, dirt, corrosion, and poor lubrication on rollers.

Fig. 10-29. Roller damage. A — Corrosion. B — Electrical pitting. C — Poor lubrication and dirt. (SKF)

SOME LOOSENESS IS NORMAL

A new bearing often feels rather loose so do not assume looseness as a sign of wear. When either raceways or rolling elements are worn enough to produce looseness, it will be evident by examining the surfaces. One or more of the conditions mentioned above will be visible.

BEARING INSPECTION

When inspecting nonseparable bearings, place the fingers of one hand through the centre ring, Fig. 10-30. Rotate the outer ring with the other. The bearing should revolve smoothly with no catching or roughness. If either condition is present, rinse and blow dry again. If the symptoms still persist, discard the bearing. Also check for signs of overheating and wear on the outer surfaces of both rings. A bearing that has been loose in the bore, or on the shaft, will have highly polished areas showing.

For separable bearings, carefully inspect the raceways and rolling elements. They should be absolutely smooth and free of heat discolouration. Inspect EACH ball or roller. Only one or two may be damaged. When satisfied as to condition, place the elements together. While forcing them together, rotate the bearing. The operation should be smooth.

When revolving bearings, do so a number of times. A single damaged ball or roller may not "catch" the first few times around. When checking thrust bearings, place one side on a solid surface. Press down on the other with the heel of your hand. While maintaining pressure, rotate. KEEP HANDS CLEAN, DRY, AND AWAY FROM RACEWAYS AND ROLLING ELEMENTS. See Fig. 10-30.

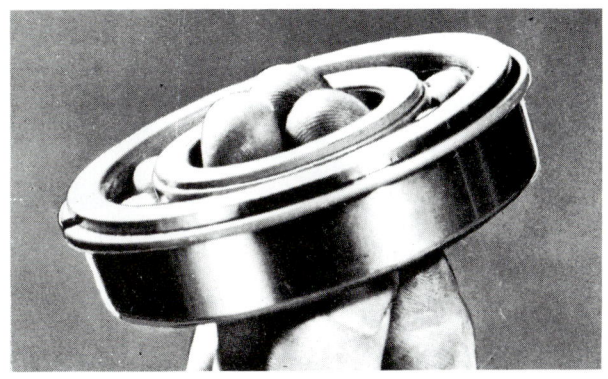

Fig. 10-30. Holding bearing for inspection.

DO NOT SAVE ONE PART

If any part, outer or inner ring, or rolling elements are damaged, discard the ENTIRE bearing. Never replace a part of a bearing.

Before discarding, write down the part number. It is a good idea to wire the parts together and keep for comparison with the replacement bearing. Mark as DEFECTIVE.

BEARING LUBRICATION

If the bearing is to be placed into service at once, it may be packed with the proper grease or it may be oiled, depending upon the need. Cover with a clean cloth until ready to install. If it is to be stored for a few days, coat with oil and place in a clean box or container. IMMEDIATELY FOLLOWING INSPECTION, COAT WITH THE DESIRED LUBRICANT TO PREVENT THE FORMATION OF RUST. Look at Fig. 10-31.

Fig. 10-31. Bearings cleaned, oiled, and placed in protective container.

If the bearing is to be stored for an extended period, coat with light grease. Wrap in oilproof paper and place in a clean box. Be sure to identify the bearings to prevent opening a number of them when looking for a specific one at some future date. When coating bearings for storage, rotate to ensure proper penetration and coverage. Fig. 10-32.

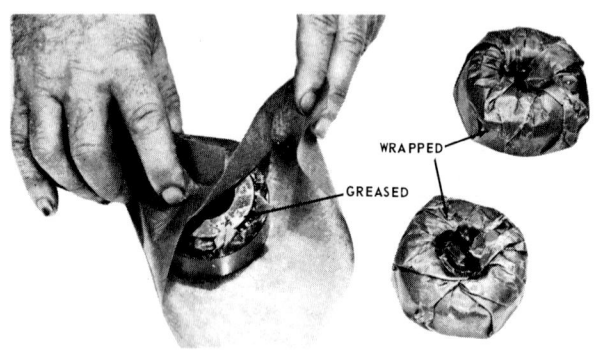

Fig. 10-32. Bearings greased and wrapped for extended storage.

PACKING WITH GREASE

When a bearing calls for grease (use the specific grease recommended by the vehicle manufacturer), use a bearing packer. If no packer is available place a quantity to grease on the palm of one hand. Your hands must be clean and dry. With the other, press the edge of the bearing into the grease (rear edge). Repeat this until grease flows out of the top. Move around to different sections until the bearing is fully packed. Separable rings should be coated also. See Fig. 10-33.

PROTECT LUBRICANTS

All grease and oil in the shop should be kept in clean containers and kept tightly covered when not in use. When opening, wipe dirt off lid and avoid dusty areas. An open can of grease near a grinder or cutting torch is an open invitation to disaster.

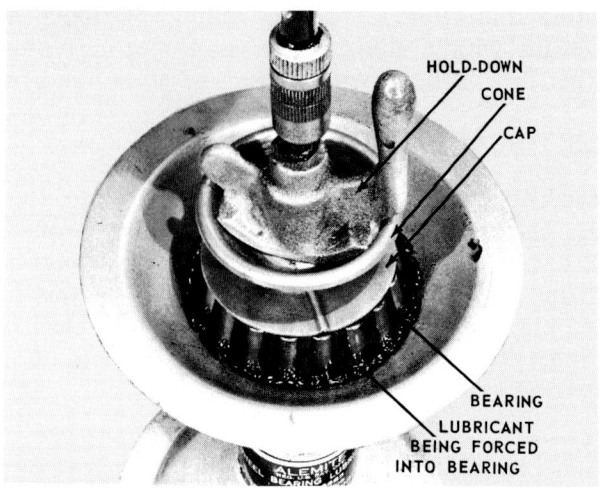

Fig. 10-33. A bearing packer is fast and efficient. (Timken)

CHECK SEALS

If any oil or grease seals are related to the job at hand, inspect them. If necessary, replace the seals at this time. In some instances, seals must be installed after the bearings.

BEARING INSTALLATION

Bearing installation calls for care and intelligent use of tools. Many otherwise good jobs have been ruined by careless installation.

MAKE CERTAIN YOU HAVE THE CORRECT BEARING

Bearings are often similar (but not exact) in type and size. Before attempting installation, make certain you are installing the correct one. Be especially careful with new replacement bearings. Check numbers and measurements.

CLEAN BORES AND SHAFTS

Clean bearing housing bores and shafts thoroughly. Remove any nicks or burrs with a fine file. Be careful not to file a flat spot. Following filing, polish with very fine emery or crocus cloth. On a shaft where the inner ring is designed to walk (creeping movement around shaft), inspect carefully. Polish if necessary. If the counterbores or press-fit shaft areas are worn from ring centre slippage, do not punch or knurl (crosshatch pattern pressed into metal) as an attempt to increase size. Such procedures will only result in failure as the bearing, under load, will quickly flatten these raised areas. The area should be built up by metal spraying (spraying molten metal onto shaft) and then ground to the correct size. Watch for dirt in threads, splines, and other areas. Look at Fig. 10-34.

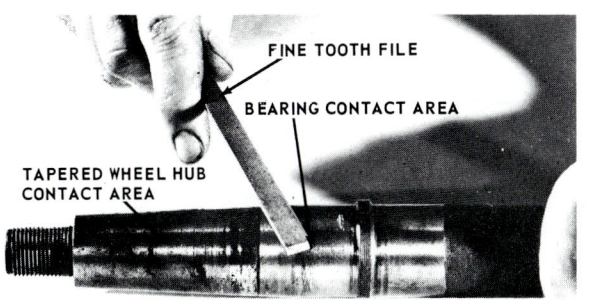

Fig. 10-34. Removing burrs from axle shaft bearing area with a fine tooth file.

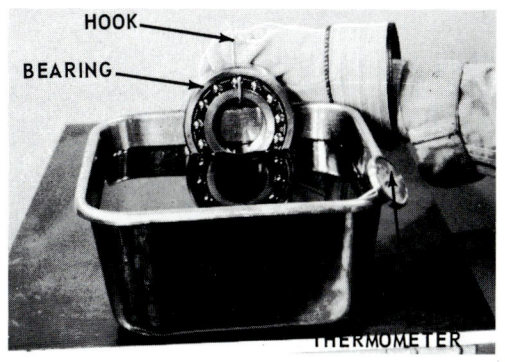

Fig. 10-36. Heating a bearing in oil. Hook keeps bearing from touching bottom of container.

USE LUBRICANT TO EASE ASSEMBLY

The use of a thin film of oil or finely powdered graphite will ease installation, prevent corrosion around ring contact area, and facilitate removal at some future date. See Fig. 10-35.

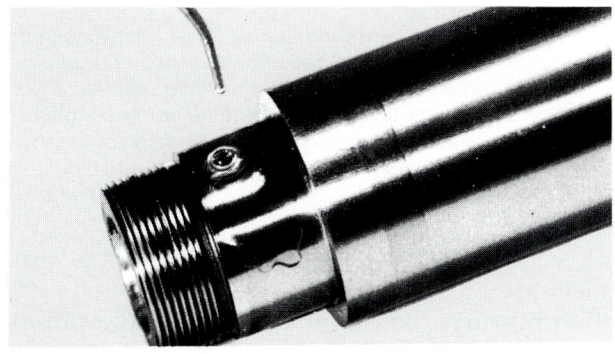

Fig. 10-35. Use lubricant to facilitate assembly.

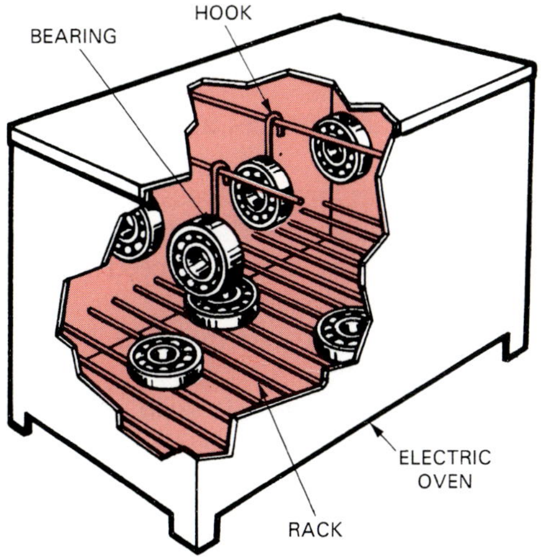

HEAT AND COLD HELPS

In difficult assembly jobs, primarily large bearings, place the outer ring in dry ice or in a deep freeze. This will reduce the diameter and help installation. Inner rings can be heated (NEVER ABOVE 135° C) in clean oil or a special electric oven. Use a thermometer. Never heat bearings with a torch. Follow manufacturer's instructions. See Figs. 10-36 and 10-37.

POSITION PROPERLY AND START SQUARELY

After determining correct installation position (do not press on backwards or fail to postition any retainers or snap rings that must go on first), start the bearing or ring with your fingers. Attach puller or set

Fig. 10-37. One type of electric bearing oven. Note that bearings are placed on racks or suspended with hooks. Do not put them in direct contact with heat source. Follow manufacturer's instructions when operating.

up in press and force bearing into place. MAKE CERTAIN IT GOES ON SQUARELY AND TO THE FULL DISTANCE REQUIRED. Apply pressure, whenever possible, on the tight ring. As in pulling, observe safety precautions.

SIMPLE TOOLS WILL OFTEN SUFFICE

If regular pressing tools are not available, simple driving tools will handle many jobs in a satisfactory manner. Make sure they are clean. Strike the tight ring only. Use soft steel tools. Brass tools tend to mushroom and chip, thus contaminating the bearings.

SHAFT AND HOUSING BORES MUST BE TRUE

A sprung shaft or bent housing will cause the bearing to operate in a distorted position, greatly shortening its life. For those jobs in which the bearing failed in a short time, despite proper installation, lubrication, and adjustment, always check shaft and housing for any warpage or other misalignment.

BEARING ADJUSTMENT

Some bearings require adjustment after installation. Proper adjustment depends on the application. Some require a specific amount of free play. Others require preloading (placing bearing under pressure so that when a driving force is applied to parts, they will not spring out of alignment). As the various service operations are described through the book, general adjustment recommendations will be given.

GENERAL RULES FOR BEARING INSTALLATION

1. Clean all contact surfaces and remove burrs and nicks.
2. Install parts that precede bearing.
3. Lubricate for easy installation.
4. If heat is required, do not exceed 135°C.
5. Start bearing squarely.
6. Align tools so that bearing will be forced on squarely.
7. For driving tools, use soft steel.
8. When possible, avoid applying pressure through balls or rollers.
9. If a vice is needed, use protective jaw covers.
10. Driving tools must have smooth, square cut ends.
11. Do not damage shaft or bore surfaces.
12. Use safety precautions.
13. Press on the full distance required.

Fig. 10-38 illustrates a few do's and dont's regarding bearing assembly.

SUMMARY

Bearings can be divided into three basic types: the BALL, ROLLER, and NEEDLE. The ball and roller bearing usually consists of an inner and outer ring with the rolling elements placed between them and positioned with a cage or separator. The needle bearing can use an outer shell, or can be placed in direct contact with a hardened and ground bore and shaft.

Bearings are designed to carry either straight thrust, radial, or combination loads.

The straight, spherical, and tapered roller, deep groove ball, angular contact ball, and self-aligning of both types are the common variations.

Bearings are normally marked with a part number.

Bearings are often sealed on one or both sides. Never wash bearings sealed on both sides.

Hydraulic and mechanical pulling or striking tools can be used to remove bearings. All must be used

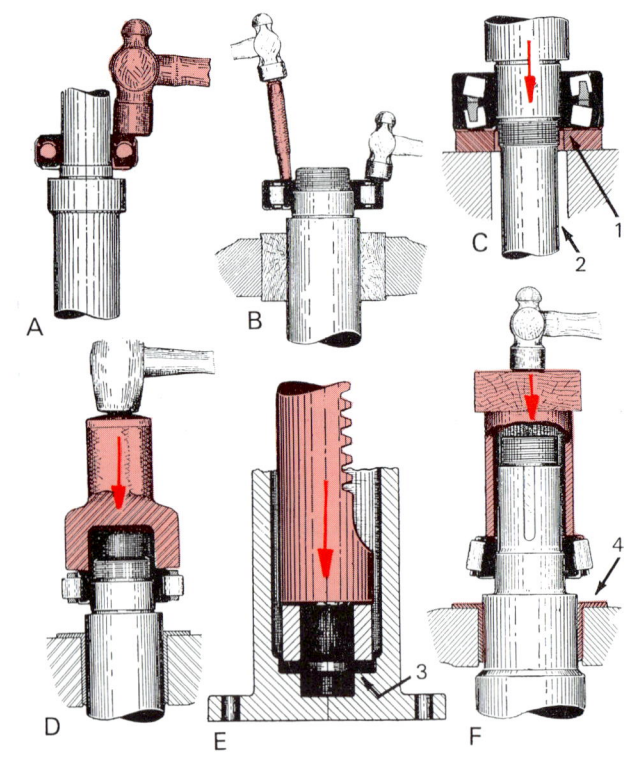

Fig. 10-38. Bearing installation hints. A — Do not strike bearing with a hammer. B — Do not use wide punches on bearings. C — Do apply force to tight ring (1) and have clearance (2) for shaft. D — Use driver with smooth, square cut ends that strike tight ring. E - Clean bearing ring recess (3) and force ring to full depth. F — Block placed on open pipe driver allows driving force to be centralised. Use protective vice jaw covers (4).

with care. If available, hydraulic and mechanical pullers are recommended.

Pull bearings, whenever possible, by the ring that is tight. Special tools are available for pulling by exerting pressure through the balls or rollers. Avoid the use of heat. Do not damage bore or shaft surfaces. When bearing is removed, if separable, keep all parts together.

Clean bearings in kerosene. Blow dry. Rinse in fresh kerosene and blow dry again. Do not spin the bearing. Air should be clean and dry. Working in a clean area, inspect bearing, Fig. 10-39. If satisfactory, oil or pack with grease at once. Keep covered until ready to install. Rejected bearings may be kept for size comparison with replacements, but MARK them as REJECTS.

Scratched, pitted, spalled, brinelled, corroded, cracked, and overheated bearings, plus those with damaged cages or dented shields MUST be RE-JECTED. Never replace one part of a bearing.

Keep all lubricants covered when not in use. A bearing packer is handy for packing bearings with grease.

Replace defective grease or oil seals. Clean bore and shaft. Remove burrs and install any parts that must precede bearing.

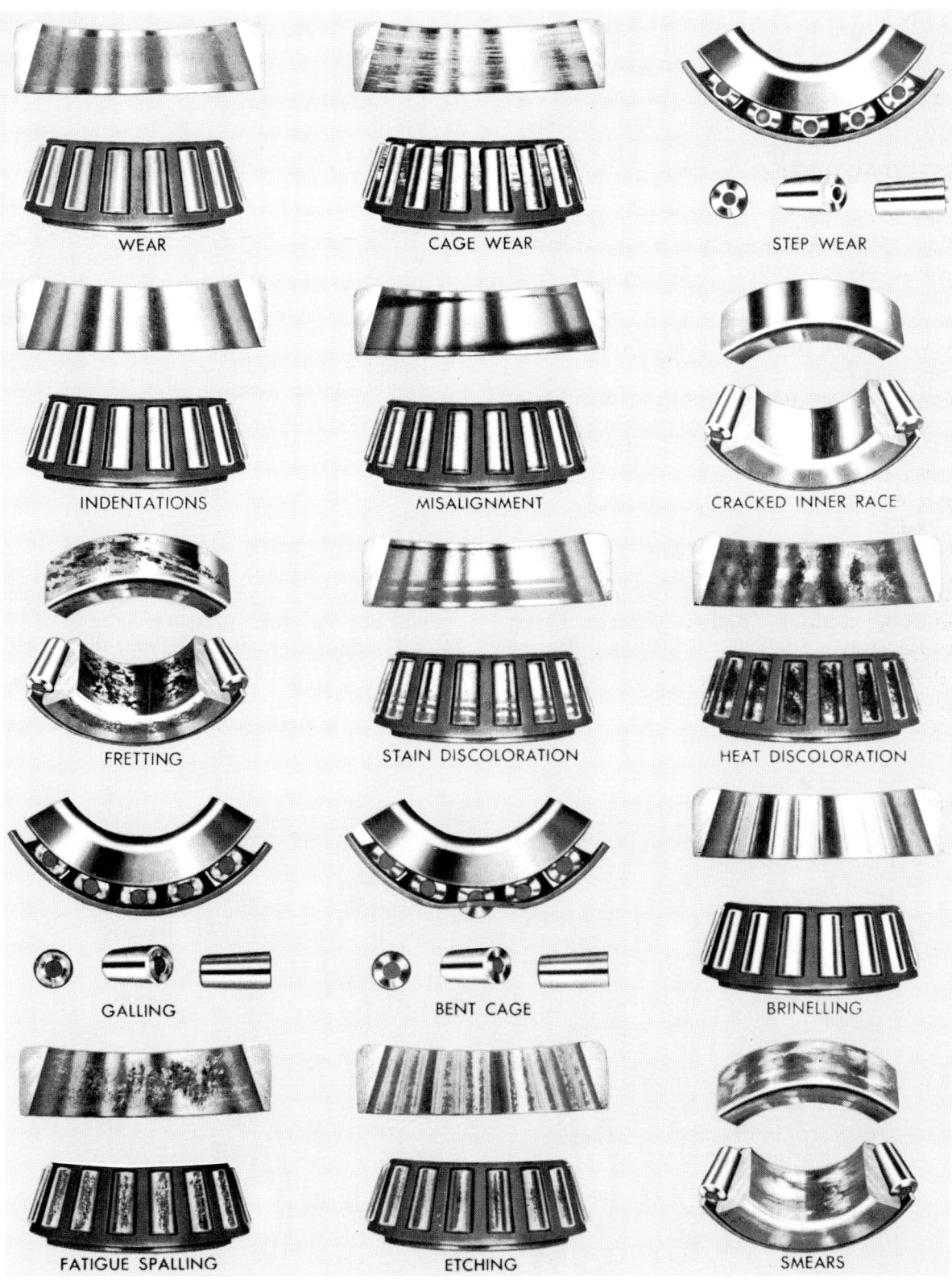

WEAR

CAGE WEAR

STEP WEAR

INDENTATIONS

MISALIGNMENT

CRACKED INNER RACE

FRETTING

STAIN DISCOLORATION

HEAT DISCOLORATION

GALLING

BENT CAGE

BRINELLING

FATIGUE SPALLING

ETCHING

SMEARS

Fig. 10-39. Some common roller bearing defects. Bearings showing these signs must be discarded.

Lubricate bearing seat area. Position bearing correctly and start by hand. Pull, press, or drive bearing fully into place, keeping square at all times. Do not damage shaft or bore. Installation tools must be spotlessly clean. In difficult assembly jobs, the use of both heat (carefully controlled) and cold will ease installation. If necessary, carefully adjust bearing.

SUGGESTED ACTIVITIES

1. Procure a number of damaged ball and roller bearings. Clean and inspect each one. Identify the cause of rejection. Try to find one good example of each typical defect.
2. Remove a bearing. Clean it properly. Inspect and pack with grease. Install the bearing following all recommendations.
3. Determine, as closely as possible, the number of ball, roller and needle bearings used in a specific car. Count them in ALL areas: clutch, transmission, drive line, rear axle, wheels, steering, pumps, motors, and other accessories.

KNOW THESE TERMS

Antifriction bearing, Cage, Ball bearing, Needle bearing, Roller bearing, Radial loading, Thrust loading, Raceway, Spalling, Brinelling, Bearing packer, Bearing oven, Etching, Fretting, Galling, Bore diameter, Outside diameter, Shell, Cup driver, Bearing axis.

REVIEW QUESTIONS

1. _____, _____, and _____ bearings are used in automotive construction.
2. Name three bearing load designs.
3. List three types of roller bearings.
4. The deep groove ball bearing will handle HEAVY thrust loads. True or False?
5. What advantage is offered by the self-aligning bearing?
6. Never _____ a bearing sealed on both sides.
7. Why are hydraulic or mechanical pullers generally superior to striking tools for bearing work?
8. Always apply pulling force to the free ring. True or False?
9. Under some circumstances, it is permissable to apply pulling pressure through the rolling element. True or False?
10. Bearings, under pulling pressure, can literally explode. True or False?
11. When heat must be applied to a bearing ring, it should not exceed _____°C.
12. Name two safety devices used when pulling bearings.
13. If a bearing is started in a "cocked" position, it will line up under pressure. True or False?
14. All pulling tools (striking type) should be of soft steel. True or False?
15. In that bearings are hardened, a little fine dirt will not hurt them. True or False?

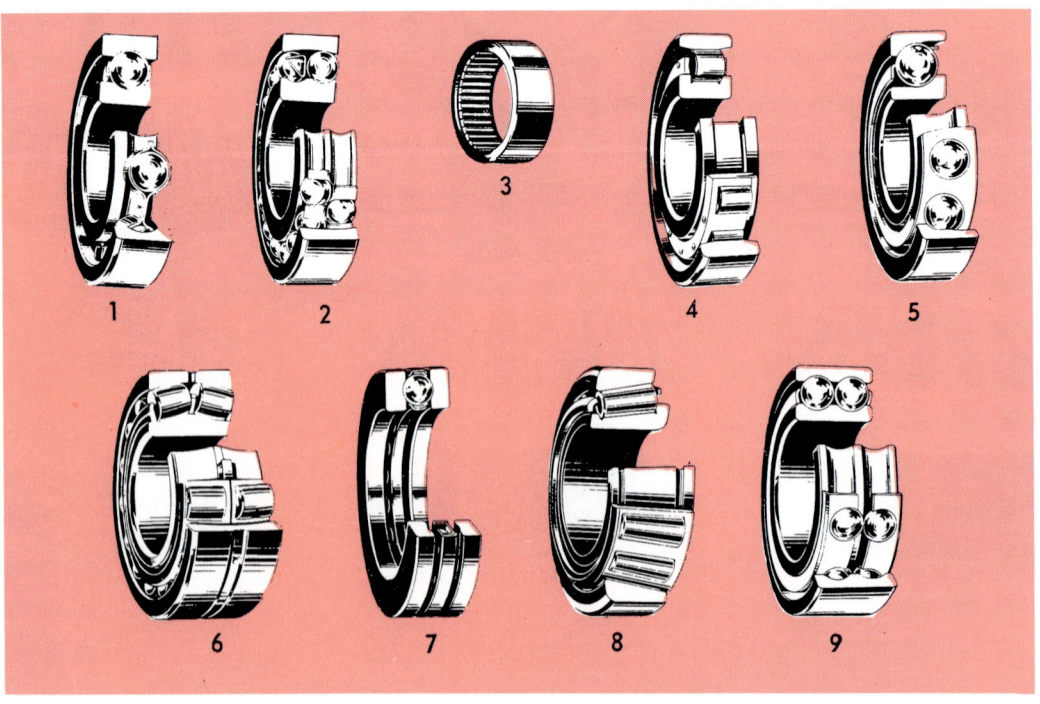

Fig. 10-40. Name these bearings. (SKF)

16. It is permissible to mix bearing parts if they are in good shape. True or False?
17. Bearings are best cleaned in _____ or _____.
18. When blowing dry, never _____ a bearing.
19. If you could only use one word to describe a proper bearing work area, that one word would be _____.
20. List six common bearing defects.
21. A bearing showing some looseness should always be rejected. True or False?
22. Oil bearings before inspecting. True or False?
23. It is important, on a separable bearing to inspect EVERY ball or roller. True or False?
24. When bearings are to be stored for some time, they should be coated with _____.
25. Always keep bearings _____ until ready to use.
26. Immediately following inspection, bearings should be _____.
27. Keep fingers away from bearing _____ elements and _____.

28. Before installing a bearing, inspect both _____ and _____ for nicks, burrs, and wear.
29. List 10 general rules regarding bearing, installation and removal.
30. Write down the numbers of the bearings illustrated in Fig. 10-40. Opposite each number, write the letter of the correct name. Some of the following names are wrong!

A. Single row, deep groove ball.
B. Self-aligning thrust.
C. Single row, tapered roller.
D. Angular contact ball.
E. Self-aligning ball.
F. Self-aligning roller.
G. Spherical roller.
H. Double row, deep groove ball.
I. Ball thrust.
J. Straight roller.
K. Needle.

Chapter 11

LUBRICATION SYSTEM FUNDAMENTALS

After studying this chapter, you will be able to:
☐ List the basic parts of a lubrication system.
☐ Summarise the operation of a lubrication system.
☐ Describe the construction of lubrication system parts.
☐ Compare different lubrication system designs.
☐ Explain the characteristics and ratings of motor oil.

The lubrication system is extremely important to engine service life because it forces oil to high friction points in the engine. Without a lubrication system, friction between parts would destroy an engine very quickly. Many of the engine parts would rapidly overheat and score from this friction. Engine bearings, piston rings, cylinder walls, and other components could be ruined.

BASIC LUBRICATION SYSTEM PARTS

A lubrication system, as shown in Fig. 11-1, basically consists of:
1. MOTOR OIL (lubricant for moving parts in engine).
2. OIL PAN (reservoir or storage area for motor oil).
3. OIL PUMP (forces oil throughout inside of engine).
4. OIL FILTER (strains out impurities in oil).
5. OIL GALLERIES (oil passages through engine).

As you will learn, other parts are added to increase system efficiency.

Lubrication system operation

With the engine running, the oil pump pulls motor oil out of the oil pan. A screen on the pickup tube removes large particles from the oil before oil enters the pump. The pump then pushes the oil through the oil filter and oil galleries.

The oil filter cleans the oil, removing very small particles. The filtered oil then flows to the camshaft, crankshaft, lifters, rocker arms, and other moving parts.

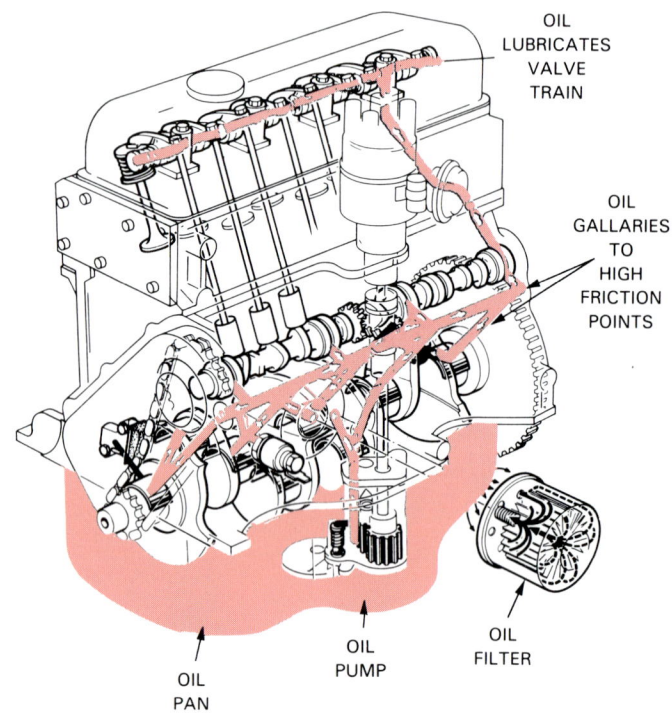

Fig. 11-1. Study the basic parts of a typical lubrication system. Also, trace flow of oil from pan through engine.

When oil leaks out of the engine bearings, it sprays on the outside of internal engine parts. For example, when oil leaks out of the connecting rod bearings, it sprays on the cylinder walls. This lubricates the piston rings, pistons, gudgeon pins, and cylinders. Oil finally drains back into the oil pan for recirculation.

FUNCTIONS OF A LUBRICATION SYSTEM

An engine lubrication system has several important functions. It:
1. Reduces friction and wear between moving parts.
2. Helps transfer heat and cool engine parts.

170

3. Cleans the inside of the engine by removing contaminants (metal, dirt, plastic, rubber, and other particles).
4. Cuts power loss and increases fuel economy.
5. Absorbs shocks between moving parts to quieten engine operation and increase engine life.

The properties of engine oil and the design of modern automotive engines allows the lubrication system to accomplish these functions.

ENGINE OIL

Engine oil, also called motor oil, is used to produce a lubricating film on the moving parts in an engine. It is commonly refined from crude oil or petroleum which is extracted from deep within the earth!

Synthetic oils (manufactured oils) are also available. They can be made from substances other than crude oil.

An oil film (thin layer of oil) separates engine parts to prevent metal-on-metal contact. This is shown in Fig. 11-2. Without the oil film, the parts would rub together and wear rapidly.

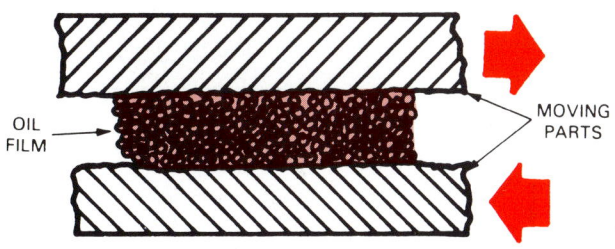

Fig. 11-2. Close-up view of clearance between moving parts shows how oil film keeps parts from touching and rubbing together.

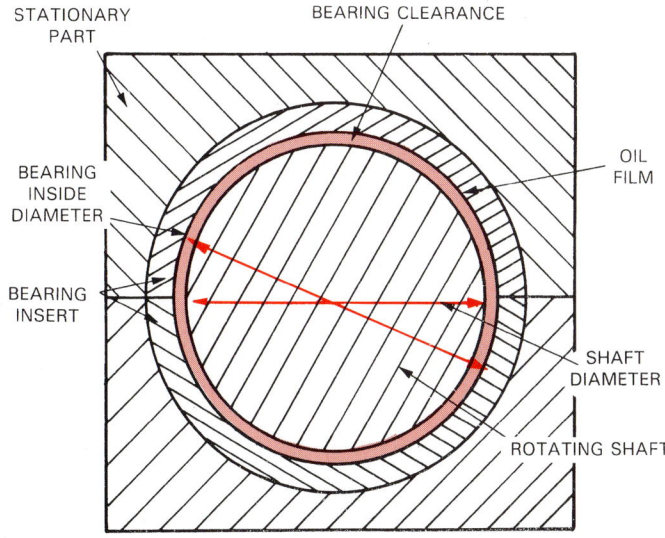

Fig. 11-3. Bearing clearance, also called oil clearance, allows oil film to hold spinning shaft away from stationary part.

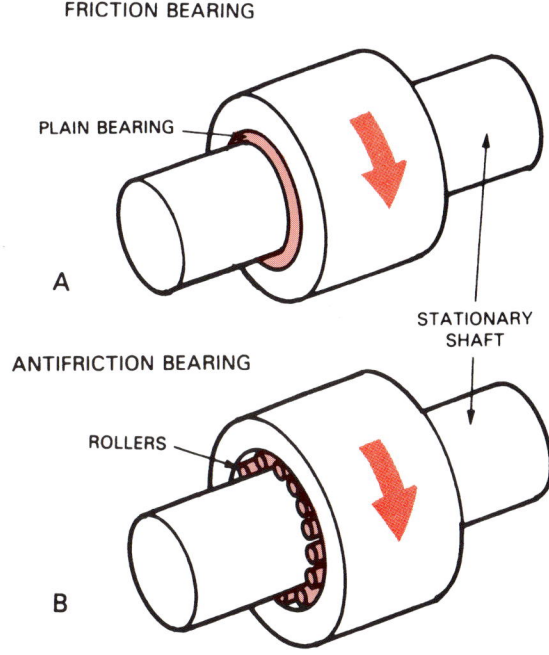

Fig. 11-4. Basic bearing types. A — Friction bearing has two smooth surfaces sliding together. B — Antifriction bearing use balls or rollers to prevent sliding (rubbing) action.

Oil clearance

Oil clearance is the small space between moving engine parts for the lubricating oil film. See Fig. 11-3.

The clearance allows oil to enter the bearing to prevent part contact.

One example, a connecting rod bearing typically has a bearing or oil clearance of about 0.05 mm. This clearance is large enough to allow oil entry. However, it is also small enough to keep the parts from hammering together during engine operation (reciprocating action).

Bearing types

There are two basic types of engine bearings: friction and antifriction types.

A friction bearing, also called plain bearings, has two smooth surfaces sliding on each other. Look at Fig. 11-4A. It is the most common type of bearing used in an engine.

Crankshaft main bearings, connecting rod bearings, cam bearings are normally friction bearings. They require a constant supply of oil under pressure for proper service life.

An antifriction bearing uses balls or rollers to avoid a sliding action between the bearing services. See Fig. 11-4B. They are only used in a few places in an engine.

A good example of an engine antifriction bearing is a roller lifter in a diesel engine. The roller cuts high friction and wear between the camshaft lobe and the bottom of the lifter.

Antifriction bearings do not require as much lubrication as a plain bearing. Usually, splash oiling is sufficient.

Oil viscosity (weight)

Oil viscosity, also called oil weight, is the thickness or fluidity (flow ability) of the motor oil. A high viscosity oil would be very thick and would resist flow, like HONEY. A low viscosity oil would be thin and running, more like WATER.

A viscosity numbering system is used to rate the thickness of engine oil. A high number would indicate thicker oil. A lower number would denote a thinner oil.

Look at Fig. 11-5. The oil's viscosity number is indicated on the oil container. The SAE (Society of Automotive Engineers) standardised this numbering system for this reason, oil viscosity is written SAE 10, SAE 20, SAE 30, etc.

Engine oil viscosities commonly range from a thin SAE 10 weight to a thick SAE 50 weight. Auto manufacturers specify an SAE number for their engines.

Temperature effects on oil

When cold, oil thickens and resists flow. When heated, oil thins and becomes runny. This can pose a problem.

The oil in a cold engine may be so thick that engine starting is difficult. The oil will not pump through the engine properly. This may increase starter drag and result in poor lubrication.

When the engine warms up, the oil film thins out. If it becomes too hot and thin, the oil film can break down and part contact can result.

It is important that the oil be thin enough for starting. It must also be thick enough to maintain lubrication when hot. Refer to Fig. 11-6.

Multi-viscosity oil

Multi-viscosity oil or multi-grade oil will exhibit operating characteristics of a thin, light oil when cold and a thicker, heavy oil when hot. A multi-grade oil can be numbered SAE 10w-30, 10w-40, 20w-50, etc.

For example, a 10w-30 multi-grade oil will flow easily (like a 10w oil) when starting a cold engine. It will then act as a thicker oil (like 30 grade) when the engine warms to operating temperature. This will make the engine start more easily in cold weather. It will also provide adequate film strength (thickness) when the engine is at full operating temperature.

Fig. 11-5. View of oil container which shows oil viscosity and service rating. This is multigrade oil that has passed strict service rating tests.

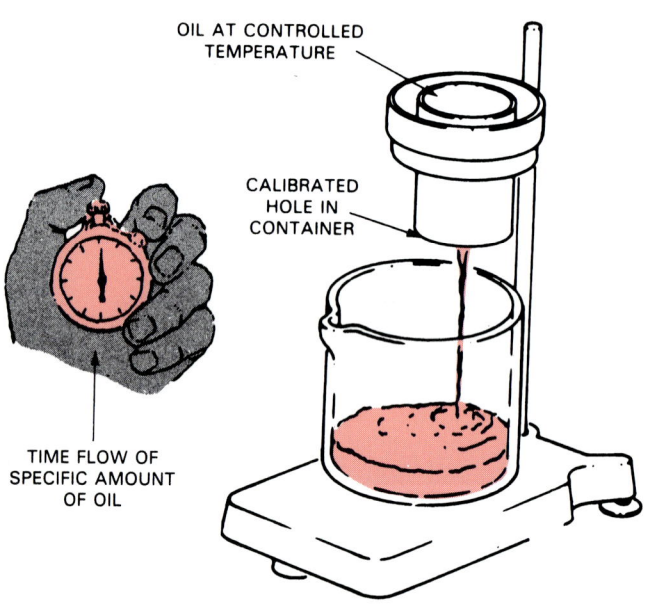

Fig. 11-6. Viscosity rating is determined by measuring how long oil takes to flow through specific opening at a specific temperature. If oil takes longer to flow into container, it would be a thicker, higher viscosity oil.

Selecting oil viscosity

Normally, you should use the oil viscosity recommended by the auto maker. However, in a very old, well-worn engine, higher viscosity oil may be beneficial. Thicker oil will tend to seal the rings and provide better bearing protection. It may also help cut engine oil consumption and smoking.

Fig. 11-7 is one auto maker's chart showing recommended SAE viscosity numbers for different temperatures.

Oil service rating

An oil service rating is a set of letters printed on the oil container to denote how well the oil will perform under operating conditions. This is a performance standard set by the American Petroleum Institute, abbreviated API. The service rating categories are:

1. SA (lowest quality straight mineral oil that should NOT be used in automotive engines).
2. SB (minimum quality for automotive petrol engines under mild service conditions, not normally recommended).
3. SC (meets oil warranty requirements for petrol engine vehicles up to 1968).
4. SD (meets oil warranty requirements for petrol engine vehicles up to 1972).
5. SE (meets oil warranty requirements for petrol engine models 1972 to 1979).
6. SF (current top grade oil recommended for severe motoring on current models beginning 1980. Covers all previous designations).
7. CA through CE (oil recommended for diesel engines).

A car owner's manual will give the service rating recommended for a specific vehicle. You can use a

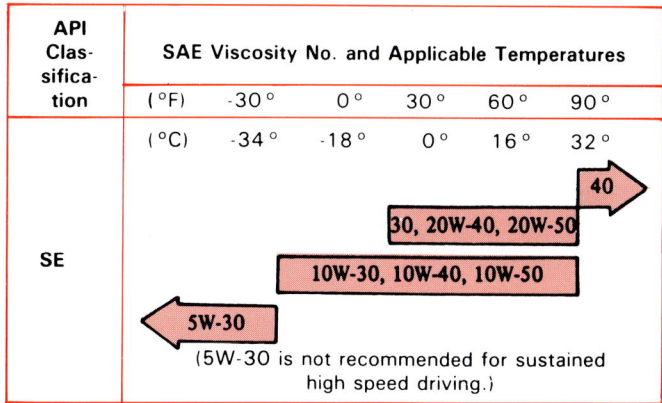

API Classification	SAE Viscosity No. and Applicable Temperatures				
	(°F) -30°	0°	30°	60°	90°
	(°C) -34°	-18°	0°	16°	32°
SE					40
			30, 20W-40, 20W-50		
			10W-30, 10W-40, 10W-50		
	5W-30				
	(5W-30 is not recommended for sustained high speed driving.)				

Fig. 11-7. Recommended SAE viscosity oil rating from one auto maker. Note how thicker oil is specified for higher outside temperatures. This maker also warns against mixing different brands of oil since different additives may be used in each. (Subaru)

better service rating than recommended, but NEVER a lower service rating!

A high service rating (SF for example) can withstand higher temperatures and loads while still maintaining a lubricating film. It will have more oil

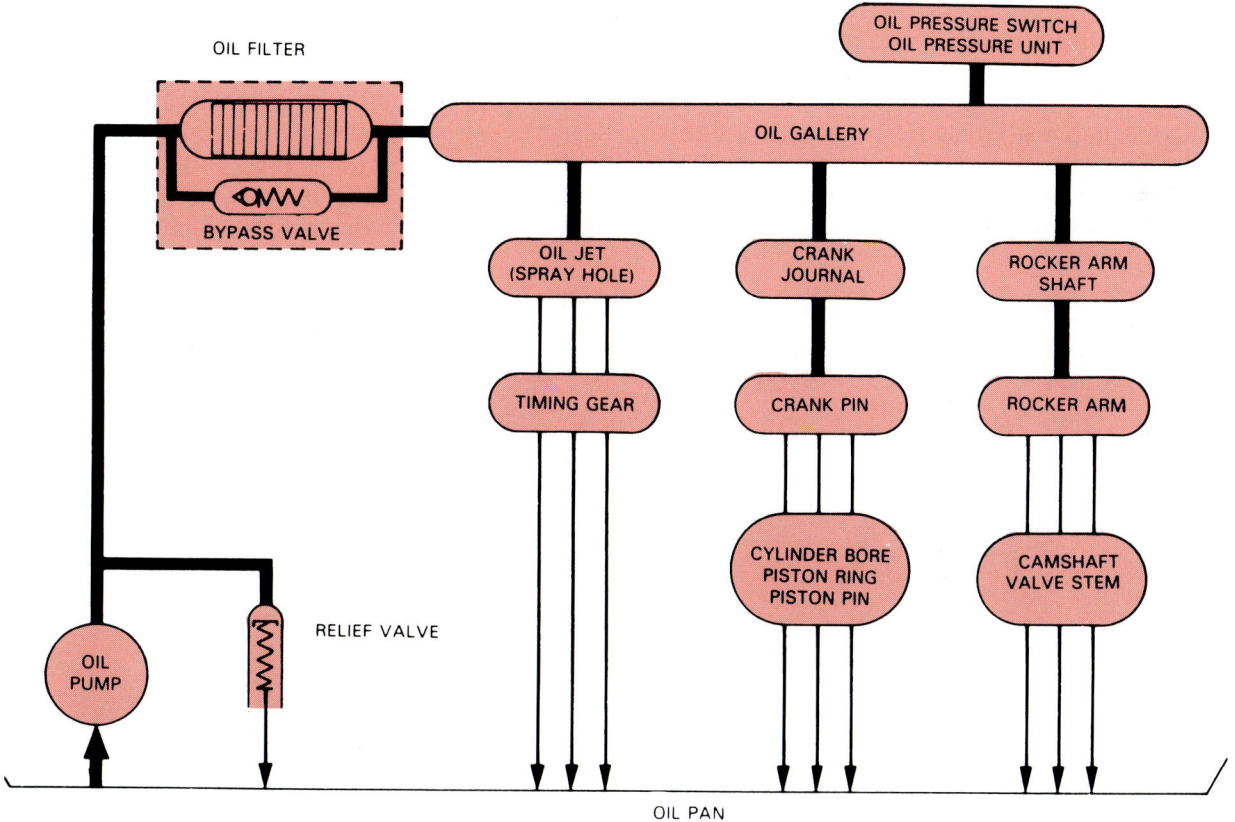

Fig. 11-8. Engine oil flow diagram. This full-flow system requires all oil to pass through filter before entering gallery. Thicker lines represent oil under pressure. Thinner lines stand for oil draining and splashing on parts. Note how filter bypass valve and pressure valve are in system.

additives (extra chemicals) to prevent oil oxidation (gumming), engine deposits (sludging), breakdown (oil changes chemicals), foaming (air bubbles form in oil), and other problems.

ENGINE OILING SYSTEMS

There are two methods for lubricating engine components: pressure fed oiling and splash oiling. See Fig. 11-8 for an oil flow diagram.

Pressure fed oiling is provided by the oil pump to the crankshaft bearings, camshaft bearings, lifters, and rocker arm assemblies. This type of oiling is needed where load and friction are very high.

Splash oiling occurs when oil sprays out and on moving parts to provide lubrication. This type of oiling is used between parts with moderate load. For instance, splash oiling is used on the piston rings, cylinders, camshaft lobes, timing chain, and many other parts.

There are two types of full pressure lubrication systems: full flow and bypass types.

Full flow lubrication system

The full flow lubrication system forces all of the oil through the oil filter before the oil reaches the parts of the engine. Refer to Fig. 11-8. It is the most common type of lubrication system for automotive engines.

Bypass lubrication system

The bypass lubrication system does NOT filter all of the oil that enters the engine bearings. It filters some of the extra oil not needed by the bearings. The bypass lubrication system is not very common. It is not as efficient as the full flow type.

OIL PAN AND SUMP

The oil pan, normally made of thin sheet metal or aluminium, bolts to the bottom of the engine block. It holds an extra supply of oil for the lubrication system. Refer to Fig. 11-9.

The oil pan is fitted with a screw-in drain plug for oil changes. Baffles may be used to keep the oil from splashing around in the pan.

The sump is the lowest area in the oil pan where oil collects, Fig. 11-10. As oil drains out of the engine, it fills the sump. Then the oil pump can pull oil out of the pan for recirculation.

OIL PICKUP AND SCREEN

The oil pickup is a tube extending from the oil pump to the bottom of the oil pan. One end of the pickup bolts or screws into the oil pump or to the engine block. The other end holds the pickup screen.

The pickup screen prevents large particles from entering the pickup tube and oil pump. See Figs. 11-9 and 10. The screen is usually part of the pickup tube. Without the screen, the oil pump could be damaged by bits of valve stem seals and other debris flushed out of the engine.

OIL PUMPS

The oil pump is the "heart" of the engine lubrication system; it forces oil out of the pan, through the engine filter, galleries, and to the engine bearings. The oil pump is frequently driven by a gear on the engine camshaft, Fig. 11-11. It may also be driven by a cogged belt or by a direct connection with the end of the camshaft or crankshaft.

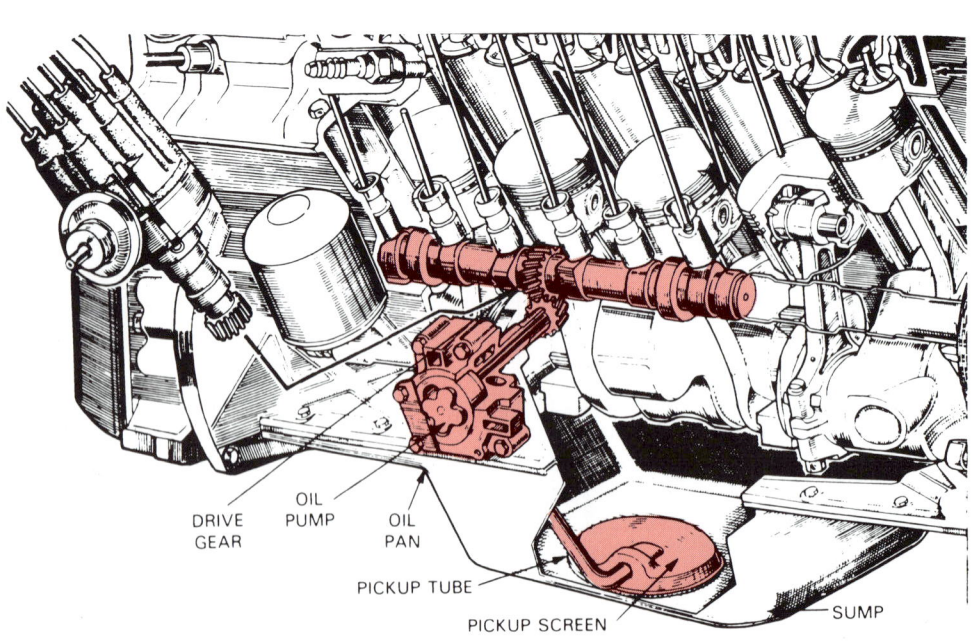

Fig. 11-9. Oil pump is commonly driven by gear on camshaft. Also note how pickup tube extends into pan.

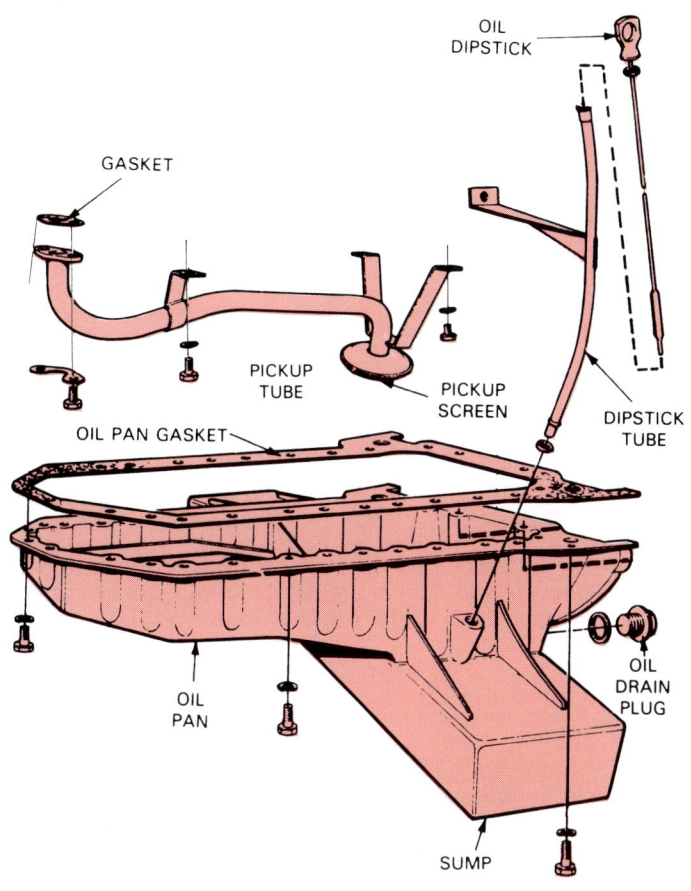

Fig. 11-10. Oil pan forms lower crankcase and sump. Gasket seals mating surface between block and pan. Also note drain plug, pickup tube and screen, and dipstick assembly. (Volvo)

Fig. 11-9 shows a cutaway view of an engine. Note how the oil pump is driven.

There are two basic types of engine oil pumps: rotary and gear. These are illustrated in Fig. 11-12.

Rotary oil pumps

A rotary oil pump uses a set of star-shaped rotors in a housing to pressurise the motor oil. Look at Fig. 11-13. Study it carefully.

As the oil pump shaft turns, the inner rotor causes the outer rotor to spin. The eccentric action of the two rotors form pockets that change in size.

A large pocket is formed on the inlet side of the pump. As the rotors turn, the oil filled pocket becomes smaller as it nears the outlet of the pump. This squeezes the oil and makes it spurt out under pressure.

As the pump spins, this action is repeated over and over to produce a relatively smooth flow of oil.

A rotary oil pump mounted on the front of the engine and driven by the timing belt is shown in Fig. 11-14. Compare it to Fig. 11-13.

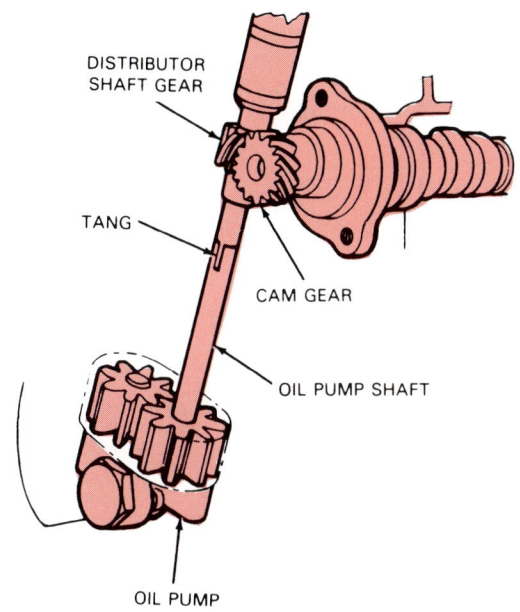

Fig. 11-11. Gear on bottom of distributor meshes with gear on camshaft. Oil pump shaft extends from distributor shaft to oil pump. With engine running, shaft turns at one-half engine speed.

Gear oil pumps

A gear oil pump uses a set of gears to produce lubrication system pressure. Figs. 11-11 and 11-12.

A shaft, usually turned by the distributor, crankshaft, or accessory shaft, rotates one of the pump gears. This gear turns the other pump gear which is supported on a very short shaft inside the pump housing.

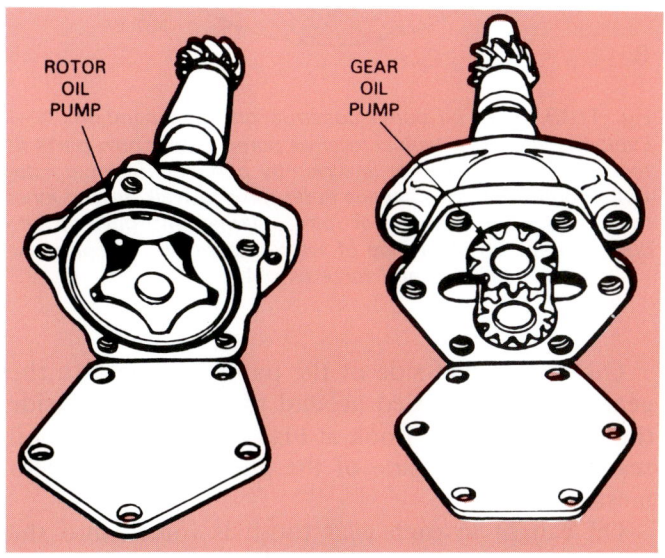

Fig. 11-12. Compare two basic types of oil pumps: rotor pump and gear pump.

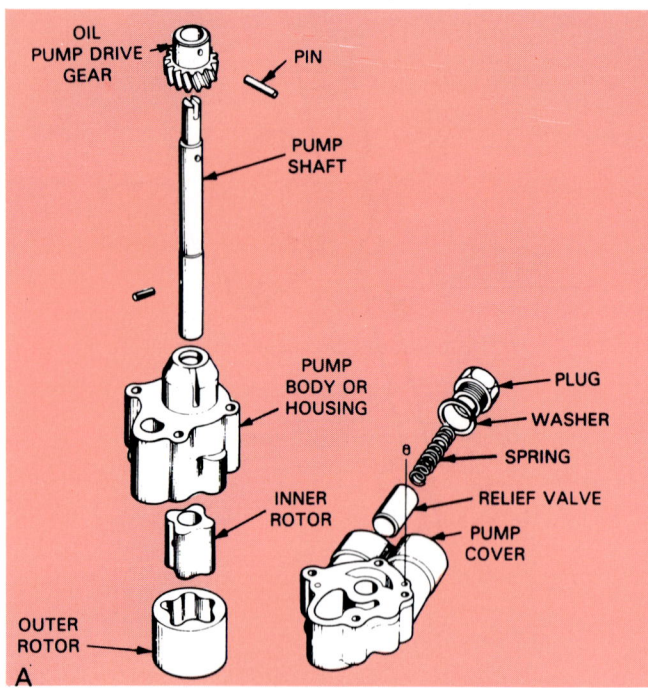

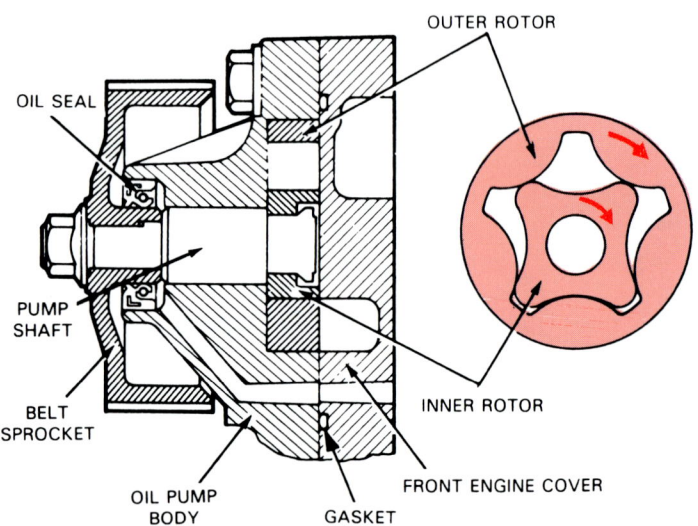

Fig. 11-14. Cutaway view of a modern belt-driven rotor oil pump. Pump bolts to front of engine. Its operation is similar to the pump in the previous illustration.

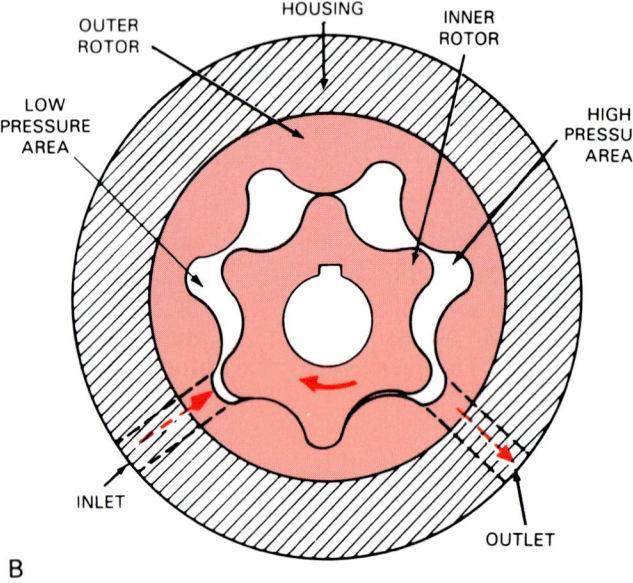

Fig. 11-13. Rotor oil pump construction and operation. A — Exploded view of a typical rotor oil pump. Study how parts fit together. B — Inner rotor is driven by pump shaft. Inner rotor turns outer rotor. This causes outer rotor to walk around inner rotor. Space on one side of rotor enlarges and pulls oil into pump. Space on other side of rotor gets smaller to compress and force oil out.

Oil on the inlet side of the pump is caught in the gear teeth and carried around the outer wall inside the pump housing. Look at Fig. 11-15. When the oil reaches the outlet side of the pump, the gear teeth mesh and seal.

Oil caught in each gear tooth is forced into the pocket at the pump outlet and pressure is formed. Oil squirts out of the pump and to the engine bearings.

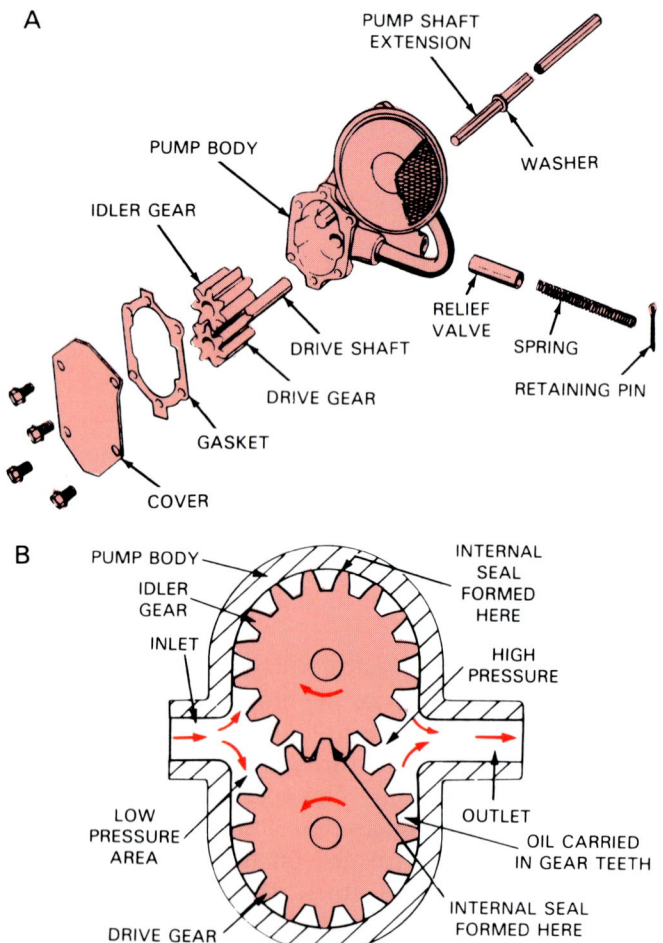

Fig. 11-15. Gear oil pump construction and operation. A — Exploded view of gear pump shows how it is like and unlike a rotor pump. Compare parts and how they fit together. B — Oil pump shaft turns one gear. That gear drives other gear. Oil is trapped in teeth of gears and carried around housing wall. When on outlet side, oil is trapped and pressurised.

An internal gear oil pump is pictured in Fig. 11-16. It uses the same general principles just discussed.

PRESSURE RELIEF VALVE

A pressure relief valve limits maximum oil pressure. It is a spring-loaded, bypass valve in the oil pump, engine block, or oil filter housing. Refer to Figs. 11-16 and 11-17.

The pressure relief valve consists of a small piston, spring, and cylinder. Look at Fig. 11-18. Under normal pressure conditions, the spring holds the relief valve closed. All of the oil from the pump flows into the oil galleries and to the bearings.

However, under abnormally high oil pressure conditions (cold, thick oil for example), the pressure relief valve opens. Oil pressure pushes the small piston back in its cylinder by overcoming spring tension. This allows some oil to bypass the main oil galleries and pour back into the oil pan. Most of the oil still flows to the bearings and preset pressure is maintained.

Some pressure relief valves are adjustable. By turning a bolt or screw or by changing spring shim thickness, the pressure setting can be altered.

OIL FILTERS

An oil filter removes small metal, carbon, rust, and dirt particles from the motor oil. It protects the moving engine parts from abrasive wear. Fig. 11-19 shows a cutaway of a modern oil filter.

An element is a paper or cotton filtering substance mounted inside the filter housing. It will allow oil flow but will block and trap small debris.

A filter bypass valve is commonly used to protect the engine from oil starvation if the filter element

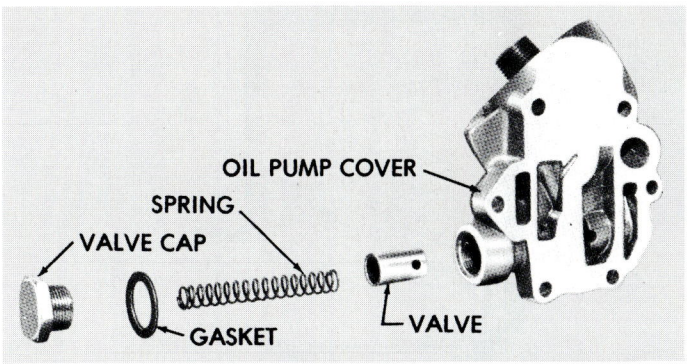

Fig. 11-17. Exploded view of pressure relief valve. Small piston or valve fits in small cylinder. Spring holds valve in normally closed position.

becomes clogged. The valve will open if too much pressure is formed in the filter. This allows unfiltered oil to flow to the engine bearings, preventing major part damage.

Oil filter types

The two classifications of engine oil filters are: spin-on filter and cartridge filter.

The spin-on oil filter is a sealed unit having the element permanently enclosed in the filter body. When it must be serviced, a new filter is simply screwed into place. This is the most common type of modern oil filter.

The cartridge oil filter has a separate element and housing. To service this type oil filter, the housing is removed. Then, a new element is installed inside the existing housing. A cartridge type oil filter is sometimes used on heavy duty or diesel applications.

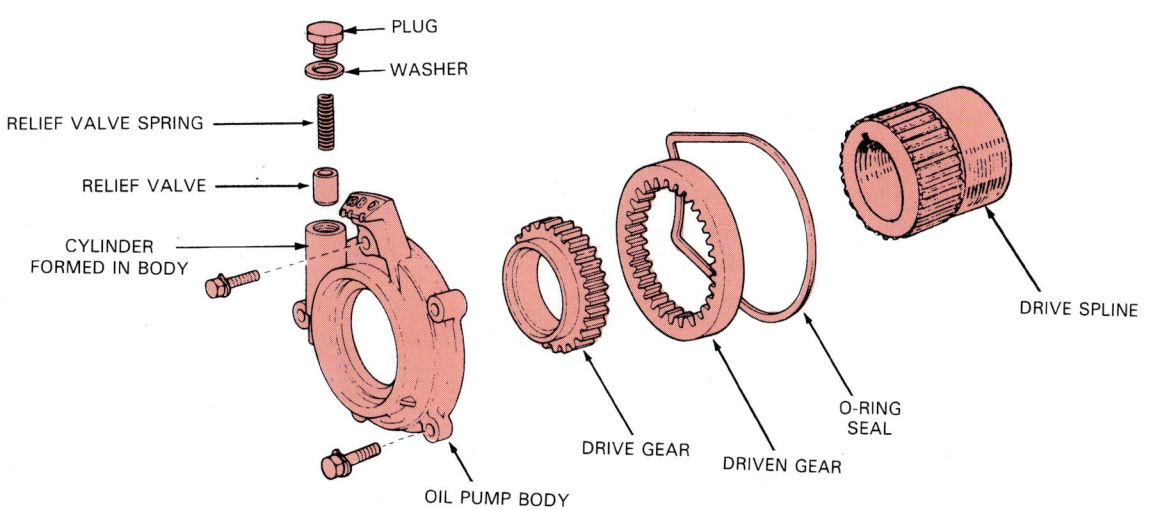

Fig. 11-16. Modern gear pump that mounts on front of engine. Drive spline on crank turns inner gear. Outer gear walks around to pump oil into block. Note pressure relief valve in pump body. (Toyota)

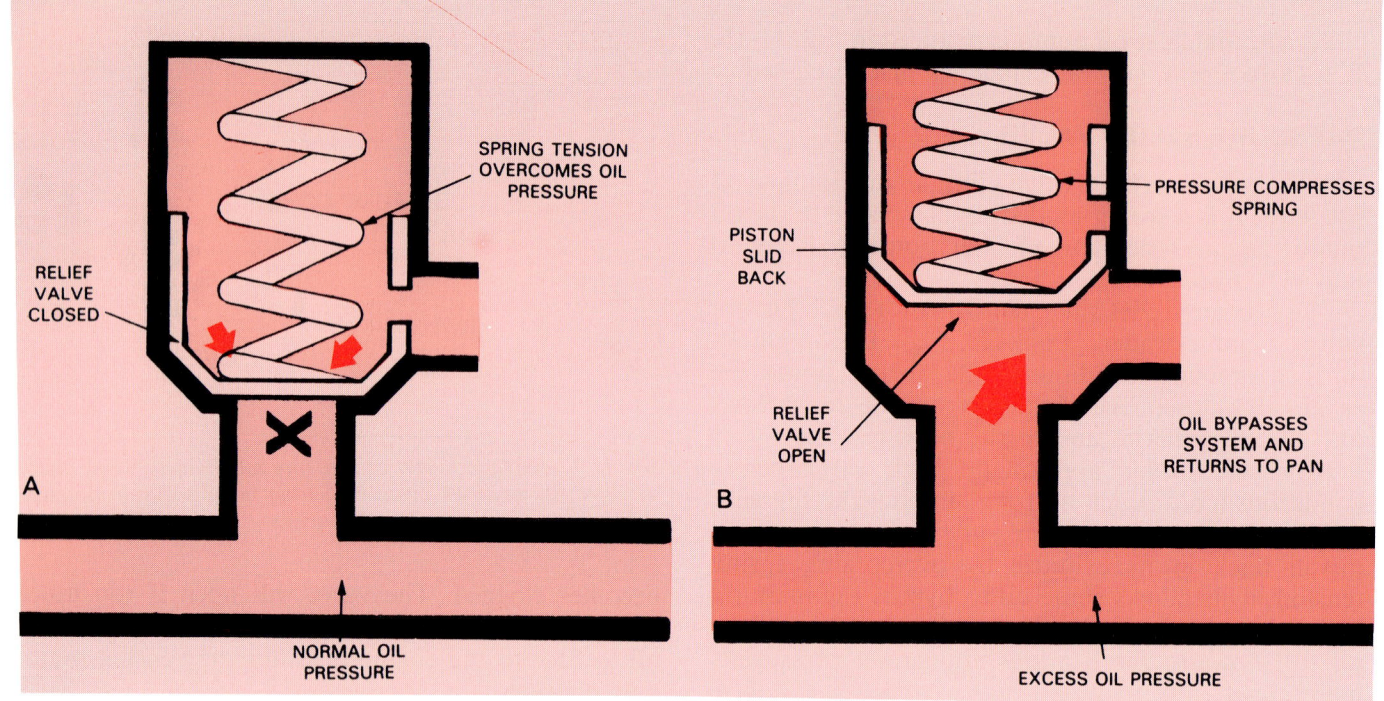

Fig. 11-18. Pressure relief valve action. A — Oil pressure normal. Spring holds relief valve closed to maintain sufficient pressure. B — Oil pressure too high, pressure compresses spring and opens valve. Excess oil bypasses system and drains into pan.

Oil filter housing

The oil filter housing is a metal part that bolts to the engine and provides a mounting place for the oil filter. The housing may also have a fitting for the oil pressure sending unit. See Fig. 11-20.

A gasket normally fits between the engine and oil filter housing to prevent leakage. Sometimes, the pressure relief valve, filter bypass valve, or oil pump are inside this housing.

OIL GALLERIES

Oil galleries are small passages through the cylinder block and head for lubricating oil. Refer to Fig. 11-21. They are cast or machined passages that allow oil to flow to the engine bearings and other moving parts.

The main oil galleries are large passages through the centre of the block. They feed oil to the crankshaft bearings, camshaft bearings, and lifters. The main galleries also feed motor oil to smaller passages running up to the cylinder heads.

OIL COOLER

An oil cooler may be used to help lower and control the operating temperature of the engine oil. Look at Fig. 11-22. It is a radiator-like device connected to the lubrication system. Oil is pumped through the cooler and back to the engine.

Airflow through the cooler removes heat and lowers the temperature of the oil.

Oil coolers are frequently used on turbocharged engines or heavy-duty applications (trailer towing package for instance).

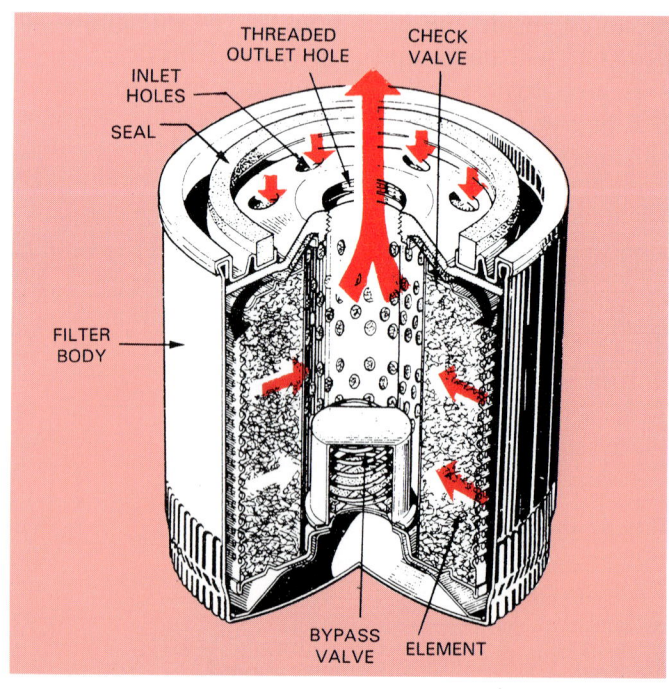

Fig. 11-19. Cutaway of modern spin-on oil filter. Oil enters small holes, passes through element, and then flows out centre hole to engine. Rubber O-ring prevents leakage between engine and filter housing. (Saab)

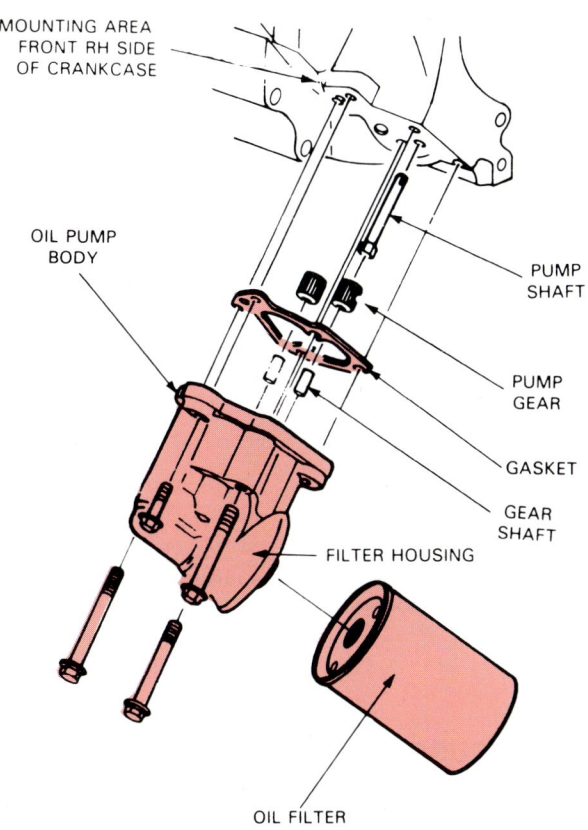

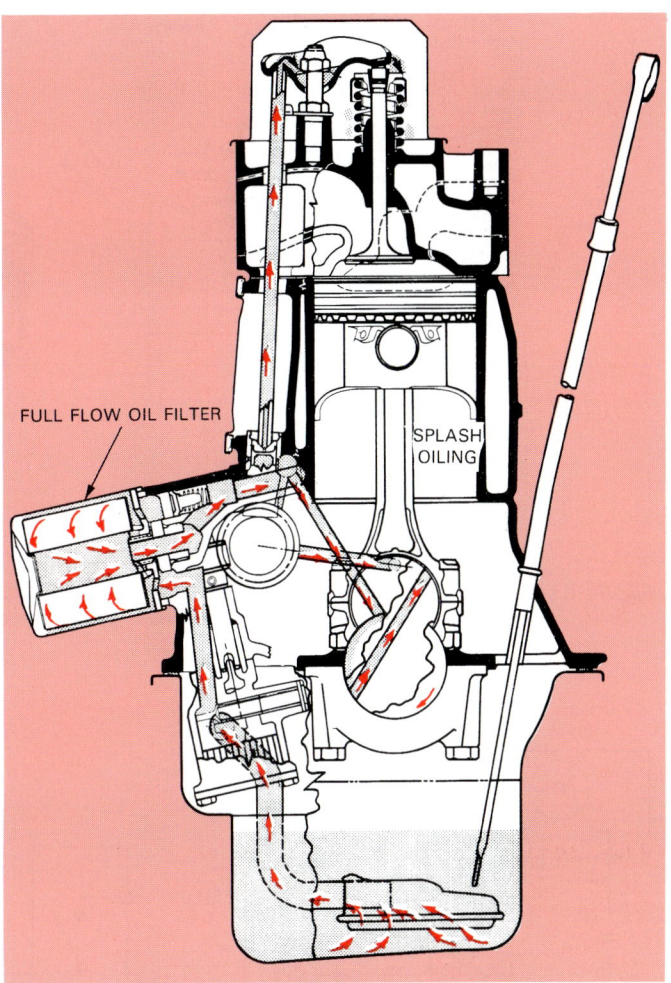

Fig. 11-20. This oil filter housing is also the oil pump housing. Gasket fits between cylinder block and housing. Filter screws on.

Fig. 11-21. Oil galleries allow oil to pass though engine. Main galleries are larger passages in block. Note how oil flows through hollow push rods to rockers.

POSITIVE CRANKCASE VENTILATION

The positive crankcase ventilation system, abreviated PCV system, draws fumes out of the engine crankcase and burns them inside the engine. This system helps prevent engine sludging (chocolate pudding-like oil formation) which could restrict oil circulation. It also prevents toxic vapours from entering and polluting the atmosphere. See Fig. 11-23.

OIL PRESSURE INDICATOR

An oil pressure indicator warns the driver of a low oil pressure problem. The circuit activates a warning light in the car's dash. A basic oil pressure light circuit is given in Fig. 11-24.

An oil pressure sending unit is a pressure sensitive switch that operates the dash indicator light. It screws into the engine and is exposed to one of the oil galleries.

When oil pressure is low (engine off or mechanical problem), the spring in the sending unit holds a pair of contacts closed. This completes the circuit and the indicator light glows.

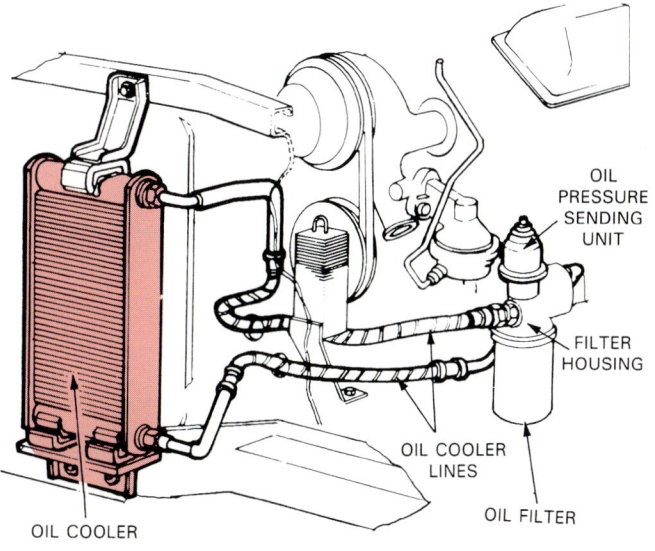

Fig. 11-22. Oil cooler allows transfer of heat out of oil and into surrounding air, like a cooling system radiator. They are used on high performance or heavy duty applications. High pressure lines carry oil to and from cooler.

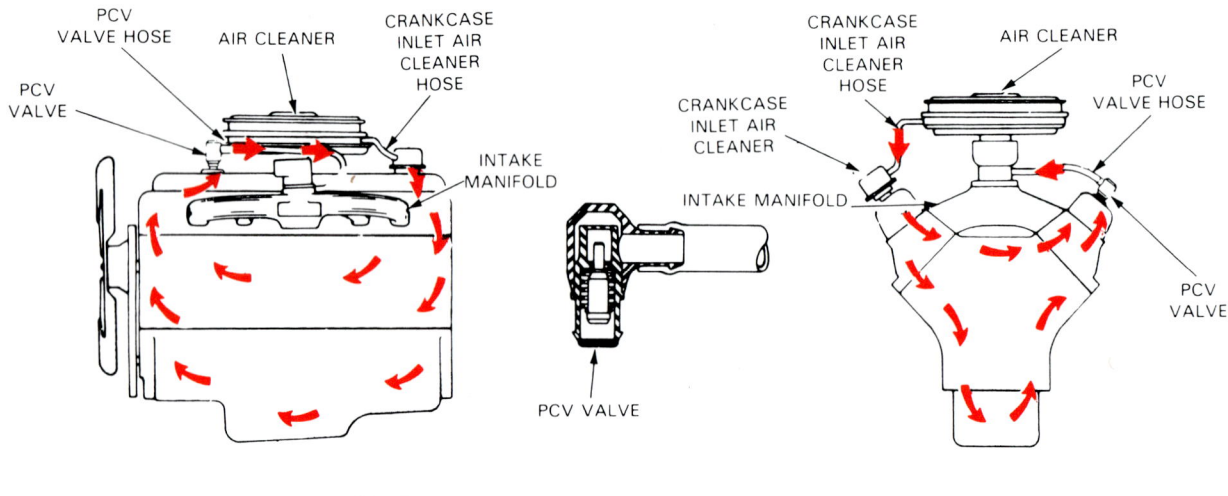

SIX CYLINDER ENGINES

Fig. 11-23. PCV system draws toxic fumes out of oil pan and other internal areas in engine. Fumes are pulled into intake manifold and combustion chambers. Fumes are then burned on power strokes.

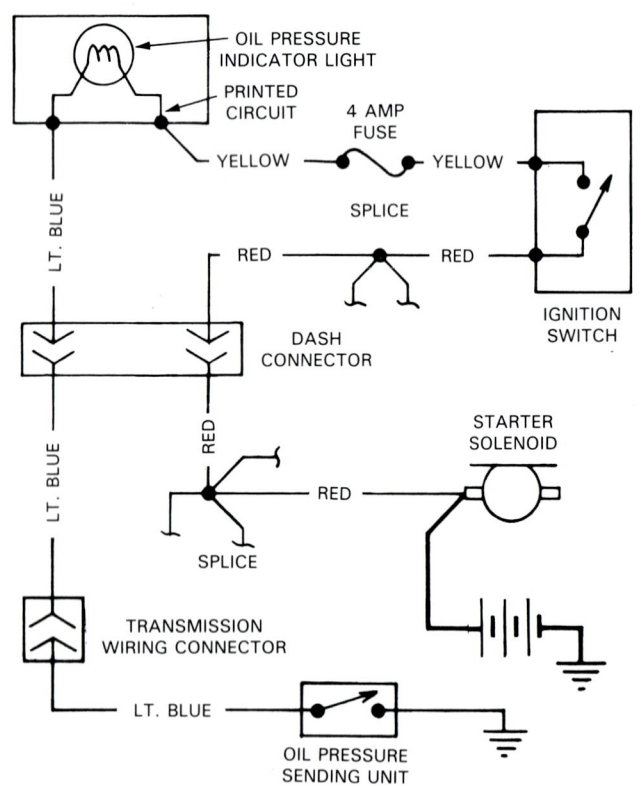

Fig. 11-24. Actual service manual wiring diagram for one make of car. Study how sending unit controls current flow through dash indicator light. Also note wire colour coding for tracing wires.

When oil pressure is normal, oil pressure acts on a diaphragm in the sending unit. Diaphragm deflection opens the contact points to break the circuit. This causes the warning light to go out, informing the driver of good oil pressure.

OIL PRESSURE GAUGE CIRCUIT

Some cars are equipped with an oil pressure gauge that registers the actual oil pressure in the engine. See Fig. 11-25. It is similar in operation to the oil pressure indicating light. However, the sending unit uses a VARIABLE RESISTANCE UNIT instead of contact points.

As more oil pressure is developed, the sending unit diaphragm is deflected a proportional amount. This causes an equal amount of sending unit resistance change.

Low pressure causes low sending unit resistance high current flow, and low oil pressure gauge readings. High engine oil pressure causes high resistance in the sending unit to allow low current flow for deflecting the pressure gauge needle to the right.

KNOW THESE TERMS

Oil film, Bearing clearance, Friction bearing, Antifriction bearing, Viscosity, Multi-grade, Oil service rating, Pressure fed oiling, Splash oiling, Full flow lubrication system, Bypass lubrication system, Oil pan, Sump, Oil pickup, Pickup screen, Oil pump, Rotary pump, Gear pump, Pressure relief valve, Spin-on oil filter, Cartridge oil filter, Oil filter housing, Oil gallery, Oil cooler, PCV, Oil pressure indicator, Oil pressure gauge.

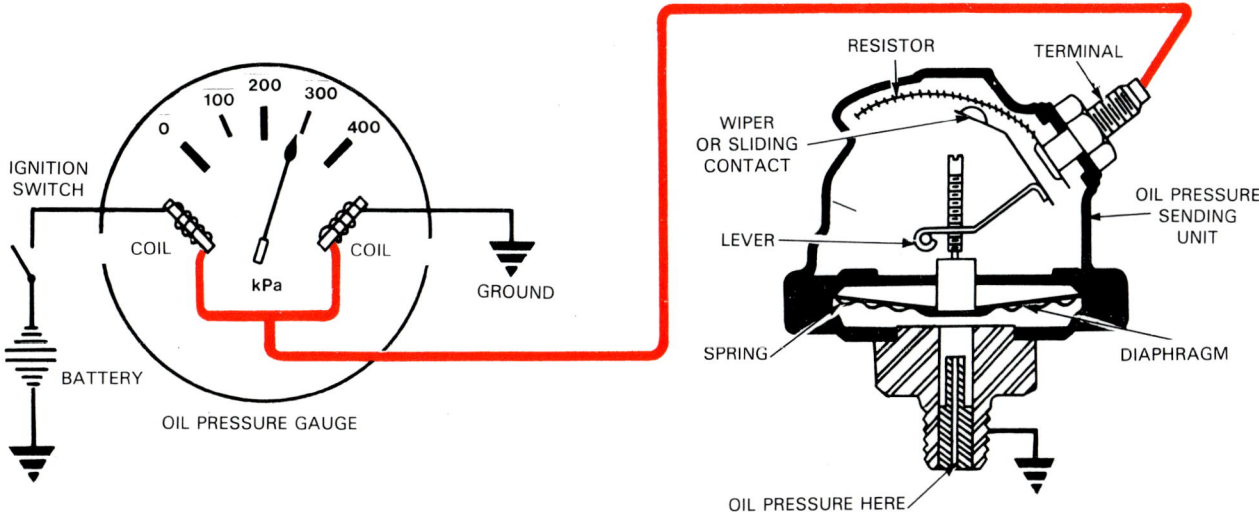

Fig. 11-25. Oil pressure gauge circuit uses variable resistance type sending unit. Changes in oil pressure causes different amounts of diaphragm deflection. This moves sliding contact on resistor. Changes in resistance and current make gauge show engine oil pressure.

REVIEW QUESTIONS

1. List and describe the five major parts of an engine lubrication system.
2. What are five functions of a lubrication system?
3. How does engine motor oil protect parts from excess wear?
4. _____ _____ is the small space between moving engine parts for the lubricating oil film.
5. Explain the difference between a friction and antifriction bearing.
6. Oil _____, also called oil _____, is the thickness or fluidity of the motor oil.
7. SAE 30 weight oil is thicker and less fluid than SAE 40 weight oil. True or False?

8. What is a multi-grade oil?
9. Pressure fed oiling is NOT used with:
   a. Crankshaft bearings.
   b. Camshaft bearings.
   c. Hydraulic lifters.
   d. Cylinder walls.
10. There are two basic types of engine oil pumps: _____ and _____ types.
11. How does a pressure relief valve work?
12. Why is an oil filter very important to engine service life?
13. _____ _____ are small passages through the cylinder head and block for lubricating oil.
14. Why is an oil cooler sometimes used?
15. Summarise the operation of an oil pressure gauge circuit.

Chapter 12

BUILDING AN ENGINE

After studying this chapter, you will be able to:
- □ Identify the basic parts of an engine.
- □ Explain engine operating principles.
- □ Describe the function of the major parts of an engine.
- □ Recall the four-stroke cycle sequence.

WHAT IS AN ENGINE?

An engine is a related group of parts assembled in a specific order. In operation, it is designed to convert the energy given off by burning fuel into a useful form.

There are many parts in a modern engine, each one being essential to the engine operation. For the time being, however, we may think of an engine as a device that allows us to pour fuel into one end and get power from the other. Fig. 12-1.

INTERNAL COMBUSTION PETROL ENGINE

Internal combustion means burning within. Petrol indicates what is being burned. Engine refers to the device in which the petrol is burned.

BUILDING A PAPER ENGINE

A excellent way to learn about engines is to build one — on paper. Make believe that the engine has not yet been invented. YOU will be the inventor. YOU will solve the problems involved, step by step.

WHAT TO USE FOR FUEL

If you are going to convert fuel into useful energy, you will need something that will ignite (burn) easily. It should burn cleanly, be reasonably inexpensive, and should produce sufficient power with good burning characteristics. It must be available in quantity, should be safe to use and easy to transport.

How about dynamite? NO, it is expensive and burns violently. In short, it would blow up the engine. Fig. 12-2.

Kerosene? NO, it is too hard to ignite and does not burn cleanly.

Petrol? Now you have found a fuel which will serve your purpose.

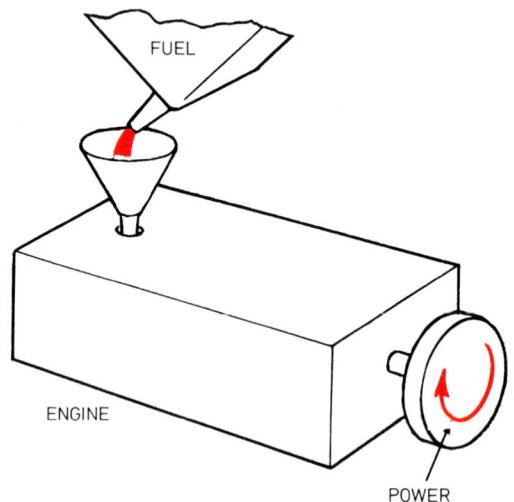

Fig. 12-1. Converting fuel into power.

Fig. 12-2. It will explode.

182

WHAT IS PETROL?

Petrol is a product obtained by refining crude oil. Figs. 12-3 and 12-4. Basically, the crude oil (obtained from oil wells in the earth) is treated in various ways to produce petrol. Petrol is only one of the many items produced from crude oil (petroleum).

Petrol used in engines is a complex mixture of basic fuels and special additives. Since petrol is a mixture of carbon and hydrogen atoms, it is termed a HYDROCARBON. Fig. 12-3.

Petrol is available in two grades. UNLEADED and SUPER. UNLEADED petrol is used in all late model cars with catalytic converters and some earlier models that are listed as suitable for running on UNLEADED petrol. SUPER petrol which contains lead additives is used in all cars which are unsuitable for running on UNLEADED petrol. Both grades are assigned an OCTANE rating. The octane rating indicates how well the petrol will resist detonation (too rapid burning) in the cylinders. The use of leaded petrol will quickly destroy the catalytic converter on cars so equipped.

Many factors enter into the quality of any petrol. It must pass exhaustive tests, both in the laboratory and in actual use. Basically, petrol must burn cleanly, ignite readily and resist vapour lock (turning too quickly from a liquid to a vapour from exposure to excessive heat). It should contain a minimum amount of harmful ingredients and be free from detonation.

WARNING: THE FOLLOWING STEPS, SHOWING HOW PETROL IS PREPARED FOR USE IN THE ENGINE, ARE FOR PURPOSES OF ILLUSTRATION ONLY AND SHOULD NOT BE ATTEMPTED BY THE STUDENT. PETROL IS DANGEROUS — TREAT IT WITH CARE AND RESPECT.

PREPARING THE FUEL FOR THE ENGINE

As you know, petrol burns readily. However, to get the most power from this fuel and, in fact, to get it to run an engine, special treatment is required.

If you were to place a small amount of petrol in a jar and drop a match in it, it would burn. Fig. 12-5.

Fig. 12-3. Petrol is a complex mixture.

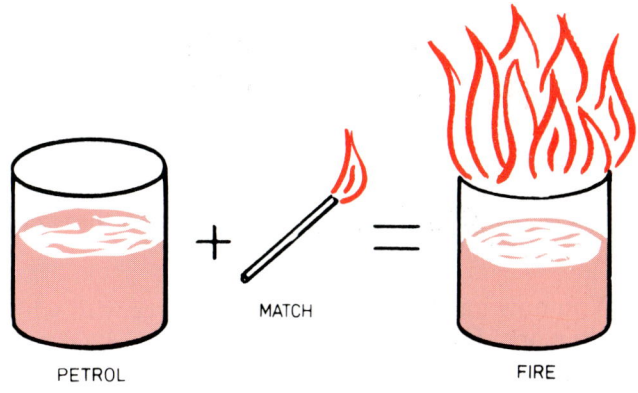

Fig. 12-5. Petrol burns.

Such burning is fine to produce heat, but it will not give us an explosive burning effect which we need to run an engine. How can we get the petrol to burn fast enough to produce an explosive force?

If you will examine Fig. 12-5, you will notice that the petrol is burning on the top side. Why is it not burning along the sides and bottom of the container? In order to burn, petrol must combine with oxygen in the air. The sides and bottom are not exposed to air so they do not burn.

BREAKING UP PETROL

For purposes of illustration, imagine having petrol in the same amount and shape, but with no

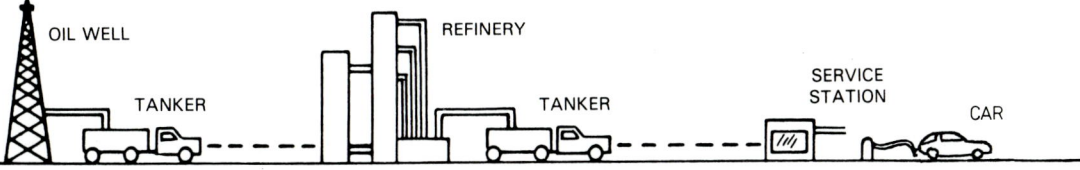

Fig. 12-4. Before reaching your car, crude oil requires much handling.

container, left, Fig. 12-6. It will then burn on all sides. This will increase the burning speed. However, it still will not burn quickly for use in an engine.

What to do? The answer: break the petrol into smaller particles. Notice that as you divide it into smaller particles, right, Fig. 12-6, you expose more and more edges to the air. If you ignite it now, it will burn considerably faster.

If you break up petrol into very tiny particles, burning is fierce. Rapid burning produces a tremendous amount of heat which, in turn, causes a rapid and powerful expansion. The burning petrol is now giving off ENERGY in the form of HEAT. Fig. 12-7.

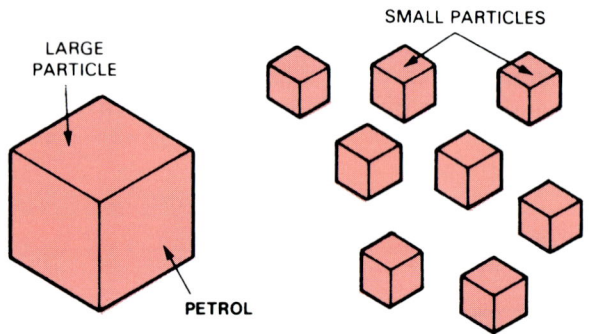

Fig. 12-6. Breaking petrol into smaller particles exposes more surface to the air, increasing the burning rate.

You now have the necessary basic force with which to do work. How do you harness it and make it work for you?

TRAPPING THE EXPLOSION

If you were to take a sturdy metal container, spray a petrol and air mixture into it, place a lid over the top and then light it, the resulting explosion would blow the lid high into the air. Fig. 12-8.

This is an example of using petrol to do work. In this case the work is blowing the lid into the air. It is obvious that a flying lid will not push a car but, think a minute, and see if the flying lid does not suggest some way of converting the power into useful motion.

A SIMPLE ENGINE

Make the same setup, only this time hook the lid to a simple crankshaft by means of a connecting rod. Then place a wheel on the other end of the crankshaft. Fig. 12-9.

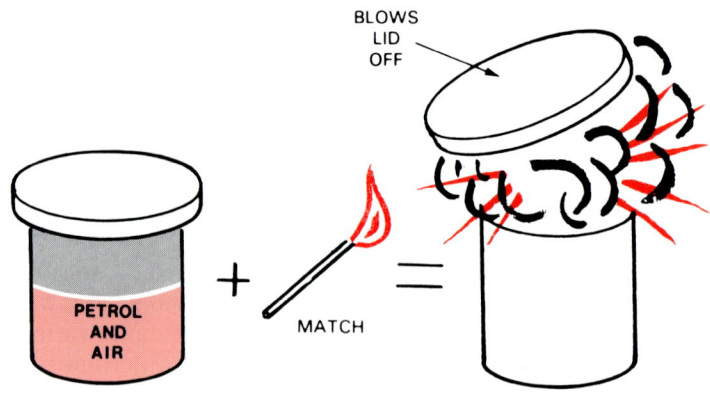

Fig. 12-8. Blowing the lid off.

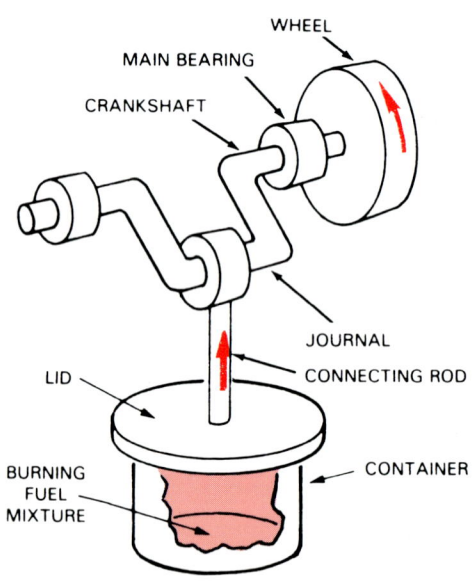

Fig. 12-9. A simple engine.

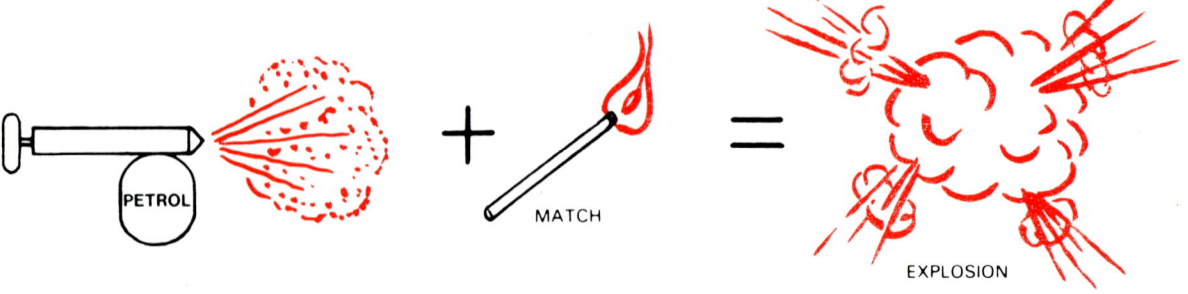

Fig. 12-7. When petrol particles are small enough they will explode.

Now if you explode the mixture, the lid, as it blows up, will give the crankshaft a sharp upward push and cause the wheel to spin. You have built a very simple engine, not practical, but it is pointing the way!

WHAT IS WRONG WITH IT?

Many things. Let us discuss them one at a time. The lid would fly up; as the wheel spun, it would be forced down. It can come down in any position, but to work, it must be over the container.

Instead of putting the lid over the container, cut it so that it just slips inside. Make the container longer so the lid can push the crank to the top of its travel, and still not get out of the container. Fig. 12-10.

If you were to bolt the container and shaft bearing so they could not change position, you would then have an engine that would spin the wheel every time you fired a fuel mixture.

In order to cause the wheel to spin in the proper direction, you would have to fire the mixture with the crank in a position similar to A, Fig. 12-10. If the crank is in the B position, the lid could not fly up without pushing the crank bearing up or the container down. If it were fired with the crank in C position, the wheel would spin backwards.

It is important that the mixture be fired when the crank is in the proper position. By studying Fig. 12-10, you can see that the crankshaft changes the RECIPROCATING (up and down) motion of the lid into ROTARY (round and round) motion.

NAME THE PARTS

At this time, it is well to name the parts you have developed. By doing so, you will learn what to call parts in a petrol engine that serve the same general purpose.

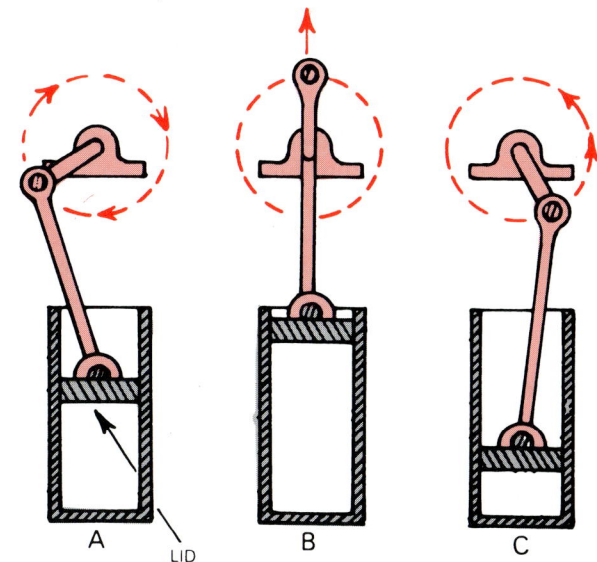

Fig. 12-10. Lid placed inside container forms a simple engine.

The container would be called the BLOCK. The hole in the container, or block, would be called the CYLINDER. The lid would be termed the PISTON. The shaft, with a section bent in the shape of a crank, would be called the CRANKSHAFT. The rod that connects the crankshaft to the piston, is called the CONNECTING ROD. The bearings that support the crankshaft are called MAIN BEARINGS. The connecting rod has an upper and lower bearing. The lower bearing (the one on the crankshaft) is called a CONNECTING ROD BEARING. The wheel is called the FLYWHEEL.

Refer back to Fig. 12-9, and see if you can substitute the correct names for those listed.

FASTENING THE PARTS

Since you are going to fasten the parts so the main bearing and cylinder cannot move, it would be wise

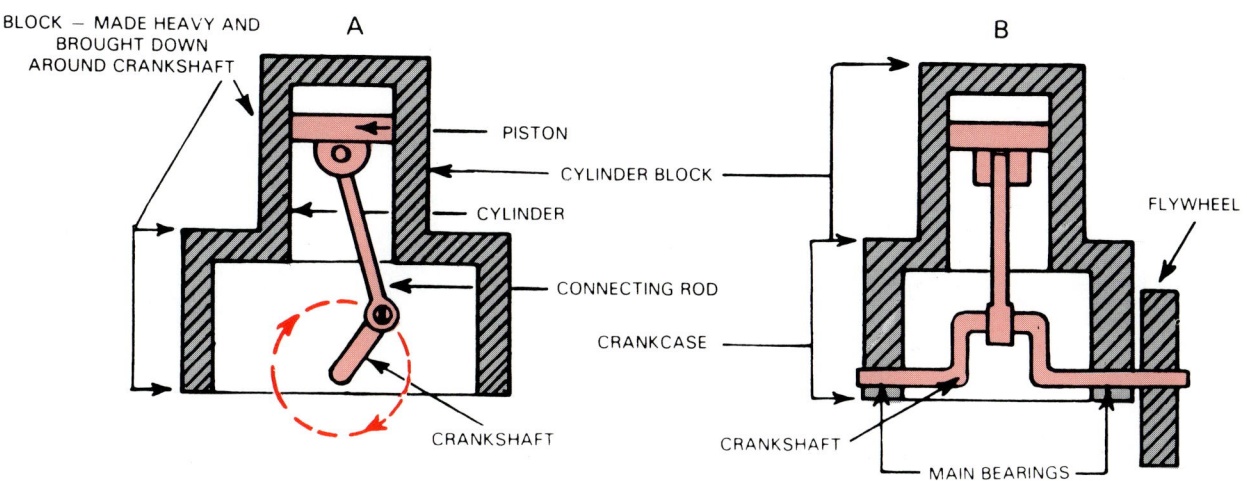

Fig. 12-11. Front, side views showing how engine is inverted, block strengthened and brought down and around crankshaft. This forms the crankcase.

to invert your engine and place the cylinder on top. Just why, will soon be obvious.

Make the block heavy to give it strength to withstand the fuel explosions. Bring it down to support the main bearing. Fig. 12-11.

By bringing the block out and down around main bearing, you now have a strong unit. Note in B, Fig. 12-11, that lower block end forms a case around the crank. It allows you to have two main bearings. This lower block end is called a CRANKCASE.

LENGTHEN THE PISTON

Now make the piston longer. This stops it from tipping sideways in the cylinder. In order to avoid a piston that is too heavy, it can be made hollow as shown in Fig. 12-12.

Tipping is illustrated in Fig. 12-13.

If the piston is to travel straight up and down, and the connecting rod is to swing back and forth in order to follow the crank, it is obvious that the connecting rod must be able to move where it is fastened to the piston. Fig. 12-14.

Now, drill a hole through the upper end of the connecting rod; also drill a hole through the piston. Line up the two and pass a pin through them. This pin is called a GUDGEON PIN or WRIST PIN. It is secured in various ways (more on this later). Fig. 12-15.

GETTING FUEL INTO THE ENGINE

You have probably noticed no way has been provided to get fuel into the upper cylinder area of your assembled engine. Your next problem is to develop a system to admit fuel, and to exhaust (blow out) the fuel after it is burned.

One opening in the upper cylinder area (called the COMBUSTION CHAMBER) will not be adequate. You cannot very well admit fuel and exhaust the burned gas out the same opening. You need TWO openings.

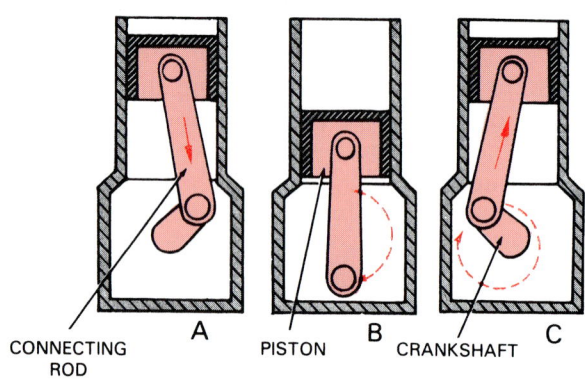

Fig. 12-14. Connecting rod must swing. Note position in A compared to C.

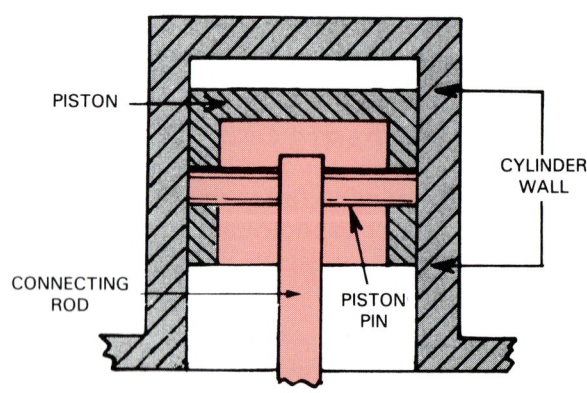

Fig. 12-12. Piston is lengthened to avoid tipping.

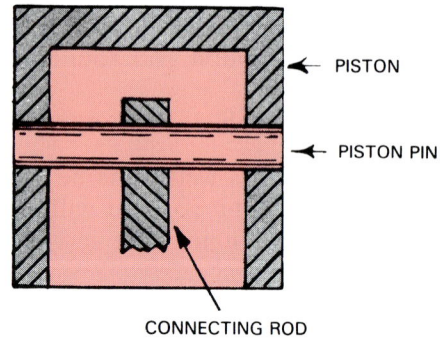

Fig. 12-15. Pin secures rod to piston.

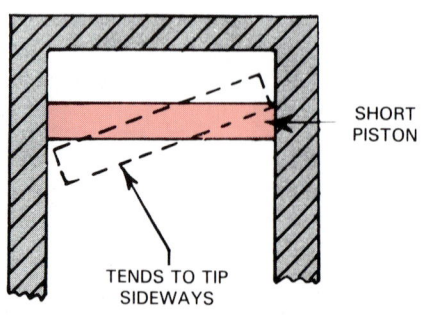

Fig. 12-13. Effect of short piston.

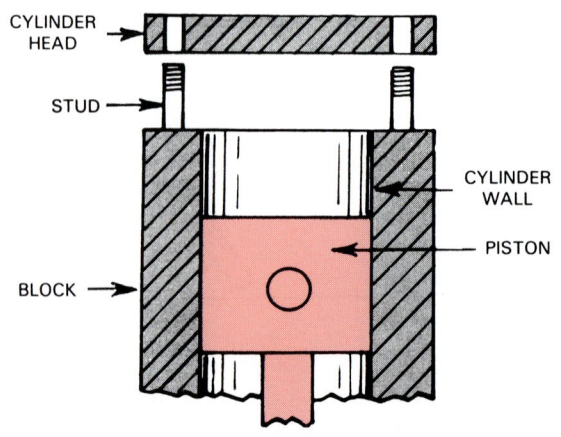

Fig. 12-16. Cylinder head. Top of cylinder has been removed.

REMOVABLE CYLINDER HEAD

Redesign your cylinder block and make the top removable. You will call this removable head a CYLINDER HEAD. It will be fastened in place with bolts or studs and nuts, Fig. 12-16.

FUEL INTAKE AND EXHAUST PASSAGES

Making your cylinder head removable has not yet solved all of the problems. You cannot take it off, and put it on, each time the engine fires.

Now make a head of much thicker metal and make two holes or passages like those shown in Fig. 12-17. This will give you a passage to take in fuel mixture, and one to exhaust it. These passages are called VALVE PORTS.

You will notice at A, Fig. 12-17, that the ports do not connect to the cylinder. To make a connection, you must cut an opening in the cylinder head that will straddle the intake and exhaust ports, plus the cylinder. Fig. 12-18.

VALVES

The next logical step is to provide a device to open and close the ports. If the ports are left open, and fuel explodes in the combustion chamber, it will blow out the openings and fail to push the piston down.

This port control device (valve) will have to be arranged so that it can be opened and closed when desired. Now arrange a valve in each opening. This may be done as shown in Fig. 12-19.

The valves may be held in a straight up and down position by a hole bored in the cylinder head metal at A, Fig. 12-19. This hole is called a VALVE GUIDE because it guides the valve up and down in a straight line.

Also to keep the valve tightly closed against its seat a spring arrangement will need to be fitted to each valve. See Fig. 12-19 which also shows a typical valve spring arrangement.

FOUR-STROKE CYCLE

At this point you have an intake valve to allow fuel to enter the cylinder and an exhaust valve to let the burned fuel out. The next problem is how to get fuel into the cylinder and out when burned.

A VACUUM IS THE CLUE

You understand that the air in which we live presses on all things. Fig. 12-20. This pressure is approximately 101kPa at sea level.

Were we to draw all of the air out of a container, we would form a vacuum. A vacuum is unnatural and atmospheric pressure will do all it can to get into the low pressure area. If there is the slightest leak in

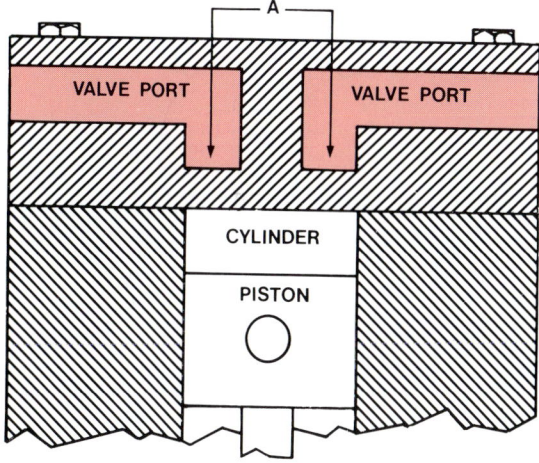

Fig. 12-17. Valve ports.

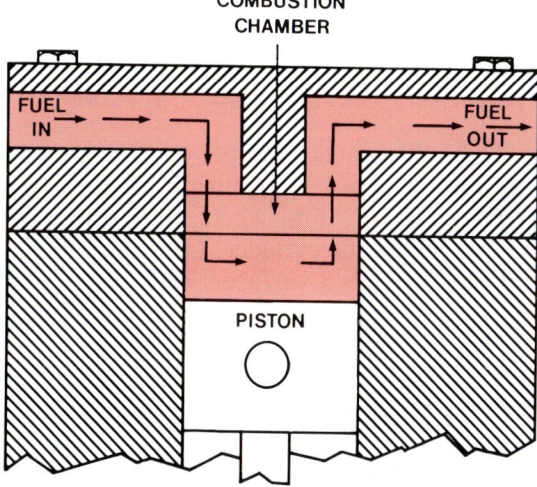

Fig. 12-18. Note opening in head allowing fuel mixture to enter and leave cylinder. This opening is called combustion chamber.

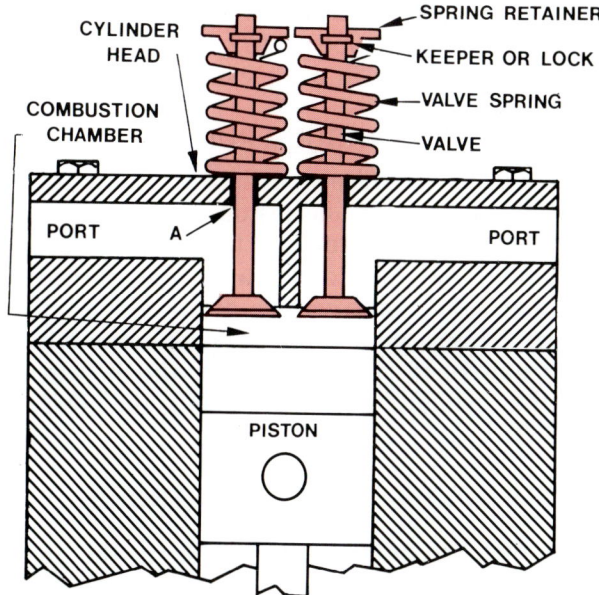

Fig. 12-19. Valves installed in ports.

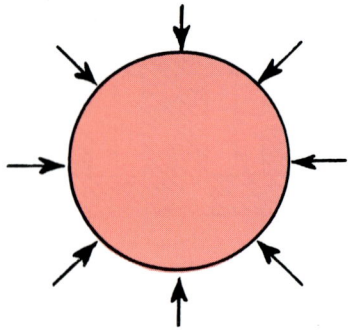

Fig. 12-20. Air pressed on all things with a pressure of approximately 101 kPa.

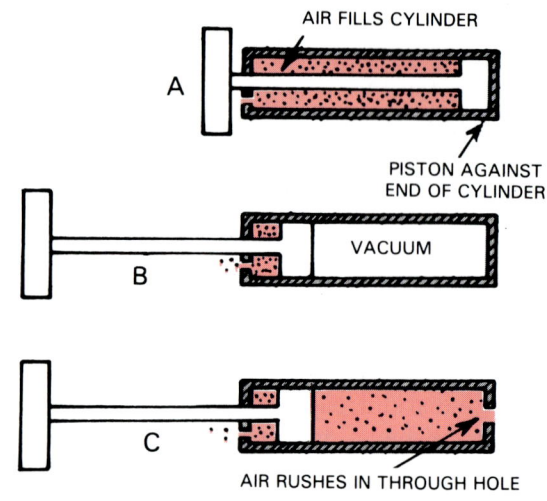

Fig. 12-21. Simple vacuum pump.

the container, air will seep in until the pressure is the same on the inside as it is on the outside.

A vacuum, then, is any area in which the air pressure is lower than that of the prevailing or outside atmospheric pressure.

YOU ALREADY HAVE A VACUUM PUMP

Fig. 12-21 illustrates a simple vacuum pump, or a cylinder into which a snug fitting piston is placed. You will see in A, Fig. 12-21, that the piston is against the end of the cylinder. Obviously there is no air between these two surfaces.

When you pull the piston through the cylinder, as in B, Fig. 12-21, you will have a larger area between the face of the piston, and the end of the cylinder. If there was no air between them before, and if the piston fits snugly against the cylinder, there still could be no air between them. This gives you a large area in which there is no air, or in other words, a vacuum.

If you were to drill a hole into the closed end of the cylinder, C, Fig. 12-21, air would immediately rush in and fill the cylinder. Any material, such as a fuel mixture, that happened to be in the surrounding air would be drawn into the cylinder.

WHERE IS YOUR VACUUM PUMP?

If the cylinder and piston, in Fig. 12-21, forms a vacuum, the cylinder and piston in your engine will do the same. If the piston is at the top of the cylinder (with both valves closed) and you turned the crankshaft, the piston will be drawn down into the cylinder. This forms a strong vacuum in the cylinder. If you now open the intake valve, the air will rush into the cylinder. This is called the INTAKE STROKE.

STROKE NO. 1 — THE INTAKE STROKE

The first stroke in your engine is the intake. Instead of opening the intake valve after you have drawn the piston down, you will find it better to open the intake valve as the piston starts down. This allows the air to draw fuel in all the time the piston is moving down. If you wait until the piston is down before opening the valve, the piston will be starting up before the cylinder can be filled with air. Fig. 12-22.

Remember: The intake stroke starts with the piston at the top of the cylinder, intake valve open and exhaust valve closed, and it stops with the piston at the bottom of its travel. This requires one-half turn of the crankshaft.

STROKE NO. 2 — THE COMPRESSION STROKE

You have discovered that the smaller the particles of petrol, mixed in air, the more powerful the explosion.

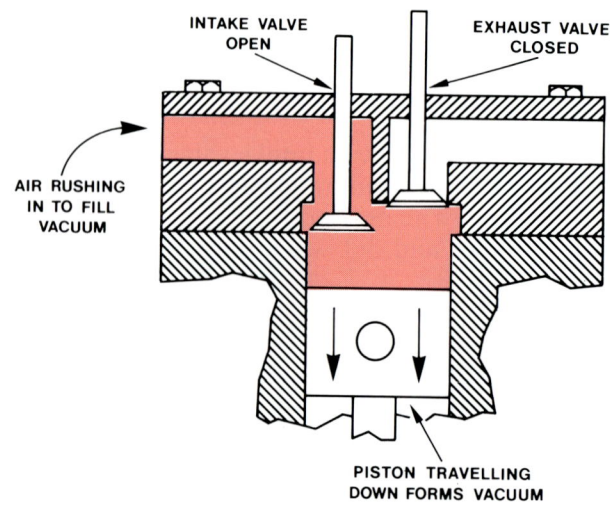

Fig. 12-22. Intake stroke.

As the crankshaft continues to move, the piston is forced up through the cylinder. If you keep both valves closed, the fuel mixture will be squeezed, or compressed, as the piston reaches the top. This is called the compression stroke. Fig. 12-23. It, too, requires one-half turn of the crankshaft.

The compression stroke serves several purposes. First, it tends to break up the fuel into even smaller particles. This happens due to the sudden swirling and churning of the mixture as it is compressed.

When engine fuel mixture is subjected to a sudden sharp compression force, its temperature rises. This increase in temperature makes the mixture easier to ignite, and causes it to explode with greater force.

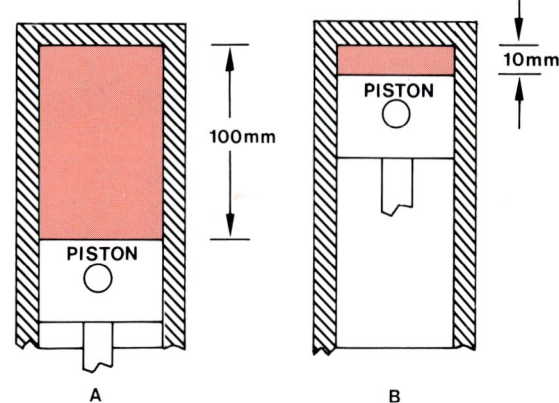

Fig. 12-24. Compression ratio.

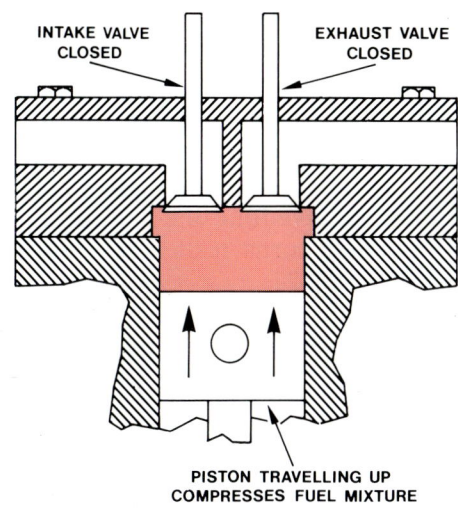

Fig. 12-23. Compression stroke.

As the piston reaches the top of its travel on the compression stroke, it has returned to the proper position to be pushed back down by the explosion.

Remember: The compression stroke starts with the piston at the bottom of the cylinder while both valves are closed, and it stops with the piston at the top of the cylinder. This requires an additional one-half turn of the crankshaft.

COMPRESSION RATIO

The amount your engine will compress the fuel mixture depends on how small a space the mixture is squeezed into. Notice in Fig. 12-24, the piston in A has travelled down 100 mm. from the top of the cylinder. This is the intake stroke. In B the piston, on the compression stroke, has travelled up to within 10 mm of the cylinder top. It is obvious that 100 mm of cylinder volume has been squeezed into 10 mm of the cylinder volume. This gives you a ratio of 10 to 1. This is termed the COMPRESSION RATIO.

STROKE NO. 3 — FIRING OR POWER STROKE

As the piston reaches the top of the compression stroke, the mixture is broken into tiny particles, and heated up. When ignited, it will explode with great force.

This is the right time to explode the mixture. A spark is provided inside the compression chamber by means of a spark plug. The spark produced at the plug is formed by the ignition system. This will be discussed later.

Just imagine that a hot spark has been provided in the fuel mixture. The mixture will explode and, in turn, force the piston down through the cylinder. This gives the crankshaft a quick and forceful push.

Both valves must be kept closed during the firing stroke or the pressure of the burning fuel will squirt out the valve ports. Fig. 12-25.

Remember: The firing stroke starts with the piston at the top of the cylinder while both valves are closed, and it stops with the piston at the bottom of the cylinder. This requires another one-half turn of the crankshaft.

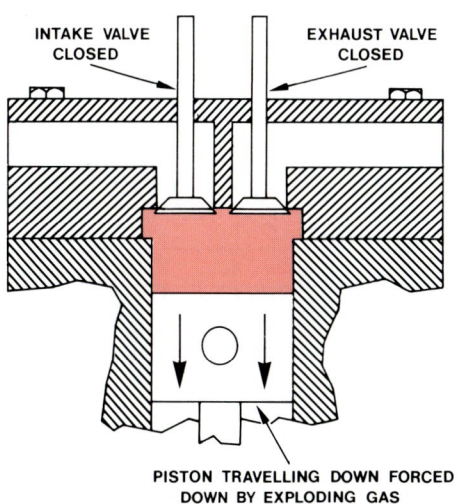

Fig. 12-25. Firing stroke.

189

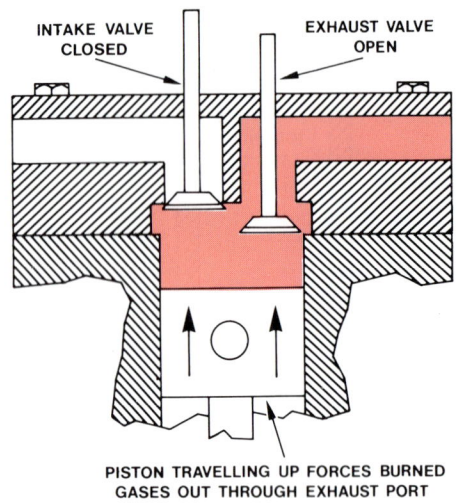

Fig. 12-26. Exhaust stroke.

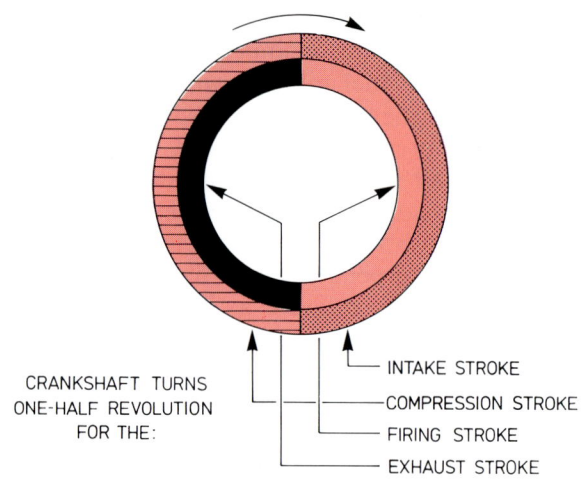

Fig. 12-27. Four strokes make two revolutions of the crankshaft.

STROKE NO. 4 — THE EXHAUST STROKE

When the piston reaches the bottom of the firing stroke the exhaust valve opens. The spinning crankshaft forces the piston up through the cylinder blowing burned gases out of the cylinder. Fig. 12-26.

Remember: The exhaust stroke starts with the piston at the bottom of the cylinder while the exhaust valve is open and intake valve is closed. It stops with the piston at the top of the cylinder. This requires one more one-half turn of the crankshaft.

COMPLETED CYCLE

If you will count the number of one-half turns in the intake, compression, firing and exhaust strokes, you will find you have four one-half turns. This gives you two complete turns (called revolutions) of the crankshaft. Fig. 12-27.

While the crankshaft is turning around twice, it is receiving power only during one-half turn, one-fourth of the time.

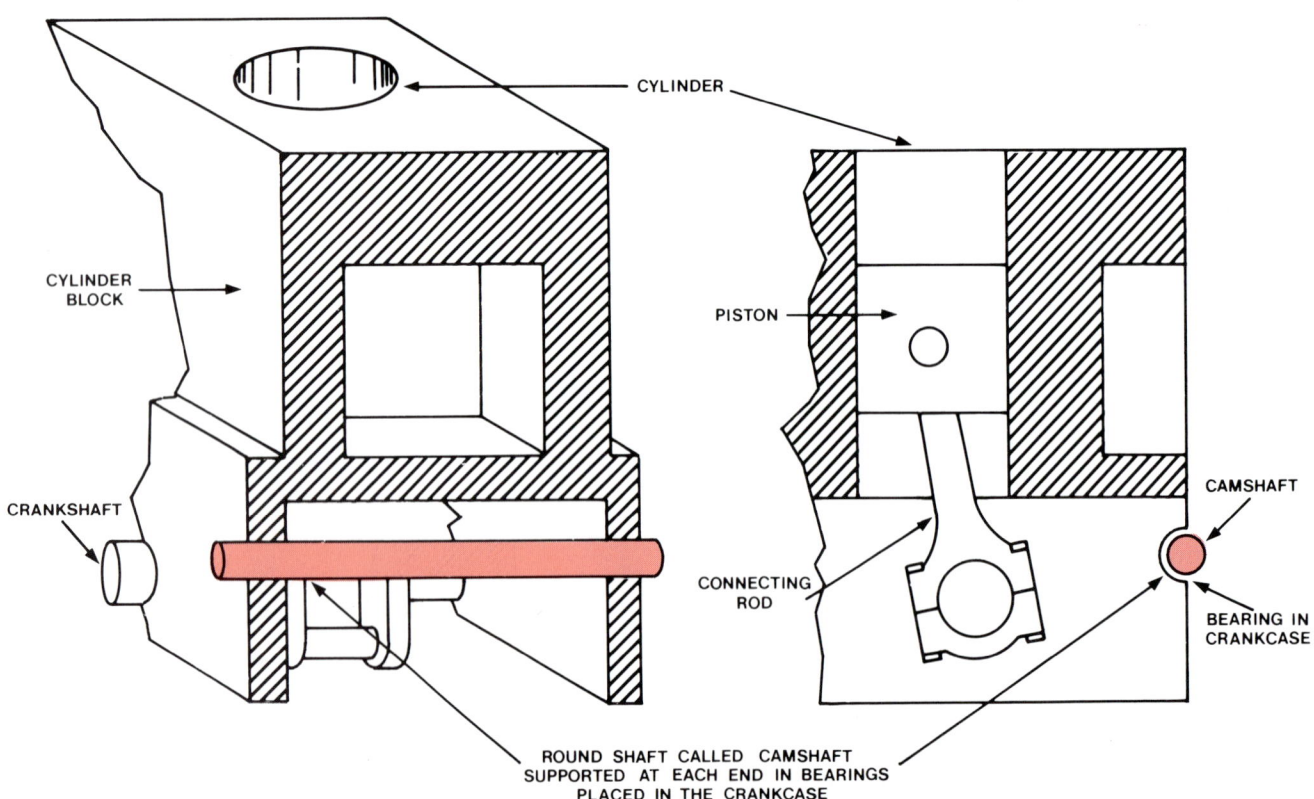

Fig. 12-28. Engine section showing how camshaft is positioned.

CYCLE REPEATED

As soon as the piston reaches the top of the exhaust stroke, it starts down on another intake, compression, firing and exhaust cycle. This is repeated time after time. Each complete cycle consists of four strokes of the piston, hence the name FOUR-STROKE CYCLE engine. There is another type, called a two-stroke cycle, that will be covered later.

WHAT OPENS AND CLOSES THE VALVES?

You have seen that the intake valve must be opened for the intake stroke. Both valves remain closed during the compression and firing strokes. The exhaust valve opens during the exhaust stroke. You must now design a device to open and close the valves.

Start by machining a round shaft that will lie beneath the valves, at some distance. This shaft is supported in bearings placed in the front and the rear of the crankcase. See Fig. 12-28.

CAMSHAFT OPENS VALVES

The shaft will have bumps called cam lobes, machined as a part of the shaft. This shaft, with the cam lobes, is called a CAMSHAFT. Fig. 12-29.

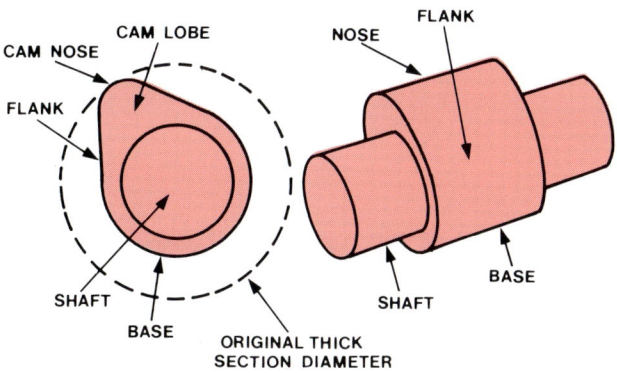

Fig. 12-29. Section of camshaft showing cam lobe. Study names of various parts of lobe.

The distance each valve will be opened, how long it will stay open, how fast it opens and closes, can all be controlled by the height and shape of the lobe. The lobe must, of course, be located directly above the valve stem. Fig. 12-30.

As you will see later, it is impractical to have the cam lobe contact the end of the valve stem itself. You have placed the camshaft some distance away from the end of the valve stem in the cylinder block. When you turn the camshaft, the lobes will not even touch the valve stem.

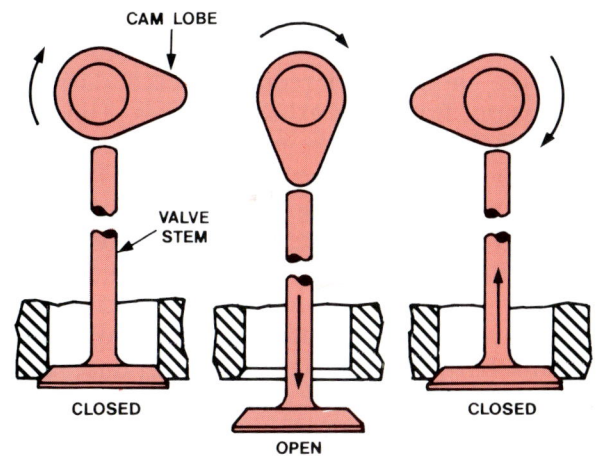

Fig. 12-30. How valve is opened.

VALVE LIFTER, PUSH ROD AND ROCKER ARM

Your next step is to construct a round bar-like unit called a cam follower or valve lifter and an arm arrangement called a rocker arm. These components with the help of a rod called a push rod are necessary to transfer camshaft action (in the cylinder block) to the ends of the valve stems (in the cylinder head).

You have now built the essential parts of the valve system. The parts, in their proper positions, are referred to as the VALVE TRAIN. Fig. 12-31.

CAMSHAFT SPEED

You have developed a method of opening and closing the valves. The next problem is how to turn the camshaft, and at WHAT SPEED.

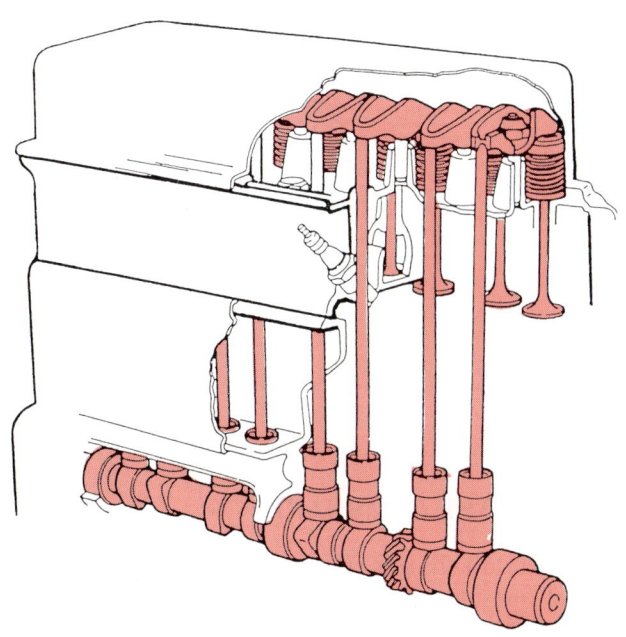

Fig. 12-31. Valve train includes valves and all mechanisms which open and close them (note coloured portion). Refer to this illustration as you review operation of valve train.

Now stop a minute and think — each valve must be open for one stroke only. The intake is open during the intake, and remains closed during the compression, firing and exhaust strokes. This would indicate that the cam lobe must turn fast enough to open the valve every fourth stroke. Fig. 12-32.

You can see, from Fig. 12-32 that it takes one complete revolution of the cam lobe for every four strokes. Remembering back, four strokes of the piston required two revolutions of the crankshaft. You can say then, that for every two revolutions of the crankshaft, the camshaft must turn once. If you are speaking of the speed of the camshaft, you can say the camshaft must turn at one-half the crankshaft speed.

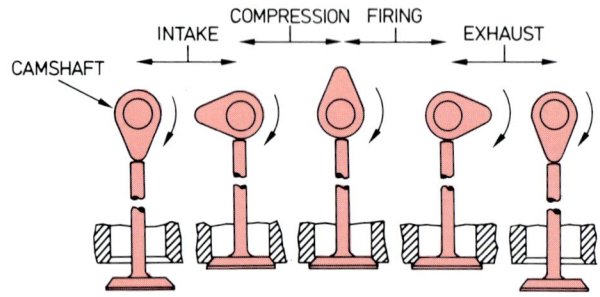

Fig. 12-32. Intake valve is opened once every fourth stroke. Each stroke turns camshaft one-quarter turn.

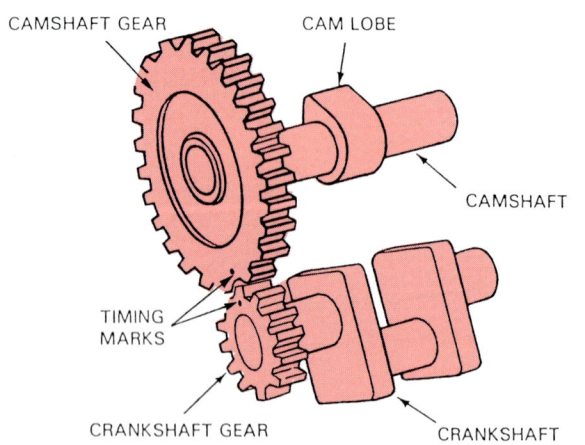

Fig. 12-33. Crankshaft turns camshaft at one-half crankshaft speed.

TURNING THE CAMSHAFT

If the crankshaft is turning, and the camshaft must turn at one-half crankshaft speed, it seems logical to use the spinning crankshaft to turn the camshaft.

One very simple way to drive the camshaft would be by means of gears, one fastened on the end of the crankshaft and the other on the end of the camshaft.

If you put a small gear on the crankshaft and a larger gear with twice as many teeth on the camshaft, the crankshaft will turn the camshaft. It will also turn it at exactly one-half crankshaft speed. Fig. 12-33.

TIMING THE VALVES

You have now developed a method of not only turning the camshaft, but also at the correct speed. The next question to be answered is how will you get the valves to open at the proper time. This is called VALVE TIMING.

Start with intake stroke. As you have discovered, the intake valve must open as the piston starts down in the cylinder. Place your piston at top dead centre (TDC — topmost point of piston travel). Fig. 12-34.

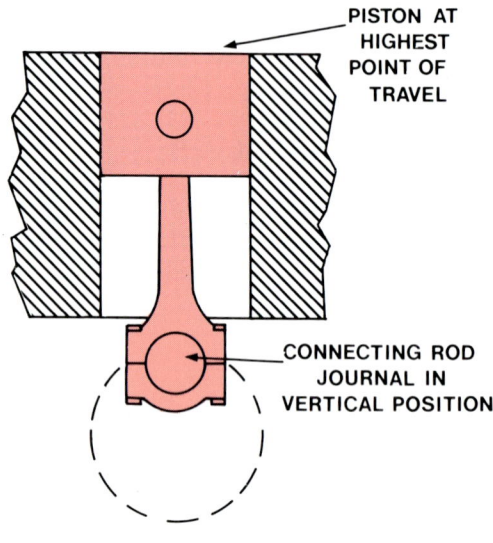

Fig. 12-34. Piston at top dead centre (TDC).

Insert the camshaft, then turn it in a counterclockwise direction until the flank of the intake cam lobe contacts the lifter. The timing gear should be meshed with the crankshaft gear. Fig. 12-35.

MARK THE GEARS

You should punch a mark on the crankshaft gear, and a mark on the camshaft gear. These are called TIMING MARKS. See Fig. 12-36.

If the camshaft is removed, it may be reinstalled by merely lining up the timing marks.

Were you to crank the engine, the crankshaft would pull the piston down on the intake stroke. The camshaft would also rotate, causing the cam lobe to turn and open and close the valve. The crankshaft would make one-half turn; camshaft one-quarter turn during this stroke. Fig. 12-37.

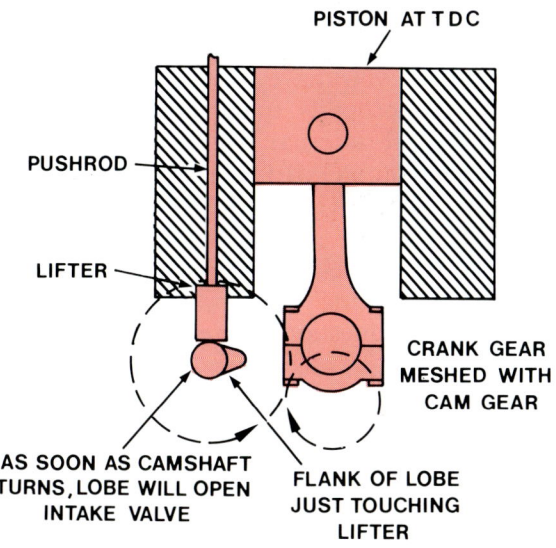

PISTON AT T D C

PUSHROD

LIFTER

CRANK GEAR
MESHED WITH
CAM GEAR

AS SOON AS CAMSHAFT
TURNS, LOBE WILL OPEN
INTAKE VALVE

FLANK OF LOBE
JUST TOUCHING
LIFTER

Fig. 12-35. Position of cam lobe at the start of the intake stroke.

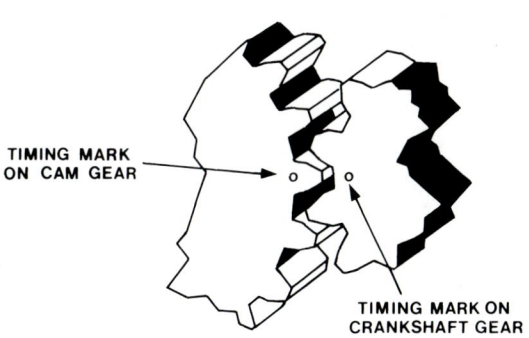

TIMING MARK
ON CAM GEAR

TIMING MARK ON
CRANKSHAFT GEAR

Fig. 12-36. Timing gears meshed and marked.

INTAKE REMAINS CLOSED

As the crank continues to turn, it will push the piston up on compression, down on firing, and up on exhaust. During these three strokes, the intake cam lobe continues to turn. When the piston reaches top dead centre (TDC) on the exhaust stroke, the flank of the cam lobe will again be touching the intake lifter.

As your next stroke will be the intake, the piston is ready to start down and the cam lobe is ready to raise the lifter to provide proper timing. Fig. 12-38.

EXHAUST TIMED THE SAME WAY

The only difference between the intake and exhaust timing is that the piston will be on bottom dead centre (BDC). On bottom dead centre the flank of the exhaust cam lobe will be contacting the lifter. When the piston begins its upward exhaust stroke the exhaust cam lobe contacts the lifter to open the exhaust valve.

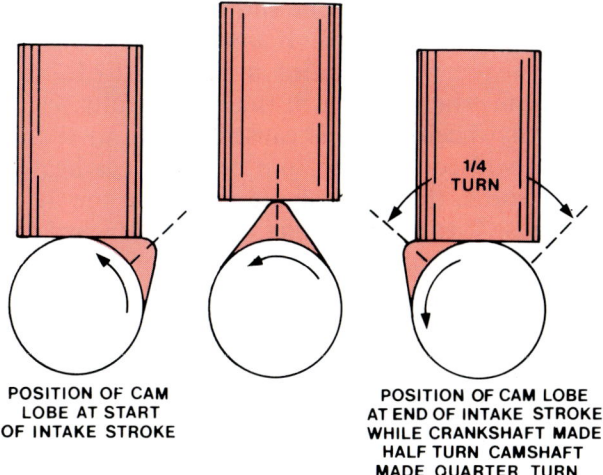

1/4
TURN

POSITION OF CAM
LOBE AT START
OF INTAKE STROKE

POSITION OF CAM LOBE
AT END OF INTAKE STROKE.
WHILE CRANKSHAFT MADE
HALF TURN CAMSHAFT
MADE QUARTER TURN

Fig. 12-37. Camshaft makes one-quarter turn during each stroke.

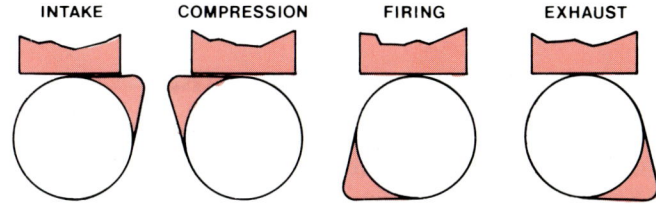

INTAKE   COMPRESSION   FIRING   EXHAUST

Fig. 12-38. Position of intake lobe at start of each stroke.

THE FLYWHEEL

When you crank an engine, it turns through all the necessary strokes. The only time the crankshaft receives power is during the firing stroke. After the firing stroke is completed, the crankshaft must continue to turn to exhaust the burned gases, take in fresh fuel and then compress it. One power stroke is not enough to keep the crankshaft turning during the four required strokes. Fig. 12-39.

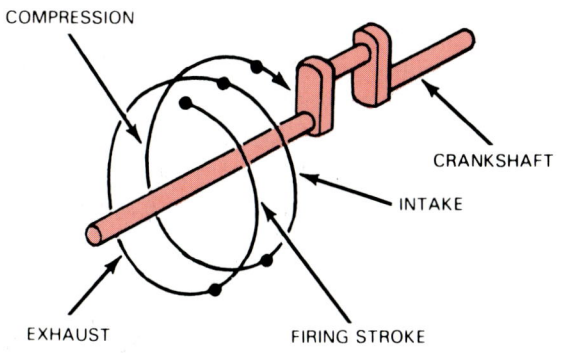

COMPRESSION

CRANKSHAFT

INTAKE

EXHAUST

FIRING STROKE

Fig. 12-39. After the firing stroke, the crankshaft must coast through the next three strokes.

Building an Engine

You will recall that one end of the crankshaft has a timing gear attached. If you will fasten a fairly large, heavy wheel, called a FLYWHEEL, to the other end, the engine will run successfully.

The flywheel, which is caused to spin by the firing stroke, will continue to spin because it is heavy. In other words, the inertia build up within the flywheel will cause it to keep turning. As it is attached to the crankshaft, it will cause the shaft to continue to turn. The shaft will now spin long enough to reach the next firing stroke. Fig. 12-40.

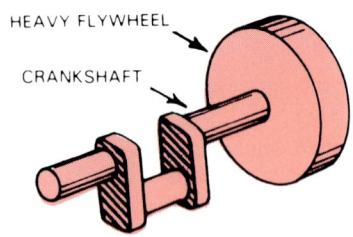

Fig. 12-40. Flywheel inertia helps engine to run.

ENGINE WILL RUN MORE SMOOTHLY

You can see that every time the engine fires, the crank receives a hard push. This will speed it up. As the power stroke is finished the shaft coasts. This slows it down. The alternating fast and slow speed would make for a very rough running engine.

With the heavy flywheel in place, the firing stroke cannot increase the crankshaft speed so quickly since it must also speed up the flywheel. When the crankshaft is coasting, it cannot slow down so fast because the flywheel wants to keep spinning. This will cause the crankshaft speed to become more constant and will, in turn, give a smooth running engine.

BASIC ENGINE COMPLETED

You have now completed a basic engine. With the addition of fuel, spark and oil, it will run. This is a one cylinder engine, and its uses would be confined to such small tasks as powering lawnmowers, scooters, generators, carts, etc.

There are many modifications and improvements that can be made to your simple engine. These will be discussed in Chapter 15.

DO NOT BE CONFUSED

Regardless of the size, number of cylinders, and horsepower a four-stroke cycle engine may have, it will contain the same basic parts that do the same job, in the same way, as the small engine you have just built. The parts may be arranged in different ways and their shape changed, but you will not find ANY different BASIC parts, just more of them.

A MOST IMPORTANT CHAPTER

The chapter you have just completed is most important. To properly understand the theory and operation of the internal combustion engine, it is essential that you COMPLETELY UNDERSTAND ALL THE MATERIAL PRESENTED IN THIS CHAPTER.

To help you check your grasp of the chapter, answer the following questions. When you CAN answer all of the questions correctly, you are on your way.

KNOW THESE TERMS

Internal combustion, Petrol, Cylinder, Piston, Crankshaft, Connecting rod, Cylinder head, Valve ports, Valves, Vacuum, Four-stroke cycle, Camshaft, Valve train, Flywheel.

REVIEW QUESTIONS

1. What is an engine?
2. Petrol is obtained from _____ _____.
3. Petrol is a mixture of _____ and _____ atoms.
4. List four features of a good petrol.
5. In order to burn petrol must have _____.
6. What must be done to petrol to speed up the burning rate?
7. Petrol, when it reaches the cylinder, must be _____ into tiny _____, and mixed with _____.
8. When the petrol explodes in the cylinder, what movable part of the engine first receives the force?
9. The connecting rod must have a bearing at _____.
10. What is a vacuum?
11. How is a vacuum developed in an engine?
12. What do you need a vacuum for?
13. What is the purpose of valve ports?
14. What is placed in each port to open and close the port?
15. List the strokes, in order, in a four-stroke cycle engine.
16. The position of each valve during the various strokes is important. Give the position of each valve for each of the strokes in question 15.
17. What opens and closes the valves?
18. The camshaft is turned by the _____ through means of _____.
19. What is the speed of the camshaft in relation to that of the crankshaft?
20. Give two uses for the flywheel.
21. Fig. 12-41 shows all of the parts of your engine. Can you identify the various parts?

194

A. _____
B. _____
C. _____
D. _____
E. _____
F. _____
G. _____
H. _____
I. _____
J. _____
K. _____
L. _____
M. _____
N. _____
O. _____

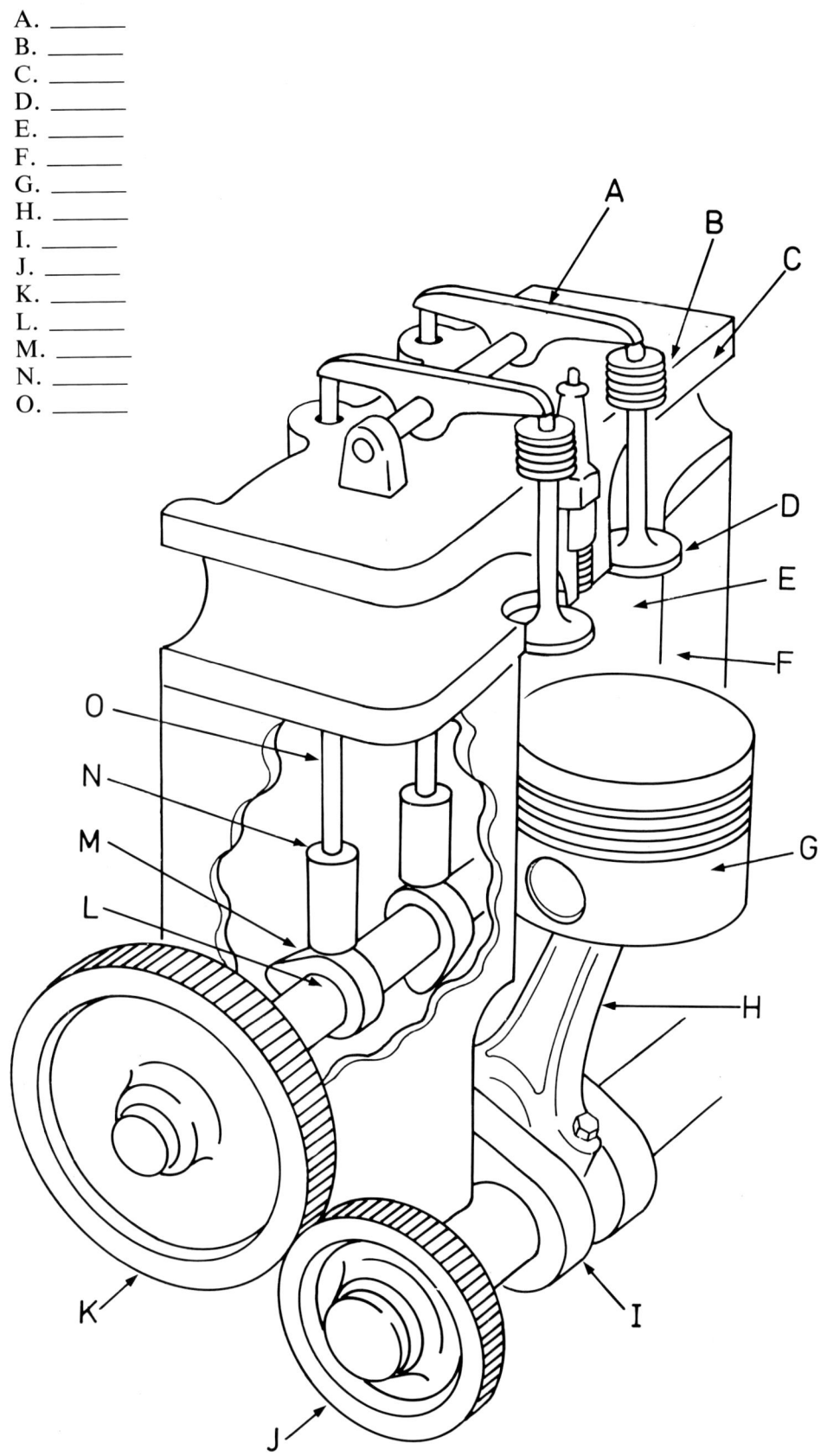

Fig. 12-41. This is the basic engine you have built. Can you identify the parts?

Chapter 13

ENGINE CLASSIFICATIONS

After studying this chapter, you will be able to:
☐ Describe basic automotive engine classifications.
☐ Compare petrol and diesel engines.
☐ Compare two and four-stroke cycle engines.
☐ Contrast combustion chamber designs.
☐ Discuss alternate engine types.

Now that you have learned about the basic parts of an engine, you should become familiar with the various engine types used in automobiles. An experienced mechanic can glance into an engine compartment and "rattle off" dozens of engine facts.

For example, you might hear a mechanic say "this is an in-line, 4-cylinder, overhead camshaft, hemi-head engine." This information has specific meaning about the particular engine.

You must understand these kinds of terms when troubleshooting and repairing car engines. Study this chapter carefully. It will help you learn the "shop talk" which all good mechanics should know.

ENGINE CLASSIFICATIONS

Even though basic parts are the same, design differences can change how engines operate and how they are repaired. For this reason, you should be able to classify engines.

Motor car engines are commonly classified by:
1. Cylinder arrangement.
2. Number of cylinders.
3. Cooling system type.
4. Valve location.
5. Camshaft location.
6. Combustion chamber design.
7. Type of fuel burned.
8. Type of ignition.
9. Number of strokes per cycle.

Cylinder arrangement

Cylinder arrangement refers to the position of the cylinders in relation to the crankshaft. There are four basic cylinder arrangements: in-line, V-type, slant, and opposed. These are shown in Fig. 13-1.

An in-line engine's cylinders are lined up in a single row. Each cylinder is located in a straight line parallel to the crankshaft. Four, five, and six-cylinder engines are commonly in-line.

Viewed from either end, a V-type engine looks like the letter "V", Fig. 13-1. The two banks of cylinders

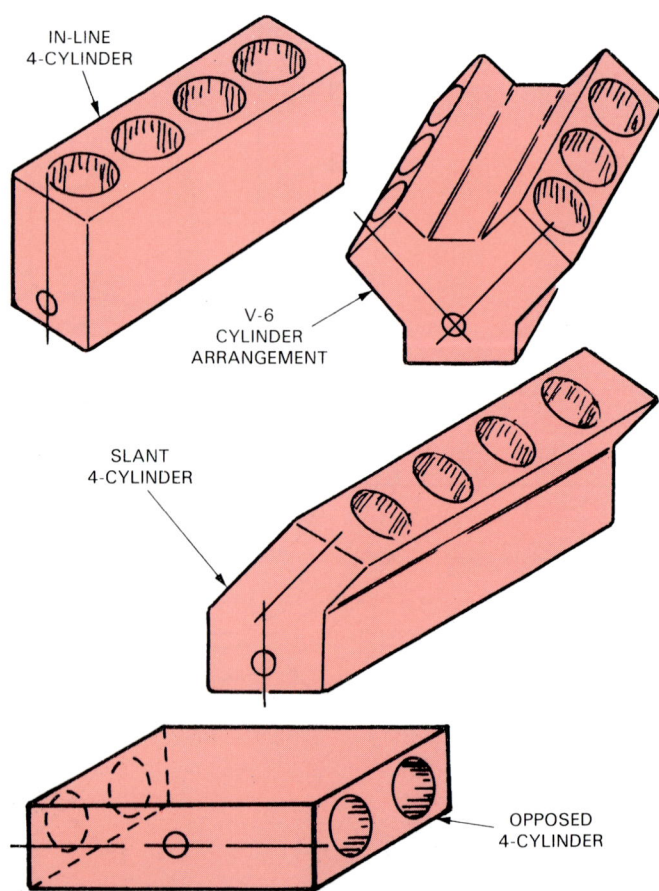

Fig. 13-1. Organisation of cylinders in block determine one engine classification. These represent four basic types used in cars.

lie at an angle to each other. This arrangement reduces the length and the height of the engine.

A slant engine has only one bank of cylinders angled or leaning to one side.

As with V-type engines, this saves space, Fig. 13-1. It allows the body and hood of the car to be much lower. A relatively large engine can be fitted into a small engine compartment.

Cylinders of an opposed engine lie flat on either side of the crankshaft, Fig. 13-1. Because of its appearance, this type is sometimes called a "flat engine". An opposed engine may be found in rear engine cars (older Volkswagens and a few late model, foreign sports cars, for example).

Number of cylinders

Normally, car engines have either 4, 6, or 8 cylinders. A few engines, however, have 5, 12, or 16 cylinders.

A greater number of cylinders generally increases engine smoothness and power. For instance, an 8-cylinder engine would produce twice as many power strokes per crank revolution as a 4-cylinder engine. This would reduce power pulsations and roughness (vibration) at idle.

Four-cylinder engines are usually inline, slant, and sometimes opposed. Six-cylinder engines can be in-line, slant, or a V-type. Five-cylinder engines are normally in-line. Eight-cylinder engines are commonly a V-type.

Cylinder numbering and firing order

Engine manufacturers number each engine cylinder to help the mechanic make repairs. The service manual will provide an illustration of the engine, as shown in Fig. 13-2.

Cylinder numbers are normally stamped on the connecting rods and sometimes, they are cast into the intake manifold. Cylinder numbering varies with different auto manufacturers.

You should keep this in mind when referring to engine classifications. Two different V-6 engines, for example, could have completely different cylinder numbering systems.

Firing order refers to the sequence in which combustion occurs in each engine cylinder. The position of the crankshaft rod journals in relation to each other determines engine firing order.

The service manual will usually have a drawing showing the firing order for the engine. This information may also be given on the engine intake manifold.

Two similar engines can have completely different firing orders. For example, a 4-cylinder, in-line engine may fire 1-3-4-2 or 1-2-4-3. Firing orders for 6 and 8-cylinder engines also vary.

A mechanic needs to know firing order when working on the engine's ignition system. It can be used when installing spark plug wires, a distributor, or when doing other tune-up related operations. Fig. 13-2.

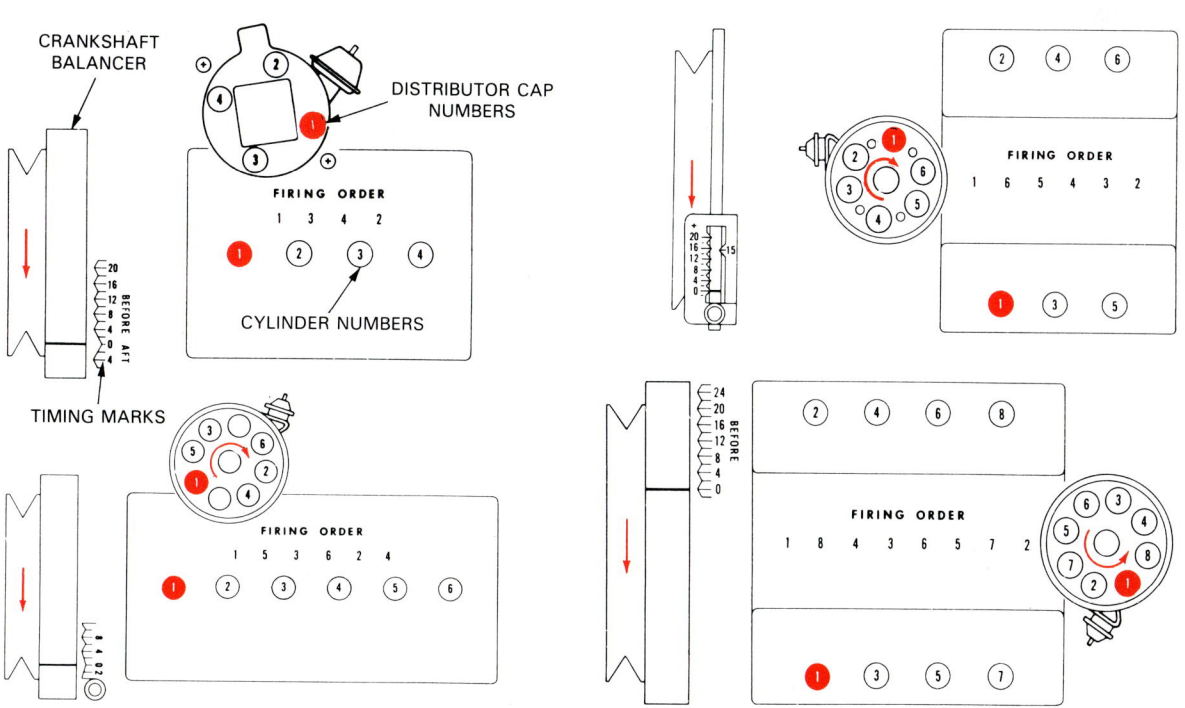

Fig. 13-2. Engine cylinder numbers, distributor cap numbers, and firing order numbers. Cylinder numbers are usually stamped on connecting rods and may be cast into intake manifold. Numbers vary from engine to engine.

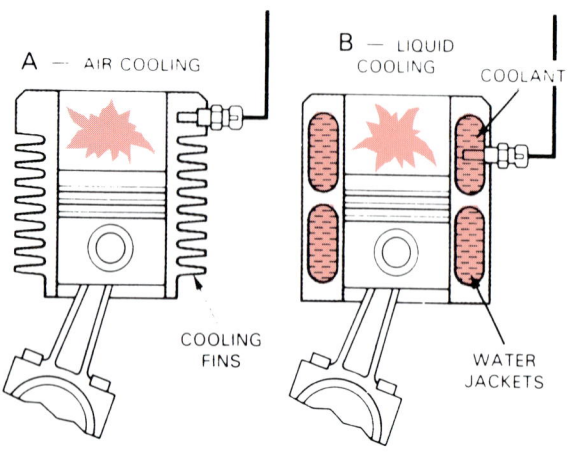

Fig. 13-3. A — Air cooling systems uses large fins around cylinder to remove heat. B — Liquid cooling system surrounds cylinder with coolant. Liquid cooling system is commonly used in cars. Motorcycles and lawnmowers use air cooling system.

Cooling system type

There are two types of cooling systems: liquid and air types. The liquid cooling system is the most common. See Fig. 13-3.

The liquid cooling system surrounds the cylinder with coolant (water and corrosion inhibitor solution). The coolant carries combustion heat out of the cylinder head and engine block to prevent engine damage.

An air cooling system circulates air over cooling fins on the cylinders. This removes heat from the cylinders to prevent overheating damage.

Air-cooled engines are no longer used in modern passenger cars. They can be found on motorcycles and lawnmowers. With strict exhaust emission regulations, auto makers have phased out air-cooled engines. They cannot maintain as constant a temperature as a liquid type cooling system. This reduces engine efficiency and increases exhaust pollution.

Fuel type

A car engine is also classified by the type of fuel it burns. A petrol engine burns petrol. A diesel engine burns diesel oil. These are the most common types of fuel for cars.

LP-gas (liquified petroleum gas), can also be used to power a car engine.

Ignition type

Two basic methods are used to ignite the fuel in an engine combustion chamber: an electric arc (spark plug) and hot air (compressed air). Look at Fig. 13-4.

A spark ignition engine uses an electric arc at the spark plug to ignite the fuel. The arc produces enough heat to start the fuel burning. Petrol engines use spark ignition, Fig. 13-4.

A compression ignition engine squeezes the air in the combustion chamber until it is hot enough to ignite the fuel. A diesel engine is a compression ignition engine. No spark plugs are used, Fig. 13-4.

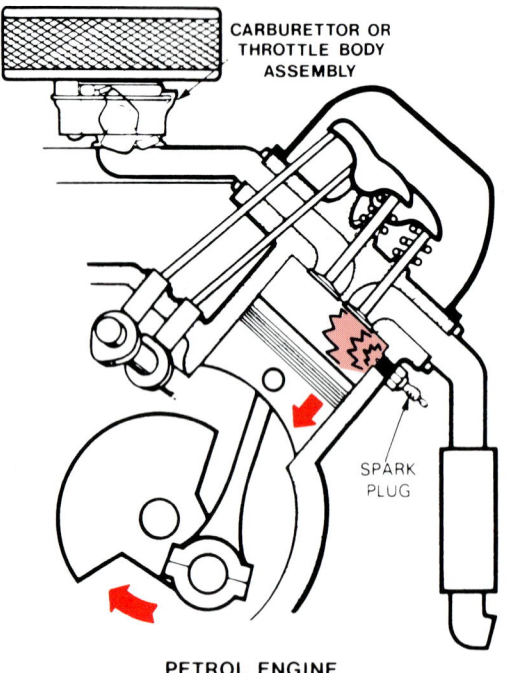

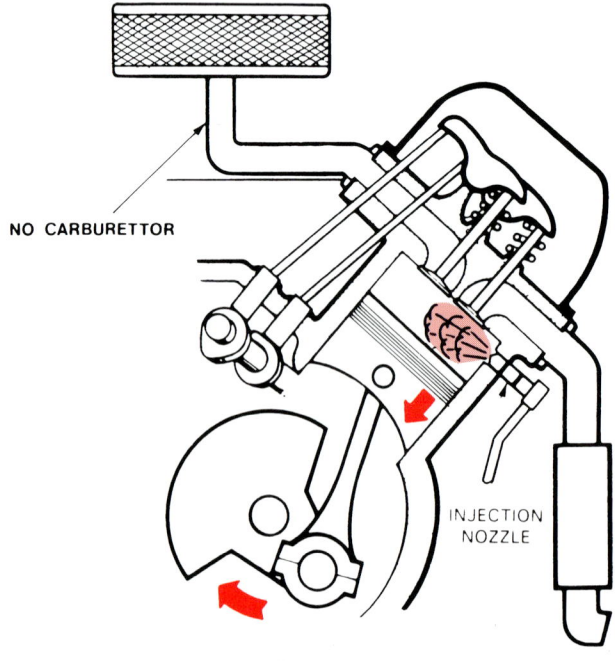

Fig. 13-4. Petrol and diesel engines use different means to ignite fuel. Petrol engines uses spark plug to start power stroke. Diesel engine compresses air in cylinder until hot. When fuel is injected into cylinder, hot air makes fuel burn.

Valve location

Another engine classification can be made by comparing the location of the valves.

An L-head engine has both the intake and exhaust valves in the block, Fig. 13-5A. Also called a flat head engine, its cylinder head simply forms a cover over the cylinders and valves. The camshaft is in the block and pushes upward to open the valves. Most four-stroke cycle lawnmower engines are L-head types. Car engines are no longer L-head types.

In an I-head engine, both valves are in the cylinder head. Another name for this design is overhead valve (OHV) engine. Fig. 13-5B.

The OHV engine has replaced the flat head engine in cars. Numerous variations of the overhead valve engine are now in use.

Other valve configurations have been used in the past. However, they are so rare that their mention is not important.

Camshaft location

There are two basic locations for the engine camshaft: in the block and in the cylinder head. Both locations are common.

A cam-in-block engine uses push rods to transfer motion to the rocker arms and valves, Fig. 13-5B. The term overhead valve (OHV) is sometimes used instead of cam-in-block.

In an overhead cam (OHC) engine, the camshaft is located in the top of the cylinder head. Push rods are NOT needed to operate the rockers and valves. This type engine is a refinement of the overhead valve engine. Refer to Fig. 13-5C.

With the cam in the head, the number of valve train parts is reduced. This cuts the weight of the valve train. Also the valves can be placed at an angle to improve breathing (airflow through cylinder head ports).

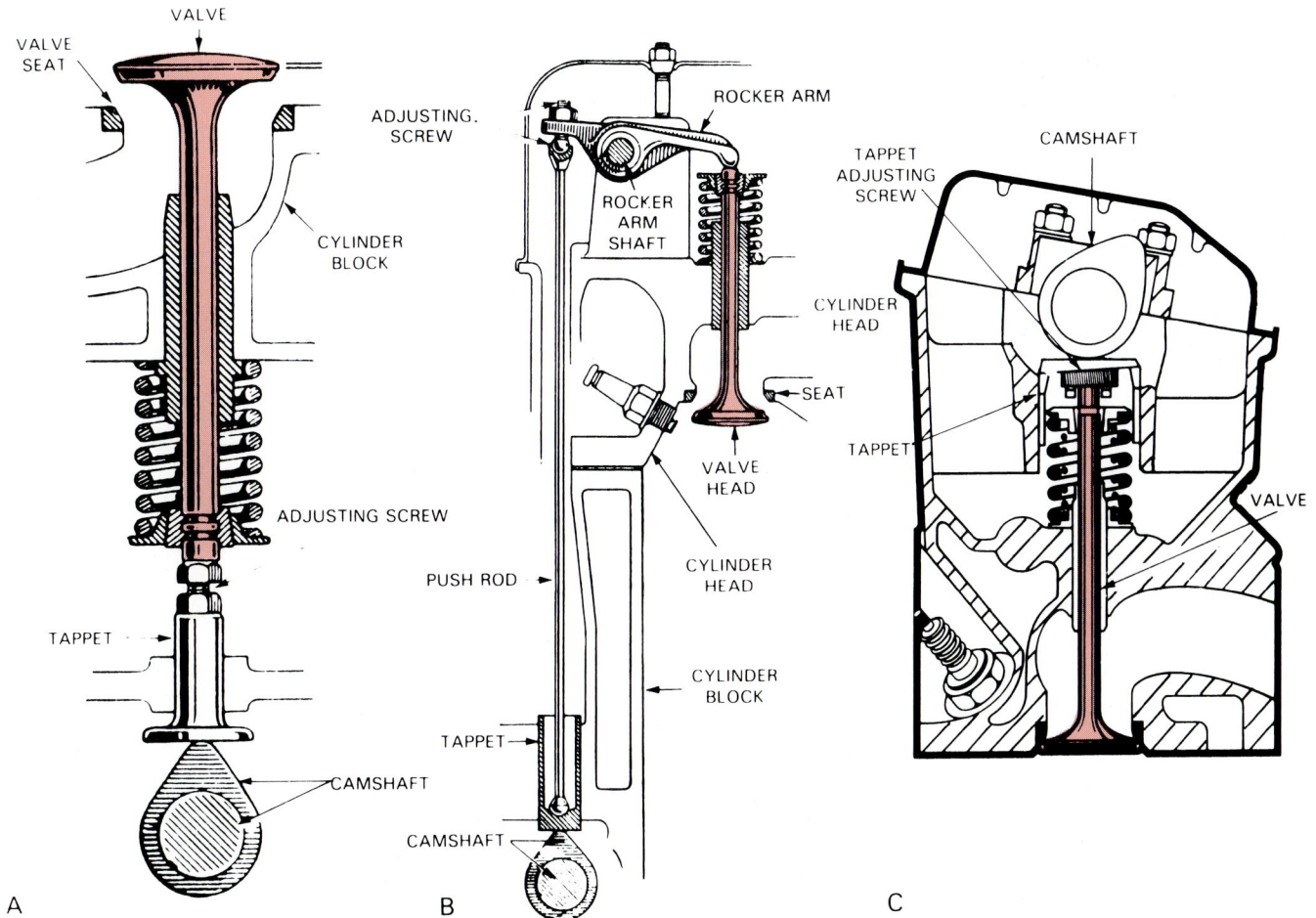

Fig. 13-5. Three common valve-camshaft locations: A — Valve in block or L-head engine is no longer used in motor cars. Small petrol engines for lawnmowers, for example, use this arrangement. B — Cam-in-block, overhead valve or I-head engine is common. C — Overhead cam engine is another form of I-head engine. It is also very common in today's cars.

OHC engines were first used in racing cars because of their high rpm (revolutions per minute) efficiency. Now they are commonly used in small, high rpm, economy car engines. Without push rods to flex, less valve train weight, and improved valve positioning. The OHC is becoming very popular.

A single overhead cam (SOHC) engine has only one camshaft per cylinder head. The cam may act directly on the valves, or rocker arms may transfer motion to the valves.

A dual overhead cam (DOHC) engine has two camshafts per cylinder head. One cam operates the intake valves, the other operates the exhaust valves. Engines of this type are used only in exotic sports cars and race car engines. A DOHC engine will be shown later in the chapter.

Combustion chamber shape

The shape of the combustion chamber provides still another method of classifying an engine. The three basic combustion chamber shapes for petrol engines are: pancake, wedge, and hemispherical. These are pictured in Fig. 13-6.

The pancake combustion chamber, also called "bath tub" chamber, has valve heads almost parallel with the top of the piston. The chamber forms a flat pocket over the piston head, Fig. 13-6A.

A wedge combustion chamber, called a wedge head, is shaped like a triangle or a wedge when viewed as in Fig. 13-6B. Valves are placed side-by-side with the spark plug next to the valves.

A squish area is commonly formed inside a wedge type cylinder head. When the piston reaches TDC, the piston comes very close to the bottom of the cylinder head. This squeezes the air-fuel mixture in that area and causes it to squirt or squish out into the main part of the chamber. Squish can be used to improve air-fuel mixing at low engine speeds.

A hemispherical combustion chamber, nicknamed hemi-head, is shaped like a dome. The valves are canted (tilted) on each side of the chamber. The spark plug is located near the centre. A hemi is shown in Fig. 13-6C. Compare it to the others.

A hemi combustion chamber is extremely efficient. There are no hidden pockets for incomplete combustion. The surface area is very small, reducing heat loss from the chamber. The centrally located spark

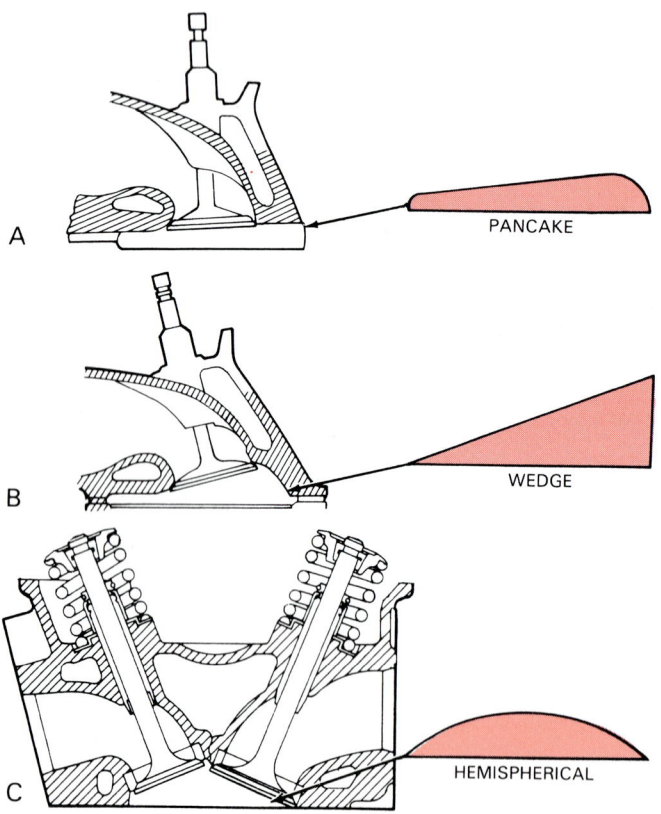

Fig. 13-6. Three basic combustion chamber shapes: A — Valve almost parallel with piston top forms pancake or bath tub shape. B — Valve at angle forms wedge-shaped combustion chamber. C — Valves at angle to each other produce hemispherical or domed chamber.

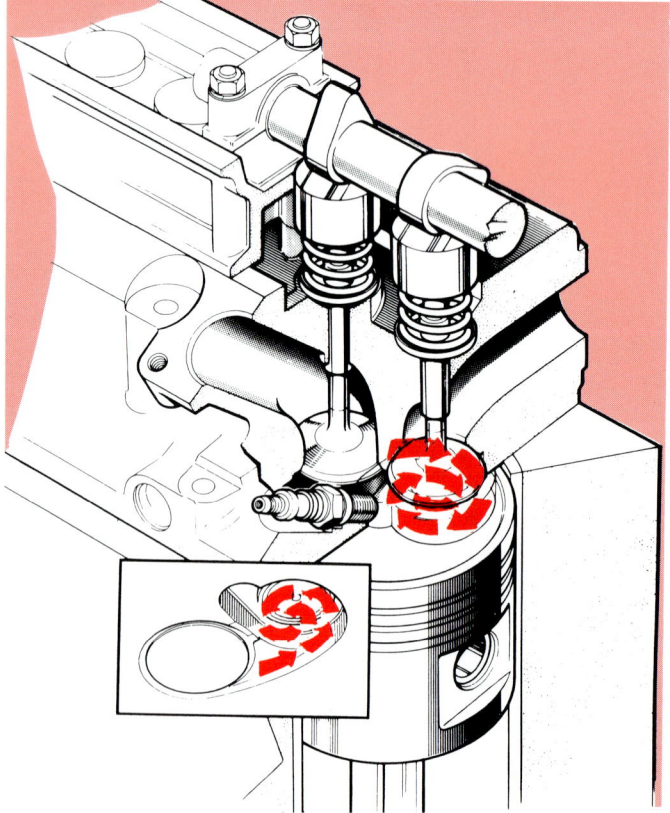

Fig. 13-7. Swirl combustion chamber has port entry designed to cause air-fuel mixture swirling. This helps stir air and fuel into finer mist for improved combustion. Many chambers use this principle. (Jaguar)

plug produces a very short flame path for combustion. The canted valves help increase breathing ability.

The hemi-head was first used in high horsepower, racing engines. It is now used in many OHC passenger car engines. It allows the engine to operate at high rpms and makes it very fuel efficient. It also produces complete burning of the fuel to reduce emissions.

Combustion chamber types

Besides the three shapes just covered, there are several other combustion chamber classifications. Each type is designed to increase the combustion efficiency and power while reducing fuel consumption and exhaust pollution.

A swirl combustion chamber causes the air-fuel mixture to twirl or spin as it enters from the intake port. Look at Fig. 13-7. This causes the air and fuel to mix into a finer mist that burns better.

A four-valve combustion chamber uses two exhaust valves and two intake valves per cylinder. This is illustrated in Fig. 13-8. The extra valves increase flow in and out of the combustion chamber. This setup is used in a few exotic, high performance engines.

A stratified charge combustion chamber uses a small combustion chamber flame to ignite and burn the fuel in the main, large combustion chamber.

A very lean mixture (high ratio of air to fuel) is admitted into the main combustion chamber. The mixture is so lean that it will not ignite and burn easily.

A richer mixture (higher ratio of fuel to air) is admitted into the small chamber by an extra valve. When the fuel mixture in the small chamber is ignited, flames blow into and ignite the fuel in the main chamber.

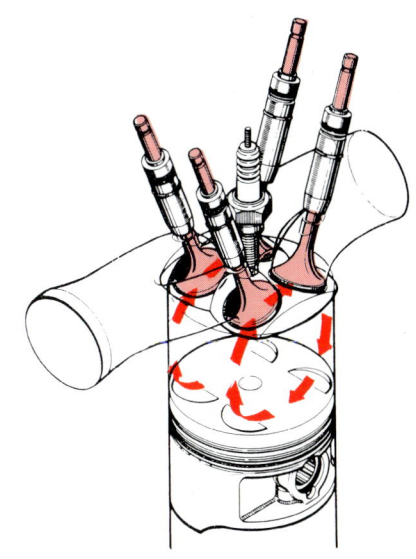

Fig. 13-8. Four-valve combustion chamber is used in a few exotic sports car and race car engines. Extra valves increase flow and engine power.

A stratified charge chamber allows the engine to operate on a lean, high efficiency air-fuel ratio. Fuel economy is increased and exhaust emission output is reduced.

An air jet combustion chamber has a single combustion chamber fitted with an extra air valve. Shown in Fig. 13-9, a passage runs from the carburettor to the combustion chamber and jet valve.

During the intake stroke, the engine camshaft opens both the conventional intake valve and the air jet valve. This allows fuel mixture to flow into the cylinder past the conventional intake valve. At the same time, a stream of air flows into the cylinder through the jet valve.

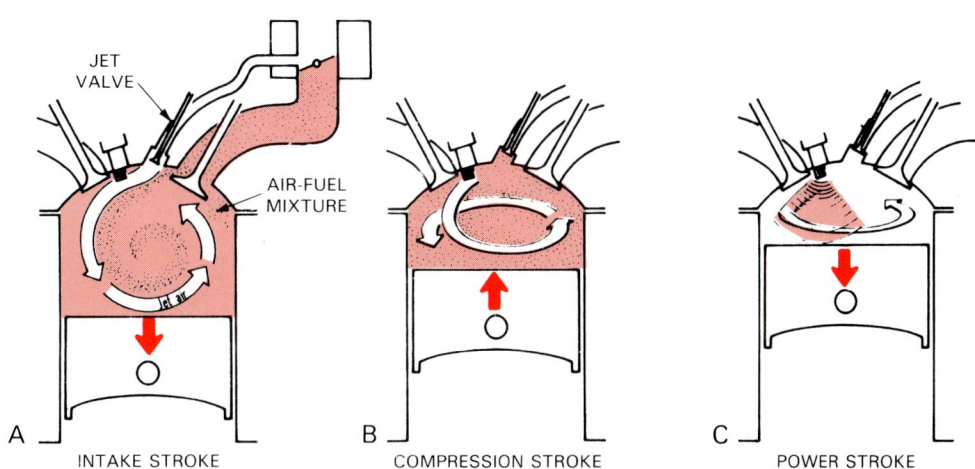

Fig. 13-9. Air jet injects stream of air into chamber at idle to improve fuel mixing and combustion.

The jet valve action causes the fuel mixture in the cylinder to swirl and mix. This increases combustion efficiency by causing more of the fuel to burn during the power stroke. The jet valve only works at idle and low engine speeds. At higher rpm, normal air-fuel mixing is adequate for efficient combustion.

A precombustion chamber is commonly used in automotive diesel engines. It is similar in shape to a stratified charge chamber for a petrol engine. Also called a DIESEL PRECHAMBER, it is used to quiet engine operation and to allow the use of a glow plug (heating element) to aid cold weather starting. Fig. 13-10 shows a cutaway view of a diesel prechamber.

During combustion, diesel oil is injected into the prechamber. If the engine is cold, the glow plug heats the air in the prechamber. This heat, along with the heat produced by compression, causes the fuel to ignite and burn. As it burns, it expands and moves into the main chamber.

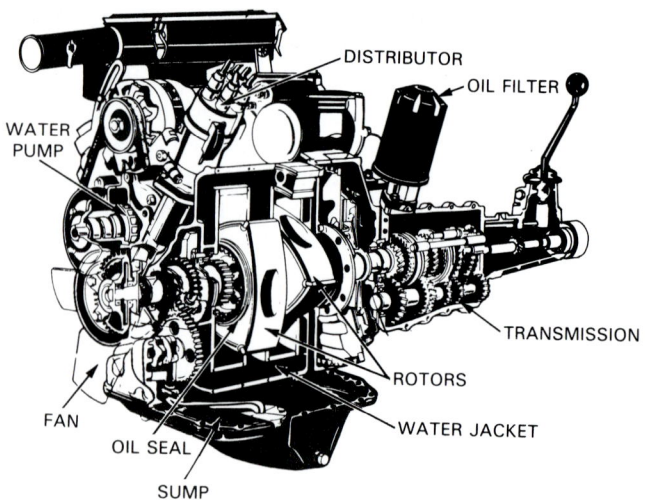

Fig. 13-11. Wankel rotary engine does not use conventional pistons that move up and down in cylinder. Rotor spins in circular motion. Note engine parts. (Mazda)

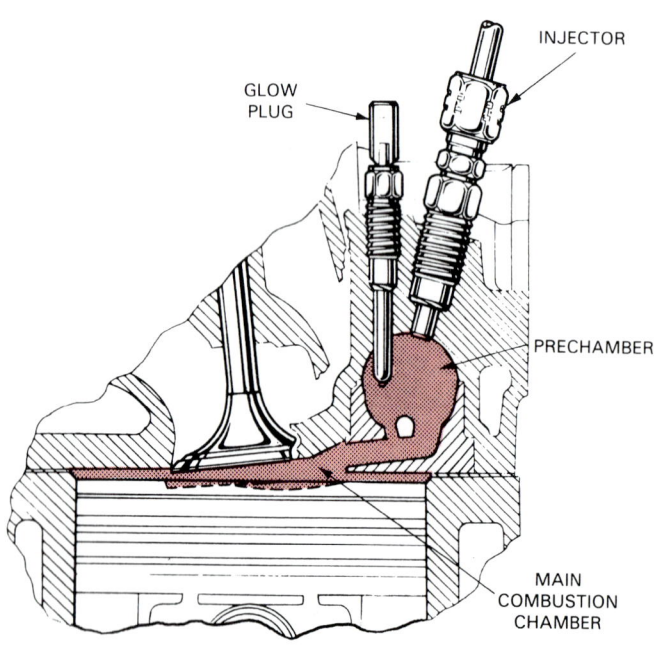

Fig. 13-10. Diesel engine prechamber should not be confused with petrol engine stratified charge chamber. Diesel prechamber quiets engine operation and allows use of glow plug. Glow plug is heating element that improves cold weather starting.

ALTERNATE ENGINES

As you have learned, cars generally use internal combustion, 4-stroke cycle, piston engines.

Alternate engines include all other engine types that may be used to power a vehicle. Various engine types have been developed, but few have been placed into production.

Wankel (rotary) engine

A wankel engine, also known as a rotary engine, uses a triangular rotor instead of conventional pistons. The rotor turns inside a specially shaped housing, as shown in Fig. 13-11.

While spinning on its own axis, the rotor orbits around a mainshaft. This eliminates the normal reciprocating (up and down) motion found in piston engines. One complete cycle (all four strokes) takes place every time the rotor turns once. Three rotor faces produce three power strokes per revolution. Fig. 13-12 illustrates the basic operation of a rotary engine.

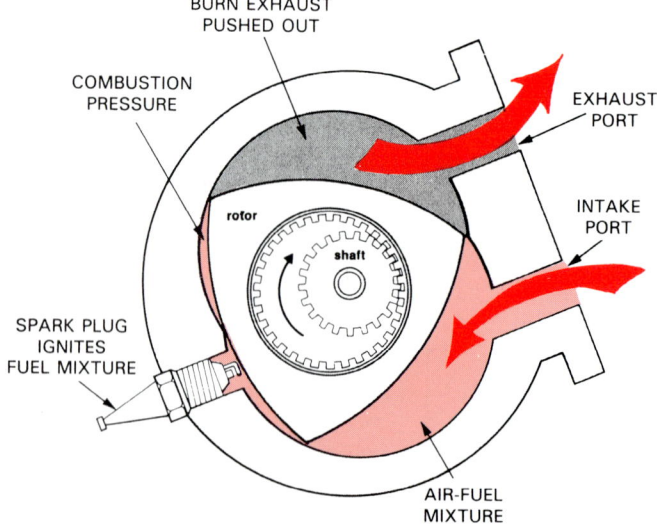

Fig. 13-12. In rotary engine, rotor movement produces low pressure area, pulling air-fuel mixture into engine. As rotor turns, mixture is compressed and ignited. As fuel burns, it expands and pushes on rotor. Rotor turns and burned gases are pushed out of engine.

A rotary engine is very powerful for its size. Also, because it spins — rather than moves up and down — engine operation is very smooth and vibration free.

A complicated emission control system is needed to make the rotary engine pass emission standards. This has limited its use. A Wankel engine is one of the few alternate engines to be mass produced and installed in production vehicles.

Electric motor

An electric motor and large storage batteries can be used to power a car. Electric cars have been produced in limited numbers by smaller companies.

They have seen some success as a means of transportation for short trips. Speed and driving distance is limited with today's technology.

Hybrid power source

A hybrid power source uses two different methods to power the car. For example, a small petrol engine and an electric motor and storage batteries may both be used to propel the car.

The batteries and electric motor supply power when the car first accelerates. This provides enough energy to accelerate the car quickly. Once cruising speeds are reached, the petrol engine takes over. It is a very small engine that can supply adequate power to keep the car moving. The petrol engine also provides enough energy to recharge the batteries.

Two-stroke cycle engine

A two-stroke cycle engine is similar to an automotive four-stroke engine, but it only requires one revolution of the crankshaft for a complete power-producing cycle. Two piston strokes (one upward and one downward) complete the intake, compression, power and exhaust events. Fig. 13-13 illustrates the basic operation of a two-stroke cycle engine.

As the piston moves up, it compresses the air-fuel mixture in the combustion chamber. At the same time, the vacuum created in the crankcase by the piston movement draws fuel and oil into the crankcase. Either a reed valve (flexible metal, flap valve) or a rotary valve (spinning, disc-shaped valve) can be used to control flow into the crankcase.

When the piston reaches the top of the cylinder, ignition occurs and the burning gases force the piston down. The reed valve or rotary valve closes, this compresses and pressurises the fuel mixture in the crankcase.

As the piston moves far enough down in the cylinder, it uncovers an exhaust port in the cylinder wall. Burned gases leave the engine through the exhaust port.

As the piston continues downward, it uncovers the transfer port. Pressure in the crankcase causes a fresh fuel charge to flow into the cylinder. Upward movement of the piston again covers the transfer and exhaust ports, compression begins and the cycle is repeated.

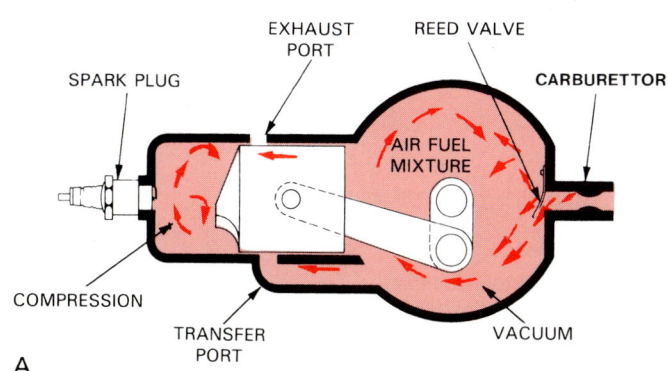

A

Piston slides in cylinder. This pulls air, petrol, and oil into crankcase. At same time, piston covers ports in cylinder and compresses fuel mixture. Spark plug fires to push piston back down.

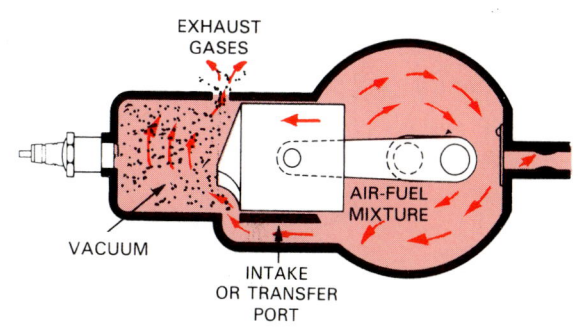

B

As piston slides down, it forms pressure in engine crankcase. This closes reed valve. When piston slides down far enough, cylinder ports open. Exhaust exits one port. Crankcase pressure pushes mixture into cylinder through other port. This prepares for another power stroke.

Fig. 13-13. Two-stroke cycle engine completes all four events in two piston movements.

Since the crankcase is used as a storage chamber for each successive fuel charge, the fuel and lubricating oil are premixed and introduced into the engine through the carburettor.

Inside the crankcase, some of the oil separates from the petrol. The oil mist lubricates and protects the moving parts inside the engine.

Generally speaking, two-stroke cycle engines are NOT used in cars because they:

1. Produce too much exhaust pollution.
2. Have poor power output at low speeds.
3. Require more service than a four-stroke.
4. Must have motor oil mixed into the fuel.
5. Are not as fuel efficient as a four-stroke engine.

TYPICAL AUTOMOTIVE ENGINES

Figs. 13-14 through 13-23 illustrate typical automotive engines. Study each of these carefully. Note the design variations between each type. Also, study the names of all of the parts. This will help you in later chapters.

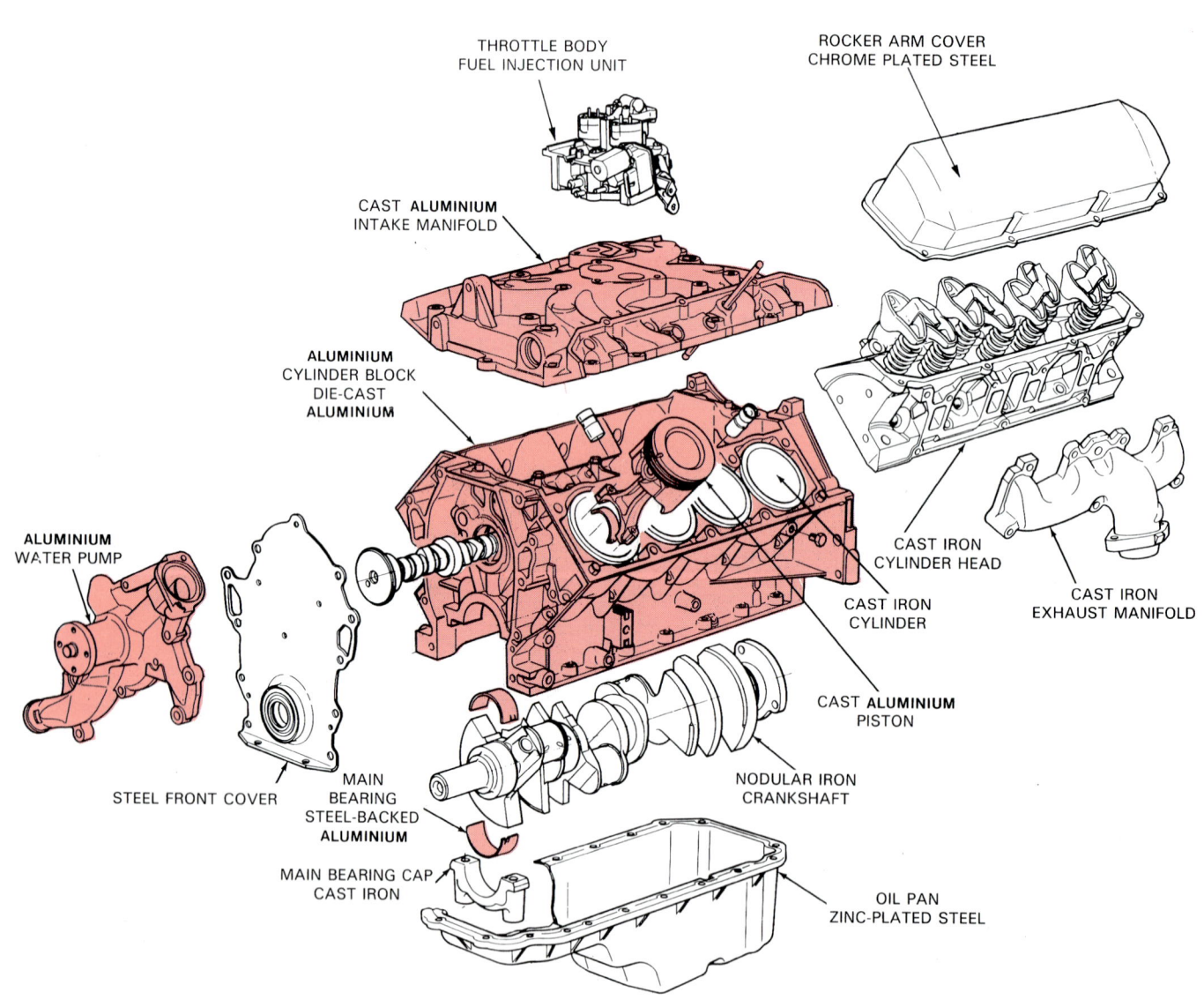

THROTTLE BODY FUEL INJECTION UNIT

ROCKER ARM COVER CHROME PLATED STEEL

CAST **ALUMINIUM** INTAKE MANIFOLD

ALUMINIUM CYLINDER BLOCK DIE-CAST **ALUMINIUM**

ALUMINIUM WATER PUMP

CAST IRON CYLINDER HEAD

CAST IRON EXHAUST MANIFOLD

CAST IRON CYLINDER

CAST **ALUMINIUM** PISTON

STEEL FRONT COVER

MAIN BEARING STEEL-BACKED **ALUMINIUM**

MAIN BEARING CAP CAST IRON

NODULAR IRON CRANKSHAFT

OIL PAN ZINC-PLATED STEEL

Fig. 13-14. Fuel injected, V-8 engine using many aluminium parts.

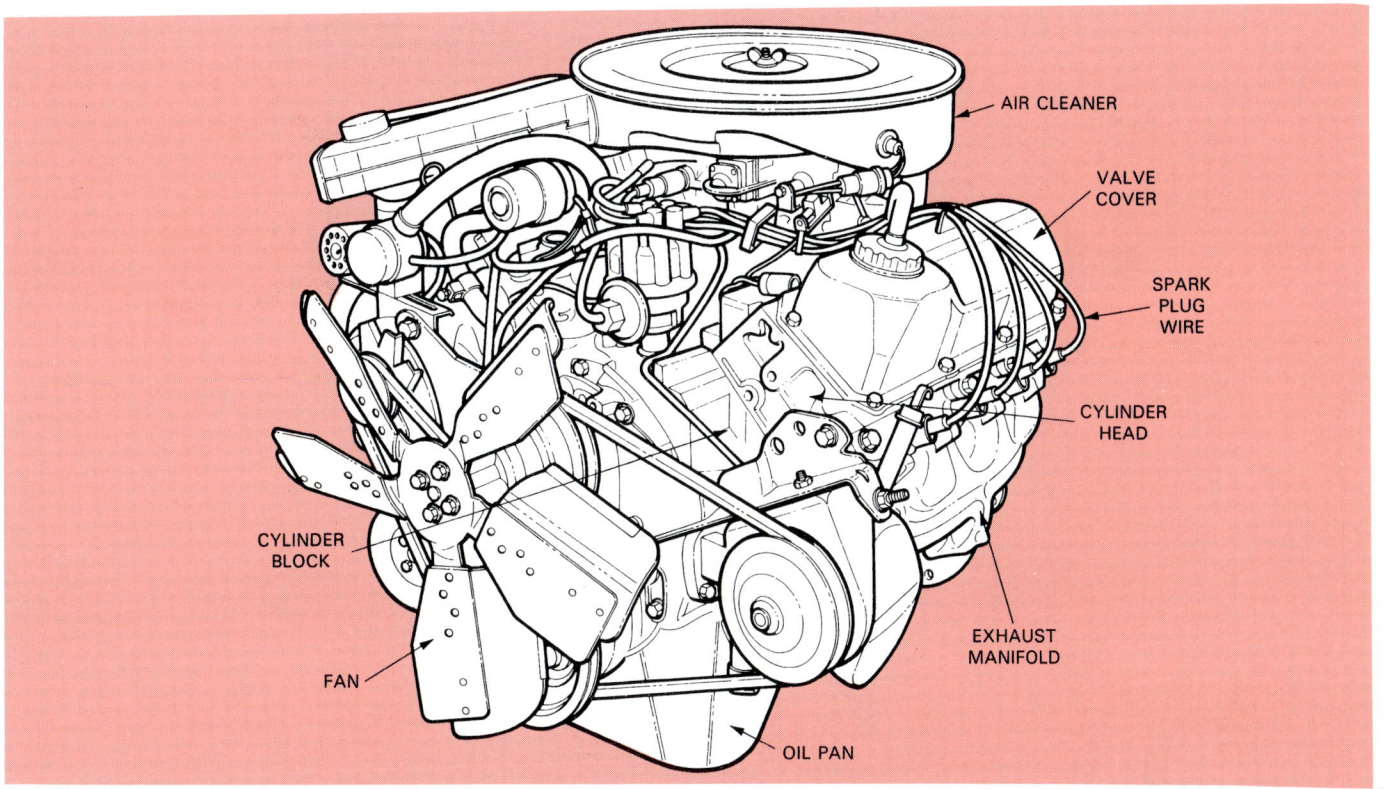

Fig. 13-15. *External view of typical V-8, petrol engine. (Ford)*

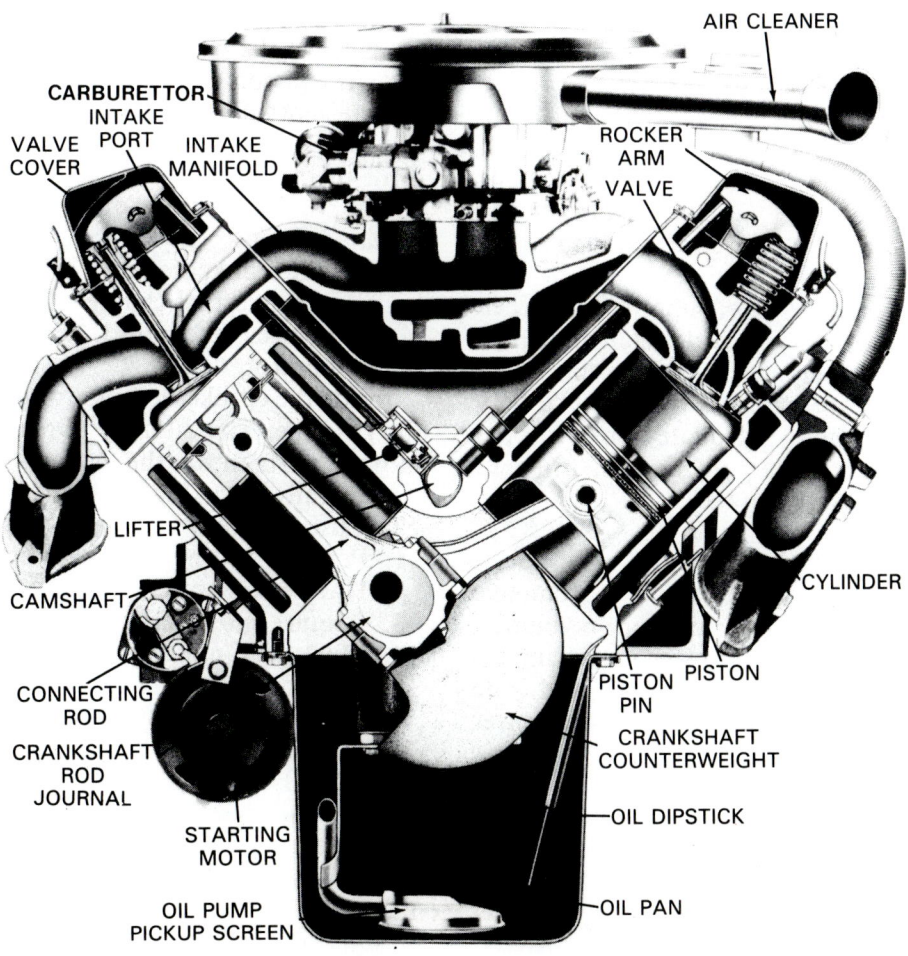

Fig. 13-16. *Cutaway view of a cam-in-block or overhead valve V-8 engine. Study relationship and names of parts.*

205

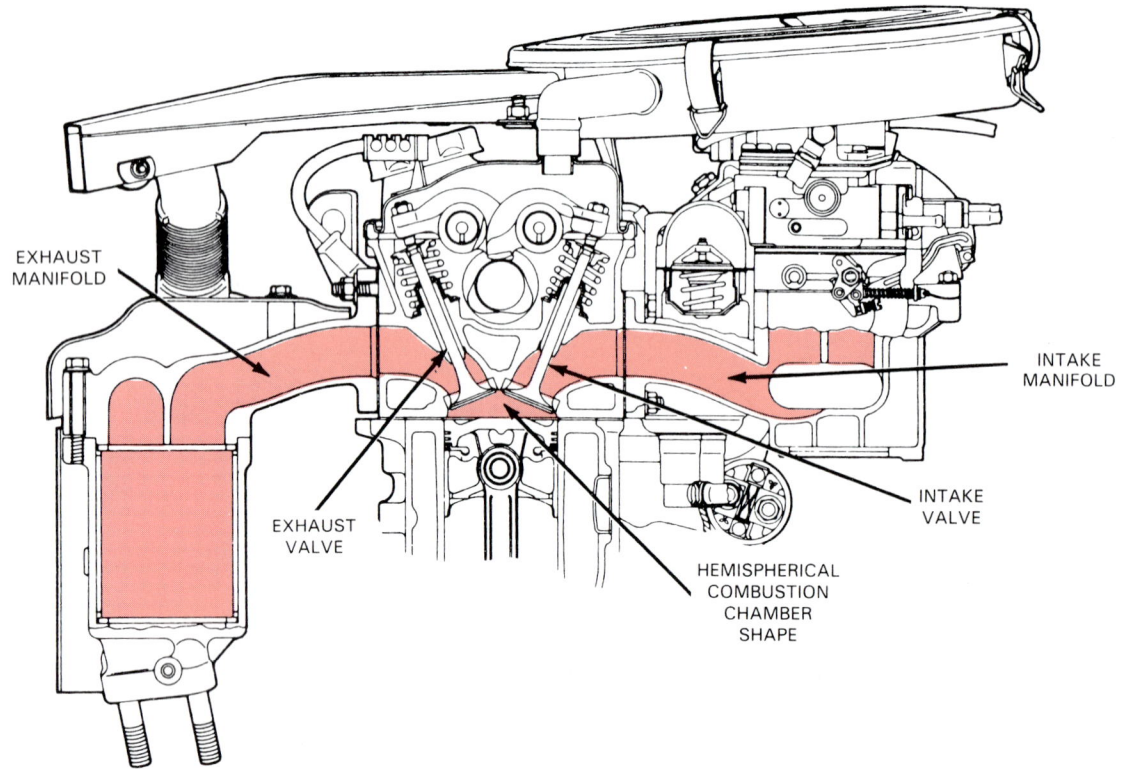

EXHAUST
MANIFOLD

INTAKE
MANIFOLD

INTAKE
VALVE

EXHAUST
VALVE

HEMISPHERICAL
COMBUSTION
CHAMBER
SHAPE

Fig. 13-17. Overhead cam, in-line using crossflow hemi head. (Chrysler)

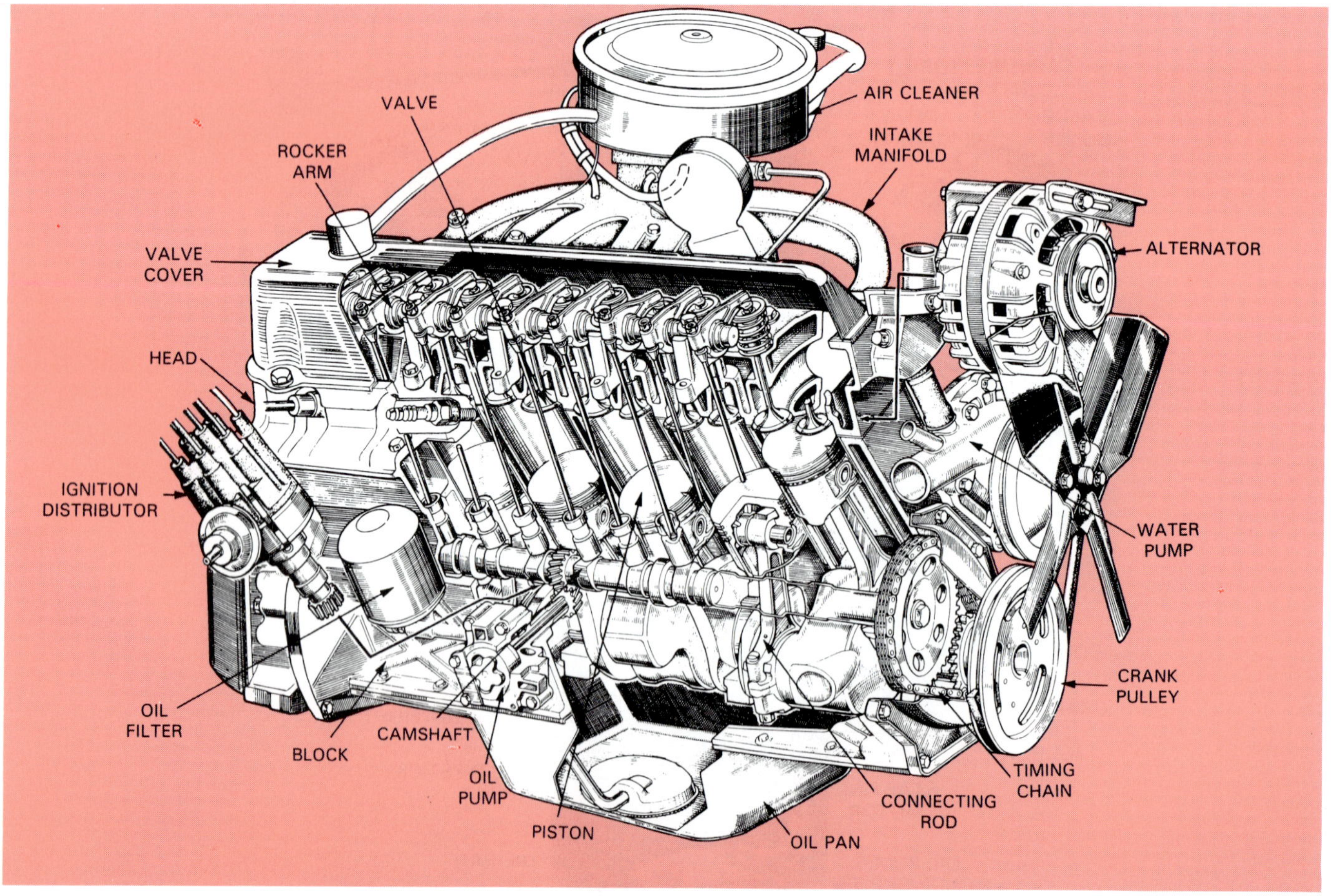

VALVE

ROCKER
ARM

VALVE
COVER

HEAD

IGNITION
DISTRIBUTOR

OIL
FILTER

BLOCK

OIL
PUMP

PISTON

OIL PAN

CONNECTING
ROD

AIR CLEANER

INTAKE
MANIFOLD

ALTERNATOR

WATER
PUMP

CRANK
PULLEY

TIMING
CHAIN

CAMSHAFT

Fig. 13-18. Slant six, in-line engine. Study parts carefully. (Chrysler)

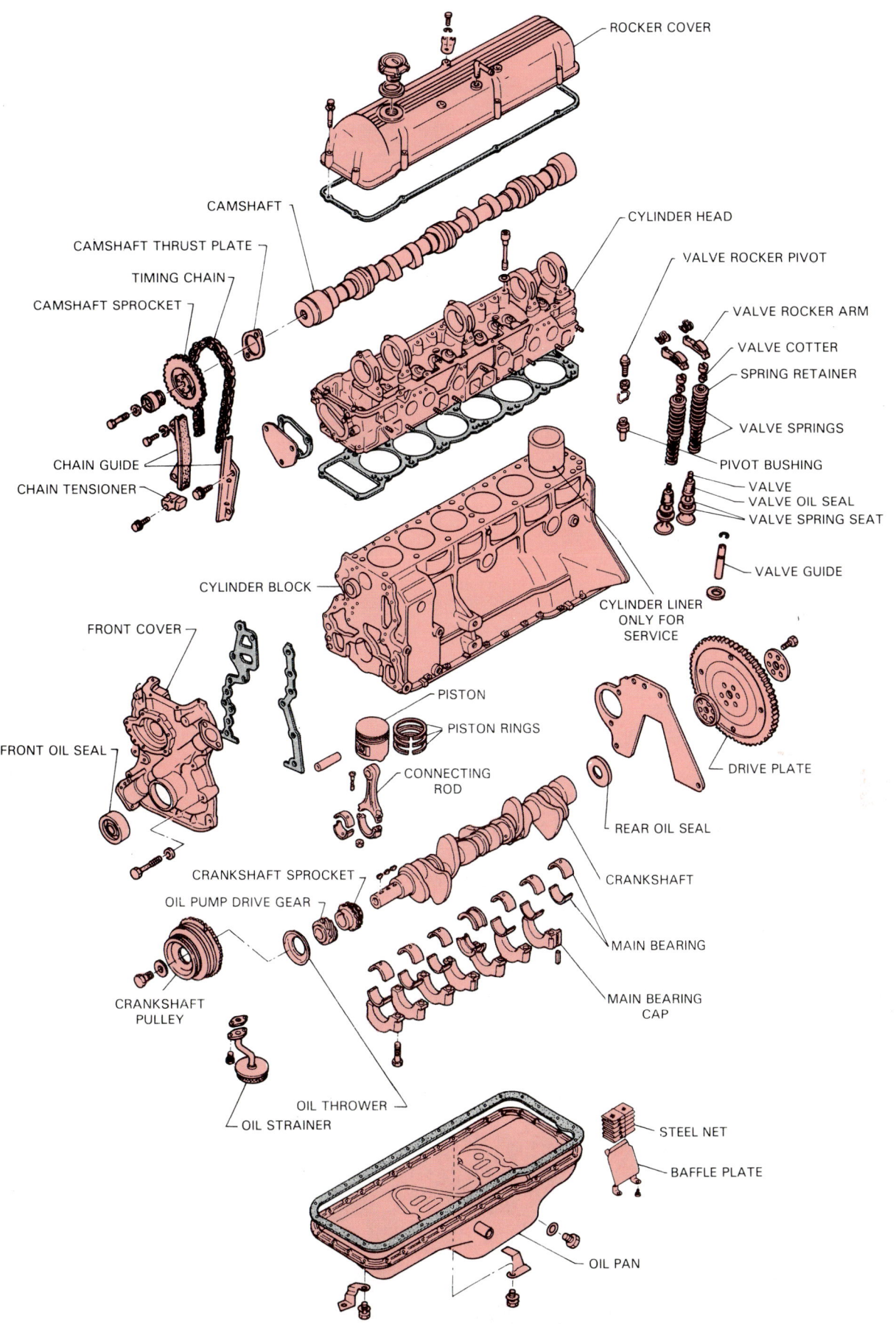

Fig. 13-19. Exploded view of 6 cylinder OHC engine and related parts. (Nissan)

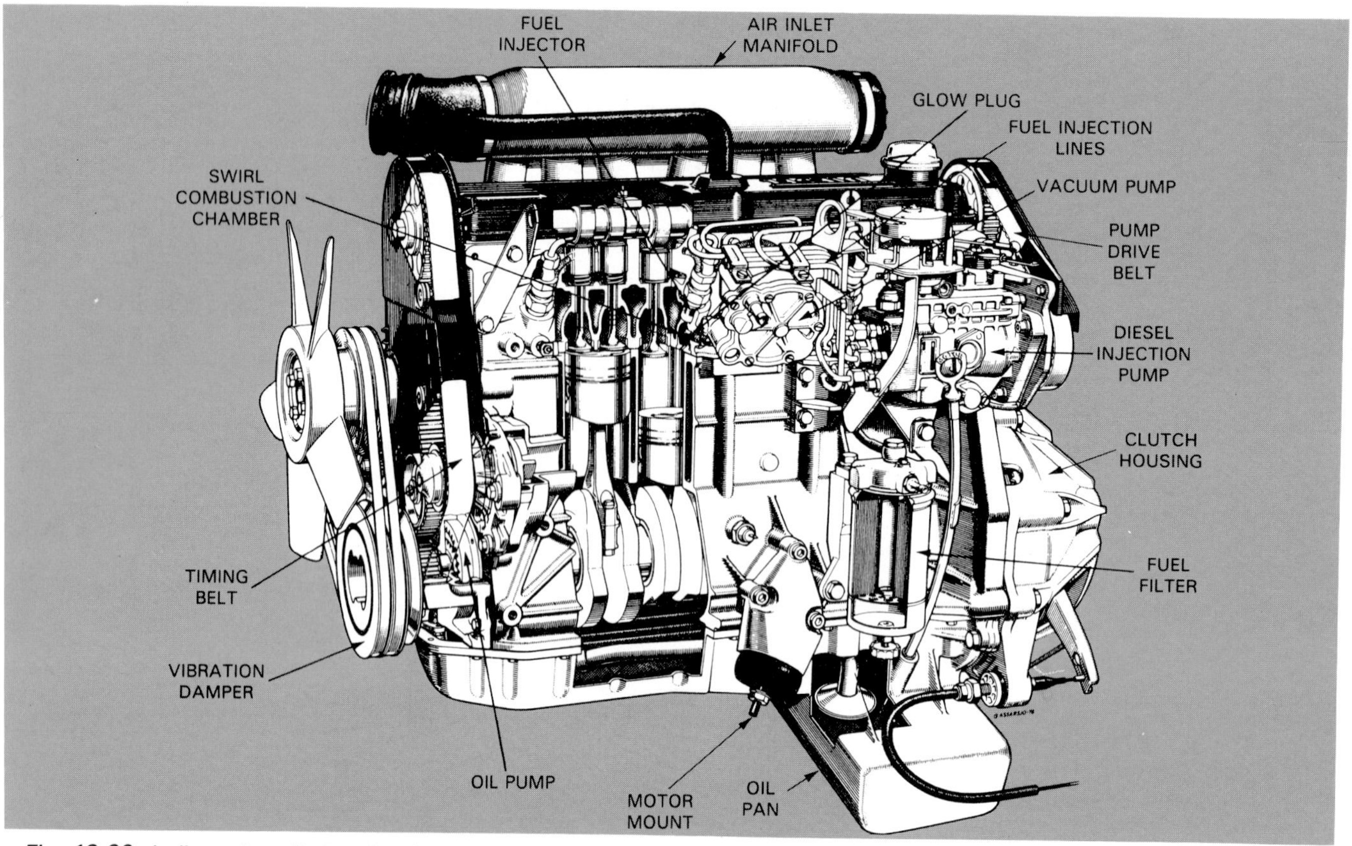

Fig. 13-20. In-line, six cylinder diesel engine. Note names of parts. Also, note rear drive belt for injection pump. (Volvo)

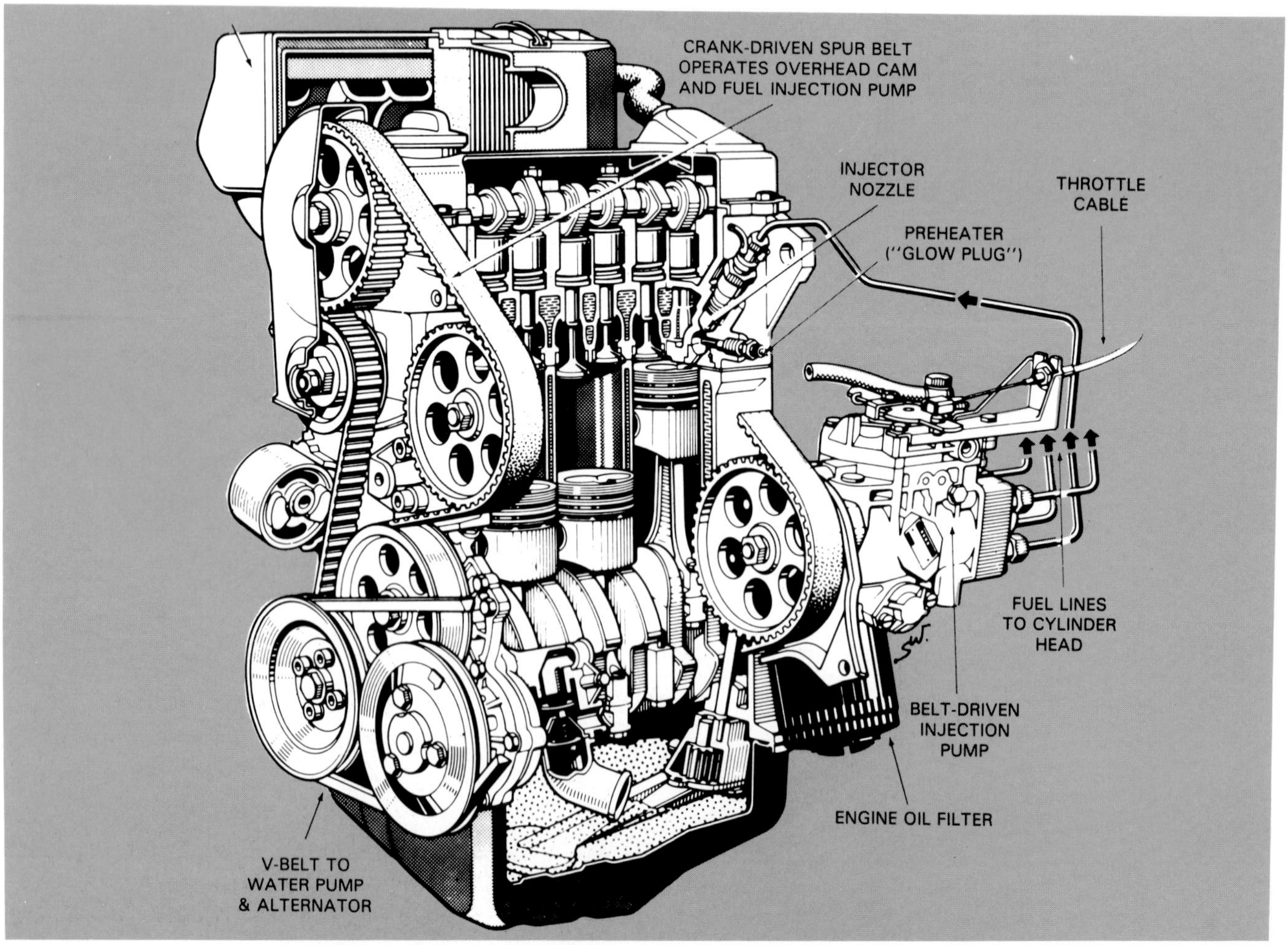

Fig. 13-21. Four cylinder diesel engine. Note use of drive belts instead of gears or chains. (VW)

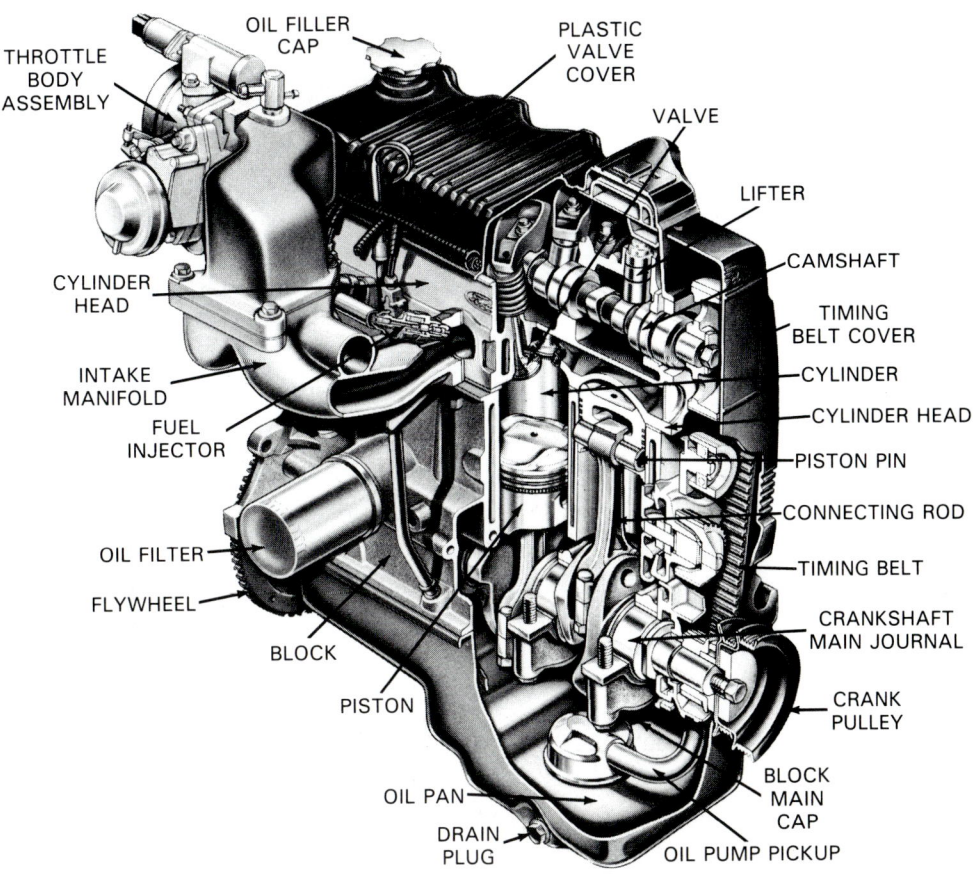

Fig. 13-22. Fuel injected, SOHC, hemi head, in-line, 4-cylinder engine. Note use of plastic valve cover. (Ford)

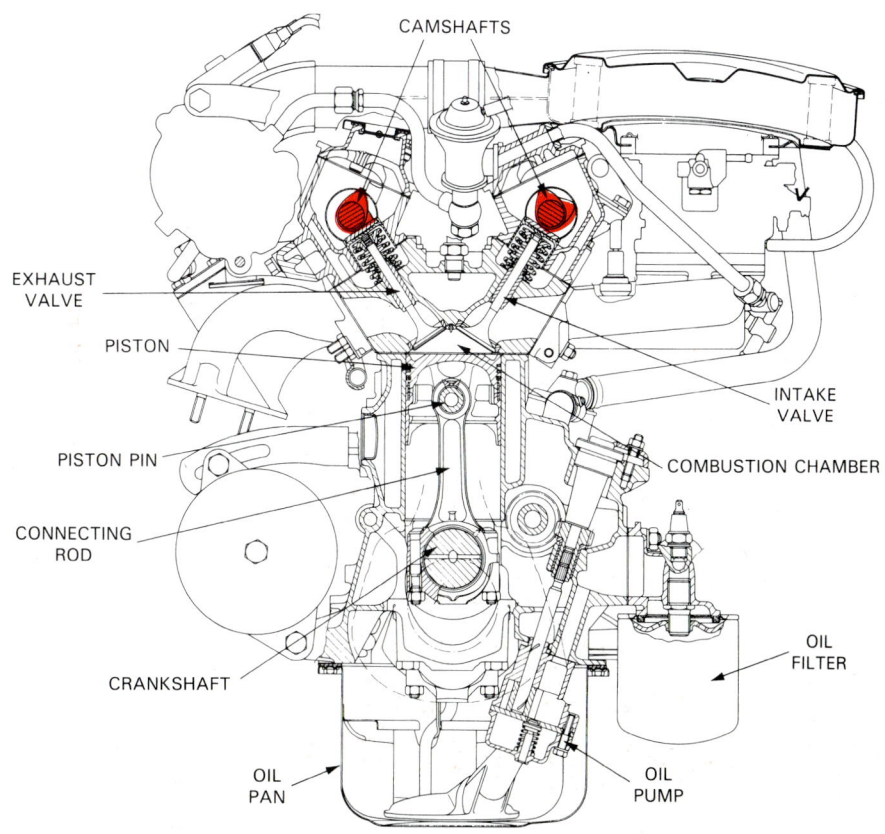

Fig. 13-23. DOHC, in-line engine. Two camshafts are located in one cylinder head. One cam operates all of the intake valves. Other operates exhaust valves. (Fiat)

Engine Classifications

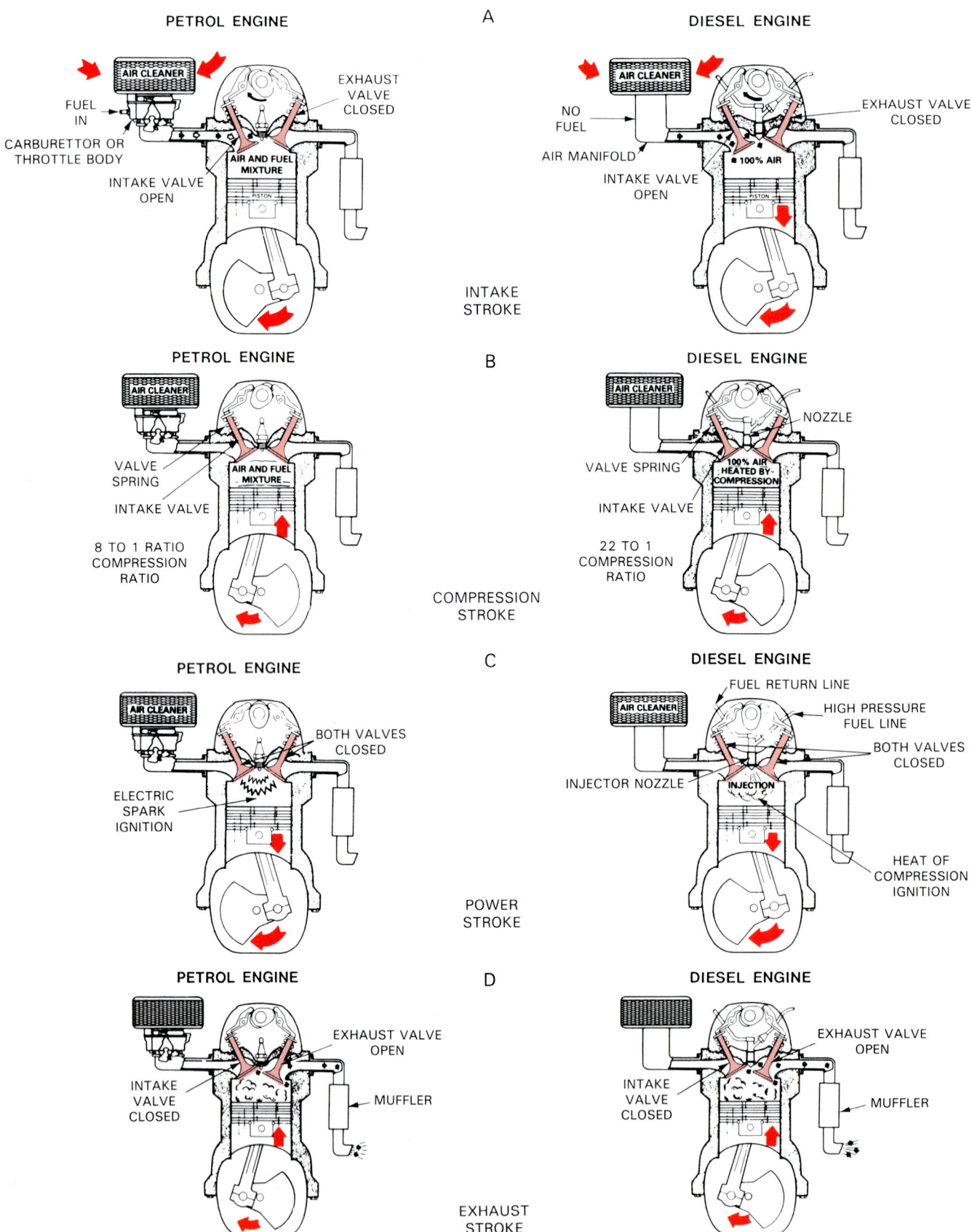

Fig. 13-24. Compare four-stroke cycle of petrol and diesel engines. A — Air-fuel mixture is drawn into petrol engine on intake stroke. Diesel only draws air into cylinder on intake stroke. B — Petrol compresses air-fuel mixture on compression stroke. Diesel compresses only air but is squeezed so much that it becomes very hot. C — On power stroke, spark at spark plug starts combustion in petrol engine. With a diesel, diesel oil is sprayed into hot, compressed air. Hot air ignites fuel. D — Exhaust strokes in petrol and diesel engines are very similar.

KNOW THESE TERMS

In-line engine, V-type engine, Slant engine, Opposed engine, Firing order, Spark ignition, Compression ignition, L-head, I-head, Overhead valve, Overhead cam, SOHC, DOHC, Pancake chamber, Wedge chamber, Hemi chamber, Squish area, Swirl chamber, Stratified charge, Air jet chamber, Precombustion chamber, Alternate engine, Wankel, Hybrid, Two-stroke engine.

REVIEW QUESTIONS

1. List nine ways to classify an automotive engine.
2. An _____ engine's cylinders are lined up in a single row.
3. Normally, the number of cylinders in car engines are _____, _____, or _____.
4. What are cylinder numbers?
5. Firing order refers to the _____ in which combustion occurs in each cylinder.
6. Most car engines have air cooling systems. True or False?
7. Explain the difference between spark ignition and compression ignition.
8. Where are the two typical locations for the engine camshaft?
9. Which of the following does NOT refer to camshaft location and design.
   a. OHC
   b. SOHC
   c. DOHC
   d. UHC.
10. A hemi-head combustion chamber is flat. True or False?
11. A few exotic, high performance engines have four valves per cylinder. True or False?
12. Explain the operation of a diesel precombustion chamber.
13. A _____ engine, also called rotary engine, does NOT use pistons that slide up and down.
14. A _____ power source uses two different methods to power the car.
15. Which of the following refers to a two-stroke engine?
   a. One crankshaft revolution completes a power stroke.
   b. Two strokes complete all four events.
   c. Uses reed valves or rotary valves.
   d. Fuel and oil are mixed together.
16. List five reasons two-stroke engines are NOT used in cars.

Chapter 14

ENGINE CAPACITY AND PERFORMANCE MEASUREMENTS

After studying this chapter, you will be able to:
☐ Describe engine capacity measurements based on bore, stroke, displacement, and number of cylinders.
☐ Explain engine compression ratio and how it affects engine performance.
☐ Explain engine torque and power ratings.
☐ Describe the different methods used to measure and rate engine performance.
☐ Explain volumetric efficiency, thermal efficiency, mechanical efficiency, and total engine efficiency.
☐ Engine capacity and performance measurements are important to the mechanic.

A service manual will list many engine capacity and performance values for specific engines. You must be able to understand this information.

ENGINE CAPACITY MEASUREMENT

Engine capacity is determined by cylinder diameter, amount of piston travel on each stroke and number of cylinders. Any of these three variables can be changed to alter engine capacity. Engine capacity information is commonly used when ordering parts or when measuring wear during major engine repairs.

Bore and stroke

Cylinder bore is the diameter of the engine cylinder. See Fig. 14-1. It is measured across the cylinder, parallel with the top of the block. Cylinder bores vary in size from about 76 to 102 mm.

Piston stroke is the distance the piston moves from TDC to BDC. Look at Fig. 14-1. The amount of offset built into the crank journal or throw controls piston stroke. Piston stroke also varies from about 76 to 102 mm.

A service manual normally gives bore and stroke specifications together. For instance, suppose a specification for bore and stroke is given as 102 mm × 76 mm. This means that the engine cylinder is 102 mm in diameter and the piston stroke

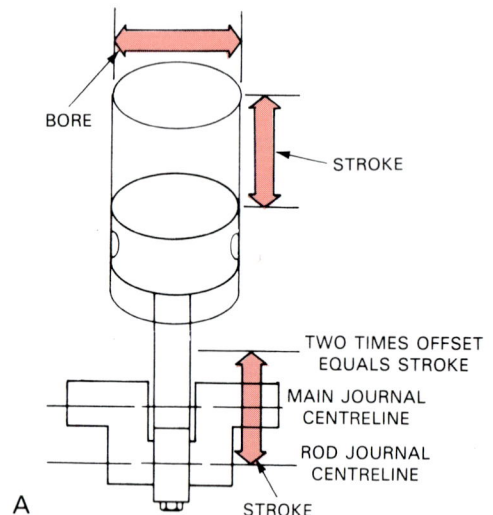

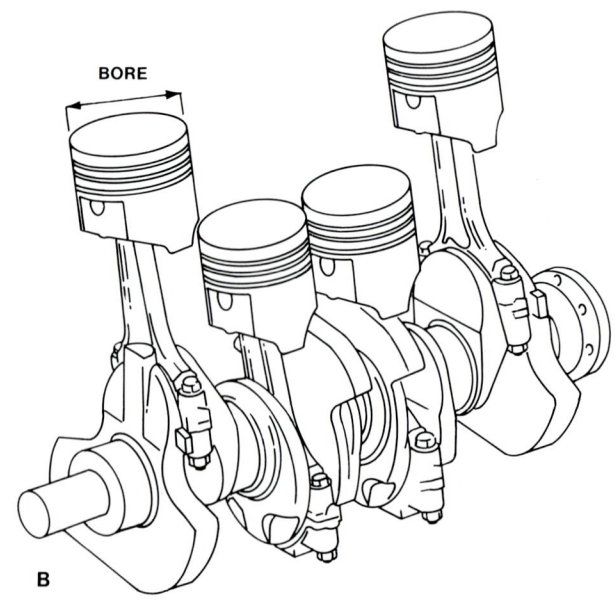

Fig. 14-1. A — Cylinder bore is measured across cylinder parallel with top face of block. Stroke is distance piston moves from BDC to TDC. Piston displacement is amount of volume piston would move in one upward stroke. B — Engine displacement is displacement for all pistons.

is 76 mm. Bore is always the FIRST VALUE GIVEN and stroke the second.

Generally, a larger bore and stroke makes the engine more powerful. It can draw in more fuel and air on each intake stroke. Then, more force is exerted on the head of the piston during the power stroke.

Piston displacement

Piston displacement is the volume the piston displaces (moves) from BDC to TDC. It is determined by comparing cylinder diameter and piston stroke. A large cylinder diameter and large piston stroke would produce a larger piston displacement.

The formula for finding piston displacement is:

Piston Displacement =

$$\frac{\text{Bore squared} \times 3.14 \times \text{Stroke}}{4}$$

If an engine has a bore of 102 mm and a stroke of 76 mm, what is its piston displacement?

To obtain an answer in cubic centimetres, it will be necessary to convert the bore and stroke dimensions into centimetres, by dividing each of them by 10.

Piston Displacement =

$$\frac{(10.2)^2 \times 3.14 \times 7.6}{4} = 621 \text{cc(ml)}$$

Note: Millilitres (ml) is the correct unit for cylinder volume but it is common trade practice to express engine displacement in cubic centimetres (cc) or (cm³). This has to be acknowledged but is not technically correct.

Engine displacement

Engine displacement or engine capacity equals piston displacement times the number of engine cylinders. For example, if one piston displaces 400 cubic centimetres and the engine has four cylinders, the engine displacement would be 1600 cubic centimetres (400 × 4 = 1600).

Cubic centimetres (cm³), litres (L) are used to state engine displacement. For example a V-6 could be a 3.3L engine. A four-cylinder engine might have a displacement of 2,300 cm³, since one litre equals 100 cm³ this may also be shown as a 2.3L engine. Prior to metrication, some engines were referred to as CID (cubic inch displacement), for example a V-8 engine might have been referred to as a 350 CID.

FORCE, WORK, AND POWER

Force

Force is a pushing and pulling action. When a spring is compressed, an outward movement or force is produced. Force is measured in newtons (N).

Work

Work is done when force causes movement. If the compressed spring moves another engine part, work has been done. If the spring does NOT cause movement, no work has been done. Work is measured in newton metres (Nm) and joules (J).

If a force of 400 Newtons is applied to an object and the object moves through a distance of 3 metres, how much work has been done?

The formula for work done is:

Work done = (force applied × distance moved).

Work done = (400 N × 3 m)

= 1200 Newton metres (Nm)

or joules (J) of work completed.

Power

Power is the rate or speed at which work is done. The metric unit for power is the watt (W), or kilowatt (kW).

Higher power output can do a large amount of work. Lower power output can only do a small amount of work. The formula for power is:

$$\text{Power} = \frac{\text{Force} \times \text{Distance}}{\text{Time in Seconds}}$$

$$= \frac{\text{Newtons} \times \text{metres}}{\text{Seconds}} = \frac{\text{Joules}}{\text{Second}} = \text{Watts}$$

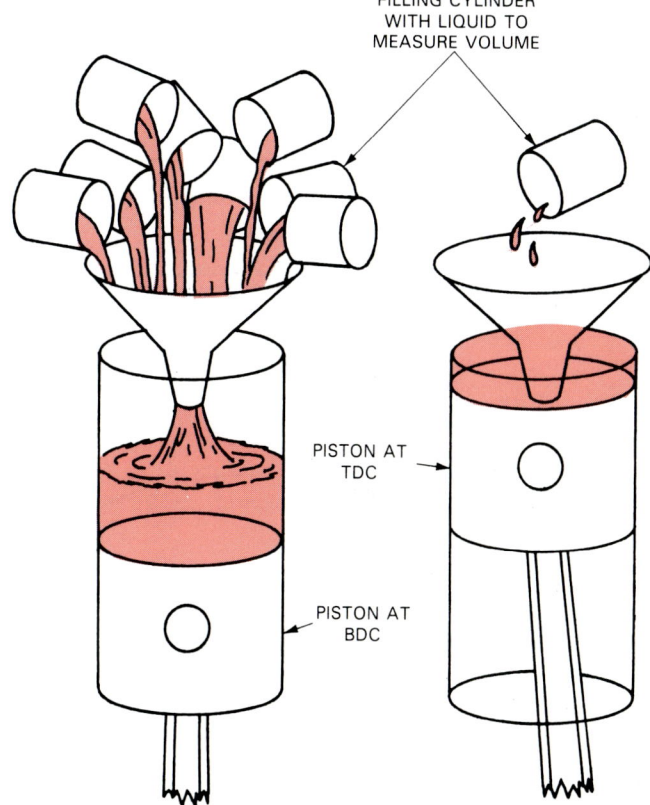

Fig. 14-2. Compression ratio is comparison of cylinder volumes with piston at TDC and BDC. This engine has eight times the volume at BDC, producing 8:1 compression ratio.

If a powered winch, moves an object 60 metres against a resistance of 800 newtons in 1 minute, how much power is needed?

$$\text{Power} = \frac{800 \times 60}{1 \times 60} = 800 \text{ W (or 0.8 kW)}$$

COMPRESSION RATIO

Engine compression ratio compares cylinder volumes with the piston at TDC and BDC. Look at Fig. 14-2. An engine's compression ratio controls how high the air-fuel mixture pressure is raised on the compression stroke.

A compression ratio is given as two numbers. For example, an engine may have a compression ratio of 9:1 (9 to 1). This means the maximum cylinder volume is nine times as large as the minimum cylinder volume. At BDC a cylinder has maximum volume, minimum volume occurs at TDC.

Fig. 14-3 illustrates two examples of compression ratio ratings. When the piston of a petrol engine is at BDC, the cylinder volume is 0.50L. When the piston moves to TDC, the cylinder volume reduces to 0.0625L. Dividing 0.50 by 0.0625 the compression ratio for this engine would be 8:1.

With the diesel (compression ignition engine), BDC cylinder volume is 20 times as large as TDC cylinder volume. The compression ratio would then be 20:1.

Older engines designed for leaded petrol had higher compression ratios. Up to a point, a high compression ratio increases engine fuel efficiency and power. However, it also causes higher exhaust emission levels.

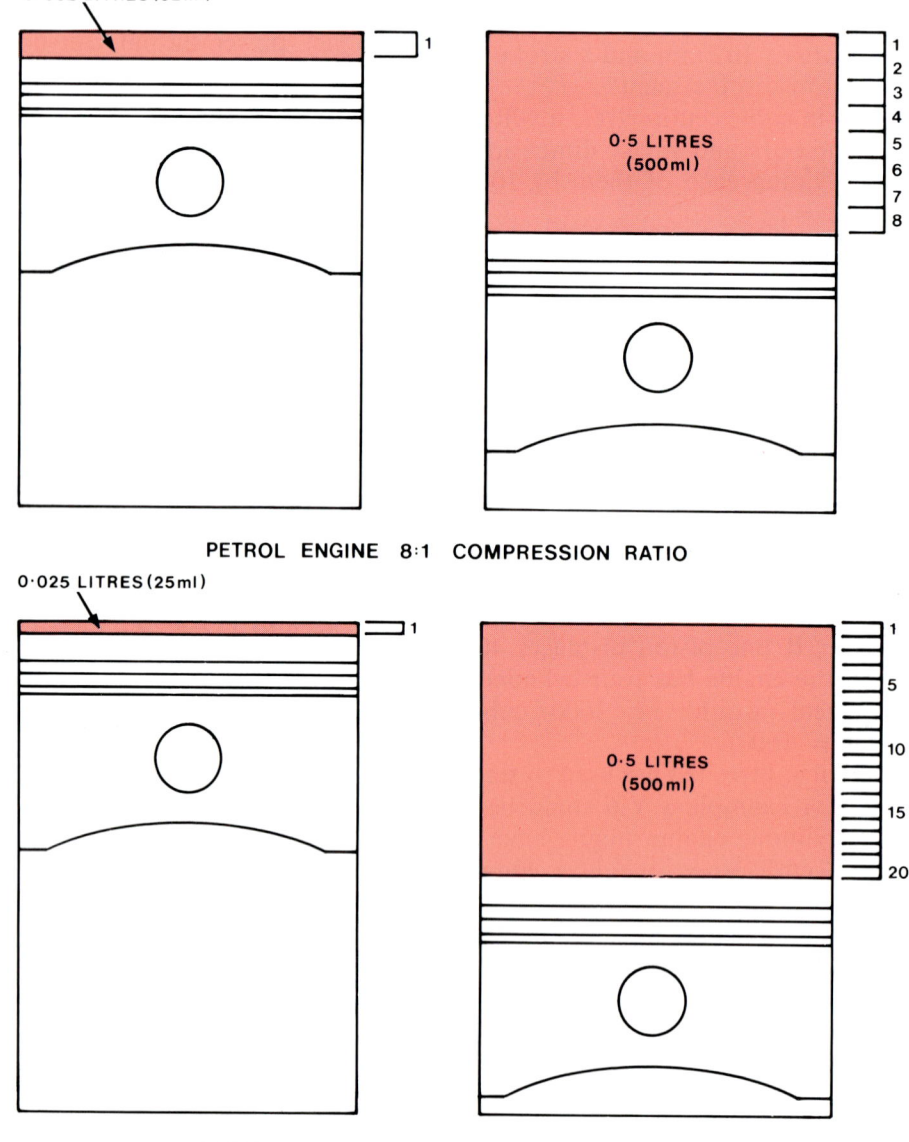

Fig. 14-3. A diesel engine has a much higher compression ratio than a petrol engine. A diesel must squeeze the fuel mixture very tight to cause combustion.

Today's petrol engines use a lower compression ratio (about 8 or 9 to 1). This allows the use of cleaner burning, unleaded fuel. There is a slight reduction in engine power and efficiency, however.

Diesel engines have a very high compression ratio (17 to 25:1 typical). The diesel compresses the air in the cylinder high enough in pressure, for it to heat up and ignite the fuel.

Note! Automotive fuels and combustion will be explained in detail in chapter 18.

Compression pressure

Compression pressure is the amount of pressure produced in the engine cylinder on the compression stroke. Compression pressure is normally measured in kilopascals (kPa) previously pounds per square inch (psi).

A petrol engine may have a compression pressure from 900 to 1300 kPa.

A diesel engine has a much HIGHER compression pressure of about 1700 to 2700 kPa.

A compression gauge is used to measure compression stroke pressure. It is secured into the spark plug, injector nozzle (or glow plug) hole. The ignition or injection system is disabled. Then, with the throttle wide open the engine is cranked over with the starter motor. The gauge will read compression stroke pressure.

Discussed in later chapters, compression stroke pressure is an indicator of engine condition. If low, there is a problem allowing air to leak out of the cylinder. The engine might have worn rings, burned valves, or a blown head gasket.

ENGINE TORQUE

Torque is a turning or twisting force. When you turn the car's steering wheel, you have applied torque to the steering wheel.

Engine torque is a rating of the turning force at the engine crankshaft. When combustion pressure pushes the piston down, a strong rotating force is applied to the crankshaft. This turning force is sent to the transmission or transaxle, drive line or drive axles, and drive wheels, propelling the car.

Engine torque specifications are given in a service manual. One example, 106 Nm at 3000 rpm is given for one particular engine. This engine would be capable of producing a maximum of 106 newton metres of torque when operating at 3000 revolutions per minute.

WHAT IS ENGINE POWER?

Power is a measure of an engine's ability to perform work.

In the SI metric system, power is measured in kilowatts, which is abbreviated to kW.

Formerly, before metrication an engine's ability to perform work was measured in horsepower (hp). At one time one horsepower was the average strength of a horse.

To measure an engine's ability to perform work (power output) an instrument or machine called a dynamometer is used. Fig. 14-4.

The engine is coupled up to the Dynamometer and drives it as it would a drive shaft.

The dynamometer is designed so that its driving resistance can be adjusted so that the engine is tested under varying loads while power readings are taken.

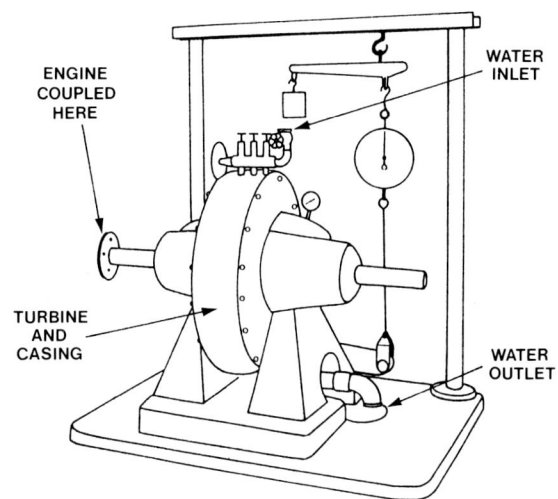

Fig. 14-4. This water brake (hydraulic) type dynamometer is used to measure engine power output.

Manufacturer's power ratings

Vehicle manufacturers rate engine power output at a specific rpm. For instance, a high performance, turbocharged engine might be rated at 225 kW at 5000 rpm. This engine power rating is normally stated in the service manual.

There are several different methods of determining engine power.

Brake power (bp) is the power available at the crankshaft for doing work. Shown in Fig. 14-5 is a prony brake, which was an equipment to measure bp.

The engine crankshaft drove the prony brake drum whilst the brake shoe mechanism was applied. From this frictional torque applied to the engine, the resulting pointer deflection on the dial could then be used to determine the bp.

An engine dynamometer (dyno) is now used to measure the brake power of modern car engines. Refer to Fig. 14-6. It functions in much the same way as a prony brake. Either an electric motor or fluid coupling is used to place a drag on the engine crankshaft. Then, power output can be determined.

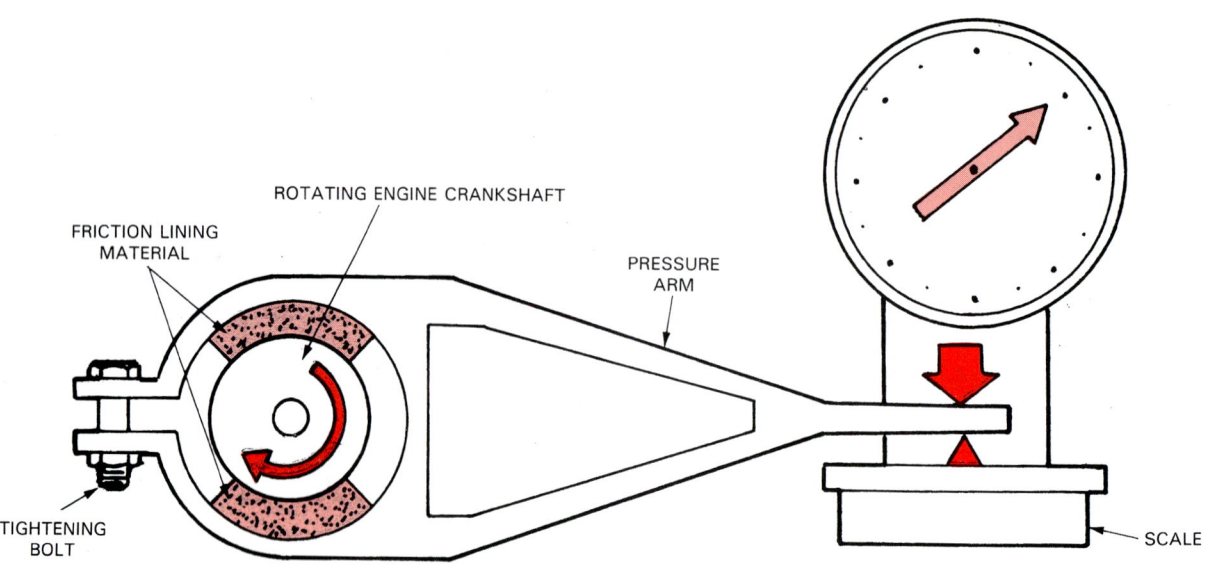

Fig. 14-5. Prony brake will measure engine brake power. Brake is applied to engine crankshaft. Amount of needle deflection can be used to find brake power.

A chassis dynamometer measures the power delivered to the vehicles driving wheels. See Fig. 14-7. It indicates the amount of power available to propel the vehicle, referred to as wheel power (wp).

Indicated power (ip) refers to the amount of power developed within the engine's combustion chambers. To determine the ip, special pressure transducers are installed to the combustion chambers. Recordings of mean effective pressure are made in order to calculate the ip.

Frictional power (fp) is the power needed to overcome friction whilst an engine is running. It is a measure of resistance to movement between engine components.

Frictional power is POWER LOST to friction. It reduces the amount of power left to drive the vehicle.

Net power is the maximum power developed when an engine is driving all auxiliaries (alternator, water pump, cooling fan, fuel pump, air injection pump, air conditioning compressor and power steering pump). It is the amount of useful power with the engine installed in the vehicle. Net power indicates the amount of power available to drive the vehicle. See Fig. 14-8.

Gross power is similar to net power, but it is the engine power available with only the basic auxiliaries installed (alternator, water pump, and cooling fan). Gross power does not include power lost to the power steering pump, air injection pump, air conditioning compressor, and/or other extra units. Refer to Fig. 14-9.

ENGINE EFFICIENCY

Engine efficiency is the ratio of power produced by the engine (brake power) and the power supplied to the engine (heat content of fuel). By comparing fuel consumption to engine power output, you can find engine efficiency.

Fig. 14-6. Technician is using engine dynamometer and elaborate computer system to monitor engine performance. Dyno loads engine to simulate driving conditions while monitoring power output and numerous other engine functions. (AC Spark Plug)

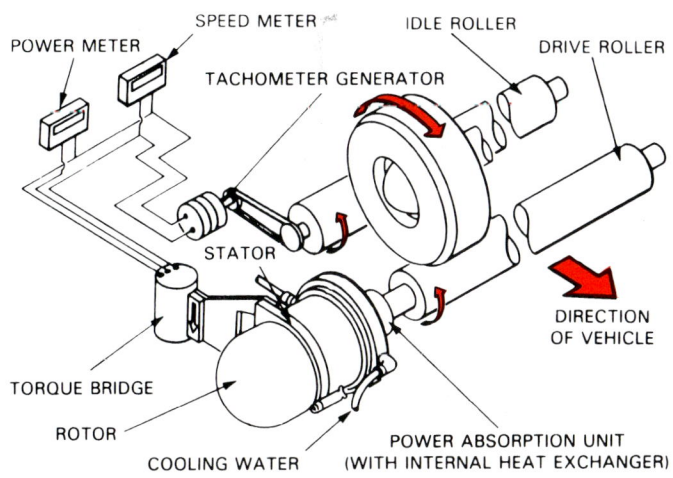

Fig. 14-7. Chassis dynamometer measures turning power at car's drive wheels. This accounts for any power consumed by driveline.

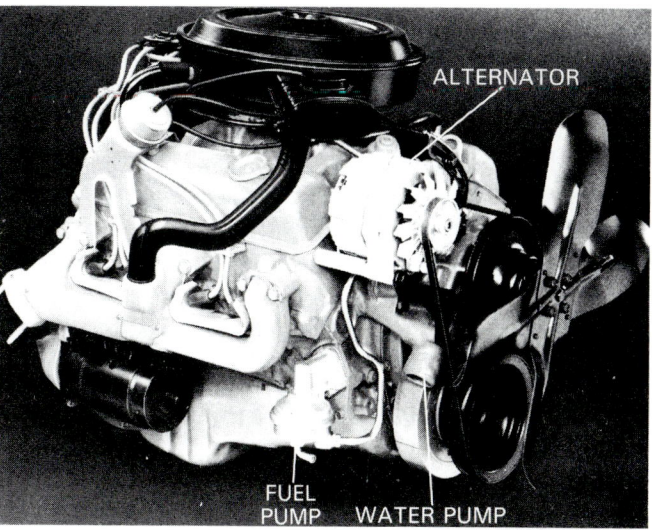

Fig. 14-9. Gross power is similar to net power rating. It does not include power lost to unneeded accessories.

Fig. 14-8. Net power is available power with engine operating all accessories such as air conditioning compressor, power steering pump, alternator, and water pump. (Ford)

If all of the heat energy in the fuel were converted into useful work, the engine would be 100 percent efficient. This much efficiency is not possible with a piston engine. Modern piston engines are only about 20 percent thermally efficient.

Fig. 14-10 illustrates how the heat energy of the fuel is used by a piston engine. About 70 percent of the fuel's heat energy enters the cooling and exhaust systems. This leaves very LITTLE HEAT ENERGY over to produce usable power.

Volumetric efficiency

Volumetric efficiency is the ratio of actual air drawn into the cylinder to the swept volume of the cylinder. It refers to how well an engine can breathe on the intake stroke.

If volumetric efficiency were 100 percent, the cylinder would completely fill with air on the intake stroke. Engines are only capable of about 80 or 90 percent volumetric efficiency. Airflow restriction in the ports and around the valves limit airflow.

High volumetric efficiency INCREASES ENGINE POWER because more fuel and air can be drawn into the cylinders and burned in the combustion chambers.

The formula for volumetric efficiency is:

Volumetric Efficiency =

$$\frac{\text{Actual volume of air taken into cylinder}}{\text{Swept volume of the cylinder}}$$

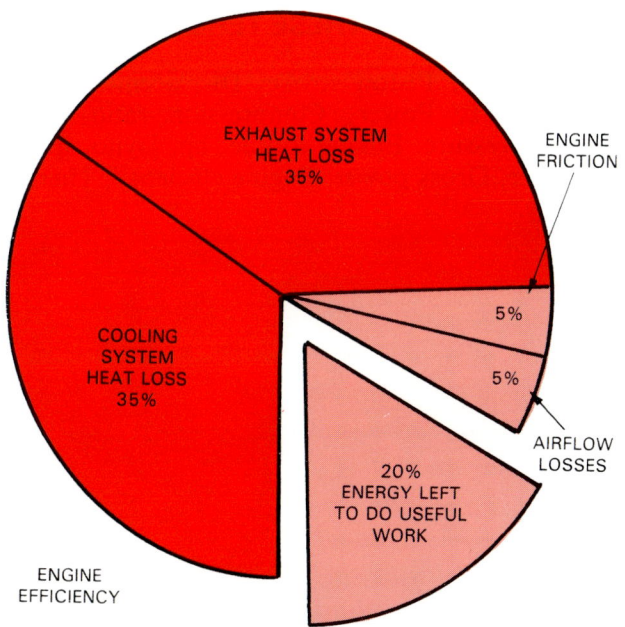

Fig. 14-10. Pie chart shows how fuel's heat energy is used by piston engine. Note that most of the heat energy is wasted.

217

Mechanical efficiency

Mechanical efficiency is the ratio of the bp to the ip. It is a measurement of mechanical friction. Remember that ip equals theoretical power produced by combustion, bp is the actual power at the engine crankshaft. The difference between the two is due to frictional power loss.

Mechanical efficiency of around 70 to 80 percent is normal. This means that about 20 to 30 percent of the engine's power is lost to friction, frictional power (fp) loss.

The friction between the piston rings and cylinder walls accounts for most of this loss.

Thermal efficiency

Thermal efficiency is the ratio of the brake power (bp) to the energy available in the fuel supplied.

Thermal efficiency =

$$\frac{\text{Output energy}}{\text{Input energy}} =$$

$$\frac{\text{Brake power (kW/sec)}}{\text{Energy in fuel (kW/sec)}}$$

Generally, engine thermal efficiency is about 20 to 30 percent. The rest of the heat energy is absorbed in the metal parts of the engine, or discharged through the exhaust system.

KNOW THESE TERMS

Cylinder bore, Piston stroke, Displacement, kW, Force, Work, Power, Compression ratio, Compression pressure, Engine torque, bp, Dyno, ip, fp, Net power, Gross power, Engine efficiency, Thermal efficiency.

REVIEW QUESTIONS

1. Engine size is determined by cylinder _____, amount of _____ on each stroke, and _____ of _____.
2. Cylinder bore is measured across the cylinder, parallel with the top of the block. True or False?
3. Piston stroke is the distance the piston moves during a complete four-stroke cycle. True or False?
4. _____ _____ is the volume the piston moves from BDC to TDC.
5. If an engine has a bore of 89 mm and a stroke of 76 mm, what is the piston displacement?
6. Define the term "engine displacement".
7. Explain the difference between force, work, and power.
8. Which of the following would NOT be a compression ratio for a car engine?
   a. 2:1
   b. 8:1
   c. 20:1
   d. All of the above are correct.
9. A petrol engine may produce a compression pressure of approximately _____ to _____ kPa.
10. A diesel engine can produce a compression pressure of about _____ to _____ kPa.
11. What is engine torque?
12. Explain the term "power".
13. Which of the following is NOT a power rating?
   a. Brake power.
   b. Fractional power.
   c. Indicated power.
   d. Frictional power.
   e. Net power.
14. Describe what the term "engine efficiency" means.
15. This is the ratio of actual air drawn into an engine to the swept volume.
   a. Mechanical efficiency.
   b. Thermal efficiency.
   c. Volumetric efficiency.
   d. Dynamometer efficiency.

Chapter 15

ENGINE FUNDAMENTALS

After studying this chapter, you will be able to:
☐ Identify the major parts of a typical automotive engine.
☐ Describe the four-stroke cycle.
☐ Define common engine terms.
☐ Explain the basic function for the major parts of an automotive engine.

In Chapter 1 THE MOTOR CAR and Chapter 12 BUILDING AN ENGINE you learned a little about engine systems and how a basic four-stroke engine operates. This chapter will build upon that information by explaining each part in a little more detail.

Note! If needed, quickly review the material in Chapters 1 and 12 as a sound understanding on basic engine theory is very important.

ENGINE OPERATION

The engine is the source of power for the car. For this reason, it is also called a power plant. An energy source (usually petrol or diesel fuel) is burned inside the engine to produce heat. The heat causes expansion (enlargement) of the fuel vapours or gases in the engine.

The burning and expansion in an enclosed space or combustion chamber produces pressure. The engine piston, connecting rod, and crankshaft converts this pressure into motion for moving the car and operating its other systems.

Fig. 15-1 illustrates how an engine converts fuel into a useful form of energy. Combustion pressure forces the piston down. By linking the piston to the crankshaft, an engine can produce a powerful spinning motion. The rotating crank can be used to drive gears, chains and sprockets, belts and sprockets, and other drive mechanisms.

Piston travel (TDC, BDC)

The distance the piston can travel up or down in the cylinder is limited by the crankshaft. The points at which the piston stops and changes direction are called TDC (top dead centre) and BDC (bottom dead centre). See Fig. 15-2.

When the piston is at the HIGHEST POINT in the cylinder, it is at TDC. When the piston slides to its LOWEST POINT in the cylinder, it is at BDC.

Piston stroke

A piston stroke is the distance the piston slides up or down from TDC to BDC. This takes one-half turn of the crankshaft. The crank rotates 180 degrees during one piston stroke. Refer to Fig. 15-2.

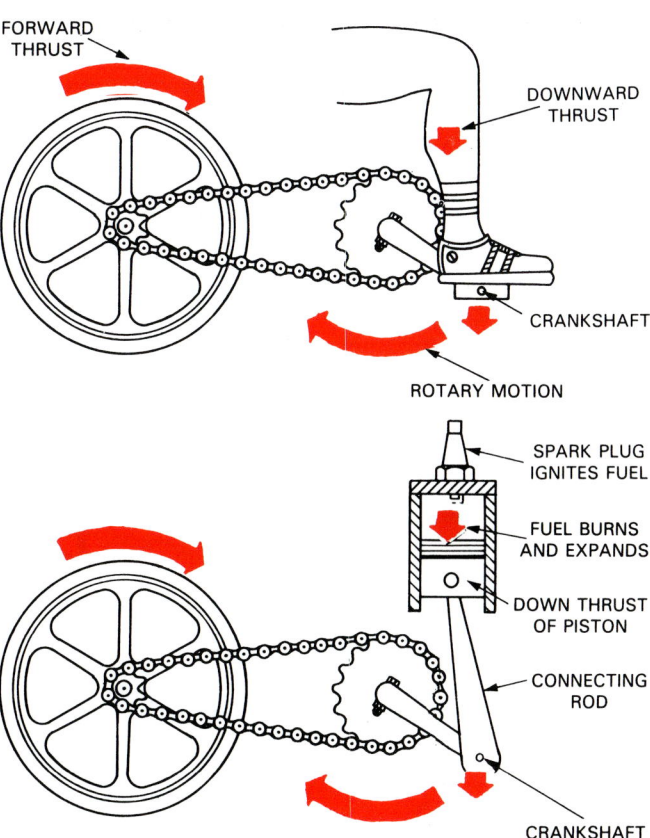

Fig. 15-1. Crank converts downward thrust of piston into useful rotating motion. Rotating motion can be used to operate drive mechanism.

Four-stroke cycle

The four-stroke cycle requires four piston strokes to complete one cycle (complete series of events). Every four strokes, the engine produces one power stroke (useful energy). Almost all automobiles use four-stroke cycle engines.

Look at Fig. 15-3 to review the four-stroke cycle.

The intake stroke of a petrol engine draws fuel and air into the engine. The intake valve is open and the exhaust valve is closed. The piston slides down and forms a low pressure area or vacuum in the cylinder. Outside air pressure then pushes the air-fuel mixture into the engine, Fig. 15-3.

The compression stroke squeezes the air-fuel mixture to prepare it for combustion (burning). The mixture is more combustible when pressurised. During this stroke, the piston slides up with both valves closed. Look at Fig. 15-3.

The power stroke burns the air-fuel mixture and pushes the piston down with tremendous force. This is the only stroke that does not consume energy. It produces energy. When the spark plug fires (petrol engine), it ignites the fuel mixture. Since both valves are still closed, pressure forms on the top of the piston. The piston is forced down, spinning the crankshaft. Refer to Fig. 15-3.

The exhaust stroke removes the burned gases from the engine and prepares the cylinder for a fresh charge of air and fuel. The piston moves up. The intake valve is closed and the exhaust valve is open. The burned gases are pushed out of the exhaust port and into the exhaust system, Fig. 15-3.

The crankshaft must rotate TWO complete revolutions to complete the four-stroke cycle. With the engine running, these series of events happen over and over very quickly.

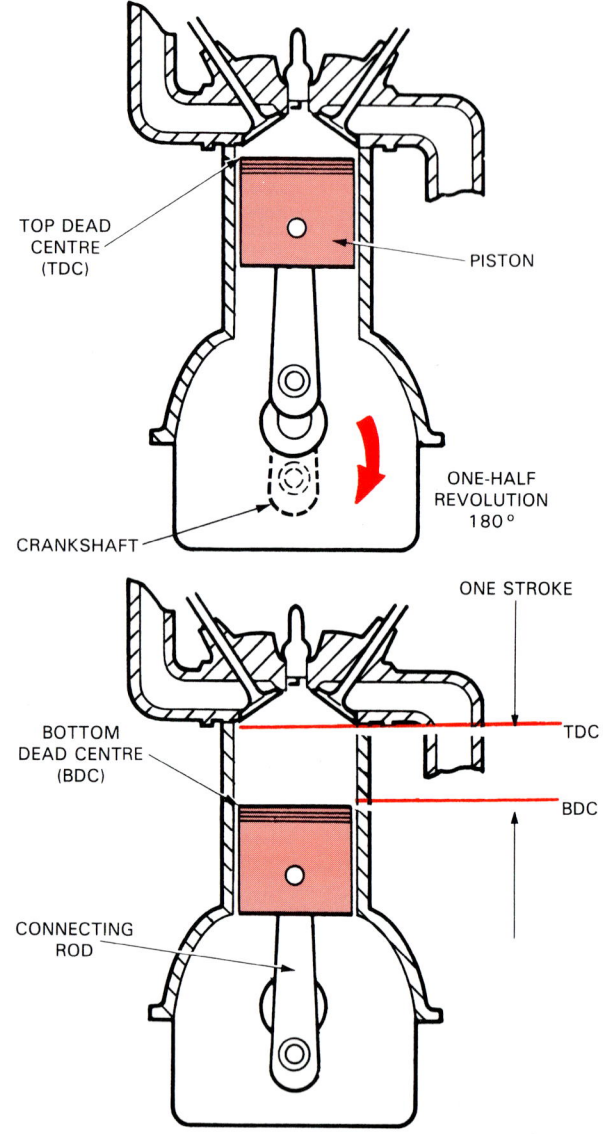

Fig. 15-2. TDC means piston is at top of stroke. BDC means piston is at bottom of stroke. Stroke is one piston movement. (Ford)

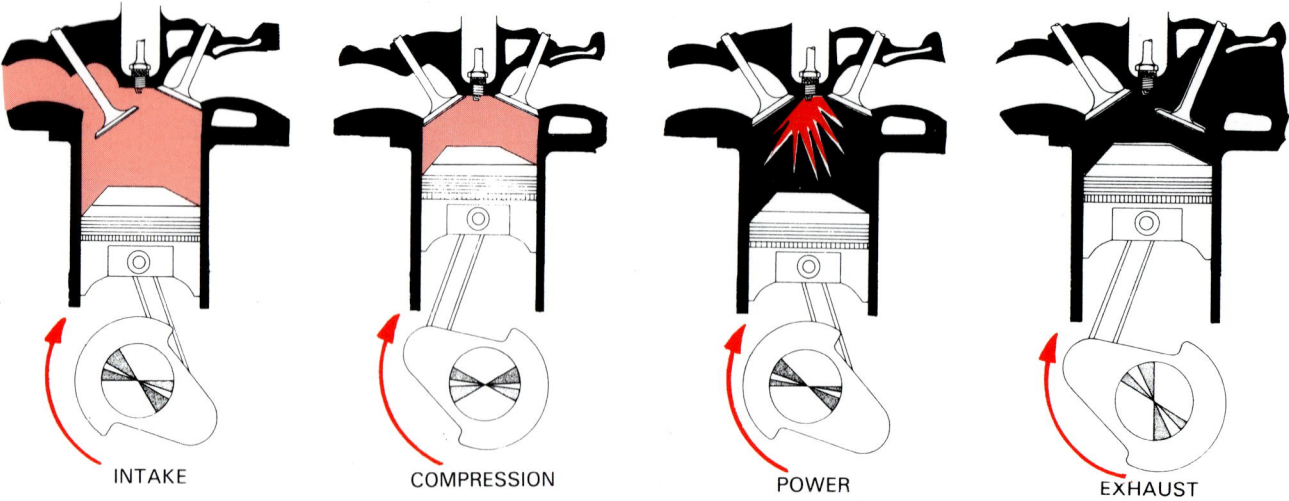

Fig. 15-3. Restudy basic four-stroke cycle.

ENGINE BOTTOM END

The term engine bottom end generally refers to the block, crankshaft, connecting rods, pistons, and related components. Another name for engine bottom end is short motor. It is an assembled engine block with the cylinder heads, intake manifold, exhaust manifold, and other external parts REMOVED.

Engine block

The engine block, also called cylinder block, forms the main body of the engine. Other parts bolt to or fit inside the block. Fig. 15-4 shows a cutaway view of a basic block with parts installed.

The cylinders, also known as cylinder bores, are large, round holes machined through the block from top to bottom. The pistons fit into the cylinders. The cylinders are slightly larger than the pistons. This allows the pistons to slide up and down freely.

The face or block face is the top of the block. It is machined perfectly flat. The cylinder head bolts to the face. Oil and coolant passages through the block face allow lubrication and cooling of the cylinder head parts.

Water jackets are coolant passages through the block. They allow a water and antifreeze solution to cool the cylinders.

Core plugs, or welch plugs, are round, metal plugs on the outside of the block. They seal holes left in the block after casting (manufacturing). The plugs prevent coolant leakage out of the water jackets.

The main bearing bores are holes machined in the bottom of the block for the crankshaft. Removable bearings fit into these bores.

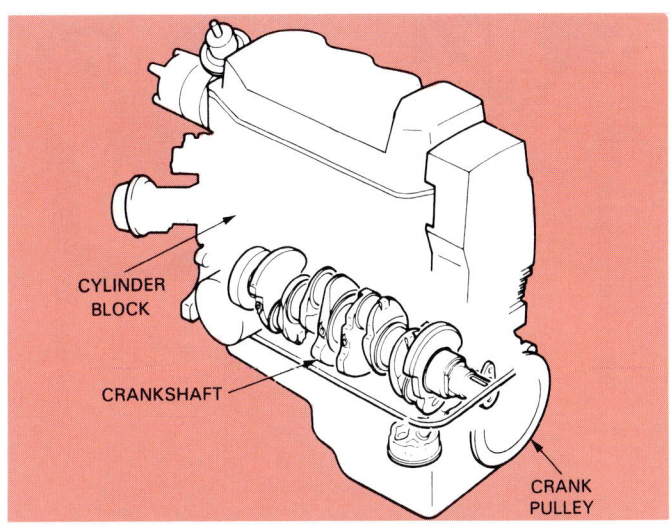

Fig. 15-5. Crankshaft fits into bottom of block. (Ford)

Main bearing caps bolt to the bottom of the block and hold the crankshaft in place. Two or four large bolts normally secure each cap to the block. The caps and the block together form the main bearing bore.

The crankcase is the lowest portion of the block. The crankshaft rotates inside the crankcase.

Crankshaft

The crankshaft harnesses the tremendous force produced by the downward thrust of the pistons. It changes the up and down motion of the pistons into a rotating motion. The crankshaft fits into the bottom of the engine block, as shown in Fig. 15-5.

Fig. 15-6 shows an engine crankshaft. Refer to this illustration as it is explained.

The crankshaft main journals are precisely machined and polished surfaces. They fit into the block main bearings.

The crankshaft connecting rod journals are also machined and polished surfaces, but they are offset from the main journals. The connecting rods bolt to the connecting rod journals. With the engine running, the connecting rod journals circle around the centreline of the crankshaft.

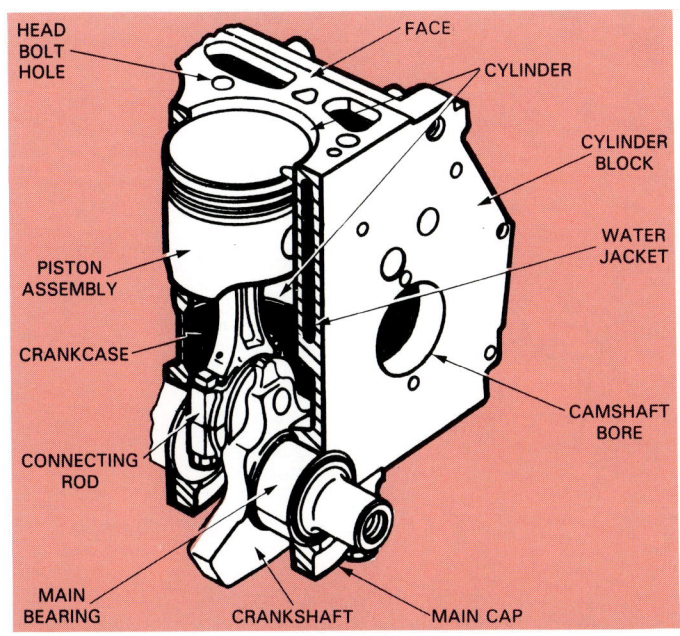

Fig. 15-4. Block is main supporting member of engine. Note how other parts fit into block. (Ford)

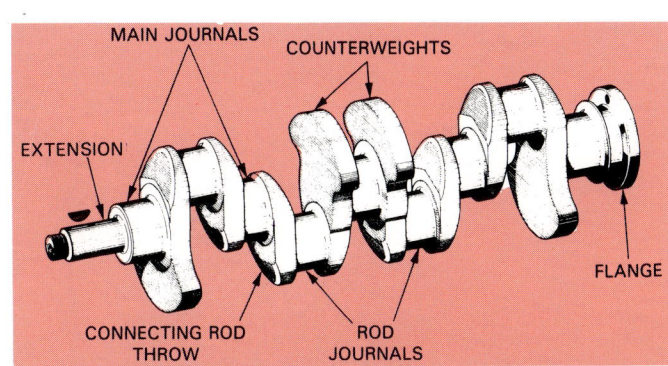

Fig. 15-6. Study basic parts of a crankshaft. Journals are very smooth surfaces for bearings. (Peugeot)

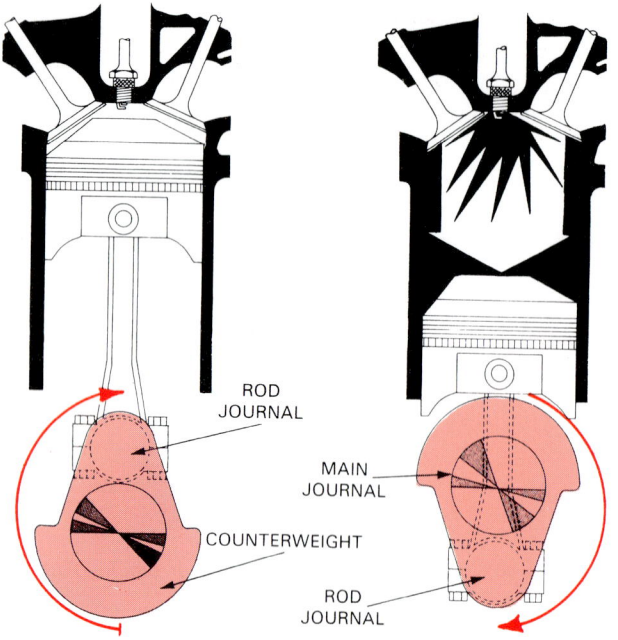

Counterweights are formed on the crankshaft to prevent vibration. The weights counteract the weight of the connecting rods, pistons, rings, and connecting rod journal offset. See Fig. 15-7.

The crankshaft extension protrudes through the front of the block. It provides a mounting place for the camshaft drive mechanism, front damper, and fan belt pulleys.

A flange for holding the flywheel is on the back of the crankshaft. The flywheel bolts to this flange. The centre of the flange has a pilot hole or bushing for the transmission torque converter or input shaft.

Motor car engines normally have 4, 6, or 8 cylinders. The crankshaft connecting rod journals are arranged so that there is always at least one cylinder on a power stroke. Then, force is always being transmitted to the crankshaft to smooth engine operation.

Engine main bearings

The engine main bearings are removable inserts that fit between the block main bore and crankshaft main journals. One-half of each insert fits into the

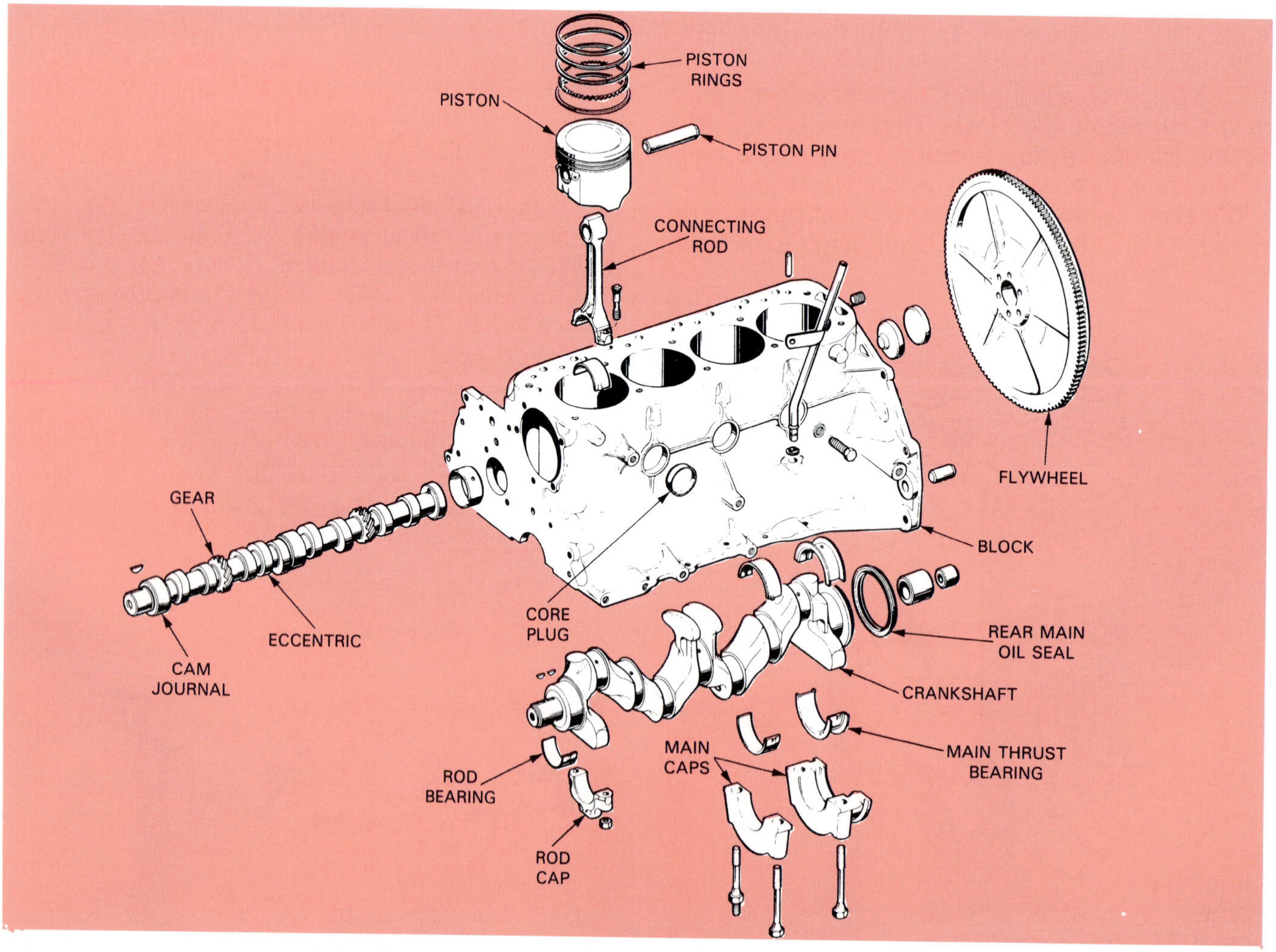

Fig. 15-7. As engine runs, connecting rod journal spins around main journal. Counterweight offsets weight of piston and rod to prevent vibration.

Fig. 15-8. Engine bottom end consists of these basic parts. Note crankshaft bearings and block main bearing caps.

block. The other half fits into the block main bearing caps. Refer to Fig. 15-8 and study the parts.

Oil holes in the upper bearing insert line up with oil holes in the block. This allows oil to flow through the block, main bearings, and into the crankshaft. The oil flows through the crankshaft to lubricate the main bearings and the connecting rod bearings. This prevents metal-on-metal contact.

A main thrust bearing limits how far the crankshaft can slide forward or rearward in the block. Flanges are formed on the main bearing. The flanges almost touch the side, thrust surfaces on the crankshaft. This limits crankshaft endplay (front-to-rear movement). See Fig. 15-9. Normally, only one of the main bearings serves as a thrust bearing.

Main bearing clearance is the space between the crankshaft main journals and the main bearing insert. The clearance allows lubricating oil to enter and separate the journal and bearing. This allows the two to rotate without rubbing on each other and wearing.

Crankshaft oil seals

Crankshaft oil seals keep oil from leaking out of the front and rear of the engine. The oil pump forces oil into the main and connecting rod bearings. This causes oil to spray out of the bearings. Seals are placed around the front and rear of the crankshaft to contain this oil.

The rear main oil seal fits around the rear of the crankshaft to prevent oil leakage, as pictured in Fig. 15-8. It can be a one-piece or a two piece seal. The seal lip rides on a smooth, machined and polished surface on the crankshaft. The front seal is covered later.

Flywheel

A flywheel is a large wheel mounted on the rear of the crankshaft. Look at Fig. 15-8. A flywheel can have several functions:

1. A flywheel for a car with a manual transmission is very heavy and can help smooth engine operation.
2. The flywheel connects the engine crankshaft to the transmission or transaxle. Either the manual clutch or the automatic transmission torque converter bolts to the flywheel, which is usually called a drive plate on automatic transmission models.
3. A large ring gear, usually on the flywheel, is used to start the engine. A small gear on the starter motor engages the flywheel ring gear and turns it.

Connecting rod

The connecting rod secures the piston to the crankshaft. It transfers piston movement and combustion pressure to the crankshaft connecting rod journals. The connecting rod also causes piston movement during the powerless strokes (intake, compression, and exhaust).

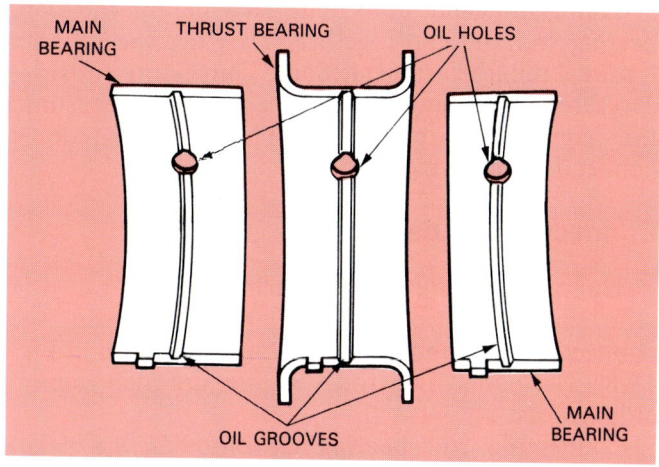

Fig. 15-9. Main bearing inserts fit between crankshaft main journals and block. One bearing has thrust surfaces to control crankshaft end play. Oil holes and grooves allow oil to lubricate bearings.

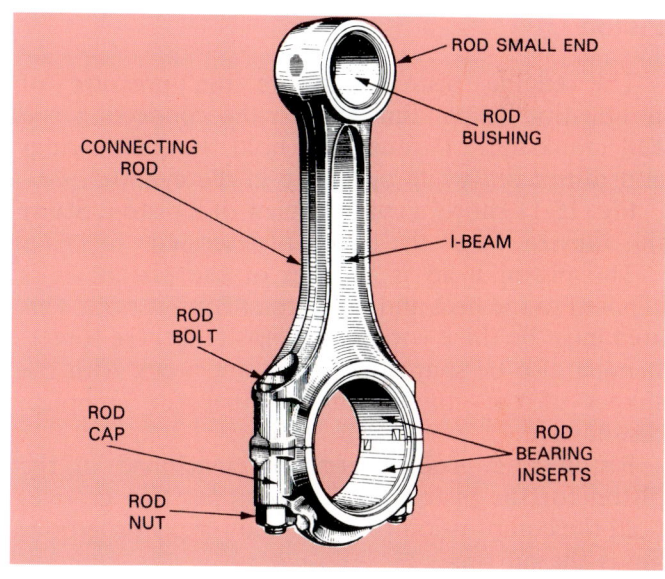

Fig. 15-10. Connecting rod is link between piston and crankshaft. (Peugeot)

Refer to Fig. 15-10 as the connecting rod is discussed.

The connecting rod small end encircles the piston pin (gudgeon pin). Also called the upper end, it contains a one-piece bushing. The bushing is pressed into the small end of the connecting rod.

The connecting rod I-beam is the centre section of the rod. The I-beam shape has a very high strength-to-weight ratio. It prevents the rod from bending, twisting, and breaking.

The connecting rod cap bolts to the bottom of the connecting rod body. It can be removed for dismantling of the engine.

The connecting rod big end or lower end is a hole machined in the rod body and cap. The connecting rod bearing fits into the big end.

Connecting rod bolts and nuts clamp the connecting rod and cap together. They are special high tensile strength fasteners. Some connecting rods use cap screws without a nut. The cap screw threads into the connecting rod itself. This design reduces connecting rod weight.

Connecting rod bearings

The connecting rod bearings ride on the crankshaft rod journals. They fit between the connecting rods and the crankshaft, as shown in Fig. 15-8. The connecting rod bearings are also removable inserts.

Connecting rod bearing clearance is the small space between the connecting rod bearing and crankshaft journals. As with the main bearings, it allows oil to enter the bearing. The oil prevents metal-to-metal contact that would wear out the crankshaft and bearings.

Piston

The engine piston transfers the pressure of combustion (expanding gas) to the connecting rod and crankshaft. It must also hold the piston rings and piston pin while operating in the cylinder.

Fig. 15-11 shows a cutaway view of a piston. Study this illustration as the piston is described.

The piston head is the top of the piston. It is exposed to the heat and pressure of combustion. This area must be thick enough to withstand these forces. It must also be shaped to match and work with the shapes of the combustion chamber for complete combustion.

Piston ring grooves are slots machined in the piston for the piston rings. The upper two grooves hold the compression rings. The lower piston groove holds the oil ring.

Oil holes in the bottom grooves allow the oil to pass through the piston. The oil then drains back into the crankcase.

The ring lands are the areas between and above the ring grooves. They separate and support the piston rings as they slide in the cylinder.

A piston skirt is the side of the piston below the last ring. It keeps the piston from tipping in its cylinder. Without a skirt, the piston could cock and jam in the cylinder.

The piston boss is a reinforced area around the piston pin hole. It must be strong enough to support the piston pin under severe loads.

A piston pin hole is machined through the piston pin boss for the piston pin. It is slightly larger than the piston pin.

Piston pin

The piston pin, also called a gudgeon pin, allows the piston to swing on the connecting rod. The piston pin fits through the hole in the piston pin boss and the connecting rod small end. This is pictured in Fig. 15-12.

Piston clearance

Piston clearance is the amount of space between the sides of the piston and the cylinder wall. Clearance is needed for a lubricating film of oil and to allow for expansion when the piston heats up. The piston must always be free to slide up and down in the cylinder block.

Piston rings

The piston rings seal the clearance between the outside of the piston and the cylinder wall. They must keep combustion pressure from entering the crankcase. They must also keep oil from entering the combustion chambers.

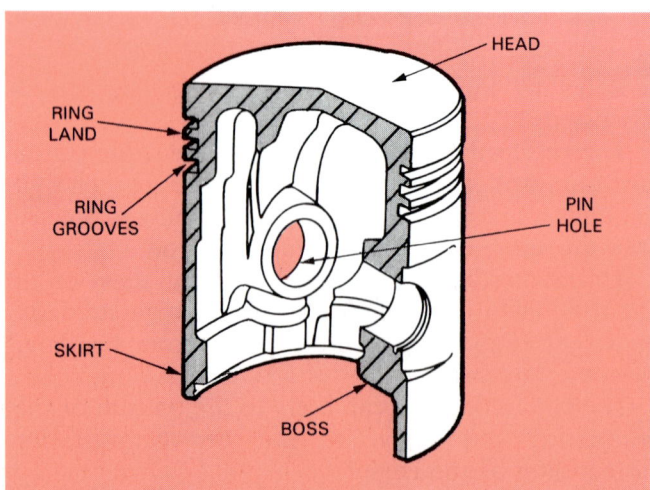

Fig. 15-11. Piston slides in cylinder and is exposed to combustion flame. It must be light, yet strong.

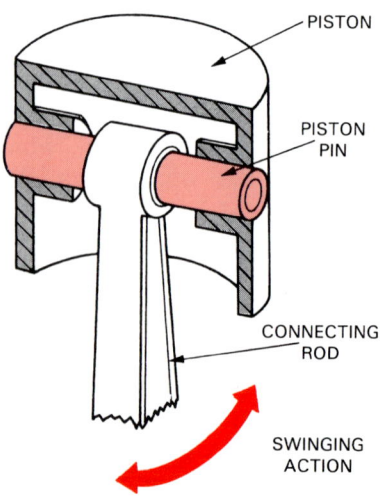

Fig. 15-12. Piston pin allows connecting rod to swing in piston. This allows crank and connecting rod bottom end movement.

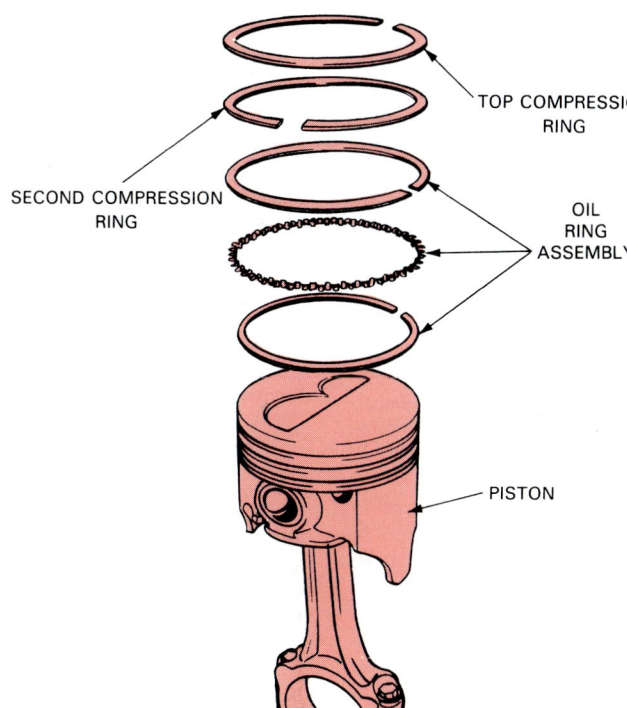

Fig. 15-13. Two top piston rings are compression rings. Bottom ring is oil ring. They fit into grooves cut in piston.

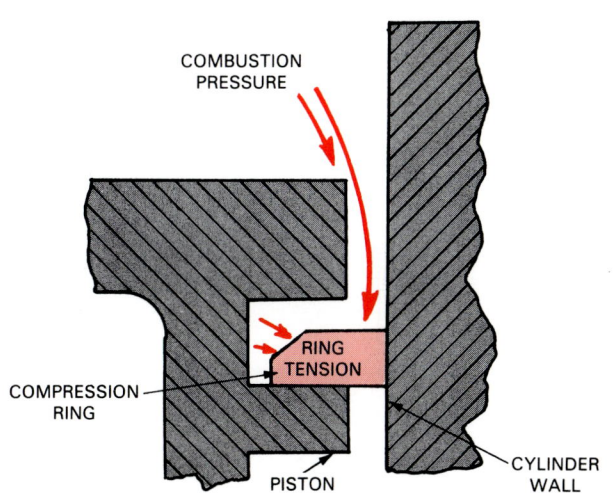

Fig. 15-14. Compression ring must prevent combustion pressure from leaking between piston cylinder wall. Pressure actually helps push ring against cylinder to aid sealing.

Most pistons use three rings: two, upper compression rings and one, lower oil ring. This is shown in Fig. 15-13. Note ring locations.

The compression rings prevent blowby (combustion pressure leaking into the engine crankcase). Fig. 15-14 shows how rings are installed to the engine.

On the compression stroke, pressure is trapped between the cylinder and piston grooves by the compression rings. Combustion pressure pushes the compression rings down in their grooves and out against the cylinder wall. This produces an almost leakproof seal.

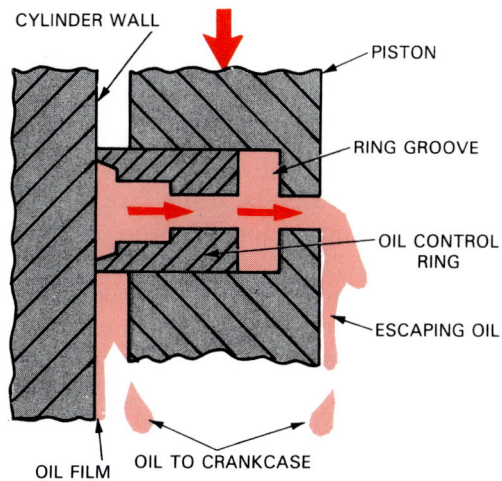

Fig. 15-15. Oil ring scrapes excess oil off cylinder wall. If oil entered combustion chamber, engine would emit blue smoke.

The main job of an oil ring is to prevent engine oil from entering the combustion chamber. It scrapes excess oil off the cylinder wall, as in Fig. 15-15.

If too much oil got into and were burned in the combustion chamber, blue smoke would come out of the car's exhaust pipe.

Ring gap is the split or space between the ends of a piston ring. The ring gap allows the ring to be spread open and installed on the piston. It also allows the ring to be made slightly larger in diameter than the cylinder. When squeezed together and installed in the cylinder, the ring spreads outward and presses on the cylinder wall. This aids ring sealing.

ENGINE TOP END

The term engine top end generally refers to the cylinder heads, valves, camshaft, and other related components. These parts work together to control the flow of air and fuel into the engine cylinders.

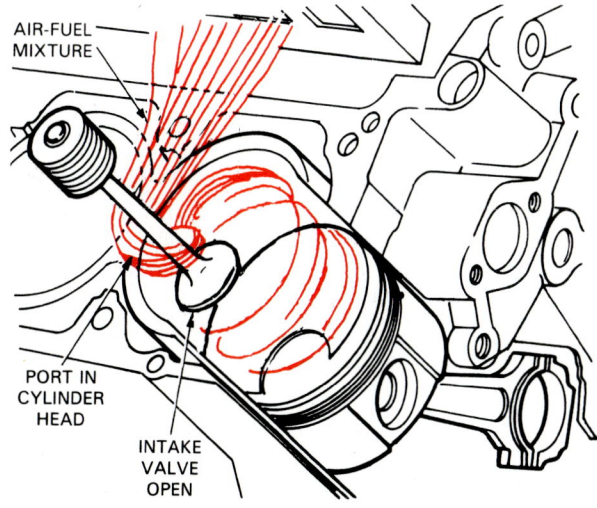

Fig. 15-16. Engine top end controls flow of mixture into cylinder. It also controls flow of exhaust out of cylinder. (Ford)

They also control the flow of exhaust out of the engine. See Fig. 15-16. It shows fuel charge entering engine.

Cylinder head

The cylinder head bolts to the face of the cylinder block. It covers and encloses the top of the cylinders. Refer to Fig. 15-17.

Combustion chambers are small pockets formed in the cylinder heads. The combustion chambers are located directly over the pistons. Combustion occurs in these areas of the cylinder head. Spark plugs (petrol engine) or injectors (diesel engine) protrude through holes and into the combustion chambers. Fig. 15-18 shows a combustion chamber.

Intake and exhaust ports are cast into the cylinder head. The intake port routes air (diesel engine) and

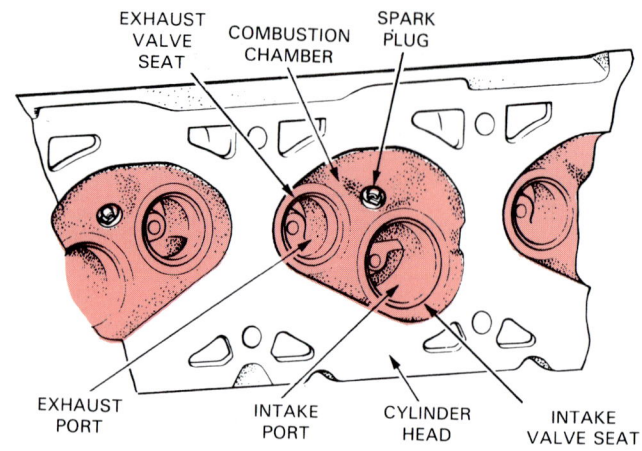

Fig. 15-18. Combustion chamber is formed in cylinder head. Valve ports enter chamber. Also, note spark plug tip and valve seats.

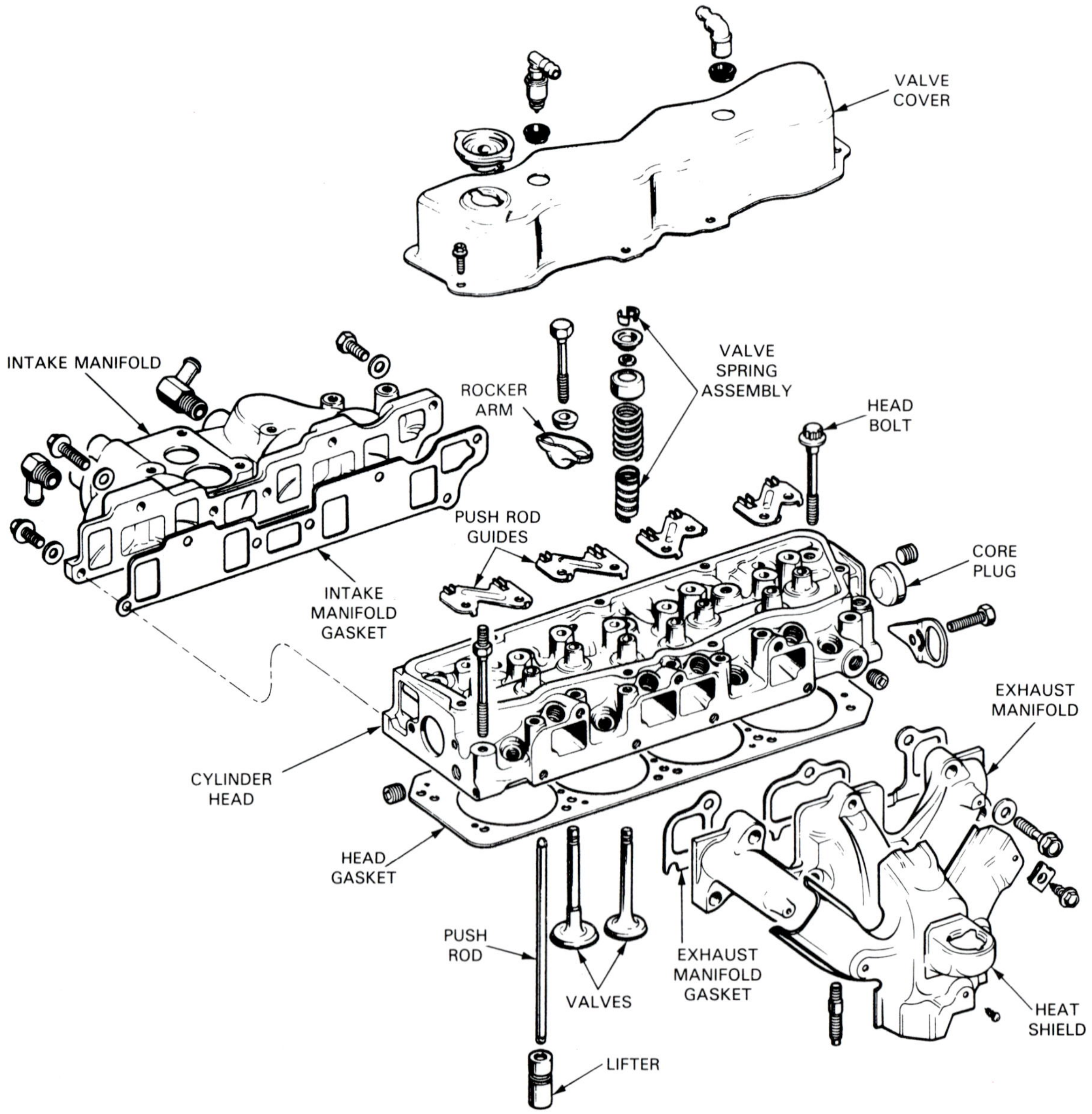

Fig. 15-17. Study basic engine top end components. Cylinder head is foundation for these parts.

air and fuel (petrol engine) into the combustion chambers. The exhaust port routes burned gases out of the engine.

Valve guides are small holes machined through the cylinder head for the valves. The valves fit into and slide in these guides.

Valve seats are round, machined surfaces in the combustion chamber port openings, Fig. 15-19. When a valve is closed, it seals against the valve seat.

Valve train

The engine valve train consists of the parts that operate the engine valves. As shown in Fig. 15-20, this includes the camshaft, lifters, push rods, rocker arms, valves, and valve spring assemblies.

The valve train must open and close the engine valves at the correct time. Note! The specific parts of a valve train vary with the engine design.

Fig. 15-21 illustrates basic valve train action.

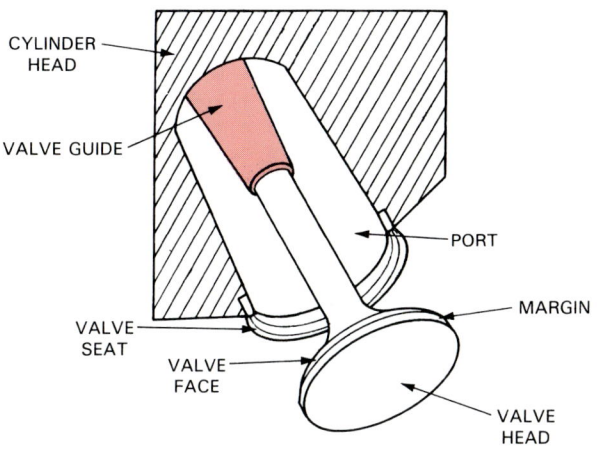

Fig. 15-19. Valve slides up and down in guide during operation. When closed, it seals against valve seat to close off port in cylinder head.

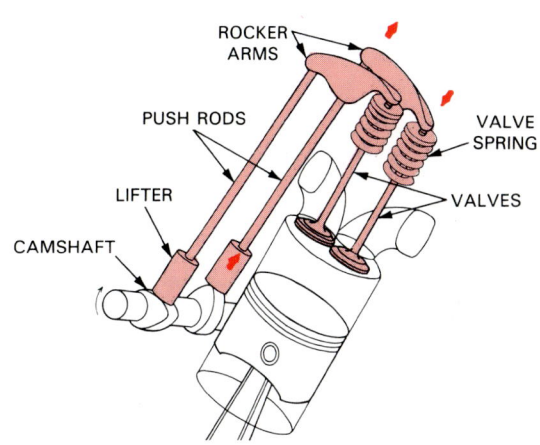

Fig. 15-20. Valve train operates engine valves. Study parts.

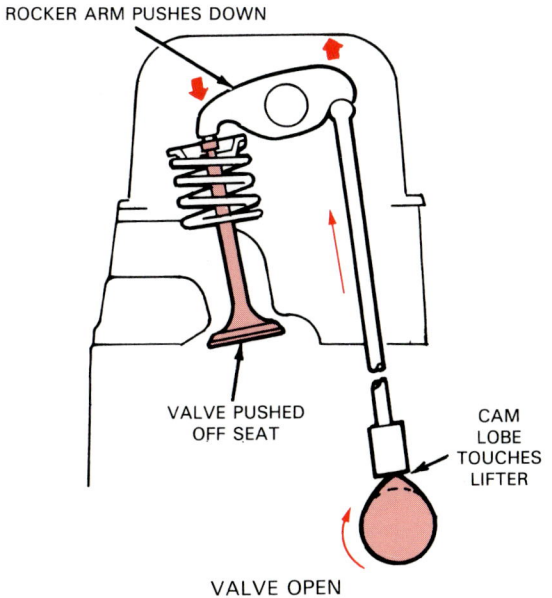

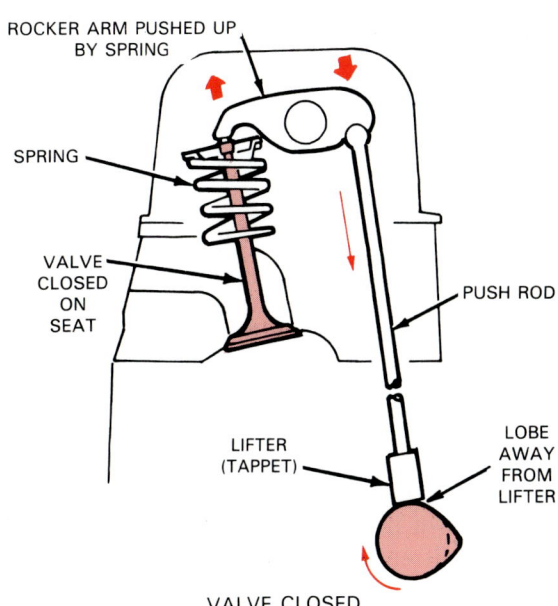

Fig. 15-21. When camshaft lobe turns into lifter, valve is pushed open. When lobe rotates away from lifter, valve spring pushes valve closed. (Ford)

227

Camshaft

The camshaft has lobes that open each valve. It can be located in the engine block or in the cylinder head. Fig. 15-22 illustrates a camshaft. Study this illustration as the camshaft is explained.

The cam lobes are egg-shaped protrusions (bumps) machined on the camshaft. One cam lobe is provided for each engine valve. A 4-cylinder engine camshaft would have eight cam lobes, a 6-cylinder twelve lobes.

The camshaft sometimes has a drive gear for operating the distributor and oil pump. A gear on the ignition system distributor may mesh with this gear.

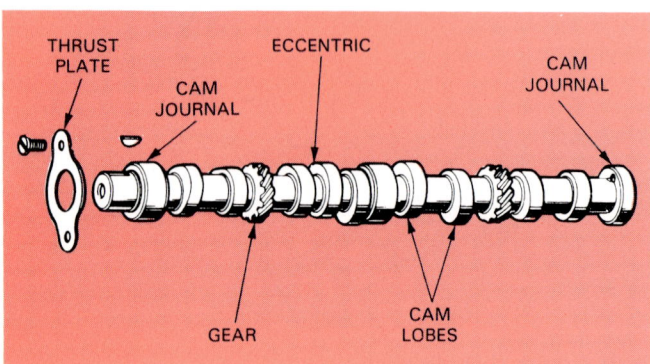

Fig. 15-22. The camshaft is a long, metal shaft with lobes, journals, and sometimes an eccentric and gears.

An eccentric (oval) may be machined on the camshaft for a mechanical (engine-driven) fuel pump. It is similar to a cam lobe but is more round. As the cam turns, the eccentric moves the fuel pump arm up and down.

Camshaft journals are precisely machined and polished surfaces for the cam bearings. Like the crankshaft, the camshaft rotates on its journals. Oil separates the cam bearings and cam journals.

Valve lifters

A valve lifter, also called a tappet, usually rides on the cam lobes and transfers motion to the rest of the valve train. The lifters can be located in the engine block or cylinder head. They fit into machined holes, termed lifter bores. Refer back to Fig. 15-20.

When the cam lobe turns into the lifter, the lifter is pushed up in its bore. This opens the valve. Then, when the lobe rotates away from the lifter, the lifter is pushed down in its bore by the valve spring. This keeps the lifter in constant contact with the camshaft.

Push rods

Push rods transfer motion between the lifters and the rocker arms, Fig. 15-21. They are needed when

the camshaft is located in the cylinder block. They are NOT needed when the camshaft is in the cylinder head.

Push rods are hollow, metal tubes with balls or sockets formed on the ends. One end of the push rod fits into the lifter. The other end fits against the rocker arm. In this way, when the lifter slides up, the push rod moves the rocker arm.

Rocker arms

Rocker arms can be used to transfer motion to the valves. They are mounted on top of the cylinder head. A pivot mechanism allows the rockers to rock up and down, opening and closing the valves. See Fig. 15-21.

Valves

Engine valves open and close the ports in the cylinder head. Two valves are normally used per cylinder: one intake valve and one exhaust valve.

The intake valve is the larger valve. It controls the flow of the fuel mixture (petrol) or air (diesel) into the combustion chamber. The intake valve fits into the port leading from the intake manifold.

The exhaust valve controls the flow of exhaust gases out of the cylinder. It is the smaller valve. The exhaust valve fits into the port leading to the exhaust manifold.

Look at Fig. 15-23 as the basic parts of a valve are introduced.

The valve head is the large, disc-shaped surface exposed to the combustion chamber. Its outside diameter determines the size of the valve.

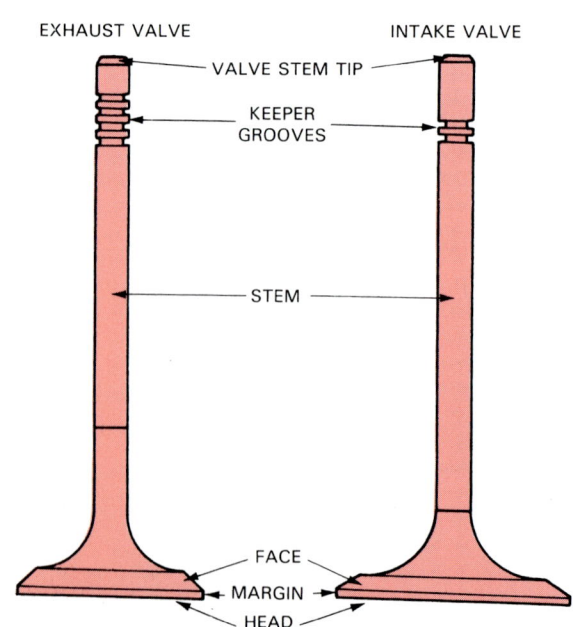

Fig. 15-23. Intake valve is larger than exhaust valve. Note parts of valve.

The valve face is a machined surface on the back of the valve head. It touches and seals against the seat in the cylinder head, Fig. 15-23.

The valve margin is the flat surface on the outer edge of the valve head. It is located between the valve head and face. The margin is needed to allow the valve to withstand the high temperatures of combustion. Without a margin, the valve head would melt and burn.

The valve stem is a long shaft extending out of the valve head. The stem is machined and polished. It fits into the guide machined through the cylinder head. Look at Fig. 15-23.

Grooves are machined into the top of the valve stem. They accept small keepers or collets that hold the spring on the valve.

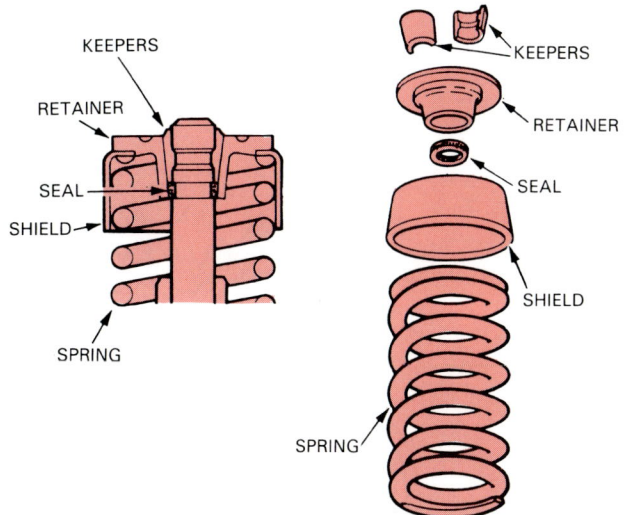

Fig. 15-25. Valve spring assembly basically consists of spring, retainer, keeper, and sometimes a shield. Note how this type seal fits on valve stem.

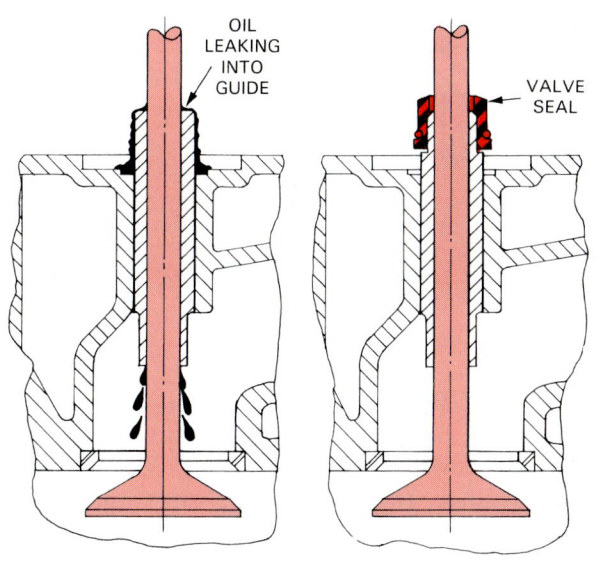

Fig. 15-24. Valve seal keeps oil from entering valve guide and combustion chamber.

Valve seals

Valve seals prevent oil from entering the combustion chambers through the valve guides. This is illustrated in Fig. 15-24.

The valve seals fit over the valve stems and keep oil from entering through the clearance between the stems and guides.

Without valve seals, oil could be drawn into the engine cylinders and burned during combustion. Oil consumption and engine smoking could result.

Valve spring assembly

The valve spring assembly is used to close the valve. It basically consists of a valve spring, retainer,

and two keepers or collets. The keepers fit into the grooves cut in the valve stem. This locks the retainer and spring on the valve. See Fig. 15-25.

Intake manifold

The engine intake manifold bolts to the side of the cylinder head or heads. The carburettor or fuel injectors (petrol engine) are mounted on the intake manifold. It contains runners (passages) going to each cylinder head port. Air (diesel engine) and fuel (petrol engine) are routed through these runners. Fig. 15-26.

Exhaust manifold

The exhaust manifold also bolts to the cylinder head; however, it fastens over the exhaust ports. During the exhaust strokes, hot gases blow into this manifold before entering the rest of the exhaust system. An engine exhaust manifold can be made of heavy cast iron or lightweight steel tubing.

Valve cover

The valve cover is a thin metal or plastic cover over the top of the cylinder head. It may also be called a rocker cover. It simply prevents valve train oil spray from leaking out of the engine. Look at Fig. 15-26.

ENGINE FRONT END

The engine front end operates the engine camshaft, and sometimes the oil pump, distributor, and diesel injection pump. Basically, the engine front end consists of a drive mechanism for the camshaft and other devices, a front cover, an oil seal, and crankshaft damper.

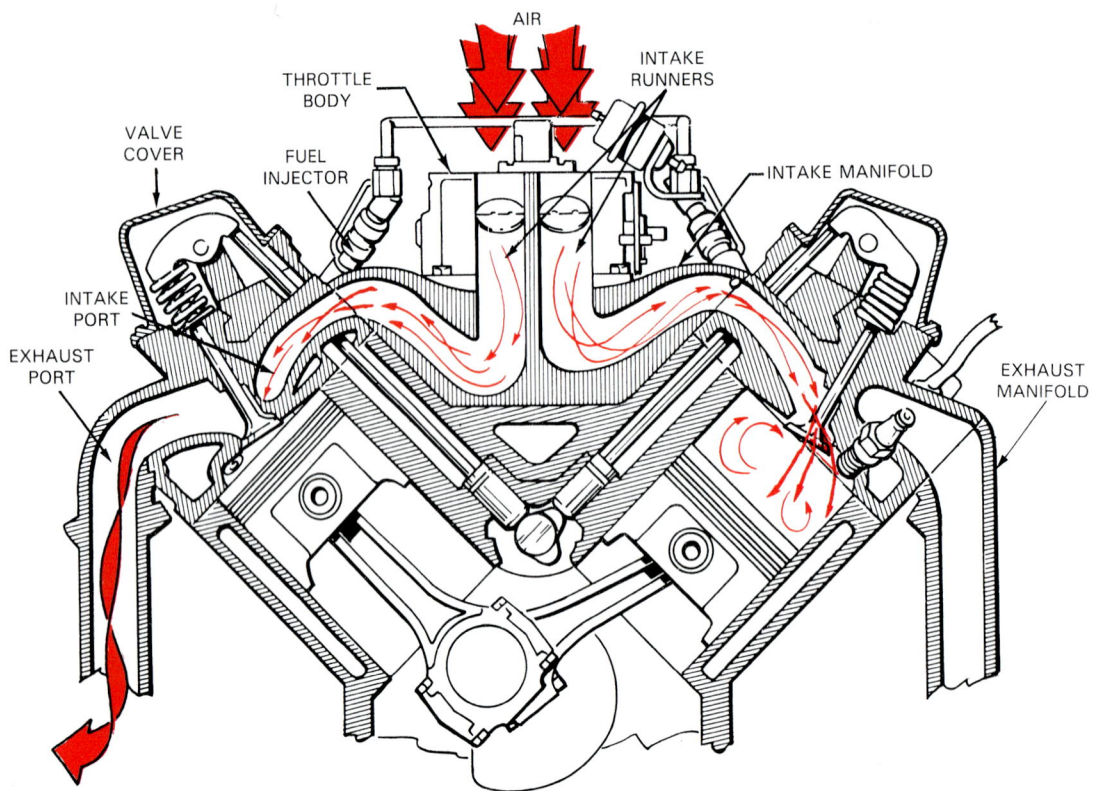

Fig. 15-26. Intake and exhaust manifolds bolt to cylinder head. Intake manifold contains runners that route fuel mixture into cylinder heads. Exhaust manifold routes burned gases into exhaust system.

Camshaft drive

A camshaft drive is needed to turn the camshaft at one-half engine speed. Gears, a chain and sprockets, or a belt and sprockets can be used to turn the camshaft. Look at Fig. 15-27.

These parts can also be called timing gears, timing chain, or timing belt because they time the camshaft with the crankshaft. See Fig. 15-28, 15-29 and 15-30.

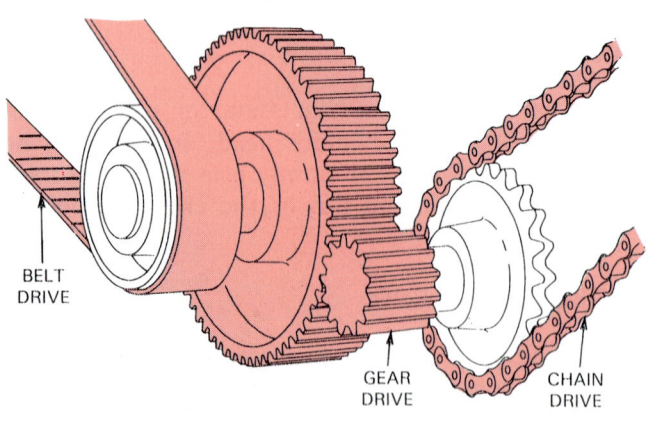

Fig. 15-27. Three types of drive mechanisms are used to turn engine camshaft: cogged belt, gears, or chain.

The camshaft must turn at one-half engine speed so that each valve will only open once for every two crankshaft revolutions.

For instance, the intake valve must only open on the intake, not the compression, power, or exhaust strokes. To do this, the camshaft gear or sprocket is twice the diameter of the gear or sprocket on the crankshaft.

Front cover

The front cover bolts over the crankshaft extension. It holds an oil seal that seals the front of the crankshaft.

When the engine uses a gear or chain type camshaft drive, the front cover can also be called the timing cover, Fig. 15-28.

With a belt drive, this cover does not enclose the cam drive or timing mechanism. A second cover is installed over the belt, Fig. 15-29.

Crankshaft damper

A crankshaft damper is a heavy wheel on the crankshaft extension. It is mounted in rubber and helps prevent crankshaft vibration and damage. This damper is also called the harmonic balancer or vibration damper.

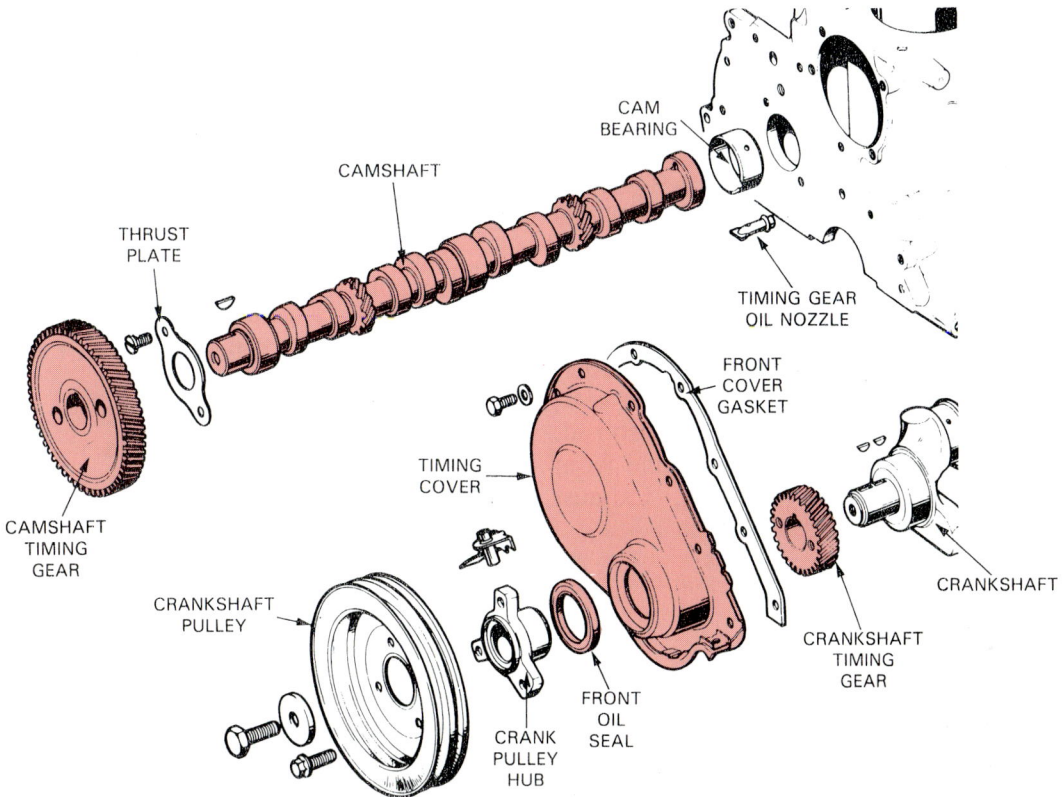

Fig. 15-28. Engine front end components primarily operate engine camshaft. This engine uses timing gears to drive camshaft at one-half engine speed. Front cover enclose gears. Front seal prevents leakage around crankshaft extension.

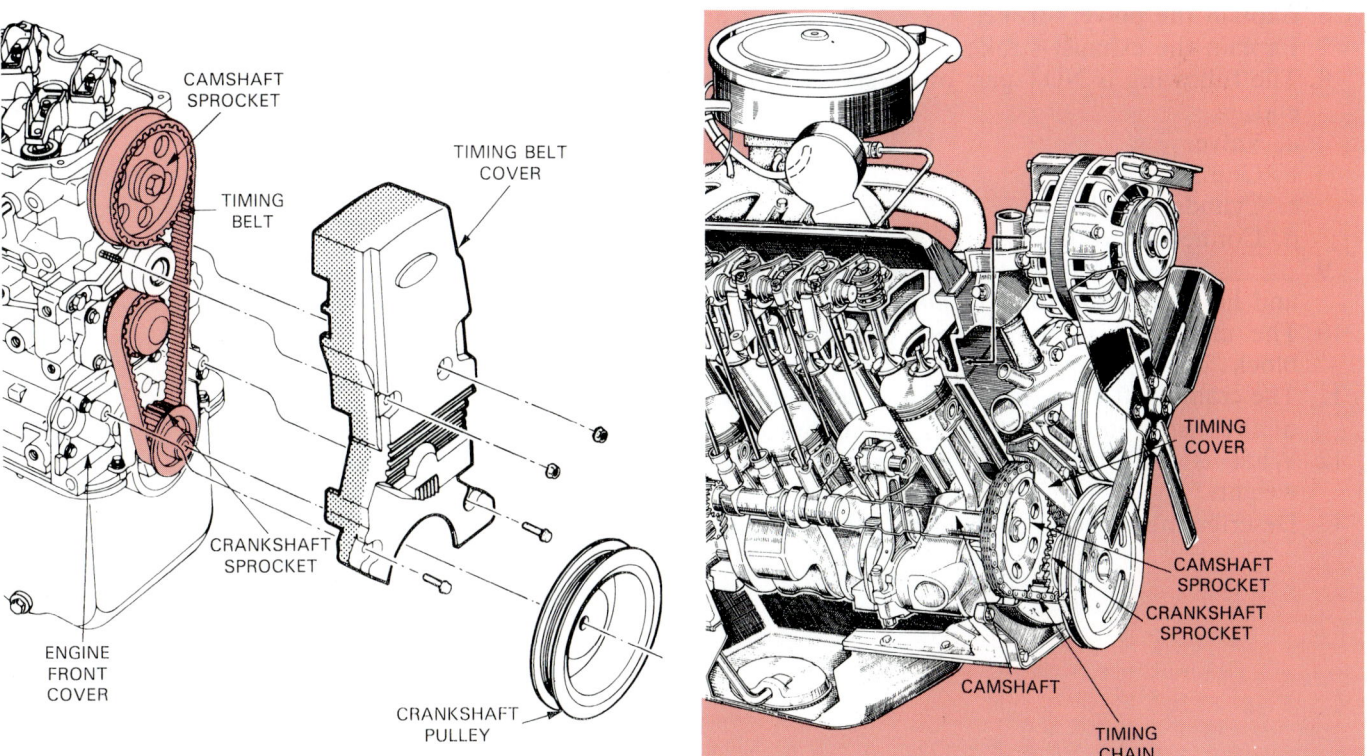

Fig. 15-29. Crankshaft sprocket turns timing belt. Timing belt turns camshaft sprocket and camshaft. Front cover simply houses front oil seal. Timing cover fits over belt. (Ford)

Fig. 15-30. Timing chain and sprockets operate camshaft in this engine. (Chrysler)

KNOW THESE TERMS

Intake stroke, Compression stroke, Power stroke, Exhaust stroke, TDC, BDC, Engine bottom end, Short motor, Cylinder block, Cylinder bore, Main bearing bores, Main bearing caps, Crankcase, Crankshaft, Main bearing journals, Connecting rod journals, Thrust bearing, Crankshaft end play, Bearing clearance, Rear main bearing oil seal, Flywheel, Connecting rod, Piston pin, Piston clearance, Compression ring, Oil ring, Ring gap, Engine top end, Cylinder head, Combustion chamber, Intake and exhaust port, Valve guide, Valve seat, Valve, Valve train, Camshaft, Valve lifter, Push rod, Rocker arm, Valve seal, Valve spring, Engine front end, Camshaft drive, Intake manifold, Exhaust manifold, Valve cover.

REVIEW QUESTIONS

1. Usually, _____ or _____ is burned inside the engine to produce heat, expansion, and resulting pressure.
2. What do TDC and BDC mean?
3. Every four strokes, the engine produces two power or energy producing strokes. True or False?
4. Explain the intake stroke.
5. Explain the compression stroke.
6. Explain the power stroke.
7. Explain the exhaust stroke.
8. The following is NOT part of an engine bottom end.
   a. Valve.
   b. Crankshaft.
   c. Cylinder block.
   d. Connecting rod.
9. _____ _____ bolt to the bottom of the block and hold the crankshaft in place.
10. The crankcase is the highest position in the block. True or False?
11. The crankshaft changes the up and down motion of the piston into a useful _____ motion.
12. What is the function of crankshaft counterweights?
13. Describe the function of the main thrust bearing.
14. The _____ _____ transfers piston movement to the crankshaft.
15. Which of the following is NOT part of a connecting rod?
   a. I beam.
   b. Lobe.
   c. Cap.
   d. Bushing.
16. Why is connecting rod bearing clearance needed?
17. Explain the function of compression and oil rings.
18. Which of the following is part of the cylinder head?
   a. Combustion chambers.
   b. Intake and exhaust ports.
   c. Valve guides.
   d. Valve seats.
   e. All of the above are correct.
19. List and explain the basic parts of a camshaft.
20. _____ open and close the ports through the cylinder head.
21. The intake valves have larger diameter heads than the exhaust valves. True or False?
22. Describe the five basic parts of an engine valve.
23. What do valve seals do and what would happen without valve seals?
24. Explain the function of the following parts.
   a. Intake manifold.
   b. Exhaust manifold.
   c. Valve or rocker cover.
25. The engine camshaft turns at one-quarter crankshaft speed. True or False?
26. Identify the engine parts in the basic engine illustration.
   A. _____.
   B. _____.
   C. _____.
   D. _____.
   E. _____.
   F. _____.
   G. _____.
   H. _____.
   I. _____.
   J. _____.
   K. _____.
   L. _____.
   M. _____.
   N. _____.

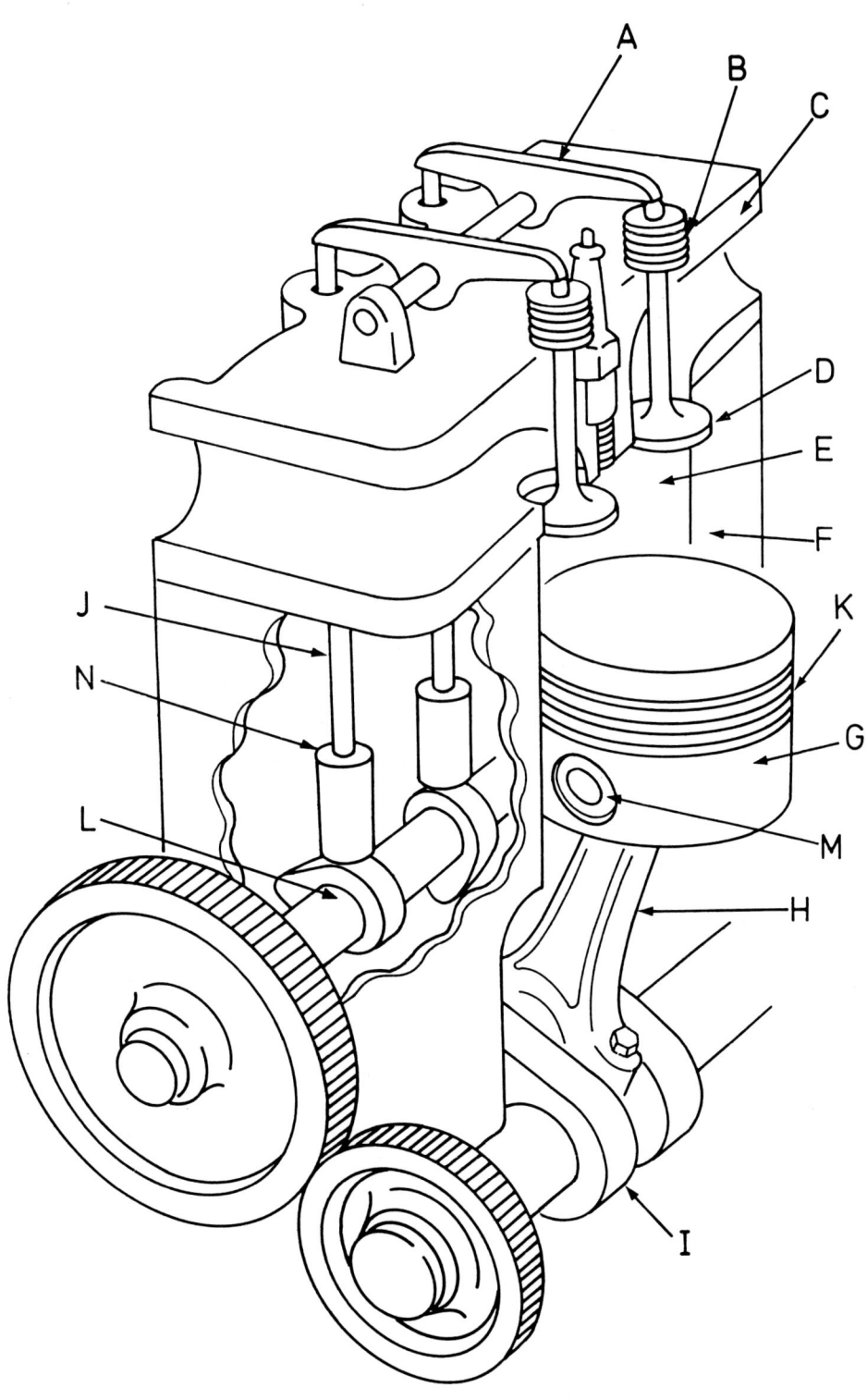

Can you identify the parts of the engine without looking at other illustrations?

Chapter 16

COOLING SYSTEM FUNDAMENTALS

After studying this chapter, you will be able to:
☐ List the basic parts of a cooling system.
☐ Describe the functions of a cooling system.
☐ Explain the operation and construction of major cooling system components.
☐ Compare cooling system design variations.
☐ Explain the importance of antifreeze.

This chapter explains the design, construction, and operation of modern cooling systems. You must fully understand today's cooling systems before learning service and repair. This is an important chapter. Study it carefully!

BASIC COOLING SYSTEM

The basic parts of a cooling system are shown in Fig. 16-1. Refer to this illustration as each part is introduced.
1. WATER PUMP (forces coolant through the engine and other system parts).
2. RADIATOR HOSES (connect the engine to the radiator).
3. RADIATOR (transfers engine coolant heat to the outside air).
4. FAN (draws air through the radiator).
5. THERMOSTAT (controls coolant flow and engine operating temperature).

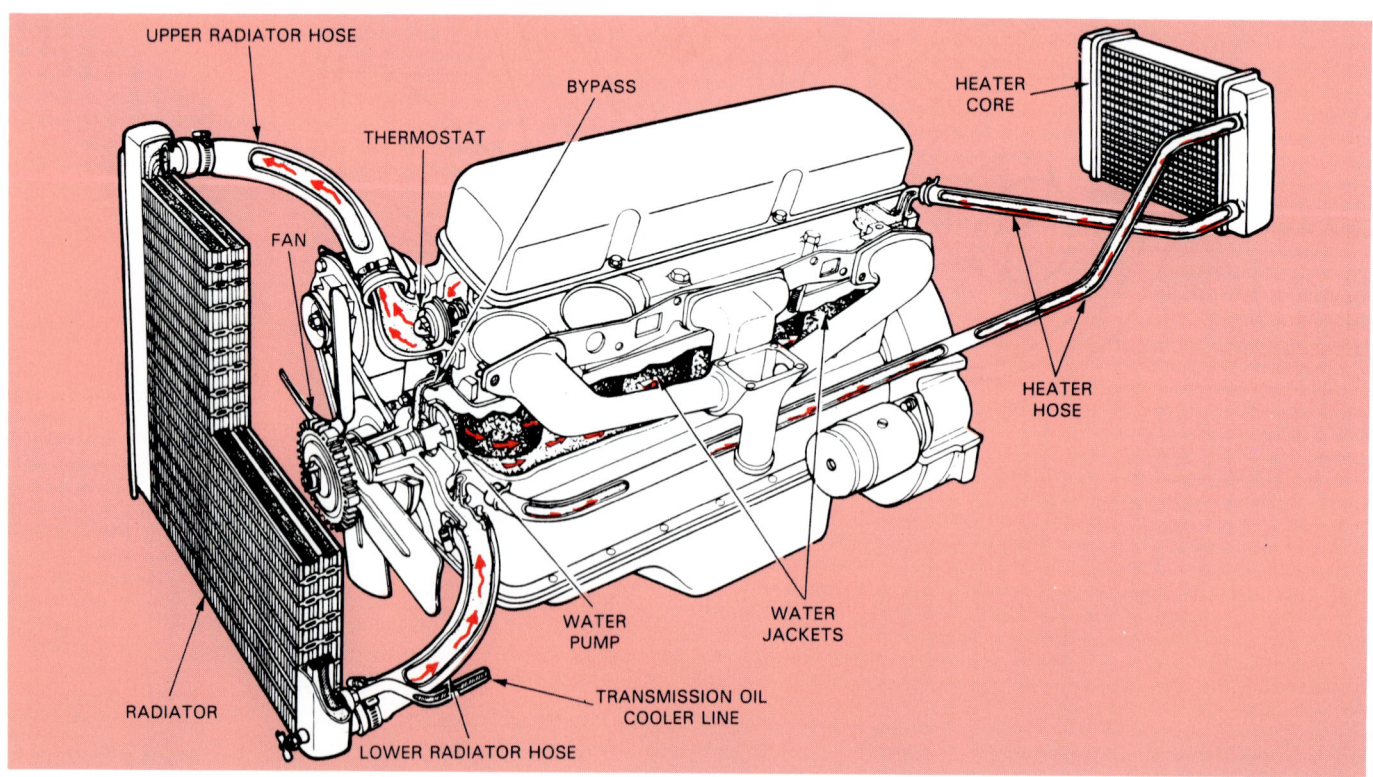

Fig. 16-1. Study basic names and location for parts of a cooling system. This will help you as each part is explained in detail. (Ford)

234

Cooling system operation

When the engine is running, a fan belt powers the water pump. The water pump forces coolant to circulate through the engine water jackets.

While the engine is cold, the thermostat remains closed. This prevents coolant from going to the radiator. Instead, it circulates around inside the engine. This helps warm the engine quickly.

When the engine reaches operating temperature, the thermostat opens. Heated coolant then flows through the radiator. Excess coolant heat is transferred to the air flowing through the radiator. This maintains a proper engine temperature.

FUNCTIONS OF A COOLING SYSTEM

A cooling system has several functions.
It must:
1. Remove excess heat from the engine.
2. Maintain a constant engine operating temperature.
3. Increase the temperature of a cold engine as quickly as possible.
4. Provide a means for heater operation (warming passenger compartment).

Removing engine heat

The burning air-fuel mixture produces a tremendous amount of heat. Combustion flame temperatures can reach 2500° C. This is enough heat to melt metal parts.

Some combustion heat is used to produce expansion and pressure for piston movement. Most combustion heat flows out the exhaust and into the metal parts of the engine. Without removal of this excess heat, the engine would be seriously damaged.

Maintain operating temperature

Engine operating temperature is the temperature the engine coolant (water and antifreeze solution) reaches under running conditions. Typically, an engine's operating temperature is between 80 and 100° C.

When an engine warms to operating temperature, its parts expand. This assures that all part clearances are correct. It also assures proper combustion, emission output levels, and engine performance.

Reaching operating temperature quickly

An engine must warm up rapidly to prevent poor combustion, part wear, oil contamination, reduced fuel economy, and other problems. A cold engine suffers from several problems.

For instance, the aluminium pistons in a cold engine will not be heat expanded (size increases from heat). This can cause too much clearance between the pistons and cylinder walls. The oil in a cold engine will be very thick. This can reduce lubrication and increase engine wear. The fuel mixture will also not vapourise and burn as efficiently in a cold engine.

Heater operation

A cooling system commonly circulates coolant to the car heater. Since the engine coolant is warm, its heat can be used to warm the passenger compartment. See Fig. 16-1.

COOLING SYSTEM TYPES

There are two major types of automotive cooling systems: liquid and air.

An air cooling system uses large cylinder cooling fins and outside air to remove excess heat from the engine. Look at Fig. 16-2.

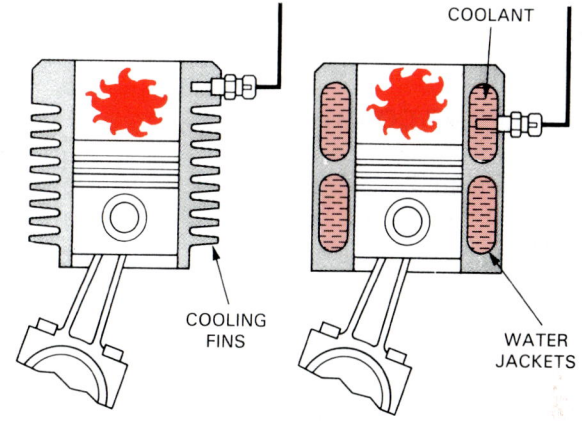

Fig. 16-2. Air-cooled engine has large fins on cylinder to dissipate heat into surrounding air. Water-cooled engine has water jackets around each cylinder to collect heat. (Robert Bosch)

The cooling fins increase the surface area of the metal around the cylinder. This allows enough heat to transfer from the cylinder.

An air cooling system commonly uses plastic or sheet metal ducts and shrouds (enclosures) to route air over the cylinder fins. Thermostatically controlled flaps regulate airflow and engine operating temperature.

Note! Air cooled automotive engines have been replaced by liquid (water) cooled engines.

A liquid cooling system circulates a solution of water and corrosion inhibitor through the water jackets (internal passages in engine). The coolant then collects excess heat and carries it out of the engine, as in Figs. 16-2 and 16-3.

A liquid cooling system has several advantages over an air type system:
1. More precise control of engine operating temperature.

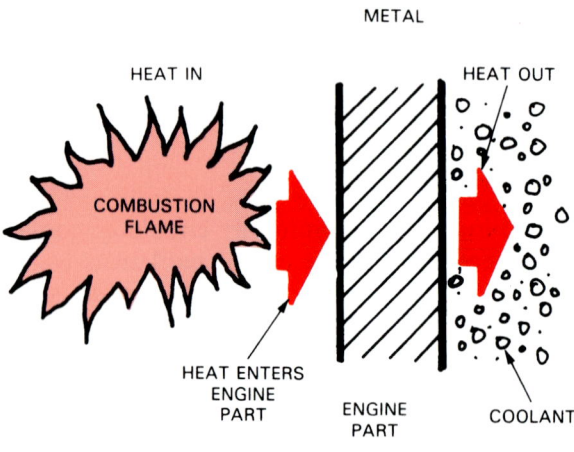

Fig. 16-3. Combustion heat transfers into cylinder wall and then into coolant. Coolant carries heat away from engine.

2. Less temperature variation inside engine.
3. Reduced exhaust emissions because of better temperature control.
4. Improved heater operation to warm passengers.

WATER PUMP CONSTRUCTION

The water pump is an impeller or centrifugal pump that forces coolant through the engine block, cylinder head, intake manifold, hoses, and radiator. It is usually driven by a fan belt running off the crankshaft pulley. Look at Fig. 16-4.

The major parts of a typical water pump includes:

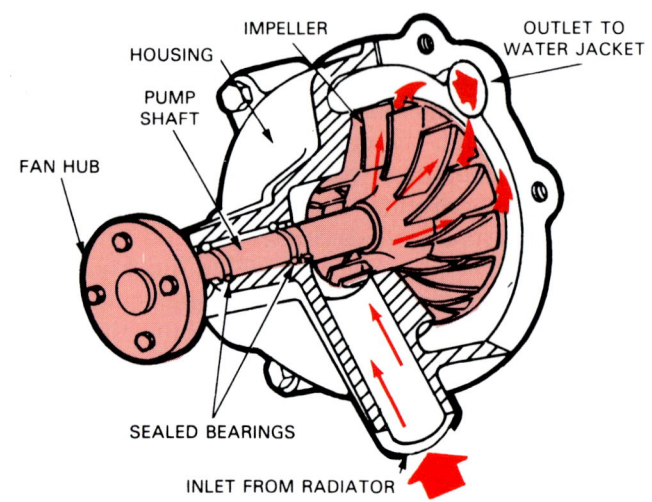

Fig. 16-5. Cutaway of simplified water pump. Note how spinning impeller throws coolant outward to produce pressure and flow.

1. WATER PUMP IMPELLER (disc with fan like blades that spins and produces pressure and flow), Fig. 16-5.
2. WATER PUMP SHAFT (steel shaft that transfers turning force from hub to impeller).
3. WATER PUMP SEAL (prevents coolant leakage between pump shaft and pump housing), Fig. 16-6.
4. WATER PUMP BEARINGS (plain or ball bearings that allows pump shaft to spin freely in housing).

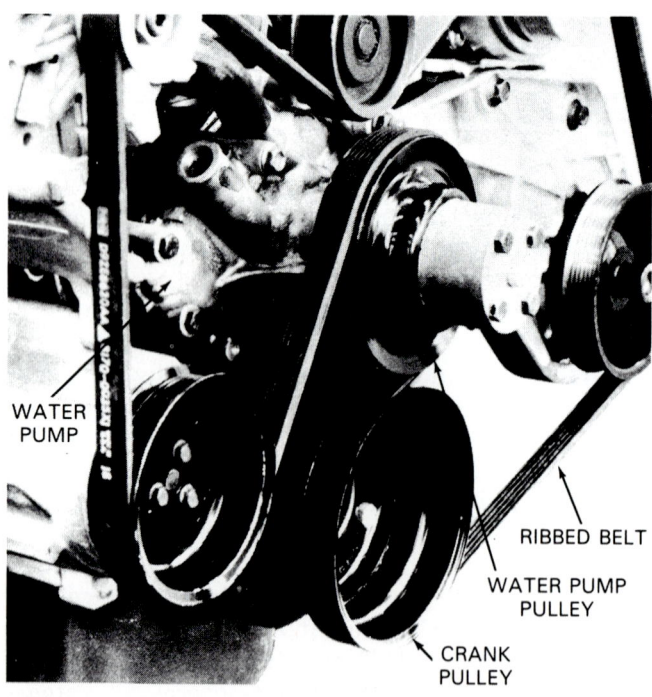

Fig. 16-4. Fan belt turns water pump pulley to operate pump. This is a modern ribbed belt that powers all accessory units. (Ford)

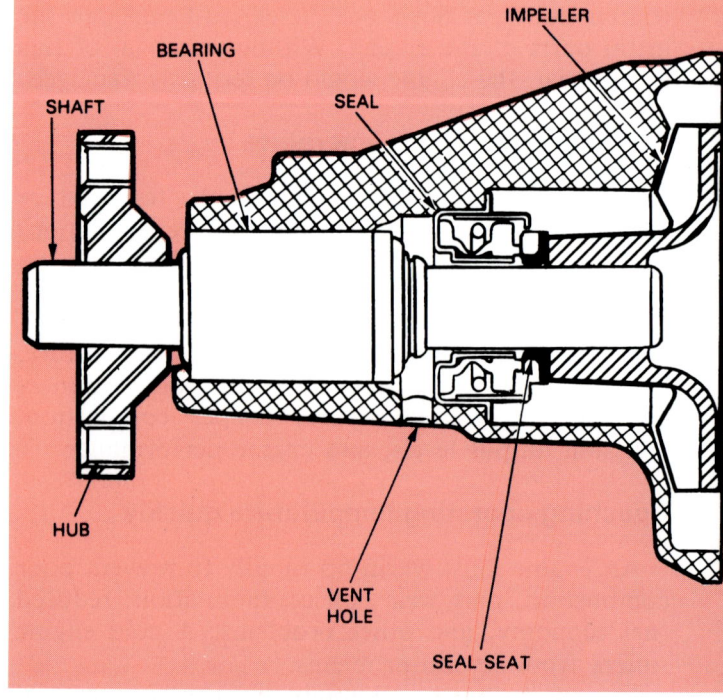

Fig. 16-6. Side view of water pump shows how seal keeps coolant from leaking out vent hole.

5. WATER PUMP HUB (provides mounting place for belt pulley and fan).

6. WATER PUMP HOUSING (iron or aluminium casting that forms main body of pump).

The water pump normally mounts on the front of the engine. With some transverse (sideways) mounted engines, it may bolt to the side of the engine and extend towards the front.

A water pump gasket fits between the engine and pump housing to prevent coolant leakage. A silicone sealer may be used instead of a gasket.

Water pump operation

Fig. 16-7 illustrates water pump action and coolant flow through an engine.

The spinning engine crankshaft pulley causes the fan belt to turn the water pump pulley, pump shaft, and impeller. The coolant trapped between the impeller blades is thrown outward. This produces suction in the central area of the pump housing. It also produces pressure in the outer area of the housing.

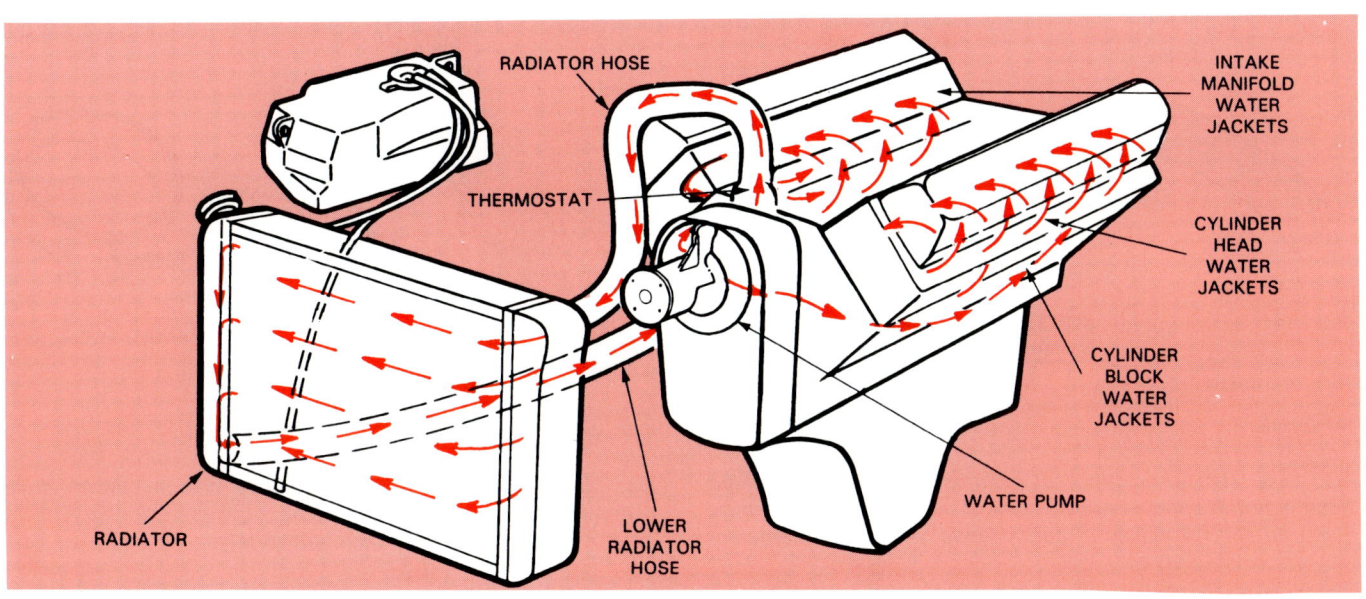

Fig. 16-7. Water pump pulls coolant out of bottom of radiator and through engine block, heads, and intake manifold. Hot coolant then re-enters radiator for cooling. (Ford)

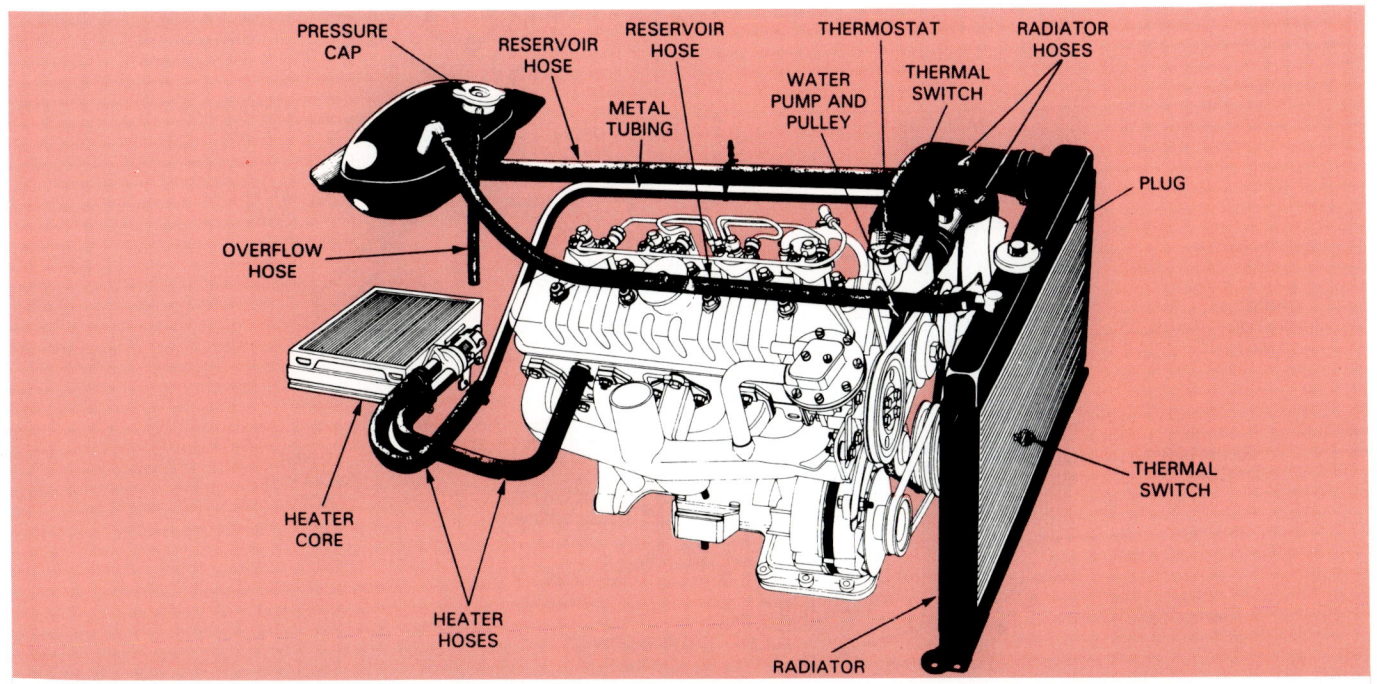

Fig. 16-8. Radiator hoses carry coolant between engine and radiator. Other hoses also contain engine coolant. (Peugeot)

Since the pump inlet opening is near the centre, coolant is pulled out of the radiator, through the lower hose, and into the engine. After being thrown outward and pressurised, the coolant flows into the engine. It circulates through the block, around the cylinders, up through the cylinder heads, and back into the radiator.

RADIATOR AND HEATER HOSES

Radiator hoses carry coolant between the engine water jackets and the radiator. Look at Fig. 16-8.

Being flexible, hoses can withstand the vibrating and rocking of the engine without breakage.

The upper radiator hose normally connects to the thermostat housing on the engine intake manifold or cylinder head. Its other end fits on the radiator. The lower hose connects the water pump inlet and the radiator.

A moulded hose is manufactured into a special shape with bends to clear parts, especially the cooling fan. It must be purchased to fit the exact year and make of vehicle. See Fig. 16-9.

A flexible hose has an accordian shape and can be bent to different angles. The pleated construction

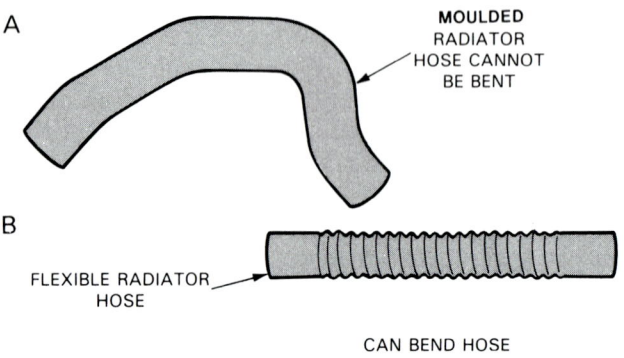

Fig. 16-9. Two basic types of radiator hoses: A — Moulded hose only fits specific applications. B — Flexible hose can be used on several makes of cars if ends are correct diameter.

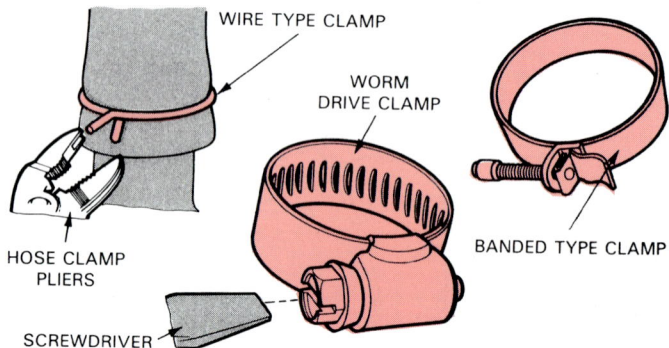

Fig. 16-10. Three basic types of hose clamps. Worm drive clamp is most common. Spring type requires hose clamp pliers with groove cut in jaws.

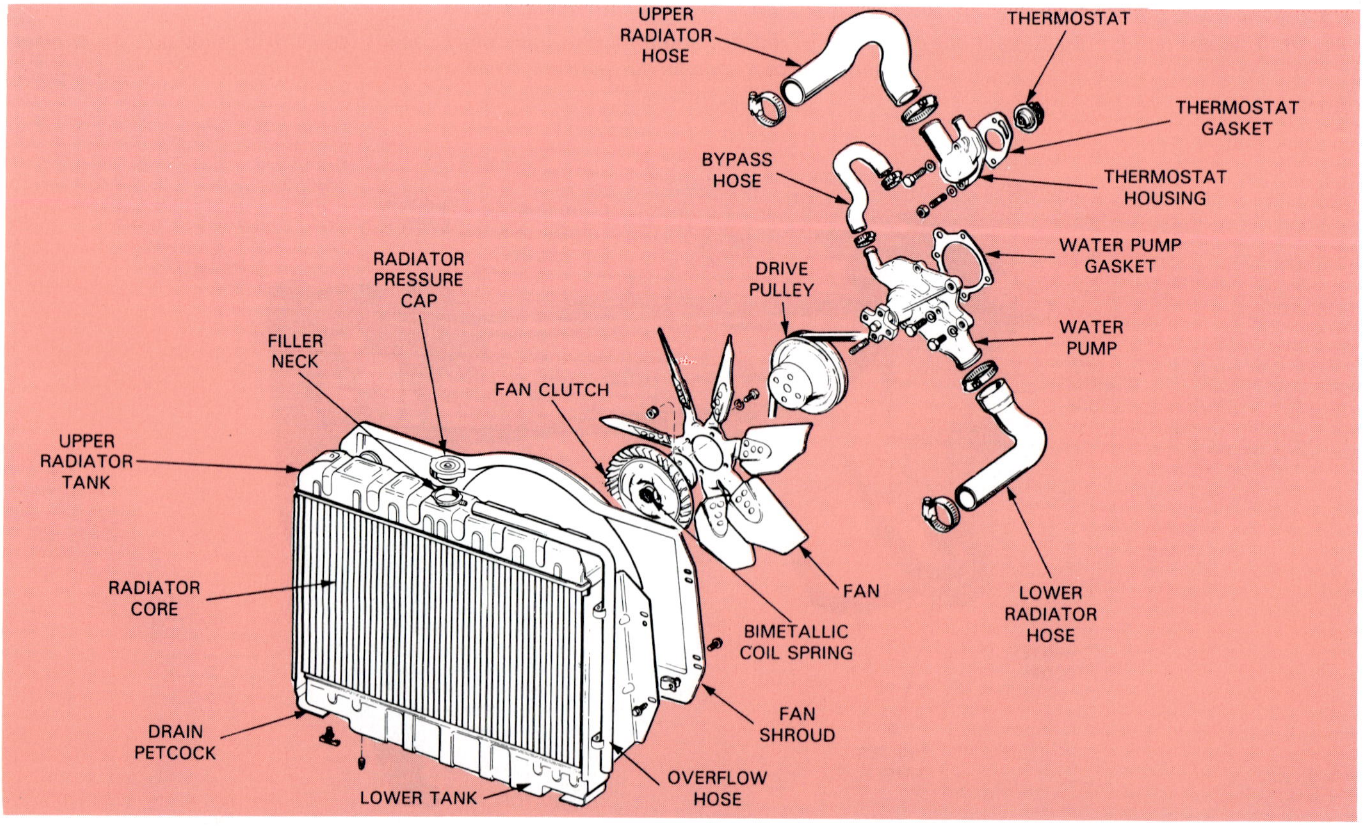

Fig. 16-11. Exploded view of major parts of cooling system. Study each component carefully!

allows the hose to bend without collapsing and blocking flow. It is also known as convoluted or universal type radiator hose.

A hose spring is frequently used in the lower radiator hose to prevent its collapse. The lower hose is exposed to suction from the water pump. The spring assures that the inner lining of the hose does NOT tear away, close up, and stop circulation.

Heater hoses are small diameter hoses that carry coolant to the heater core (small radiator-like device under car dash). Refer to Fig. 16-8.

Hose clamps hold the radiator and heater hoses on their fittings. Three types of hose clamps are pictured in Fig. 16-10.

RADIATOR CONSTRUCTION

The radiator transfers coolant heat to the outside air. See Fig. 16-11. The radiator is normally mounted in front of the engine. Cool outside air can flow freely through it.

A radiator typically consists of:
1. CORE (centre section of radiator made up of tubes and cooling fins).
2. TANKS (metal or plastic ends that fit over core tube ends to provide storage for coolant and fittings for hoses).
3. FILLER NECK (opening for adding coolant, also holds radiator cap and overflow tube).
4. OIL COOLER (inner tank for cooling automatic transmission or transaxle fluid).

Radiator action

Under normal operating conditions, hot engine coolant circulates through the radiator tanks and core tubes. Heat transfers into the core's tubes and fins. Since cooler air is flowing over and through the radiator fins, heat is removed from the radiator. This reduces the temperature of the coolant before it flows back into the engine.

Radiator types

The two types of radiators are the crossflow and downflow. Both are shown in Fig. 16-12.

A downflow radiator's tanks are on the top and bottom and the core tubes run vertically. See Fig. 16-12. Hot coolant from the engine enters the top tank. The coolant flows downward through the core tubes. After cooling, coolant flows out the bottom tank and back into the engine.

A crossflow radiator is a more modern design that has its tanks on the side of the core. The core tubes are arranged for horizontal coolant flow. Look at Fig. 16-12. The tank with the radiator cap is normally the outlet tank. A crossflow radiator can be shorter, allowing for a lower car bonnet.

Transmission oil cooler

A transmission oil cooler is often placed in the radiator on cars with automatic transmissions. It is a small tank enclosed in one of the main radiator tanks, Fig. 16-13. Since the transmission fluid is hotter than the engine coolant, heat is removed from the fluid as it passes through the radiator and cooler.

In downflow radiators, the transmission oil cooler is located in the lower tank. In crossflow radiators, it is in the tank having the radiator cap. Both tanks are cooler outlet tanks.

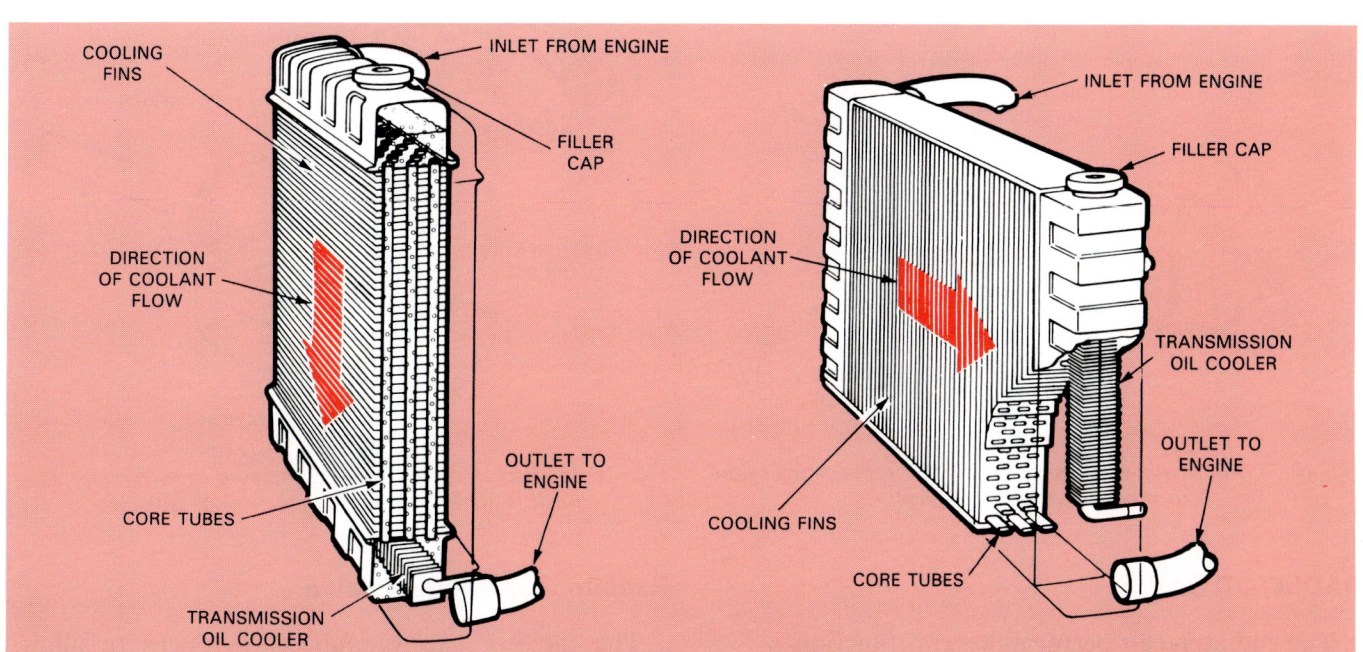

Fig. 16-12. Two types of radiators. Left. Downflow radiator has core tubes running up and down. Right. Crossflow radiator has cooling tubes running horizontally. Crossflow is common on late model cars.

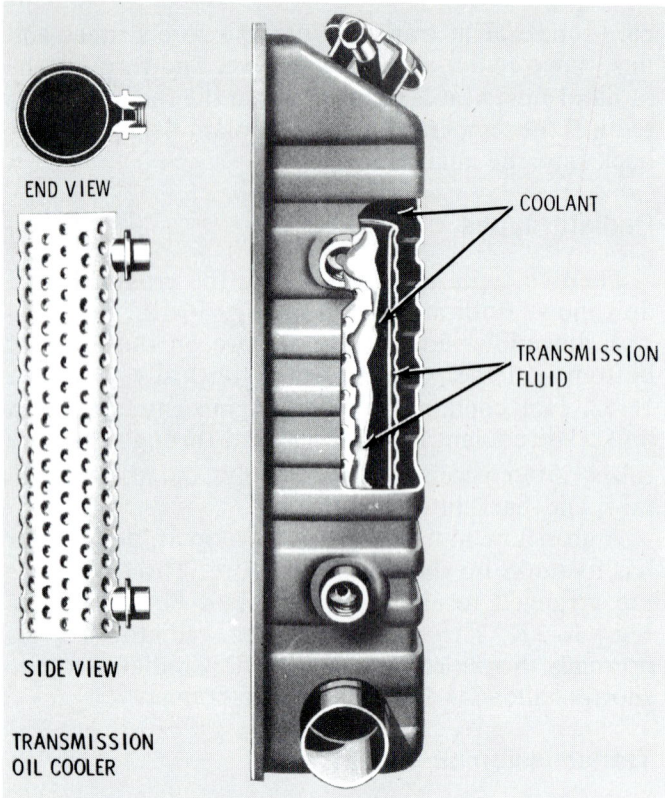

Fig. 16-13. *Transmission oil cooler prevents overheating of automatic transmission fluid. It is small tank inside radiator tank. Note transmission line fittings.*

Line fittings from the cooler extend through the radiator tank to the outside. Metal lines from the automatic transmission connect to these fittings. Fig. 16-14. The transmission oil pump forces the fluid through the lines and cooler.

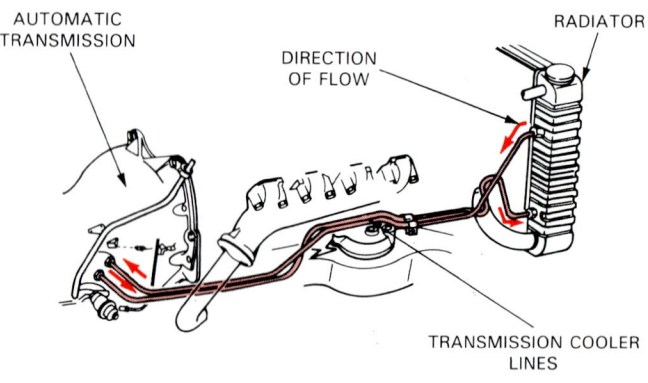

Fig. 16-14. *Automatic transmission lines run from transmission to radiator oil cooler fittings.*

RADIATOR CAP

The radiator cap performs several functions:
1. Seals top of radiator filler neck to prevent leakage.

2. Pressurises system to raise boiling point of coolant. This keeps coolant from boiling and turning to steam.
3. Relieves excess pressure to protect against system damage.
4. In a closed system, it allows coolant flow into and from coolant reservoir.

The radiator cap locks onto the radiator tank filler neck. Rubber or metal seals make the cap-to-neck joint airtight.

Radiator cap pressure valve

The radiator cap pressure valve, Fig. 16-15, consists of a spring-loaded disc that contacts the filler neck. The spring pushes the valve into the neck to form a seal.

Under pressure, water's boiling point increases. Normally, water boils at 100° C. However, for every 10 kPa of pressure increase, the boiling point goes up about 2° C. The radiator cap works on this principle.

Typical radiator cap pressure is 90 to 110 kPa. This raises the boiling point of the engine coolant to about 120 to 130° C. Many surfaces inside the water jackets can be above 100° C.

If the engine overheats and pressure exceeds the cap rating, the pressure valve opens. Excess pressure forces coolant out the overflow tube and into the reservoir or onto the ground. This prevents high pressure from rupturing the radiator, gaskets, seal, or hoses.

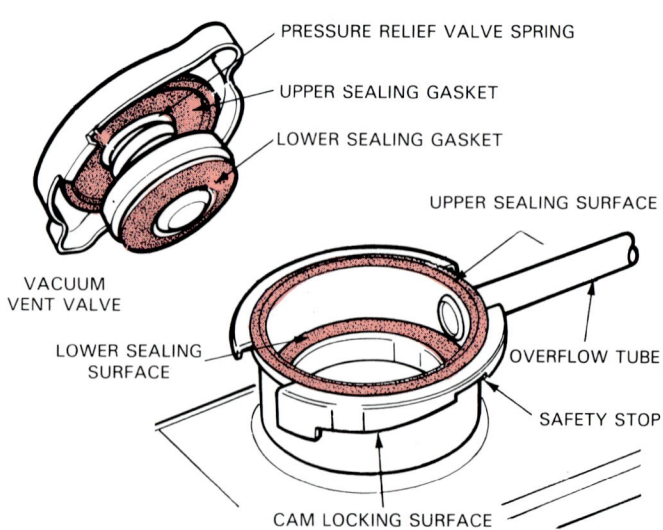

Fig. 16-15. *Radiator pressure cap screws onto radiator filler neck. Rubber or metal seals prevent leakage.*

Radiator cap vacuum valve

The radiator cap vacuum valve opens to allow reverse flow back into the radiator when the coolant temperature drops after engine operation. Look at

Fig. 16-16. It is a smaller valve located in the centre, bottom of the cap.

The cooling and contraction of the coolant and air in the system could decrease coolant volume and pressure. Outside atmospheric pressure could then crush inward on the hoses and radiator. Without a cap vacuum or vent valve, the radiator hoses and radiator tanks could collapse.

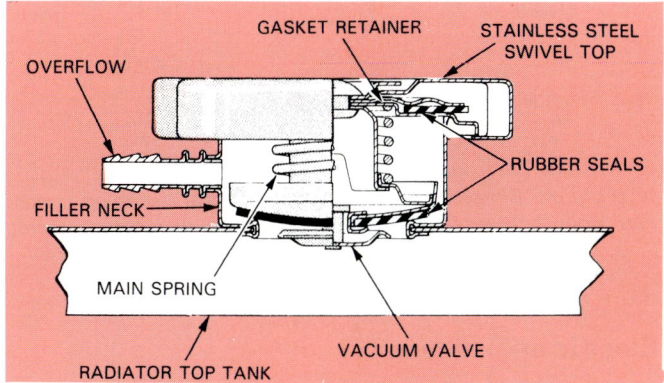

Fig. 16-16. Cutaway view shows how pressure cap installs and seals on radiator filler neck.

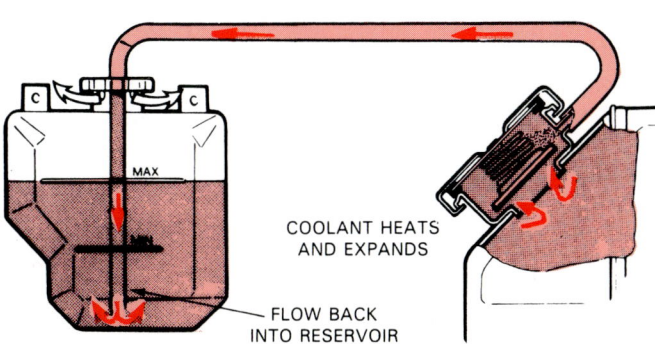

A

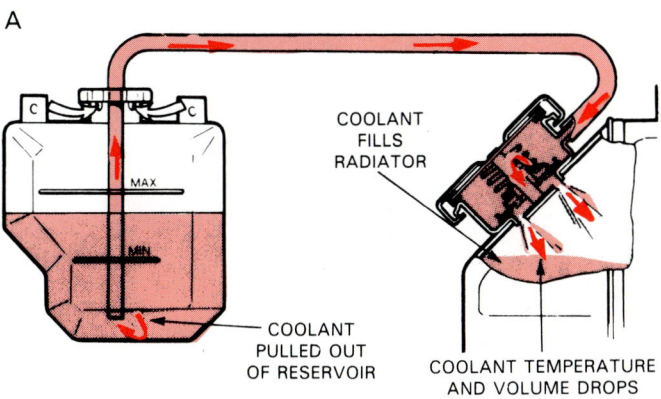

B

Fig. 16-17. Pressure cap operation: A — When engine heats up, coolant expands. Excess fluid opens pressure valve and coolant enters reservoir. B — When engine is shut off, coolant temperature drops. This causes coolant to reduce in volume. Cap vent valve opens to let coolant flow back into radiator. (Ford)

CLOSED COOLING SYSTEM

A closed cooling system uses an expansion tank or reservoir and a special closed system radiator cap. The overflow tube is routed into the bottom of the reservoir tank. Pressure and vacuum valve action pull coolant in and out of the reservoir tank as needed. This keeps the cooling system correctly filled at all times.

Fig. 16-17 shows the operation of a closed cooling system. In A, the engine is heating up. The coolant has expanded and opened the cap pressure valve. Instead of leaking onto the ground, the coolant flows into the reservoir.

In B, the engine has been shut off. As the coolant temperature drops, its volume decreases. This causes the vacuum valve to open. Atmospheric pressure (system suction) can then force coolant back into the radiator. This will compensate for any small system leak, keeping the system properly filled.

OPEN COOLING SYSTEM

An open cooling system does NOT use a coolant reservoir. The overflow tube allows excess coolant to leak onto the ground. Also, it does not provide a means of adding fluid automatically as needed.

The open cooling system is no longer used on modern cars. It has been replaced by the closed system which requires less maintenance.

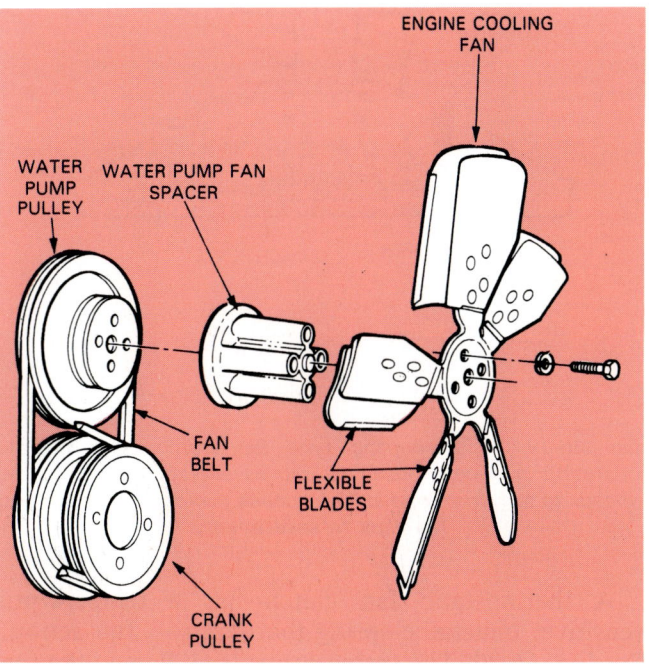

Fig. 16-18. Flex type radiator fan. High rpm causes fan blades to flex or bend and reduce blowing action. Note how spacer is used to move fan closer to radiator. (Ford)

COOLING SYSTEM FANS

A cooling system fan pulls air through the core of the radiator and over the engine to help remove heat. It increases the volume of air flowing through the radiator, especially when the car is standing still. The fan is driven by a fan belt or an electric motor.

Engine powered fans

An engine powered fan normally bolts to the water pump hub and pulley. Refer to Fig. 16-18. Sometimes, a spacer fits between the fan and pulley to move the fan closer to the radiator.

A flex fan has thin, flexible blades that alter airflow with engine speed. At low speeds, the fan blades remain curved and pull air through the radiator. At higher engine speeds, the blades flex until they are almost straight. This reduces fan action and saves engine power. Fig. 16-18.

A fluid coupling fan clutch is designed to slip at higher engine speeds. It performs the same function as a flexible fan. The clutch is filled with silicone-based oil, Fig. 16-19. At a specific fan speed, there is enough load to make the clutch slip.

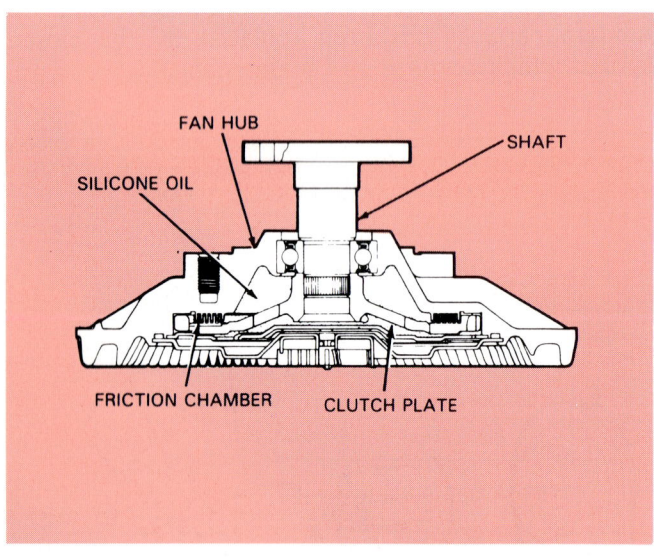

Fig. 16-19. Fluid coupling type fan clutch. Clutch plate operating in silicone-based oil causes enough friction at low speeds to turn fan. Load at high speeds overcomes friction and fan slips to save energy.

A thermostatic fan clutch has a temperature sensitive, bimetallic spring that controls fan action. See Fig. 16-20. The spring controls oil flow in the fan clutch. When cold, the spring causes the clutch to slip, speeding engine warmup. After reaching operating temperature, it locks the clutch, providing forced air circulation.

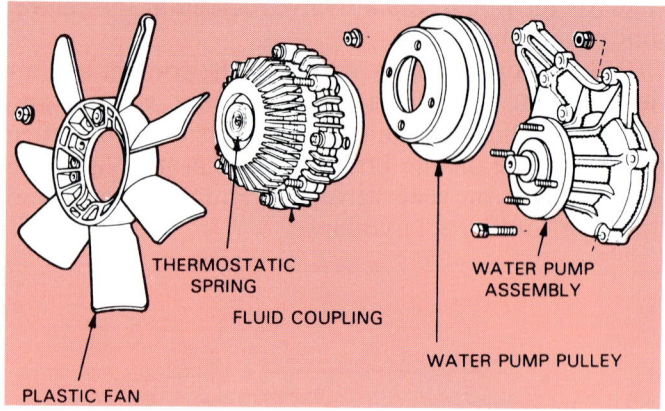

Fig. 16-20. Thermostatic fan clutch is similar to fluid coupling type clutch. Bimetal spring is used to control clutching action. Fan only operates when engine is hot and when spring activates clutch mechanism. (Toyota)

Electric engine fans

An electric engine fan uses an electric motor and a thermostatic switch to provide cooling action. Look at Fig. 16-21. An electric fan is needed on front-wheel-drive cars having transverse (sideways) mounted engines. The water pump is normally located away from the radiator.

The fan motor is a small, DC (direct current) motor. It mounts on a bracket secured to the radiator. A metal or plastic fan blade mounts on the end of the motor shaft.

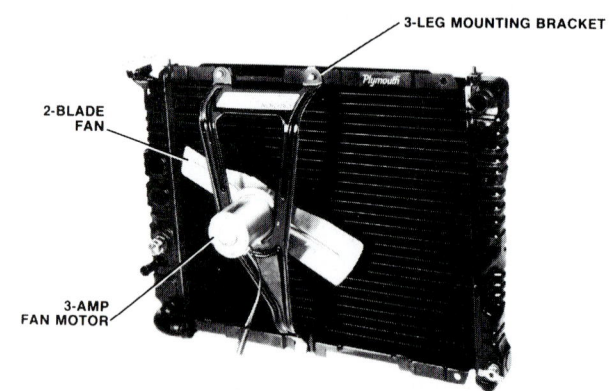

Fig. 16-21. Electric cooling fan uses battery or alternator current for power. DC motor spins metal or plastic fan blade.

The fan switch or thermo switch is a temperature sensitive switch that controls fan motor operation. When the engine is cold, the switch is open. This keeps the fan from spinning and speeds engine warmup. After warmup, the switch closes to operate the fan and provide cooling. This is illustrated in Fig. 16-22.

An electric engine fan saves energy and increases cooling system efficiency. It only functions when needed. By speeding engine warmup, it reduces emissions and fuel consumption. In cold weather, the electric fan may shut off at highway speeds. There may be enough cool air rushing through the car's grille to provide adequate cooling.

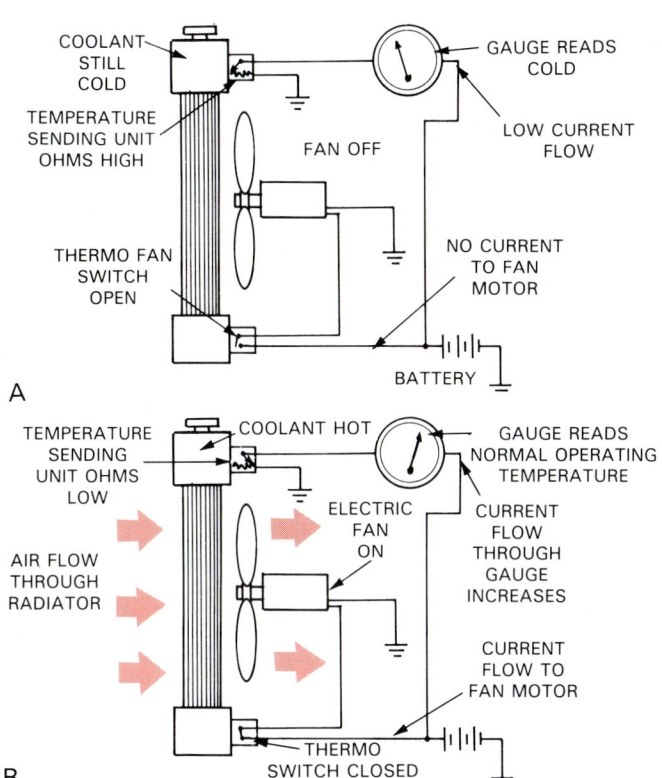

A

B

Fig. 16-22. Electric cooling fan operation. A — Engine is cold. Thermo-switch is open to prevent electric fan operation. This speeds engine warmup. B — Engine at full operating temperature. Thermo-switch closes. Current then flows to fan motor.

RADIATOR SHROUD

The radiator shroud helps assure that the fan pulls air through the radiator. See Fig. 16-23. It fastens to the rear of the radiator and surrounds the area around the fan.

When the fan is spinning, the plastic shroud keeps air from circulating between the back of the radiator and the front of the fan. As a result, a huge volume of air flows through the radiator core. Without a fan shroud, the engine could overheat.

THERMOSTAT

The thermostat senses engine temperature and controls coolant flow through the radiator. It reduces coolant flow when the engine is cold and increases flow when the engine is hot. The thermostat normally

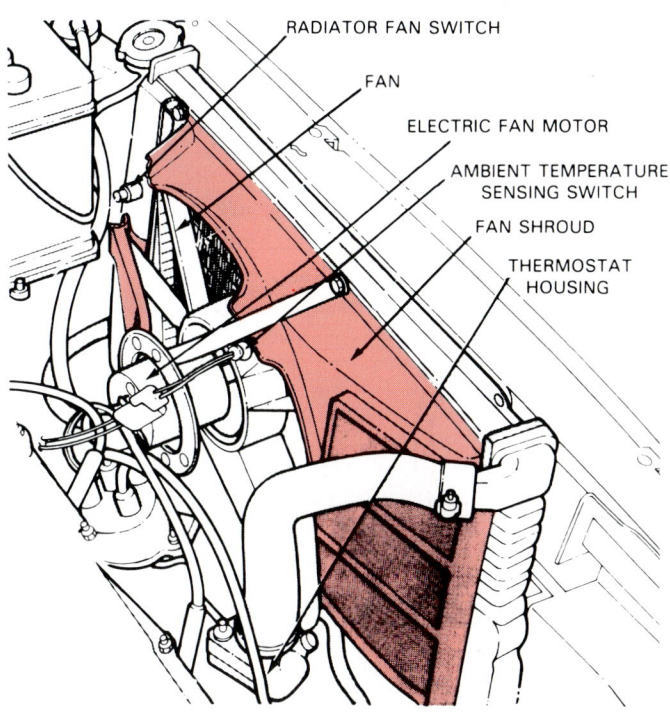

Fig. 16-23. Fan shroud assures that fan pulls air through radiator core. Without shroud, air could circulate between fan and back of radiator. Engine overheating could result.

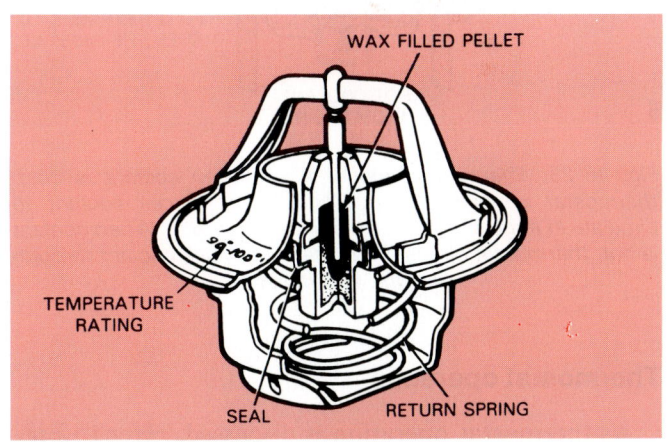

Fig. 16-24. Thermostat is temperature sensitive valve. Note pellet of wax enclosed in cylinder-piston chamber. When heated, pellet expands and pushes against spring tension.

fits under a thermostat housing between the engine and the end of the upper radiator hose.

The thermostat has a wax-filled pellet, as shown in Fig. 16-24. The pellet is contained in a cylinder and piston assembly. A spring holds the piston and valve in a normally closed position.

When the thermostat is heated, the pellet expands and pushes the valve open. As the pellet and thermostat cool, spring tension overcomes pellet expansion and the valve closes. Refer to Fig. 16-25.

Thermostat rating is stamped on the thermostat to indicate the operating (opening) temperature of the thermostat. Normal ratings are between 82 and 91° C.

High thermostat heat ranges are used in modern cars because they reduce exhaust emissions and increase combustion efficiency.

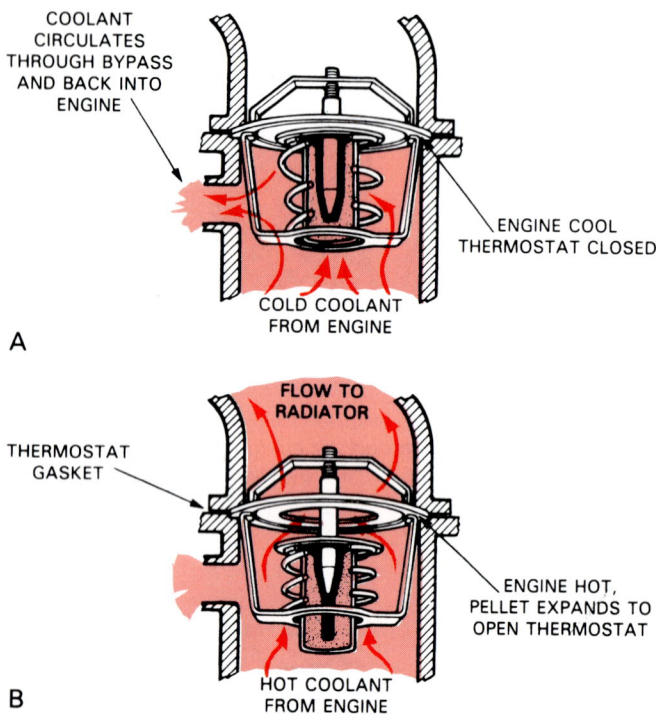

Fig. 16-25. Thermostat action. A — When coolant is cold, thermostat remains closed. Water pump forces coolant to circulate in engine, but not through radiator. B — When coolant is hot, thermostat opens. Pump can then push coolant through engine and radiator.

Thermostat operation

As thermostat operation is discussed, refer to Fig. 16-26. It shows thermostat action.

When the engine is cold, the thermostat will be closed and coolant cannot circulate through the radiator. Instead, the coolant circulates around inside the engine block, cylinder head, and intake manifold until the engine is warm.

As the heat range of the thermostat is reached, the hot engine coolant causes the pellet inside the thermostat to expand. The thermostat gradually opens and allows coolant to flow through the system.

Since the amount of thermostat opening is dependent upon engine temperature, the exact

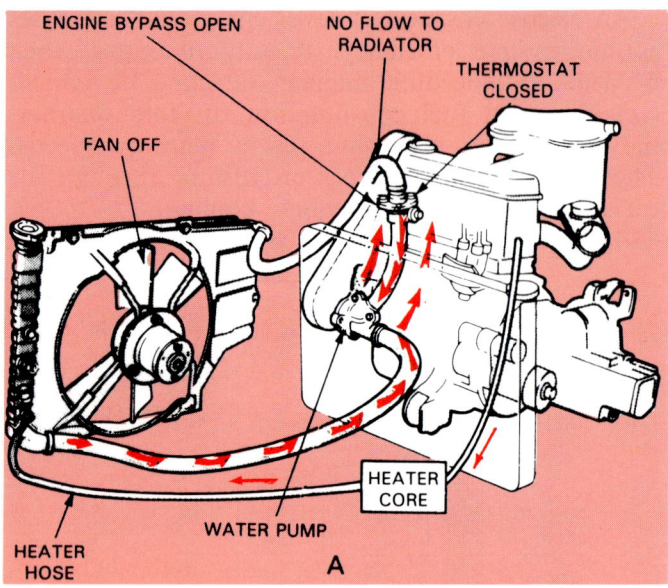

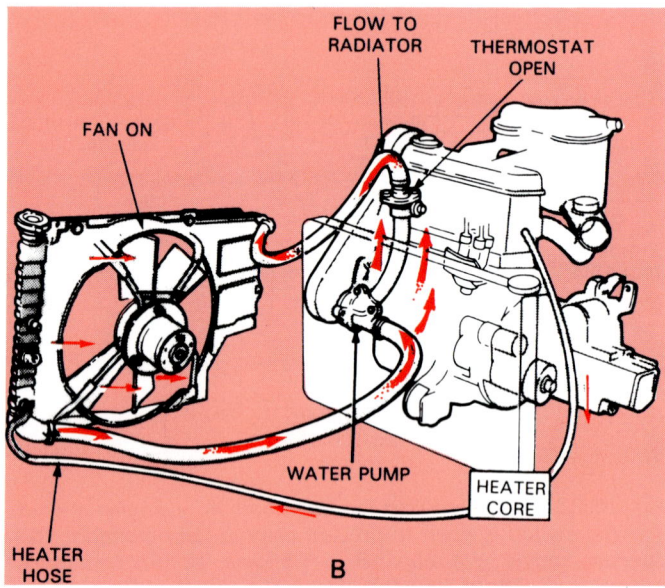

Fig. 16-26. A — Thermostat does not allow coolant to enter radiator when engine is below operating temperature. B — When at operating temperature, thermostat opens and allows flow into radiator. Thermostat moves open and closed different amounts to maintain correct temperature.

operating temperature of the engine can be precisely controlled.

A bypass valve, Fig. 16-27, or a bypass hose permits coolant circulation through the engine when the thermostat is closed. If the coolant could NOT circulate, hot spots could develop inside the engine.

TEMPERATURE WARNING LIGHT

A temperature warning light informs the driver when the engine is overheating, Fig. 16-28. When the engine coolant becomes too hot, a temperature sending unit (switch) in the engine cylinder head or

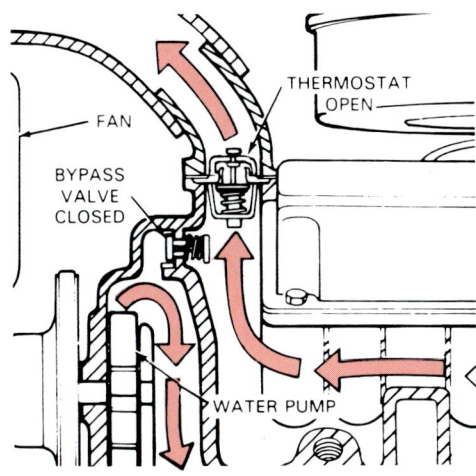

Fig. 16-27. Bypass valve is sometimes used to allow circulation in the engine. It only opens when thermostat is closed and when pressure is stronger than bypass valve spring.

block closes. This completes the circuit and the dash indicating light glows.

When the engine is cold or at normal operating temperature, the sending unit circuit is open and the light remains OFF.

ENGINE TEMPERATURE GAUGE

An engine temperature gauge shows the exact operating temperature of the engine coolant. It is similar to the circuit in Fig. 16-28. However, a variable resistance sending unit and gauge are used in the circuit.

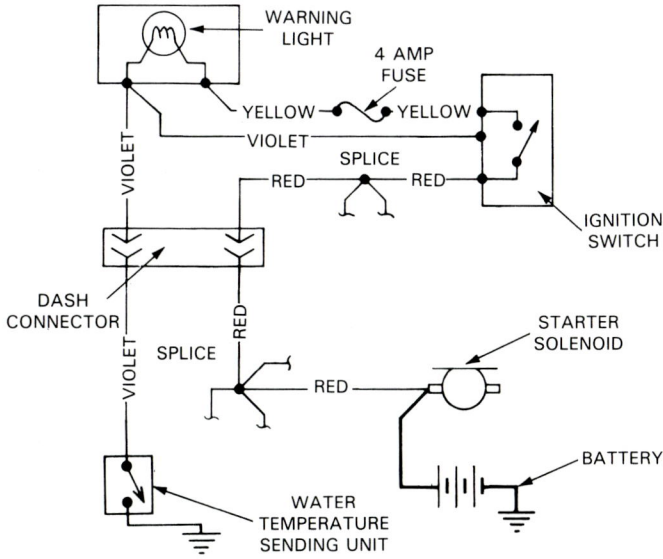

Fig. 16-28. Circuit diagram for an engine temperature warning light. Sending unit screws into engine water jacket. It closes when engine overheats to light indicator bulb. Ignition switch lights bulb when turned to start. This lets driver know bulb is not burned out.

When the engine is cold, the gauge sending unit has high resistance and current does NOT flow to the gauge. The temperature gauge reads cold.

As engine temperature increases, the resistance in the sending unit drops. Current increases in the gauge circuit. Current causes the gauge needle to deflect to the right, showing engine temperature.

ANTIFREEZE (ETHYLENE GLYCOL BASED ENGINE COOLANT)

Antifreeze, or inhibited ethylene glycol, is mixed with water to produce the engine coolant. Antifreeze has several functions.

Prevent winter freeze up

Antifreeze keeps the coolant from freezing in very cold weather (outside temperature below 0° C).

Coolant freezing can cause serious cooling system or engine damage. As ice forms, it expands. This expansion can produce tonnes of force. The water pump housing, cylinder head, engine block, radiator, or other parts could be cracked and ruined.

Prevent rust and corrosion

Antifreeze also prevents rust and corrosion inside the cooling system. It provides a protective film on part surfaces.

Lubricates water pump

Antifreeze acts as a lubricant for the water pump. It increases the service life of the water pump bearings and seals.

Cools the engine

Antifreeze conducts heat better than plain water and, therefore, cools the engine better. It is normally recommended in hot weather.

For example, air conditioning increases the temperature of the air flowing through the radiator. Antifreeze can help prevent overheating in very hot weather when the air conditioning is on.

Antifreeze/water mixture

For ideal cooling and winter protection from freeze up, a 50/50 mixture of water and antifreeze is usually recommended. It will provide protection from ice formation to about −35° C. Higher ratios of antifreeze may produce even lower freezing temperatures but this much protection is not normally needed.

NOTE! Plain water should NEVER be used in a cooling system or the four antifreeze functions just discussed will NOT be provided.

KNOW THESE TERMS

Engine operating temperature, Air cooling system, Liquid cooling system, Water pump, Impeller, Radiator hoses, Heater hoses, Radiator, Downflow, Crossflow, Transmission oil cooler, Radiator cap, Pressure valve, Vacuum valve, Cap pressure rating, Closed system, Open system, Flex fan, Fluid coupling fan clutch, Thermostatic fan clutch, Electric engine fan, Shroud, Thermostat rating, Bypass valve, Antifreeze.

REVIEW QUESTIONS

1. List and explain the five major parts of a cooling system.
2. What are the four functions of a cooling system?
3. Typically, an engine's operating temperature is between _____ and _____ ° C.
4. Not using a thermostat in hot weather is acceptable because the engine would run cooler. True or False?
5. Why has the liquid cooling system replaced the air types?
6. List and explain the six major parts of a water pump.
7. Which of the following does NOT relate to radiator construction?
   a. Core.
   b. Filler neck.
   c. Tanks.
   d. Impeller.
8. Explain the differences between downflow and crossflow radiators.
9. How does an automatic transmission oil cooler work?
10. Describe the four functions of a radiator cap.
11. Typical radiator cap pressure is _____ to _____ kPa, which raises the boiling point of the coolant to about _____ to _____ ° C.
12. How do closed and open cooling systems differ?
13. A _____ _____ is commonly used to turn an electric engine cooling fan on and off.
14. Why is a radiator shroud used?
15. Summarise the operation of a cooling system thermostat.
16. A temperature _____ _____ (switch) on the engine is used to operate the temperature warning light.
17. List and explain four reasons why inhibited antifreeze should be used in the cooling system.
18. For ideal cooling, this mixture of water and antifreeze is typical.
   a. 30 percent water, 70 percent antifreeze.
   b. 50 percent water, 50 percent antifreeze.
   c. 100 percent antifreeze.
   d. 70 percent water, 30 percent antifreeze.
19. Should plain water (no antifreeze) be used in a cooling system during warm weather? Why?

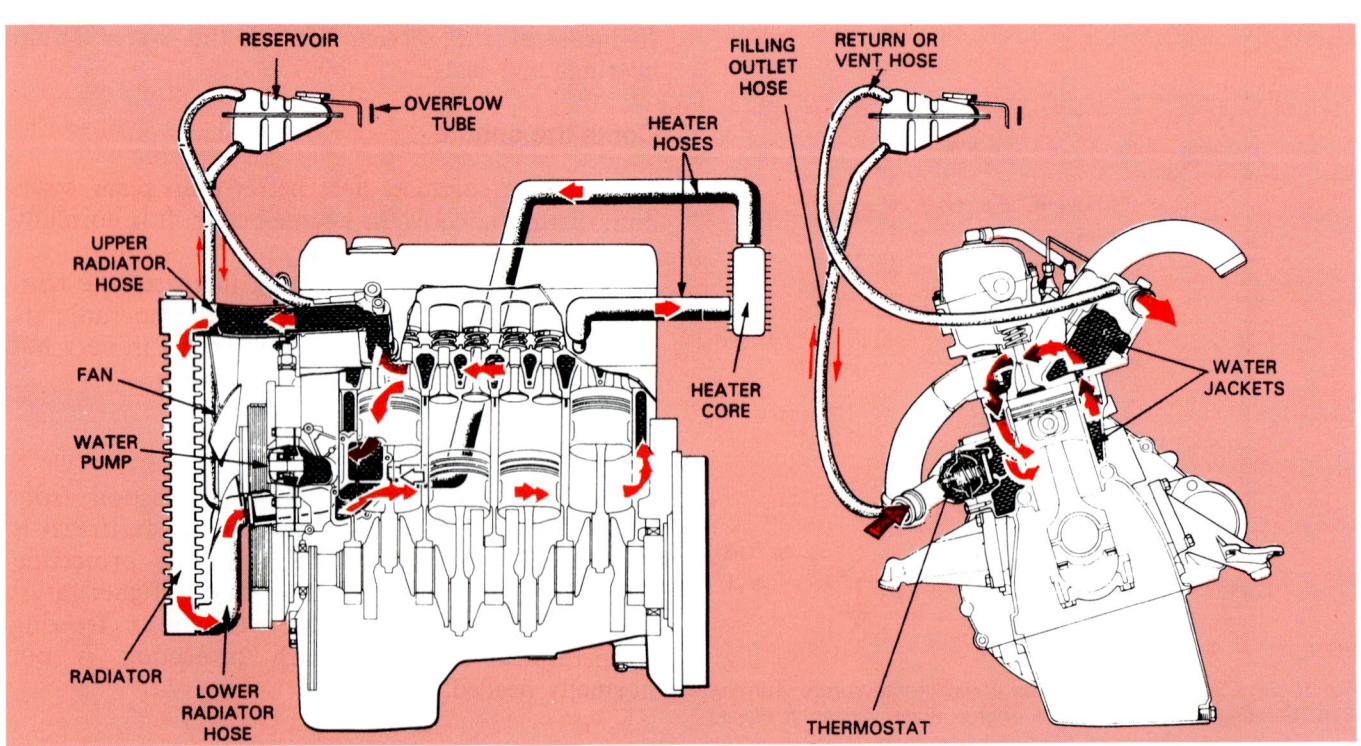

Can you explain the purpose of the major parts of a cooling system? If not, review the chapter! (Mercedes Benz)

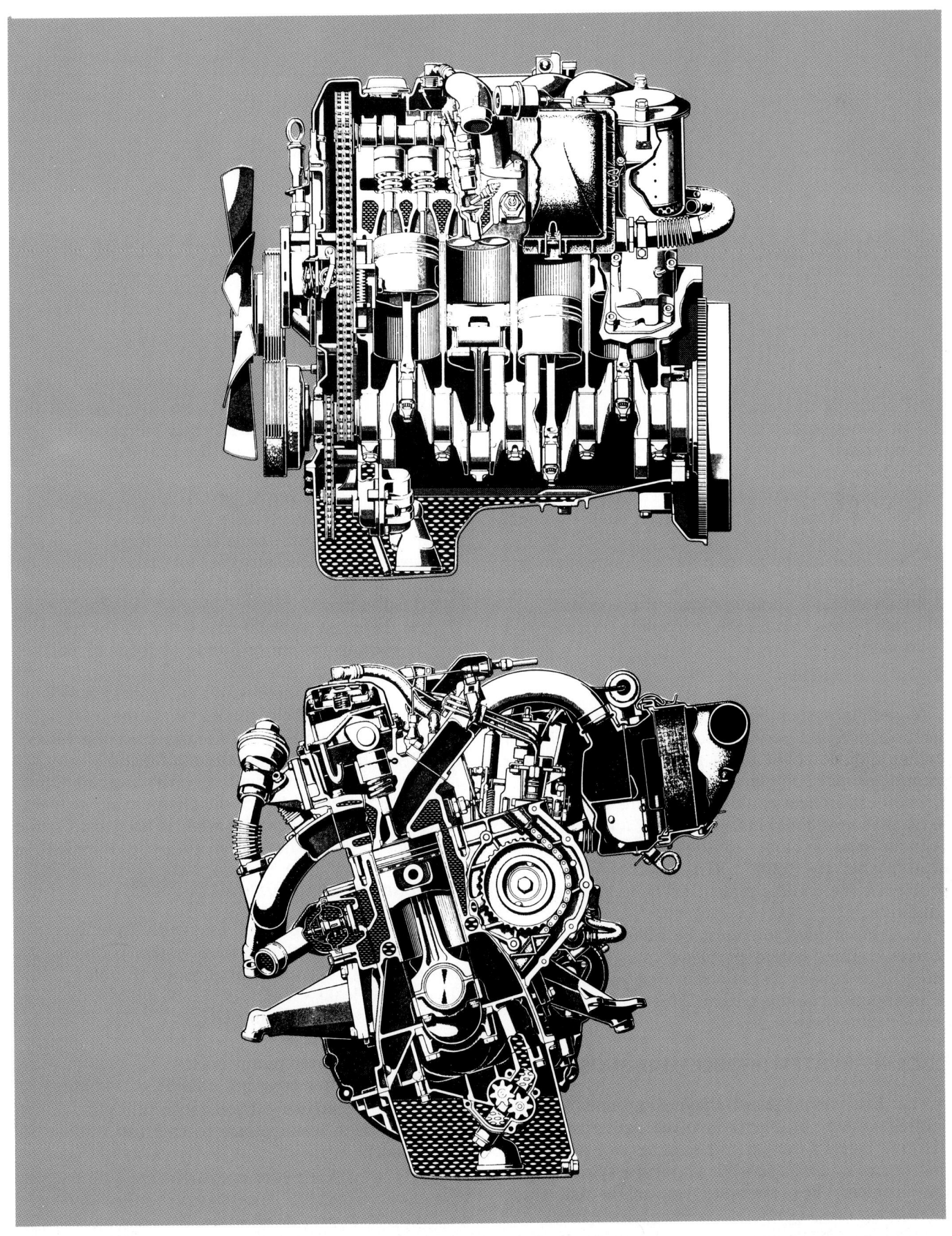

Study side and front view of this modern four-cylinder diesel engine. It uses overhead camshaft to operate valves. Also, note cooling system water jackets in cylinder head and cylinder block. Thermostat is located at front, centre of engine. Can you find it? (Mercedes Benz)

Chapter 17

COOLING SYSTEM TESTING, MAINTENANCE, REPAIR

After studying this chapter, you will be able to:
- List common cooling system problems and their symptoms.
- Describe the most common causes of system leakage, overheating, and overcooling.
- Perform a combustion leak test and a system pressure test.
- Check the major parts of a cooling system for proper operation.
- Replace faulty cooling system components.
- Drain, flush, and refill a cooling system with coolant.

A cooling system is extremely important to the performance and service life of an engine. Major engine damage could result in a matter of minutes without proper cooling.

Combustion heat could collect in the metal engine parts. The heat could melt pistons, crack or warp the cylinder head or block, cause valves to burn, or the head gasket to ''blow''. To prevent these costly problems, the cooling system must be kept in good condition.

As a mechanic, you must be able to locate and correct cooling system problems quickly and accurately. It is equally as important that you know how to maintain a cooling system. This chapter will help you develop these skills.

COOLING SYSTEM PROBLEM DIAGNOSIS

The first step toward diagnosing and locating cooling system problems involves gathering information. Talk to the car owner or service adviser to find out as much as possible about the symptoms of the problem. For example, you might ask these questions:
1. Can you describe the cooling system problem (engine temperature light ON, overheating, coolant loss)?
2. When does the problem seem to occur (all the time, at highway speeds, when idling only)?
3. How long have you had the problem?
4. When was the last time the coolant was replaced (one year, two years, never)?
5. Have any other repairs been performed (new thermostat, hoses, or engine repairs)?
6. Have you noticed any coolant leaks (puddles on ground, wetness around engine), or have you added coolant?
7. Are there any unusual noises that might relate to the cooling system (grinding at front of engine, hissing)?

The answers to these kinds of questions will be very useful. It will help you eliminate the least likely sources so that you can concentrate on the MOST PROBABLE CAUSES of the malfunction.

After gathering information, verify the complaint. Test drive the car. Inspect the engine compartment. Listen to engine noises. Do what is needed to make sure the symptoms have been properly described.

Inspecting cooling system

A visual inspection will frequently be enough to find the source of the cooling system problem. As shown in Fig 17-1 look for obvious troubles:
1. Coolant leaks.
2. Loose or missing fan belts.
3. Low coolant level.
4. Abnormal water pump noises.
5. Foreign matter covering outside of radiator.
6. Coolant in engine oil (oil looks milky).
7. Combustion leakage into coolant (air bubbles in coolant).

CAUTION! Keep your hands and tools away from a spinning engine fan. Wear eye protection and stand behind, not over, the spinning fan blade. Then, if tools are dropped into the fan or a fan blade breaks off, you are not likely to be hit and injured by flying parts.

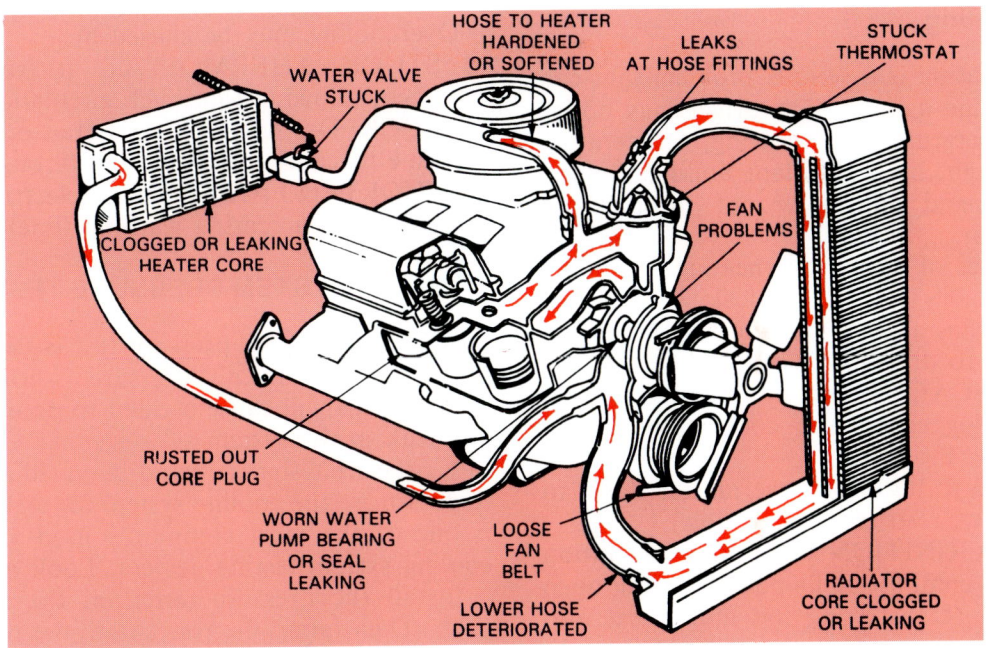

Fig. 17-1. These are common problem areas in a cooling system. Note leakage points.

COOLING SYSTEM PROBLEMS

Cooling system problems can be grouped into three general categories:
1. COOLANT LEAKS (crack or rupture, allowing pressure cap action to push coolant out of system).
2. OVERHEATING (engine operating temperature too high, warning light on, temperature gauge shows hot, or coolant and steam blowing out overflow).
3. OVERCOOLING (engine fails to reach full operating temperature, engine performance poor or sluggish).

COOLANT LEAKS

Coolant leaks show up as a wet, discoloured (darkened or rust coloured) areas in the engine compartment or on the ground. The leaking fluid will smell like antifreeze and have the same general colour. Leaks can occur almost anywhere in the system. Fig. 17-1.

A low coolant level may indicate a leak. If not visible, the leak may be internal (cracked engine part, blown head gasket), as illustrated in Fig. 17-2.

Remember to check the coolant level in modern systems at the overflow or reservoir tank. DO NOT remove the radiator cap. Only on older, open systems (no reservoir), do you need to remove the pressure cap to check coolant level.

DANGER! Never remove a radiator cap when the engine is hot. The pressure release can make the coolant begin to boil and expand. Boiling coolant could spurt out of the filler neck — causing SEVERE BURNS!

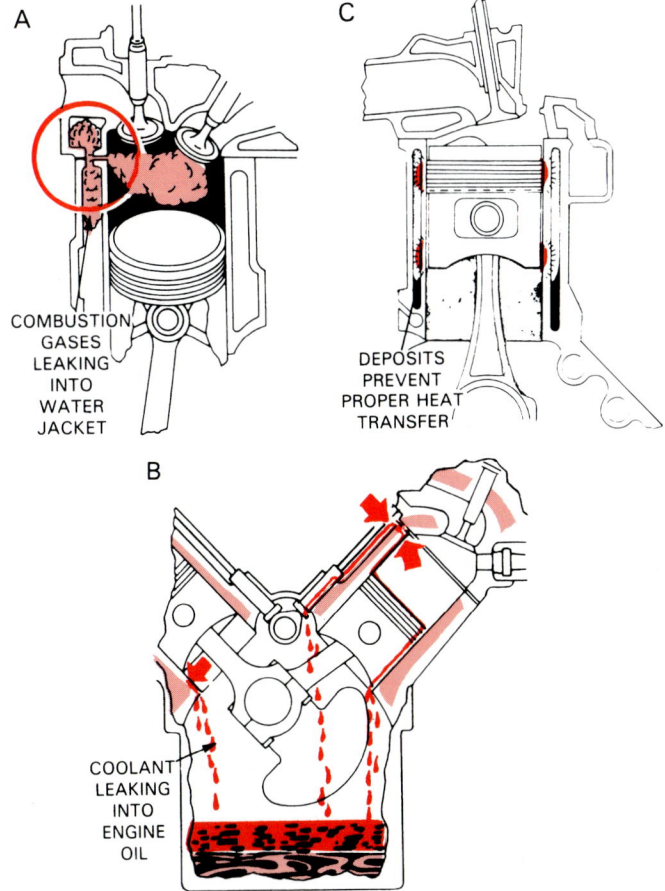

Fig. 17-2. Engine problems can affect cooling system. A — Blown head gasket allowing combustion gases to enter coolant. B — Cracked part or blown head gasket allowing coolant to leak into engine oil. C — Mineral deposits prevent proper heat transfer. Overheating can result.

249

Engine overheating

Engine overheating is a serious problem that can cause major engine damage. The driver may notice the engine temperature light glowing, temperature gauge reading high, or the coolant boiling. Boiling coolant will expand and blow out through the overflow as steam.

Common causes of engine overheating are:
1. LOW COOLANT LEVEL (leak or lack of maintenance has allowed coolant level in engine and radiator to drop too low).
2. RUST OR SCALE (mineral accumulations in system has clogged radiator core or built up in water jackets, Fig. 17-2).
3. STUCK THERMOSTAT (thermostat fails to open normally, restricts coolant flow).
4. RETARDED IGNITION TIMING (late ignition timing allows combustion flame to blow out open exhaust valve, transferring too much heat into exhaust valves, ports, and manifold).
5. LOOSE FAN BELT (water pump drive belt slips under load and reduces coolant circulation).
6. BAD WATER PUMP (broken pump shaft or damaged impeller blades prevent normal pumping action).
7. COLLAPSED LOWER HOSE (suction from water pump may flatten hose if spring is missing or hose is badly deteriorated).
8. MISSING FAN SHROUD (air circulates between fan and back of radiator, reducing airflow through radiator).
9. ICE IN COOLANT (coolant frozen from lack of antifreeze can block circulation and cause overheating).
10. ENGINE FAN PROBLEMS (fan clutch or electric fan troubles can prevent adequate airflow through radiator).

Any of these, or other troubles, can make the engine overheat. You must use your knowledge of system operation and basic testing methods to find the problem's source. Methods for locating specific troubles will be covered later in this chapter.

Overcooling

Overcooling may be indicated by slow engine warm up, insufficient warmth from the heater, low fuel economy, and sluggish engine performance.

Overcooling can cause increased part wear. Because parts are not at full operating temperature, their clearances will be too great. The parts will not expand enough to produce the correct fit.

Overcooling also reduces fuel economy because more combustion heat transfers into the metal parts of the engine. Less heat remains to produce expansion of gases and pressure on the pistons.

Overcooling may be caused by:
1. STUCK THERMOSTAT (thermostat stuck open, allowing too much circulation).
2. LOCKED FAN CLUTCH (fan operates all the time to cause excess airflow through radiator).
3. SHORTED FAN SWITCH (electric fan runs all the time, increasing warmup time).

COOLING SYSTEM PRESSURE TEST

A cooling system pressure test is used to quickly locate leaks. Low air pressure is forced into the system. This will cause coolant to pour or drip from any leak in the system.

A pressure tester is a hand-operated air pump used to pressurise the cooling system for leak detection. It is one of the most commonly used and important cooling system testing devices. Look at Fig. 17-3.

Install the pressure tester on the radiator filler neck. Then pump the tester until the pressure gauge reads radiator cap pressure (around 100 kPa).

CAUTION! Do not pump too much pressure into the cooling system or part (radiator, hose, or gasket) damage may result. Only equal radiator cap pressure when testing.

With pressure in the system, inspect all parts for coolant leakage. Check at all hose fittings, at gaskets, under the water pump, around the radiator, and at engine expansion (core) plugs. If a leak is found, tighten, repair, or replace parts as needed.

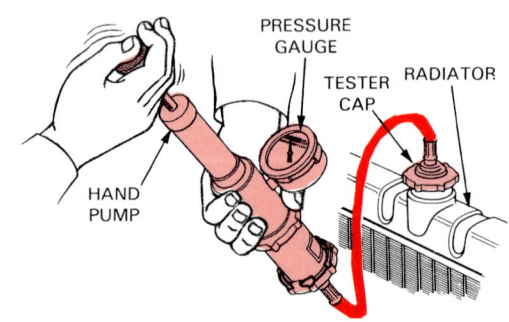

Fig. 17-3. Use pressure tester to pump cap-rated pressure into system. This will cause coolant to drip from any leak. (Toyota)

COMBUSTION LEAK TEST

A combustion leak test checks for the presence of combustion gases in the engine coolant. It should be performed when signs (overheating, bubbles in coolant, rise in coolant level upon starting) point to a blown head gasket, cracked block, or cracked cylinder head. Refer back to Fig. 17-2.

A block tester, sometimes called a combustion leak tester, is placed in the radiator filler neck. The engine is started and the testing bulb is squeezed and then released. This will pull air from the radiator through the test fluid, as in Fig. 17-4.

The fluid in the block tester is normally blue. The chemicals in exhaust gases cause a reaction in the test fluid, changing its colour.

A combustion leak will turn the fluid YELLOW. If the fluid REMAINS BLUE, there is no combustion leakage.

If combustion leakage is indicated, short out spark plugs one at a time. Test the cooling system with each plug shorted. When the fluid does not change colour, the cylinder being checked (shorted) has a combustion leak.

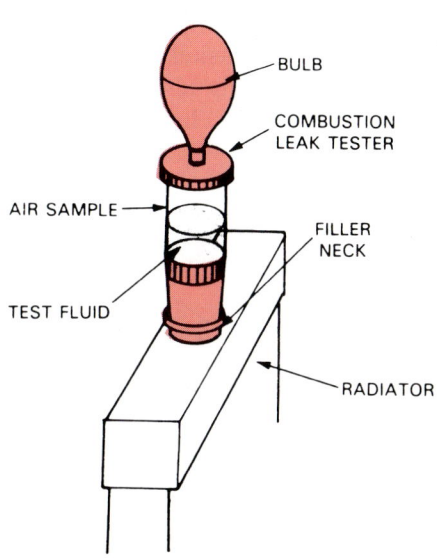

Fig. 17-4. Use bulb to draw radiator air sample into combustion leak tester. If test fluid turns yellow, engine problems are allowing combustion gas leakage into cooling system. Combustion leakage can make engine overheat.

Combustion leakage into the cooling system is very damaging. Exhaust gases mix with the coolant and form very corrosive acids. The acids can eat holes in the radiator and etch or corrode other components.

Fig. 17-5 shows how an exhaust gas analyser will also check for combustion gases in the cooling system.

COOLANT IN OIL

When engine coolant, and oil mix, the solution turns milky white in colour. See Fig. 17-2. If found in the engine oil or in valve covers, it is an indication of a coolant leak. The cause may be one of the same problems that create combustion leakage (blown head gasket, cracked head or block, or leaking intake manifold gasket on V-type engines).

It is possible to have both combustion leakage into the coolant and coolant leakage into the engine oil. Always correct an engine problem causing internal leakage.

Coolant containing ethylene glycol (antifreeze), when mixed with engine oil, can cause engine damage. The ethylene glycol can collect on the cylinder walls where it burns, causing piston and cylinder gumming or scoring.

DIAGNOSIS CHARTS

A cooling system diagnosis chart should be used when problems are difficult to locate and correct. A service manual will give a chart for the particular type engine and cooling system. It will be very accurate and will help you decide what tests and repairs are needed.

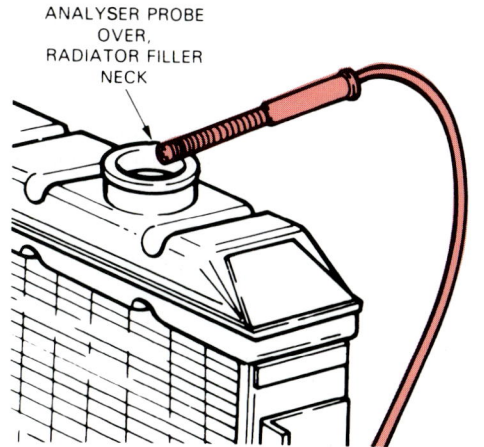
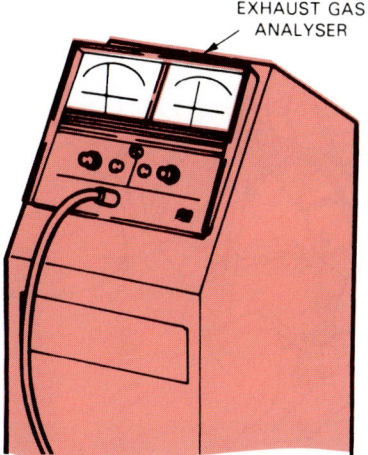

Fig. 17-5. An exhaust gas analyser will also detect combustion pressure leakage into coolant. Place probe over filler neck and accelerate engine. HC or hydrocarbon reading indicates leakage.

WATER PUMP SERVICE

A bad water pump may leak coolant (worn seal), fail to circulate coolant (broken shaft or damaged impeller), or it may produce a grinding sound (faulty pump bearings).

Rust in the cooling system or lack of rust and corrosion inhibitor are common reasons for pump failure. These conditions could speed up seal, shaft, and bearing wear. An over-tightened fan belt is another common cause for premature water pump failure.

Checking water pump

To check for a bad water pump seal, pressure test the system and watch for leakage at the pump. Coolant will leak out of the small drain hole at the bottom of the pump or at the end of the pump shaft. Replace or rebuild a leaking pump.

To check for worn water pump bearings, try to wiggle the fan or pump pulley up and down. Look at Fig. 17-6. If the pump shaft is loose in the bearing case, the pump bearings are badly worn. Pump replacement would usually be necessary. A stethoscope can also be used to listen for worn, noisy water pump bearings.

To check water pump action, warm the engine. Squeeze the top radiator hose while someone starts the engine. You should feel a pressure surge (hose swelling) if the pump is working. If not, pump shaft or impeller problems are indicated. You can also watch for coolant circulation in the radiator with the engine at operating temperature.

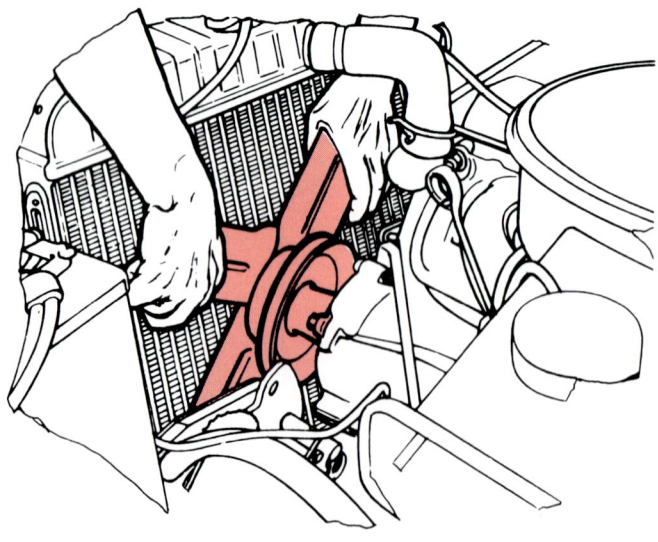

Fig. 17-6. Wiggle engine fan up and down to check for water pump bearing wear. Pump shaft should not wiggle.

Removing a water pump

To remove the water pump, unbolt all brackets and other components (air conditioning compressor, power steering pump, alternator, shroud, fan, etc) preventing pump removal. Then, unscrew the bolts holding the pump to the engine. Note the location of each bolt to aid reassembly.

Scrape off old gasket or sealer material. The engine-to-pump mating surfaces must be perfectly clean to prevent coolant leakage. On soft aluminium parts, be careful not to gouge or scratch the sealing surfaces.

Installing water pump

To install a water pump gasket, use approved sealer to stick the new gasket to the pump. This will keep the gasket in alignment over the bolt holes during pump installation. Look at Fig. 17-7.

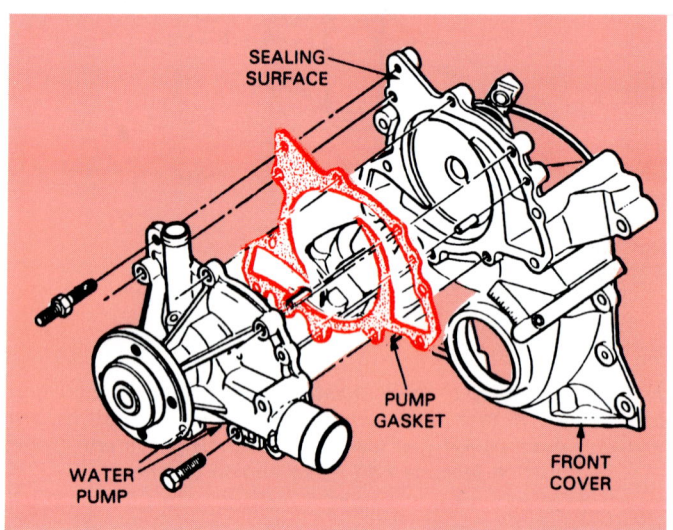

Fig. 17-7. Gasket is used to seal water pump-to-engine sealing surfaces. Make sure gasket is correct one and aligned.

To use a chemical silicone gasket (sealer used in place of fibre gasket), squeeze out a bead of approved sealer around the pump sealing surface. Form a continuous bead of consistent width about 3 mm. This is illustrated in Fig. 17-8.

Fit the pump onto the engine. Move it straight into place. Do not shift the gasket or break the sealant bead. Start ALL of the bolts by hand. Screw them in about two turns. Check that all bolt lengths are correct. Each bolt should be sticking out the same amount.

Torque all of the fasteners a little at a time in a crisscross pattern. Go over the bolts several times to assure correct tightening. Install the other components and tighten the belt properly.

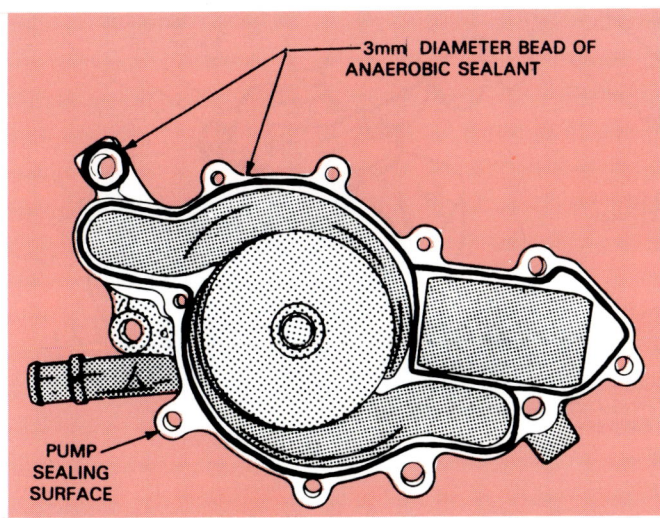

Fig. 17-8. Using sealer instead of water pump gasket. Form continuous bead and do not break bead when installing pump.

If needed, refer to a service manual. It will give detailed directions on pump service for the exact make and model of car. You may need to remove the radiator in some cars, but not with others. It will give this kind of essential information.

Water pump rebuild

A water pump rebuild involves pump disassembly, cleaning, part inspection, worn part replacement, and reassembly. Few mechanics rebuild water pumps. They purchase new or factory rebuilt pumps. Rebuilding takes too much time and would not usually be cost effective. Fig. 17-9 shows an exploded view of a water pump.

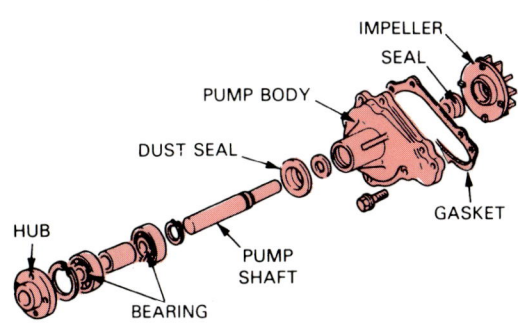

Fig. 17-9. Exploded view shows major parts of water pump. Pump rebuild typically involves replacing pump bearings, seals, shaft, and sometimes impeller. (Mazda)

THERMOSTAT SERVICE

A stuck thermostat can either cause engine overheating or engine overcooling.

If the thermostat is stuck shut, coolant will not circulate through the radiator. As a result, overheating could make the coolant boil.

If the thermostat is stuck open, too much coolant may circulate through the radiator. The engine may not reach proper operating temperature. The engine may run poorly for extended periods in cold weather. Engine efficiency (power, fuel economy, and driveability) will be reduced.

Thermostat testing

To check thermostat action, watch the coolant through the radiator neck. When the engine is cold, coolant should NOT flow through the radiator. When the engine warms, the thermostat should open. Coolant should then begin to circulate through the radiator. If this action does not occur, the thermostat may be defective.

In some instances, the thermostat may have to be removed from the engine for testing. As pictured in Fig. 17-10, the thermostat is placed in a container of water and heated on a hot plate or stove. The opening temperature of the thermostat is observed and compared to specifications.

If the thermostat does NOT open at the correct temperature, it is defective and should be replaced.

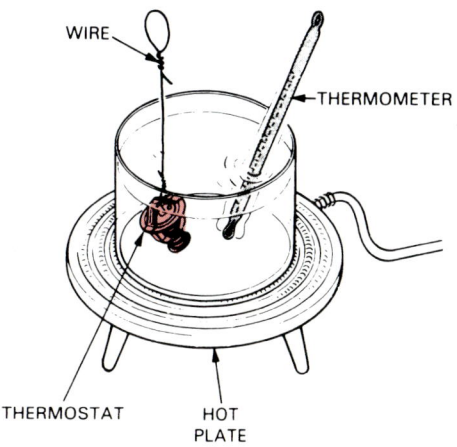

Fig. 17-10. Using a hot plate to check thermostat action. Place thermostat in container and heat. Note opening temperature of thermostat using thermometer. Replace thermostat if not within specs. (Mazda)

Thermostat replacement

The thermostat is normally located under the thermostat housing (fitting for upper radiator hose).

To remove the thermostat, drain the coolant and remove the upper radiator hose from the engine. Unscrew the bolts holding the thermostat housing to the engine. Tap the housing free with a rubber hammer. Lift off the housing and thermostat. Fig. 17-11A.

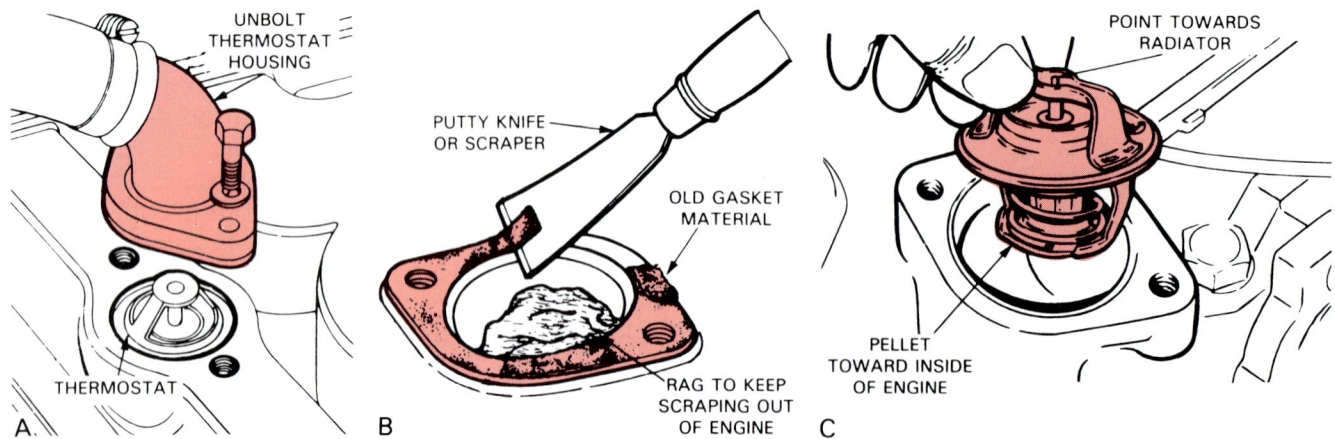

Fig. 17-11. A — Thermostat is removed by unbolting thermostat housing. B — Scrape off all old gasket or sealer from engine and thermostat housing. C — Install thermostat with pellet towards inside of engine. (Ford)

Scrape all of the old gasket material off the thermostat housing and the sealing surface on the engine. See Fig. 17-11B.

Make sure the thermostat housing is not warped. Place it on a flat surface and check for gaps between the housing and surface. If warped, file the surface flat. This will prevent coolant leakage.

Make sure the temperature rating is correct. Then place the new thermostat into the engine, Fig. 17-11C. Normally, the rod (pointed end) on the thermostat should face the radiator hose. The pellet chamber should face the inside of the engine.

Position the new gasket with approved sealer. Start the fasteners by hand. Then torque them to specs in a crisscross pattern. Do NOT overtighten the thermostat housing bolts or warpage may result. Most housings are made of soft aluminium.

COOLING SYSTEM HOSE SERVICE

Old radiator hoses and heater hoses are frequent causes of cooling system problems. After a few years of use, hoses deteriorate. They may become soft and mushy, or hard and brittle. Cooling system pressure can rupture the hoses and result in coolant loss.

A softened lower radiator hose can be collapsed by the suction of the water pump. The collapsed hose will restrict coolant circulation and cause overheating. The spring inside the lower radiator hose normally prevents hose collapse and it should never be removed.

Checking cooling system hoses

Inspect the radiator and heater hoses for cracks, bulges, cuts, or any other sign of deterioration or damage. Look at Fig. 17-12.

SQUEEZE the hoses to check whether they are hardened or softened and faulty. Flex or bend the

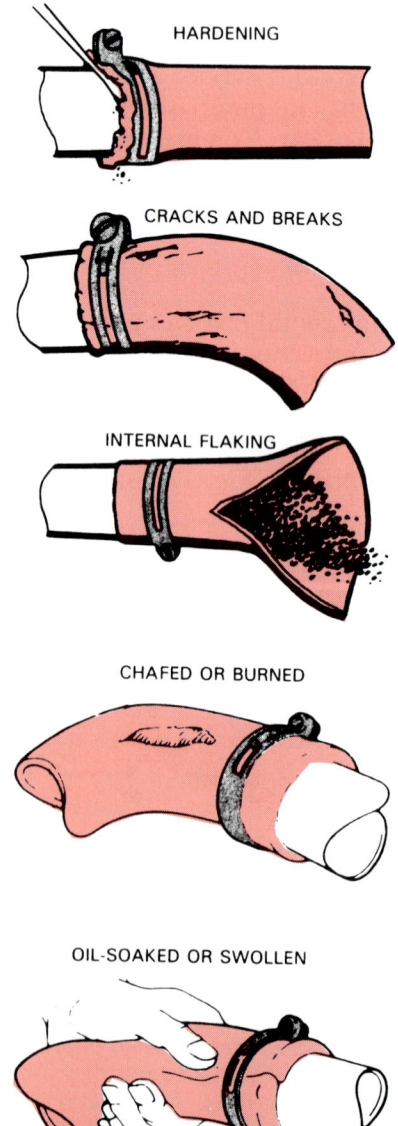

Fig. 17-12. Check cooling system hoses for these kinds of problems. (Ford)

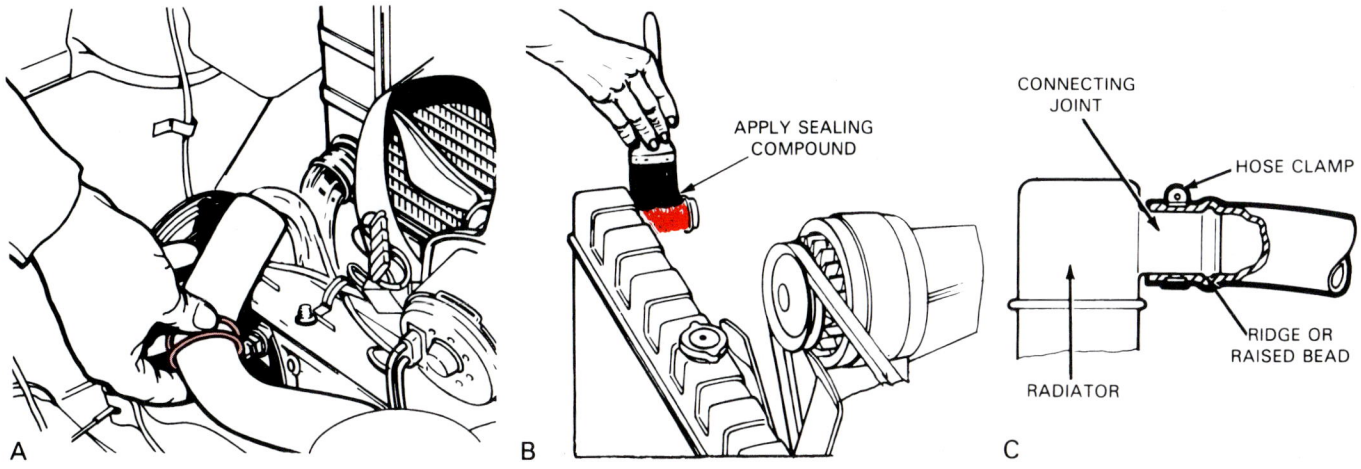

Fig. 17-13. Hose replacement. A — Loosen hose clamp. Twist and pull hose off fitting. B — Clean fitting and coat with nonhardening sealer. C — Slide on new hose and clamp. Make sure clamp is positioned inside bead and fitting. Tighten clamp. (Ford)

heater hoses and watch for surface cracks. If any problem is detected, the affected hoses should be replaced.

Hose replacement

To remove a hose, loosen the hose clamps. Twist the hose while pulling it. Fig. 17-13A. If a new hose is to be installed, you may cut a slit in the end of the old hose to aid removal.

Clean the metal hose fittings. Coat them with a non-hardening sealer, Fig. 17-13B. Install the new hose.

Position the hose clamps so that they are fully over the metal hose fittings, Fig. 17-13C. Then, tighten the clamps.

Fill the engine with coolant and pressure test the system. Check all fittings for leaks.

RADIATOR AND PRESSURE CAP SERVICE

When overheating problems occur and the system is NOT leaking, check the radiator and the pressure cap. They are common sources of overheating. The pressure cap could have bad seals, allowing pressure loss. The radiator may be clogged and not permitting adequate air or coolant flow.

Inspecting radiator and pressure cap

Inspect the outside of the radiator for foreign matter such as insects, leaves and road dirt. Also, make sure the radiator shroud is in place and unbroken. These troubles could limit air circulation through the core.

If needed, use a water hose to wash foreign matter out of the core. Spray water from the back to push the matter out of the front of the radiator. You may also use compressed air if pressure is low enough not to damage the core.

Inspect the radiator cap and filler neck. Check for cracks or tears in the cap seal. Check the filler neck sealing surfaces for nicks or dents. Replace the cap or have the neck repaired as needed.

Pressure testing radiator cap

A radiator cap pressure test measures cap opening pressure and checks the condition of the sealing washer. The cap is installed on a cooling system pressure tester, as pictured in Fig. 17-14.

Pump the tester to pressurise the cap. Watch the pressure gauge. The cap should release air at its rated pressure (pressure stamped on cap). It should also hold that pressure for at least one minute. If not, install a new cap.

Radiator repair

A radiator repair shop specialises in radiator repairs. It has the facilities to properly disassemble,

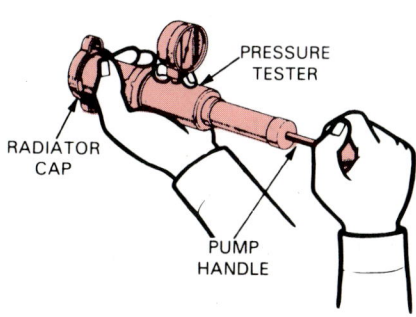

Fig. 17-14. A pressure tester can be used to check opening value of radiator cap. It should open within specs and hold pressure without leaking. (Toyota)

bayonet out (clean), solder (reassemble or repair), and pressure test a radiator. Few mechanics attempt radiator repairs.

A radiator repair shop can solder pin hole leaks. It has special cleaning tanks for loosening and removing scale built up inside the radiator. A radiator repair shop can also remove the tanks and solder in a new core, if needed.

FAN BELT SERVICE

A loose fan belt will slip and may rotate the water pump and fan too slowly. The engine may overheat. Always inspect the condition and tension (tightness) of fan belts when servicing a cooling system. If a fan belt is cracked, frayed, glazed (hard, shiny surface), or oil soaked, it should be replaced.

DANGER! Keep your hands away from a moving engine belt. The belt can pull your fingers into the pulleys, causing severe hand injuries.

ENGINE FAN SERVICE

A faulty engine fan can cause overheating, overcooling, vibration, and water pump wear, or damage. Always check the fan for bent blades, cracks, and other problems. A flexible fan is especially prone to these problems. If any troubles are found, replace the fan.

DANGER! A fan with cracked or bent blades is extremely dangerous. Broken blades can be thrown out with great force, causing severe lacerations or death.

Testing a fan clutch

To test a thermostatic fan clutch, start the engine. The fan should slip when cold. When the engine warms, the clutch should engage. Air should begin to flow through the radiator and over the engine. You will be able to hear and feel the air when the fan clutch locks up.

If the fan clutch is locked all the time (cold and hot), it is defective and must be replaced. Excessive play or oil leakage also indicates fan clutch failure.

Electric cooling fan service

Most electric cooling fans are controlled by a heat sensitive switch located somewhere in the cooling system (radiator, engine block, thermostat housing). When the engine is cold, the switch keeps the electric fan motor OFF to speed engine warmup. Then, when a predetermined temperature is reached, the switch closes and the fan begins to cool the engine.

Testing an electric cooling fan

To test an electric cooling fan, observe whether the fan turns ON when the engine is warm. Make sure

the fan motor is spinning at NORMAL SPEED and is forcing enough air through the radiator.

If the fan does NOT function, check the fuse, electrical connections, and supply voltage to the motor. Refer to Fig. 17-15.

If the fan motor fails to operate with voltage applied, the motor should be repaired or replaced.

If the engine is warm and no voltage is supplied to the fan motor, check the action of the fan switch. Use either a voltmeter or test light. The switch should have almost zero resistance (pass current and voltage) when engine is warm. Resistance should be infinite (stop current and voltage) when the engine is cold.

If these tests do not locate the trouble, refer to a service manual for instructions. There may be a defective relay, connection, or other problem.

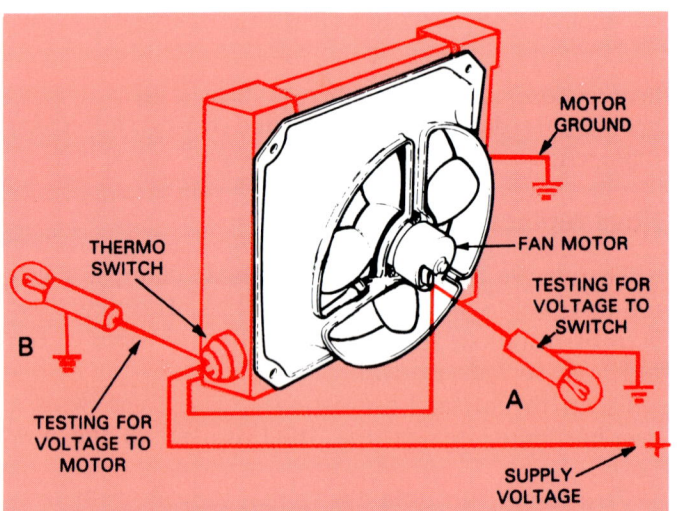

Fig. 17-15. Testing electric fan circuit. A — Check for power to fan with engine warm. Light should glow. B — With no power to fan, check action of fan switch. Switch should be open when cold and closed when hot. (Honda)

EXPANSION PLUG SERVICE

A leaking engine expansion plug (core or freeze plug) is a frequent cause of coolant loss and overheating. Since the engine plugs are thinner than the metal in the engine block or head, they will RUST through before the other parts of the engine.

Replacing expansion plugs

To replace an expansion plug, drive a drift or large screwdriver (full shank type) through the plug, Fig. 17-16A. Pry sideways, without scraping the engine block or cylinder head. The plug should pop out.

Sand the core plug hole in the engine and wipe it clean. Coat the plug hole and plug with nonhardening sealant. Then, as shown in Fig. 17-16B, drive the expansion plug squarely into position.

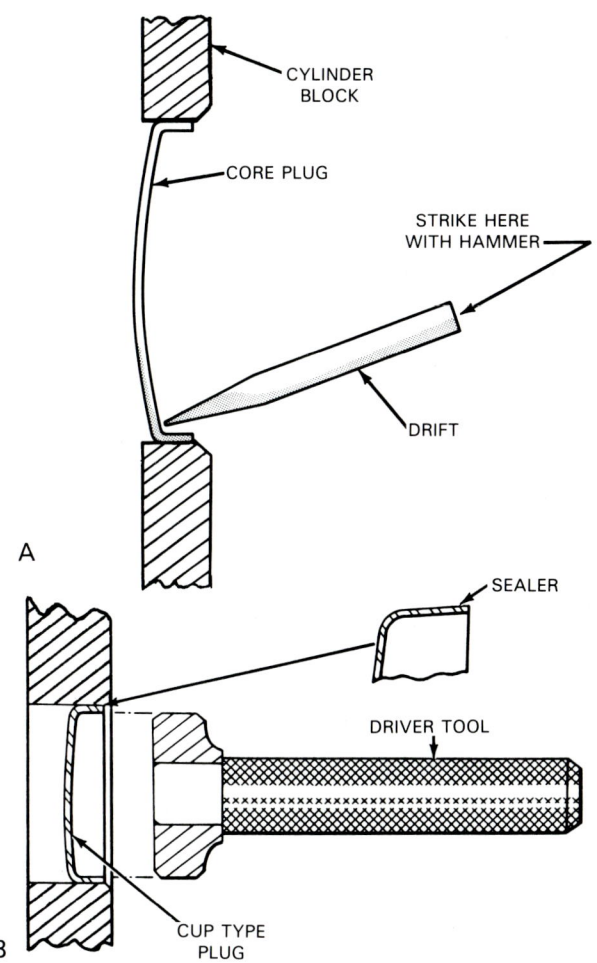

Fig. 17-16. Freeze or core plug replacement. A — Drive drift or full shank screwdriver through old plug. Pry out. B — After cleaning and coating hole with sealer, drive new plug into place. Drive plug in squarely and to proper depth. (Ford)

Special expansion plugs are available for tight quarters. They are installed by tightening a nut which causes the plug to expand and lock into the hole. This allows the plug to be installed without hammering.

ANTIFREEZE (ETHYLENE GLYCOL BASED COOLANT) SERVICE

Antifreeze (inhibited ethylene glycol coolant) should be checked and changed at regular intervals. After prolonged use, antifreeze will break down and become very corrosive. It can lose its rust and corrosion preventive properties and the cooling system can rapidly fill with rust and corrosion.

Inspecting antifreeze

A visual inspection of the antifreeze will help determine its condition. Rub your finger inside the radiator filler neck, as in Fig. 17-17. Check for rust, oil (internal engine leak), scale, or transmission fluid

(leaking oil cooler). Also, find out how long the antifreeze solution has been in service.

If contaminated or too old, replace the antifreeze. If badly rusted, you may also need to flush (clean) the system, as described shortly.

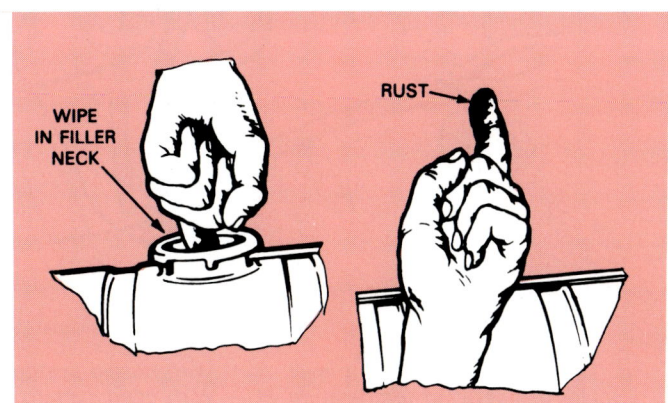

Fig. 17-17. To check for antifreeze contamination, wipe finger inside filler neck. Badly rusted coolant would require replacement of antifreeze and possibly system flushing.

Changing antifreeze

Antifreeze should be changed when contaminated or when two years old. Check in a service manual for exact change schedules.

With the pressure cap removed, loosen the drain tap on the bottom of the radiator. Allow the old coolant to drain into a pan, Fig. 17-18. If a drain tap

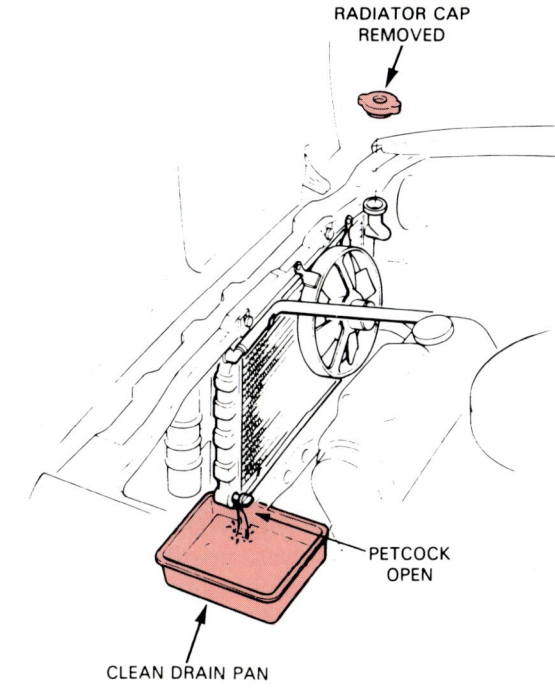

Fig. 17-18. To drain coolant, remove radiator cap. Place pan under drain fitting. Then, turn petcock correct direction. Many have left-hand threads and must be turned clockwise for opening. (Honda)

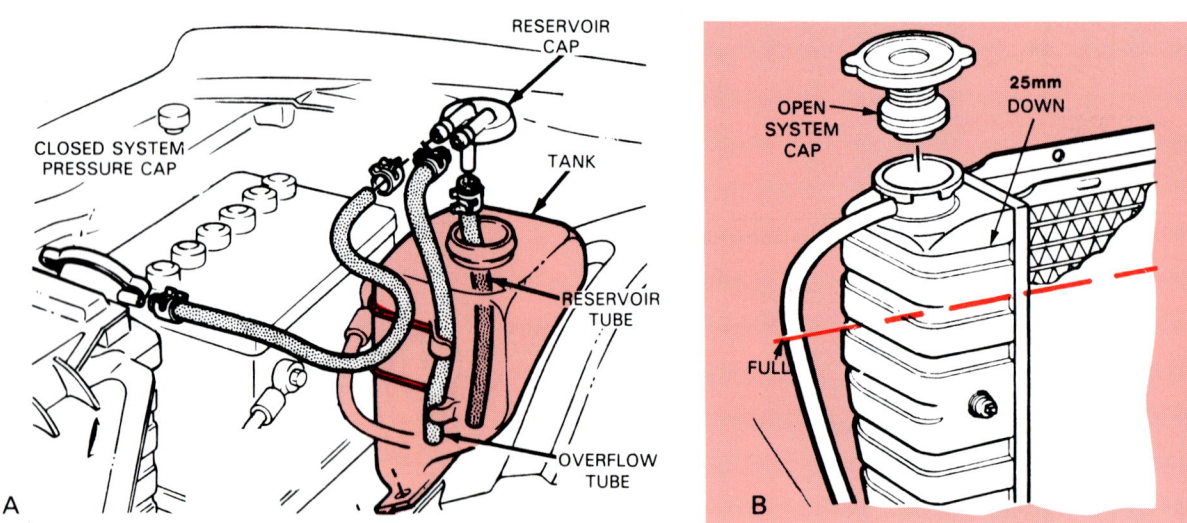

Fig. 17-19. Checking coolant level. A — With closed system, coolant should be even with correct marking on reservoir with coolant at operating temperature. B — With open system, coolant should be about 25 mm down from top of tank. (Ford)

or plug is not fitted to the radiator it will be necessary to disconnect the bottom radiator hose from the radiator.

If the antifreeze is not contaminated with rust, you may then refill the system. Tighten the drain tap or reconnect the hose. Pour in the needed amount and strength of antifreeze (coolant).

Start and warm the engine. The coolant level may drop when the thermostat opens. Add more water, if needed. Then install the radiator cap.

Fig. 17-19 shows how to tell when the cooling system is full. Note the difference between checking closed and open cooling systems.

Testing antifreeze strength

Antifreeze strength is a measurement of the concentration of antifreeze compared to water. It determines the freeze up protection of the solution.

A cooling system hydrometer can be used to measure the freezing point of the cooling system antifreeze solution. One type is shown in Fig. 17-20.

Minimum antifreeze strength

Minimum antifreeze strength should be several degrees lower than the lowest possible temperature for the climate of the area. For example, if the lowest normal temperature for the area is −10° C, the antifreeze should test to −20° C.

A 50/50 mix of antifreeze and water is commonly used to provide protection for most weather conditions. Generally, between 4 and 8 litres of antifreeze is mixed with enough water to fill the cooling system.

Corrosion of aluminium

Note! Many late model cars use aluminium cooling system and engine parts. Radiators, water pumps, cylinder heads, blocks, and intake manifolds can be made of aluminium. Only use antifreeze formulated for aluminium components.

Aluminium can be corroded by some types of antifreeze. Check the car's service manual or the antifreeze label for details.

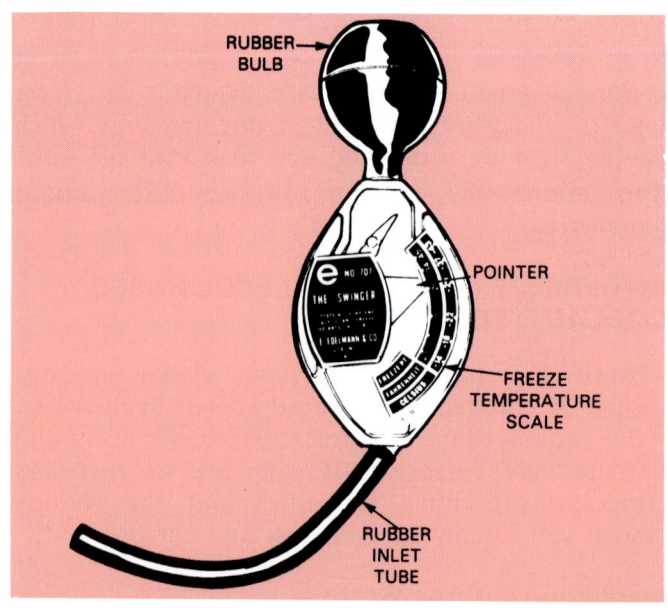

Fig. 17-20. Cooling system hydrometer. Squeeze and release bulb to draw coolant into tester. Needle will float and show freeze protection. Many testers must be corrected for coolant temperature differences.

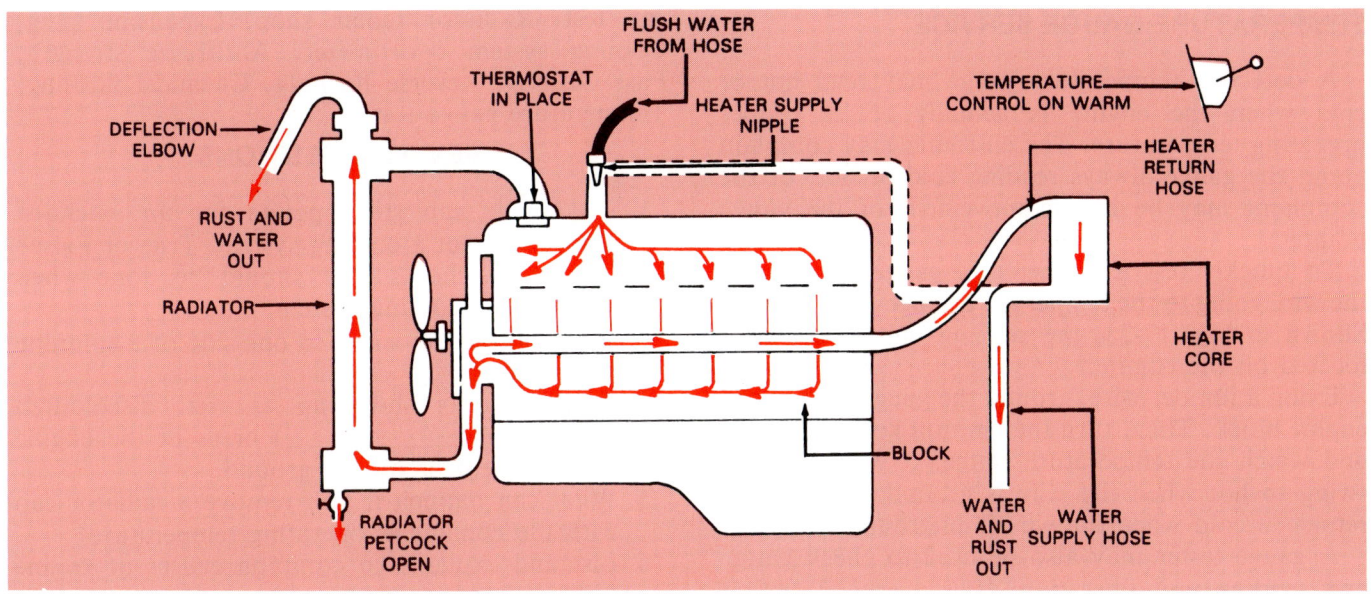

Fig. 17-21. Method of fast flushing cooling system. Water hose is connected to heater hose fitting. This will force water and rust out heater hose and top of radiator.

FLUSHING A COOLING SYSTEM

Flushing (cleaning) of a cooling system should be done when rust or scale is found in the system. Flushing involves running water or a cleaning chemical through the cooling system. This washes out contaminants.

Rust is very harmful to a cooling system. It can cause premature water pump wear. Rust can also collect in and clog the radiator or heater core tubes.

Fast flushing is a common method of cleaning a cooling system because the thermostat does not have to be removed from the engine. Look at Fig. 17-21.

A water hose is connected to a heater hose fitting. The radiator cap is removed and the drain tap is opened. When the water hose is ON and water flows into the system, rust and loose scale are removed.

Reverse flushing of a radiator requires a special adapter that is connected to the radiator outlet tank by a piece of hose. Another hose is attached to the inlet tank. Compressed air, under low pressure, is used to force water through the core backwards. This can be done on the engine block as well. See Fig. 17-22.

Chemical flushing is needed when a scale build up in the system is causing engine overheating. A chemical cleaner is added to the coolant. The engine is operated for a specific amount of time to allow the chemical to act on the scale. Then the system is flushed with water.

DANGER! Always follow manufacturer's instructions when using a cooling system cleaning agent. The chemical may cause eye and skin burns.

After flushing, always add the recommended type and amount of antifreeze. Antifreeze has rust inhibitors and lubricants for the water pump. Never leave plain water in the system.

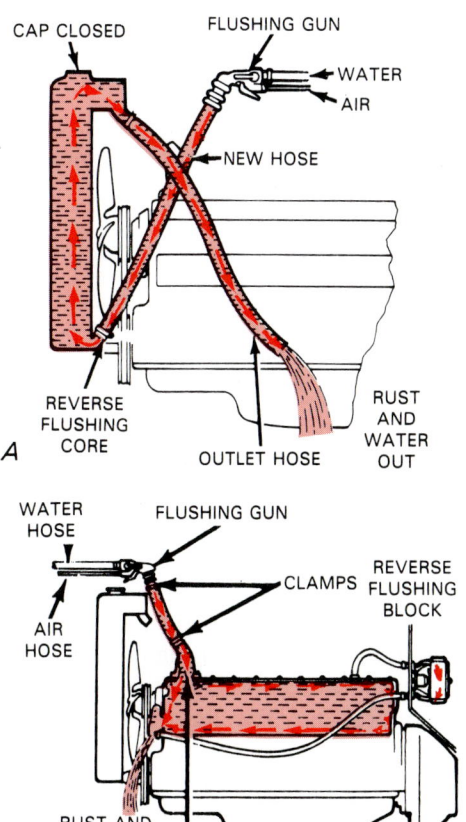

Fig. 17-22. A — Reverse flushing radiator. B — Reverse flushing engine block.

TEMPERATURE GAUGE SERVICE

A defective temperature gauge may read hot or cold when the engine is actually at its proper operating temperature. The customer may complain about the gauge always reading cold or hot, or the complaint may be erratic movement of the gauge pointer.

To quickly test a temperature gauge, disconnect the wire going to the temperature gauge sending unit. Shown in Fig. 17-23, the sending unit is normally located on the engine.

Using a jumper wire, ground the gauge wire to the engine block. Then, turn the ignition key switch ON and watch the temperature gauge. It will normally swing to hot when the wire is grounded. It should return to cold when the wire is ungrounded.

A gauge tester may also be used to check gauge and sending unit operation. It is a special testing device with a variable resistor. Set the tester to a specified ohms value and the temperature gauge should read as specified.

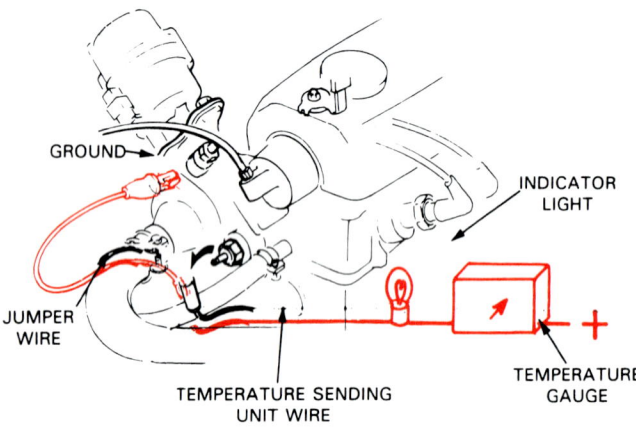

Fig. 17-23. To check action of temperature gauge or indicator light, ground wire to temperature sending unit. This should cause gauge to read hot or light to glow. If not, circuit before sending unit is faulty. If gauge or light function, sending unit may be faulty. (Honda)

If available, use a gauge tester. Some temperature gauges could be damaged by grounding the sending unit wire.

If the gauge begins to function when grounded, the sending unit is defective and should be replaced. If the gauge does NOT function when grounded, either the gauge circuit or the gauge is faulty.

To test a temperature indicating light, perform the same basic operation. The light should glow when the sending unit wire is grounded. It should go out when the wire is ungrounded.

KNOW THESE TERMS

Cooling system pressure test, Combustion leak test, Milky or white oil, Water pump rebuild, Pressure cap test, Radiator repair shop, Expansion plug, Cooling system hydrometer, Antifreeze strength, Fast flushing, Reverse flushing, Chemical flushing, Temperature gauge tester.

REVIEW QUESTIONS

1. An engine can still operate for an extended period without a cooling system. True or False?
2. List seven checks that should be done when inspecting a cooling system.
3. Why should you stand to one side of a spinning engine fan?
4. Coolant leaks show up as wet, discoloured (_____ or _____ _____) areas in the engine compartment or on the ground.
5. What can happen if you remove a radiator cap with the coolant at operating temperature?
6. List and explain ten common causes of engine overheating.
7. Which of the following is NOT a typical cause of engine overcooling?
   a. Stuck thermostat.
   b. Locked fan clutch.
   c. Ice in cooling system.
   d. Shorted electric fan switch.
8. A cooling system _____ _____ is used to quickly find leaks in the system.
9. A _____ _____ test checks for the presence of combustion gases in the engine coolant, indicating an engine problem.
10. When water, antifreeze, and oil mix, the solution turns _____ _____ in colour.
11. A leaking water pump will usually leak coolant:
    a. At the pump-to-block gasket.
    b. At a drain hole in the bottom of the housing.
    c. Out the front of the housing.
    d. Out the bypass connection.
12. A customer complains of sluggish engine performance and a lack of adequate warmth from the heater.
    Mechanic A says that this could not be caused by the cooling system. There may be separate problems with the engine and heating system.
    Mechanic B says that a missing or stuck open thermostat might cause these symptoms. The thermostat should be checked first before checking other possible components.
    Who is correct?
    a. Mechanic A.
    b. Mechanic B.
    c. Both Mechanic A and B are correct.
    d. Neither Mechanic A nor B are correct.
13. How do you replace an engine expansion or core plug?
14. How often should you change antifreeze and how do you check antifreeze strength?
15. How can you quickly determine if a dash temperature gauge is functioning?

Chapter 18

AUTOMOTIVE FUELS, PETROL AND DIESEL COMBUSTION

After studying this chapter, you will be able to:
□ Summarise how crude oil is converted into petrol, diesel fuel, LP-gas, and other products.
□ Describe properties of petrol and diesel fuel.
□ Explain octane and cetane ratings.
□ Describe normal and abnormal combustion of petrol and diesel fuel.
□ Summarise the properties of alternate fuels.

A car engine burns a fuel as a source of energy. Various types of fuel will burn in an engine: petrol, diesel fuel, alcohol, LP-gas, and other alternate fuels. As a mechanic, you must understand how fuel burns inside an engine. Combustion (burning) is a primary factor controlling fuel economy, power, emissions, and the service life of an engine.

PETROLEUM (CRUDE OIL)

Petroleum, also called crude oil, is oil taken directly out of the ground. It is used to make petrol, diesel fuel, liquefied petroleum gas, and many other non-fuel materials. Fig. 18-1 shows some of the products made out of petroleum.

Natural crude oil is a mixture of semi-solids (neither solid nor liquid), liquids, and gases. Chemically, crude oil consists of highly flammable hydrocarbons.

Hydrocarbons are chemical mixtures of about 12 percent hydrogen (light gas vapour) and 82 percent carbon (heavy black solid). Crude oil also contains impurities (sulphur, nitrogen, metals, and salt water) that must be removed.

Processing crude oil

Oil deposits are hidden deep inside the earth. Oil companies perform many tests (seismic studies, surface mapping, test drilling), called exploration tests, to help find oil.

After finding where oil might be located, a drilling crew bores a hole thousands of feet into the ground. A huge steel derrick is used for the drilling operation.

It has a cutting bit capable of passing through dirt, sand, and rock. See Fig. 18-2.

Once the oil deposit has been reached, the oil is pumped to the surface.

Next, oil is sent to the refinery. The refinery converts the crude oil into more useful substances.

Distillation is the first process. It uses a fractioning tower to break the crude oil down into different parts or fractions (LP-gas, petrol, kerosene, fuel oil, lubricating oils). Look at Fig. 18-3. After distillation, other processes purify these products.

PETROL

Petrol is the most common type of automotive fuel. It is an abundant and highly flammable part of crude oil. Extra chemicals, called additives (lead, detergents, antioxidants, etc.), are mixed into petrol to improve its operating characteristics.

Antiknock additives, usually tetraethyl lead (TEL) or tetramethyl lead (TML), are used to slow down the ignition and burning of petrol. This helps prevent engine ping or knock (knocking sound produced by abnormal and excessively rapid combustion).

Leaded and unleaded petrol

Leaded petrol has lead antiknock additives. The lead allows a higher engine compression ratio to be used without the fuel igniting prematurely.

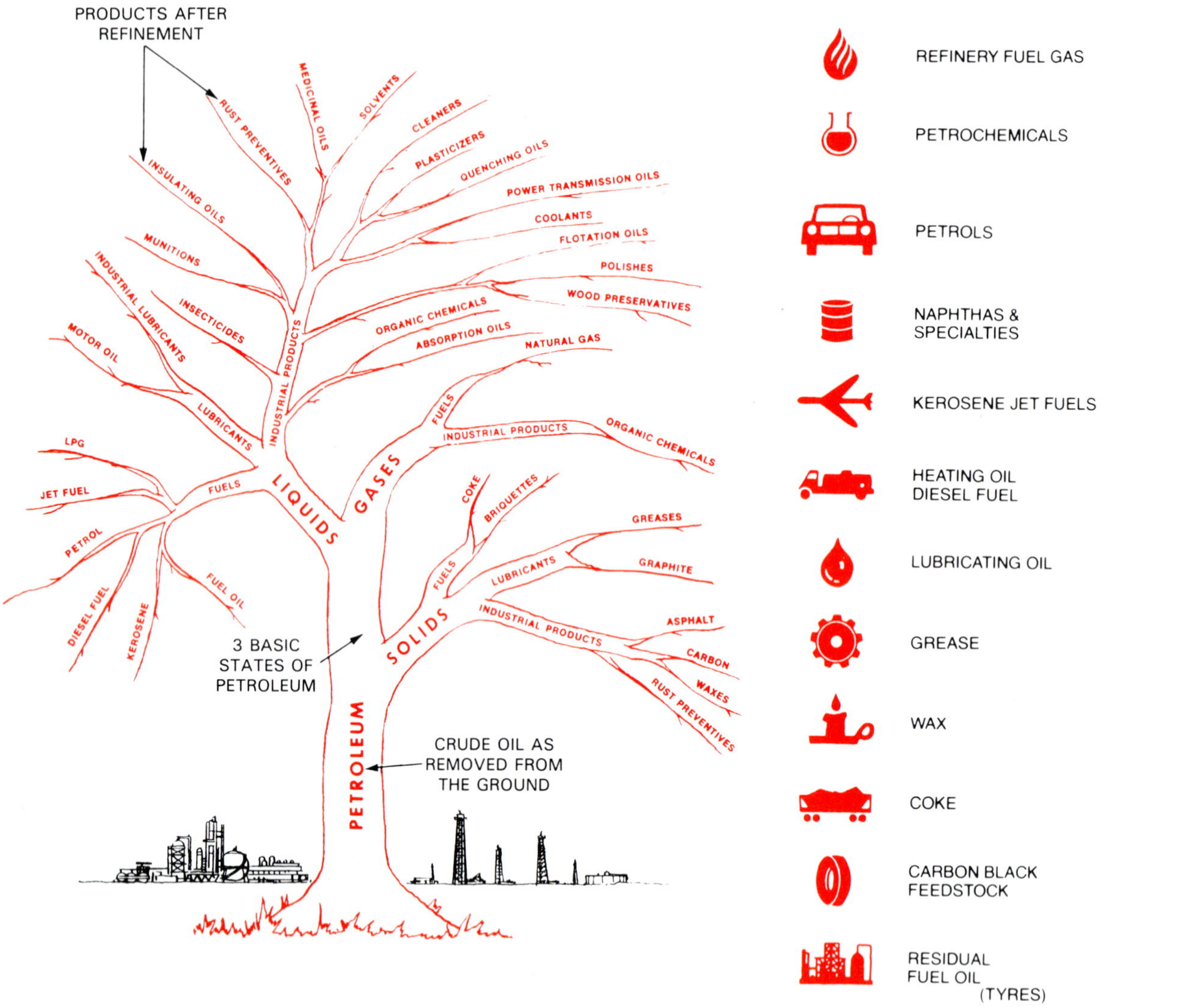

PRODUCTS AFTER REFINEMENT

RUST PREVENTIVES
MEDICINAL OILS
SOLVENTS
CLEANERS
PLASTICIZERS
QUENCHING OILS
INSULATING OILS
POWER TRANSMISSION OILS
COOLANTS
FLOTATION OILS
POLISHES
MUNITIONS
WOOD PRESERVATIVES
INDUSTRIAL LUBRICANTS
INSECTICIDES
ORGANIC CHEMICALS
ABSORPTION OILS
MOTOR OIL
NATURAL GAS
INDUSTRIAL PRODUCTS
LUBRICANTS
ORGANIC CHEMICALS
LPG
FUELS
INDUSTRIAL PRODUCTS
COKE
BRIQUETTES
GREASES
JET FUEL
FUELS
GRAPHITE
PETROL
FUELS
LUBRICANTS
ASPHALT
INDUSTRIAL PRODUCTS
DIESEL FUEL
FUEL OIL
CARBON
KEROSENE
WAXES
RUST PREVENTIVES

LIQUIDS
GASES
SOLIDS
PETROLEUM

3 BASIC STATES OF PETROLEUM

CRUDE OIL AS REMOVED FROM THE GROUND

REFINERY FUEL GAS
PETROCHEMICALS
PETROLS
NAPHTHAS & SPECIALTIES
KEROSENE JET FUELS
HEATING OIL DIESEL FUEL
LUBRICATING OIL
GREASE
WAX
COKE
CARBON BLACK FEEDSTOCK
RESIDUAL FUEL OIL (TYRES)

Fig. 18-1. Petroleum is used to make many products besides petrol and diesel fuel. (See also back page of book which shows listing of common automotive lubricants and applications).

Leaded petrol is designed to be used in older cars which have little or no emission controls. They were not required to pass strict air pollution standards.

Note! The lead additives in petrol also act as a LUBRICANT. The lead coats the face of the valves and seats in the engine, helping to prevent wear.

Unleaded petrol, also called lead-free petrol does NOT contain lead-based additives, however such fuels are not strictly lead-free as a very small allowable tolerance of lead is permitted. State Governments passed laws making cars meet strict emission levels. As a result, car makers began using catalytic converters (device in exhaust system for treating exhaust emissions) and unleaded fuel.

Lead additives cannot be used in a car with a catalytic converter. The lead will coat the inside of the converter and prevent it from working.

If unleaded fuel is used in a car designed for leaded fuel, it can increase engine valve and seat wear.

Only use the type of fuel recommended by the car maker.

Fig. 18-2. Large cutter can penetrate dirt and rock to drill hole deep into ground. When found, oil can be pumped to surface.

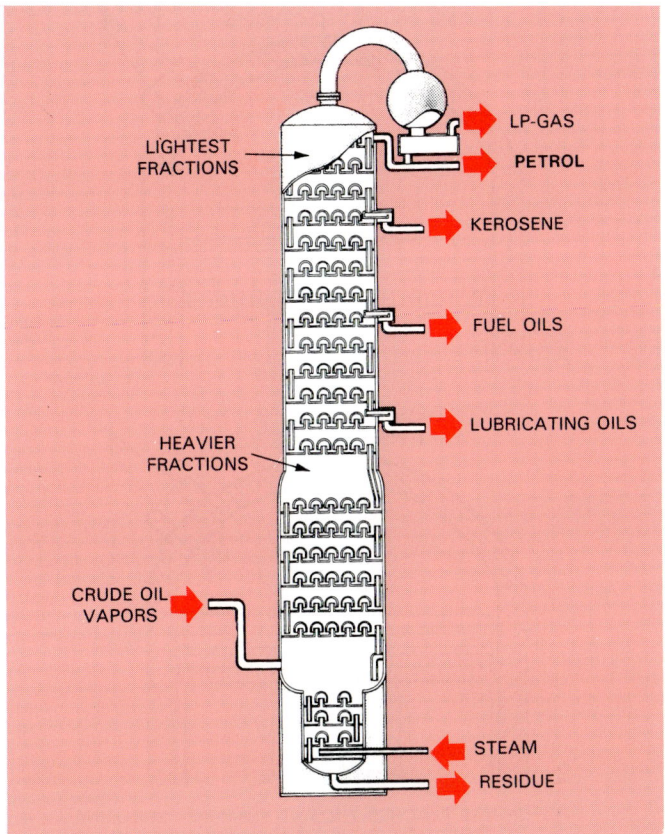

Fig. 18-3. Fractioning tower allows crude oil vapours to condense and separate into trays as shown. (Ford)

Petrol octane rating

The octane rating of petrol is a measurement of the fuel's ability to resist knock or ping. A high octane rating indicates the fuel will NOT knock or ping easily. It should be used in a high compression engine. A low octane petrol is suitable for low compression engines.

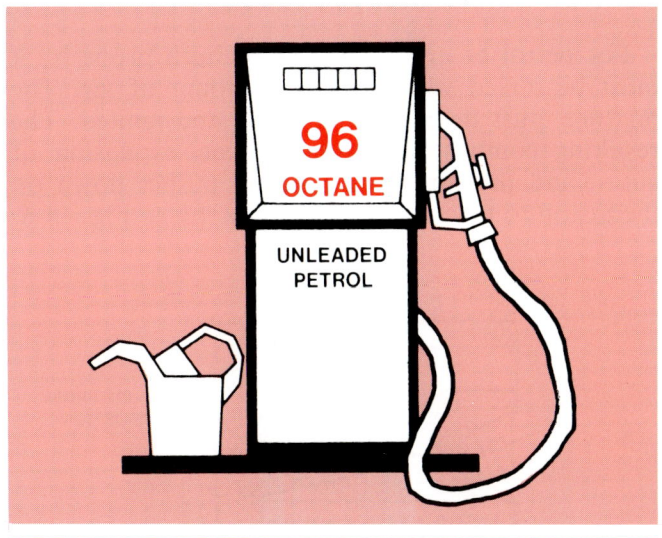

PETROL GRADES	
HIGH OCTANE HIGH ANTIKNOCK	LOW OCTANE LOW ANTIKNOCK
96 OCTANE	91 OCTANE

Fig. 18-4. Octane rating indicates antiknock value of petrol.

Octane numbers give the antiknock value of petrol. A high octane number (96, for example) will resist ping better than petrol with a low octane number (92, for example). See Fig. 18-4.

Here in Australia unleaded petrol is now available in two different grades or octane ratings. One popular oil company for example call their two unleaded grades Ultra High which is 96 octane and Ultra which is 92-93 octane.

Always use an octane rating as high or higher than the car maker recommends.

Fig. 18-5 summarises several factors that control engine octane requirements.

OCTANE REQUIREMENT FACTORS	
Octane number requirement tends to go UP when: 1. Ignition timing is advanced. 2. Air density rises due to supercharging, a larger throttle opening, or higher barometric pressures. 3. Humidity or moisture content of air decreases. 4. Inlet air temperature goes up. 5. Lean fuel-air ratios. 6. Compression ratio is increased. 7. Coolant temperature is raised. 8. Antifreeze (glycol) engine coolant is used. 9. Combustion chamber design provides little or no quench area. 10. Vehicle weight is increased. 11. Engine loading is increased such as climbing a grade, pulling a trailer or increasing wind resistance with a car-top carrier.	Octane number requirement tends to go DOWN when: 1. Car is operated at higher altitudes (lower barometric pressure). 2. Fuel-air ratio is richer or leaner than that producing maximum knock. 3. Spark plug location in combustion chamber provides shortest path of flame travel. 4. Combustion chamber design gives maximum turbulence of fuel-air charge. 5. Compression ratio is lowered. 6. Exhaust gas recycle system operates at part-throttle. 7. Ignition timing retard devices are used. 8. Humidity of the air increases. 9. Ignition timing is retarded. 10. Inlet air temperature is decreased. 11. Reduced engine loads are employed.

Fig. 18-5. Note various factors controlling octane requirements.

PETROL COMBUSTION

For petrol or any other fuel to burn properly, it must be mixed with the right amount of air. The mixture must then be compressed and ignited. The resulting combustion produces heat, expansion of gases, and pressure. The pressure pushes down on the piston to turn the crankshaft. Refer to Fig. 18-6.

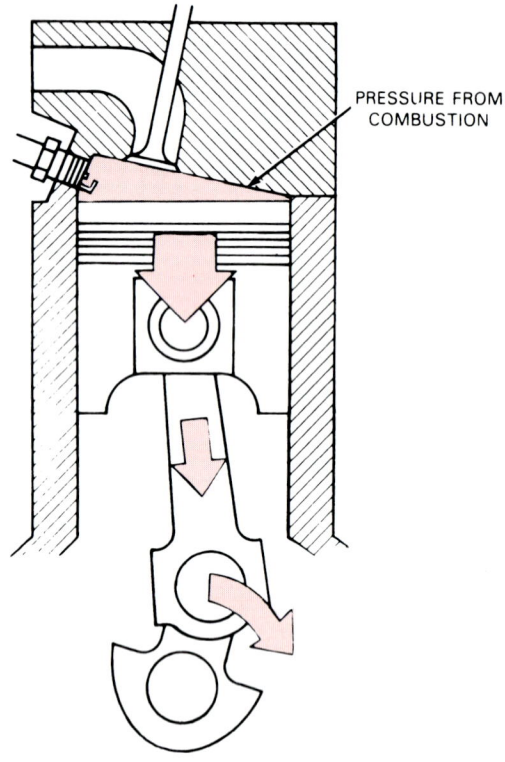

PRESSURE FROM COMBUSTION

Fig. 18-6. Combustion produces heat. Heat causes gases to expand. Expansion causes pressure. Pressure pushes piston down on power stroke. (Ford)

Normal petrol combustion

Normal petrol combustion occurs when the spark plug ignites the fuel and burning progresses smoothly through the fuel mixture. Maximum cylinder pressure should be produced a few degrees of crankshaft rotation after piston TDC on the power stroke.

Fig. 18-7 illustrates normal petrol combustion. In A, a spark at the spark plug starts the fuel burning. A small ball of flame forms around the tip of the spark plug. The piston is moving up in the cylinder, compressing the fuel mixture.

In B, the flame spreads faster and moves about halfway through the mixture. Generally, the flame is moving evenly through the fuel mixture. The piston is nearing TDC, causing increased pressure.

In C, the piston reaches TDC. The flame picks up more speed.

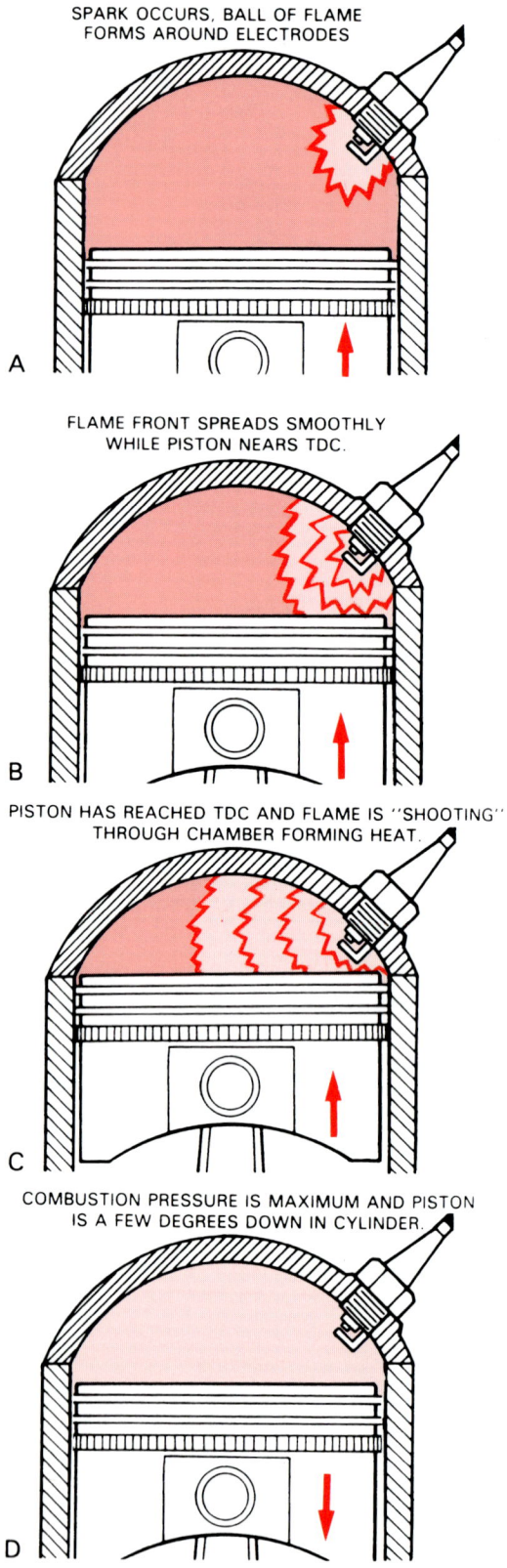

SPARK OCCURS, BALL OF FLAME FORMS AROUND ELECTRODES

A

FLAME FRONT SPREADS SMOOTHLY WHILE PISTON NEARS TDC.

B

PISTON HAS REACHED TDC AND FLAME IS "SHOOTING" THROUGH CHAMBER FORMING HEAT.

C

COMBUSTION PRESSURE IS MAXIMUM AND PISTON IS A FEW DEGREES DOWN IN CYLINDER.

D

Fig. 18-7. Study action of normal combustion. Single flame moves smoothly through air-fuel mixture.

In D, the flame shoots out to consume the rest of the fuel in the chamber. Combustion is complete with the piston only a short distance down in the cylinder.

Normal combustion only takes about 3/1000 of a second. This is much slower than an explosion. Dynamite explodes in about 1/50,000 of a second.

Under some undesirable conditions, however, petrol can be made to burn too quickly, making part of combustion like an explosion. This is detailed later.

AIR-FUEL MIXTURE RATIO

For proper combustion and engine performance, the right amounts of air and fuel must be mixed. If too much fuel or too much air is used, engine power, fuel economy, and efficiency will suffer.

Stoichiometric Fuel Mixture

A stoichiometric fuel mixture is a chemically correct or perfect air-fuel mixture or ratio. For petrol, it is a mixture ratio of about 14.9:1 (14.9 parts air to 1 part fuel by weight). Under constant engine conditions, this ratio can help assure that all of the fuel is burned during combustion.

As you will learn in later chapters, the conditions in an engine are not always ideal. The fuel system must change the air-fuel ratio with changes in engine operating conditions.

Lean air-fuel mixture

A lean air-fuel mixture contains a large amount of air. Look at Fig. 18-8. For petrol 20:1, for example, would be a very lean mixture.

A slightly lean mixture is desirable for good fuel economy and low exhaust emissions. Extra air in the cylinder assures that all of the fuel is burned. Too lean of a mixture, however, can cause poor engine performance (lack of power, missing, and even engine damage).

Rich air-fuel mixture

A rich air-fuel mixture is the opposite of a lean mixture; a little more fuel is mixed with the air. For petrol 8:1 (8 parts air to one part fuel) would be a very rich fuel mixture. Refer to Fig. 18-8.

A slightly rich mixture tends to increase engine power. However, it also increases fuel consumption and exhaust emissions. An over-rich mixture will reduce engine power, foul spark plugs, and cause incomplete burning (black smoke at engine exhaust).

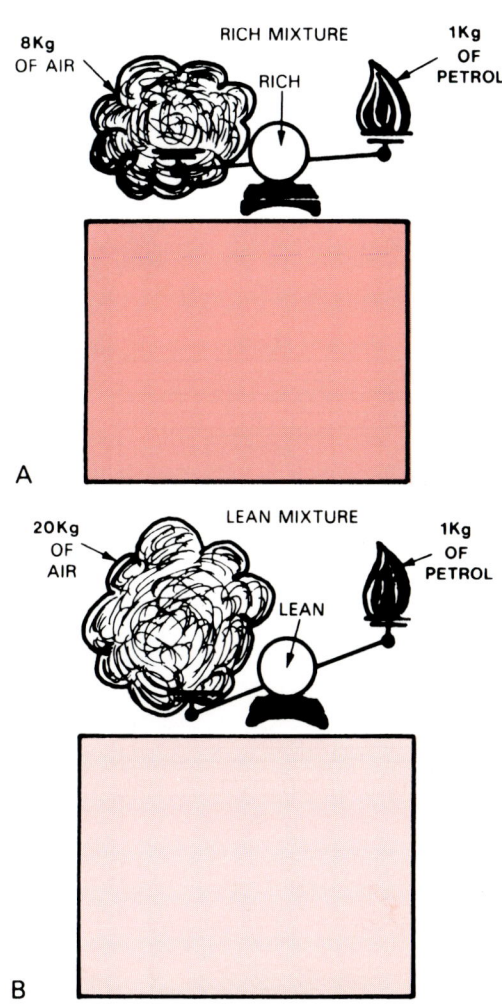

Fig. 18-8. A — Rich fuel mixture has more fuel mixed into air. B — Lean fuel mixture has less fuel mixed with air.

ABNORMAL COMBUSTION

Abnormal combustion occurs when the flame does NOT spread evenly and smoothly through the combustion chamber. The lean air-fuel mixtures, high operating temperatures, and low octane, unleaded fuels of today make abnormal combustion a major problem.

Detonation

Detonation results when part of the unburned fuel mixture explodes violently. This is the most severe and engine-damaging type of abnormal combustion.

Engine knock is a symptom of detonation. The pressure rises so quickly that parts of the engine vibrate. Detonation sounds like a hammer hitting the side of the engine.

Fig. 18-9 shows what happens during detonation. Study the four phases.

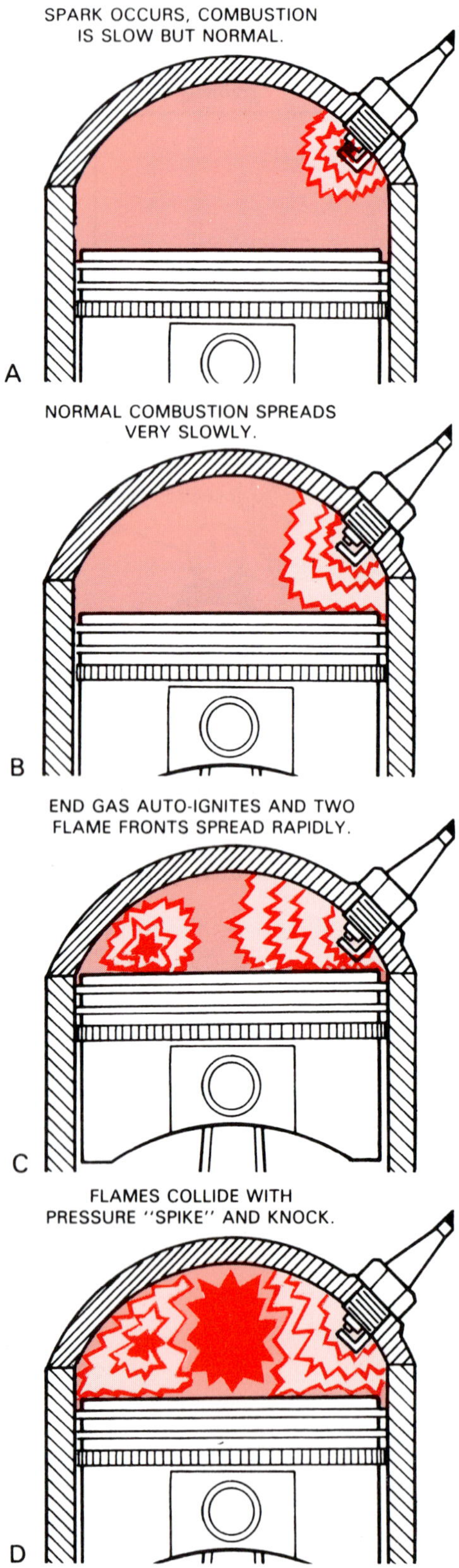

As you can see, the end gas or residual gas (unburned portion of mixture) is heated and pressurised for an extended period. Normal combustion is too slow because of a fuel mixture problem, lack of turbulence, or fuel distribution problem. This causes the end gas to explode with a "bang" (knock).

Detonation can greatly increase the pressure and heat in the engine combustion chamber. It can crack cylinder heads, blow head gaskets, burn pistons, and shatter spark plugs. See Figs. 18-10 and 18-11.

SPARK OCCURS, COMBUSTION
IS SLOW BUT NORMAL.

A

NORMAL COMBUSTION SPREADS
VERY SLOWLY.

B

END GAS AUTO-IGNITES AND TWO
FLAME FRONTS SPREAD RAPIDLY.

C

FLAMES COLLIDE WITH
PRESSURE "SPIKE" AND KNOCK.

D

Fig. 18-9. Detonation is caused by normal combustion being too slow. End gas or unburned fuel mixture ignites and two flame fronts collide with a loud knock.

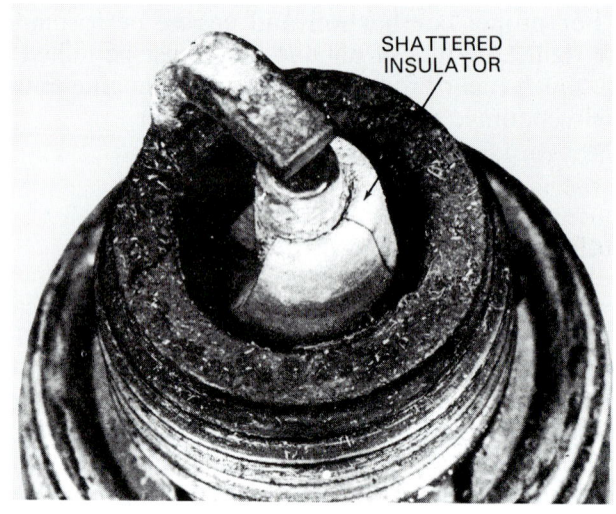

Fig. 18-10. Detonation has shattered insulator on this spark plug. (Champion Spark Plugs)

Fig. 18-11. Detonation has blown a hole in head of piston. (Champion Spark Plugs)

Preignition

Preignition results when an overheated surface in the combustion chamber ignites the fuel mixture. Termed surface ignition, a "hot spot" (overheated bit of carbon, sharp edge, hot exhaust valve) causes the mixture to burn prematurely.

A ping or a mild knock is a light tapping noise that can be heard during preignition. It is NOT as loud nor as harmful as detonation knock.

Study Fig. 18-12. Preignition is similar to detonation, but the actions are reversed. Detonation begins AFTER the start of normal combustion. Preignition begins BEFORE the start of normal combustion.

Preignition, ping, or mild knock is very common to the modern car. Some car makers say that some preignition is normal, especially when accelerating under a load.

NOTE! Prolonged preignition can produce harmful detonation. If an engine pings or knocks excessively, serious engine damage can result. Correct the problem right away.

Dieseling

Dieseling, also called after-running or run-on, is a problem where the engine keeps running after the key is turned OFF. A knocking, coughing, or fluttering noise may be heard as the fuel ignites and the crankshaft spins.

When dieseling, the petrol engine ignites the fuel from heat and pressure, somewhat like a diesel engine. With the ignition key off, the engine runs without voltage to the spark plugs.

The most common causes of dieseling are a high idle speed, carbon deposits in combustion chambers, low octane fuel, overheated engine, or spark plugs with too high of a heat range.

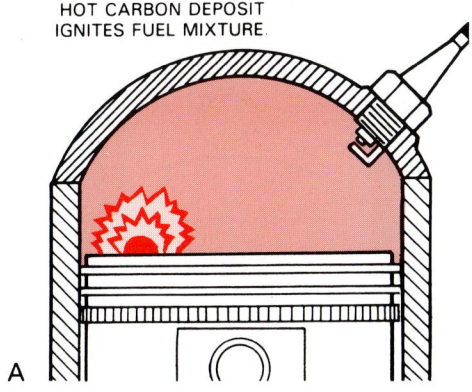

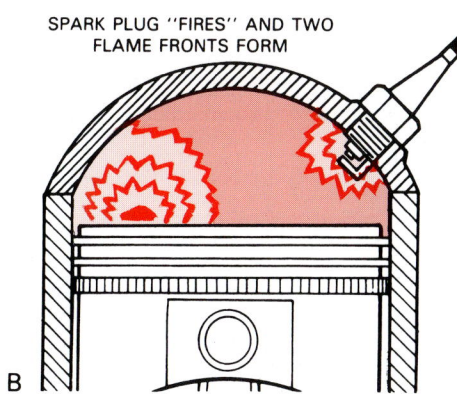

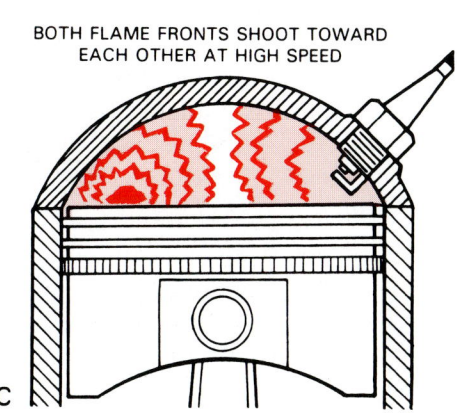

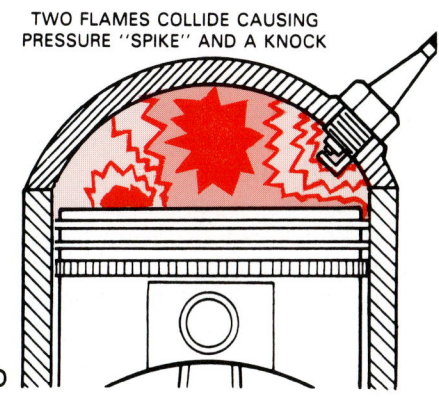

Fig. 18-12. Preignition is caused by early, abnormal ignition of fuel mixture. Abnormal and normal flames collide, producing, a pinging noise.

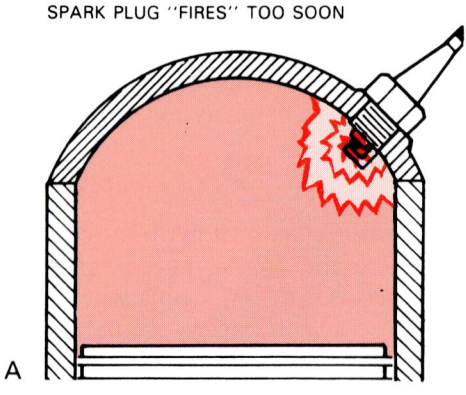

SPARK PLUG "FIRES" TOO SOON

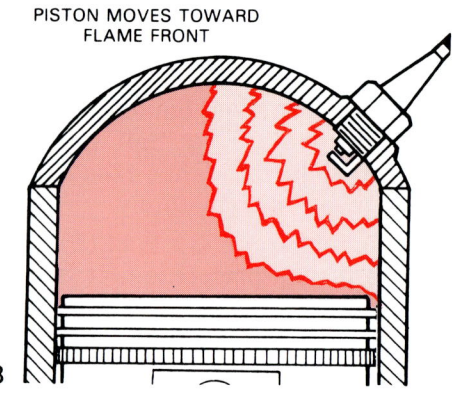

PISTON MOVES TOWARD
FLAME FRONT

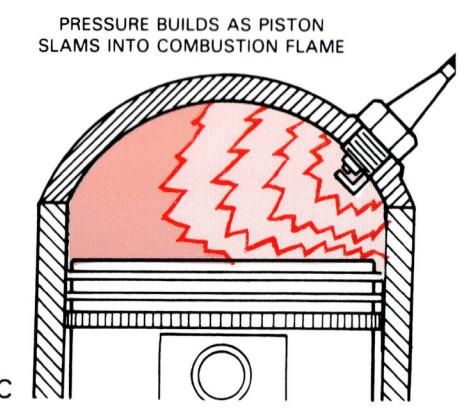

PRESSURE BUILDS AS PISTON
SLAMS INTO COMBUSTION FLAME

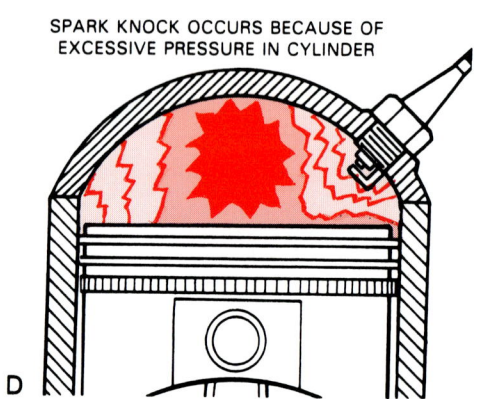

SPARK KNOCK OCCURS BECAUSE OF
EXCESSIVE PRESSURE IN CYLINDER

*Fig. 18-13. Spark knock is generally understood to mean ping
or knock caused by an ignition timing problem.*

Spark knock

Spark knock is another engine combustion problem caused by the spark plug firing too soon in relation to the position of the piston. The spark timing is advanced too far and is causing combustion to slam into the upward moving piston. This causes maximum cylinder pressure to form before TDC, not after TDC as it should.

Fig. 18-13 shows what happens during spark knock. Spark knock can also lead to preignition and more damaging detonation.

Spark knock and preignition produce about the same symptoms — pinging under load. To find its cause, first check the ignition timing. If timing is correct, check other possible causes.

DIESEL FUEL

Diesel fuel which is also known as distillate or diesel oil, is the second most popular type of automotive fuels. Diesels are becoming more common because of their high fuel economy.

A litre of diesel fuel contains more heat energy than a litre of petrol. It is a thicker fraction (part) of crude oil. Diesel fuel can produce more cylinder pressure and vehicle movement than an equal amount of petrol.

Since it is thicker and has different burning characteristics, a high pressure injection system must be used to spray the fuel directly into the combustion chambers. A carburettor or low pressure injection system would not meter the thicker diesel fuel properly. Look at Fig. 18-14.

Diesel fuel will NOT vapourise (change from a liquid to a gas) as easily as petrol. If diesel fuel entered the intake manifold of an engine, it would collect on the inner walls of the manifold. This would upset engine operation.

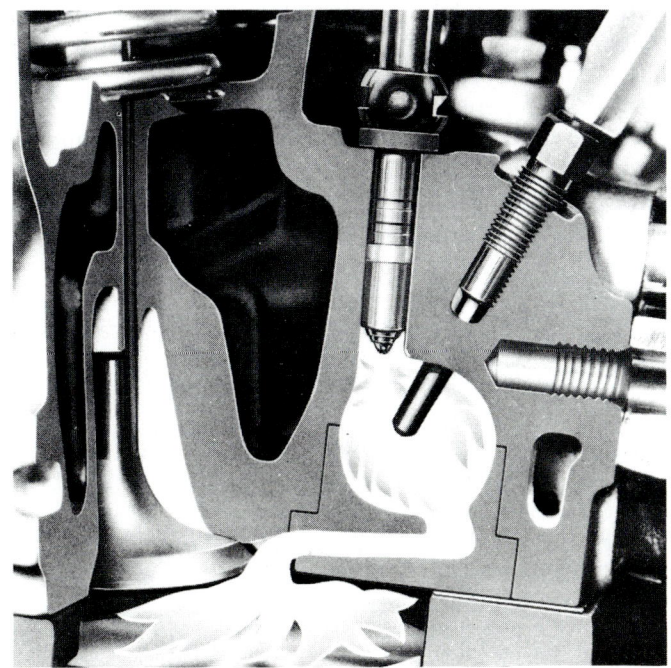

Fig. 18-14. Diesel engine is a compression ignition engine. High compression pressure heats air in cylinder. When fuel is sprayed into hot air, combustion begins.

Combustion requires fuel to be in a vapour state. Diesel engines inject the diesel oil directly into the combustion chamber. The compressed, hot air vapourises and burns the fuel.

WARNING! Diesel fuel or diesel oil should NOT be confused with FUEL OIL or HOME HEATING OIL. Diesel fuel has less impurities in it than fuel oil. Fuel oil should never be used in a diesel engine or damage will result.

Diesel fuel cloud point

One of the substances in diesel fuel is paraffin (wax). At very cold temperatures, this wax can separate from the other parts of the fuel. When this happens, the fuel will turn cloudy or milky.

Cloud point is the temperature at which wax separates out of the fuel. At cloud point, the wax can clog fuel filters and prevent diesel engine operation.

Diesel fuel water contamination

Water contamination is a common problem with diesel engines. Water, when mixed with diesel fuel, can clog filters and corrode components. The parts in diesel injection pumps and nozzles are very precise. They can be easily damaged by water.

Many later model diesel injection systems have water separators to prevent water damage. These will be covered in later chapters.

Diesel fuel cetane rating

A cetane rating indicates the cold starting ability of diesel fuel. A numbering system is used. The higher the cetane number, the easier the engine will start and run in cold weather.

Most car makers recommend a cetane rating of about 45. In Australia the refining oil companies produce only one grade of diesel fuel for automotive use, which usually has a cetane rating of between 47 and 50.

A cetane number, in some ways, is the opposite of a petrol octane number. This is shown in Fig. 18-15. A high cetane number means that the fuel will ignite easily from heat and pressure. In a diesel, fuel should ignite and burn as soon as it touches the hot air in the combustion chamber.

CETANE AND OCTANE COMPARISON

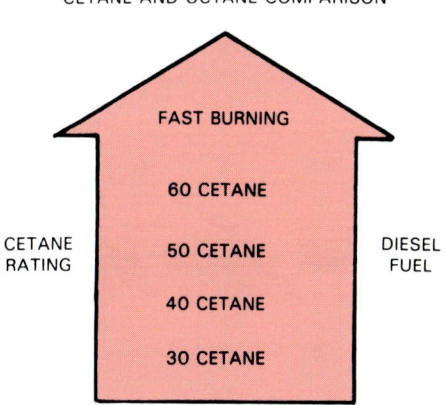

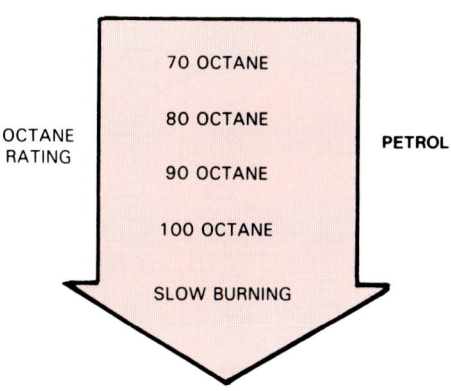

Fig. 18-15. Diesel oil cetane rating is opposite of petrol octane rating.

DIESEL COMBUSTION

A diesel engine is a compression ignition engine. It compresses air until the air is hot enough to ignite the fuel. A spark plug would NOT ignite diesel fuel properly. If petrol were used in a diesel engine it would detonate on the compression stroke and not produce useful energy.

Since diesel fuel is thick and hard to ignite, a high pressure, mechanical pump and nozzles force the fuel into the engine combustion chambers. An extremely high compression ratio heats the air in the cylinder. Then, when fuel is sprayed into the hot air, it begins to burn.

Fig. 18-16 shows the phases of normal diesel combustion. Study this illustration closely.

In A, the piston moves up to compress and heat the air in the cylinder. Note that this is different than a petrol engine that compresses both fuel and air.

In B, diesel fuel is injected directly into the combustion chamber. The hot air makes the fuel begin to burn and expand.

In C, more fuel is sprayed into the chamber. More pressure is developed and the piston begins to move down in the cylinder.

In D, the rest of the fuel is injected into the chamber. Pressure continues to form, pushing the piston down on the power stroke.

Note that fuel was injected into the engine for several degrees of crankshaft rotation. This caused a smooth, steady build up of pressure for quiet diesel engine operation.

Diesel combustion knock

Diesel engines, when compared to petrol engines, knock almost all of the time. You have probably heard a diesel engine running. It clatters or rattles as the diesel fuel ignites in the combustion chambers.

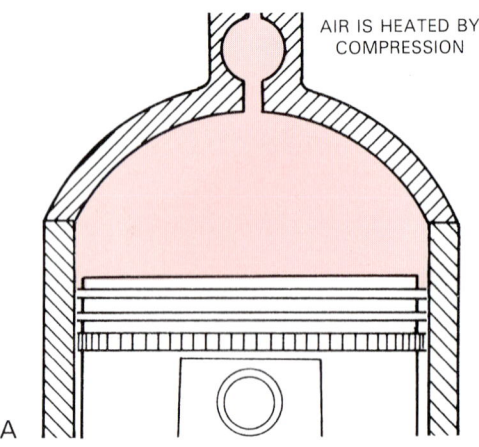
AIR IS HEATED BY COMPRESSION

A

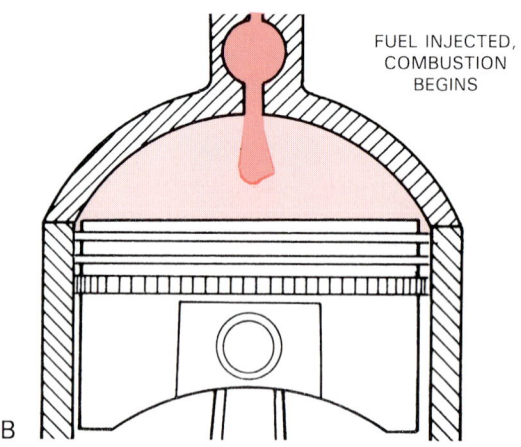
FUEL INJECTED, COMBUSTION BEGINS

B

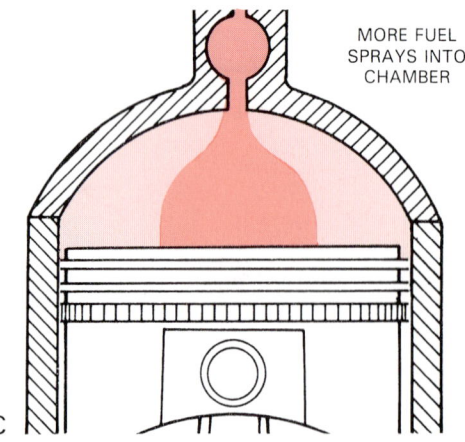
MORE FUEL SPRAYS INTO CHAMBER

C

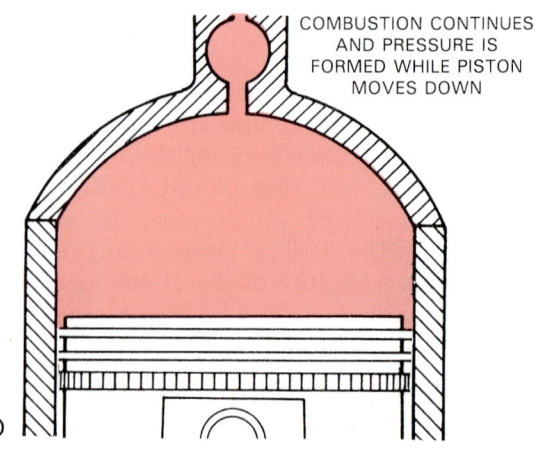
COMBUSTION CONTINUES AND PRESSURE IS FORMED WHILE PISTON MOVES DOWN

D

Fig. 18-16. Study action during normal diesel combustion.

Diesel knock occurs when too much fuel ignites at one time, producing a louder then normal knocking noise. Excessive diesel knock can reduce engine power, fuel economy, and engine life.

Ignition lag is the time it takes diesel fuel to heat up, vapourise, and begin to burn. It is the time lapse between initial fuel injection and actual ignition (burning).

Ignition lag is a major controlling factor of diesel knock. If lag time is too long, a large amount of fuel can ignite, producing a louder than normal knock. A high cetane fuel, with a short lag time, reduces the chances of diesel knock.

Fig. 18-17 shows the basic phases of diesel knock. Study this illustration.

As you can see, diesel knock is caused by too much fuel burning at one time. This can be due to a cold engine, low cetane fuel, improper fuel spray pattern, or incorrect injection timing. This will be discussed in a later chapter.

ALTERNATE FUELS

Alternate fuels are generally understood to include any fuel other than petrol and diesel fuel. LP-gas, alcohol, and hydrogen are examples of alternate fuels.

LP-gas

LP-gas or liquefied petroleum gas (LPG) is sometimes used as a fuel for cars and trucks. It is one of the lightest fractions of crude oil. Refer again to Fig. 18-3.

Chemically, LP-gas is similar to petrol. However, at room temperature and pressure, LP-gas is a vapour, NOT a liquid.

LP-gas is commonly used in taxi cabs and industrial equipment, such as fork lifts. Being a gas, LP-gas burns cleanly, producing few exhaust emissions.

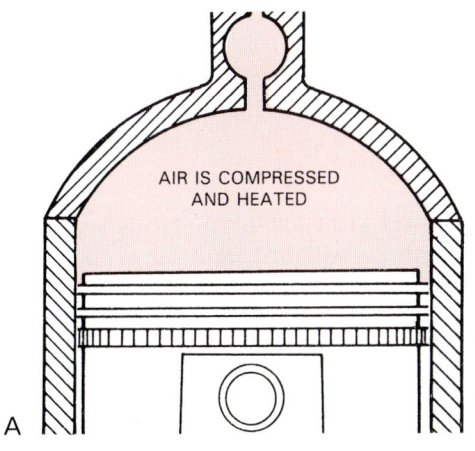
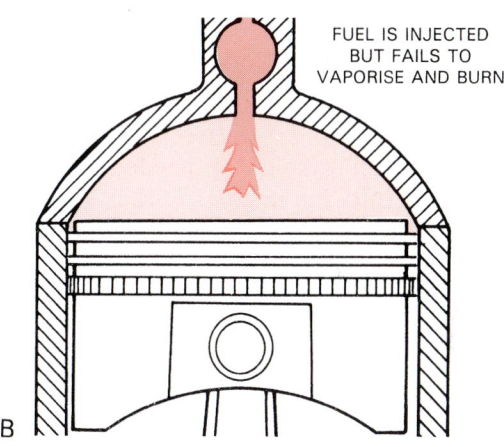

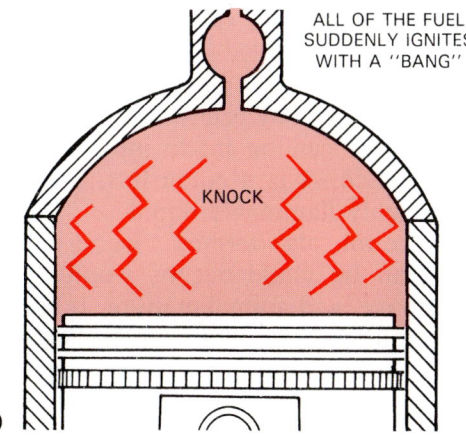

Fig. 18-17. Basically, diesel knock is caused by too much fuel igniting at one time. Fuel does not ignite quick enough when injection begins.

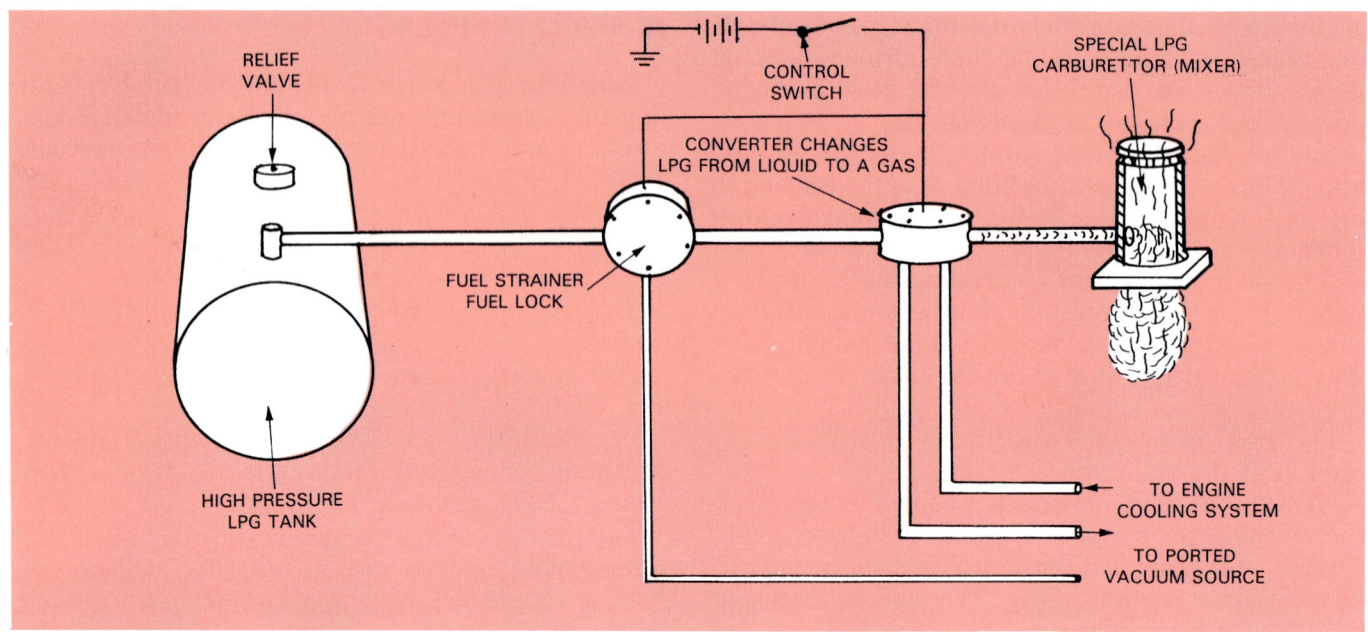

Fig. 18-18. LP-gas fuel system uses a high pressure storage tank. Fuel strainer-fuel lock cleans and prevents leakage when engine is not running. Converter uses engine coolant heat to change liquid LPG into a gas. Special carburettor meters LP-gas into engine.

A special fuel system is needed to meter the gaseous LPG into the engine. One is shown in Fig. 18-18. Note the names and construction of the basic parts.

Alcohol

Alcohol is an excellent alternate fuel for auto engines. It is especially desirable because it is made from sources other than crude oil. The two types of alcohol that are suitable for use in motor vehicles are ethyl alcohol and methyl alcohol.

Ethyl alcohol, also called grain alcohol or ethanol, is made from farm crops. Grain, wheat, sugar cane, potatoes, fruit, oats, soy beans, and other crops rich in carbohydrates can be made into ethyl alcohol. This type of alcohol is a colourless, harsh tasting, toxic, and highly flammable liquid.

Methyl alcohol, also termed "wood alcohol" or methanol, can be made out of wood chips, petroleum, garbage, and animal manure. It has a strong odour, is colourless, poisonous, and very flammable.

Alcohol is a clean burning fuel for motor vehicles. It is not commonly used because it is expensive to use and produce. Also the car's fuel system requires modification before it can burn straight alcohol. Almost TWICE as much alcohol must be burned, compared to petrol. This reduces fuel economy by 50 percent.

Synthetic fuels

Synthetic fuels are fuels made from coal, shale oil (rock filled with petroleum), and tar sand (sand filled with petroleum). See Fig. 18-19.

Synthetic fuels are synthesised (changed) from a solid hydrocarbon (petroleum) state into a liquid or a gaseous state. Synthetic fuels are now being experimented with as a means of supplementing crude oil. As crude oil-based fuels become more expensive, synthetic fuels will become more practical.

Hydrogen

Hydrogen is a highly flammable gas that is a promising alternate fuel of the future. Hydrogen is one of the most abundant elements on our planet. It can be produced through the electrolysis of water (sending electric current through salt water to produce hydrogen gas). This is illustrated in Fig. 18-20.

Hydrogen is an ideal fuel. The sun is burning hydrogen. Hydrogen burns almost perfectly, leaving only water and harmless carbon dioxide as by-products.

At present, hydrogen is too expensive to make and store. However, as we use up our supply of crude oil, we may someday make hydrogen a major source of automotive fuel.

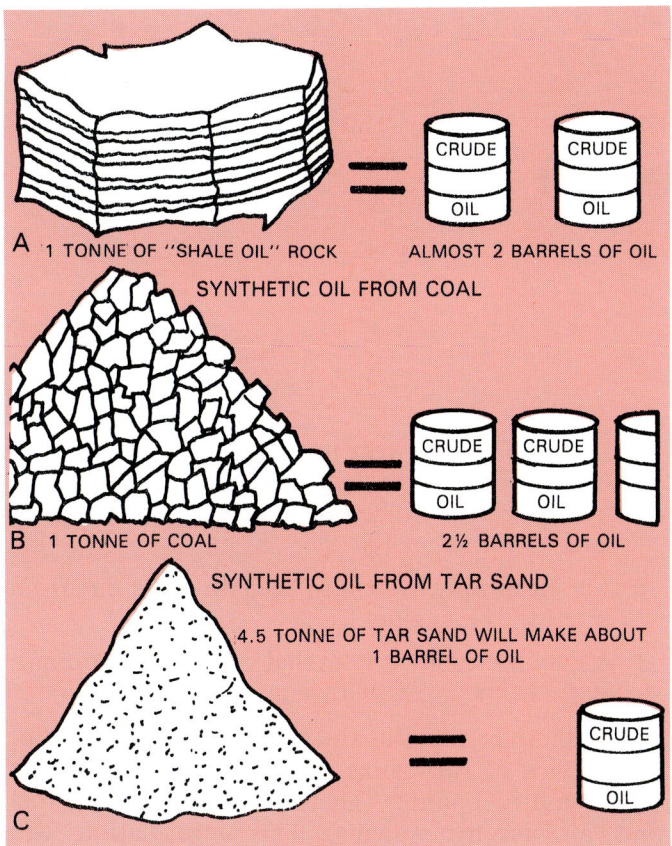

Fig. 18-19. Synthetic sources of oil. A — Shale rock can be converted into oil. B — Coal can produce about two and one-half barrels of oil per tonne. C — Four and one-half tonnes of tar sand can be changed into about one barrel of oil.

HYDROGEN PRODUCTION BY ELECTROLYSIS

HYDROGEN GAS VAPOURS →

SOURCE OF ELECTRICITY

WATER

DC CURRENT

Fig. 18-20. Hydrogen gas can be made through electrolysis of water. Solar cells, wind energy, or ocean thermal energy may make production feasible some day.

KNOW THESE TERMS

Petroleum, Hydrocarbon, Leaded petrol, Unleaded petrol, Octane number, Stoichiometric fuel mixture, Lean air-fuel ratio, Rich air-fuel ratio, Detonation, Preignition, Knock, Ping, Spark knock, Dieseling, Diesel fuel, Cloud point, Cetane number, Compression ignition, Ignition lag, Alternate fuel, LPG, Ethyl alcohol, Methyl alcohol, Synthetic fuels.

REVIEW QUESTIONS

1. _____, also called _____ _____, is oil taken directly out of the ground.
2. What are hydrocarbons?
3. Explain the difference between leaded and unleaded petrol.
4. The lead in leaded petrol acts as a lubricant for the engine valves and valve seats. True or False?
5. The _____ _____ of petrol is a measurement of the fuel's ability to resist knock or ping.
6. If a car maker recommends an octane number of 96, is it ok to use 92. True or False?
7. Which of the following is NOT needed for proper combustion?
   a. air.
   b. compression.
   c. condensation.
   d. ignition.
8. Describe normal petrol combustion.
9. Define the term "stoichiometric fuel mixture".
10. A _____ air-fuel mixture ratio contains a large amount of air.
11. A _____ air-fuel mixture ratio contains a large amount of fuel.
12. What are the results of lean and rich air-fuel mixtures?
13. Explain detonation, preignition, spark knock, and dieseling in a petrol engine.
14. How does diesel fuel differ from petrol?
15. Explain how ignition lag affects diesel combustion.
16. _____ _____ is made from farm crops and _____. _____ can be made out of wood chips, petroleum, garbage, and animal manure.
17. What is LPG?
18. LPG can be used without fuel system modifications. True or False?
19. Which of the following does NOT pertain to hydrogen as an alternate fuel?
   a. Made from most abundant element.
   b. Produced through electrolysis.
   c. Burns without toxic emissions.
   d. Economical or inexpensive to make.
20. Hydrogen burns almost perfectly, leaving only _____ and harmless _____ as byproducts.

Chapter 19

FUEL SUPPLY SYSTEM

After studying this chapter, you will be able to:
- List the components of a fuel supply system.
- Describe the operation of mechanical and electric fuel pumps.
- Describe the construction and action of air filters.
- Explain the tests used to diagnose problems with fuel pumps, fuel filters, and fuel lines.
- Repair a fuel line or replace a fuel hose.
- Locate and replace fuel filters in both petrol and diesel fuel systems.

A fuel system provides a combustible air-fuel mixture to power the engine.

There are several different types of fuel systems. Today's cars commonly use carburettors, petrol injection systems, and diesel injection systems.

A modern fuel system has three subsystems:
1. A Fuel Supply System which provides filtered fuel to carburettor, injection pump, and injectors.
2. An Air Supply System to keep dust and dirt from entering the engine.
3. A Fuel Metering System that controls the amount of fuel that mixes with air.

This chapter explains the construction, operation, and service of fuel tanks, fuel lines, fuel filters, air filters, and fuel pumps. These parts make up the fuel and air supply systems. This information will prepare you for later chapters on fuel metering systems. Study this chapter carefully!

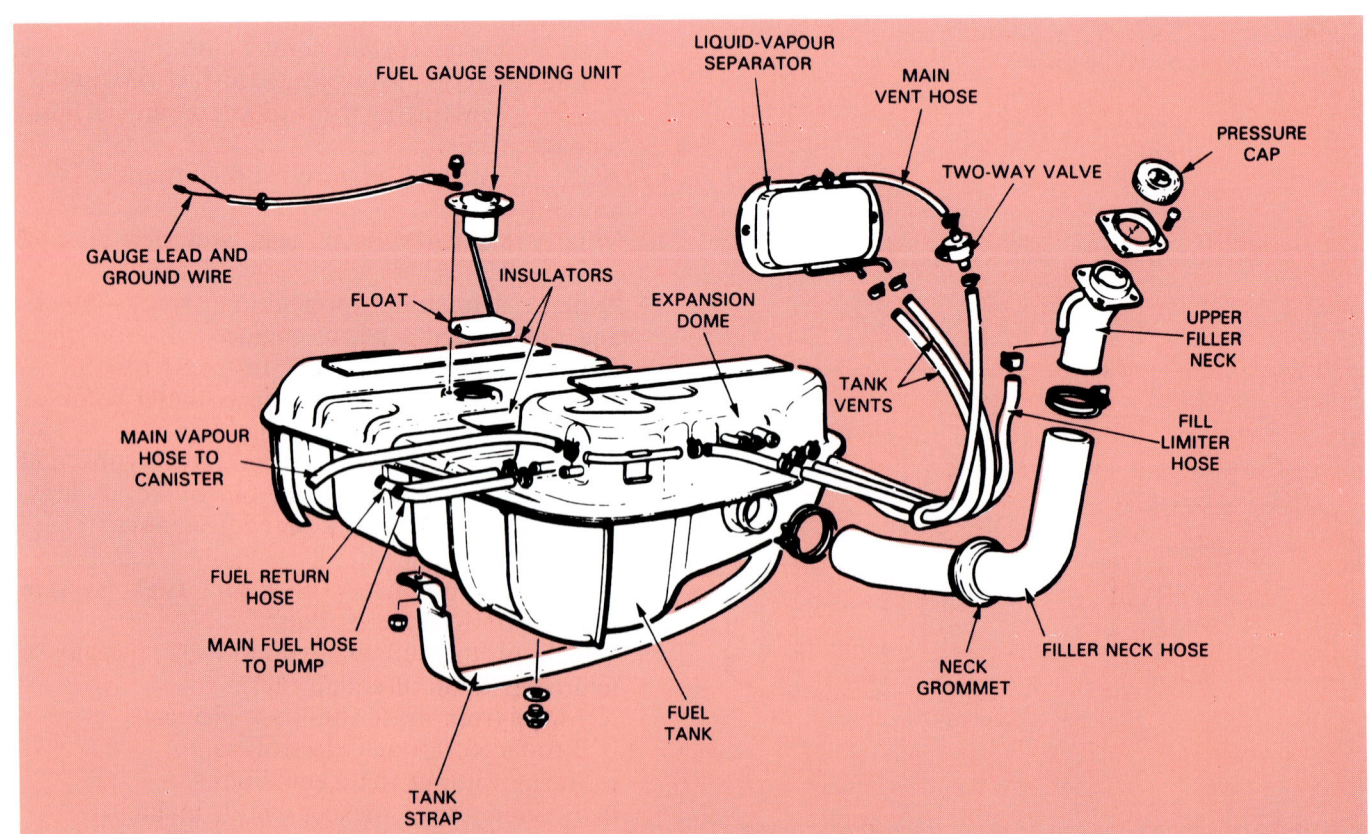

Fig. 19-1. Note basic parts of fuel tank assembly.

274

FUEL SUPPLY SYSTEM

A fuel supply system draws fuel from the fuel tank and forces it into the fuel metering device (carburettor, petrol injectors, or diesel injection pump). One type uses a mechanical (engine driven) fuel pump. Another type uses an electric fuel pump.

The basic parts of a fuel supply system include:
1. FUEL TANK (which stores petrol, diesel oil, or LP-gas).
2. FUEL LINES (which carry fuel between tank, pump and other parts).
3. FUEL PUMP (which draws fuel from tank and forces it to engine or fuel metering device).
4. FUEL FILTERS (which removes contaminants from the fuel).

FUEL TANKS

An automotive fuel tank must safely hold an adequate supply of fuel for prolonged engine operation. It is normally mounted in the rear of the car, under the luggage compartment or the rear seat. See Fig. 19-1.

The size of a fuel tank determines, in part, a car's driving range (greatest distance car can be driven without stopping for fuel).

Fuel tank capacity is the rating of how much fuel a fuel tank can hold when full. An average fuel tank capacity is around 40 to 80 litres.

Fuel tank construction

Fuel tanks are usually made of thin sheet metal or plastic. The main body of a metal tank is made by soldering or welding two formed pieces of sheet metal together. As shown in Fig. 19-1, other parts (filler neck, baffles, vent tubes, expansion chamber) are added to form the complete fuel tank assembly. A lead-tin alloy is normally plated to the sheet metal to keep the tank from rusting.

The fuel tank filler neck is the extension on the tank for filling the tank with fuel. The filler cap fits on the end of the filler neck, Fig. 19-2. The neck extends from the tank through the body of the car. A flexible hose is normally used as part of the filler neck. It allows tank vibration without part breakage.

A filler neck restrictor is used inside the tank filler neck on cars requiring unleaded fuel. It prevents the accidental use of leaded fuel in an engine designed for unleaded petrol. The restrictor is too small to accept the larger leaded fuel type petrol station pump nozzle. Look at Fig. 19-2.

WARNING! If the neck restrictor is removed and leaded fuel is used in a car designed for unleaded fuel, the catalytic converter will be damaged. This is also a violation of state laws. NEVER remove a neck restrictor!

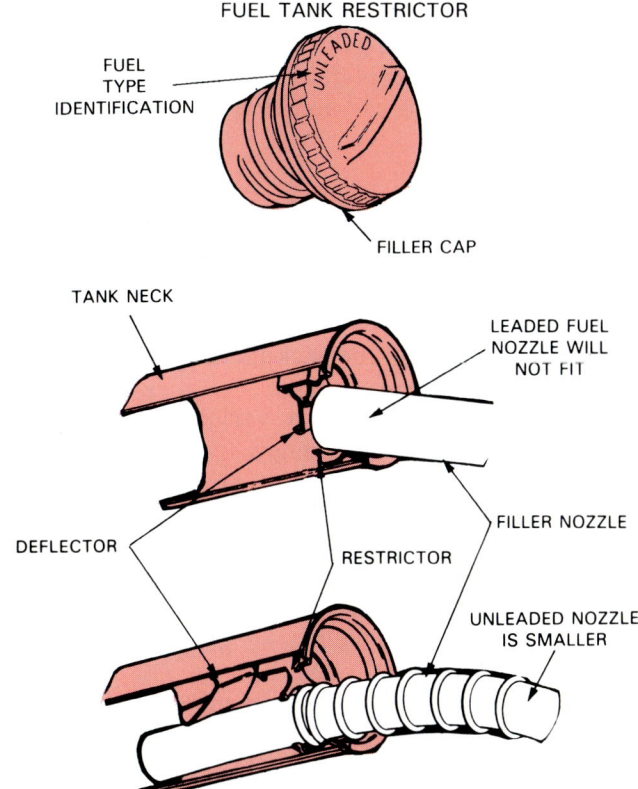

Fig. 19-2. Neck restrictor prevents leaded petrol from being used in a car designed for unleaded petrol. Never remove restrictor.

Modern fuel tank caps are sealed to prevent the escape of fuel and fuel vapours (emissions) from the tank. Fig. 19-3 shows a cutaway view of a fuel cap. This cap has pressure and vacuum valves that only open under abnormal conditions of high pressure or vacuum. Normally it is NOT vented to atmosphere.

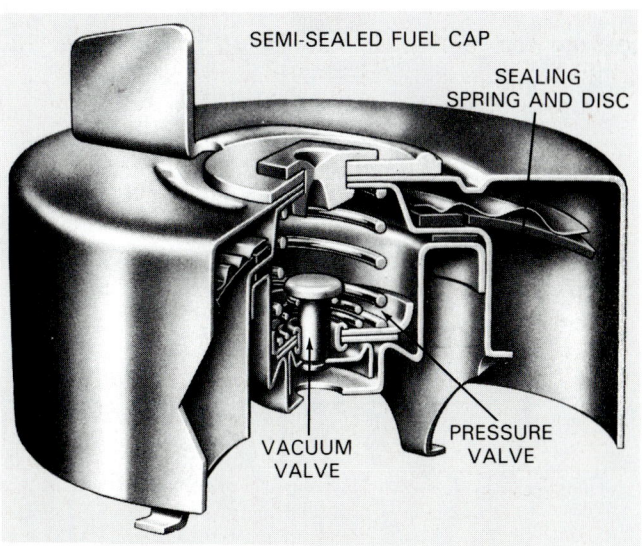

Fig. 19-3. Modern fuel tank caps prevent fuel vapours from escaping into and polluting atmosphere. Valves only release under extreme conditions.

Fuel tank baffles are placed inside the fuel tank to keep fuel from sloshing or splashing around in the tank. The baffles are metal plates that restrict fuel movement when the car accelerates, decelerates, or turns a corner.

Tank pickup-sending unit

A tank pickup-sending unit extends down into the tank to draw out fuel and to operate the fuel gauge. One is shown in Fig. 19-4. A coarse filter is usually placed on the end of the pickup tube to strain out larger pieces of foreign matter.

A tank sending unit is a variable resistor. Its resistance changes with changes in fuel level. This causes it to control the amount of current reaching the fuel gauge in the instrument panel. Fig. 19-5 shows the basic action of a fuel tank sending unit.

When the fuel level in the tank is low (float down), the tank unit has a high resistance. Only a small amount of current flows to the gauge. The gauge shows a low fuel level.

When the tank is full, the float moves up, moving the variable resistor in the tank unit. This causes the tank unit to have a low resistance. More current can then flow to the gauge. The gauge needle moves to full.

FUEL SUPPLY TUBE
LOCATING TANG
FILTER
TERMINAL
LOCK RING
RUBBER SEAL
FUEL GAUGE (TANK UNIT)
FLOAT

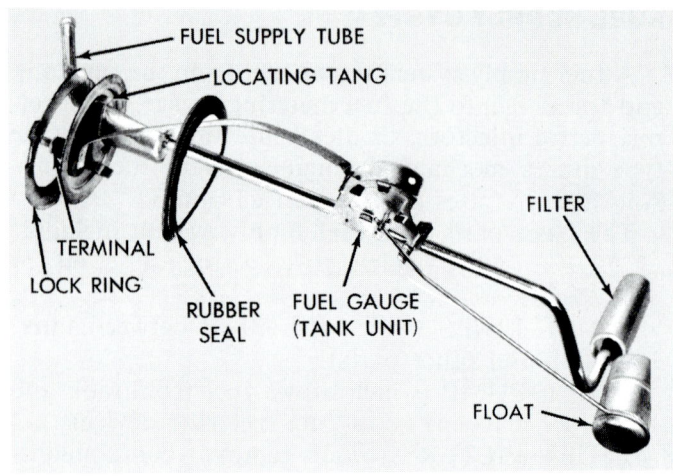

Fig. 19-4. Typical fuel tank pickup-sending unit. "Sock" Filter on end of pickup strains out debris in tank. Tank sending unit operates instrument panel fuel gauge.

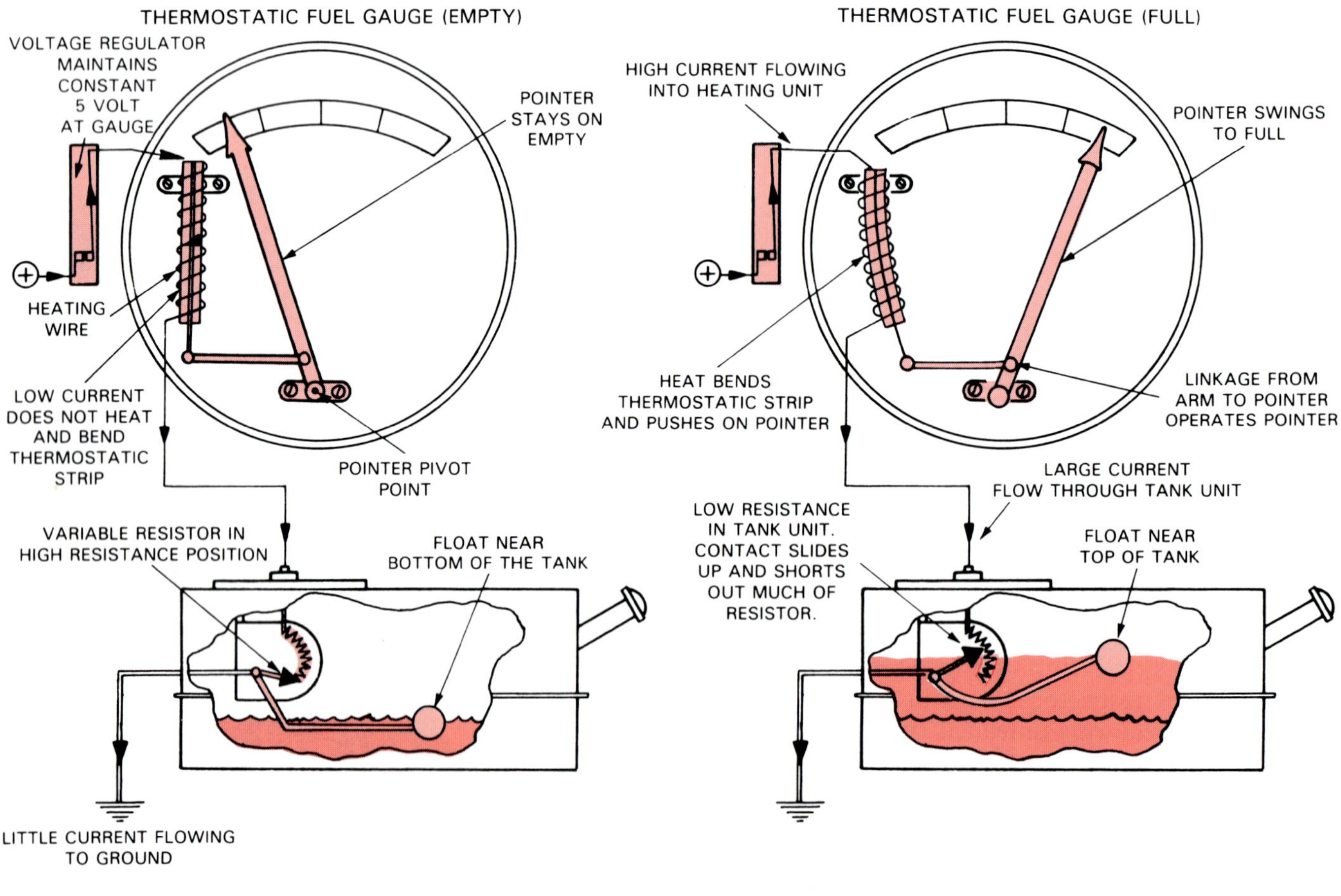

THERMOSTATIC FUEL GAUGE (EMPTY)
VOLTAGE REGULATOR MAINTAINS CONSTANT 5 VOLT AT GAUGE
POINTER STAYS ON EMPTY
HEATING WIRE
LOW CURRENT DOES NOT HEAT AND BEND THERMOSTATIC STRIP
POINTER PIVOT POINT
VARIABLE RESISTOR IN HIGH RESISTANCE POSITION
FLOAT NEAR BOTTOM OF THE TANK
LITTLE CURRENT FLOWING TO GROUND
THERMOSTATIC FUEL GAUGE (FULL)
HIGH CURRENT FLOWING INTO HEATING UNIT
POINTER SWINGS TO FULL
HEAT BENDS THERMOSTATIC STRIP AND PUSHES ON POINTER
LINKAGE FROM ARM TO POINTER OPERATES POINTER
LARGE CURRENT FLOW THROUGH TANK UNIT
LOW RESISTANCE IN TANK UNIT. CONTACT SLIDES UP AND SHORTS OUT MUCH OF RESISTOR.
FLOAT NEAR TOP OF TANK

Fig. 19-5. Fuel tank sending unit and fuel gauge operation. A — Low fuel level causes float to move down. Tank unit has high resistance. Low current flow does not heat bi-metal strip and gauge shows low. B — Full tank moves float up. This moves resistor to low resistance position. High current flow through gauge tank unit heats bi-metal strip. Strip bends and moves needle to full.

FUEL LINES AND HOSES

Fuel lines and fuel hoses carry fuel from the tank to the engine. A main fuel line allows a fuel pump to draw fuel out of the tank. The fuel is drawn through this line to the pump and then into the carburettor or metering section of the injection system.

Fig. 19-6 shows a complete set of fuel lines, including the fuel vapour lines for the evaporation control (emission control) system. Study the routing of lines. Note how they connect to the fuel system components.

Fuel lines are normally made of strong, double-wall steel tubing. For fire safety reasons, a fuel line must be able to withstand the constant and severe vibration produced by the engine and road surface.

rosion, dirt) from entering the carburettor, throttle body, injectors, injection pumps, and any other part that could be damaged by foreign matter. A fuel filter is normally located on the fuel tank pickup tube. A second fuel filter is located in the main fuel line or inside the carburettor or fuel pump. See Fig. 19-9.

Fuel filter construction

Some fuel filters use pleated paper elements to trap dirt in the fuel, Fig. 19-9. Others use a very fine metal gauze or sintered bronze (porous metal) element, Fig. 19-9. Both types are capable of stopping very small particles. Some filters are also capable of trapping water.

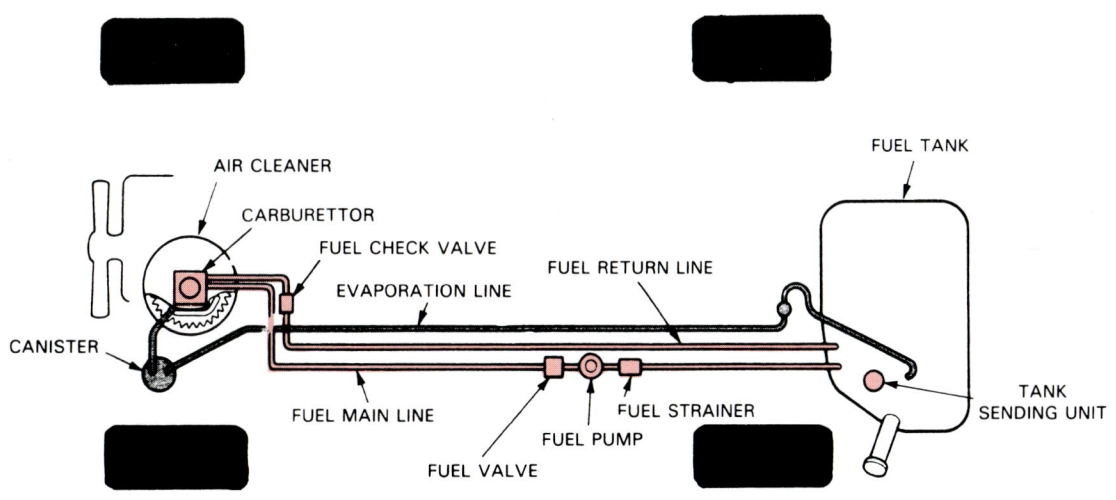

Fig. 19-6. Study fuel and emission lines. Note location of fuel pump, filters, and other devices. (Mazda)

Fuel hoses, made of synthetic rubber, are needed where severe movement occurs between parts. For example, a fuel hose is used between the main fuel line and the engine. The engine is mounted on rubber engine mountings. The soft mountings allow the engine some movement in the car frame or body. Fig. 19-7. A flexible hose can absorb this movement without breakage.

Hose clamps secure the fuel hoses to the fuel line or to metal fittings.

Fuel return system

A fuel return system helps cool the fuel and prevent vapour lock (bubbles which form in overheated fuel and stop fuel flow). Fig. 19-8 shows a fuel return system. A second, return fuel line is used to carry excess fuel back to the tank. This keeps cool fuel constantly circulating through the system.

FUEL FILTERS

Fuel filters stop contaminants (rust, water, cor-

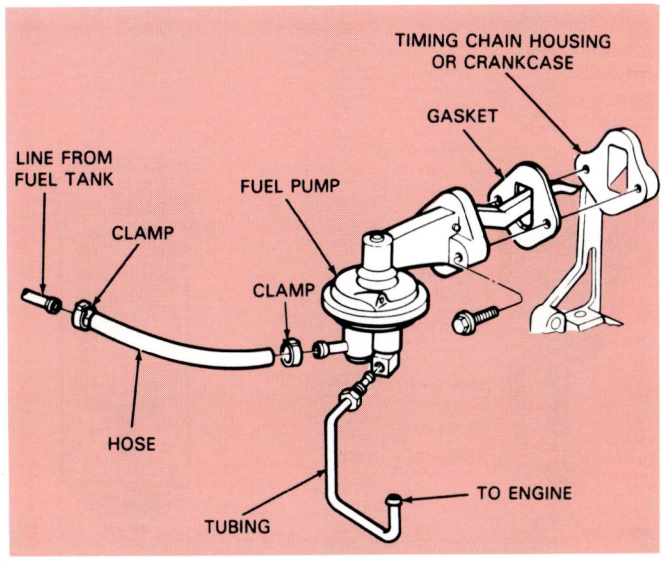

Fig. 19-7. Hose is used between pump and main fuel line to allow for engine movement on mounts. Metal tubing is used where there is no movement between two connections.

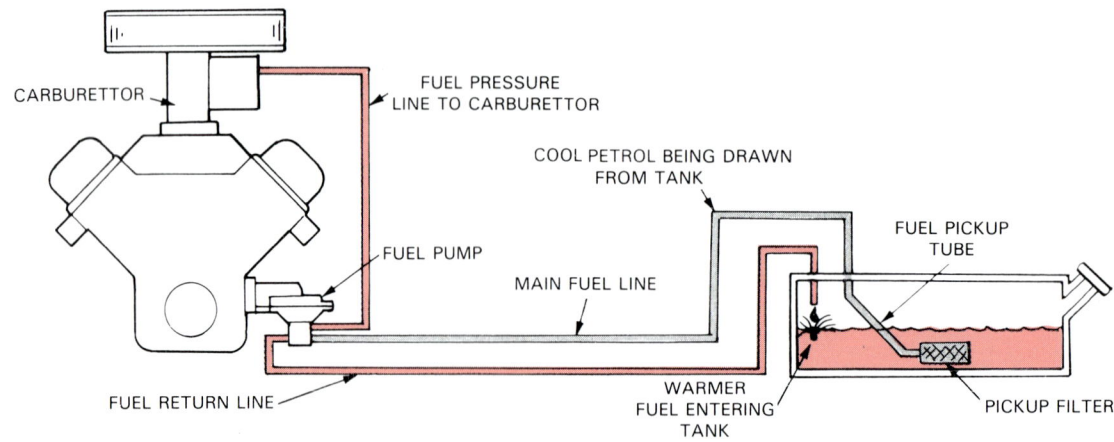

CARBURETTOR

FUEL PRESSURE
LINE TO CARBURETTOR

COOL PETROL BEING DRAWN
FROM TANK

FUEL PICKUP
TUBE

FUEL PUMP

MAIN FUEL LINE

FUEL RETURN LINE

WARMER
FUEL ENTERING
TANK

PICKUP FILTER

Fig. 19-8. Basic action of a fuel return system. Constant flow of fuel to and from tank cools fuel to help prevent vapour lock and other similar problems.

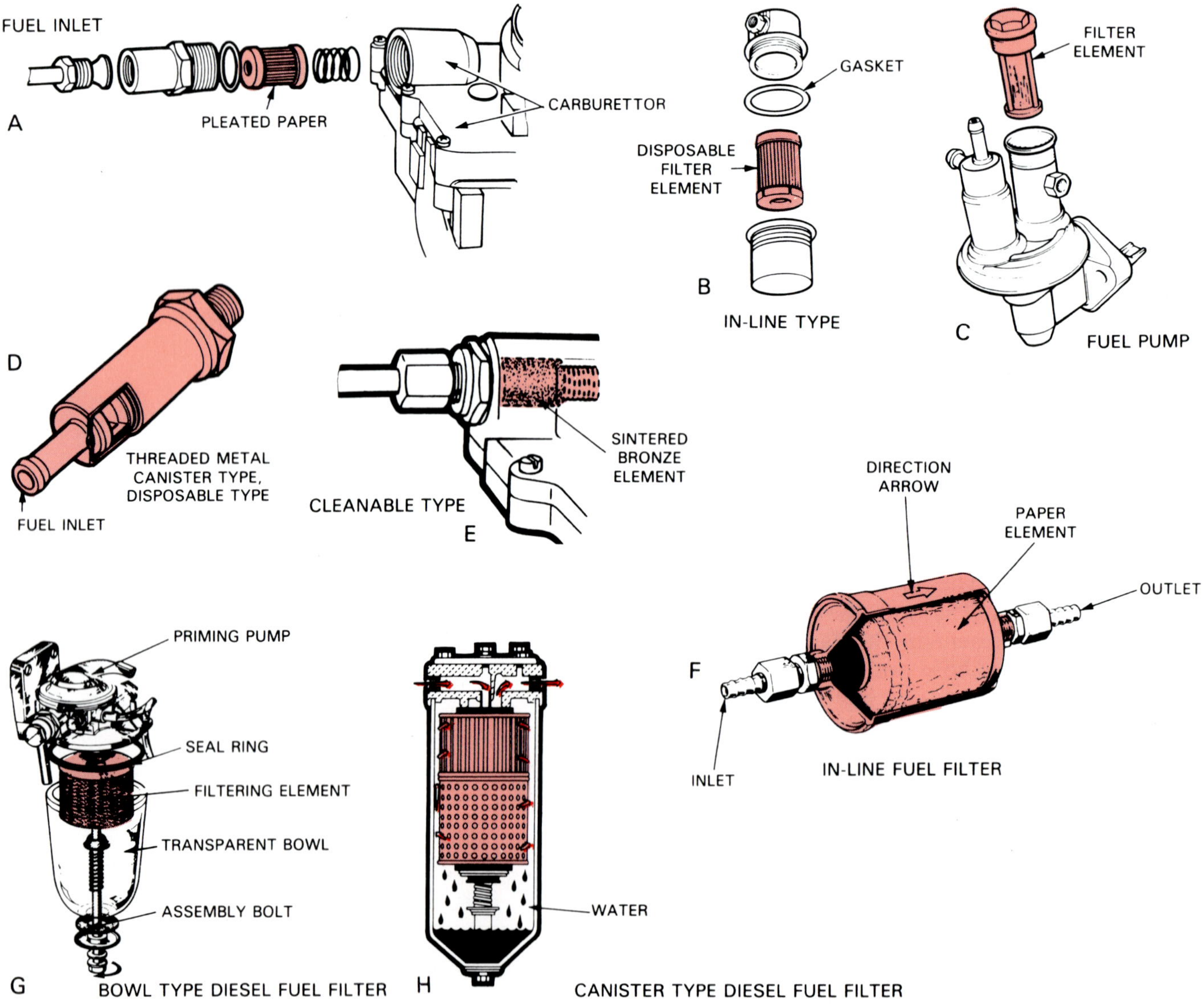

FUEL INLET

A — PLEATED PAPER — CARBURETTOR

GASKET

DISPOSABLE
FILTER
ELEMENT

B — IN-LINE TYPE

FILTER
ELEMENT

C — FUEL PUMP

D — THREADED METAL
CANISTER TYPE,
DISPOSABLE TYPE

FUEL INLET

SINTERED
BRONZE
ELEMENT

CLEANABLE TYPE

E

DIRECTION
ARROW

PAPER
ELEMENT

OUTLET

F

INLET

IN-LINE FUEL FILTER

PRIMING PUMP

SEAL RING

FILTERING ELEMENT

TRANSPARENT BOWL

ASSEMBLY BOLT

G — BOWL TYPE DIESEL FUEL FILTER

WATER

H — CANISTER TYPE DIESEL FUEL FILTER

Fig. 19-9. Variations of automotive fuel filters. A — Pleated paper fuel filter in carburettor. B — In-line canister fuel filter. C — Filter element in fuel pump. D — Threaded metal canister fuel filter. E — Sintered bronze fuel filter. F — In-line paper fuel filter. G — Bowl type fuel filter for diesel system. Bowl types are also used on some petrol fuel systems. H — Large, in-line canister fuel filter for diesel systems. (Peugeot, Fram, Saab, Ford, Chrysler)

A bowl fuel filter is sometimes used on both petrol and diesel engines. Look at Fig. 19-9. Many types use a glass bowl that allows the mechanic to see the filter element. When the bowl and element become dirty, the mechanic can remove the bowl for service.

A canister fuel filter is sometimes used on diesel engines. Refer to Fig. 19-9. The filter element is housed inside a metal container. When the filter requires replacement, a new element can be installed inside the container.

Vapour separator-filter

One type of fuel filter or strainer also serves as a vapour separator, as in Fig. 19-10. If too much engine heat transfers into the fuel, bubbles can form in the fuel. The bubbles collect at the top of the vapour separator-filter. They are then carried back to the fuel tank with excess fuel.

FUEL PUMPS

A fuel pump forces fuel out of the tank and to the engine under pressure. There are two basic types of fuel pumps: mechanical fuel pump and electric fuel pump. Both types are used on today's cars.

Mechanical fuel pumps

A mechanical fuel pump is usually powered by an eccentric (egg shaped lobe) on the engine camshaft. See Fig. 19-11 .The mechanical fuel pump bolts to the

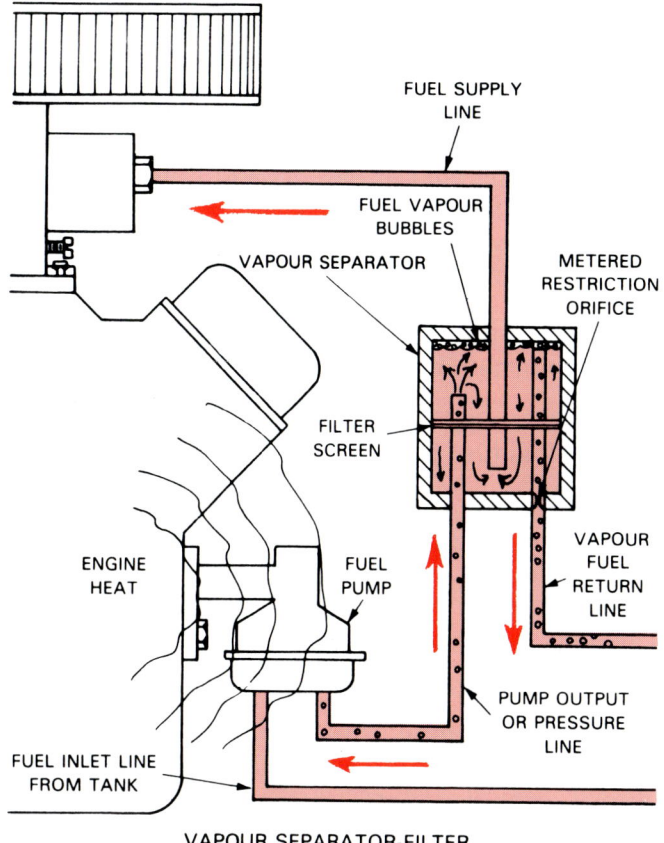

VAPOUR SEPARATOR-FILTER

Fig. 19-10. Study action of fuel filter-vapour separator. Screen traps dirt in fuel. Bubbles or vapours in fuel collect at top of separator and are carried back to fuel tank through return line.

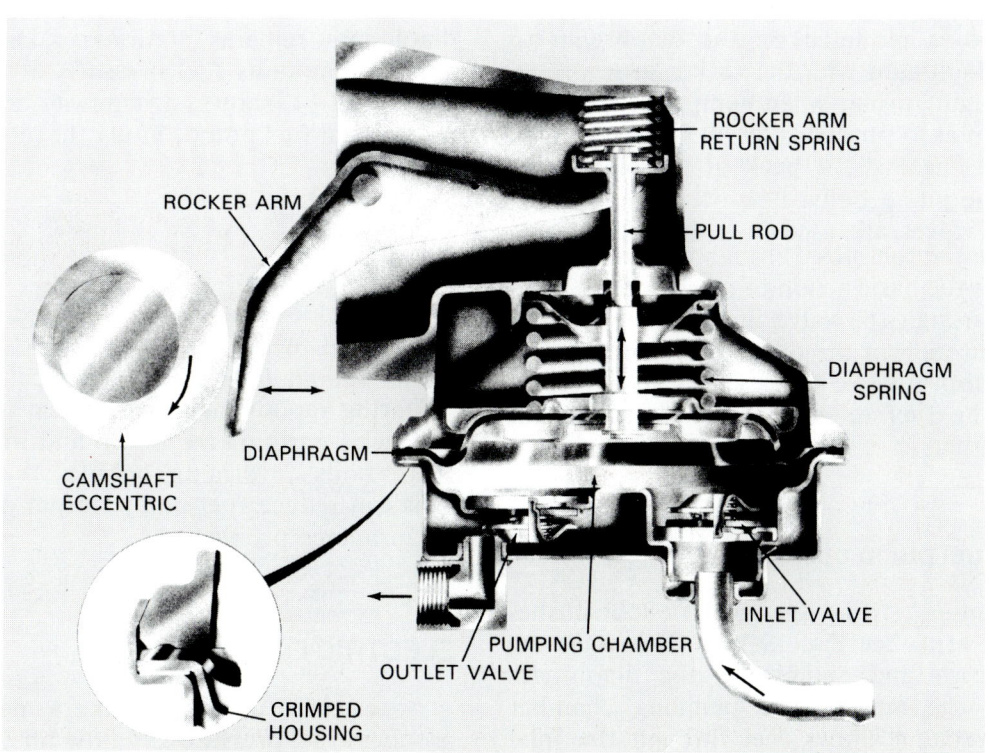

Fig. 19-11. Cutaway view of a typical mechanical fuel pump. Note part names and locations. (AC-Delco)

side of the cylinder block or in some cases the cylinder head or timing cover. A gasket prevents oil leakage between the pump and engine.

Mechanical fuel pumps are commonly used with carburettor type fuel systems. They are the oldest type of fuel pump, but they are still found on many cars. Since the mechanical pump uses a back and forth motion, it is a RECIPROCATING type pump.

The parts of a basic mechanical fuel pump are shown in Fig. 19-11. Note how the rocker arm, diaphragm, springs, and check valves are positioned in the pump body.

Mechanical fuel pump construction

The rocker arm, also called an actuating lever, is a metal arm hinged in the middle. A small pin passes through the arm and fuel pump body. The outer end of the rocker arm rides on the camshaft eccentric. The inner end operates the diaphragm.

Sometimes, a push rod fits between the pump rocker arm and the eccentric. In any case, the rocker arm moves up and down as the eccentric turns.

The fuel pump return spring keeps the rocker arm pressed against the eccentric. Without a return spring, the rocker arm would make a loud CLATTERING SOUND as the eccentric lobe hits the rocker arm.

The diaphragm is a synthetic rubber disc clamped between the halves of the pump body. Refer to Fig. 19-11 once more. The core of the diaphragm is usually cloth which adds strength and durability. A metal pull rod is mounted on the diaphragm to connect the diaphragm with the rocker arm.

The diaphragm spring, when compressed, pushes on the diaphragm to produce fuel pressure and flow. This spring fits against the back of the diaphragm and against the pump body.

Two check valves are used in a mechanical fuel pump to make the fuel flow through the pump. Fig. 19-12 illustrates the basic action of a check valve. Fuel flows easily through the valve in one direction but cannot flow through in the other direction.

In a fuel pump, the two check valves are reversed. This causes the fuel to enter one valve and exit through the other.

Mechanical fuel pump operation

During the intake stroke, the eccentric lobe pushes on the rocker arm. See Fig. 19-13A. This pulls the diaphragm down and compresses the diaphragm spring. Since the area in the pumping chamber increases, a vacuum draws fuel through the inlet check valve. This fills the pump with fuel.

On the output stroke, the eccentric lobe rotates

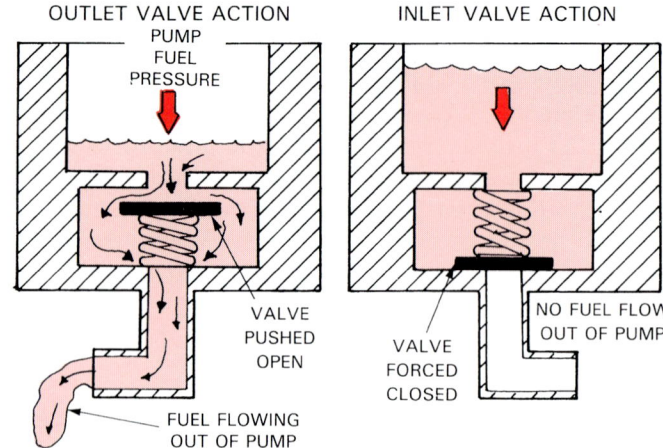

Fig. 19-12. Check valves only allow fuel in one direction. A — Check valve is positioned to allow fuel flow out of pressure chamber. Pressure acts on valve, compresses spring, and fuel moves through. B — Check valve is positioned to stop flow. Pressure helps close valve to seal pressure chamber.

away from the pump rocker arm. This releases the diaphragm, as in Fig. 19-13B. The diaphragm spring then pushes on the diaphragm and pressurises the fuel in the pumping chamber. The amount of spring tension controls fuel pressure. The fuel flows out the outlet check valve.

When the engine is running slowly, the fuel pump would produce more fuel than the engine consumes. For this reason, the fuel pump is made to idle when fuel is not needed. See 19-13C. The diaphragm pull rod is free to slide through the rocker arm. This lets the rocker arm move up and down while the diaphragm remains stationary. Diaphragm spring tension maintains fuel pressure.

Fig. 19-14 shows a cutaway view of another mechanical fuel pump. Study the names of the parts.

Vapour lock

Vapour lock is a problem created when bubbles in overheated fuel reduce or stop fuel flow. Fig. 19-15 shows vapour lock in a mechanical fuel pump.

During vapour lock, engine heat transfers through the metal parts of the pump and into the fuel. The fuel "boils", forming bubbles that displace fuel. This can reduce fuel pump output and cause engine performance problems.

ELECTRIC FUEL PUMPS

An electric fuel pump, like a mechanical pump, produces fuel pressure and flow for the fuel metering section of a fuel system. Electric fuel pumps are commonly used in petrol injection type fuel systems.

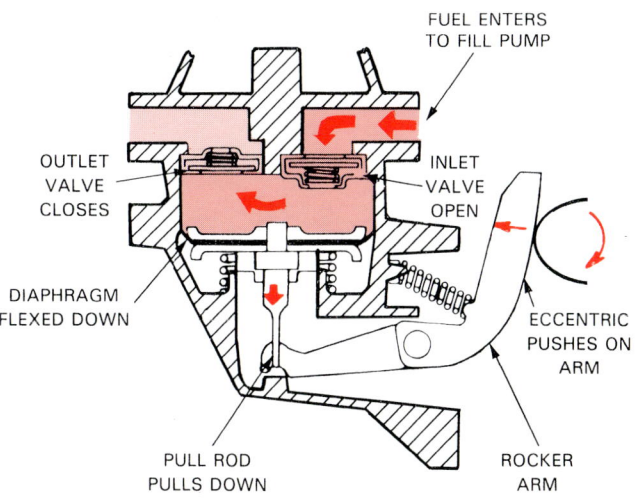

OUTLET VALVE CLOSES

FUEL ENTERS TO FILL PUMP

INLET VALVE OPEN

DIAPHRAGM FLEXED DOWN

ECCENTRIC PUSHES ON ARM

PULL ROD PULLS DOWN

ROCKER ARM

A — During intake stroke, eccentric pushes on rocker arm. Rocker arm pulls down on link and diaphragm. This pulls fuel into pump and compresses diaphragm spring.

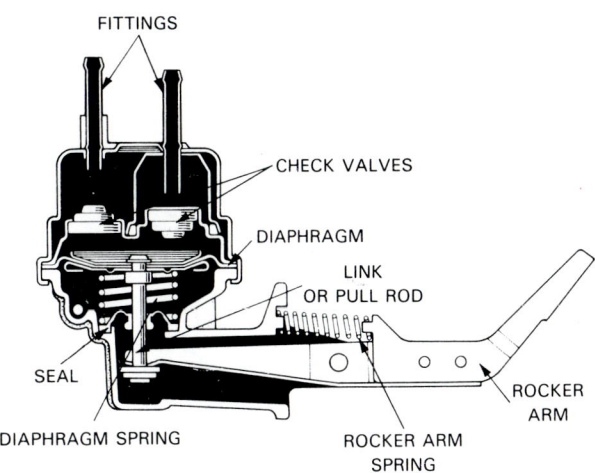

FITTINGS

CHECK VALVES

DIAPHRAGM

LINK OR PULL ROD

SEAL

DIAPHRAGM SPRING

ROCKER ARM SPRING

ROCKER ARM

Fig. 19-14. Cutaway view of another mechanical fuel pump. (Toyota)

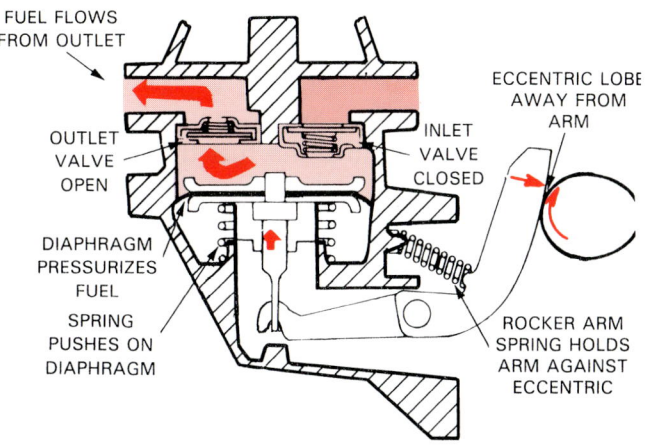

FUEL FLOWS FROM OUTLET

ECCENTRIC LOBE AWAY FROM ARM

OUTLET VALVE OPEN

INLET VALVE CLOSED

DIAPHRAGM PRESSURIZES FUEL

SPRING PUSHES ON DIAPHRAGM

ROCKER ARM SPRING HOLDS ARM AGAINST ECCENTRIC

B — During output stroke, eccentric releases rocker arm. Spring then pushes on diaphragm to pressurize fuel in pump chamber. Fuel then flows out pump and to engine.

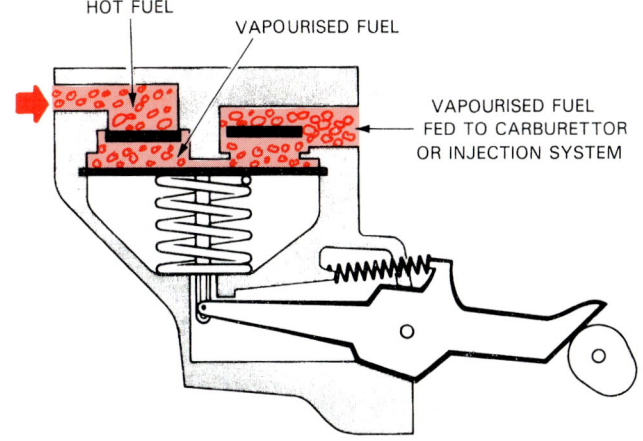

HOT FUEL

VAPOURISED FUEL

VAPOURISED FUEL FED TO CARBURETTOR OR INJECTION SYSTEM

Fig. 19-15. Vapour lock is caused by too much engine heat transferring into fuel. Bubbles form and displace fuel. This can prevent fuel from flowing through system. (Ford)

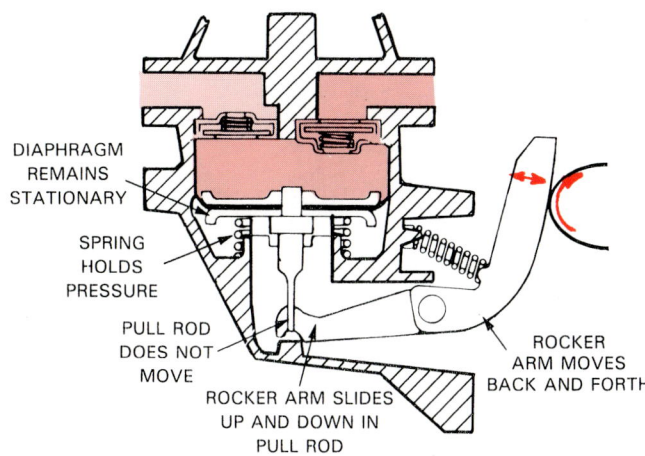

DIAPHRAGM REMAINS STATIONARY

SPRING HOLDS PRESSURE

PULL ROD DOES NOT MOVE

ROCKER ARM SLIDES UP AND DOWN IN PULL ROD

ROCKER ARM MOVES BACK AND FORTH

C — When idling, rocker slides in link. Diaphragm is stationary. Spring maintains pressure.

Fig. 19-13. Mechanical fuel pump operation.

An electric fuel pump can be located inside the fuel tank as part of the fuel pickup-sending unit. It can also be located in the fuel line between the tank and engine. See Fig. 19-16.

An electric fuel pump has some possible advantages over a mechanical fuel pump. An electric pump can produce almost instant fuel pressure. A mechanical pump slowly builds pressure as the engine is cranked for starting. Also, most electric fuel pumps are a rotary type. They produce a smoother flow of fuel (less pressure pulsations) than a reciprocating, mechanical pump.

Since most electric fuel pumps are located away from the engine, they help prevent vapour lock. An electric fuel pump pressurises all of the fuel lines near engine heat. This helps avoid vapour lock because pressure makes it more difficult for bubbles to form in the fuel.

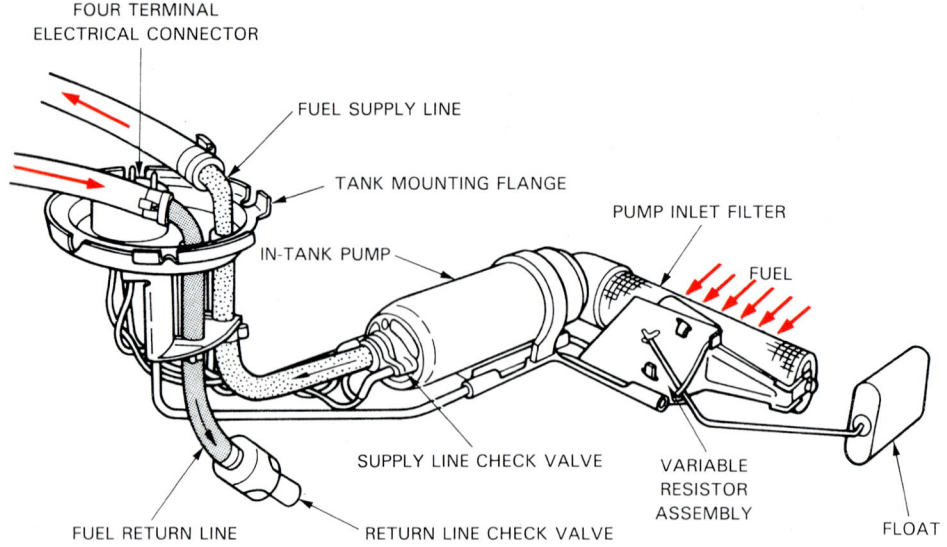

Fig. 19-16. In-tank electric fuel pump sending unit. This particular pump feeds fuel to a higher pressure pump for petrol injection system.

Rotary fuel pumps

Rotary fuel pumps include the impeller, roller vane, and sliding vane types. They use a circular or spinning motion to produce pressure.

An impeller electric fuel pump is a centrifugal type pump. Normally, it is located inside the fuel tank. See Fig. 19-17. This pump uses a small DC motor to spin the impeller (fan blade). The impeller blades cause the fuel to fly outward due to centrifugal force (spinning matter flys outward). This produces enough pressure to move the fuel through the fuel lines.

A roller vane electric fuel pump is a positive displacement pump (each pump rotation moves a specific amount of fuel). It is normally located in the main fuel line. Look at Fig. 19-18. Small rollers and an offset mounted rotor disc produce fuel pressure.

When the rotor disc and rollers spin, they draw fuel in on one side. Then, the fuel is trapped and

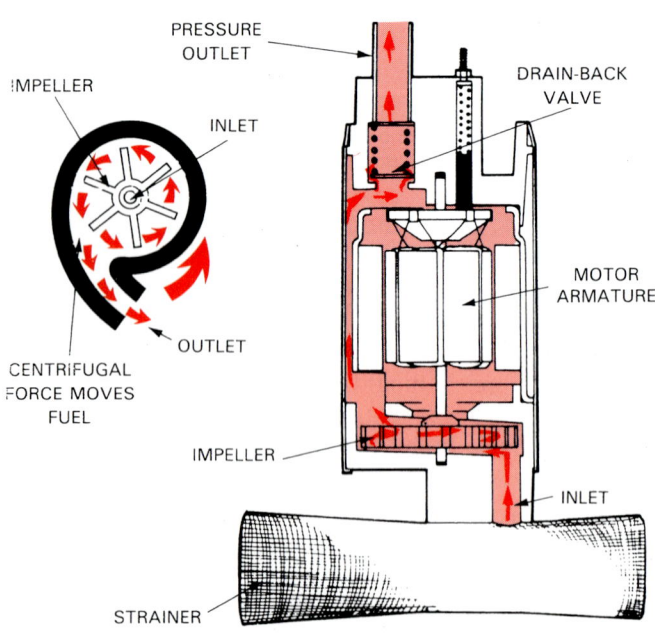

Fig. 19-17. In-tank, impeller type electric fuel pump. Impeller produces very smooth flow of fuel through system. Check valve prevents fuel from draining out of lines and back into tank when pump is not running. (Volvo, Ford)

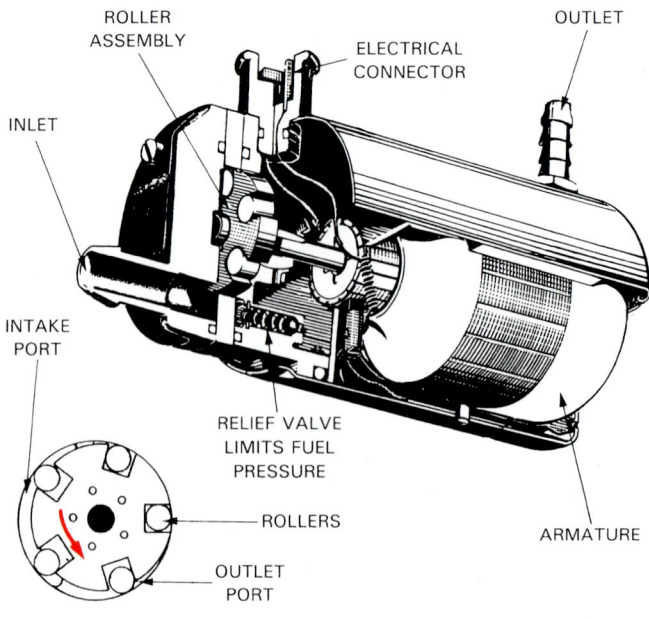

Fig. 19-18. Roller vane electric fuel pump. This type pump is usually capable of producing higher pressure and more volume than an impeller type pump. Relief valve limits fuel pressure. (Volvo)

pushed to a smaller area on the opposite side of the pump housing. This squeezes the fuel between the rollers and the fuel flows out under pressure.

A sliding vane electric fuel pump is similar to a roller vane pump. Vanes (blades) are used instead of rollers.

As shown in Figs. 19-17 and 19-18, most rotary electric fuel pumps also have check valves and relief valves.

The electric fuel pump check valves keep fuel from draining out of the fuel line when the pump is not running. A relief valve limits the maximum output pressure.

A reciprocating electric fuel pump has the same basic action as a mechanical fuel pump. However, it uses a solenoid instead of a rocker arm to produce a pumping motion.

A reciprocating pump commonly uses either a bellows or a plunger. The solenoid turns on and off to force the bellows or plunger up and down. This pushes fuel through the check valves and fuel system. This type pump is not very common.

Electric Fuel Pump Circuit

A circuit for an electric fuel pump is shown in Fig. 19-19. Study the electrical connections. This circuit has a switch controlled by air flow.

The fuel pump switch which is built into the air flow meter is a safety feature that prevents the fuel pump from working if the engine is not running.

Many modern electric fuel pumps are controlled by an ON-BOARD COMPUTER.

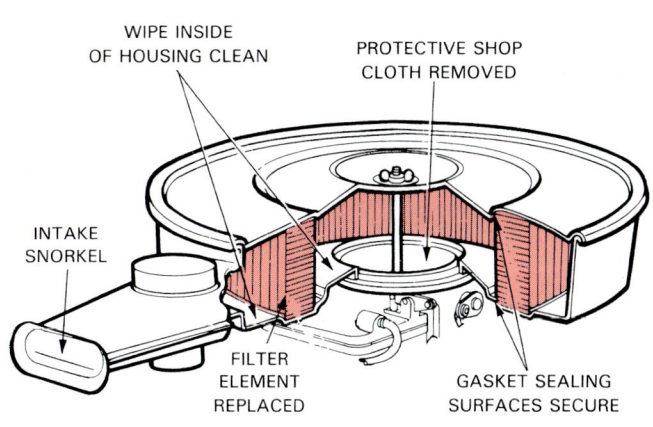

Fig. 19-20. Typical automotive air filter. Most air filters are made of pleated paper. Metal housing encloses paper element. Housing seals against top of carburettor, throttle body, or air inlet on diesel engines.

AIR FILTERS

An air filter, or air cleaner, removes foreign matter (dirt and dust) from the air entering the engine intake manifold. See Fig. 19-20. Most air filters use a paper element (filter material). For heavy duty operation, ie if the vehicle is operating in extremely dusty conditions some cars use a polyurethane (foam) filtering element in conjunction with the paper element. The element fits inside a metal or plastic housing.

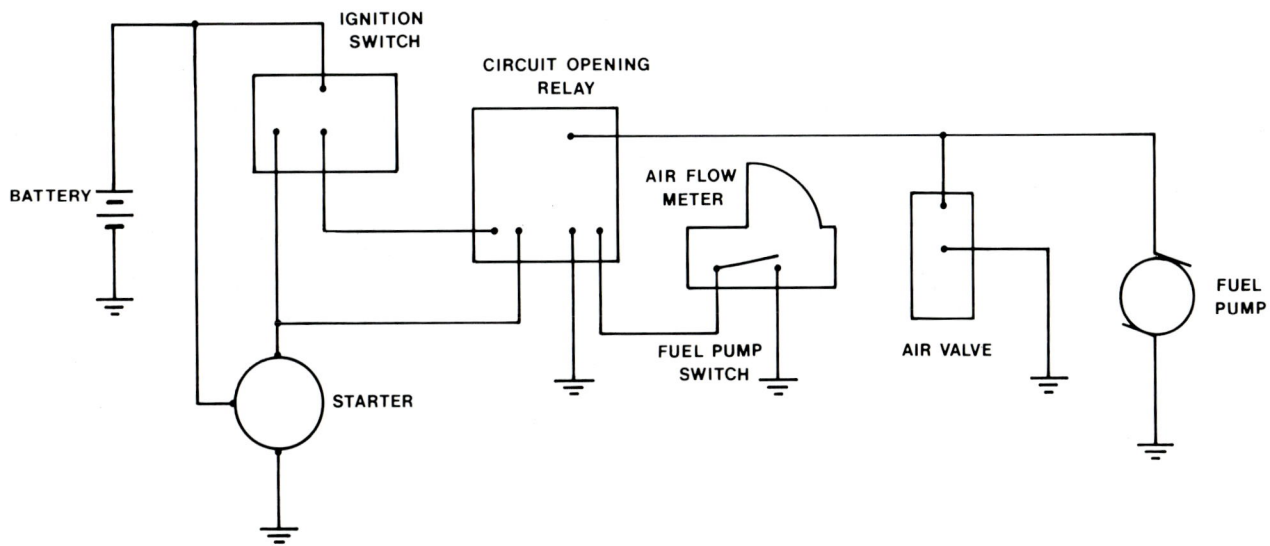

Fig. 19-19. Typical electric fuel pump circuit.

FUEL SUPPLY SYSTEM SERVICE

Now that you have a basic understanding of how a fuel supply system functions, you are ready to learn about problems, tests, and repairs common to the fuel pump, lines, filters, and tank.

IMPORTANT! Always keep a fire extinguisher handy when working on a car's fuel system. During a fire, a few minutes of time can be a LIFETIME!

FUEL TANK SERVICE

Typical fuel tank problems include fuel leakage, physical damage (accidents), and contamination by foreign matter (rust, dirt, and water). Vibration and rusting can cause a fuel tank to develop pinhole leaks. Rocks can fly up and puncture the tank. Foreign matter can get in the fuel tank from internal tank deterioration or from the petrol station pump.

DANGER! Do NOT weld or solder a fuel tank. Send the leaking tank, if not badly rusted, to a well trained specialist. Even an empty tank can EXPLODE when fuel gum melts, vapourises, and ignites from the heat of soldering or welding.

Fuel tank removal and replacement

A fuel tank can be located under the luggage compartment, in a body panel, or under the rear seat. It may be held in the vehicle by large metal straps or by bolts passing through the tank flange.

CAUTION! Before servicing a fuel tank, empty the tank. A full tank is very heavy and can rupture if dropped. Fire could result!

To remove fuel from the tank, unscrew the drain plug and drain the fuel into an approved safety can. If a drain plug is NOT provided, use an approved pumping or syphoning method to draw the fuel out of the tank. After the tank is empty, you can remove it from the car.

DANGER! Wipe up fuel spills immediately with a workshop cloth or rag. Do NOT spread sawdust or similar material on fuel spills because the sawdust will become extremely flammable.

When installing a fuel tank, make sure you replace rubber insulators. Check that all fuel lines are properly secured. Replace the fuel in the tank and check for leaks. If needed, a service manual for the particular vehicle will detail exact tank installation procedures.

FUEL TANK SENDING UNIT SERVICE

A faulty fuel tank sending unit can make the fuel gauge reading inaccurate. Usually, the variable resistor in the sending unit fails. However, you should remember that the fuel gauge or the gauge circuit may be at fault.

First, test the action of the fuel gauge. Fig. 19-21 shows a fuel gauge tester. It is connected to the wire going to the fuel tank sending unit.

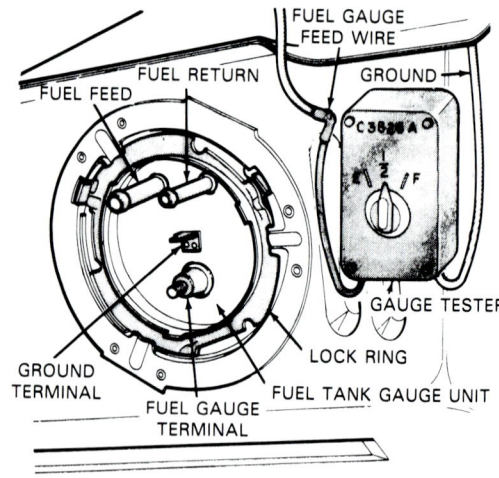

Fig. 19-21. Special tester will quickly check condition of fuel gauge and circuit. If circuit and gauge are working well, problem may be in tank sending unit. Note how sending unit is held in fuel tank by lock ring.

When the tester is set on full, for example, the fuel gauge should read full.

If the gauge does NOT function, either the gauge or the gauge circuit is faulty. If the fuel gauge begins to work with the tester in place, the tank sending unit is probably faulty.

If your tests indicate an inoperative fuel tank sending unit, remove the unit after draining the tank. Unscrew the cam lock holding the sending unit in the fuel tank, Fig. 19-22. Lift the unit out of the tank.

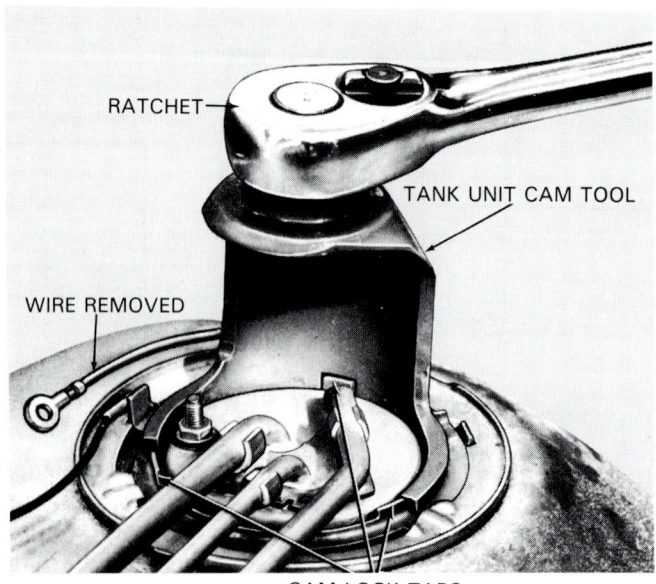

Fig. 19-22. To remove tank sending unit, turn cam lock ring. Special tool makes this easy. Light taps with a hammer on a full shank screwdriver will also work.

With the sending unit removed, measure its resistance with an ohmmeter, Fig. 19-23. If the resistance is not within factory specs, install a new sending unit.

Also check the sending unit float for leakage. Shake the float next to your ear. If you can hear liquid splashing, replace the float.

If the tank unit resistance is good, check the tank ground. A poor ground could prevent operation.

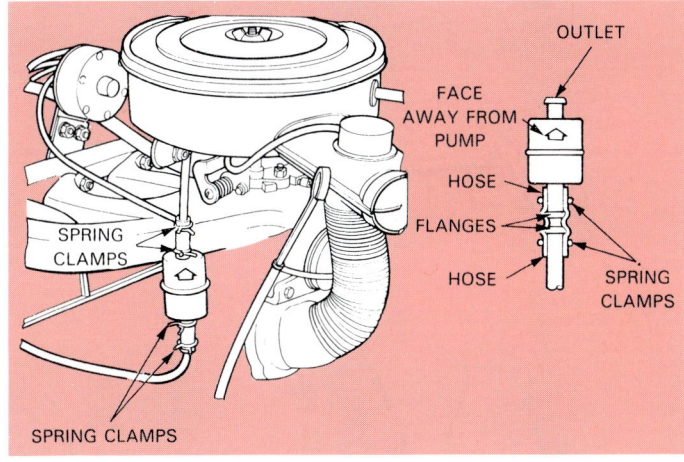

Fig. 19-24. Always check all fuel hoses for signs of deterioration. When replacing a hose, make sure it is pushed fully over fittings.

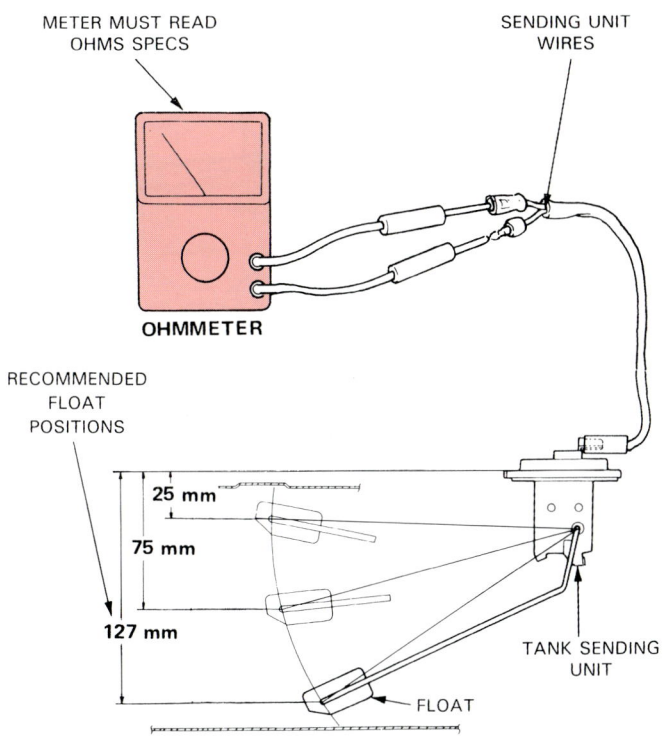

Fig. 19-23. Ohmmeter can be used to check condition of tank sending unit. Ohms should be within specs with float in prescribed positions. (Honda)

FUEL LINE AND HOSE SERVICE

Faulty fuel lines and hoses are a common source of fuel leaks. See Fig. 19-24. Fuel hoses can become hard and brittle after being exposed to engine heat and the elements. Engine oil can soften and swell them. Always inspect hoses closely and replace any in poor condition.

Metal fuel lines seldom cause problems. However, they should be replaced when squashed, kinked, rusted, or leaking.

Fuel line and hose service rules

Remember these rules when working with fuel lines and hoses:

1. Place a workshop cloth around the fuel line fitting during removal. This will keep fuel from spraying on you or on the hot engine. Use a flare nut or tubing wrench on fuel system fittings.
2. Only use approved double-wall steel tubing for fuel lines. Never use copper or plastic tubing.
3. Make smooth bends when forming a new fuel line. Use a bending spring or bending tool.
4. Form double-lap flares on the ends of the fuel line. A single lap flare is not approved for fuel lines. Refer to Fig. 19-25.
5. Reinstall fuel line hold-down clamps and brackets. If not properly supported, the fuel line can vibrate and fail.
6. Route all fuel lines and hoses away from hot or moving parts. Double-check clearance after installation.
7. Only use approved synthetic rubber hoses in a fuel system. If vacuum type rubber hose is accidentally used, the fuel can chemically attack and rapidly ruin the hose. A dangerous leak could result.
8. Make sure a fuel hose fully covers its fitting or line before installing the clamps. Pressure in the fuel system could force the hose off if not installed properly.
9. Double-check all fittings for leaks. Start the engine and inspect the connections closely.

DANGER! Most fuel injection systems have very high fuel pressure. Follow recommended procedures for bleeding or releasing pressure before disconnecting a fuel line or fitting. This will prevent fuel spray from possibly causing injury or a fire!

FUEL FILTER SERVICE

Fuel filter service involves periodic replacement or cleaning of system filters. It may also include

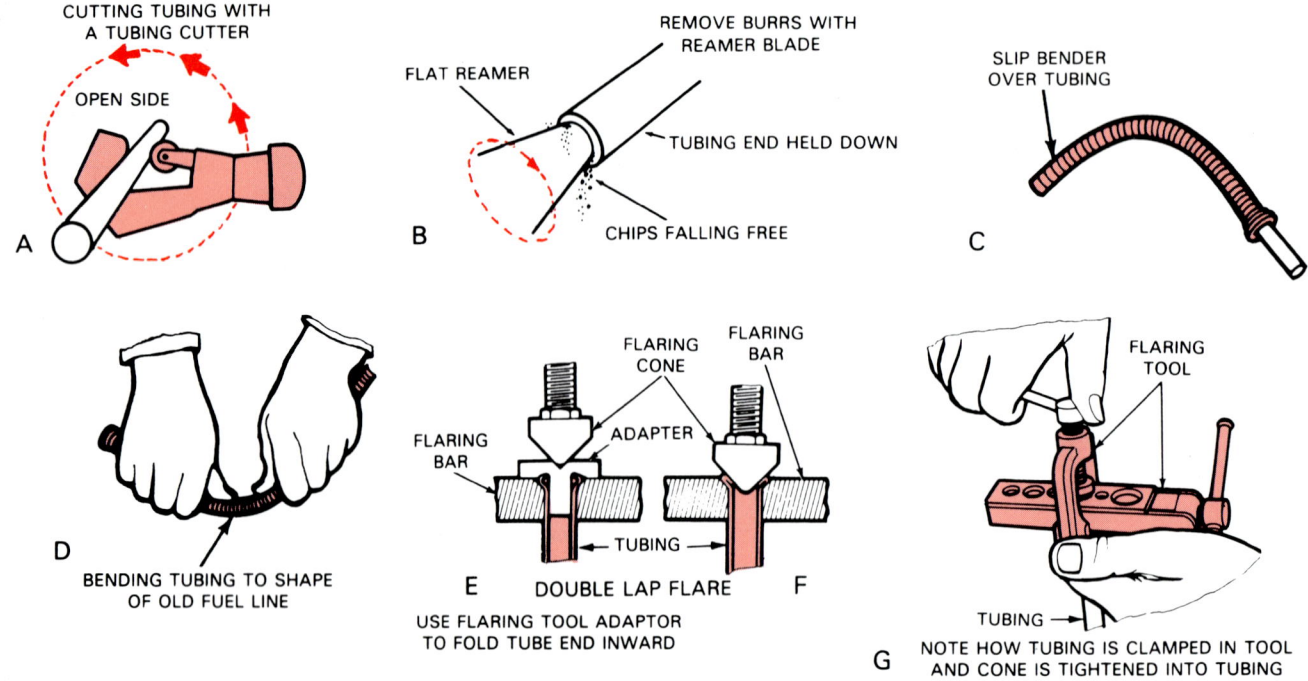

Fig. 19-25. Fuel lines need double-lap flares. Study basic steps for making a new fuel line. This procedure also applies to other lines (brake lines, vacuum lines, etc.)

locating clogged fuel filters that are upsetting fuel system operation. Paper element filters must be replaced when clogged or after prolonged service. Sintered bronze or metal gauze fuel filters can usually be cleaned and reinstalled.

A clogged fuel filter can restrict the flow of fuel to the carburettor, petrol injectors, or diesel injection pump. Engine performance problems will usually show up at higher cruising speeds.

For example, the engine may temporarily loose power or stall when a specific engine speed is reached.

A partially clogged filter may pass enough fuel at low engine speeds. However, when engine speed and fuel flow increase, the engine may "starve" for fuel.

On older cars, a clogged in-tank strainer is a common and hard to diagnose problem. The in-tank filter, when clogged, can collapse and stop all fuel flow. Then, after the engine stalls, the strainer can open back up, leaving no trace of a restriction.

Some fuel filters have a check valve that opens when the filter becomes clogged. This will allow fuel contaminants to flow into the system. When contaminants are found in the filters and system, the tank, pump, and lines should be flushed with clean fuel.

Fuel filter locations

Fuel filters can be located in the following locations:

1. In the fuel line before the carburettor, fuel injectors, or diesel injection pump.
2. Inside the fuel pump.
3. In the fuel line right after the electric fuel pump.
4. Under the fuel line fitting in the carburettor.
5. A fuel strainer is also located in the fuel tank on the end of the fuel pickup tube.

When in doubt about fuel filter locations, refer to a service manual. It will give information about service intervals, cleaning, and replacement of all system filters.

If a fitting must be loosened when changing a fuel filter, use a flare nut wrench, as in Fig. 19-26. Do not overtighten and strip the fitting when tightening. Double-check that the fuel hose, if used, is fully installed over the fitting barbs.

FUEL PUMP SERVICE

Fuel pump problems usually show up as low fuel pressure, inadequate fuel flow, abnormal pump noise, or fuel leakage from the pump. Both mechanical and electric fuel pumps can fail after prolonged operation.

Low fuel pump pressure can be caused by a weak diaphragm spring, ruptured diaphragm, leaking check valves, or physical wear of moving parts. Low fuel pump pressure can make the engine starve for fuel at higher engine speeds.

High fuel pump pressure, more frequent with electric pumps, indicates an inoperative pressure

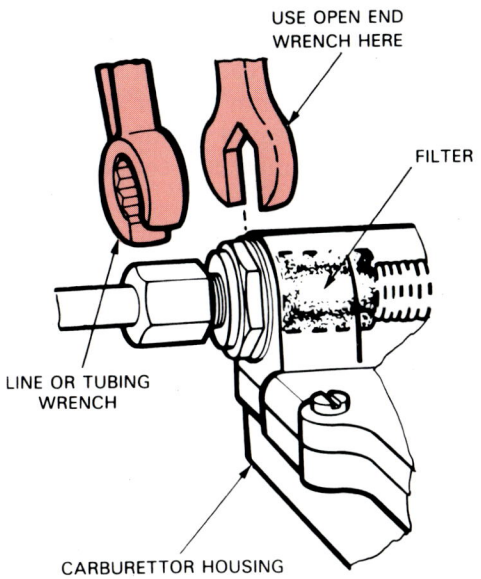

Fig. 19-26. When loosening or tightening a fuel line fitting, use a line wrench. This will help prevent rounding off soft fitting nut. Do not overtighten fittings. Their threads will strip easily.

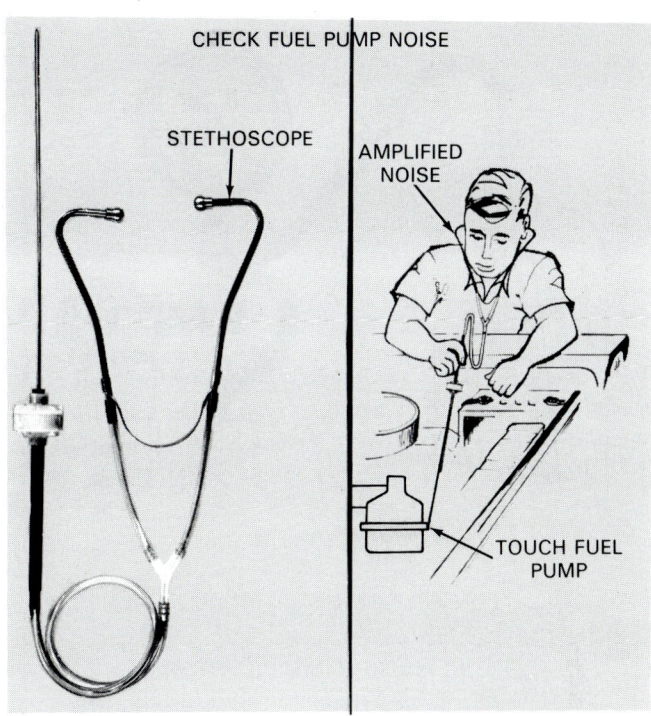

Fig. 19-27. A stethoscope will quickly locate a noisy mechanical or electric fuel pump. A clattering mechanical pump will sound like a faulty engine lifter.

relief valve. If the relief valve fails to open, both pressure and volume can be above normal. High fuel pressure can produce a rich fuel mixture or even flood the engine.

Mechanical fuel pump noise (clacking sound from inside pump) is commonly caused by a weak or broken rocker arm return spring or by wear of the rocker arm pin or arm itself. Mechanical fuel pump noise can be easily confused with valve or tappet clatter. Both sound very similar. To verify mechanical fuel pump noise, use a stethoscope, Fig. 19-27.

NOTE! Most electric fuel pumps make some noise (buzz or whirr sound) when running. Only when the pump noise is abnormally loud should an electric fuel pump be considered faulty.

A clogged tank strainer can cause excessive pump noise. Pump speed can increase because fuel is not entering the pump properly.

Fuel pump leaks are caused by physical damage to the pump body or deterioration of the diaphragm or gaskets. Most mechanical fuel pumps have a small vent hole in the pump body. When the diaphragm is ruptured, fuel will leak out of this hole.

It is possible for a ruptured mechanical fuel pump diaphragm to contaminate the engine oil with petrol. Fuel can leak through the diaphragm and pump body into the side of the block, and down into the engine sump. When a petrol smell is noticed in the oil, correct the pump problem and change the oil and filter.

Fuel pump tests

Fuel pump testing commonly involves measuring fuel pump pressure and volume. Since exact procedures vary, depending upon fuel system type, refer to a service manual for exact testing methods. Sometimes, fuel pump vacuum is measured as another means of determining pump and supply line condition.

WARNING! Most petrol injection systems operate on very high fuel pressure. Make sure you tighten all test connections and follow prescribed procedures when testing fuel pump output.

Always remember that there are several other problems that can produce symptoms similar to those caused by a faulty fuel pump. Before testing a fuel pump, check for:
1. Restricted fuel filters.
2. Squashed or kinked fuel lines or hoses.
3. Air leak into vacuum side of pump or line.
4. Carburettor or injection system troubles.
5. Ignition system problems.
6. Low engine compression.

Measuring fuel pump pressure

To measure fuel pump pressure, connect a pressure gauge to the outlet line of the fuel pump. This is illustrated in Fig. 19-28.

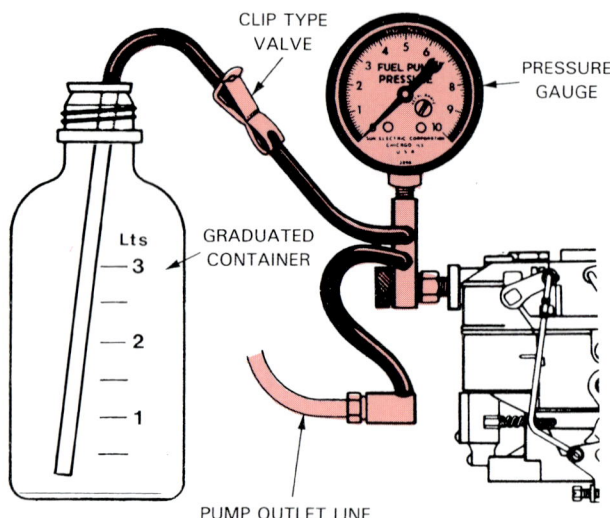

FUEL PUMP SPECIFICATIONS		
	VOLUME (Litres/min @ rpm)	PRESSURE (kPa @ rpm)
FOUR CYLINDER	1.6 @ 1800 rpm	20-27 @ 1800
SIX CYLINDER	0.6 @ 1200 rpm	24-32 @ variable
EIGHT CYLINDER	1.5 @ 1200 rpm	48-59 @ slow idle

Fig. 19-28. To test fuel pump pressure and volume, make connections of test equipment as shown. Connect gauge to line before carburettor, fuel manifold, or throttle body. Close clip and start engine to measure pressure. Open clip to measure volume over prescribed time span. Chart shows fuel pump specs for one make and model car.

Typically, you would start and idle the engine at spec rpm with a mechanical fuel pump. With an electric fuel pump, you may only need to activate (supply voltage to) the pump motor. Compare your pressure readings to specifications.

Fuel pressure for a carburettor type fuel system should be about 20 to 50 kPa pressure. A petrol injection system will usually have a higher pressure output. Fuel pressure can run from 100 to 280 kPa. A diesel supply pump should produce around 40 to 160 kPa. It feeds fuel to the very high pressure injection pump.

Always remember to use factory values when determining fuel pump condition. Pressures can vary considerably from system to system.

If fuel pump pressure is NOT within specs, check pump volume and the lines and filters before replacing the fuel pump.

Measuring fuel pump volume

Fuel pump volume also called capacity, is the amount of fuel the pump can deliver in a specific amount of time. It is measured by allowing fuel to pour into a graduated (marked) container for a certain time period.

To check fuel pump volume, use a setup similar to that in Fig. 19-28. Route the outlet line from the fuel pump into a special container. For safety, a valve or clip should be used to control fuel flow into the container.

With the engine idling at a set speed, allow fuel to pour into the container for the prescribed amount of time (normally one minute). Close off the clip or valve. Compare volume output to specs.

Fuel pump volume output should be a minimum of about ONE LITRE in one minute for carburettor systems. Fuel injection systems typically have a slightly higher volume output from the supply pump. Refer to a service manual or specifications sheet for values for the particular fuel pump and car.

Measuring fuel pump vacuum

Fuel pump vacuum should be checked when a fuel pump fails pressure and volume tests. A vacuum test will eliminate possible problems in the fuel lines, hoses, filters, and pickup screen in the tank.

For example, a clogged fuel pickup screen could make the fuel pump fail the volume test. If the same pump passes a vacuum test, you need to check the lines and filters for problems.

To measure fuel pump vacuum, connect a vacuum gauge to the inlet side of the pump. Leave the fuel hose in your graduated container from the volume test. Open the valve and activate the pump (start engine and allow to run on fuel in carburettor or connect voltage to electric pump). Compare your reading to specs.

Typically, fuel pump vacuum should be about 25 to 35 kPa. A good vacuum reading indicates a good fuel pump. If the pump failed the pressure or volume test but passed the vacuum test, the fuel supply lines and filter may be at fault.

Electric fuel pump circuit tests

Many electric fuel pump problems are caused by circuit problems. Broken wires, faulty relays, shorts, blown fuses, computer malfunctions, and other troubles can affect electric fuel pump operation.

If an electric fuel pump does NOT pass its pressure or volume tests, measure the amount of voltage being fed to the pump motor. Look at Fig. 19-29. If supply voltage is low, there is a problem in the electrical circuit to the pump.

When circuit problems must be found, use your knowledge of basic testing and a service manual.

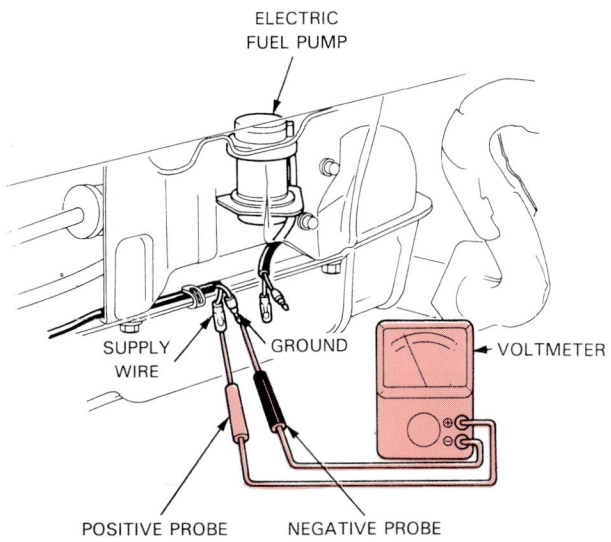

Fig. 19-29. Before condemning an electric fuel pump, make sure circuit is in good condition. Measure amount of voltage being supplied to pump and compare to specs. If voltage is low, repair circuit. (Honda)

Generally, test at various points until the source of the trouble is found. Fig. 19-30 shows a wiring diagram for one electric fuel pump circuit. Note that there are only a few components and electrical connections that could upset pump operation.

FUEL PUMP REMOVAL AND REPLACEMENT

When a fuel pump does NOT pass its performance tests, it must be removed for replacement or overhaul.

To remove a mechanical fuel pump, simply disconnect the fuel lines and unbolt the pump from the engine, Fig. 19-31. If needed, tap the side of the pump with a plastic hammer to free the gasket.

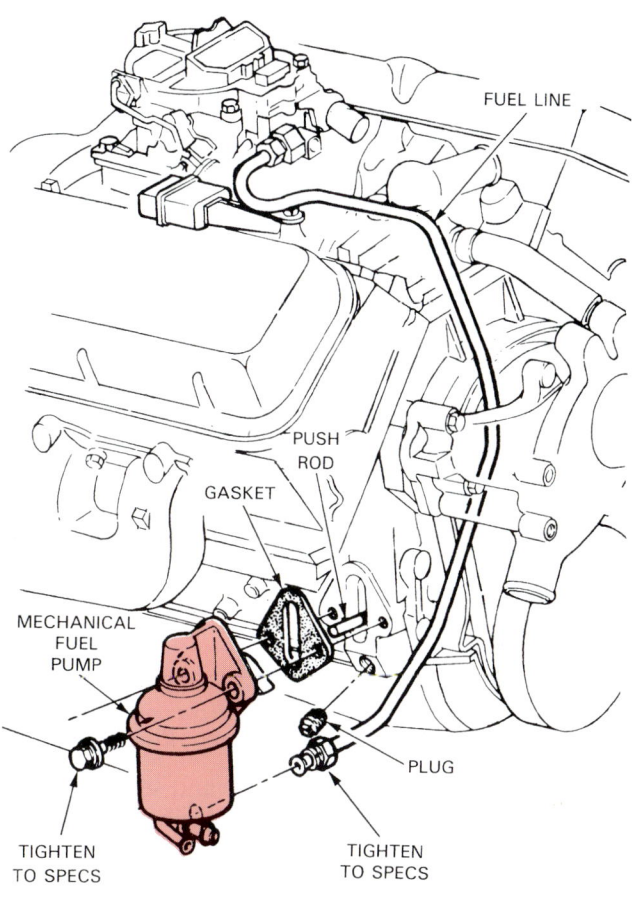

Fig. 19-31. To service mechanical fuel pump, remove lines and hoses. Then remove pump-to-block fasteners. Use new gasket when installing. Torque fasteners to specs.

An electric fuel pump may be located in the main fuel line or in the fuel tank. With an in-line pump, simply disconnect the fuel fittings, and remove the pump. An in-tank pump will usually need to be

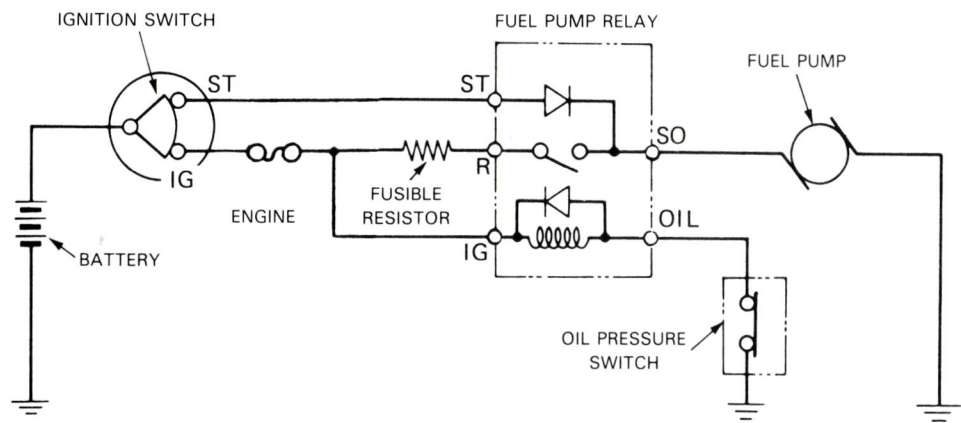

Fig. 19-30. Typical fuel pump control circuit. Note fusible resistor, relay, oil pressure switch, and pump. Any defective part could upset pump operation. (Toyota)

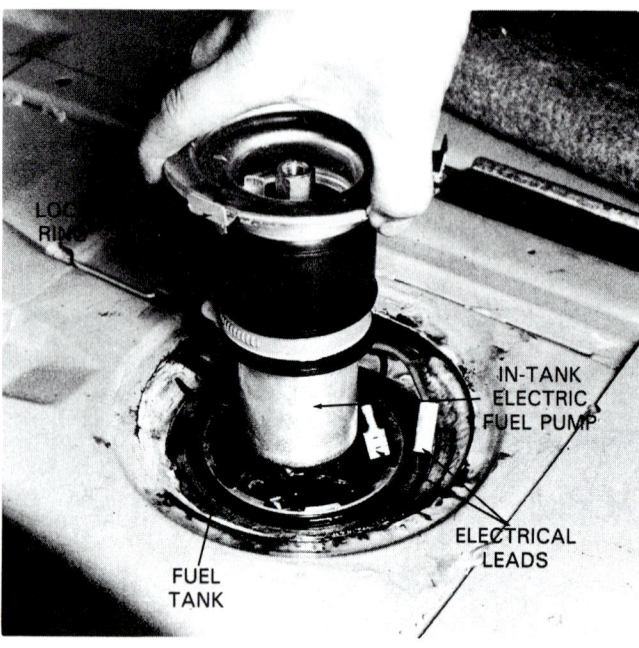

Fig. 19-32. In-tank electric fuel pump is removed like a tank pickup-sending unit. Unscrew lock and pull out. Use new O-ring seal when installing pump. (Saab)

removed as part of the tank pickup and sending unit, Fig. 19-32. This procedure was described earlier in the chapter.

Fuel pump overhaul

A large percentage of modern mechanical fuel pumps and almost all electric fuel pumps are sealed at the factory and CANNOT be overhauled. See Fig. 19-33.

A. SCREW	M. RETAINER
B. FUEL PUMP CAP	N. BALL
C. CAP GASKET	O. DIAPHRAGM ASSEMBLY
D. OUTLET CONNECTOR	P. BODY LOWER (COMPLETE)
E. INLET CONNECTOR	Q. ROCKER ARM SPRING
F. BODY LOWER (COMPLETE)	R. ROCKER ARM
G. PACKING	S. GASKET
H. VALVE ASSEMBLY	T. SPACER
I. RETAINER	U. ROCKER PIN
J. SCREW	V. PLAIN WASHER
K. DIAPHRAGM ASSEMBLY	W. SPRING WASHER
L. DIAPHRAGM SPRING	X. NUT

Fig. 19-34. Exploded view of a mechanical pump. Rebuilding involves replacing parts suffering wear or deterioration: diaphragm, check valves, gaskets, springs, and any other part in poor condition. These parts usually come in a rebuild kit. (Nissan)

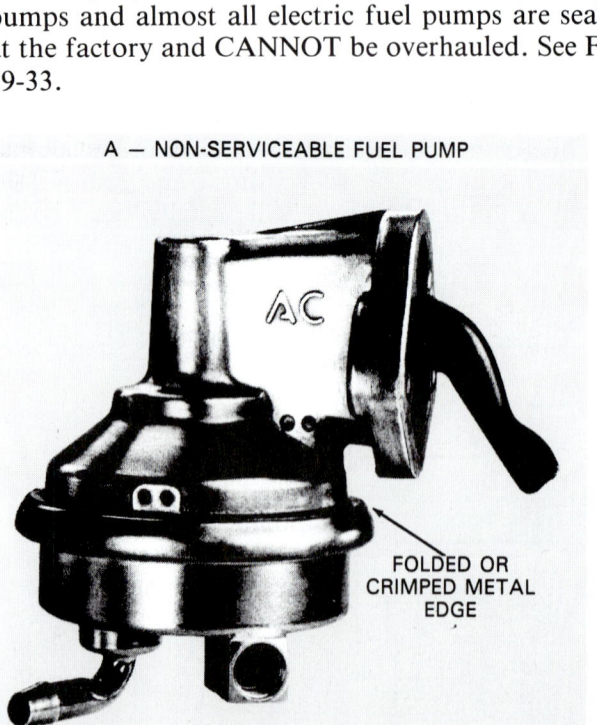

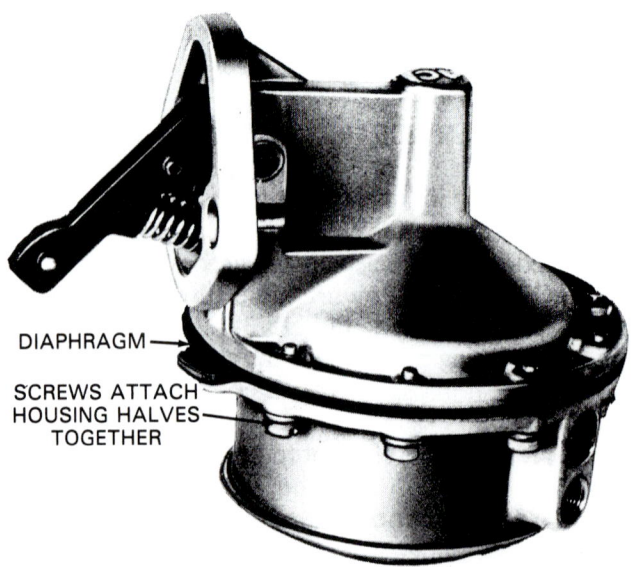

Fig. 19-33. Most mechanical fuel pumps are not serviceable. They must be replaced when faulty. A — Non-serviceable pump cannot be taken apart. B — Serviceable pump is held together by screws. Check for parts availability before doing too much work on old pump. (AC-Delco)

Some mechanical fuel pumps are held together with screws and can be overhauled if parts are available. Normally, when either a mechanical or electric fuel pump is faulty, it is replaced with a new pump.

To overhaul a mechanical pump, you must install the new parts provided in the repair kit. This normally includes a new diaphragm, check valves, springs, and other parts that suffer wear. Fig. 19-34 shows an exploded view of a mechanical fuel pump.

HINT! When installing a mechanical fuel pump, position the camshaft eccentric AWAY from the pump rocker arm. This will make it much easier to hold the pump against the engine while starting the bolts.

If a push rod is used, Fig. 19-35, coat it with heavy grease. Then, push it up into place. This will hold the pushrod out of the way.

the most common, are usually replaced with a new unit. Look at Fig. 19-36. Polyurethane filter elements can be cleaned and reused, Fig. 19-37.

When replacing a filter element, you should also wipe out the filter housing. Dirt can collect in the bottom of the housing.

CAUTION! Be careful not to drop anything into the air inlet opening in the carburettor or fuel injection system. As a precaution, place a clean cloth over the engine's air inlet.

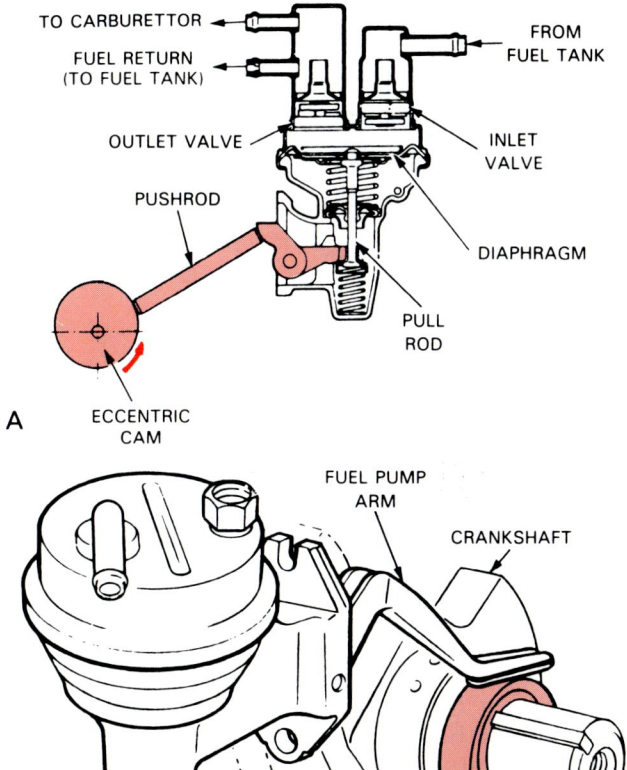

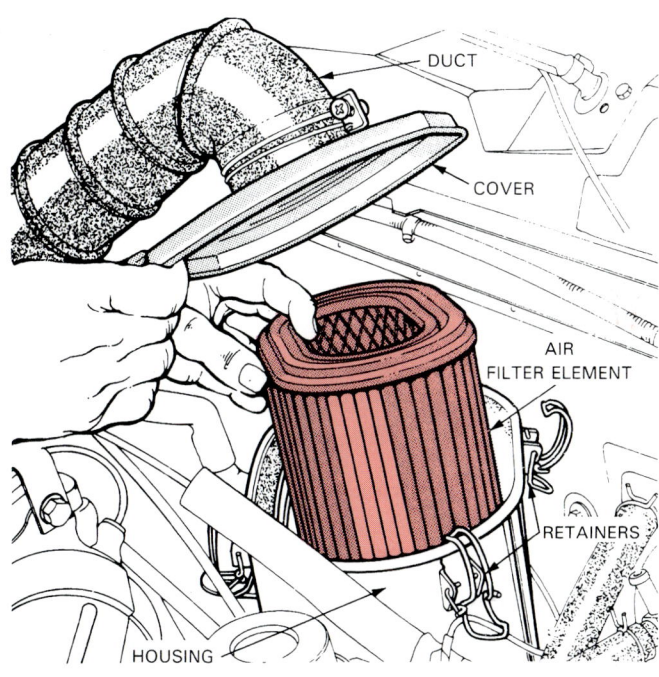

Fig. 19-36. Paper filter elements are normally replaced when dirty. Some auto makers allow using compressed air to blow dirt out of element in reverse direction. Make sure housing is secure and not leaking.

Fig. 19-35. A — This mechanical fuel pump uses a push rod between eccentric cam and rocker arm. When installing pump, use heavy grease, hacksaw blade, or special tool to hold push rod up in block. B — This supply pump works off an eccentric on diesel engine crankshaft.

AIR FILTER SERVICE

Air filter service usually involves replacement or cleaning of the filter element. Paper filter elements,

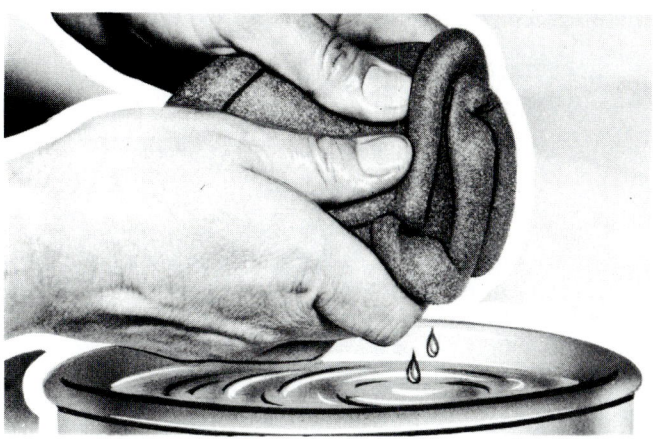

Fig. 19-37. Foam type air filters can be cleaned in solvent and reused. Check in service manual when in doubt.

KNOW THESE TERMS

Driving range, Fuel tank capacity, Filler neck restrictor, Tank sending unit, Fuel return system, Pleated paper filter, Sintered bronze filter, Vapour separator-filter, Mechanical fuel pump, Electric fuel pump, Check valve, Vapour lock, Double-lap flare, Fuel pump pressure, Fuel pump volume, Fuel pump vacuum.

REVIEW QUESTIONS

1. List and describe the three subsystems of a modern fuel system.
2. A _____ _____ system draws fuel from the fuel tank and forces it to the fuel metering device.
3. An average fuel tank capacity is around 40 to 80 litres. True or False?
4. What is the purpose of a filler neck restrictor in a fuel tank assembly?
5. Explain what can happen if leaded fuel is used in a car designed for unleaded fuel.
6. Which of the following is NOT part of a fuel tank pickup-sending unit?
   a. Vapour separator.
   b. In-tank fuel strainer or filter.
   c. Variable resistor.
   d. Pickup tube.
7. Fuel lines are normally made of single-wall steel tubing. True or False?
8. When are fuel hoses needed?
9. Explain the operation of a fuel return system.
10. The following could NOT be a fuel filter.
    a. Pleated paper filter.
    b. Sintered bronze filter.
    c. Bowl filter.
    d. Canister filter.
    e. All of the above are correct.
11. The two basic types of fuel pumps are the _____ and _____ types.
12. Explain the difference between the two types of fuel pumps in Question 11.
13. The _____ _____ controls fuel pressure in a mechanical fuel pump.
14. During the intake stroke of a mechanical fuel pump, the camshaft's eccentric pushes on the pump rocker arm. True or False?
15. Can a mechanical fuel pump idle? Explain.
16. Define the term "vapour lock".
17. Where can electric fuel pumps be located?
18. Which of the following is an advantage of an electric fuel pump?
    a. Instant fuel pressure.
    b. Very little pressure pulsations.
    c. Helps prevent vapour lock.
    d. All of the above are correct.
19. Rotary type electric fuel pumps include the _____, _____ _____, and _____ _____ types.

20. A _____ _____ limits the maximum output pressure of an electric fuel pump.
21. Why is a fuel pump switch sometimes used in an electric fuel pump circuit?
22. List nine service rules for fuel lines and hoses.
23. A car engine repeatedly stalls or stops running when driven at a constant speed higher than approximately 60 km/h.
    Mechanic A says that the electric fuel pump may not be running due to a blown fuse.
    Mechanic B says that a fuel filter may be clogged and limiting fuel flow at higher engine speeds.
    Who is correct?
    a. Mechanic A.
    b. Mechanic B.
    c. Both are correct.
    d. Neither A nor B are correct.
24. Name five typical fuel filter locations.
25. An engine is producing a clacking sound from the lower left side of the engine. It is a constant rap noise that speeds up as engine speed increases.
    Mechanic A says that a stethoscope should be used to check for mechanical fuel pump noise. There may be a weak or broken rocker arm spring.
    Mechanic B says that it may be a faulty hydraulic lifter because the speed of the noise changes with engine speed. The problem could NOT be fuel system related.
    Who is correct?
    a. Mechanic A.
    b. Mechanic B.
    c. Both are correct.
    d. Neither A nor B are correct.
26. Fuel pump pressure for a carburettor type fuel system should be approximately _____ to _____ kPa.
27. A petrol injection system can have fuel pressure much higher than a carburettor type fuel system. True or False?
28. What is a fuel pump volume test?
29. An engine fuel pump has failed the pressure and volume tests.
    Mechanic A says that the fuel pump is faulty and should be replaced or overhauled.
    Mechanic B says that a fuel pump vacuum test should be done to make sure the lines and filters before the pump are in good condition.
    Which mechanic is correct?
    a. Mechanic A.
    b. Mechanic B.
    c. Both are correct.
    d. Both are wrong.
30. If an electric fuel pump fails all performance tests, check the supply voltage to the pump before pump replacement. True or False?

Chapter 20

CARBURETTOR FUNDAMENTALS

After studying this chapter, you will be able to:
☐ Describe and identify the basic parts of a carburettor.
☐ Compare carburettor design differences.
☐ List and explain the fundamental carburettor systems.
☐ Explain special carburettor devices.

The principles of supplying an engine with the right amounts of fuel and air have not changed over the years. However, stricter exhaust emission laws and the need for improved fuel economy have changed carburettor requirements. Today's carburettors use numerous devices to alter the air-fuel ratio with changes in engine speed, temperature, and load.

This chapter introduces the fundamental principles of carburation. It discusses carburettor systems, design differences and auxiliary control devices. It will prepare you to later study the service and repair of carburettors.

BASIC CARBURETTOR

A carburettor is basically a device for mixing air and fuel in the correct proportions (amounts) for efficient combustion. The carburettor bolts to the engine intake manifold. The air cleaner fits over the top of the carburettor to trap dust and dirt. See Fig. 20-1.

When the engine is running, downward moving pistons on their intake strokes produce a suction in the intake manifold. Air rushes through the carburettor and into the engine to fill this low pressure. The airflow through the carburettor is used to meter fuel and mix it with the air.

Atmospheric pressure

Atmospheric pressure is the pressure formed by the air surrounding the earth. At sea level, atmospheric pressure exerts about 101 kPa on everything. This pressure is caused by the weight of the air, as shown in Fig. 20-2.

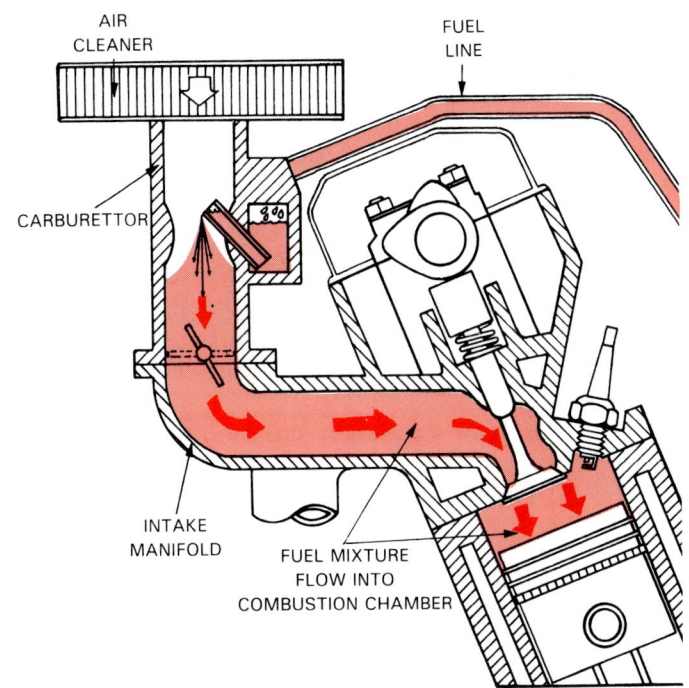

Fig. 20-1. Carburettor bolts to engine intake manifold. It meters and mixes fuel with incoming air.

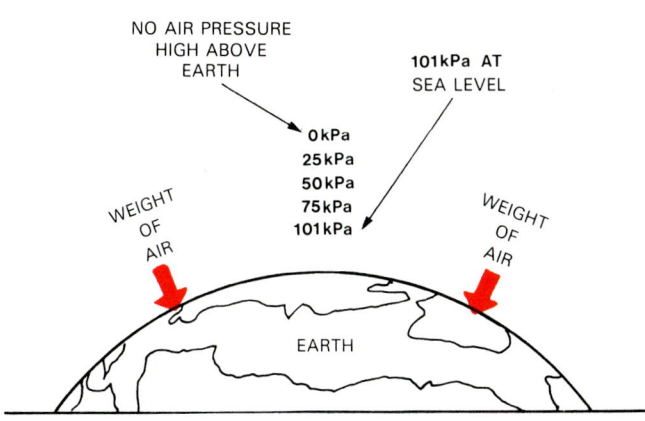

Fig. 20-2. Atmospheric pressure is produced by weight of air above earth. Pressure changes with altitude above sea level. This affects carburation, as you will learn.

Vacuum

A vacuum is lower than atmospheric pressure in an enclosed area. Suction is another word for vacuum. Any space with less than 101 kPa at sea level has a vacuum.

Differences in pressure cause flow

A difference in pressure inside two areas can be used to cause flow. For instance when you suck on a straw, atmospheric pressure pushes down on the liquid in the glass. This causes the liquid to flow through the straw and into the vacuum in your mouth.

A carburettor uses differences in pressure to force fuel into the engine. The engine acts as a vacuum pump, producing a low pressure area (vacuum) in the intake manifold. The carburettor itself also produces a vacuum, as you will learn shortly.

Basic carburettor parts

Fig. 20-3 shows the basic parts of a carburettor. Refer to this illustration as these parts are introduced. A basic carburettor consists of:
1. CARBURETTOR BODY (main carburettor housing).
2. BARREL (air passage containing venturi, throttle valve, and end of main discharge tube).
3. THROTTLE VALVE (airflow control valve in barrel).
4. VENTURI (restriction or narrowed area formed in barrel).
5. MAIN DISCHARGE TUBE (fuel passage between fuel bowl and barrel).

6. FUEL BOWL (fuel storage area in body).

Carburettor body

The carburettor body is a cast metal housing for the other carburettor components, Fig. 20-3. It contains cast and drilled passages for air and fuel. (In an actual carburettor the main discharge tube, venturi, fuel bowl and throttle valve, are normally made as part of the carburettor body.) A flange on the bottom of the body allows the carburettor to be bolted to the engine.

Barrel

The carburettor barrel, routes outside air into the engine intake manifold, Fig. 20-3. It contains the throttle valve, venturi, and outlet end of the main discharge tube.

Throttle valve

The carburettor throttle valve is the disc shaped valve that controls airflow through the barrel. Look at Fig. 20-3. It is mounted on a shaft in the lower part of the barrel.

When closed, the throttle valve restricts the flow of air and fuel into the engine. When the throttle is opened, airflow, fuel flow, and engine power increase.

Fig. 20-4 shows how a car's accelerator pedal and throttle cable control the throttle valve. When the driver presses the accelerator pedal, the throttle cable slides inside its housing. This swings the throttle valve open to increase engine power and speed.

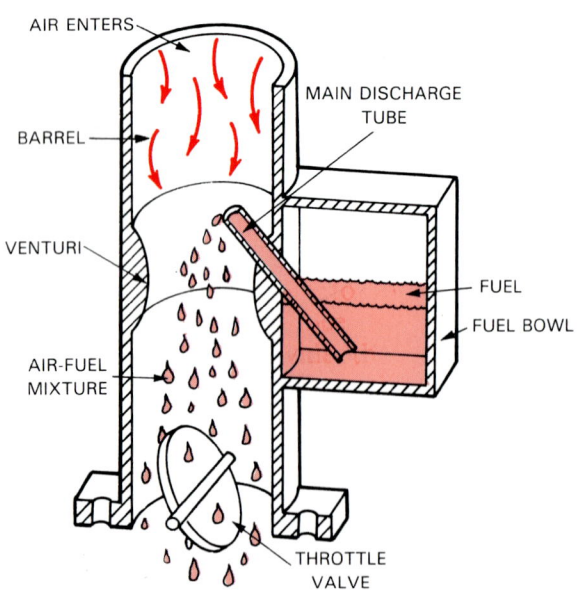

Fig. 20-3. A carburettor controls amount of air and fuel entering engine. Study basic parts.

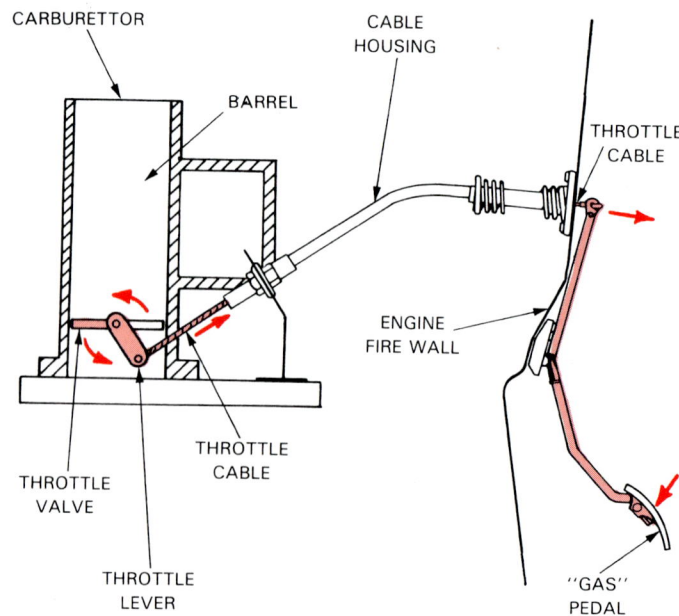

Fig. 20-4. Driver's accelerator pedal is connected to carburettor throttle valve. Valve controls airflow and engine power output.

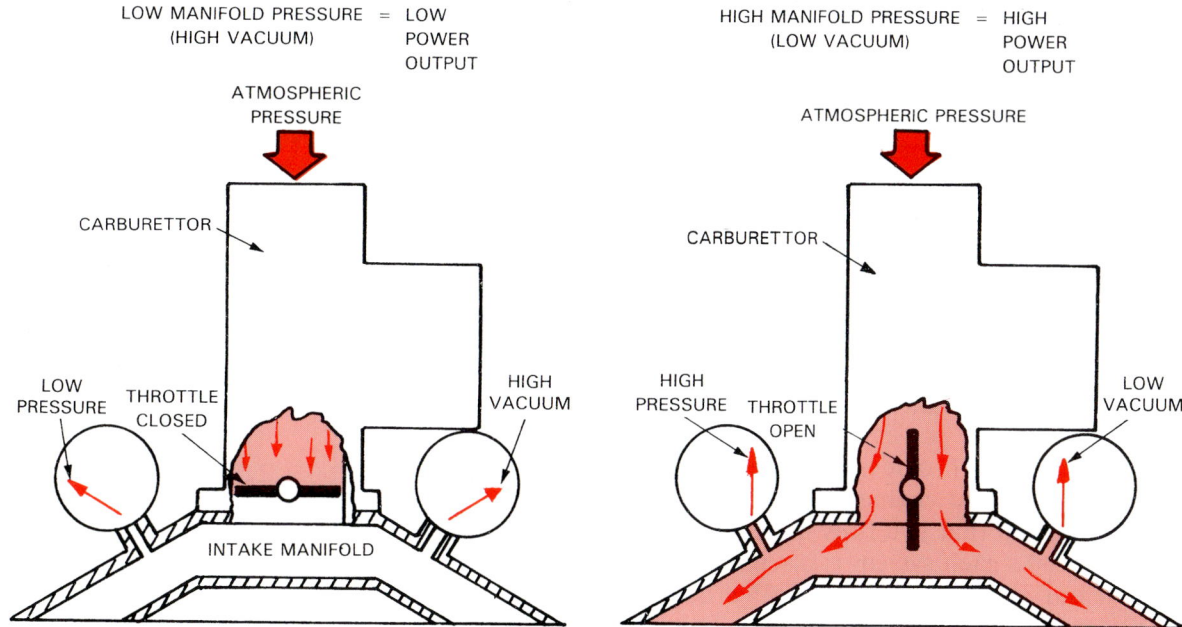

LOW MANIFOLD PRESSURE = LOW
(HIGH VACUUM)   POWER
                       OUTPUT

ATMOSPHERIC
PRESSURE

CARBURETTOR

LOW   THROTTLE   HIGH
PRESSURE   CLOSED   VACUUM

INTAKE MANIFOLD

HIGH MANIFOLD PRESSURE = HIGH
(LOW VACUUM)   POWER
                       OUTPUT

ATMOSPHERIC PRESSURE

CARBURETTOR

HIGH   THROTTLE   LOW
PRESSURE   OPEN   VACUUM

Fig. 20-5. Throttle valve position controls air flow and amount of vacuum in intake manifold. A — Throttle valve closed produces high vacuum in manifold. Engine tries to draw air through carburettor, but cannot. B — Throttle opening allows airflow, reducing vacuum in intake manifold.

When the accelerator pedal is released, a throttle return spring pulls the throttle valve closed. This returns the engine to a slow idle speed. Look at Fig. 20-5.

Venturi

A venturi produces sufficent suction to draw fuel out of the main discharge tube. Venturi action is illustrated in Fig. 20-6. Note how vacuum is highest inside the venturi. The narrowed airway increases air velocity, forming a low pressure area in the barrel.

Main discharge tube

The main discharge tube uses venturi vacuum to feed fuel into the barrel and engine. Also called MAIN FUEL NOZZLE, it is a passage in the

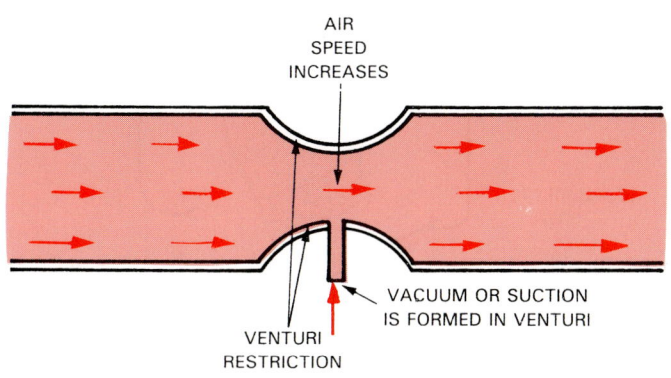

AIR
SPEED
INCREASES

VACUUM OR SUCTION
IS FORMED IN VENTURI

VENTURI
RESTRICTION

Fig. 20-6. Venturi is used to produce vacuum from airflow. Note how vacuum is highest inside venturi. (Ford)

carburettor body that connects the fuel bowl to the centre of the venturi. Refer to Fig. 20-3. Note how main discharge tube is located in carburettor body.

Fuel bowl

The carburettor fuel bowl holds a supply of fuel that is NOT under fuel pump pressure. Several additional carburettor parts are mounted in the fuel bowl. These will be discussed later.

BASIC CARBURETTOR SYSTEMS

A carburettor system is a network of passages and related parts that help control the air-fuel ratio under a specific engine operating condition. Also called a CARBURETTOR CIRCUIT, each system applies a predetermined air-fuel mixture as the temperature, speed, and load of the engine change.

For example, a petrol engine's air-fuel mixture may vary from a rich 8:1 ratio to a lean 18:1 ratio. An automotive carburettor, using its various systems, must be capable of providing air-fuel ratios of approximately:

1. 8:1 for cold engine starting.
2. 16:1 for idling.
3. 15:1 for part throttle.
4. 13:1 for full acceleration.
5. 18:1 for normal cruising at highway speeds.

Note! Older cars, not subject to strict emission control regulations, have a slightly richer air-fuel ratio. Late model cars have leaner carburettor settings to help reduce exhaust pollution.

The seven basic carburettor systems are the:

1. FLOAT SYSTEM (maintains supply of fuel in carburettor bowl).
2. IDLE SYSTEM (provides a small amount of fuel for low speed engine operation).
3. OFF-IDLE SYSTEM (provides correct air-fuel mixture slightly above idle speeds).
4. ACCELERATION SYSTEM (squirts fuel into barrel when throttle valve opens and engine speed increases).
5. HIGH SPEED SYSTEM (supplies lean air-fuel mixture at cruising speeds).
6. FULL POWER SYSTEM (enriches fuel mixture slightly when engine power demands are high).
7. CHOKE SYSTEM (provides extremely rich air-fuel mixture for cold engine starting).

It is very important that you fully understand each of these systems. As each system is discussed, try to draw a "mental picture" of how a carburettor operates under the conditions described. This will help you when diagnosing and repairing carburettor problems.

FLOAT SYSTEM

The float system must maintain the correct level of fuel in the carburettor bowl. Since the carburettor uses differences in pressure to force fuel into the barrel, the fuel in the bowl must be kept at atmospheric pressure. The float system keeps the fuel pump from forcing too much petrol into the carburettor bowl.

Float system parts

Look at Fig. 20-7. The basic parts of a carburettor float system are the fuel bowl, float, needle valve, needle seat, bowl vent, and hinge assembly. Study the relationship of each part.

The carburettor float rides on top of the fuel in the bowl to open and close the needle valve. It is normally made of thin brass or plastic. One end of the float is hinged to the side of the carburettor body. The other end is free to swing up and down. Fig. 20-7.

The needle valve in the fuel bowl regulates the amount of fuel passing through the fuel inlet and needle seat. See Fig. 20-7. The needle valve is usually made of steel. Sometimes, the end of the needle valve will have a soft (synthetic rubber) tip. The soft tip seals better than a metal tip, especially if dirt gets caught in the needle seat.

The carburettor float needle seat works with the needle valve and float to control fuel flow into the bowl. It is normally a brass fitting that threads into the carburettor body, Fig. 20-7.

A bowl vent prevents a pressure or vacuum buildup in the carburettor fuel bowl. Refer to Fig. 20-7 again. Without venting, pressure could form in

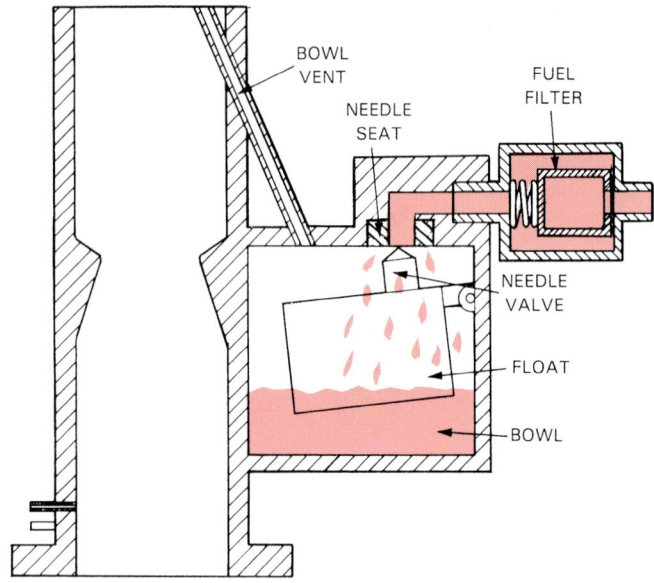

Fig. 20-7. Basic parts of a float system. Float opens and closes needle valve as fuel level falls and rises. Study part names.

the bowl as the fuel pump fills the carburettor. This could also cause vacuum to form in the bowl as fuel is drawn out of the carburettor and into the engine.

Fig. 20-8 shows a bowl vent mechanism on a car equipped with an evaporation control type emission system. Instead of venting the fuel bowl into the outside air, it is vented into a hose going to a charcoal canister. The canister stores toxic fuel vapours and prevents them from entering and polluting the atmosphere.

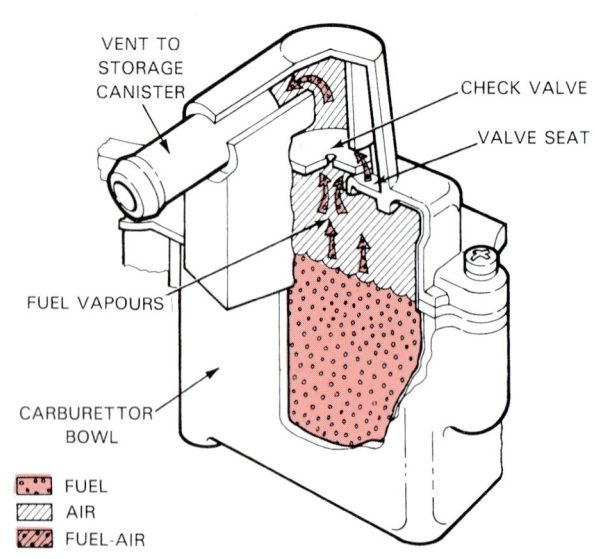

Fig. 20-8. This fuel bowl is vented to emission controlling charcoal canister. Canister stores fuel vapours until engine is started. (Ford)

Float system operation

When engine speed or load increases, fuel is rapidly drawn out of the fuel bowl and into the venturi. Illustrated in Fig. 20-9, this makes the fuel level and float drop in the bowl. The needle valve also drops away from its seat. The fuel pump can then force more fuel into the bowl.

As the fuel level in the bowl rises, the float pushes the needle valve back into the seat. When the fuel level is high enough, the float closes the opening between the needle valve and seat.

With the engine running, the needle valve usually lets some fuel leak into the bowl. As a result, the float system maintains a stable quantity of fuel in the bowl. This is very important because the fuel level in the bowl can affect the air-fuel ratio.

IDLE SYSTEM

A carburettor idle system provides the engine's air-fuel mixture at speeds below approximately 800 rpm or 30 km/h.

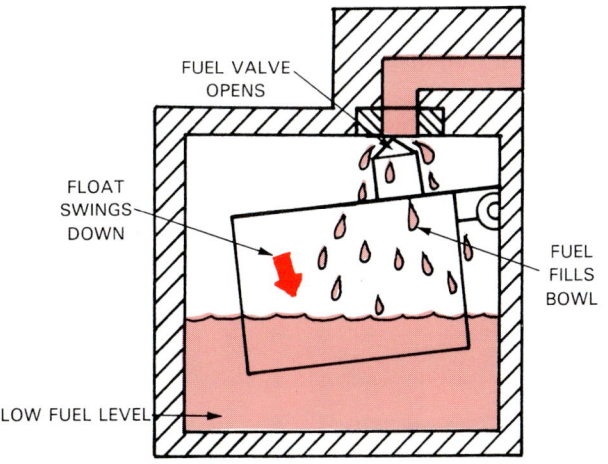

Fuel level low. Float drops and opens needle valve. Fuel flows through needle seat to refill bowl.

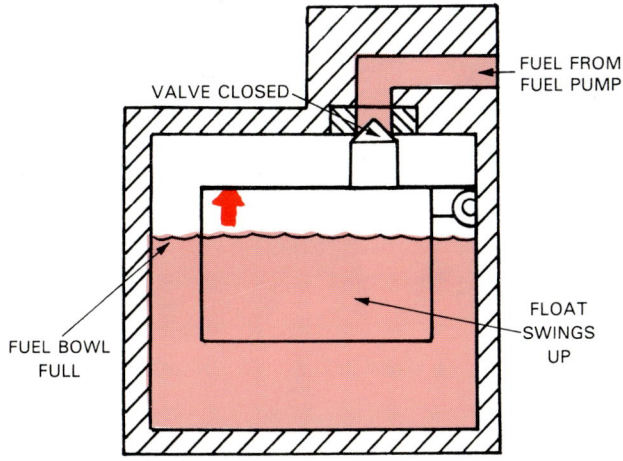

Fuel level up to specs. Float rises and pushes needle valve into needle seat. This blocks fuel entry from pump.

Fig. 20-9. Basic float operation.

When an engine is idling, the throttle valve is almost closed. Airflow through the barrel is too restricted to produce enough vacuum in the venturi. Venturi vacuum cannot draw fuel out of the main discharge tube. Instead, the high intake manifold vacuum BELOW the throttle valve and a separate idle circuit are used to feed fuel into the barrel.

Idle system parts

The fundamental parts of a carburettor idle system include a section of the main discharge tube, a low speed jet, idle air bleed, bypass, idle passage economiser, idle port, and an idle mixture screw. These parts are illustrated in Fig. 20-10.

The low speed jet is a restriction in the idle passage that limits maximum fuel flow in the idle circuit. It is placed in the fuel passage before the idle air bleed and economiser.

The idle air bleed works with the economiser and bypass to add air bubbles to the fuel flowing to the idle port. As shown in Fig. 20-10, the air bubbles help break up or atomise the fuel. This makes the air-fuel mixture burn more efficiently once in the engine.

The idle passage carries the mixture of liquid fuel and air bubbles to the idle screw port.

The idle screw port is an opening into the barrel below the throttle valve.

The idle mixture screw allows adjustment of the size of the opening in the idle screw port, Fig. 20-10.

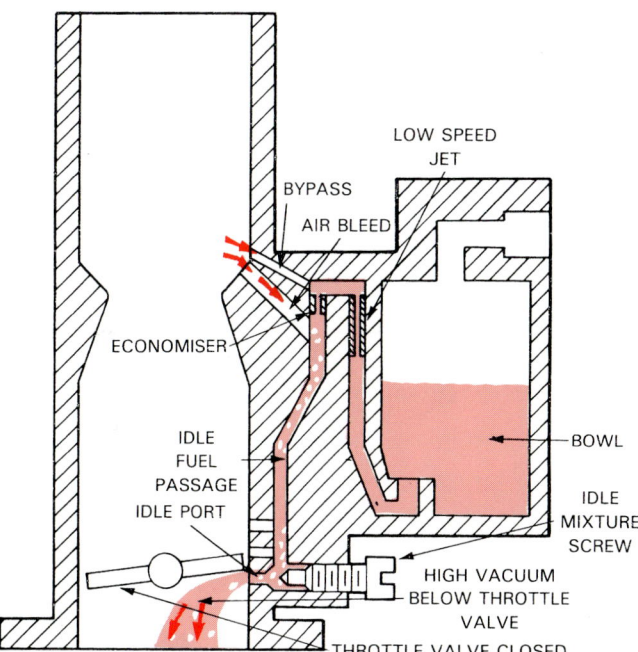

Fig. 20-10. Idle system feeds fuel when throttle is closed for low engine speed operation. High vacuum below throttle pulls fuel out idle port. Mixture screw allows adjustment of mixture at idle. Air bleed helps premix air and fuel.

Turning the idle screw IN reduces the size of the idle port and amount of fuel entering the barrel. Turning the idle screw OUT usually increases fuel flow and enriches the fuel mixture at idle.

Most modern carburettors have sealed idle mixture screws that are NOT normally adjusted. The idle mixture screws are covered with metal plugs, as pictured in Fig. 20-11. This prevents tampering with the factory setting of the idle mixture.

Sometimes, plastic limiter caps are pressed over the heads of the idle mixture screws. They restrict how far the screws can be turned toward rich or lean settings.

The idle screw adjustment of today's carburettors is very critical to exhaust emissions.

Idle system operation

For the idle system to function, the throttle valve must be closed. Then, high intake manifold vacuum can draw fuel out of the idle circuit. Refer to Fig. 20-10.

At idle, fuel flows out of the fuel bowl, through the main discharge, and into the low speed jet. The low speed jet restricts maximum fuel flow.

At the bypass, outside air is drawn into the idle system. This partially atomises the fuel. As the fuel and air bubbles pass through the economiser, the air bubbles are reduced in size to further improve mixing.

The fuel and air mixture then enters the idle screw port. The setting of the idle screw controls how much fuel enters the barrel at idle.

OFF-IDLE SYSTEM

The off-idle system, often termed the PART THROTTLE CIRCUIT, feeds more fuel into the barrel when the throttle valve is partially open. Look at Fig. 20-12. It is an extension of the idle system. It functions ABOVE approximately 800 rpm (30 km/h).

Without the off-idle system, the fuel mixture would become too lean slightly above idle. The idle circuit alone is not capable of supplying enough fuel to the airstream passing through the carburettor. The off-idle circuit helps supply fuel during transition (change) from idle to high speed. Refer to Fig. 20-12.

Off-idle system operation

The off-idle system begins to function when the driver presses lightly on the accelerator pedal and

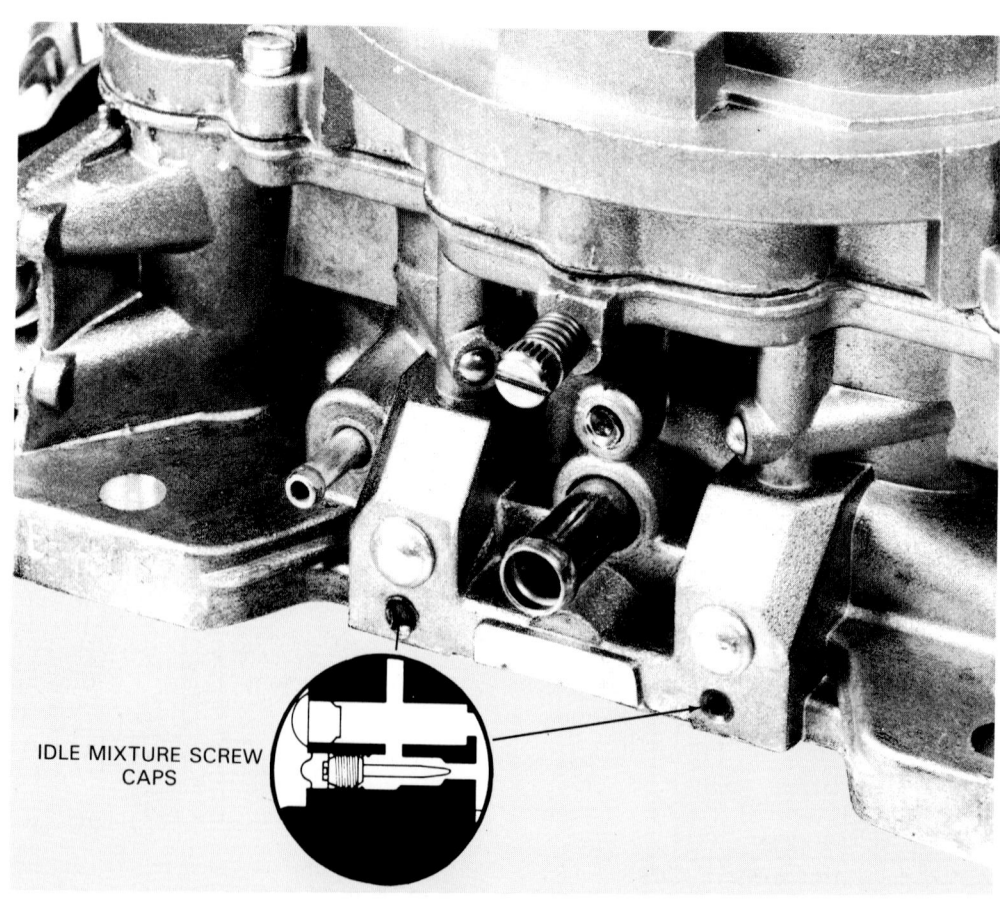

IDLE MIXTURE SCREW CAPS

Fig. 20-11. Modern idle mixture screws are covered with metal plugs. This prevents tampering which would upset mixture and increase exhaust emissions. (Carter Carburettor Div.)

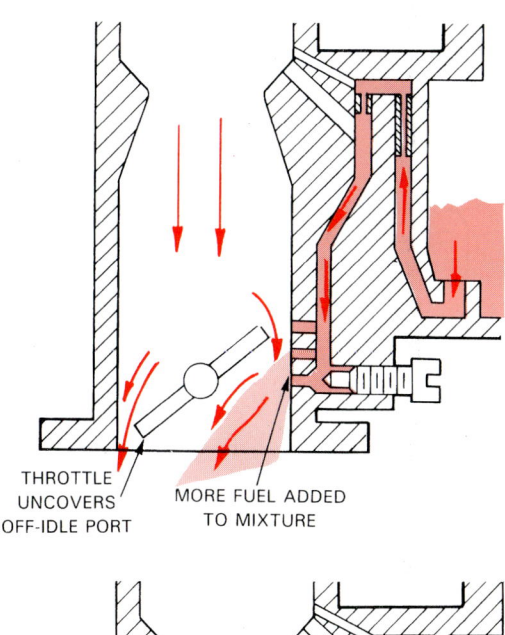

THROTTLE
UNCOVERS
OFF-IDLE PORT

MORE FUEL ADDED
TO MIXTURE

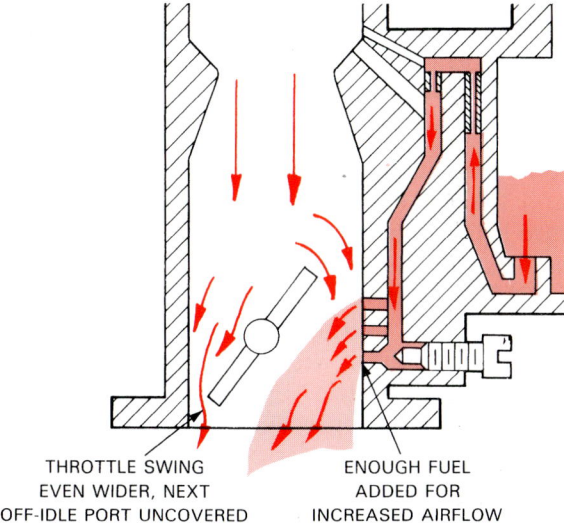

THROTTLE SWING
EVEN WIDER, NEXT
OFF-IDLE PORT UNCOVERED

ENOUGH FUEL
ADDED FOR
INCREASED AIRFLOW

Fig. 20-12. Off-idle system feeds fuel when throttle is opened slightly. It adds a little extra fuel to the extra air flowing around throttle valve.

cracks open the throttle valves. As the throttle valves swing open, they expose the off-idle ports to intake manifold vacuum. Vacuum then begins too draw fuel out of idle screw port and the off-idle ports. This provides enough extra fuel to mix with the additional air flowing around the throttle valves.

Fig. 20-13 pictures the bottom of a carburettor barrel. Notice the idle screw port, idle screw tip, and off-idle ports. Study how the throttle plate exposes all of the ports to vacuum when partially opened.

ACCELERATION SYSTEM

The carburettor's acceleration system, like the off-idle system, provides extra fuel when changing from the idle circuit to the high speed circuit (main discharge).

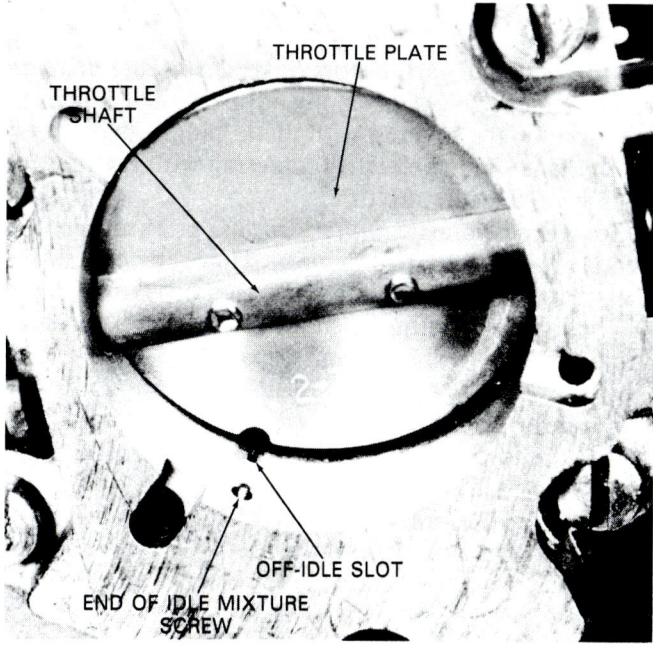

THROTTLE
SHAFT

THROTTLE PLATE

OFF-IDLE SLOT

END OF IDLE MIXTURE
SCREW

Fig. 20-13. Bottom view of actual carburettor shows idle mixture screw tip, idle port opening, and off-idle slot. (Carter Carburettor Div.)

The acceleration system SQUIRTS a stream of extra fuel into the barrel whenever the accelerator pedal is pressed (throttle valves swing open). This is illustrated in Fig. 20-14.

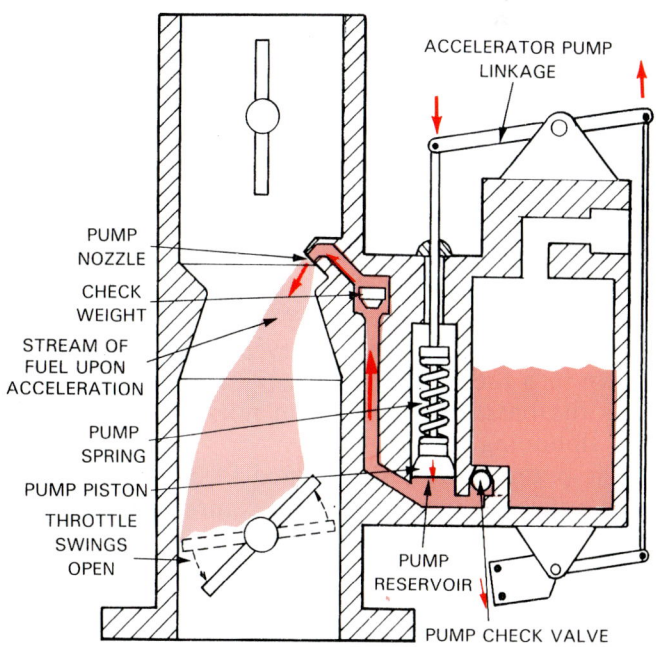

ACCELERATOR PUMP
LINKAGE

PUMP
NOZZLE

CHECK
WEIGHT

STREAM OF
FUEL UPON
ACCELERATION

PUMP
SPRING

PUMP PISTON

THROTTLE
SWINGS
OPEN

PUMP
RESERVOIR

PUMP CHECK VALVE

Fig. 20-14. Accelerator pump system squirts fuel into air horn every time throttle is opened. This adds fuel to rush of air entering engine and prevents temporary lean condition. Study part names.

Without the acceleration system, too much air would rush into the engine as the throttle is quickly opened. The mixture would become too lean for combustion and the engine would HESITATE or STALL. The acceleration system prevents a lean air-fuel mixture from upsetting a smooth increase in engine speed.

Acceleration system parts

The basic parts of a carburettor acceleration system are the pump linkage, accelerator pump, check ball, pump reservoir, pump check weight, and pump nozzle. These parts are given in Fig. 20-14.

The accelerator pump develops the pressure to force fuel out of the pump nozzle and into the barrel. There are two types of accelerator pumps: piston and diaphragm. See Figs. 20-15 and 20-16.

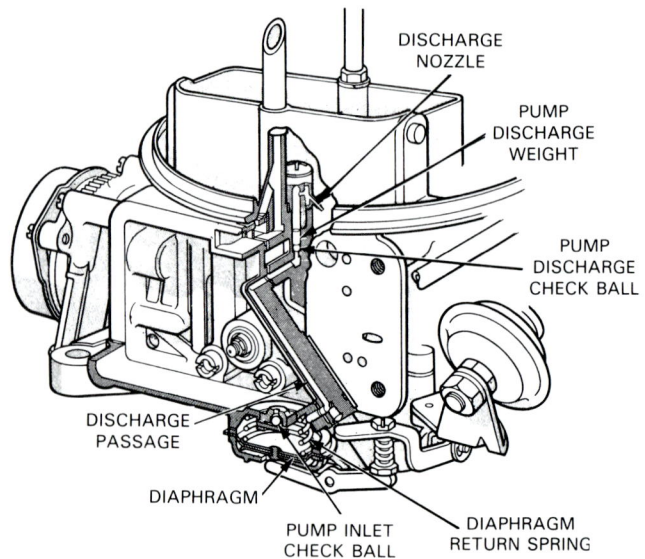

Fig. 20-16. Cutaway view of carburettor using a diaphragm type accelerator pump. (Holley)

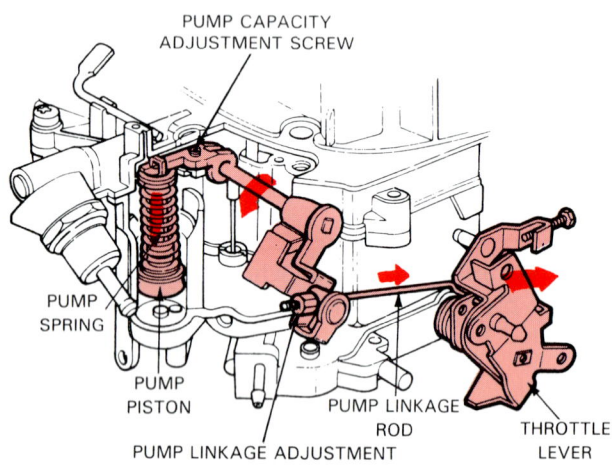

Fig. 20-15. Most accelerator pump systems use mechanical linkage from throttle lever. When driver presses accelerator pedal for acceleration, both the throttle valve and pump are actuated. (Ford)

The pump check ball only allows fuel to flow into the pump reservoir. It stops fuel from flowing back into the fuel bowl when the pump is actuated.

The pump check weight prevents fuel from being drawn into the barrel by venturi vacuum. Its weight seals the passage to the pump nozzle and prevents fuel siphoning.

The pump nozzle, also termed PUMP JET, has a fixed orifice (opening) that helps control fuel flow out of the pump circuit. It also guides the fuel stream into the centre of the barrel, Fig. 20-14.

Acceleration system operation

When the driver presses the accelerator pedal, the throttle valves swing open. This causes the accelerator pump piston or diaphragm to compress the fuel in the pump reservoir.

Accelerator pump pressure closes the pump check ball and fuel flows toward the pump check weight. Pressure lifts the pump check weight off its seat and fuel squirts into the carburettor barrel, as if from a TOY SQUIRT GUN.

A spring is used on the accelerator pump assembly to produce a smooth, steady flow of fuel out the pump nozzle. Throttle opening compresses the spring. Then the compressed spring pushes on the pump piston to pressurise the fuel and produce prolonged fuel flow.

As the accelerator pedal is released, the pump piston or diaphragm retracts. This closes the discharge check weight and opens the pump check ball. Fuel flows out of the bowl to refill the accelerator pump reservoir. The system is then ready to spray another stream of fuel into the barrel when the car accelerates.

Fig. 20-17 shows an auxiliary acceleration system. It supplements the main acceleration system when the engine is cold.

HIGH SPEED SYSTEM

The carburettor's high speed system, also called MAIN METERING SYSTEM, supplies the engine's air-fuel mixture at normal cruising speeds, Fig. 20-18. This circuit begins to function when the throttle valves are open wide enough for venturi action. Airflow through the carburettor must be relatively high for venturi vacuum to draw fuel out of the main discharge tube.

The high speed system provides the leanest, most fuel efficient air-fuel ratio. It functions from about 30 to 90 km/h or 2000 to 3000 rpm.

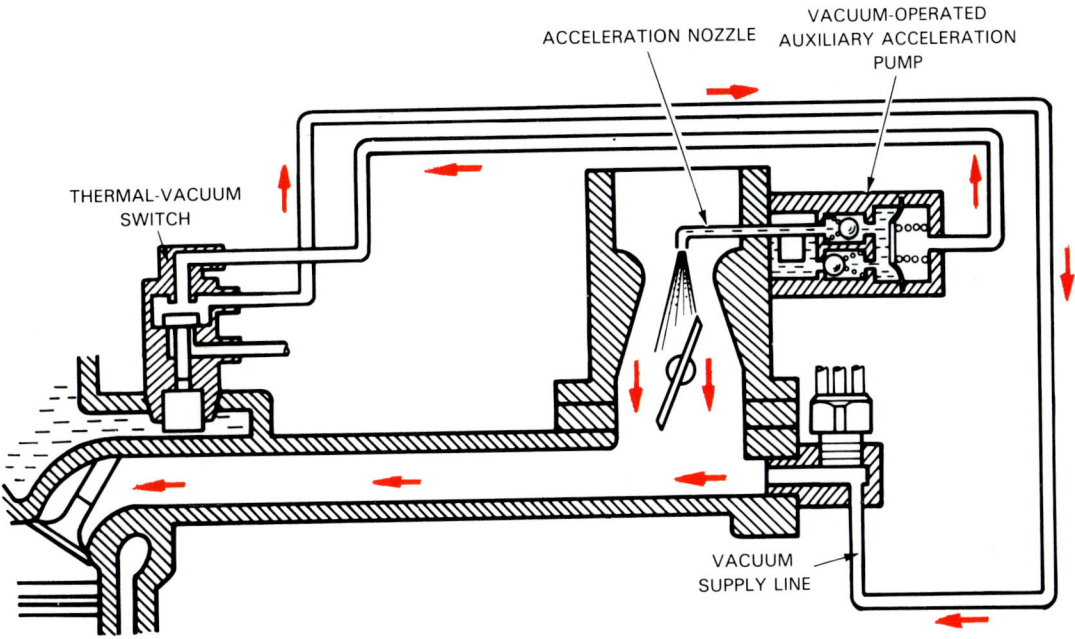

Fig. 20-17. Auxiliary accelerator pump system is sometimes used to aid conventional mechanical pump system. Thermal-vacuum valve is open when engine is cold. This allows engine vacuum to operate vacuum-operated accelerator pump.

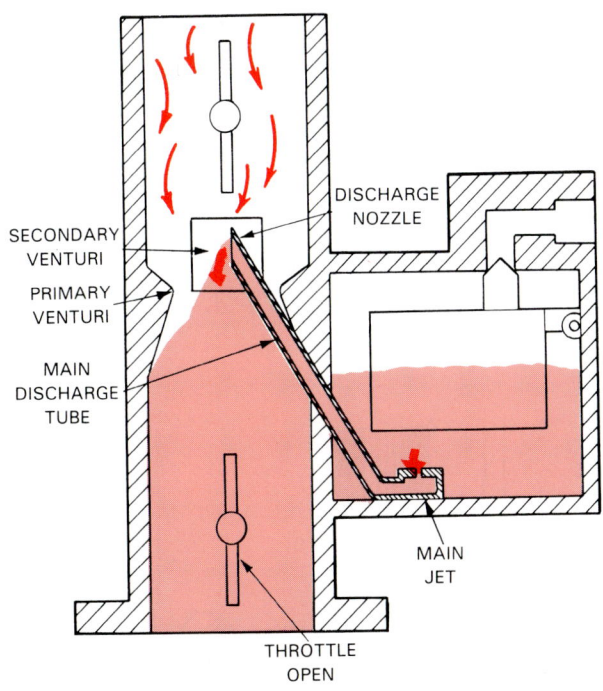

Fig. 20-18. High speed system is simple. Main jet controls fuel flow and mixture. At higher speeds, there is enough airflow through venturi to produce vacuum. This pulls fuel through main discharge. Study part names.

High speed system parts

The high speed system is the simplest carburettor circuit. It consists of a high speed jet, main discharge passage, emulsion tube, air bleed, and venturi.

The high speed jet is a fitting with a precision hole drilled in the centre. This jet screws into a threaded hole in the fuel bowl, Fig. 20-19. One jet is used for each barrel.

The hole size in the main jet determines how much fuel flows through the circuit. A number is usually stamped on a high speed jet to denote the diameter of the hole in the jet. Since jet numbering systems vary, refer to the carburettor manufacturer's manual for information on jet size.

The emulsion tube and the air bleed add air to the fuel flowing through the main discharge tube. See Fig. 20-19. The premixing of air with fuel helps the fuel atomise as it discharges into the barrel.

The primary venturi is the venturi formed in the side of the carburettor barrel.

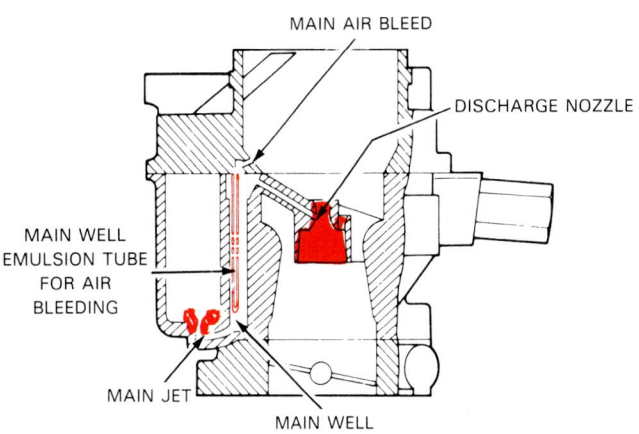

Fig. 20-19. Emulsion tube is commonly used to premix air and fuel before entry into air horn. Note screw-in main jet. (Holley)

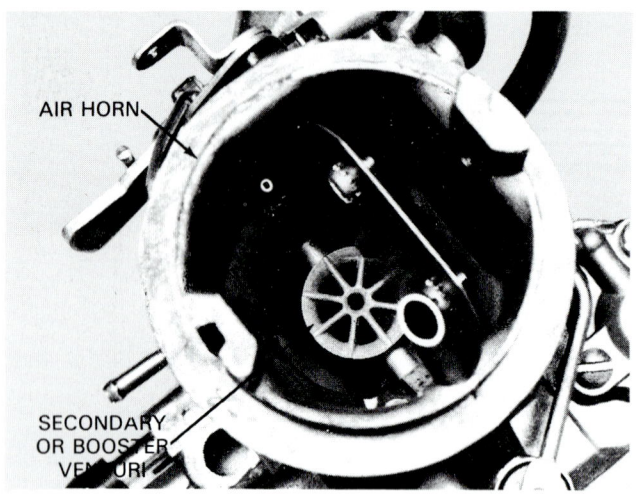

Fig. 20-20. Secondary venturi, also termed booster venturi, is place inside primary venturi. It helps produce venturi vacuum at lower engine speeds. (Holley)

One or two booster venturis can be added inside the primary venturi to increase vacuum at lower engine speeds. This is illustrated in Fig. 20-20.

High speed system operation

When engine speed is high enough, airflow through the carburettor forms a high vacuum in the venturi. The vacuum draws fuel through the main metering system.

Fuel flows through the main jet which meters the amount of petrol entering the circuit. Then, the fuel flows into the main discharge tube and emulsion tube.

The emulsion tube causes air from the air bleed to mix with the fuel. The fuel, mixed with air, is finally drawn out of the main nozzle and into the engine.

FULL POWER SYSTEM

The carburettor full power system provides a means of enriching the fuel mixture for high speed, high power conditions. This circuit operates, for example, when the driver presses the accelerator pedal to pass another vehicle or to climb a steep hill. A simplified illustration of a full power system is given in Fig. 20-21.

The full power system is usually an addition to the main metering system. Either a metering rod or a power valve (jet) can be used to provide a variable, high speed air-fuel ratio.

Metering rod action

A metering rod is a stepped rod that moves in and out of the main jet to alter fuel flow.

As shown in Fig. 20-22, when the metering rod is down inside the jet, flow is restricted and a leaner

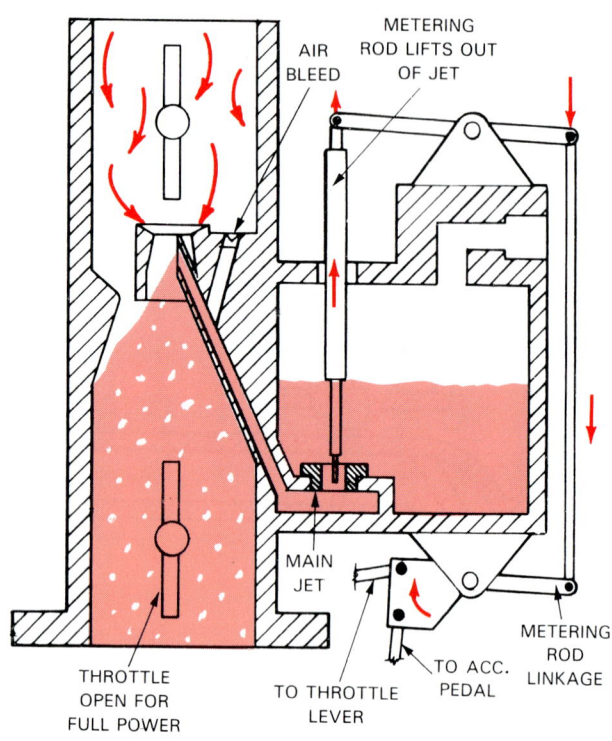

Fig. 20-21. High speed-full power system enriches high speed circuit when needed. When accelerator pedal is pushed down for full power, throttle linkage acts on metering rod linkage. Metering rod is lifted out of main jet to add more fuel to the mixture.

fuel mixture results. When the metering rod is pulled out of the jet, more fuel can flow through the system to enrich the mixture for more power output.

Either mechanical linkage or engine vacuum can be used to operate a metering rod.

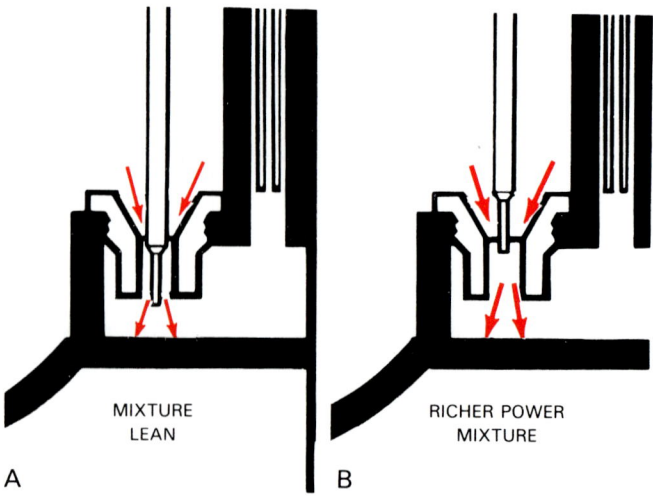

Fig. 20-22. Metering rod action. A — Metering rod lowered into jet. Less fuel can flow through jet, leaning mixture. B — Metering rod pulled out of jet. This allows more fuel flow through jet, enriching mixture.

The metering rod can be linked to the throttle lever. Then, whenever the throttle is opened wide, the linkage lifts the metering rod out of the jet.

A metering rod controlled by engine vacuum is connected to a diaphragm. At steady speeds, power demands are low and engine vacuum is high. The opposite is true under heavy power demands (wide open throttle); intake manifold vacuum drops. This vacuum-load relationship is ideal for controlling a metering rod or power valve.

Power valve action

A power valve, also known as an ECONOMISER VALVE, performs the same function as a metering rod; it provides a variable high speed fuel mixture. A power valve consists of a fuel valve, a vacuum diaphragm, and a spring.

Look at Fig. 20-23. The spring holds the power valve in the normally open position. A vacuum passage runs to the power valve diaphragm. When the power valve is open, it serves as an extra jet that feeds fuel into the high speed circuit.

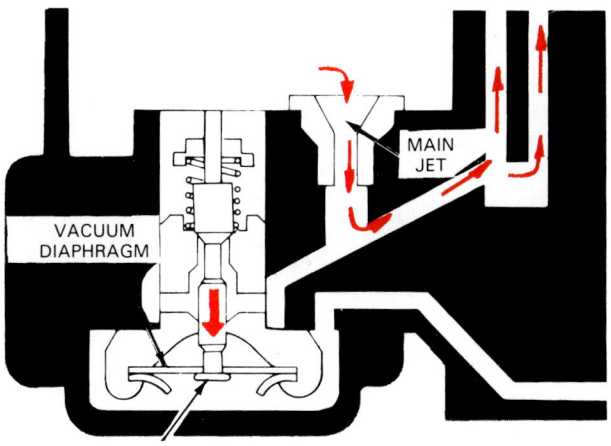

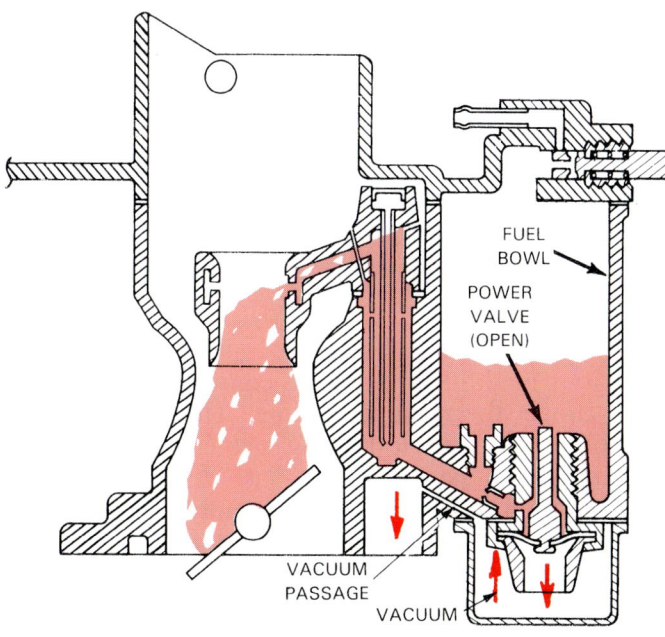

Fig. 20-23. Power valve serves same function as metering rod. It enriches mixture under high load, low intake manifold vacuum conditions. When vacuum is low, spring opens power valve. Extra fuel can then flow through valve and into main discharge.

When the engine is cruising at normal highway speeds, engine intake manifold vacuum is high. This vacuum acts on the power valve diaphragm and pulls the fuel valve closed, Fig. 20-24. No additional fuel is added to the main metering system under normal driving conditions.

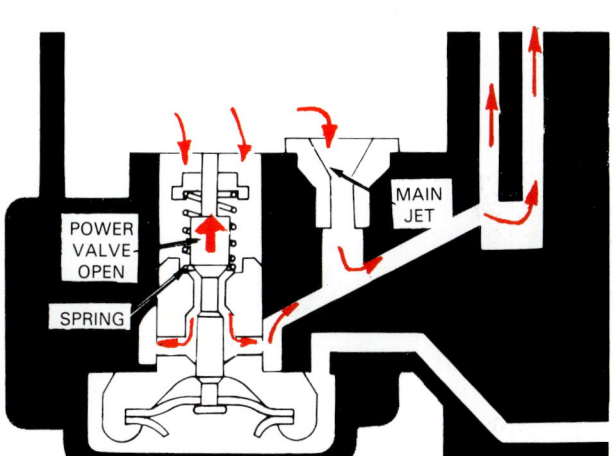

Fig. 20-24. Power valve action. A — High vacuum, low power output closes power valve by pulling on diaphragm. No extra fuel enters main system. B — Engine power output is high, causing intake manifold vacuum to drop. This allows spring to open power valve for more fuel.

However, when the throttle valves are swung open for passing or climbing a hill, engine manifold vacuum drops. Then, the spring in the power valve can push the fuel valve open. Fuel flows through the power valve and into the main metering system. This adds more fuel for more engine power.

CHOKE SYSTEM

The choke system is designed to supply an extremely rich air-fuel ratio to aid cold engine starting.

For the fuel mixture to burn properly, the fuel entering the intake manifold must atomise and vapourise. When the engine is cold, the fuel entering the intake tends to condense into a liquid. As a result, not enough fuel vapours enter the combustion chambers and the engine could miss or stall when cold. A choke is used to prevent this lean condition.

Choke system parts

A choke system has a choke valve (plate), thermostatic spring, and other parts depending upon choke design. See Fig. 20-25.

The choke valve is a butterfly (disc) type valve located near the top of the carburettor barrel.

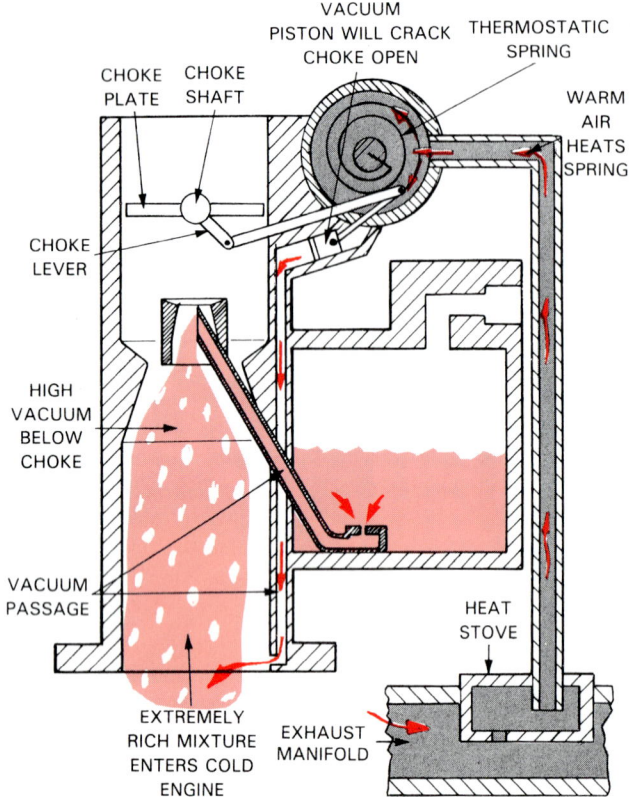

Fig. 20-25. Basic choke system parts. Thermostatic spring is main control of choke operation. When engine is cold, spring closes choke. High vacuum below choke pulls large amount of fuel out of main discharge. When engine warms, hot air causes spring to open choke. Vacuum piston cracks choke upon engine starting to prevent flooding.

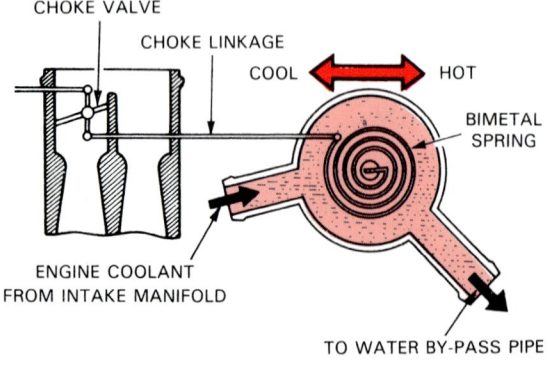

Fig. 20-26. Instead of hot air, this thermostatic choke spring is warmed by engine coolant. (Toyota)

When the choke valve is closed, it blocks normal airflow through the carburettor. This causes high intake manifold vacuum to form below the choke valve. Vacuum pulls on the main discharge tube, even though air is not flowing through the venturi. Fuel is drawn out to prime the engine with extra fuel.

A thermostatic spring may be used to open and close the choke. Refer to Fig. 20-26.

The thermostatic spring is a bimetal spring (spring made of two disimilar metals). The two metals have different rates of expansion that make the spring coil tighter when cold. It uncoils when heated. This coiling-uncoiling action is used to operate the choke.

Basic choke action

Before the engine starts, the cold thermostatic spring holds the choke closed. When the engine is started, the closed choke causes high vacuum in the carburettor barrel. This draws a large amount of fuel out of the main discharge. The rich mixture helps keep the cold engine running.

As the engine and thermostatic spring warm, the spring uncoils and opens the choke. This produces a leaner mixture. A warm engine would not run properly if the choke were to remain closed.

Manual choke

A manual choke simply uses a cable mechanism that allows the driver to open or close the choke valve. Normally, when the driver pulls on the choke knob, a cable pulls the choke valve closed for cold engine starting. The driver must open the choke valve after the engine starts and warms slightly.

Automatic chokes

Various methods are used to control the warming of a choke thermostatic spring. Hot air, engine coolant, or an electric heating element can operate the thermostatic spring.

An integral hot air choke is mounted on the side of the carburettor. It uses WARM AIR from the engine to heat the thermostatic spring. One is shown in Fig. 20-25. An integral choke may also use engine coolant instead of warm air, Fig. 20-26.

A nonintegral choke mounts the thermostatic spring in the top of the intake manifold. Then, as the engine and manifold warm, the thermostatic spring uncoils to open the choke plate. See Fig. 20-27.

An electric assist choke uses both hot air and an electric heating element to operate the thermostatic spring. Look at Fig. 20-27. The electric assist choke system uses a temperature sensitive switch to operate a choke heating element.

When the engine is started cold, the choke switch is open and current does not flow to the heating element. The choke thermostatic spring is only

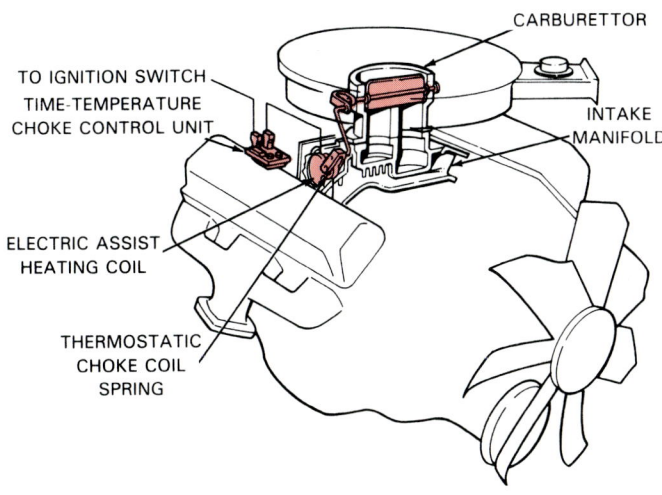

Fig. 20-27. Electric assist choke uses engine heat and an electric heating element to warm choke.

warmed by hot air and the choke remains partially closed to aid cold engine operation.

When the engine warms, the temperature sensing switch closes and current flows to the heating element. This speeds thermostatic spring action and the choke opens more quickly. As a result, the control of choke opening is more precise, reducing fuel consumption and exhaust emissions.

An all electric choke uses neither hot air nor coolant to aid thermostatic spring action. Instead, a two-stage heating element provides full control of choke operation. See Fig. 20-28.

When the engine is cold, only the first stage of the heating element is activated. The thermostatic spring

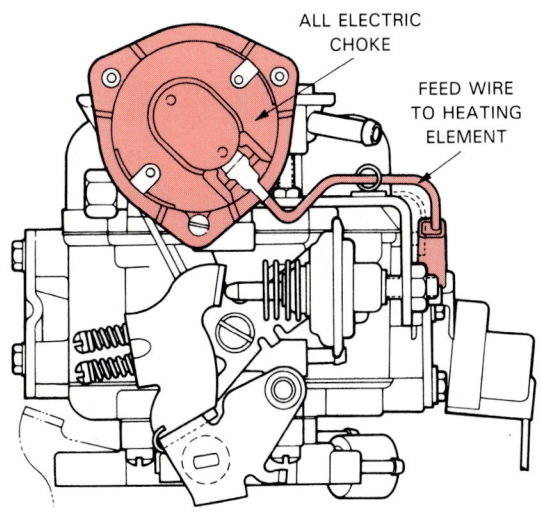

Fig. 20-28. All electric choke has two stage heating element. One stage warms thermostatic spring when engine is cold. When engine is partially warm, both heating stages function. (Ford)

warms slowly to keep the choke partially closed. When the engine warms, both stages of the heating element operate. The element heats up very quickly to make the thermostatic spring open the choke.

Most choke systems use other components to aid choke action. You will learn about them next.

Mechanical choke unloader

A mechanical choke unloader physically opens the choke valve whenever the throttle swings fully open. Look at Fig. 20-29.

A mechanical choke unloader uses a metal lug on the throttle lever. When the throttle lever moves to the fully open position, the lug pushes on the choke linkage (fast idle linkage). This gives the driver a means of opening the choke. Air can then enter the barrel to help clear a flooded engine (engine with too much liquid fuel in cylinders and intake manifold).

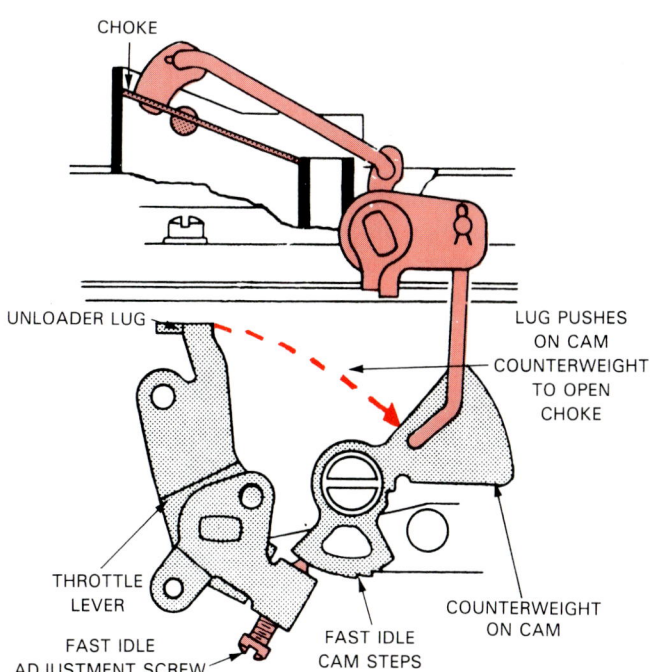

Fig. 20-29. Choke unloader physically opens choke when accelerator pedal is pushed to floor. The throttle lever lug moves the choke linkage. This lets driver clean out flooded engine.

Vacuum choke unloader

A vacuum choke unloader, also called a CHOKE BREAK, uses engine vacuum to crack open the choke valve as soon as the engine starts. It automatically prevents the engine from flooding with too much fuel.

A vacuum choke unloader consists of a manifold vacuum fitting, vacuum hose, vacuum diaphragm

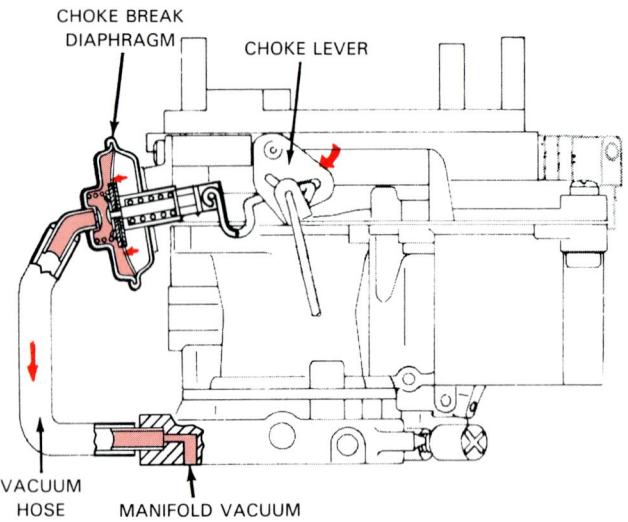

Fig. 20-30. *Vacuum choke break or unloader cracks choke open as soon as engine is started. Engine vacuum pulls on diaphragm. Diaphragm pulls on choke linkage.*

(choke break), and linkage connected to the choke lever. These parts are shown in Fig. 20-30.

Before the engine starts, the choke spring holds the choke valve almost completely closed. This primes the engine with enough fuel for starting.

Then, as the engine starts intake manifold vacuum acts on the choke break diaphragm. The diaphragm pulls on the choke linkage and lever to swing the choke valve open slightly. This helps avoid an overrich mixture and improves cold engine driveability.

FAST IDLE CAM

A fast idle cam increases engine idle speed when the choke is closed. It is a stepped cam fastened to the choke linkage, Fig. 20-31.

When the choke closes, the fast idle cam swings around in front of a fast idle screw. The fast idle screw is mounted on a throttle lever. As a result, the fast idle cam and fast idle screw prevent the throttle valves from closing. Engine idle speed is increased to smooth cold engine operation and prevent stalling.

As soon as the engine warms and the choke opens, the fast idle cam is deactivated. When the throttle is opened, the choke linkage swings the cam away from the fast idle screw and the engine returns to curb idle (normal, hot idle speed).

FAST IDLE SOLENOID

A fast idle solenoid opens the carburettor throttle valves during engine operation but allows the throttle valves to close as soon as the engine shuts off.

In this way, a faster idle speed can be used while still avoiding dieseling (engine keeps running even though ignition key is turned off).

Sometimes termed an anti-dieseling solenoid, the fast idle solenoid is mounted on the carburettor so that it contacts the throttle lever, as in Fig. 20-32.

When the engine is running, current flows to the fast idle solenoid. This causes the plunger in the solenoid to move outward. The throttle valves are held open to increase engine speed.

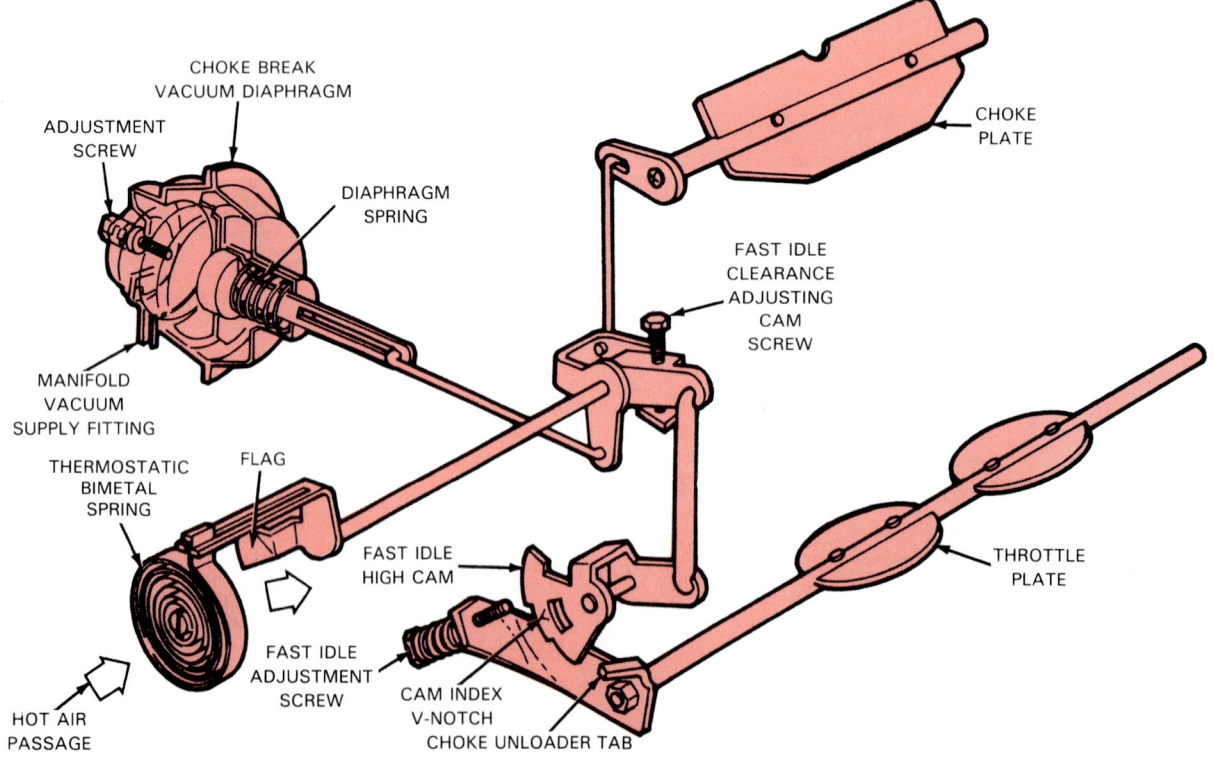

Fig. 20-31. *Note relationship of these parts. They work together to provide smooth engine operation when engine is cold. (Ford)*

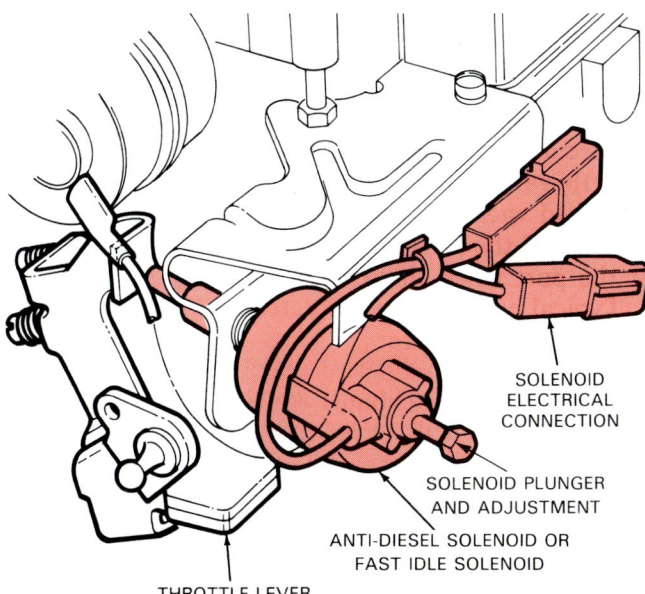

Fig. 20-32. Fast idle solenoid holds throttle open when engine is running. When ignition key is turned off, solenoid allows throttle to close more. This keeps engine from dieseling or after-running. (Ford)

When the engine is shut off, current flow to the solenoid stops. The solenoid plunger retracts and the throttle valves are free to swing almost closed.

THROTTLE RETURN DASHPOT

A throttle return dashpot causes the carburettor throttle valves to close slowly, Fig. 20-33. Frequently called an ANTI-STALL DASHPOT, it is commonly used on carburettors for automatic transmission equipped cars.

Without a throttle return dashpot, the engine could stall when the engine quickly returns to an idle. The drag of the automatic transmission could ''kill'' the engine.

The throttle return dashpot works something like a shock absorber. It uses a spring-loaded diaphragm mounted in a sealed housing. A small hole is drilled in the diaphragm housing. The small air hole prevents rapid movement of the dashpot plunger and diaphragm. Air must bleed out of the small hole slowly.

When the car is travelling down the road (throttle valves open), the spring pushes the dashpot plunger outward. Then, when the engine returns to idle, the throttle lever strikes the extended dashpot plunger. As air leaks out of the throttle return dashpot, the engine slowly returns to curb idle. This gives the automatic transmission enough time to disconnect (torque converter releases) from the engine, without engine stalling.

HOT IDLE COMPENSATOR

A hot idle compensator is a carburettor device that prevents engine stalling or a rough idle under high engine temperature conditions. As pictured in Fig. 20-34, it is a temperature sensitive valve that admits extra air into the engine to increase idle speed and smoothness.

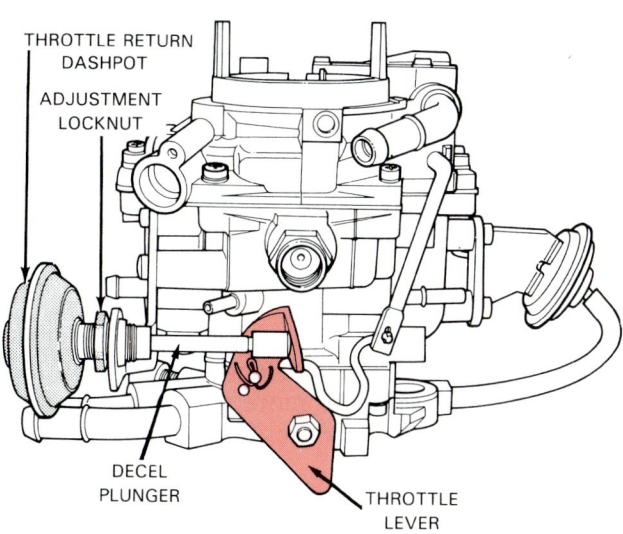

Fig. 20-33. Throttle return dashpot keeps engine from stalling when engine is quickly returned to an idle. It is normally used on cars with automatic transmission. Dashpot makes throttle plates close slowly.

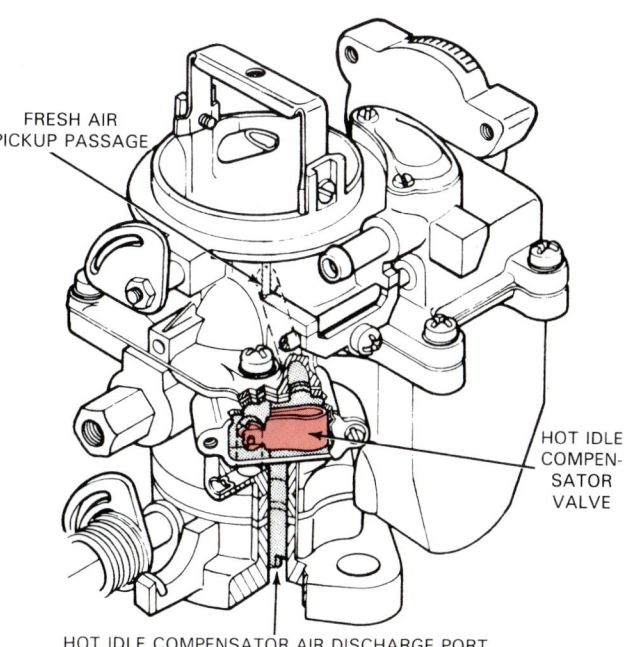

Fig. 20-34. Hot idle compensator adds extra air during high temperature conditions. More air is needed to offset extra fuel vapours caused by heat. (Ford)

With normal engine temperatures, the hot idle compensating valve remains closed. The engine idles normally.

When temperatures are high (prolonged idling periods for example), fuel vapours can enter the barrel and enrich the air-fuel mixture. At this time, the hot idle compensator opens to allow extra air to enter the intake manifold. This compensates for the extra fuel vapours and the correct fuel mixture is maintained.

ALTITUDE COMPENSATOR

An altitude compensator can be used to change the carburettor's air-fuel mixture with changes in the car's height above or below sea level. An altitude compensator normally has an aneroid (bellows device that expands and contracts with changes in atmospheric pressure), Fig. 20-35.

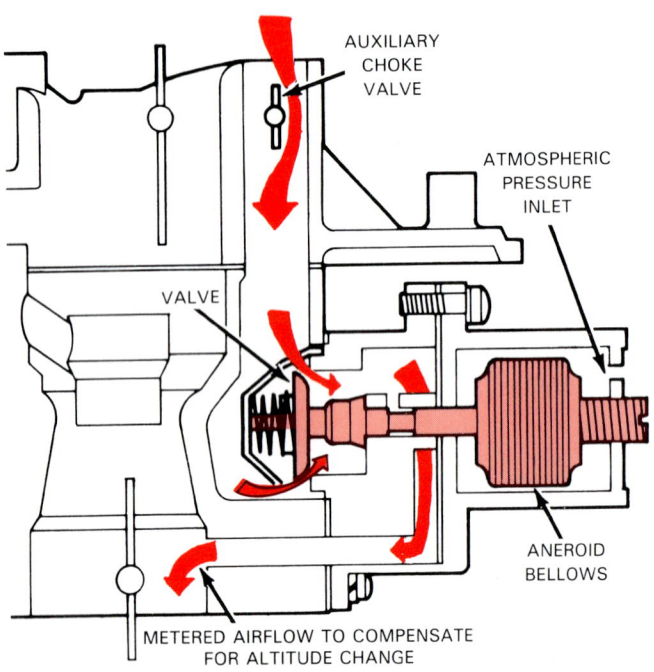

Fig. 20-35. Altitude compensation circuit used aneroid bellows. Bellows expands and contracts with changes in altitude and atmospheric pressure. This increases or decreases air flow through circuit, maintaining correct air-fuel ratio.

When a car is driven up a mountain, for example, the density of the air around the car decreases. This tends to make the air-fuel mixture richer. The reduced air pressure causes the aneroid to expand, opening an air valve. Extra air flows into the air horn and the air-fuel mixture becomes leaner as needed.

The opposite occurs when the car's height above sea level decreases. The greater air density and pressure tends to make the carburettor mixture too lean. The increased air pressure collapses the aneroid and the air valve closes. This enriches the mixture enough to compensate for the lower altitude.

CARBURETTOR VACUUM CONNECTIONS

A modern carburettor has numerous vacuum connections. Look at Fig. 20-36. When the vacuum connection or port is BELOW the carburettor throttle valve, the port ALWAYS receives full intake manifold vacuum. However, when the vacuum port is ABOVE the throttle valve, vacuum is only present at the port when the THROTTLE IS OPENED.

Typical components operated off carburettor vacuum connections are:

1. EGR VALVE (exhaust emission control device).
2. DISTRIBUTOR VACUUM ADVANCE (diaphragm for advancing ignition timing).
3. CHARCOAL CANISTER (emission control container for storing fuel vapours).
4. CHOKE BREAK (diaphragm for partially opening choke when engine is running).

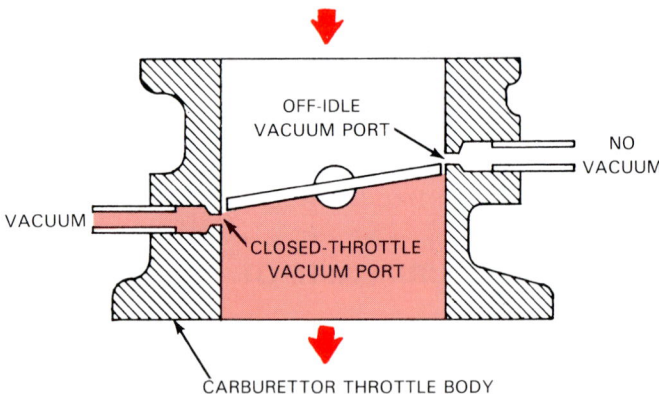

Fig. 20-36. Position of vacuum port in carburettor controls when vacuum is present. When below throttle valve, vacuum is present at idle. When above throttle, vacuum is present above idle.

TWO AND FOUR-BARREL CARBURETTORS

So far, this chapter has discussed mainly the single barrel carburettor (carburettor body with one barrel). Automotive carburettors also are available in two-barrel (two barrels in single body) and four-barrel (four barrels in one body) designs. All are illustrated in Fig. 20-37. Compare them.

Multiple barrel carburettors are used to provide increased "engine breathing" (air intake). The amount of fuel and air that enter the engine is a factor limiting engine horsepower output. Extra carburettor barrels allow more air and fuel into the engine at wide open throttle. This allows the engine to develop more power.

Carburettor primary and secondary

In-line, two-barrel and all four-barrel carburettors are divided into two sections: the primary and the secondary.

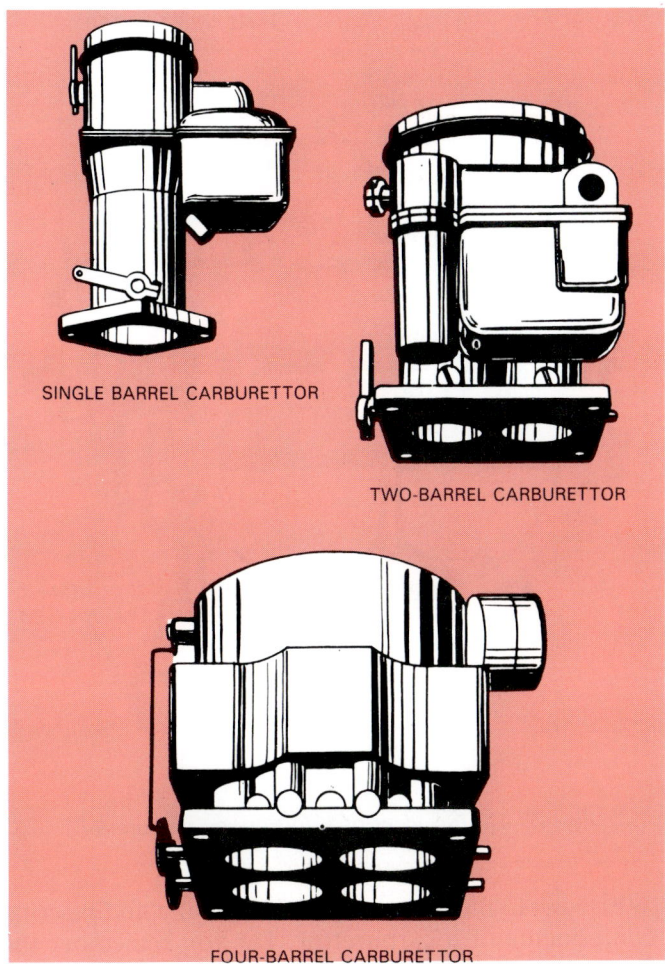

Fig. 20-37. One, two, and four-barrel carburettors are common. More barrels or air horns are used with larger engines.

The primary of a carburettor includes the components that operate under normal driving conditions. In a four-barrel carburettor for example, it consists of the two front throttle valves and related components. Refer to Fig. 20-38.

The secondary of a carburettor consists of the components or circuits that function under high engine power output conditions, Fig. 20-38. In a four-barrel, this would be the two rear barrels. They only function when more fuel is needed for added power.

A secondary diaphragm is normally used to open the secondary throttle valves, causing the secondary circuits to function. Illustrated in Fig. 20-38, a diaphragm is connected to the secondary throttle lever. A vacuum passage runs from this diaphragm to the venturi in the primary throttle bore.

Under normal driving conditions, vacuum in the primary is NOT high enough to actuate the secondary diaphragm and throttles. The engine would run using only the primary of the carburettor.

If the driver passes another car, for example, increased airflow in the primary produces enough vacuum to operate the secondary diaphragm. Vacuum pulls on the diaphragm and compresses the diaphragm spring. This opens the secondary throttle valves for increased engine horsepower.

AUXILIARY AIR VALVE

Some four-barrel and a few two-barrel carburettors use an auxiliary butterfly valve over the secondary throttle valves. One type of air valve is

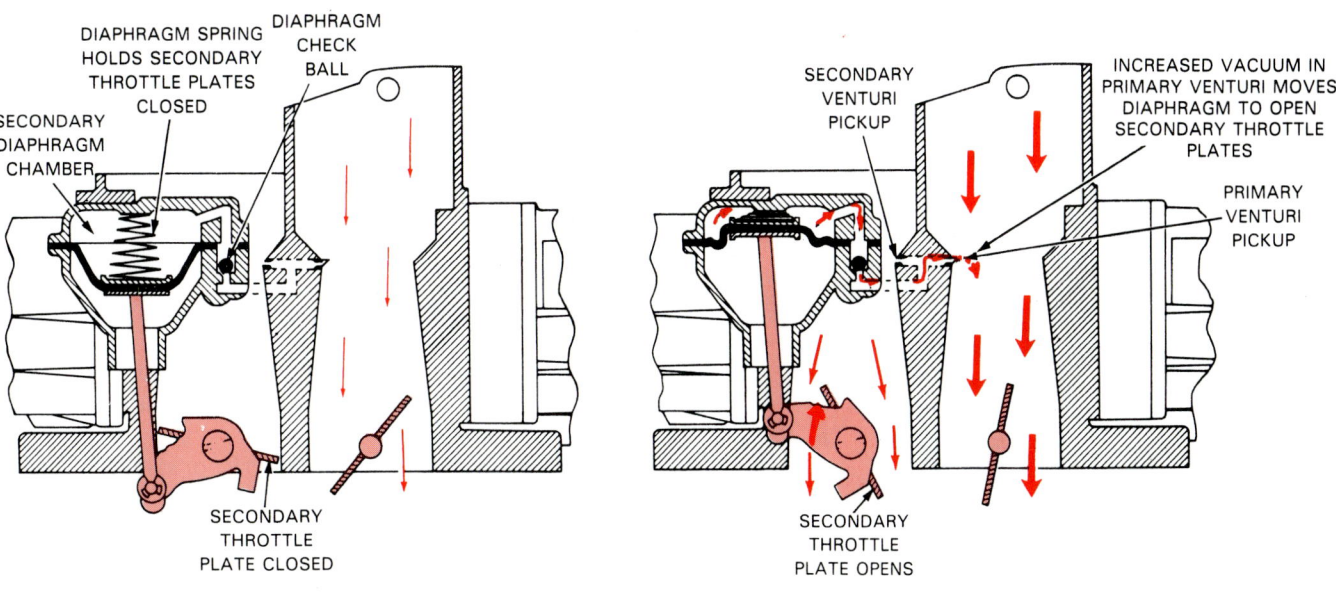

Fig. 20-38. Primary is front barrel or barrels of carburettor. Secondary is rear of carburettor. Note how secondary diaphragm opens rear throttle plates when engine power output is high. (Holley)

shown in Fig. 20-39. It should not be confused with a choke valve. They look very similar.

An auxiliary throttle valve is designed to keep secondary barrel air velocity high enough to assure complete fuel mixing and atomisation. It stops secondary throat operation until primary air speed is high enough to allow efficient operation.

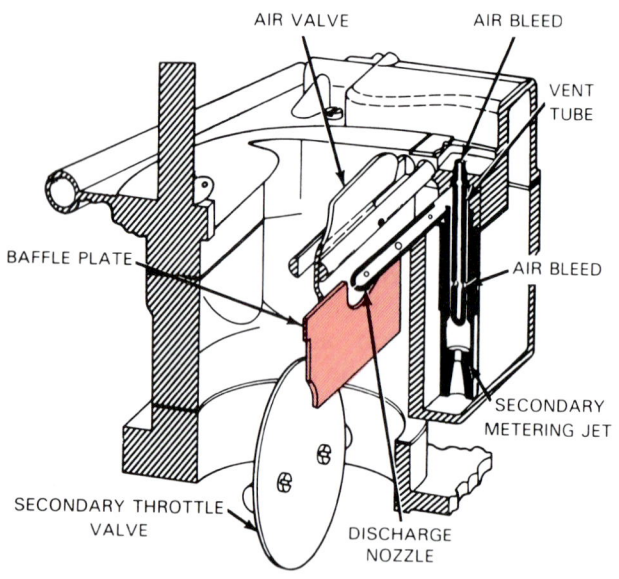

Fig. 20-39. Air valve over secondary throttles only opens if air flow is high enough for proper secondary operation.

The auxiliary air valve is normally closed and covers the barrel entrance. It is held shut by a light spring or counterweight.

Figs. 20-40 and 20-41 show two and four-barrel carburettors. Study the part names and their locations.

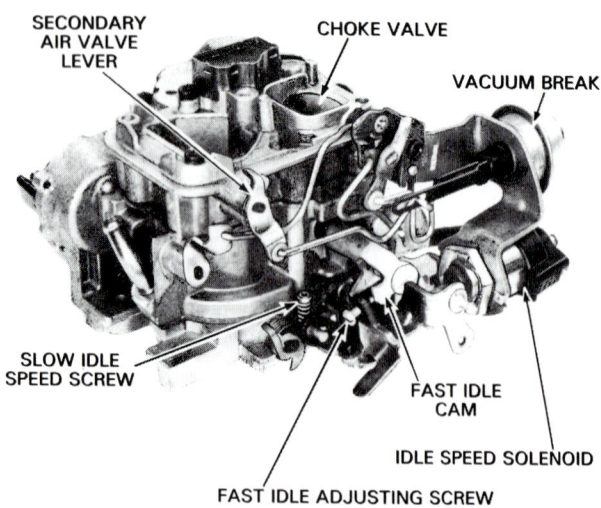

Fig. 20-40. Modern two-barrel carburettor. Note parts.

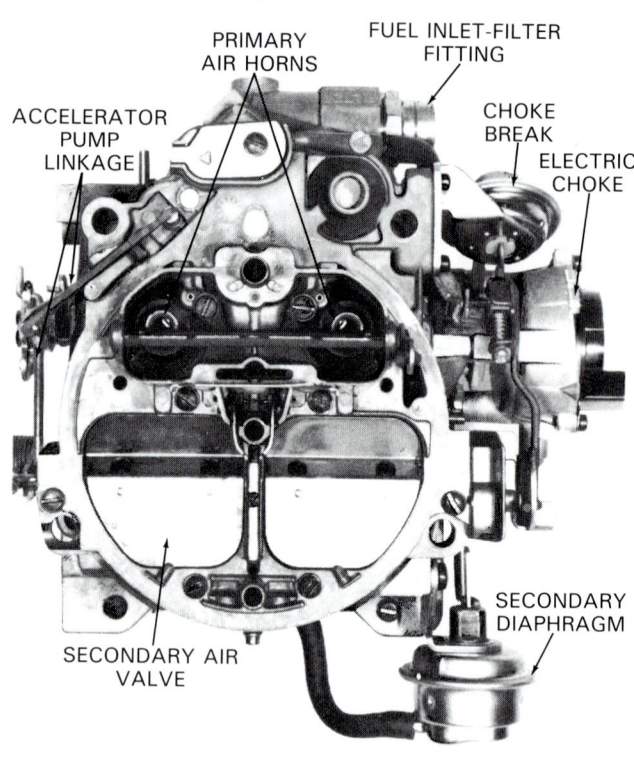

Fig. 20-41. Top view of a four-barrel carburettor.

Fig. 20-42 illustrates all of the circuits and internal components of a carburettor. Do you remember the functions of each part?

KNOW THESE TERMS

Atmospheric pressure, Vacuum, Barrel, Throttle valve, Venturi, Main discharge tube, Carburettor system, Float, Needle valve, Bowl vent, Idle mixture screw, Accelerator pump, Main jet, Metering rod, Power valve, Thermostatic spring, Mechanical choke unloader, Vacuum choke break, Fast idle cam, Fast idle solenoid, Throttle return dashpot, Hot idle compensator, Altitude compensator, Primary, Secondary.

REVIEW QUESTIONS

1. A carburettor is basically a device for _____ air and fuel in the correct _____ (amounts) for efficient combustion.
2. Define the term "atmospheric pressure".
3. Define the term "vacuum".
4. List and explain the six major parts of a carburettor.
5. The barrel is the main fuel passage in the carburettor. True or False?
6. The _____ _____ control airflow through the carburettor to allow the driver to control engine power.

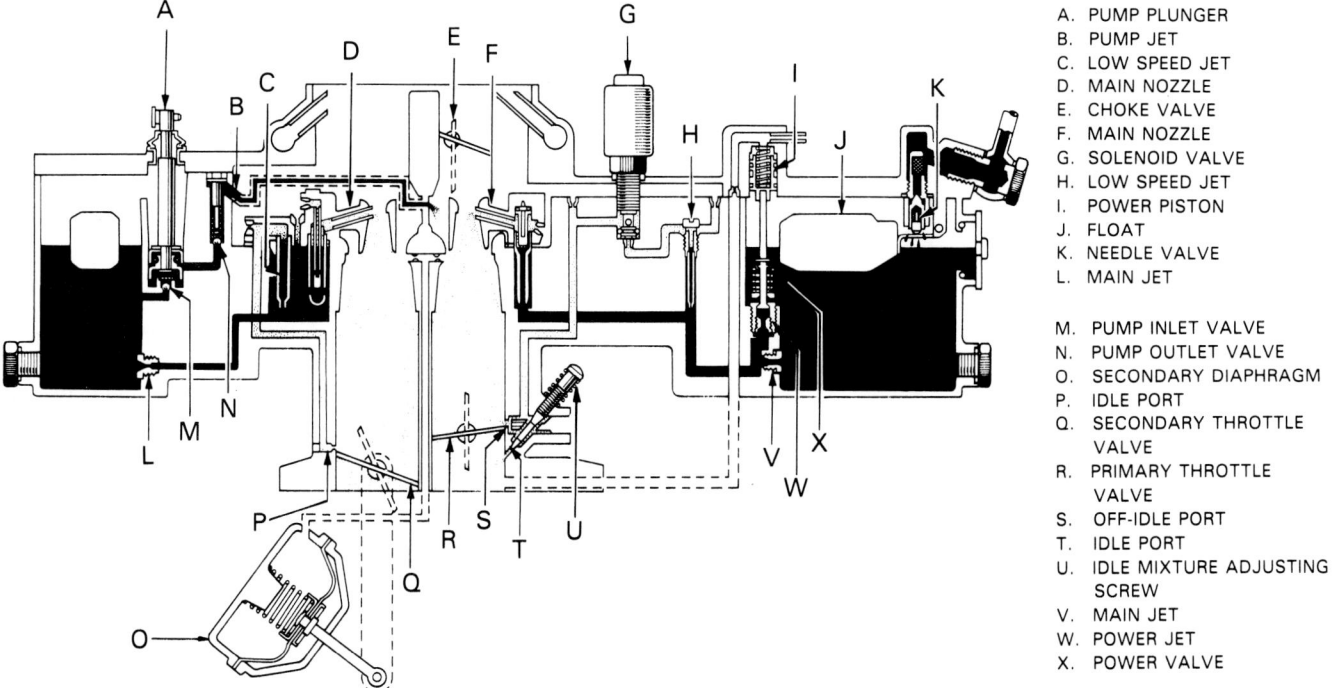

A. PUMP PLUNGER
B. PUMP JET
C. LOW SPEED JET
D. MAIN NOZZLE
E. CHOKE VALVE
F. MAIN NOZZLE
G. SOLENOID VALVE
H. LOW SPEED JET
I. POWER PISTON
J. FLOAT
K. NEEDLE VALVE
L. MAIN JET

M. PUMP INLET VALVE
N. PUMP OUTLET VALVE
O. SECONDARY DIAPHRAGM
P. IDLE PORT
Q. SECONDARY THROTTLE
   VALVE
R. PRIMARY THROTTLE
   VALVE
S. OFF-IDLE PORT
T. IDLE PORT
U. IDLE MIXTURE ADJUSTING
   SCREW
V. MAIN JET
W. POWER JET
X. POWER VALVE

Fig. 20-42. Try to identify and explain carburettor parts before looking at listed names. (Toyota)

7. A venturi is used to produce vacuum in the carburettor barrel. True or False?
8. Which of the following is NOT a typical air-fuel ratio for a petrol engine?
   a. 8:1
   b. 16:1
   c. 3:1
   d. 18:1
9. List and explain the seven major carburettor systems or circuits.
10. The carburettor _____ rides on top of the fuel to open and close the needle valve as needed.
11. A bowl vent is used to allow fuel vapours to enter the outside air. True or False?
12. The carburettor _____ system provides the engine's air-fuel mixture at speeds below about _____ rpm 30 km/h.
13. What is the purpose of an idle air bleed?
14. What happens when you turn the idle mixture screw in and out?
15. Why are modern idle mixture screws covered with a metal plug?
16. A carburettor accelerator pump works like a toy squirt gun to squirt fuel into the barrel when the throttle is opened for acceleration. True or False?

17. The _____ _____ carburettor system provides the leanest, most fuel efficient air-fuel mixture at normal cruising speeds.
18. What is a high speed jet?
19. Which of the following could NOT be included in the full power system?

   a. Power valve.
   b. Metering rod.
   c. Economiser valve.
   d. Choke valve.

20. What is the function of a carburettor choke?
21. List and explain four types of automatic choke.
22. A mechanical choke unloader uses a metal lug on the throttle lever to physically open the choke plate at full throttle. True or False?
23. How does a vacuum choke break work?
24. What carburettor component is used to prevent dieseling?
25. A throttle return dashpot is used on cars with manual transmissions. True or False?
26. Define the term "aneroid".
27. Describe the difference between the primary and secondary of a carburettor.

Chapter 21

PETROL INJECTION FUNDAMENTALS

After studying this chapter, you will be able to:
□ List some of the possible advantages of petrol injection.
□ Describe the classifications of petrol injection.
□ Explain the operation of electronic single-point (throttle body) petrol injection.
□ Explain the operation of electronic multi-point (port) petrol injection.
□ Summarise the operation of electronic air-flow sensing, hydraulic-mechanical (continuous), and manifold pressure sensing petrol injection systems.
□ Compare the various types of petrol injection systems.

This chapter introduces the operating principles of modern petrol injection systems. Specific systems vary, but many of the parts (sensors, fuel injectors, computer) are very similar. This chapter offers you a broad background in the many petrol injection systems found on today's cars.

PETROL INJECTION FUNDAMENTALS

A modern petrol injection system uses pressure from an electric fuel pump to spray fuel into the engine intake manifold. See Fig. 21-1. Like a carburettor, it must provide the engine with the correct air-fuel mixture for specific operating conditions. Unlike a carburettor however, PRESSURE, not engine vacuum (suction), is used to feed fuel into the engine. This makes a petrol injection system very efficient.

Petrol injection advantages

A petrol injection system has several possible advantages over a carburettor type fuel system. A few of these include:

1. Improved atomisation (fuel is forced into intake manifold under pressure which helps break fuel droplets into a fine mist).
2. Better fuel distribution (more equal flow of fuel vapours into each cylinder).
3. Smoother idle (lean fuel mixture can be used without rough idle because of better fuel distribution and low speed atomisation).
4. Improved fuel economy (high efficiency because of more precise fuel metering, atomisation, and distribution).
5. Lower emissions (lean, efficient air-fuel mixture reduces exhaust pollution).
6. Better cold weather driveability (injection provides better control of mixture enrichment than a carburettor choke).
7. Increased engine power (precise metering of fuel to each cylinder and increased airflow can result in more horsepower output).

PETROL INJECTION CLASSIFICATIONS

There are many types of petrol injection systems. Before studying the most common ones, you should have a basic knowledge of the different classifications of petrol injection. This will help you relate the similarities and differences between systems.

A petrol injection system is commonly called a fuel injection system. To avoid confusion, remember that a diesel injection system is also a fuel injection system. The two are quite different, however.

Single and multi-point injection

The point (location) of fuel injection is one way to classify a petrol injection system.

A single-point injection system, also called throttle body injection (TBI), has the injector nozzle(s) in a

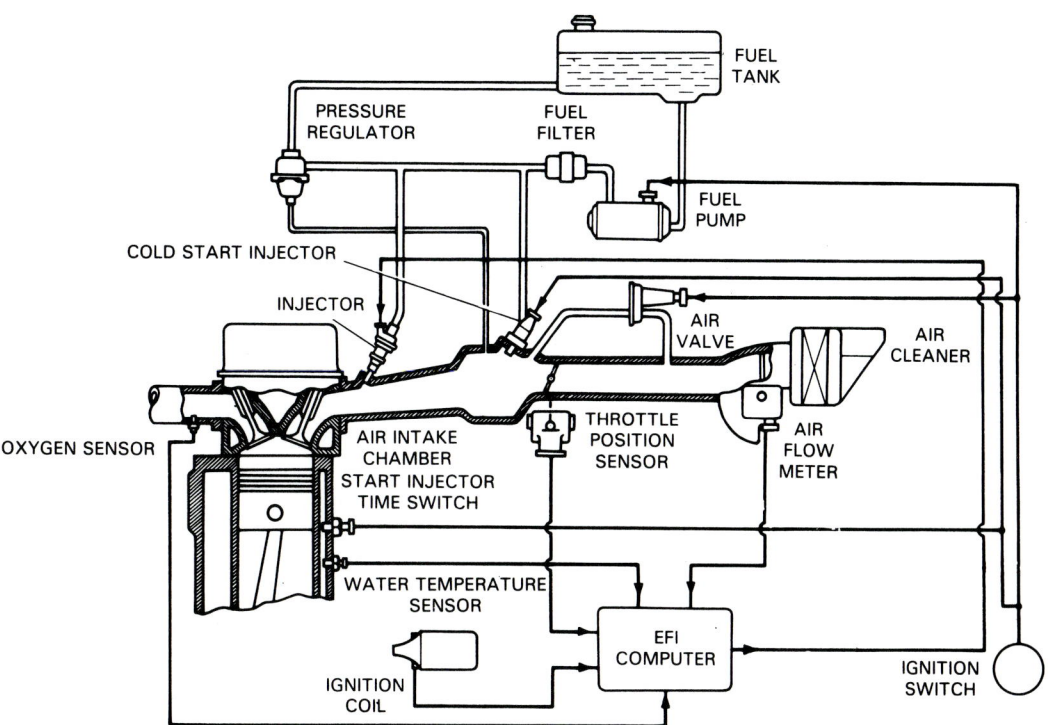

Fig. 21-1. Note general location of electronic fuel injection parts. (Toyota)

throttle body assembly on top of the engine. Fuel is sprayed into the top, centre of the intake manifold. Single-point (one location) injection is illustrated in Fig. 21-2A.

A multi-point injection system, also called a port injection has a fuel injector in the port (air-fuel runner or passage) going to each cylinder. See part B in Fig. 21-2. Petrol is sprayed into each intake port and toward each engine intake valve. Hence, the term multi-point (more than one location) fuel injection is used.

Both single-point and multi-point injection are used on today's cars. Australian made cars have used both single-point and multi-point injection. Foreign cars commonly have multi-point injection. Multi-point is a more commonly used system.

Indirect and direct injection

An indirect injection system sprays fuel into the engine intake manifold. Most petrol injection systems are this type.

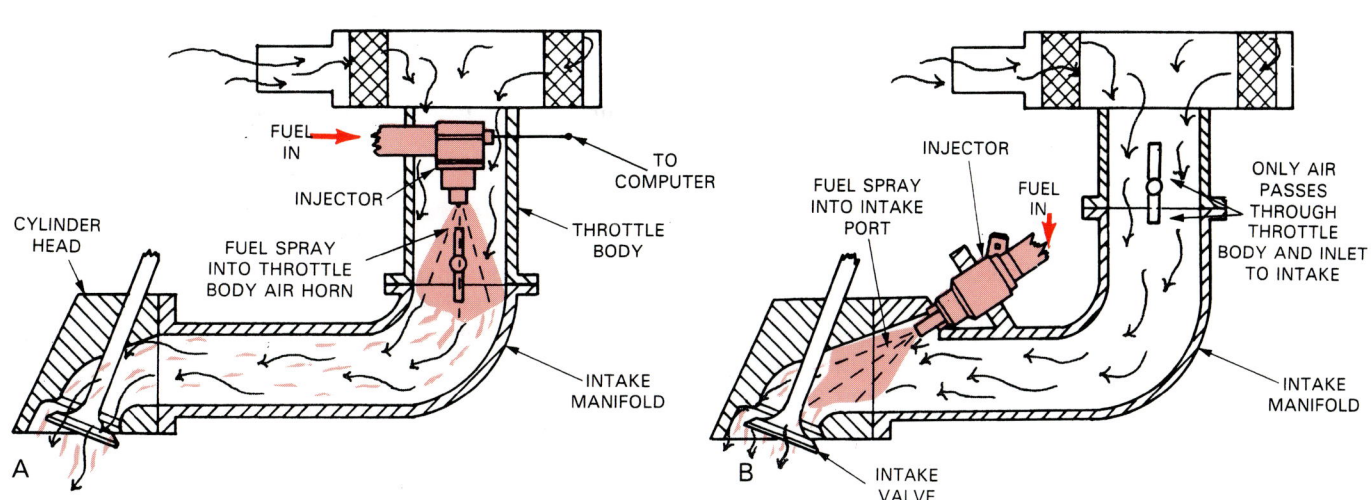

Fig. 21-2. A — Throttle body injection has injector inside throttle body. Like a carburettor, fuel enters air stream in air horn. B — Port injection has one injector for each engine cylinder. Fuel sprays toward engine intake valve.

A direct injection system forces fuel into the engine combustion chambers. All diesel injection systems are a direct type. Indirect and direct injection systems are shown in Fig. 21-3.

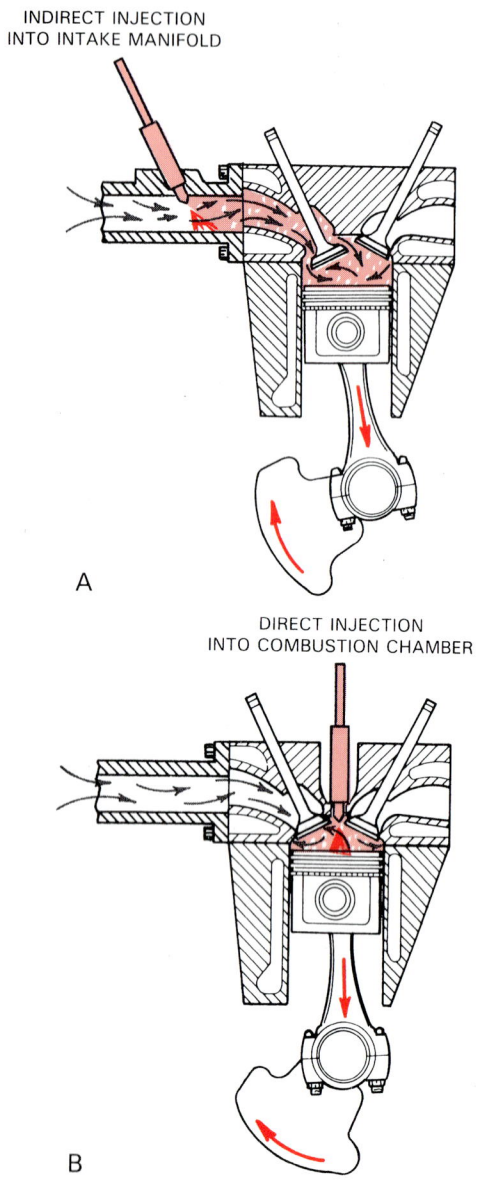

INDIRECT INJECTION
INTO INTAKE MANIFOLD

A

DIRECT INJECTION
INTO COMBUSTION CHAMBER

B

Fig. 21-3. Indirect injection sprays fuel into intake manifold. B — Direct injection sprays fuel into combustion chambers. Petrol injection systems are usually indirect type. Diesels are direct type.

PETROL INJECTION CONTROLS

There are three common methods used to control the amount of petrol injected into the engine: electronic controls, hydraulic controls, and mechanical controls. Older petrol injection systems use a combination of each.

Electronic fuel injection control uses various engine sensors and a computer to control the opening and closing of the injection valves. Look at Fig. 21-4. This is the most modern and common type of

petrol injection system. It will be covered in detail.

Hydraulic fuel injection control refers to hydraulically (air or fuel pressure) moved control devices. Hydraulic control uses an airflow sensor and a fuel distributor (hydraulic valve mechanism) to meter petrol into the engine. Covered later, it is a system used on several European cars.

Mechanical fuel injection control uses throttle linkage, a mechanical pump, and a governor speed device to control injection quantity. This is a very old, seldom used type injection system for high performance or racing cars. Diesel injection systems are mechanical type injection systems. Diesel systems are covered in a later chapter.

Petrol injection timing

The timing of a petrol injection system relates engine valve action to the time when fuel is sprayed into the engine intake manifold. There are three basic classifications of petrol injection timing: intermittent, timed and continuous.

An intermittent petrol injection system opens and closes the injection valves independently of the engine intake valves. This type of injection system may spray fuel into the engine when the valves are open or when they are closed.

Another name for an intermittent injection system is MODULATED injection system. This is one of the most common types of petrol injection.

A timed injection system squirts fuel into the engine right before or as the intake valves open. It is timed to the opening of the engine intake valves. The best example of timed injection is a diesel injection system.

A continuous petrol injection system sprays fuel into the intake manifold all of the time. Anytime the engine is running, some fuel is forced out of the injector nozzles and into the engine.

The air-fuel ratio is controlled by increasing or decreasing fuel pressure at the injectors. This increases or decreases fuel flow out of the injectors. A continuous type injection system is frequently used on European cars. This type is discussed near the end of the chapter.

Injector opening relationship

Simultaneous injection means all of the injectors open at the same time. The injectors are pulsed ON and OFF together.

Sequential injection has the injectors open one after the other. One opens and then another.

Group injection has several, but not all injectors opening at the same time. For example, A V-8 engine might have four injectors open at once and then the other four open next. A six cylinder engine would have two groups (three injectors in each group) open at different times.

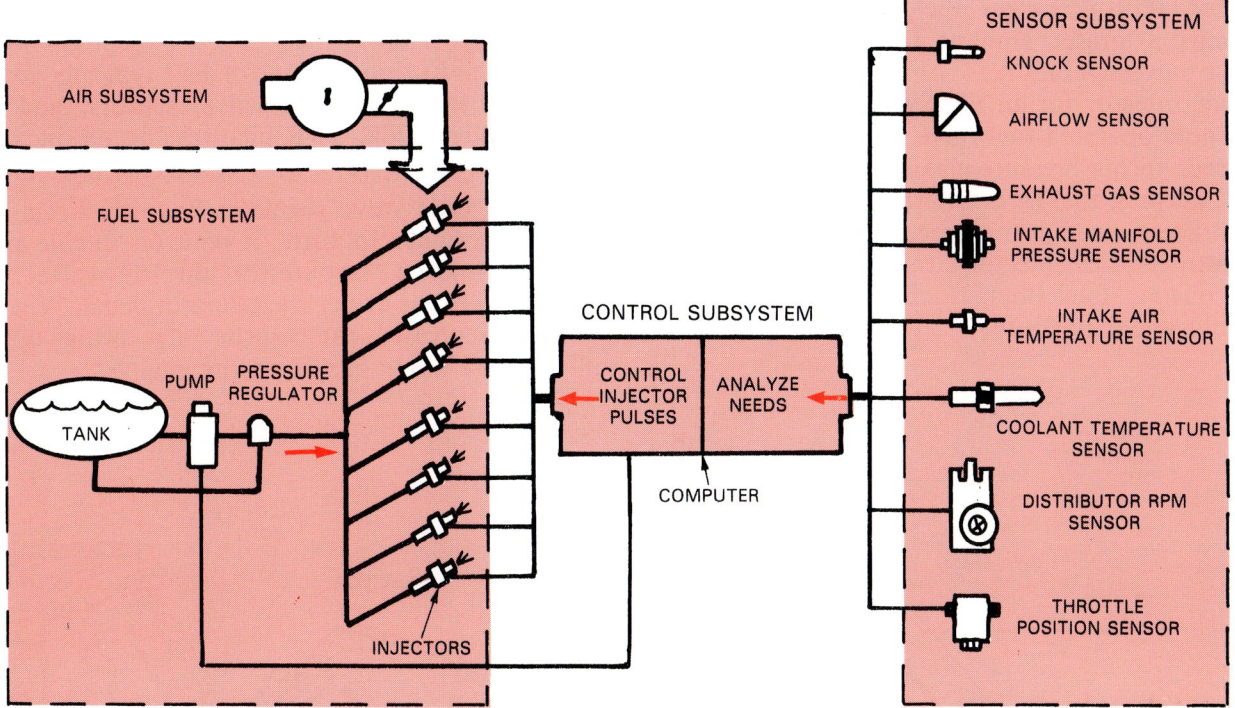

Fig. 21-4. Note four subsystems of an electronic petrol injection system (dashed line boxes). Sensor systems feeds data to computer. Computer uses this data to operate fuel delivery system. Parts of air system can also be controlled by computer.

ELECTRONIC FUEL INJECTION (EFI)

An electronic fuel injection system, abbreviated EFI, can be divided into four subsystems:

1. Fuel delivery system.
2. Air induction system.
3. Sensor system.
4. Computer control system.

These four subsystems are illustrated in Fig. 21-4.

Fuel delivery system

The fuel delivery system of an EFI system includes an electric fuel pump, fuel filter, pressure regulator, injector valves, and connecting lines and hoses. This is illustrated in Fig. 21-5.

The electric fuel pump draws petrol out of the tank and forces it into the pressure regulator.

The fuel pressure regulator controls the amount of pressure entering the injector valves. Look at Fig. 21-6. When sufficient pressure is attained, the

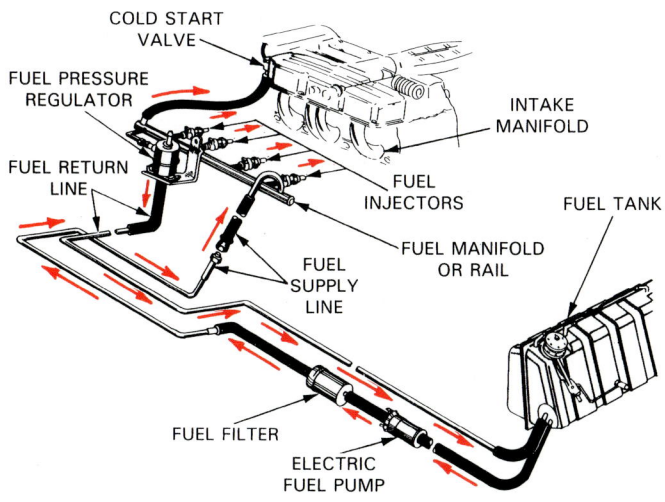

Fig. 21-5. Fuel delivery system typically consists of these parts. Pressure regulator maintains constant fuel pressure for injectors. Injectors are fuel valves that spray fuel into engine intake manifold. (Fiat)

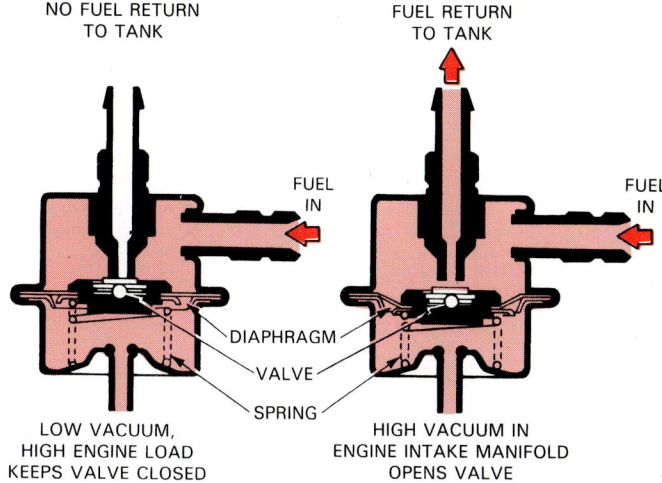

Fig. 21-6. Study pressure regulator action. A — Low engine vacuum would indicate high engine load. Spring could then hold regulator return closed to increase fuel pressure for more power. B — High engine vacuum would indicate low load. Vacuum would act on diaphragm, opening regulator return to tank. This would reduce or limit fuel pressure. (Lancia)

regulator returns excess fuel to the tank. This maintains a preset amount of fuel pressure for injector valve operation.

A fuel injector for an EFI system is simply a coil or solenoid operated fuel valve, Fig. 21-7. When not energised, spring pressure makes the injector remain closed, keeping fuel from entering the engine. When current flows through the injector coil, the magnetic field attracts the injector armature. The injector valve opens. Fuel then squirts into the intake manifold under pressure.

Air induction system

An air induction system for EFI typically consists of an air filter, throttle valve, sensors, and connecting ducts. Fig. 21-8 pictures the air induction system for one type of EFI equipped engine.

The throttle valve regulates how much air flows into the engine. In turn, it controls engine power output. Like a carburettor throttle valve, it is connected to the driver's accelerator pedal. When the pedal is depressed, the throttle valve swings open to allow more air to rush into the engine.

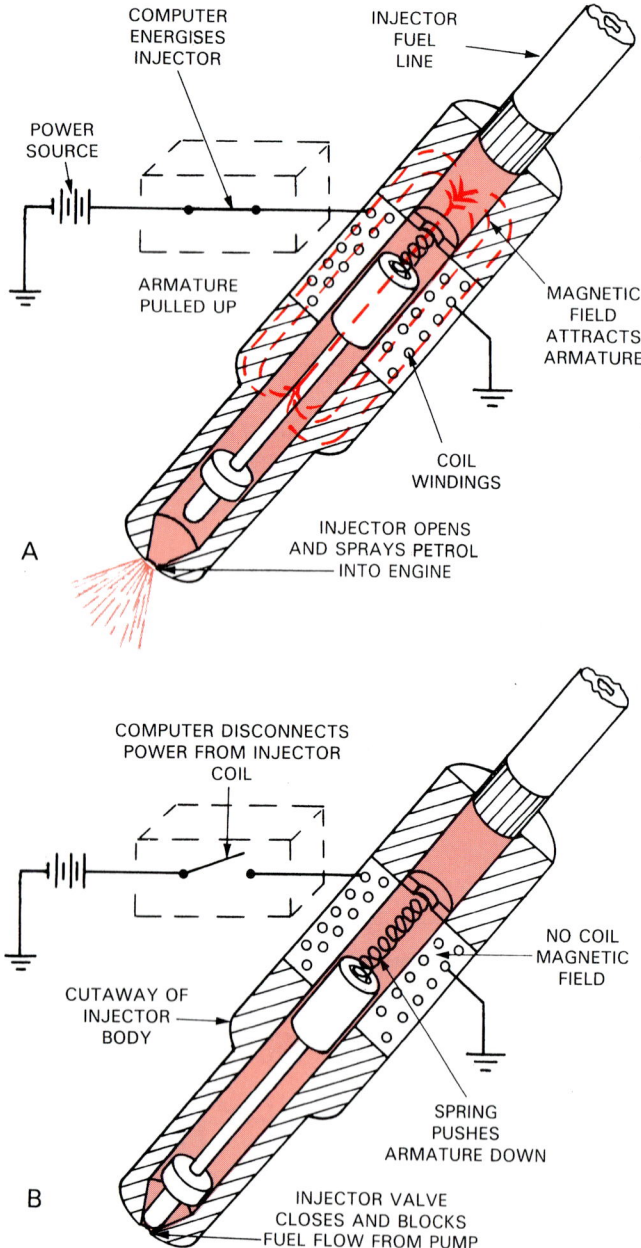

Fig. 21-7. EFI injector operation. A — Current flow through injector coil builds magnetic field. Field attracts and pulls up on armature to open injector. Fuel then sprays out injector. B — When computer breaks circuit, spring can push injector valve closed to stop fuel spray.

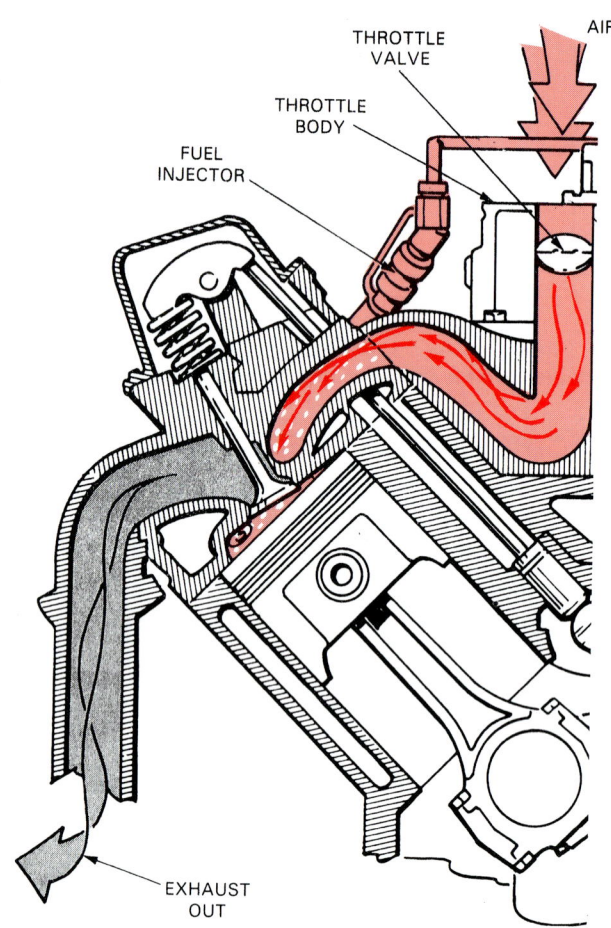

Fig. 21-8. Air induction system mainly consists of throttle body. It contains throttle plates that control airflow into engine.

Sensor system

The EFI sensor system monitors engine operating conditions and reports this information to the computer. See Fig. 21-9. A typical EFI sensor system consists of an oxygen (exhaust gas) sensor, engine coolant temperature sensor, air inlet temperature sensor, throttle position sensor, intake manifold pressure (vacuum) sensor, engine speed sensor, and other sensors.

An engine sensor is an electrical device that changes circuit resistance or voltage with a change in a condition (temperature, pressure, position of parts,

Petrol Injection Fundamentals

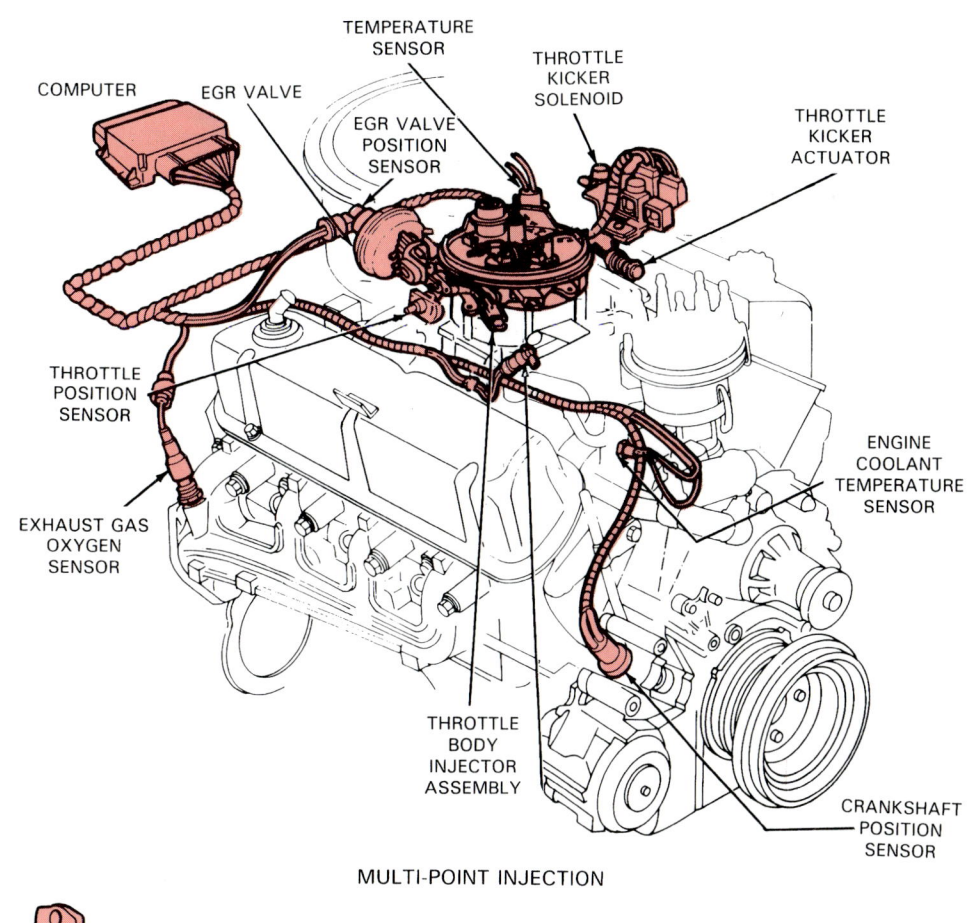

THROTTLE BODY INJECTION

COMPUTER

EGR VALVE

TEMPERATURE SENSOR

EGR VALVE POSITION SENSOR

THROTTLE KICKER SOLENOID

THROTTLE KICKER ACTUATOR

THROTTLE POSITION SENSOR

EXHAUST GAS OXYGEN SENSOR

ENGINE COOLANT TEMPERATURE SENSOR

THROTTLE BODY INJECTOR ASSEMBLY

CRANKSHAFT POSITION SENSOR

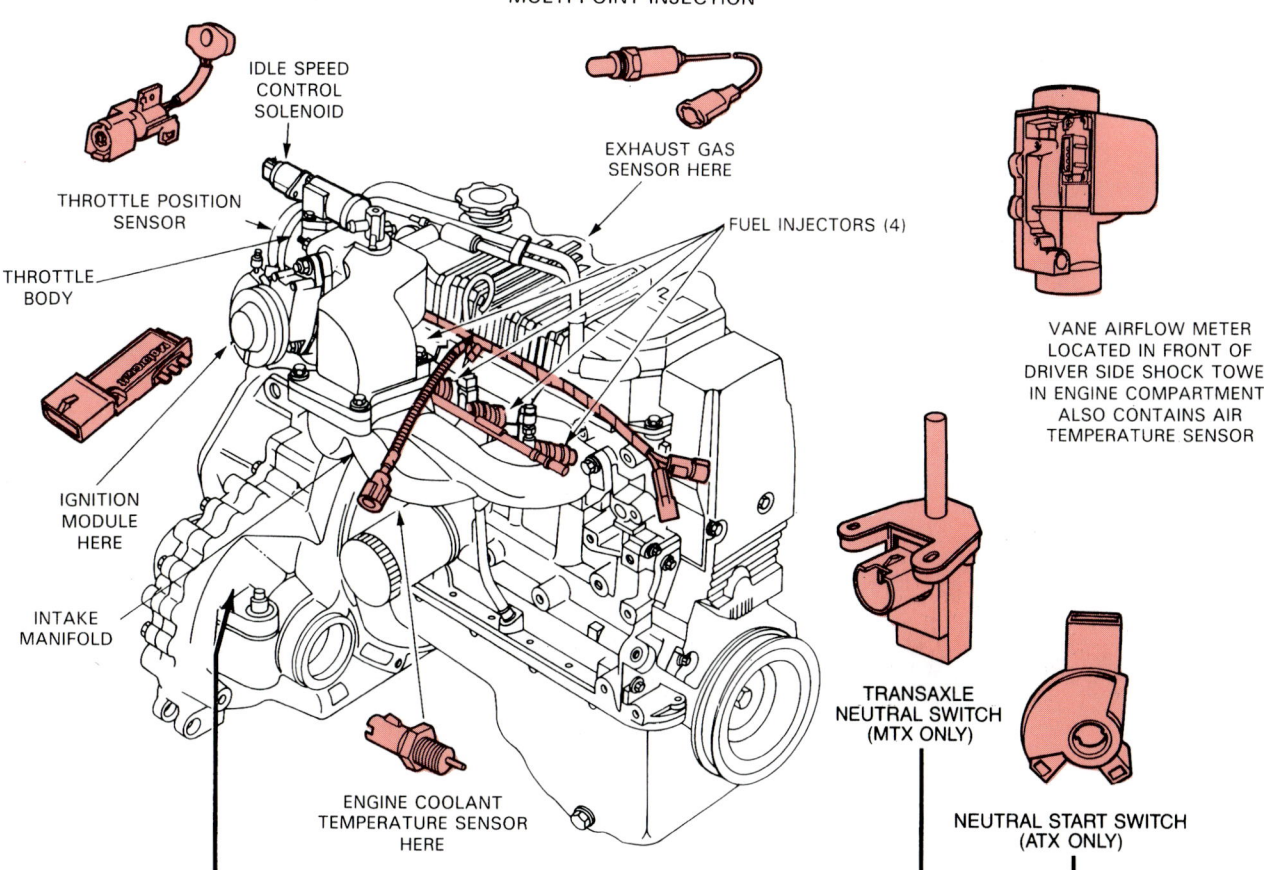

MULTI-POINT INJECTION

IDLE SPEED CONTROL SOLENOID

THROTTLE POSITION SENSOR

THROTTLE BODY

IGNITION MODULE HERE

INTAKE MANIFOLD

EXHAUST GAS SENSOR HERE

FUEL INJECTORS (4)

VANE AIRFLOW METER LOCATED IN FRONT OF DRIVER SIDE SHOCK TOWER IN ENGINE COMPARTMENT. ALSO CONTAINS AIR TEMPERATURE SENSOR

TRANSAXLE NEUTRAL SWITCH (MTX ONLY)

ENGINE COOLANT TEMPERATURE SENSOR HERE

NEUTRAL START SWITCH (ATX ONLY)

Fig. 21-9. Study sensors for single-point throttle body and multi-point or port injection systems. These are typical. (Ford)

etc). For example, a temperature sensor's resistance may decrease as temperature increases. The computer can use the increased current flow through the sensor to calculate any needed change in injector valve opening.

Computer control system

The computer control system uses electrical data from the sensors to control the operation of the fuel injectors. Refer to Fig. 21-10. A wiring harness connects the engine sensors to the input of the computer. Another wiring harness connects the output of the computer to the fuel injectors.

The computer, also called an ELECTRONIC CONTROL UNIT (ECU), is the "brain" of the electronic fuel injection system. Refer to Fig. 21-11. It is a preprogrammed microcomputer (preset, miniature electronic circuit). The ECU uses sensor output to calculate when and how long to open the fuel injectors.

To open an injector, the computer connects the injector coil to battery voltage. To close the injector, the computer opens the circuit between the battery and the injector coil.

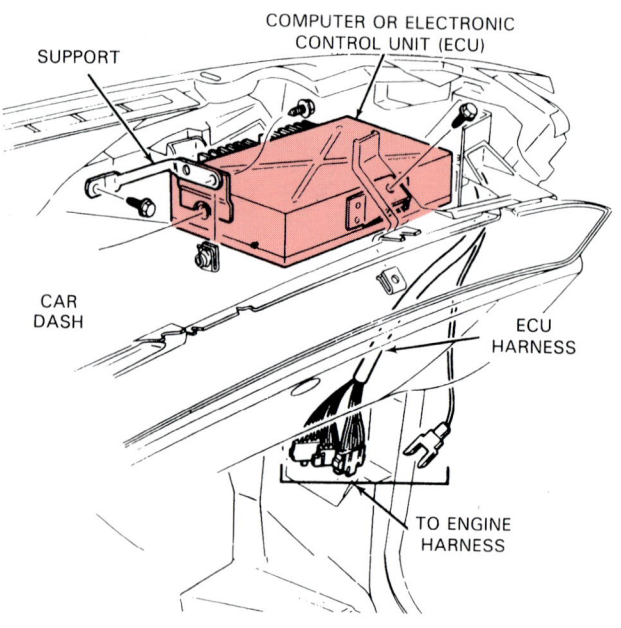

Fig. 21-10. Computer or electronic control unit is commonly mounted behind car instrument panel (dash). This keeps it away from damaging engine heat and vibration. Some systems, however, mount control unit on air cleaner or elsewhere in engine compartment.

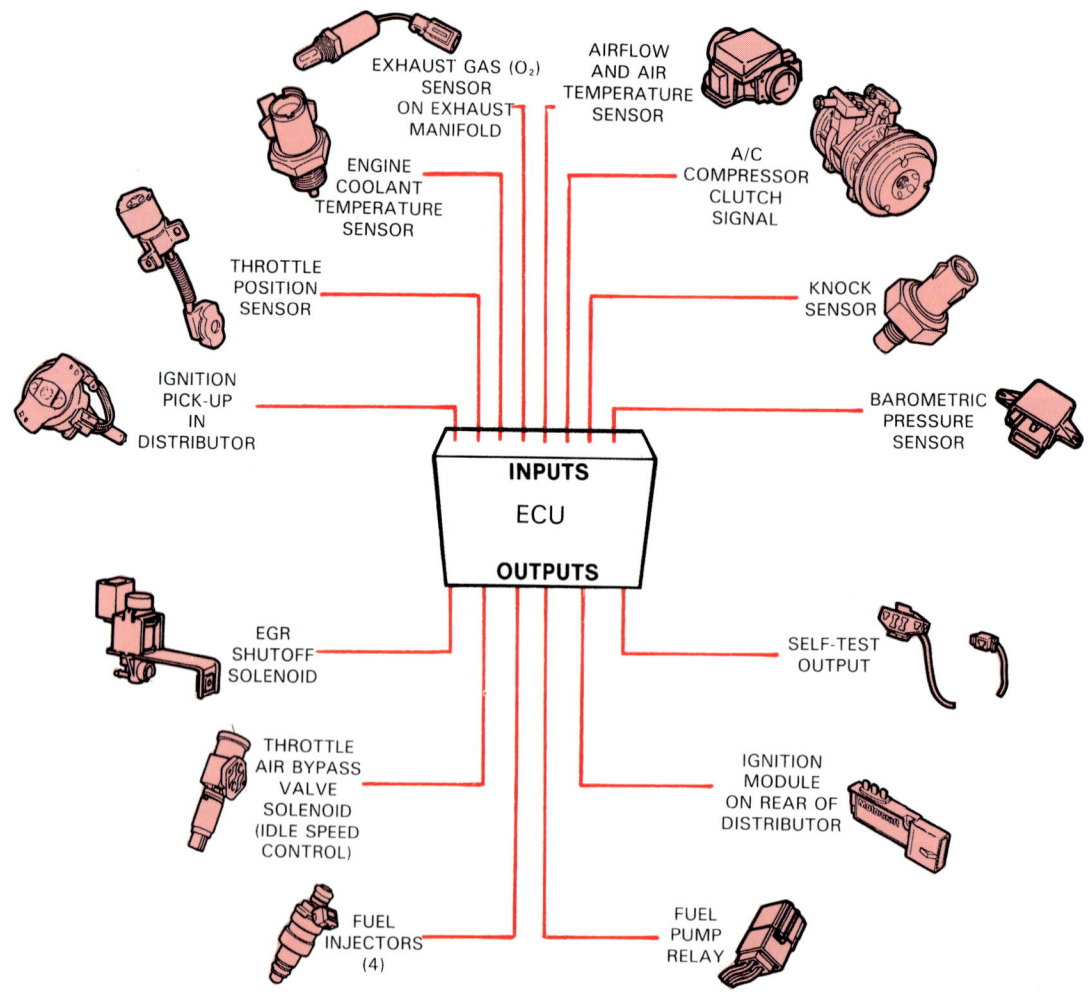

Fig. 21-11. Sensors feed information to computer. Computer uses this data to operate other components of system. (Ford)

ENGINE SENSORS

Typical sensors for an EFI system include:
1. An exhaust gas or oxygen sensor.
2. A manifold pressure sensor.
3. A throttle position sensor.
4. An engine temperature sensor.
5. An airflow sensor.
6. An inlet air temperature sensor.
7. A crankshaft position sensor.

An exhaust gas sensor, also called an oxygen sensor, measures the oxygen content in the engine's exhaust system as a means of checking combustion efficiency. It fits into the exhaust manifold or pipe at a point before the catalytic converter. Look at Figs. 21-12. and 21-13.

The oxygen sensor voltage output changes with any change in the content of the exhaust. For example, an increase in oxygen (lean mixture) might make the sensor output voltage decrease. A decrease in oxygen (rich mixture) might cause the sensor output to increase.

In this way, the sensor supplies data (different current levels) to the computer. The computer can then alter the opening and closing of the injectors to maintain a correct air-fuel ratio for maximum efficiency.

A manifold pressure sensor (abbreviated MAP for manifold absolute pressure) measures the pressure (vacuum) inside the engine intake manifold. Discussed in an earlier chapter, engine manifold pressure is an excellent indicator of engine load.

High pressure (low intake vacuum) indicates a high load, requiring a rich mixture. Low manifold pressure (high intake vacuum) indicates very little load, requiring a leaner mixture.

The manifold pressure sensor changes resistance with changes in engine load, Fig. 21-14. This data is used by the computer to alter the fuel mixture.

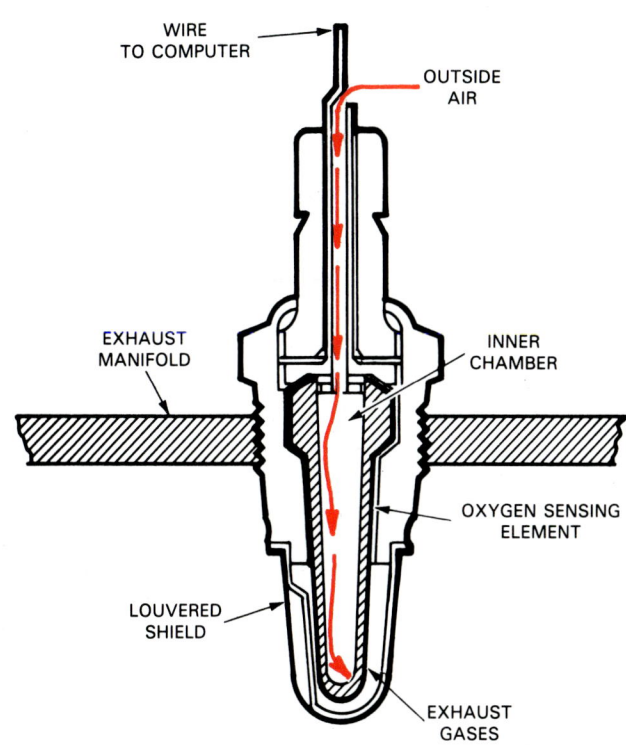

Fig. 21-13. Exhaust gas oxygen sensor compares amount of oxygen in engine exhaust with oxygen in outside air. Note passage to outside air.

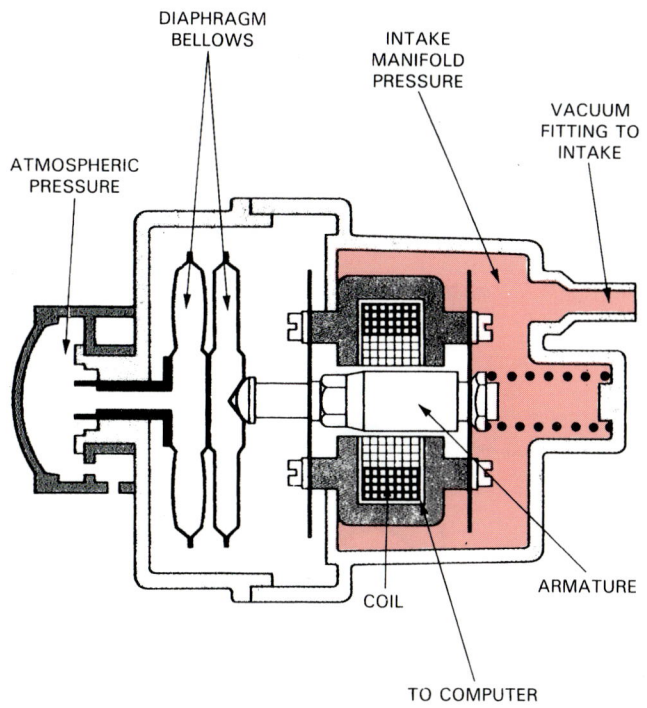

Fig. 21-14. Engine intake manifold pressure sensor changes resistance with changes in vacuum (pressure). High engine intake vacuum indicates a low load condition. Low engine intake manifold vacuum indicates a high load or power output condition. Pressure sensor reports this information to computer as a change in current flow. (Robert Bosch)

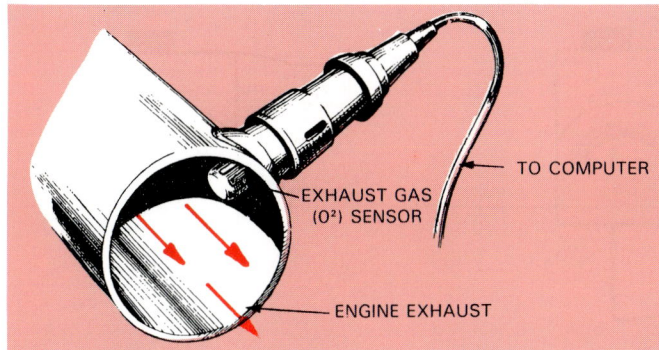

Fig. 21-12. Exhaust gas oxygen (EGO) sensor is very important sensor commonly used in EFI systems. It allows system to self-test air-fuel mixture setting by measuring oxygen in engine exhaust.

A throttle position sensor is a variable resistor connected to the throttle plate shaft. Look at Fig. 21-15. When the throttle swings open for more power or closed for less power, the sensor changes resistance and signals the computer. The computer can then enrichen or lean the mixture as needed.

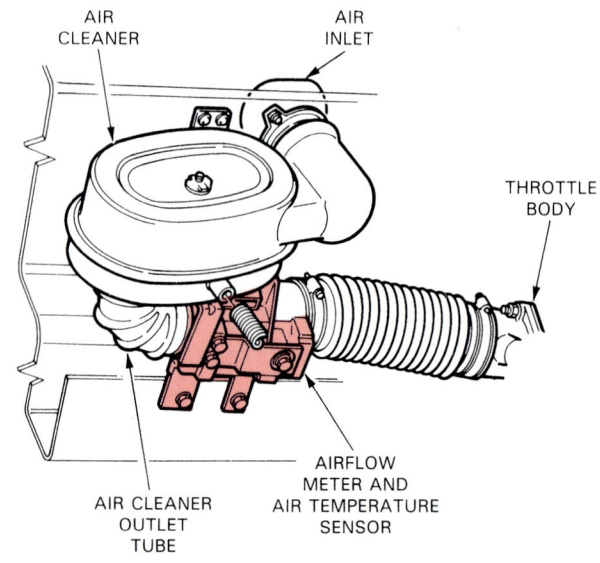

Fig. 21-16. Airflow meter and air inlet temperature sensor are housed under air cleaner on this engine. (Ford)

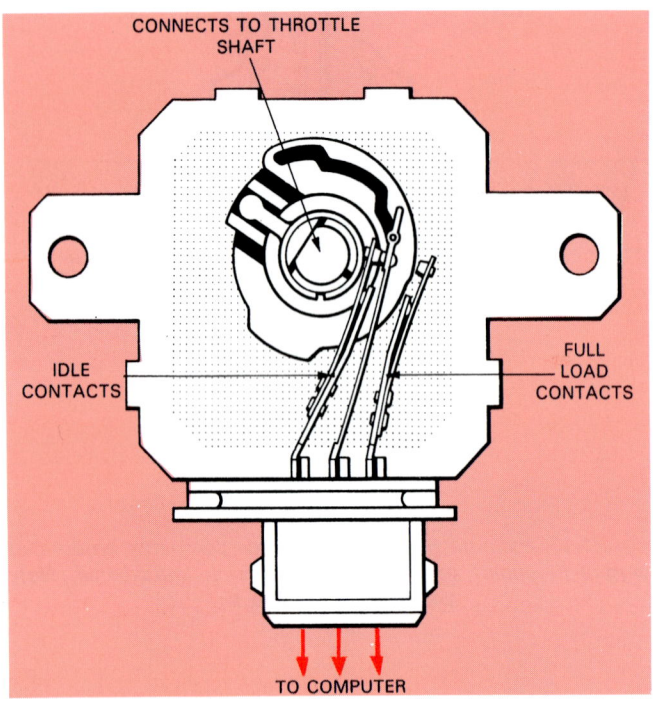

Fig. 21-15. Throttle position sensor uses contacts to report amount of throttle opening to computer. Throttle shaft rotation causes different contacts to close. Each set of contacts is connected to circuit resistor of different value. In this way, different current levels are produced for different throttle positions. Computer can then alter fuel mixture for idle and wide open throttle positions. (Robert Bosch)

An engine temperature sensor monitors the operating temperature of the engine. It is mounted so that it is exposed to the engine coolant.

When the engine is cold, the sensor might provide a high current flow (low resistance). The computer would then enrichen the air-fuel mixture for cold engine operation. When the engine warms, the sensor would supply information (high resistance for example) so that the computer could make the mixture more lean.

An airflow sensor is used in many EFI systems to measure the amount of outside air entering the engine. This helps the computer determine how much fuel is needed. Refer to Fig. 21-16.

Pictured in Fig. 21-17, it is usually an air flap or door that operates a variable resistor. Increased

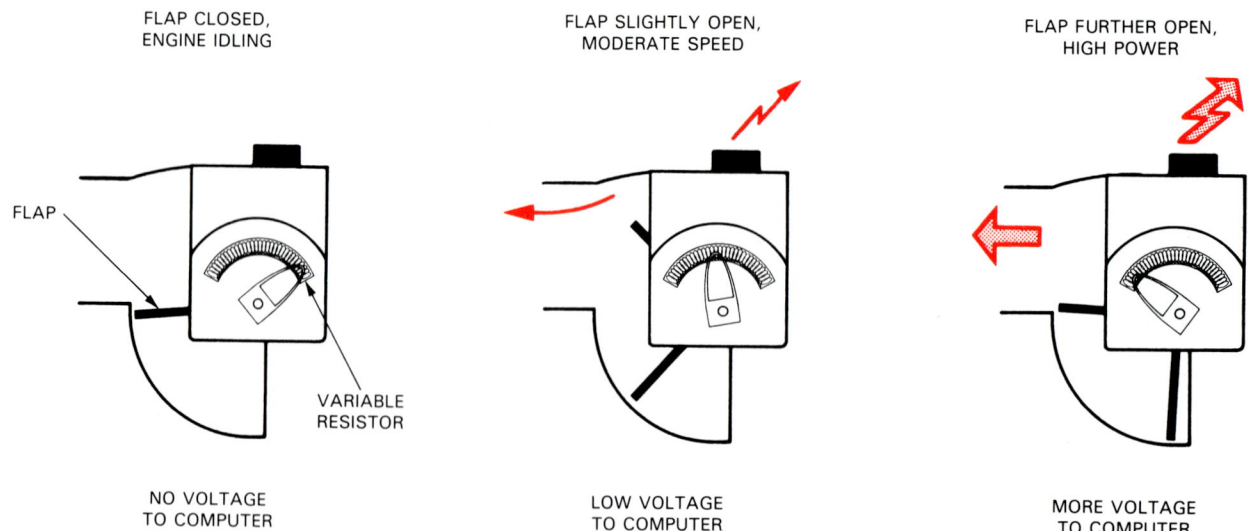

Fig. 21-17. Airflow sensor operates variable resistor. Low airflow at idle would not open sensor flap and resistance might stay high. As airflow increases, flap swings open, decreases sensor resistance and increases current flow to computer. (VW)

airflow opens the flap more to change the position of the variable resistor. Information is then sent to the computer indicating air inlet volume.

Note! An airflow sensor will be covered in more detail later in the chapter.

An inlet air temperature sensor measures the temperature of the air entering the engine. Look at Fig. 21-16 again.

Cold air is more dense than warm air, requiring a little more fuel. Warm air is NOT as dense as cold air, requiring a little less fuel. The air temperature sensor helps the computer compensate for changes in outside air temperature and maintain an almost perfect air-fuel mixture ratio.

A crankshaft position sensor is used to detect engine speed. Refer to A in Fig. 21-9. It allows the computer to change injector opening with changes in engine rpm. Higher engine speeds generally require more fuel.

Other sensors include an A/C compressor sensor, transmission sensor, EGR sensor, and engine knock sensor. They provide additional data about operating conditions affecting engine fuel needs.

Analog and digital signals

The signal from the engine sensors can be either a digital or analog type output.

Sensor digital signals are on-off signals. An example of a sensor providing a digital signal is the crankshaft position sensor which shows engine rpm. Voltage output or resistance goes from maximum to minimum, like a switch.

An analog signal changes in strength to let the computer know about a change in a condition. Sensor internal resistance may smoothly increase or decrease with temperature, pressure, or part position. The sensor acts as a variable resistor.

Open loop and closed loop

When in open loop, the electronic injection system does NOT use engine exhaust gas content as a main control of the air-fuel mixture. Illustrated in Fig. 21-18A, the system operates on information stored in the computer.

With the engine cold, the exhaust gas sensor cannot accurately provide data for the computer. The computer is set to ignore this data when the engine is cold. The system would then function in an open loop mode.

When in a closed loop mode, the EFI system uses information from the exhaust gas sensor and other sensors to control the air-fuel mixture. Shown in Fig. 21-18B, a complete loop (circle) is formed in theory as data flows from the sensor to the computer, and back through the system.

Under most operating conditions, an electronic petrol injection system functions in closed loop. This lets the computer double-check the fuel mixture it is providing to the engine.

INJECTOR PULSE WIDTH

The injector pulse width indicates the amount of time each injector is energised and kept open. The computer controls the injector pulse width.

As an example, under full acceleration, the computer would sense a wide open throttle, high intake manifold pressure, and high inlet airflow. The computer would then increase injector pulse width to enrichen the mixture for more power.

Under low load conditions, the computer would shorten the injector pulse width. With the injectors closed a large percentage of time, the air-fuel mixture would be leaner for better fuel economy.

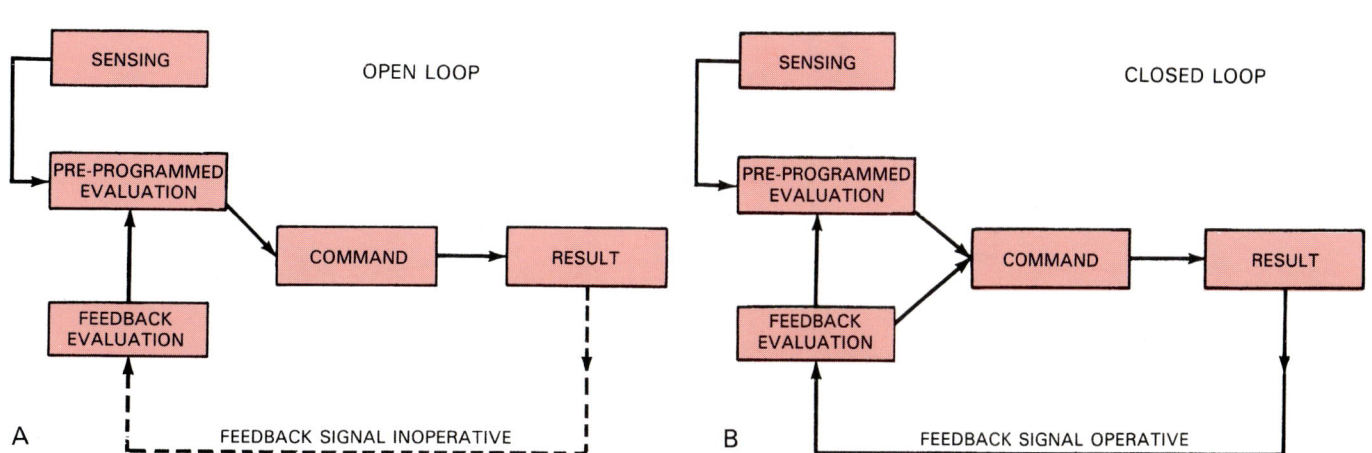

Fig. 21-18. Note basic flow of information with EFI system in open loop (A) and closed loop (B).

Fig. 21-19 shows how pulse width controls injector output. Study it carefully!

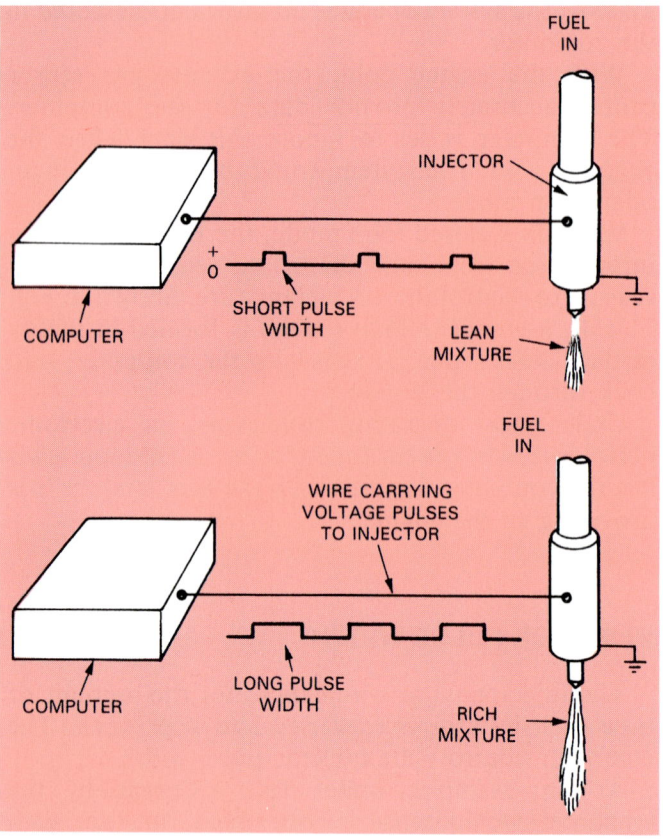

Fig. 21-19. Pulse width is used to control amount of fuel injected into engine. A longer pulse width would richen mixture. A shorter pulse width would lean mixture.

THROTTLE BODY INJECTION (TBI)

A throttle body injection system, abbreviated TBI, uses one or two injector valves mounted in a throttle body assembly. A diagram of an injector and its control circuits is shown in Fig. 21-20. Note fuel spray in the illustration.

The injectors spray fuel into the top of the throttle body air horn, Fig. 21-21. The TBI fuel spray mixes with the air flowing through the air horn. The mixture is then drawn into the engine by intake manifold vacuum.

Fig. 21-21. One or two injectors may be mounted in TBI unit. Fuel sprays into air horn, just as fuel is pulled into air horn of carburettor by vacuum. Pressure, not vacuum, forces fuel out of injector.

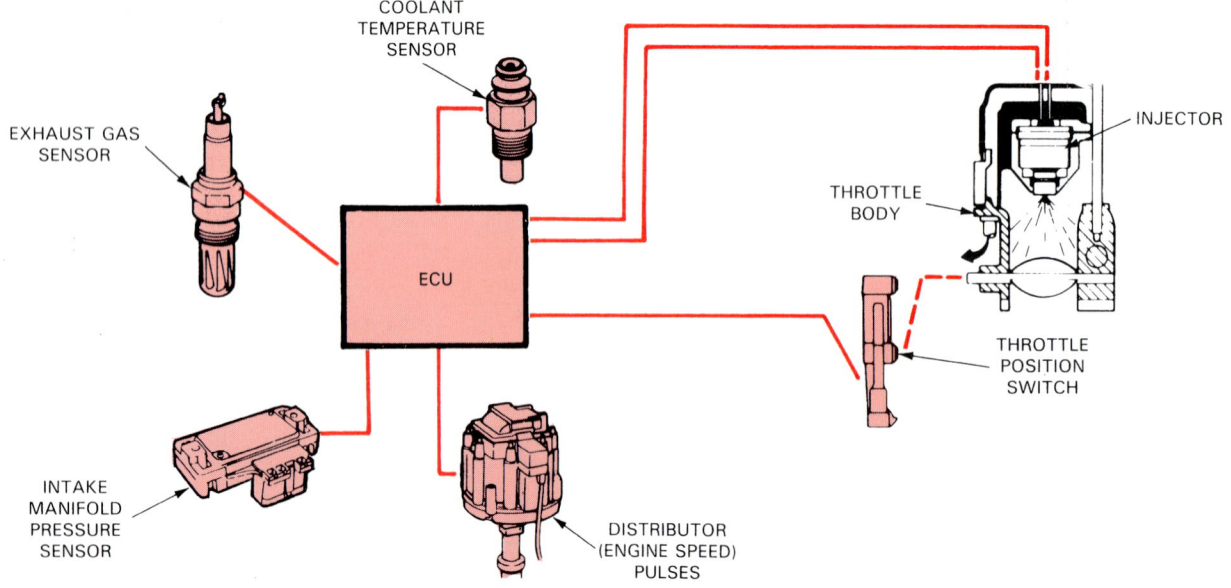

Fig. 21-20. Throttle body injection has sensors and computer operated injector mounted inside throttle body. It uses many of the basic components already introduced.

TBI assembly

The TBI assembly, Fig. 21-22, typically consists of:

1. THROTTLE BODY HOUSING (metal castings that hold injector(s), fuel pressure regulator, throttle valves, and other parts).
2. FUEL INJECTORS (solenoid operated fuel valves mounted in upper section of throttle body assembly).
3. FUEL PRESSURE REGULATOR (springloaded bypass valve that maintains constant pressure at injectors).
4. THROTTLE POSITIONER (motor assembly that opens or closes throttle valves to control engine idle speed).
5. THROTTLE POSITION SENSOR (variable resistor that senses opening or closing of throttle valves).

6. THROTTLE VALVES (butterfly valves that control airflow through throttle body).

TBI throttle body

The TBI throttle body, like a carburettor body, bolts to the pad on the intake manifold, Fig. 21-23. Throttle valves are mounted in the lower section of the throttle body. A linkage mechanism or cable connects the throttle valves with the driver's accelerator pedal. An inlet fuel line connects to one fitting on the throttle body. An outlet return line to the tank connects to another fitting on the throttle body.

Throttle body injector

A throttle body injector consists of an electric solenoid coil, armature or plunger, ball or needle valve, ball or needle seat, and injector spring. These parts are pictured in Fig. 21-24 and on the right in Fig. 21-25.

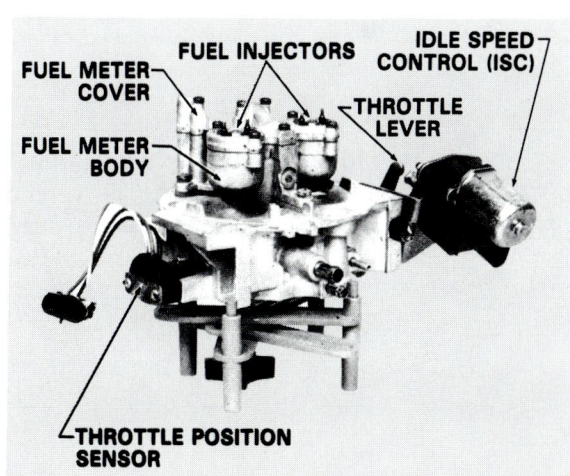

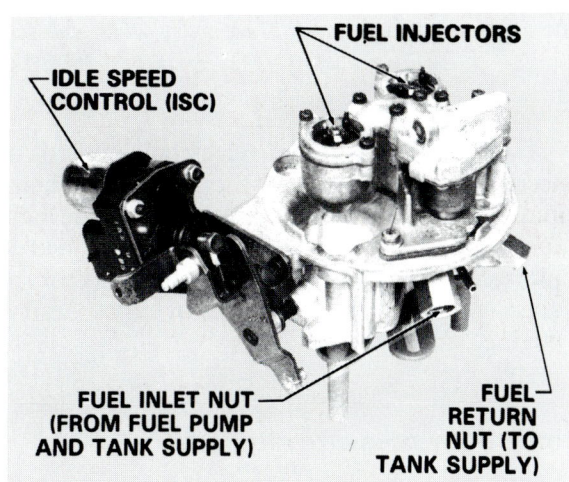

Fig. 21-22. External views of typical TBI assembly showing main parts. Study carefully.

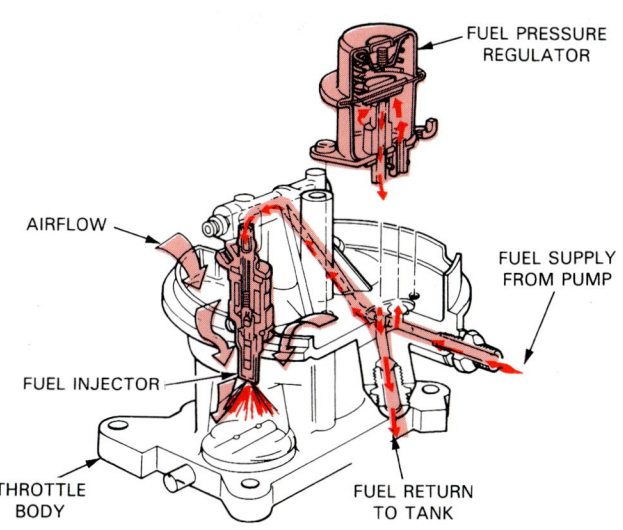

Fig. 21-23. Fuel enters throttle body from pump. It then enters pressure regulator before passing into injector. Fuel spray out of injector mixes with air entering air horn. (Ford)

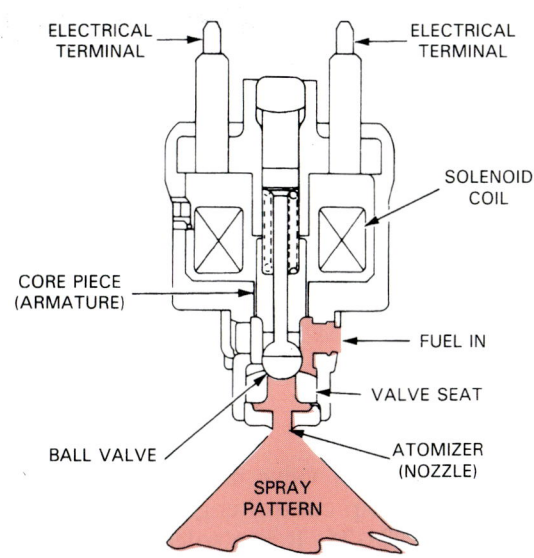

Fig. 21-24. This TBI injector uses a ball type valve, instead of a pointed needle. Note part names.

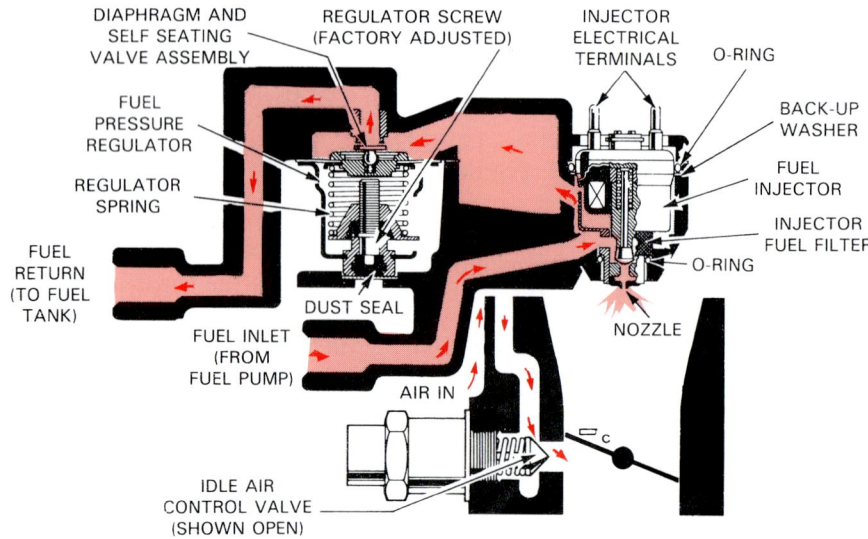

DIAPHRAGM AND SELF SEATING VALVE ASSEMBLY

REGULATOR SCREW (FACTORY ADJUSTED)

INJECTOR ELECTRICAL TERMINALS

O-RING

FUEL PRESSURE REGULATOR

BACK-UP WASHER

REGULATOR SPRING

FUEL INJECTOR

FUEL RETURN (TO FUEL TANK)

INJECTOR FUEL FILTER

O-RING

DUST SEAL

NOZZLE

FUEL INLET (FROM FUEL PUMP)

AIR IN

IDLE AIR CONTROL VALVE (SHOWN OPEN)

Fig. 21-25. Cutaway illustration shows basic action inside a typical throttle body assembly. Regulator limits maximum pressure inside injector to preset level. Then, pulse width will accurately control air-fuel ratio. Idle air control valve is used to increase or decrease idle speed as needed.

Wires from the computer (electronic control unit) connect to the terminals on the injector. When the computer energises the injector, a magnetic field is produced in the injector coil. The magnetic field pulls the plunger and valve up to open the injector. Fuel can then squirt through the injector nozzle and into the engine.

Throttle body pressure regulator

The throttle body pressure regulator consists of a fuel valve, diaphragm, and spring. When fuel pressure is low (initial engine starting), the spring holds the fuel valve closed. This causes pressure to build as fuel flows into the regulator from the electric fuel pump. Refer to Fig. 21-23.

When a preset pressure is reached, pressure acts on the diaphragm. The diaphragm compresses the spring and opens the fuel valve. Fuel can then flow back to the fuel tank. This limits the maximum fuel pressure at the injectors.

Fig. 21-25 shows how the pressure regulator functions in one type of throttle body assembly.

Idle air control valve

An idle air control valve may be used in a TBI throttle body to help control engine idle speed. One is shown in Fig. 21-25. It is a solenoid operated air bypass valve. The system computer or built-in thermostat normally opens and closes the idle air control valve.

When open, the idle air control valve allows more air to enter the intake manifold. This tends to increase idle rpm. When closed the valve decreases bypass air and idle speed. The valve can be used to control both slow and fast idle speeds. It is comparable to a carburettor's fast idle cam.

Throttle positioner

A throttle positioner (in addition to an idle air control valve) is sometimes used on throttle body assemblies to control engine idle speed. The computer actuates the positioner to open or close the throttle valves.

In this way, the computer can maintain a precise idle speed with changes in engine temperature, load (air conditioning ON for example), and other conditions.

ELECTRONIC MULTI-POINT INJECTION

Electronic multi-point injection systems use a computer, engine sensors, and one solenoid injector for each engine cylinder. This is a very common type system on late model cars. Look at Fig. 21-26.

The operation of an electronic multi-point type system is similar to a modulated, single-point injection system covered earlier. However, fuel is injected into each intake port, instead of the top centre of the intake manifold.

Petrol Injection Fundamentals

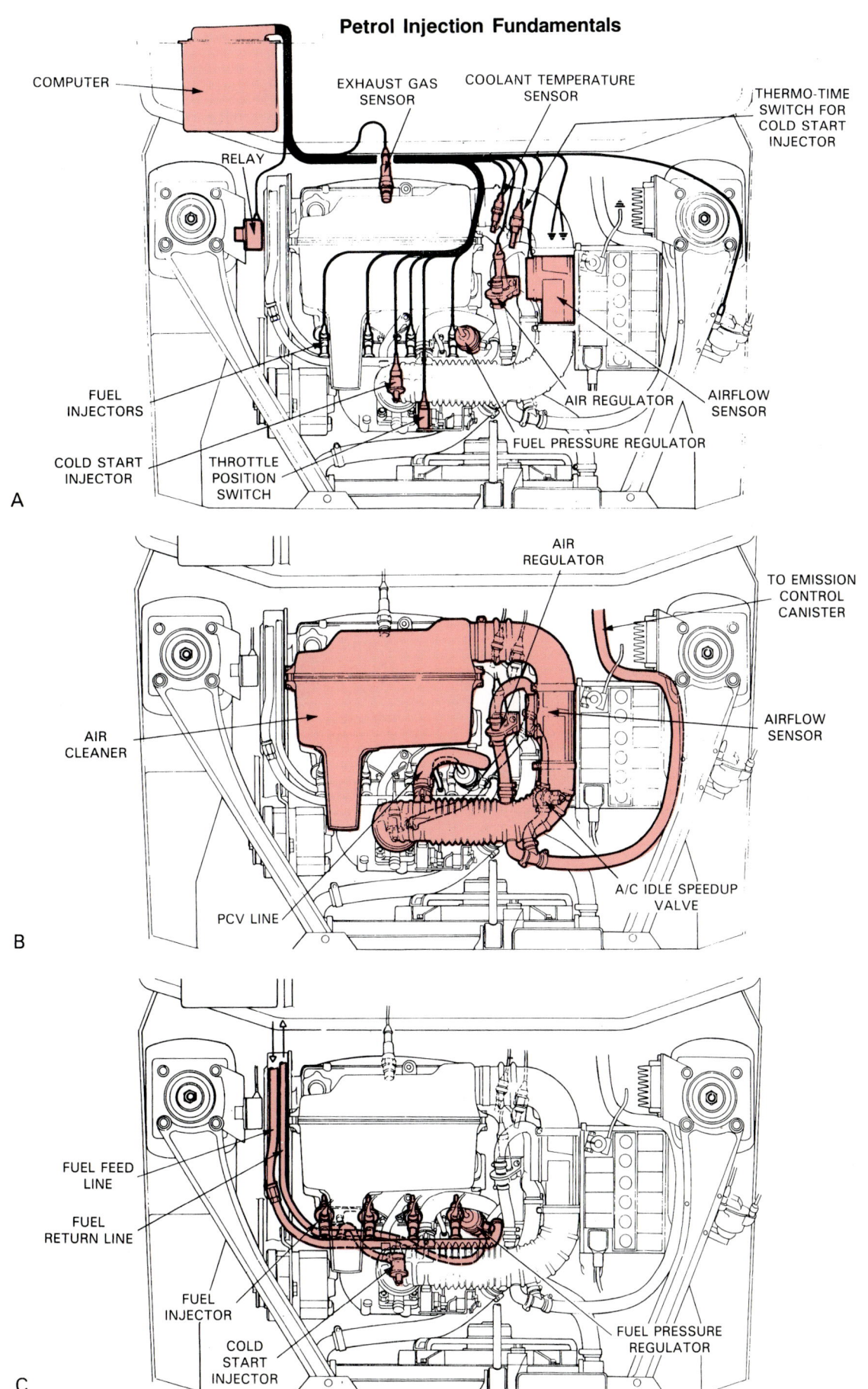

A

COMPUTER

EXHAUST GAS SENSOR

COOLANT TEMPERATURE SENSOR

THERMO-TIME SWITCH FOR COLD START INJECTOR

RELAY

FUEL INJECTORS

COLD START INJECTOR

THROTTLE POSITION SWITCH

AIR REGULATOR

AIRFLOW SENSOR

FUEL PRESSURE REGULATOR

B

AIR REGULATOR

TO EMISSION CONTROL CANISTER

AIRFLOW SENSOR

AIR CLEANER

A/C IDLE SPEEDUP VALVE

PCV LINE

C

FUEL FEED LINE

FUEL RETURN LINE

FUEL INJECTOR

COLD START INJECTOR

FUEL PRESSURE REGULATOR

Fig. 21-26. Note systems of multi-point electronic injection system. A — Sensor and control systems of EFI multi-point system. Sensors feed data to computer. Computer can then operate injectors and other components for maximum efficiency. B — Air delivery system parts. Note airflow sensor that monitors air volume entering engine. C — Fuel delivery system basically includes injectors, pressure regulator, lines, and hoses. Some systems use a cold start injector. (Lancia)

A multi-point throttle body assembly contains the throttle valves, throttle position sensor, but does NOT contain the injector valves. See upper right in Fig. 21-27. Its main function is to control airflow into the engine.

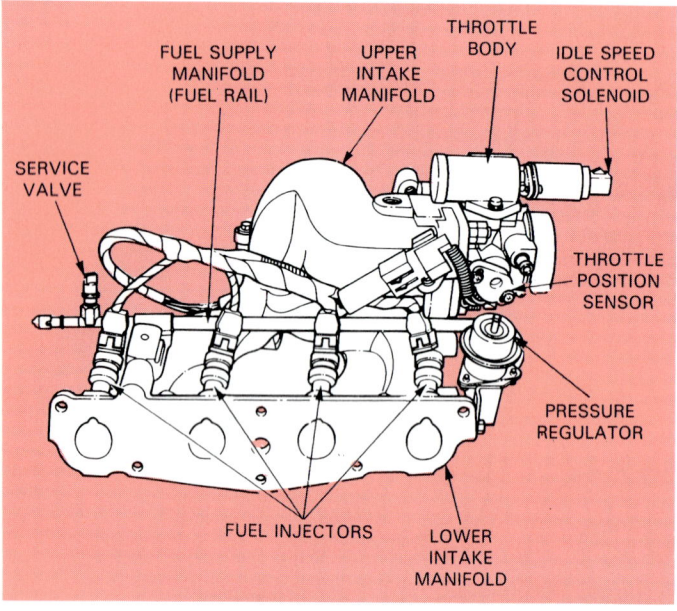

Fig. 21-27. A throttle body assembly is part of this multi-point injection system. Note how multi-point injectors install in intake runners or ports. Also note other parts. (Ford)

A multi-point pressure regulator is mounted in the fuel line before or after the injectors, Fig. 21-28. It performs the same function as the pressure regulator covered earlier. It maintains a constant pressure at the inlet to the injector valves by acting as a bypass branch.

A fuel rail feeds fuel to several or all of the injectors. It is a tubing assembly that connects the main fuel line to the inlet of each injector. Look at Figs. 21-27 and 21-28. Locate the fuel rail.

EFI multi-point injector

An EFI multi-point injector valve is usually a pressfit into the runner (port) in the intake manifold. Each injector is aimed to spray towards an engine intake valve. It is constructed something like an intermittent, throttle body injector.

An EFI multi-point injector typically consists of:
1. ELECTRIC TERMINALS (electrical connection for completing circuit between injector coil and computer).
2. INJECTOR SOLENOID (armature and coil that opens and closes valve).
3. INJECTOR SCREEN (screen filter for trapping foreign matter before it can enter injector nozzle).
4. NEEDLE VALVE (end of armature that seals on needle seat).
5. NEEDLE SEAT (round hole in end of injector that seals against needle valve tip).

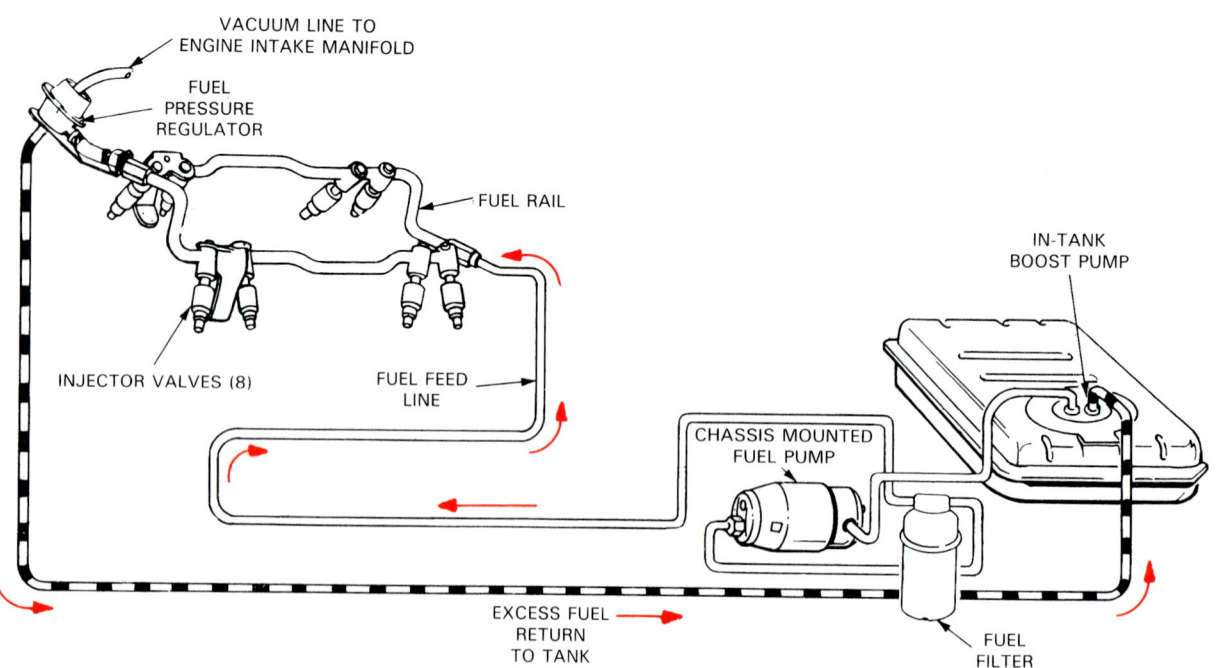

Fig. 21-28. Pressure regulator action. Fuel pump forces fuel into fuel rail, injectors, and regulator. Regulator allows excess fuel to flow back to fuel tank. Vacuum line to regulator causes fuel pressure to increase and decrease with changes in engine vacuum and load.

6. INJECTOR SPRING (small spring that returns needle valve to closed position).

7. O-RING SEAL (rubber seal that fits around outside of injector body and seals in intake manifold).

8. INJECTOR NOZZLE (outlet of injector that produces fuel spray pattern).

Fig. 21-29 shows an EFI multi-point injector. Study this illustration carefully.

There are several variations of electronic multi-point injection. It is important that you understand the primary differences between each type system.

EFI (airflow sensing, multi-point)

An airflow sensing multi-point EFI uses an airflow sensor as a main control of the system. As shown in Fig. 21-30, an airflow sensor is placed at the inlet to the intake manifold. It and other engine sensors provide electrical data to the computer.

The airflow sensor is a flap-operated variable resistor, Fig. 21-31. Airflow through the sensor causes an air door (flap) to swing to one side. Since the air door is connected to a variable resistor, the amount of airflow into the engine is converted into an electrical signal for the computer.

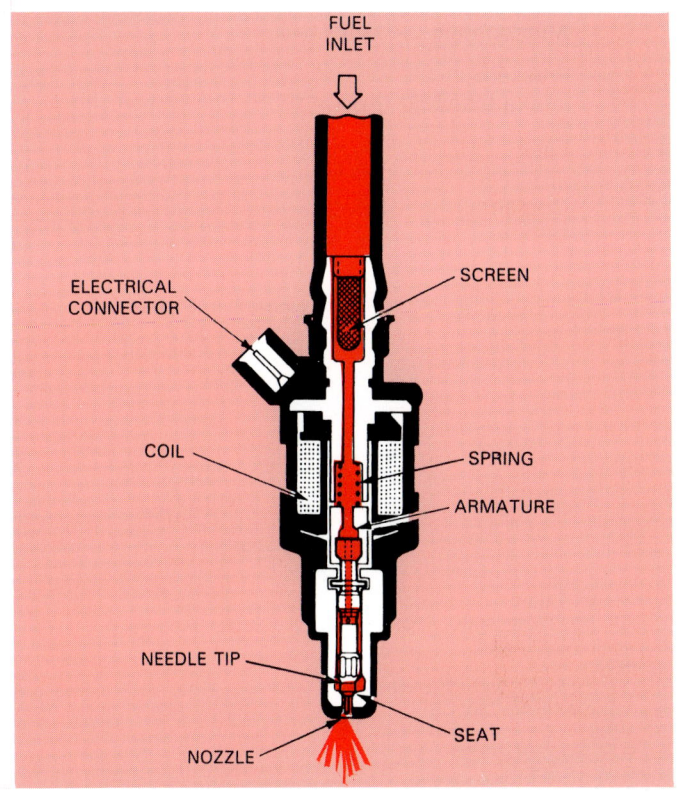

Fig. 21-29. Note basic parts of this electronic multi-point or port fuel injector. Solenoid opens injector when current flow through coil builds magnetic field and acts on armature. (Lancia)

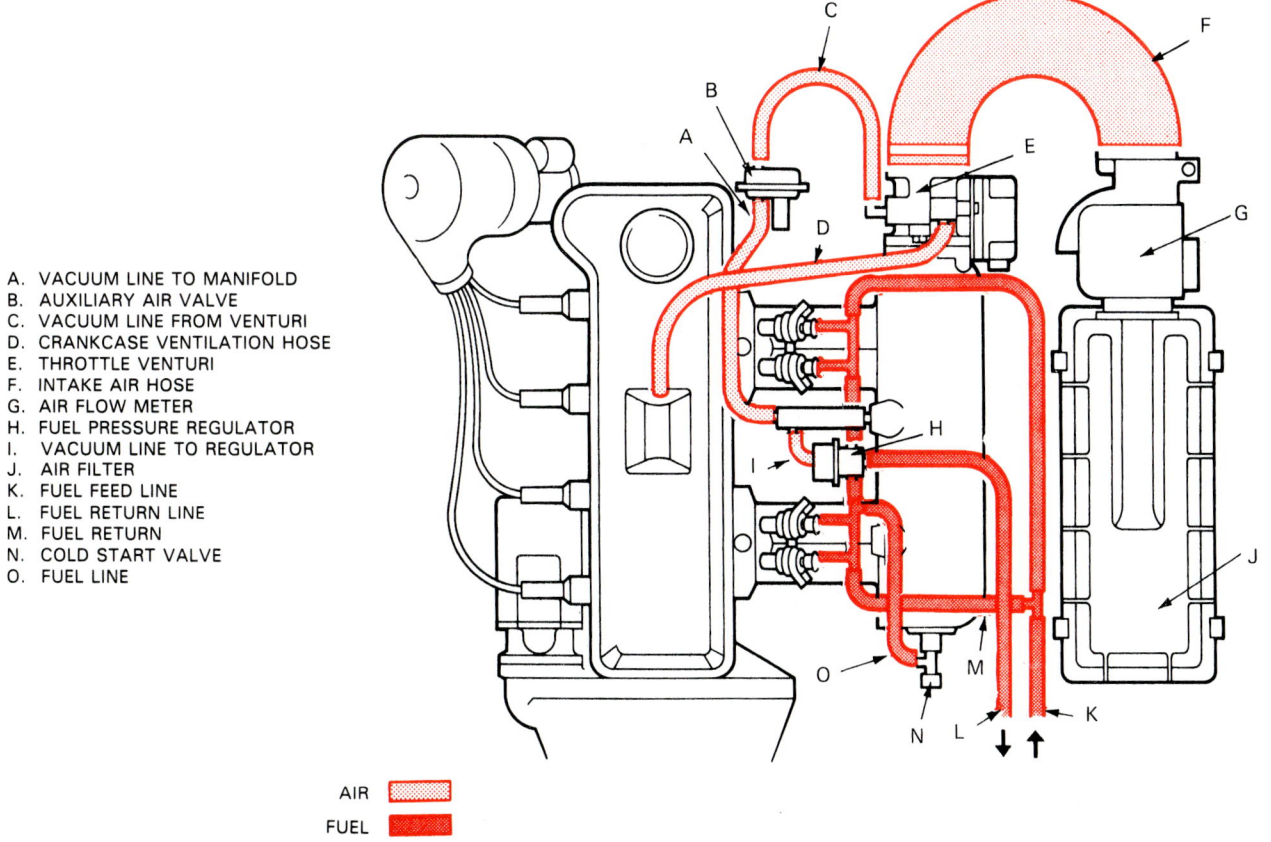

A. VACUUM LINE TO MANIFOLD
B. AUXILIARY AIR VALVE
C. VACUUM LINE FROM VENTURI
D. CRANKCASE VENTILATION HOSE
E. THROTTLE VENTURI
F. INTAKE AIR HOSE
G. AIR FLOW METER
H. FUEL PRESSURE REGULATOR
I. VACUUM LINE TO REGULATOR
J. AIR FILTER
K. FUEL FEED LINE
L. FUEL RETURN LINE
M. FUEL RETURN
N. COLD START VALVE
O. FUEL LINE

AIR
FUEL

Fig. 21-30. Top view of an airflow sensing multi-point petrol injection system. Study location of basic parts. (Robert Bosch)

327

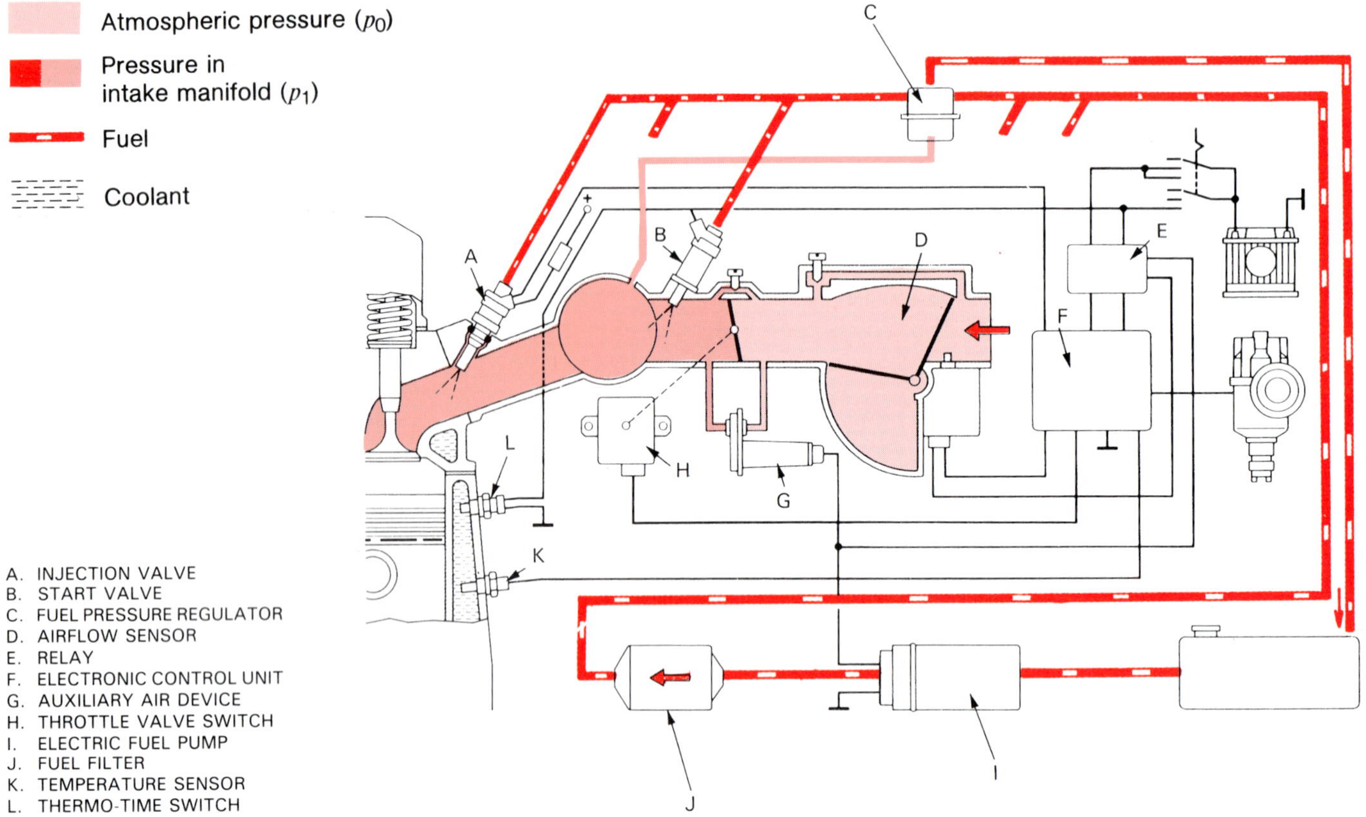

Atmospheric pressure (p_0)

Pressure in intake manifold (p_1)

Fuel

Coolant

A. INJECTION VALVE
B. START VALVE
C. FUEL PRESSURE REGULATOR
D. AIRFLOW SENSOR
E. RELAY
F. ELECTRONIC CONTROL UNIT
G. AUXILIARY AIR DEVICE
H. THROTTLE VALVE SWITCH
I. ELECTRIC FUEL PUMP
J. FUEL FILTER
K. TEMPERATURE SENSOR
L. THERMO-TIME SWITCH

Fig. 21-31. Diagram shows how each part is connected in airflow sensing EFI system. Air flow sensor is primary sensor. Also note use of cold start injector. (Robert Bosch)

Airflow sensor operation

When the throttle valve is closed (engine idling), the airflow meter door almost remains closed. See Fig. 21-32. The computer then produces a short injector pulse width. Only a small amount of fuel is injected into the intake ports.

When the driver presses the accelerator pedal and swings open the throttle plate, airflow increases. The airflow door is pushed out of the way, changing sensor resistance. The computer then increases injector pulse width for a richer mixture.

EFI (pressure sensing, multi-point)

Pressure sensing multi-point injection uses intake manifold pressure (vacuum) as a primary control of the system. Look at Fig. 21-33. A pressure sensor is

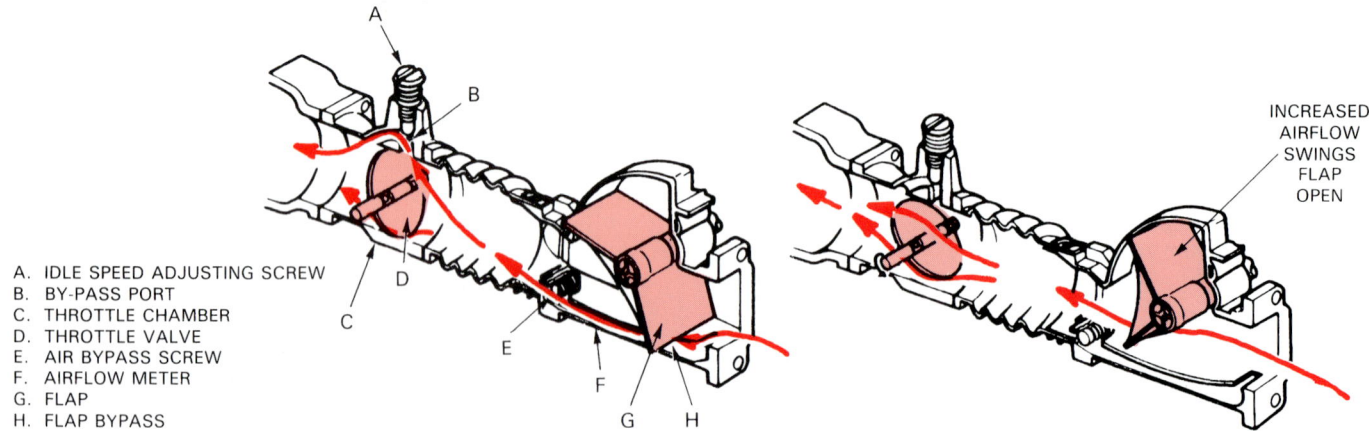

A. IDLE SPEED ADJUSTING SCREW
B. BY-PASS PORT
C. THROTTLE CHAMBER
D. THROTTLE VALVE
E. AIR BYPASS SCREW
F. AIRFLOW METER
G. FLAP
H. FLAP BYPASS

INCREASED AIRFLOW SWINGS FLAP OPEN

Fig. 21-32. Throttle valve controls engine speed and power output. Left — Throttle almost closed. Engine running slowly. Airflow sensor would detect little air flow. Computer would produce short injection pulse width for small injection quantity. Right — Throttle moved open for more power. Increased flow pushes sensor flap open. Computer would know to increase pulse width for richer mixture. (Nissan)

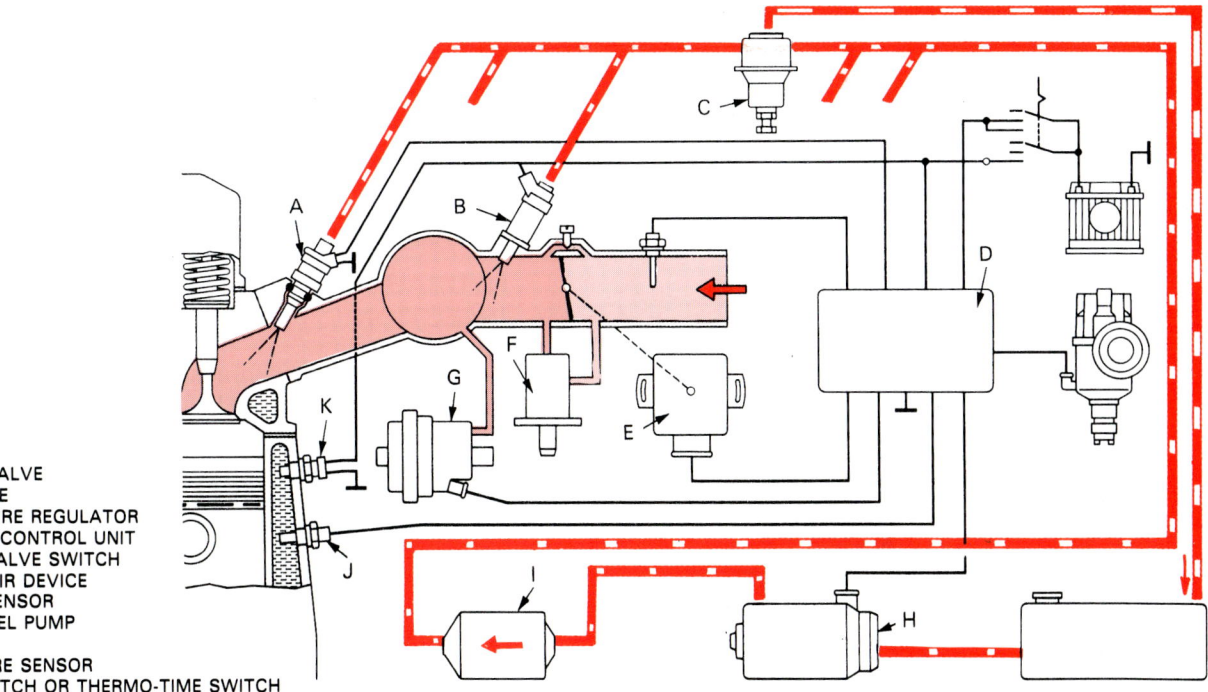

A. INJECTION VALVE
B. START VALVE
C. FUEL PRESSURE REGULATOR
D. ELECTRONIC CONTROL UNIT
E. THROTTLE VALVE SWITCH
F. AUXILIARY AIR DEVICE
G. PRESSURE SENSOR
H. ELECTRIC FUEL PUMP
I. FUEL FILTER
J. TEMPERATURE SENSOR
K. THERMO-SWITCH OR THERMO-TIME SWITCH

Fig. 21-33. Pressure sensing petrol injection system uses intake manifold vacuum as main source of computer information. High intake manifold vacuum indicates a low load condition, needing a lean air-fuel mixture. Low intake vacuum would indicate a high load condition, requiring a richer mixture. Compare this system to the one in Fig. 21-31.

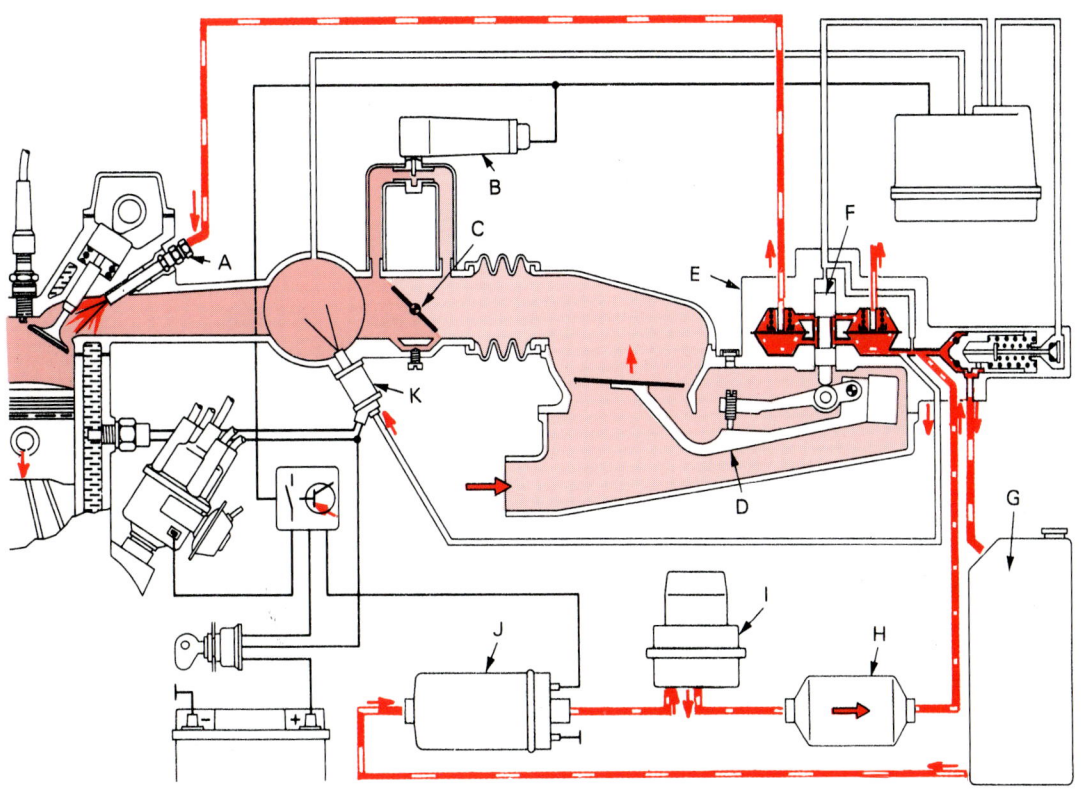

A. CONTINUOUS INJECTOR
B. AIR BYPASS VALVE
C. THROTTLE VALVE OR PLATE
D. AIRFLOW PLATE AND LEVER
E. FUEL DISTRIBUTOR
F. FUEL CONTROL PLUNGER
G. FUEL TANK
H. FUEL FILTER
I. FUEL ACCUMULATOR
J. FUEL PUMP
K. COLD START INJECTOR

Fig. 21-34. Hydraulic-mechanical injection system uses mechanical airflow sensor to operate hydraulic fuel distributor assembly. Note that continuous injector is used to spray fuel into engine any time the engine is running. (Robert Bosch)

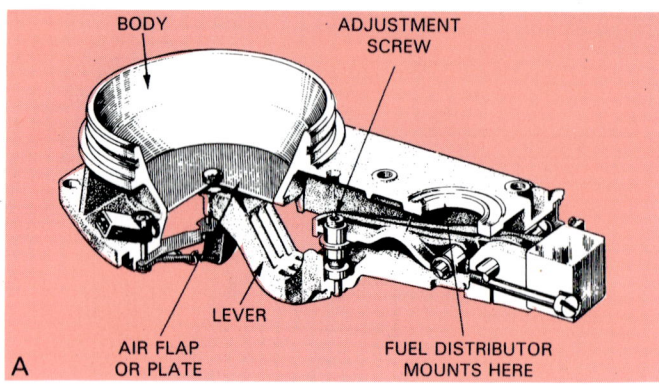

connected to a passage going into the intake manifold.

The pressure sensor converts changes in manifold pressure into changes in electrical resistance or current flow. The computer uses this electrical data to calculate engine load and air-fuel ratio requirements.

HYDRAULIC-MECHANICAL, CONTINUOUS INJECTION

A hydraulic-mechanical continuous injection system (CIS) uses a MIXTURE CONTROL UNIT (airflow sensor-fuel distributor assembly) to operate the injectors. Fig. 21-34 shows a diagram of this system. This is not an electronic type system.

The airflow sensor for this type injection system is pictured in Fig. 21-35. A round air sensor plate is hinged inside the centre of the sensor housing. The sensor plate is mounted on a sensor arm.

When airflow into the engine increases, the sensor plate is pushed up. This also pulls up on the sensor arm. The sensor arm then operates the fuel distributor. Refer to Fig. 21-36.

CIS fuel distributor

A fuel distributor is a hydraulic operated valve mechanism that controls fuel flow (pressure) to each CIS injector. See Figs. 21-35 and 21-36. The fuel

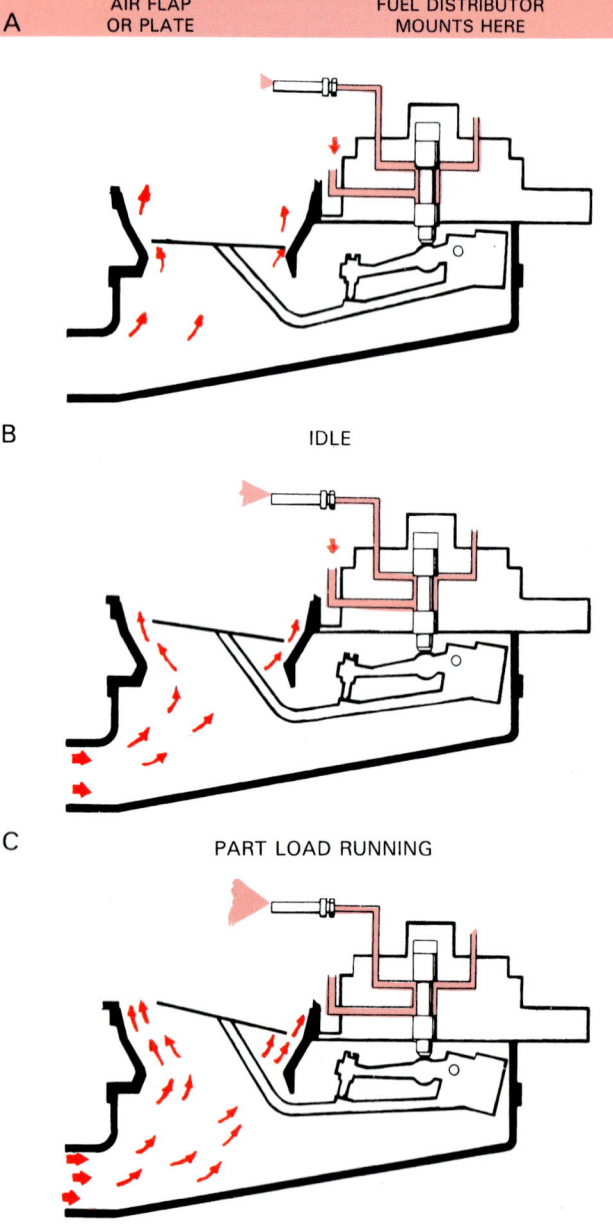

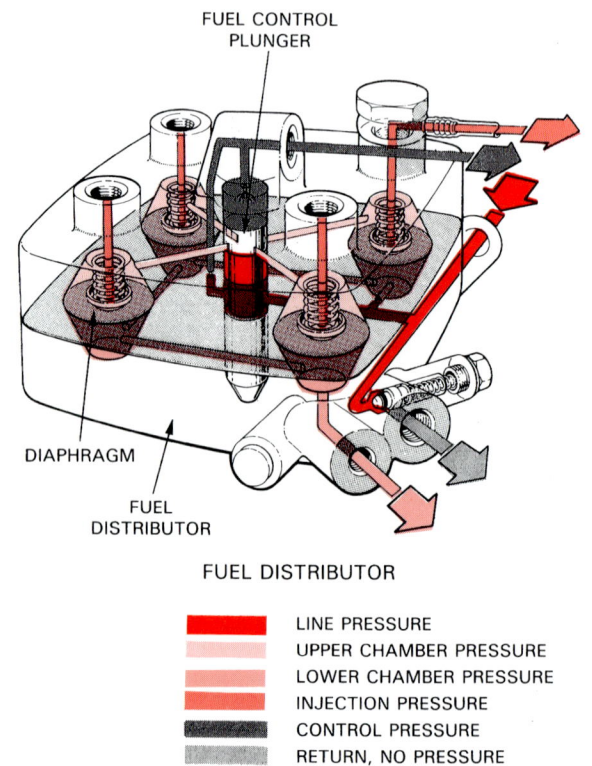

Fig. 21-35. A — Airflow sensor has large disc shaped flap hinged in airhorn. Disc operates lever arm. B — At idle, low air flow only moves sensor plate a little. Lever arm pushes up lightly on fuel control plunger for small injection quantity. C — At part load, more airflow moves sensor plate and control plunger up more. More fuel sprays out injectors. D — Fuel load condition and high air flow pushes sensor plate up high. This opens fuel control plunger fully for maximum injection pressure and volume. (Robert Bosch)

Fig. 21-36. Fuel distributor is complex set of pressure differentiating diaphragm valves. They assure that the same amount of fuel is sent to each injector. (SAAB)

control plunger is located in the centre of the distributor. Fuel is fed from the plunger to spring-loaded diaphragms. The diaphragms compensate for pressure differences in each injection line. They help assure that the same amount of fuel is sent to each injector.

Note! A fuel distributor is only used in one type of CIS system. However, this is a common system found on many European cars.

Continuous fuel injector

A continuous fuel injector is simply a spring-loaded valve. It injects fuel ALL the TIME when the engine is running. See Fig. 21-37.

A spring holds the valve in a normally closed position. A filter is placed in the injector to trap dirt. The injector is usually a push-fit into plastic bushings in the cylinder head or intake manifold.

With the engine off, the injector spring holds the injector valve closed. This keeps fuel from dripping into the engine.

When the engine starts, fuel pressure builds and pushes the injector valve open. A steady stream of petrol then sprays toward each engine intake valve. The fuel is drawn into the engine when the intake valves open.

With CIS injectors, injection quantity (air-fuel ratio) is controlled by increasing or decreasing fuel pressure to the injectors.

Cold start injector

A cold start injector is an additional fuel injector valve used to supply extra petrol for cold engine starting. Refer back to Figs. 21-33 and 21-34. Either a thermo-time switch or the system computer can be used to operate the cold start injector.

A cold start injector can be used in electronic airflow sensing, electronic pressure sensing, and hydraulic mechanical type multi-point systems. It is constructed like a conventional, solenoid type injector. One is pictured in Fig. 21-38.

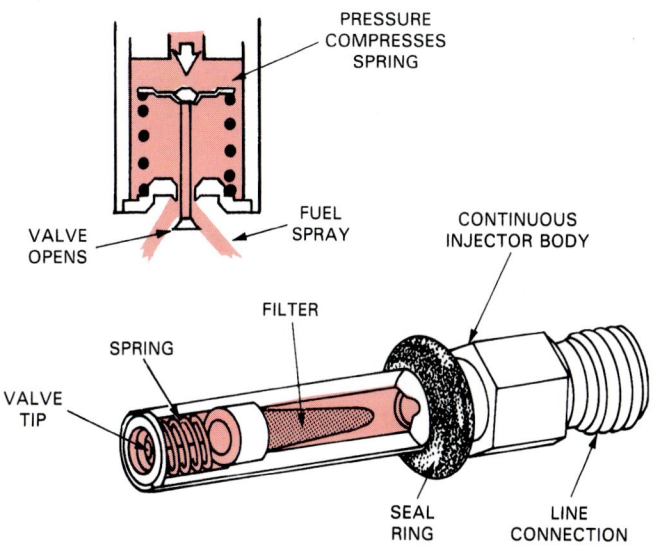

Fig. 21-37. Continuous type injector is simply spring-loaded valve. With enough fuel pressure, injector valve opens and fuel sprays into intake port of engine.

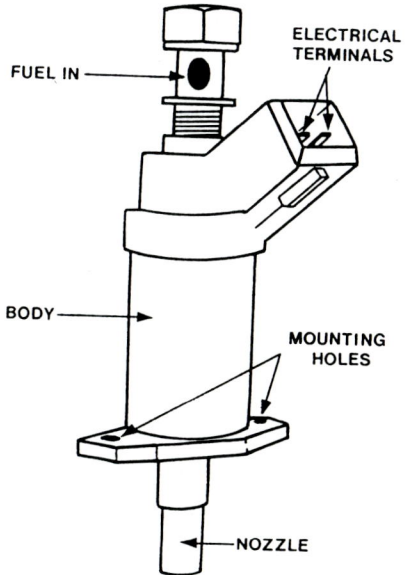

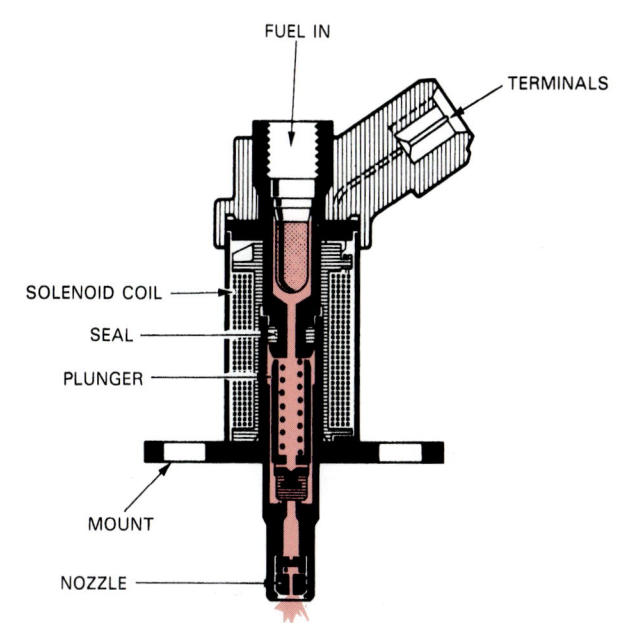

Fig. 21-38. Cold start injector is like injector for EFI system. It is a solenoid type injector valve. Note parts and construction. (SAAB)

Cold start injector operation

Fig. 21-39 shows a basic cold start injector-thermo-time switch circuit. It illustrates cold start valve action.

When the sensor detects a cold engine, the switch closes to energise the cold start injector. The cold start injector and the other injectors all spray fuel into the intake manifold. Like a carburettor choke, this provides a very rich mixture to sustain cold engine operation.

FUEL ACCUMULATOR

A fuel accumulator can be used in an injection system to dampen pressure pulses. One is shown in Fig. 21-40. The accumulator may also maintain pressure when the system is shut down. This aids engine restarting.

OTHER INJECTION SYSTEMS

There are other fuel injection systems besides those discussed in this chapter. They use the same basic principles that have already been explained.

For complete details of a particular petrol injection system, refer to a vehicle service manual. It will detail the operation of the specific system. Many fuel injection systems vary from year to year and model to model.

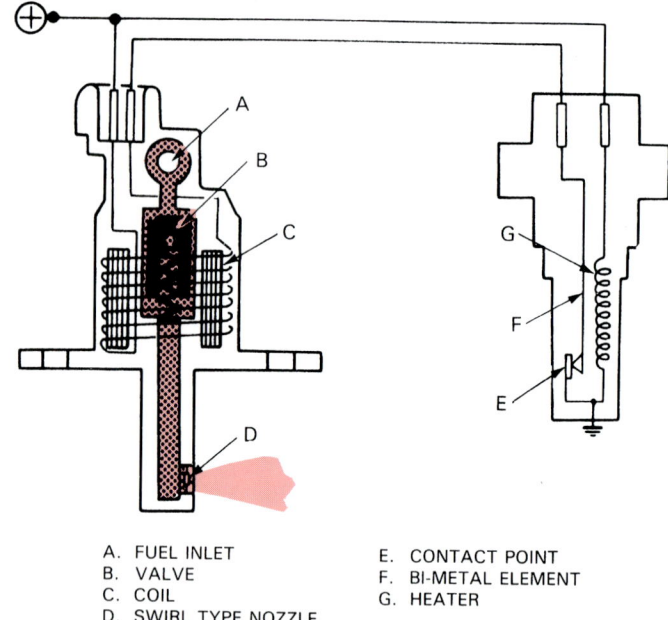

A. FUEL INLET
B. VALVE
C. COIL
D. SWIRL TYPE NOZZLE
E. CONTACT POINT
F. BI-METAL ELEMENT
G. HEATER

Fig. 21-39. Basic cold start injector, thermo-time switch circuit. Thermo-time switch energises injector when engine temperature is low enough. Cold start injector then sprays extra fuel into engine to help keep the cold engine running smoothly. When engine warms enough, thermo-time switch opens to shut off injector. Heating element in switch assures that injector stays on a short period of time, even if engine is very cold.

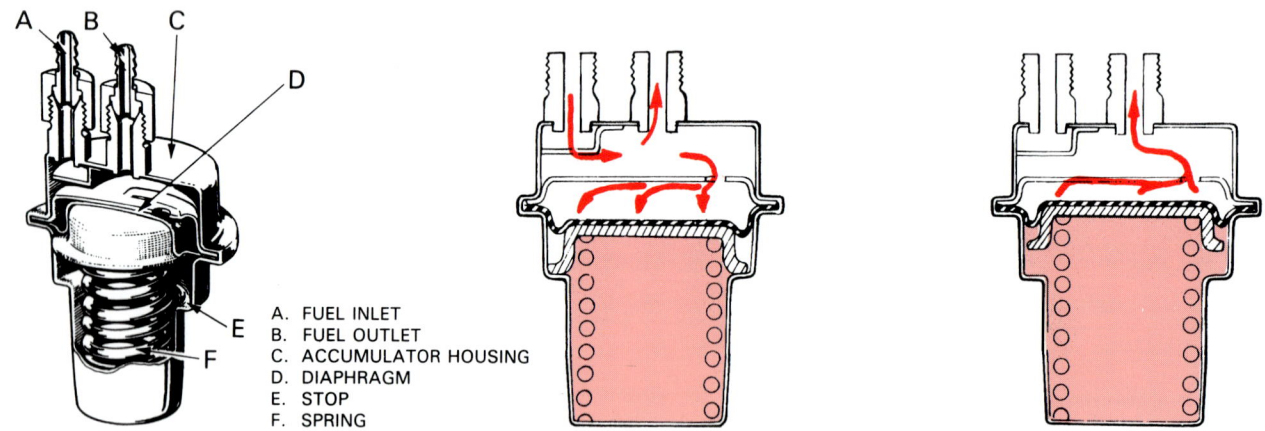

A. FUEL INLET
B. FUEL OUTLET
C. ACCUMULATOR HOUSING
D. DIAPHRAGM
E. STOP
F. SPRING

Fig. 21-40. Fuel accumulator is simply a spring-loaded diaphragm. It dampens pressure pulsations in system. It also maintains fuel pressure when engine is shut off. Left — Note basic parts of fuel accumulator. Centre — With engine running, fuel pressure compresses diaphragm spring. Right — When engine is shut off, spring pushes up on diaphragm to hold pressure in system. (Volvo)

KNOW THESE TERMS

Petrol injection system, Fuel injection system, Single-point, TBI, Multi-point, Port injection, EFI, Timed injection, Modulated injection, Continuous injection, Group injection, Fuel pressure regulator, Fuel injector, Throttle valve, Computer, ECU, Exhaust gas or oxygen sensor, Manifold pressure sensor, Throttle position sensor, Engine Temperature sensor, Airflow sensor, Inlet air temperature sensor, Crankshaft position sensor, Digital signal, Analog signal, Open loop, Closed loop, Pulse width, Idle air control valve, Throttle positioner, Fuel rail, Fuel distributor, Cold start injector, Thermo-time switch, Fuel accumulator.

REVIEW QUESTIONS

1. A petrol injection system uses pressure from an _____ _____ _____ to spray fuel into the engine _____ _____.

2. List seven possible advantages of a petrol injection system over a carburettor system.

3. Explain the difference between single-point (throttle body) and multi-point (port) injection systems.

4. Petrol injection systems use direct injection; diesels use indirect injection. True or False?

5. This is the most common and modern type of petrol injection system.

   a. Mechanical fuel injection.
   b. Hydraulic fuel injection.
   c. Electronic fuel injection.
   d. Pneumatic fuel injection.

6. This type of petrol injection pulses the injectors open and closed independently of the engine valve action.

   a. Timed injection.
   b. Intermittent injection.
   c. Continuous injection.
   d. Bank injection.

7. List the parts typically included in an EFI fuel delivery system.

8. How does an EFI injector open and close?

9. Explain the action of an EFI system throttle valve.

10. An engine _____ is an electrical device that changes circuit resistance or voltage with a change in a condition (temperature, pressure, position of part, etc.).

11. Define the term "EFI computer".

12. An _____ _____ _____, also called _____ _____ measures the oxygen content in the engine's exhaust system as a means of checking _____ _____.

13. When the intake manifold pressure sensor detects high pressure (low vacuum), the computer would know that a _____ mixture is needed for load conditions.

14. A throttle position sensor is a _____ _____ connected to the _____ _____ _____.

15. Which of these is NOT a typical EFI system sensor?

   a. Exhaust back pressure sensor.
   b. Throttle position sensor.
   c. Engine temperature sensor.
   d. Air inlet temperature sensor.

16. Explain the difference between sensor analog and digital signals.

17. When an EFI system is in open loop, the computer uses stored information to operate the system. True or False?

18. When an EFI system is in closed loop, the computer uses engine sensor information to control the system. True or False?

19. Define the term "injector pulse width".

20. List and explain the six major parts of a TBI unit.

21. What are the main differences between the throttle body for multi-point injection and single-point injection?

22. An EFI multi-point injector fits into the _____ _____ in the _____ manifold.

23. List and explain the eight major parts of an EFI multi-point injector.

24. Describe the mixture control unit in a hydraulic-mechanical, CIS.

25. A continuous fuel injector is a spring-loaded fuel valve that does NOT use an electrical coil. True or False?

Chapter 22

EXHAUST SYSTEMS, TURBOCHARGING

After studying this chapter, you will be able to:
☐ Describe the basic parts of an exhaust system.
☐ Compare exhaust system design differences.
☐ Perform exhaust system repairs.
☐ Explain the fundamental parts of a turbocharging system.
☐ Describe the construction and operation of a turbocharger and waste gate.
☐ Remove and replace a turbocharger and waste gate.

This chapter begins by covering the basic parts of an exhaust system. It then explains how to repair the system by replacing rusted or damaged components. The second part of the chapter covers turbocharging. With smaller, high efficiency engines, turbocharging is commonly used to increase engine power output.

BASIC EXHAUST SYSTEMS

An exhaust system quietens engine operation and carries exhaust fumes to the rear of the car. The parts of a typical exhaust system are shown in Fig. 22-1. They include:
1. EXHAUST MANIFOLD (connects cylinder head exhaust ports to engine).
2. ENGINE PIPE (steel tubing that carries exhaust gases from exhaust manifold to catalytic converter or muffler).
3. CATALYTIC CONVERTER (device for removing pollutants from engine exhaust).
4. INTERMEDIATE PIPE (tubing sometimes used between engine pipe and muffler or catalytic converter and muffler).

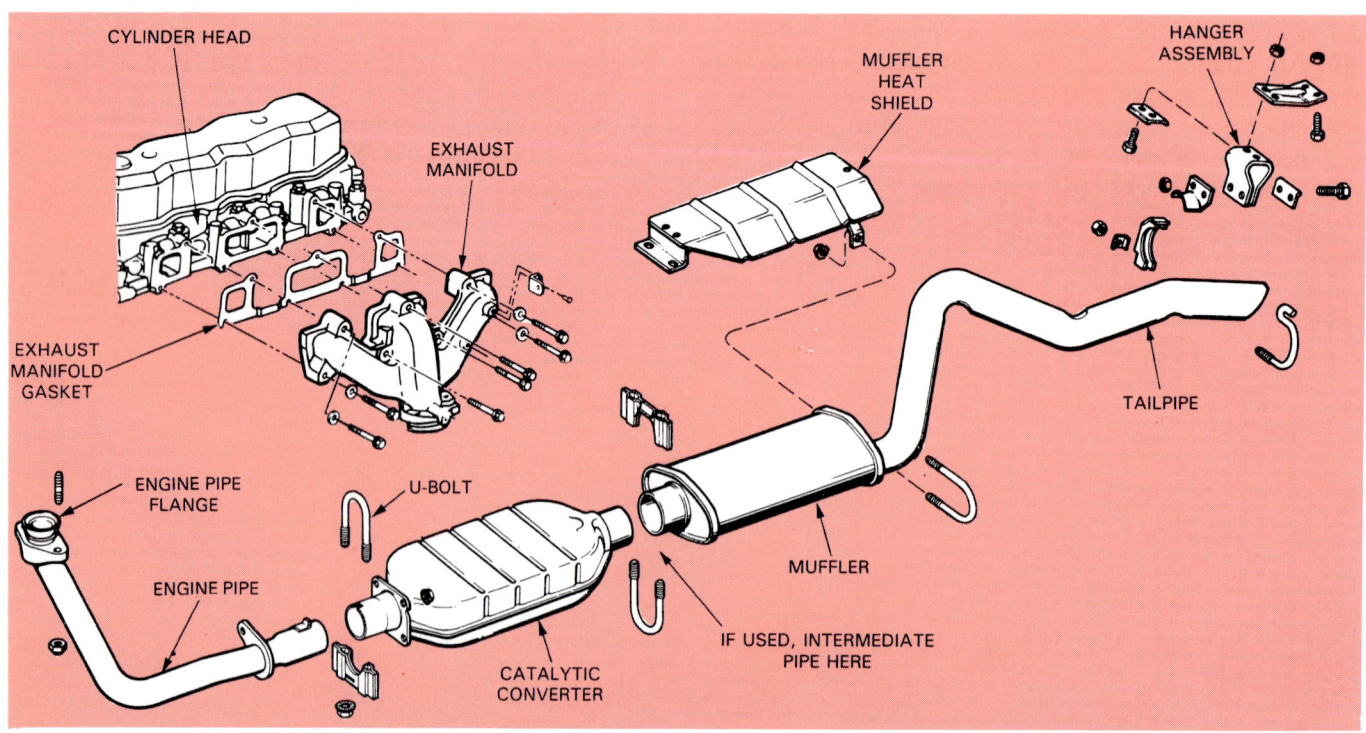

Fig. 22-1. Note parts of typical exhaust system. Exhaust comes out of cylinder head, into manifold, and then through system.

5. MUFFLER (metal chamber for damping pressure pulsations to reduce exhaust noise).
6. TAILPIPE (tubing that carries exhaust from muffler to rear of car body).
7. HANGERS (devices for securing exhaust system to underside of car body).
8. HEAT SHIELDS (metal plates that prevent exhaust heat from transferring into another object).
9. MUFFLER CLAMPS (U-bolts for connecting parts of exhaust system together).

When an engine is running, extremely hot gases blow out of the cylinder head exhaust ports. The gases enter the exhaust manifold. They flow through the engine pipe, catalytic converter, intermediate pipe, muffler, and out of the tailpipe.

Exhaust back pressure

Exhaust back pressure is the amount of pressure developed in the exhaust system when the engine is running. High back pressure reduces engine power. A well designed exhaust system should have LOW back pressure.

The restriction of the exhaust pipes, catalytic converter, and muffler contribute to exhaust back pressure. Larger pipes and a free-flowing muffler, for example, would reduce back pressure.

Single and dual exhaust systems

A single exhaust system has one path for exhaust flow through the system. Typically, it has only one engine pipe, main catalytic converter, muffler, and tailpipe. The most common type, it is used from the smallest four-cylinder engines, on up to large V-8 engines.

A dual exhaust system has two separate exhaust paths to reduce back pressure. It is two single exhaust systems combined into one. A dual exhaust system is sometimes used on high performance cars with large V-6 or V-8 engines. It allows the engine to "breathe" better at high rpm.

A crossover pipe normally connects the right and left side engine pipes to equalise back pressure in a dual system. This also increases engine power slightly.

EXHAUST MANIFOLD

An exhaust manifold bolts to the cylinder head to enclose the exhaust port openings, Fig. 22-2. The manifold is usually made of cast iron. It is sometimes made of stainless steel or lightweight steel tubing. The cylinder head mating surface is machined smooth and flat. An exhaust manifold gasket is commonly used between the cylinder head and manifold to help prevent leakage.

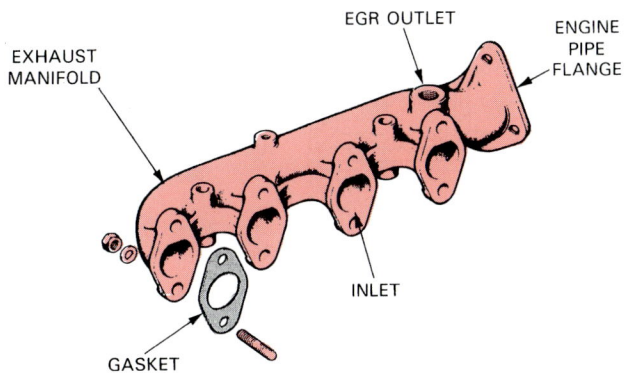

Fig. 22-2. Exhaust manifold is metal casting that bolts over exhaust ports on side of cylinder head. An exhaust manifold gasket prevents leakage between head and manifold.

The outlet end of the exhaust manifold has a single round opening with provisions for studs, bolts or cap screws. A gasket or O-ring (doughnut) seals the connection between the exhaust manifold outlet and engine pipe to prevent leakage.

Exhaust manifold heat valve (heat riser)

An exhaust manifold heat valve allows hot exhaust gases to flow into the intake manifold to aid cold weather starting. Look at Fig. 22-3.

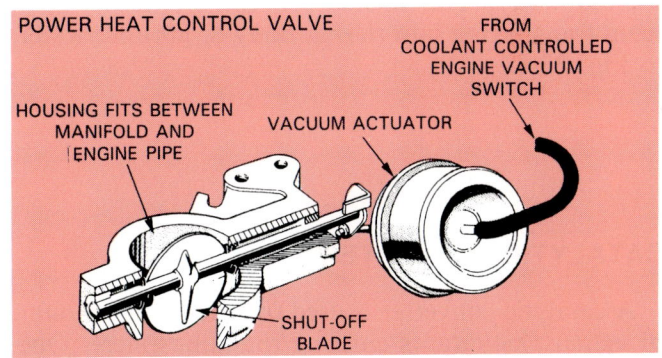

Fig. 22-3. Heat control valve, sometimes called heat riser, allows hot exhaust gases to flow into the intake manifold. This helps engine run smoothly. Valve opens as engine warms up.

A butterfly valve may be located in the outlet of the exhaust manifold. A heat sensitive spring or a vacuum diaphragm and temperature sensing vacuum switch may operate the valve.

When the engine is cold, the heat valve is closed. This increases EXHAUST BACK PRESSURE. Hot gases blow into an exhaust passage in the intake manifold, Fig. 22-4. This warms the floor of the intake manifold to hasten fuel vapourisation. The heat valve opens as the engine warms up.

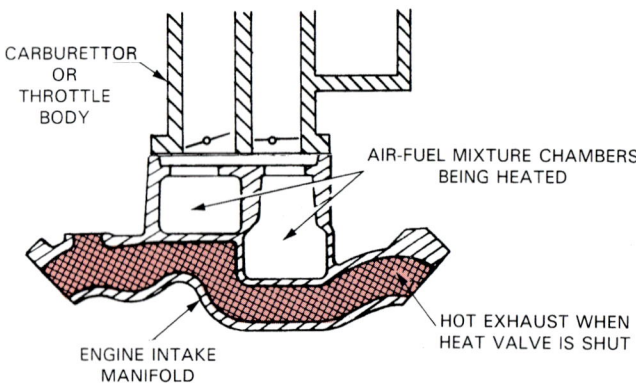

Fig. 22-4. Heat control valve causes back pressure in exhaust system. This directs a large amount of hot exhaust into chamber in bottom of intake manifold. This action warms and helps vapourise fuel.

EXHAUST PIPES

The exhaust pipes (engine pipe, intermediate pipe, and tailpipe) are usually made of rust resistant steel tubing. The inlet end of the engine pipe has a flange for securing the pipe to the exhaust manifold studs. Fig. 22-1. One end of each pipe may be enlarged to fit over the end of the next pipe.

HEAT SHIELDS

Heat shields are located where the exhaust system (especially catalytic converter and muffler) is close to the car body. The shields prevent too much heat from transferring into the car body or ground. Refer to Fig. 22-1.

DANGER! Always reinstall all exhaust system heat shields. Without a heat shield, car undercoating, carpeting, dry leaves on the ground, and other flammable materials could catch on fire!

CATALYTIC CONVERTER

A catalytic converter is used to reduce the amount of exhaust pollutants entering the atmosphere. One or more catalytic converters can be located in the exhaust system, Fig. 22-1.

MUFFLERS

A muffler reduces the pressure pulses and resulting noises produced by the engine exhaust. When an engine is running, the exhaust valves are rapidly opening and closing. Each time an exhaust valve opens, a blast of hot gas shoots out of the engine. Without a muffler, these exhaust gas pulsations are very loud.

Fig. 22-5 shows the inside of a muffler. Note how various chambers, tubes, holes, and baffles are arranged to cancel out the pressure pulsations in the exhaust.

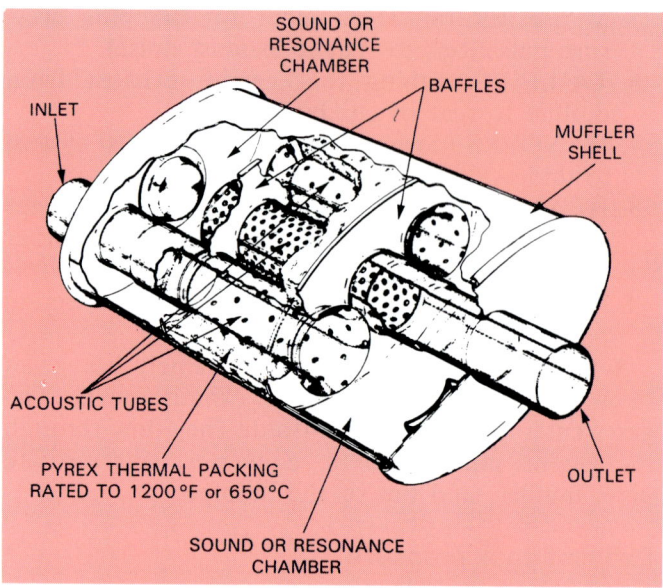

Fig. 22-5. Basic muffler contains baffles, resonance chambers, and acoustic tubes to reduce exhaust noise.

EXHAUST SYSTEM SERVICE

Exhaust system service is usually needed when a component in the system rusts and begins to leak. Because engine combustion produces water and acids, an exhaust system can fail in a relatively short time.

DANGER! A leaking exhaust system could harm the passengers of a car. Since engine exhaust is poisonous, a leaky exhaust can allow toxic gases to flow through any opening in the body and into the passenger compartment.

Exhaust system inspection

To inspect an exhaust system, raise the car on a hoist. Using a lead light, closely inspect the system for problems (rusting, loose connections, leaks). In particular, check around the muffler, all pipe connections, gaskets, and pipe bends.

DANGER! Parts of the exhaust system especially the catalytic converter, can be VERY HOT. Remember not to touch any component until after it has cooled.

Exhaust system repairs

Faulty exhaust system components must be removed and renewed. If only the muffler is rusted, a new muffler can be installed in the existing system. After prolonged service, several parts or ALL of the exhaust system may require renewal.

When repairing an exhaust system, remember the following:

1. Use RUST PENETRANT on all threaded fasteners that will be reused, Fig. 22-6. This is especially important on the exhaust manifold flange nuts or bolts.

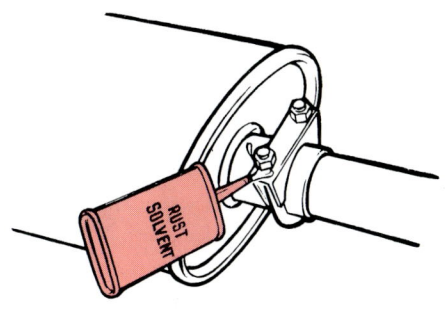

Fig. 22-6. Rust penetrant or solvent will ease removal of badly rusted fasteners.

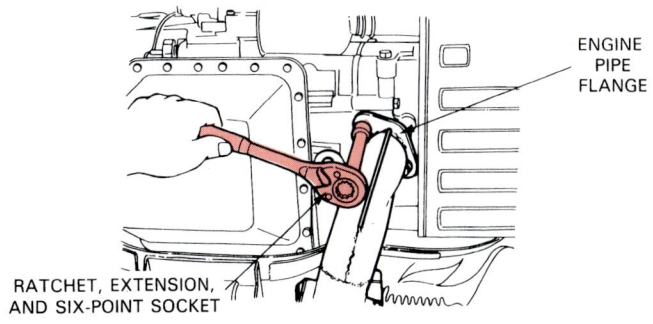

Fig. 22-8. Engine pipe fasteners can be difficult to remove. Use rust penetrant, six-point socket, extension, and ratchet. This will usually remove fasteners. (Subaru)

2. Use an air chisel, cut-off tool, cutting torch, or hacksaw to remove faulty parts. Make sure you do NOT damage parts that will be reused. See Fig. 22-7 for some examples.

3. A SIX-POINT SOCKET and ratchet or air impact will usually allow quick fastener removal without rounding off the fastener heads. Refer to Fig. 22-8.

4. Wear SAFETY GLASSES or goggles to keep rust and dirt from entering your eyes.

5. Obtain the correct replacement parts.

6. A pipe expander should be used to enlarge pipe ends as needed, Fig. 22-9. A pipe shaper can be used to straighten dented pipe ends, Fig. 22-10.

7. Make sure all pipes are fully inserted. Position all clamps properly, as in Fig. 22-11.

8. Double-check the routing of the exhaust system. Keep adequate clearance between it, the car body, and chassis. See Fig. 22-12.

9. Tighten all clamps and hangers evenly. Torque the fasteners only enough to hold the parts. Over-tightening will squash and deform the pipes, possibly causing leakage.

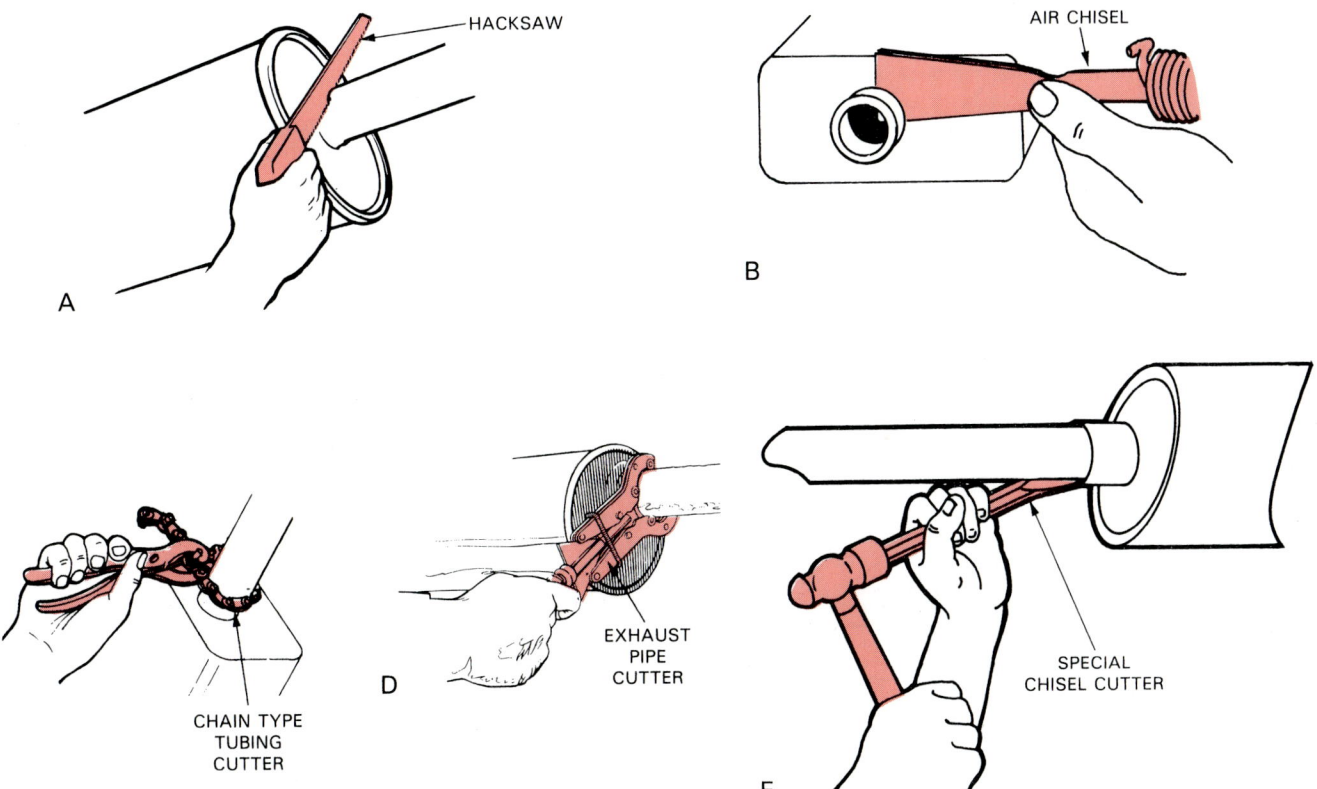

Fig. 22-7. Several methods of removing old exhaust system parts. A — Hacksaw. B — Air chisel. C — Chain type cutting tool. D — Exhaust pipe cutter. E — Specially shaped, hand type cutter or chisel.

337

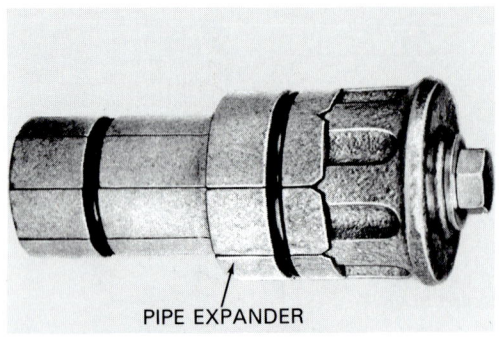

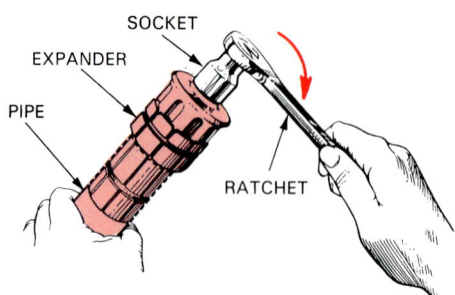

Fig. 22-9. Pipe expander will enlarge ID (inside diameter) of pipes. Then one pipe will fit over another.

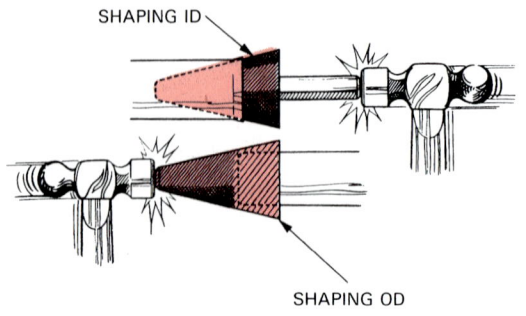

Fig. 22-10. Pipe shaper will round dented pipe ends.

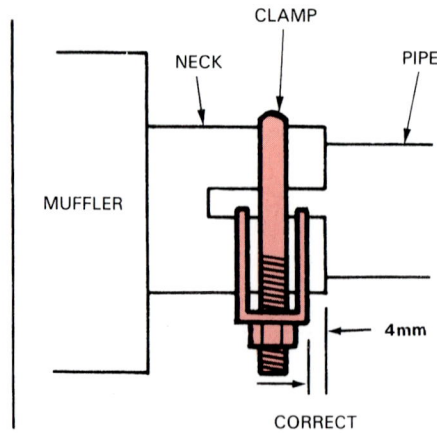

Fig. 22-11. Make sure muffler clamps are installed correctly. Clamps must be positioned around both pipes. If not, one pipe can pull out of other.

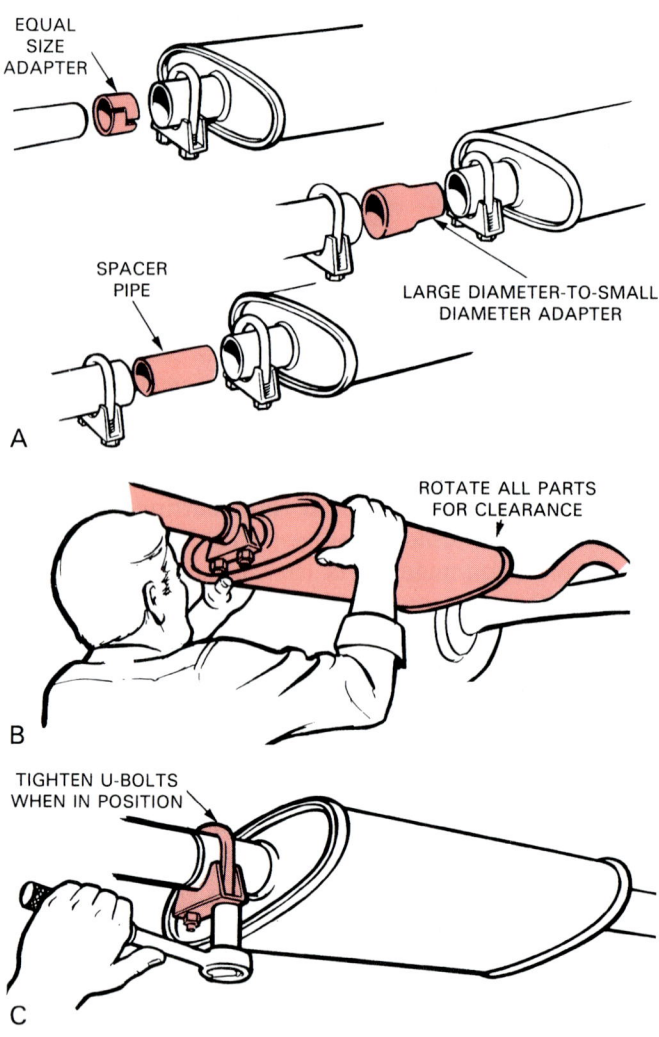

Fig. 22-12. A — Adaptors are sometimes needed to make muffler work on existing system. B — Double-check exhaust system-to-car clearance carefully. C — After checking clearance, tighten all clamps evenly and properly.

10. When replacing an exhaust manifold, use a gasket and check sealing surface flatness. If the manifold is warped, it must be machined flat. Torque the exhaust manifold bolts to specs, Fig. 22-13.
11. Always use new gaskets and O-rings.
12. Check heat riser operation using the information in a service manual.
13. Install all heat shields.
14. Check the system for leaks and rattles after repairs.

SUPERCHARGERS

A supercharger is an air pump that increases engine power by pushing a denser air-fuel charge into the combustion chambers. With more fuel and air, combustion produces more heat energy and pressure to push the piston down in the cylinder.

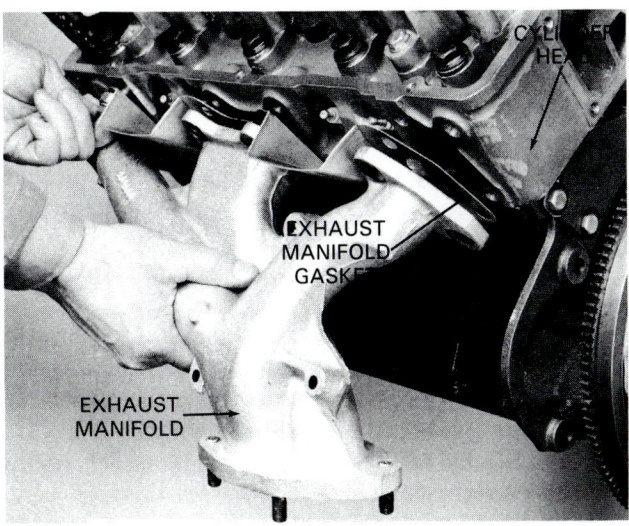

Fig. 22-13. Exhaust gasket is normally recommended. Gasket is held in position as all fasteners are started by hand. Torque fasteners to recommended value in a crisscross pattern. (Saab)

Sometimes termed a blower, the supercharger raises the air pressure in the engine intake manifold. Then, when the intake valves open, more air-fuel mixture (petrol engine) or (diesel engine) can flow into the cylinders.

A normally aspirated engine uses atmospheric pressure (101 kPa at sea level) to push air into the engine. It is a nonsupercharged engine. With only outside air pressure as a moving force, only a limited amount of fuel can be burned on each power stroke.

Supercharger types

There are three basic types of superchargers:
1. Centrifugal supercharger.
2. Rotor (Rootes) supercharger.
3. Vane supercharger.

These types are shown in Fig. 22-14. Note the differences in construction and operation.

In the field, the term "supercharger" generally refers to a blower driven by a belt, gears, or chain. Superchargers are used on large diesel truck engines and racing engines. They are not commonly found on passenger cars.

The term turbocharger or "turbo" refers to a blower driven by engine exhaust gases. Turbochargers are commonly used on passenger cars, trucks, and competition engines.

TURBOCHARGERS

A turbocharger is an exhaust driven supercharger (fan or blower) that forces air into the engine under pressure. Turbochargers are frequently used on small petrol and diesel engines to increase power output. By harnessing engine exhaust energy, a turbocharger can also improve engine efficiency (fuel economy and

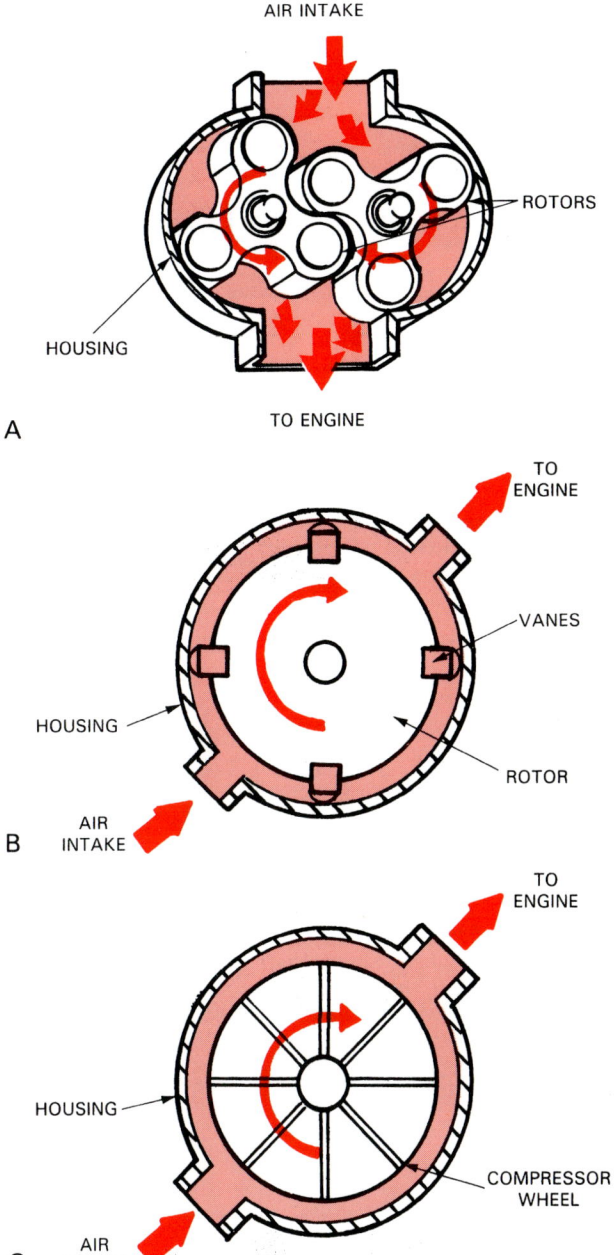

Fig. 22-14. Three basic types of superchargers. A — Rotor or Rootes type supercharger. B — Vane type supercharger. C — Centrifugal supercharger, more commonly called a turbocharger.

emission levels). This is especially true with a diesel engine.

As shown in Fig. 22-15, the basic parts of a turbocharger are:

1. TURBINE WHEEL (exhaust driven fan that turns turbo shaft and compressor wheel).
2. TURBINE HOUSING (outer enclosure that routes exhaust gases around turbine wheels).
3. TURBO SHAFT (steel shaft that connects turbine and compressor wheels. It passes through centre of turbo housing).

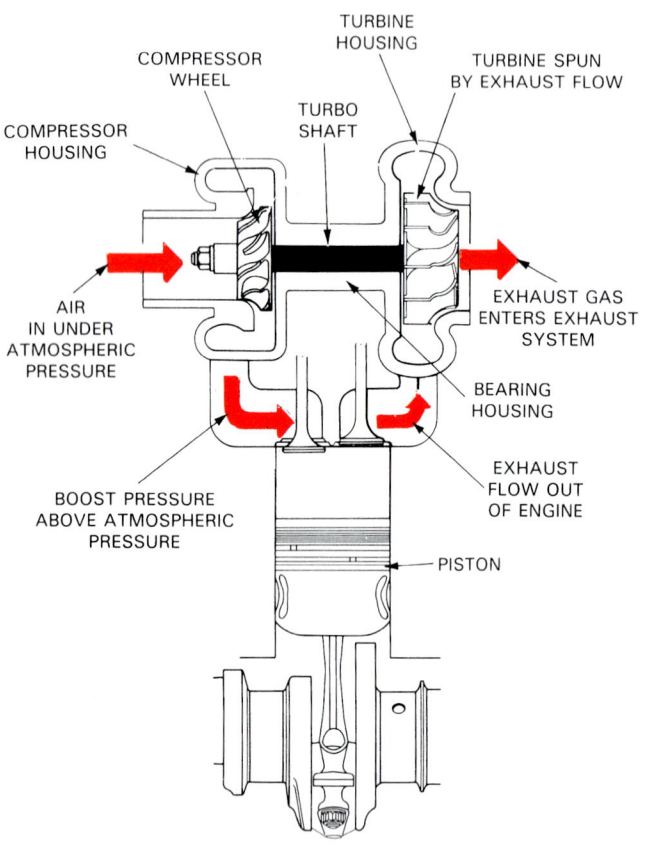

4. COMPRESSOR WHEEL (driven fan that forces air into engine intake manifold under pressure).

5. COMPRESSOR HOUSING (part of turbo housing that surrounds compressor wheel. Its shape helps pump air into engine).

6. BEARING HOUSING (enclosure around turbo shaft that contains bearings, seals, and oil passages).

Turbocharger operation

When the engine is running, hot exhaust gases blow out of the open exhaust valves and into the exhaust manifold. The exhaust manifold and connecting tubing route these gases into the turbine housing. Refer to Fig. 22-16.

As the gases pass through the turbine housing, they strike the fins or blades on the turbine wheel. When engine load is high enough, there is enough exhaust gas flow to rapidly spin the turbine wheel, Fig. 22-16.

Since the turbine wheel is connected to the compressor wheel by the turbo shaft, the compressor wheel rotates with the turbine. Compressor wheel rotation pulls air into the compressor housing. Centrifugal force throws the spinning air outward. This causes air to flow out of the turbocharger and into the engine cylinder under pressure.

Turbocharger location

A turbocharger is usually located on one side of the engine. See Fig. 22-17. An exhaust pipe connects

Fig. 22-15. Turbocharger uses exhaust gas flow to spin turbine wheel. Turbine wheel spins shaft and compressor wheel. Compressor wheel then pressurises air entering engine for more power output. (Mercedes Benz)

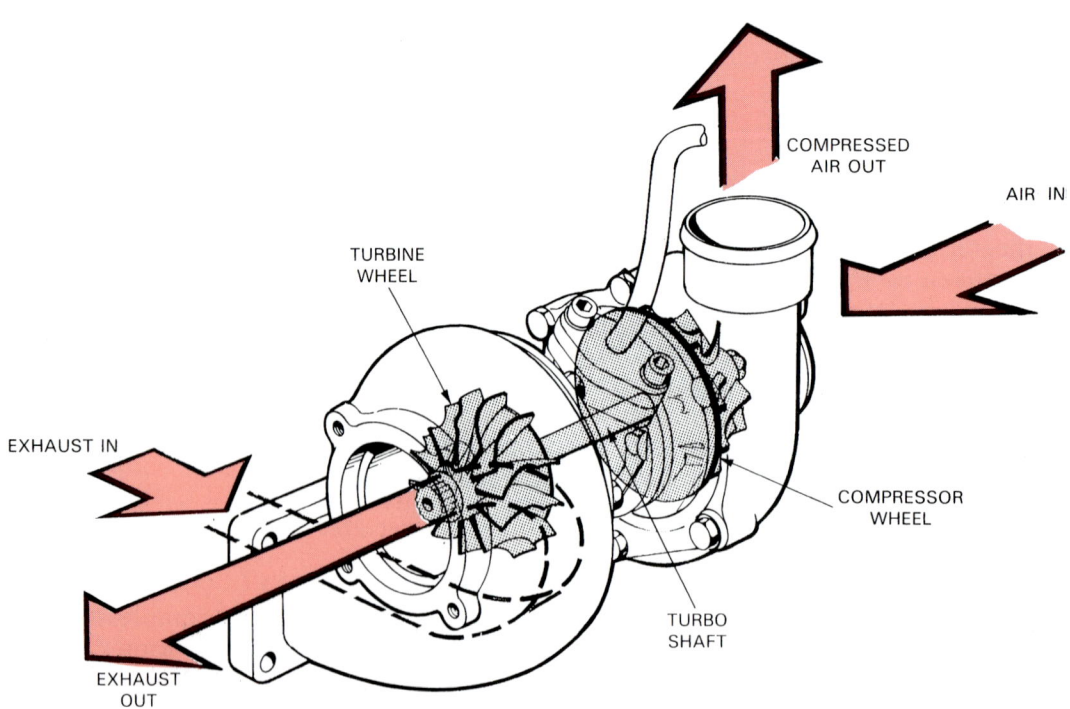

Fig. 22-16. Exhaust flow spins turbine wheel, shaft, and compressor wheel. Normally wasted energy in exhaust is used to increase compression stroke pressure in cylinders for more violent combustion. (Saab)

Fig. 22-17. Turbocharger normally bolts to one side of engine. Pipes route exhaust through turbine housing. Compressed air leaves turbo and enters intake tract and engine. (Ford)

the engine exhaust manifold to the turbine housing. The exhaust system engine pipe connects to the outlet of the turbine housing.

A blow-through turbo system has the turbocharger located before the carburettor or throttle body. The turbo compressor wheel only pressurises air. Fuel is mixed with the air after leaving the compressor.

A draw-through turbo system locates the turbocharger after the carburettor or throttle body assembly. As a result, both air and fuel (petrol engine) pass through the compressor housing.

Theoretically, the turbocharger should be located as close to the engine exhaust manifold as possible. Then, a maximum amount of exhaust heat will enter the turbine housing. When the hot gases move past the spinning turbine wheel, they are still expanding and help rotate the turbine.

Turbocharger lubrication

Turbocharger lubrication is needed to protect the turbo shaft and bearings from damage. A turbocharger can operate at speeds up to 100,000 rpm. For this reason, the engine lubrication system forces motor oil into the turbo shaft bearings. Look at Fig. 22-18.

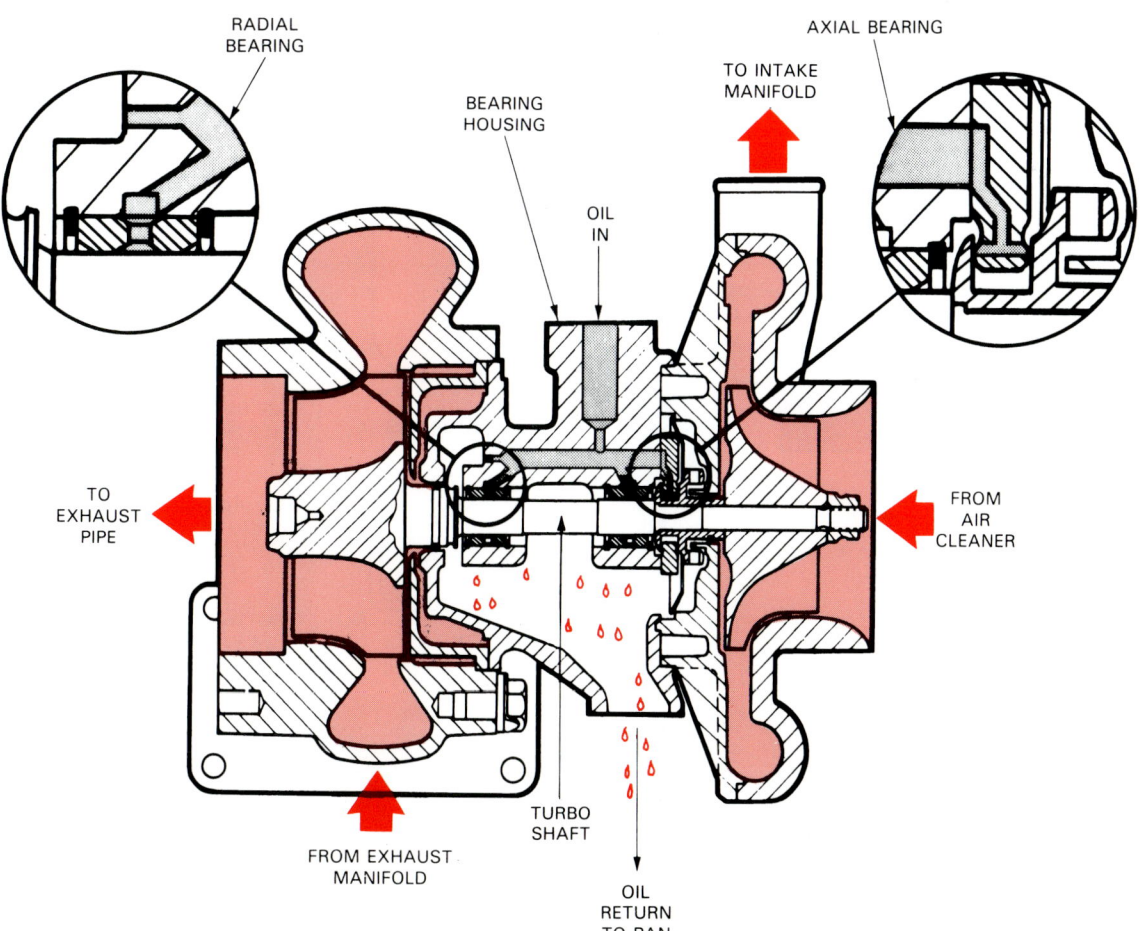

Fig. 22-18. High turbo speeds require pressure lubrication. Engine oil is fed to turbo through oil line. Oil flows through bearings and then drains into oil pan through drain line. (Saab)

Oil passages are provided in the turbo housing and bearings. An oil supply line runs from the engine to the turbo. With the engine running, oil enters the turbo under pressure. See Fig. 22-19.

Sealing rings (piston type rings) are placed around the turbo shaft, at each end of the turbo housing. See Fig. 22-18. They prevent oil leakage into the compressor and turbine housings.

A drain passage and drain line allow oil to return to the engine oil pan after passing through the turbo bearings.

Turbo lag

Turbo lag refers to a short delay before the turbo develops sufficient boost (pressure above atmospheric pressure).

When the car's accelerator pedal is pressed down

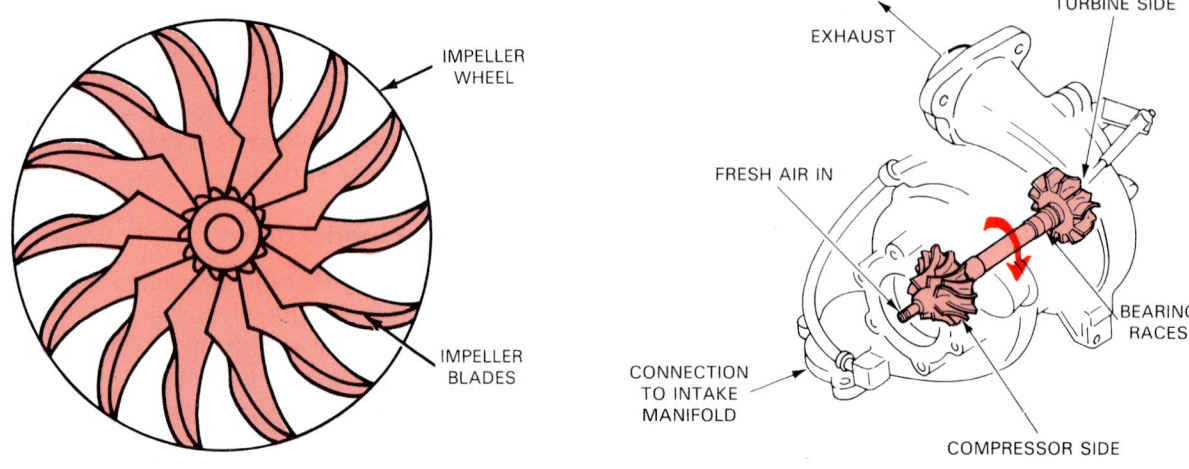

Fig. 22-19. Turbine and compressor wheels have specially shaped fins or blades that act as fans. Exhaust flow spins turbine fan. Spinning compressor fan blows air into engine under pressure.

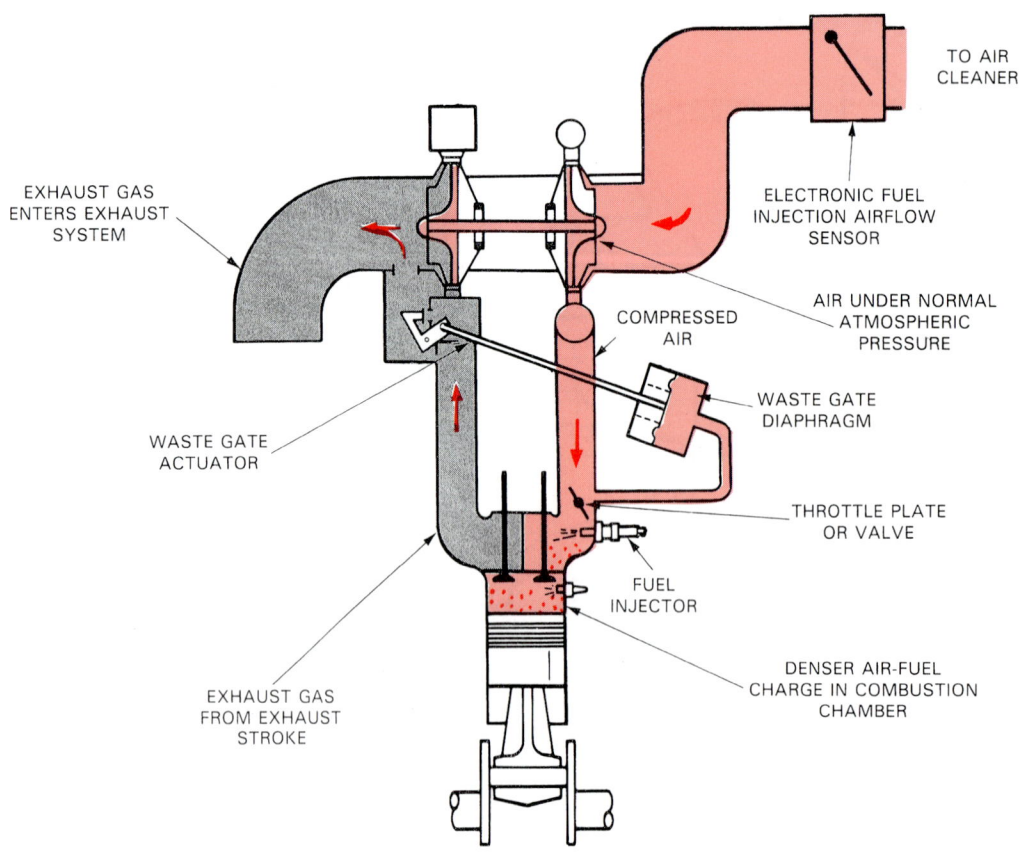

Fig. 22-20. Study basic exhaust and inlet airflow through complete turbo system. (Ford)

for rapid acceleration, the engine may lack power for a few seconds. This is caused by the compressor and turbine wheels not spinning fast enough. It takes time for the exhaust gases to bring the turbo up to operating speed.

Modern turbo systems suffer very little from turbo lag. Their turbine and compressor wheels are very light so that they can accelerate up to rpm quickly.

Turbocharger intercooler

A turbocharger intercooler is an air-to-air heat exchanger that cools the air entering the engine. It is a radiator like device mounted at the pressure outlet of the turbocharger.

Outside air flows over and cools the fins and tubes of the intercooler. Then, when the air flows through the intercooler, heat is removed.

By cooling the air entering the engine, engine power is increased because the air is more dense (contains more oxygen by volume). Cooling also reduces the tendency for engine detonation.

WASTE GATE

A waste gate limits the maximum amount of boost pressure developed by the turbocharger. It is a butterfly or poppet type valve that allows exhaust to bypass the turbine wheel. See Fig. 22-20.

Without a waste gate, the turbo could produce too much pressure in the combustion chambers. This could lead to detonation (spontaneous combustion) and engine damage.

Basically, a waste gate is a valve operated by a diaphragm assembly, Fig. 22-21. Intake manifold pressure acts on the diaphragm to control waste gate valve action. The valve controls the opening and closing of a passage around the turbine housing. Fig. 22-22.

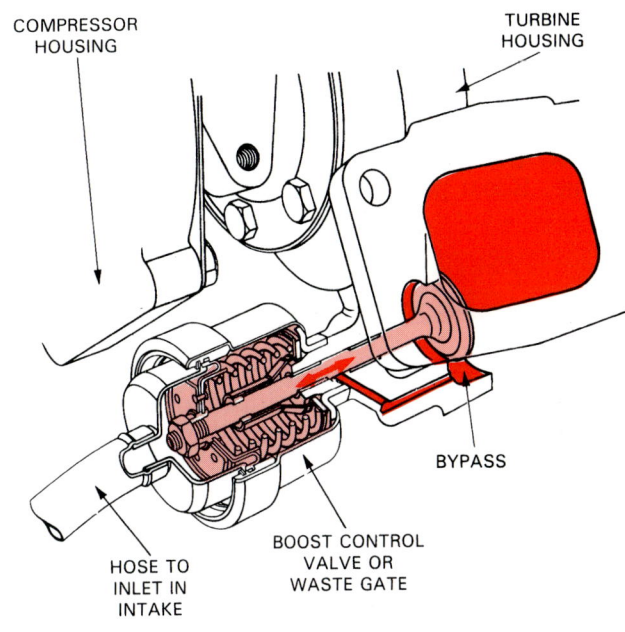

Fig. 22-22. Note hose that connects waste gate with compressor housing. When boost pressure is too high, pressure acts on waste gate diaphragm to open valve and reduce boost. (Garrett)

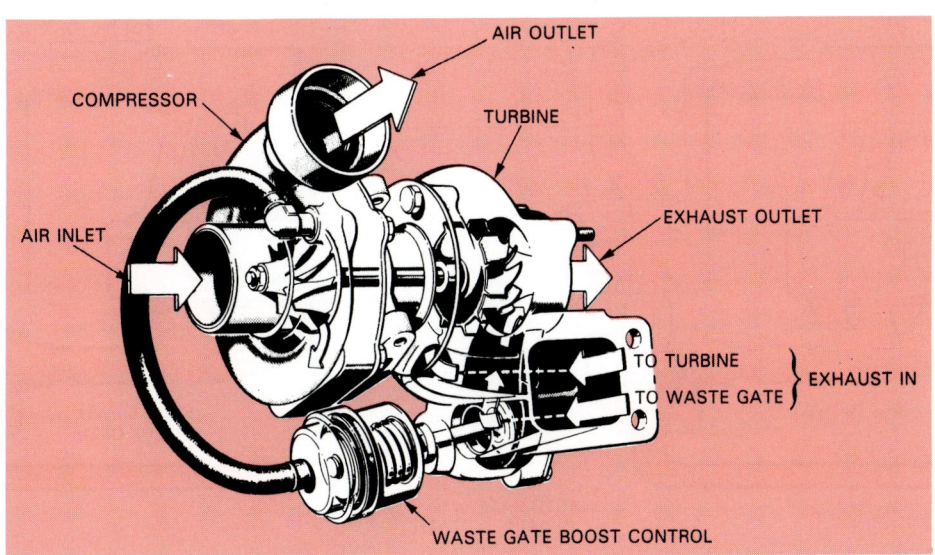

Fig. 22-21. Waste gate or boost control is a valve in turbine housing. When needed, it can open to limit boot pressure by reducing amount of exhaust acting on turbine wheel. (Mercedes-Benz)

Waste gate operation

Fig. 22-23 illustrates the basic operation of a turbocharger waste gate. Under partial load, the system routes all of the exhaust gases through the turbine housing. The waste gate is closed by the diaphragm spring. This assures that there is adequate boost to increase engine power.

Under full load, boost may become high enough to overcome diaphragm spring pressure. Manifold pressure compresses the spring and opens the waste gate valve. This permits some of the exhaust gases to flow through the waste gate passage and into the exhaust system. Less exhaust is left to spin the turbine. Boost pressure is limited to a preset value.

TURBOCHARGED ENGINE MODIFICATIONS

A turbocharged engine normally has several modifications to make it withstand the increased power output. A few of these are shown in Fig. 22-24 and include:

1. Lower compression ratio.
2. Stronger connecting rods, pistons, and crankshaft.
3. Higher volume oil pump and an oil cooler.
4. Larger cooling system radiator.
5. O-ring type head gasket.
6. Heat resistant valves.
7. Knock sensor (ignition retard system).

KNOCK SENSOR

A knock sensor is used to retard ignition timing if the engine begins to knock (detonate or ping). The sensor is mounted on the engine. It works something like a microphone. When it "hears" a knocking sound, an electrical signal is sent to the onboard computer. The computer then retards the timing until the knock stops.

A knock sensor helps the computer keep the ignition timing advanced as much as possible. This improves engine power and fuel consumption. It also protects the engine from detonation damage.

TURBOCHARGING SYSTEM SERVICE

Turbocharging system problems usually show up as inadequate boost pressure (lack of engine power), leaking shaft seals (oil consumption), damaged turbine or compressor wheels (vibration and noise), or excessive boost (detonation).

Refer to a vehicle service manual for a detailed troubleshooting chart if needed. It will list the common troubles for the particular turbo system.

To protect a turbocharger from damage, some auto makers recommend that the oil in a turbocharged engine be changed more frequently. The turbo bearings and shaft, because of the high rotating speeds, are very sensitive to oil contaminants. Engine oil must be kept clean to assure long turbocharger life.

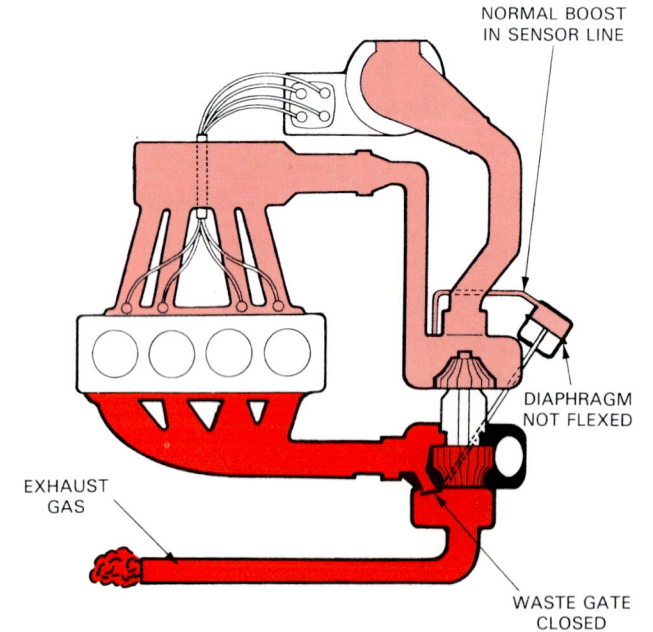
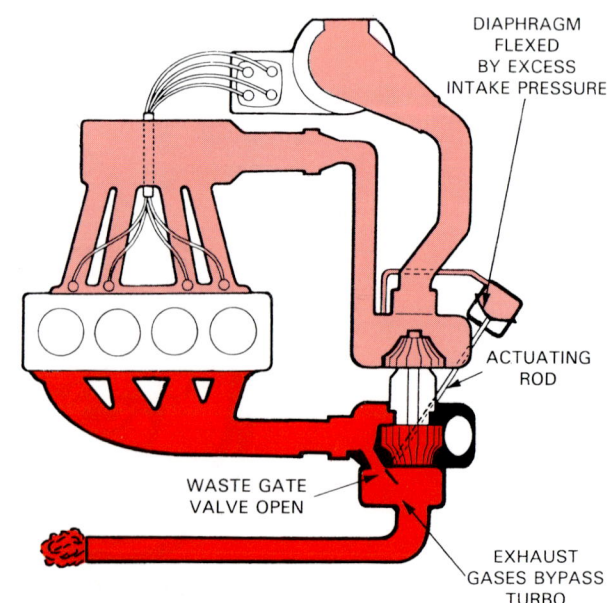

Fig. 22-23. Study operations of complete turbo system. A — Under part throttle, normal boost conditions, waste gate remains closed. All exhaust flow is directed over turbine wheel. B — Under full load, boost pressure may increase too much. High intake pressure deflects waste gate diaphragm to open waste gate valve. This allows some of the exhaust to bypass turbine wheel. Turbine wheel slows down and reduces boost pressure. (Saab)

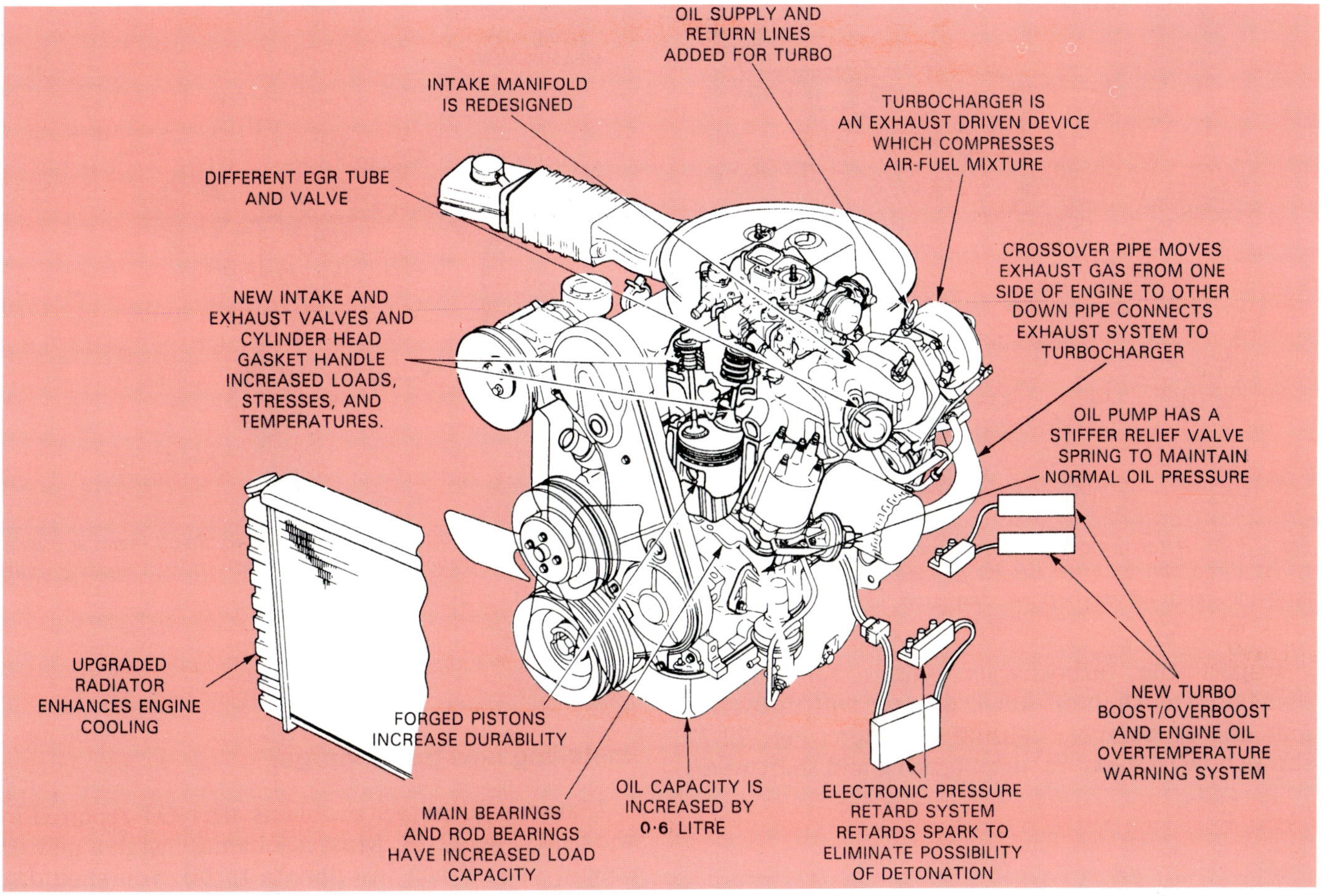

OIL SUPPLY AND
RETURN LINES
ADDED FOR TURBO

INTAKE MANIFOLD
IS REDESIGNED

TURBOCHARGER IS
AN EXHAUST DRIVEN DEVICE
WHICH COMPRESSES
AIR-FUEL MIXTURE

DIFFERENT EGR TUBE
AND VALVE

CROSSOVER PIPE MOVES
EXHAUST GAS FROM ONE
SIDE OF ENGINE TO OTHER
DOWN PIPE CONNECTS
EXHAUST SYSTEM TO
TURBOCHARGER

NEW INTAKE AND
EXHAUST VALVES AND
CYLINDER HEAD
GASKET HANDLE
INCREASED LOADS,
STRESSES, AND
TEMPERATURES.

OIL PUMP HAS A
STIFFER RELIEF VALVE
SPRING TO MAINTAIN
NORMAL OIL PRESSURE

UPGRADED
RADIATOR
ENHANCES ENGINE
COOLING

NEW TURBO
BOOST/OVERBOOST
AND ENGINE OIL
OVERTEMPERATURE
WARNING SYSTEM

FORGED PISTONS
INCREASE DURABILITY

MAIN BEARINGS
AND ROD BEARINGS
HAVE INCREASED LOAD
CAPACITY

OIL CAPACITY IS
INCREASED BY
0·6 LITRE

ELECTRONIC PRESSURE
RETARD SYSTEM
RETARDS SPARK TO
ELIMINATE POSSIBILITY
OF DETONATION

Fig. 22-24. Note many engine modifications commonly used with turbocharging. Turbocharging increases demands on engine. (Ford)

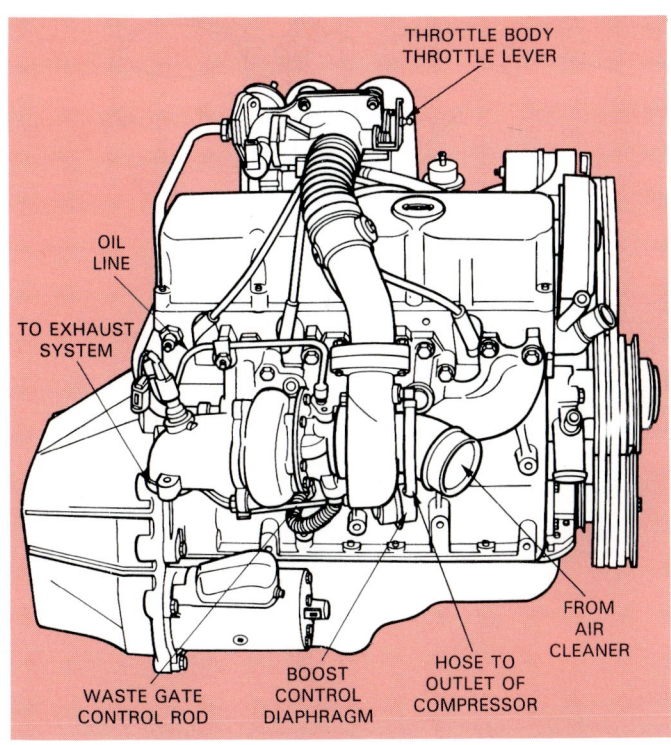

THROTTLE BODY
THROTTLE LEVER

OIL
LINE

TO EXHAUST
SYSTEM

FROM
AIR
CLEANER

WASTE GATE
CONTROL ROD

BOOST
CONTROL
DIAPHRAGM

HOSE TO
OUTLET OF
COMPRESSOR

Fig. 22-25. Side view of engine shows location and mounting of turbocharger. (Ford)

Checking turbocharging system

There are several checks that can be made to determine turbocharging system condition. These include:

1. Check connection of all vacuum lines to waste gate and oil lines to turbo, Fig. 22-25.
2. Use a regulated, low pressure air hose to check for waste gate diaphragm leakage and operation.
3. Use the dash gauge or a test gauge to measure boost pressure (pressure developed by turbo under a load). If needed, connect the pressure gauge to an intake manifold fitting. Compare to specs.
4. Use a stethoscope to listen to bad turbocharger bearings.

Checking turbocharger

To check the internal condition of a turbo, remove the unit from the engine, as in Fig. 22-26. Unbolt the connections at the turbo. Remove the oil lines and take the unit to your workbench.

345

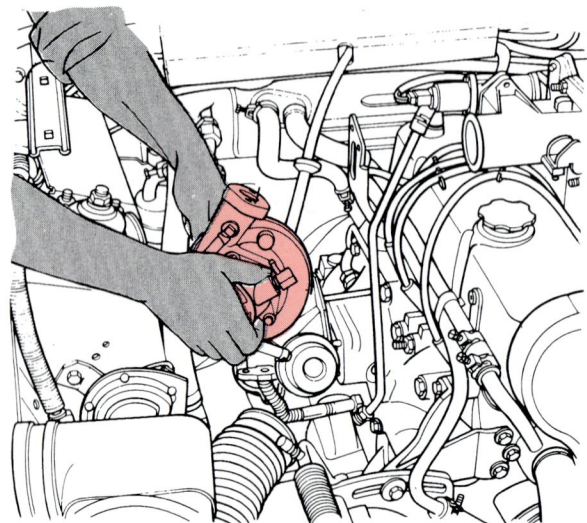

Fig. 22-26. After removing mounting fasteners and any other parts, turbo can be lifted off for replacement. A turbo is not normally repaired in-shop. A new unit is usually installed. (Ford)

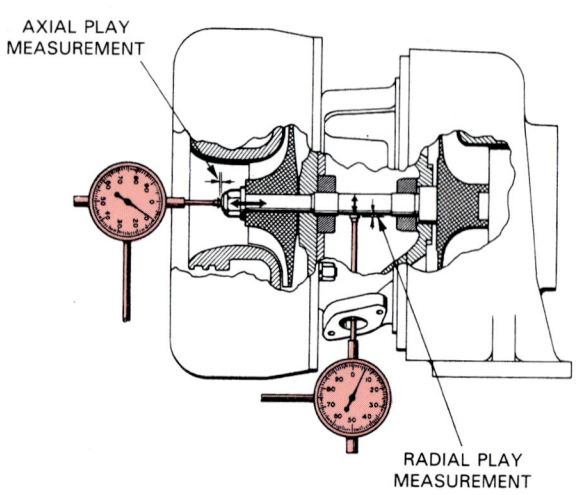

Fig. 22-27. Dial indicator can be used to check radial and axial play of turbo shaft. If not within specs, replace unit. (Waukesha)

Inspect the turbocharger wheels for physical damage. The slightest knick or dent will throw the unit out of balance, causing vibration. Fig. 22-27 shows how to measure turbo bearing and shaft wear.

WARNING! Never use a hard metal object or sandpaper to remove carbon deposits from the turbine wheel. If you gouge or remove metal, the wheel can vibrate and destroy the turbo. Only use a soft wire brush and solvent to clean the turbo wheels.

Installing new turbocharger

Many turbocharger problems are NOT repaired in the field, Fig. 22-28. Most mechanics install a new or rebuilt unit. When installing a turbo, you should:

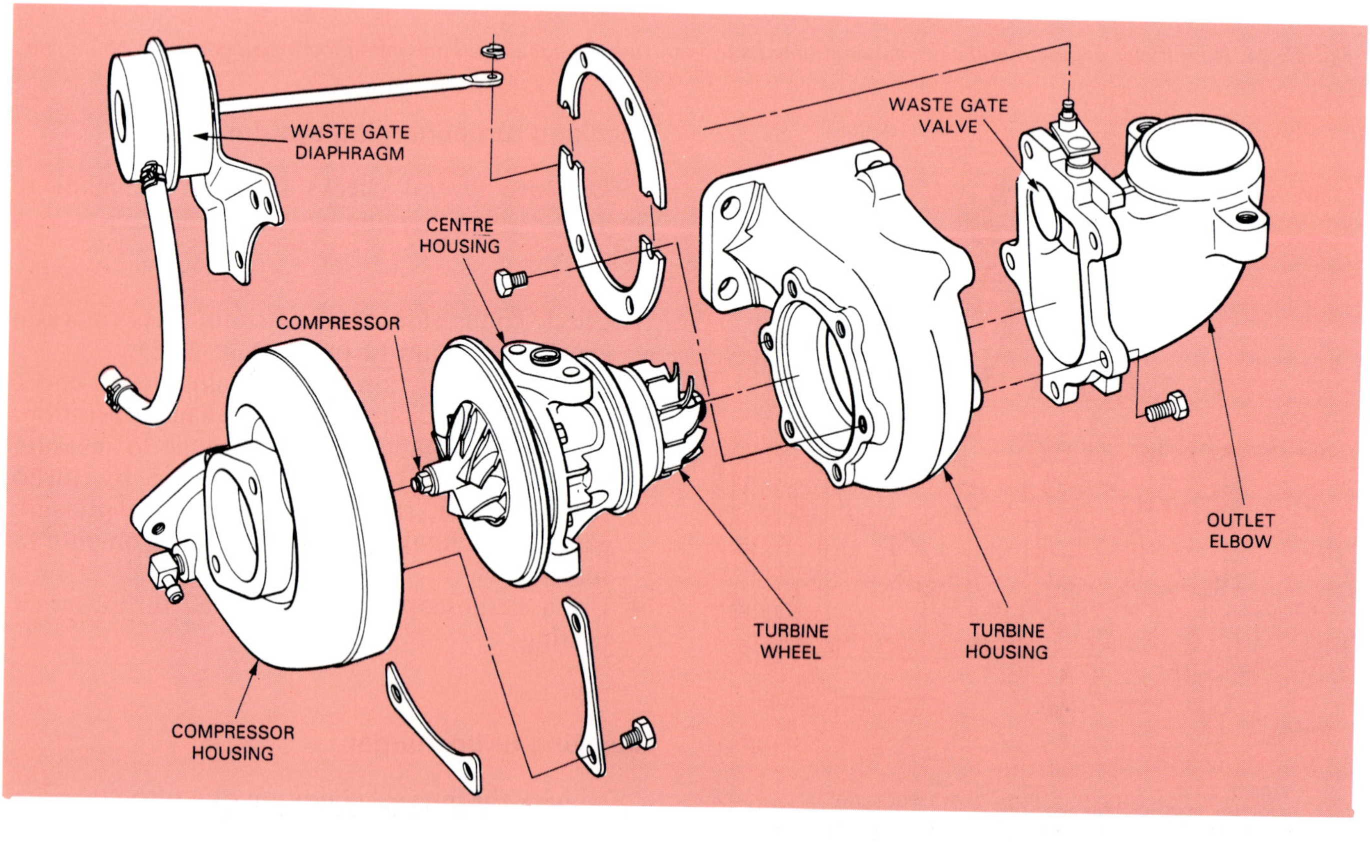

Fig. 22-28. Exploded view of modern turbocharger. Only external parts are serviceable. Turbine-compressor wheel is very precise, balanced assembly. Slightest nick or chip on blade can cause unit to explode in service. (Ford)

1. Make sure the new turbo is the correct type. Compare part numbers.
2. Use new gaskets and seals.
3. Torque all fasteners to specs.
4. If needed, change engine oil and flush oil lines before starting engine.
5. If the failure was oil related, check oil supply pressure in feed line to turbo.

Fig. 22-30 shows a turbocharged engine. Can you identify all of the parts. Trace flow of fuel charge into engine and exhaust out of engine.

KNOW THESE TERMS

Exhaust manifold, Engine pipe, Catalytic converter, Muffler, Tailpipe, Hangers, Heat shields,

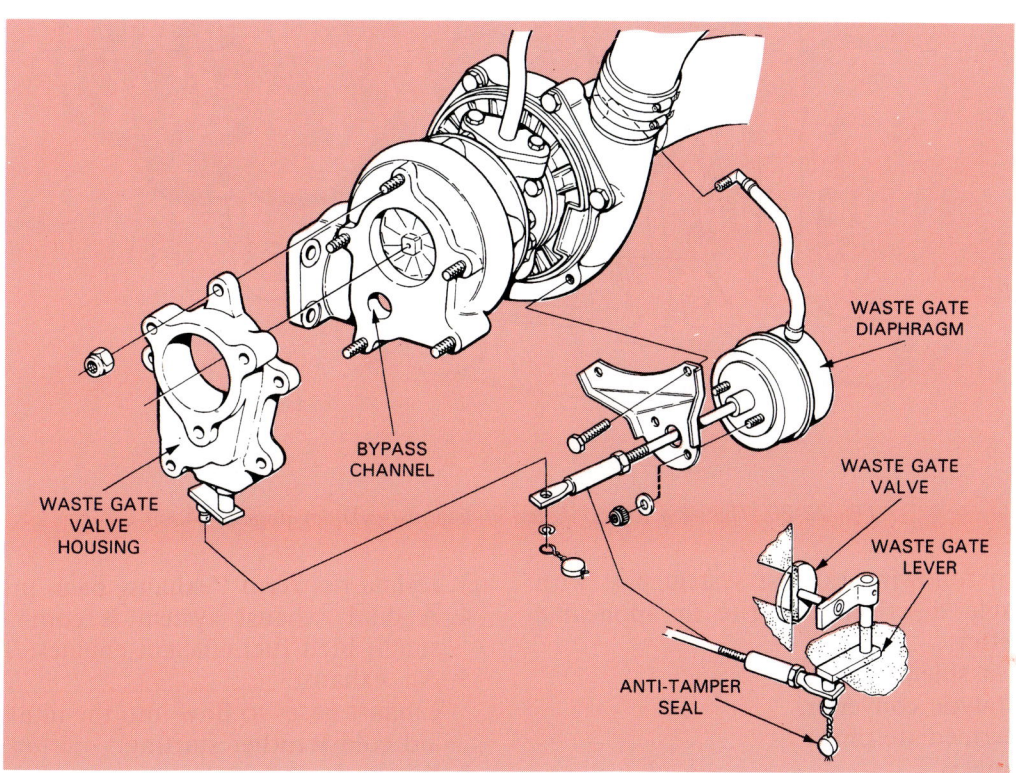

Fig. 22-29. Waste gate normally bolts to side of turbo housing. Linkage rod connects diaphragm with valve mechanism. Also note seal that prevents tampering with boost setting. Although overboost will increase power, it can also cause engine damage. (Saab)

Waste gate service

An inoperative waste gate can either cause too much or too little boost pressure. If stuck open, the turbo will not produce boost pressure and the engine will lack power. If stuck closed, detonation and engine damage can result from excessive boost.

Before condemning the waste gate, always check other parts. Check the knock sensor (spark retard system if used) and the ignition timing. Make sure the vacuum-pressure lines are all connected properly.

Follow service manual instructions when testing or replacing a waste gate. As shown in Fig. 22-29, waste gate removal is simple. Unbolt the fasteners. Remove the lines and lift the unit off of the engine. Many manuals recommend waste gate renewal, rather than repairs.

Muffler clamps, Back pressure, Crossover pipe, Exhaust manifold heat valve, Rust penetrant, Air chisel, Pipe expander, Pipe shaper, Supercharger, Turbocharger, Normal aspiration, Turbine wheel, Turbine housing, Turbo shaft, Compressor wheel, Compressor housing, Bearing housing, Turbo bearing, Blowthrough turbo, Draw-through turbo, Sealing rings, Turbo lag, Waste gate, Knock sensor, Boost pressure.

REVIEW QUESTIONS

1. An exhaust system _____ engine operation and carries _____ _____ to the rear of the car.

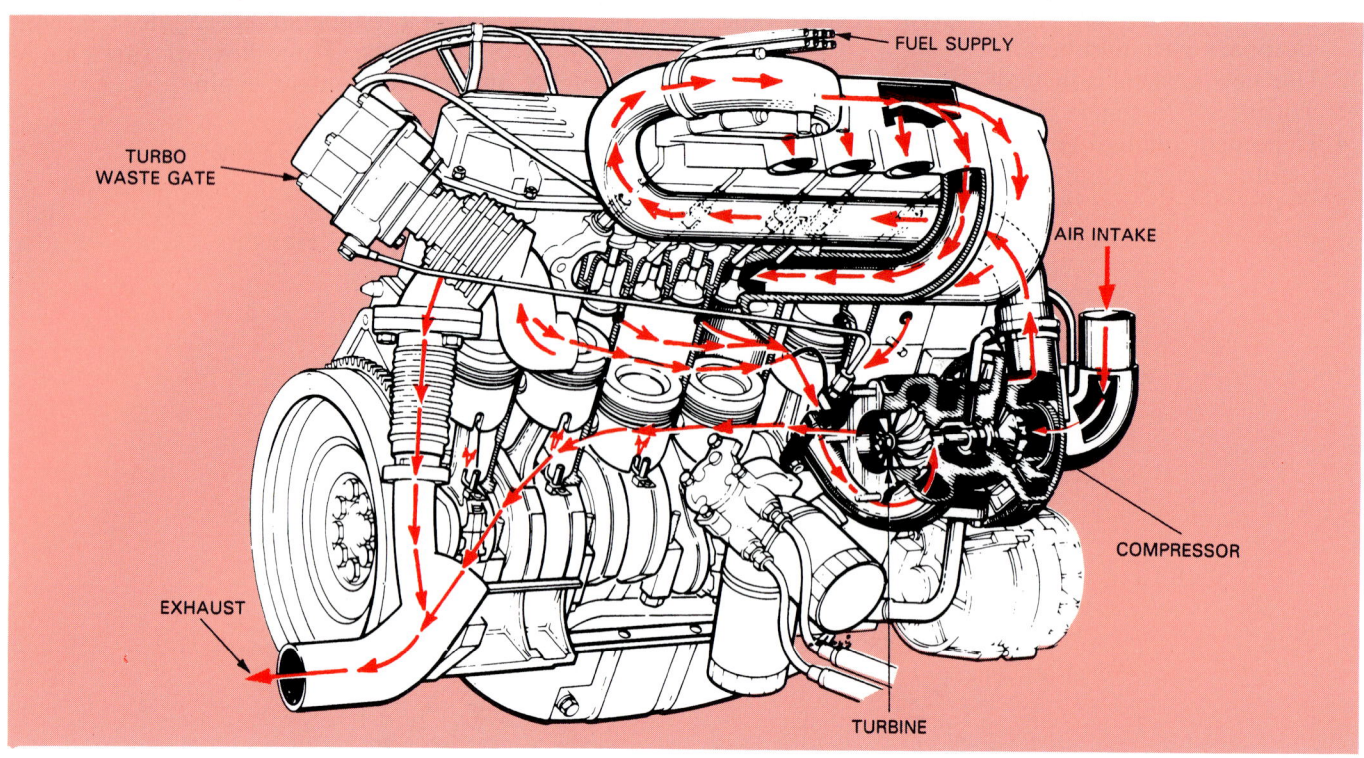

Fig. 22-30. This is a turbocharged, in-line, six-cylinder engine. (Audi)

2. Match each following exhaust system part with its applicable description. Write in applicable alphabet letters.
   _____ Heat shields.
   _____ Catalytic converter.
   _____ Intermediate pipe.
   _____ Hangers.
   _____ Engine pipe.
   _____ Tailpipe.
   _____ Muffler clamp.
   _____ Exhaust manifold.
   _____ Muffler.

   a. U-bolt for connecting parts of exhaust system.
   b. Tubing that connects exhaust manifold to rest of system.
   c. Chamber for damping out pressure pulsations.
   d. Carries exhaust from muffler to rear of car body.
   e. Connects cylinder head exhaust ports to engine pipe.
   f. Prevents heat from transferring into other objects.
   g. Connects exhaust manifold to tailpipe.
   h. Device for removing pollutants from exhaust.
   i. Pipe between catalytic converter and muffler.
   j. Connects exhaust system to underside of car body.

3. Define the term "exhaust back pressure".
4. A dual exhaust system is commonly used on small, high fuel efficient engines. True or False?
5. An exhaust _____ _____ _____ allows hot exhaust gases to flow into the intake manifold to aid cold weather starting.
6. When is exhaust system service commonly needed?
7. List fourteen rules to remember when servicing an exhaust system.
8. A _____ is an air pump that increases engine power by pushing a denser air-fuel charge into the combustion chambers.
9. What is a normally aspirated engine?
10. Which of the following is NOT a type of supercharger?
    a. Vane.
    b. Gear.
    c. Rotor.
    d. Centrifugal.

11. Explain the term "Turbocharger".
12. In the field, the term "supercharger" generally refers to a blower driven by a _____, _____ or _____.
13. List and explain the six basic parts of a turbocharger.
14. A _____ turbocharger has the turbo located before the carburettor or throttle body.
15. A turbocharger can operate at speeds up to _____ rpm.

16. _____ _____ refers to the short delay before the turbo develops sufficient _____ (pressure above atmospheric pressure).

17. A waste gate limits the minimum amount of boost produced by the turbocharger. True or False?

18. What could happen if a waste gate did not open?

19. List seven engine modifications commonly found on a turbocharged engine.

20. Which of the following is NOT a recommended practice when servicing a turbocharging system?
    a. Inspect vacuum and oil lines to turbo and waste gate.
    b. Remove carbon from turbo compressor with gasket scraper.
    c. Use regulated, low pressure air hose to check waste gate diaphragm for leakage and operation.
    d. Use pressure gauge to measure boost pressure.
    e. Use stethoscope to listen for faulty turbo bearings.

21. A turbocharger was badly damaged because of excess bearing and shaft wear.

Mechanic A says that a new unit should be installed and that oil pressure to the turbo should be checked. Engine oil should be changed after a short break-in period.

Mechanic B says that the oil should be drained and all lines should be flushed before installing the new turbo. Oil pressure to the unit should also be checked.

Who is correct?
a. Mechanic A.
b. Mechanic B.
c. Both are correct.
d. Neither A nor B is correct.

22. An inoperative turbocharger waste gate can cause:
a. High boost pressure.
b. Low boost pressure.
c. Low engine power.
d. Detonation.
e. All of the above are correct.

Chapter 23

DIESEL INJECTION FUNDAMENTALS

After studying this chapter, you will be able to:
□ Explain the operating principles of a diesel injection system.
□ Summarise the differences between petrol and diesel engines.
□ Describe the major parts of a diesel injection system.
□ Compare variations in the design of diesel injection systems.

A diesel fuel injection system is a super-high pressure, mechanical system that delivers fuel directly into the engine combustion chambers. It is unlike a lower pressure, petrol system that meters fuel into the engine intake manifold. Diesel fuel injection is relatively simple. It uses a mechanical pump as the main control of the engine's air-fuel mixture.

Note! Several earlier chapters discuss information essential to this chapter. If needed, use the index to locate applicable chapters which discuss diesel engines, diesel fuel, fuel pumps, fuel filters, compression ratios, and diesel combustion.

BASIC DIESEL INJECTION SYSTEM

Fig. 23-1 illustrates a basic diesel injection system. Refer to it as each part is explained.
1. INJECTION PUMP (high pressure, mechanical pump that meters the correct amount of fuel and delivers it to each injector nozzle at the right time).
2. INJECTION LINES (high strength, steel tubing that carries fuel to each injector nozzle).
3. INJECTOR NOZZLES (spring-loaded valves that spray fuel into each combustion chamber).
4. GLOW PLUGS (electric heating elements that warm air in pre-combustion chambers to aid starting of cold engines).
The fuel supply system feeds fuel to the injection pump, Fig. 23-2. The engine-driven injection pump

then controls when and how much fuel is forced to the injector nozzles. The high pressure injection lines carry fuel to the injectors. The spring-loaded nozzles are normally closed. However, when the injector pump produces enough pressure, each nozzle opens and squirts a fuel charge into the engine to start combustion. A return line carries excess fuel back to the tank.

DIESEL AND PETROL ENGINE DIFFERENCES

Before covering the parts of a diesel injection system in detail, you should review the major differences between a petrol engine and a diesel engine. This will help you understand diesel injection, Fig. 23-3.

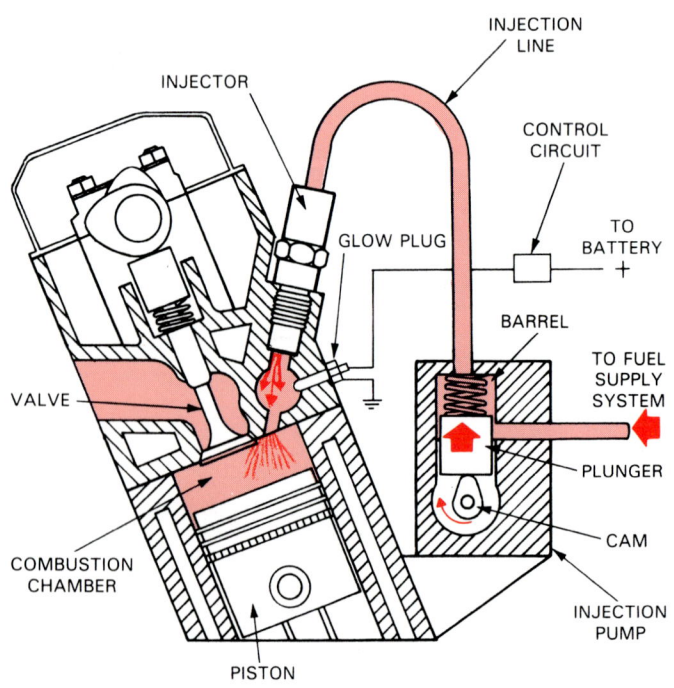

Fig. 23-1. Basic parts of a simplified diesel injection system. Injection pump plunger produces very high pressure to open injector valve. Glow plug warms air in precombustion chamber when engine is cold.

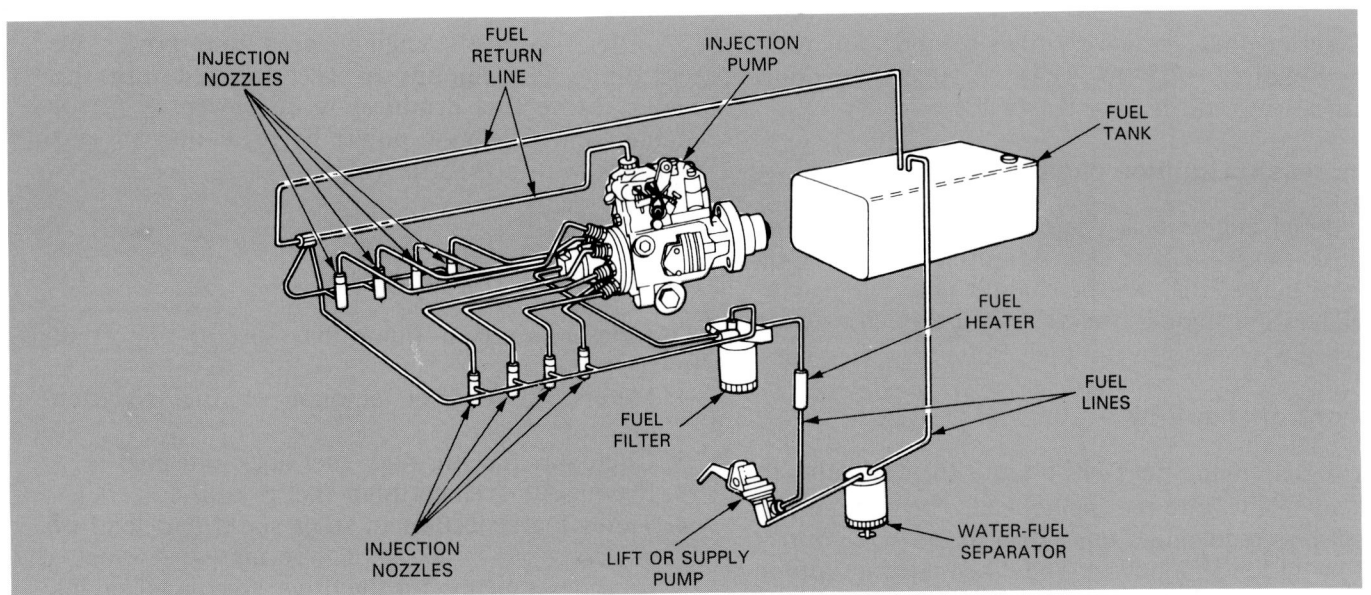

Fig. 23-2. Schematic of complete fuel injection system. Lift pump feeds fuel into injection pump. Injection pump produces high pressure for injection into combustion chambers. (Ford)

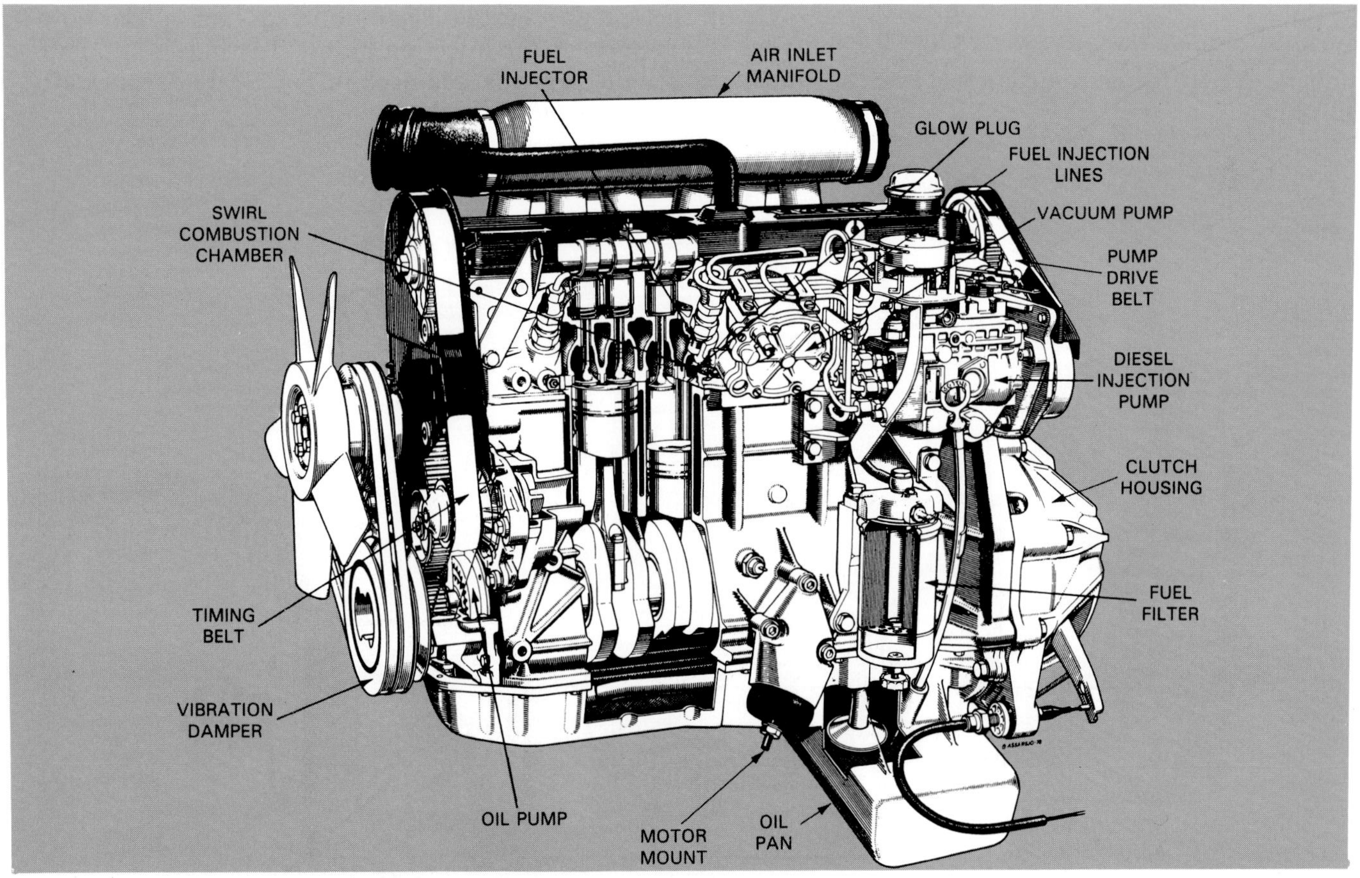

Fig. 23-3. In-line, six cylinder diesel engine. Note names of parts. Also, note rear drive belt for injection pump. (Volvo)

Higher compression ratio

Diesel engines use a very high compression ratio (approximately 17:1 to 23:1). A petrol engine's compression ratio is only 8:1 or 9:1.

Compression ignition engine

A diesel engine is a compression ignition engine because it uses the heat from compressed air to ignite and burn the fuel. Petrol engines are classified as spark ignition engines because they use an electric arc (spark plug) to ignite the fuel.

No control of airflow

A diesel engine does NOT use a throttle valve to control airflow into the engine. Both carburetted and petrol injected engines use a throttle valve to control airflow and engine power. The diesel injection pump controls engine power output, as you will learn.

Compresses only air

A diesel engine compresses only air on its compression stroke. A petrol engine compresses a mixture of air and fuel.

Injects fuel into combustion chambers

A diesel engine forces fuel into the combustion chambers. A petrol engine meters fuel into the intake manifold.

Fuel controls engine speed and power

A diesel controls engine speed and power by controlling the amount of fuel injected into the engine. More fuel produces more power. A petrol engine controls engine power by regulating air and fuel flow with a throttle valve.

DIESEL INJECTION PUMPS

A diesel injection pump has several important functions. It:
1. Meters the correct amount of fuel to each injector.
2. Circulates fuel through fuel lines and nozzles.
3. Produces extremely high fuel pressure.
4. Times fuel injection to meet speed and load of engine.
5. Provides a means for the driver to control engine power output.
6. Controls engine idle speed and maximum engine speed.
7. Helps close injector nozzles after injection.
8. Provides a means of shutting off engine.

The diesel injection pump is normally bolted to the side or top of the engine. See Fig. 23-3. The fuel supply system (fuel tank, lines, filters, conventional fuel pump) pushes clean, filtered fuel, under low pressure, to the injection pump. The injection pump uses the principle pictured in Fig. 23-1. A camshaft acts on a plunger.

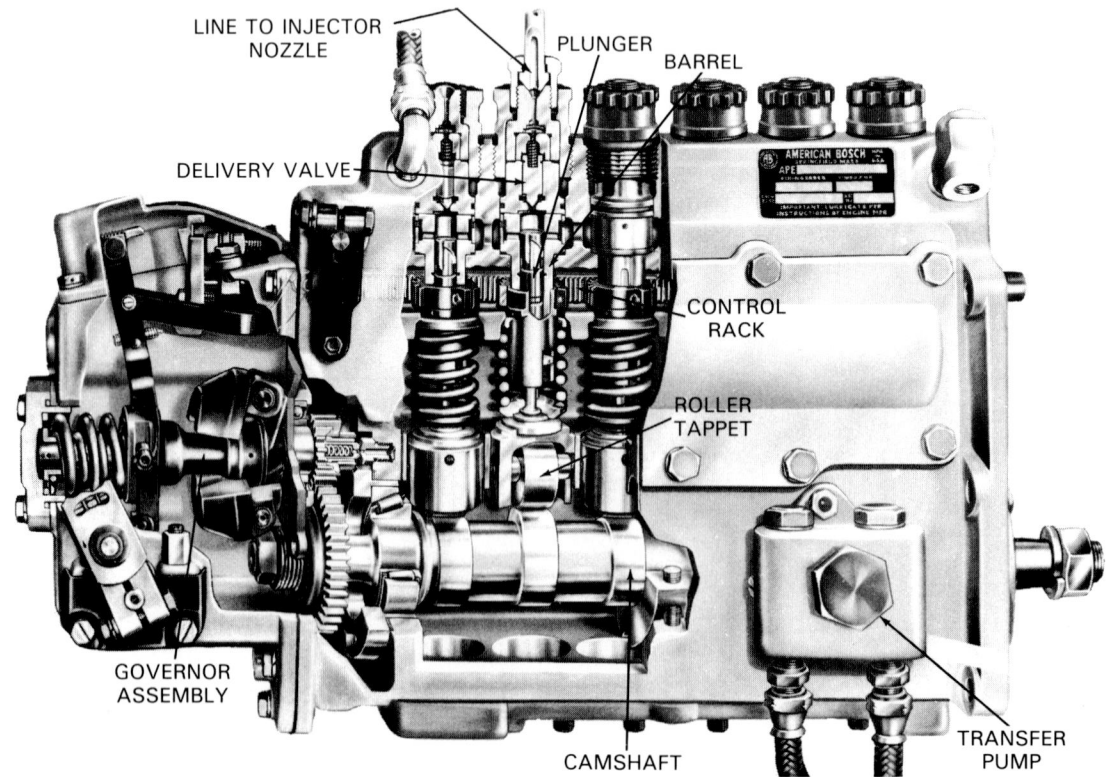

Fig. 23-4. Cutaway shows basic parts of an in-line diesel injection pump. (American Bosch)

The plunger slides up in its barrel (cylinder), compressing and pressurising the fuel. The fuel then flows through the injection line and out of the injector nozzle.

A diesel injection pump is powered by the engine. Power may be transferred by a set of gears, a chain, or toothed belt. There are two common types of automotive diesel injection pumps: in-line type and distributor type.

IN-LINE DIESEL INJECTION PUMPS

An in-line diesel injection pump has one pumping plunger (piston) for each engine cylinder. The pumping plungers are lined up in a row, like pistons of an in-line engine.

The major parts of an in-line injection pump are shown in Figs. 23-4 and 23-5. Refer to these illustrations as the parts are introduced.

The in-line injection pump camshaft operates the pumping plungers. It has lobes like an engine camshaft. When the engine turns the pump camshaft, the lobes push on the roller tappets to move them up and down.

The injection pump roller tappets transfer camshaft action to the pumping plungers. Like roller lifters in an engine, the rollers reduce friction and wear on the cam lobes.

In-line pump plungers are small pistons that push on and pressurise the diesel fuel. When the cam lobe

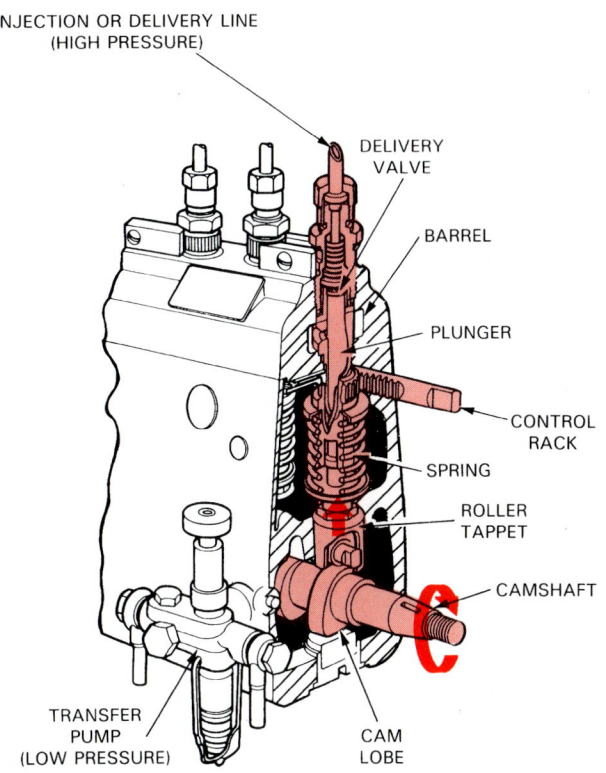

Fig. 23-5. Camshaft acts on roller tappet. Tappet pushes up on plunger to push fuel out delivery valve and into injection line. Control rack can be used to change injection quantity.

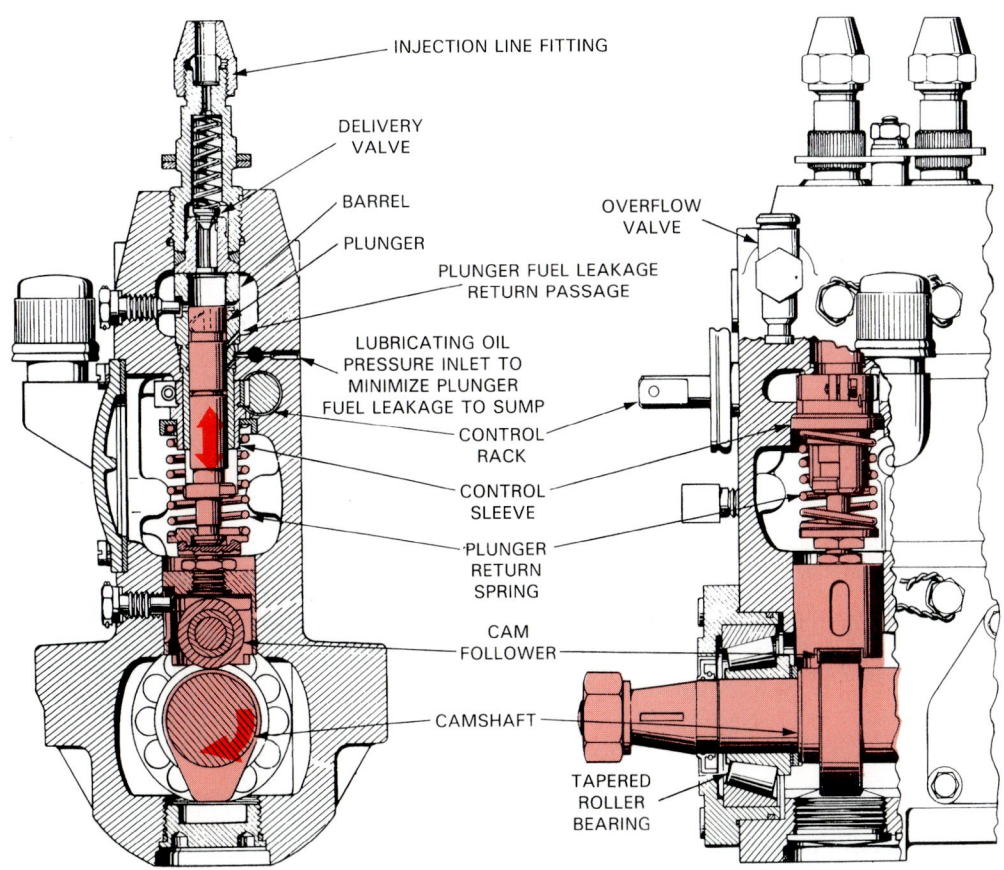

Fig. 23-6. Another view of an in-line diesel injection pump. Compare this illustration to previous ones. (Waukesha)

acts on the roller tappet, both the tappet and the plunger are pushed upward.

The barrels are small cylinders that hold the pumping plungers. When the plunger slides upwards in its barrel, extremely high fuel pressure is obtained.

The plunger return springs keep a downward pressure on the pumping plungers and roller tappets. This action holds the tappets against the camshaft when the lobes rotate away from the rollers.

Control sleeves turn on the pumping plungers to alter how much fuel is delivered to each injector nozzle. See Fig. 23-6 and locate the control sleeve.

A control rod or rack is a toothed shaft that acts as a throttle to control diesel engine speed and power. It rotates the control sleeves to increase or decrease injection pump output and engine power.

Delivery valves are spring-loaded valves in the outlet fittings to the injection lines. They help to ensure quick, leak-free closing of the injector nozzles.

In-line injection pump operation

When the engine is running, the injection pump camshaft rotates at one-half engine speed. While the cam lobe is not pushing on the roller tappet, the lift pump (supply pump) fills the barrel with fuel. Then, as the cam lobe puts pressure on the roller tappet, the plunger is forced upward in its barrel. This forces fuel through the delivery valve, through the injection line, out of the nozzle, and into the engine. See Fig. 23-7.

As the plunger reaches the end of its stroke, pressure drops and the delivery valve closes. The delivery valve action helps reduce injection line pressure rapidly, preventing fuel from dripping out of the injector nozzle.

When the cam lobe moves away from the roller tappet, the plunger return spring pushes the plunger down. Fuel can again flow into and fill the barrel. This prepares the plunger to supply fuel for another power stroke.

In-line injection pump fuel metering

To control the amount of fuel injected into the engine, the control rod or rack slides across the control sleeves to rotate them. Look at Fig. 23-8.

A helix (spiral) shaped groove and a slot are cut into the slide of the plunger. When the helix is aligned with the port (hole) in the side of the barrel, the plunger CANNOT develop pressure. Fuel will flow down the slot through the helix groove, and out of the port.

The effective plunger stroke is the amount of

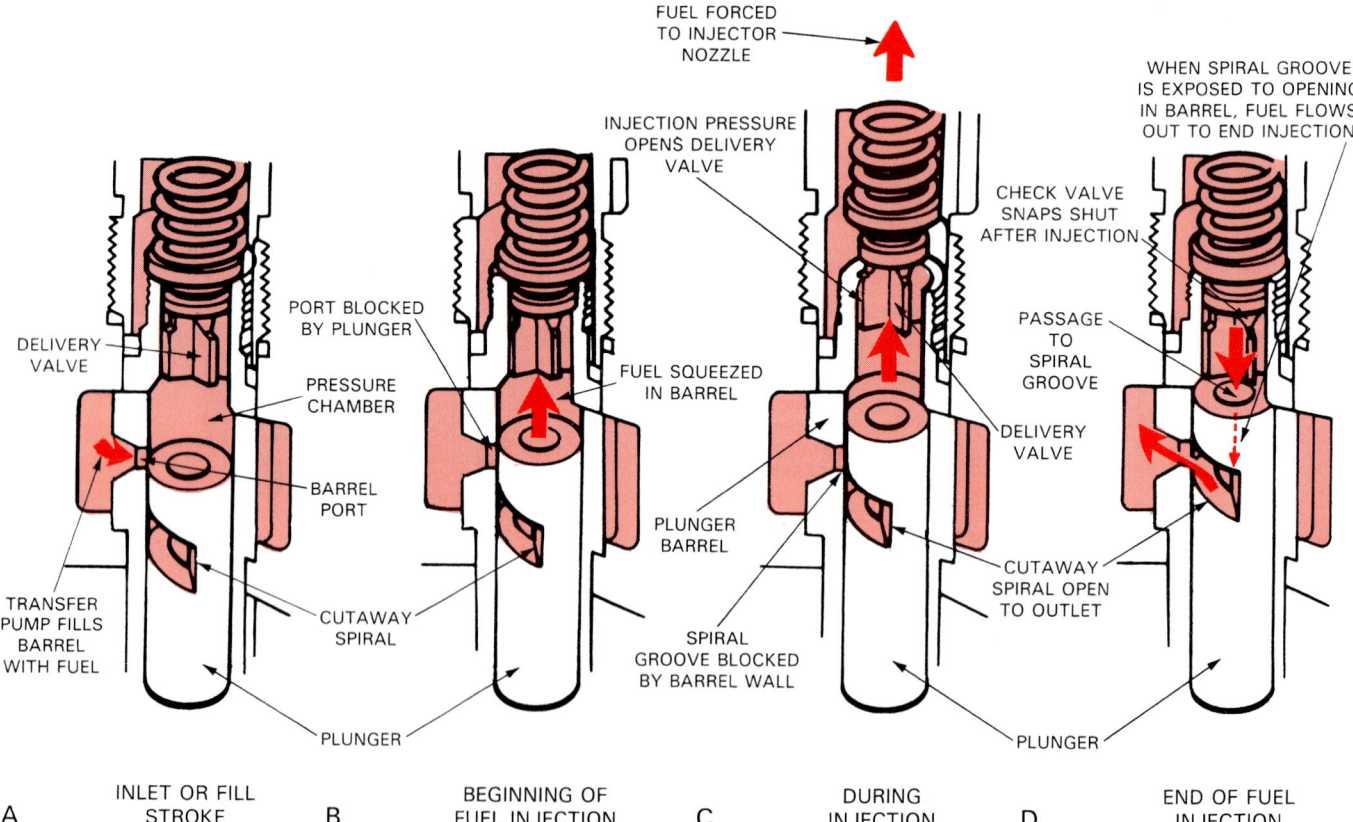

Fig. 23-7. Pumping plunger action in an in-line injection pump. A — Cam lobe away from tappet. Plunger is down. Supply pump fills barrel with fuel. B — Cam lobe acts on tappet. Plunger is moved up. This causes plunger to cover port in barrel. Fuel is pressurised. C — More movement of plunger builds enough pressure to open delivery valve. Fuel flows out to injector. D — Even more movement of plunger has exposed spiral groove in plunger. This allows fuel to pass through plunger and out port, ending fuel injection.

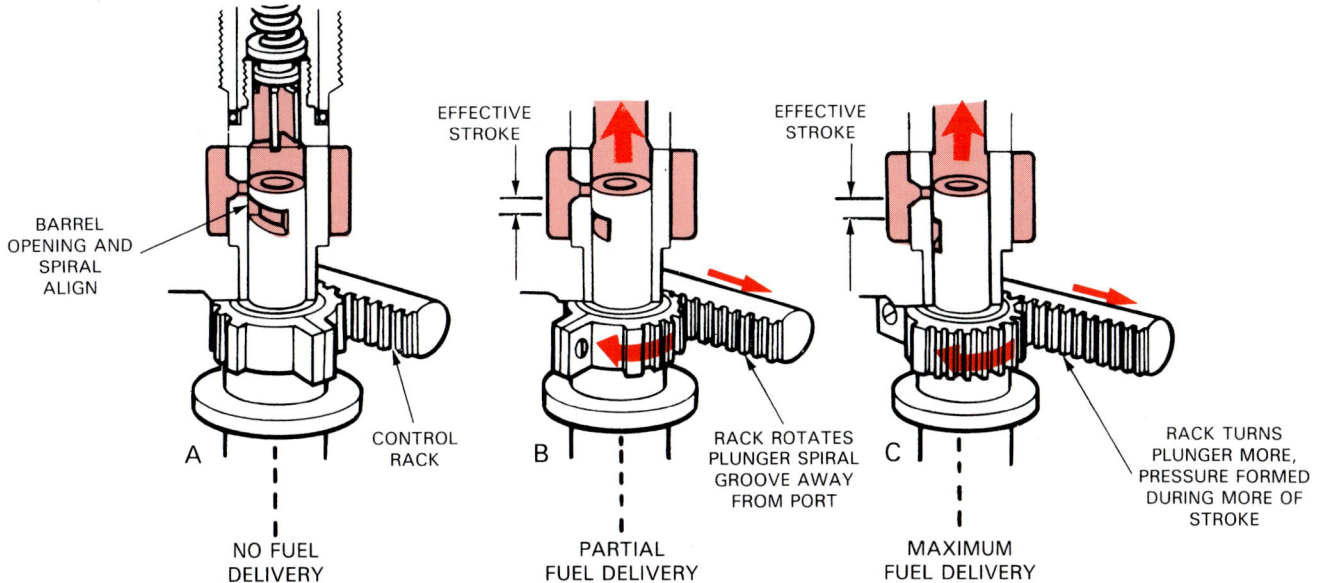

Fig. 23-8. Control rack is linked to driver's accelerator pedal and governor. Two forces work together to control rack position and amount of fuel injection. A — Control rack moved so that spiral groove is always open to port. This would prevent injection. B — Movement of control rack has turned plunger spiral groove away from port. This causes fuel delivery. C — More movement of rack and plunger increases plunger effective stroke. Plunger can move further before spiral groove is open to port.

plunger movement that pressurises fuel. It controls the amount of fuel delivered to the injectors.

When the plunger moves up and the helix is NOT aligned with the barrel port, fuel is trapped and pressurised. In this way, rotation of the sleeve can be used to regulate how much fuel is injected into the engine's combustion chambers.

In-line injection pump governor

A governor is used on an in-line injection pump to control engine idle speed and also to limit maximum engine speed. Look at Fig. 23-9. A diesel engine can be DAMAGED if allowed to run too fast.

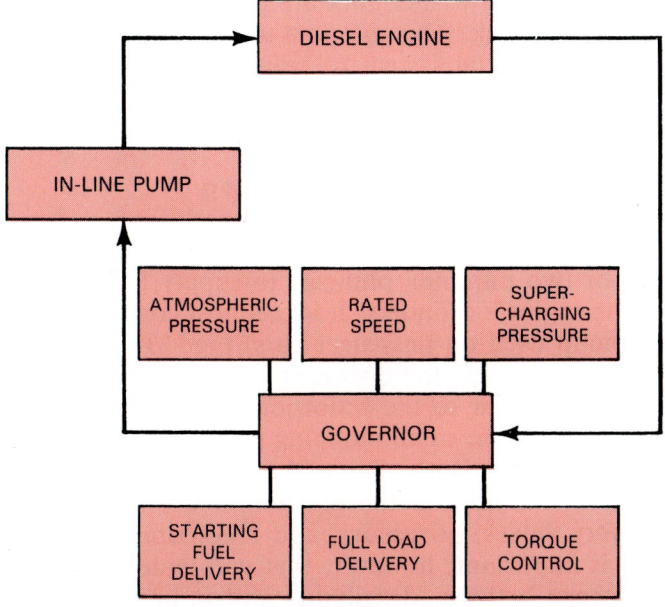

Fig. 23-9. Diagram illustrates governor operation.

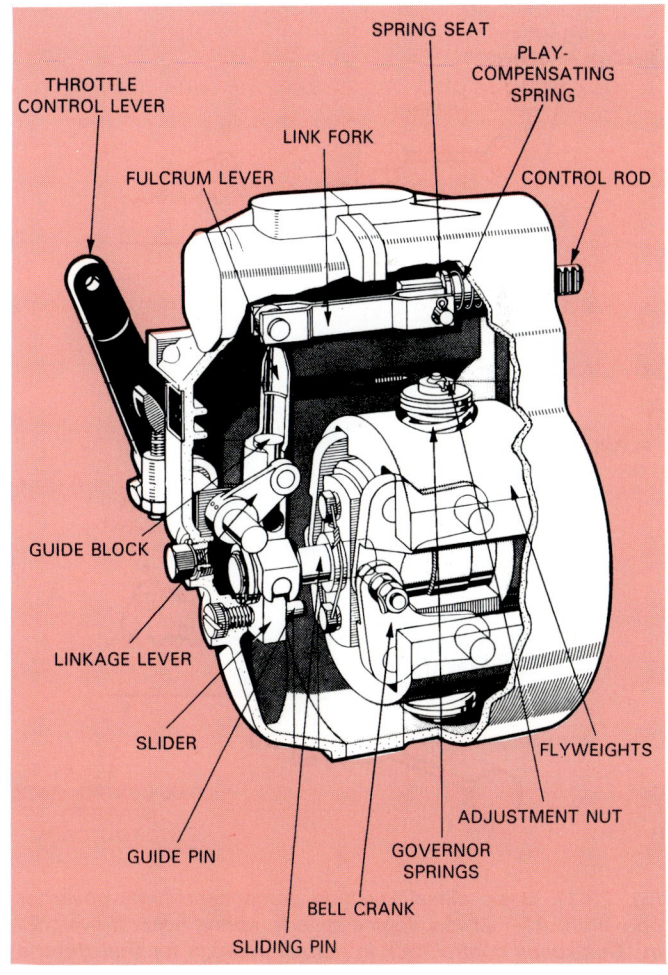

Fig. 23-10. Governor for an in-line injection pump. Note that throttle control lever and governor are both connected to control rack. (Robert Bosch)

Fig. 23-10 shows a cutaway view of an in-line injection pump governor.

Notice how the governor uses centrifugal (spinning) weights, springs and levers, Figs. 23-10 and 23-11. The levers are connected to the control rack or rod. If engine speed increases too much, the governor weights are thrown outward. This moves the levers and control rack to reduce the effective stroke of the plungers. Engine speed and driver output are limited.

When the driver presses the accelerator pedal for more power, it moves a control or throttle lever on the side of the injection pump governor. See Fig. 23-12. This causes throttle lever spring pressure to overcome governor spring pressure. The control rack is moved to increase fuel delivery and engine power. Only when engine rpm reaches a preset level does the governor overcome the full throttle lever position.

Injection timing

Injection timing refers to when fuel is injected into the combustion chambers in relation to the engine's

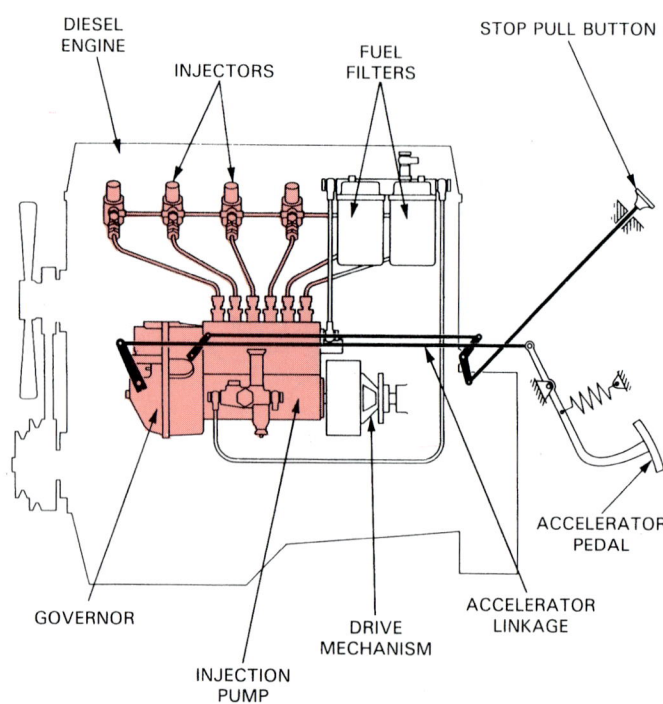

Fig. 23-12. Basic accelerator pedal-to-injection pump linkage arrangement. (Robert Bosch)

piston position. It is similar to ignition timing in a petrol engine.

Injection timing with an in-line injection pump is usually controlled by spring-loaded weights. As engine speed increases, the weights fly outward to advance injection timing. This gives the diesel fuel enough time to ignite and burn completely.

In-line injection pump fuel flow

Look at Fig. 23-13. It illustrates the flow of fuel through a diesel injection system using an in-line type pump. Note how fuel lines are provided that return to the fuel tank. This allows a steady flow of excess fuel through the system to help cool and lubricate moving parts.

DISTRIBUTOR INJECTION PUMPS

A distributor injection pump normally uses only one or two pumping plungers to supply fuel for all of the engine's cylinders. It is the most common type of pump used on passenger cars, Fig. 23-14.

In many ways, the operating of a distributor type pump is similar to the action of an in-line injection pump. Both use small pumping plungers to trap and pressurise fuel. Both align and misalign fuel ports to control fuel flow to the injector nozzles. Both use delivery valves, governors, and other similar parts.

It is important, however, that you understand the difference between distributor and in-line injection pumps.

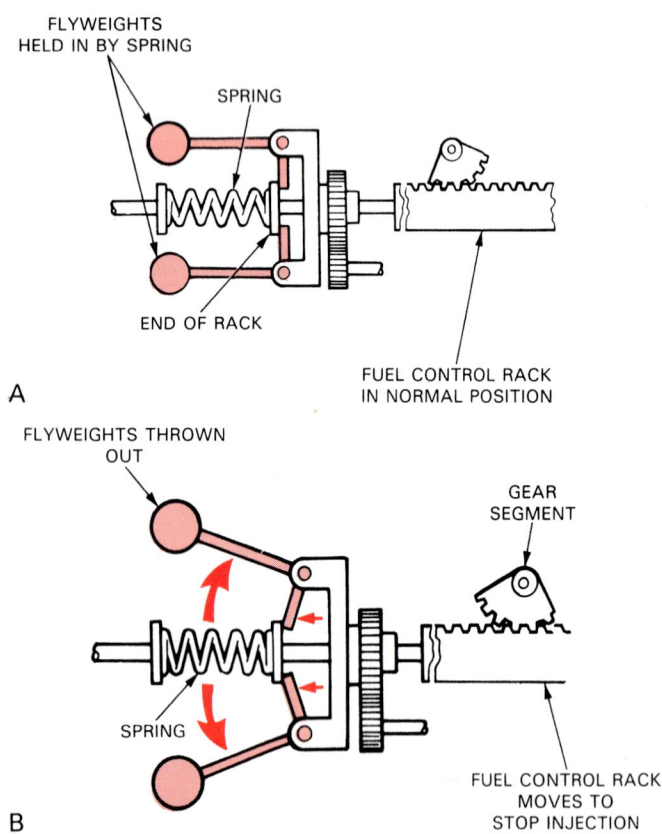

Fig. 23-11. Basic diesel injection pump centrifugal governor operation. A — At low engine speeds, spring holds flyweights in. This keeps control rack in normal position for that throttle lever position. B — When maximum speed is reached, flyweights are spinning fast enough for centrifugal force to compress spring. This causes lever action to move control rack into no injection position. This shuts off diesel engine power until rpm drops.

OVERFLOW CIRCUIT
LEAK-OFF CIRCUIT
INJECTION LINE
MAIN FILTER
INJECTION NOZZLE
FUEL FEED PUMP
INJECTION PUMP
PRE-FILTER
FUEL TANK
STRAINER

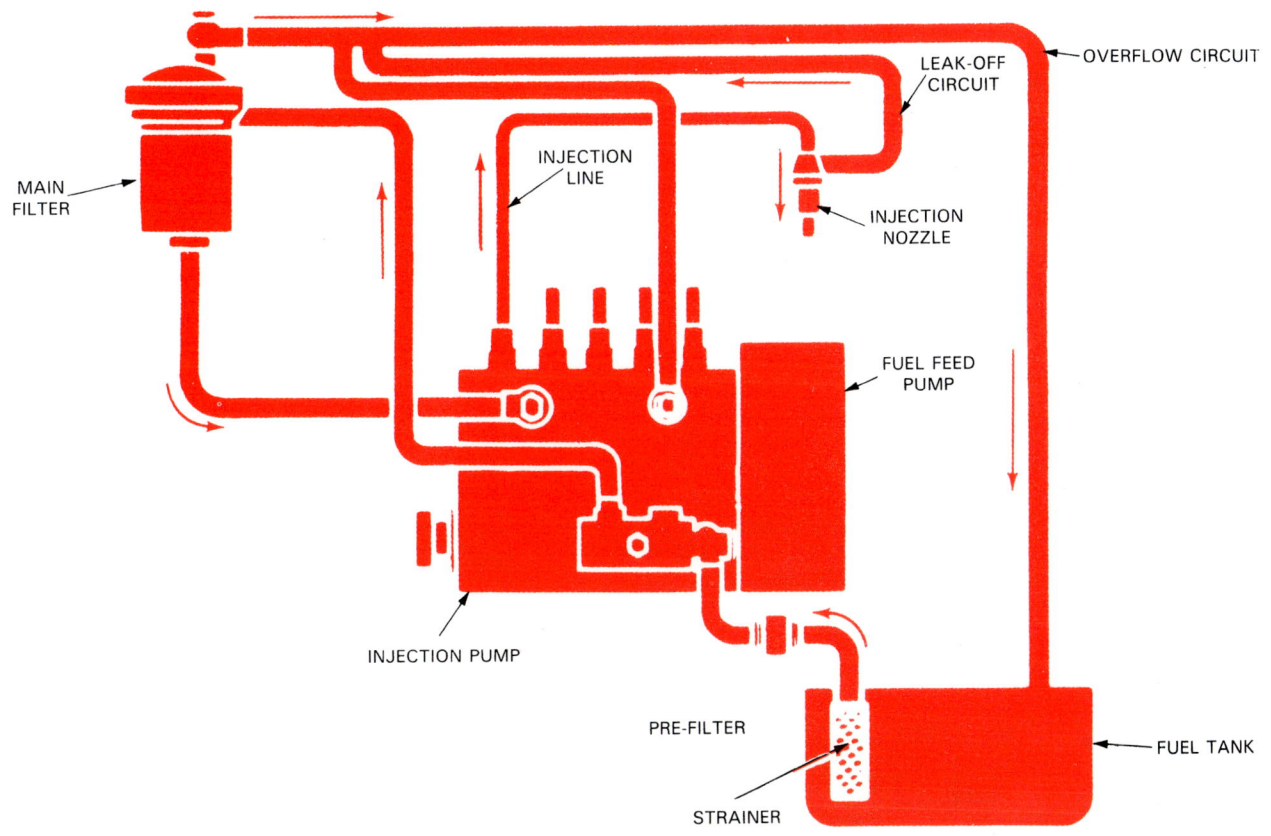

Fig. 23-13. Trace flow of fuel through injection pump and lines. (Mercedes-Benz)

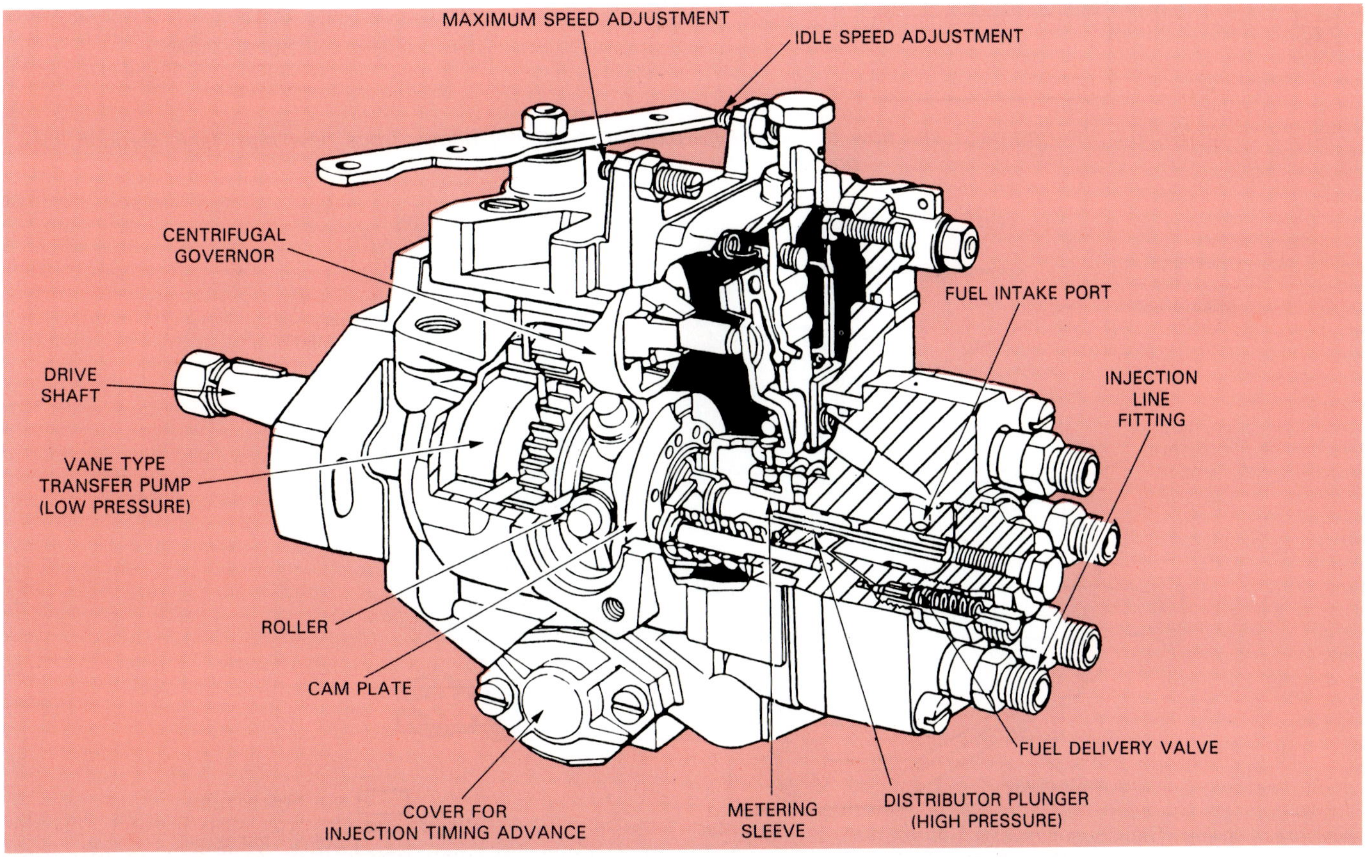

MAXIMUM SPEED ADJUSTMENT
IDLE SPEED ADJUSTMENT
CENTRIFUGAL GOVERNOR
FUEL INTAKE PORT
DRIVE SHAFT
INJECTION LINE FITTING
VANE TYPE TRANSFER PUMP (LOW PRESSURE)
ROLLER
CAM PLATE
FUEL DELIVERY VALVE
COVER FOR INJECTION TIMING ADVANCE
METERING SLEEVE
DISTRIBUTOR PLUNGER (HIGH PRESSURE)

Fig. 23-14. Typical single-plunger distributor diesel injection pump. Study part names.

There are two common variations of the distributor injection pump: single-plunger and two-plunger. Both will be discussed.

Single-plunger distributor injection pump

The major parts of single-plunger distributor injection pump are shown in Figs. 23-15 and 23-16.

Refer to these illustrations as the parts are discussed.

The drive shaft uses engine power to operate the parts in the injection pump. The outer end of the shaft holds either a gear, chain sprocket, or a belt sprocket. This provides a drive mechanism for the pump.

A transfer pump is a small pump that forces diesel fuel into and through the injection pump. This

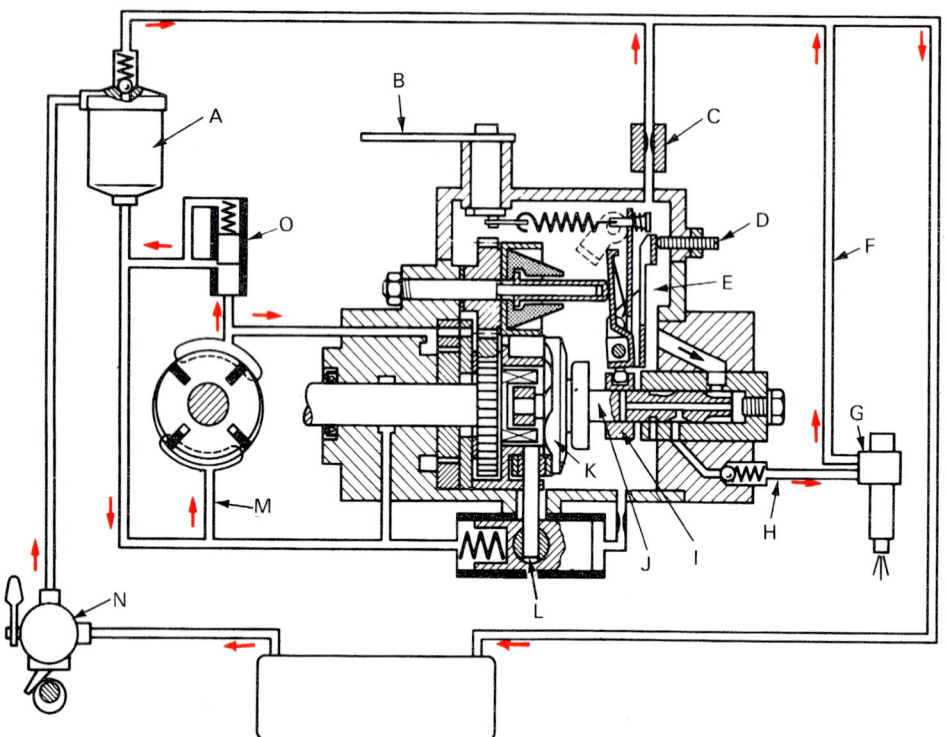

A. FINE FUEL FILTER
B. CONTROL OR THROTTLE LEVER
C. FUEL RETURN RESTRICTIONS
D. FUEL LOAD OR MAXIMUM SPEED ADJUSTMENT
E. INJECTION PUMP INTERIOR
F. FUEL RETURN LINE
G. INJECTOR
H. INJECTION LINE
I. REGULATING COLLAR OR METERING SLEEVE
J. PUMP PLUNGER
K. CAM PLATE
L. INJECTION TIMING DEVICE
M. INLET LINE TO VANE PUMP
N. SUPPLY PUMP
O. SUPPLY PRESSURE REGULATING VALVE

Fig. 23-15. Schematic showing parts and flow through single-plunger distributor pump. Trace flow from tank, through lines, pump, injector, and back to tank.

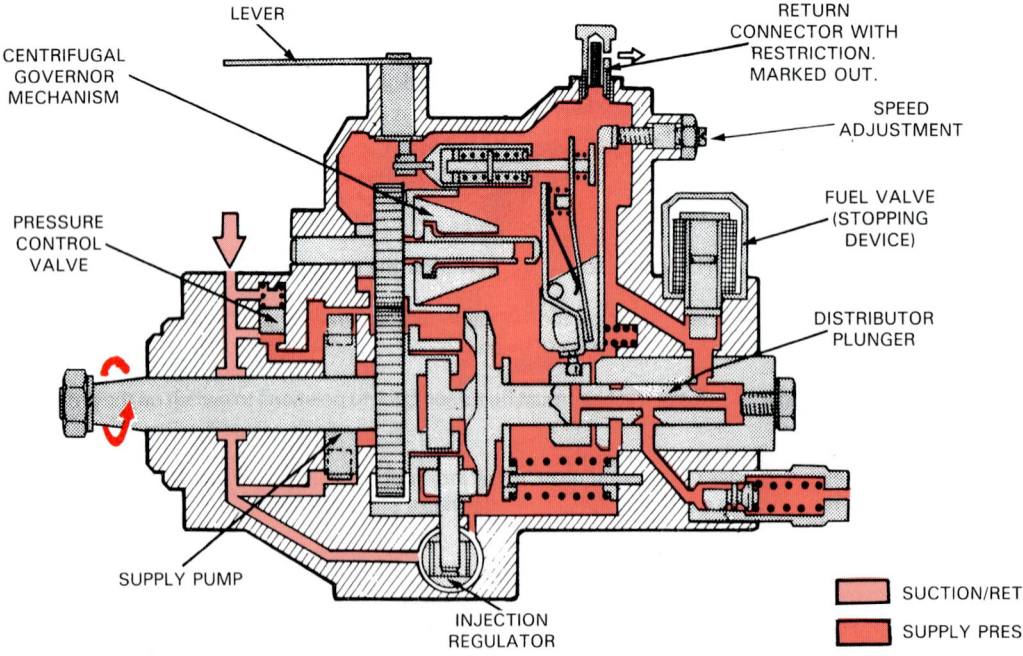

Fig. 23-16. Main parts of single plunger distributor pump. Note arrows. They show fuel flow. (Volvo)

lubricates the pump and fills the pumping chambers. Most transfer pumps for distributor pumps are a VANE TYPE.

A pumping plunger for a distributor type injection pump is a small piston that produces high fuel pressure. It is comparable to an in-line plunger.

A cam plate is a rotating lobed disc that operates the pumping plunger. Like an in-line pump camshaft, it forces the pumping plunger to move and develop injection pressure.

A fuel metering sleeve can be slid sideways on the pumping plunger to change the effective plunger stroke (plunger movement that causes fuel pressure). It surrounds the pumping plunger. The fuel metering sleeve performs the same function as the sleeve and control rack in an in-line pump. The sleeve controls

injection quantity, engine speed, and power output.

The hydraulic head, is the housing around the pumping plunger. It contains passages for filling the plunger barrel with fuel and for allowing fuel to be injected into the delivery valves.

A centrifugal governor helps to control the amount of fuel injected and engine speed. Flyweight action moves the metering sleeve to limit top rpm.

Single-plunger distributor pump operation

As the injection pump shaft rotates, the fill port in the hydraulic head lines up with the port in the plunger. At this point, transfer pump can force fuel into the high pressure chamber in front of the plunger. Refer to Fig. 23-17.

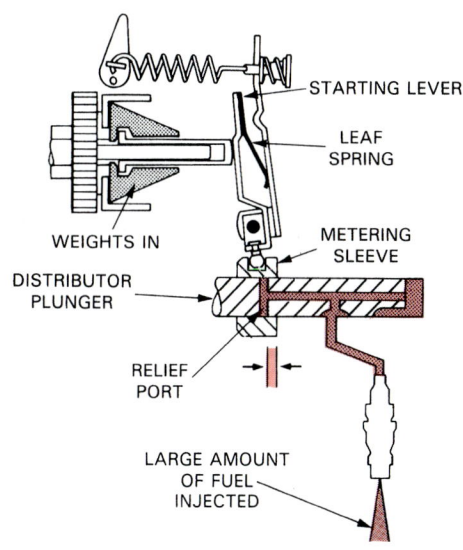

A

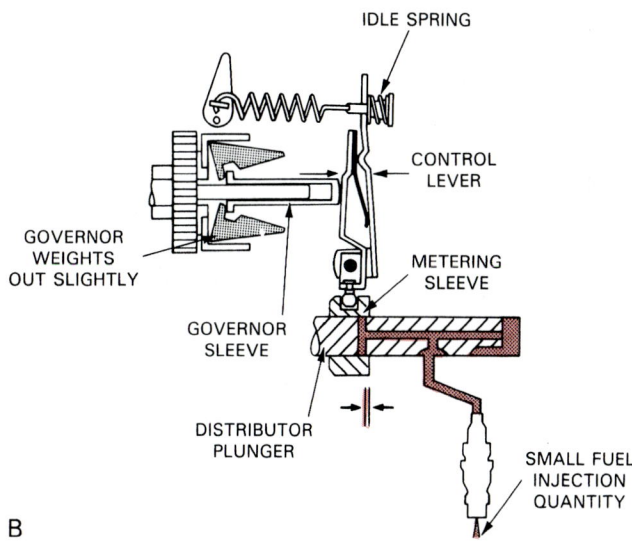

B

Starting — Leaf spring presses starting lever to left so metering sleeve moves to right. Distributor plunger moves further before relief port is exposed. Injection lasts longer.

Idle — Weights in centrifugal governor are partly expanded so governor sleeve moves to right. Metering sleeve moves to left. Distributor plunger now moves a short distance before relief port is uncovered.

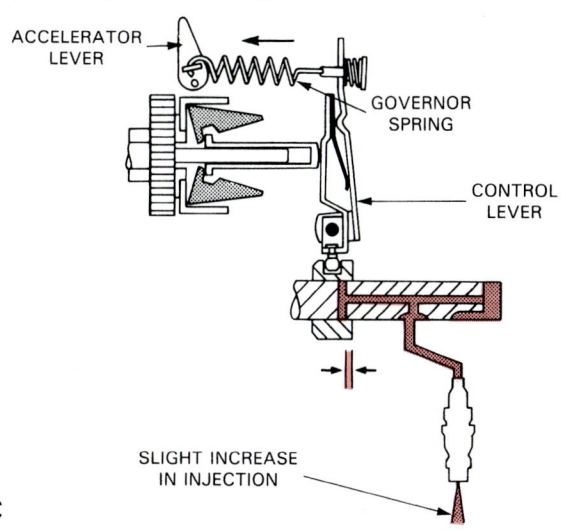

C

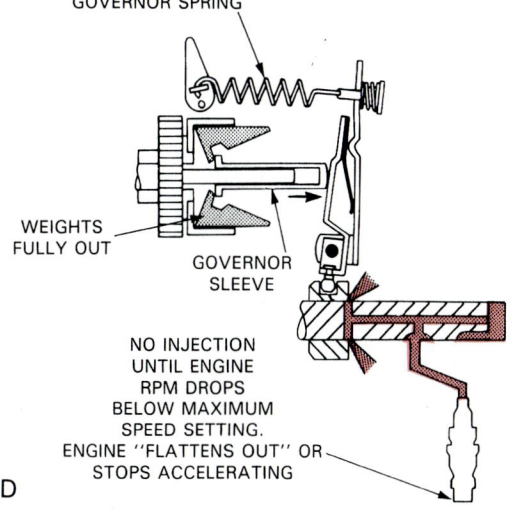

D

Acceleration — Control lever is pulled to left by linkage from accelerator pedal. Metering sleeve is moved to right. Engine speed increases until governor "neutralizes" effect of pedal linkage.

Maximum Speed — Governor is spinning with enough centrifugal force for governor sleeve to stretch governor spring. Metering sleeve uncovers relief port at beginning of distributor plunger stroke.

Fig. 23-17. Basic operation of a single-plunger distributor diesel injection pump.

With more shaft and plunger rotation, the fill port moves out of alignment and an injection port lines up. At this instant, the cam plate lobe pushes the plunger sideways. Fuel is forced out of the injection port to the correct injector nozzle.

This process is repeated several times during each rotation of the injection pump drive shaft. Fuel injection must be timed to occur at each nozzle as that engine piston nears TDC on its compression stroke.

Single-plunger distributor pump fuel metering

In a single-plunger distributor injection pump, the amount of fuel injected is controlled by movement of the sleeve on the pumping plunger. This is illustrated in Fig. 23-17. The sleeve slides one way to increase fuel delivery by covering the spill port. The sleeve moves the other way to reduce delivery by uncovering the spill port.

Single-plunger distributor pump injection timing

At the end of the engine compression stroke, diesel fuel must be injected directly into the precombustion chamber. Injection must continue past TDC to make sure all of the fuel burns and adequate power is developed.

As engine speed increases, injection must occur sooner to ensure peak combustion pressure right after TDC. Fig. 23-18 shows how one type of injection pump advances injection timing with an increase in engine speed.

Increased engine rpm causes the transfer (vane) pump to spin faster. This increases the pressure output of the transfer pump. The pressure is used to move an injection advance piston. The piston, in turn, causes the cam plate ramps (lobes) to engage the pumping plunger sooner, advancing the injection timing.

Two-plunger distributor pump

A two-plunger distributor injection pump is pictured in Fig. 23-19. Besides many of the basic parts already covered, this injection pump consists of:

1. TWO PUMPING PLUNGERS (two small pistons that move in and out to force fuel to each injector nozzle).
2. DISTRIBUTOR ROTOR (slotted shaft that controls fuel flow to each injector nozzle).
3. INTERNAL CAM RING (lobed collar that acts as a cam to force plungers inward for injection of fuel).
4. FUEL METERING VALVE (rotary valve that regulates fuel injection quantity by controlling how two pumping plungers move apart on fill stroke).

The other parts of a two-plunger distributor pump (transfer pump, hydraulic head, delivery valve) are almost the same as in a single-plunger type pump. Fig. 23-20 gives a circuit diagram for a two-plunger distributor pump. Study this illustration carefully.

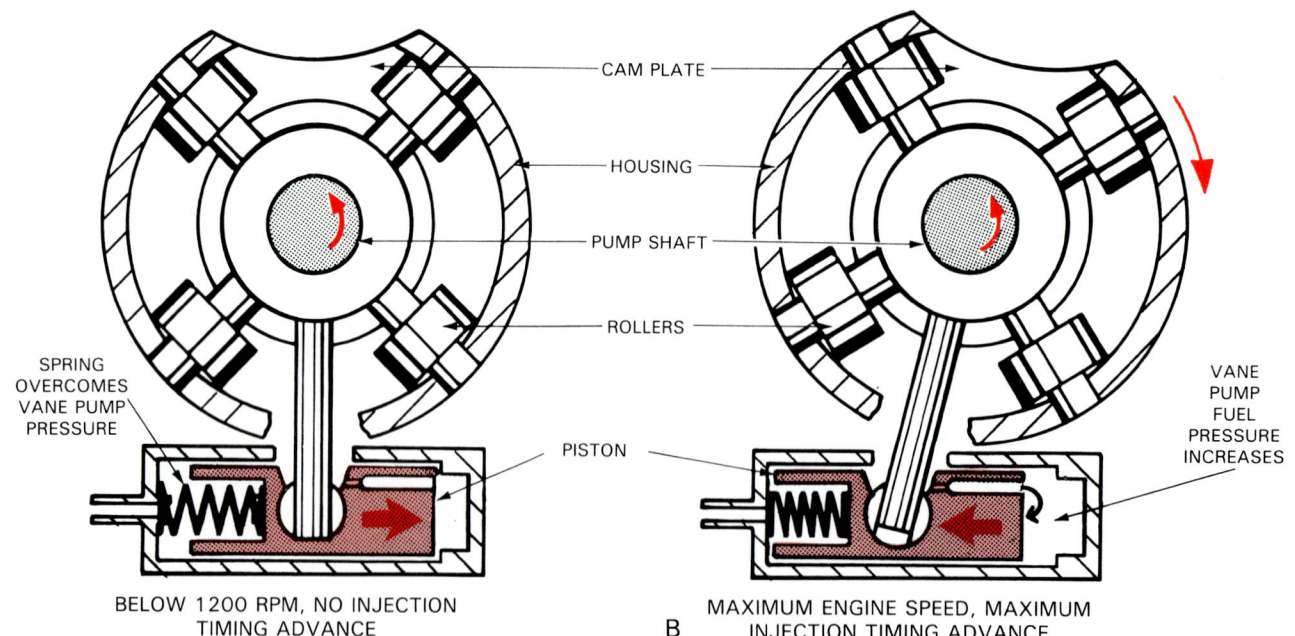

Fig. 23-18. Vane pump pressure can be used to control injection timing. A — At low engine speeds, lower vane pump pressure cannot compress spring. Injection timing remains retarded. B — At higher engine speeds, vane pump spins faster and develops more pressure. This causes piston to compress spring and rotate roller housing. Since cam plate motion is opposite piston movement, plate ramps or lobes engage rollers sooner to cause injection advance. This gives more time for fuel to burn at higher engine rpm. (VW)

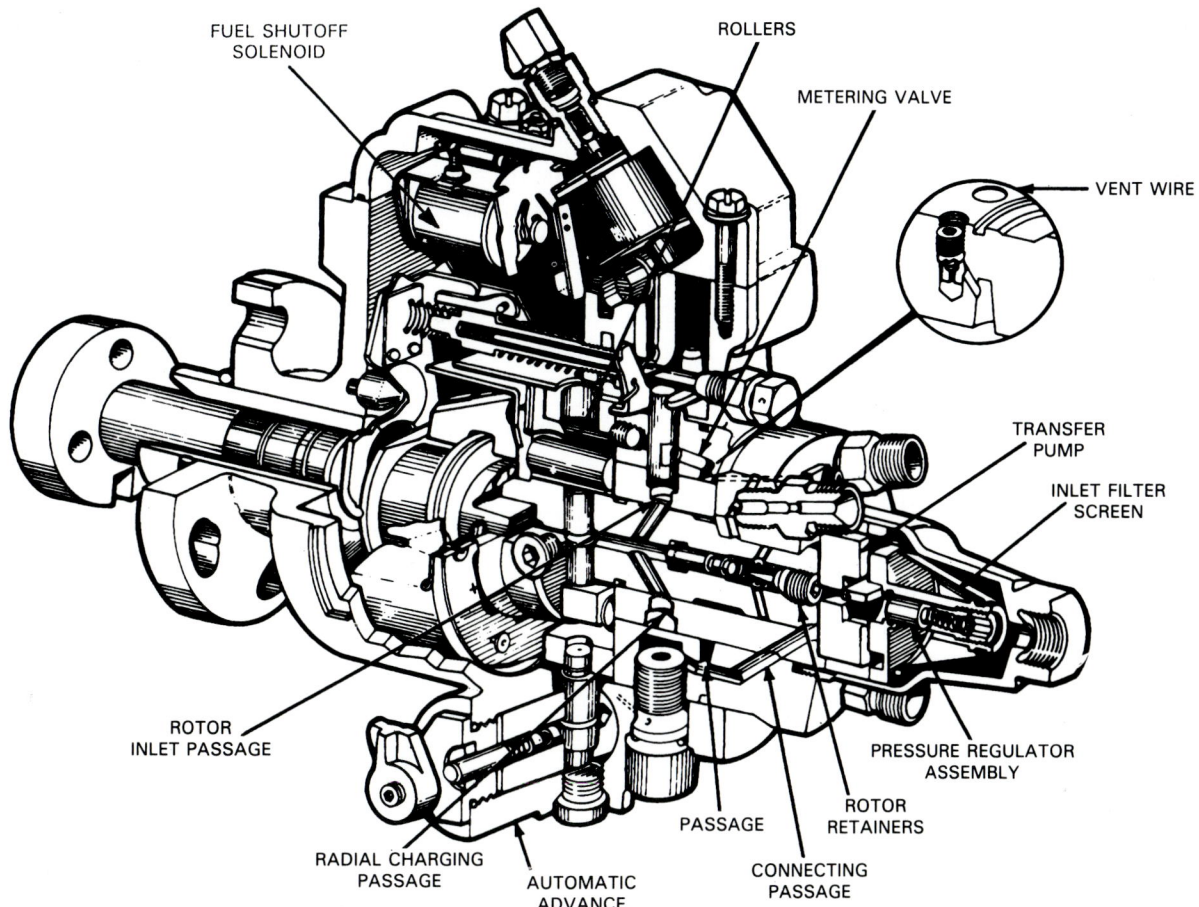

Fig. 23-19A. Two-plunger distributor pump. It is similar to single-plunger type pump. Can you find the major differences and similarities?

Two-plunger distributor pump operation

When the engine is running, the drive shaft turns the transfer (vane) pump. This pulls fuel into the injection pump. When the charging ports line up, fuel fills the high pressure pumping chamber, Fig. 23-19A.

As the shaft continues to turn, the charging ports move out of alignment and the discharging ports line up. At this instant, the plungers are forced inward by the cam ring. This pressurises the fuel and pushes it out of the hydraulic head to the injector nozzles. Look at Fig. 23-19B.

OTHER DIESEL INJECTION PUMP FEATURES

There are numerous design variations of the injection pumps that were just explained. For details of each pump design, refer to the pump manufacturer's service manual. It will explain the construction and operation of the particular pump.

A few additional features with many automotive diesel injection pumps include:

1. An ELECTRICAL FUEL SHUT-OFF is a solenoid that stops fuel injection when the ignition key is turned off. See Fig. 23-19A.

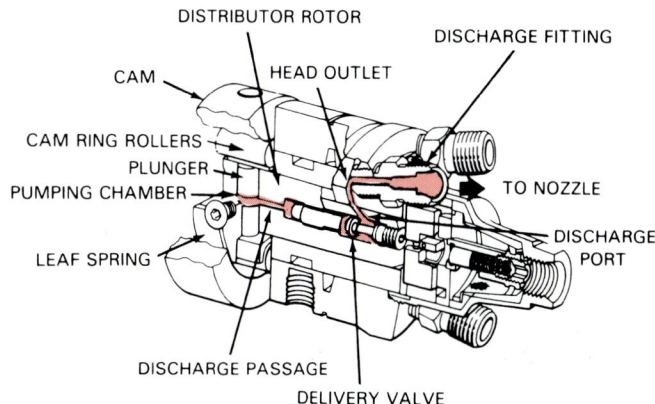

Fig. 23-19B. Pump in discharge cycle position. Cams force plungers together. Fuel is pumped through delivery valve to proper injector line.

2. A VISCOSITY COMPENSATING VALVE allows for different fuel weights or thicknesses and temperatures.

3. An INJECTION PUMP VENT is a small passage that allows fuel to return to the fuel tank. This helps bleed air out of system, Fig. 23-20.

DIESEL INJECTOR NOZZLES

Diesel injector nozzles are spring-loaded valves that spray fuel directly into the engine precombustion chambers. See Fig. 23-21.

The injector nozzles are threaded into the cylinder head. One injector is provided for each engine cylinder. The inner tip of the injector nozzle is exposed to the heat of combustion, like a spark plug in a petrol engine.

Diesel injector parts

The basic parts of a diesel injector are pictured in Figs. 23-22 and 23-23. Look at these illustrations as the parts are explained.

A diesel injector heat shield helps protect the injector from engine heat. It also helps make a good seal between the injector and the cylinder head.

The injector body is the main section of the injector that holds the other parts. The body threads into the heat shield. The heat shield threads into the cylinder head. Fuel passages are provided in the injector body. A needle seat is formed by the lower opening in the injector body.

The diesel injector needle valve opens and closes the nozzle (fuel opening). It is a precisely machined rod with a specially shaped tip. The tip of the needle seals against the injector body when closed.

The injector spring holds the injector needle in the normally closed position. It fits around the needle and against the injector body. Spring tension helps control injector opening pressure.

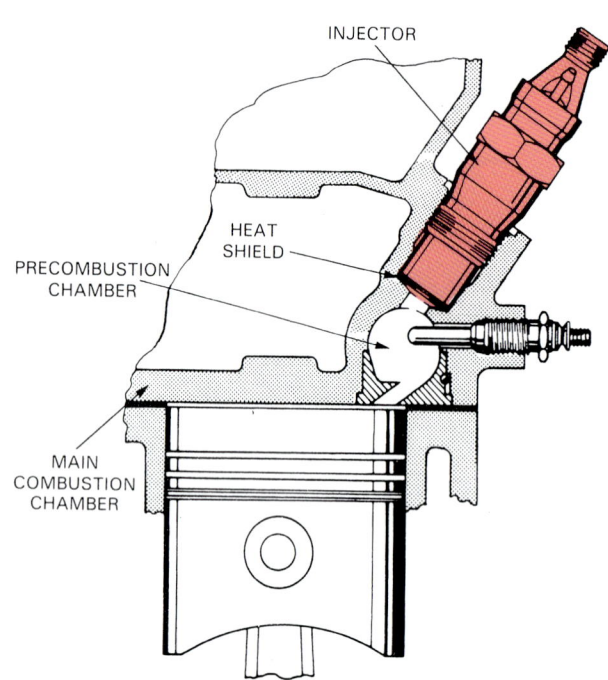

Fig. 23-21. Diesel injector screws into cylinder head. Heat shield fits between injector body and head. Fuel is injected into precombustion chamber. (VW)

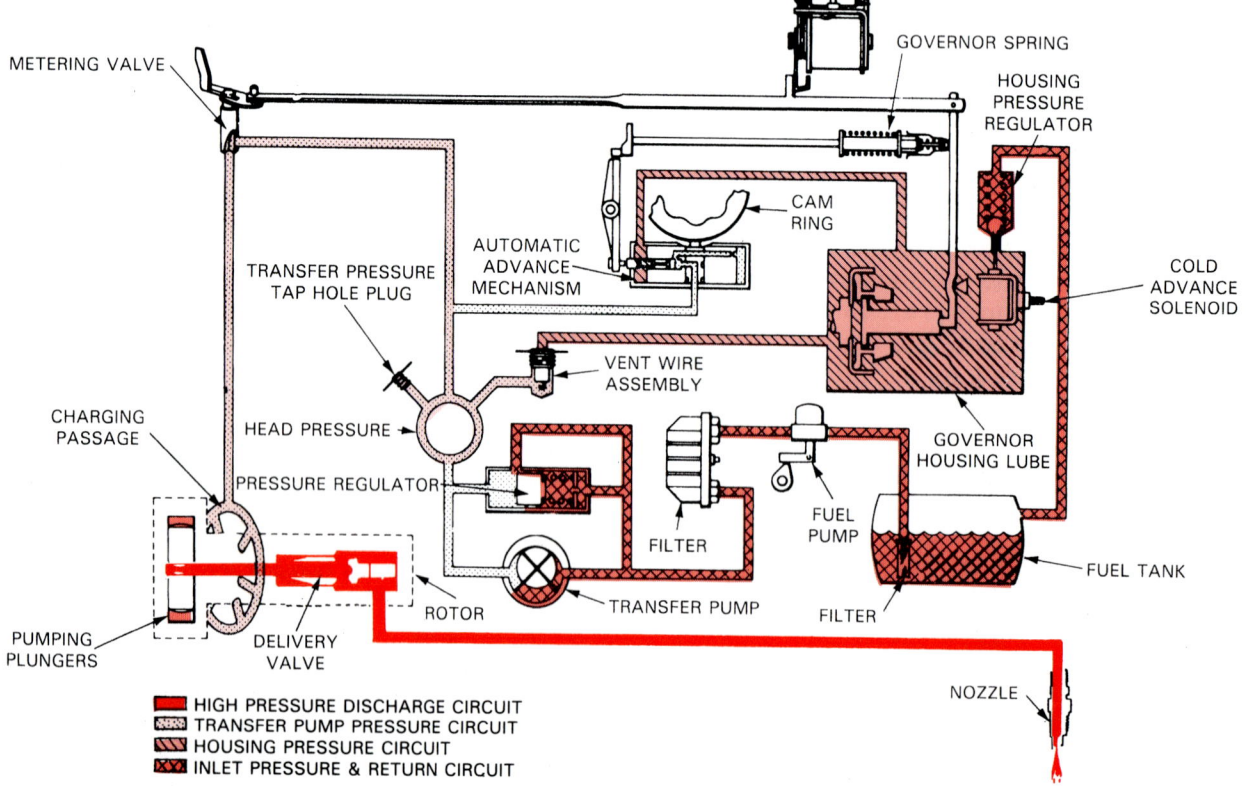

Fig. 23-20. Typical fuel circuit diagram for a two-plunger type distributor injection pump. Can you trace fuel through system?

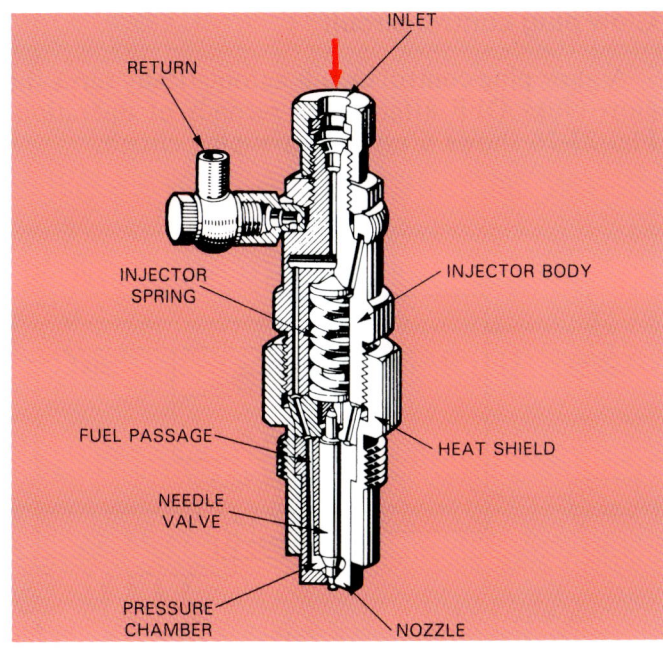

Fig. 23-22. Basic diesel injector construction. Note part names, shapes, and locations. (Robert Bosch)

An injector pressure chamber is formed around the tip of the injector needle and inner cavity in the injector body. Injection pump pressure forces fuel into this chamber to push the needle valve open.

Diesel injector nozzle operation

When the injection pump produces high pressure, fuel flows through the injection line and into the inlet of the injector nozzle. Look at Fig. 23-23. Fuel then

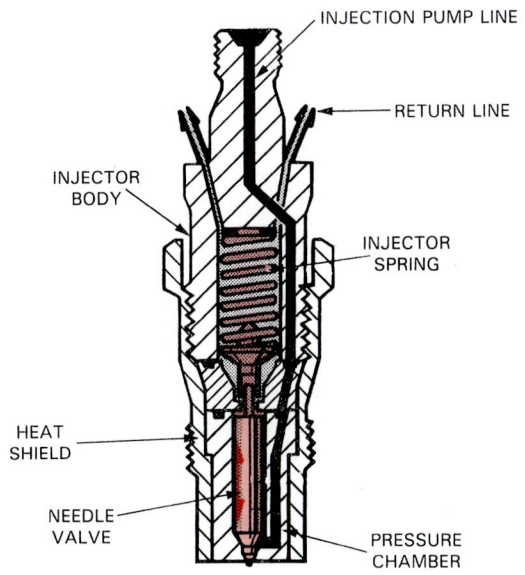

Fig. 23-23. High pressure from injection pump enters top of injector valve. Fuel flows through body passage to pressure chamber. With enough injection pressure, needle valve is pushed up and fuel sprays into precombustion chamber. (VW)

flows down through the fuel passage in the injector body and into the pressure chamber.

The high fuel pressure in the pressure chamber forces the needle upward, compressing the injector spring. This allows diesel fuel to spray out forming a CONE-SHAPED spray pattern. Some fuel leaks past the injector needle and returns to the fuel tank through the return lines.

Diesel injector nozzle types

Several types of injector nozzles are used in diesel engines. The most common of these are the:

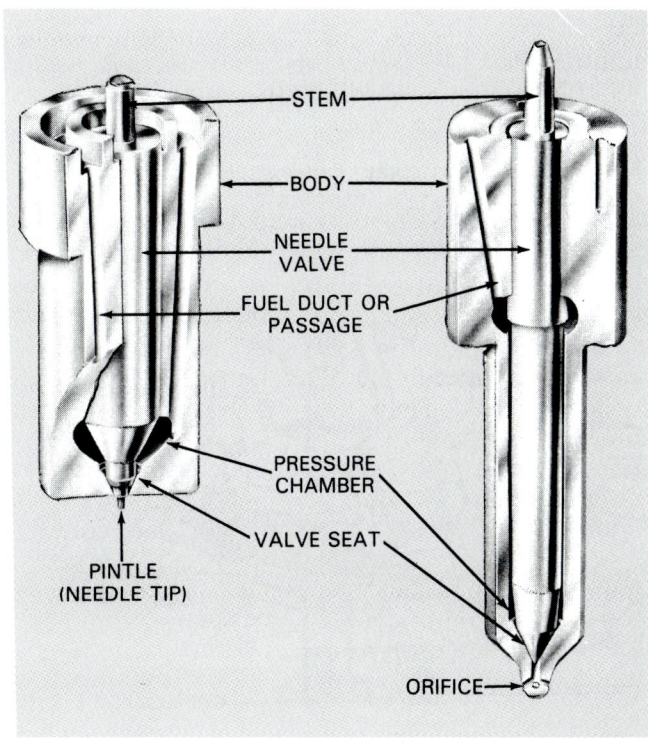

Fig. 23-24. Inward opening injector nozzles. Pintle type injector is very common in automotive diesel engine. Right — Hole type injector is less common. (American Bosch)

1. Inward opening injector nozzle, Fig. 23-24.
2. Outward opening injector nozzle, Fig. 23-25.
3. Pintle injector nozzle, Fig. 23-24.
4. Hole injector nozzle, Fig. 23-24.

Most automotive diesels use an inward opening, pintle type injector nozzle. Fig. 23-25 shows some typical spray patterns.

GLOW PLUGS

Glow plugs are heating elements that warm the air in precombustion chambers to help start a cold diesel engine. Refer to Figs. 23-26 and 23-27.

The glow plugs are threaded into holes in the cylinder head. The inner tip of the glow plug extends into the precombustion chamber.

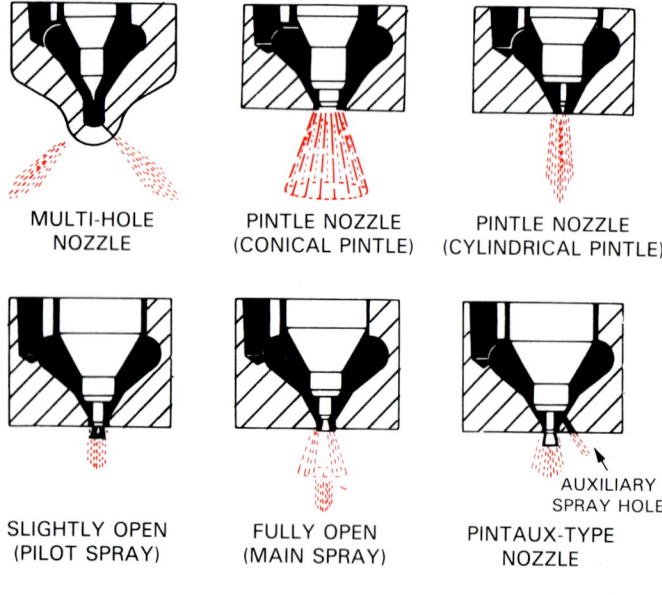

MULTI-HOLE NOZZLE

PINTLE NOZZLE (CONICAL PINTLE)

PINTLE NOZZLE (CYLINDRICAL PINTLE)

SLIGHTLY OPEN (PILOT SPRAY)

FULLY OPEN (MAIN SPRAY)

PINTAUX-TYPE NOZZLE

AUXILIARY SPRAY HOLE

Fig. 23-25. Study spray patterns for different types of injectors.

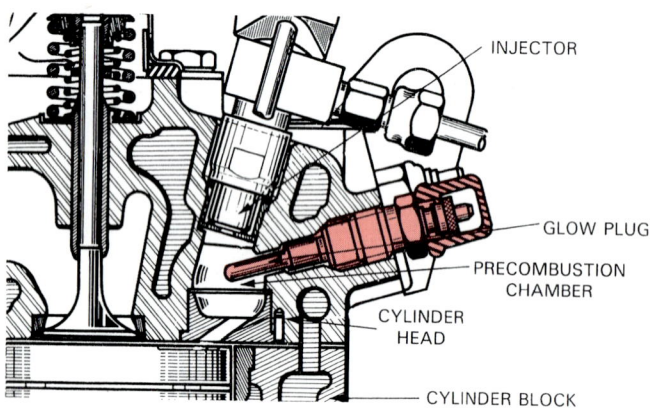

INJECTOR

GLOW PLUG

PRECOMBUSTION CHAMBER

CYLINDER HEAD

CYLINDER BLOCK

Fig. 23-26. Glow plug screws into cylinder head next to injector. Its tip protrudes into precombustion chamber. (Peugeot)

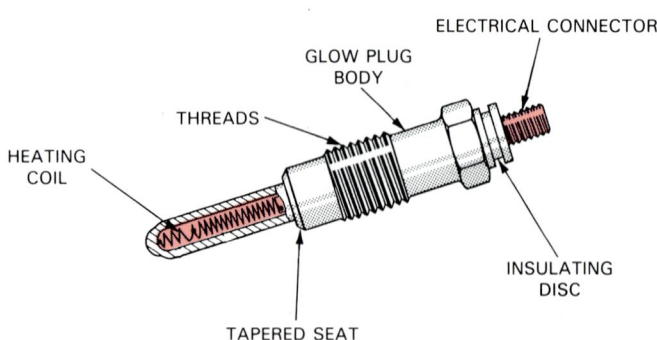

ELECTRICAL CONNECTOR

GLOW PLUG BODY

THREADS

HEATING COIL

INSULATING DISC

TAPERED SEAT

Fig. 23-27. Glow plug is simply an electric heating element. Current flow through plug heats element to warm air in precombustion chamber. This aids starting of cold engine. (Mercedes-Benz)

Glow plug control circuit

A glow plug control circuit automatically turns the glow plugs OFF after a few seconds of operation. Fig. 23-28 shows a typical glow plug circuit.

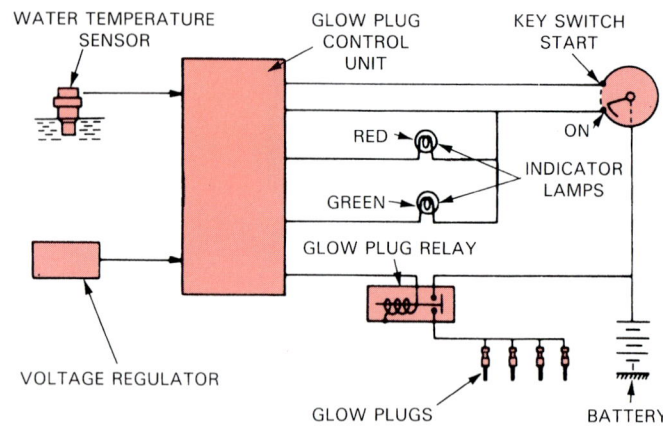

WATER TEMPERATURE SENSOR

GLOW PLUG CONTROL UNIT

KEY SWITCH START

RED

ON

GREEN

INDICATOR LAMPS

GLOW PLUG RELAY

VOLTAGE REGULATOR

GLOW PLUGS

BATTERY

Fig. 23-28. Basic circuit for glow plugs. Control unit monitors engine temperature and informs driver whether glow plugs have been on long enough for engine starting.

A sensor checks the temperature of the engine coolant. It feeds this electrical data to a control unit. Thus, if the engine is already warm, the control unit will not turn the glow plugs on.

Indicator lights, also operated by the control unit, inform the driver whether or not the engine is ready to start. The glow plugs need a few seconds to heat up.

Glow plug operation

When the engine is cold and the driver turns the ignition switch to run, a large current flows from the battery to the glow plugs. In a few seconds, the glow plug tips will heat to a dull red glow, Fig. 23-27.

When the glow plug indicator light goes out, the driver can start the engine. The compression stroke pressure and heat, along with the heat from the glow plugs, causes the engine to start easily.

WATER DETECTOR

A water detector may be used to warn the driver of water in the diesel fuel. Such contamination is very harmful to a diesel fuel system. The water mixes with the fuel and can cause corrosion of the precision parts in the injection pump and injectors. Fig. 23-29 shows a circuit using an in-tank water detector.

FUEL HEATER

A fuel heater is sometimes used to warm the diesel fuel, preventing the fuel from waxing (turning into a semisolid). An optional device, it is needed in very

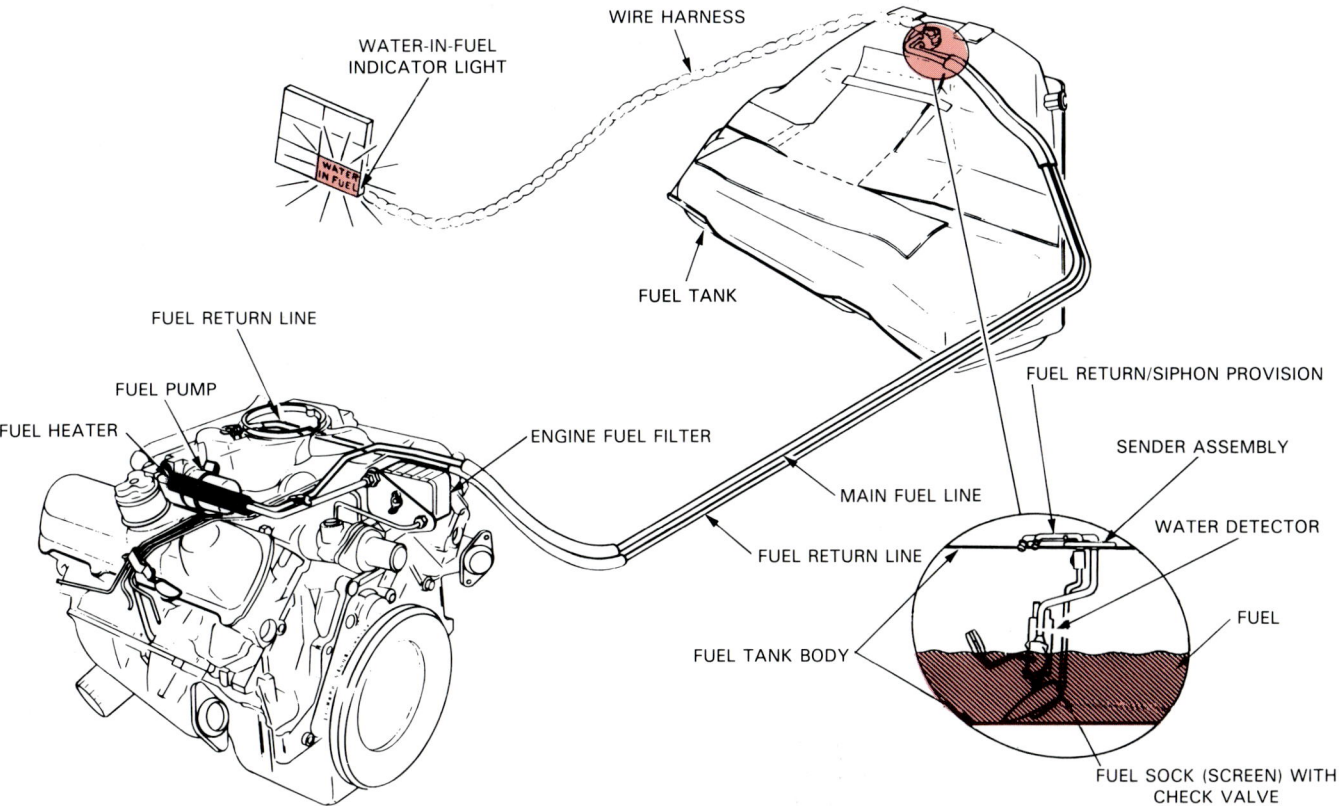

Fig. 23-29. This diesel injection system has a water detector-warning light system and a fuel heater. Indicator light glows if there is an excessive amount of water in tank. Fuel heater warms fuel in line before fuel enters injection pump. Heat is only needed in cold weather.

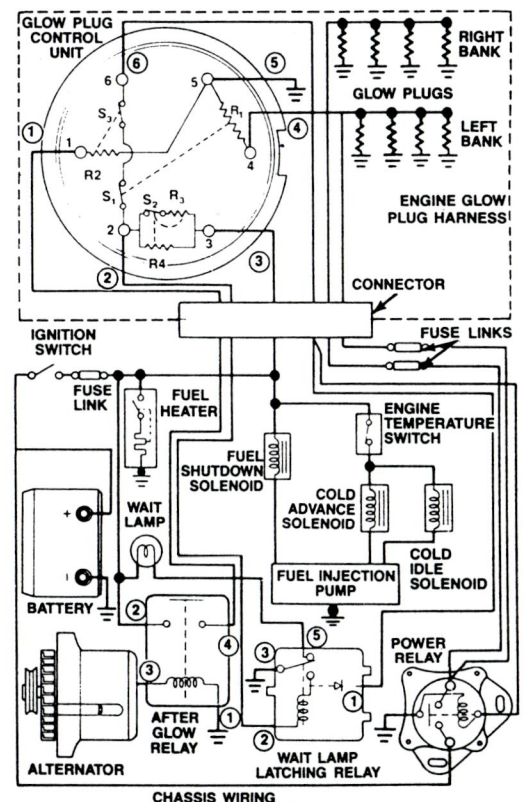

Fig. 23-30. Complete electrical circuit for a typical diesel injection system. Note parts and electrical connections. (Ford)

cold climates. The heater is simply an electric heating element in the fuel line ahead of the injection pump. See Figs. 23-29 and 23-30.

VACUUM PUMP

A vacuum pump is frequently used on a diesel engine to provide a source of vacuum (suction) for the vehicle. Vacuum is needed for the power brakes, A/C-ventilation system, and emission control devices.

A petrol engine has a natural source of vacuum in the intake manifold. A diesel engine, because it does NOT have a throttle valve restriction, has very little vacuum in the intake manifold. For this reason, a vacuum pump is needed.

A vacuum pump on a diesel engine may be a reciprocating diaphragm type bolted to the front or rear of the engine, Fig. 23-31. It may also be a rotary type pump at the rear of the alternator, Fig. 23-32.

A vacuum pump uses the same principle as a fuel pump (see Chapter 19). However, it pumps air, not liquid fuel, out of an enclosed area to reduce pressure and form a vacuum. The diesel engine provides power for the vacuum pump, usually through a belt.

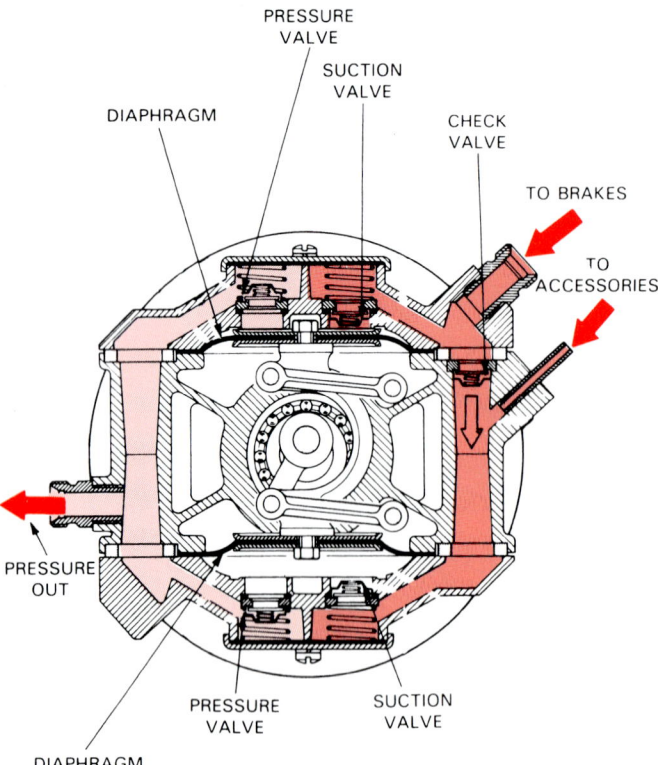

Fig. 23-31. Construction of a diaphragm vacuum pump. (Mercedes-Benz)

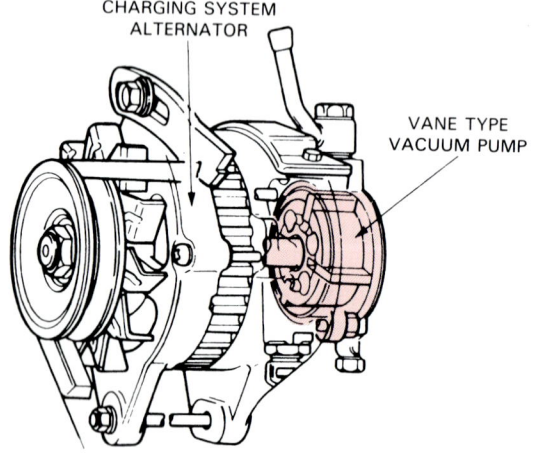

Fig. 23-32. Vane type vacuum pump is mounted in back of charging system alternator. It produces vacuum for power brakes and other vacuum operated devices. (Chrysler)

KNOW THESE TERMS

Compression ignition, Injection pump, In-line pump, Pumping plunger, Barrel, Control sleeve, Control rod, Delivery valve, Effective plunger stroke, Governor, Injection timing, Distributor injection pump, Transfer pump, Hydraulic head, Distributor rotor, Internal cam ring, Injector, Needle valve, Pintle, Pressure chamber, Glow plug, Water detector, Fuel heater, Vacuum pump.

REVIEW QUESTIONS

1. A diesel fuel injection system is a super-high pressure, mechanical system that delivers fuel directly to the engine combustion chambers. True or False?
2. List and explain the four major components of a diesel injection system.
3. Which of the following is NOT related to a diesel engine?
   a. Compression ignition.
   b. No throttle valves for air control.
   c. Fuel quantity controls engine speed.
   d. All of the above are correct.
4. There are two common types of diesel injection pumps: _____ and _____ types.
5. What is the function of a pumping plunger in an injection pump?
6. Roller tappets are commonly used in an in-line injection pump. True or False?
7. Explain how an in-line injection pump alters the amount of fuel forced to the injector nozzles.
8. _____ _____ are spring-loaded valves in the outlet fittings to the injection lines for assuring quick, leak-proof closing of the injector nozzles.
9. Define the term "effective plunger stroke".
10. Why is a governor needed on a diesel engine?
11. _____ _____ refers to when fuel is injected into the combustion chambers in relation to piston position.
12. A _____ _____ _____ normally uses only one or two pumping plungers and is the most common type diesel injection pump.
13. Which of the following is NOT part of a single-plunger distributor injection pump?
   a. Drive shaft.
   b. Fuel metering sleeve.
   c. Hydraulic head.
   d. Roller tappet.
14. How does a single-plunger distributor pump develop injection pressure?
15. How does a single-plunger distributor pump control injection quantity?
16. Which of the following is NOT part of a two-plunger distributor injection pump?
   a. External camshaft.
   b. Internal cam ring.
   c. Distributor rotor.
   d. Fuel metering valve.
17. A vane type transfer pump is commonly used to pull fuel into the two-plunger distributor injection pump. True or False?
18. An electric fuel shut-off is only used on in-line type injection pumps. True or False?
19. A diesel _____ _____ is a spring-loaded valve that sprays fuel into the engine precombustion chamber.

366

20. List and explain the five major parts of a diesel injector.
21. Why are glow plugs needed in a diesel engine?
22. Explain the purpose of a water detector.
23. A _____ _____ can be used to help keep the diesel fuel from waxing in cold weather.
24. Explain why a diesel engine, unlike a petrol engine, needs a vacuum pump.

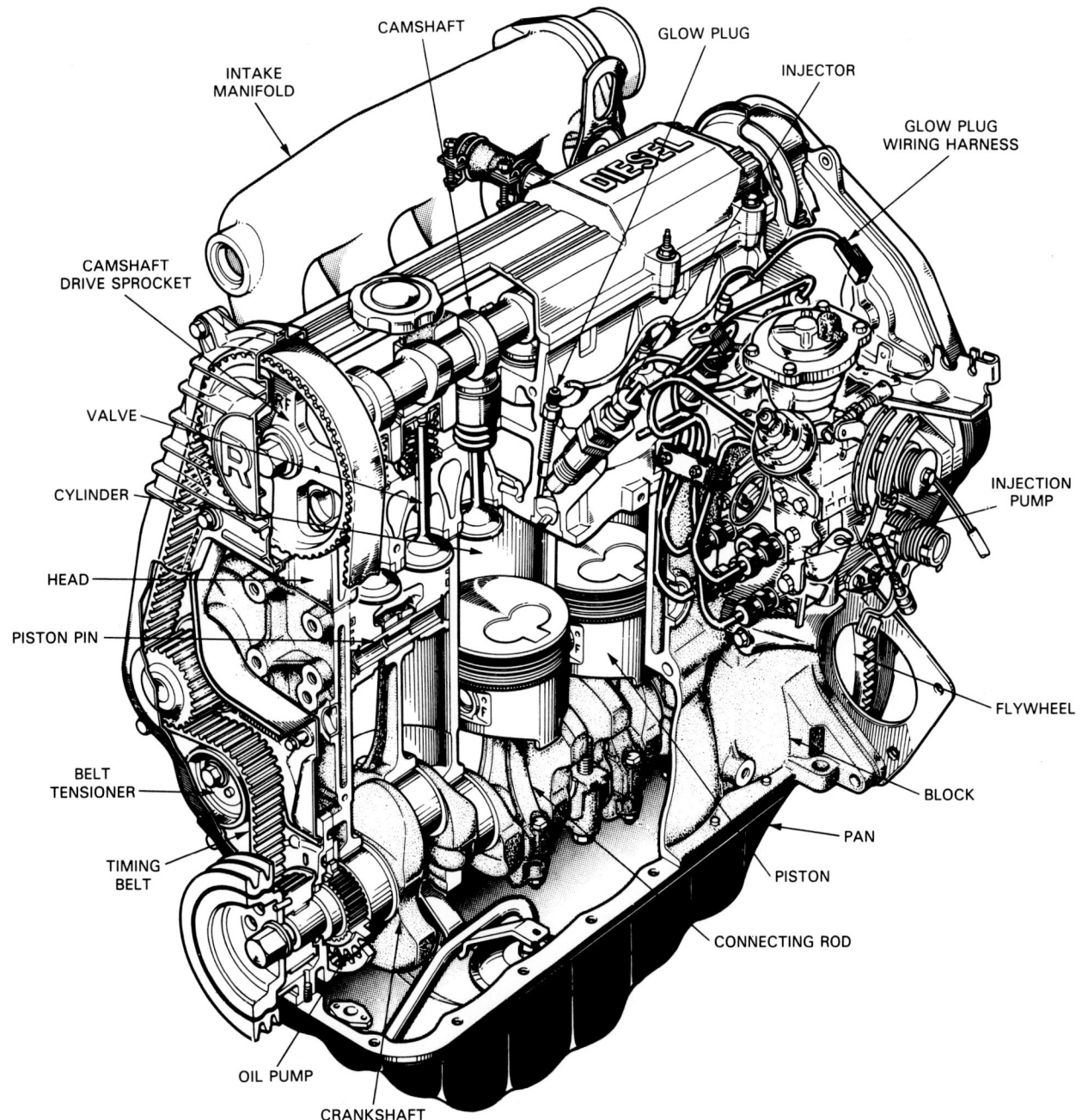

INTAKE MANIFOLD

CAMSHAFT

GLOW PLUG

INJECTOR

GLOW PLUG WIRING HARNESS

DIESEL

CAMSHAFT DRIVE SPROCKET

VALVE

CYLINDER

HEAD

PISTON PIN

INJECTION PUMP

BELT TENSIONER

FLYWHEEL

BLOCK

PAN

PISTON

TIMING BELT

CONNECTING ROD

OIL PUMP

CRANKSHAFT

Cutaway view of a four-stroke cycle, 4-cylinder diesel engine. This 1998 cc engine has a compression ratio of 22.7. It also features an aluminium alloy cylinder head. (Mazda)

Chapter 24

CLUTCH FUNDAMENTALS

After studying this chapter, you will be able to:

☐ List the basic parts of an automotive clutch.
☐ Explain the operation of a clutch.
☐ Describe the construction of major clutch components.
☐ Compare clutch design differences.
☐ Explain the different types of clutch release mechanisms.

This chapter begins your study of a car's drive train. The clutch is the first drive train component powered by the engine crankshaft. The clutch lets the driver control power flow between the engine and transmission or transaxle (transmission-differential unit).

Fig. 24-1 shows the basic parts of drive trains for both rear and front-wheel drive cars. Review the relationship of the parts.

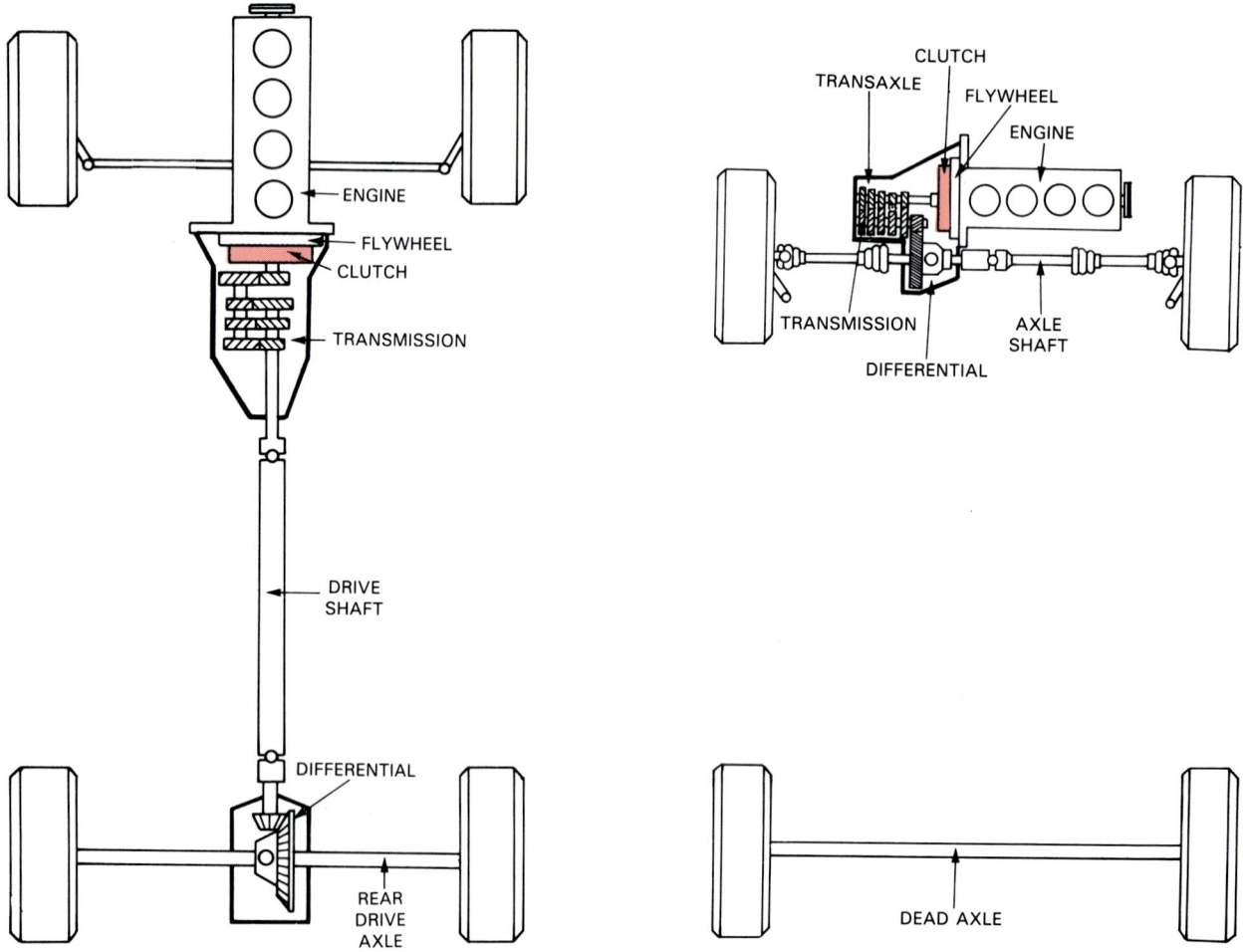

Fig. 24-1. Compare drive trains for front and rear-wheel drives. Clutch mounts on engine flywheel.

The next chapters cover transmissions, drive shafts, transfer cases, differentials, transaxles, and other drive train units. To fully understand these later chapters, you must first understand clutches.

Each succeeding chapter builds upon the knowledge gained in the previous chapter. For this reason, it is very important that you learn everything about to be discussed. Study carefully!

PURPOSE OF THE CLUTCH

An automotive clutch is used to connect and disconnect the engine and manual (hand-shifted) transmission or transaxle. The clutch is located between the back of the engine and the front of the transmission.

Only cars with manual transmissions require a clutch. Cars with automatic transmissions do NOT need a clutch. They have a fluid coupling or torque converter that automatically disengages the engine and transmission at low engine rpm.

CLUTCH PRINCIPLES

Power flow from one unit to another can be controlled with a drive disc and a driven disc. Look at Fig. 24-2. Relating to an automotive clutch, one disc is fastened to the rear of the engine crankshaft. The other disc is attached to the input shaft of the transmission.

When the discs do NOT touch, the crankshaft can rotate while the transmission input shaft remains stationary. However, when the transmission disc is forced into the spinning disc on the crankshaft, both spin together. Power flow is then transferred out of the engine and into the transmission. This principle is used in a car's clutch.

BASIC CLUTCH PARTS

Fig. 24-3 shows the major parts of an automotive

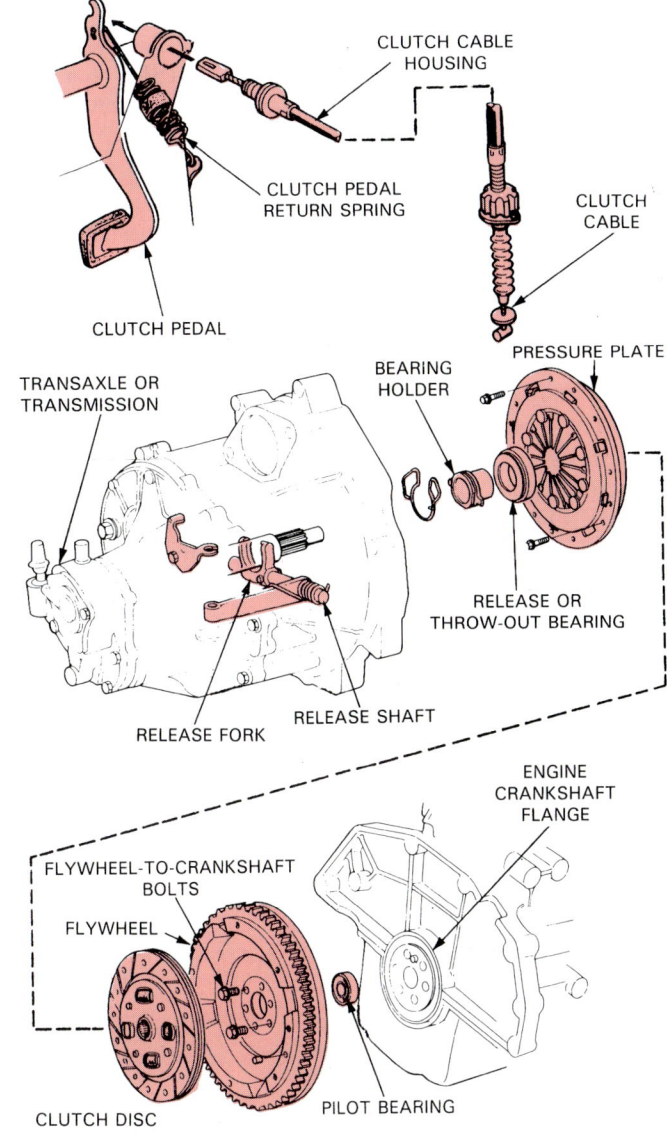

Fig. 24-3. Fundamental components of a clutch. Note how parts are located in relation to each other and with rear of engine. (Honda)

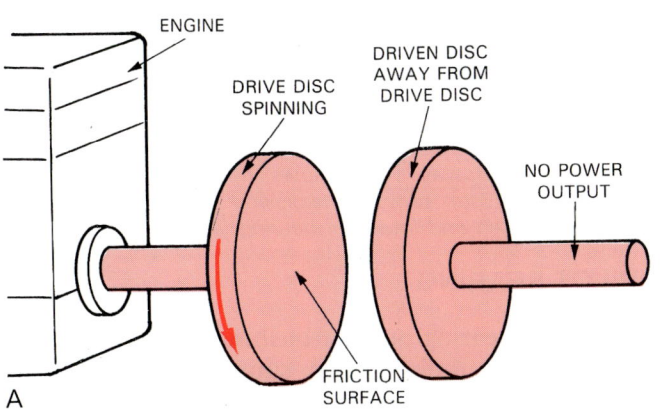

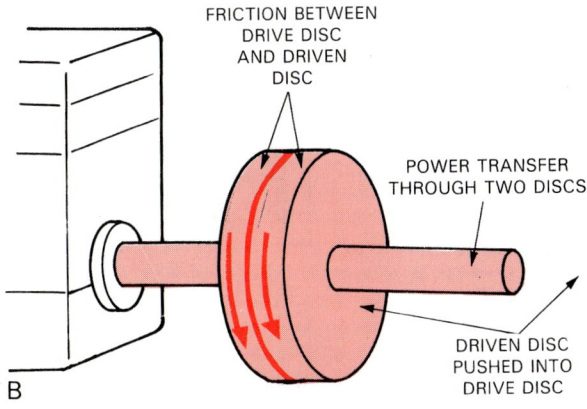

Fig. 24-2. Rotating discs demonstrate action of automotive clutch. A — Crankshaft spins drive disc on left. Other disc is not in contact with drive disc. No power transfers. B — Two discs are pushed together. Friction causes crankshaft disc to turn other disc connected to transmission input shaft. Power is transferred through clutch.

clutch. Study the names, locations, and relationship of the components.

1. CLUTCH RELEASE MECHANISM (cable, linkage, or hydraulic system allows driver to disengage clutch with foot pedal).
2. CLUTCH FORK (lever that forces release bearing into pressure plate).
3. RELEASE BEARING (bearing that cuts friction between clutch fork and pressure plate).
4. PRESSURE PLATE (spring-loaded device that clamps clutch disc against flywheel).
5. CLUTCH DISC (friction disc splined [fastened] to transmission input shaft and pressed against face of flywheel).
6. FLYWHEEL (provides mounting place for clutch and friction surface for clutch disc).
7. PILOT BEARING (bushing or bearing that supports forward end of transmission input shaft).

Clutch action

When the driver presses the clutch pedal, the clutch release mechanism pulls or pushes on the clutch release lever or fork. See Fig. 24-4.

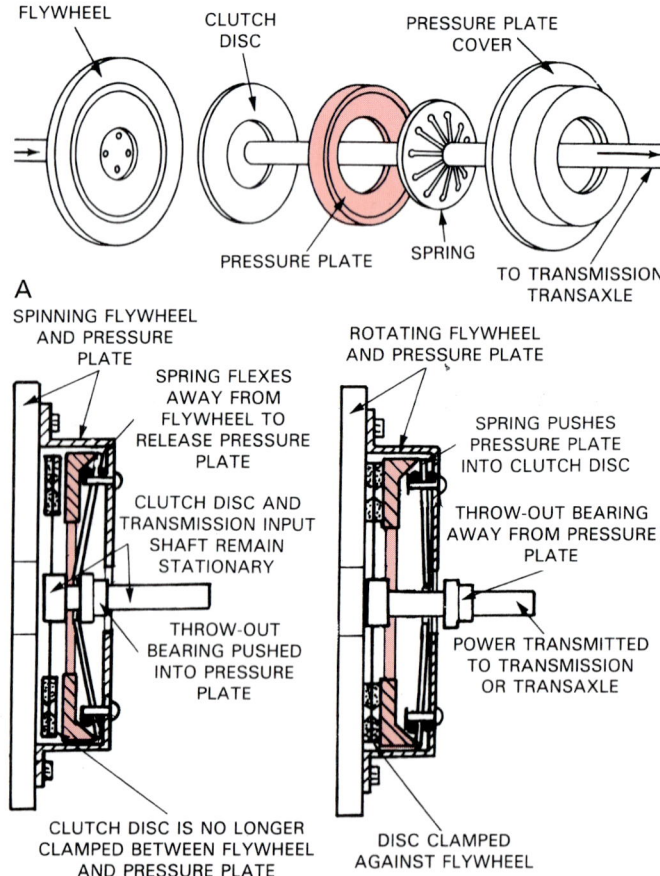

Fig. 24-4. A — Basic clutch parts. B — Clutch disengaged. Pressure plate does not clamp friction disc against flywheel. No power flow through clutch. C — Clutch engaged. Pressure plate spring action clamps friction disc and flywheel together. Clutch disc then turns transmission input shaft.

The fork moves the release bearing into the centre of the pressure plate. This causes the pressure plate face to pull away from the clutch disc, releasing the disc from the flywheel. The engine crankshaft can then turn without turning the clutch disc and transmission input shaft.

When the clutch pedal is released by the driver, spring pressure inside the pressure plate pushes forward on the clutch disc. It locks the flywheel, disc, pressure plate and transmission input together. The engine again rotates the transmission input shaft, transmission gears, drive train, and car's wheels.

CLUTCH CONSTRUCTION

Now that you understand the basic action of a clutch, we will discuss, in more detail, how each part is made. This information will be useful when learning to diagnose and repair a clutch.

Refer to Fig. 24-5. It shows a side view of an assembled clutch.

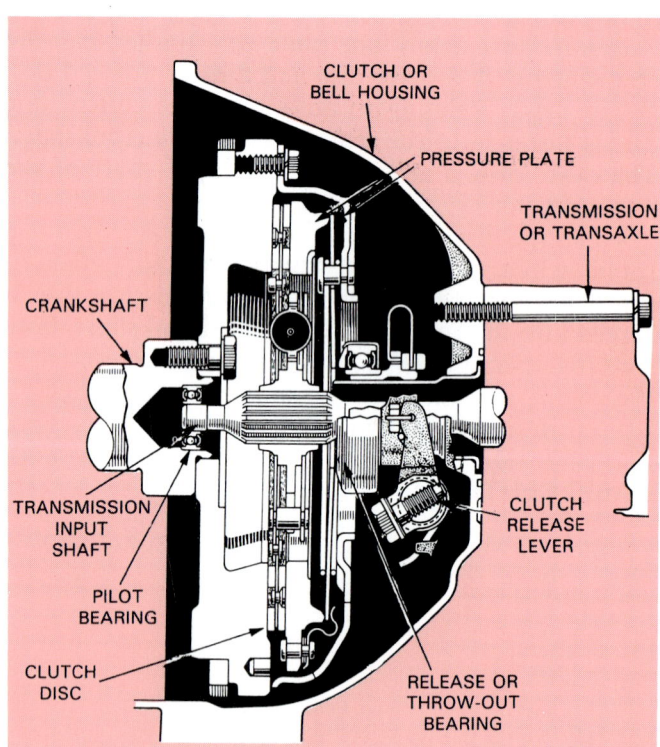

Fig. 24-5. Cutaway view shows how clutch looks when assembled. Note how transmission input shaft extends through clutch and into pilot bearing in crankshaft. (Toyota)

PILOT BEARING

A pilot bearing or pilot bushing is pressed into the end of the crankshaft to support the end of the transmission input shaft. Usually, the pilot is a solid bronze bushing. It may also be a roller or ball bearing.

As shown in Fig. 24-5, the end of the transmission

input shaft has a small journal machined on its end. This journal slides inside the pilot bearing. The pilot prevents the transmission shaft and clutch disc from wobbling up and down when the clutch is released. It helps the input shaft centre the disc on the flywheel.

FLYWHEEL

The flywheel is the mounting place for the clutch. The pressure plate bolts to the flywheel face. The clutch disc is pinched and held against the flywheel by the spring action of the pressure plate. Look at Figs. 24-5 and 24-6.

The face of the flywheel is precision machined to a smooth surface. Normally, the face of the flywheel, that touches the clutch disc is made of iron. Even if the flywheel is aluminium, the face is iron because it wears well and dissipates heat better.

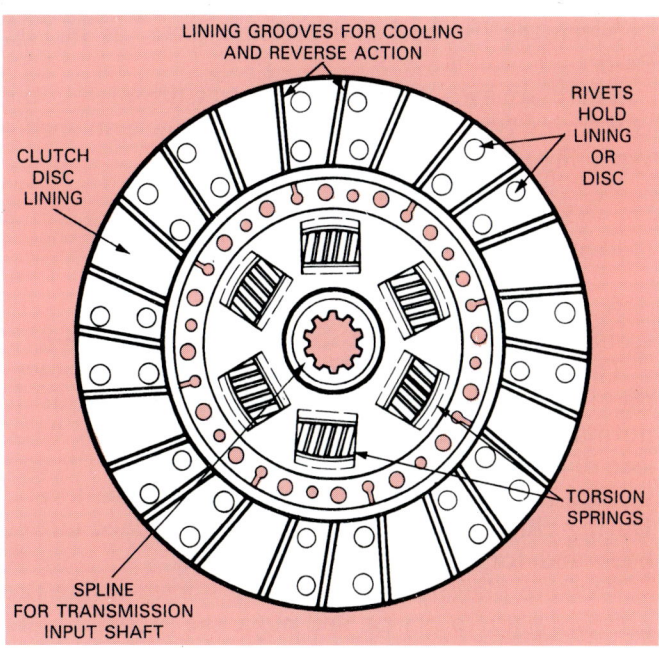

Fig. 24-7. Clutch disc construction. Asbestos lining or friction material is held on metal disc by rivets. Splines in centre of disc fit over splines on transmission input shaft.

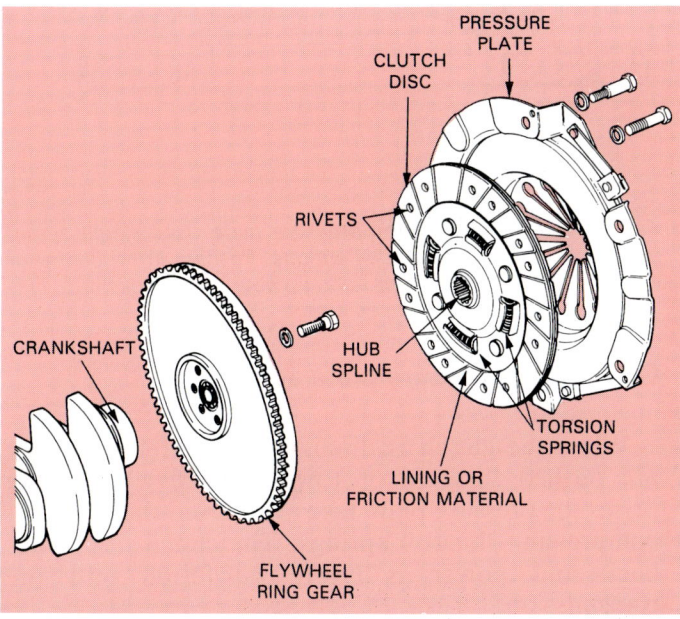

Fig. 24-6. Flywheel bolts to crankshaft flange. Pressure plate bolts to flywheel. Clutch disc is clamped between flywheel and pressure plate. (Mazda)

CLUTCH DISC

The clutch disc, also called friction disc, consists of a splined hub and a round metal plate covered with friction material (lining). One is pictured in Fig. 24-7.

The splines (grooves) in the centre of the clutch disc mesh with splines on the transmission input shaft. This makes the input shaft and disc turn together. However, the disc is free to slide back and forth on the shaft.

Clutch disc torsion springs

Clutch disc torsion springs, also termed damping springs, help absorb some of the vibration and shock

produced by clutch engagement. They are small coil springs located between the clutch disc splined hub and the friction disc assembly.

When the clutch is engaged, the pressure plate jams the stationary disc against the spinning flywheel. The torsion springs compress and soften the shock as the disc first begins to turn with the flywheel, Fig. 24-7.

Clutch disc facing springs

Clutch disc facing springs, also called cushioning springs, are flat, metal springs located under the disc's friction material. These springs have a slight wave or curve. They allow the friction material to flex inward slightly during initial clutch engagement. This also smoothes engagement.

Clutch disc friction material

The clutch disc friction material, also called disc lining or facing, is usually made of heat resistant asbestos, cotton fibres, and copper wires woven or moulded together. Refer to Fig. 24-7.

Grooves are cut in the friction material to aid cooling and release of the clutch disc. Rivets are used to bond the friction material to both sides of the metal body of the disc.

PRESSURE PLATE

The pressure plate is a spring-loaded device that can either lock or unlock the clutch disc and the flywheel. It bolts to the flywheel. The clutch disc fits between the flywheel and pressure plate, as in Fig. 24-6.

There are two basic types of pressure plates: the coil spring type, Fig. 24-8, and the diaphragm spring type, Fig. 24-9.

Coil spring pressure plate

A coil spring pressure plate uses small, coil springs, similar to valve springs. Study Fig. 24-8.

The pressure plate face is a large ring that contacts the friction disc during clutch engagement. It is normally made of iron. Refer to Fig. 24-8. The back side of the pressure plate face has pockets for the coil springs and brackets for hinging the release levers. During clutch action, the pressure plate face moves back and forth inside the clutch cover.

Pressure plate release levers are hinged inside the pressure plate to pry on and move the pressure plate face away from the disc and flywheel. Small clip-type springs fit around the release levers to keep them from rattling and in the fully retracted (released) position.

The pressure plate cover fits over the springs, release levers, and pressure plate face. Its main purpose is to hold the parts of the pressure plate assembly together. Holes around the outer edge of the cover are for bolting the pressure plate to the flywheel.

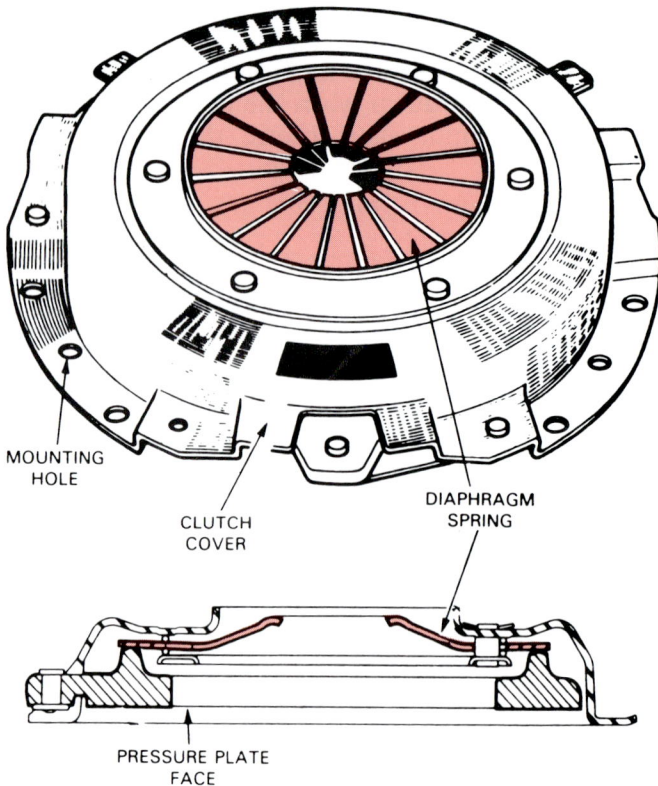

Fig. 24-9. Diaphragm type pressure plate uses single spring instead of several small, coil springs. Pushing in on centre of spring bends outer edge of spring away from crankshaft. This releases clutch disc.

Coil spring pressure plate action

When the clutch is disengaged, the release levers are pushed forward or toward the flywheel. This prises the pressure plate face away from the flywheel, compressing the coil springs. The clutch disc slides back and power is NOT transferred into the transmission.

When the clutch is engaged, the release bearing moves away from the release levers. Then, the pressure plate springs force the face and disc forward, into the rotating flywheel. The disc and transmission input shaft begin to spin and transmit power.

Centrifugal action of pressure plate

A semi-centrifugal pressure plate uses weighted release levers or rollers, and the resulting centrifugal force, to increase clamping pressure on the clutch disc. The weights on the release levers or the rollers are positioned so that, as engine speed increases, their outward thrust acts on the release levers. The extra force helps keep the clutch from SLIPPING. It also permits the use of slightly weaker coil springs to reduce the amount of foot pedal pressure required to disengage the clutch.

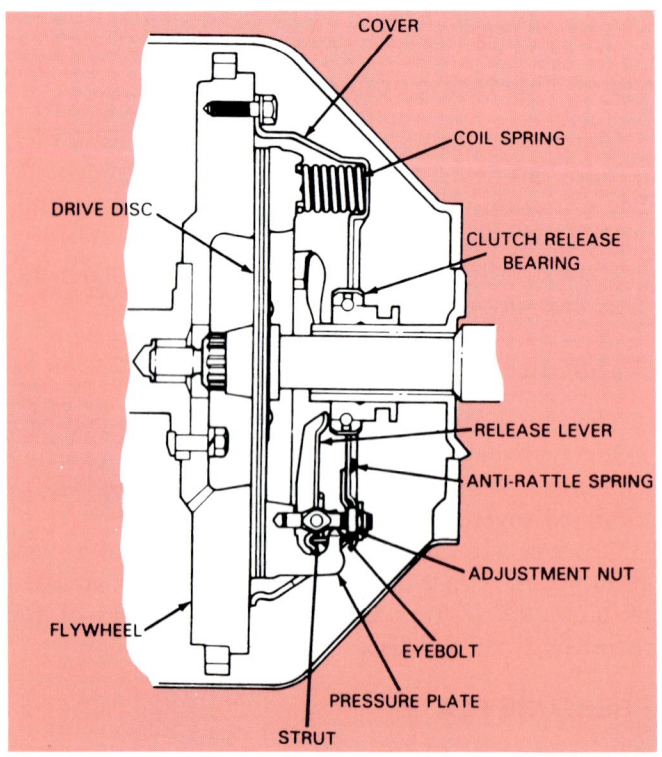

Fig. 24-8. Cutaway view of coil spring type pressure plate. Study parts.

Diaphragm pressure plate

A diaphragm pressure plate uses a single diaphragm spring instead of several coil springs. One is pictured in Fig. 24-9. This type pressure plate functions almost like a coil spring pressure plate.

The diaphragm spring is a large, round disc of spring steel. The spring is bent or dished and has pie-shaped segments (splits) running from the outer edge to the centre opening. The diaphragm spring is mounted in the pressure plate with the outer edge touching the back of the pressure plate face.

A pivot ring mounts behind the diaphragm spring. It is located partway in from the outer edge of the diaphragm spring. See Fig. 24-9.

Diaphragm pressure plate action

When the centre of the diaphragm spring is pushed toward the engine, its outer edge bends back away from the engine. This lets the pressure plate face and clutch disc slide away from the spinning flywheel.

When the centre of the diaphragm spring is released, the spring trys to return to its normal dished shape. As a result the outer edge of the spring pushes the pressure plate face into the clutch disc.

RELEASE BEARING (THROW-OUT BEARING)

The release bearing, also called a throw-out bearing, is usually a ball bearing and collar assembly.

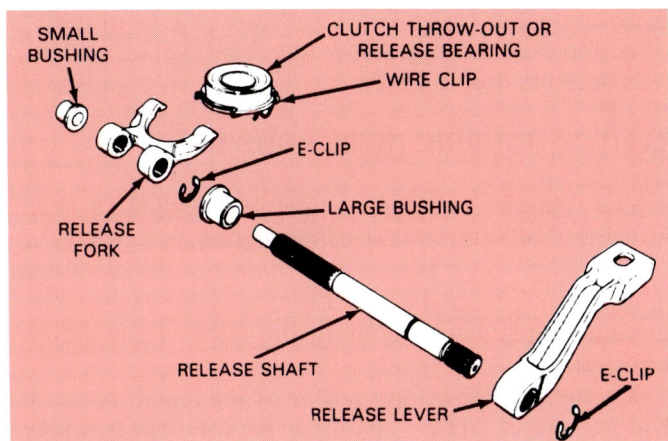

Fig. 24-10. Throw-out or release bearing acts on centre of pressure plate. It is an antifriction bearing that cuts down rubbing contact between clutch fork and pressure plate. Note other parts that operate this throw-out bearing.

It reduces friction between the pressure plate levers and the clutch fork. The release bearing is a sealed unit packed with grease. It slides on a hub or sleeve extending out from the front of the manual transmission or transaxle. Refer to Fig. 24-10.

A few cars, especially foreign, use a graphite type throw-out bearing. The ring-shaped block of friction resistant graphite presses on a smooth flat plate on the clutch release levers.

The throw-out bearing usually snaps over the end of the clutch fork. Small spring clips hold the

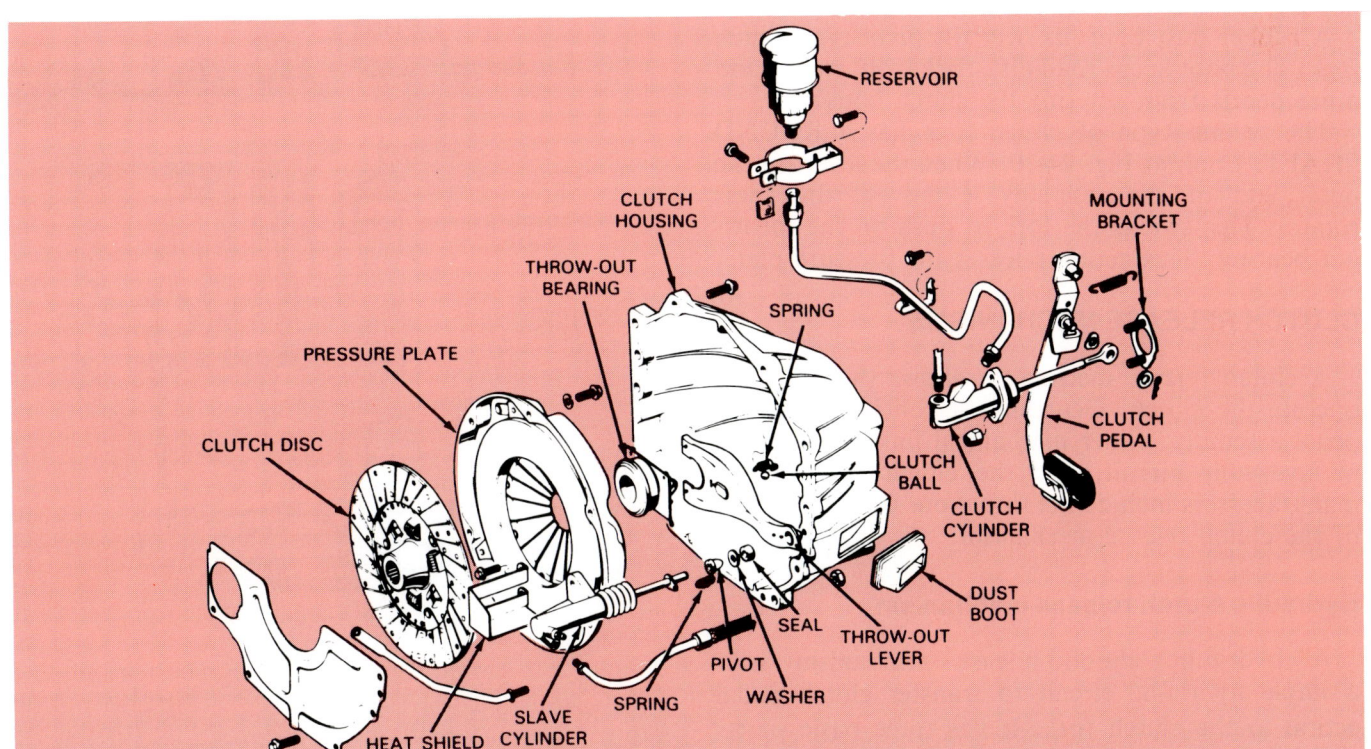

Fig. 24-11. This diaphragm type clutch is operated by a hydraulic release mechanism. When driver presses clutch pedal, clutch master cylinder develops pressure in system. Pressure actuates slave cylinder piston and operates clutch fork to release clutch.

373

bearing on the fork. Then, fork movement in either direction slides the throw-out bearing along the transmission hub sleeve.

CLUTCH HOUSING (BELL HOUSING)

The clutch housing, sometimes called a bell housing, bolts to the rear of the engine, enclosing the clutch assembly. It can be made of aluminium, magnesium, or cast iron. Look at Fig. 24-11. The manual transmission bolts to the back of the clutch housing.

A hole is provided in the side of the clutch housing for the clutch fork. The fork or fork shaft sticks through the housing. A bracket or ball is needed to hold an arm type fork, Fig. 24-11.

The lower front of the clutch housing usually has a thin, sheet metal cover, Fig. 24-11. It can be removed for flywheel ring gear inspection or when the engine must be separated from the clutch assembly.

CLUTCH FORK (CLUTCH ARM)

The clutch fork, also called a clutch arm or release arm, transfers motion from the clutch release mechanism to the throw-out bearing and pressure plate. There are two basic types of clutch forks.

A lever type clutch fork projects through a square hole in the bell housing and mounts on a pivot. Refer to Fig. 24-11. When moved by the release mechanism, the clutch fork PRISES on the throw-out bearing to disengage the clutch.

A rubber dust boot fits over the pivot type clutch fork. The boot keeps road dirt, rocks, oil, water, and other debris from entering the clutch housing.

The second type of clutch fork has a ROUND SHAFT. Look at Fig. 24-10. When the lever on the outer end of the assembly is moved, the shaft rotates. This swings the fork to push on the throw-out bearing. This action releases the pressure plate.

CLUTCH RELEASE MECHANISMS

A clutch release mechanism allows the driver to operate the clutch. Generally, it consists of clutch pedal assembly, either mechanical linkage, a cable, or hydraulic circuit, and the clutch fork. Many manufacturers include the throw-out bearing as part of the clutch release mechanism.

Hydraulic clutch release mechanism

A hydraulic clutch release mechanism uses a simple hydraulic circuit to transfer clutch pedal action to the clutch fork. It has three basic parts: clutch cylinder, hydraulic line, and slave cylinder. Refer to Fig. 24-11.

The clutch cylinder, sometimes called the clutch

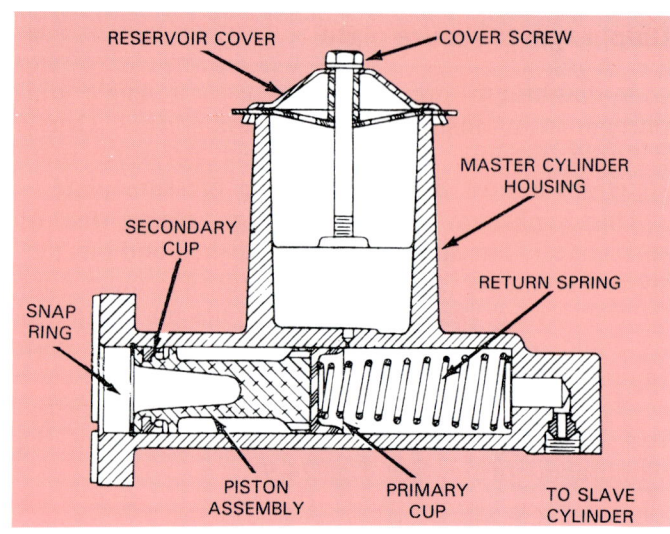

Fig. 24-12. Cutaway view shows inside of clutch master cylinder. Clutch pedal and linkage pushes piston and cup into cylinder producing hydraulic pressure.

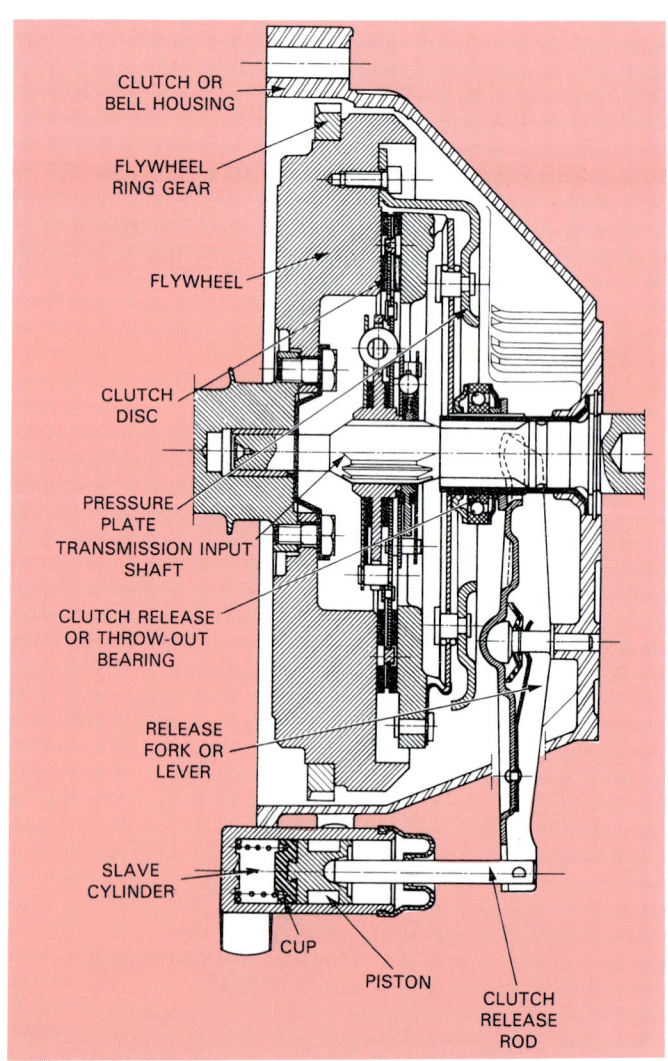

Fig. 24-13. Slave cylinder releases this clutch. Pressure from master cylinder enters cylinder, moving small piston towards clutch fork. Study other components. (Peugeot)

master cylinder, produces the hydraulic pressure for the system. Look at Fig. 24-12. It contains a piston mounted in a cylinder. The piston has rubber cups that produce a leakproof seal between the piston and cylinder wall.

A fluid reservoir is mounted above or on top of the clutch cylinder to hold extra fluid. See Figs. 24-11 and 24-12. Most hydraulic clutch systems use BRAKE FLUID as the medium for pressure transfer.

A cap and seal are threaded onto the reservoir to keep fluid from leaking out and to keep road dirt, and water, from entering the system.

The clutch cylinder usually mounts on the bulkhead. A push rod links the clutch pedal and the cylinder piston.

When the clutch pedal is pressed, the push rod moves the piston to produce pressure in the cylinder.

The hydraulic line is a high pressure, rubber hose-metal line assembly that moves fluid from the clutch cylinder to the slave cylinder. When pressure is produced in the clutch cylinder, fluid flows through the hydraulic line, Fig. 24-11.

The slave cylinder uses the system's hydraulic pressure to cause clutch fork movement. It contains a piston assembly inside a cylinder, as in Fig. 24-13.

When the master cylinder forces fluid into the slave cylinder, pressure pushes the piston outward.

Hydraulic clutch action

When the clutch pedal is depressed, linkage pushes on the piston in the clutch cylinder. A check valve in the clutch cylinder keeps fluid from entering the reservoir. As a result, fluid flows into the hydraulic line and slave cylinder. Pressure forms in the system and the slave cylinder piston is slid outward. The slave cylinder piston and push rod then act on the clutch fork to disengage the clutch.

When the clutch pedal is released, a spring on the clutch pedal pulls it back. Other springs inside the two cylinders push the pistons back into their retracted positions. Brake fluid flows back through the line and into the reservoir.

Clutch linkage mechanism

A clutch linkage mechanism uses levers and rods to transfer motion from the clutch pedal to the clutch fork. One configuration is shown in Fig. 24-14. When the pedal is pressed, a push rod pushes on the bellcrank.

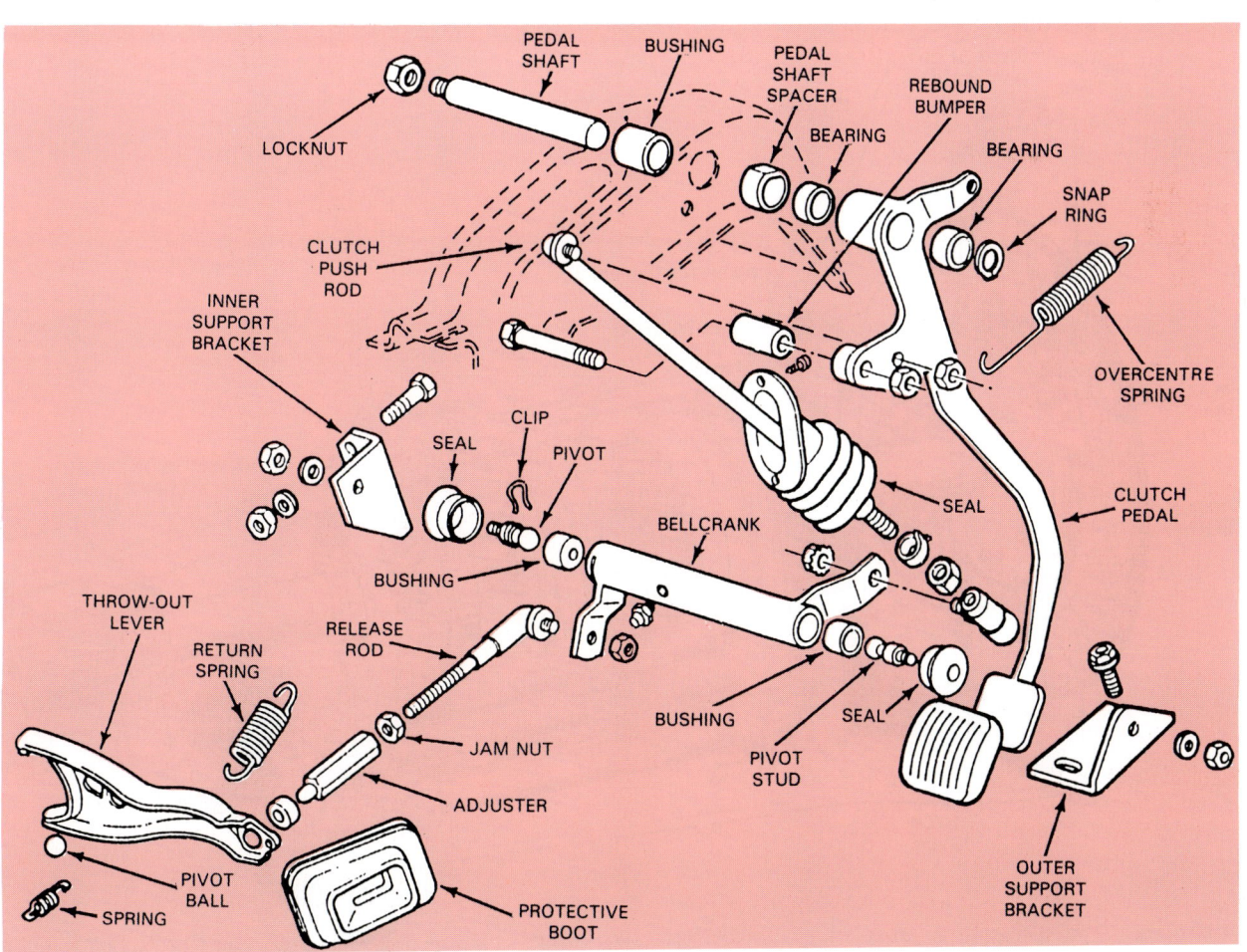

Fig. 24-14. Typical clutch linkage release mechanism. Arms and rods transfer clutch pedal action to clutch fork.

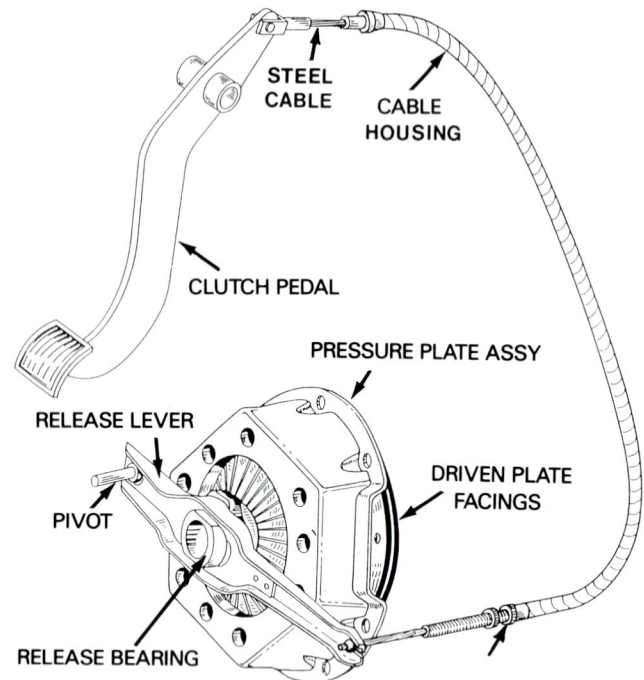

Fig. 24-15. Typical clutch cable mechanism. Steel cable runs through stationary housing. When clutch pedal is pressed, cable slides in housing to operate release lever and throw-out bearing. Also note cable adjuster at bottom end of housing.

The bellcrank reverses the forward movement of the clutch pedal. The other end of the bellcrank is connected to a release rod.

The release rod transfers bellcrank movement to the fork and usually provides a method of adjustment.

Study the clutch linkage parts and their locations in Fig. 24-14 very carefully. This is a typical arrangement.

Clutch cable mechanism

A clutch cable mechanism uses a steel cable inside a flexible housing to transfer pedal movement to the clutch fork. This is a simple mechanism.

Shown in Fig. 24-15, the cable is usually fastened to the upper end of the clutch pedal. The other end of the cable connects to the clutch fork. The cable housing is mounted in a stationary position. This causes the cable to slide inside the housing whenever the clutch pedal is moved.

When the clutch pedal is depressed, the clutch pulls on the clutch fork to disengage the clutch. When the clutch pedal is released, a strong spring pulls back on the pedal, cable, and fork to engage the clutch.

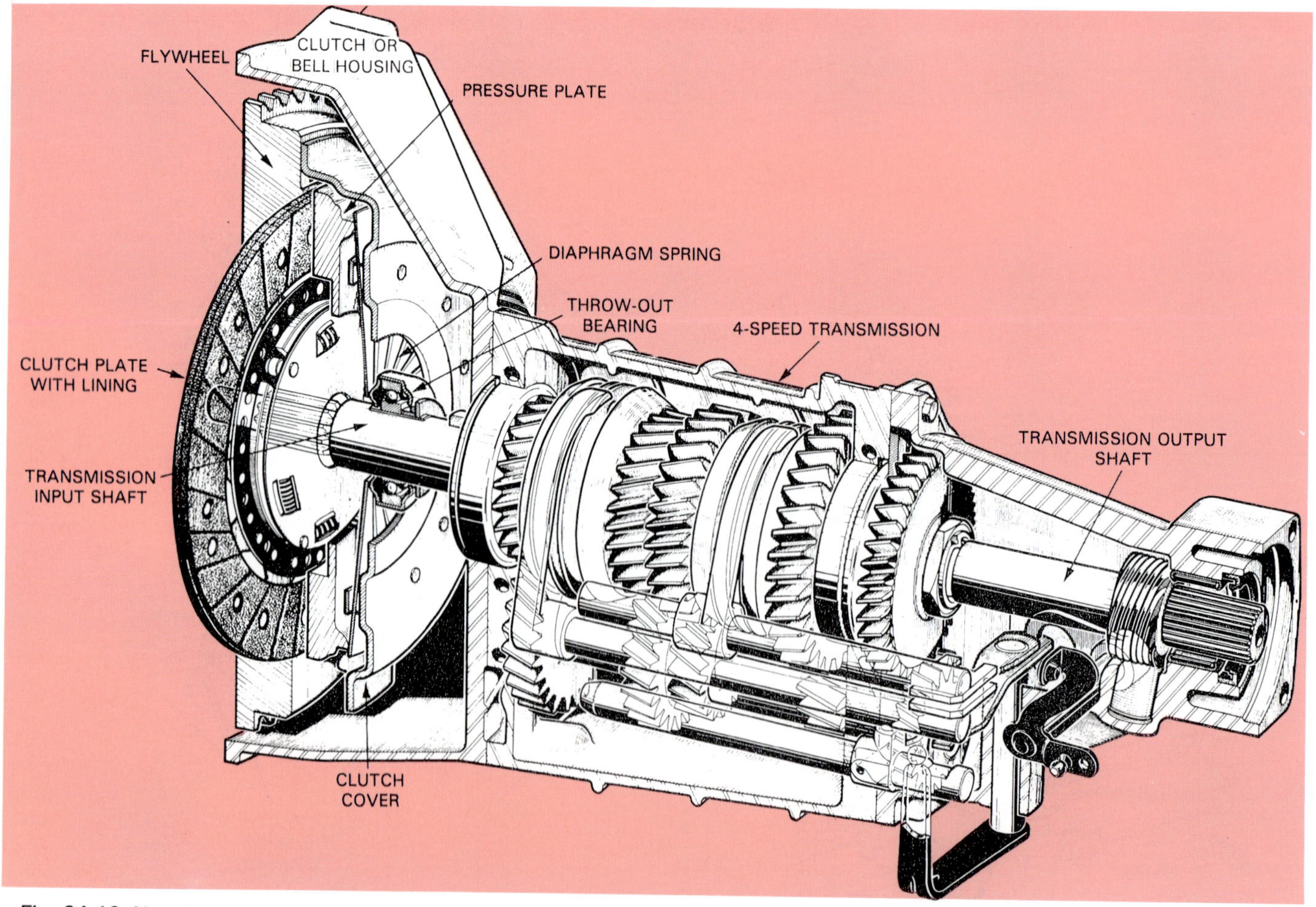

Fig. 24-16. Note how clutch installs in bell housing. Transmission bolts to rear of bell housing. Can you explain function of clutch parts? Transmissions are covered in Chapter 25. (Peugeot)

One end of the clutch cable housing usually has a threaded sleeve for CLUTCH ADJUSTMENT.

REVIEW IF NEEDED

The information you just covered on clutches will be useful when studying many of the following chapters on drive train components.

Look at Figs. 24-16 and 24-17. They show the location of clutches for transmission (rear-wheel drive) and transaxle (front-wheel drive) equipped cars. Locate the clutch components. If you cannot describe the functions of each part, quickly review the chapter.

KNOW THESE TERMS

Automotive clutch, Clutch release mechanism, Clutch fork, Throw-out bearing, Pressure plate, Clutch disc, Pilot bearing, Clutch lining, Torsion springs, Cushioning springs, Pressure plate face, Pressure plate release levers, Pressure plate cover, Semi-centrifugal clutch, Diaphragm spring clutch, Bell housing, Clutch master cylinder, Slave cylinder, Clutch linkage, Clutch cable.

REVIEW QUESTIONS

1. An automotive _____ connects and disconnects the engine and manual transmission or transaxle.

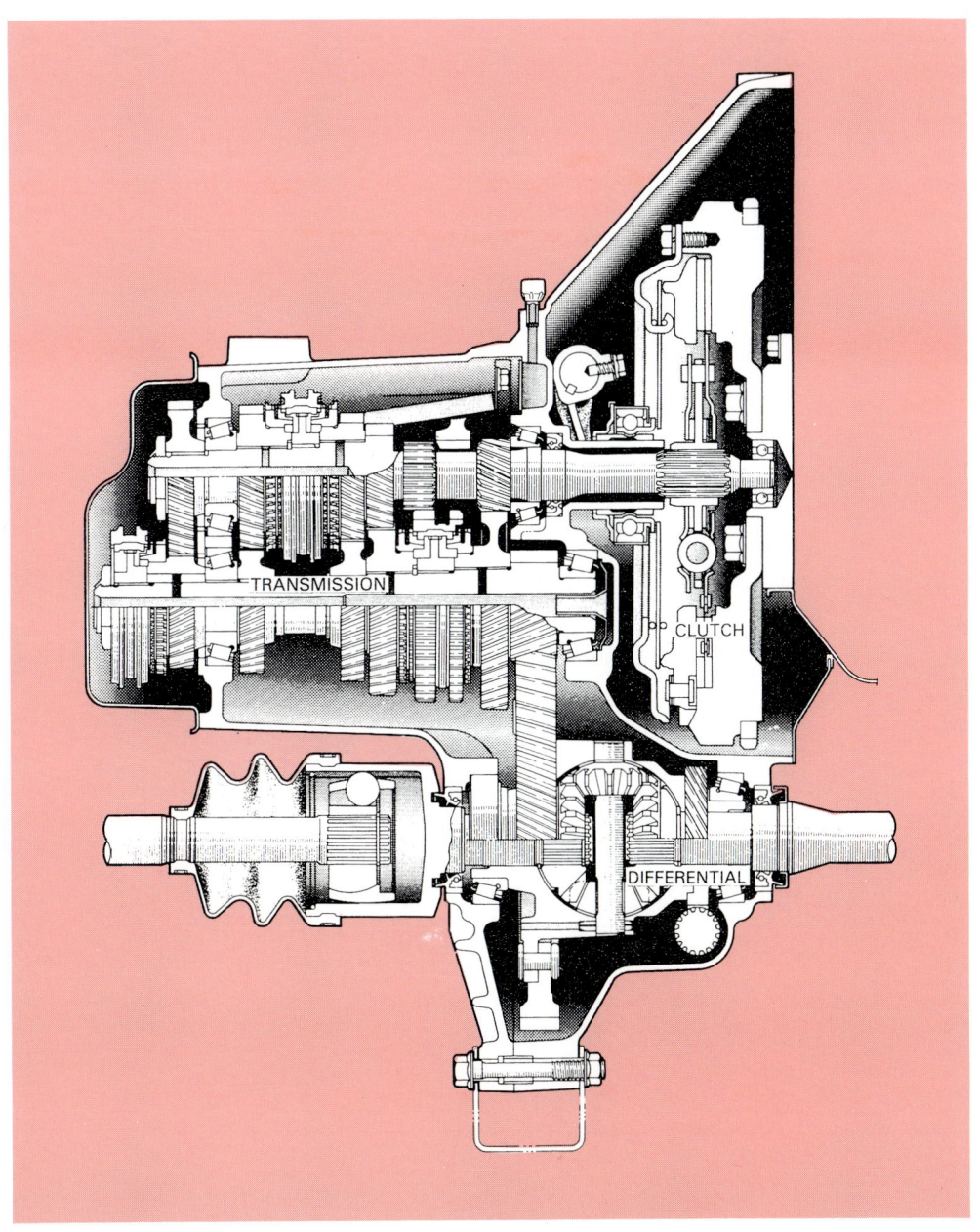

Fig. 24-17. See how clutch is located in relation to manual transaxle for this front-wheel drive vehicle. Transaxles are covered in Chapter 29.

2. List and explain the seven basic parts of an automotive clutch.
3. What is the purpose of the pilot bearing or bushing?
4. The _____ is the mounting place for the clutch.
5. The clutch _____, also called _____, _____, consists of a splined hub, and round metal plate covered with friction material (lining).
6. Clutch disc torsion springs help absorb some of the vibration and shock produced by clutch engagement. True or False?
7. What part of a clutch is commonly made of asbestos?

8. The _____ _____ is a spring-loaded device that can either lock or unlock the clutch disc and flywheel.
9. The two most common types of automotive pressure plates are, _____ _____ and _____.
10. How does the release or throw-out bearing work?
11. What is the clutch or bell housing?
12. Explain clutch fork action.
13. A _____ clutch release mechanism uses a clutch cylinder to operate a slave cylinder.
14. Clutch linkage and clutch cable release mechanisms are both common. True or False?

Chapter 25

MANUAL TRANSMISSION FUNDAMENTALS

After studying this chapter, you will be able to:
☐ Describe gear operating principles.
☐ Identify and define all of the major parts of a transmission.
☐ Explain the fundamental operation of a manual transmission.
☐ Trace the power flow through transmission gears.
☐ Compare the construction of different types of manual transmissions.
☐ Explain the purpose and operation of a transmission overdrive ratio.

A manual transmission must be shifted by hand. It is normally bolted to the clutch housing at the rear of the engine. See Fig. 25-1. The clutch disc rotates the transmission input shaft. Gears inside the transmission transfers engine power to the drive shaft and rear wheels. A column or floor shift lever allows the driver to select which set of transmission gears are engaged.

A manual transmission should not be confused with an automatic transmission or automatic transaxle. A manual transmission is shifted by hand. It is

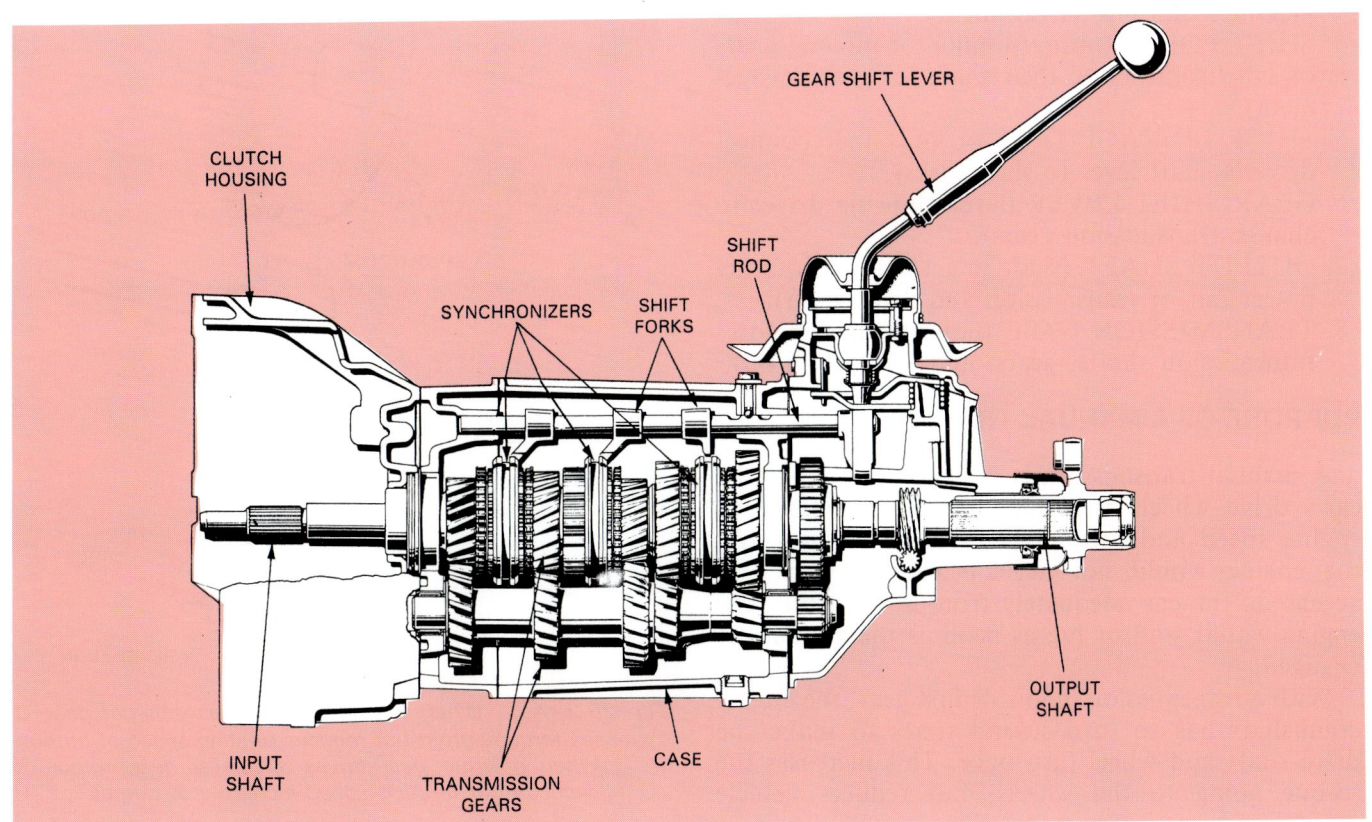

Fig. 25-1. Study basic names and locations of manual transmission parts. This will help you as you learn about each part in more detail. (Fiat)

379

normally used in a front engine, rear-wheel drive vehicle. A foot-operated friction clutch is needed.

An automatic transmission, covered in Chapter 26 uses hydraulic pressure and sensing devices to shift gears. It detects engine speed and load to determine shift points. An automatic transmission also uses a fluid coupling, instead of a dry friction clutch.

A transaxle combines both the transmission and the differential into a single housing. It is commonly used with front-wheel drive cars. A transaxle can contain either a manual or automatic transmission. Transaxles are covered in Chapter 29.

BASIC TRANSMISSION PARTS

To help you understand later sections of the chapter, study the parts of the transmission in Fig. 25-1. Learn to identify and locate the fundamental components. Then, you will be prepared for more specific details of transmission construction and operation.

1. TRANSMISSION INPUT SHAFT (shaft, operated by clutch, that turns gears inside transmission).
2. TRANSMISSION GEARS (provide a means of changing output torque and speed leaving transmission).
3. SYNCHRONIZERS (devices for meshing, or locking gears into engagement).
4. SHIFT FORKS (pronged units for moving gears or synchronizers on their shaft for gear engagement).
5. SHIFT LINKAGE (arms or rods that connect driver's shaft lever to shift forks).
6. GEAR SHIFT LEVER (lever allowing driver to change transmission gears).
7. OUTPUT SHAFT (shaft that transfers rotating power out of transmission and drive shaft).
8. TRANSMISSION CASE (housing that encloses transmission shafts, gears, and lubricating oil).

PURPOSE OF A MANUAL TRANSMISSION

A manual transmission is designed to change the car's drive wheel speed and torque in relation to engine speed and torque. Without a transmission, the engine would not develop enough power to accelerate the car adequately from a standstill. The engine would stall or lug as soon as the clutch was engaged.

With a transmission in low or first gear, the engine crankshaft has to turn several times to make the drive shaft and wheel turn once. This increases the torque going to the wheels, but reduces vehicle speed.

Then, as the transmission is shifted through the gears and into high, the engine and drive shaft begin to turn at approximately the same speed. Wheel and vehicle speed increases, while engine speed drops.

If in proper operating condition, a manual transmission should:
1. Be able to increase torque going to the drive wheels for quick acceleration.
2. Supply different gear ratios to match different engine load conditions.
3. Have a reverse gear for moving the car backwards.
4. Provide the driver with an easy means of shifting transmission gears.
5. Operate quietly, with minimum power loss.

GEAR FUNDAMENTALS

Gears are round wheels with teeth machined on their outer diameter. They are commonly used to transmit turning effort from one shaft to another. Basically, one size gear is used to turn another size gear to change output speed and torque (turning power). This is illustrated in Fig. 25-2.

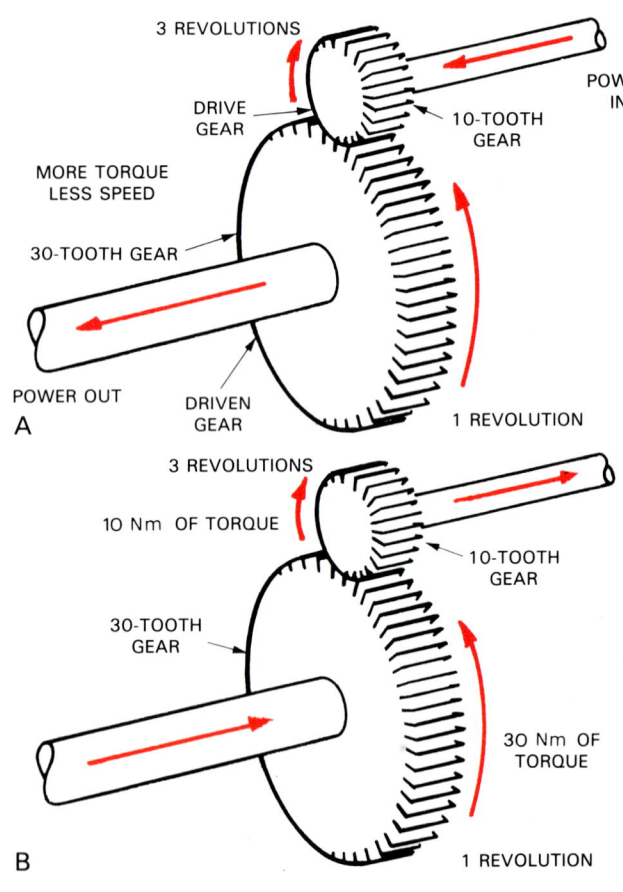

Fig. 25-2. A — When a small gear drives a larger gear, it increases torque output but reduces rotating speed of output. B — When a larger gear drives a smaller gear, torque is reduced but rotating speed increases at output.

Gear ratios

A gear ratio is the number of turns a driving gear must turn before the driven gear turns one complete revolution. Gear ratio is calculated by dividing the number of teeth on the driven gear by the number of teeth on the driving gear.

For example, look at Fig. 25-3. If the drive gear has 12 teeth and the driven gear 24 teeth (24 divided by 12), the gear ratio would be TWO TO ONE, written 2:1.

In this example, the drive gear would have to revolve two times to turn the other gear once. As a result, the speed of the larger, driven gear would be half as fast as the drive gear. However, the torque on the shaft of the larger gear would be twice that of the input shaft.

Various sizes of drive and driven gears can be used to produce any number of gear ratios. As the number of teeth on the driven gear increase in relation to the number of teeth on the drive gear, the gear ratio increases. A gear ratio of 10:1 would be larger than a ratio of 5:1, for example.

An overdrive ratio results when a larger gear drives a small gear. As shown in Fig. 25-2, the speed of the output gear increases, but torque drops.

Gear types

Manual transmissions commonly use two types of gears: spur gears and helical gears.

Spur gears have their teeth cut parallel to the centreline of the gear shaft. As shown in Fig. 25-4A, they are sometimes called straight-cut gears.

Spur gears are somewhat noisy and are no longer used as the main drive gears in a transmission. They may be used for sliding reverse gear, however.

Helical gears have their teeth machined at an angle to the centreline of gear rotation. Modern transmissions commonly use helical gears as the main drive gears. See Fig. 25-4B. Helical gears are quieter and stronger than spur gears.

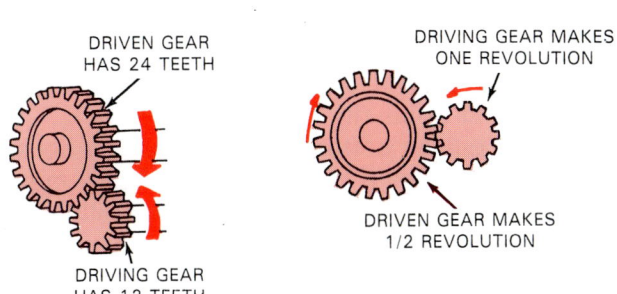

Fig. 25-3. Gear ratio is determined by number of teeth on drive and driven gears. If drive gear has half as many teeth as driven gear, a two to one ratio would be produced.

Transmission gear ratios

Transmission gear ratios vary with the manufacturer. However, approximate gear ratios average 3:1 for first gear, 2:1 for second gear, 1:1 for third or high gear, and 3:1 for reverse gear.

In first or low gear, there would be a high gear ratio. A small gear would drive a large gear. This would reduce output speed but increase output torque. The car would accelerate easily, even with low engine rpm and low power conditions.

In high gear, the transmission frequently has a 1:1 ratio. The transmission output shaft would spin at the same speed as the engine crankshaft. There would be NO torque multiplication (increase), but the car would travel faster. Very little torque is needed to propel a car at a constant speed on level ground.

Gear reduction and overdrive

Gear reduction occurs when a small gear drives a large gear to increase turning force. Gear reduction is used in the lower transmission gears. Fig. 25-2.

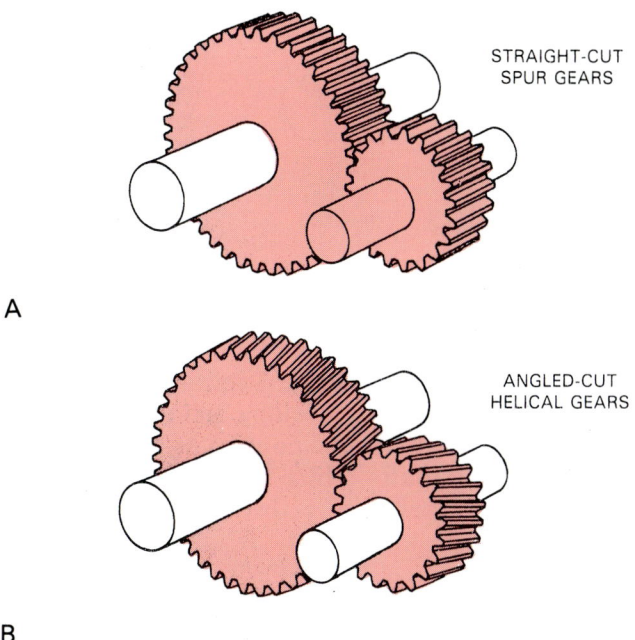

Fig. 25-4. Two basic types of gears used in manual transmission are straight-cut spur gears and angled-cut helical gears.

Gear backlash

Gear backlash is the small clearance between the meshing gear teeth. Clearance allows lubricating oil to enter the high friction area between the gear teeth. This reduces friction and wear. Backlash also allows the gears to heat up and expand during operation without binding or being damaged.

MANUAL TRANSMISSION LUBRICATION

The bearings, shafts, gears, and other moving parts in a transmission are lubricated by oil throw-off or splash. As the gears rotate, they sling oil around inside the transmission case.

Typically, 80 or 90W gear oil is recommended for use in a manual transmission. However, follow manufacturer's recommendations.

TRANSMISSION BEARINGS

Manual transmissions normally use three basic types of bearings: ball bearings, roller bearings, and needle bearings. These three types are shown in Fig. 25-5. Bearings are used to reduce the friction between the surfaces of rotating parts in the transmission.

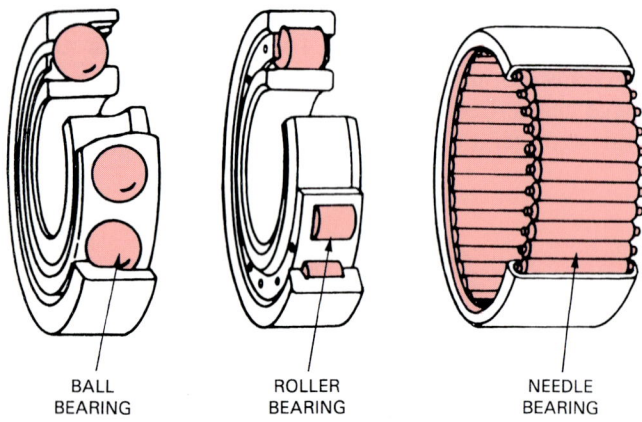

BALL BEARING   ROLLER BEARING   NEEDLE BEARING

Fig. 25-5. Three types of antifriction bearings found in transmissions: ball, roller, needle.

The bearings are lubricated by oil spray off of the spinning transmission gears. Typically, antifriction bearings (bearing using a rolling action) fit between the transmission shafts and housing or between some of the gears and shafts. These are high friction points that must be capable of withstanding the engine's power.

MANUAL TRANSMISSION CONSTRUCTION

Now that you have a general grasp of gear and transmission principles, we will assemble each part of a working transmission. We will start out with the case, then install the shafts, gears, bearings, and other parts.

Transmission case

The transmission case must support the transmission bearings and shafts and provide an enclosure for gear oil. Refer to Fig. 25-6. A manual transmission case is usually cast of either iron or aluminium. Aluminium is becoming more common because of its lightness.

A drain plug and a fill plug are usually provided in the transmission case. The drain plug is on the bottom of the case. The fill plug is on the side of the case.

The fill plug also serves as a means of checking the oil level in the transmission. Typically, the oil should be level with the fill plug when the transmission is at operating temperature.

Extension housing and front bearing hub

The extension housing, also termed the tailshaft or propeller shaft housing, bolts to the rear of the transmission case. It encloses the transmission output shaft and holds the rear oil seal. See Fig. 25-6.

A flange on the bottom of the extension housing provides a base for the rubber transmission mount, also called rear engine mount. A gasket usually seals the mating surfaces between the transmission case and extension housing.

A front bearing hub, sometimes called front bearing cap or retainer, covers the front transmission bearing and acts as a sleeve for the clutch throw-out bearing. It bolts to the transmission case. A gasket fits between the front hub and case to prevent oil leakage.

Transmission shafts

Basically, a manual transmission has four steel shafts mounted inside its case. It normally has an input shaft, countershaft, reverse idle shaft, and an output shaft. Figs. 25-7 and 25-8 show the general location and shape of these shafts.

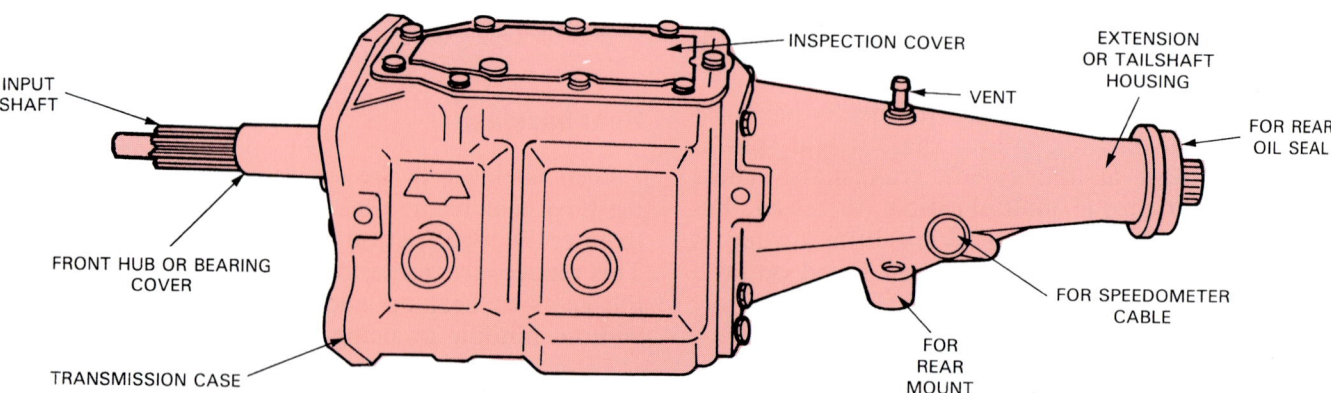

INSPECTION COVER

EXTENSION OR TAILSHAFT HOUSING

INPUT SHAFT

VENT

FOR REAR OIL SEAL

FRONT HUB OR BEARING COVER

TRANSMISSION CASE

FOR SPEEDOMETER CABLE

FOR REAR MOUNT

Fig. 25-6. Case is centre section on transmission. Extension housing bolts to rear of case. Front bearing hub bolts to front of case. It encloses front input shaft bearing and supports clutch throw-out bearing. This transmission also has a sheet metal inspection cover bolted to top of case.

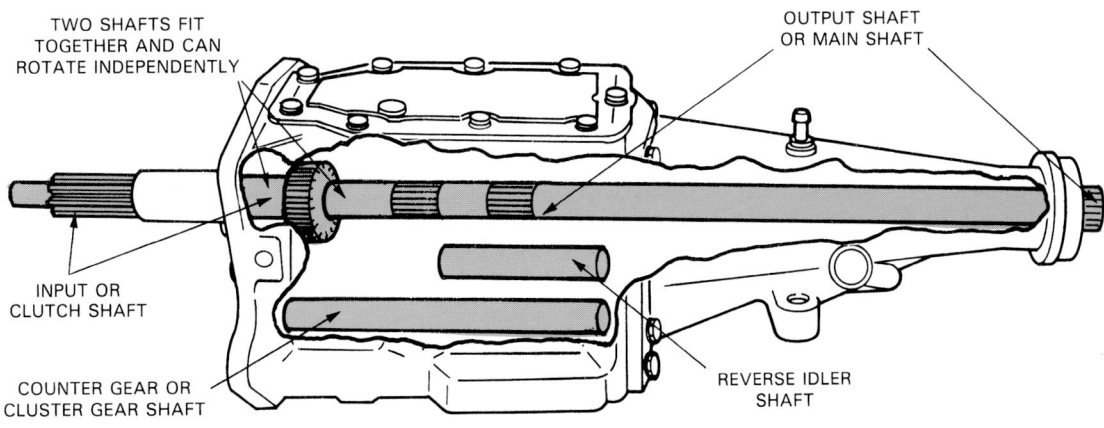

Fig. 25-7. Note how transmission shafts are located in transmission case. Input shaft is driven by clutch. Output shaft is on same centreline as input shaft. Countershaft and reverse idler shafts mount below and to one side in case.

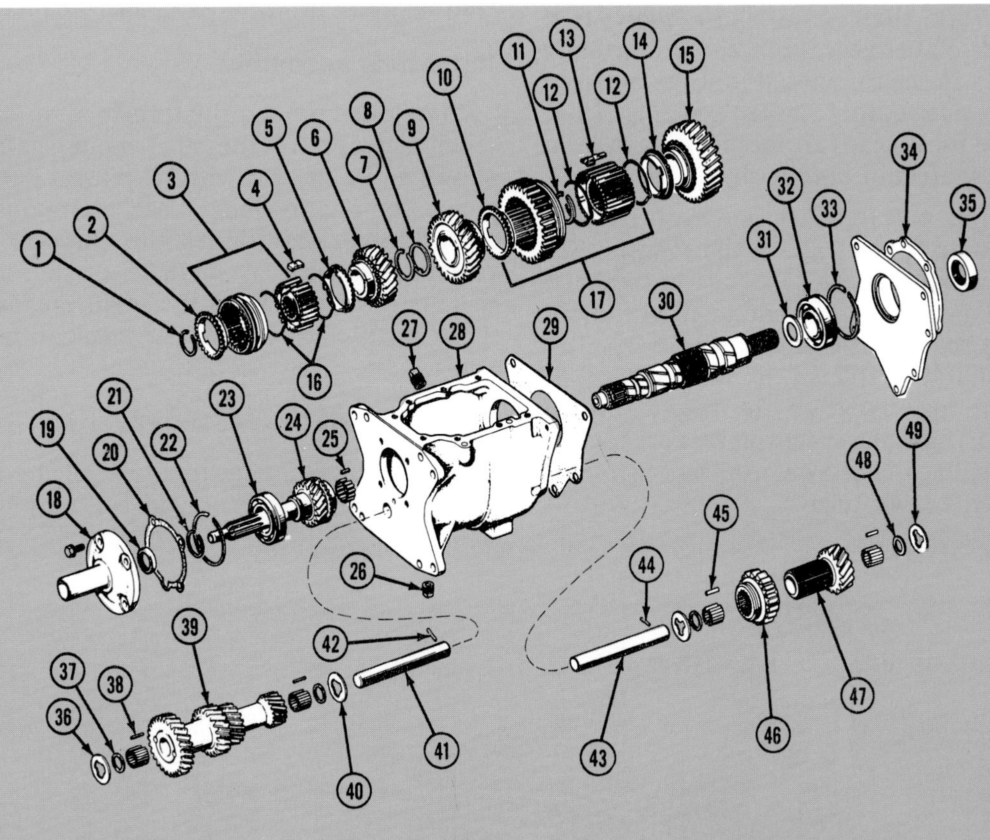

Fig. 25-8. Exploded view shows major parts of a typical transmission. Note four shafts and how components fit on shafts.

1. 3rd-4th GEAR SNAP RING
2. 4th GEAR SYNCHRONIZER RING
3. 3rd-4th GEAR CLUTCH ASSEMBLY
4. 3rd-4th GEAR PLATE
5. 3rd GEAR SYNCHRONIZER RING
6. 3rd SPEED GEAR
7. 2nd GEAR SNAP RING
8. 2nd GEAR THRUST WASHER
9. 2nd SPEED GEAR
10. 2nd GEAR SYNCHRONIZER RING
11. MAIN SHAFT SNAP RING
12. 1st-2nd SYNCHRONIZER SPRING
13. LOW-2nd PLATE
14. 1st GEAR SYNCHRONIZER RING
15. 1st GEAR
16. 3rd-4th SYNCHRONIZER SPRING
17. 1st-2nd GEAR CLUTCH ASSEMBLY

18. FRONT BEARING CAP
19. OIL SEAL
20. GASKET
21. SNAP RING
22. LOCK RING
23. FRONT BALL BEARING
24. CLUTCH SHAFT
25. ROLLER BEARING
26. DRAIN PLUG
27. FILL PLUG
28. CASE
29. GASKET
30. SPLINE SHAFT
31. 1st GEAR THRUST WASHER
32. REAR BALL BEARING
33. SNAP RING
34. ADAPTER PLATE

35. ADAPTER SEAL
36. FRONT COUNTERSHAFT GEAR THRUST WASHER
37. ROLLER WASHER
38. REAR ROLLER BEARING
39. COUNTERSHAFT GEAR
40. REAR COUNTERSHAFT THRUST WASHER
41. COUNTERSHAFT
42. PIN
43. IDLER GEAR SHAFT
44. PIN
45. IDLER GEAR ROLLER BEARING
46. REVERSE IDLER SLIDING GEAR
47. REVERSE IDLER GEAR
48. IDLER GEAR WASHER
49. IDLER GEAR THRUST WASHER

The input shaft, often termed clutch shaft, transfers rotation from the disc to the countershaft gears in the transmission. The outer end of the shaft is splined. The inner end of the shaft has a gear machined on it. See Fig. 25-8.

A bearing in the transmission case supports the input shaft in the case. Anytime the clutch disc turns, the input shaft gear and gears on the countershaft turn.

The countershaft, also called cluster gear shaft, holds the countershaft gear into mesh with the input gear and other gears in the transmission. It is located slightly below and to one side of the clutch shaft, Fig. 25-9.

Normally, the counter shaft does NOT turn in the transmission case. It is locked in the case by either a steel pin, force fit, or locknuts. Refer to Fig. 25-8.

The reverse idler shaft is a short shaft that supports the reverse idler gear, Fig. 25-8. It normally mounts stationary in the case about midway between the countershaft and output shaft. See Fig. 25-7. Then, the reverse idler gear can mesh with gears on both the countershaft and output shaft.

The transmission output shaft, also called main shaft, holds the output gears and synchronizers. See Fig. 25-8. The rear of this shaft extends to the back of the extension housing. It connects to the drive shaft to turn the rear wheels of the car.

The output shaft is splined in the centre. In modern transmissions, the gears are free to revolve on the output shaft, but the synchronizers are locked on the shaft by splines. The synchronizers will only turn when the shaft itself turns.

TRANSMISSION GEARS

Transmission gears can be typically classified into four groups: input shaft gear, countershaft gears, output shaft gears, and reverse idle gear. Illustrated in Fig. 25-9, the input shaft gear turns the countershaft gears. The countershaft gears turn the output shaft gears and reverse idle gear.

In low gear, a small gear on the countershaft drives a larger gear on the output shaft, Fig. 25-10A. This provides a high gear ratio for accelerating. Then, in high gear, a larger countershaft gear drives an equal or smaller size output shaft gear. Fig. 25-10B. This reduces the gear ratio and the car moves faster.

When in reverse, power flows from the countershaft gear, to the reverse idler gear, and to the engaged gear on the output shaft. This reverses output shaft rotation as shown in Fig. 25-10C.

Input gear assembly

Mentioned briefly, the transmission input gear is a machined part of the steel input shaft. Fig. 25-11 shows an input gear with its related parts. Study the shape and relationship of each component carefully.

The input gear drives the forward gear on the countershaft gear. A small set of spur gear teeth are usually located next to the main, helical drive gear. This small gear is for engagement of the synchronizer.

Countershaft gear assembly

The countershaft gear, also called countergear or laygear, turns the gears on the output shaft. This

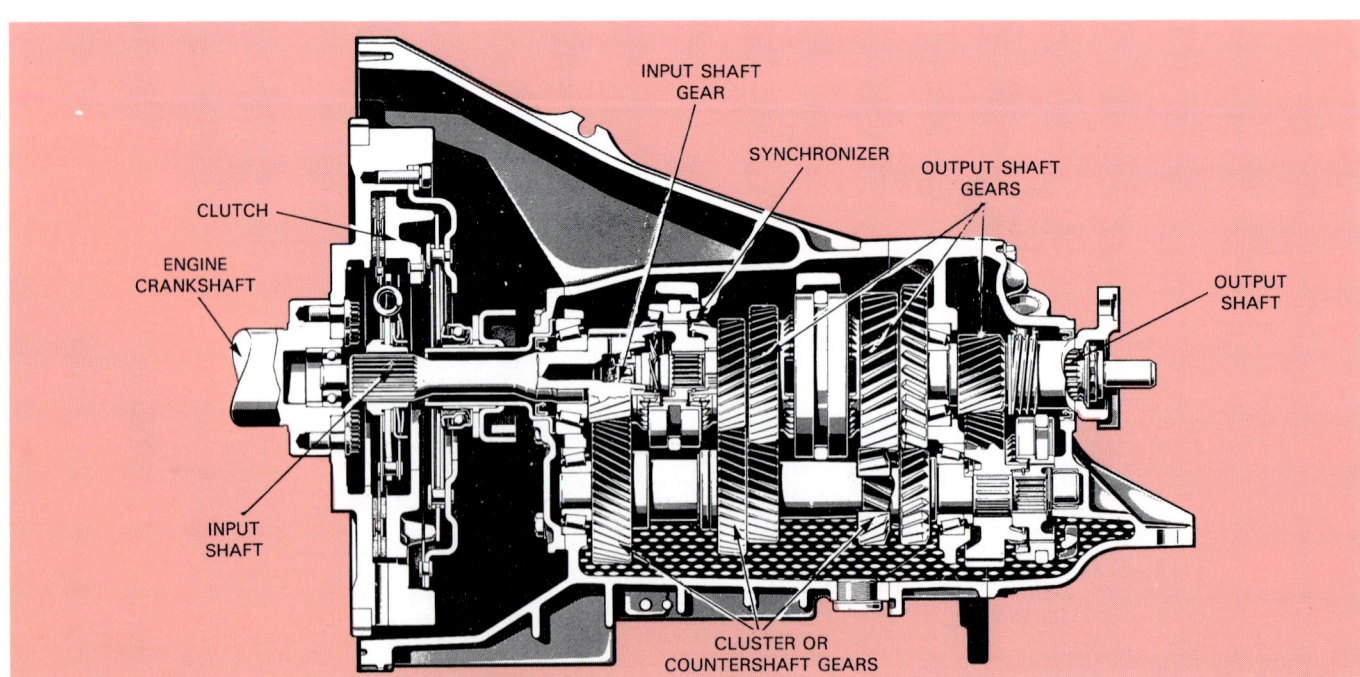

Fig. 25-9. Cutaway of modern transmission shows gears assembled on their shafts. Gear on clutch or input shaft drives countershaft gears. Countershaft gears turn output gears on main or output shaft. (Mercedes-Benz)

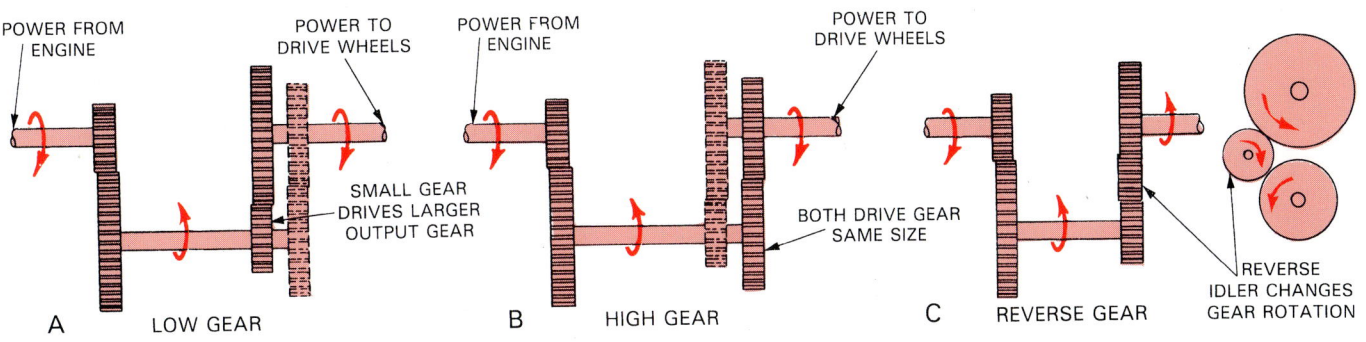

Fig. 25-10. Simplified transmission action. A — Low gear, input shaft gear turns gears on countershaft. Small countershaft gear drives larger output shaft gear to produce gear reduction. B — High gear. Engaged gears are same size. Output shaft turns faster than when in low gear. Less torque increase is needed. C — Reverse. Reverse idler gear is used between countershaft gear and output shaft gear. This reverses direction of rotation at output shaft.

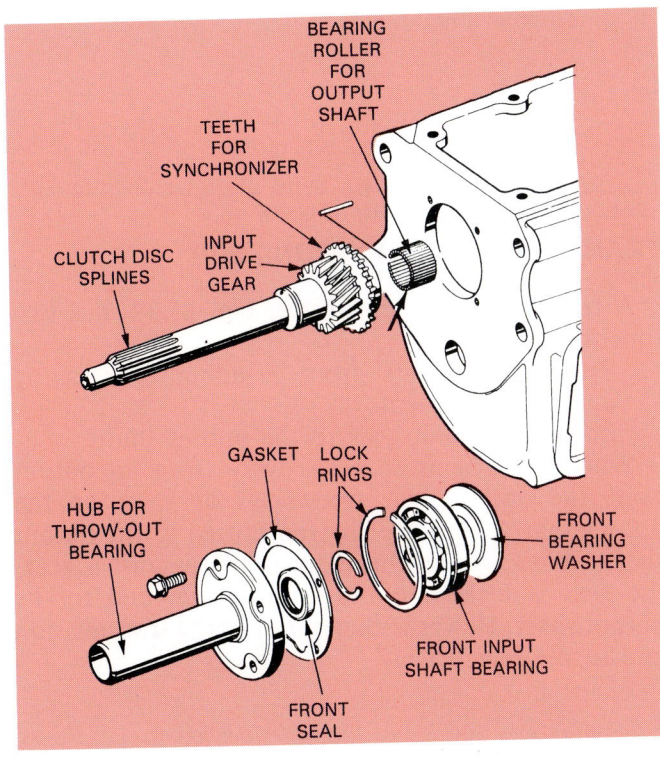

Fig. 25-11. Exploded view of input shaft and gear. Gear is normally machined part of shaft. Large bearing supports shaft in front of transmission case. Individual roller bearings support rear of shaft. Snap rings secure assembly in case.

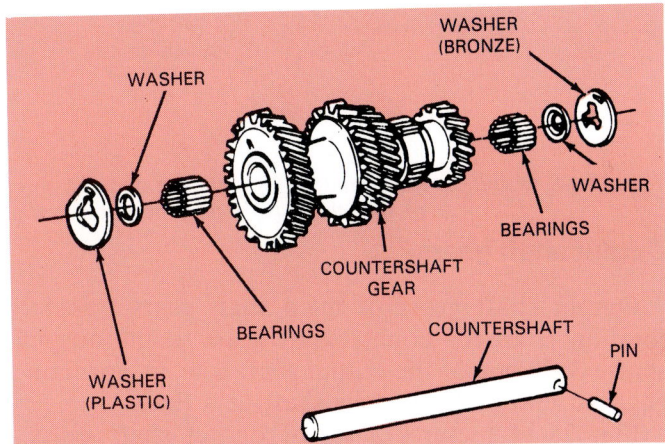

Fig. 25-12. Countershaft assembly. Countershaft gear has several gears formed as a single unit. They mount on roller bearings and countershaft. Washers control end play of unit in case.

Reverse idler gear assembly

A reverse idler gear assembly is shown in Fig. 25-13. Note how it is constructed like the other transmission shaft-gear assemblies just discussed.

gear is actually several gears machined out of a single piece of steel. Hence, it is often called the cluster gear, Fig. 25-12.

When the input gear drives the matching countershaft gear, all of the countershaft gears turn as a single unit. However, since each forward gear is a different size, the countershaft gear unit is capable of providing several gear ratios.

Note in Fig. 25-12 how the countergear rides on roller bearings. Thrust washers fit on each end of the gear to set end play or case-to-gear clearance.

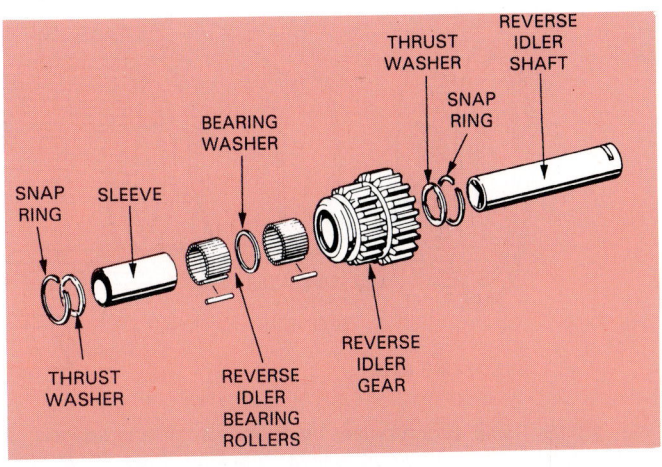

Fig. 25-13. Reverse idler gear and shaft assembly.

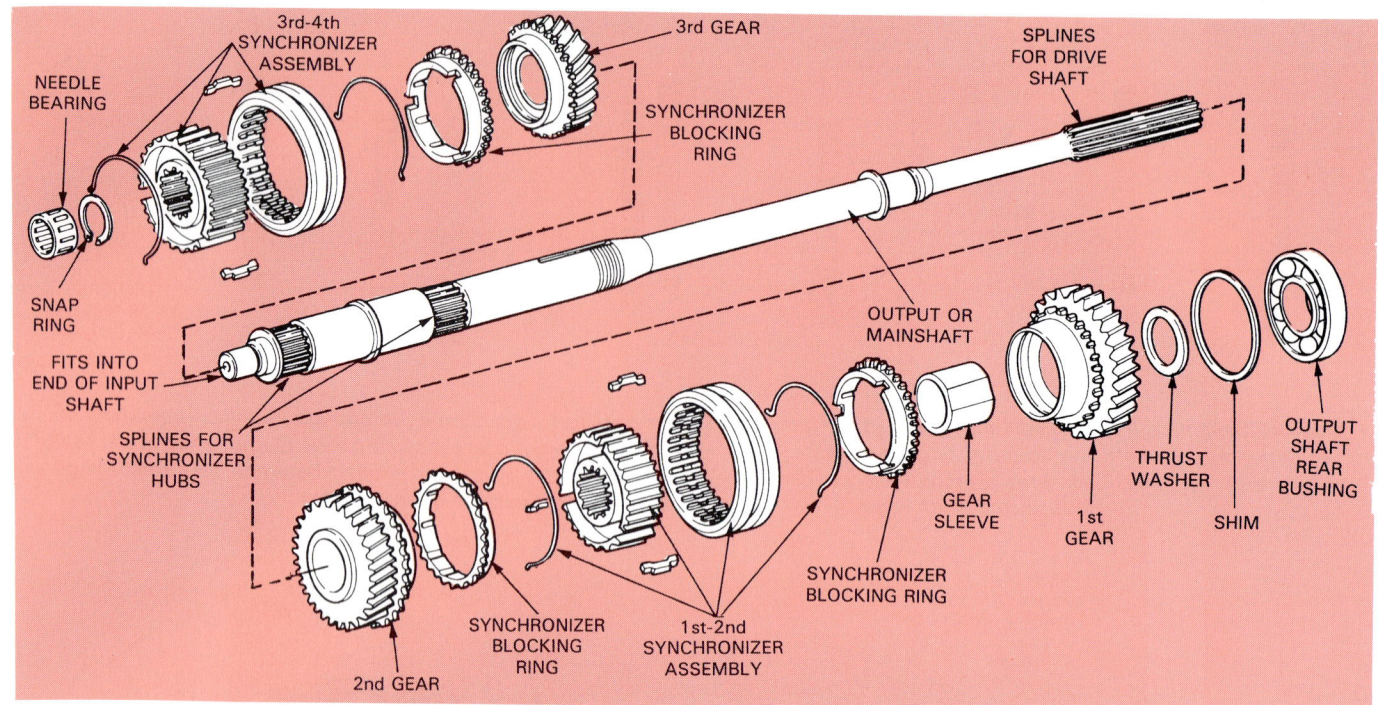

Fig. 25-14. Output or main shaft is long shaft extending through transmission tailshaft housing. Drive shaft is splined to rear of this shaft. Note how output gears and synchronizers install on shaft. (Mazda)

Output shaft gears

Output shaft gears or main shaft gears transfer rotation from the countershaft gears to the output shaft. Only one of the output shaft gears is normally engaged and locked to the shaft at a time.

Fig. 25-14 pictures a set of output shaft gears. Notice how it has a main drive gear (helical gear) and a smaller synchronizer gear (spur gear).

The inside bore of each output shaft gear is smooth so that it can spin freely on its shaft when not engaged. Normally, one output shaft gear will be provided for each transmission speed, including reverse.

TRANSMISSION SYNCHRONIZERS

A transmission synchronizer, Fig. 25-14, has two functions. It must:

1. Prevent the gears from grinding or clashing during engagement.
2. Lock the output gear to the output shaft.

When the synchronizer is away from an output gear, the output gear freewheels or spins on the output shaft. No power is transmitted to the output shaft. When the synchronizer is slid against a gear, the gear is locked to the synchronizer and to the output shaft. Power is then sent out the transmission and to the rear wheels.

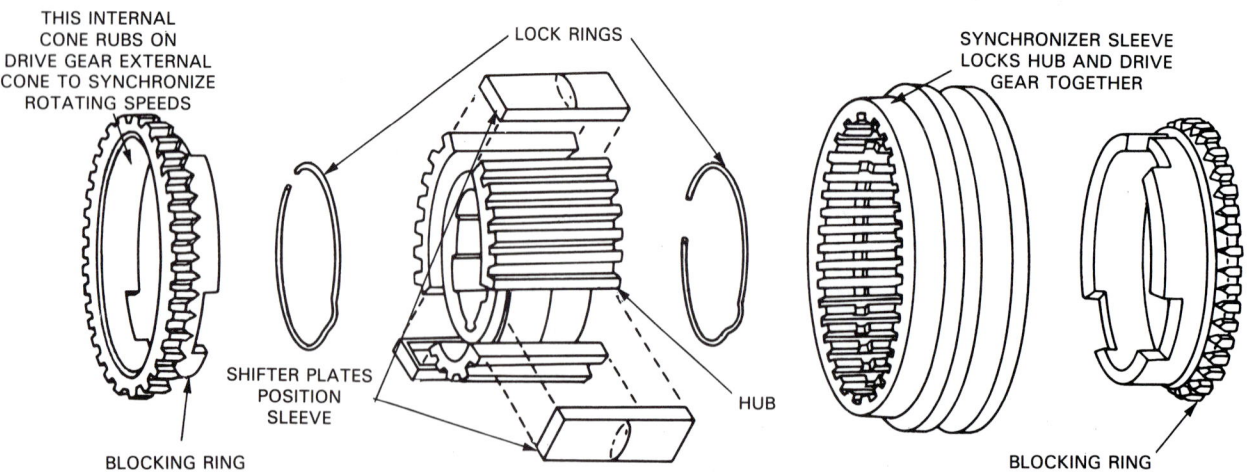

Fig. 25-15. Basic synchronizer components. Hub is splined to output shaft. It will slide but not turn on shaft. Sleeve fits over hub. Shifter plates position sleeve. Blocking rings allow sleeve to slide into and mesh with output gear without clashing or grinding.

Synchronizer construction

The most popular type synchronizer consists of an inner splined hub, shift plates, shift plate springs, an outer sleeve, and blocking rings. See Fig. 25-15.

The synchronizer hub is splined on the output shaft. It is held in a stationary position between the transmission gears. Shift plates fit between the hub and sleeve. The springs push the shift plates into the sleeve. This helps hold and centre the sleeve on its hub. The blocking rings fit on the outer ends of the hub and sleeve.

Synchronizer operation

When the driver shifts gears, the synchronizer's sleeve slides on its splined hub toward the main drive gear.

First, the blocking ring cone rubs on the side of the gear cone, setting up friction between the two, Fig. 25-16. This causes the gear, synchronizer, and output shaft to begin to spin at the SAME SPEED.

As soon as the speed is equalized or synchronized, the sleeve can slide completely over the blocking ring and over the small, spur gear teeth on the drive gear. This locks the output gear to the synchronizer hub and to the shaft. Power then flows through that gear and to the rear wheels of the car.

FULLY SYNCHRONIZED TRANSMISSION

Fully synchronized means that all of the forward output gears use a synchronizer. This allows the driver to downshift into any lower gear (except reverse) with the car moving. Most modern manual transmissions are fully synchronized.

Many older three-speed transmissions did NOT have first gear synchronized. The driver had to wait until the car came to a complete stop before downshifting into first. Trying to shift into first with the car in motion would cause first gear to grind and clash.

SHIFT FORKS

The shift forks fit around the synchronizer sleeves to transfer movement to the sleeves from the gear shift linkage. This is illustrated in Fig. 25-17.

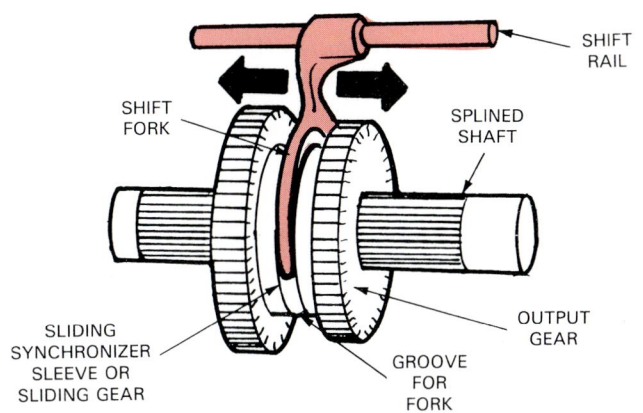

Fig. 25-17. Shift fork is used to move synchronizer sleeve or sliding gear on splined shaft. Shift linkage or rail operates shift fork.

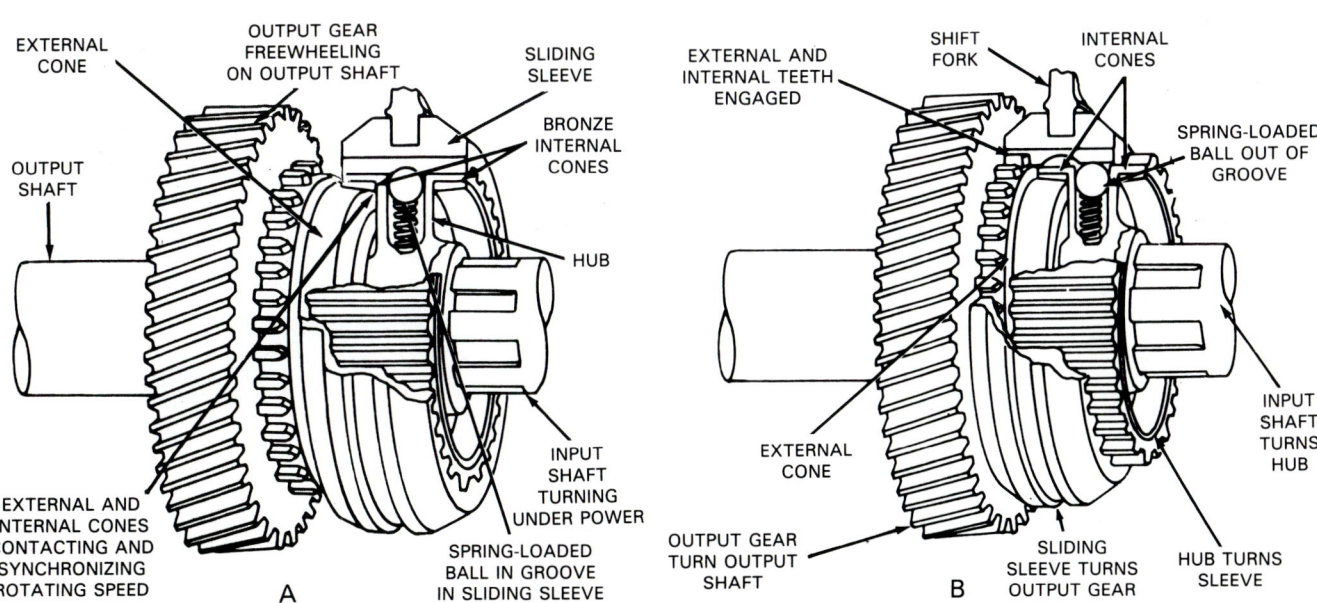

Fig. 25-16. Synchronizer operation. A — As synchronizer sleeve moves into output gear, cone on sleeve rubs against cone on gear. Friction makes gear and sleeve begin to turn at same speed. B — When at same speed, sleeve can slide over and mesh with small, spur teeth on side of output gear. This locks output gear, sleeve, hub, and input shaft together. Output gear is then engaged to output shaft.

The shift fork fits into a groove cut into the synchronizer sleeve. A linkage rod or shift rail connects the fork to the driver's shift lever. When the lever moves, the linkage or rail moves the shift fork and sleeve to engage the correct transmission gear.

Fig. 25-18 pictures a typical, complete shift fork assembly. Study the parts and how they fit together.

TRANSMISSION SHIFT LINKAGE AND SHIFT LEVER

There are two general types of transmission linkage: EXTERNAL ROD type and INTERNAL SHIFT RAIL type. Both perform the same function. They connect the shift lever with the shift fork mechanism.

Look at Fig. 25-19. It shows the components of an external shift rod type linkage. The rods fit into levers on the shift mechanism and fork assembly. Spring clips hold the rods in the levers. One end of each linkage rod is threaded so that the linkage can be adjusted.

An internal rail type is shown in Fig. 25-20.

When the driver shifts gears, the bottom of the shift lever catches in one of the gates (notched unit

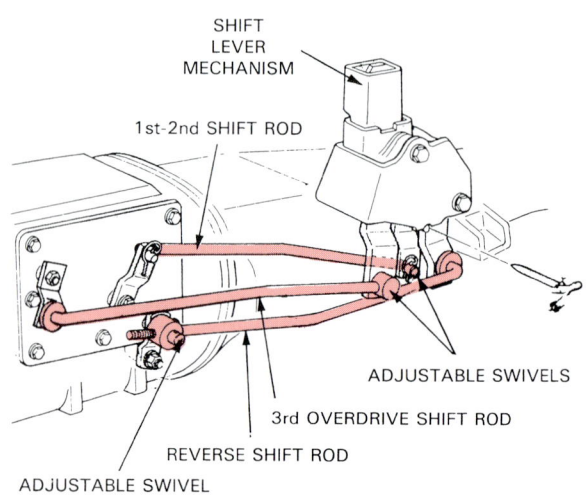

Fig. 25-19. Side view of transmission showing shift linkage and lower part of shift lever. Study parts.

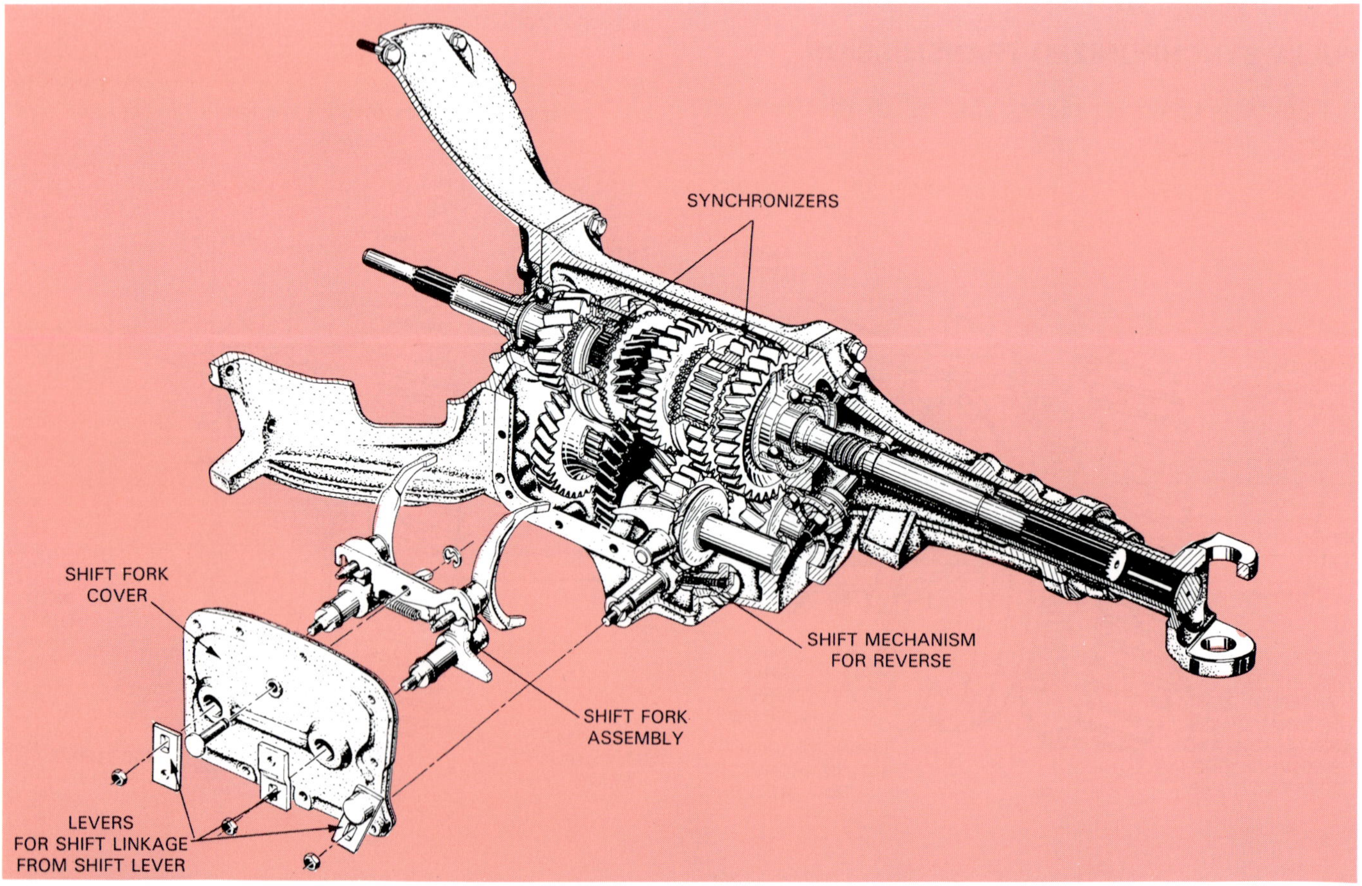

attached to shift rail), Fig. 25-21. Each gate is mounted on a shift rail. As a result, movement of the lever places a prying action on the rail. Since the fork is located on the rail, it is also moved to change transmission gears. Spring-loaded balls are some-

Fig. 25-18. Shift fork fits over groove machined in centre of synchronizer sleeve. Movement of shift linkage to transmission moves fork forward or rearward in transmission to engage different output gear.

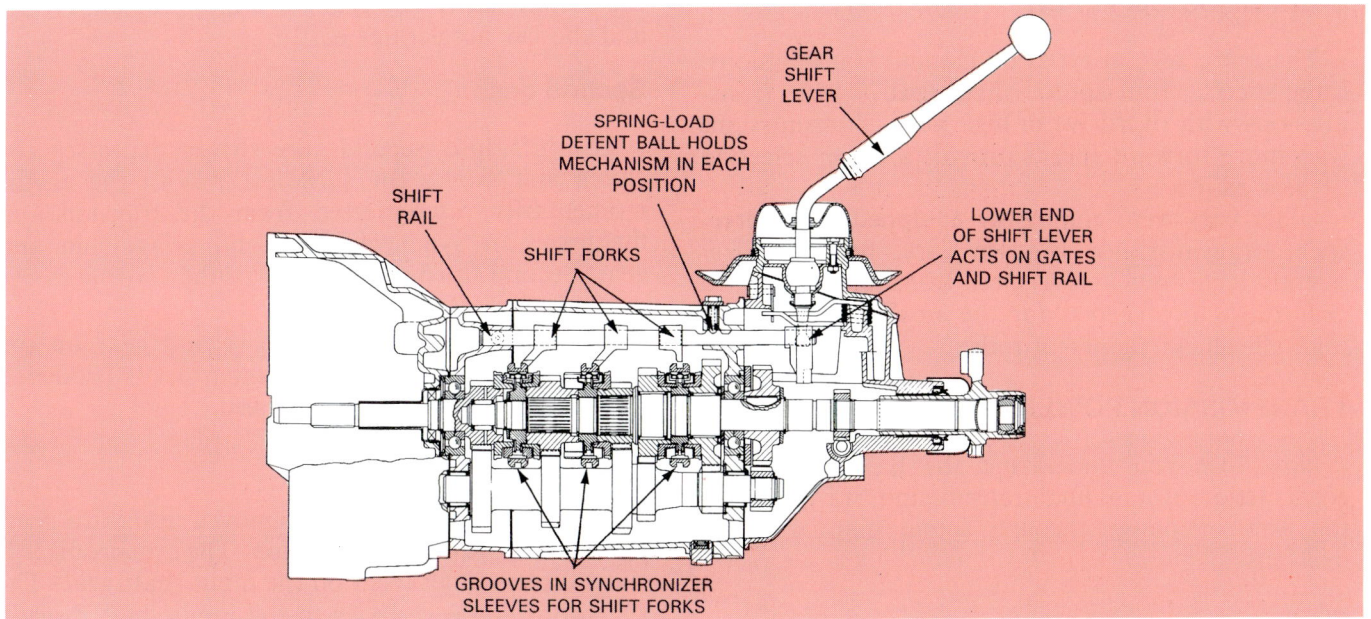

Fig. 25-20. This transmission uses an internal shift rail mechanism instead of external shift rods. Shift lever acts on rail. Rail then operates shift forks and synchronizer sleeves. (Fiat)

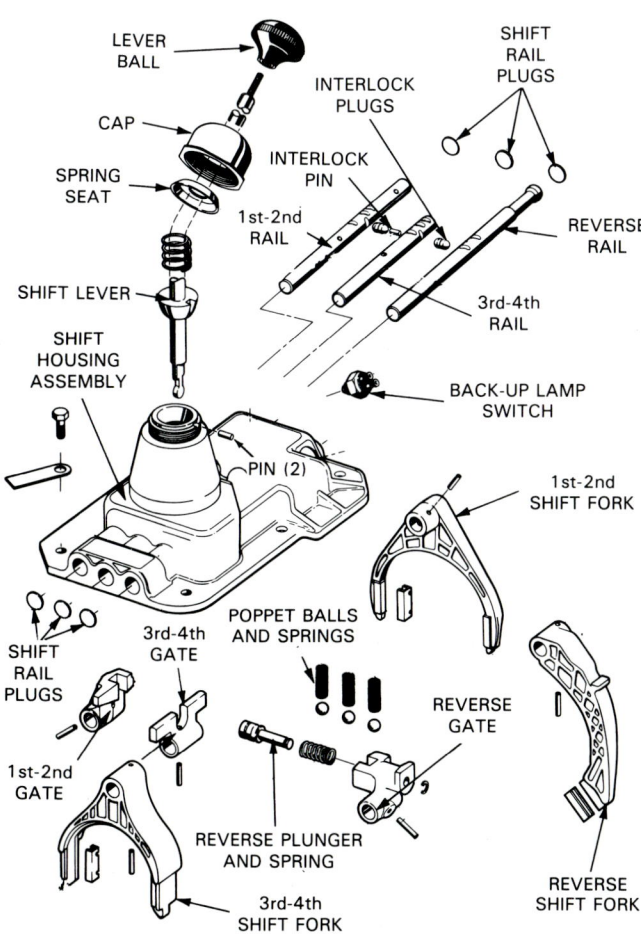

Fig. 25-21. Exploded view of shift mechanism for late model transmission. Shift lever acts on shift gates. Shift gates are attached to shift rails. Rails move forks for gear changes. Spring-loaded balls and plunger help position shift forks during each change. Also note back-up lamp switch.

times used to lock the shift rail(s) into position when in neutral or in gear.

Variations in shift rail linkages are also available. However, their basic construction and operation is almost the same.

The transmission shift lever assembly can be moved to cause movement of the shift linkage or rail, shift forks, and synchronizers.

The parts of a floor mounted shift lever assembly are shown in Figs. 25-19, 20 and 21. One has external shift rods and the others internal rails. Fig. 25-22 pictures a steering column mounted shift lever. Study the parts and how they function.

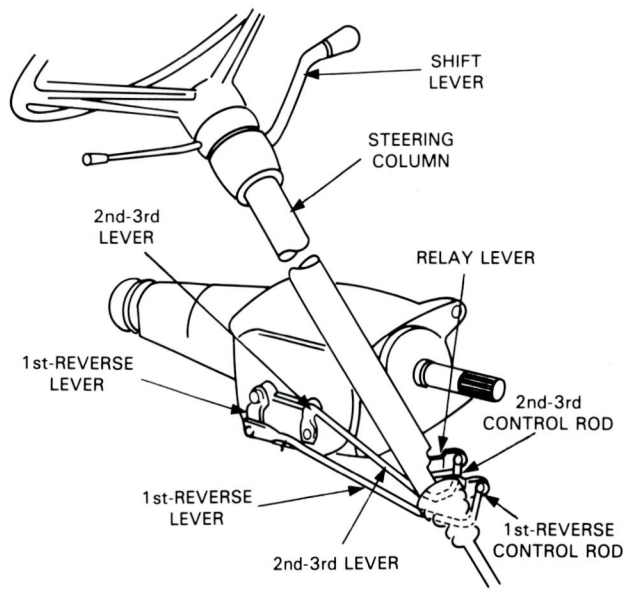

Fig. 25-22. Typical column type shift mechanism. Gear shift lever operates levers on bottom of column. Rods then transfer movement to transmission.

TRANSMISSION TYPES

There are several types of manual transmissions: three-speed, four-speed, five-speed, and transmissions with overdrive in high gear. Transmissions with more forward speeds provide a better selection of gear ratios.

Older cars were commonly equipped with three-speed transmissions. Modern cars, however, frequently have four or five-speed transmissions. Extra gear ratios are needed for the smaller, low power, high efficiency engines of today.

TRANSMISSION POWER FLOW

Now that you understand the basic parts and construction of a manual transmission, we will cover the flow of power through actual manual transmissions.

First gear

To get the car moving from a standstill, the driver moves the gear shift lever into first. The clutch pedal must be pressed to stop power flow into the transmission. The linkage rods move the shift forks so that first gear synchronizer is engaged to first output gear. The other output gears are in neutral. Look at Fig. 25-23.

As the driver releases the clutch pedal, the clutch shaft gear begins to spin the countershaft gears. Since only first gear is locked to the output shaft, a small gear on the countershaft drives a larger gear on the output shaft. The gear ratio is approximately 3:1 and the car accelerates easily.

Second gear

To shift into second, the driver depresses the clutch and moves the shift lever. With the engine momentarily disconnected from the transmission, the first gear synchronizer is slid away from first gear. Second-third synchronizer is then engaged. See Fig. 25-24.

Now power flow is through second gear on the output shaft. A gear ratio of about 2:1 is produced to give the car a little more speed.

Third gear

When the gear shift lever is moved into third gear, there is no torque multiplication. The synchronizer is slid over the small teeth on the input shaft gear. The synchronizer sleeve locks in the input shaft directly to the output shaft. Refer to Fig. 25-25.

A 1:1 gear ratio results with no torque increase. All of the output shaft gears freewheel or spin on their shaft. Power flow is straight through the transmission. The car travels at highway speeds while the engine rpm is relatively low.

Reverse

When shifted into reverse, a synchronizer is moved into the reverse gear on the output shaft. This locks the gear to the output shaft. Power flows through the countershaft, reverse idler gear, reverse gear, and to the drive shaft, as in Fig. 25-26.

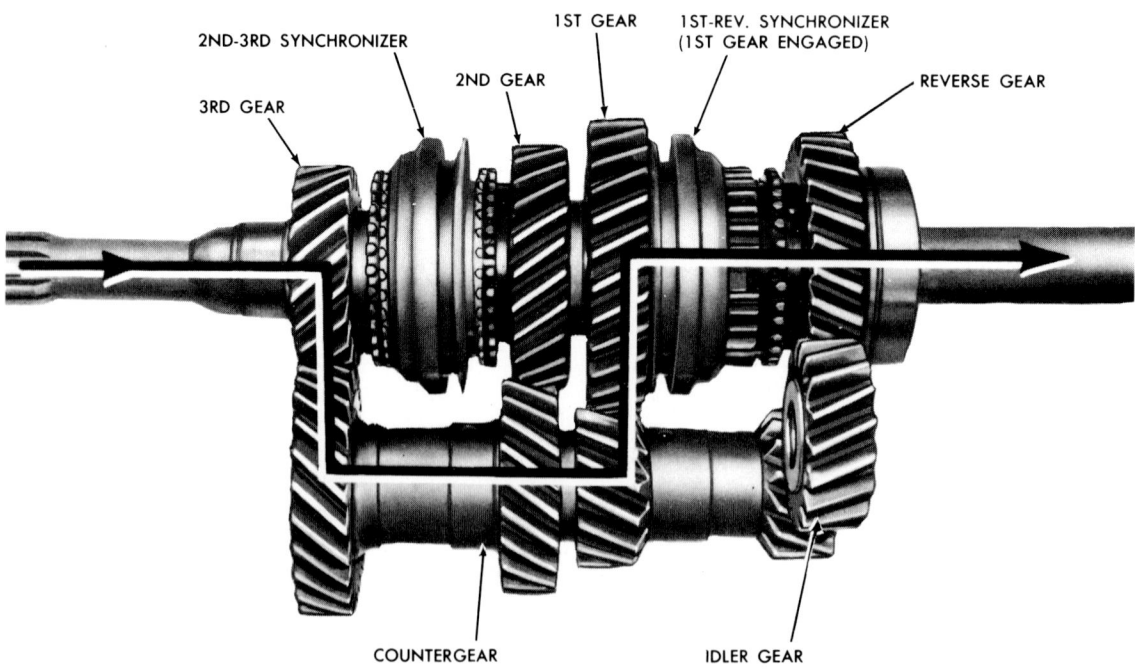

Fig. 25-23. Transmission in first gear. First-reverse synchronizer engaged with first output gear. Other synchronizer is in neutral position. First output gear is locked to output shaft and transfers high torque to drive shaft.

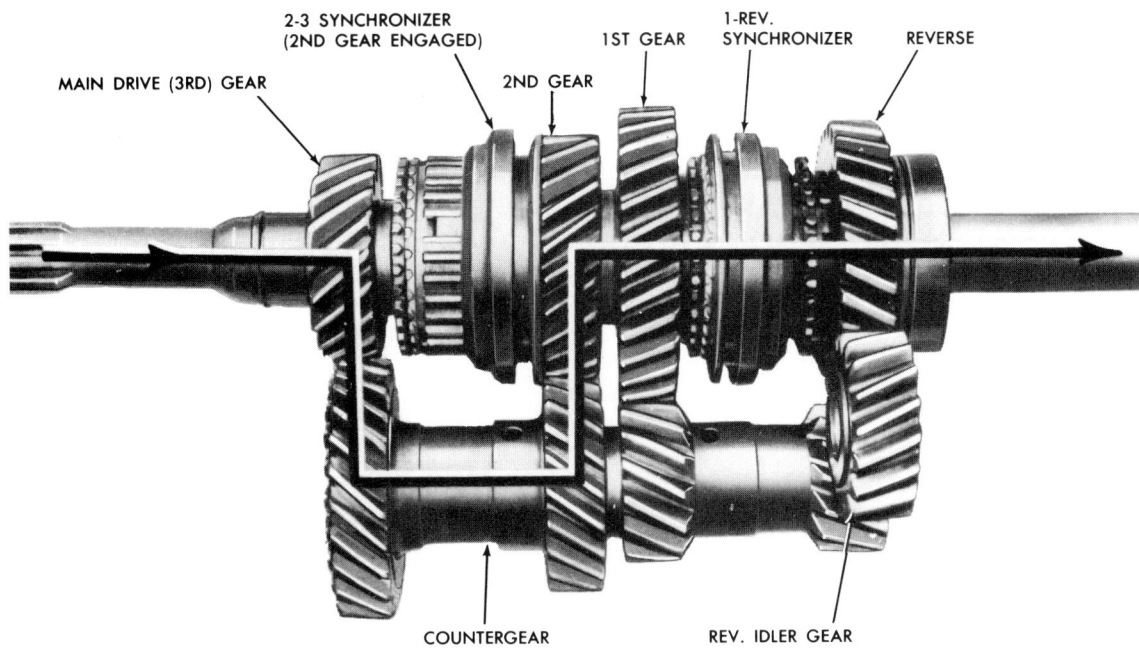

Fig. 25-24. Transmission in second gear. First-reverse synchronizer is moved into neutral. Second-third synchronizer is engaged with second output gear, locking it to its shaft.

Neutral

In neutral, all of the synchronizer sleeves are located in the centre of their hubs, Fig. 25-27. This allows all of the output shaft gears to freewheel on the output shaft. No power is transmitted to the output shaft.

Overdrive gear

When in high gear, many modern transmissions have overdrive. Either fourth (4-speed) or fifth (5-speed) has a ratio of less than 1:1 (0.87:1 for example) to increase fuel economy.

Fig. 25-28 shows the power flow through a late model, 5-speed, overdrive transmission. It is designed for a low power, diesel engine. Trace power flow through the transmission in each gear. The first four forward speeds allow the diesel to accelerate quickly. The overdrive high gear keeps engine rpm down at highway speeds to increase fuel economy and engine service life.

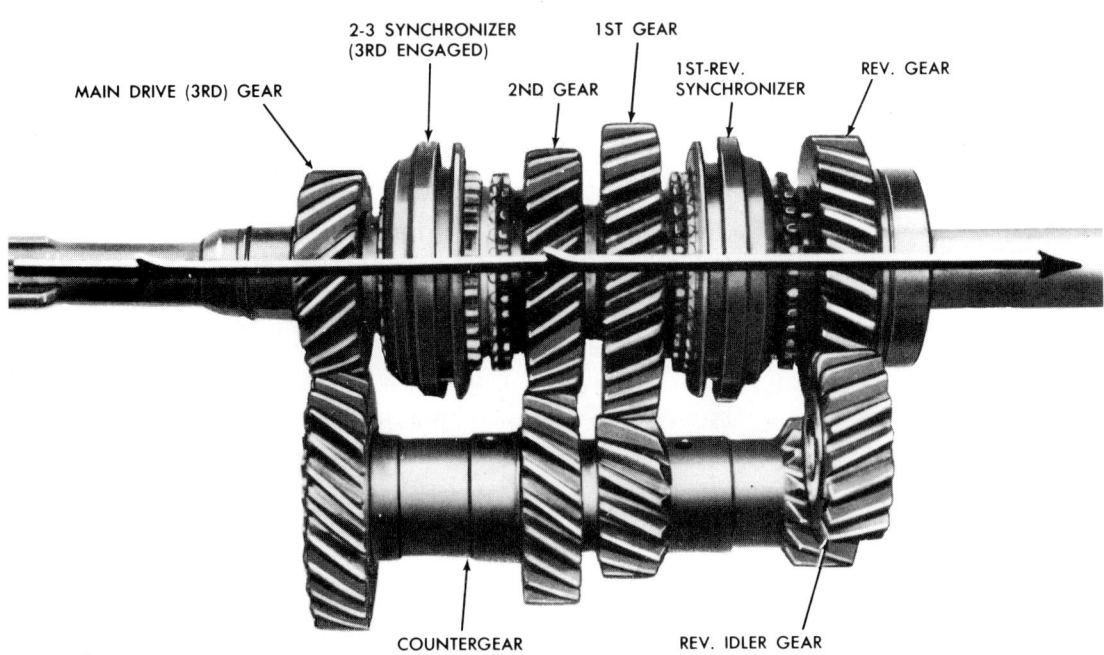

Fig. 25-25. Transmission in third gear. Second-third gear synchronizer is slid to engage gear on input shaft. This locks input shaft directly to output shaft. Both shafts turn at same speed for no gear reduction.

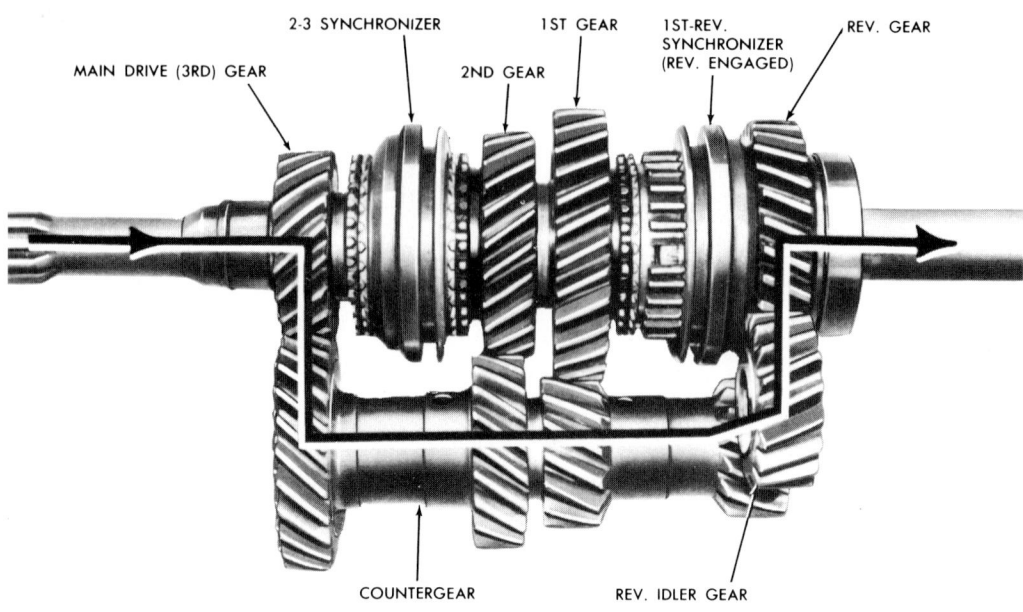

Fig. 25-26. Transmission in reverse. Second-third synchronizer moved into neutral. First-reverse synchronizer slid into mesh with reverse output gear. Countershaft gear drives reverse idler. Idler drives output shaft backwards.

OTHER TRANSMISSION DESIGNS

Many transmission design variations are used by the numerous car manufacturers. However, all transmissions use the basic operation and construction principles just explained.

Review the transmission parts in Figs. 25-29 and 25-30. Can you explain the basic function of each part?

SPEEDOMETER DRIVE

Normally, a manual transmission has a worm gear on the output shaft that drives the speedometer gear and cable. See Fig. 25-29. The gear on the output shaft turns a plastic gear on the end of the speedometer cable. The cable runs through a housing up to the speedometer head (speed indicator assembly) in the dash. A retainer and bolt hold the cable

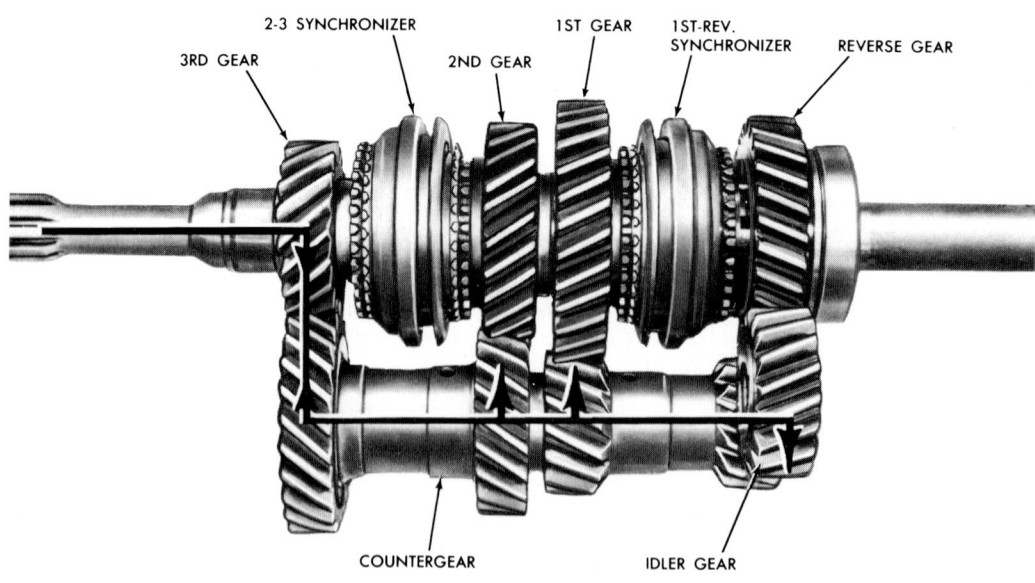

Fig. 25-27. In neutral, both synchronizers are in centre positions. No output gears are locked to output shaft. Gears freewheel and do not transfer power to drive shaft.

Manual Transmission Fundamentals

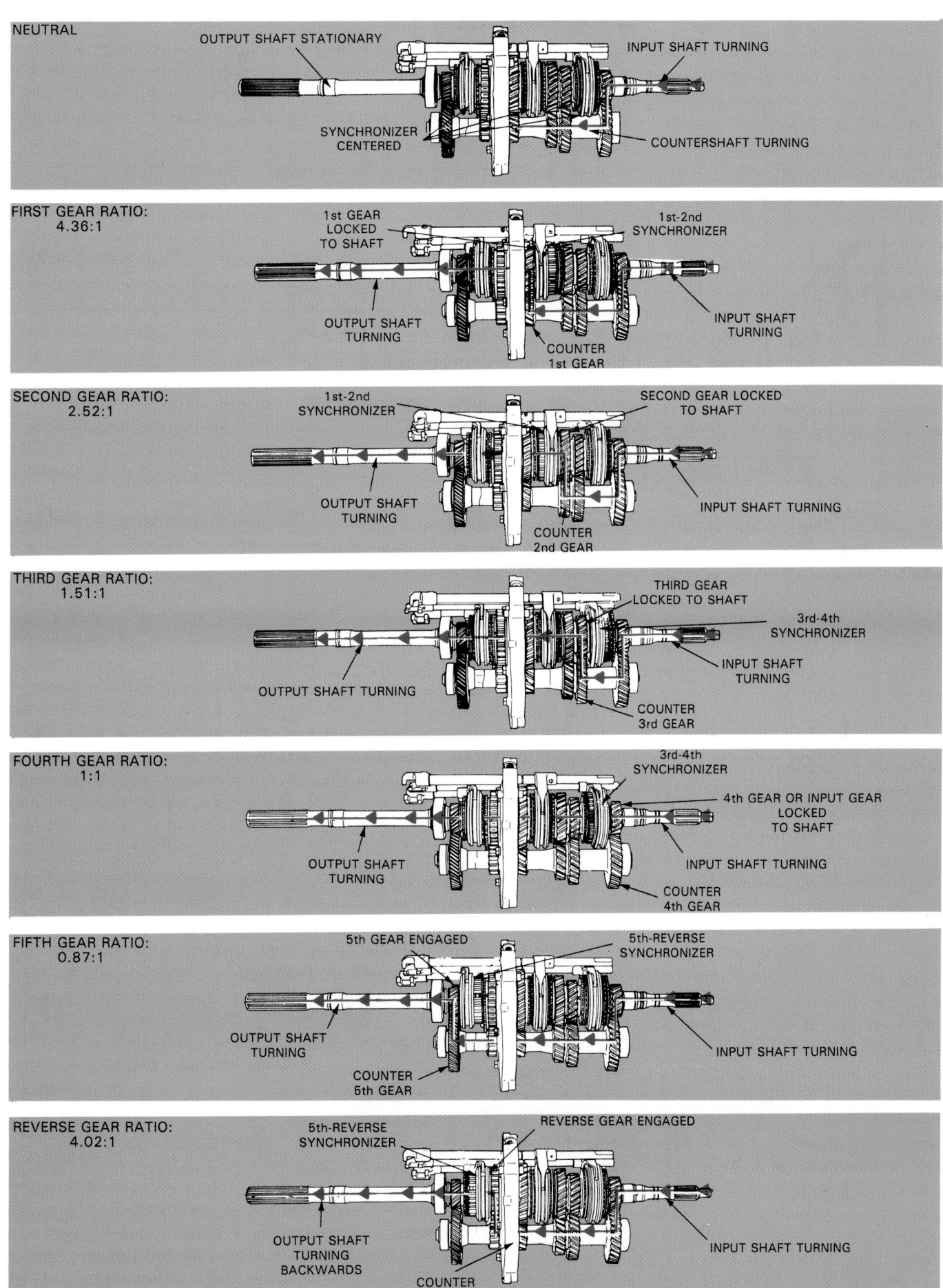

NEUTRAL

OUTPUT SHAFT STATIONARY

INPUT SHAFT TURNING

SYNCHRONIZER CENTERED

COUNTERSHAFT TURNING

FIRST GEAR RATIO: 4.36:1

1st GEAR LOCKED TO SHAFT

1st-2nd SYNCHRONIZER

OUTPUT SHAFT TURNING

INPUT SHAFT TURNING

COUNTER 1st GEAR

SECOND GEAR RATIO: 2.52:1

1st-2nd SYNCHRONIZER

SECOND GEAR LOCKED TO SHAFT

OUTPUT SHAFT TURNING

INPUT SHAFT TURNING

COUNTER 2nd GEAR

THIRD GEAR RATIO: 1.51:1

THIRD GEAR LOCKED TO SHAFT

3rd-4th SYNCHRONIZER

INPUT SHAFT TURNING

OUTPUT SHAFT TURNING

COUNTER 3rd GEAR

FOURTH GEAR RATIO: 1:1

3rd-4th SYNCHRONIZER

4th GEAR OR INPUT GEAR LOCKED TO SHAFT

OUTPUT SHAFT TURNING

INPUT SHAFT TURNING

COUNTER 4th GEAR

FIFTH GEAR RATIO: 0.87:1

5th GEAR ENGAGED

5th-REVERSE SYNCHRONIZER

OUTPUT SHAFT TURNING

INPUT SHAFT TURNING

COUNTER 5th GEAR

REVERSE GEAR RATIO: 4.02:1

5th-REVERSE SYNCHRONIZER

REVERSE GEAR ENGAGED

OUTPUT SHAFT TURNING BACKWARDS

INPUT SHAFT TURNING

COUNTER REVERSE GEAR

Fig. 25-28. Power flow through a five speed transmission with overdrive in high gear. Study each carefully.

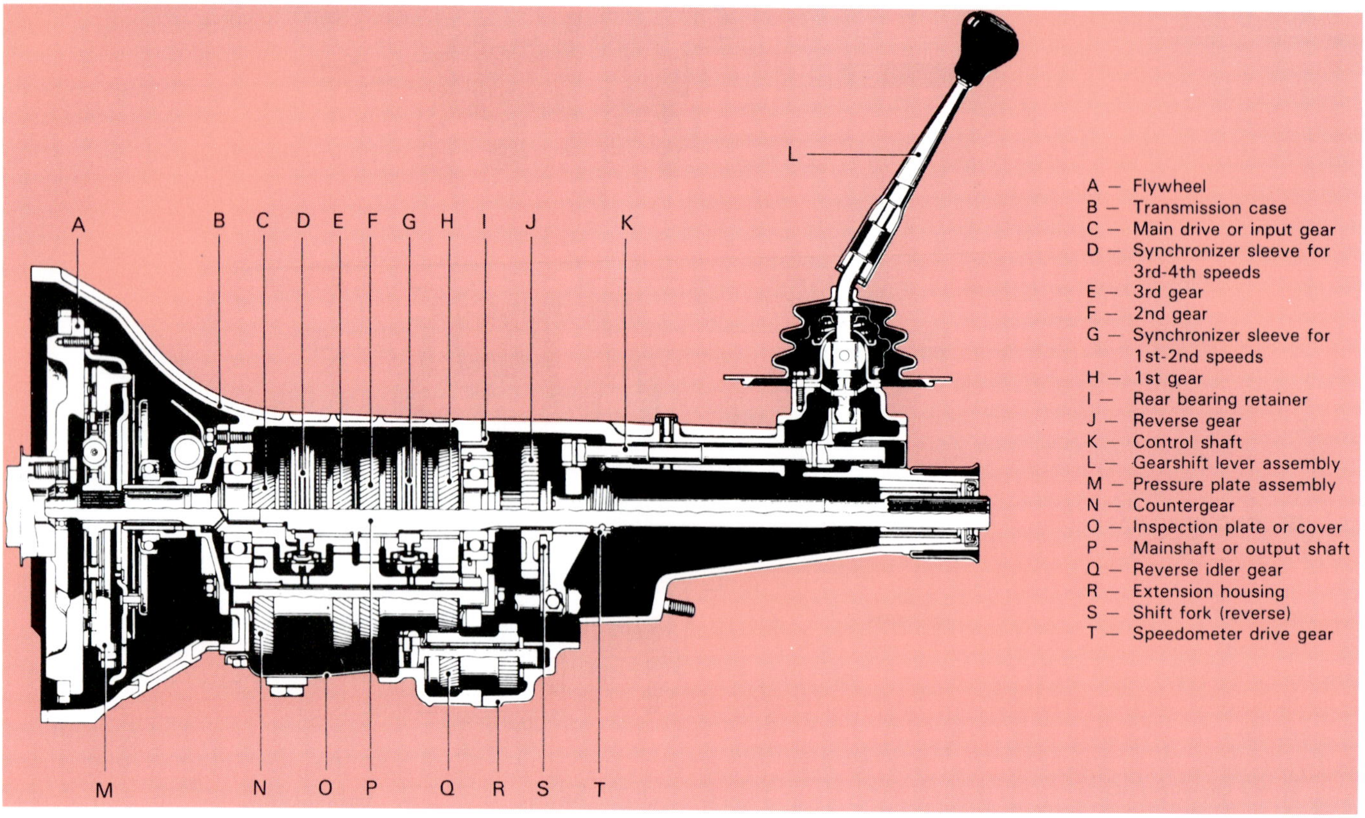

Manual Transmission Fundamentals

A — Flywheel
B — Transmission case
C — Main drive or input gear
D — Synchronizer sleeve for
    3rd-4th speeds
E — 3rd gear
F — 2nd gear
G — Synchronizer sleeve for
    1st-2nd speeds
H — 1st gear
I — Rear bearing retainer
J — Reverse gear
K — Control shaft
L — Gearshift lever assembly
M — Pressure plate assembly
N — Countergear
O — Inspection plate or cover
P — Mainshaft or output shaft
Q — Reverse idler gear
R — Extension housing
S — Shift fork (reverse)
T — Speedometer drive gear

Fig. 25-29. Cutaway view of late model four speed transmission. Note part names and locations.

assembly in the transmission extension housing.

Whenever the output shaft turns, the speedometer cable turns. This makes the speedometer head register the road speed of the car.

MANUAL TRANSMISSION SWITCHES

Two types of electric switches are sometimes mounted on the manual transmission: the back-up light switch and the ignition spark switch.

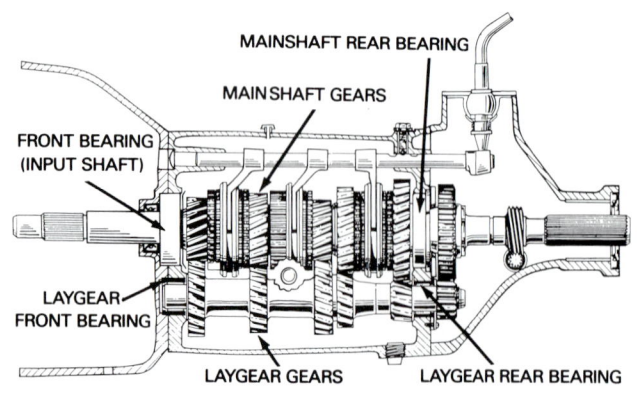

Fig. 25-30. Five-speed, manual transmission. Can you describe function of major parts?

The back-up light switch is an electric switch closed by the action of the reverse gear shift linkage. Refer to Fig. 25-21. When shifted into reverse, the linkage closes the switch to connect the backup lamps to the battery. This warns others that the car is being moved backwards and also helps the driver see behind the car.

A few manual transmissions have an ignition spark switch which only allows distributor vacuum advance in high gear. The switch usually mounts in the side of the transmission. It is normally closed, until activated in high gear. This retards the ignition timing in lower gears to reduce exhaust pollution.

KNOW THESE TERMS

Manual transmission, Gear ratio, Gear reduction, Overdrive ratio, Spur gears, Helical gears, Gear backlash, Gear oil, Transmission case, Extension housing, Input shaft, Countershaft, Reverse idler shaft, Output shaft, Countershaft gear, Output shaft gears, Synchronizer, Fully synchronized transmission, Shift fork, Transmission linkage, Shift rail, Shift lever, Backup light switch, Ignition spark switch.

REVIEW QUESTIONS

1. List and explain the eight major parts of a manual transmission.
2. Define the term "gear ratio".
3. How do you find the gear ratio of two gears?
4. Approximate manual transmission gear ratios are _____ for first, _____ for second, _____ for high, and _____ for reverse.
5. A gear reduction results when a small gear drives a larger gear to increase turning force. True or False?
6. This would be an overdrive ratio.
   a. 1:1.
   b. 0.87:1.
   c. 1:0.87.
   d. 3:1.
7. _____ _____ is the small clearance between the meshing gear teeth for lubrication and heat expansion.
8. Typically, _____ or _____ gear oil is used in a manual transmission.
9. What is the transmission extension housing?
10. Name and describe the four shafts in a manual transmission.
11. List and explain the general gear classifications found in a manual transmission.
12. A manual transmission synchronizer is used to:
   a. Prevent gear clashing or grinding.
   b. Lock output gear to output shaft.
   c. Both of the above are correct.
   d. None of the above are correct.
13. What is a fully synchronized transmission?
14. Describe the two major types of transmission shift linkages.
15. Why is an overdrive ratio used?

Chapter 26

AUTOMATIC TRANSMISSION FUNDAMENTALS

After studying this chapter, you will be able to:
- Identify the basic components of an automatic transmission.
- Describe the function and operation of the major parts of an automatic transmission.
- Trace the flow of power through automatic transmissions.
- Explain how an automatic transmission shifts gears.
- Compare different types of automatic transmissions.

An automatic transmission performs the same functions as a standard transmission. However, it "shifts gears" and "releases the clutch" automatically. A majority of modern cars use an automatic transmission (or transaxle) because it saves the driver from having to move a shift lever and depress a clutch pedal.

As you will learn, an automatic transmission normally senses engine rpm (speed) and engine load (engine vacuum or throttle position) to determine

gear shift points. It then uses internal oil pressure to shift gears. Computers can also be used to sense or control automatic transmission shift points.

BASIC AUTOMATIC TRANSMISSION

Before detailing the construction and operation of each individual part, it is important for you to have a general idea of how an automatic transmission works. Then, you will be able to relate the details of each part to the complete transmission assembly.

Refer to Fig. 26-1 as the following parts of an automatic transmission are introduced.

1. TORQUE CONVERTER (fluid coupling that connects and disconnects engine and transmission).
2. INPUT SHAFT (transfers power from torque converter to internal drive members and gearsets).
3. OIL PUMP (produces pressure to operate hydraulic components in transmission).

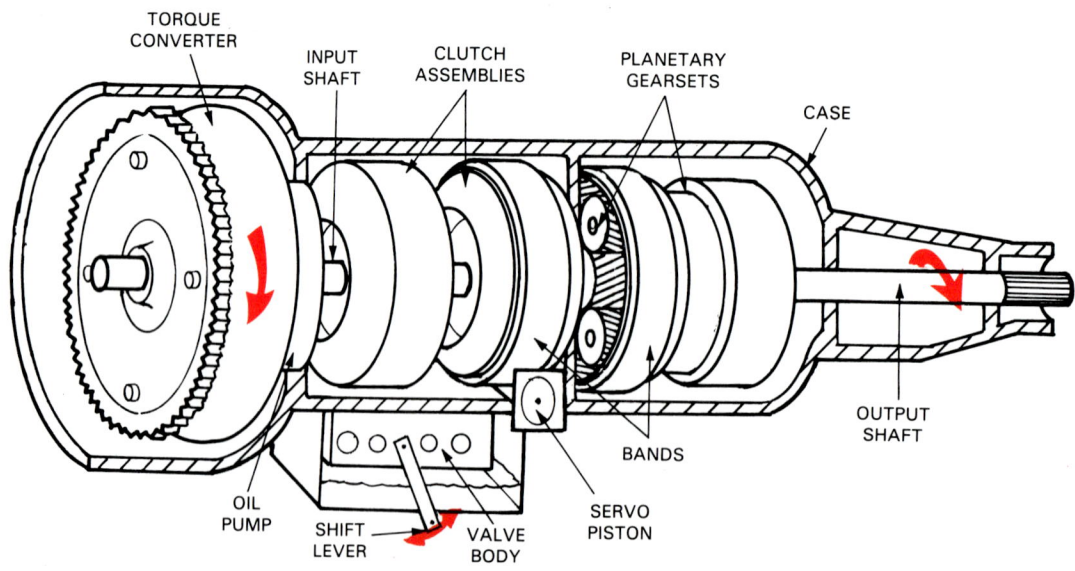

Fig. 26-1. Study basic parts of simplified automatic transmission. Note general shape and location of components. This will help prepare you to learn details of each part.

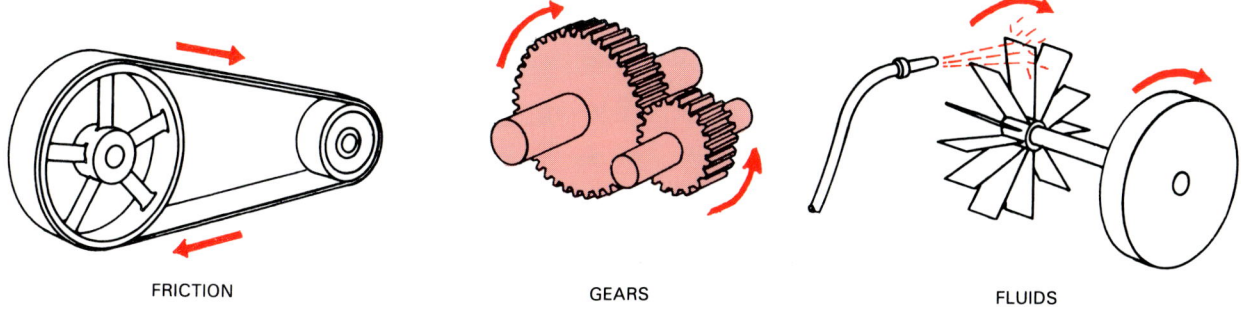

FRICTION GEARS FLUIDS

Fig. 26-2. An automatic transmission uses these methods of transmitting power.

4. VALVE BODY (operated by shift lever and sensors, controls oil flow to pistons and servos).
5. PISTONS AND SERVOS (actuate bands and clutches).
6. BANDS and CLUTCHES (apply clamping or driving pressure on different parts of gearsets to operate them).
7. PLANETARY GEARSETS (provide different gear ratios and reverse gear).
8. OUTPUT SHAFT (transfers engine torque from gearsets to drive shaft, and rear wheels).

TRANSMITTING POWER

As you will see, an automatic transmission uses three methods to transmit power: fluids, friction, and gears. This is illustrated in Fig. 26-2.

The torque converter uses FLUID to transfer power. The bands and clutches use FRICTION. The transmission GEARS, not only transmit power, they can increase or decrease speed and torque.

TRANSMISSION HOUSINGS AND CASE

An automatic transmission is normally constructed with four main components: converter housing, case, pan, and rear extension housing. These parts support and enclose all of the other components in the transmission. Refer to Fig. 26-3.

The converter housing or bell housing surrounds the converter and holds the transmission against the engine. It is usually made of aluminium.

Bolts fit through holes in the converter housing and attach to the engine block. The converter housing also keeps road dirt, rocks, and other debris away from the rotating torque converter and flywheel, Fig. 26-3.

The transmission case encloses the clutches, bands,

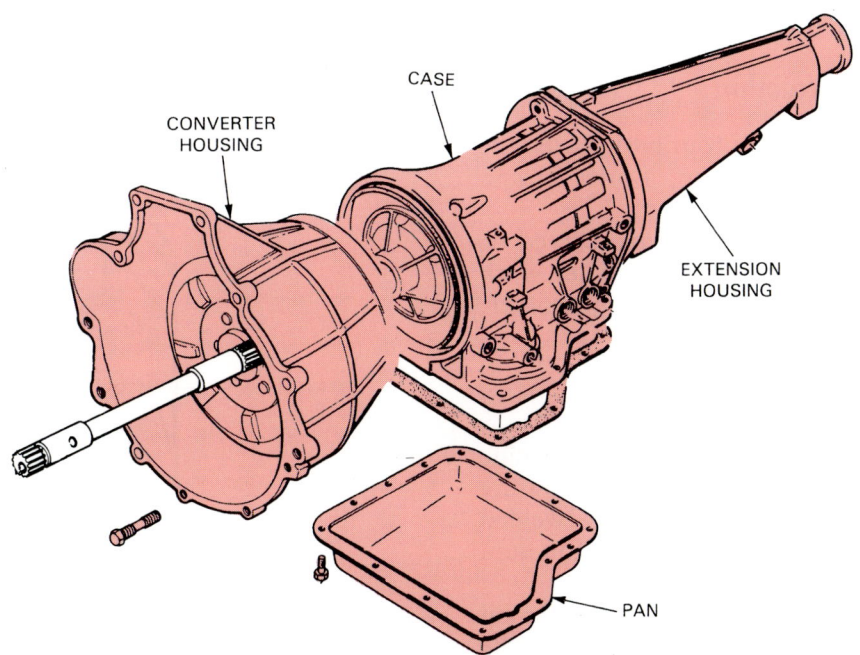

CONVERTER HOUSING
CASE
EXTENSION HOUSING
PAN

Fig. 26-3. Most automatic transmissions are constructed with front converter housing, central case, rear extension housing, and lower pan. Most parts fit inside case.

gearsets, and inner ends of the transmission shafts. The converter housing bolts to the front of the case. The extension housing bolts to the rear of the case. The valve body and pan bolt to the bottom of the case. It may be made of aluminium or cast iron, Fig. 26-3.

The oil pan, also called transmission pan, collects and stores a supply of transmission fluid. It is usually made of thin, stamped steel. The pan fits over the valve body. A gasket or sealant prevents leakage between the case and oil pan, Fig. 26-3.

The extension housing slides over and supports the output shaft. The housing uses a gasket on the front and a seal on the rear to prevent oil leakage. It is often made of aluminium, or sometimes cast iron, Fig. 26-3.

TORQUE CONVERTER

The torque converter is a fluid clutch that performs the same basic function as a manual transmission's dry friction clutch. It provides a means of uncoupling the engine for stopping the car in gear. It also provides a means of coupling the engine for acceleration.

Torque converter principles

Two house fans can be used to demonstrate the basic action inside a torque converter. Look at Fig. 26-4. One fan is plugged in and is rotating. The other fan is NOT plugged into electrical power.

Since the whirling fan is facing the other, it can be used to rotate the unplugged fan, transferring power through a liquid (air). This same principle applies inside a torque converter, but oil is used instead of air.

Torque converter construction

A torque converter consists of four basic parts: the outer housing, an impeller or pump, a turbine, and a stator. These are shown in Fig. 26-5.

The impeller, stator, and turbine have curved or curled fan blades, as shown in Fig. 26-6. They work like our simple example of one fan driving another. The impeller drives the turbine.

Converter housing

The impeller, stator, and turbine are housed inside a doughnut-shaped housing. The converter housing is normally made of two pieces of steel welded together. The housing is filled with transmission fluid (oil).

The impeller is the driving fan that produces oil movement inside the converter whenever the engine is running. It is sometimes called the converter pump, Figs. 26-6 and 26-7.

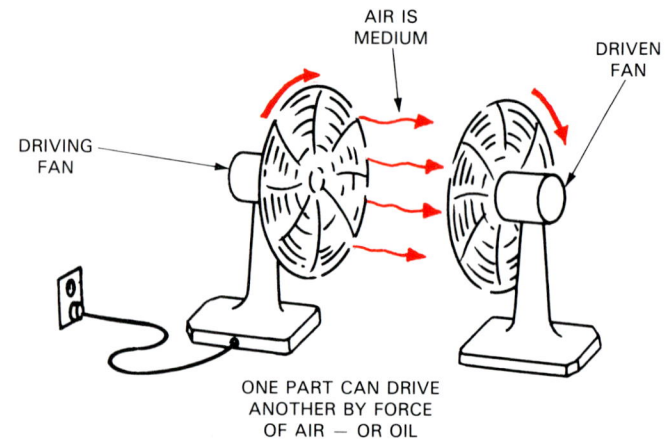

Fig. 26-4. Two fans demonstrate principle of fluid coupling or torque converter.

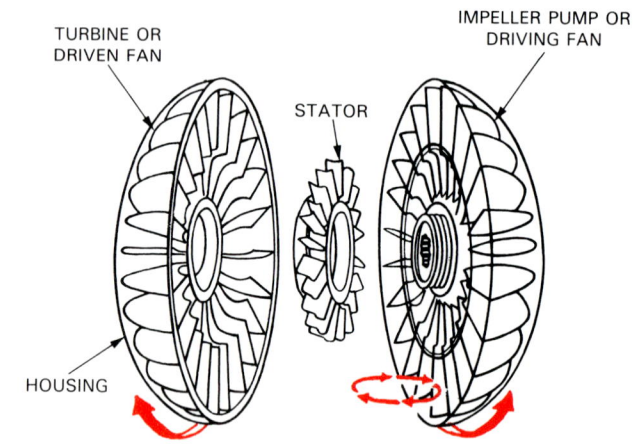

Fig. 26-5. Four major parts of torque converter: housing, impeller or pump, stator, and turbine.

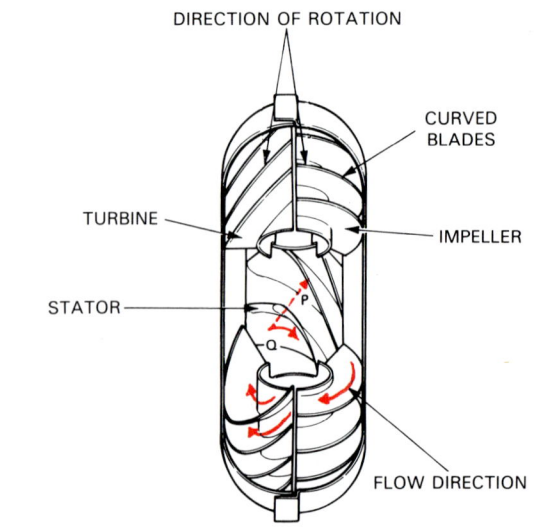

Fig. 26-6. Blades on impeller and stator direct oil circulation onto blades of turbine. Impeller is driven by engine. Turbine is driven by impeller. (Subaru)

The turbine is a driven fan splined to the input shaft of the automatic transmission. It fits in front of the stator and impeller in the housing. The turbine is NOT fastened to the impeller, but is free to turn independently. Oil is the only connection between the two.

The stator is designed to improve oil circulation inside the torque converter. It increases efficiency and torque by causing the oil to swirl around inside the converter housing. This makes use of all the force produced by the moving oil.

Flywheel action

The transmission torque converter is very large and heavy. This allows it to serve as a flywheel to smooth out engine power pulses. Its inertia reduces vibration entering the transmission and drive line.

An automatic transmission uses a very thin and light flywheel, which is called a drive plate. It is simply a stamped disc with a ring gear for the starter motor.

If the ring gear is on the torque converter, a drive plate, without a ring gear, can be used to connect the crankshaft to the torque converter.

The crankshaft bolts to one side of the drive plate. The converter bolts to the other side.

Torque converter operation

With the ENGINE IDLING, the impeller rotates slowly. Only a small amount of oil is thrown into the stator and turbine. Not enough force is developed inside the torque converter to drive the turbine. The car would remain stationary with the transmission in gear.

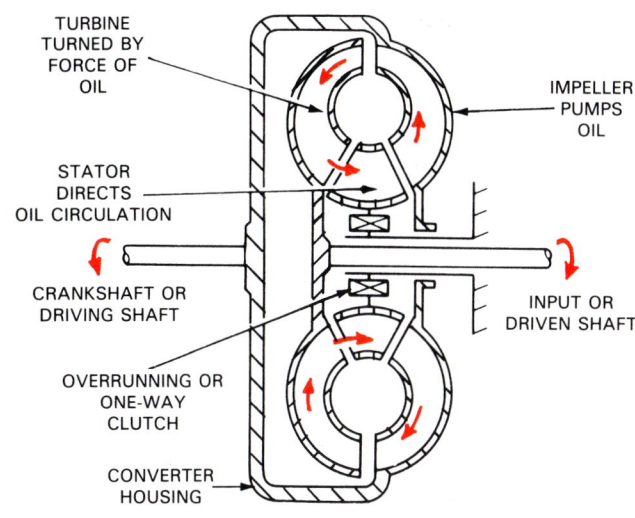

Fig. 26-7. Crankshaft is fastened to converter housing and impeller. Stator is mounted on one-way clutch. When engine crankshaft rotates fast enough, oil movement rotates turbine and transmission input shaft. (Subaru)

During ENGINE ACCELERATION, the engine crankshaft, converter housing, and impeller begin to rotate faster. More oil is thrown out by centrifugal force. This makes the turbine begin to turn. As a result, the transmission input shaft and car would start to move, but with some slippage, Fig. 26-7.

At CRUISING SPEEDS, the impeller and turbine rotate at almost the same speed, with very little slippage. When the impeller is rotating fast enough, centrifugal force throws the oil out hard enough to almost lock the impeller and turbine. See Fig. 26-7.

Converter one-way clutch

A one-way clutch allows the stator to turn in one direction but not the other. See Fig. 26-8. The stator

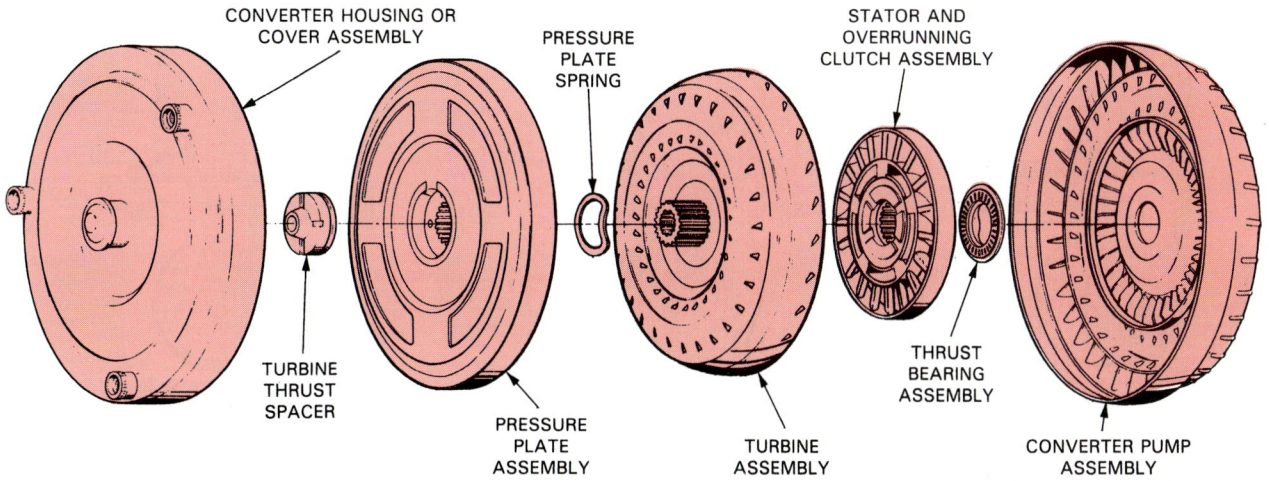

Fig. 26-8. Lock-up torque converter is conventional converter with a friction pressure plate added. The pressure plate can be used to lock turbine to converter housing, eliminating slippage and increasing fuel economy.

mounts on the clutch mechanism. Stator action is only needed when the impeller and turbine are turning at different speeds.

The one-way clutch locks the stator when the impeller is turning faster than the turbine. This causes the stator to route oil flow over the turbine vanes properly. Then, when turbine speed almost equals impeller speed, the stator can freewheel on its shaft, so not to obstruct oil flow.

Torque multiplication

Torque multiplication refers to the ability of a torque converter to increase the amount of engine torque applied to the transmission input shaft. Just as a small gear driving a large gear increases torque, a torque converter can act as several different gear ratios to alter torque output. Torque can be doubled by the converter under certain conditions.

Torque multiplication occurs when the impeller is rotating FASTER than the turbine. For instance, if the engine is accelerated quickly, the engine and impeller rpm might increase rapidly while the turbine is almost stationary. At this time, torque multiplication would be maximum. When the turbine speed nears impeller speed, torque multiplication drops off.

Torque is increased in the converter by sacrificing motion. The turbine rotates slower than the impeller during torque multiplication.

Torque converter stall speed

The stall speed of a torque converter basically occurs when the impeller is at maximum speed without rotation of the turbine. This causes the oil to be thrown off the stator vanes at tremendous speeds. The greatest torque multiplication occurs at stall speed.

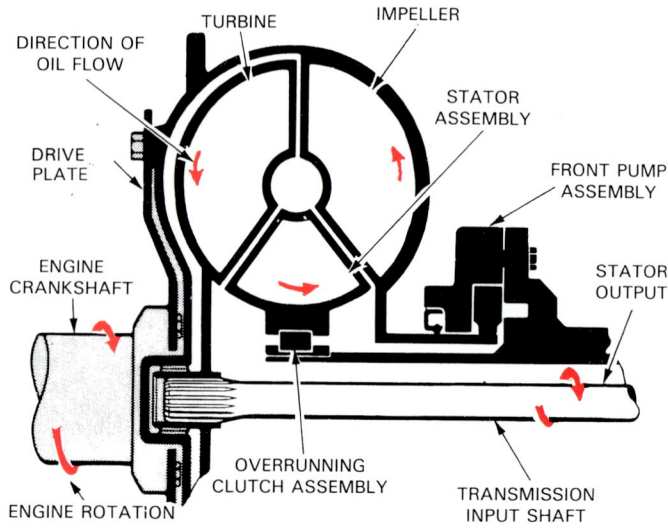

Fig. 26-10. Transmission input shaft extends through stator support. Shaft is splined to turbine. Also note how stator mounts on one-way clutch. (Ford)

Lock-up torque converters

A lock-up torque converter has an internal friction clutch mechanism for locking the impeller to the turbine in high gear. In a conventional converter, there is always some slippage between the impeller and turbine. By locking these components with a friction clutch, the torque converter does not slip. This improves fuel economy.

Typically, a lock-up mechanism in a torque converter consists of a hydraulic piston, torsion springs, and clutch friction material. See Figs. 26-8 and 26-9.

In lower transmission gears, the converter clutch is released. The torque converter operates normally, allowing slippage and torque multiplication.

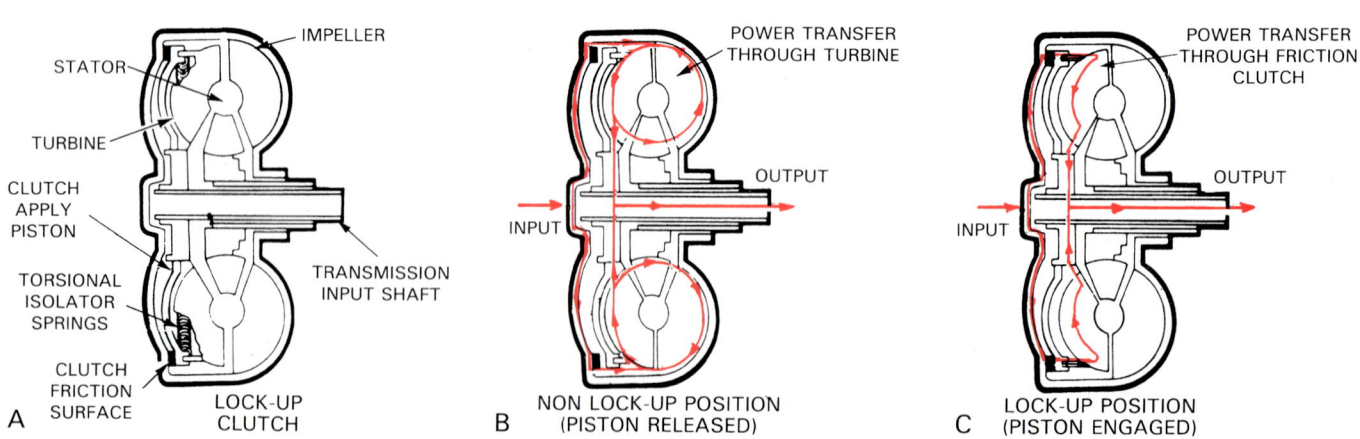

Fig. 26-9. Lock-up torque converter operation. A — Parts of lock-up converter. B — In lower gears, no oil pressure acts on clutch apply piston. Torque converter operates like conventional unit, impeller drives turbine. C — In high gear, oil is transferred into piston chamber. Clutch apply piston forces friction surfaces together. Turbine is mechanically locked to converter housing and impeller. Crankshaft drives transmission input shaft directly, without slippage.

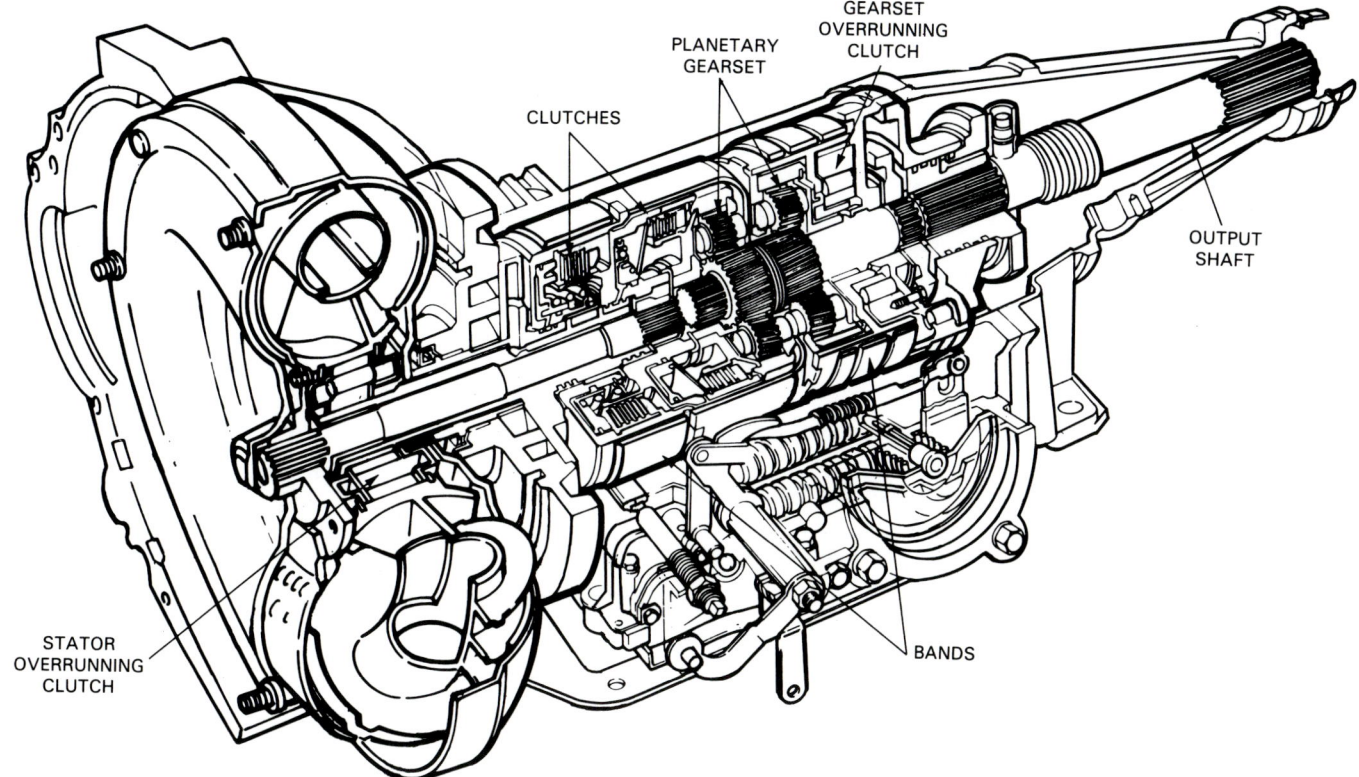

Fig. 26-11. Study location of gearsets and holding devices. Clutches, bands, and rear one-way clutch operate gearsets. Other one-way clutch operates torque converter stator. (Ford)

Then, when shifted into high or direct drive, oil is channelled to the converter piston. The piston pushes the friction discs together to lock the converter. The torsion springs help dampen engine power pulses entering the drive train. Refer to Fig. 26-9.

AUTOMATIC TRANSMISSION SHAFTS

Typically, an automatic transmission has two main shafts: the input shaft and output shaft.

An automatic transmission input shaft or turbine shaft connects the torque converter with the driving components in the transmission. Look at Fig. 26-10.

Each end of the input shaft has male (external) splines. These splines fit into splines in the torque converter turbine and a driving unit in the transmission. The input shaft rides on bushings. Transmission fluid lubricates the shaft and bushings.

The output shaft connects the driving components in the transmission with the drive shaft. Refer to Fig. 26-11. This shaft runs on the same centreline as the input shaft. Its front end almost touches the input shaft.

STATOR SUPPORT

The stator support, also called stator shaft, is usually a stationary shaft splined to the torque converter stator assembly. As pictured in Fig. 26-10, it is a tube that extends forward from the front of the transmission. It surrounds the input shaft.

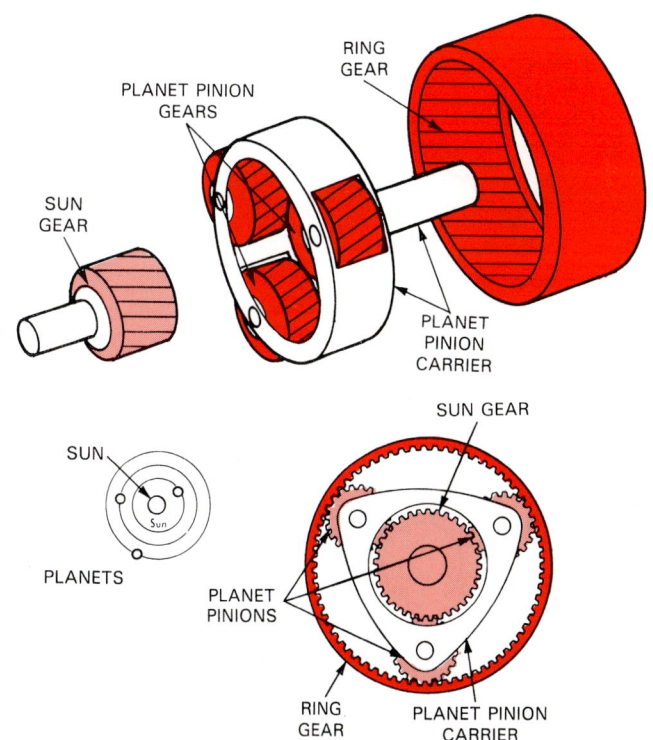

Fig. 26-12. Simplified planetary gearset. Planet gears fit between ring gear and sun gear. Planet gears are mounted on planet carrier. Gears are always in mesh, making a compact, strong, and dependable assembly. Name (planetary gears) is derived from how planet gears revolve around sun gear, like solar system planets.

401

PLANETARY GEARS

A planetary gearset consists of a sun gear, several planet gears, a planet gear carrier, and a ring gear. A simple planetary gearset is shown in Fig. 26-12.

The name planetary gearset is easy to remember because it refers to our solar system. Just as our planets (Earth, Jupiter, Mars) circle the sun, the planet gears revolve around the sun gear.

As you can see, a planetary gearset is always in mesh. It is very strong and compact. An automatic transmission will commonly use two or more planetary gearsets.

By holding or releasing the components of a planetary gearset, it is possible to:

1. Reduce output speed and increase torque (gear reduction).
2. Increase output while lowering torque (overdrive).
3. Reverse output direction (reverse gear).
4. Serve as a solid unit to transfer power (one to one ratio).
5. Freewheel to stop power flow (park or neutral).

Planetary reduction

One method of obtaining a gear reduction and torque increase is to hold the sun gear (stop it from turning) while driving the ring gear. This makes the planet carrier the output member. Refer to Fig. 26-13A.

When power turns the ring gear, the planet pinion gears "walk" (rotate) around the locked sun gear. The planet gears move in the same direction as the ring gear, but NOT as fast. As a result, more torque is applied to the output member (planet carrier) and output shaft.

Gear reduction can also be produced in the planetary gearset by turning the sun gear and holding the ring gear.

Planetary overdrive

Driving the carrier while holding the ring gear achieves an overdrive ratio in a planetary gearset. Look at Fig. 26-13B.

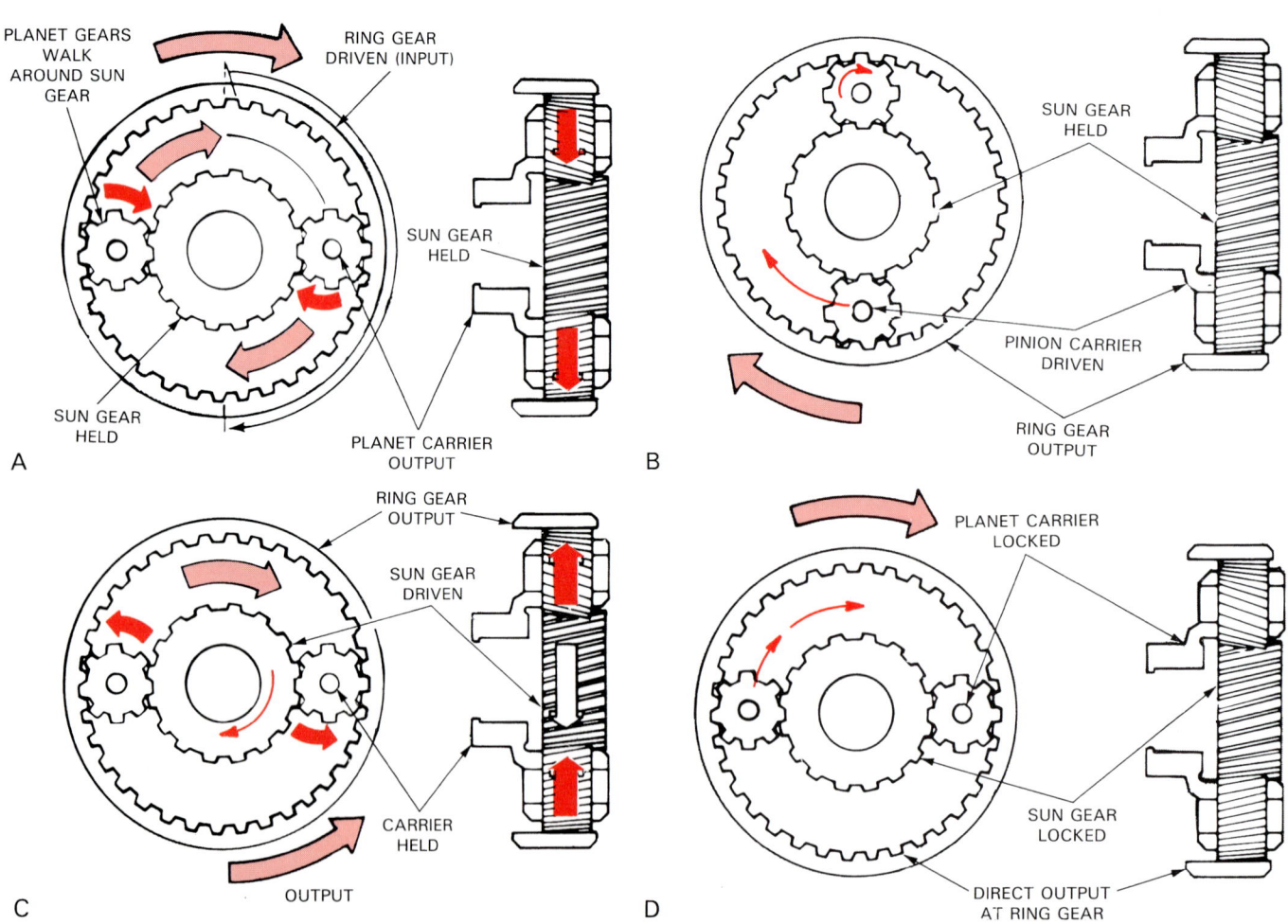

Fig. 26-13. Study how different planetary gearset members can be held to provide different gear ratios and reverse. A — Simple gear reduction. Sun gear is stationary. Ring gear is driven. Planet carrier is output. Input torque increases and speed decreases. B — Overdrive. Sun gear is held stationary. Pinion carrier is driven. Ring gear is output and turns faster than input. C — Simple reverse gear. Pinion carrier is held. Sun gear is driven. Ring gear turns backwards as output. D — Direct drive results when any two members of planetary gearset are held, or by driving any two members from same input.

The input shaft powers the planet carrier. The sun gear is the output member driving the output shaft. The planet gears "walk" in the ring gear and power the sun gear. The sun gear rotates faster than the carrier. Torque is lost but speed is increased.

Planetary reverse

A planetary gearset can also reverse output direction. The input shaft drives the sun gear, as in Fig. 26-13C. The carrier is held and the ring gear turns the output shaft. The planet pinion gears simply act as idler gears. They reverse the direction of rotation between the sun gear and ring gear.

Planetary direct drive

A planetary gearset will act as a solid unit when TWO of its members are held. This causes the input and output members to turn at the same speed, Fig. 26-13D.

Planetary neutral

When none of the planetary members are held, the unit will NOT transfer power. This freewheeling condition is used when an automatic transmission is placed in neutral or park.

Compound planetary gearset

A compound planetary gearset combines two planetary units in one housing or ring gear. It may have two sun gears or a long sun gear to operate two sets of planet pinion gears, Fig. 26-14.

With this design, short planet gears engage forward sun gear. Long planet gears mesh with the rear sun gear. The ring gear engages both sets of planet gears.

Another design, called Simpson Compound Gearset, uses a long sun gear to operate two sets of planet gears on the same ring gear. This type is common.

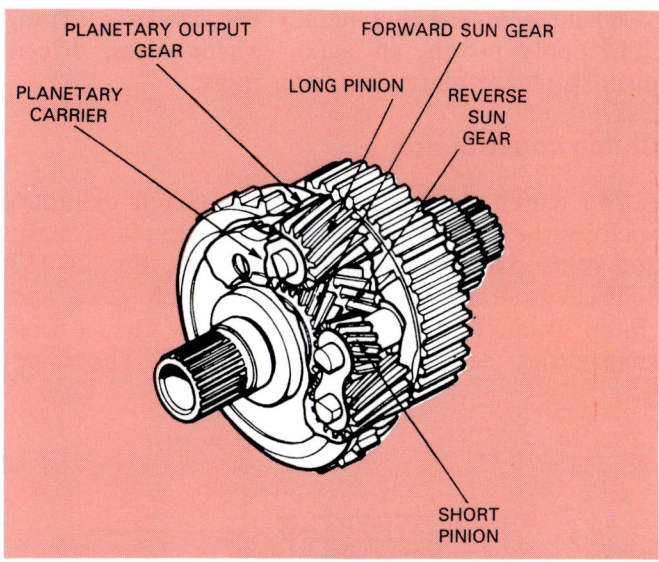

Fig. 26-14. Compound planetary gearset acts like two gearset assemblies mounted together. Normally, a common ring gear is used for two separate sets of planet gears. (Subaru)

A compound planetary gearset is used because it can provide more forward gear ratios than a simple planetary gearset.

CLUTCHES AND BANDS

Automatic transmission clutches and bands are friction devices that drive or lock planetary gearset members. They are used to enable the gearsets to transfer power. Refer again to Figs. 26-1 and 26-11.

Multiple disc clutches

A multiple disc clutch has several clutch discs that can be used to couple or hold planetary gearset members. As shown in Fig. 26-15, the front clutch assembly usually drives a planetary sun gear. The next clutch transmits power to the planetary ring gear when engaged. This can vary, however.

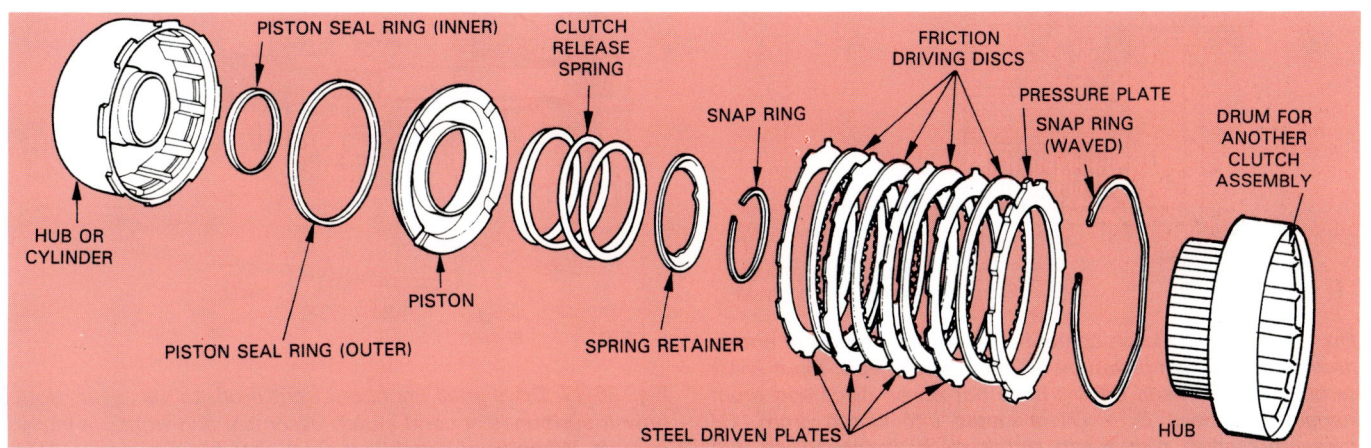

Fig. 26-15. Study construction of clutch assembly from automatic transmission. Clutch parts and hub fit inside clutch drum. Also note difference in clutch discs. Driving discs are splined to hub. Driven discs are locked in drum by tabs.

A clutch assembly generally consists of a drum, hub, apply piston, spring(s), driving discs, driven discs, pressure plate, and snap rings.

Clutch construction

The clutch drum, also called a clutch cylinder, encloses the apply piston, discs, pressure plate, seals, and other parts of the clutch assembly, Fig. 26-15.

The clutch hub fits inside the clutch discs and clutch drum. It has teeth on its outer surface that engage the teeth on the driving discs. The front

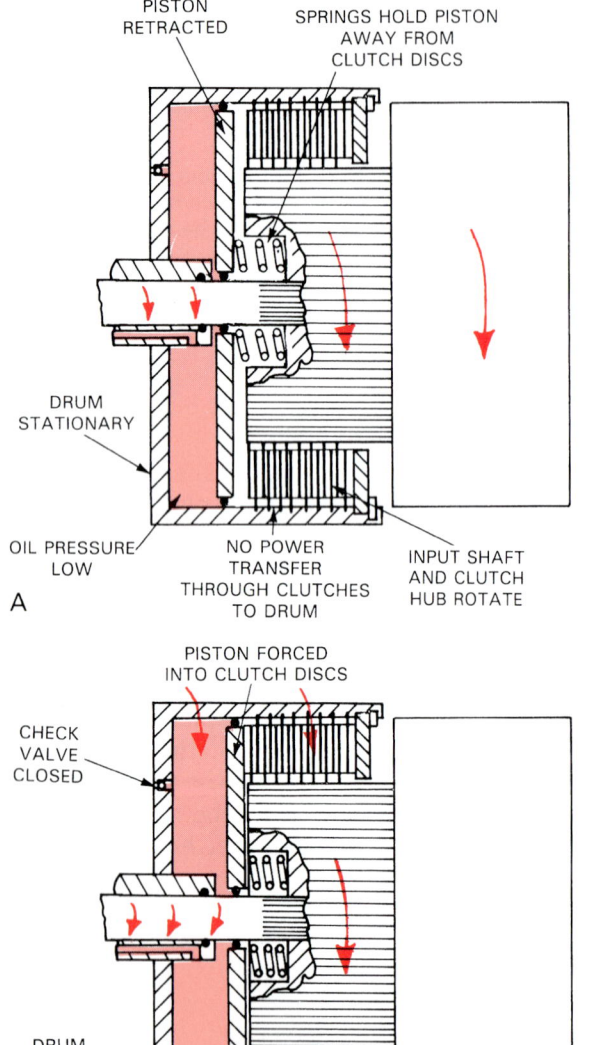

Fig. 26-16. Basic clutch operation. A — No oil pressure enters drum. Springs hold piston away from clutch discs. Input shaft turns clutch hub and driving discs but driven discs and drum remain stationary. B — Oil is routed into clutch drum. Oil pressure pushes piston into clutch discs, forcing discs into pressure plate. This locks discs, hub, and drum together. Power is then transferred from input to drum.

clutch hub is also splined to the transmission input shaft.

The driving discs are usually covered with friction lining. They have teeth on their inside diameter that engage the clutch hub. See Fig. 26-15.

The driven discs are steel plates that have outer tabs that lock into the clutch drum. A driven disc fits between each driving disc. This enables the hub and friction discs to turn the steel discs and drum when the clutch is activated.

The clutch apply piston slides back and forth inside the clutch drum to clamp the driving and driven discs together. Seals fit on the piston to prevent fluid leakage during clutch application.

The pressure plate serves as a stop for clutch discs when the piston is applied. The piston pushes the discs against the pressure plate. Look at Fig. 26-15.

A clutch spring or springs are used to push the apply piston away from the clutch discs during clutch disengagement. One is shown in Fig. 26-15.

Clutch operation

During clutch engagement, oil pressure is routed into the clutch drum. Shown in Fig. 26-16A, oil pressure acts on the large piston. The piston is then forced into the clutch discs. Friction locks the driving and driven discs together to transfer power through the clutch assembly.

When oil pressure is blocked from the piston, the return spring pushes the clutch discs apart, Fig. 26-16B. Power is no longer transferred through the clutch. The driving and driven discs are free to turn, independently.

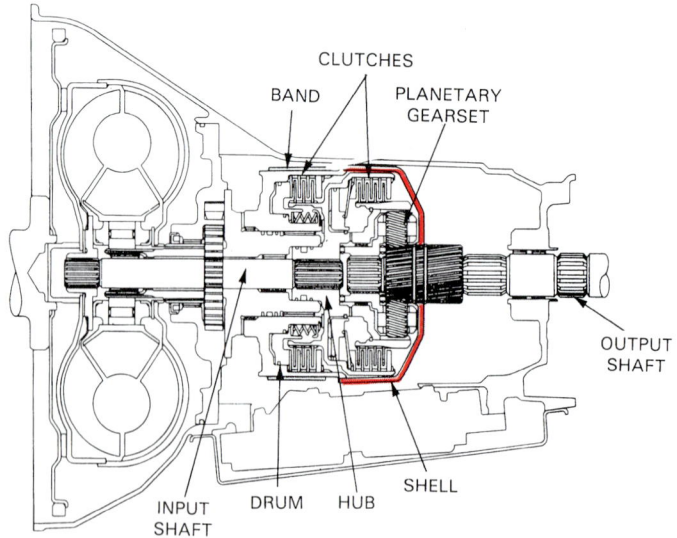

Fig. 26-17. Drive shell connects front drum to sun gear. Note how it surrounds second clutch assembly and front planetary gearset. When front clutch is locked, shell turns sun gear. Also note band used to hold front drum and sun gear stationary. (Ford)

Driving shell

A driving shell or clutch shell is commonly used to transfer power to one of the planetary sun gears, Fig. 26-17. It is a thin, metal cylinder-shaped part that frequently connects the front clutch drum and sun gear.

The shell may surround the second clutch assembly and forward planetary gearset. Tabs on the shell fit into notches on the front clutch drum. This makes the shell, drum, and sun gear turn together.

Bands and servos

Automatic transmission bands are also friction devices for holding members of the planetary gearsets. Two or three bands are commonly used in modern transmissions. Bands are shown in Figs. 26-1, 26-11, 26-17, and 26-18.

Servos are apply pistons that operate the bands. Fig. 26-19 shows the parts of a band and a servo assembly.

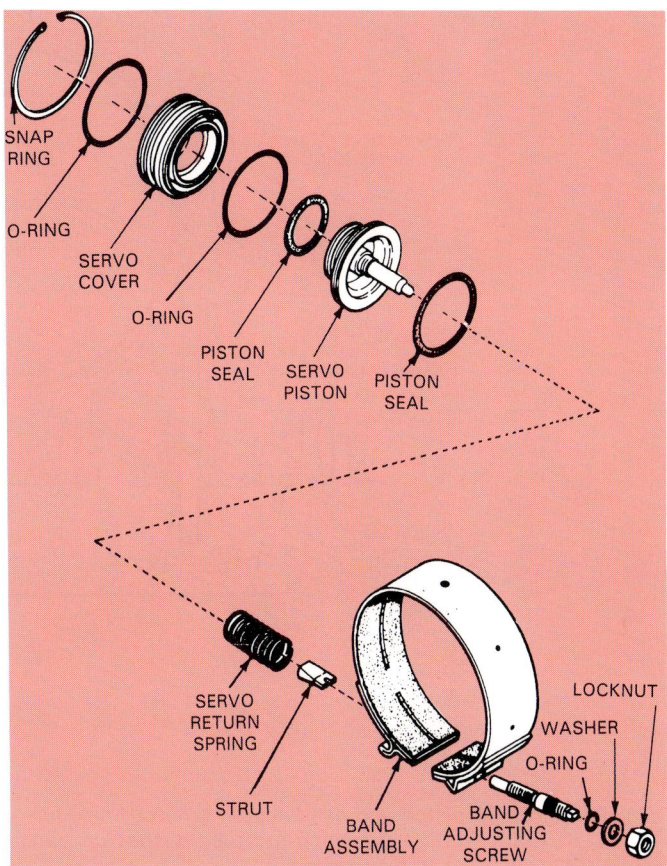

Fig. 26-19. Exploded view of band and servo assembly. Note relationship between parts. (Subaru)

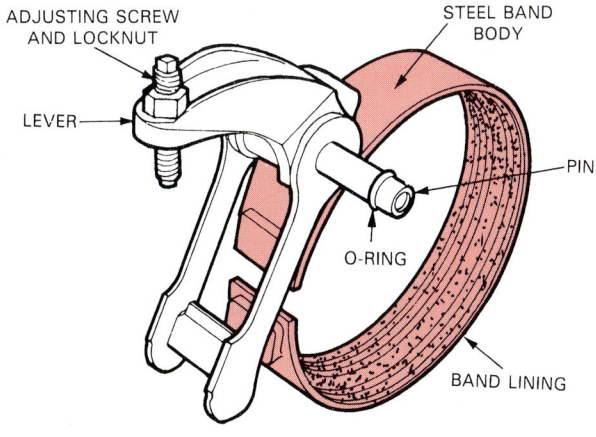

Fig. 26-18. Band is steel strap with friction lining on its inner surface. One end of band is anchored in case.

Band and servo construction

A band is a steel strap with lining (friction material) on its inner surface. The band's lining can be clamped around a clutch drum to stop drum rotation.

The friction material on the inside of the band is designed to operate in automatic transmission fluid. It resists the lubricating qualities of the fluid.

The servo piston is a metal plunger that operates in a cylinder machined in the transmission case. Rubber seals fit around the outside of the piston to prevent fluid leakage. See Fig. 26-20.

A rod on the servo piston attaches to one end of the brake band. The other end of the brake band is anchored to the transmission case.

A band adjustment screw provides a means of adjusting the band-to-drum clearance. It moves the band closer to the drum as the friction material wears.

Servo seals prevent fluid leakage around the servo piston and cylinder. Snap rings hold the piston in its cylinder.

Transmission band operation

To activate a brake band, oil pressure is sent into the servo cylinder. Pressure acts on the servo piston. The piston then slides in the cylinder and pushes on one end of the brake band, as in Fig. 26-20.

Since the other end of the band is anchored, the band tightens or squeezes around the drum. The friction material rubs on the drum and stops it from turning. This keeps one of the planetary components from revolving.

When the oil flow to the piston servo is blocked, the servo spring pushes on the piston. This slides the piston rod away from the band. The band then releases the drum and planetary gearset member.

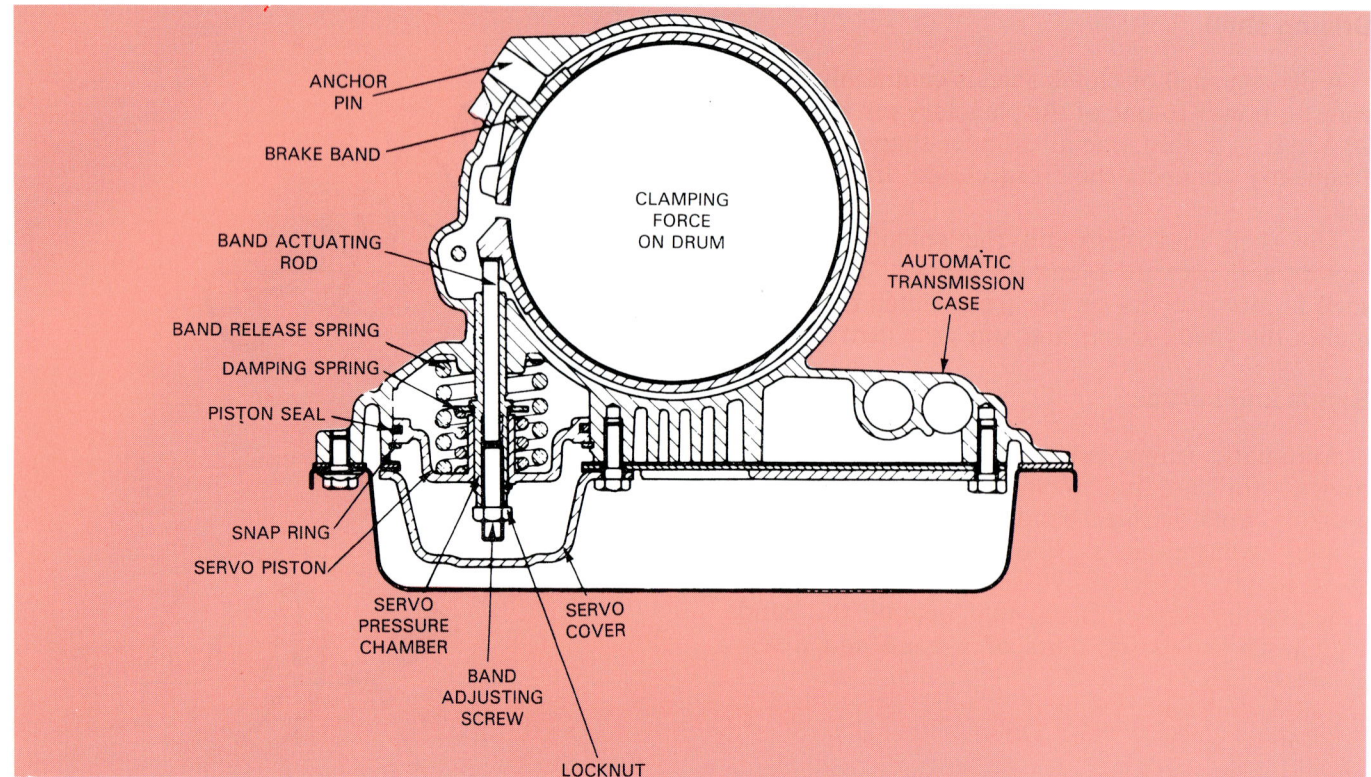

Fig. 26-20. Servo piston and band action. When oil pressure enters servo pressure chamber, servo piston slides up in cylinder. Actuating rod then pushes on band to squeeze band inward on drum. (Fiat)

Accumulator

An accumulator is used in the apply circuit of a band or clutch to cushion initial application. It temporarily absorbs some of the oil pressure to cause slower movement of the apply piston.

OVERRUNNING CLUTCHES

Besides the bands and clutches, an overrunning clutch can be used to hold a planetary gearset member. It is a one-way, roller clutch that locks in one direction and freewheels in the other.

An overrunning clutch for the planetary gears is similar to the ones in a torque converter stator or an electric starter motor drive gear. The typical locations of automatic transmission overrunning clutches (stator clutch and gearset clutch) are illustrated in Fig. 26-11.

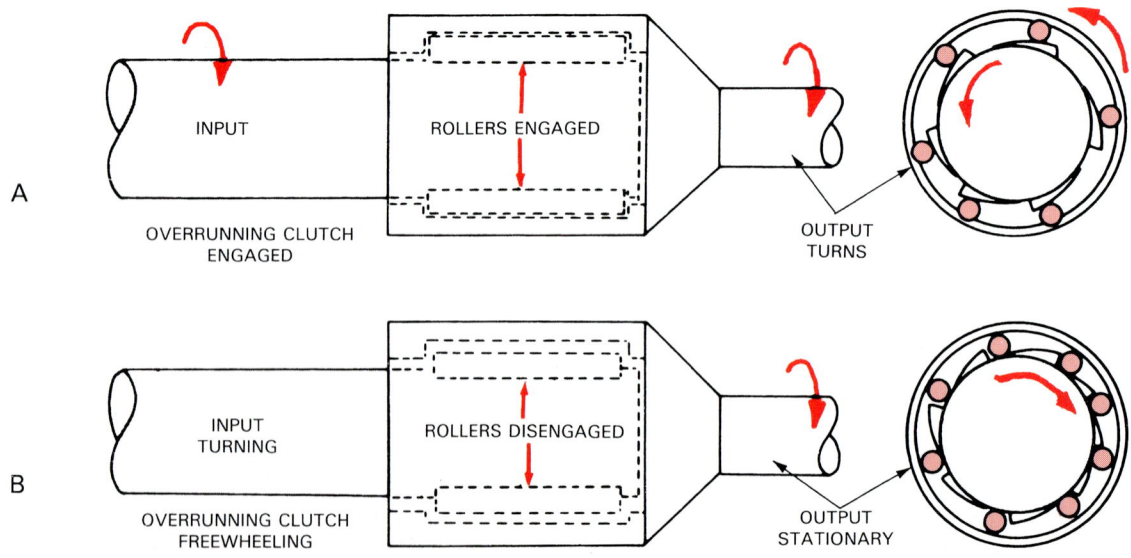

Fig. 26-21. Overrunning or one-way clutch action. A — When driven in one direction, rollers lock between ramps on inner race and on outer race. Both races turn together. This action can also be used to stop movement of planetary member, for example. B — When turned in other direction, rollers walk off ramps. Two races are free to turn independently.

A planetary gearset overrunning clutch consists of an inner race, set of springs, rollers, and an outer race. Fig. 26-21 shows overrunning clutch operation.

HYDRAULIC VALVE ACTION

The basic action of a hydraulic valve and a piston are illustrated in Fig. 26-22. Oil pump pressure causes oil to flow through the spool valve and pressure lines to the left end of the piston cylinder. This pushes the piston to the right.

When the spool valve is moved the other way, pump pressure is not sent to the piston. The piston is then forced back into its cylinder.

Valves like this are used to operate the band servos and clutch pistons.

HYDRAULIC SYSTEM

The hydraulic system for an automatic transmission typically consists of a pump, pressure regulator valve, manual valve, vacuum modulator valve, governor valve, shift valves, servos, pistons, and valve body. These parts work together to form the "brain" (sensing) and "muscles" (control) of an automatic transmission. See Fig. 26-23.

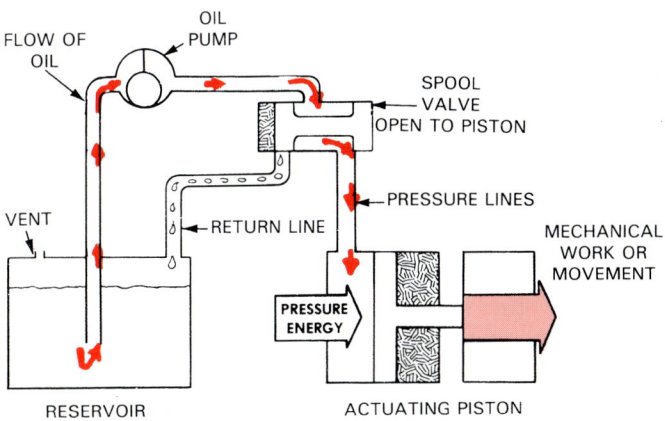

Fig. 26-22. Basic hydraulic circuit. Pump draws oil out of reservoir and forces it through spool valve. In this position, spool valve routes oil to piston. Piston uses oil pressure to produce movement or clamping pressure.

The hydraulic system also forces oil to high friction points in the transmission. This prevents wear and overheating by lubricating the moving parts.

Hydraulic pump (oil pump)

The hydraulic pump, also called the oil pump, produces the pressure to operate an automatic

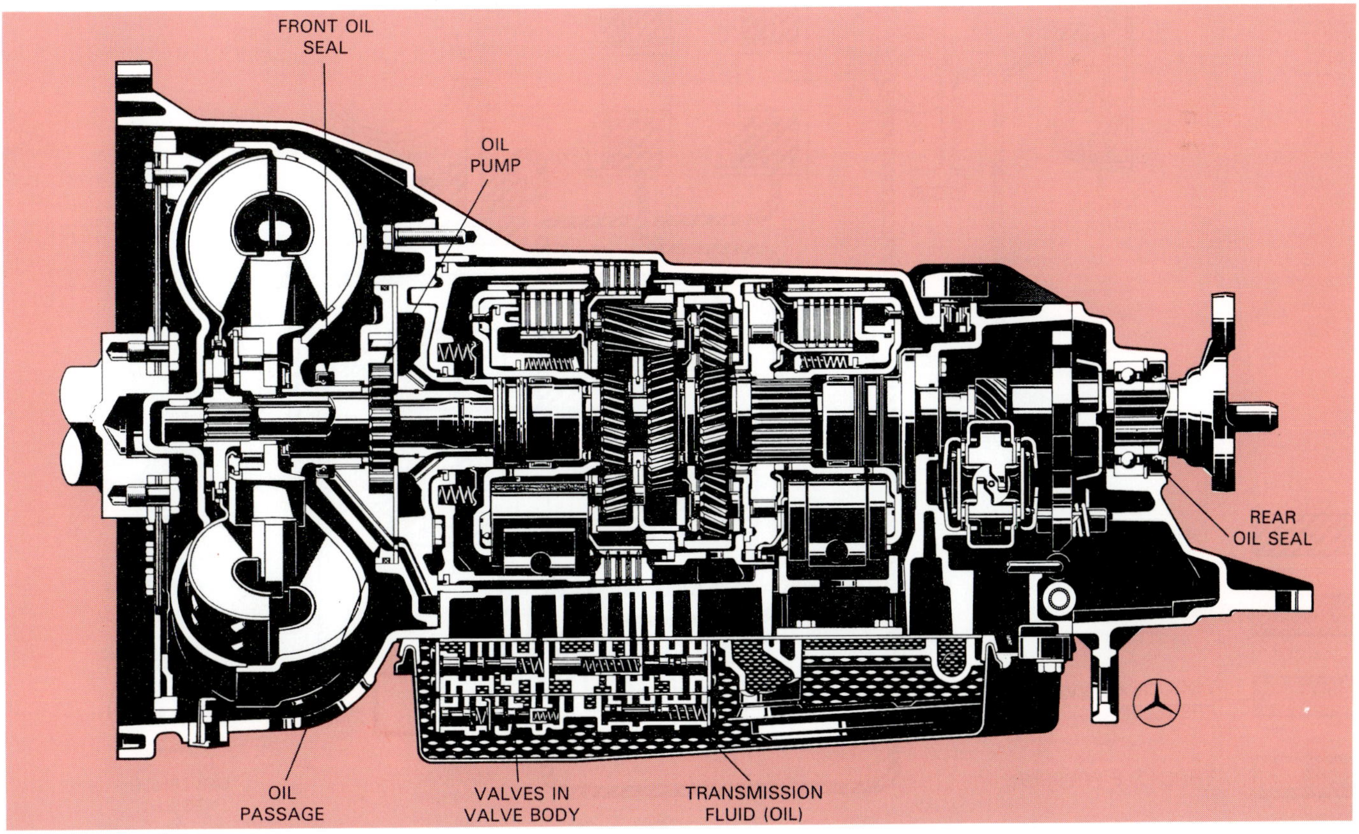

Fig. 26-23. Oil pump is normally located in front of case. Oil is drawn out of pan, circulated through passages and to hydraulic components. Also note location of oil seals. (Mercedes-Benz)

transmission. Modern transmissions only have one pump located directly behind the torque converter.

Look at Fig. 26-23. The sleeve or collar on the rear of the torque converter drives the pump.

The automatic transmission oil pump has several basic functions:

1. Produces pressure to operate the clutches, bands, and gearsets.
2. Lubricates the moving parts in the transmission.
3. Keeps the torque converter filled with oil for proper operation.
4. Circulates oil through the transmission and cooling tank (radiator) to transfer heat.
5. Operates hydraulic valves in the transmission.

There are two commonly used oil pumps: The gear type and the rotor type, Fig. 26-24.

When the torque converter rotates the oil pump, transmission fluid is drawn into the pump from the pan. The pump compresses the oil and forces it to the pressure regulator. This is illustrated in Fig. 26-25.

Pressure regulator

The pressure regulator limits the maximum amount of oil pressure developed by the oil pump,

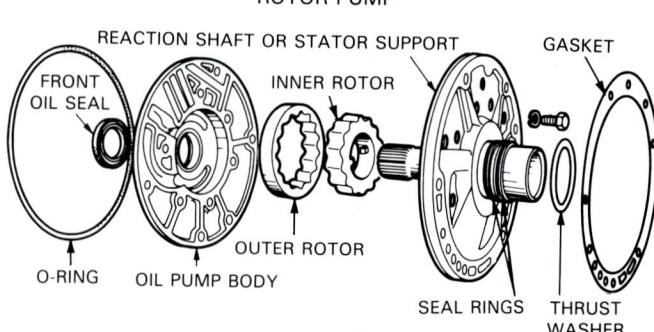

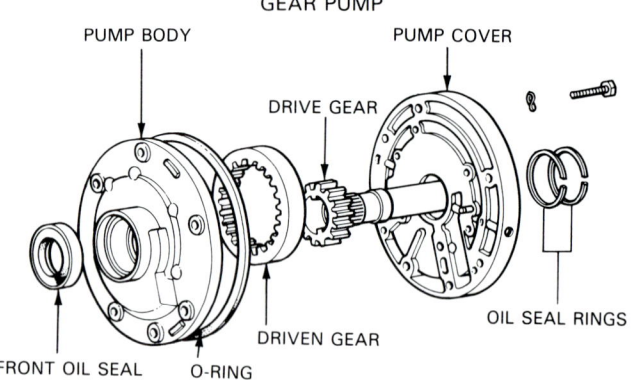

Fig. 26-24. Two basic types of automatic transmission pumps, rotor and gear. Study similarities and differences. Torque converter normally drives pump. (Toyota)

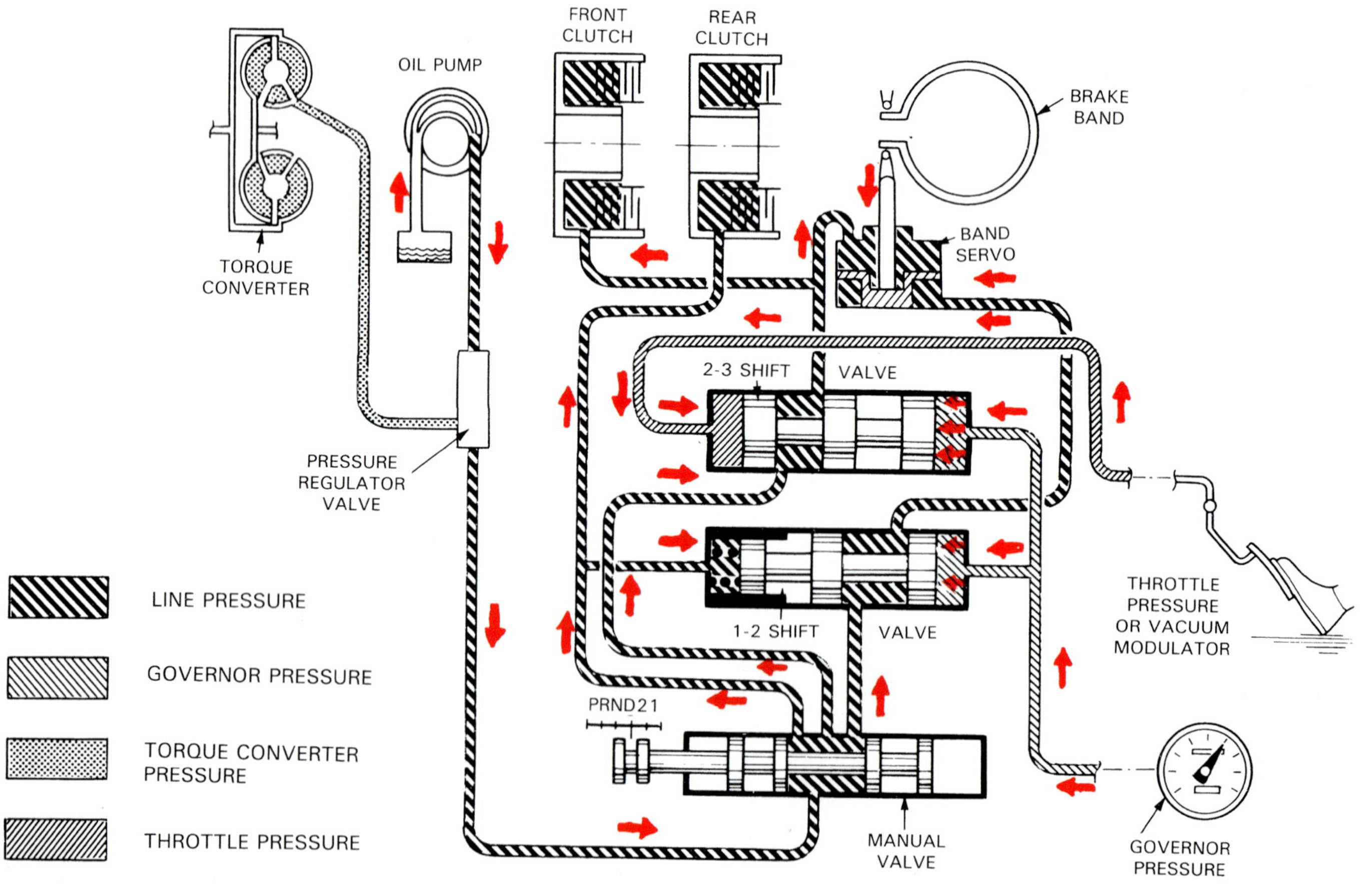

LINE PRESSURE

GOVERNOR PRESSURE

TORQUE CONVERTER PRESSURE

THROTTLE PRESSURE

Fig. 26-25. Simplified circuit showing hydraulic action in automatic transmission. Manual valve pressure, throttle valve pressure, and governor valve pressure operate balance or shift valves. Shift valves then direct oil pressure to correct clutch or band pistons. Study this diagram carefully. (Nissan)

Fig. 26-25. It is a spring-loaded valve that routes excess pump pressure out of the hydraulic system. This assures proper transmission operation.

Manual valve

A manual valve, operated by the shift mechanism, allows the driver to select park, neutral, reverse, or different drive ranges. When the gear shift lever is moved, the shift linkage moves the manual valve. As a result, the valve routes oil pressure to the correct components in the transmission. Look at Fig. 26-25.

Vacuum modulator valve

The vacuum modulator valve, also termed throttle valve, senses engine load (vacuum) and determines when the transmission should shift to a higher gear. Refer to Figs. 26-25 and 26-26. A vacuum line runs from the engine intake manifold to this valve.

As engine vacuum (load) rises and falls, it moves the diaphragm inside the vacuum modulator. This, in turn, moves the rod and hydraulic valve to change throttle control pressure in the transmission. In this way, the vacuum modulator can match transmission shift points to engine loading.

For example, if a car is climbing a steep hill (under a heavy pull), engine vacuum will be very low. This will allow the spring in the modulator to slide the modulator valve further into the transmission. The valve then directs oil pressure to delay the upshift. The transmission stays in a lower gear longer to allow the car to accelerate up the hill. Look at Fig. 26-26.

Governor valve

The governor valve senses engine speed (transmission output shaft speed) to help control gear shifting, Fig. 26-25. The vacuum modulator and governor work together to determine shift points.

Illustrated in Fig. 26-27 is one type of governor assembly. It consists of a drive gear, centrifugal weights, springs, hydraulic valve, and shaft. The governor gear is usually meshed with a gear on the transmission output shaft. Whenever the car and output shaft are moving, the centrifugal weights rotate.

When the output shaft and weights are rotating slowly, The weights are held IN by the governor springs. This causes a low pressure output and the transmission remains in a low gear ratio.

As engine and shaft speed increase, the weights are thrown out further and governor pressure increases. This moves the shift valve and causes the transmission to shift to a higher gear.

Other types of governor valves are also used. However, they do the same job.

Shift valves (balanced valves)

Shift valves, also called balanced valves, use control pressure (oil pressure from regulator, governor, throttle, and manual valves) to operate the bands, servos, and gearsets. Fig. 26-25 shows how the shift valves are connected to the other transmission components.

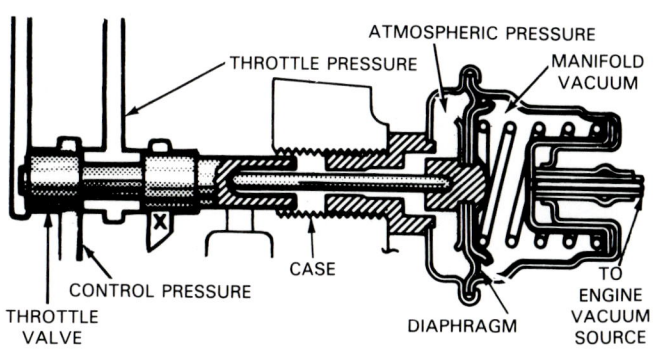

Fig. 26-26. Vacuum modulator operates throttle valve. Engine vacuum allows modulator to sense engine load. For example, with engine acceleration and high load, vacuum drops. The vacuum modulator spring could then overcome vacuum pull on the diaphragm. The spring would push the valve to the left. This would alter throttle oil pressure, keeping the transmission in a lower gear. (Ford)

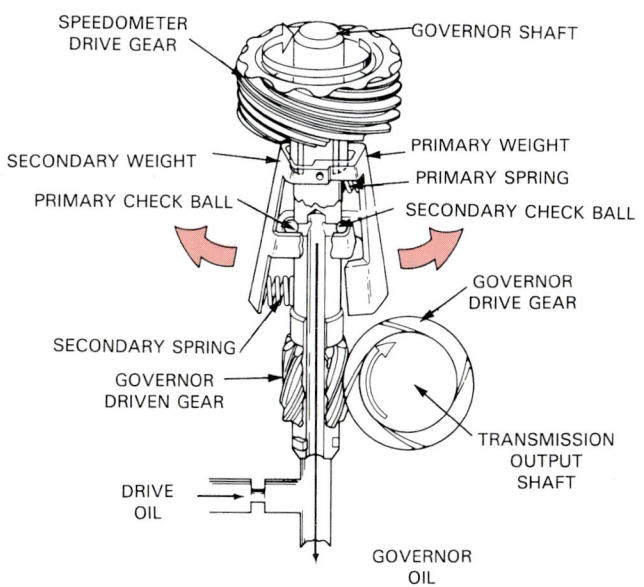

Fig. 26-27. Governor senses engine vehicle speed. Gear on transmission output shaft rotates governor. As speed increases, centrifugal weights are thrown outward. This opens the governor valve enough to change governor pressure and cause an upshift.

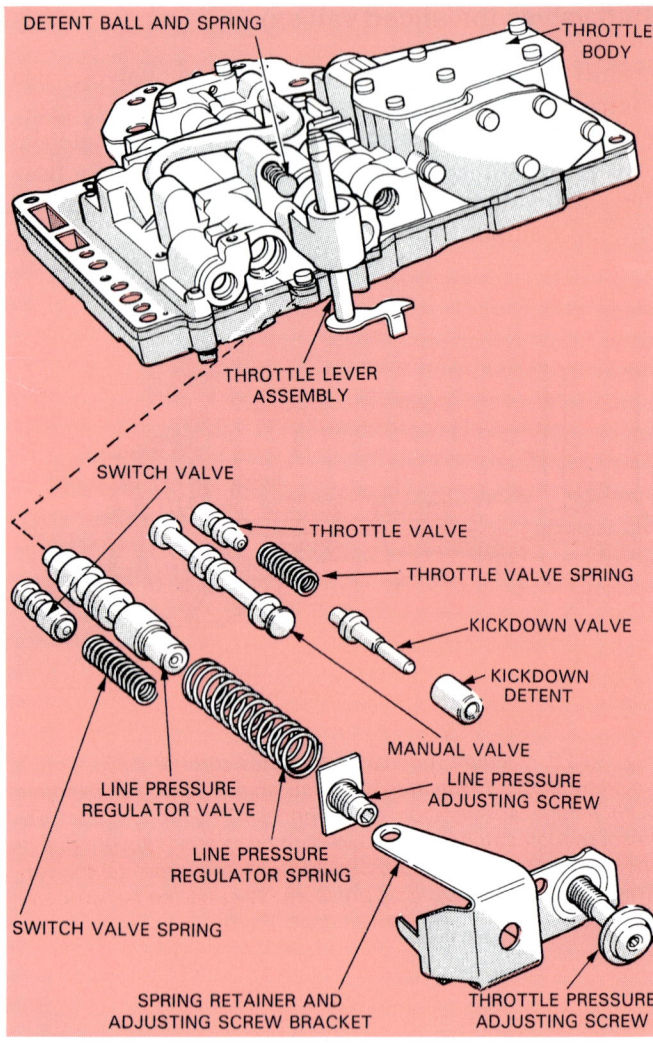

DETENT BALL AND SPRING

THROTTLE BODY

THROTTLE LEVER ASSEMBLY

SWITCH VALVE

THROTTLE VALVE

THROTTLE VALVE SPRING

KICKDOWN VALVE

KICKDOWN DETENT

MANUAL VALVE

LINE PRESSURE REGULATOR VALVE

LINE PRESSURE ADJUSTING SCREW

LINE PRESSURE REGULATOR SPRING

SWITCH VALVE SPRING

SPRING RETAINER AND ADJUSTING SCREW BRACKET

THROTTLE PRESSURE ADJUSTING SCREW

Fig. 26-28. Throttle body bolts to bottom of transmission case. It houses manual valve, pressure regulator valve, kickdown valve, and other valves.

Oil pressure from the other transmission valves act on each end of the shift valves. For example, if the pressure from the governor is high and the pressure from the throttle and manual valves are low, the shift valves will be moved sideways in their cylinder.

In this way, the shift valves are sensitive to engine load (throttle valve oil pressure), engine speed (governor valve oil pressure) and gear shift position (manual valve oil pressure). The shift valves move according to these forces and keep the transmission shifted into the correct gear ratio for the driving conditions.

Kickdown valve

A kickdown valve causes the transmission to shift into a lower gear during fast acceleration. A rod or cable links the carburettor or fuel injection throttle body to a lever on the transmission.

When the driver presses down on the accelerator pedal, the lever moves the kickdown valve. This causes hydraulic pressure to override normal shift control pressure and the transmission downshifts, Fig. 26-28.

Valve body

The valve body contains many of the hydraulic valves (pressure regulating valve, shift valves, manual valve, etc.) of an automatic transmission. See Fig. 26-28.

The valve body bolts to the bottom of the transmission case. It is housed in the transmission pan. A filter or screen is usually attached to the bottom of the valve body, Fig. 26-29.

Passages in the valve body route fluid from the pump to the valves and then into the transmission case. Passages in the case carry fluid to the other hydraulic components.

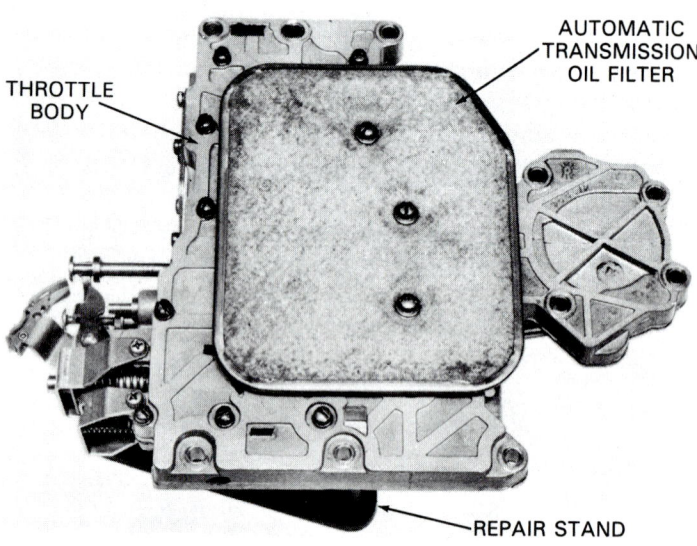

THROTTLE BODY

AUTOMATIC TRANSMISSION OIL FILTER

REPAIR STAND

Fig. 26-29. Oil filter is fastened to bottom of throttle body. It removes particles before they can enter hydraulic circuit.

Automatic transmission fluid

Automatic transmission fluid is a special type of oil having several additives that make it compatible with the friction clutches and bands in the transmission. Different types of automatic transmission fluids are available for different transmissions.

Transmission oil cooling

A tremendous amount of heat is developed inside an automatic transmission. When the torque converter slips, friction heats the fluid. This heat must be removed or transmission failure could result.

Many transmissions have an oil cooling system which includes external oil lines and a cooling tank inside the engine radiator. Look at Fig. 26-30.

When the engine is running, the transmission pump forces oil through the cooling lines and into the radiator tank. Since transmission oil is hotter than the engine coolant, oil temperature drops. The cooled oil returns to the transmission through the other line.

Some cars, especially those used to pull a heavy load (caravans, boats) use an auxiliary transmission oil cooler. It is a small radiator, separate from the engine radiator. Air passes over the radiator to cool the transmission fluid.

PARKING PAWL

A parking pawl is used to lock the transmission output shaft and keep the car from rolling when not in use. Fig. 26-31 shows its basic action.

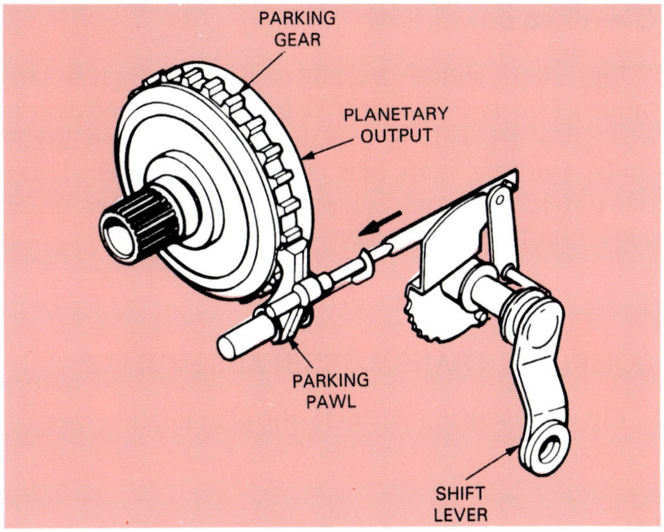

AUTOMATIC TRANSMISSION POWER FLOW

The flow of power through an automatic transmission depends on its specific design. However, you should have a GENERAL understanding of how power is transmitted through the major parts of modern transmissions.

Fig. 26-31. Parking pawl is simply a latch that locks into large teeth on parking gear. Since pawl is mounted on case, this locks parking gear and output shaft so car will not roll. (Subaru)

Fig. 26-32 shows how torque moves from the input shaft to the output shaft. This is a typical three-speed transmission. Study each illustration carefully, noting which clutches, bands and gearset members are activated.

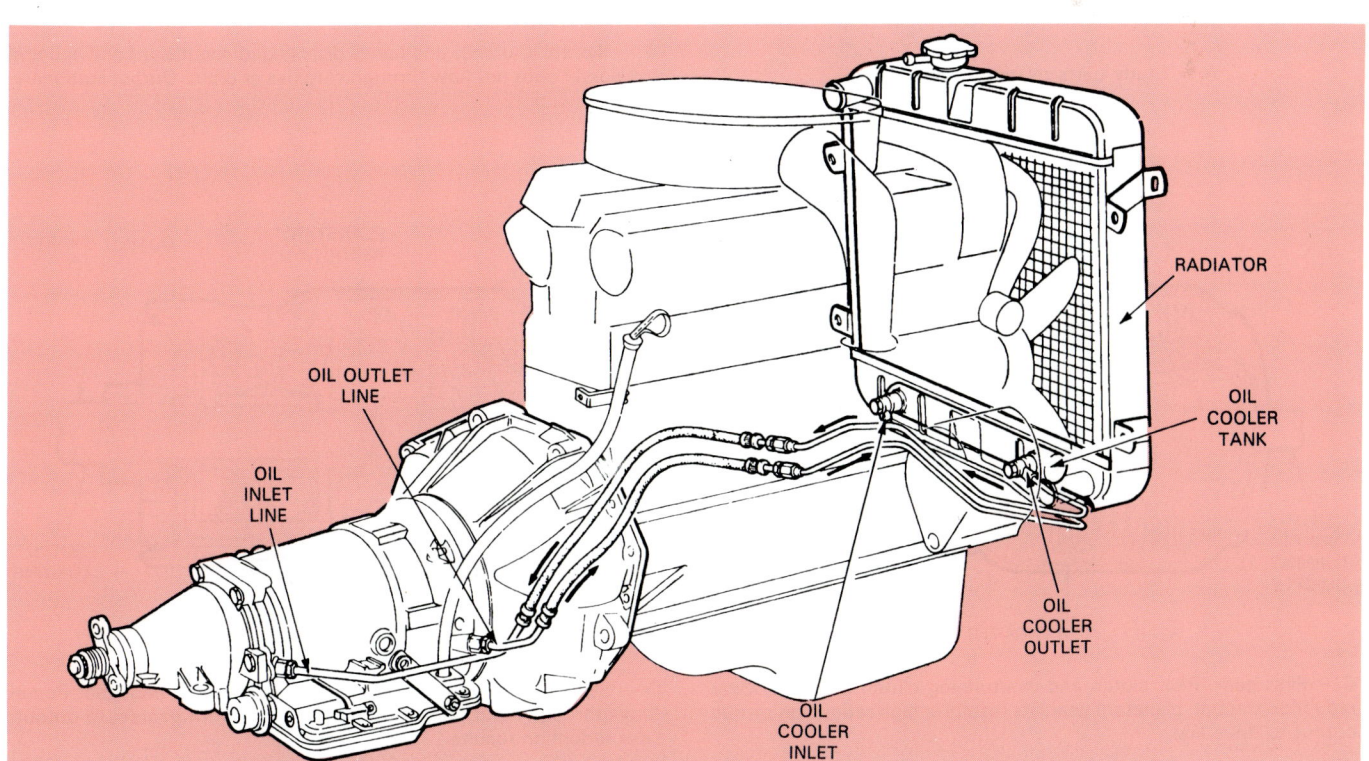

Fig. 26-30. Oil cooler tank is commonly used in transmission. Oil pump pushes oil through lines and cooler tank to maintain acceptable oil temperature. (Fiat)

Overdrive power flow

Fig. 26-33 shows the power flow through a late model, four-speed, overdrive automatic transmission in high gear. This is a design that uses two input shafts (turbine shaft and direct input shaft). Trace the power flow and compare it to the other more conventional transmissions covered earlier.

COMPLETE TRANSMISSION ASSEMBLIES

Fig. 26-34 through 26-36 show different types of

automatic transmissions. Study each closely. As you look at each part, try to remember its function.

Refer to service manuals for more information on a particular automatic transmission. The manual will give hydraulic circuit diagrams, specific illustrations, as well as detailed operating and construction descriptions for major components.

AUTOMATIC TRANSMISSION ELECTRONIC CONTROLS

Some new cars use the on-board computer to help control transmission shift points and monitor trans-

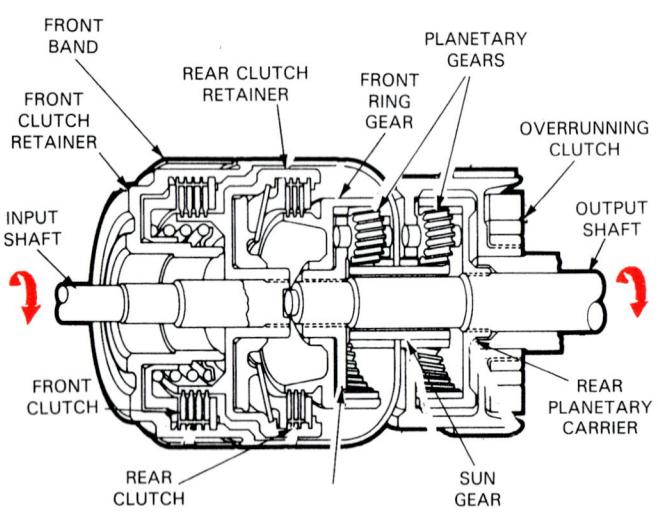

A — Study parts relating to power flow.

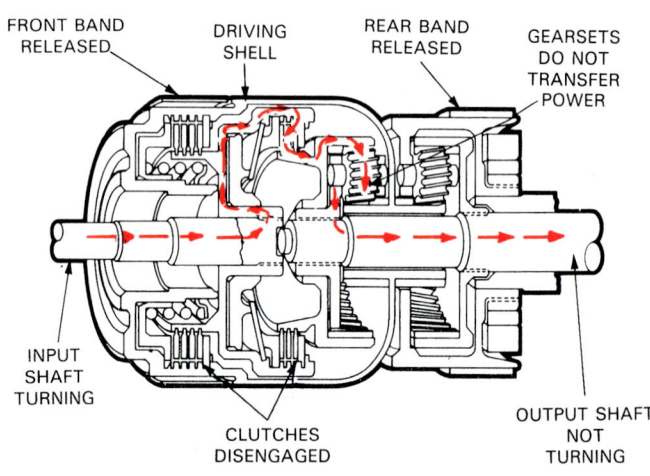

B — Neutral. Clutches and bands disengaged. Input shaft and hub turn but power does not flow through clutches or drum. Output stationary.

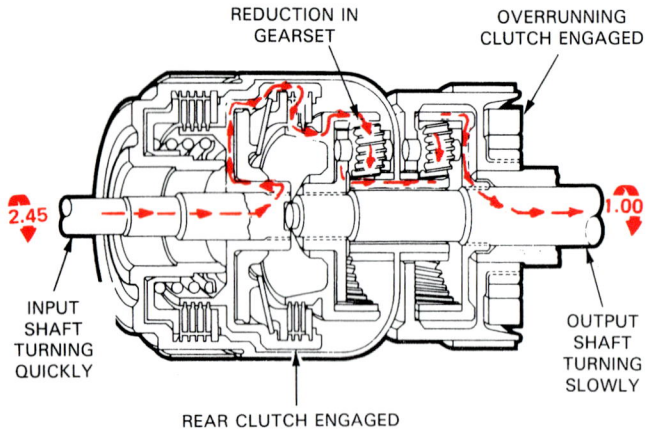

C — First gear. Rear clutch and overrunning clutch engaged. Gear reduction through planetary gearsets results in high ratio, high torque output to drive line.

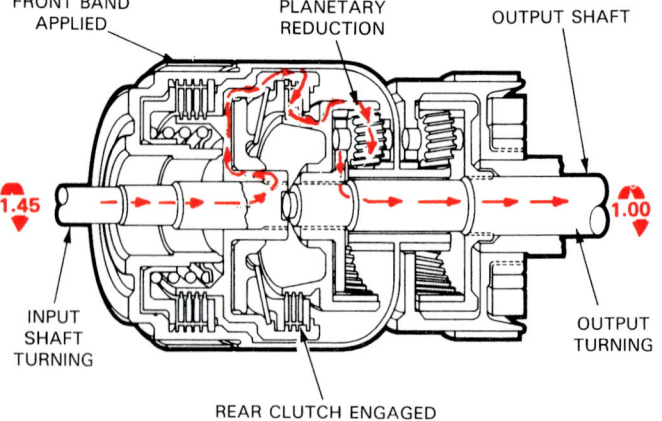

D — Second gear. Front band applied. Rear clutch engaged. Power flows through input, hub, clutch, drum, and front gearset to output. Less reduction results.

Fig. 26-32. Power flow through a typical automatic transmission.

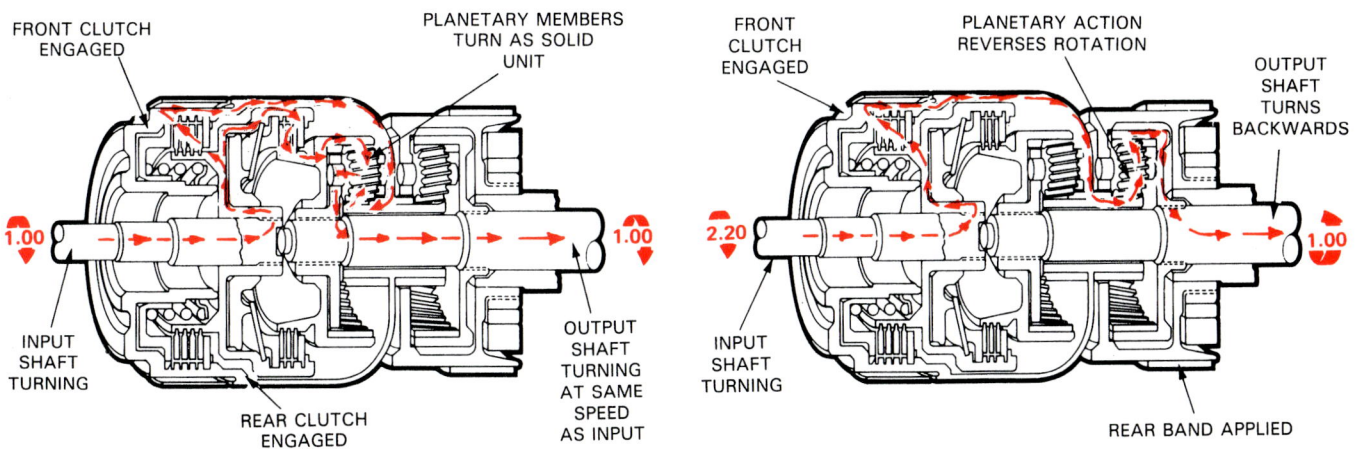

E — Third gear. Front and rear clutches engaged. Planetary members locked for direct drive. One-to-one ratio for higher vehicle speeds results.

F — Reverse. Front clutch and rear band applied. Power flows through clutch, shell, and sun gear to rear planetary gearset which reverses rotation.

(Fig. 26-32 continued)

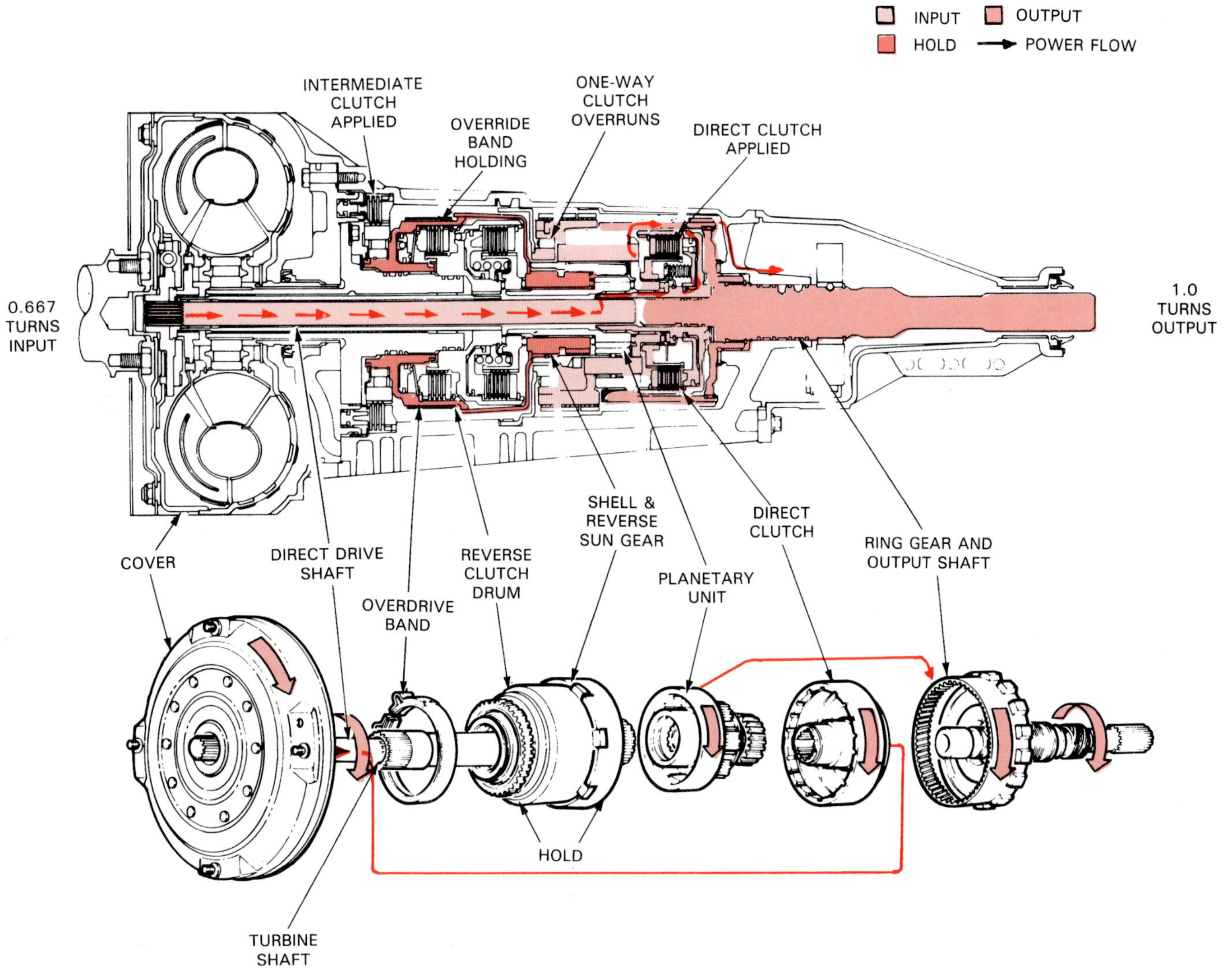

INPUT   OUTPUT
HOLD   POWER FLOW

INTERMEDIATE CLUTCH APPLIED

OVERRIDE BAND HOLDING

ONE-WAY CLUTCH OVERRUNS

DIRECT CLUTCH APPLIED

0.667 TURNS INPUT

1.0 TURNS OUTPUT

COVER

DIRECT DRIVE SHAFT

OVERDRIVE BAND

REVERSE CLUTCH DRUM

SHELL & REVERSE SUN GEAR

PLANETARY UNIT

DIRECT CLUTCH

RING GEAR AND OUTPUT SHAFT

HOLD

TURBINE SHAFT

Fig. 26-33. Power flow in high gear of modern four-speed automatic with overdrive. Study differences with transmissions already covered in chapter. (Ford)

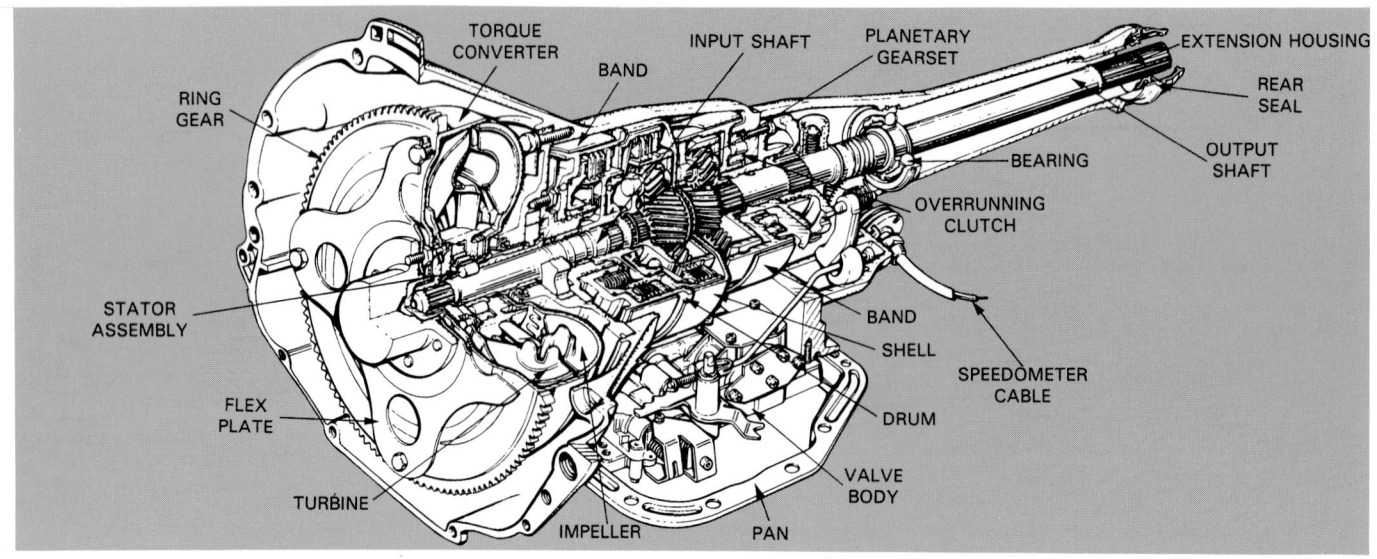

Fig. 26-34. This is a three-speed automatic transmission. Study part locations. Can you recall their function?

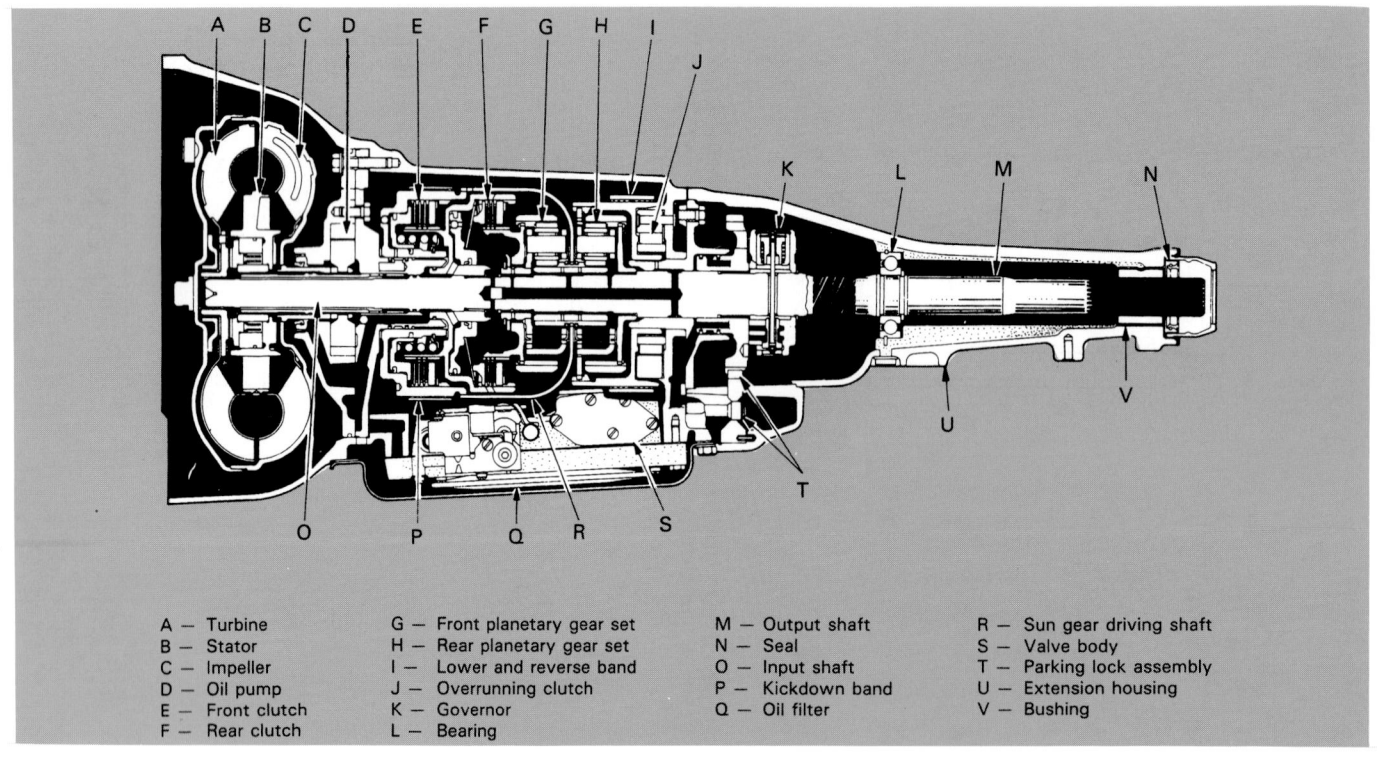

A — Turbine	G — Front planetary gear set	M — Output shaft	R — Sun gear driving shaft
B — Stator	H — Rear planetary gear set	N — Seal	S — Valve body
C — Impeller	I — Lower and reverse band	O — Input shaft	T — Parking lock assembly
D — Oil pump	J — Overrunning clutch	P — Kickdown band	U — Extension housing
E — Front clutch	K — Governor	Q — Oil filter	V — Bushing
F — Rear clutch	L — Bearing		

Fig. 26-35. Cutaway of Chrysler Torqueflite transmission. Compare it to the transmission in previous illustration.

mission operation. A simplified diagram of an electronic control system is given in Fig. 26-37.

The computer typically monitors engine speed, load, throttle position, transmission output shaft speed, gear shift position, and other variables. It can then provide control for transmission shift points, torque converter lockup, ignition timing, fuel injection timing, emission control system operation, and other functions. This keeps the transmission and other engine systems functioning at maximum efficiency.

CONTINUOUSLY VARIABLE TRANSMISSION

A continuously variable transmission, abbreviated CVT, has an infinite number of driving ratios NOT three, four, or five forward speeds, as with conventional transmissions. It uses centrifugally operated

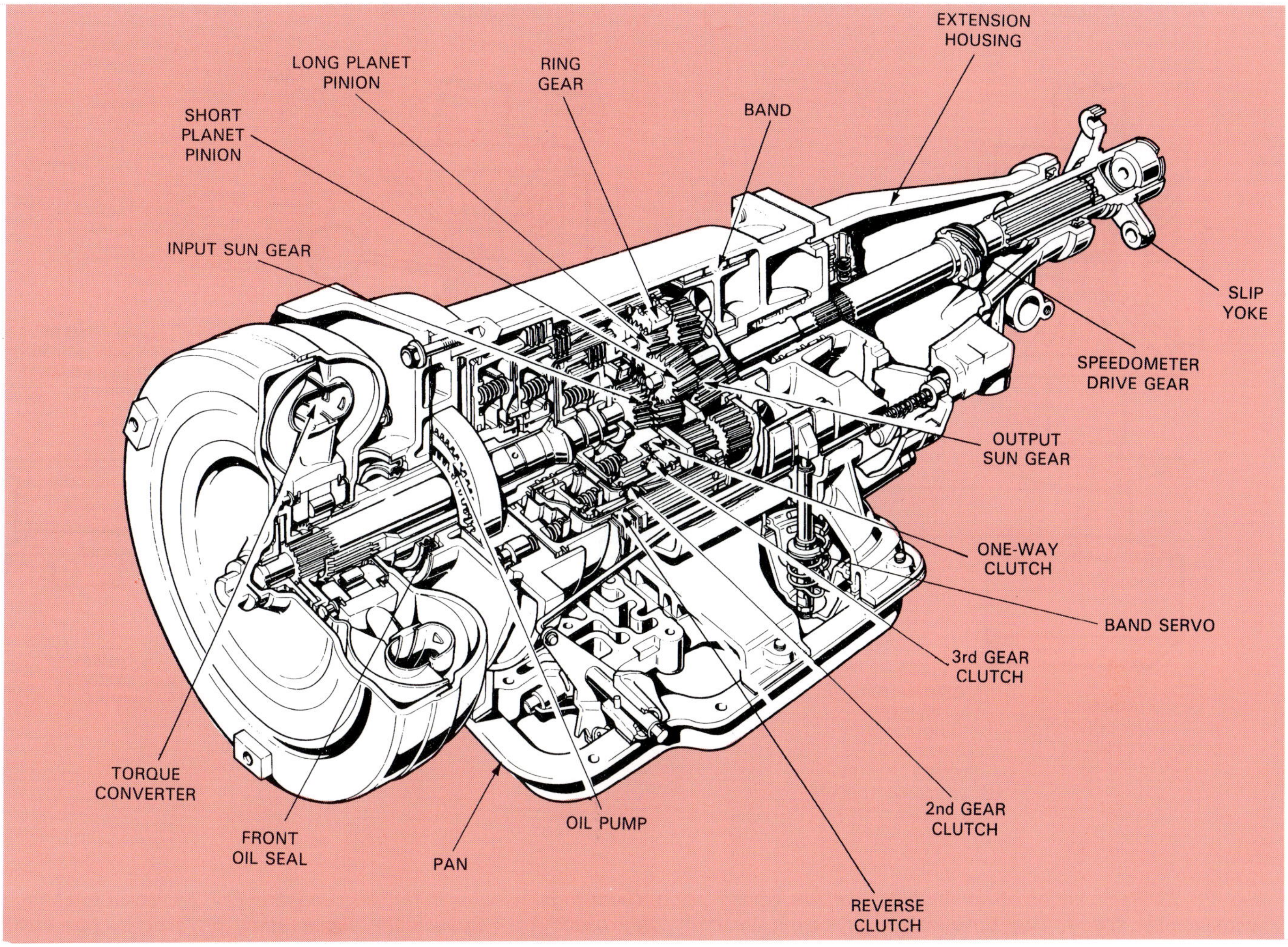

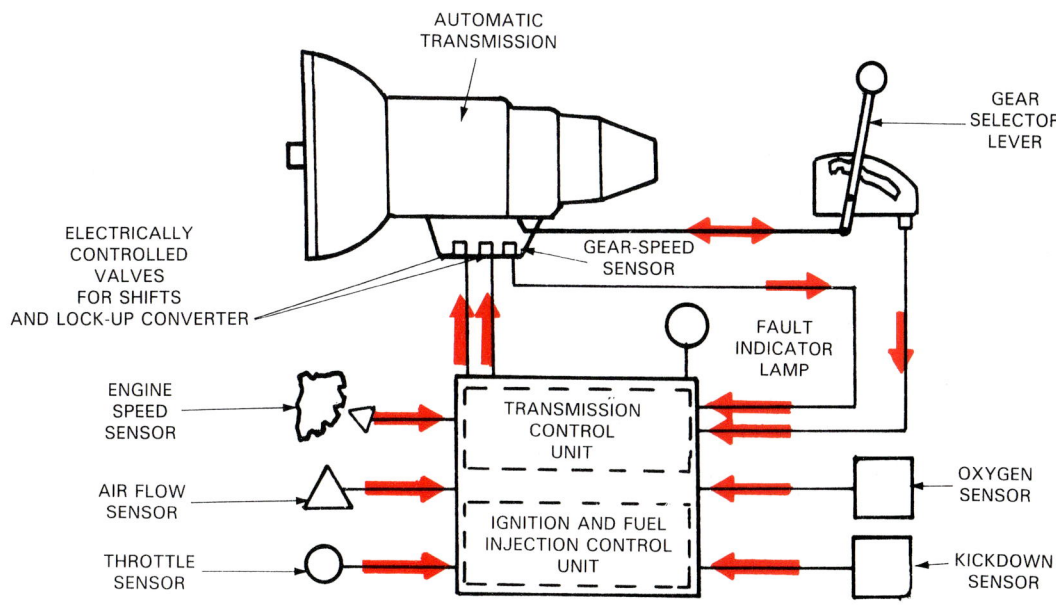

EXTENSION
HOUSING

LONG PLANET
PINION

RING
GEAR

BAND

SHORT
PLANET
PINION

INPUT SUN GEAR

SLIP
YOKE

SPEEDOMETER
DRIVE GEAR

OUTPUT
SUN GEAR

ONE-WAY
CLUTCH

BAND SERVO

3rd GEAR
CLUTCH

TORQUE
CONVERTER

FRONT
OIL SEAL

PAN

OIL PUMP

2nd GEAR
CLUTCH

REVERSE
CLUTCH

Fig. 26-36. Cutaway view shows internal parts of another automatic transmission. Note how servo piston is attached to band. (Fiat)

AUTOMATIC
TRANSMISSION

GEAR
SELECTOR
LEVER

ELECTRICALLY
CONTROLLED
VALVES
FOR SHIFTS
AND LOCK-UP CONVERTER

GEAR-SPEED
SENSOR

FAULT
INDICATOR
LAMP

ENGINE
SPEED
SENSOR

TRANSMISSION
CONTROL
UNIT

AIR FLOW
SENSOR

OXYGEN
SENSOR

THROTTLE
SENSOR

IGNITION AND FUEL
INJECTION CONTROL
UNIT

KICKDOWN
SENSOR

Fig. 26-37. Some late model automatic transmissions use a computer to help control shift points. Note flow of data to and from computer and transmission.

A — ACCELERATION

B — CRUISING SPEEDS

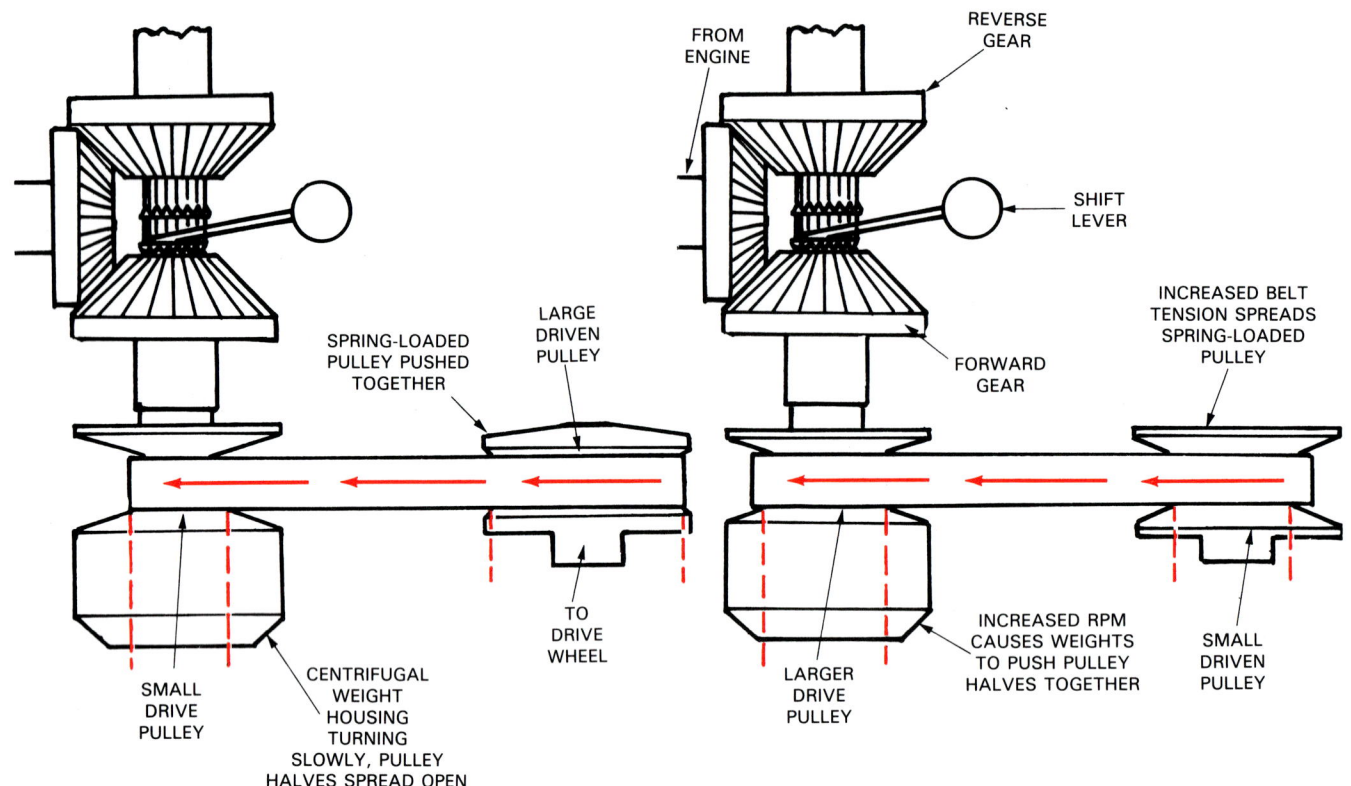

Fig. 26-38. Basic action of continuously variable transmission. Centrifugal weights in housing cause pulley diameters to change with vehicle speed. Two drive belt mechanisms are commonly used. A — Upon initial acceleration, drive pulley has small diameter and driven pulley has larger diameter. This provides gear reduction for rapid acceleration. B — As car and pulley speed increases, centrifugal weights push one pulley together, increasing its diameter. This increases belt tension, pulling other pulley apart. As a result, ratio constantly decreases with increase in speed.

pulleys and usually two V-belts instead of planetary gearsets.

Fig. 26-38 shows a simplified drawing of a CVT. During initial acceleration, a small drive pulley turns a larger pulley, drive reduction results.

As speed increases, centrifugal force pushes the halves of the drive pully together. The belt rides out in the pulley, increasing the pulley's effective diameter. As a result, a larger pulley drives a smaller pulley for more vehicle speed.

A CVT transmission is used in some European cars and is being experimented with by Japanese car manufacturers. It is capable of increasing fuel economy approximately 25 percent because it keeps the engine at its most efficient operating speed. Engine rpm can be kept relatively constant. The engine does NOT have to accelerate through each gear, resulting in an almost perfectly smooth increase in vehicle speed.

KNOW THESE TERMS

Converter housing, Case, Extension housing, Torque converter, Impeller, Stator, Turbine, Lock-up converter, Overrunning clutch, Torque multiplication, Stall speed, Stator support, Planetary gearset, Multiple disc clutch, Clutch piston, Drum, Hub, Shell, Band, Servo, Oil pump, Pressure regulator, Manual valve, Kickdown valve, Valve body, Accumulator, Automatic transmission fluid, Transmission cooler, Parking pawl, Automatic transmission electronic controls, Continuously variable transmission.

REVIEW QUESTIONS

1. List and explain the eight major parts of an automatic transmission.
2. An automatic transmission uses the following methods of transferring power.
   a. Friction.
   b. Fluids.
   c. Gears.
   d. All of the above.
   e. None of the above.
3. Describe the four major housings or components of an automatic transmission.
4. A _____ _____ is a fluid clutch that provides a means of coupling and uncoupling the engine and transmission.
5. Which of the following is NOT part of a torque converter?
   a. Band.
   b. Stator.
   c. Impeller.
   d. Turbine.
6. _____ _____ refers to the ability of a torque converter to increase the amount of engine torque applied to the transmission's input shaft.
7. Define the term "stall speed".
8. Why do many late model cars use a lock-up torque converter?
9. A planetary gearset consists of a _____ gear, several _____ gears, _____ gear _____, and a _____ gear.
10. List five functions of a planetary gearset.
11. Automatic transmission _____ and _____ are friction devices that drive and lock planetary gearset members.
12. Explain the operation of a clutch apply piston.
13. What is a servo?
14. An _____ is used in the apply circuit of a band or clutch to cushion initial application.
15. An overrunning clutch locks in one direction and freewheels in the other. True or False?
16. List and describe the major parts of the hydraulic system in an automatic transmission.
17. List five functions of the oil pump in an automatic transmission.
18. This valve senses engine speed (transmission output shaft rpm) to help control gear shifting.
   a. Vacuum modulator valve.
   b. Governor valve.
   c. Regulator valve.
   d. Manual valve.
19. How do the shift or balanced valves work?
20. Engine oil is compatible with the friction material in an automatic transmission. True or False?

Chapter 27

DIFFERENTIAL, REAR AXLE FUNDAMENTALS

After studying this chapter, you will be able to:
☐ Identify the major parts of a rear drive axle assembly.
☐ List the functions of a rear axle assembly.
☐ Describe the operation of a differential.
☐ Explain differential design variations.
☐ Compare different types of axles.
☐ Describe the principles of a limited-slip differential.
☐ Relate rear axle ratios to vehicle performance.

After engine power flows through the transmission and drive shaft, it enters the rear axle assembly. The rear axles transfer torque to the rear wheels of the car.

When the car's engine is in the front and drive wheels are at the rear, a rear drive axle and differential assembly is needed. Today's high performance cars and older cars commonly use this setup.

Many new cars are front engine, front-wheel drive and they use a transaxle. A rear drive axle is NOT needed on these vehicles. However, many of the operating principles in a transaxle and a differential are the same. This chapter, as a result, will prepare you for the later chapter covering front-wheel drive vehicles.

BASIC REAR DRIVE AXLE ASSEMBLY

A simple rear drive axle assembly, Fig. 27-1, consists of:
1. PINION DRIVE GEAR (transfers power from drive shaft to ring gear).

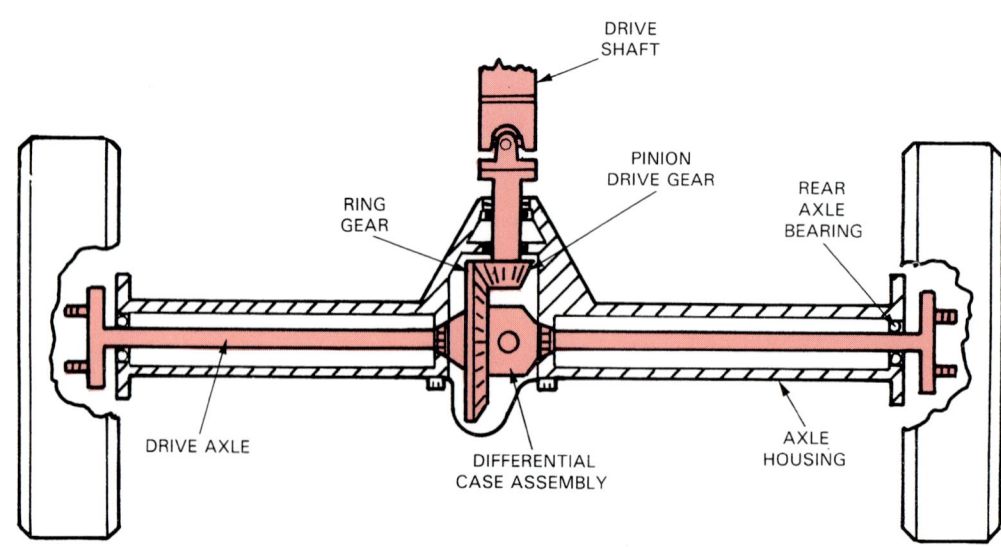

Fig. 27-1. Study fundamental components of rear drive axle. Drive shaft turns pinion gear. Pinion gear turns ring gear and differential case assembly. Differential transfers power to drive axle and wheels.

2. RING GEAR (transfers turning power to differential case assembly).

3. DIFFERENTIAL CASE ASSEMBLY (holds ring gear and other components that drive rear axles).

4. REAR DRIVE AXLES (steel shafts that transfer torque from differential assembly to drive wheels).

5. REAR AXLE BEARINGS (ball or roller bearings that fit between axles and inside of axle housing).

6. AXLE HOUSING (metal body that encloses and supports parts of rear axle assembly).

Rear axle power flow

Power enters the rear axle assembly from the drive shaft, Fig. 27-1. The drive shaft spins the pinion drive gear. The pinion gear turns the larger ring gear to produce a gear reduction.

Since the ring gear is bolted to the differential case, the case rotates with the ring gear. Small gears, inside the differential case, send torque to each axle.

The axles extend out of the axle housing. The axles normally hold and turn the rear wheels and tyres to propel the car.

FUNCTION OF A REAR DRIVE AXLE

A rear-wheel axle assembly has several functions. These include:

1. Send power from the drive shaft to the rear wheels.
2. Provide a final gear reduction.
3. Transfer torque through a 90° angle.
4. Split the amount of torque going to each wheel.
5. Allow for different wheel rotating speeds when the car is turning corners.
6. Support the rear axles, brake assemblies, suspension components, and chassis.

DIFFERENTIAL CONSTRUCTION

A differential assembly uses drive shaft rotation to transfer power to the axle shafts. The term "differential" can be remembered by thinking of the words "different" and "axle". The differential must be capable of providing torque to both AXLES, even when they are turning at DIFFERENT speeds (car turning corner, for example).

Look at Fig. 27-2. It shows the major parts of a rear drive axle.

Pinion gear

The pinion gear turns the ring gear when the drive shaft is rotating. One is shown in Fig. 27-2.

The outer end of the pinion drive gear is splined to the rear U-joint companion flange or yoke. The inner end of the pinion gear meshes with the teeth on the ring gear. The pinion gear is normally mounted on tapered roller bearings. They allow the pinion gear to

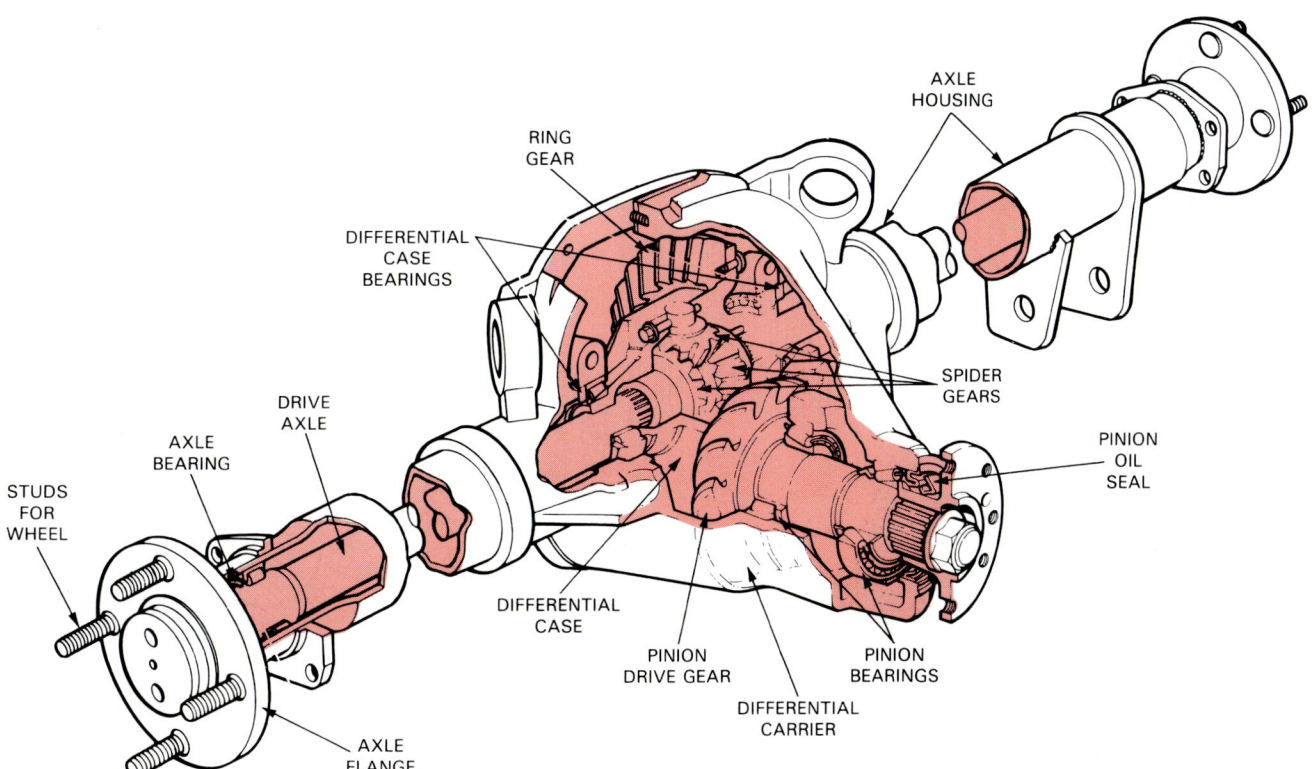

Fig. 27-2. Cutaway shows detailed view inside axle housing. Note relationship of parts. Axle fastens to suspension components. (Ford)

revolve freely in the carrier. Either a crushable sleeve or shims are used to preload the pinion gear bearings.

With some differentials, the extreme inner end of the pinion gear is supported by a pinion pilot bearing. It is usually a straight roller bearing. The pinion pilot bearing helps the two tapered roller bearings support the pinion gear during periods of heavy load. See Fig. 27-3.

Ring gear

The ring gear is driven by the pinion gear; it transfers rotating power through an angle change of 90 degrees. The ring gear has more teeth than the pinion gear. Bolts hold the ring gear securely to the differential case. Refer to Fig. 27-4.

The ring and pinion drive gears are commonly a matched set. They are lapped (meshed and spun together with abrasive compound on teeth) at the factory. Then one tooth on each gear is marked to show correct teeth engagement. Lapping produces quieter operation and assures longer gear life.

Hunting and non-hunting gears

In a hunting-type gear set, any one pinion gear tooth comes into contact with all drive gear teeth. In this type, several revolutions of the ring gear are required to make all possible gear combinations.

In a non-hunting type gear set, any one pinion gear tooth comes into contact with only a few ring gear teeth. In this type, only one revolution of the ring gear is required to make all possible tooth contact combinations.

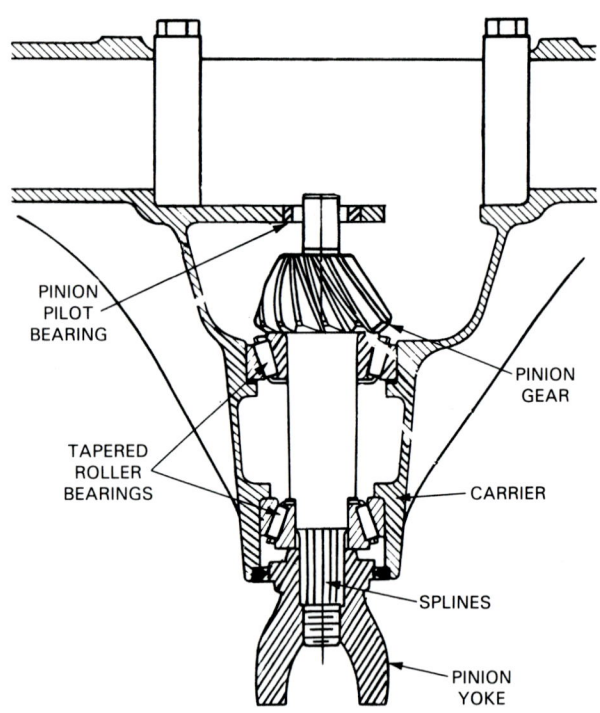

Fig. 27-3. Pinion gear is mounted on tapered roller bearings and sometimes pinion pilot bearing. Drive shaft yoke is splined to pinion drive gear. (Ford)

Hypoid and spiral bevel gears

Hypoid gears have the driving pinion centreline offset or lowered from the centreline of the ring gear. Modern differential ring and pinion gears are hypoid type. Refer to Fig. 27-5.

Spiral bevel gears have curved gear teeth with the

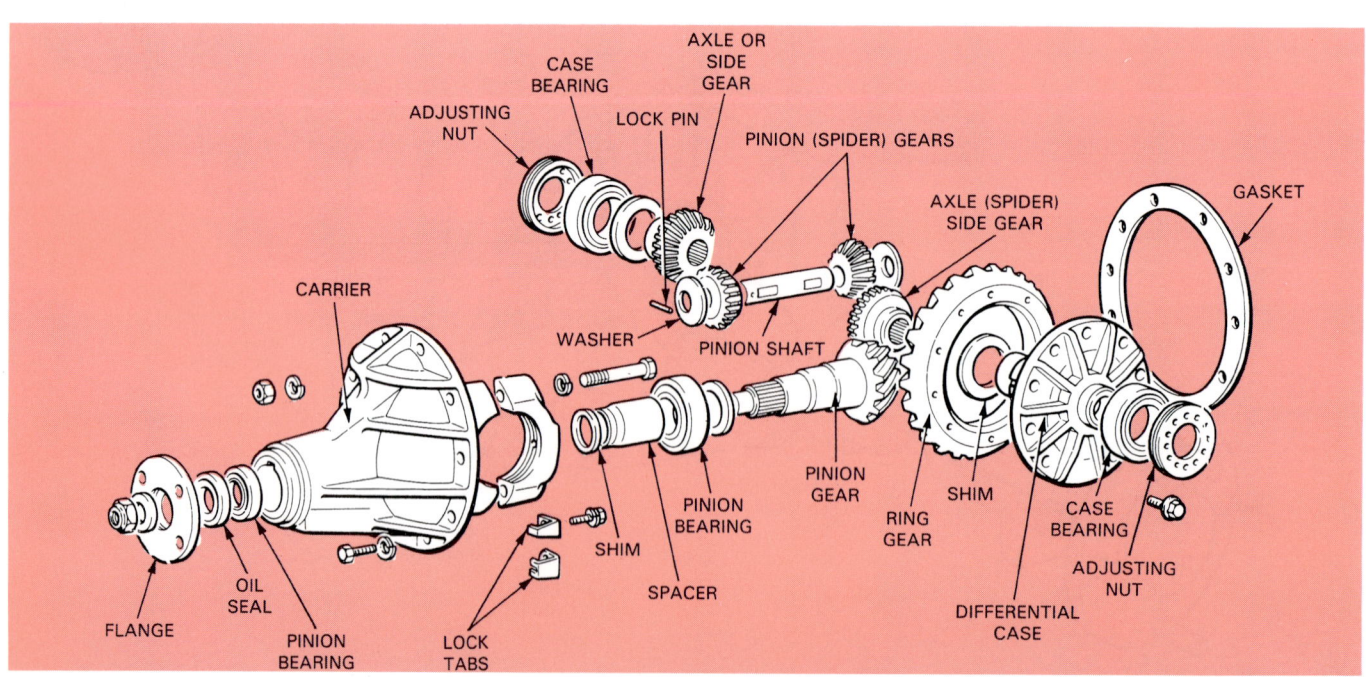

Fig. 27-4. Exploded view shows all of the major parts of a differential assembly. Ring gear bolts to differential. Differential case mounts in bearings which are adjusted using large nuts. Study part names and how they fit together.

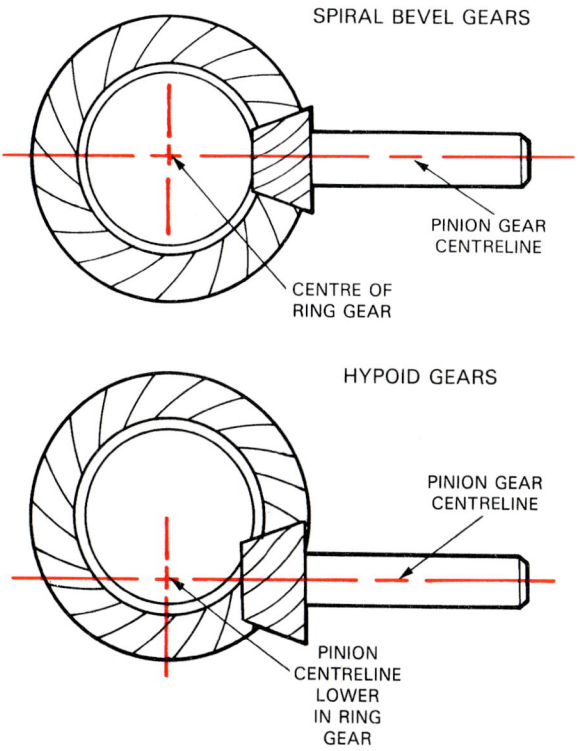

SPIRAL BEVEL GEARS

PINION GEAR CENTRELINE

CENTRE OF RING GEAR

HYPOID GEARS

PINION GEAR CENTRELINE

PINION CENTRELINE LOWER IN RING GEAR

Fig. 27-5. Modern ring and pinion gears are hypoid type. Note how hypoid lowers centreline of drive pinion gear. This improves gear tooth contact and lowers drive shaft hump in floor of car.

pinion and ring gears on the same centreline. This type gear setup is no longer used.

Hypoid gears have replaced spiral bevel gears because they lower the hump in the car floor and improve gear meshing action. With more than one gear tooth in contact, a hypoid design increases gear life and reduces gear noise.

Rear axle ratio

Rear axle ratio, also termed differential ratio, is determined by comparing the number of teeth on the pinion drive gear and on the ring gear.

To calculate rear axle ratio, count the number of teeth on each gear. Then divide the number of pinion teeth into the number of ring gear teeth.

For example, if the drive pinion gear has 10 teeth and the ring gear 30 teeth (30 divided by 10), the rear axle ratio would be 3:1.

Generally, auto makers install a rear axle ratio that provides a compromise between performance and economy. An average ratio is 3.50:1.

A higher axle (numerical) ratio, 4.11:1 for instance, could increase acceleration and pulling power but it would also decrease fuel economy. The engine would have to run at a higher rpm to maintain an equal cruising speed.

A lower axle (numerical) ratio, 3:1, would reduce acceleration and pulling power but it could improve fuel economy. The engine would run at a lower rpm while maintaining the same speed.

Differential carrier

The differential carrier provides a mounting place for the drive pinion gear, differential case, and other differential components. There are two basic types of differential carriers: the removable type and the integral (unitised) type.

A removable carrier bolts to the front of the axle housing, as in Fig. 27-6A. Stud bolts are installed in the housing to provide proper carrier alignment.

A gasket fits between the carrier and housing to prevent oil leakage.

An integral carrier is constructed as part of the axle housing, Fig. 27-6B. A stamped metal or cast aluminium cover bolts to the rear of the integral carrier.

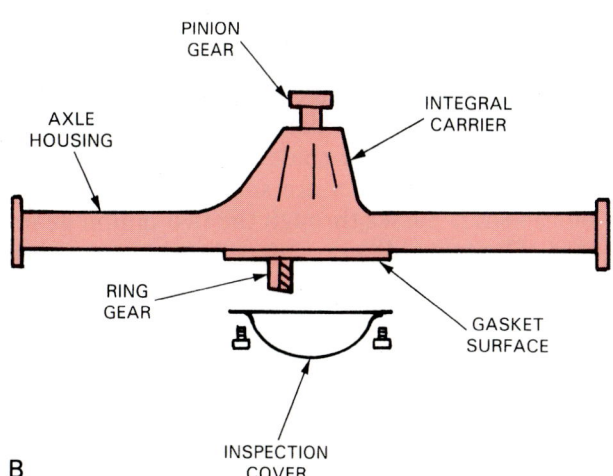

PINION GEAR

REMOVABLE CARRIER

RING GEAR

AXLE HOUSING

A

PINION GEAR

AXLE HOUSING

INTEGRAL CARRIER

RING GEAR

GASKET SURFACE

INSPECTION COVER

B

Fig. 27-6. Two basic types of rear axle carriers: A — Removable carrier is handy because it can be serviced at workbench. B — Integral or unitised carrier is formed as part of axle housing. It must be serviced in-car or in housing.

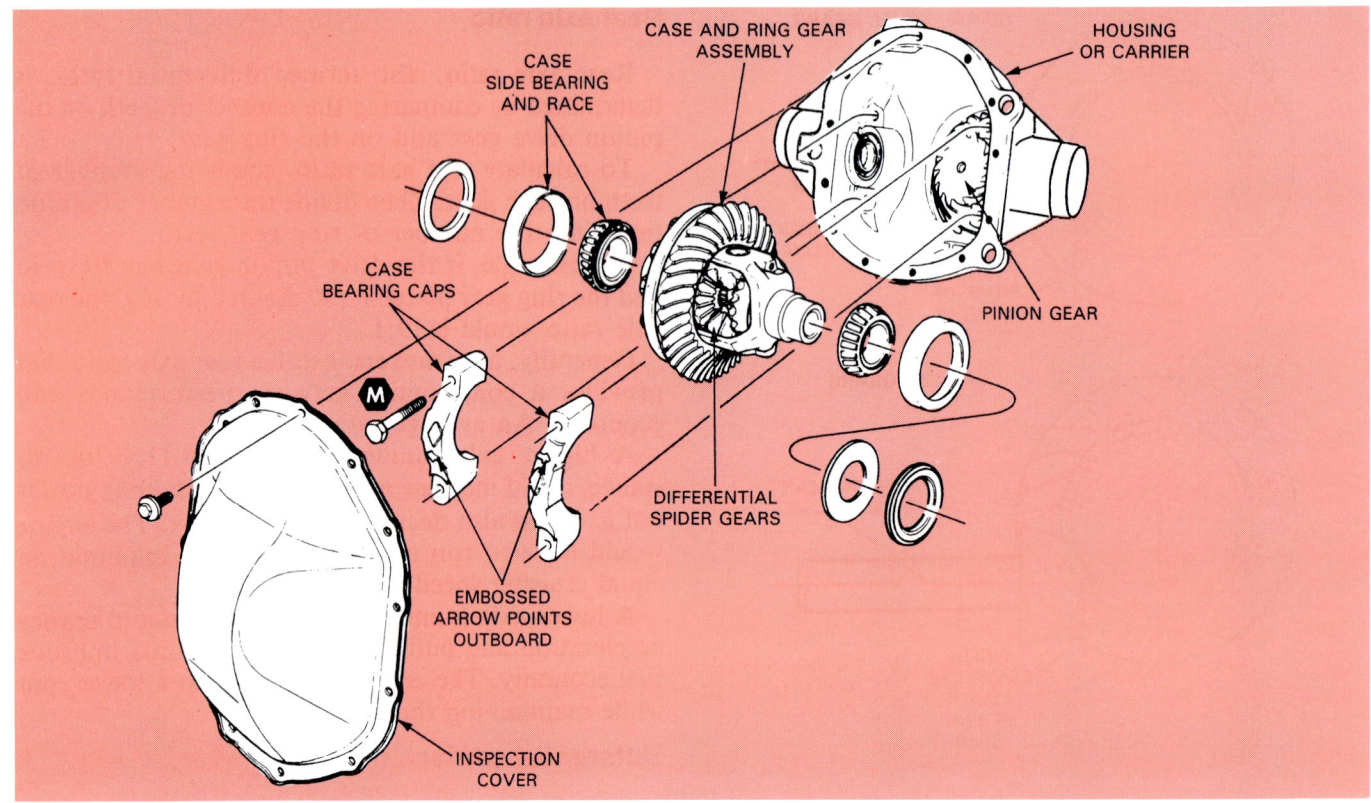

Fig. 27-7. Differential case mounts in carrier on tapered roller bearings. Races fit between bearings and carrier. Large caps secure bearing assemblies. Shims are used to adjust bearings and ring gear on this unit.

Differential case

The differential case holds the ring gear, spider gears, and inner ends of the axles. It mounts and rotates in the carrier, Fig. 27-7.

Case bearings, also called carrier bearings, fit between the outer ends of the differential case and the carrier. Refer to Figs. 27-4 and 27-7.

Spider or differential gears

The spider gears or differential gears basically include two axle gears (differential side gears) and two pinion gears (differential idler gears). Refer to Figs. 27-4 and 27-7. The spider gears mount inside the differential case. They are small bevel gears.

A pinion shaft passes through the two pinion gears and case. The two side gears are splined to the inner ends of the axles. This is shown in Fig. 27-8A.

DIFFERENTIAL ACTION

The rear wheels of a car do not always turn at exactly the same speed. When the car is turning, or when tyre diameters differ slightly, the rear wheels must rotate at different speeds.

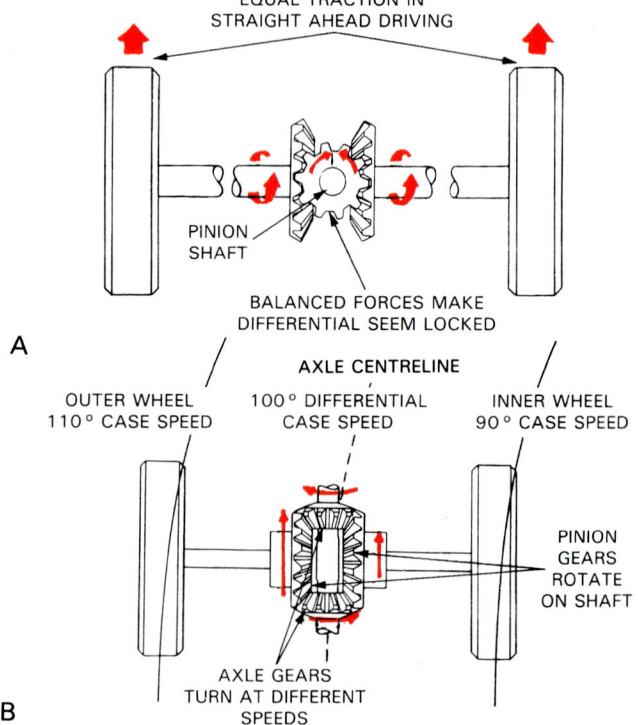

Fig. 27-8. Differential action allows wheels to turn at different speeds. A — Car travelling straight ahead. Differential spider gears inactive. Spider gears seem locked. B — With car turning corner, outer wheel must travel farther and turn faster than inner wheel. Pinion spider gears can rotate on their shaft, allowing axle side gears to turn at different speeds.

If there were a solid connection between each rear axle and the differential case, the tyres would tend to slide, squeal, and wear whenever the driver turned the steering wheel of the car. A differential is designed to prevent this problem.

Driving straight ahead

Look at view A, Fig. 27-8. Both rear wheels are turning at the same speed. The spinning case and pinion shaft rotate the differential pinion gears. The teeth on pinion gears apply torque to the axle side gears and axles. Balanced forces make the differential seem to be locked.

Turning corners

Fig. 27-8, view B, illustrates the action of a differential when the car is rounding a corner. Note how the outer wheel is turning faster than the inside wheel. The outer wheel must travel farther than the inner wheel.

The action of the spider gears allows each axle to change speed while still transferring torque to propel the car. Without a differential (solid case holding axles), you could break an axle or wear out tyres because of the different turning speeds of the rear wheels.

LIMITED SLIP DIFFERENTIALS

With a conventional differential, there may NOT be adequate traction on slippery pavement, in mud, or during rapid acceleration.

When one wheel of a conventional rear axle assembly lacks traction (on ice for example), the other wheel will NOT propel the car. Torque will flow

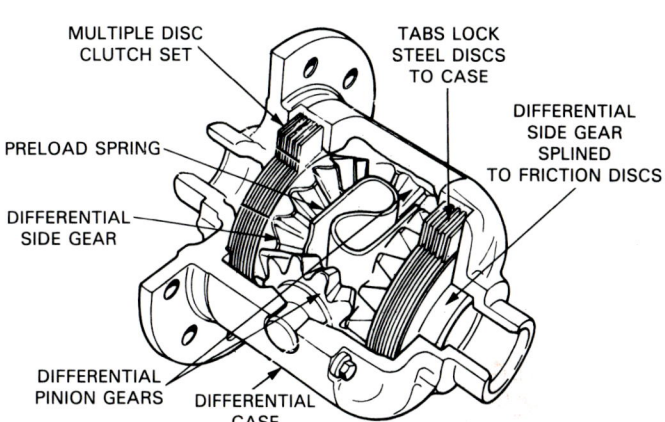

Fig. 27-9. Section view of limited slip differential shows how clutch discs and spider gears fit in case. Spring pushes side gears and clutch discs together. This sets up predetermined amount of friction that makes both axles drive car. (Ford)

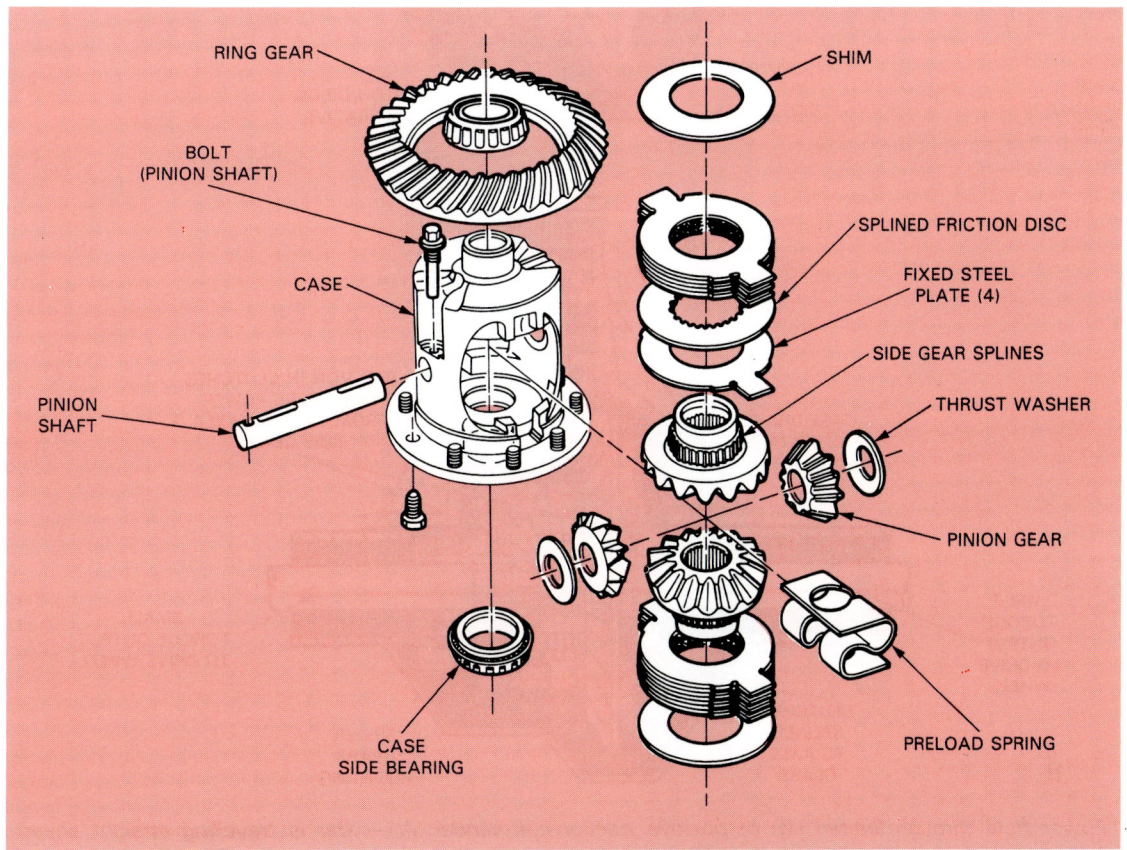

Fig. 27-10. This limited slip differential uses belleville or diaphragm springs to preload clutch discs. Friction discs are splined to axle side gears. Steel discs are locked to case by large tabs.

through the spider gears and to the axle that turns easiest.

A limited slip differential provides driving force to both rear wheels at all times. It transfers a portion of the driving torque to both the slipping wheel and the driving wheel. This will help prevent the car from being stuck in mud or snow.

Other names for a limited slip differential are positraction, trac-lock, sure-grip, equal-lock, or no-spin.

Clutch pack differential

The most popular type of limited slip differential uses a clutch pack (set of friction discs and steel plates). Look at Fig. 27-9. The friction discs are sandwiched between the steel plates inside the differential case.

The friction discs are usually splined to the differential side gears, Fig. 27-10. The steel plates have tabs which lock into notches in the differential case. The friction discs turn with the axle side gears. The plates turn with the case.

Springs (bellville springs, coil springs, or leaf spring) force the friction discs and steel plates

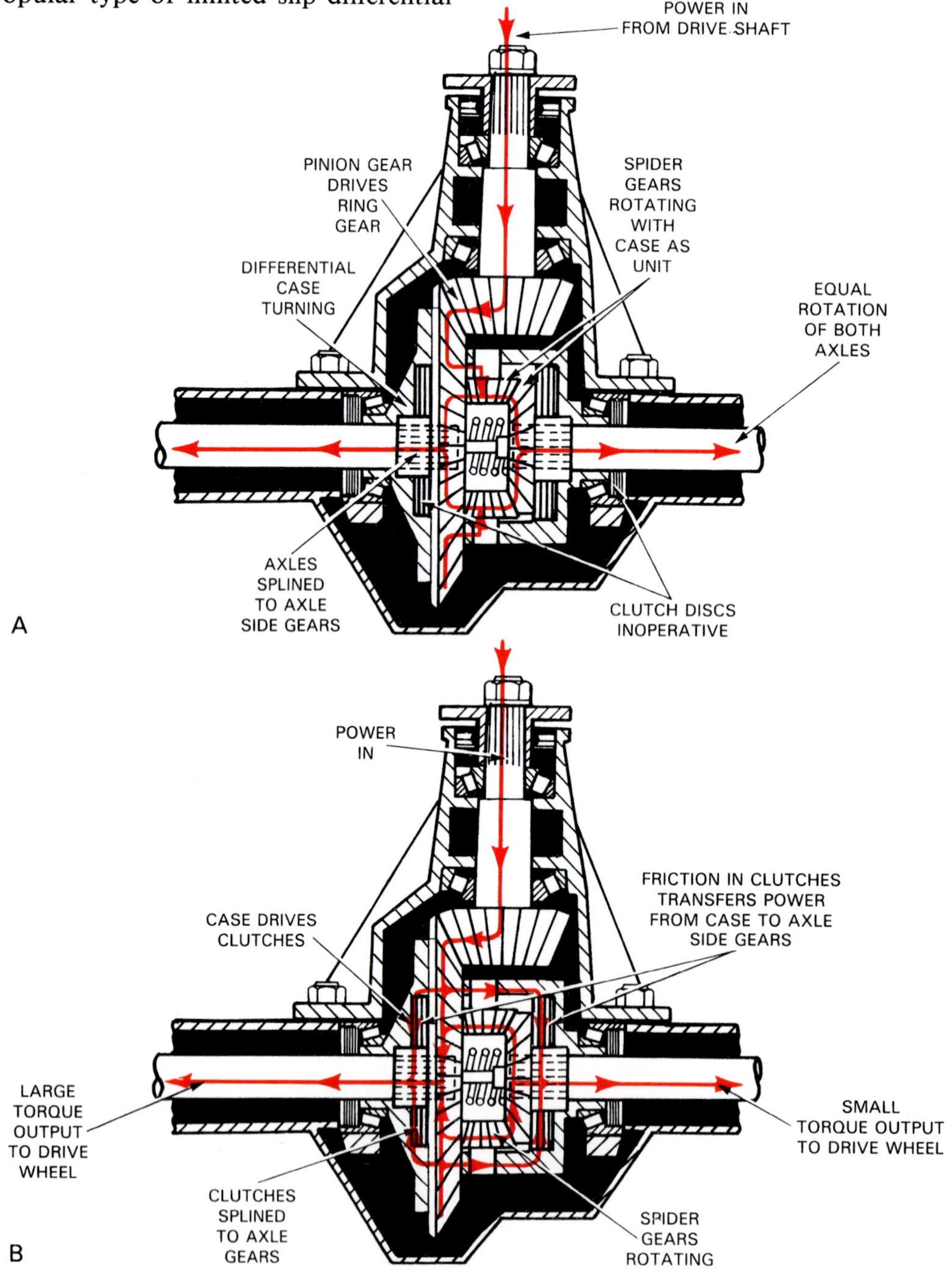

Fig. 27-11. Power flow through limited slip or positive traction differential. A — Car is travelling straight ahead or on dry pavement. Spider gears transfer power normally. Clutch discs turn together and do not slip. B — One wheel is on slippery pavement and it spins, friction between clutches still transfers torque to other axle. This gives car more traction than with conventional differential.

together. As a result, both rear axles try to turn with the differential case. Look at Figs. 27-10 and 27-11.

The thrust action of the spider gears normally helps the clutch spring(s) apply the clutch pack. Under high torque conditions, the rotation of the differential pinion gears PUSHES OUT on the axle side gears. The axle side gears then push on the clutch discs. This action helps lock the discs and keep both rear wheels turning.

However, when driving normally, the car can turn a corner without both wheels rotating at the same speed. The clutch pack will slip in turns, Fig. 27-11.

Cone clutch differential

A cone clutch limited slip differential uses the friction produced by cone shaped axle gears to provide improved traction. See Fig. 27-12.

Springs are used to force the cones against the ends of the differential case. With the axles splined to cone gears, the axles tend to rotate with the case.

Under rapid acceleration, the differential pinion gears, as they drive the cone gears, push outward on the cone gears. This increases friction between the cones and case even more and the drive wheels are turned with even greater torque.

REAR DRIVE AXLES

The rear drive axles connect the differential side gears to the drive wheels. They usually support the weight of the vehicle, Fig. 27-13. Rear axles are

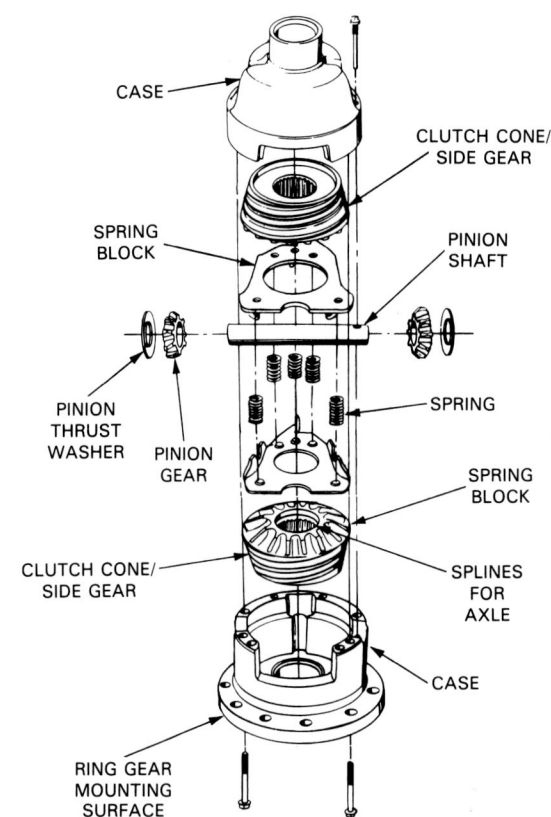

Fig. 27-12. Cone type limited slip differential. Cone surface on axle side gears serves as friction surface to drive both axles. Increased engine torque pushes side gears and cones outward to lock axles. In turns, side thrust on axles helps release one axle.

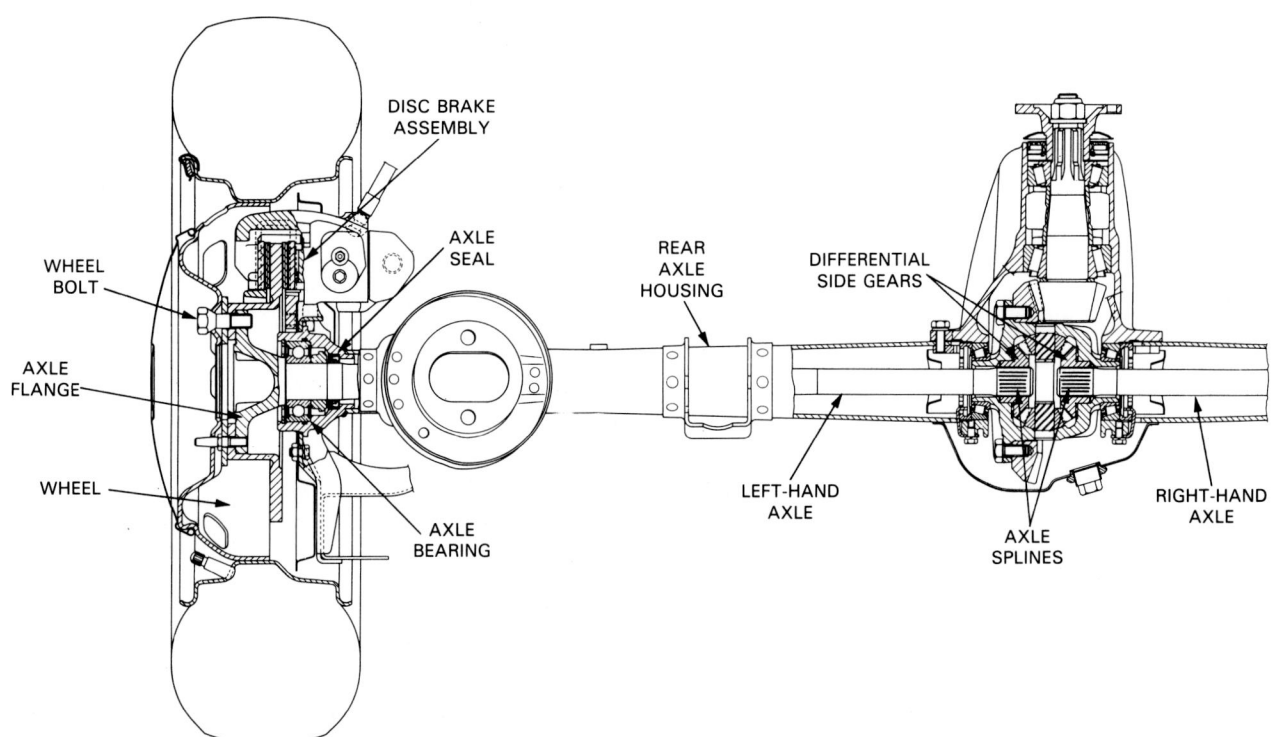

Fig. 27-13. Solid steel, case hardened rear drive axle extends from differential to outside of axle housing. Case and side gear supports inner end of axle. Rear wheel bearing supports outer end of axle. Also note flange on axle for wheel.

usually induction hardened for increased strength. There are several types of rear axle designs: semi-floating, three quarter floating, full floating, and swing axles.

Modern cars normally use semi-floating and swing types. Four-wheel drive vehicles and some light commercial vehicles use three-quarter and full floating axles. Compare the construction of the axle types in Fig. 27-14.

Semi-floating axle

The semi-floating axle turns the drive wheel and supports the weight of the car. It is the most common type of axle found on cars. See Fig. 27-14A and B.

A ball or roller bearing fits between the axle shaft and the axle housing. Splines on the inner end of the axle fit into matching splines in the differential side gears. A flange is usually machined on the outer end of the axle shaft. A collar may be used to hold the axle bearing on the axle.

A variation of the semi-floating axle is shown in Fig. 27-14B. It has a tapered end that accepts a wheel hub. A key and large nut lock the hub to the axle. Fig. 27-14C shows a full floating axle.

The rear wheel bearings reduce friction between the axle and axle housing. They allow the axle to turn freely. The inner bearing race fits against the axle. The outer bearing race fits into the machined end of the axle housing. Ball or roller bearings can be used.

The rear axle seals usually press into the axle housing, as shown in Fig. 27-14A. The seal lips contact the axles or axle collars to prevent lubricant leakage from the housing.

Axle shaft retainers can bolt to the outside of the axle housing to keep the axles from sliding out. They are normally used with a removable carrier type differential. See Fig. 27-15.

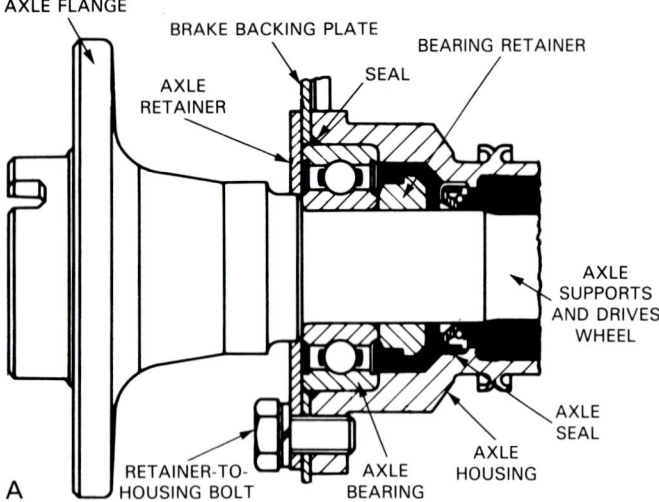

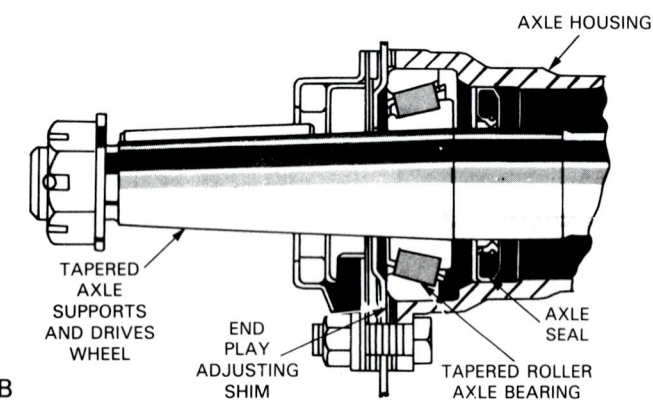

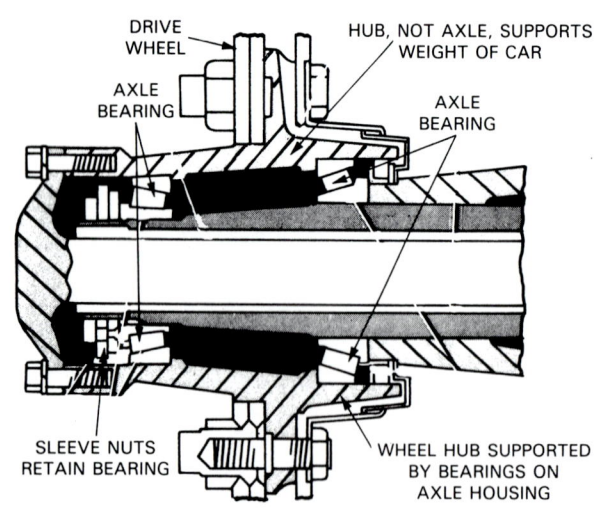

Fig. 27-14. Three common rear wheel bearing variations: A — Semi-floating, ball bearing type. B — Semi-floating, roller bearing type. C — Full-floating axle, used on heavy duty, light commercial and large truck applications. Other types of wheel bearings for swing axle or front-wheel drive cars are also used. These are covered in a later chapter.

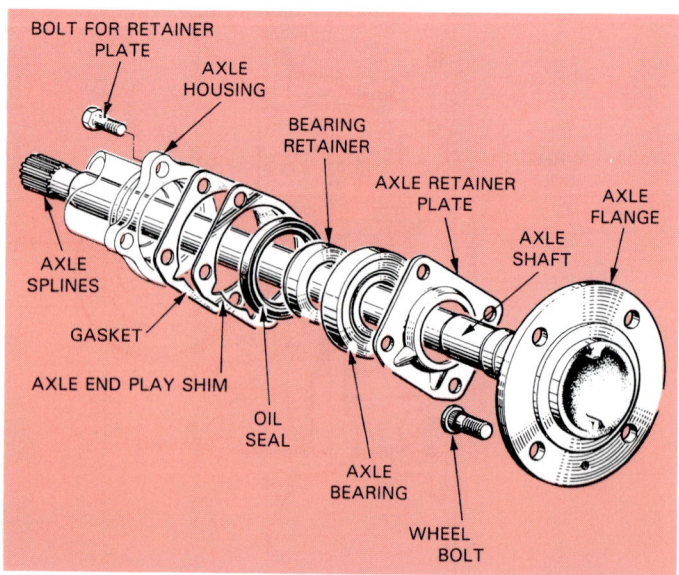

Fig. 27-15. Most rear axles with removable carriers have an axle retainer plate that bolts to axle housing. Shims are used to adjust axle end play. Gasket and oil seal prevent leakage of differential fluid. (Toyota)

Special bolts usually fit through holes in the housing flange and retainer. Nuts screw on these bolts to secure the axle into the housing.

Axle shims are frequently used between the axle shaft retainer and the housing to limit axle end play (in and out movement). A thicker shim can be used to reduce end play. A thinner shim will increase axle end play. Refer to Figs. 27-15 and 27-16.

DIFFERENTIAL LUBRICANT

Differential lubricant, usually SAE 80W-90 gear oil, is used to reduce friction between the moving parts in the rear axle assembly. Ring gear rotation splashes the oil on all moving parts to prevent wear.

A limited slip differential usually requires a SPECIAL GEAR LUBRICANT. It is needed for the

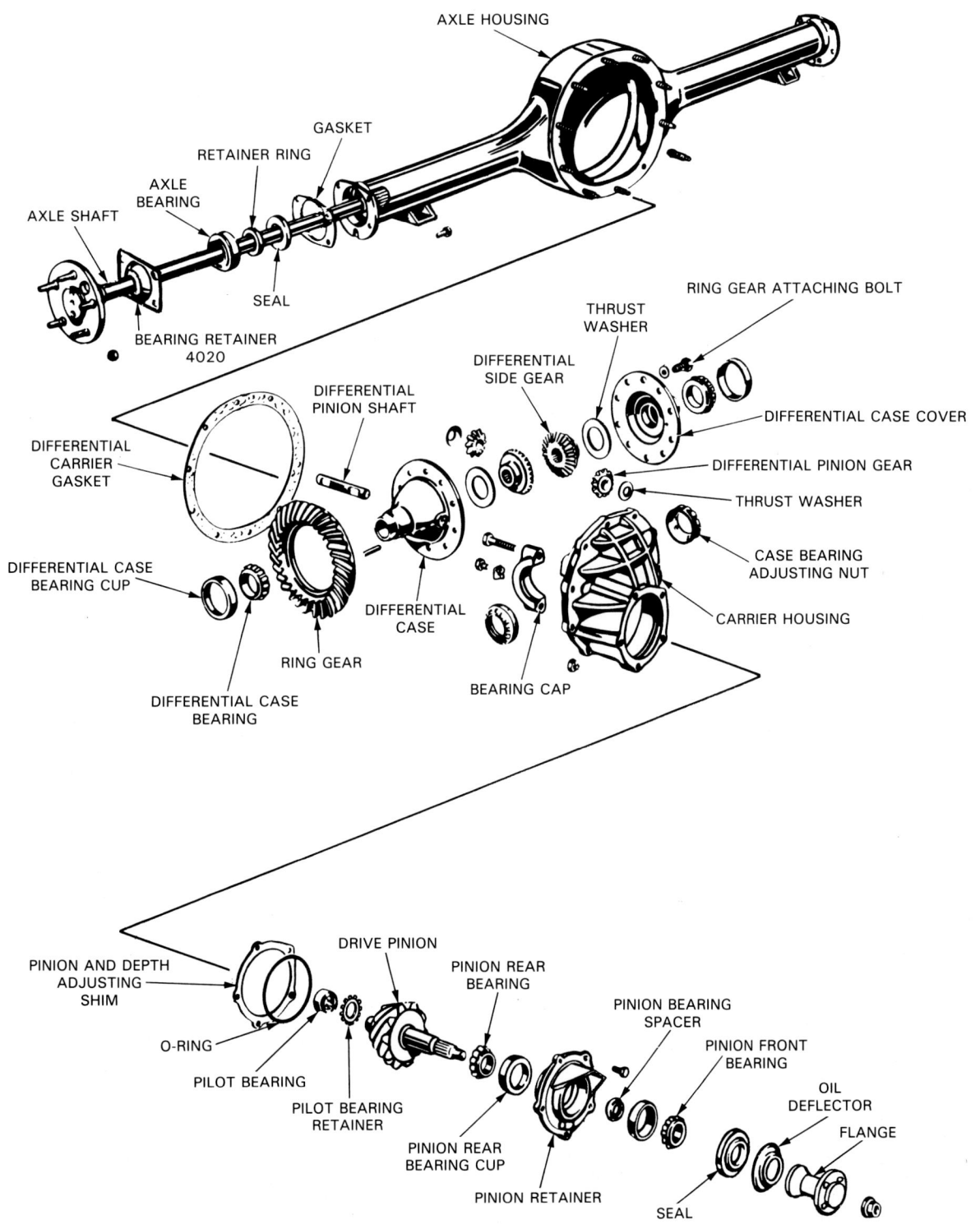

Fig. 27-16. This rear axle assembly has removable carrier, semi-floating axles, and conventional differential. (Ford)

clutch pack. The friction discs will NOT function properly with regular gear oil.

Differential breather tube

A differential breather tube vents pressure or vacuum in or out of the rear axle housing with changes in temperature. Look at Fig. 27-17.

Without a breather tube, pressure could build as the differential lubricant warmed to operating temperature. Lubricant could blow out the axle seals or pinion drive gear seal.

FRONT, FOUR-WHEEL DRIVE AXLE

A front, four-wheel drive axle assembly is similar to a rear drive axle, however, provisions must be made for steering the front wheels. Look at Fig. 27-18. Note how the outer ends of the axles have universal joints. The U-joints let the front wheels and hubs swivel while still transferring driving power to the hubs and wheels.

Locking hubs transfer power from the driving axles to the driving wheels on a four-wheel drive vehicle. There are three basic types of locking hubs:

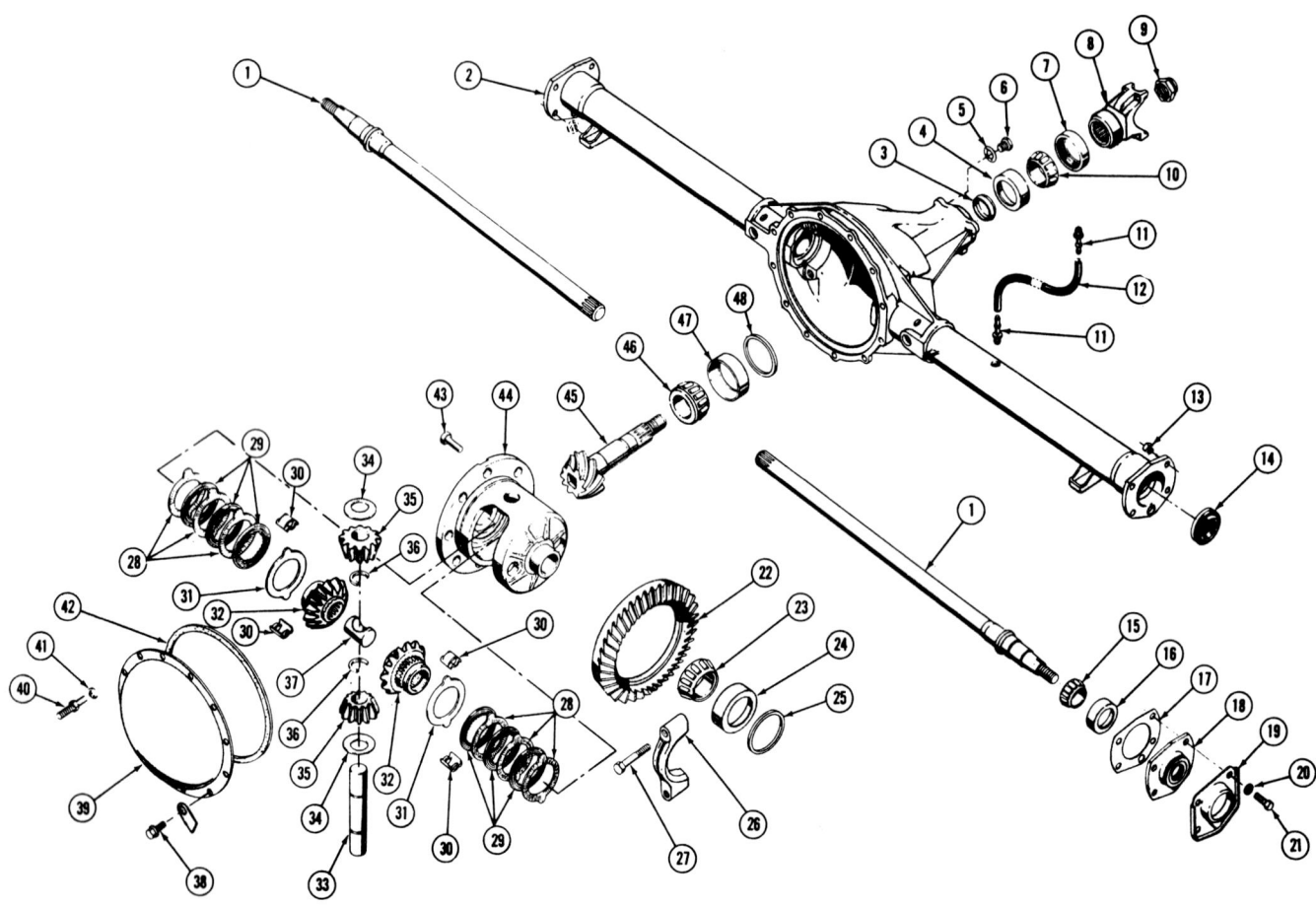

1. Axle Shaft	18. Axle Shaft Oil Seal	34. Differential Pinion Thrust Washer
2. Rear Axle Housing	19. Axle Shaft Oil Seal Retainer	35. Differential Pinion
3. Pinion Front Bearing Cup	20. Washer	36. Differential Pinion Shaft Snap Ring
4. Collapsible Spacer	21. Bolt	37. Differential Pinon Shaft Thrust Block
5. Filler Plug Gasket	22. Ring Gear	
6. Filler Plug	23. Differential Bearing	38. Bolt
7. Pinion Oil Seal	24. Differential Bearing Cup	39. Housing Cover
8. Universal Joint Yoke	25. Differential Bearing Shim	40. Stud
9. Pinion Nut	26. Differential Bearing Cap	41. Washer
10. Front Pinion Bearing	27. Bolt	42. Housing Cover Gasket
11. Breather (2)	28. Clutch Plates	43. Bolt
12. Breather Hose	29. Clutch Discs	44. Differential Case
13. Nut	30. Clutch Retainer Clip	45. Pinion Gear
14. Axle Shaft Inner Oil Seal	31. Clutch Belleville Spring	46. Rear Pinion Bearing
15. Axle Shaft Bearing	32. Differential Side Gear	47. Rear Pinion Bearing Cup
16. Axle Shaft Bearing Cup	33. Differential Pinion Shaft	48. Pinion Depth Adjusting Shim
17. Axle Shaft Bearing Shim		

Fig. 27-17. Disassembled view of complete rear axle assembly. Study parts carefully. Note that axles use tapered roller bearings that ride on separate races. Differential is limited slip. Carrier is integral (one piece) with axle housing.

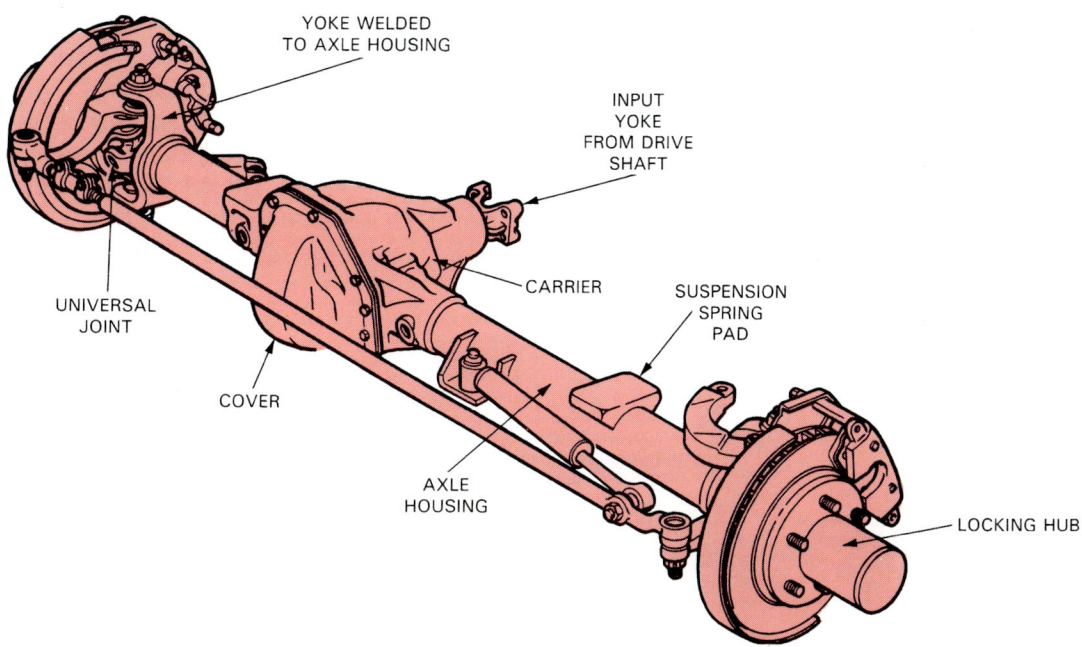

YOKE WELDED
TO AXLE HOUSING

INPUT
YOKE
FROM DRIVE
SHAFT

CARRIER

SUSPENSION
SPRING
PAD

UNIVERSAL
JOINT

COVER

AXLE
HOUSING

LOCKING HUB

Fig. 27-18. A typical front drive axle for four-wheel drive vehicle. It is conventional differential with U-joints on outer end of axles to allow for steering action. Special hubs lock drive axle to hub and wheel when in four-wheel drive.

1. Manual locking hub (driver must turn latch on hub to lock hub for four-wheel drive action).

2. Automatic locking hub (hub locks front wheel to axles when driver shifts into four-wheel drive).

3. Full time hub (front hubs are always locked and drive front wheels).

Manual and automatic locking hubs are common. Used with part-time, four-wheel drive, they enable the drive line to be in two-wheel drive for vehicle use on dry bitumen. The front wheels can turn without turning the front axles. This increases fuel economy and reduces drive line wear. Fig. 27-19 shows the basic parts of a manual locking hub.

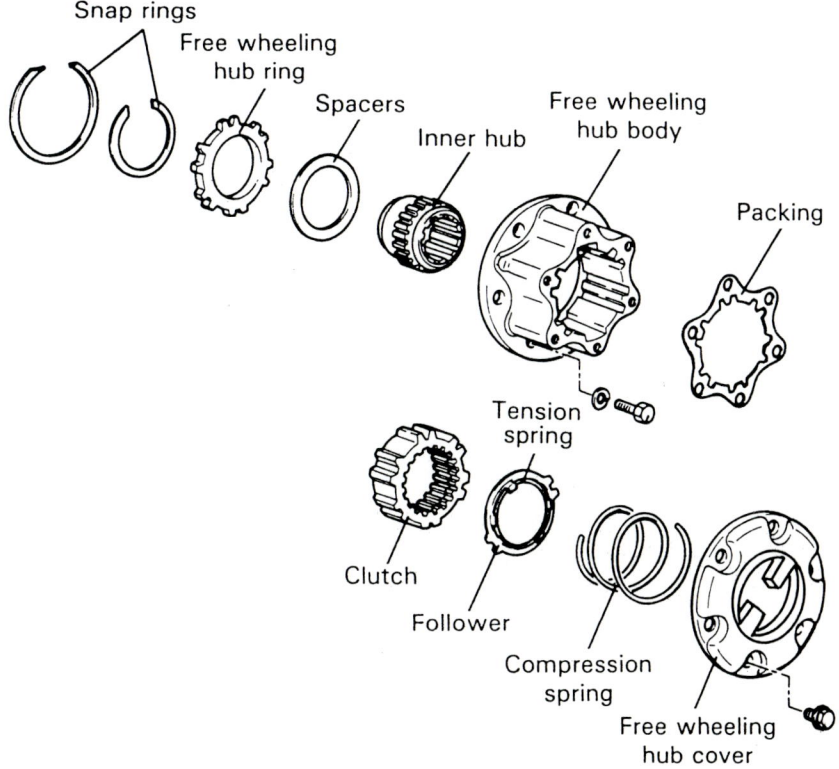

Snap rings

Free wheeling
hub ring

Spacers

Inner hub

Free wheeling
hub body

Packing

Tension
spring

Clutch

Follower

Compression
spring

Free wheeling
hub cover

Fig. 27-19. Dismantled view of a manual locking hub.

SWING AXLES (REAR-WHEEL DRIVE)

Swing axles are used when the differential is mounted solid on the car's frame. Universal joints in the axles are needed to allow for up and down suspension action. Fig. 27-20 illustrates a rear drive axle assembly using swing axles.

The differential works like a conventional unit. However, the drive axles are not solid, steel shafts. They are flexible. Each has two U-joints, one on each end. Fig. 27-21 shows another view of a swing axle.

For more information on swing axles, refer to Chapter 29. It covers front-wheel drive transaxles which use similar drive axles.

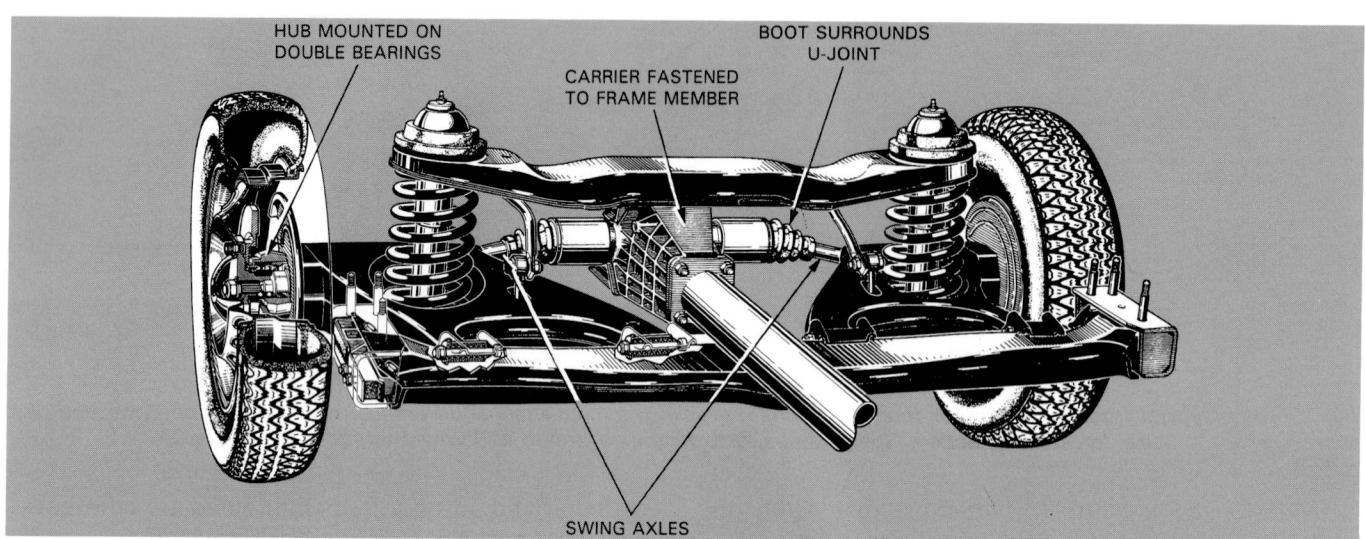

Fig. 27-20. Swing axle has differential assembly mounted on frame member. Universal joints allows axles and wheels to move up and down with suspension action. (Peugeot)

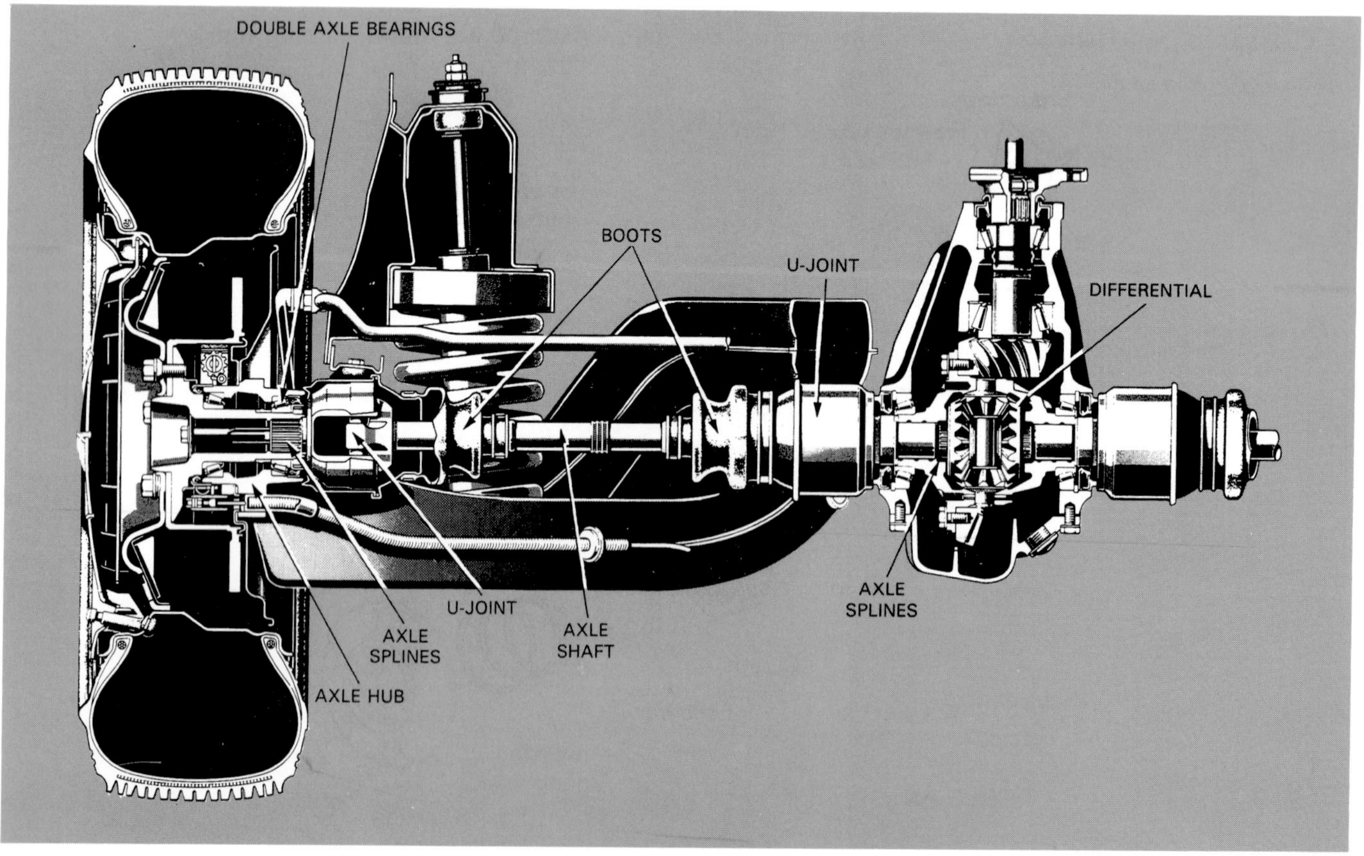

Fig. 27-21. Section view shows internal parts of differential and swing axles. Double roller bearings support hub. (Mercedes-Benz)

KNOW THESE TERMS

Rear drive axle assembly, Differential, Pinion gear, Ring gear, Pinion pilot bearing, Rear axle ratio, Hypoid gears, Differential carrier, Differential case, Spider gears, Pinion shaft, Limited slip differential, Rear drive axle, Semi-floating axle, Rear wheel bearing, Axle shaft retainer, Axle shims, Differential lubricant, Locking hub, Swing axle.

REVIEW QUESTIONS

1. Which of the following are basic parts of the rear axle assembly?
   a. Differential case assembly.
   b. Axle housing.
   c. Pinion drive gear.
   d. Rear axle bearings.
   e. Rear drive axles.
   f. Ring gear.
   g. All of the above.
2. The _____ must be capable of providing torque to both axles when turning corners.
3. The purpose of the pinion gear is to transfer power from the ring gear to the axle. True or False?
4. Explain the difference between a hunting gearset and a non-hunting gearset.
5. Rear axle ratio is determined by comparing the number of teeth on the _____ drive gear to the number of teeth on the _____ _____.
6. An integral carrier is constructed as part of the axle housing. True or False?
7. What major problem is a differential designed to prevent?
8. Which of the following statements best describes a clutch pack for a limited slip differential?
   a. Set of friction discs and steel plates usually splined to the differential side gears.
   b. Friction-producing cone shaped axle gears that are splined to the axles.
9. A limited slip differential usually uses 80W-90 gear oil as a lubricant. True or False?
10. Swing axles are found on a differential which is mounted solid on the auto frame. True or False?

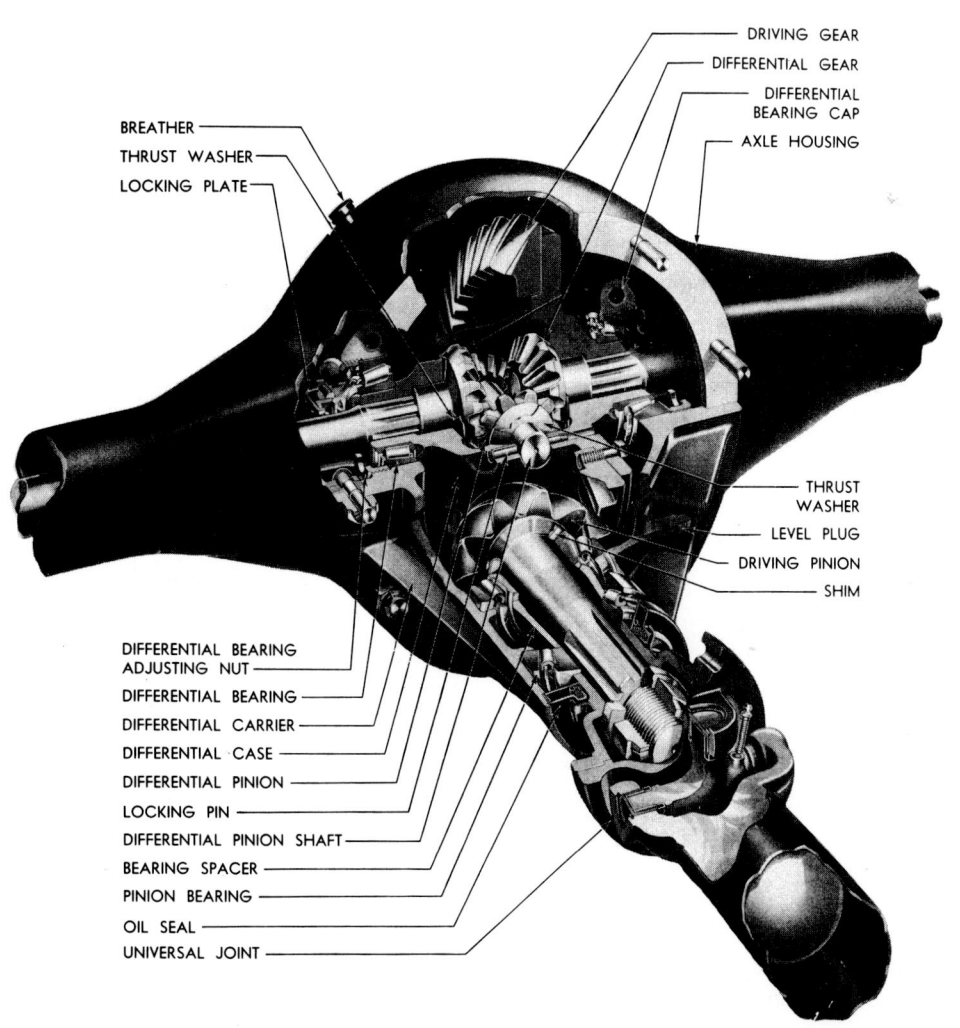

DRIVING GEAR
DIFFERENTIAL GEAR
DIFFERENTIAL BEARING CAP
AXLE HOUSING
BREATHER
THRUST WASHER
LOCKING PLATE
THRUST WASHER
LEVEL PLUG
DRIVING PINION
SHIM
DIFFERENTIAL BEARING ADJUSTING NUT
DIFFERENTIAL BEARING
DIFFERENTIAL CARRIER
DIFFERENTIAL CASE
DIFFERENTIAL PINION
LOCKING PIN
DIFFERENTIAL PINION SHAFT
BEARING SPACER
PINION BEARING
OIL SEAL
UNIVERSAL JOINT

Can you explain action of each part in this differential? If you cannot, review chapter as needed. (Ford)

Chapter 28

DRIVE SHAFTS AND TRANSFER CASES

After studying this chapter, you will be able to:
☐ Identify the parts of a modern drive shaft assembly.
☐ Explain the functions of a drive shaft.
☐ Define the major drive shaft parts.
☐ List the different types of drive lines.
☐ Describe the different types of universal joints.
☐ Identify the major parts of four-wheel drive type drive line.
☐ Explain the basic operation of a transfer case.

The term drive line generally refers to the parts that transfer power from the transmission to the drive wheels. Cars with the engine in the front and drive axle assembly in the rear have a drive line with a long drive shaft. Fig. 28-1 shows these basic parts.

Front-wheel drive and rear or mid-engine cars do NOT use a drive line with a single, long drive shaft.

Instead, they use a transaxle (transmission-differential assembly) and two drive axle shafts or swing axles. The short axle shafts extend directly out of the transaxle to power the drive wheels.

Chapter 29 covers front-drive axles and transaxles. Refer to this chapter for more information.

DRIVE SHAFT ASSEMBLY

A drive shaft assembly typically consists of a front slip yoke, two universal joints, drive or propeller shaft, and rear yoke. This is illustrated in Fig. 28-1.
1. SLIP YOKE (connects transmission output shaft to front universal joint).
2. FRONT UNIVERSAL JOINT (swivel connection that fastens slip yoke to drive shaft).
3. DRIVE SHAFT (hollow metal tube that transfers turning power from front universal joint to rear universal joints).

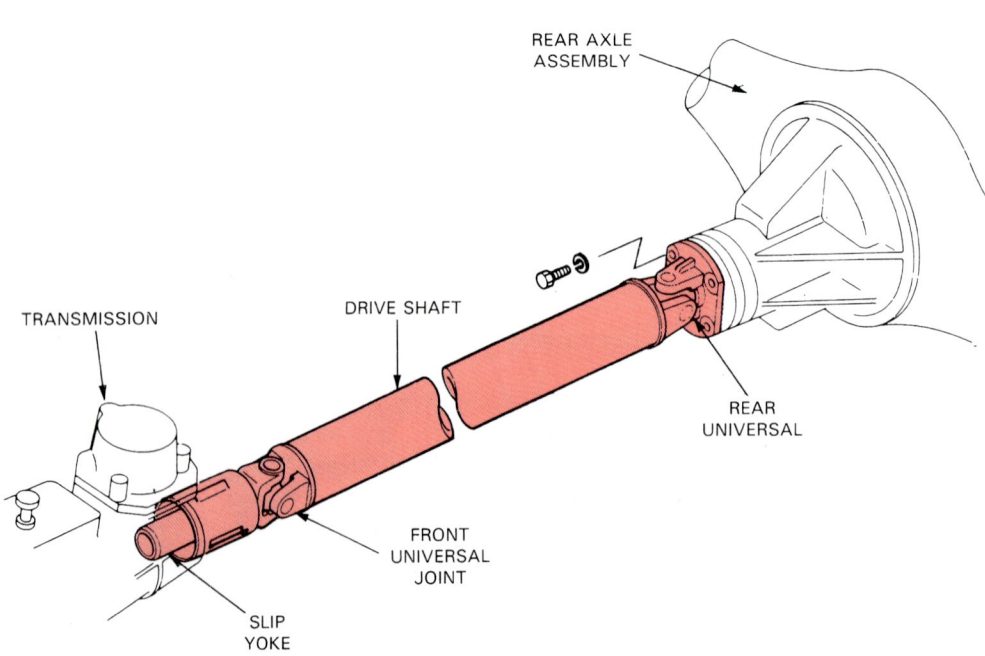

Fig. 28-1. Drive shaft assembly connects transmission output shaft with rear axle assembly. Note major parts. (Mazda)

4. REAR UNIVERSAL JOINT (another flex joint connecting drive shaft to differential yoke).

5. REAR YOKE (it holds rear universal and transfers torque to gears in rear axle assembly).

This drive shaft, Fig. 28-1, is the most common type used on front engine, rear-wheel drive automobiles. Discussed shortly, a few variations are sometimes used to satisfy special applications or to improve the smoothness of power transfer.

FUNCTIONS OF DRIVE SHAFT

The drive shaft assembly has several important functions. It must:

1. Send turning power from the transmission to the rear axle assembly.
2. Flex and allow up and down movement of the rear axle assembly.
3. Provide a sliding action to adjust for changes in drive line length.
4. Provide smooth power transfer.

DRIVE SHAFT OPERATION

With the car moving, the transmission output shaft turns the slip yoke. Refer to Fig. 28-2. The slip yoke then turns the front universal, drive shaft, rear universal, and rear yoke on the differential. The differential contains gears that transfer power to the rear drive axles. The axles rotate the wheels to propel the car.

Drive line flex

When the car tyres strike a bump in the road, the rear suspension and springs are compressed. This pushes the rear axle upward in relation to the car body. Suspension movement smooth's the car's ride.

The universal joints let the drive line flex without damaging the drive shaft. This is illustrated in Fig. 28-2.

Changes in drive line length

The movement of the rear axle assembly also causes the distance between the rear axle and transmission to change. The slip yoke allows for this change of length. Look at Fig. 28-2.

SLIP YOKE (SLIP JOINT) CONSTRUCTION

The slip yoke or slip joint, splined to the transmission output shaft, allows for any changes in drive line length by sliding in and out of the transmission. Cutaway views of a slip joint are shown in Fig. 28-3.

Note how the inside of the slip joint has splines that fit over the transmission output shaft splines.

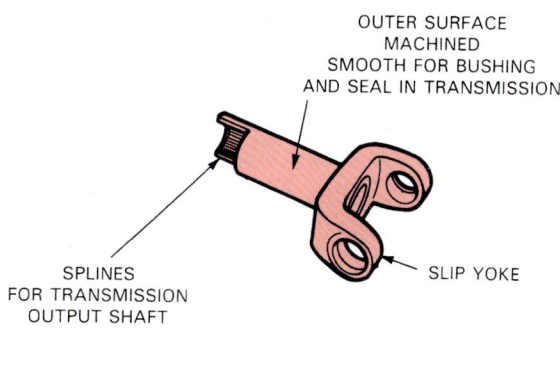

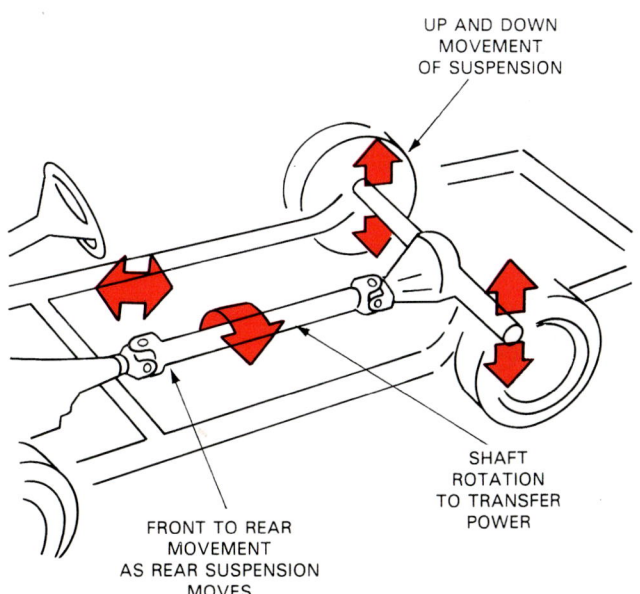

Fig. 28-2. Drive shaft universal joints allows drive line bend or flex as rear axle moves up and down over bumps in road. Most types also allow for length changes to allow suspension action.

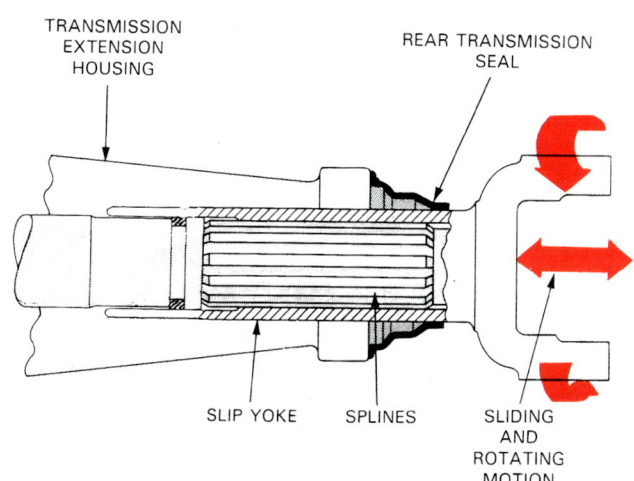

Fig. 28-3. Slip yoke is splined to transmission output shaft and fits inside transmission extension housing. Transmission seal contacts slip yoke. Yoke rides on bushing in extension housing. Slip yoke rotates with output shaft but is free to slide in and out of transmission. (Ford)

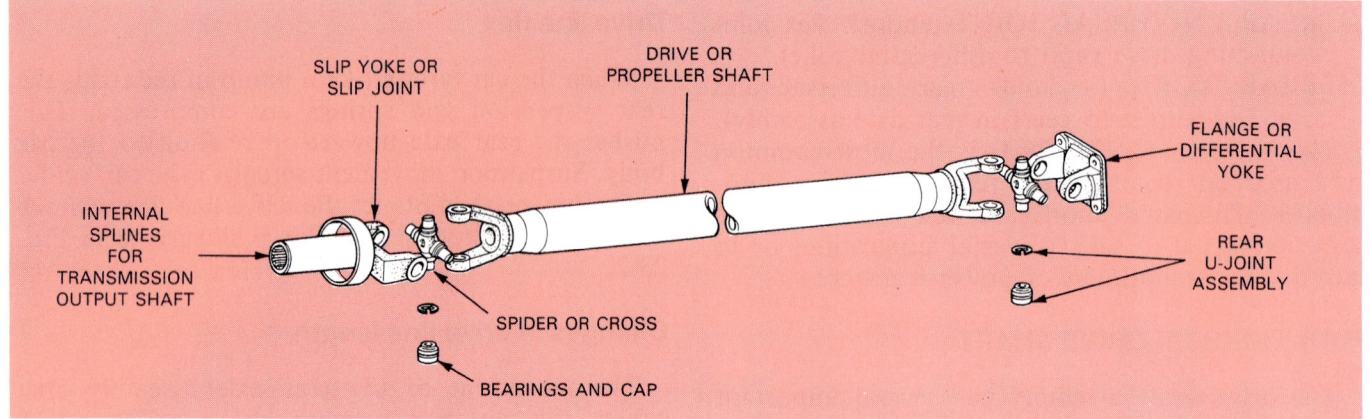

Fig. 28-4. Typical drive or propeller shaft assembly. Note basic components. (Toyota)

This causes the two to rotate together. However, it also permits the yoke to slide on the splines.

The outer diameter of the yoke is machined smooth. This smooth surface provides a bearing surface for the bushing and oil seal in the transmission.

The extension housing bushing supports the slip yoke as it rotates in the transmission. Refer to Fig. 28-3.

The transmission rear seal rides on the slip yoke and prevents fluid leakage out of the rear of the transmission. The seal also keeps road dirt out of the transmission and off the slip yoke.

Normally the outside of the slip joint is lubricated by the transmission fluid. Transmission lubricant prevents bushing, yoke, and seal wear.

Some types of yokes, however, require special heavy grease on their splines. The splines are sealed from the transmission lubricant. This keeps the oil or fluid from washing the grease off the splines.

DIFFERENTIAL YOKE CONSTRUCTION

The differential yoke is the yoke bolted to the outer end of the pinion (drive) gear on the rear axle assembly. The rear universal is held by this yoke, Fig. 28-4.

DRIVE SHAFT (PROPELLER SHAFT)

The drive shaft, also called a propeller shaft, is commonly a hollow steel tube with permanent yokes welded on each end. See Figs. 28-4 and 28-5. A tubular design makes the drive shaft very strong and light. Since the drive shaft rotates much faster than the wheels and tyres, it must be straight and perfectly balanced.

Most cars use a single, one-piece drive shaft. However, a few large passenger cars and some pickup trucks have a two-piece drive shaft. This cuts the length of each shaft to avoid drive line vibration.

Drive shaft balance

Since a drive shaft can rotate at full engine rpm in high gear, it must be perfectly balanced (weight evenly distributed around centreline of shaft). If NOT balanced, the shaft could vibrate violently.

Drive shaft balancing weights are frequently welded to the shaft, as in Fig. 28-5. The drive shaft is rotated on a balancing machine at the factory. If needed, small, metal weights are attached to the shaft on the light side. This counteracts the heavy side to smooth operation.

Sometimes the drive shaft has a large ring-shaped weight mounted on rubber. This ring, called a drive shaft vibration damper, also helps keep shaft spinning smoothly by absorbing torsional (twisting) vibrations.

UNIVERSAL JOINTS

A universal joint, also called U-joint, is a swivel connection capable of transferring power or turning force through an angle. A simple universal joint is

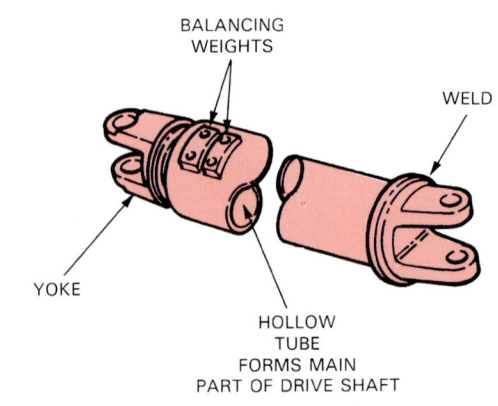

Fig. 28-5. Drive shaft is normally hollow tube with yokes welded to each end. Note balancing weights welded to shaft to prevent vibration. (Ford)

made of two Y-shaped yokes or knuckles connected by a cross or spider. See Fig. 28-6. Bearings on each end of the cross allows the two yokes to swing into various angles while turning.

Today's drive shafts use two or more U-joints. A majority only use two. Extra universals are sometimes needed in very long drive shafts.

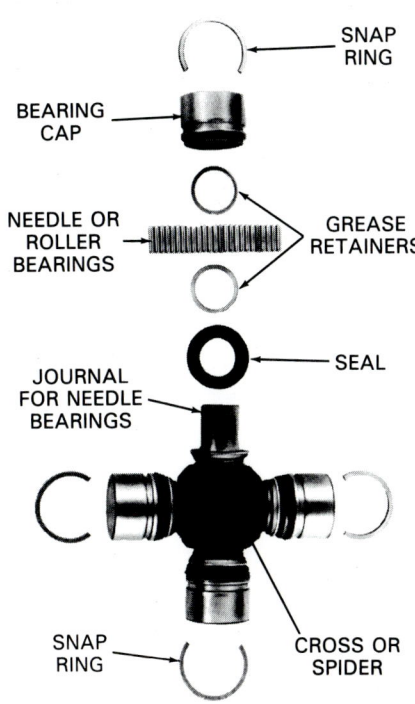

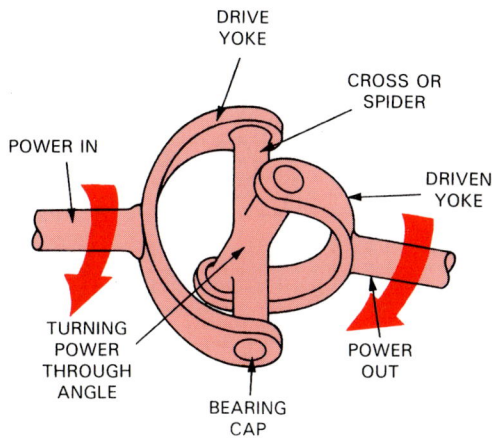

Fig. 28-6. Universal joint or U-joint will swivel to allow for changes in transmission-to-differential alignment. Two yokes are connected to central spider. Needle bearings fit between yokes and spider. Each yoke can be swivelled in relation to the other.

Fig. 28-7. Partially disassembled view of cross and roller type U-joint. Needle bearings are packed with grease. Snap rings hold bearing caps in their yokes. Rubber cup or boot keeps grease inside joint.

Cross and roller universal joint

The cross and roller, also called a cardan universal joint, is the most common type of drive shaft U-joint. Pictured in Fig. 28-7, it consists of four bearing caps, four needle roller bearings, a spider or cross, grease seals, grease retainers, and snap rings.

The bearing caps are held stationary in the drive shaft yokes. Roller bearings fit between the caps and cross to reduce friction. The cross is free to rotate inside the caps and yokes.

Snap rings usually fit into grooves cut in the caps or the yoke bores. There are several other methods of securing the bearing caps in the yokes. These are pictured in Fig. 28-8. Sometimes, bearing covers,

U-bolts, or injected plastic rings keep the caps and rollers from flying out of the rotating drive shaft assembly.

Fig. 28-9 shows a dismantled view of a cross and roller drive shaft assembly. Note how the universal joint components fit together.

Hotchkiss and torque tube drives

A hotchkiss drive line has an open drive shaft which operates a rear axle assembly mounted on springs. This is the most common rear-wheel drive type. It usually has cross and roller U-joints.

A hotchkiss type drive line is pictured in Fig. 28-9. It has almost totally replaced the torque tube setup.

The torque tube drive line uses a solid steel drive shaft enclosed in a large hollow tube. Only one

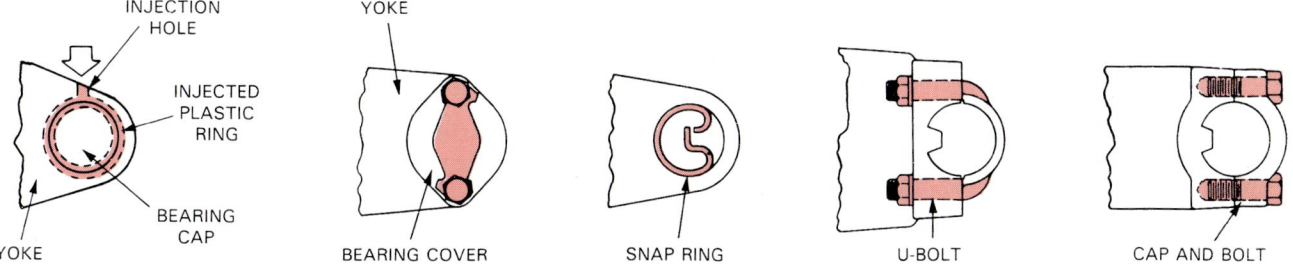

Fig. 28-8. Several methods can be used to hold U-joint caps in yoke. Study each.

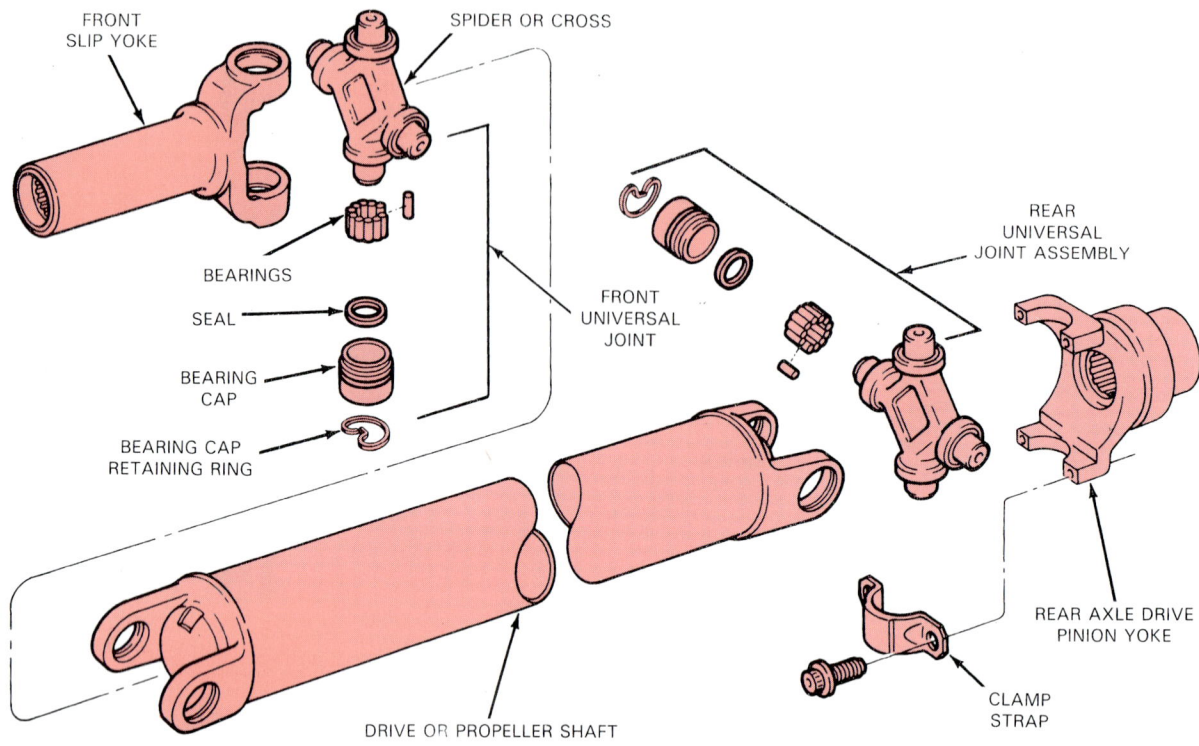

Fig. 28-9. Dismantled view of drive shaft using cross and roller universals. Note how parts fit together.

swivel joint is used at the front. The rear of the torque tube is formed as a rigid part of the rear axle housing.

CENTRE SUPPORT BEARING

A centre support bearing is needed to hold the middle of a two-piece drive shaft. See Fig. 28-10. The centre bearing bolts to the car frame or underbody. It supports the centre of the drive shaft, where the two shafts come together.

Commercial vehicles commonly use a centre support bearing. A two-piece drive shaft is required because of the great distance between the transmission and rear axle.

A cutaway view of a centre support bearing is shown in Fig. 28-11. A sealed ball bearing allows the drive shaft to rotate freely. The outside of the ball bearing is held by a thick, rubber, doughnut-shaped mount. The rubber mount prevents noise and vibration from transferring into the driver's compartment.

TRANSFER CASES

A transfer case sends power to both the front and rear axle assemblies in a four-wheel drive vehicle. Look at Fig. 28-12. The transfer case usually mounts behind and is driven by the transmission. Two drive

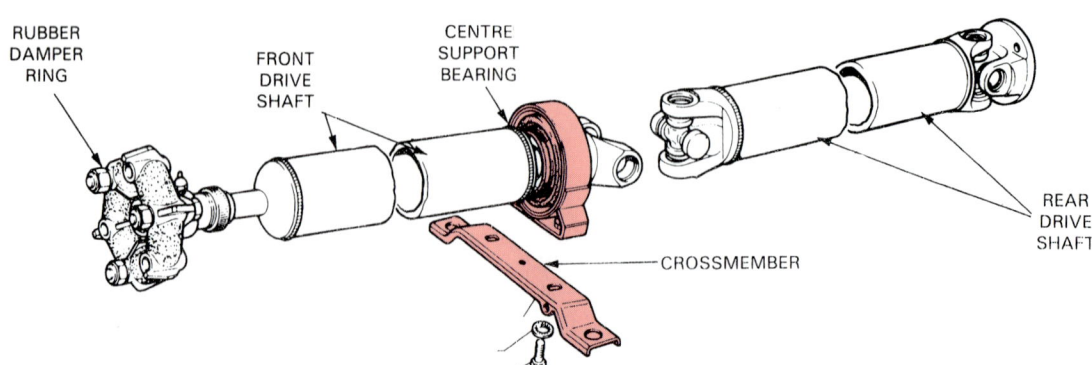

Fig. 28-10. Centre support bearing holds centre of two-piece drive shaft. It is roller bearing mounted in rubber. Also note rubber torsion damping ring. (Fiat)

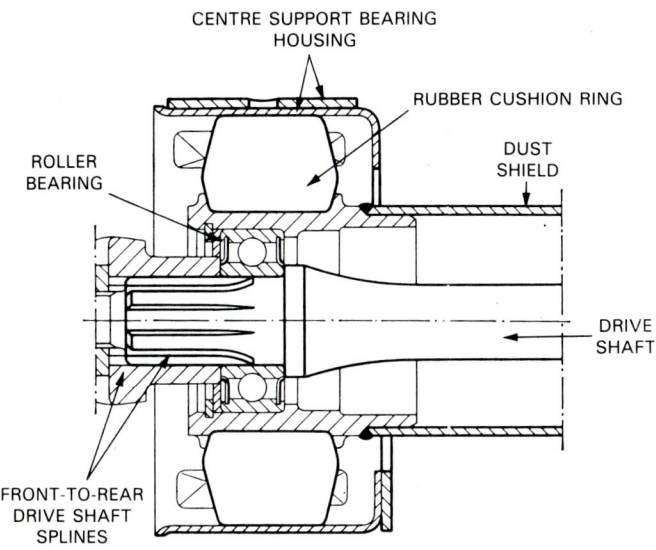

Fig. 28-11. Cutaway view of centre support bearing. Study parts. (Fiat)

shafts normally run from the transfer case, one to each drive axle.

Most modern transfer cases provide a 2H (two-wheel drive, high range), a 4H (four-wheel drive, high range), and a 4L (four-wheel drive, low range).

High range normally has a gear ratio of 1:1 low range typically has a gear ratio of 2:1 for climbing steep hills or pulling heavy loads. A 2H is provided for highway driving, when four-wheel drive traction is not needed.

Transfer case operation

Fig. 28-13 shows the major parts of a transfer case. Fig. 28-14 shows how power flows through a transfer case in different shifter positions. Study and compare the parts in both illustrations.

Notice that this unit uses a planetary gearset to produce the two gear ratios. A hand-shifted sliding clutch regulates power transfer for two and four-wheel drive.

Two-wheel drive, high range (2H)

In 2H, torque flows from the input gear, through the locked planetary gearset, and annulus gear which rotate as a single unit. Torque is transferred to the mainshaft through the planetary carrier splined to the mainshaft. Power finally flows out of the rear yoke, through the rear drive shaft, and to the rear differential. Refer to Fig. 28-14.

In 2H, the sliding clutch remains in the neutral position. As a result, torque is NOT transferred to the front axle assembly.

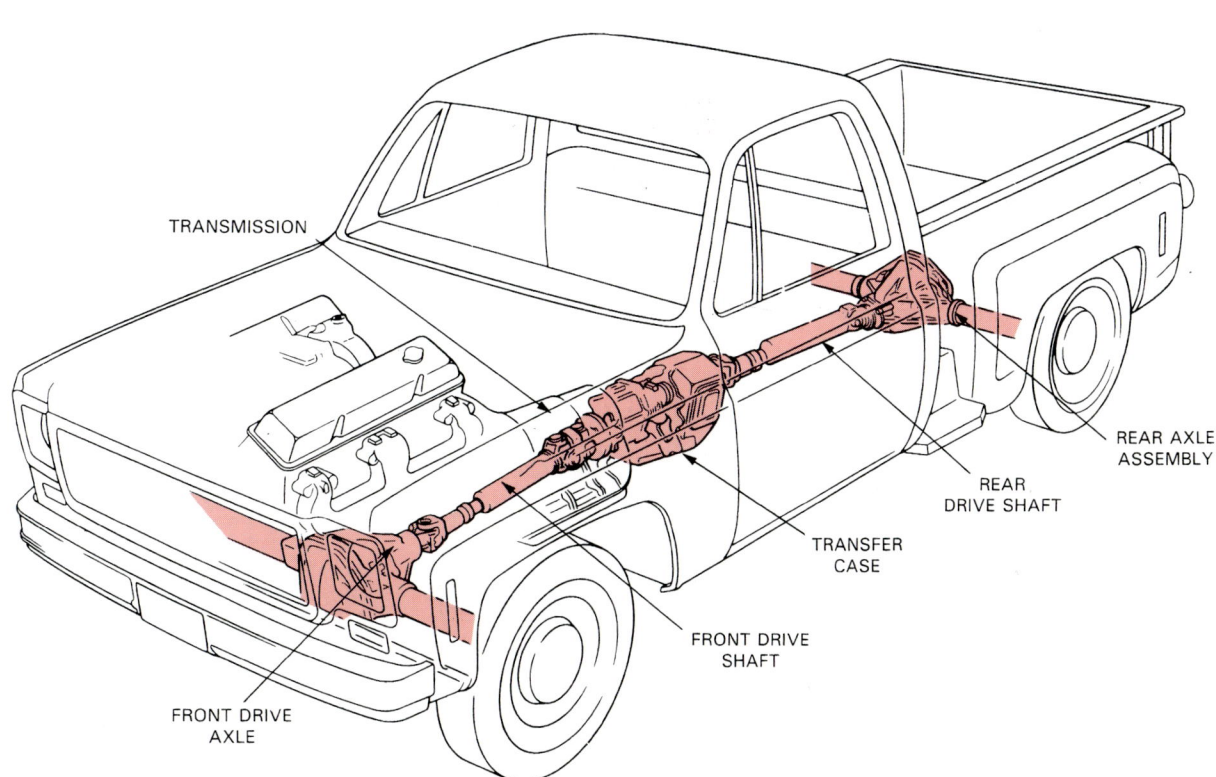

Fig. 28-12. Light commercial vehicles and passenger cars commonly use four-wheel drive. Transfer case is power takeoff unit that sends power to both front and rear drive axle assemblies. Drive shafts extend out of front and rear of transfer case.

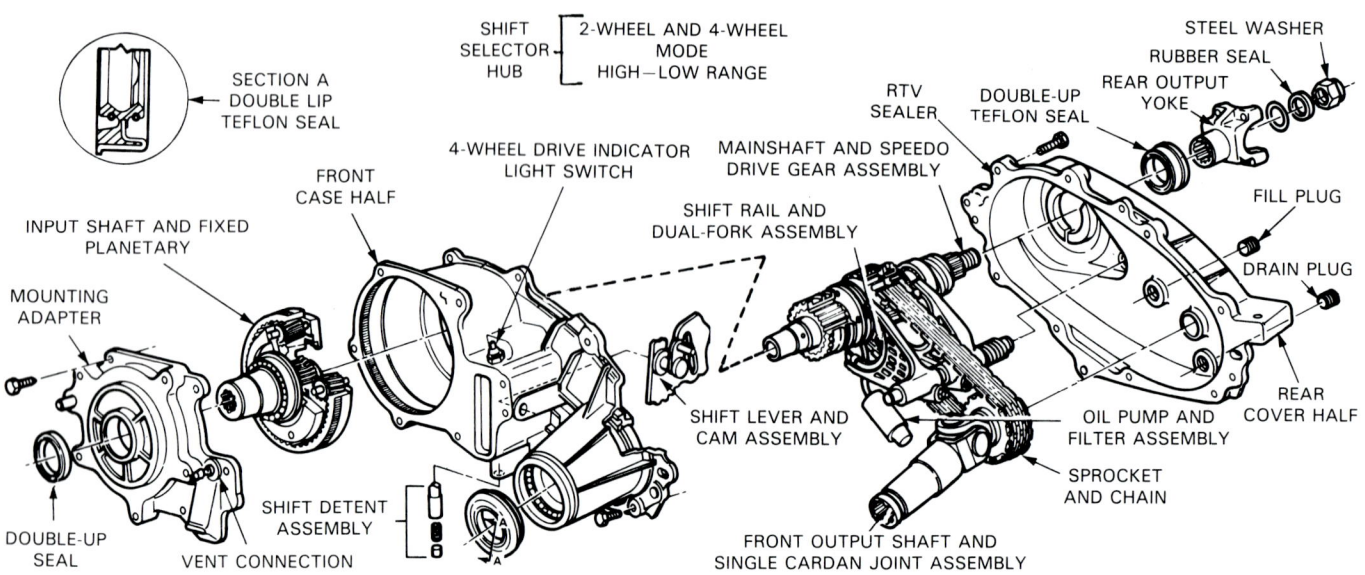

SECTION A
DOUBLE LIP
TEFLON SEAL

SHIFT SELECTOR HUB — 2-WHEEL AND 4-WHEEL MODE / HIGH—LOW RANGE

STEEL WASHER
RUBBER SEAL
REAR OUTPUT YOKE

RTV SEALER

DOUBLE-UP TEFLON SEAL

MAINSHAFT AND SPEEDO DRIVE GEAR ASSEMBLY

4-WHEEL DRIVE INDICATOR LIGHT SWITCH

FRONT CASE HALF

SHIFT RAIL AND DUAL-FORK ASSEMBLY

FILL PLUG

DRAIN PLUG

INPUT SHAFT AND FIXED PLANETARY

MOUNTING ADAPTER

SHIFT LEVER AND CAM ASSEMBLY

OIL PUMP AND FILTER ASSEMBLY

REAR COVER HALF

DOUBLE-UP SEAL

SHIFT DETENT ASSEMBLY

VENT CONNECTION

SPROCKET AND CHAIN

FRONT OUTPUT SHAFT AND SINGLE CARDAN JOINT ASSEMBLY

Fig. 28-13. Major parts of modern transfer case. Planetary gearset provides high and low ranges. Large chain sends power to front output shaft. Shift rail and fork assembly is activated to control two-wheel or four-wheel drive mode. (Ford)

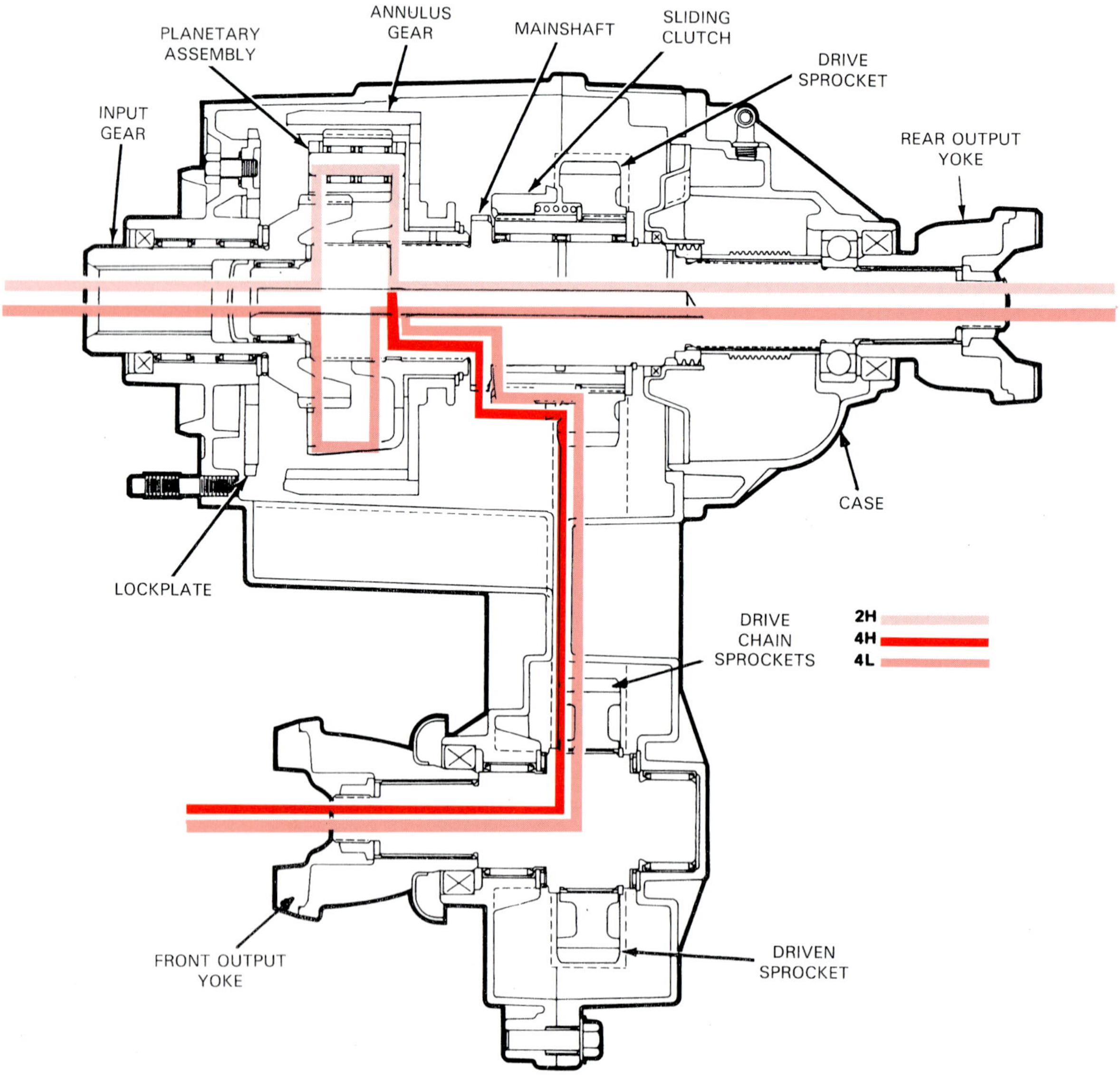

PLANETARY ASSEMBLY

ANNULUS GEAR

MAINSHAFT

SLIDING CLUTCH

DRIVE SPROCKET

INPUT GEAR

REAR OUTPUT YOKE

CASE

LOCKPLATE

DRIVE CHAIN SPROCKETS

2H
4H
4L

FRONT OUTPUT YOKE

DRIVEN SPROCKET

Fig. 28-14. Trace power flow through transfer case in 2H, 4H, and 4L modes. Note parts transferring power in each mode.

Four-wheel drive, high range (4H)

In 4H, torque flows from the input gear, through the planetary gear and annulus (ring) gear in the same fashion as in 2H. However, the sliding clutch is shifted into the mainshaft clutch gear. Torque then flows through the drive chain, front output yoke, and to the front drive axle assembly, Fig. 28-14. Both the front and rear axles drive the vehicle.

Four-wheel drive, low range (4L)

In 4L, torque transfer is almost the same as in 4H. However, the annulus gear is shifted forward into the lock plate. This holds the annulus gear stationary. As a result, the planetary pinions walk inside the annulus gear, producing a gear reduction.

Transfer case construction

A transfer case is constructed something like a transmission. It uses shift forks, splines, gears, shims, bearings, and other components found in manual and automatic transmissions. Look at Fig. 28-13.

A transfer case has an outer case made of cast iron or aluminium. It is filled with lubricant (oil) that cuts friction on all moving parts. Seals hold the lubricant in the case and prevent leakage around shafts and yokes. Shims set up the proper clearances between the internal components and the case.

All-wheel drive

All-wheel drive refers to a four-wheel drive train that does not use a conventional transfer case. It is a relatively new system designed from a front-wheel drive transaxle. Shown in Fig. 28-15, it is a simple system using the principles covered earlier.

KNOW THESE TERMS

Drive line, Drive shaft assembly, Slip yoke, Differential yoke, Drive shaft, Universal joint, Hotchkiss drive, Torque tube, U-joint, Centre support bearing, Transfer case.

REVIEW QUESTIONS

1. The term _____ _____ generally refers to the parts that transfer power from the transmission to the drive wheels.
2. List and explain the five major parts of a drive shaft.
3. What are four functions of a drive shaft?
4. The movement of the rear axle assembly also causes the distance between the rear axle and transmission to change. True or False?

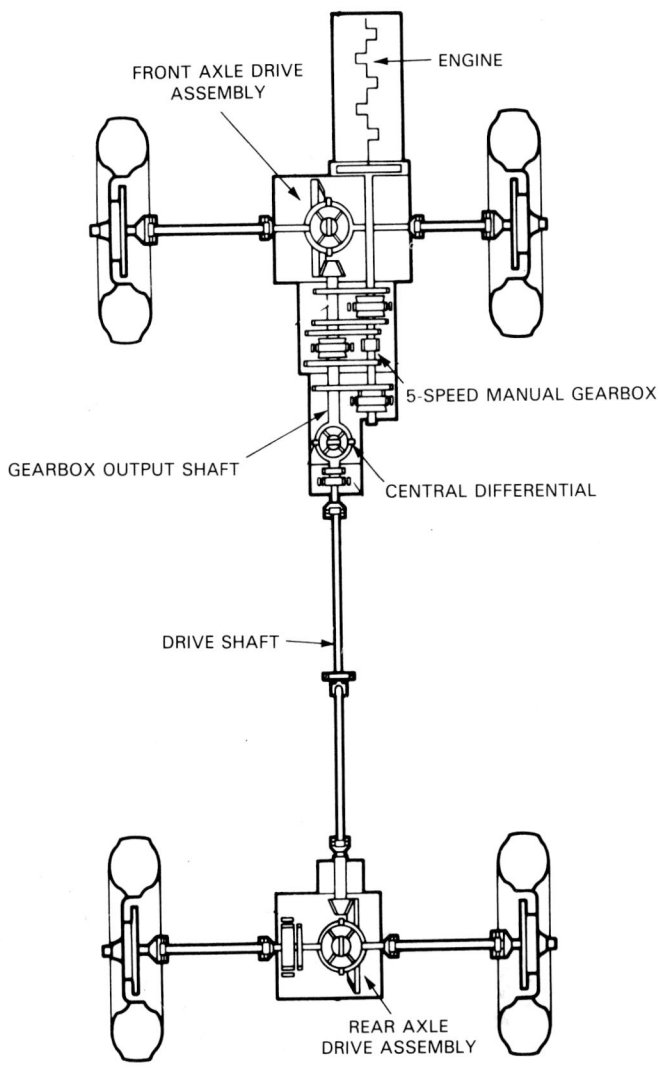

Fig. 28-15. All-wheel drive uses design variation of transmission and transaxle. Main gearbox shaft drives front differential directly. Rear of same gearbox shaft turns drive shaft going to rear axle assembly. A conventional transfer case is not needed. (Porsche-Audi)

5. The _____ _____ or _____ _____ is splined to the transmission output shaft.
6. Describe the construction of a typical drive shaft, not including the universal joints or other parts.
7. Which of the following does NOT attach to or touch an assembled drive shaft?
   a. Balance weights.
   b. Yokes.
   c. Universal joints.
   d. All of the above are correct.
   e. None of the above are correct.
8. When is a centre support bearing needed and why?
9. Explain the basic operation of a transfer case.

Chapter 29

TRANSAXLE, FRONT WHEEL DRIVE FUNDAMENTALS

After studying this chapter, you should be able to:
☐ Identify the major parts of a transaxle assembly.
☐ Explain the operation of a manual transaxle.
☐ Explain the operation of an automatic transaxle.
☐ Trace the flow of power through manual and automatic transaxles.
☐ Describe design differences in transaxles.
☐ Identify the parts of front drive axles.
☐ Compare design differences in front drive axle CV-joints.

This chapter describes the construction and operating principles of both manual and automatic transaxles. It relys and builds upon the information given in previous text chapters on clutches, manual transmissions, automatic transmissions, and differentials.

TRANSAXLE

A transaxle is a transmission and a differential combined in a single assembly. See Fig. 29-1. A transaxle is commonly used in late model, front-wheel drive cars. However, a few rear or mid-engine sports cars and some older, rear engine economy cars also use a transaxle.

A transaxle allows the wheels next to the engine to propel the vehicle. Short drive axles can be used to connect the transaxle output to the hubs and drive wheels.

Auto makers claim that a car having a transaxle and front-wheel drive has several advantages over a front engine car with rear-wheel drive. A few of these advantages are:
1. Reduced drive train weight and improved efficiency.
2. Improved traction on slippery road because of more weight on the drive wheels.
3. Increased passenger compartment space (no hump in floor for drive shaft to rear axle).

4. Smoother ride because of less unsprung weight (weight that must move with suspension action).
5. Quieter operation since engine and drive train noise is centrally located in engine compartment (no transmission, drive shaft, and rear axle under passenger compartment).
6. Improved safety because of increased mass in front of passengers.

Both manual and automatic transaxles are available. A manual transaxle uses a friction clutch and

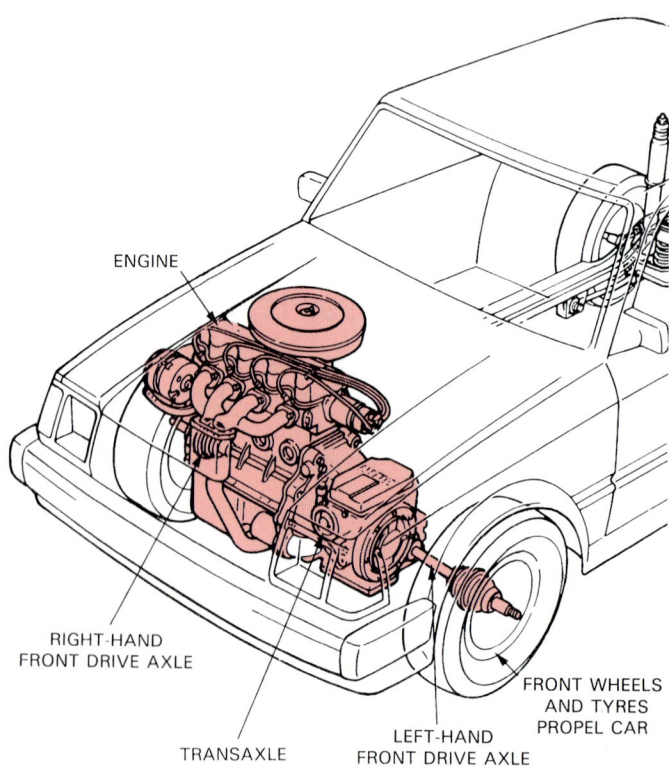

Fig. 29-1. Transaxle is used in modern front-wheel drive cars. It has both a transmission and differential that drives front axles. (Ford)

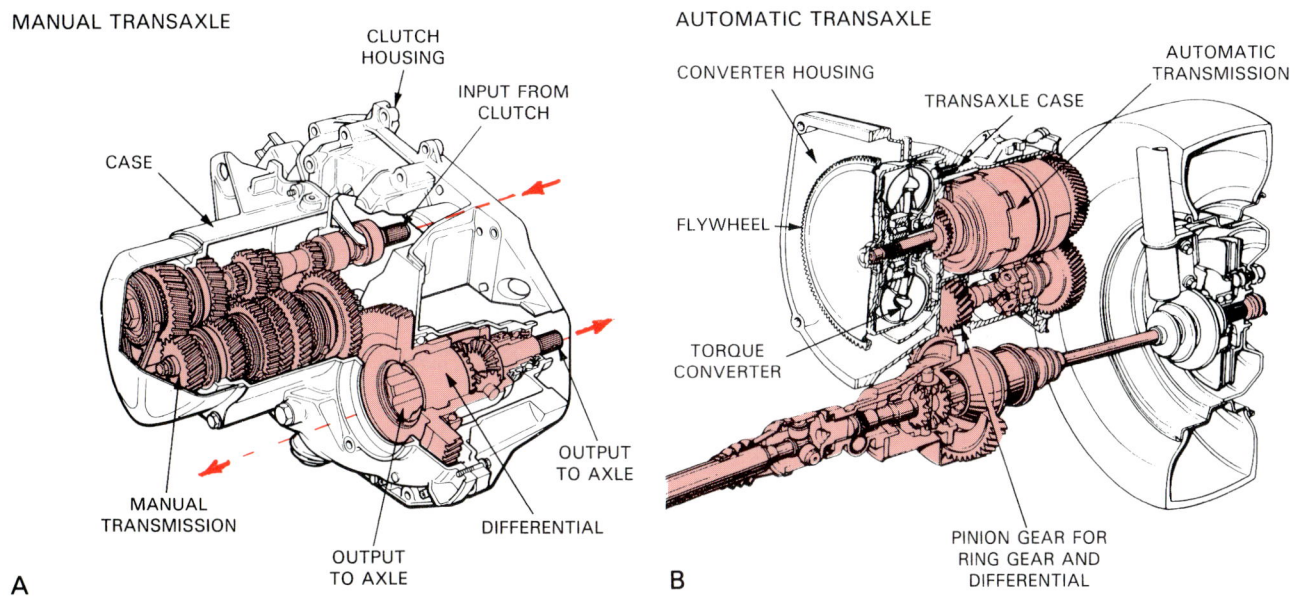

MANUAL TRANSAXLE

CLUTCH HOUSING

INPUT FROM CLUTCH

CASE

MANUAL TRANSMISSION

OUTPUT TO AXLE

DIFFERENTIAL

OUTPUT TO AXLE

A

AUTOMATIC TRANSAXLE

CONVERTER HOUSING

TRANSAXLE CASE

AUTOMATIC TRANSMISSION

FLYWHEEL

TORQUE CONVERTER

PINION GEAR FOR RING GEAR AND DIFFERENTIAL

B

Fig. 29-2. A — Typical manual transaxle is a manual transmission and a differential in single assembly. B — Typical automatic transaxle is automatic transmission and differential combined.

standard transmission type gearbox. An automatic transaxle uses a torque converter and a hydraulic system to control gear engagement. Compare the manual and automatic transaxles in Fig. 29-2.

Transaxle for transverse engine

Most transaxles are designed so that the engine can be transverse (sideways) mounted in the engine compartment. Look at Fig. 29-3A. Study the basic arrangement of this drive system. The engine crankshaft centreline points at both drive wheels. The transaxle bolts to the rear of the engine. This produces a very compact unit.

Engine torque enters the clutch and transaxle's transmission. The transmission transfers power into the differential. Then, the differential turns the drive axles which rotate the front wheels, Fig. 29-4A.

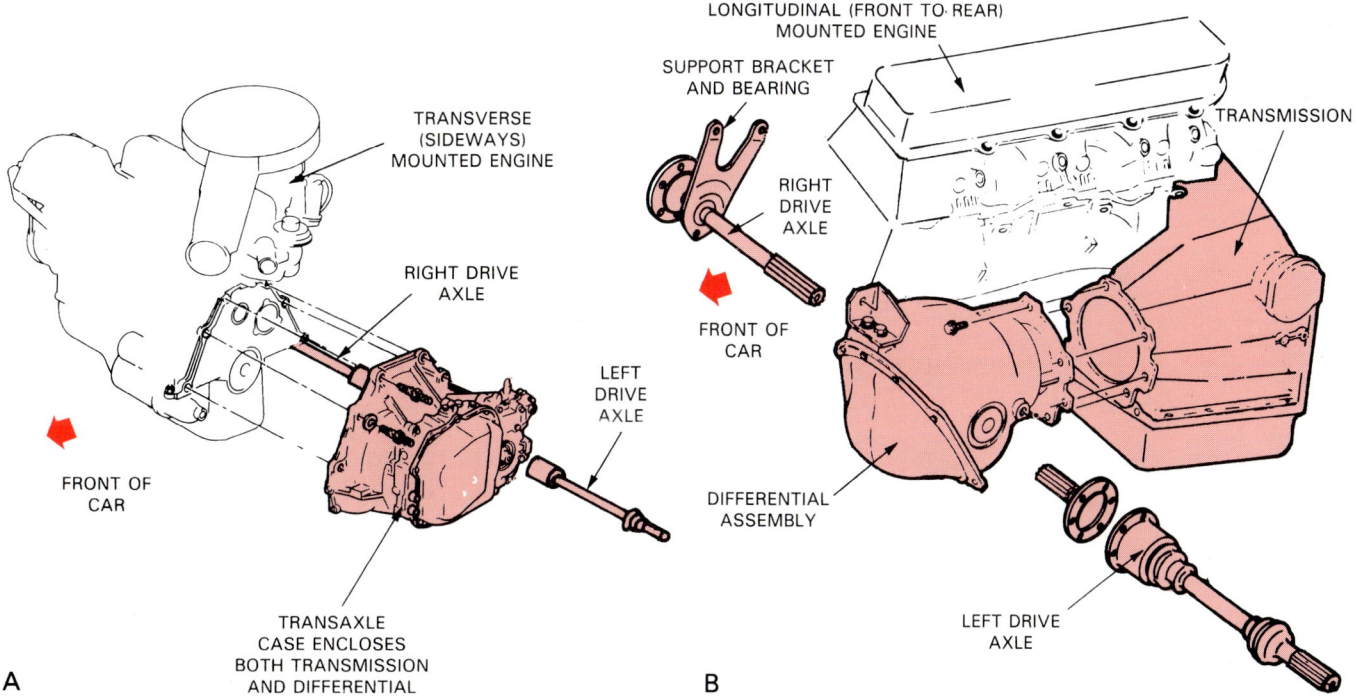

TRANSVERSE (SIDEWAYS) MOUNTED ENGINE

RIGHT DRIVE AXLE

LEFT DRIVE AXLE

FRONT OF CAR

TRANSAXLE CASE ENCLOSES BOTH TRANSMISSION AND DIFFERENTIAL

A

LONGITUDINAL (FRONT TO REAR) MOUNTED ENGINE

SUPPORT BRACKET AND BEARING

RIGHT DRIVE AXLE

TRANSMISSION

FRONT OF CAR

DIFFERENTIAL ASSEMBLY

LEFT DRIVE AXLE

B

Fig. 29-3. There are two basic transaxle differential design variations: A — With transverse mounted engine, engine crankshaft centreline and axle centreline are on same plane. B — With longitudinal mounted engine, differential must change power flow 90 degrees, as with rear-wheel drive.

Transaxle, Front Wheel Drive Fundamentals

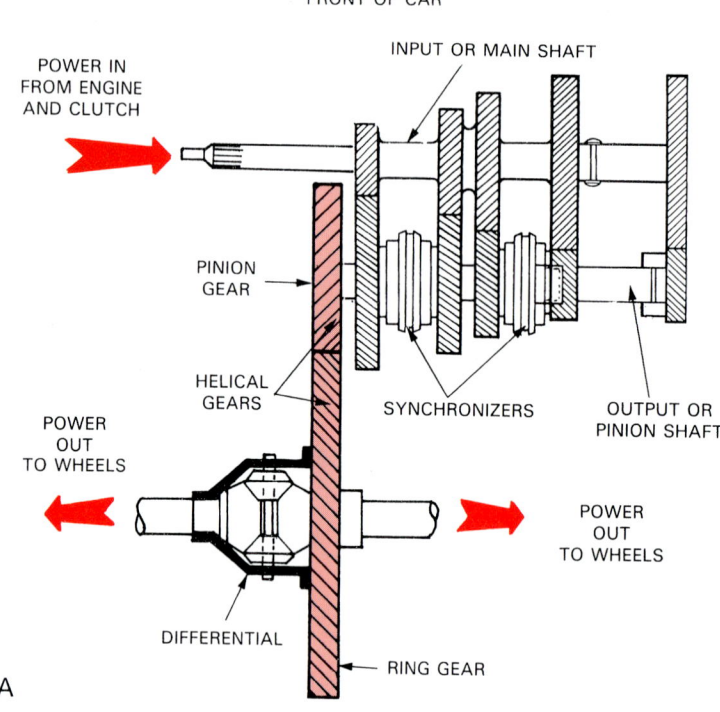

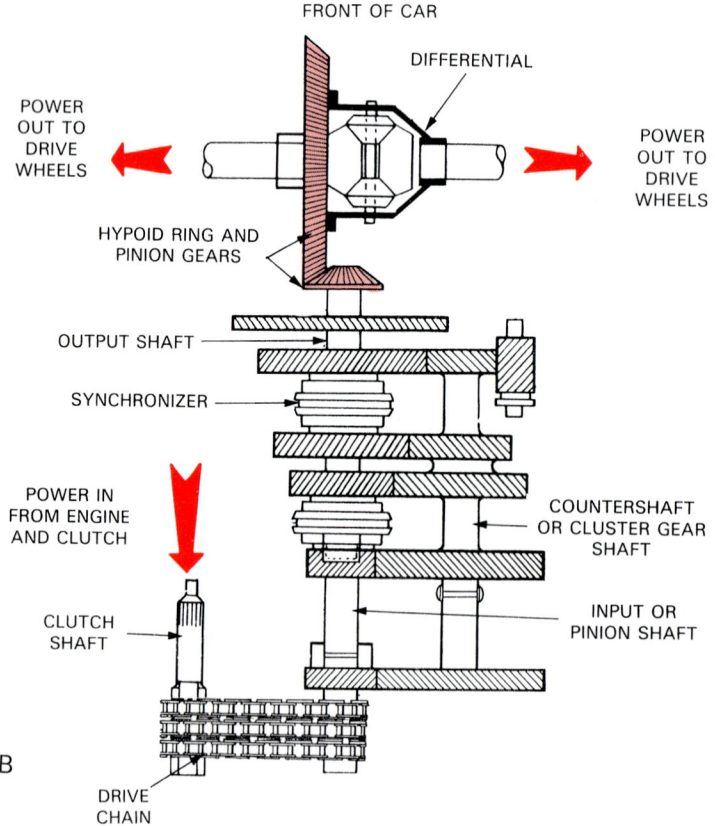

Fig. 29-4. Study transmission-to-differential locations. A — With transverse mounted engine, pinion and ring gears are helical gears. They are positioned in same direction as transmission gears. B — With longitudinal engine, differential uses hypoid gears to change direction of output. Also note drive link or chain that transfers power from crankshaft and clutch to input shaft. (Saab)

Transaxle for longitudinal engine

A few transaxles are made so that the engine is mounted longitudinally (lengthwise). The crankshaft centreline points toward the front and rear of car. Look at Fig. 29-3B.

A transaxle for a longitudinally mounted engine frequently uses a more conventional differential with helical ring and pinion gears. See Fig. 29-4B. The flow of engine torque must be changed 90 degrees in order to turn the drive axles.

As you can see from Figs. 29-3 and 29-4, a transaxle uses the same principles as a conventional transmission and rear axle assembly. However, the parts are located differently.

MANUAL TRANSAXLE

A manual transaxle uses a standard or manual clutch and transmission. A foot-operated clutch engages and disengages the engine and transaxle. A hand-operated shift lever allows the driver to change gear ratios.

As the basic parts relating to manual transaxle are introduced, Refer to Fig. 29-5.
1. TRANSAXLE INPUT SHAFT (main shaft splined to clutch disc; turns gears in transaxle).
2. TRANSAXLE INPUT GEARS (either freewheeling or fixed gears on input shaft; mesh with output gears).
3. TRANSAXLE OUTPUT GEARS (either freewheeling or fixed gears driven by input gears).
4. TRANSAXLE OUTPUT SHAFT (pinion shaft that transfers torque to ring and pinion gears and differential).
5. TRANSAXLE SYNCHRONIZERS (splined hub assemblies that can be used to lock freewheeling gears to their shafts for engagements).
6. TRANSAXLE DIFFERENTIAL (transfers gearbox torque to driving axles and allows axles to turn at different speeds).
7. TRANSAXLE CASE (aluminium housing that encloses and supports parts of transaxle).

Manual transaxle clutch

A manual transaxle clutch is almost identical to a clutch used with a manual transmission for a rear-wheel drive car. It uses a friction disc and spring-loaded pressure plate bolted to a heavy flywheel, Fig. 29-6.

Some transaxles use a conventional clutch release mechanism (throw-out bearing and fork). Others use a long push rod passing through the input shaft, as shown in Fig. 29-6.

Manual transaxle transmission

A manual transaxle transmission provides several (usually four or five) forward gear ratios and reverse.

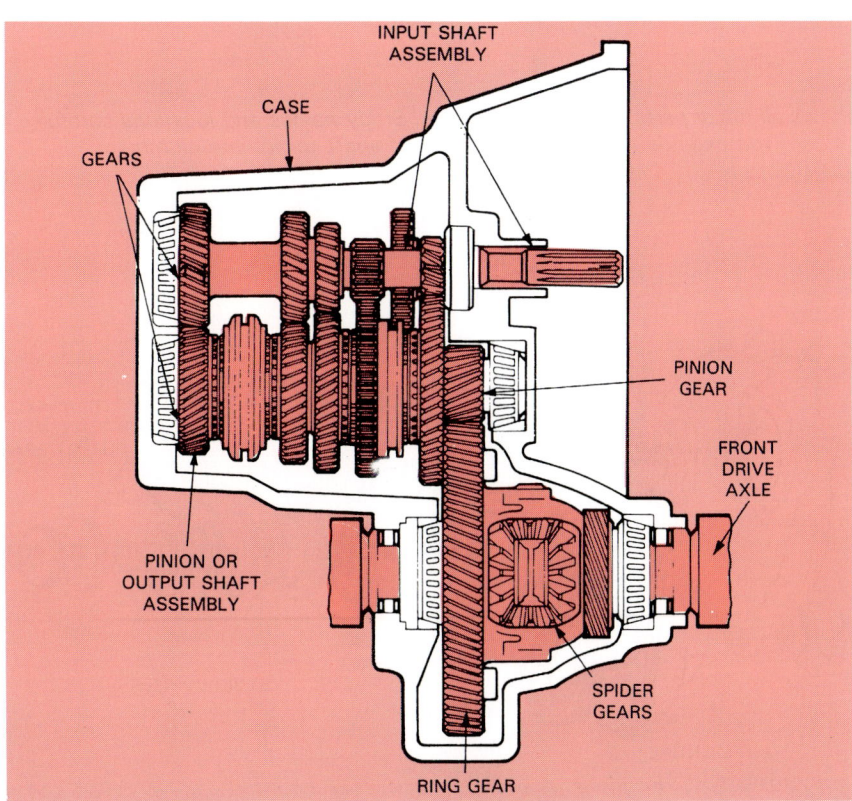

Fig. 29-5. Note basic parts of typical transaxle assembly. (Ford)

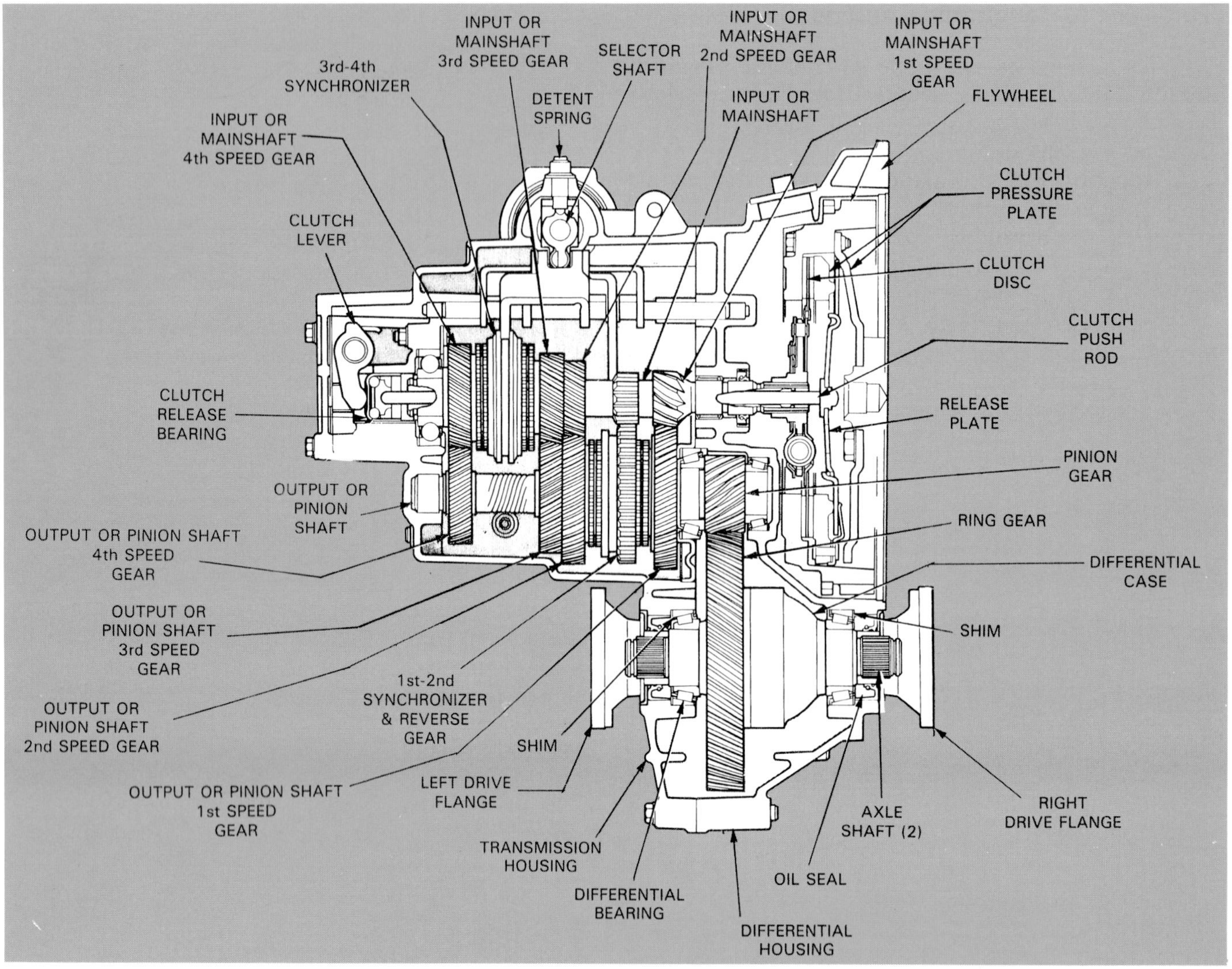

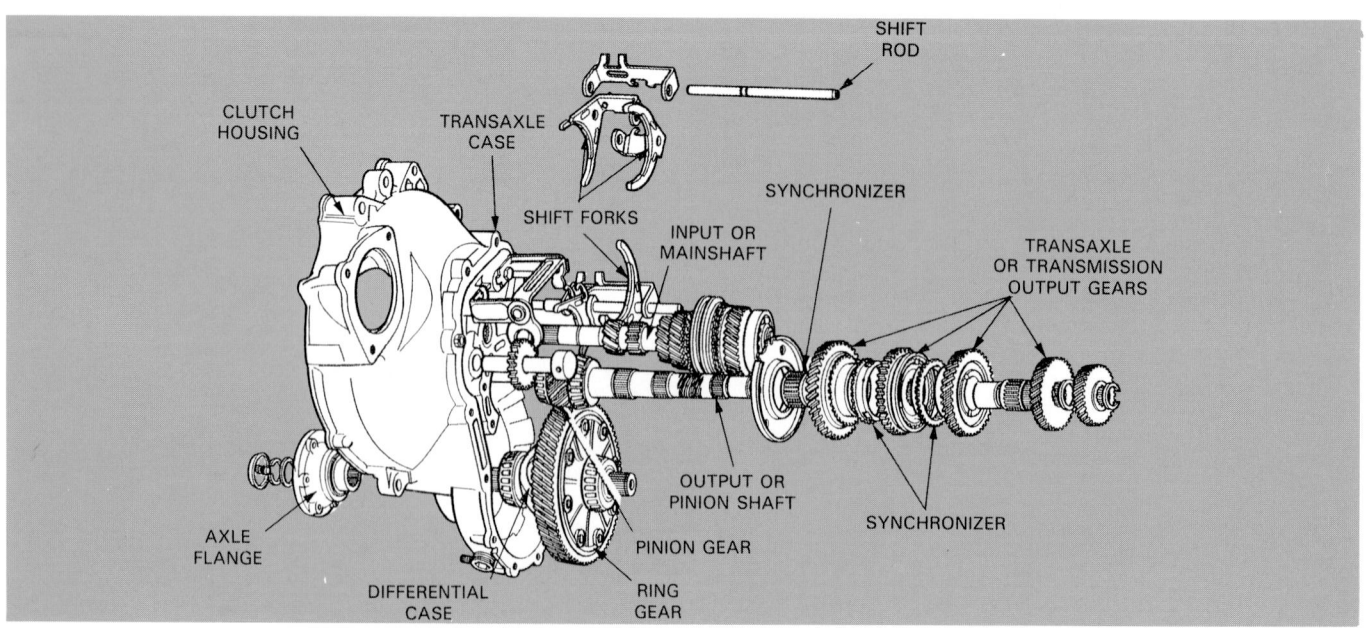

Fig. 29-6. Detailed view shows all major parts of manual transaxle. Study names and locations carefully. In this unit, long push rod extends through centre of input shaft to actuate clutch.

Fig. 29-7. As you can see, the inside of manual transaxle is very similar to manual transmission. However, this transaxle has freewheeling gears and synchronizers on both the input and output shafts. Shift rods and forks operate synchronizers.

Sometimes, high gear can provide an overdrive ratio for increased fuel economy.

As you read service manuals, you will find that the names of the shafts, gears, and other parts in a transaxle will vary. This will depend upon the location and function of the components. For example, look at Figs. 29-6, and 29-7.

The input shaft can sometimes be called the main shaft. The output shaft may be called the pinion shaft because it drives the ring and pinion gears.

Sometimes, the input or output shaft gears are called the CLUSTER GEAR or COUNTERSHAFT GEAR assembly. Like a manual transmission cluster or countershaft gear, several gears in the transaxle can be machined together as a unit. Refer to Figs. 29-5, 29-6 and 29-7.

The transaxle shafts are normally mounted in either tapered roller or ball bearings. The shaft bearings fit into the transaxle case. As in Fig. 29-7, the output shaft usually has a gear or sprocket for driving the differential ring gear.

The transaxle synchronizers are almost identical to those used in many manual transmissions. See Fig. 29-8. The inside hub of the synchronizer is splined to a transaxle shaft. The outer sleeve is free to slide on the hub.

When a shift fork moves a synchronizer into one of the freewheeling gears, the outer sleeve or ring of the synchronizer meshes with the small, outer teeth on the gear, Fig. 29-8. This locks the gear to the shaft.

Transaxle differential

A transaxle differential, like a rear axle differential, transfers power to the axles and wheels while allowing one wheel to turn at a different speed than the other. Look at Figs. 29-9 and 29-10. A small

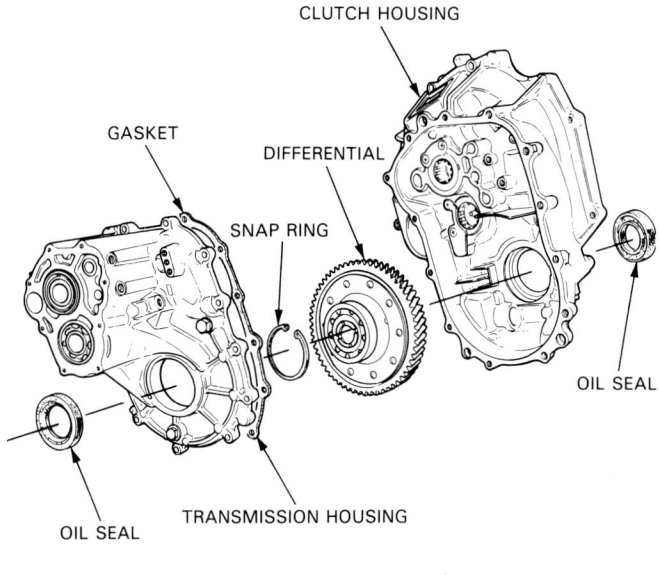

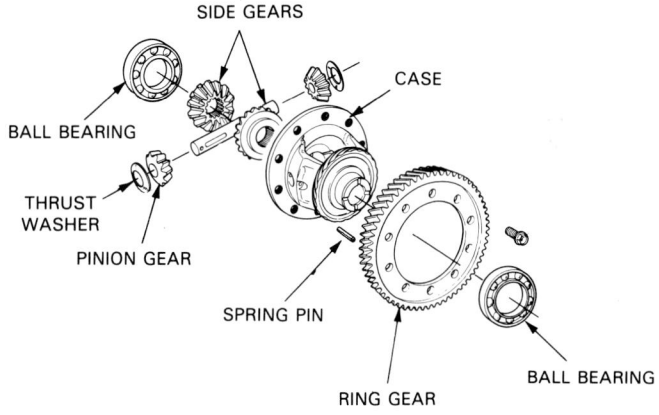

Fig. 29-9. Differential in transaxle uses spider gears, pinion shaft, and case to provide turning power to drive axles. Note that this unit for transverse engine uses helical ring and pinion gears instead of hypoid gears.

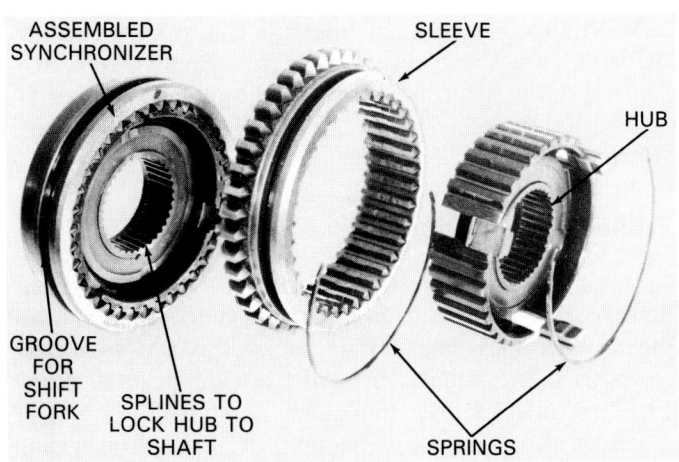

Fig. 29-8. Synchronizer for manual transaxle is also similar to one in manual transmission. Inner hub is splined to shaft. Outer sleeve can slide on hub to engage small teeth on side of freewheeling gear. This locks gear to hub and shaft for engagement.

pinion gear on the gearbox output shaft or countershaft turns the differential ring gear.

The differential ring gear is fastened to the differential case. The case holds the spider gears (pinion gears and axle side gears) and a pinion shaft. The axle shafts are splined to the differential side gears.

When the gearbox rotates the differential ring gear and case, the pinion shaft and pinion gears also revolve. The pinion gear teeth transfer power to the side gears. This causes the axle shafts to rotate. See Fig. 29-10.

Fig. 29-9 illustrates how the transmission and differential are positioned in the transaxle case. The transaxle case is normally made of cast aluminium. The shaft and differential bearing races are press fits into the case. The case also has provisions for mounting the shift forks and other accessory components.

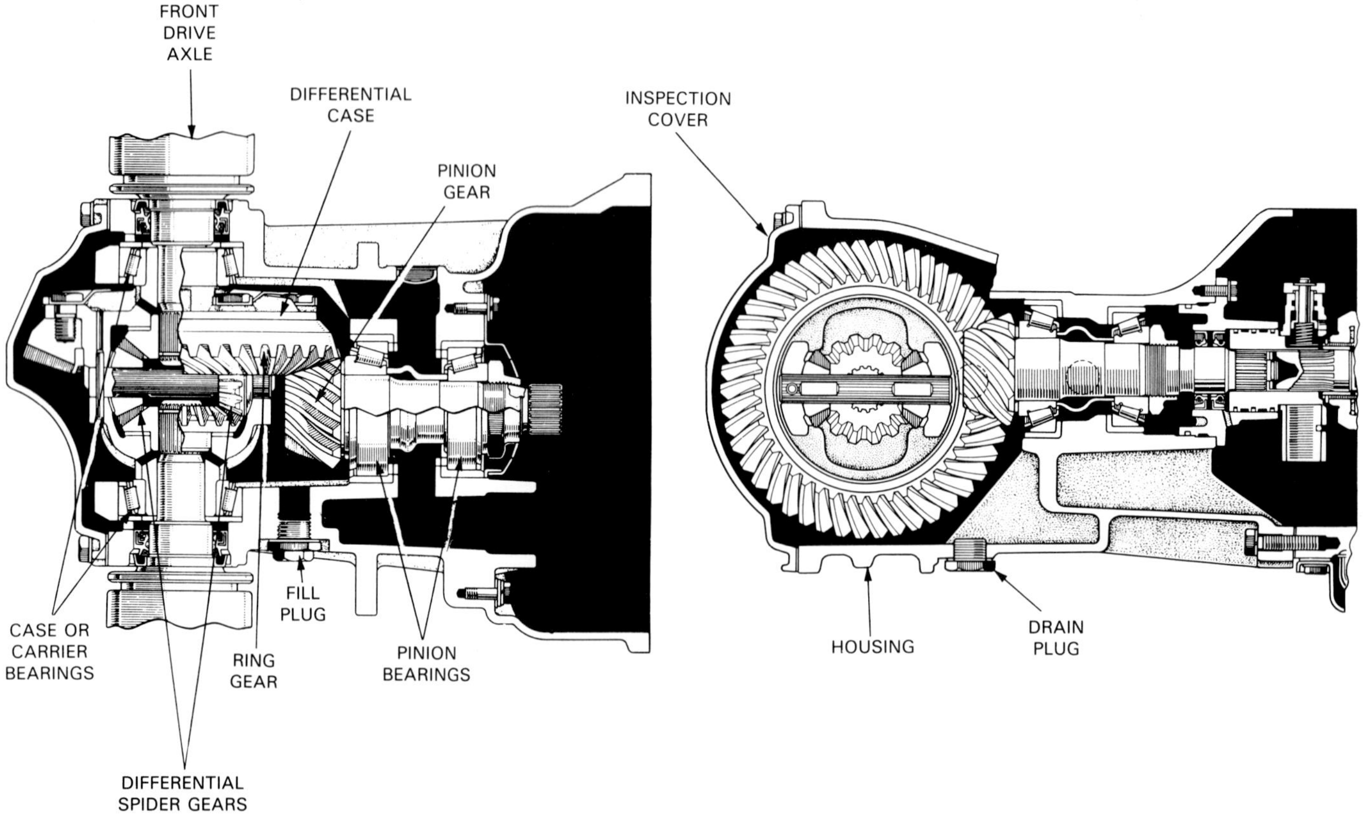

FRONT DRIVE AXLE

DIFFERENTIAL CASE

PINION GEAR

INSPECTION COVER

CASE OR CARRIER BEARINGS

RING GEAR

FILL PLUG

PINION BEARINGS

DIFFERENTIAL SPIDER GEARS

HOUSING

DRAIN PLUG

Fig. 29-10. Transaxle for longitudinal engine is almost identical to differential in rear-wheel drive axle. Hypoid gears transfer driving power. (Toyota)

Drive link (chain)

A drive link or chain is sometimes used to transfer crankshaft power to the transaxle gearbox or transmission. Pictured in Fig. 29-11, it is used with a longitudinally mounted engine.

One sprocket is mounted on a shaft connected to the engine crankshaft. The other sprocket drives the transaxle input shaft. This allows the transaxle to be located below and to one side of the engine. The differential and drive axles are under approximately the centre of the engine.

DRIVE SPROCKET

DRIVEN SPROCKET

LINK ASSEMBLY

Fig. 29-11. Drive link or chain transfers power from crankshaft input shaft to main shaft in transaxle. This allows transaxle to be mounted to one side and below engine.

MANUAL TRANSAXLE POWER FLOW

As the operation of a manual transaxle is explained, refer to Fig. 29-12. This illustration shows how power flows through a modern four-speed transaxle.

Transaxle in neutral

With the transaxle in neutral, the engine rotates the input shaft. However, since the synchronizers are centred away from any freewheeling gears, power is NOT transferred to the output shaft and differential. See Fig. 29-12A.

Transaxle in first gear

When the driver shifts the transaxle into first gear, a shift fork slides the first-second synchronizer into mesh with first gear. This locks the synchronizer teeth with the small teeth on the side of first gear. The first gear is now locked to its shaft.

Power flows through the input shaft first gear, output shaft first, gear pinion gear, ring gear, and to the drive axles through the differential spider gears, Fig. 29-12B.

Since the input shaft first gear is much smaller than the first gear on the output, a gear reduction is

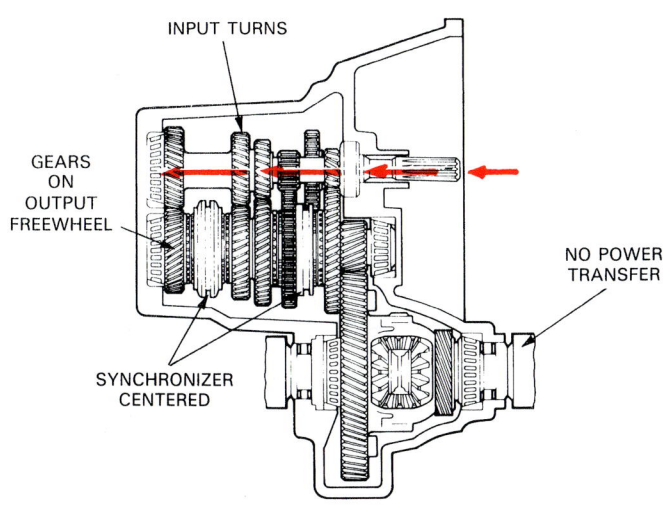

A — In neutral, no synchronizers are engaged with gears. Input shaft spins but gears freewheel and do not transfer power to output.

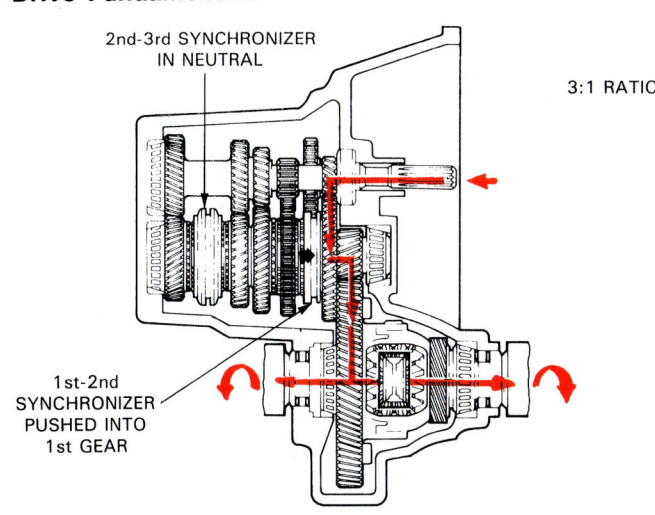

B — In first gear, first-second synchronizer slides to left, engaging first output gear. This locks gear to shaft and power flows to output shaft and differential.

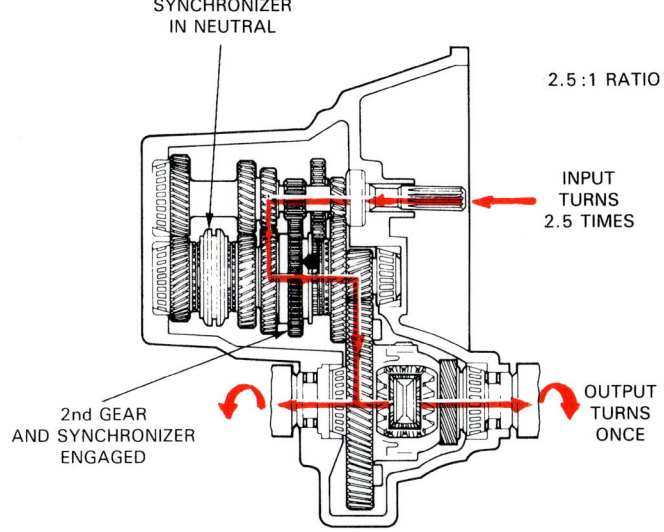

C — In second gear, same synchronizer slides to left. This locks output second gear to shaft. Power flows through second gears and to differential.

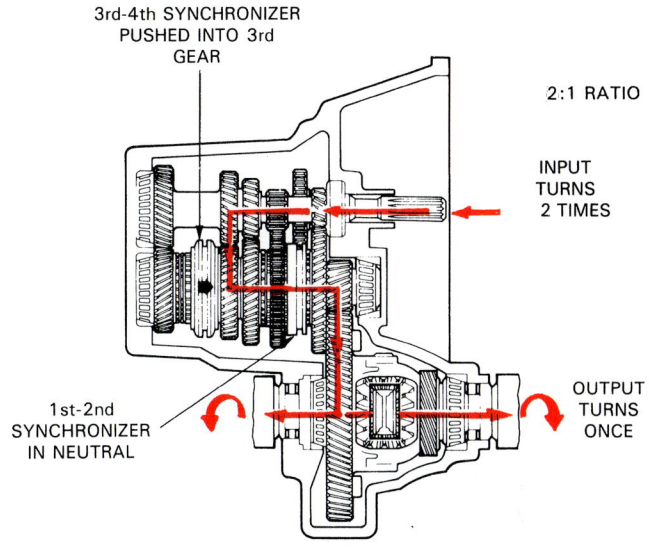

D — In third gear, first-second synchronizer is centered. Third-fourth synchronizer moves to right, engaging third output gear to its shaft.

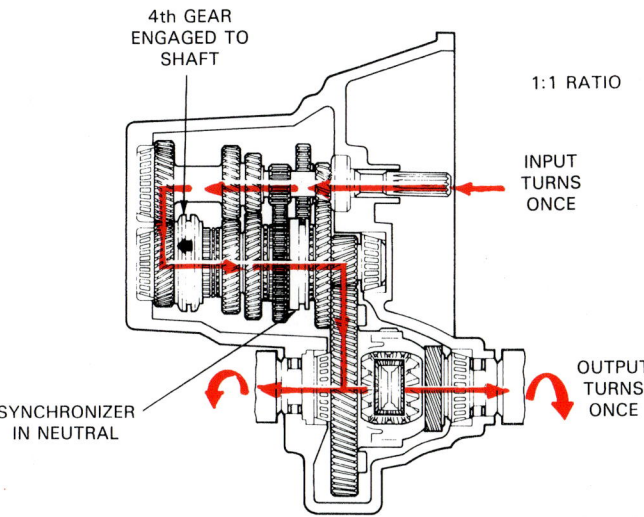

E — When shifted into fourth, the same synchronizer is slid the other way. Fourth output gear is locked to shaft and transmits power.

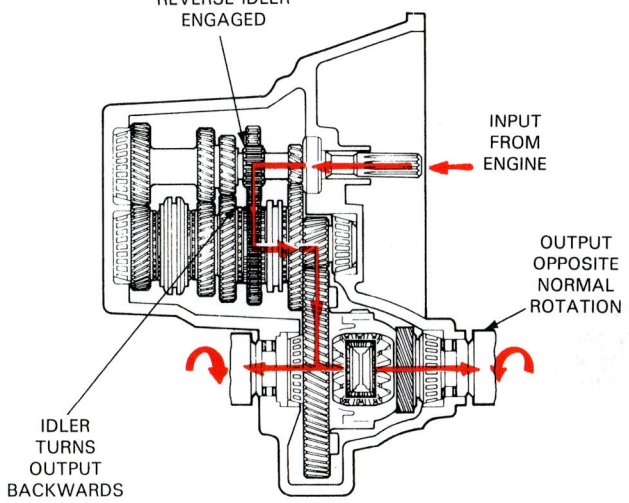

F — In reverse, the reverse idler gear is engaged. It causes the output shaft and differential to turn backwards.

Fig. 29-12. Tracing power flow through typical manual transaxle. (Ford)

produced. Torque is increased since the engine must rotate several times to produce one axle shaft rotation.

Transaxle in second gear

When the transaxle is shifted into second, the first-second synchronizer is moved into mesh with second gear on the input shaft. The second gear can no longer freewheel on its shaft. Power then flows through the second gears and into the differential. From the differential, power flows to the axle shafts and drive wheels. Refer to Fig. 29-12C.

Transaxle in third gear

With the transaxle in third gear, the first-second synchronizer is shifted into neutral so that the first and second gears freewheel. At the same time, the third-fourth synchronizer is slid into mesh with third gear. Third gear is then locked to its shaft. Power flows through the transaxle, Fig. 29-12D.

Transaxle in fourth gear

When the transaxle is in fourth gear, the three-four synchronizer is moved into contact with fourth gear. Power flows through the two fourth gears, into the differential, and to the front wheels. Fig. 29-12E.

Since the fourth gear on the input shaft and the fourth gear on the countershaft are almost the same size, the gear ratio is reduced. Compared to its rotation in the other gears, the engine turns slowly while the differential case and axles rotate at a relatively high speed. This allows the car to cruise at highway speeds with the engine running at low rpm.

Transaxle in reverse

When the transaxle is shifted into reverse, the reverse sliding gear is moved into mesh with the reverse gears on the input shaft and output shaft. The sliding gear reverses the direction of rotation. As a result, the differential and axle shafts are turned backwards and the car moves in reverse. Look at Fig. 29-12F.

Figs. 29-13 and 29-14 show two more manual transaxles. Compare these to the ones shown earlier.

AUTOMATIC TRANSAXLE

An automatic transaxle is a combination auto-

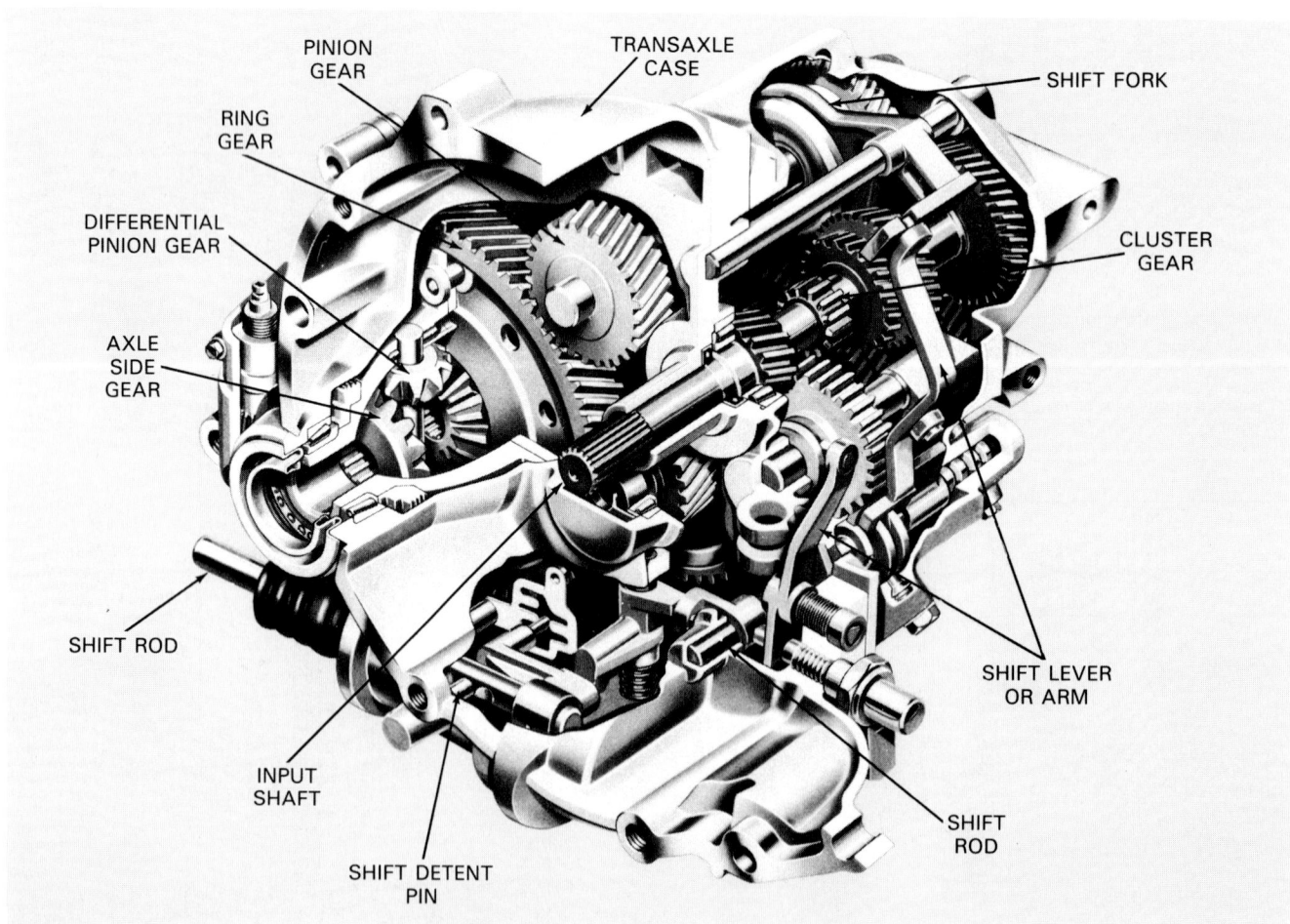

Fig. 29-13. Cutaway view shows inside of modern 5-speed manual transaxle with overdrive in high gear. Study how shift levers and rods connect to shift forks. (Ford)

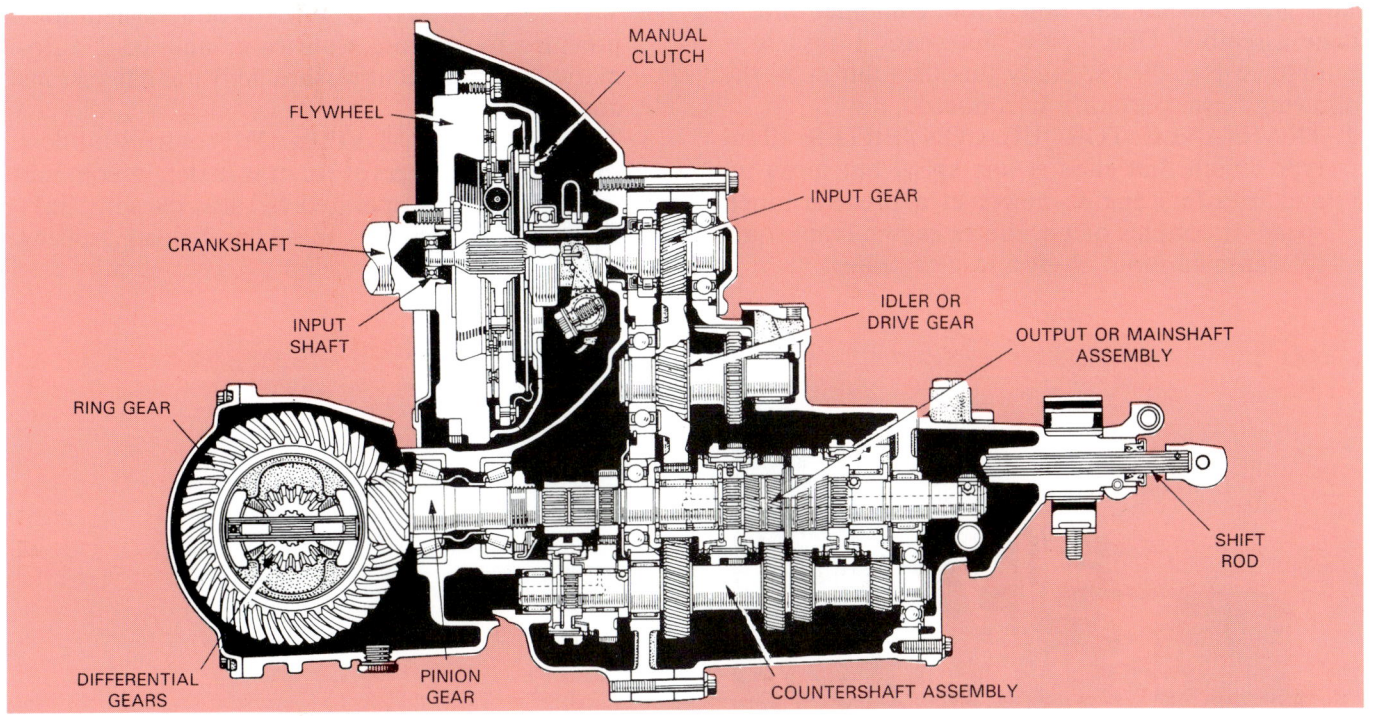

Fig. 29-14. Late model manual transaxle for longitudinally mounted engine. Note how gears, instead of drive chain, transfer power to transaxle gearbox.

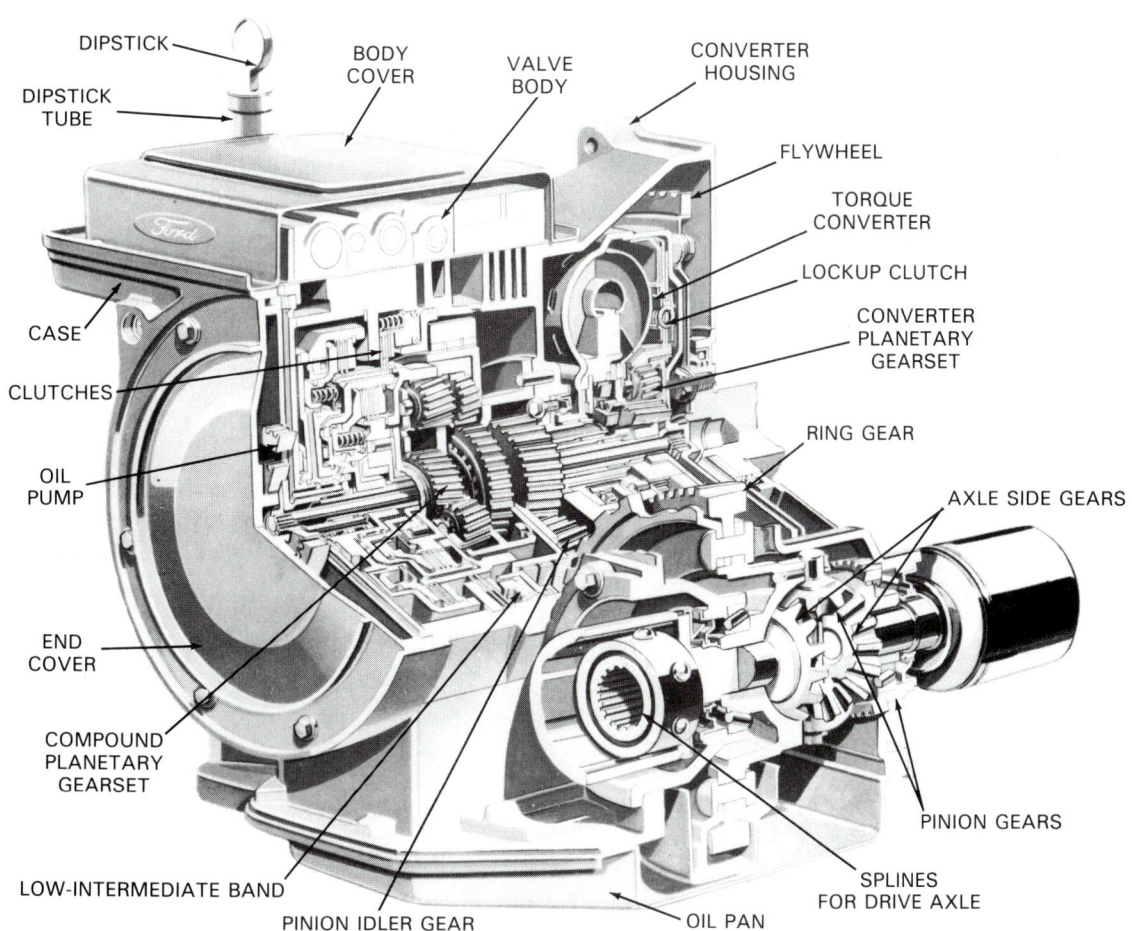

Fig. 29-15. Study basic parts of automatic transaxle. It uses same parts covered in chapter on automatic transmissions.

matic transmission and differential combined into a single assembly. One type of automatic transaxle is pictured in Fig. 29-15. Study this illustration as the following basic parts are introduced.

1. TRANSAXLE TORQUE CONVERTER (fluid type clutch that slips at low speed but locks up and transfers engine power at a predetermined speed; it couples or uncouples engine crankshaft to transaxle input shaft and gear train).

2. TRANSAXLE OIL PUMP (produces hydraulic pressure to operate, lubricate, and cool automatic transaxle; its pressure activates the pistons and servos).

3. TRANSAXLE VALVE BODY (controls oil flow to pistons and servos in transaxle; it contains hydraulic valves operated by driver's shift linkage and by engine speed and load sensing devices).

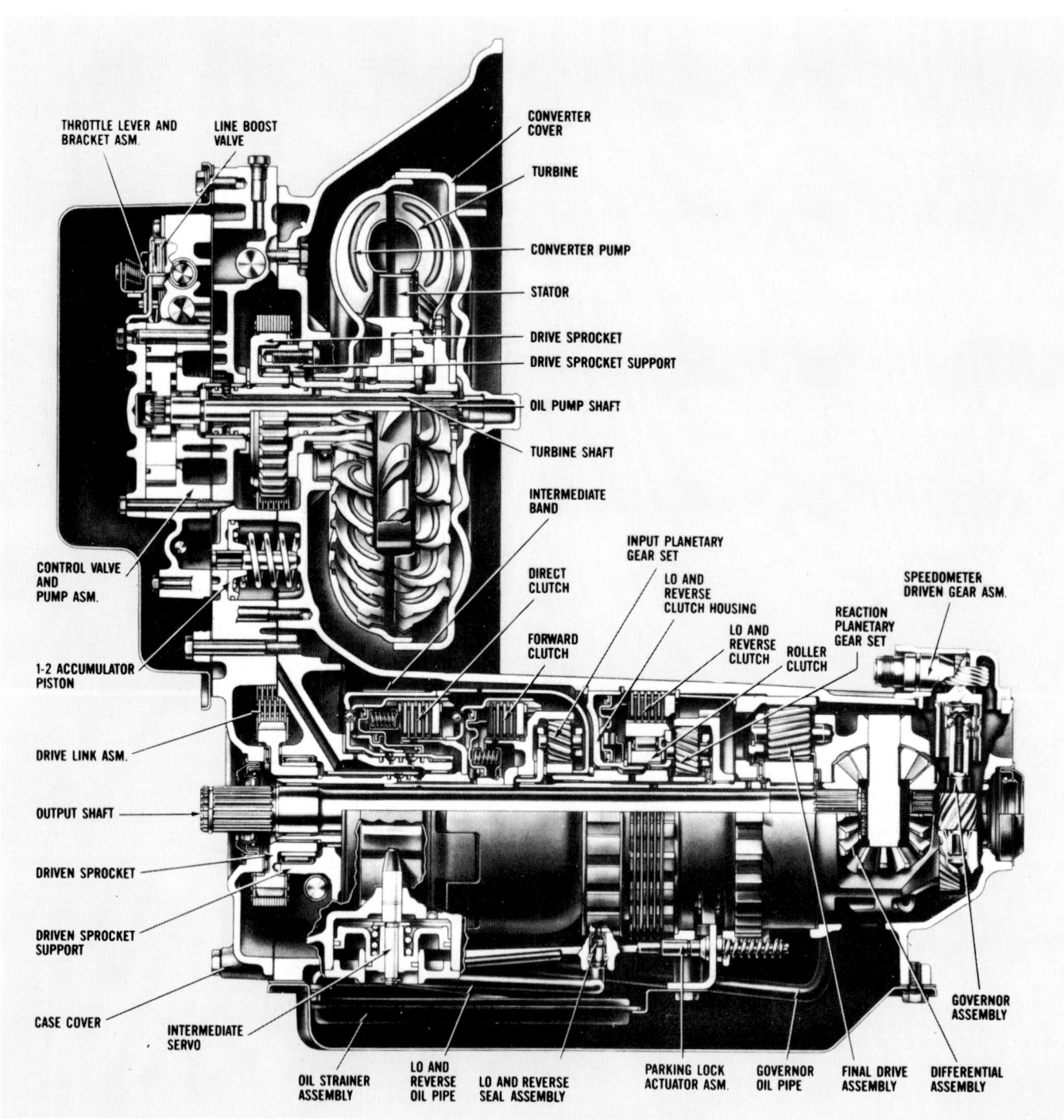

Fig. 29-16. This automatic transaxle uses drive link or chain between torque converter input and transmission proper. Output shaft connects to pinion gear in conventional type differential.

4. TRANSAXLE PISTONS and SERVOS (operate clutches and bands when activated by oil pressure from valve body).

5. TRANSAXLE BANDS and CLUTCHES (apply planetary gears in transaxle; different bands and clutches can be activated to operate different units in gearsets).

6. TRANSAXLE PLANETARY GEARSETS (provide different gear ratios and reverse gear in automatic transaxle).

7. TRANSAXLE DIFFERENTIAL (transfers power from transmission components to axle shafts).

As you can see, an automatic transaxle uses many of the same parts found in an automatic transmission.

For a review of the operating principles of these components, turn back to Chapter 26, Automatic Transmission Fundamentals.

Fig. 29-16 shows another variation of an automatic transaxle. Study how the major parts are located differently than in the transaxle given in Fig. 29-15.

Fig. 29-17 shows some of the parts of a transaxle hydraulic system. Note that the valve body on this particular unit bolts to the top of the transaxle case. The oil filter and sump are located on the bottom of the case. Gaskets seal the valve body cover and oil pan to the case.

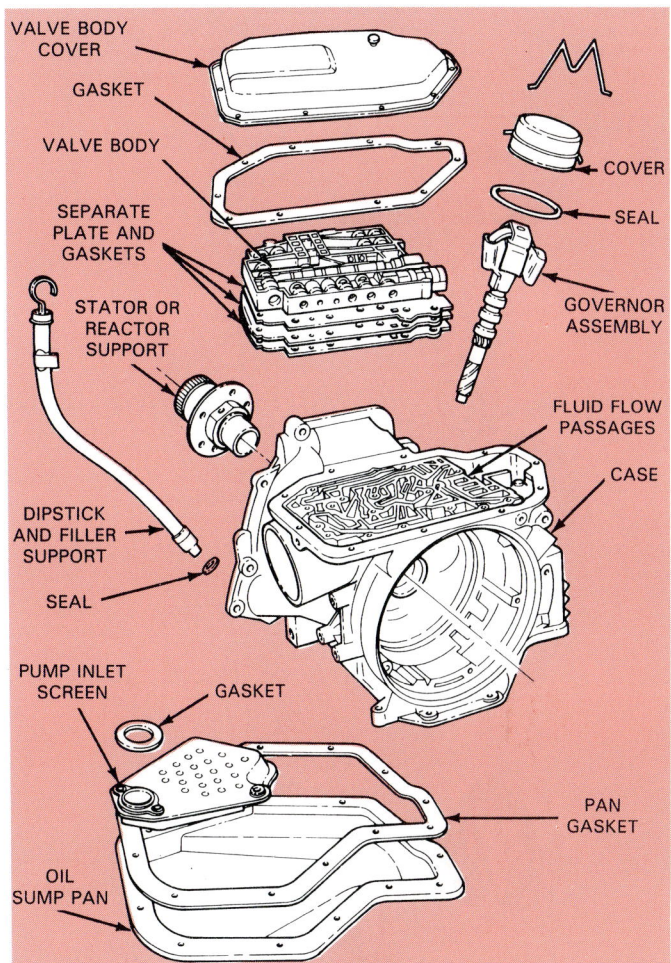

Fig. 29-17. General arrangement of major hydraulic components in automatic transaxle. With this particular unit, valve body is on top and sump and pan are on bottom. Case construction is similar to automatic transmission. (Ford)

AUTOMATIC TRANSAXLE POWER FLOW

The flow of power through an automatic transaxle is similar to power flow through an automatic transmission. Engine torque enters the torque converter. The torque converter then turns the input shaft and planetary gearsets.

Depending upon which bands and clutches hold the gearset members, power flows through the planetary gearsets to the ring and pinion gears. The differential powers the axle shafts and front wheels.

Fig. 29-18 shows the power flow through one make of automatic transaxle.

In A, the transaxle is in first gear. The band holds the forward sun gear. The one-way clutch sends turbine shaft torque to the low-reverse sun gear. This produces a gear reduction in the planetary gearset for initial acceleration.

In B, the transaxle is in second gear. The band remains applied and holds the forward sun gear. The intermediate clutch is applied, locking the intermediate shaft to the ring gear. This causes a slightly less gear reduction in the planetary gearset.

In C, the transaxle is in third gear. Both clutches are applied. This locks both members of the planetary gearset and the unit turns as a single member for direct drive to the differential.

In D, the transaxle is in reverse. The reverse clutch holds the planetary ring gear stationary. The direct clutch locks the turbine shaft to the low-reverse sun gear. The one-way clutch allows the turbine shaft to turn low-reverse sun gear clockwise. Then, the output from the planetary gearset is reversed.

Figs. 29-19 and 29-20 illustrate other automatic transaxle design variations. Compare these transaxles. Make sure you can identify all of the major components.

FRONT DRIVE AXLES (AXLE SHAFTS)

Front drive axles, also called axle shafts or front drive shafts transfer power from the differential to the hubs and wheels of the car, Fig. 29-21.

Most modern front drive axles consists of two or three separate shafts and two universal joints. This enables the drive axle to transfer power smoothly as the front wheels move up and down over bumps and to the left or right for steering.

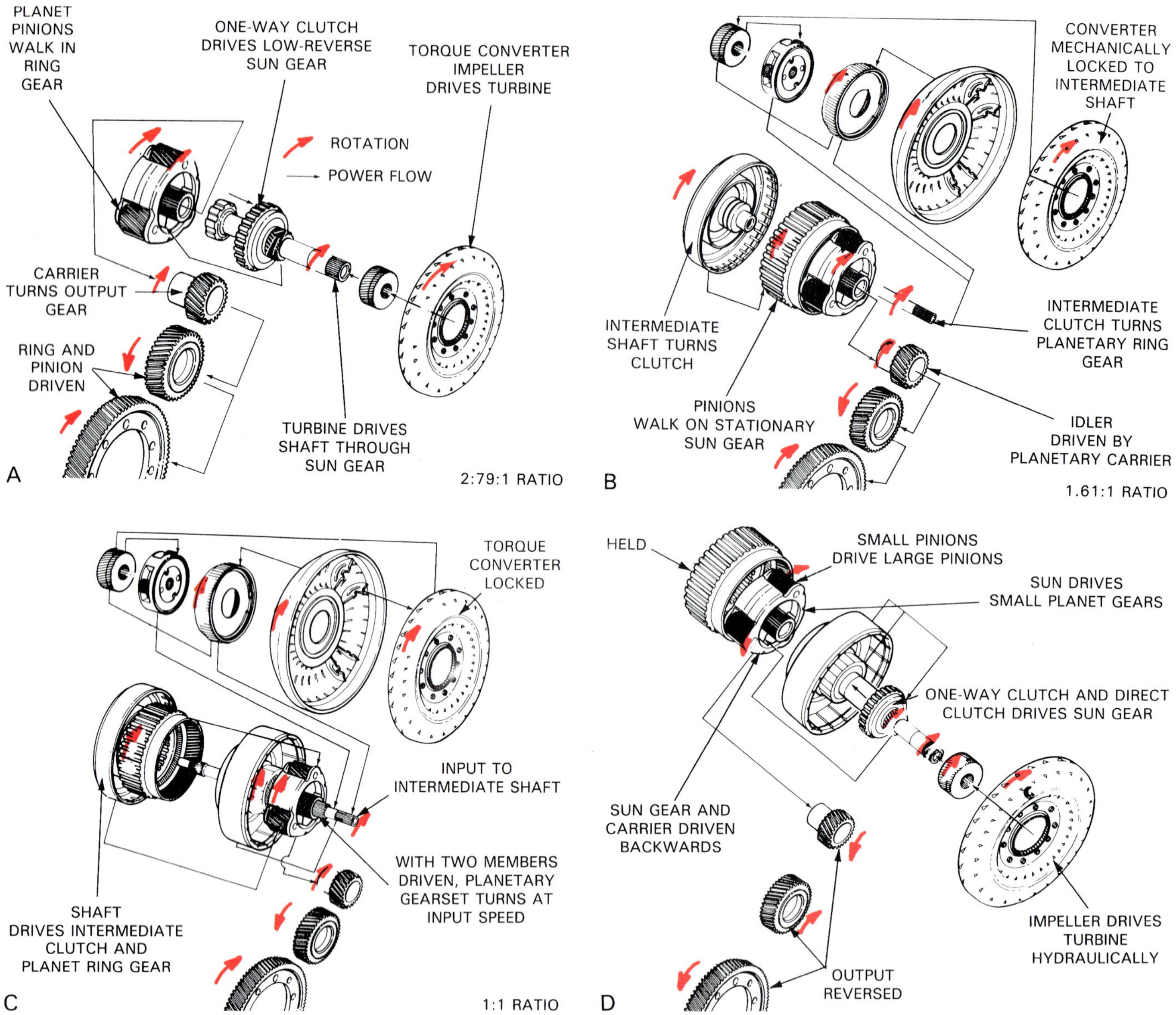

A

PLANET PINIONS WALK IN RING GEAR

ONE-WAY CLUTCH DRIVES LOW-REVERSE SUN GEAR

TORQUE CONVERTER IMPELLER DRIVES TURBINE

→ ROTATION

→ POWER FLOW

CARRIER TURNS OUTPUT GEAR

RING AND PINION DRIVEN

TURBINE DRIVES SHAFT THROUGH SUN GEAR

2:79:1 RATIO

B

CONVERTER MECHANICALLY LOCKED TO INTERMEDIATE SHAFT

INTERMEDIATE SHAFT TURNS CLUTCH

PINIONS WALK ON STATIONARY SUN GEAR

INTERMEDIATE CLUTCH TURNS PLANETARY RING GEAR

IDLER DRIVEN BY PLANETARY CARRIER

1.61:1 RATIO

C

TORQUE CONVERTER LOCKED

INPUT TO INTERMEDIATE SHAFT

WITH TWO MEMBERS DRIVEN, PLANETARY GEARSET TURNS AT INPUT SPEED

SHAFT DRIVES INTERMEDIATE CLUTCH AND PLANET RING GEAR

1:1 RATIO

D

HELD

SMALL PINIONS DRIVE LARGE PINIONS

SUN DRIVES SMALL PLANET GEARS

ONE-WAY CLUTCH AND DIRECT CLUTCH DRIVES SUN GEAR

SUN GEAR AND CARRIER DRIVEN BACKWARDS

IMPELLER DRIVES TURBINE HYDRAULICALLY

OUTPUT REVERSED

Fig. 29-18. Basic power flow through modern automatic transaxle. (Ford)

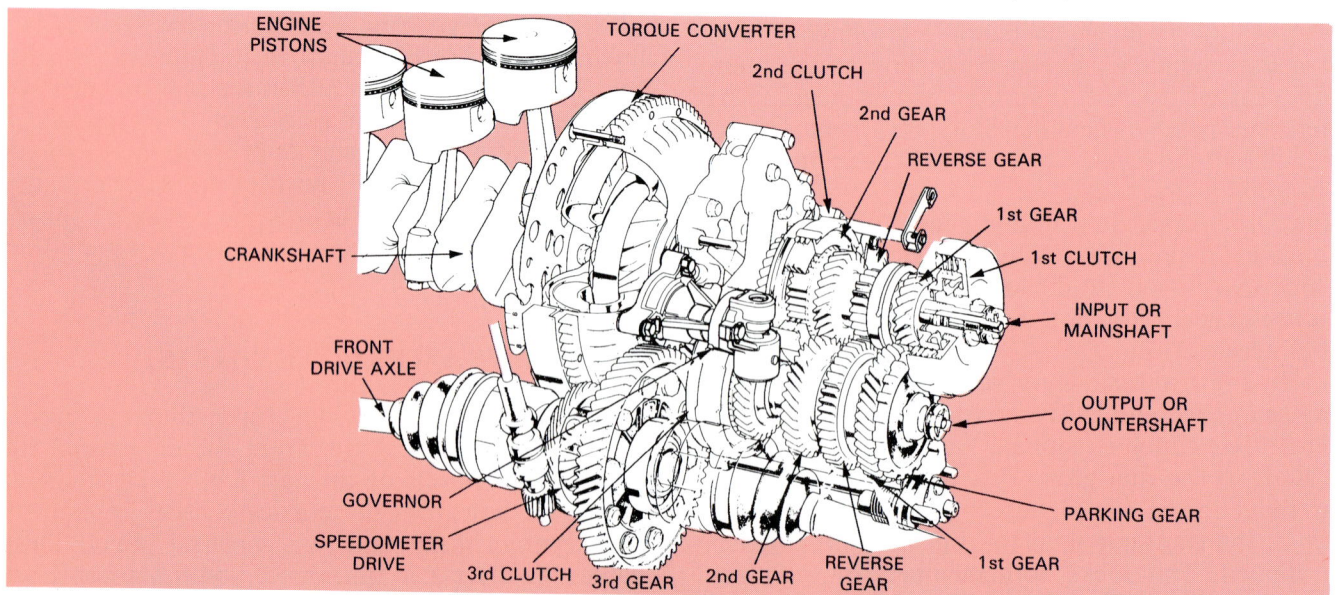

ENGINE PISTONS

TORQUE CONVERTER

2nd CLUTCH

2nd GEAR

REVERSE GEAR

1st GEAR

1st CLUTCH

INPUT OR MAINSHAFT

OUTPUT OR COUNTERSHAFT

PARKING GEAR

1st GEAR

REVERSE GEAR

2nd GEAR

3rd GEAR

3rd CLUTCH

SPEEDOMETER DRIVE

GOVERNOR

FRONT DRIVE AXLE

CRANKSHAFT

Fig. 29-19. This automatic transaxle does not use planetary gears. It uses hydraulically operated clutches to activate helical gears. Study its construction. (Honda)

452

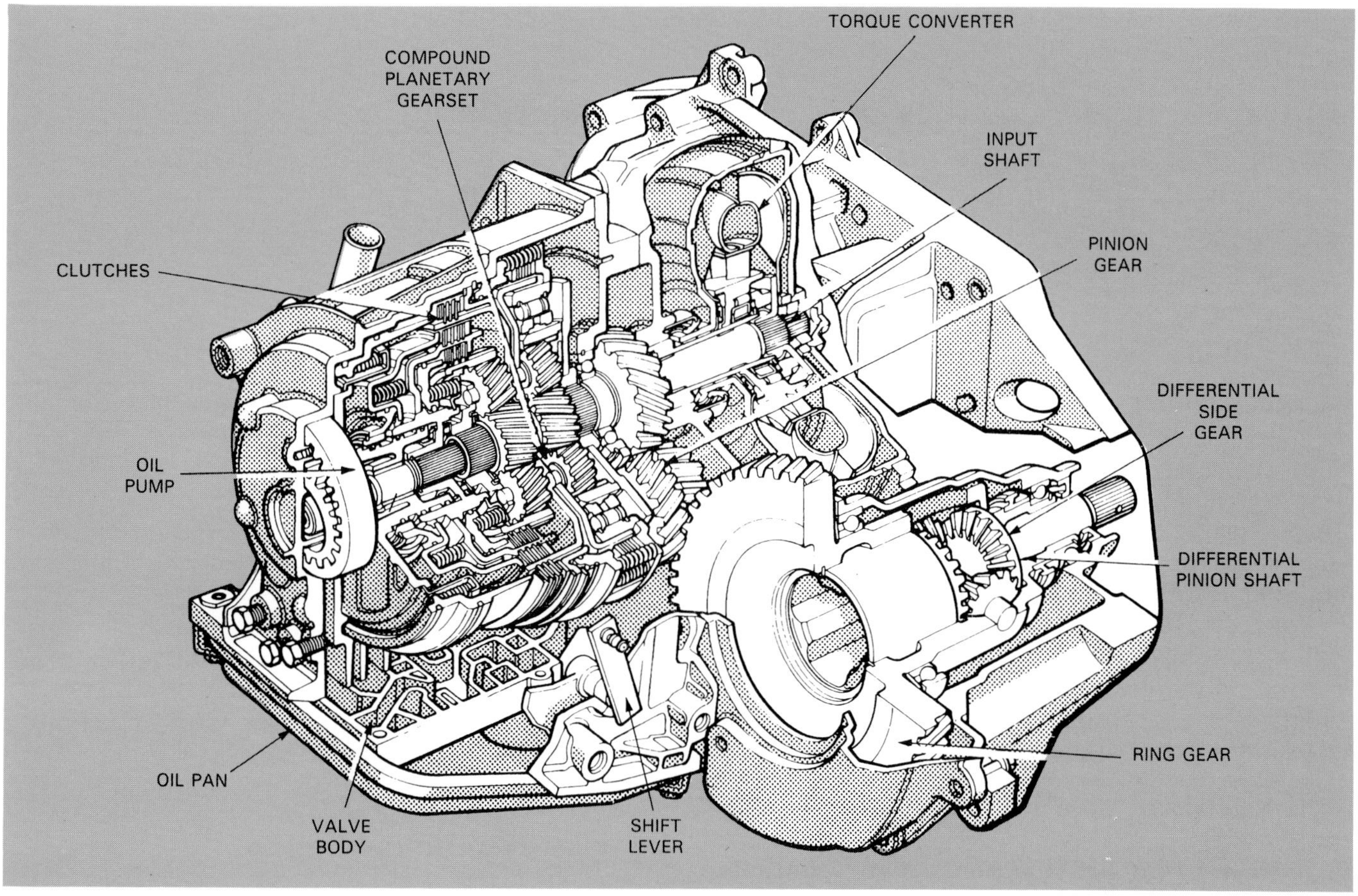

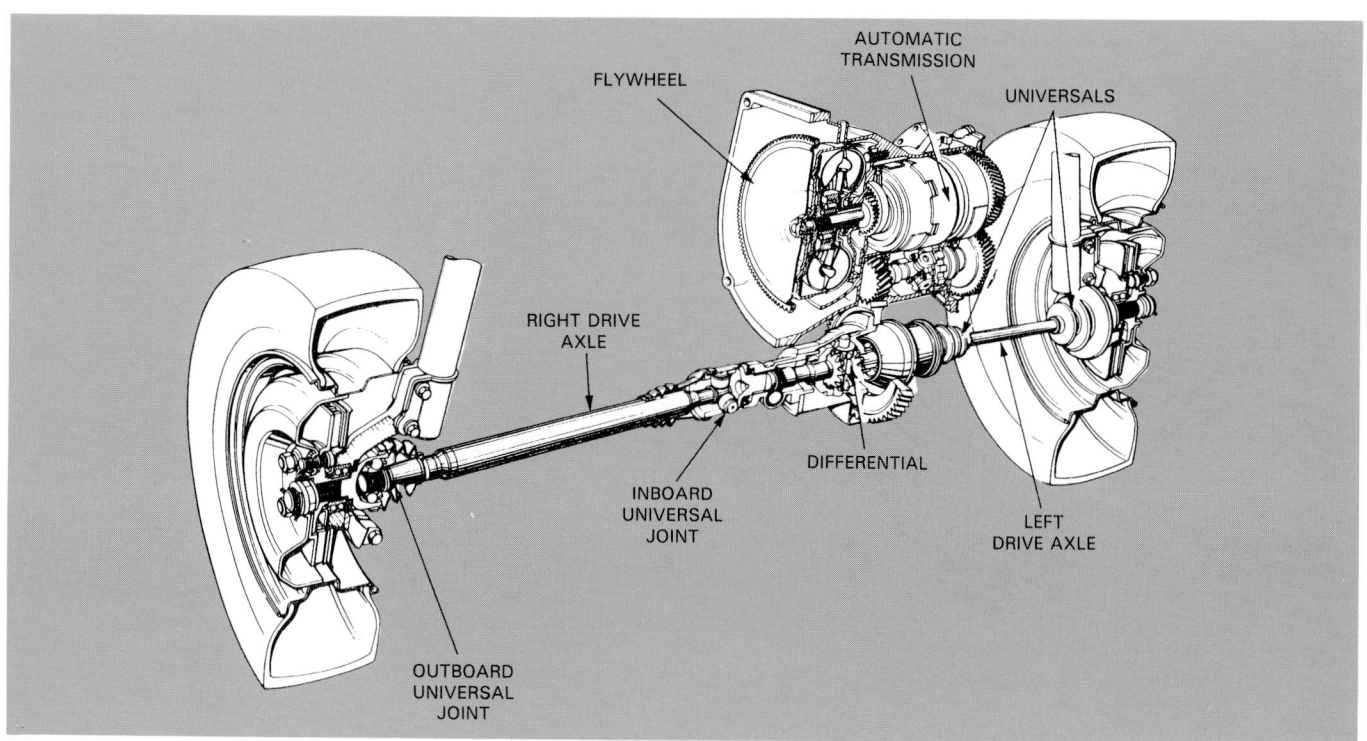

Fig. 29-20. *Another automatic transaxle. This is a typical unit that uses compound planetary gearset, bands, and clutches. (Renault)*

Fig. 29-21. *Front drive axles connect differential side gears to wheel hubs. When differential side gears turn, axles rotate hubs and front wheels to propel car. (Typical)*

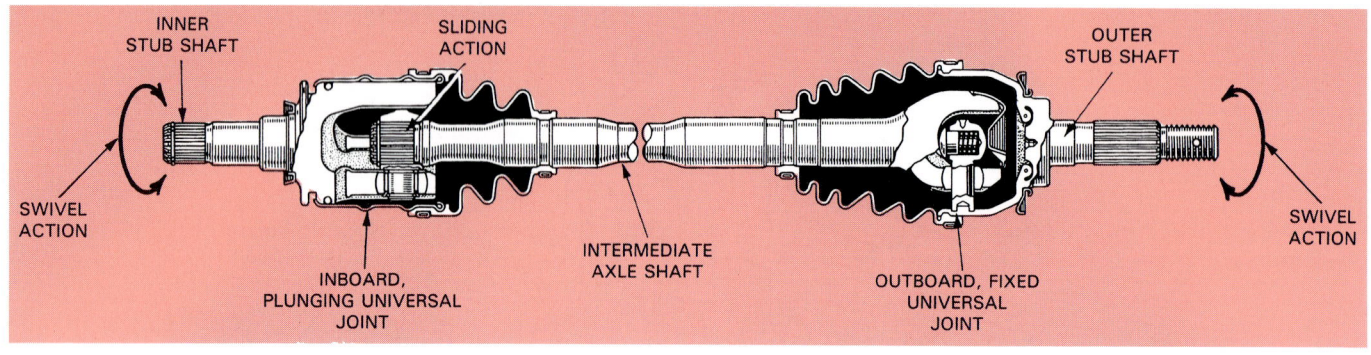

Fig. 29-22. Constant velocity universal joints, commonly called CV-joint, allow drive axle to swivel into various angles. This is a three-piece drive axle: inner stub shaft, intermediate shaft, and outer stub shaft. Inboard CV-joint is normally a sliding joint to allow for length changes with suspension and steering action. (Toyota)

Front drive axles turn much SLOWER than a drive shaft for a rear-wheel drive car. They turn about one-third slower than a rear drive shaft. They are connected directly to the drive wheels and do NOT have to act through the reduction of rear axle ring and pinion gears.

Axle shafts (front-wheel drive)

The axle shafts of a front drive axle typically consists of:

1. INNER STUB SHAFT (short shaft splined to side gears in differential and connected to inner universal joints), Fig. 29-22.
2. OUTER STUB SHAFT (short shaft connected to outer universal joint and front wheel hub).

3. INTERCONNECTING SHAFT (centre or intermediate shaft that fits between two universal joints), Fig. 29-22.

The outer end of each shaft are machined. They may have splines for meshing with splines on mating parts. They may also have a portion of a universal joint machined as an integral part. Grooves are also cut in the shafts for snap rings, boots, and other components.

Universal joints (front-wheel drive)

Universal joints in the front drive axle assemblies allow the shafts to operate through an angle without damage. They are normally the CONSTANT VELOCITY (abbreviated CV) type. Normally, either

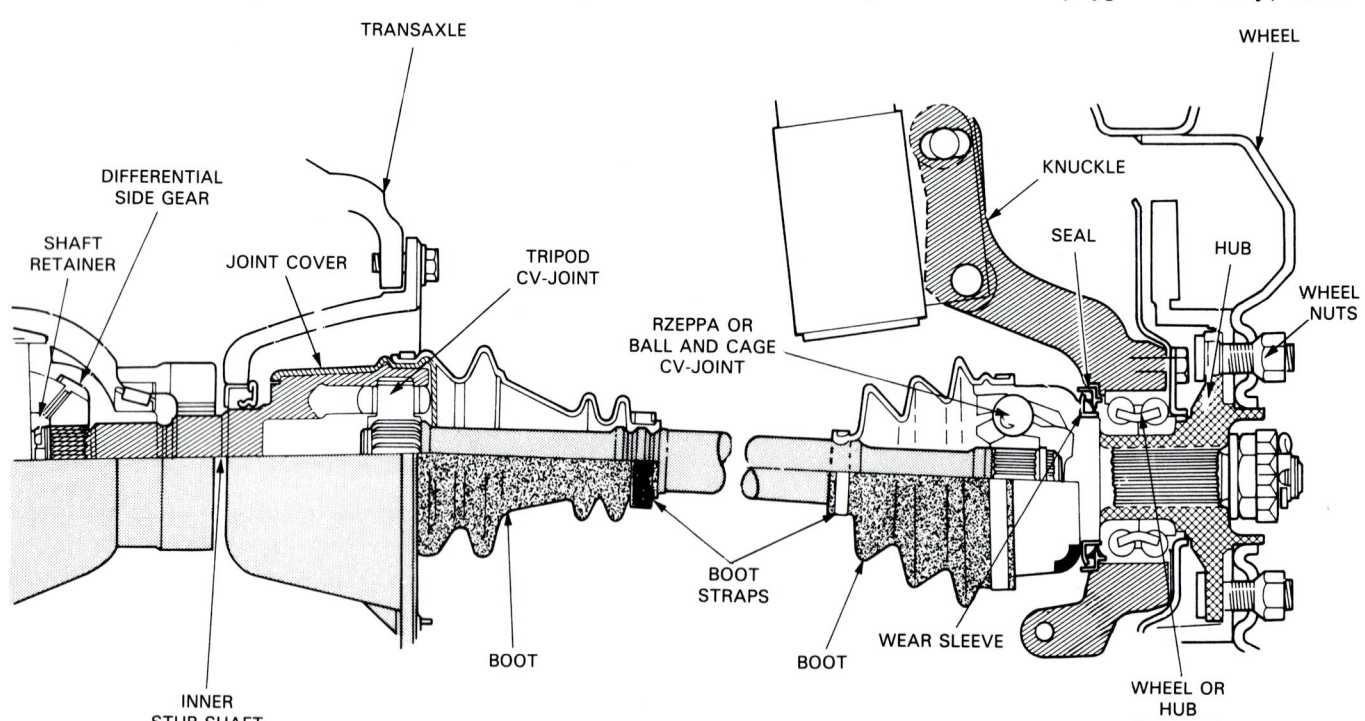

Fig. 29-23. Study how inner end of front drive axle is splined to axle side gear in transaxle differential. Outer end of drive axle extends through and is splined to front wheel hub. Outer CV-joint is fixed Rzeppa or ball and cage type. Inboard CV-joint is plunging tripod type.

Rzeppa (ball and cage) or tripod (ball and housing) type CV-joints are used in front drive axles. A Cardan (cross and roller) joint, however, may sometimes be used.

The outboard CV-joint (outer universal) is normally a FIXED (nonsliding) ball and cage or Rzeppa type joint, Fig. 29-23. Sometimes, it is a fixed tripod type, Fig. 29-22. The outboard CV-joint transfers rotating power from the axle shaft to the hub assembly.

The inboard CV-joint (inner universal) is commonly a PLUNGING (sliding) ball and housing or tripod joint. It acts like a slip joint in a drive shaft for a rearwheel drive car.

The plunging action of the inner CV-joint allows for a change in distance between the transaxle and wheel hub. As the front wheels move up and down over bumps in the road, the length of the drive axle (inner joint) must change. Look at Figs. 29-22 and 29-23.

CV-joint construction (front-wheel drive)

A Rzeppa or ball and cage CV-joint consists of a star shaped inner race, several balls, ball cage, outer race or housing, and a rubber boot. Refer to Fig. 29-24A.

The inner race of this type joint is normally splined to the axle shaft. The outer race can be as part of the axle or it may be splined and held on the axle with snap rings.

The balls fit between the inner and outer races, Fig. 29-24B. When the axle turns, power is trans-

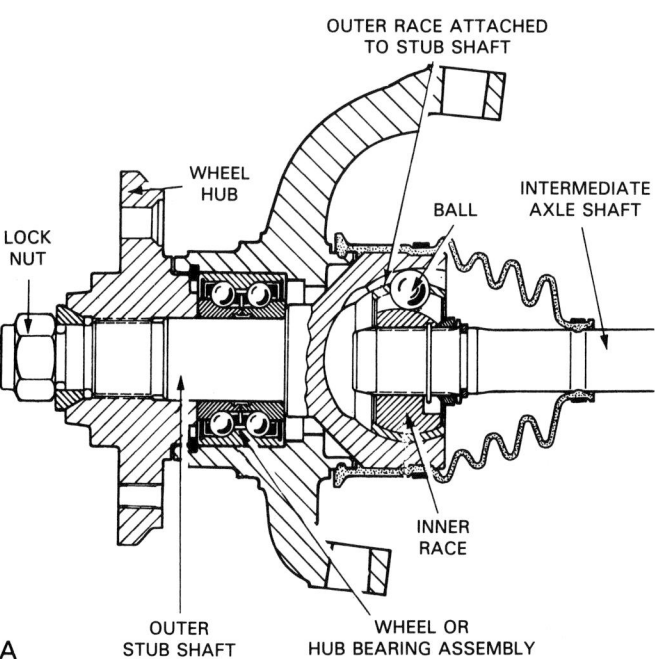

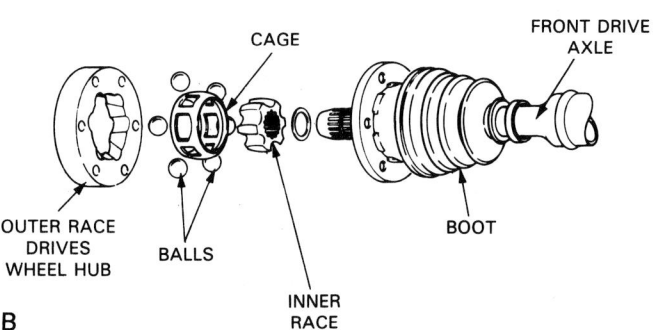

Fig. 29-24. Study construction of Rzeppa or ball and cage CV-joint. A — Drive axle turns inner race. Inner race turns balls. Balls transfer turning force to outer race and hub, rotating wheels to propel car. B — Exploded view of ball and cage joint. Study how parts fit together. When inner joint, this type of joint is plunging type. When outer joint, it is fixed type. (Saab)

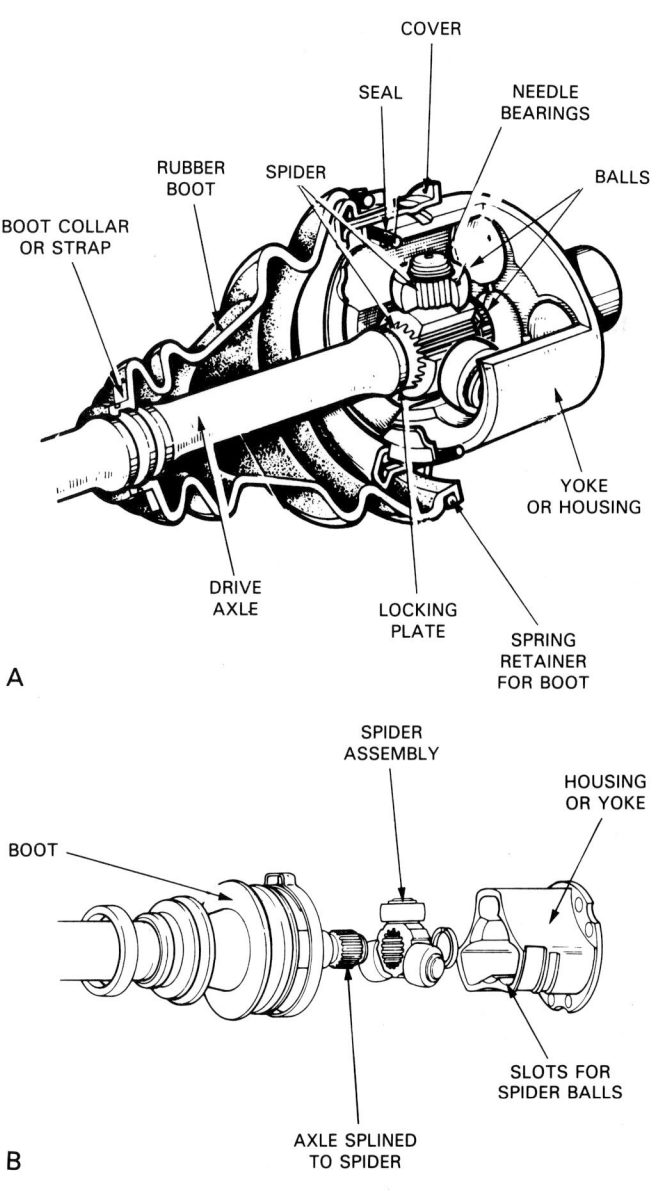

Fig. 29-25. Tripod type CV-joint construction. A — Cutaway view shows how tripod fits together. Axle is splined to spider. Spider rotates balls and housing. Balls turn on needle bearing to allow swivelling action. B — Exploded view of tripod joint. When inboard CV-joint, it serves as sliding or plunging joint. When used as outboard joint, tripod is fixed and does not slide in and out. Tripod is commonly inboard plunging joint. (Renault)

ferred through inner race, balls, outer race, and to the wheel hub.

A tripod or ball and housing CV-joint consists of a spider, usually three balls, needle bearings, outer yoke, and boot. A cutaway view of a modern tripod joint is shown in Fig. 29-25A.

The inner spider is normally splined to the axle shaft. The needle bearings and three balls fit around the spider. The yoke or housing then slides over the balls. Slots in the yoke allow the balls to slide in and out and swivel.

During operation, the axle shaft turns the spider and ball assembly. The balls transfer power to the outer housing. Since the housing is connected to the axle stub shaft or hub, power is sent through the joint to propel the car.

In Fig. 29-25B, study how the balls would allow the axle shaft to swivel and slide in the yoke.

CV-joint boots

Boots are used to keep road dirt out of the CV-joints on a front drive axle. They also prevent the loss of lubricant (grease). Shown in Figs. 29-24 and 29-25, they are accordion-shaped or pleated boots that flex with movement of the CV-joint.

Retaining collars or straps secure the boots to the drive axle. They are usually plastic straps or metal spring clamps that hold or squeeze in on the ends of the boot, providing a tight seal.

KNOW THESE TERMS

Transaxle, Transverse, Longitudinal, Manual transaxle input shaft, Transaxle output shaft, Transaxle differential, Transaxle gearbox, Drive chain, Front drive axles, Inner stub shaft, Interconnecting shaft, Outer stub shaft, CV, Outboard CV-joint, Inboard CV-joint, Rzeppa CV-joint, Tripod CV-joint, Boot.

REVIEW QUESTIONS

1. Define the term "transaxle".
2. List six possible advantages of front-wheel drive.
3. Both manual and automatic transaxles are available. True or False?
4. Summarise the differences between transaxles for transverse and longitudinally mounted engines.
5. Name and explain the seven major parts of a manual transaxle.
6. Which of these parts is NOT found in a manual transaxle?
   a. Synchronizers.
   b. Differential.
   c. Transmission.
   d. Fluid coupling.
   e. All of the above.
   f. None of the above.
7. A drive _____ or _____ is sometimes used to send crankshaft power to the transaxle with a longitudinally mounted engine.
8. List and explain the six major components of an automatic transaxle.
9. Describe the two common types of CV-joints used on front-wheel axles.
10. Which of the following is NOT part of a front-drive axle assembly?
    a. Boots.
    b. CV-joints.
    c. Stub shafts.
    d. Interconnecting shaft.
    e. Pivot shaft.

Chapter 30

SUSPENSION SYSTEM FUNDAMENTALS

After studying this chapter, you will be able to:
☐ Identify the major parts of a suspension system.
☐ Describe the basic function of each suspension system component.
☐ Explain the operation of the four common types of springs.
☐ Compare the various types of suspension systems.
☐ Explain automatic suspension levelling systems.

The suspension system works with the tyres, frame or unitised body, wheels, wheel bearings, brake system, and steering system. All of the parts in these systems work together to provide a safe and comfortable means of transportation. For this reason, make sure you learn all of the material in this chapter. You will then be prepared to study later chapters.

FUNCTIONS OF A SUSPENSION SYSTEM

A suspension system has several important functions:
1. Support the weight of the frame, body, engine, transmission, drive train, and passengers.
2. Provide a smooth, comfortable ride by allowing the wheels and tyres to move up and down with minimum movement of the car body.
3. Allow rapid cornering without extreme body roll (car leans to one side).
4. Keep the tyres in firm contact with the road, even after striking bumps or holes in the road.
5. Prevent excessive body squat (body tilts down in rear) when accelerating or when carrying a heavy load.
6. Prevent excessive body dive (body tilts down in front) when braking.
7. Allow the front wheels to turn from side-to-side for steering.
8. Work with the steering system to help keep the wheels in correct alignment.

As you will learn, a suspension system uses springs, swivel joints, damping devices, movable arms, and other components to accomplish these functions.

BASIC SUSPENSION SYSTEM

Before discussing each component in detail, you should be able to visualise each major part and how it functions in relation to the other parts. Look at Fig. 30-1 as each component is introduced.
1. CONTROL ARM (movable lever that fastens steering knuckle to the car frame or unitised body).
2. STEERING KNUCKLE (provide spindle or bearing support for mounting wheel hub, bearings, and wheel assembly).

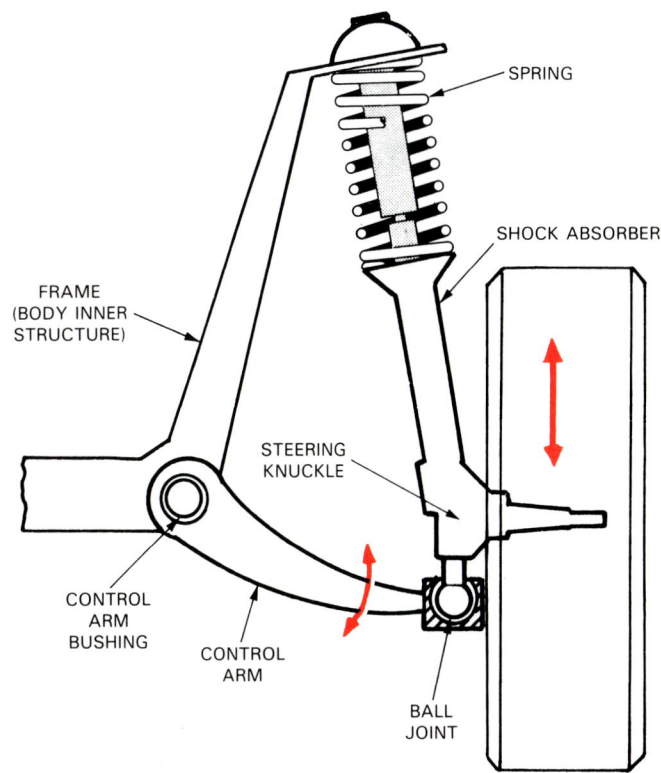

Fig. 30-1. Elementary parts of suspension system. Study basic motion of components.

3. BALL JOINT (swivel joint that allows control arm and steering knuckle to move up or down and from side to side).
4. SPRING (supports weight of car; flexes to permit control arm and wheel to move up and down).
5. SHOCK ABSORBER or DAMPER (keeps suspension from continuing to bounce up and down after spring compression and extension).
6. CONTROL ARM BUSHING (sleeve that allows control arm to swing up and down on frame).

INDEPENDENT AND NONINDEPENDENT SUSPENSION

Suspension systems may be grouped into two broad categories: independent and nonindependent.

Independent suspension

Independent suspension allows one wheel to move up and down with a minimum effect on the other wheels. Look at Fig. 30-2A.

Since each wheel is attached to its own suspension unit, movement of one wheel does NOT cause direct movement of the wheel on the other side of the car.

Detailed later, there are many types of independent suspension. It is the most popular type for modern passenger cars.

Nonindependent suspension

Nonindependent suspension has both the right and left wheels attached to the same, solid axle, Fig. 30-2B. When one tyre hits a bump in the road, its upward movement causes a slight upward TILT of the other wheel. Hence, neither wheel is independent of the other.

SUSPENSION SYSTEM SPRINGS

Suspension system springs must jounce (compress) and rebound (extend) with bumps and holes in the road surface. They support the weight of the car while still allowing suspension travel (movement).

The most common types of springs are the coil springs, leaf springs, air springs and torsion bar.

Coil spring

A coil spring is a length of spring steel rod wound into a spiral, Fig. 30-3A. This is the most common type of spring found on modern suspension systems. Coil springs may be used on either the front or rear of the car, as shown in Fig. 30-4.

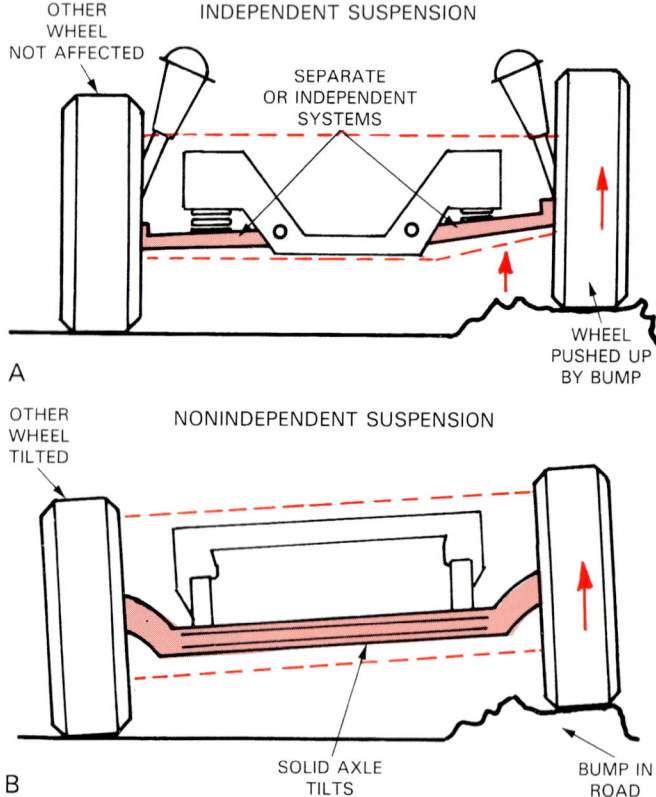

Fig. 30-2. Comparison of independent and nonindependent suspensions. A — Independent suspension allows one wheel to roll over bump with minimum effect on other wheel. B — Nonindependent suspension causes action of one wheel to tilt and affect other wheel.

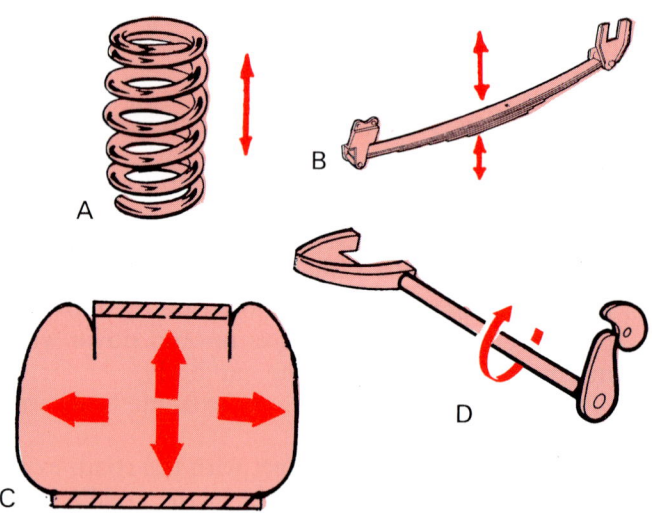

Fig. 30-3. Study different types of suspension system springs.

Leaf spring

A leaf spring is commonly made of flat plates or strips of spring steel bolted together. A few are made of fibreglass. Although leaf springs were once used on front suspension systems, they are now limited to the rear of some cars. Fig. 30-3B illustrates a simple leaf spring rear suspension.

Fig. 30-5 illustrates an exploded view of a leaf spring assembly.

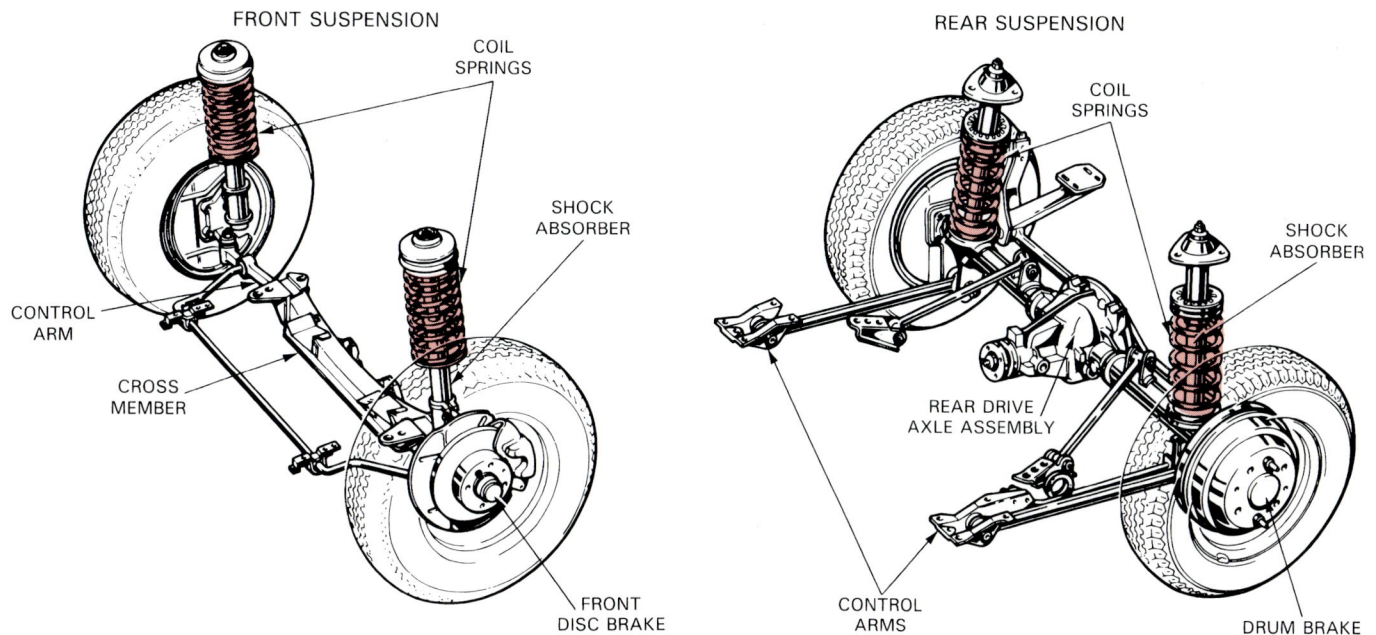

Fig. 30-4. Both front and rear of car can use coil springs. Coil springs are becoming more common with today's suspension system designs. (Fiat)

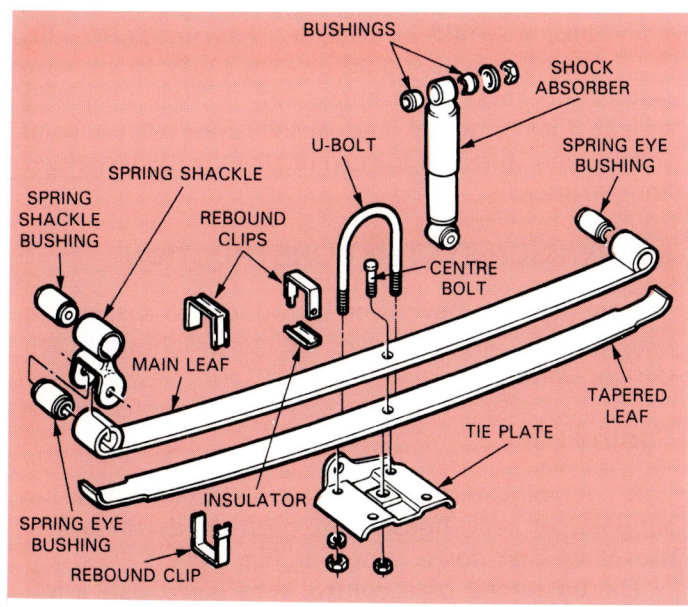

Fig. 30-5. Exploded view of simple leaf spring assembly.

Insulators are placed between the springs to prevent squeaks and rattles.

Each end of the leaf spring has an eye (cylinder-shaped hole) which holds a bushing.

A shackle fastens the rear leaf spring eye to the car frame. It allows the spring to change length when bent.

The front spring eye normally bolts directly to the frame structure. Two large U-bolts secure the axle or axle housing to the leaf springs.

Leaf spring windup is a condition causing the rear leaf springs to flex when driving or braking force is applied to the suspension system. Fig. 30-6 illustrates spring windup. The twisting and distortion of the spring can cause body squat and dive.

Air spring

An air spring is typically a two-ply rubber cylinder filled with air. End caps are formed on the air spring for mounting. Air pressure in the rubber cylinder

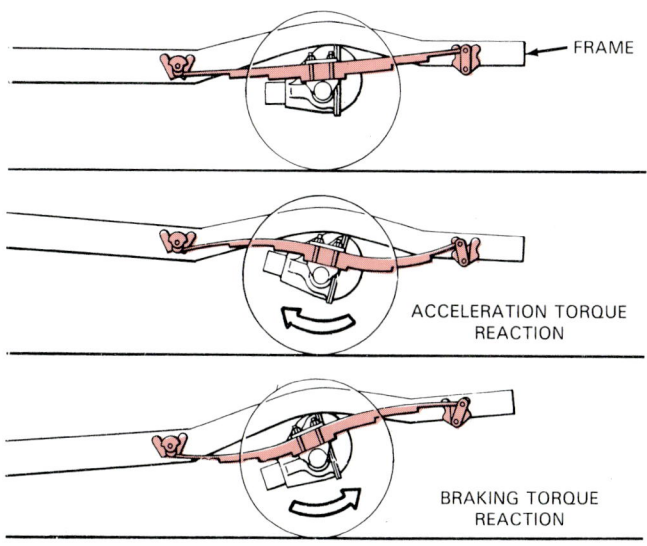

Fig. 30-6. Leaf spring windup is problem when leaves support driving axle. Torque tends to twist spring. (Ford)

makes the unit have a spring action, like a coil spring. Refer to Fig. 30-7.

An air spring is lighter than a coil spring. This gives it the potential to produce a smoother ride than a coil spring. Special synthetic rubber compounds must be used so the air spring can operate properly in cold weather. Low temperatures tend to harden or stiffen rubber.

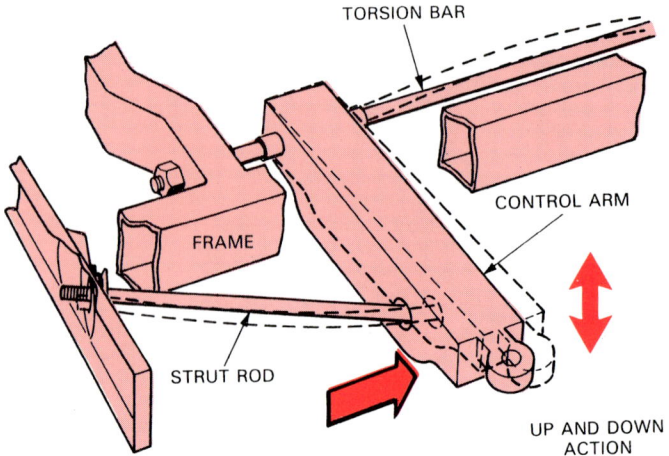

Fig. 30-8. Torsion bar is twisted with control arm movement. Bar resists twisting action and acts like conventional spring. (Typical)

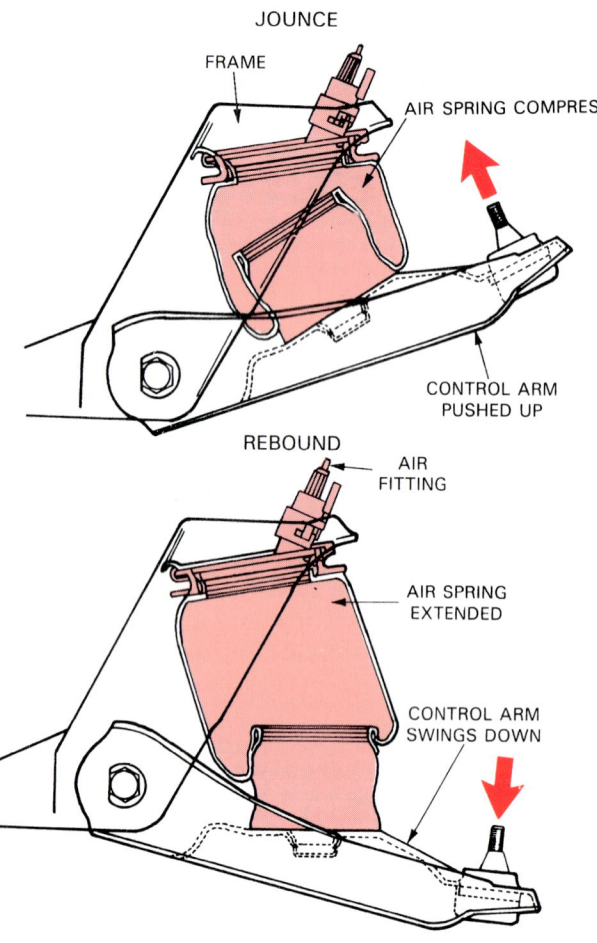

Fig. 30-7. Air spring is used on some late model cars. They are especially adaptable to automatic levelling systems.

Torsion bar (spring)

A torsion bar is another type of spring made of a large, spring steel rod. See Fig. 30-8. One end of the torsion bar is attached to the frame. The other end is fastened to the suspension system control arm.

Up and down movement of the suspension system twists the steel bar. It will then try to flex back into its original shape, moving the suspension arm back into place.

Spring terminology

Spring rate refers to the stiffness or tension of a spring. The rate of a spring is determined by the weight needed to bend it.

Sprung weight refers to the weight of the parts that are supported by the springs and suspension system. Sprung weight should be kept HIGH in proportion to unsprung weight.

The unsprung weight of a car is the weight of the parts that are NOT supported by the springs. The tyres, wheels, wheel bearings, steering knuckles, or axle housing would be considered unsprung weight.

Unsprung weight should be kept LOW to improve ride smoothness. Movement of a high unsprung weight (heavy wheel and suspension components) would tend to transfer movement into the passenger compartment.

SUSPENSION SYSTEM CONSTRUCTION

Now that you have been introduced to suspension system basics, we will cover the construction of each part in detail.

Control arms

A control arm holds the steering knuckle, bearing support, or axle housing in position as the wheel moves up and down. Look at Fig. 30-8.

The outer end of a control arm has a ball joint. The inner end has bushings. A rear suspension control arm may have bushings on both ends.

Control arm bushings act as bearings, allowing the arm to swing up and down on a shaft bolted to the frame or suspension unit. Refer to Fig. 30-9. These bushings may either be pressed or screwed into the holes in the control arm.

Strut rod

A strut rod fastens to the outer end of the lower control arm and to the frame. See Fig. 30-8. It keeps the control arm from swinging toward the rear or front of the car.

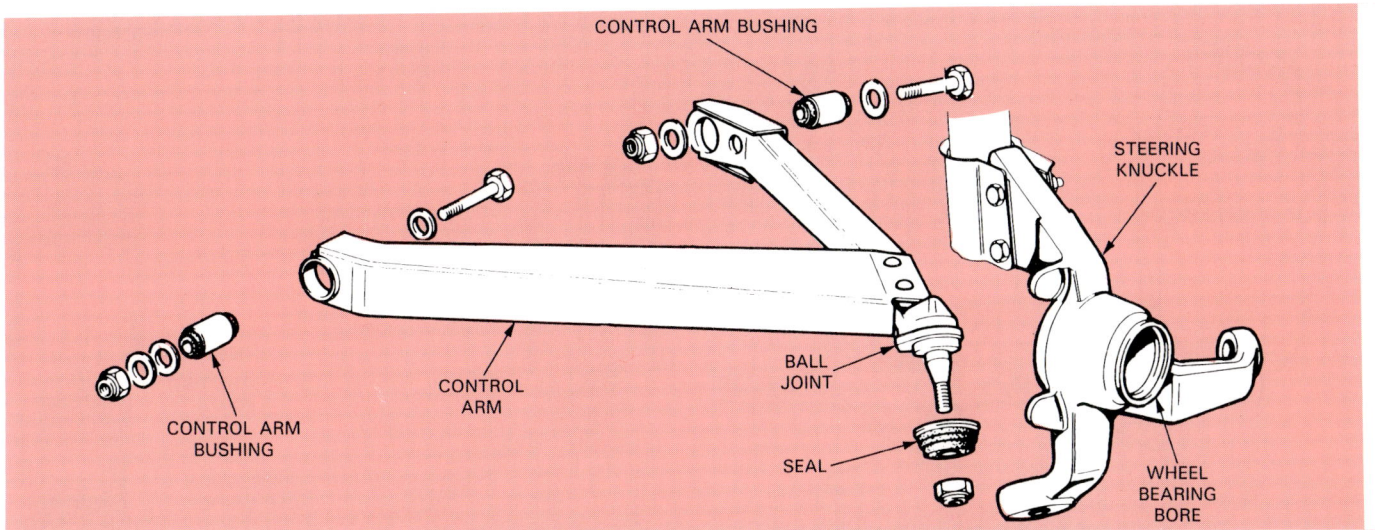

Fig. 30-9. Study basic parts of control arm. Bushings fit into inner ends of arm. Ball joint fits into outer end of control arm. Ball joint connects to steering knuckle. (Fiat)

The front of the strut rod has rubber bushings that soften the action of the strut rod. They allow a controlled amount of lower control arm movement while allowing full suspension travel.

Ball joints

Ball joints are swivel connections mounted in the outer ends of the control arms, Fig. 30-9. They may be pressed, bolted, riveted, or screwed into the control arm.

Shown in Fig. 30-10, a ball joint is a ball stud mounted inside a socket.

Since the ball joint must be filled with grease, a grease fitting and grease seal are normally placed on the joint. Fig. 30-11. The end of the stud on the ball joint is threaded for a large nut. When the nut is tightened, it force fits the tapered stud in the steering knuckle or bearing support.

SHOCK ABSORBERS

Shock absorbers limit spring oscillations (compression-extension movements) to smooth the car's ride. Without shock absorbers, the car would continue to bounce up and down after striking a dip or hump in the road. This would make the ride uncomfortable and unsafe.

Fig. 30-12 shows the basic parts of a shock absorber. They include a piston rod, rod seal, piston, reservoir, compression cylinder, extension cylinder, and flow control valves. Most shocks are filled with oil. Some are air or gas and oil filled.

When the shock is compressed or extended, the oil causes resistance to movement. The rod tends to drag slowly in or out. This dampens spring and suspension system action.

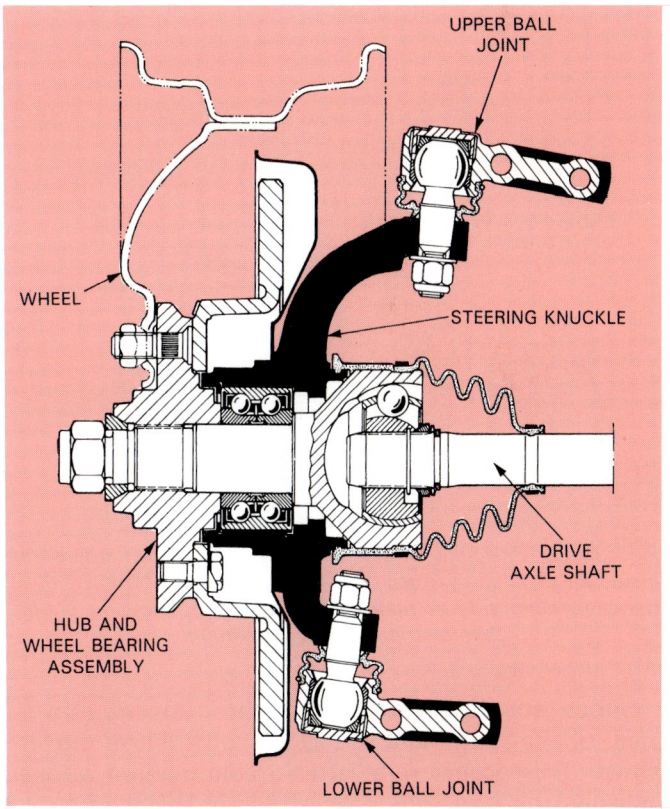

Fig. 30-10. Cutaway view shows ball joints, steering knuckle, and driving hub for front-wheel drive vehicles. Study construction of parts. (Chrysler)

One end of the shock absorber connects to a suspension component, usually a control arm. The other end of the shock fastens to the frame. In this way, the shock rod is pulled in and out to restrain movement.

Fig. 30-13 shows shock absorber operation.

Shock absorber compression occurs when the car's tyre is forced upward upon hitting a bump.

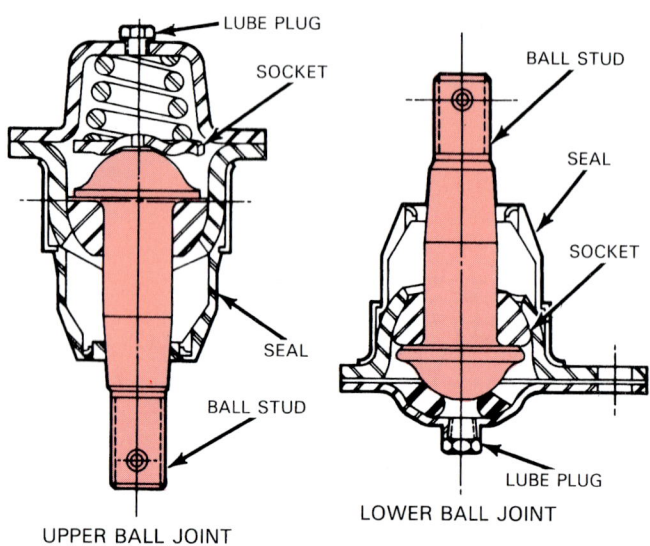

Fig. 30-11. Ball joint is simply a swivel socket. Ball stud is free to swing or turn in housing. This allows control arm and steering knuckle to move up and down freely.

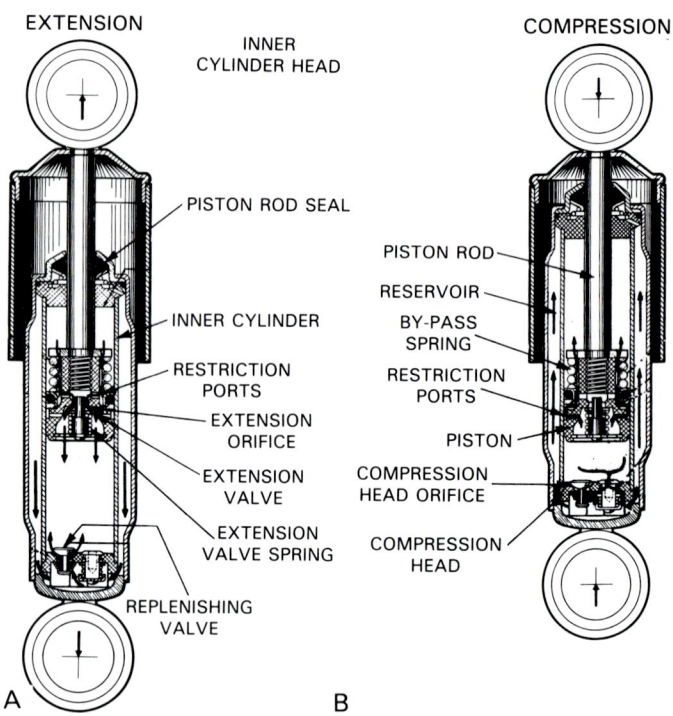

Fig. 30-13. Cutaway view of shock absorber in action. A — Extension stroke causes oil to be pulled back into lower area. Note valve action. B — Compression stroke forces piston down in cylinder. Oil is forced into upper area of shock.

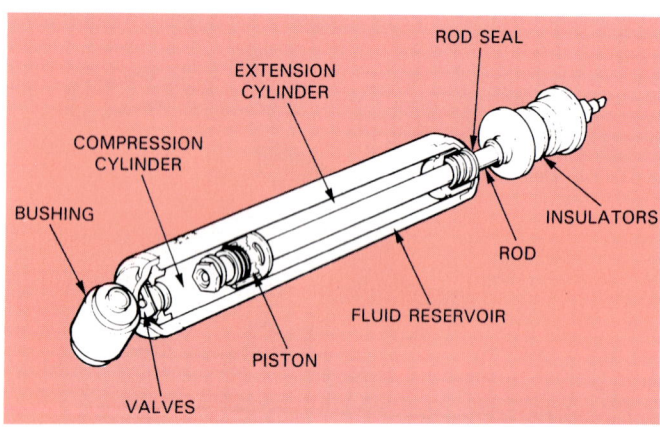

Fig. 30-12. Basically, shock absorber is a piston operating inside an oil-filled cylinder. Small holes cause oil to resist flow between each side of piston. This produces damping action that restricts spring oscillations.

Shock absorber extension is the outward movement of the piston and rod as the control arm moves down. This occurs right after a compression stroke or when the tyre rolls over a hole in the road.

Gas-charged shock absorbers

Gas-charged shock absorbers use a low-pressure gas to help keep the oil in the shock from foaming. See Fig. 30-14. Usually, hydrogen gas is enclosed in a chamber separate from the main oil cylinder. The shock piston operates in the oil. The gas maintains constant pressure on the oil to stop air bubbles from forming. This increases shock performance during rapid jounce and rebound.

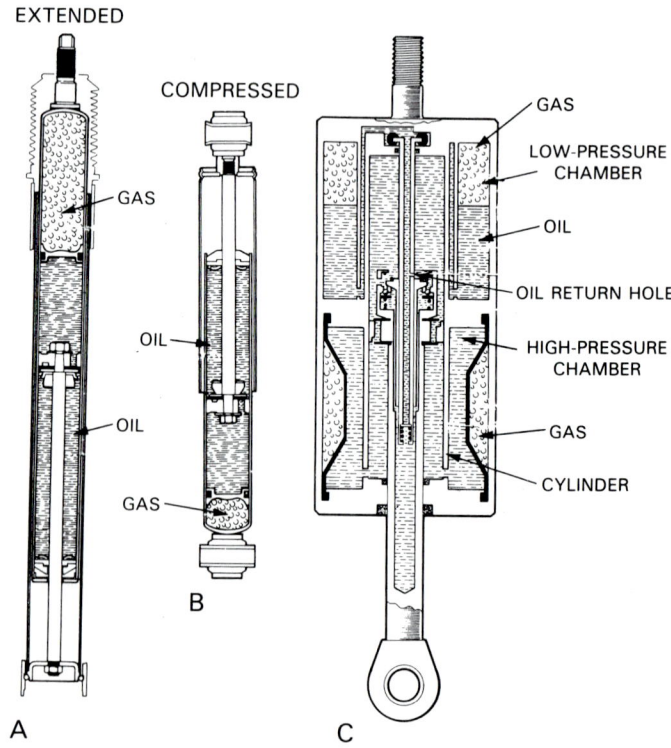

Fig. 30-14. Gas-pressure damped shocks operate like conventional oil filled shocks. Gas is used to keep oil pressurised, which reduces oil foaming and increases efficiency under severe conditions. A — Gas-charged shock for strut. B — Conventional style gas-charged shocks. C — Gas-charged, self-levelling shock. (Volvo)

Self-levelling shock absorbers

A self-levelling shock absorber uses a special design that causes a hydraulic lock action to help maintain normal vehicle curb height. One is shown in Fig. 30-14C. Study its construction and operation.

Adjustable shocks

Adjustable shock absorbers provide a means of changing shock stiffness. Usually, by turning the shock outer body or an adjustment knob, you can set the shock soft for a smooth ride or stiff for better handling.

Strut assembly

A strut assembly consists of a shock absorber, coil spring (most types), and upper damper unit. The strut replaces the upper control arm. Only the lower control arm and strut are needed to support the front wheel assembly. Look at Fig. 30-15.

The basic parts of a typical strut assembly are shown in Fig. 30-16. They include:

1. STRUT SHOCK ABSORBER (piston operating in oil-filled cylinder to prevent coil spring oscillations).
2. DUST SHIELD (metal shroud or rubber boot that keeps road dirt off shock absorber rod).
3. LOWER SPRING SEAT (lower mount formed around shock body for coil spring).
4. COIL SPRING (supports weight of car and allows suspension action).
5. UPPER SPRING SEAT (holds upper end of coil spring, contacts strut bearing).
6. STRUT BEARING (ball bearing that allows shock and spring assembly to rotate for steering action; only used on front of car).
7. RUBBER BUMPERS (jounce and rebound bumpers that prevent metal on metal contact during extreme suspension compression and extension).
8. RUBBER ISOLATORS (prevent noise from transmitting into body structure of car).

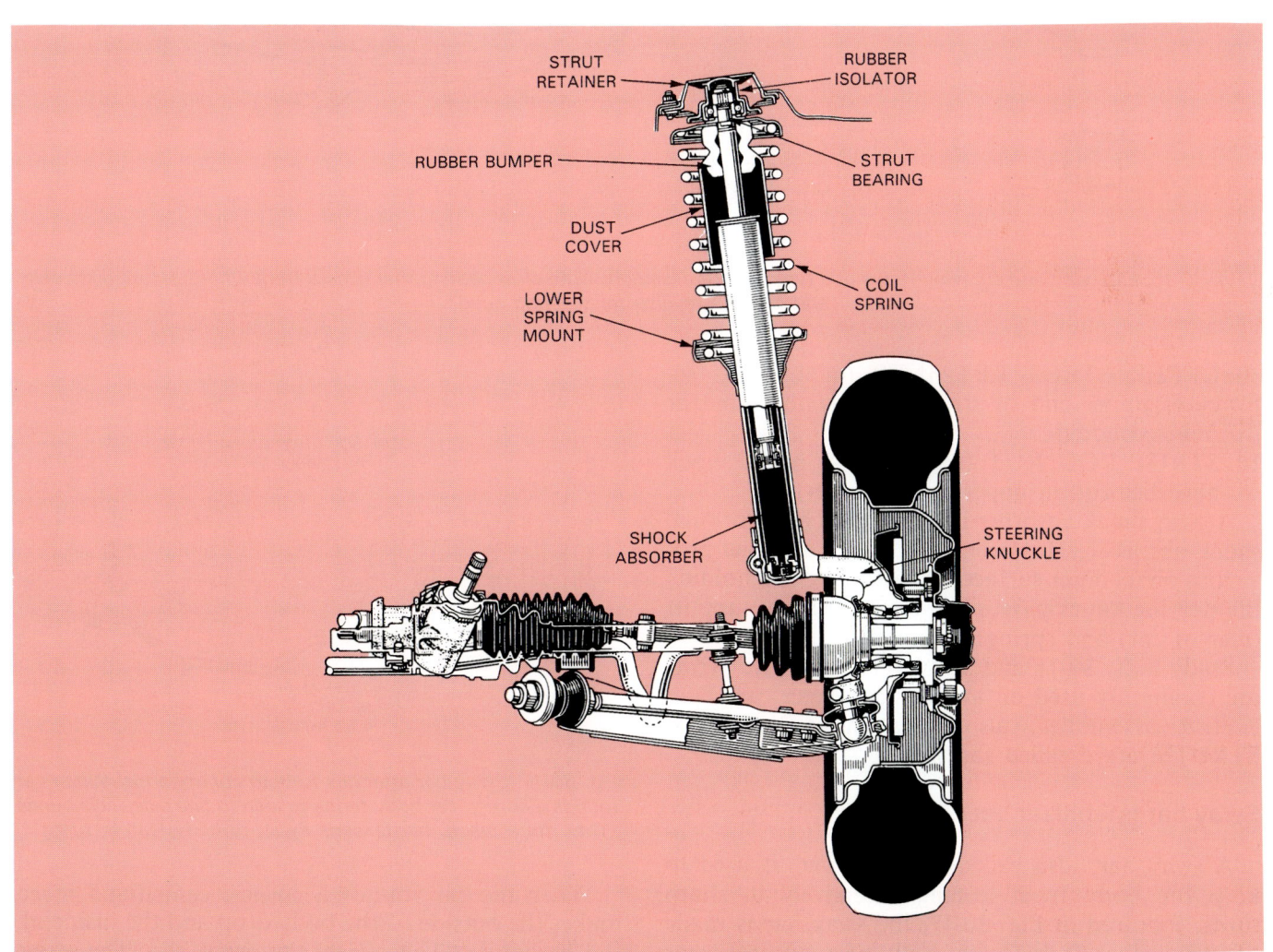

Fig. 30-15. Study parts of strut assembly very closely. This is one of the most modern suspension systems and is commonly used on today's cars. (Toyota)

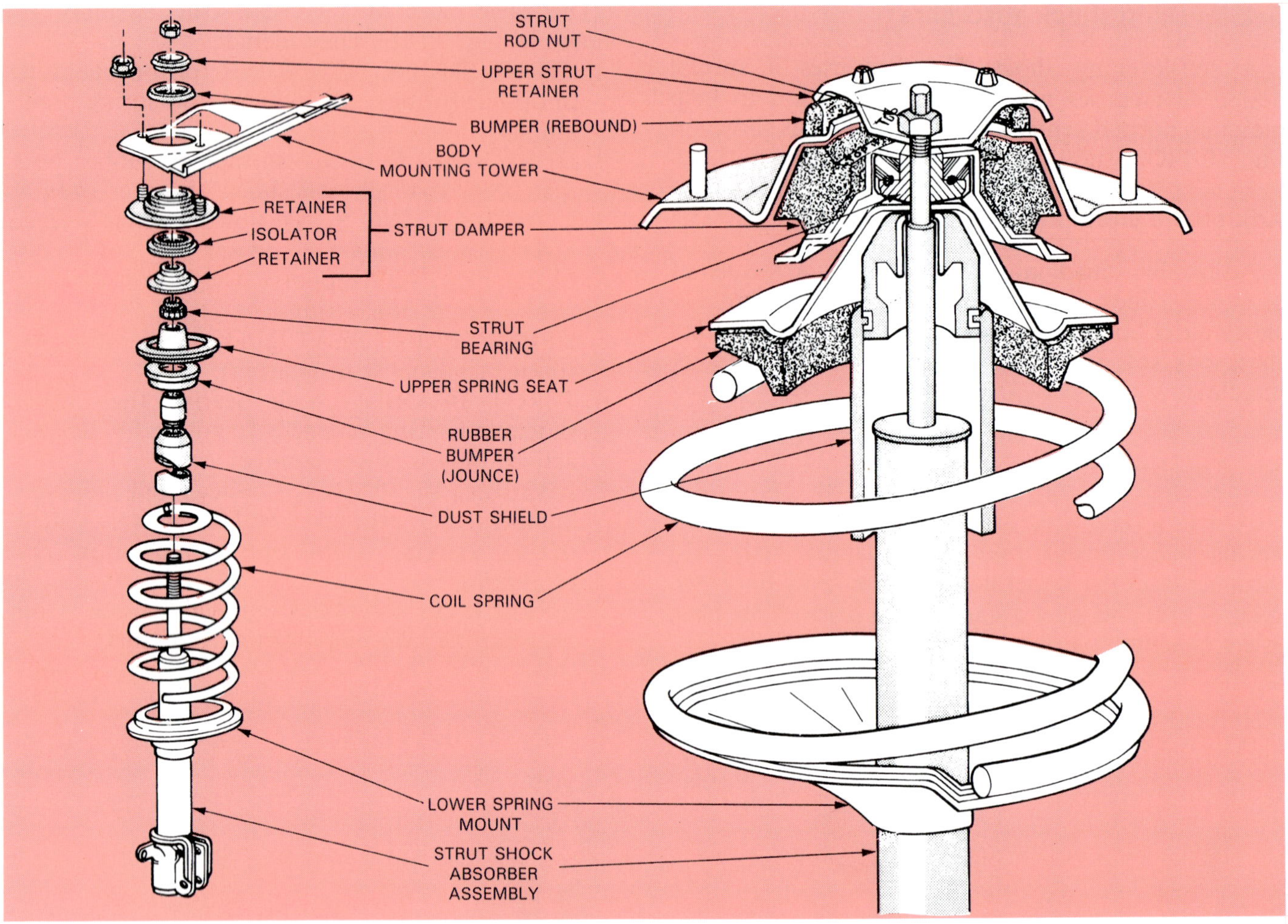

STRUT
ROD NUT

UPPER STRUT
RETAINER

BUMPER (REBOUND)

BODY
MOUNTING TOWER

RETAINER
ISOLATOR   STRUT DAMPER
RETAINER

STRUT
BEARING

UPPER SPRING SEAT

RUBBER
BUMPER
(JOUNCE)

DUST SHIELD

COIL SPRING

LOWER SPRING
MOUNT

STRUT SHOCK
ABSORBER
ASSEMBLY

Fig. 30-16. Exploded and cutaway views of a typical strut assembly. Note strut bearing that allows front wheel, steering knuckle, and strut to revolve for steering action.

9. UPPER STRUT RETAINER (mount that secures upper end of strut assembly to frame or unitised body).
10. STRUT ROD NUT (hex nut that holds shock absorber rod in upper strut retainer).

A strut shock absorber is similar to a conventional shock absorber. However, it is longer and has provisions on its outer surface for mounting and holding the steering knuckle (front of car) or bearing support (rear of car) and spring.

Study Figs. 30-15 and 30-16 very carefully. Struts are commonly used on modern passenger cars.

Strut assemblies, also called MACPHERSON STRUTS, are detailed shortly.

Sway bar (stabilizer bar)

A sway bar, also called stabilizer bar, is used to keep the body from leaning excessively in sharp turns. Pictured in Fig. 30-17, the sway bar is made of spring steel. It fastens to both lower control arms and to the frame. Rubber bushings fit between the bar, control arms, and frame.

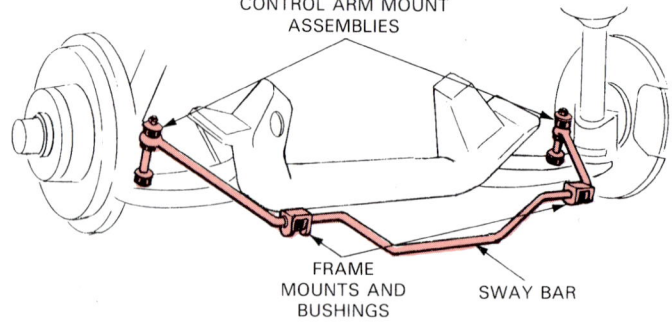

Fig. 30-17. Sway bar attaches to both control arms. When car rounds a corner, car body tends to lean to one side. This bends bar. As a result, bar lessens sway or body lean in turns.

When the car rounds a corner, centrifugal force makes the outside of the body drop and the inside of the body rise. This twists the sway bar. The sway bar's resistance to this twist limits body lean in corners.

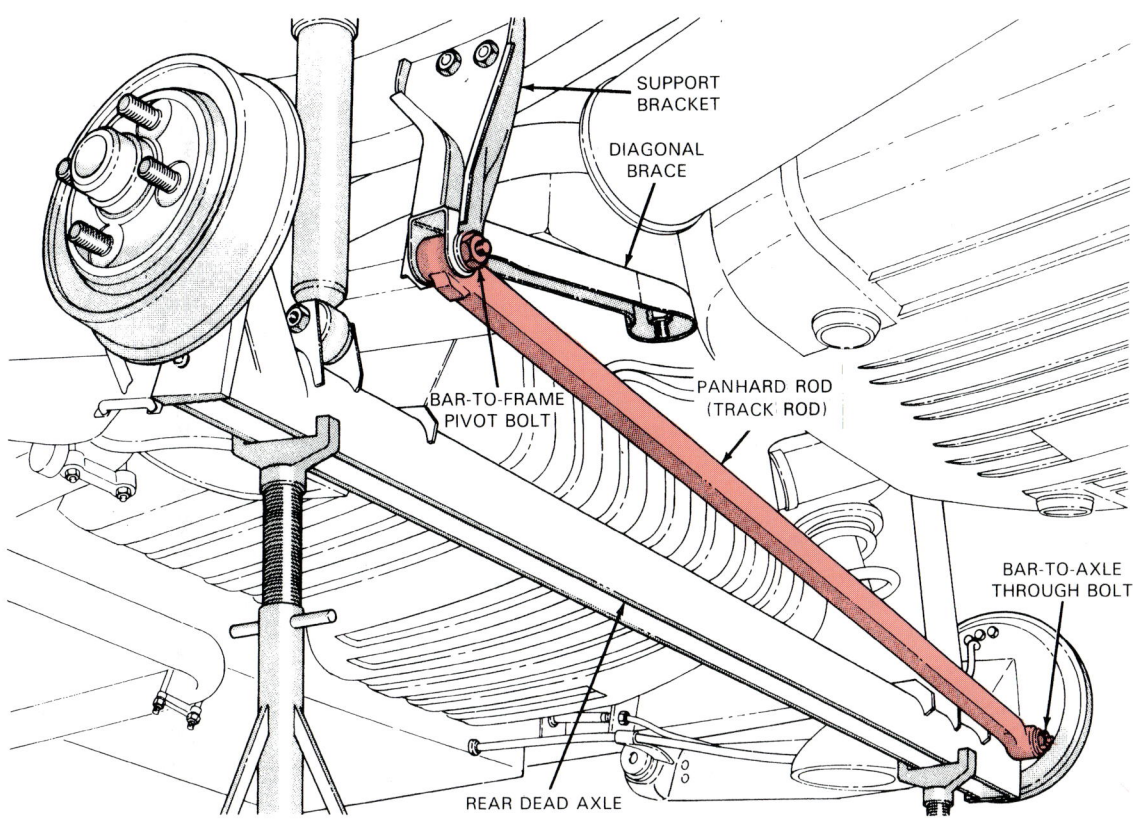

Fig. 30-18. Panhard rod, or track rod, is commonly used on rear axle to prevent side-to-side movement. Note how rod connects to frame and axle.

Panhard rod (track rod)

A panhard rod, also termed track rod or track bar, is sometimes used on rear suspension systems to prevent axle side-to-side movement when cornering, Fig. 30-18. The panhard rod runs almost parallel with the rear axle. It fastens to the axle and to the frame or body structure.

Jounce bumpers

Jounce bumpers are blocks of hard rubber that keep the suspension system parts from hitting the frame when the car hits large bumps or holes, Fig. 30-19.

LONG-SHORT ARM SUSPENSION

A long-short arm suspension uses control arms of different lengths to keep the tyres from tilting with suspension action. The upper control arms are made shorter than the lower arms, as in Fig. 30-19.

If the control arms were the same length, the front tyres would pivot outward at the top when the car hits a bump. This would cause undue tyre scuffing and wear. See Fig. 30-19.

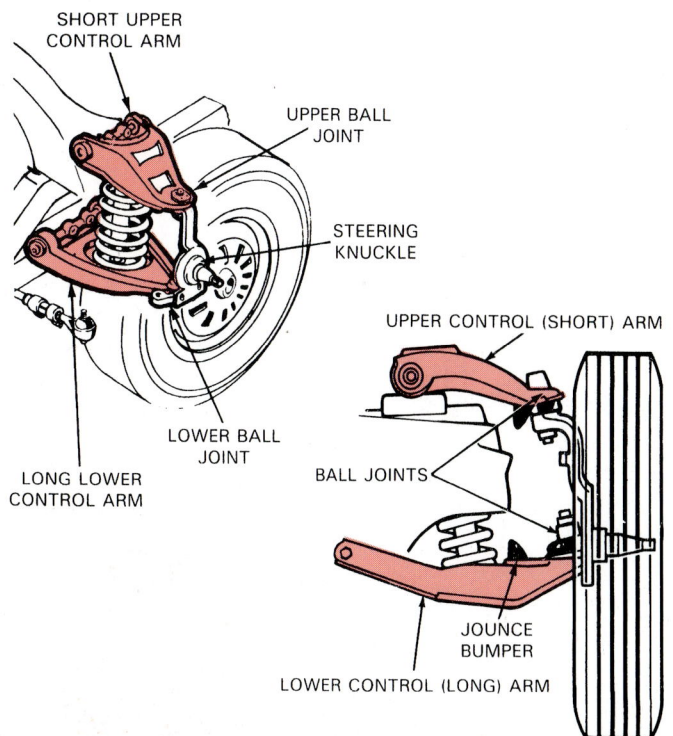

Fig. 30-19. Long-short arm suspension has different control arm lengths. This helps keep steering knuckle in alignment with suspension travel.

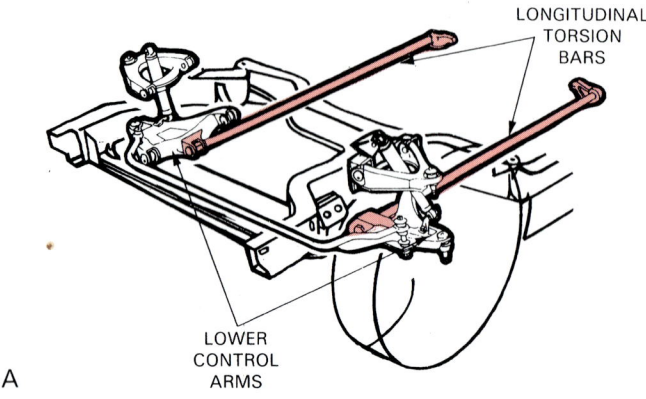

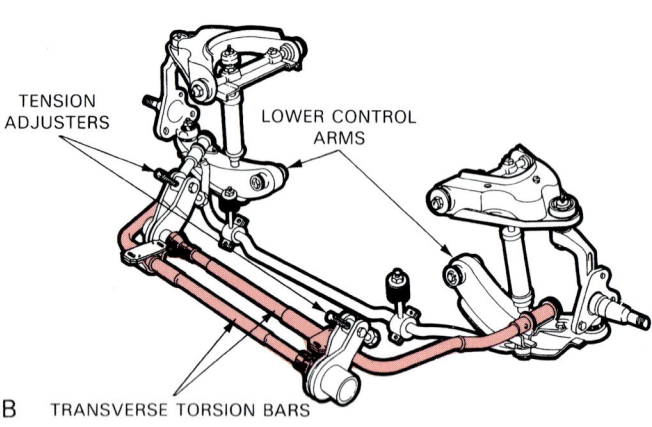

Fig. 30-20. A — Torsion bar suspension using bars mounted parallel with frame. B — Torsion bars on this modern suspension mount crosswise in car. Both can be adjusted to change vehicle height.

TORSION BAR SUSPENSION

A torsion bar suspension is a long-short arm suspension with torsion bar springs replacing the coil springs. Fig. 30-20 illustrates two types.

Most torsion bar suspensions allow easy adjustment of curb height (distance from road up to specific point on car). By turning an adjustment bolt, you can increase or decrease the tension on the torsion bar. This will either raise or lower the body and frame of the car.

MACPHERSON STRUT SUSPENSION

A MacPherson strut suspension uses only ONE control arm and a strut (spring, damper, and shock absorber unit) to support each wheel assembly, Fig. 30-21.

A conventional lower control arm attaches to the frame and to the lower ball joint. The ball joint holds the control arm to the steering knuckle or bearing support. The top of the steering knuckle or bearing support is bolted to the strut. The top of the strut is fastened to the reinforced body structure.

A MacPherson strut is the most common type of suspension found on late model cars, Fig. 30-22. It may be used on both the front or rear wheels. It reduces the number of moving parts in the suspension system to decrease unsprung weight, and ride smoothness.

A modified strut suspension has the coil spring mounted on top of the lower control arm, not around the strut. This type is illustrated in Fig. 30-23.

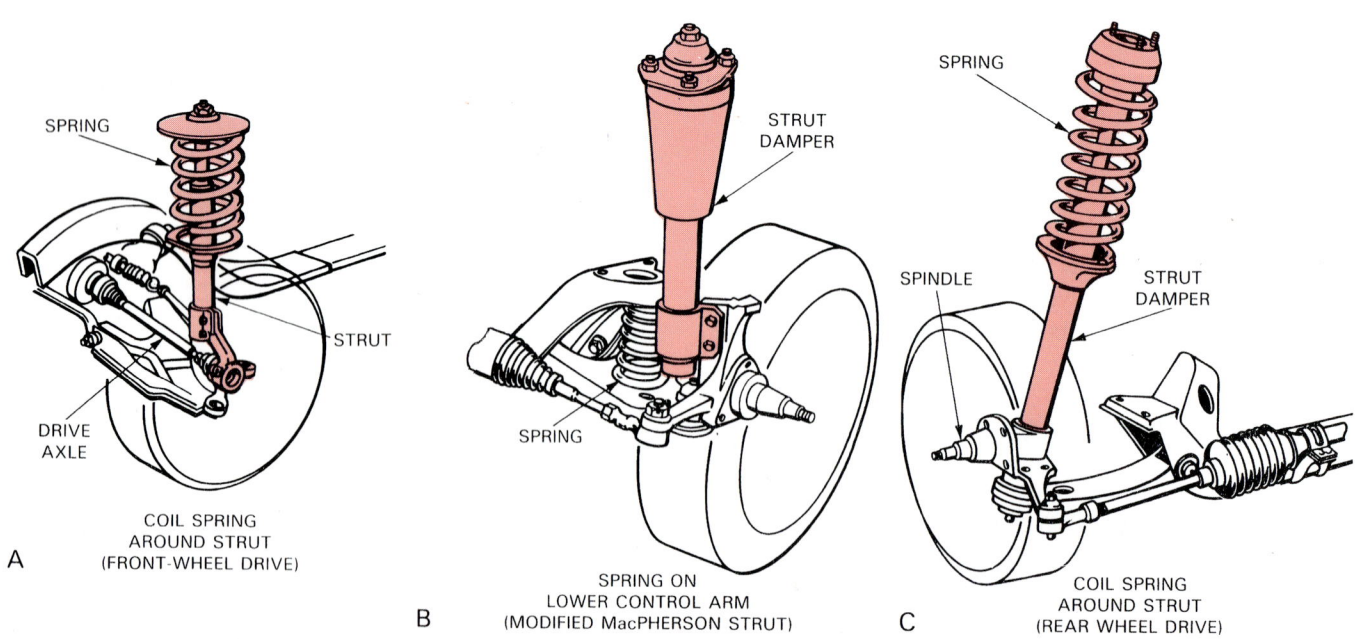

Fig. 30-21. MacPherson strut suspensions. A — Coil spring around strut, front-wheel drive. B — Modified strut has coil spring mounted on control arms. C — Same as A, but without front-wheel drive.

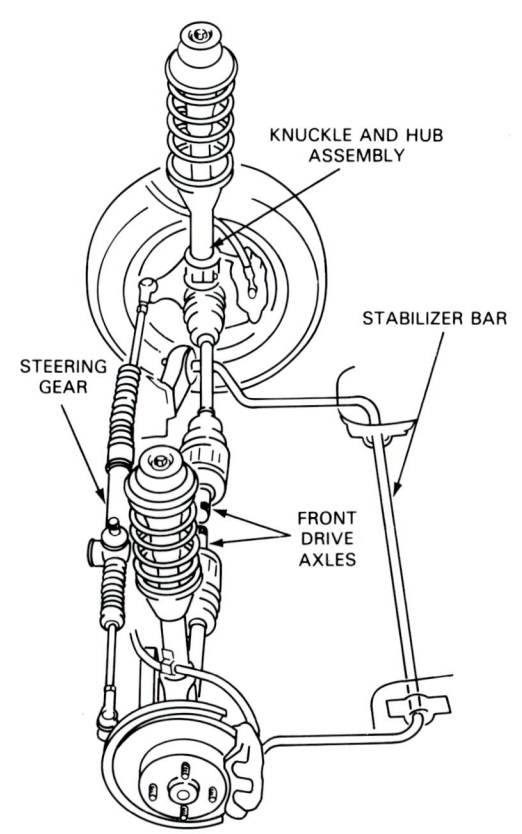

Fig. 30-22. Another view of MacPherson strut suspension. (Honda)

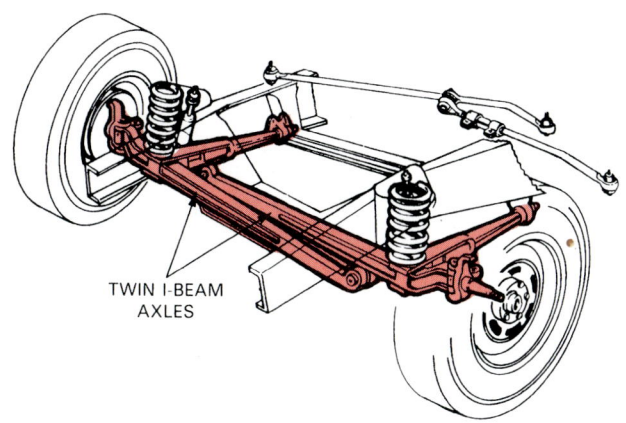

Fig. 30-24. Twin I-beam suspension is used on some light commercials. (Ford)

LIGHT COMMERCIAL VEHICLE SUSPENSION SYSTEMS

Light commercial vehicles use numerous suspension system designs: long-short control arm, MacPherson strut, solid axle, twin axle or twin I-beam suspension. The control arm and strut type are basically the same as those used on passenger cars.

Fig. 30-24 shows a twin-axle or TWIN-I-BEAM suspension.

A four-wheel drive vehicle can have a solid axle housing and differential in the front. The steering

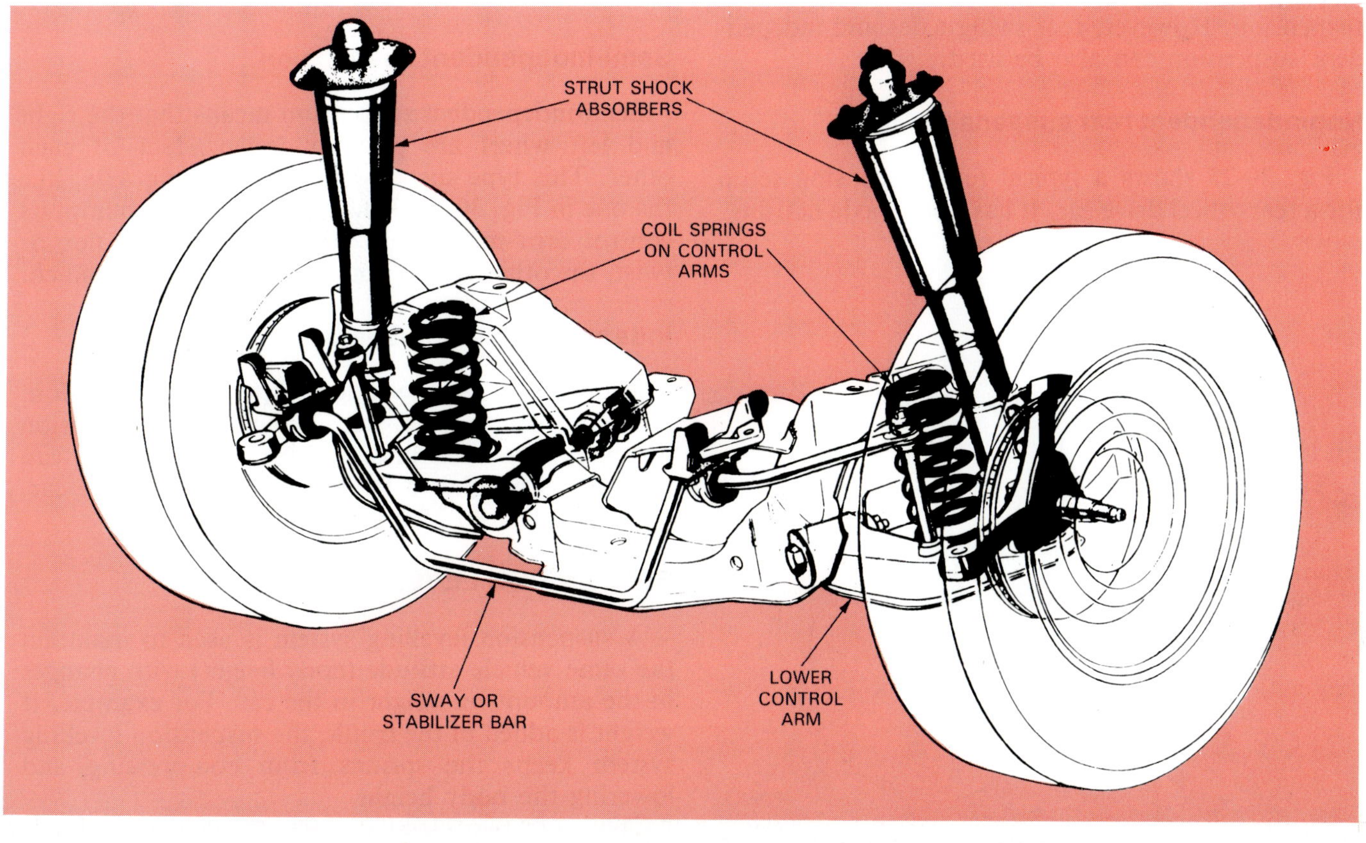

Fig. 30-23. Modified strut for front of rear-wheel drive car. (Typical)

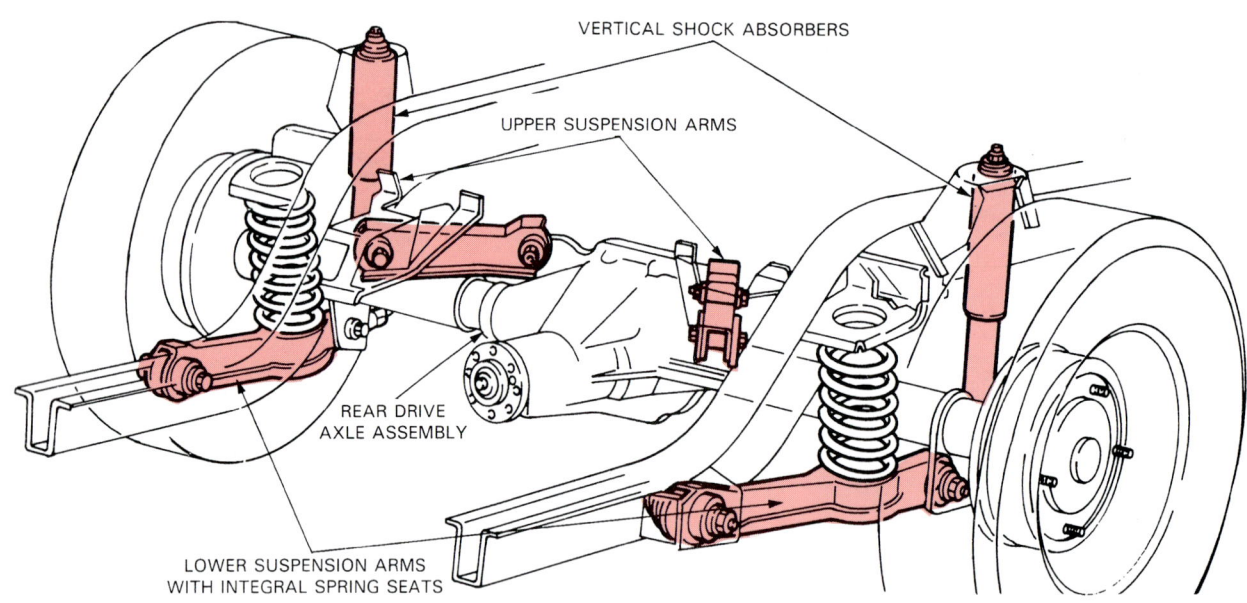

Fig. 30-25. Solid axle, rear suspension for rear-wheel drive car. Study parts. (Typical)

knuckles are mounted on the axle housing so that they will turn from left to right for cornering.

REAR SUSPENSION SYSTEMS

Rear suspension systems are similar to front suspension systems but they do NOT have to provide for steering. With a rear-wheel drive car, the rear axle housing may be solid, resulting in nonindependent suspension. However, rear swing axles and independent suspension can also be used.

Nonindependent rear suspension

Fig. 30-25 shows a typical rear suspension setup for a rear-wheel drive car. It has a solid axle housing.

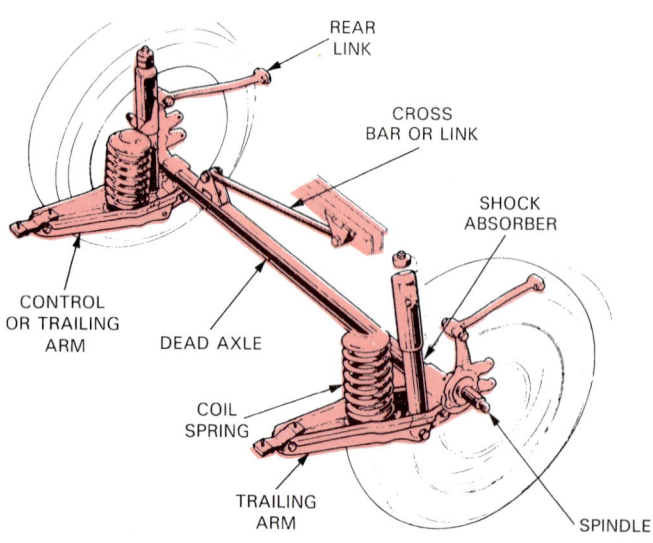

Fig. 30-26. Dead axle generally refers to solid axle that does not have drive wheels. (Saab)

Note how the coil springs are mounted between the control arms and frame.

Dead axle

A dead axle is a term used to describe a solid rear axle on a front-wheel drive vehicle. Fig. 30-26. Since the front wheels transfer driving power to the road, the rear axle is simply a straight or solid type axle.

Semi-independent suspension

Semi-independent suspension means that the right and left wheel are partially independent of each other. This type suspension uses a flexing axle, like the one in Fig. 30-26. When one tyre hits a bump, its control arm moves up. Since the axle can flex or twist, the other tyre is not affected or tilted as much.

Independent rear suspension

Many new cars use independent rear suspension. As with front suspension, independent suspension increases ride smoothness. This type suspension can be used with either a front or rear-wheel drive car. Refer to Figs. 30-27 and 30-28.

SUSPENSION LEVELLING SYSTEMS

A suspension levelling system is used to maintain the same vehicle attitude (body height) with changes in the amount of weight in the car. For example, if weight is added in the trunk, the suspension levelling system keeps the springs from compressing and lowering the body height.

There are two classifications of suspension levelling system: manual and automatic.

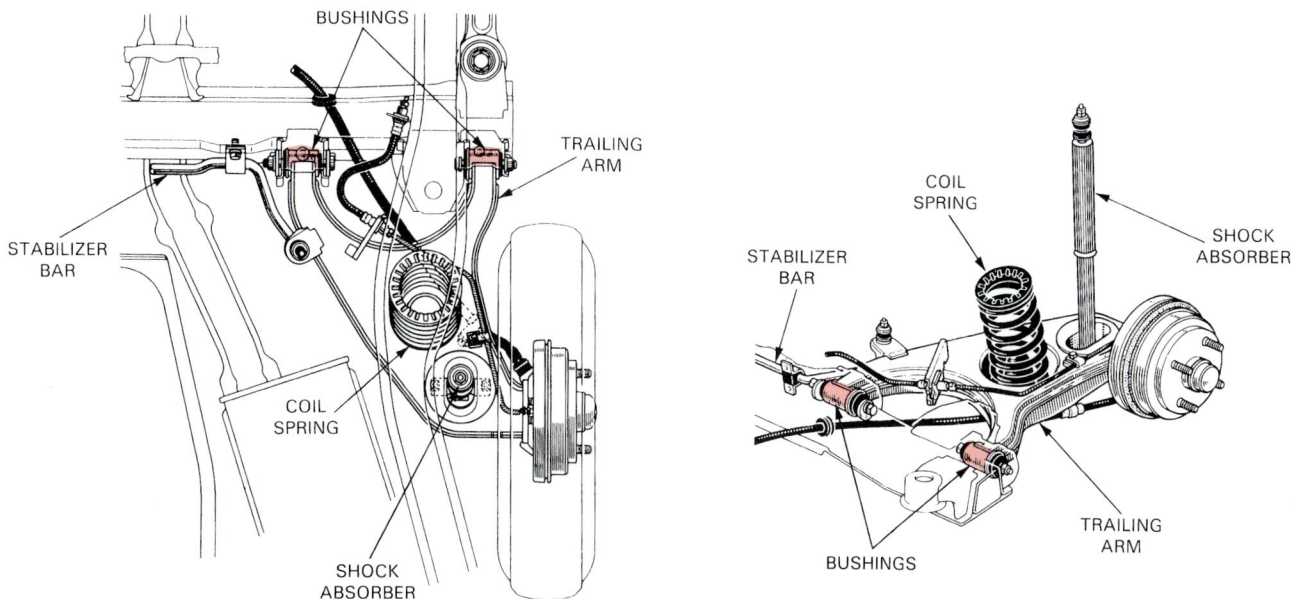

Fig. 30-27. Top and side views of trailing arm, independent rear suspension. Note location of bushings, spring, and shock absorber. (Toyota)

Manual suspension levelling system

A manual suspension levelling system uses air shocks and an electric compressor to counteract changes in passenger and luggage weight. A manual switch can be used to activate the compressor to alter air shock pressure and body height.

Automatic suspension levelling system

Automatic suspension levelling systems use air shocks or air springs, height sensors, and a compressor to maintain curb height. System designs vary.

Fig. 30-29 shows an automatic suspension levelling

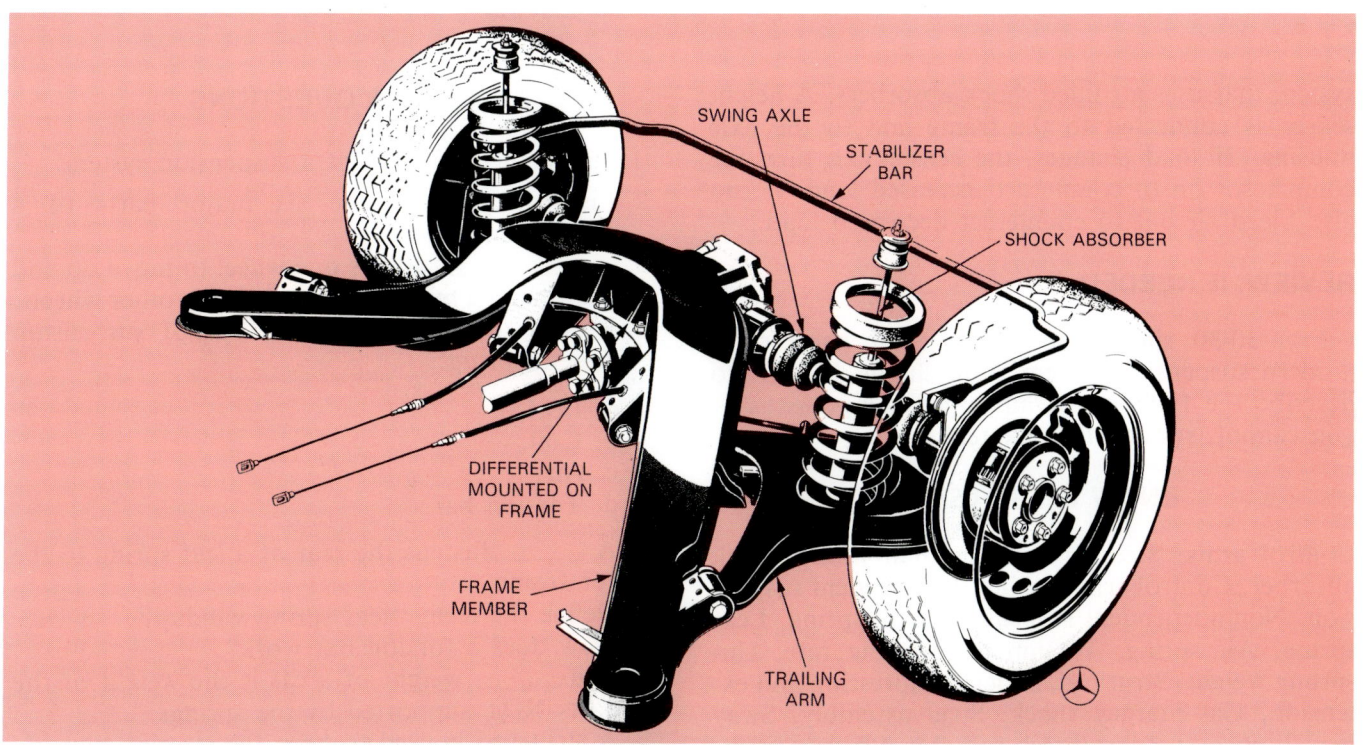

Fig. 30-28. This rear drive axle uses a differential mounted solid on frame. Swing axles extend out to drive wheels. Note trailing arms and other components. (Mercedes-Benz)

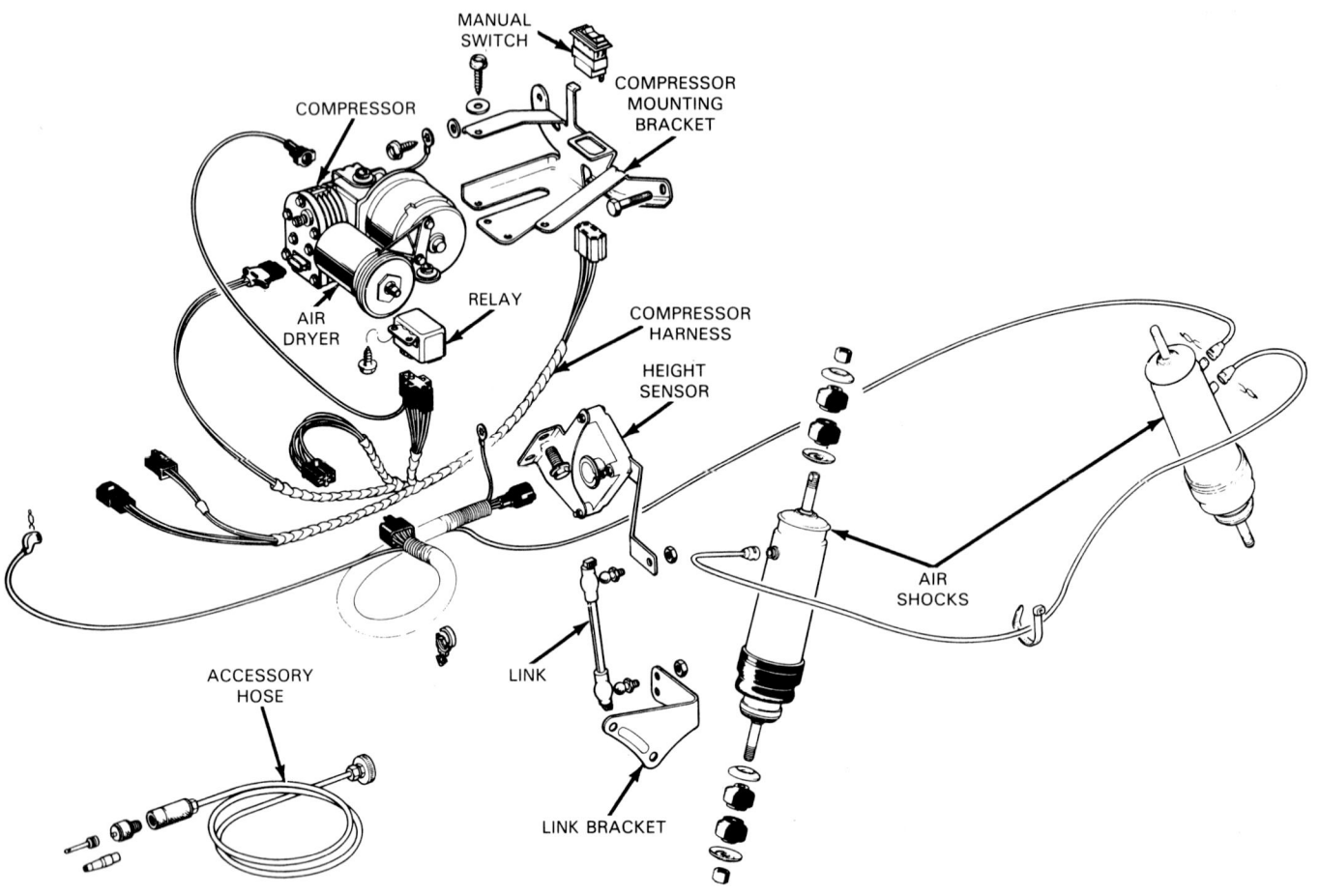

Fig. 30-29. Illustration showing major parts of a typical suspension levelling system. Basically, height sensor operates air pump. Air pump operates air shocks to maintain correct vehicle height.

system that uses air filled shock absorbers. A height sensor is connected to the frame and to the axle housing. If load changes, the sensor can turn the compressor on to counteract increased load. It can also bleed air out to counteract decreased load.

REVIEW, IF NEEDED

Fig. 30-30 shows the fundamental parts of a modern suspension system. You should be able to describe the function of all major components. If you cannot, restudy the chapter.

KNOW THESE TERMS

Control arms, Steering knuckle, Ball joint, Shock absorber, Control arm bushing, Independent suspension, Nonindependent suspension, Coil spring, Leaf spring, Air spring, Torsion bar, Spring rate, Unsprung weight, Strut rod, Shock compression and extension, Gas-charged shock, Strut assembly, Sway bar, Jounce bumper, Track rod, MacPherson strut, Twin I-beam, Dead axle, Suspension levelling system.

REVIEW QUESTIONS

1. List eight functions of a suspension system.
2. List and explain the six major parts of a suspension system.
3. _____ _____ allows one wheel to move up and down with a minimum effect on the other wheels.
4. This is the most common type of suspension system spring:

   a. Leaf.
   b. Coil.
   c. Air.
   d. Torsion bar.

5. A _____ fastens the rear of a leaf spring to the car frame.
6. Define the term "leaf spring windup".
7. How does a torsion bar work?
8. The _____ weight of a car is the weight of the parts NOT supported by the springs.
9. A strut rod is used to keep the steering knuckle from swivelling. True or False?
10. Why are ball joints needed?

470

11. Summarise the basic operation of a conventional shock absorber.
12. What is the advantage or purpose of gas-charged shocks?
13. List and explain the ten major parts of a strut assembly.
14. This part is used to keep the car body from rolling or leaning excessively in turns or corners.
    a. Strut rod.
    b. Jounce bumper.
    c. Track rod.
    d. Sway bar.
15. In your own words, describe a MacPherson strut suspension.

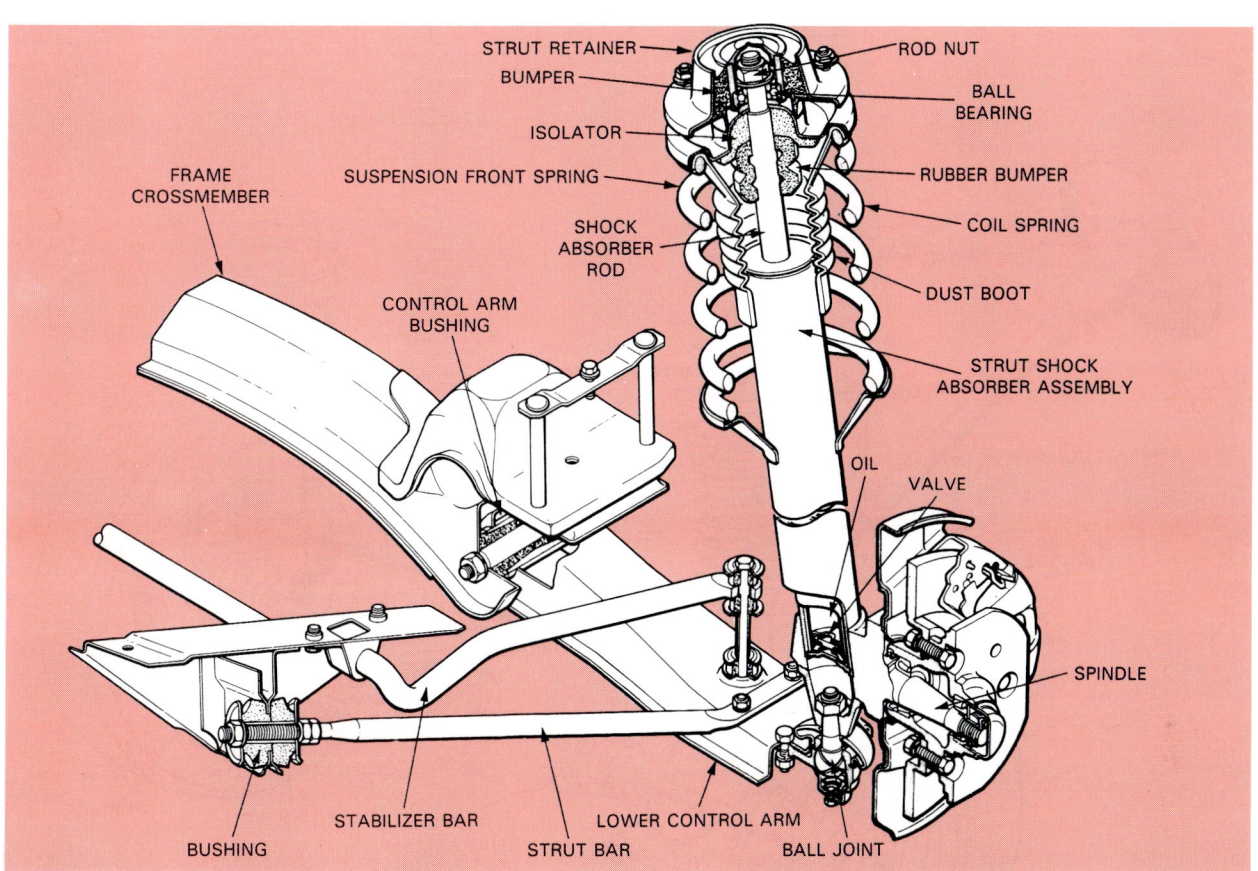

Fig. 30-30. Can you identify all of the components in this illustration? If not, review chapter as needed.

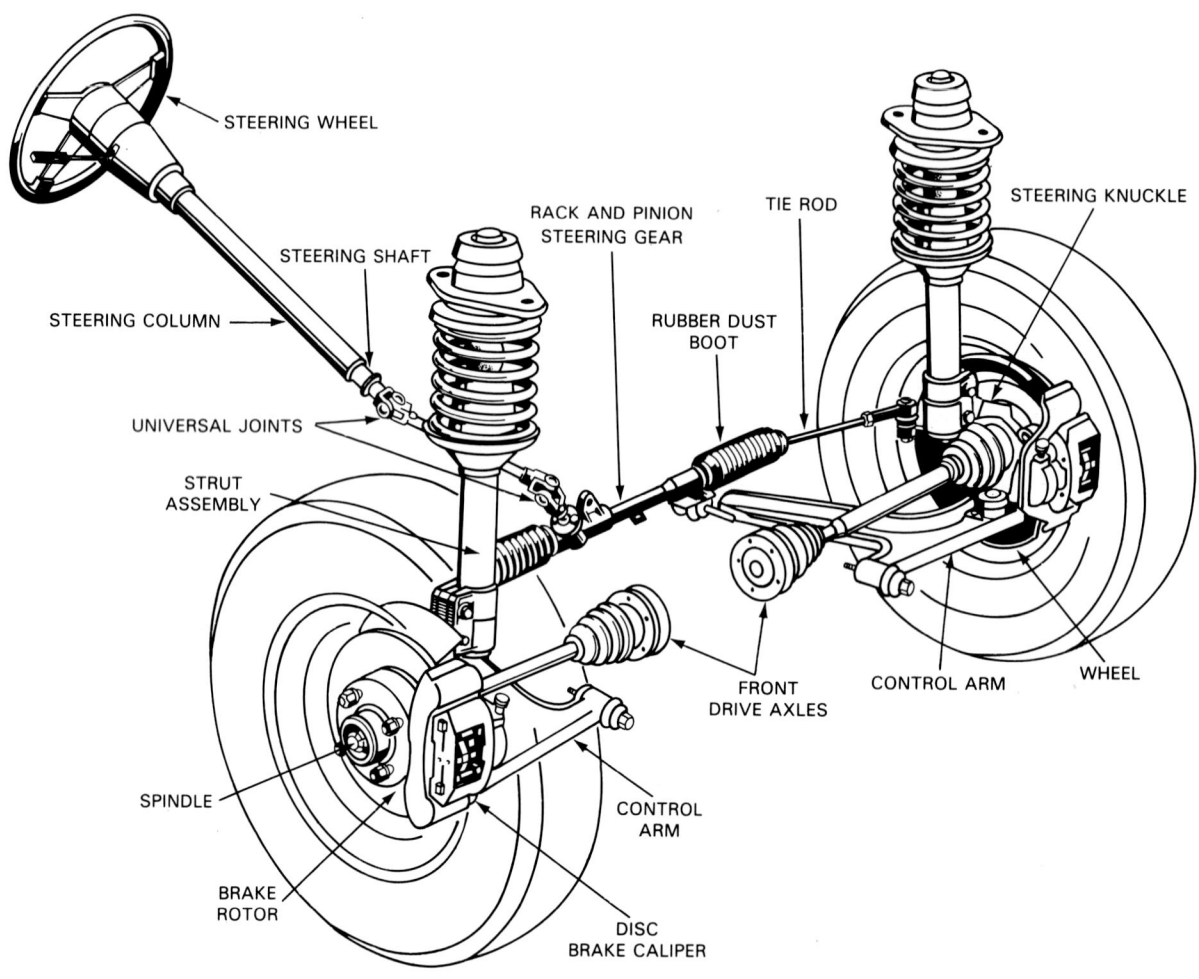

STEERING WHEEL

RACK AND PINION
STEERING GEAR

TIE ROD

STEERING KNUCKLE

STEERING SHAFT

RUBBER DUST
BOOT

STEERING COLUMN

UNIVERSAL JOINTS

STRUT
ASSEMBLY

FRONT
DRIVE AXLES

CONTROL ARM

WHEEL

SPINDLE

CONTROL
ARM

BRAKE
ROTOR

DISC
BRAKE CALIPER

Can you find all parts included in steering system? Note how they work with suspension and braking systems on this front-wheel drive setup.

Chapter 31

STEERING SYSTEM FUNDAMENTALS

After studying this chapter, you will be able to:
☐ Identify the major parts of a steering system.
☐ Explain the operating principles of steering systems.
☐ Compare the differences between linkage and rack and pinion type steering.
☐ Describe the operation of power steering systems.

This chapter will build upon your knowledge of auto mechanics by introducing modern steering systems. The steering mechanism works with the suspension system to provide a safe handling car.

There are two basic kinds of steering systems in wide use today: linkage (worm gear) steering and rack and pinion steering. See Fig. 31-1. They may be operated manually or with power assistance.

FUNCTIONS OF A STEERING SYSTEM

The steering system must perform several important functions.
1. Provide precise control of front wheel direction.
2. Maintain correct amount of effort needed to turn the front wheels.
3. Transmit road feel (slight steering wheel pull caused by road surface) to the driver's hands.
4. Absorb most of the shock going to the steering wheel as the tyres hit bumps and holes in road.
5. Allow for suspension action.

BASIC STEERING SYSTEM PARTS

Before studying each part, you should have a basic understanding of both linkage and rack and pinion type steering systems. This will allow you to develop a better "picture" of how each component operates.

Basic Linkage steering system

A linkage steering system, Fig. 31-1A, consists of the following parts:

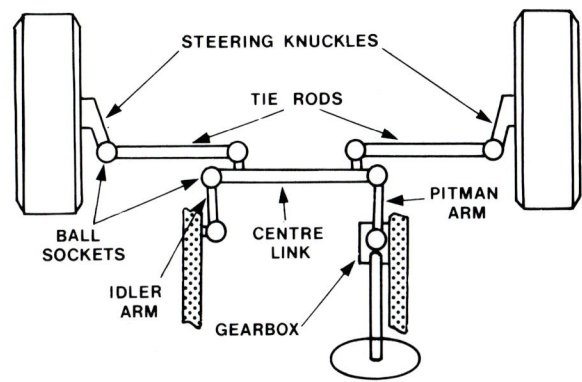

LINKAGE STEERING

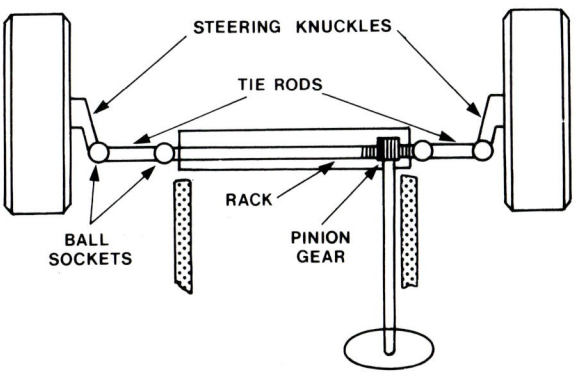

RACK AND PINION STEERING

Fig. 31-1. There are two types of steering systems. Both are found on today's cars.

1. STEERING WHEEL (used by driver to rotate steering shaft that passes through steering column).
2. STEERING SHAFT (transfers turning motion from steering wheel to steering gearbox).
3. STEERING COLUMN (supports steering wheel and steering shaft).
4. STEERING GEARBOX (changes turning motion into straight line motion to the left or right).

5. STEERING LINKAGE (connects steering gearbox to steering knuckles and wheels).
6. BALL SOCKETS (allow linkage arms to swivel up and down for suspension action and from left to right for turning).

Basic rack and pinion steering system

A rack and pinion steering system also uses a steering wheel, steering column, and steering shaft. See Fig. 31-1B. These components transfer the driver's turning effort to gears in the steering gear assembly.

Other than the parts just covered the major components of a rack and pinion steering system are:
1. PINION GEAR (rotated by steering wheel and steering shaft; has teeth that mesh with teeth on rack).
2. RACK (long steel bar with teeth along one section, slides sideways as pinion gear turns).
3. GEAR HOUSING (holds pinion gear and rack).
4. TIE RODS (connect rack with steering knuckles).

STEERING COLUMN ASSEMBLY

The steering column assembly consists of the steering wheel, steering shaft, column (outer housing), ignition key mechanism, and, sometimes a flexible coupling and universal joint. Look at Fig. 31-2.

The steering column normally bolts to the underside of the dash. The column projects through the bulkhead and fastens to the steering gear assembly.

Bearings fit between the steering shaft and column. They permit the shaft to rotate freely. The steering wheel is locked to the shaft by splines. A large nut holds the steering wheel on the shaft splines.

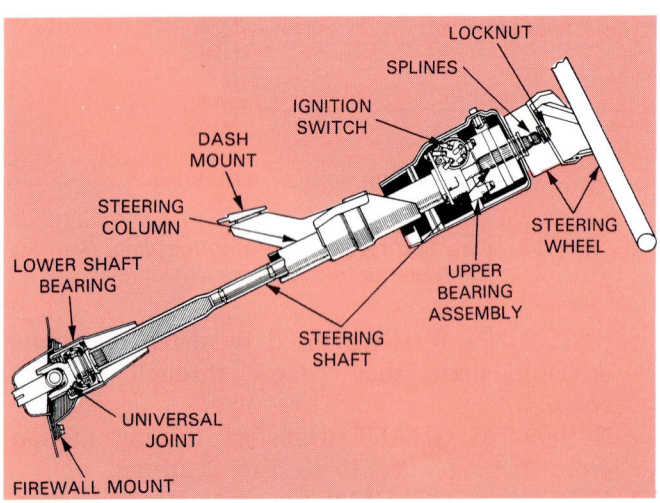

Fig. 31-2. Note steering column components. Steering wheel is splined to shaft that extends through column and down to steering gearbox. (Toyota)

Ignition lock and switch

Most modern cars have the ignition lock and switch mechanism mounted on the steering column. The key mechanism is normally on the top, right-hand side of the column. The ignition switch is usually bolted to the upper end of the steering column.

For more information on ignition switches, refer to Chaper 35.

Locking steering wheel

To help prevent theft, late model cars also have a locking steering wheel. When the ignition key is off, the steering wheel cannot be turned. Fig. 31-3 shows a common method of locking the steering wheel.

A rack and a sector are used to slide a steel pin into mesh with a slotted disc at the steering wheel hub. Since the disc is splined to the steering shaft, the steering wheel will NOT turn.

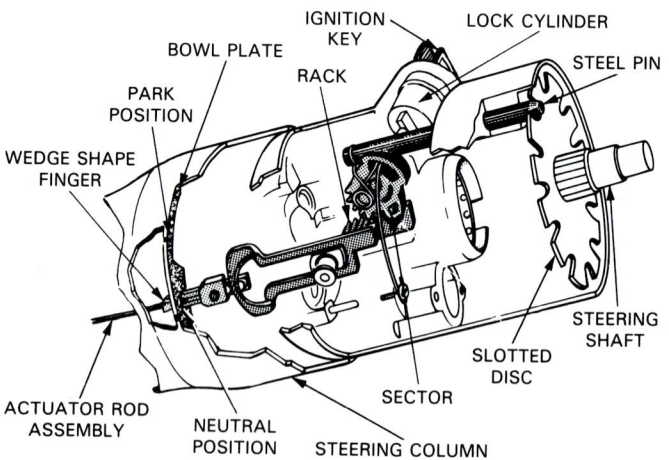

Fig. 31-3. Cutaway view shows how locking steering column functions. Steel pin slides into slot in disc to keep wheel from turning with key removed. Also note small rod that extends down to ignition switch. (Typical)

Collapsible steering column

Today's cars use a collapsible steering column to help prevent driver injury during an auto accident. The column is designed to crumple or slide together when forced. Look at Fig. 31-4.

When a car hits a stationary object, the engine and front body structure can be pushed rearward, into the steering column. At the same time, the driver could be thrown forward into the steering wheel. With a rigid steering column, the driver's chest could be injured.

There are several types of collapsible steering columns: steel mesh (crushing) type, tube and ball (sliding) type, and the shear capsule (break and slide) type. In all types, the column is two piece. The two sections move toward or into each other upon severe impact.

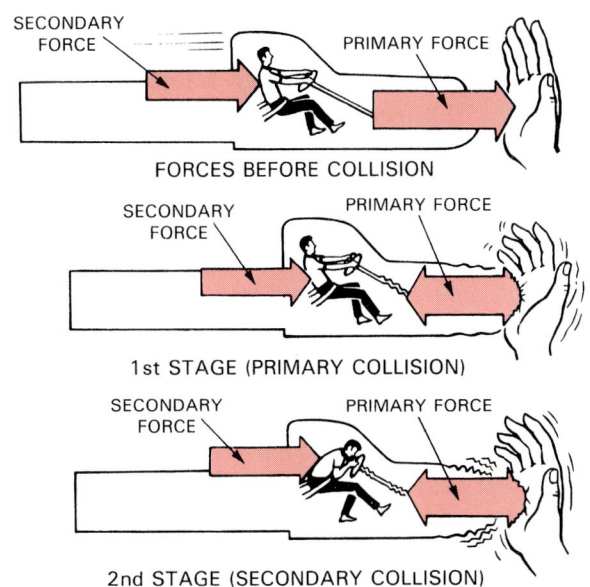

Fig. 31-4. Collapsible steering column crushes and protects driver.

STEERING GEAR PRINCIPLES

As was mentioned briefly, some steering systems use a worm type steering gear assembly. Others use a pinion gear and a rack. These two gear principles are illustrated in Figs. 31-5 and 31-6.

Recirculating ball gearbox

The recirculating ball gearbox is the most common type used with a linkage steering system. Pictured in Fig. 31-7, it has small steel balls that circulate between the gear members.

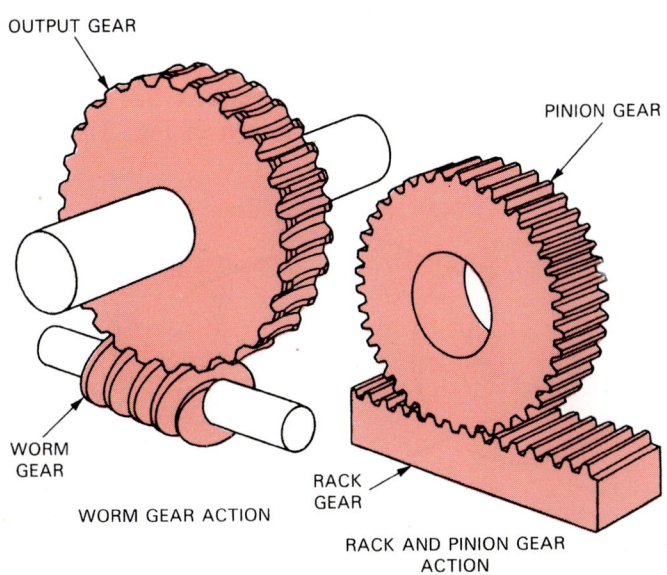

Fig. 31-5. Two basic types of gear mechanisms found in steering gearboxes: worm gear and rack and pinion gearset.

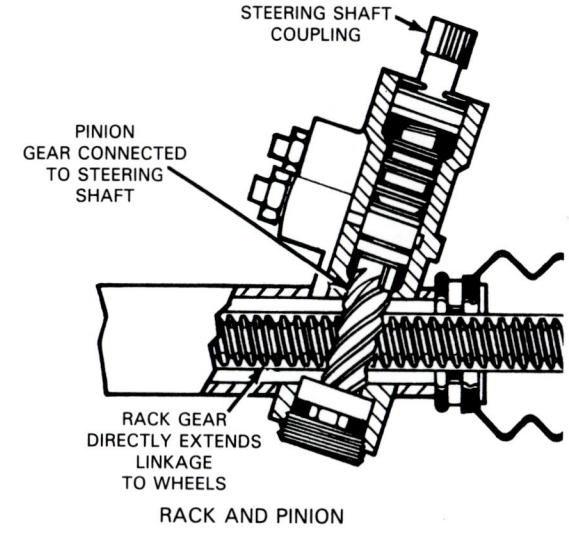

A   RACK AND PINION

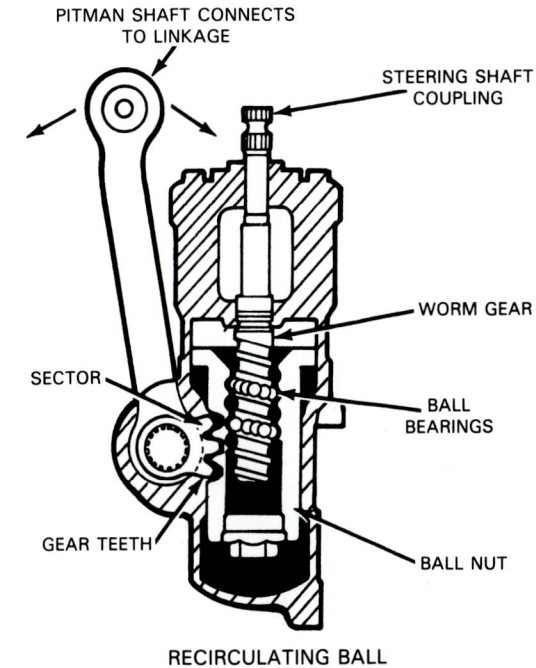

B   RECIRCULATING BALL

Fig. 31-6. Another view of two types of steering gears. A — Rack and pinion steering gear. B — Worm steering gearbox.

A worm shaft is the input connected to the steering column shaft. The balls fit and ride in the grooves in the worm gear.

The sector shaft is the output gear from the steering gearbox. It transfers motion to the steering linkage. Fig. 31-7. A sector gear is machined on the inner end of the sector shaft.

A ball nut rides on the steel balls and worm gear. See Fig. 31-8. Grooves are cut in the ball nut to match the shape of the worm gear. Since the ball nut cannot rotate, it slides up and down as the worm gear rotates.

Ball guides route extra balls in and out from between the worm and ball nut. Refer to Fig. 31-8.

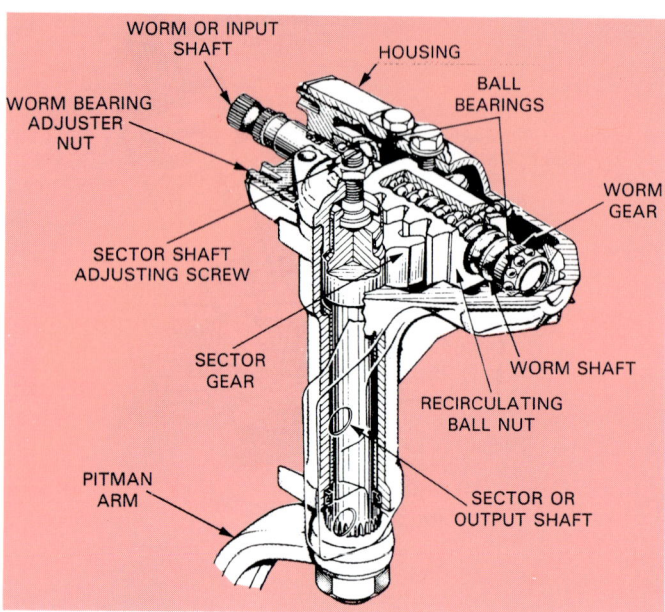

Fig. 31-7. Cutaway view of recirculating ball steering gearbox. Study part relationships.

The worm shaft is mounted in either ball or roller bearings. The sector shaft is also mounted on antifriction bearings. See Fig. 31-8.

A bearing adjuster nut is usually provided to set worm shaft bearing preload.

An adjusting screw is used to set the sector shaft clearance.

The gearbox housing provides an enclosure for the other components, Fig. 31-8. Seals are pressed into the housing to prevent lubricant leakage at the worm

and sector shafts. The shaft bearings also press into the gearbox housing.

The housing bolts to the car frame or reinforced area on the unitised body. An end cover normally bolts on the housing to cover the end of the sector gear. It can be removed for gearbox service.

Gearbox ratio (steering gear reduction)

Gearbox ratio, also termed steering gear reduction, is basically a comparison between steering wheel rotation and sector shaft rotation. Steering gearbox ratios range from 15:1 down to 24:1. With a 15:1 ratio, the worm shaft turns 15 times to turn the sector shaft once.

A manual gearbox will have a high ratio to reduce the amount of effort needed to turn the steering wheel.

Power steering gearboxes have a lower ratio.

Constant and variable ratio steering

A variable ratio gearbox changes the internal gear ratio as the front wheels are turned from the centre position. Most modern recirculating ball gearboxes are variable ratio. Refer to Fig. 31-9.

Variable ratio steering is faster when cornering, requiring fewer turns of the steering wheel from full right to full left. It also provides better control and response when manoeuvring.

Variable ratio steering is accomplished by changing the length of the gear teeth on the sector shaft gear. This changes the effective LEVER ARM action between the gears. Many manual steering gearboxes and most power steering gearboxes are variable ratio.

A constant ratio gearbox has the same gear reduction from full left to full right. The sector gear teeth are the same length.

Worm and roller steering gearbox

A worm and roller steering gearbox has a roller meshed with the worm gear. Shown in Fig. 31-10, a

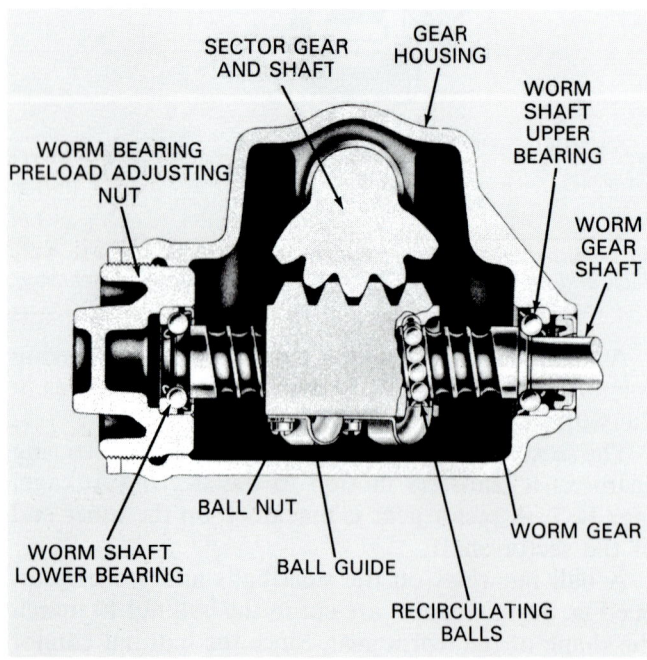

Fig. 31-8. When steering wheel is turned, shaft and worm gear rotate. This causes balls and ball nut to walk on worm. As a result, ball nut turns sector gear and shaft.

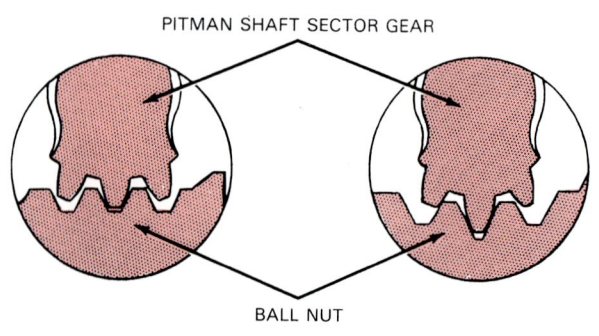

Fig. 31-9. Note difference in teeth with constant and variable ratio gearboxes. (Typical)

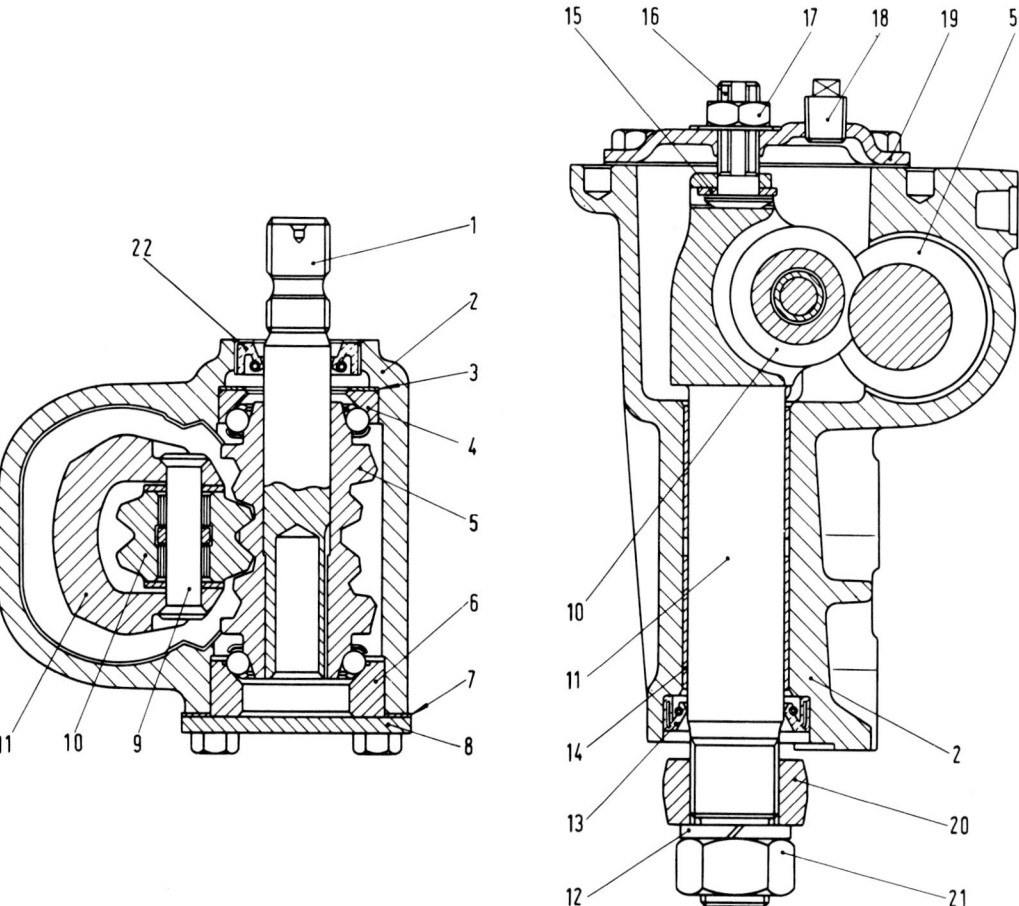

1. Steering shaft. - 2. Steering box. - 3. Worm upper bearing shims. - 4. Rear ball bearing. - 5. Worm. - 6. Front ball bearing. - 7. Worm lower bearing shims. - 8. Worm thrust cover. - 9. Roller pin. - 10. Roller. - 11. Roller shaft. - 12. Spring washer under pitman arm nut. - 13. Roller shaft oil seal. - 14. Roller shaft bushing. - 15. Roller shaft adjusting disc. - 16. Roller shaft adjusting screw. - 17. Locknut. - 18. Plug. - 19. Steering box cover. - 20. Pitman arm. - 21. Pitman arm nut. - 22. Steering shaft oil seal.

Fig. 31-10. Worm and roller gearbox is similar to recirculating ball type. Single roller replaces balls and ball nut. (Fiat)

roller, instead of balls, is used to reduce internal friction.

STEERING LINKAGE (WORM TYPE GEARBOX)

The steering linkage is a series of arms, rods, and ball sockets that connect the steering gearbox to the steering knuckles. Fig. 31-11 shows these parts.

The steering linkage used with a worm type gearbox typically includes a pitman arm, centre link, idler arm, and two tie rod assemblies.

Pitman arm

The pitman arm transfers gearbox motion to the steering linkage. The pitman arm is splined to the gearbox sector (output) shaft. A large nut and lock washer secure the arm to its shaft. Refer to Fig. 31-7.

The outer end of the pitman arm normally uses a ball socket (swivel joint), Fig. 31-11.

Centre link (relay rod)

The centre link, also called a relay rod, is simply a steel bar that connects the right and left sides of the steering linkage. As shown in Fig. 31-11, it has holes that accept the pitman arm, tie rod ends, and idler arm.

Idler arm

The idler arm supports the end of the centre link on the passenger side of the car. As pictured in Fig. 31-11, the idler arm bolts to the car frame or subframe.

Ball sockets

Ball sockets are like small ball joints; they provide a swivel connection between two parts. See Fig. 31-11. Ball sockets are needed so that the steering linkage is

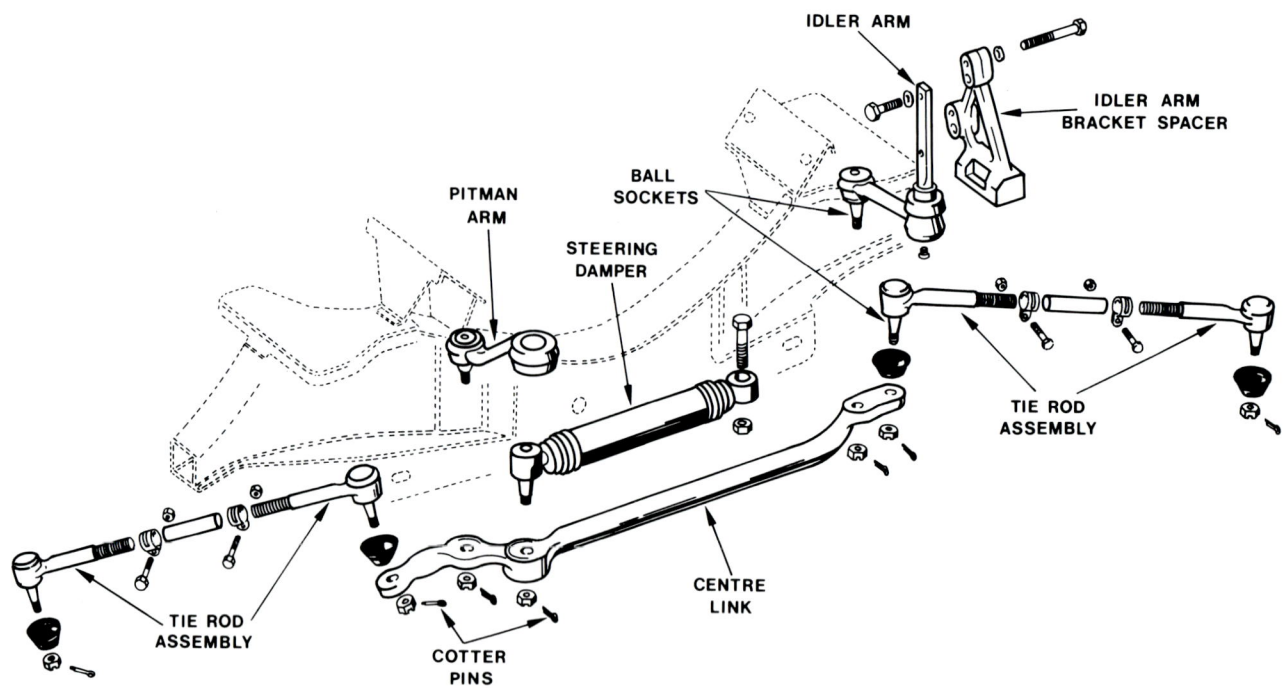

Fig. 31-11. Study arrangement of linkage type steering. Pitman arm connects to steering gearbox. Arm swings right or left and moves other linkage components.

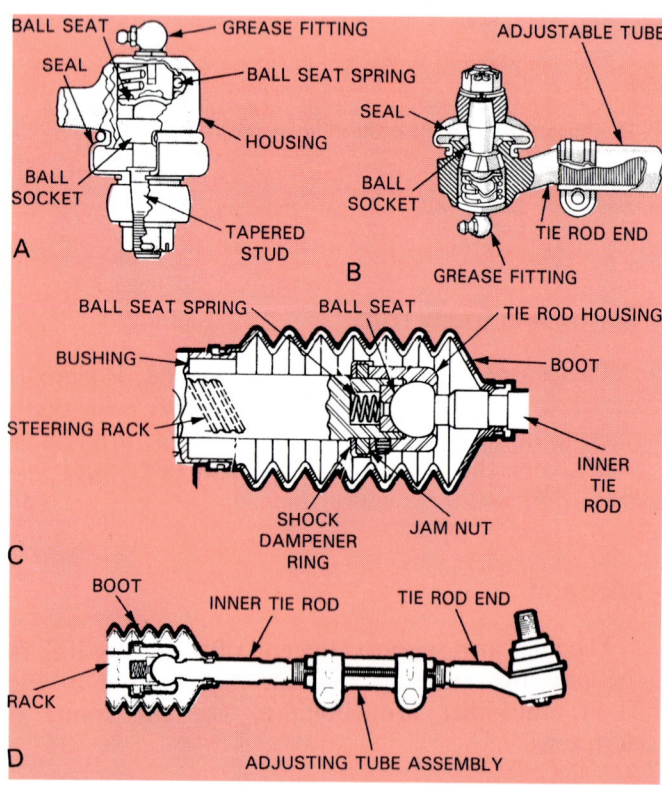

Fig. 31-12. Ball sockets allow linkage components to swivel freely when steering car. They are commonly used on end of pitman arm, idler arm, and tie rods. A — Ball socket for idler arm. B — Ball-socket for tie rod end. C — Ball socket for inner end of tie rod on rack and pinion assembly. D — Ball sockets on both ends of tie rod for rack and pinion steering. (Ford)

NOT bent when the wheels turn or move up and down over rough road surfaces.

Cutaway views for various ball sockets are given in Fig. 31-12. Note how a ball stud fits into a socket.

Ball sockets are filled with grease to prevent friction and wear. Some ball sockets are sealed. Others have a grease fitting that allows chassis grease to be inserted with a grease gun.

Tie rod assemblies

Two tie rod assemblies are used to fasten the centre link to the steering knuckles. Ball sockets are normally used on both ends of both tie rods. Look at Figs. 31-11 and 31-12. An adjustment tube assembly is provided for changing the length of the tie rod during wheel alignment.

MANUAL RACK AND PINION STEERING

Rack and pinion steering is the most popular type of steering system on today's cars. For this reason, it is important that you understand its parts and operating principles.

Fig. 31-13 pictures the external parts of a manual rack and pinion steering mechanism. Note that the steering gear bolts to the frame crossmember.

Steering shaft, flexible coupling, and U-joint

Many steering systems have a flexible coupling or a

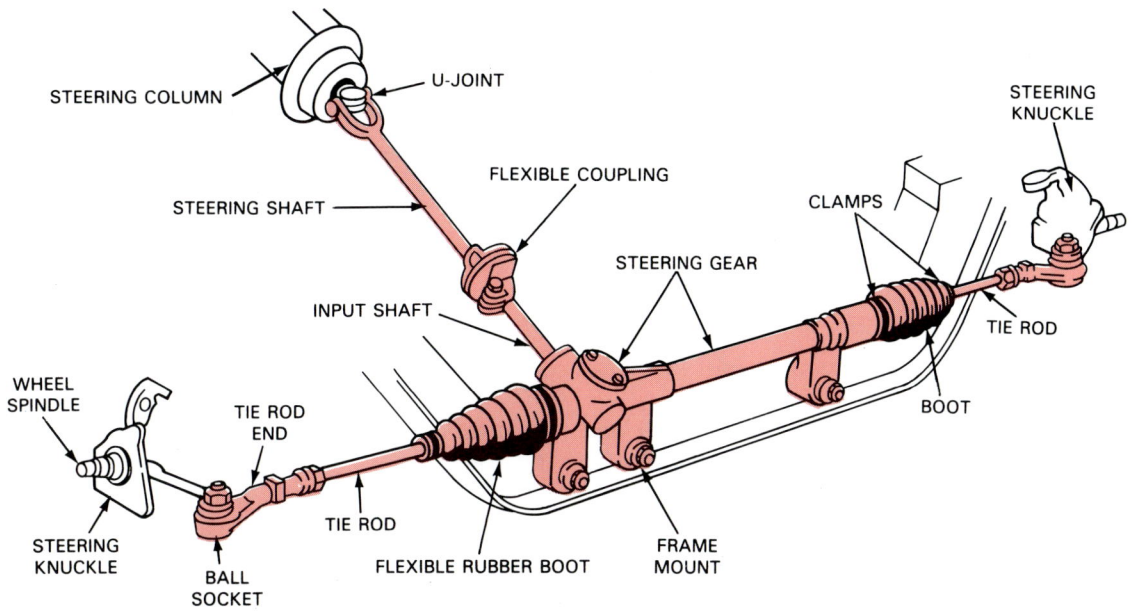

Fig. 31-13. Study components of manual rack and pinion steering gear. It uses fewer parts than linkage system. (Ford)

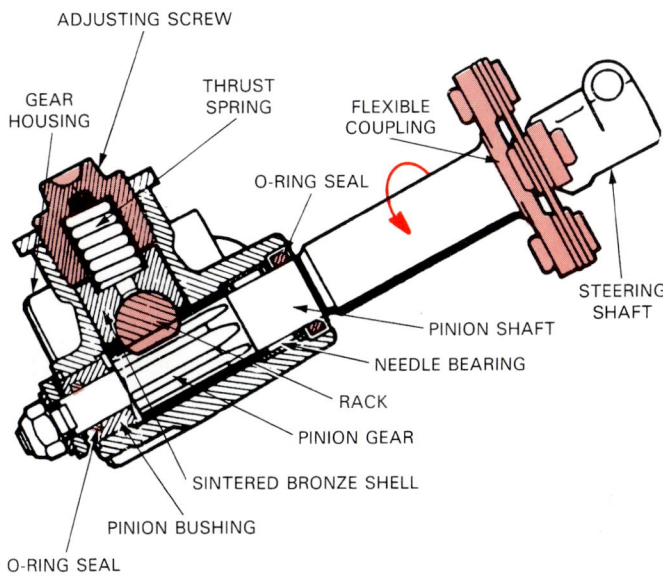

Fig. 31-14. Cutaway shows how pinion shaft gear rotates in housing. Pinion gear teeth mesh and act on rack gear teeth to slide rack left or right for steering action. Thrust spring holds rack and pinion gears in contact. (Ford)

universal joint in the steering shaft. Look at Figs. 31-13 and 31-14.

The flexible coupling helps keep road shock from being transmitted to the steering wheel. It also allows for slight misalignment of the steering shaft and steering gear input shaft.

A universal joint allows for a change in angle between the steering column and steering gear input shaft. One is shown in Fig. 31-13.

Rack and pinion steering gear

A manual rack and pinion steering gear basically consists of a pinion shaft, rack, thrust spring, bearings, seals, and gear housing. A cutaway view of one is shown in Fig. 31-14.

When the steering shaft turns the pinion shaft, the pinion gear acts on the rack gear. The rack then slides sideways inside the gear housing, Fig. 31-15.

The thrust spring preloads the rack and pinion gear teeth to prevent excessive gear backlash (play), Fig. 31-15. Adjustments screws or shims may be used for setting thrust spring tension.

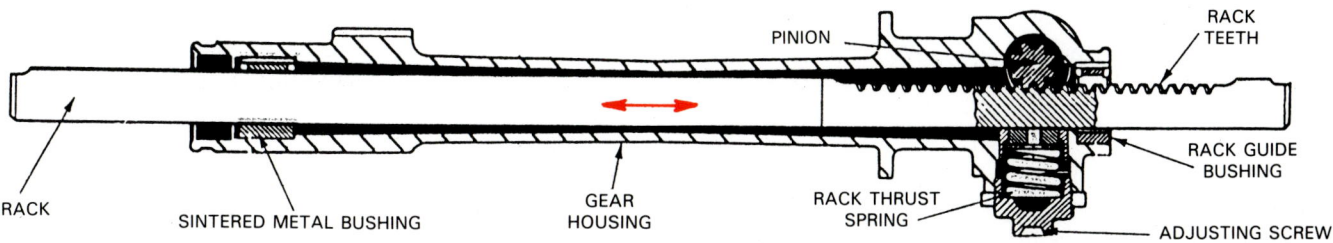

Fig. 31-15. Rack slides sideways and pushes or pulls on tie rods. This rotates steering knuckles and front wheels. (Typical)

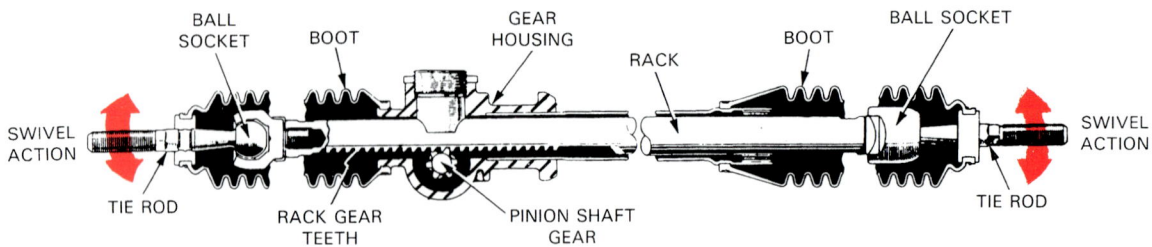

Fig. 31-16. Note how tie rods attach to rack. (Toyota)

Either bushings or roller bearings may be used on the pinion shaft and rack. Frequently, the pinion shaft uses roller bearings and the rack uses plain bushings. This is pictured in Figs. 31-15 and 31-16.

Rack and pinion tie rods

Tie rod assemblies for rack and pinion steering connect the ends of the rack with the steering knuckles. Look at Fig. 31-16.

Note the unique construction of the inner ends of the tie rods used on rack and pinion steering, Fig. 31-17. They have a very large ball formed on one end of each tie rod. These balls fit into separate sockets that normally screw onto the ends of the rack. Con-

ventional ball joints are used on the outer ends of the tie rods.

Rubber dust boots fit over the inner ball sockets to keep out road dirt, water, and to hold in lubricating grease. See Fig. 31-17. Clamps secure each end of the dust boots.

POWER STEERING SYSTEMS

Power steering systems normally use an engine driven pump and hydraulic system to assist steering action. The schematic in Fig. 31-18 illustrates a power steering system.

Pressure from an oil pump is used to operate a piston and cylinder assembly. When the control valve

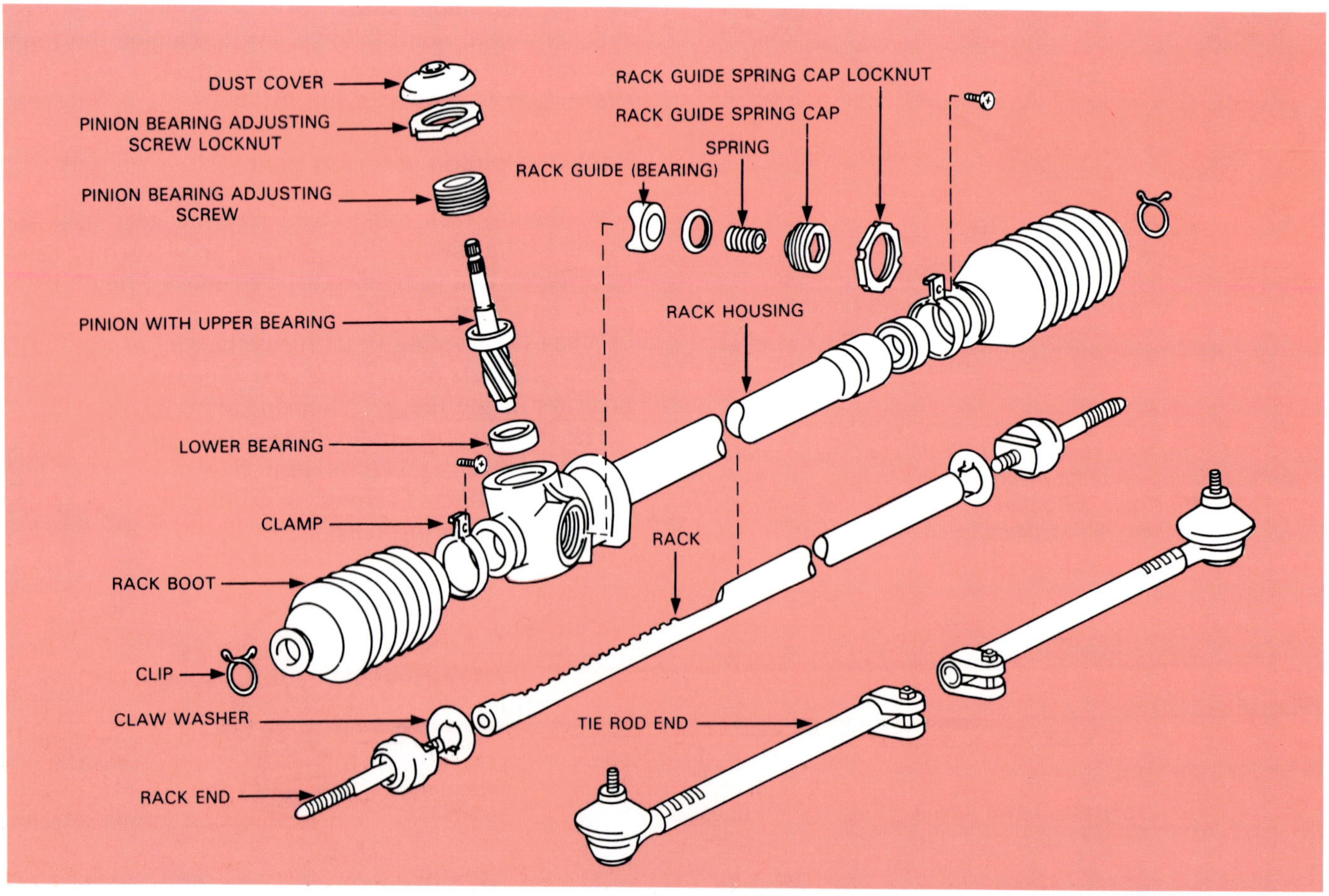

Fig. 31-17. Note how parts of rack and pinion gear system fit together. Study this carefully! (Ford)

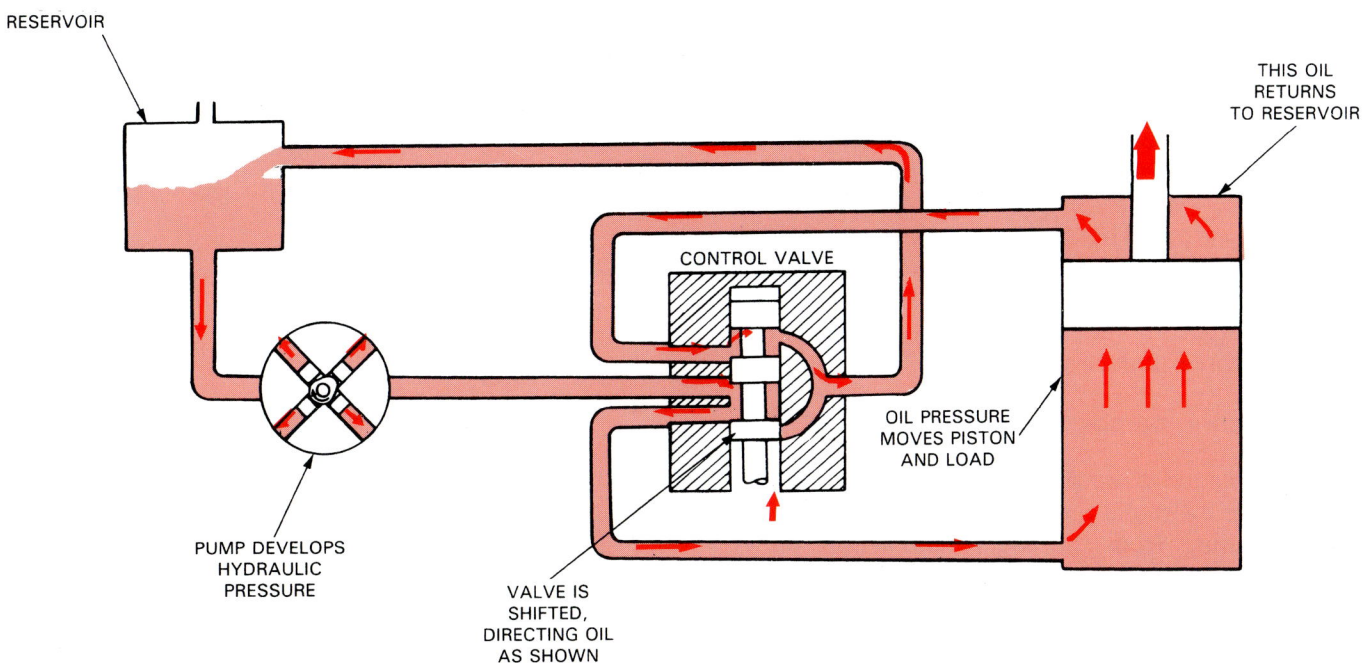

RESERVOIR

THIS OIL
RETURNS
TO RESERVOIR

CONTROL VALVE

OIL PRESSURE
MOVES PISTON
AND LOAD

PUMP DEVELOPS
HYDRAULIC
PRESSURE

VALVE IS
SHIFTED,
DIRECTING OIL
AS SHOWN

Fig. 31-18. Basic components of power steering system. Hydraulic or oil pump pressurises system. Control valve routes oil into either side of piston. Piston action can then be used to aid steering of front wheels.

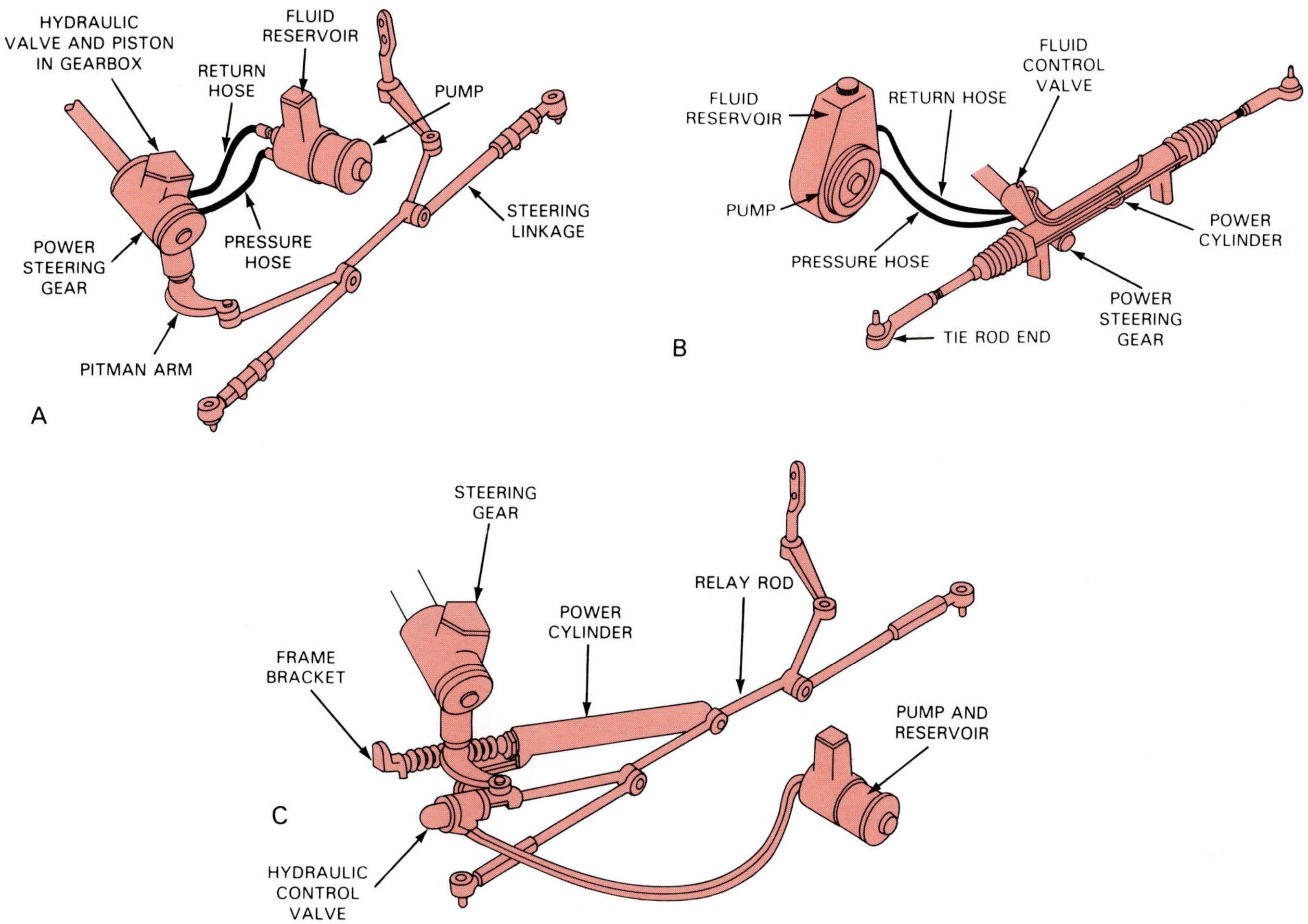

HYDRAULIC
VALVE AND PISTON
IN GEARBOX

FLUID
RESERVOIR

RETURN
HOSE

PUMP

STEERING
LINKAGE

POWER
STEERING
GEAR

PRESSURE
HOSE

PITMAN ARM

A

FLUID
RESERVOIR

FLUID
CONTROL
VALVE

RETURN HOSE

PUMP

PRESSURE HOSE

POWER
CYLINDER

POWER
STEERING
GEAR

TIE ROD END

B

STEERING
GEAR

RELAY ROD

POWER
CYLINDER

FRAME
BRACKET

PUMP AND
RESERVOIR

C

HYDRAULIC
CONTROL
VALVE

Fig. 31-19. The three major power steering systems. A — Integral piston-linkage type. B — Rack and pinion type. C — External piston-linkage type.

routes oil pressure into one end of the piston, the piston slides in its cylinder. Piston movement can then be used to help move the steering system components and front wheels of the car.

There are three major types of power steering systems used on modern autos: integral piston-linkage type, external piston-linkage type, and rack and pinion type. The rack and pinion system is further divided into the integral and the external power piston. The integral rack and pinion power steering system is the most common. Fig. 31-19 shows the three main types.

Power steering pumps

The power steering pump is engine driven and produces the hydraulic pressure for system operation. A belt running from the engine crankshaft pulley normally powers the pump. Refer to Fig. 31-20.

Fig. 31-21 shows four basic types of power steering pumps: roller, vane, slipper, and gear types.

During pump operation, the drive belt turns the pump shaft and pumping elements. Oil is pulled into one side of the pump by vacuum. The oil is then trapped and pushed to a smaller volume inside the pump. This pressurises the oil at the output and oil flows to the rest of the power steering system.

A cutaway view of a modern power steering pump is in Fig. 31-22.

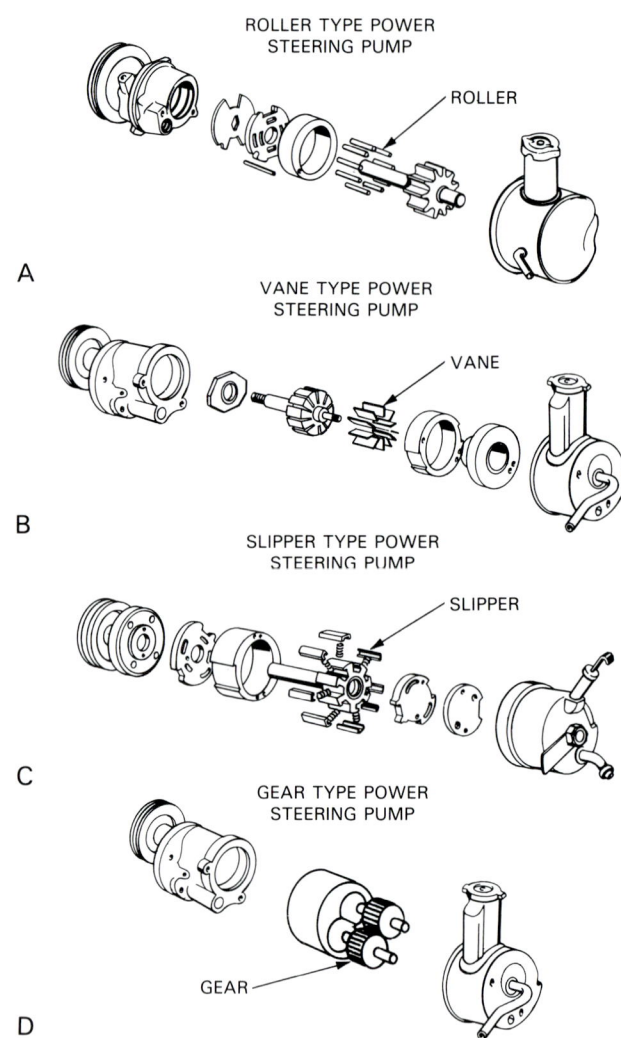

Fig. 31-21. Four basic types of power steering pumps.

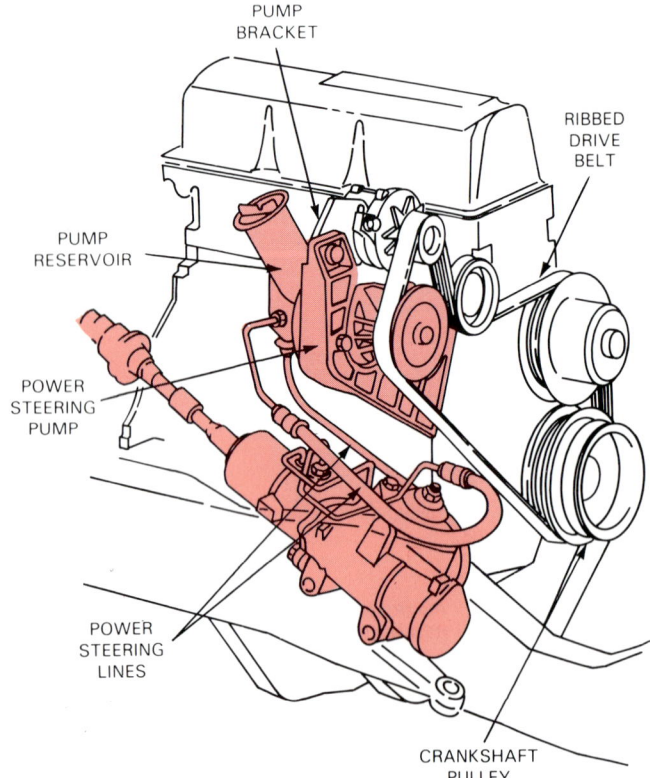

Fig. 31-20. Power steering pump bolts to front of engine. Belt spins pump. Power steering lines connect pump to control valve on gearbox or linkage. (Ford)

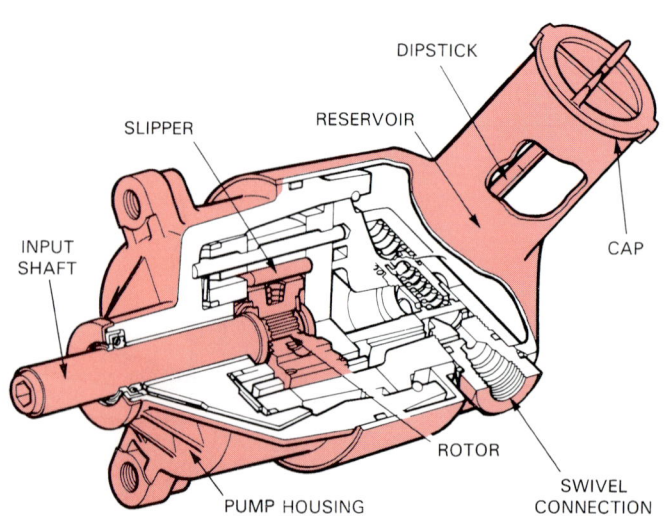

Fig. 31-22. Rotation of input shaft turns rotor and slippers. Power steering fluid is forced out under pressure. (Ford)

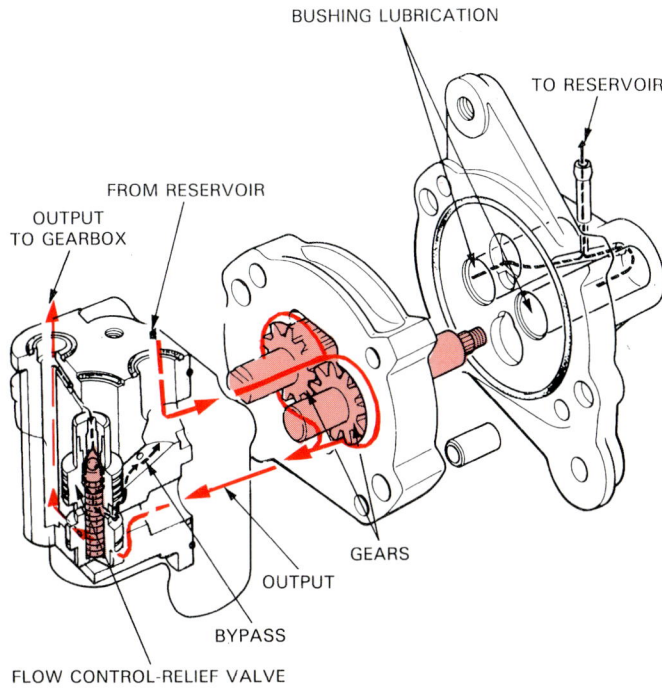

BUSHING LUBRICATION

TO RESERVOIR

FROM RESERVOIR

OUTPUT
TO GEARBOX

GEARS

OUTPUT

BYPASS

FLOW CONTROL-RELIEF VALVE

Fig. 31-23. Note action inside this gear type power steering pump. Flow control-relief valve is built into body of pump.

Pressure relief valve

A pressure relief valve is used in a power steering system to control maximum oil pressure. It prevents system damage by limiting pressure when needed. Fig. 31-23 shows the fundamental operation of a pressure relief valve in a modern power steering pump.

Power steering hoses

Power steering hoses are high pressure, hydraulic, rubber hoses that connect the power steering pump and the integral gearbox or power cylinder. Refer to Fig. 31-20. One line serves as the pressure feed line. The other acts as a return line to the reservoir.

INTEGRAL POWER STEERING SYSTEM (LINKAGE TYPE)

An integral power steering system has the hydraulic piston mounted inside the steering gearbox housing. It is a very common type of linkage power steering system. Basically, it consists of a power steering pump, hydraulic lines, and a special integral power-assist gearbox.

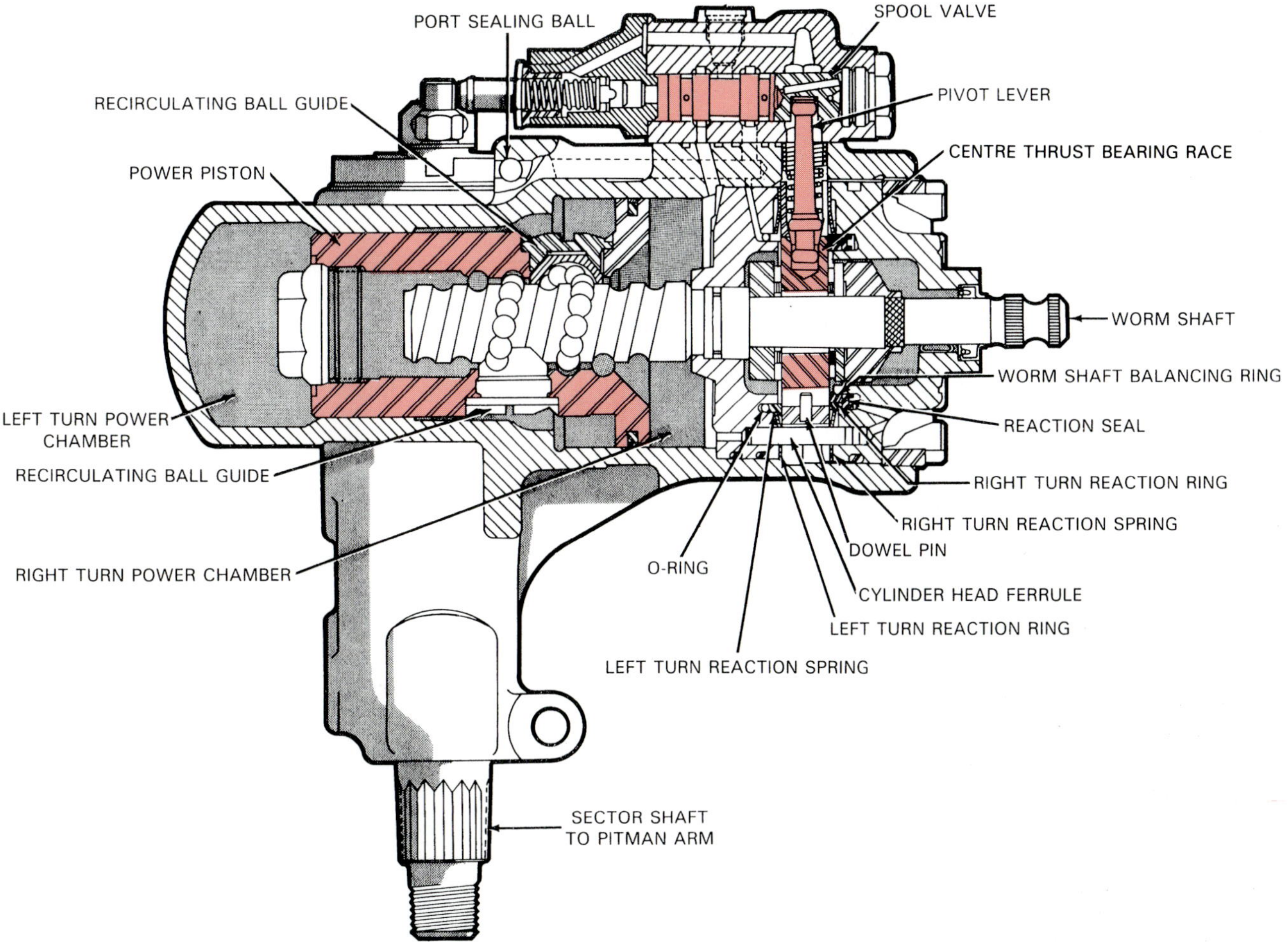

PORT SEALING BALL

SPOOL VALVE

RECIRCULATING BALL GUIDE

PIVOT LEVER

POWER PISTON

CENTRE THRUST BEARING RACE

WORM SHAFT

WORM SHAFT BALANCING RING

LEFT TURN POWER CHAMBER

REACTION SEAL

RECIRCULATING BALL GUIDE

RIGHT TURN REACTION RING

RIGHT TURN REACTION SPRING

DOWEL PIN

RIGHT TURN POWER CHAMBER

O-RING

CYLINDER HEAD FERRULE

LEFT TURN REACTION RING

LEFT TURN REACTION SPRING

SECTOR SHAFT
TO PITMAN ARM

Fig. 31-24. Cutaway of power steering gearbox for linkage steering system. Spool valve controls pressure on each side of power piston.

The integral power steering gearbox contains a conventional worm and sector gear, a hydraulic piston, and a flow direction valve. One type of integral power steering gearbox uses a spool valve. Another popular type has a rotary valve.

Fig. 31-24 shows a spool valve type power steering gearbox. Note that it uses a small spool valve to control pressure entering the two power chambers on each side of the piston.

When the steering wheel is turned to the right, the pivot lever moves the spool valve so that pressure enters the right turn chamber. This forces the power piston to the left and helps turn the sector shaft for a right turn. Pressure enters the opposite chamber when the steering wheel is turned to the left.

A rotary valve type power steering gearbox has a small torsion bar to detect steering wheel turning direction and turning effort. See Fig. 31-25.

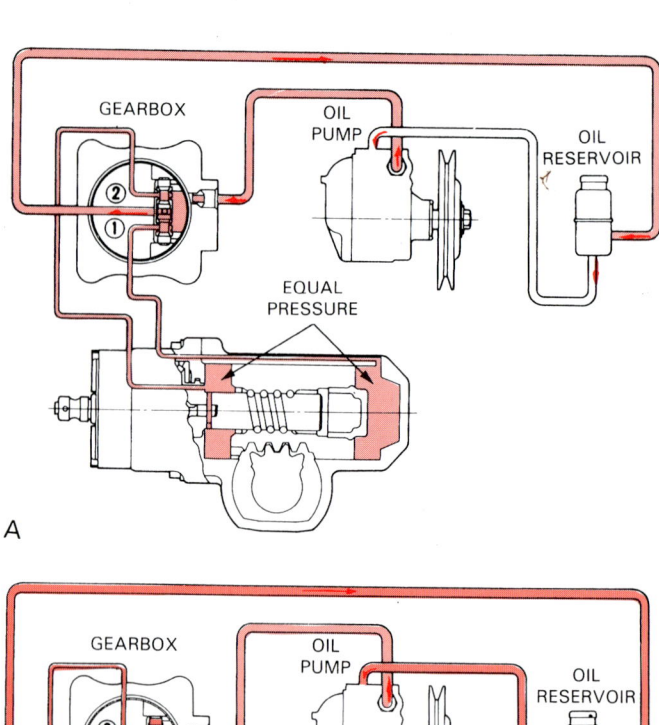

A

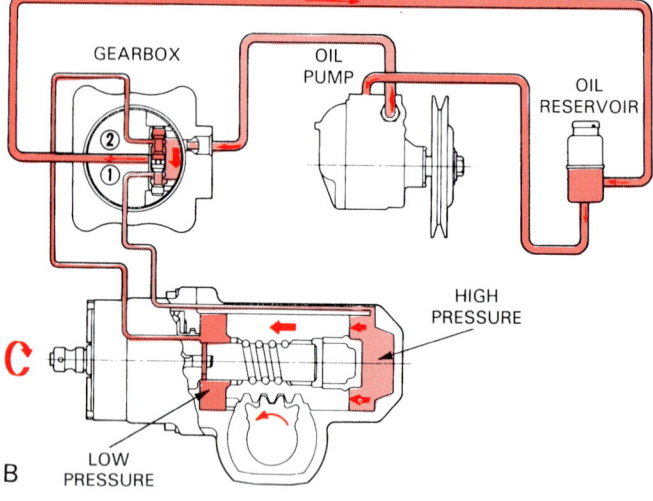

B

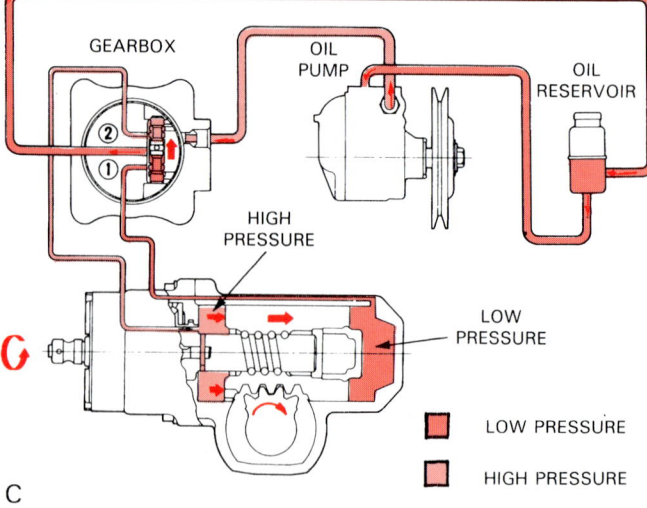

C

Fig. 31-26. Integral power steering gear operation. A — Steering wheel held straight ahead or neutral. Control valve balances pressure on both sides of power piston. Oil returns to pump reservoir from valve. B — With right turn, control valve routes oil to one side of power piston. Piston is pushed in cylinder to aid pitman shaft rotation. C — With left turn, control valve routes oil to other nonpressure side of piston back through control valve and to pump.

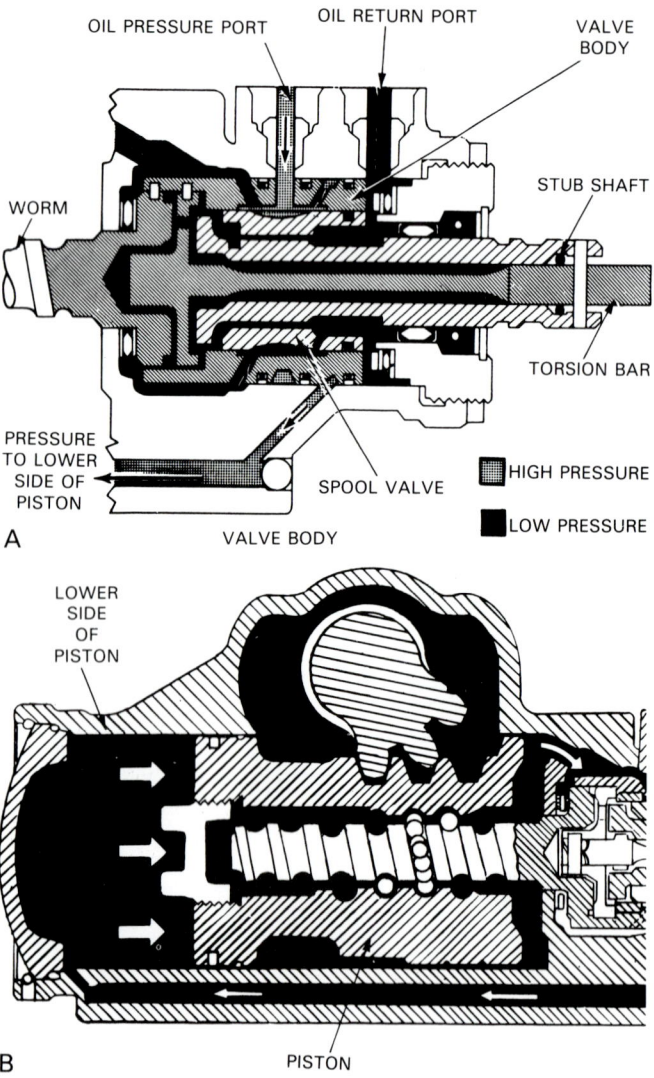

Fig. 31-25. This power steering gearbox uses a rotary valve and torsion bar to control pressure to power piston. Twist of torsion bar causes rotary valve to open port to right or left piston chamber.

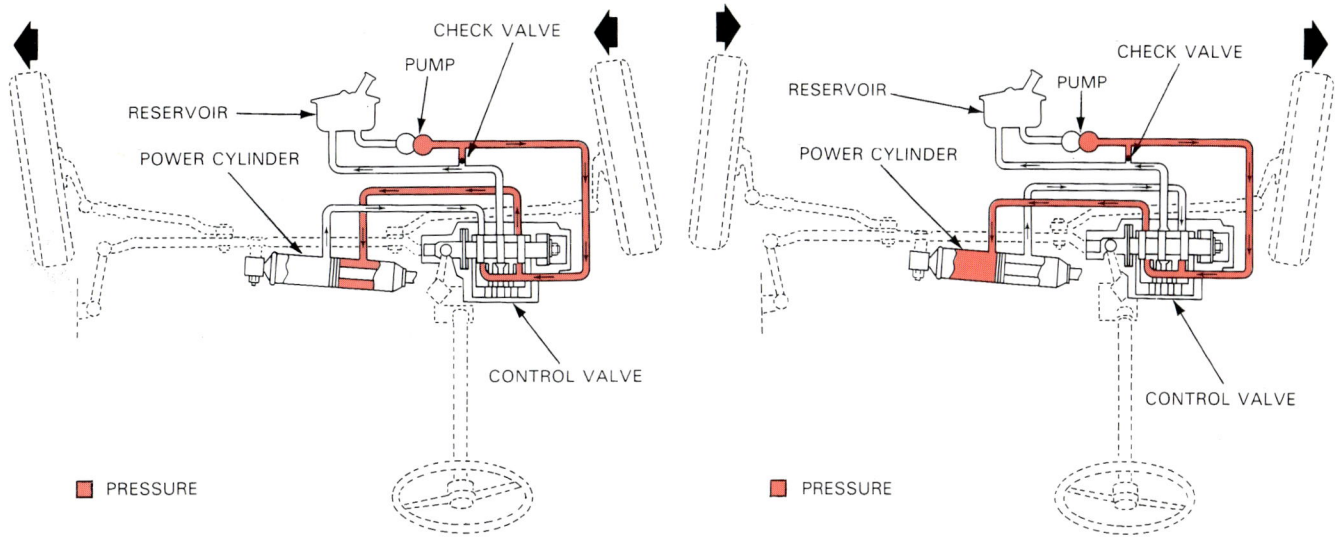

Fig. 31-27. Power steering system using power cylinder mounted on steering linkage. Note system pressures for right and left turns. (Ford)

When the steering wheel is turned, the torsion bar twists and turns the rotary valve. The rotary valve then directs hydraulic pressure to the correct side of the power piston.

Rotary valve action is detailed later when discussing power rack and pinion steering.

Study the flow of oil in Fig. 31-26. Notice how the pressure is used to slide the power piston. The piston pushes on and turns the sector shaft and pitman arm.

EXTERNAL CYLINDER POWER STEERING (LINKAGE TYPE)

An external cylinder power steering system commonly has the power piston bolted to the frame and to the centre link. This is shown in Fig. 31-27. The control valve may be located in the gearbox or on the steering linkage.

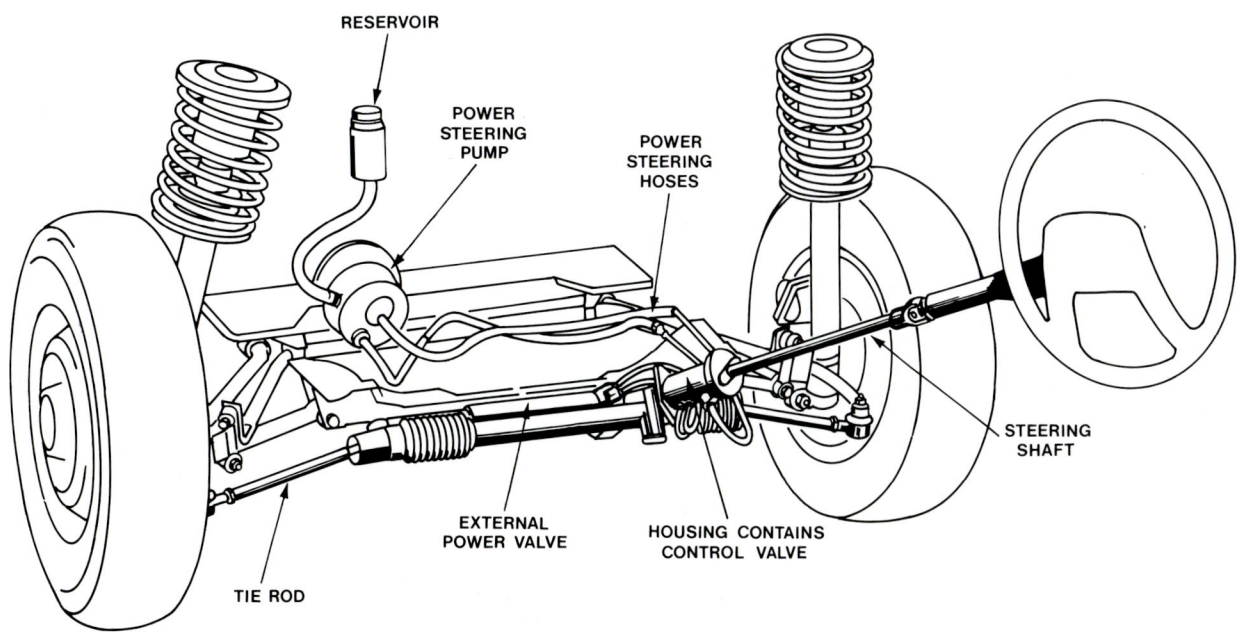

Fig. 31-28. External view of power rack and pinion setup. (Peugeot)

POWER RACK AND PINION STEERING

Power rack and pinion steering uses hydraulic pump pressure to assist the driver in moving the rack and front wheels. Fig. 31-28 shows this type of system.

The power steering pump normally mounts on the front of the engine. A belt powers the pump. Power steering hoses and metal lines connect the pump with the rack and pinion gear.

A power rack and pinion assembly basically consists of:

1. POWER CYLINDER (hydraulic cylinder machined inside rack or gear housing). See Fig. 31-29.
2. POWER PISTON (hydraulic, double-acting piston formed on rack).
3. HYDRAULIC LINES (steel tubing connecting control valve and power cylinder).
4. CONTROL VALVE (either a rotary or spool type hydraulic valve that regulates pressure entry into each end of the power piston).

The other parts of the assembly are similar to those found on a manual rack and pinion assembly. Note in Fig. 31-29 how routing oil pressure into either end of the power cylinder causes piston operation.

Power cylinder and piston

A power cylinder for rack and pinion steering is precisely machined to accept the power piston. Provisions are made for the hydraulic lines. The cylinder housing bolts to the car frame member, just like a manual unit.

The power piston is formed by attaching a hydraulic piston to the centre of the rack. A rubber seal fits around this piston. Seals are also used on each end of the piston to keep fluid from leaking out. This is shown in Fig. 31-30.

Power rack and pinion control valves

There are two types of control valve mechanisms used on power rack and pinion gears: rotary valve and spool valve. The rotary valve is more common.

The rotary valve uses a torsion shaft connected to the pinion gear to operate the hydraulic control valve.

Study Fig. 31-30 closely. It illustrates the operation of a power rack and pinion gear using a rotary type control valve. Today's cars frequently use this design.

A spool valve uses the thrust action of the pinion shaft to shift the control valve. The control valve can then route oil to the power cylinder.

Fig. 31-31 shows a simplified view of a power rack and pinion gear with a spool type control valve.

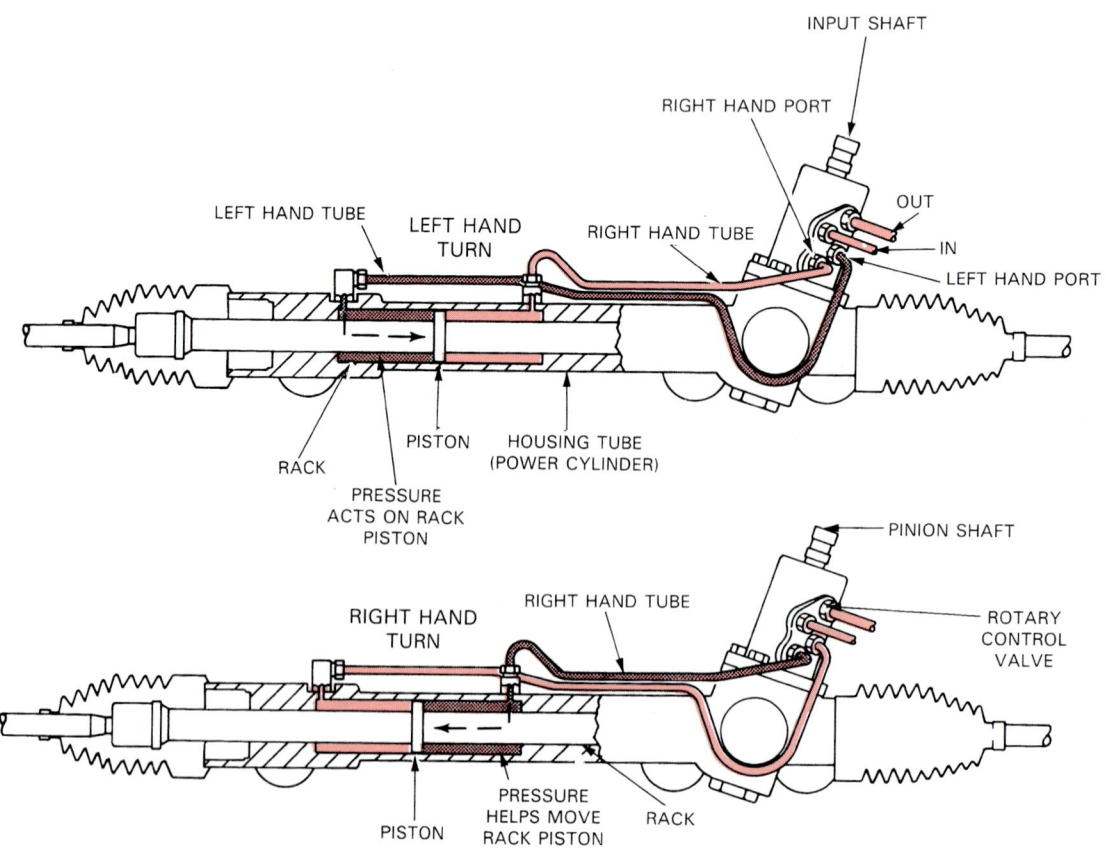

Fig. 31-29. Power cylinder is formed around rack. Pressure acts on rack piston to help slide rack in its housing. (Ford)

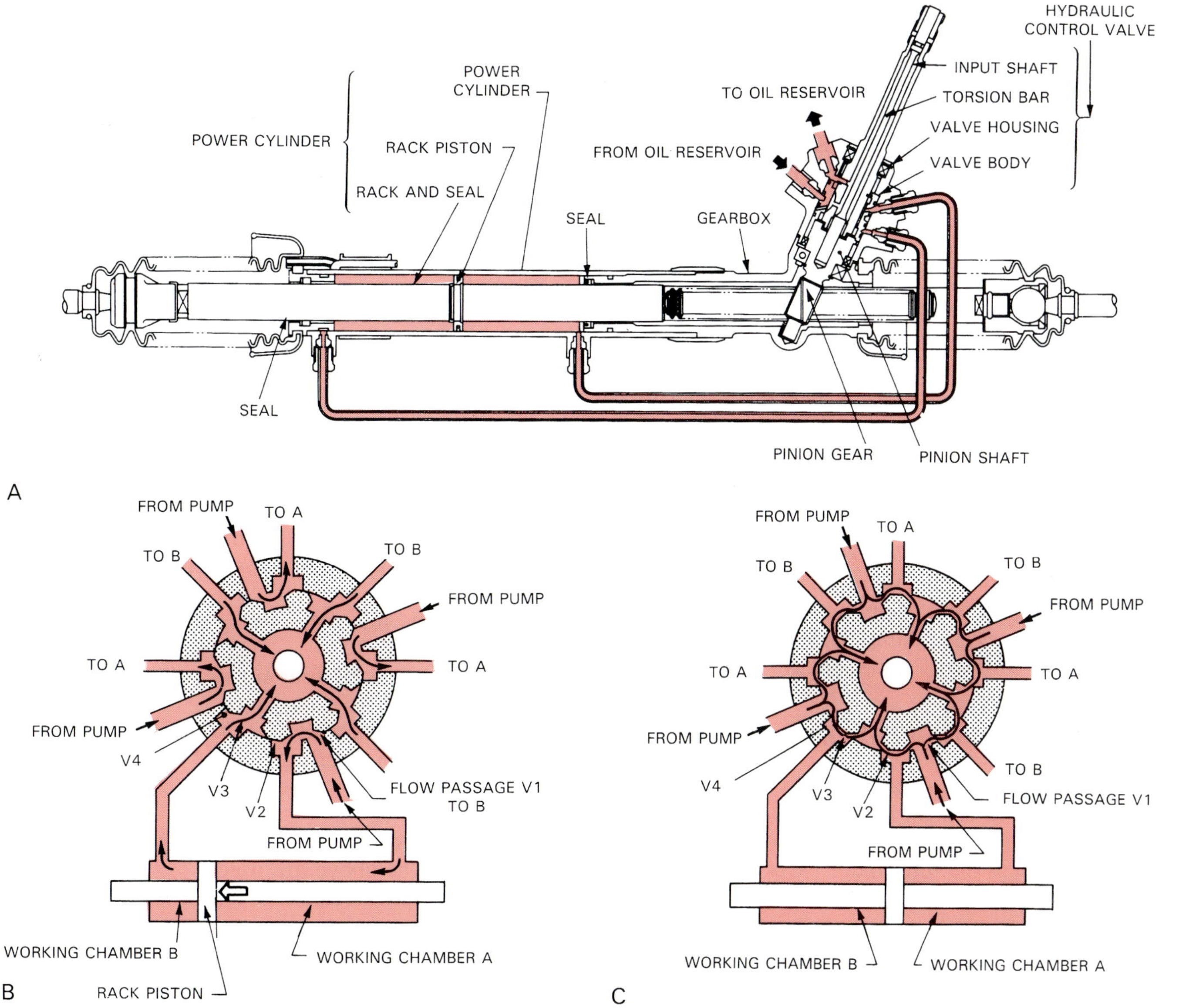

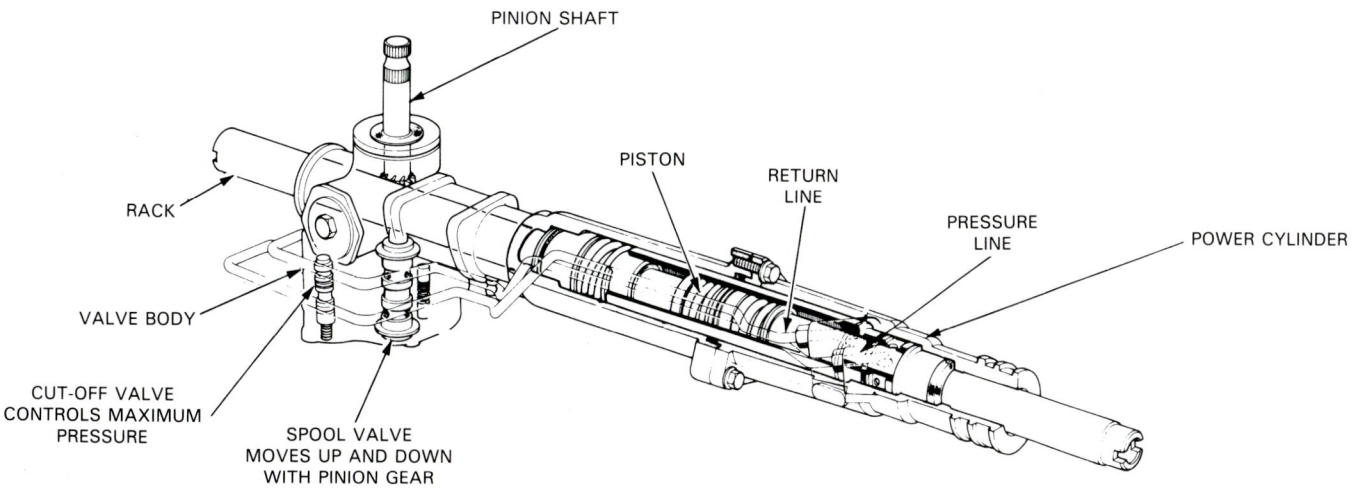

Fig. 31-30. A — Study parts. B — Steering effort applied. C — No effort applied. (Subaru)

Fig. 31-31. This power rack and pinion steering assembly uses a spool valve that detects thrust action of the helical pinion gear. It can then control oil pressure to the rack piston. (Honda)

Power rack and pinion operation

When the steering wheel is turned, the weight of the car causes the front tyres to resist turning. This twists a torsion bar (rotary valve mechanism) or thrusts the pinion shaft (spool valve mechanism) slightly. This makes the control valve move and align specific oil passages.

Pump pressure then flows through the control valve, hydraulic line, and into the power cylinder. Pressure acts on the power piston. The piston helps push the rack and front wheels for turning.

Since the steering gear is filled with oil (usually automatic transmission fluid), the internal parts of the system are always lubricated. They slide or turn easily with little friction and wear.

KNOW THESE TERMS

Steering shaft, Steering gearbox, Steering linkage, Ball sockets, Pinion gear, Rack, Tie rod, Steering column, Recirculating ball, Worm shaft, Sector shaft, Ball nut, Gearbox ratio, Pitman arm, Centre link, Idler arm, Power steering pump, Relief valve, Integral power steering.

REVIEW QUESTIONS

1. Name five functions of a steering system.
2. List and explain the six major parts of a linkage type steering system.
3. List and explain the four major parts of a manual rack and pinion steering system.
4. Today's cars commonly use a _____ steering column to help prevent driver injury during an auto accident.
5. Which of the following is NOT a part in a recirculating ball steering gearbox.
   a. Roller.
   b. Worm shaft.
   c. Sector shaft.
   d. Ball nut.
6. Define the term "gearbox ratio".
7. The idler arm supports the pitman arm on the passenger side of the car. True or False?
8. _____ and _____ is the most popular type steering system on modern cars.
9. What is the purpose of tie rods on a rack and pinion steering system?
10. Power steering systems normally use an engine driven _____ and _____ system to assist steering action.
11. A pressure relief valve is used in a power steering system to control maximum system pressure. True or False?
12. Describe an integral, linkage power steering system.
13. List and explain the four major parts of a power rack and pinion steering system.
14. Which of the following is NOT used in a power rack and pinion steering system?
    a. Spool valve.
    b. Rotary valve.
    c. Poppet valve.
    d. None of the above are used.
15. Define the term "road feel".

Chapter 32

WHEEL ALIGNMENT

After studying this chapter, you will be able to:
☐ Explain the principles of wheel alignment.
☐ List the purpose of each wheel alignment setting.
☐ Perform a pre-alignment inspection of tyres, steering, and suspension systems.
☐ Describe how to adjust caster, camber, and toe.
☐ Explain toe-out on turns, steering axis inclination, and tracking.
☐ Describe the different types of equipment used during wheel alignment service.

The term alignment means ''to position in a straight line''. Relating to motor vehicles, alignment means to position the four tyres so that they roll freely and evenly over the road surface.

Correct wheel alignment is essential to motor vehicle safety, sure handling, maximum fuel economy, and long tyre life. This chapter introduces both the principles and the basic procedures for wheel alignment.

WHEEL ALIGNMENT PRINCIPLES

The main purpose of wheel alignment is to make the tyres roll without scuffing, slipping, or dragging under all operating conditions. Six fundamental angles or specifications are needed for proper wheel alignment:

1. CASTER.
2. CAMBER.
3. TOE.
4. STEERING AXIS INCLINATION.
5. TOE-OUT ON TURNS.
6. TRACKING.

CASTER

Caster is basically the forward or rearward tilt of the steering knuckle (spindle support) when viewed from the side of the car. You are probably familiar with the term caster from furniture casters, Fig. 32-1.

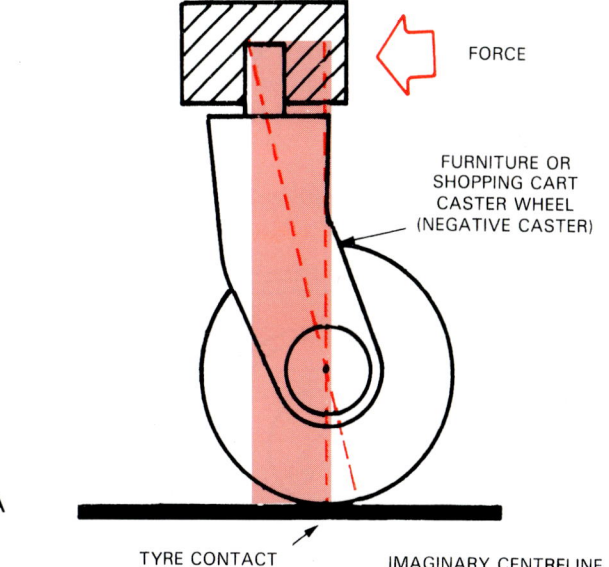

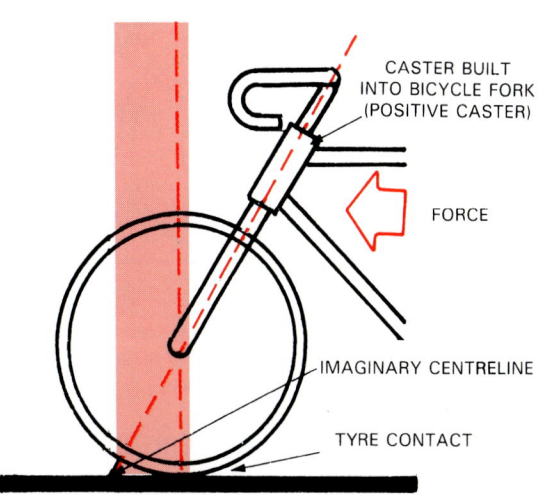

Fig. 32-1. Caster is determined by contact point of tyre and imaginary centre-line through spindle support. A — Negative caster on shopping cart causes wheel to follow irregular surfaces in floor. B — Positive caster, like on bike, makes front wheel travel straight ahead.

489

Caster controls where the tyre touches the road in relation to an imaginary centreline drawn through the spindle support. It is NOT a tyre wearing angle.

The basic purposes of caster are:
1. To aid directional control of the vehicle.
2. To cause the wheels to return to the straight-ahead position.
3. To offset road crown pull (steering wheel pull caused by hump in centre of road).

Positive caster tilts the top of the steering knuckle toward the rear of the car. See Fig. 32-2A. Positive caster helps to keep the car's wheel travelling in a straight line. When you turn the wheels, it lifts the car. Since this takes extra turning force, the wheels resist turning and try to return to the straight-ahead position.

Negative caster tilts the top of the steering knuckle toward the front of the car. Look at Fig. 32-2B. It is the opposite of positive caster. With negative caster, the wheels will be easier to turn. However, the wheels will tend to swivel and follow imperfections in the road.

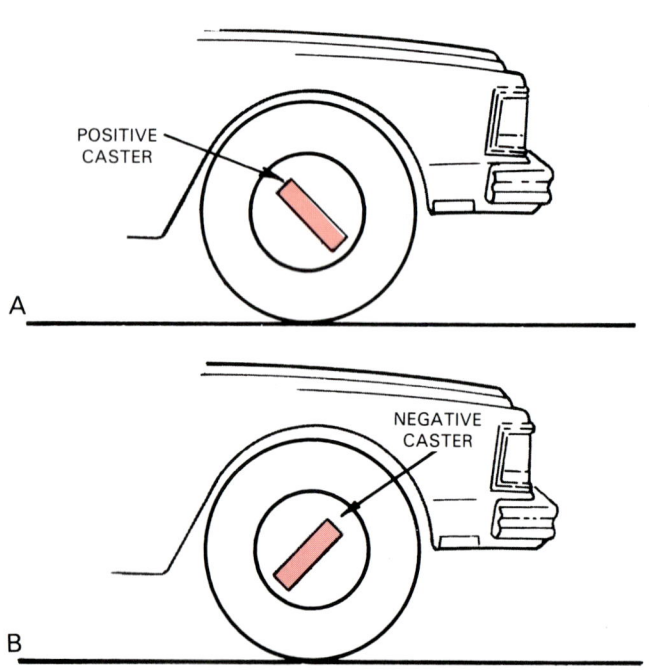

Fig. 32-2. A — Positive caster tilts steering knuckle towards front of car. B — Negative caster tilts steering knuckle towards rear of car.

Caster is measured in DEGREES starting at the true vertical (plumb line). Illustrated in Fig. 32-3, auto manufacturer's give specifications for caster as a specific number of degrees positive or negative.

Typically, specifications list more positive caster for cars with power steering. More negative caster is recommended for cars with manual steering to ease steering effort.

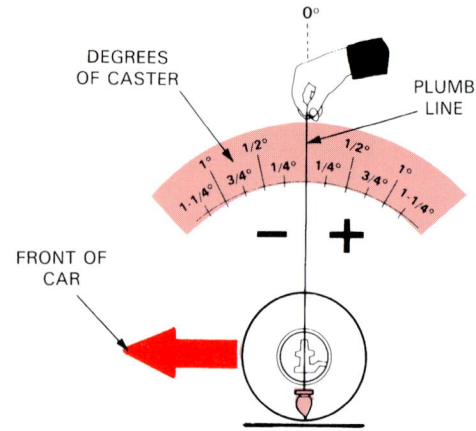

Fig. 32-3. Caster is measured in degrees, as shown.

Caster-road crown effect

Caster is a DIRECTIONAL CONTROL angle. It determines whether the car travels straight or pulls (steering wheel tries to turn) to the right or left.

Road crown is the normal slope of the road surface, Fig. 32-4. Most road surfaces angle downward from the centre. This helps keep rain water from collecting on the road. If the caster of both front wheels were the same, the road crown could make the car pull to the left. The car would want to steer off the outside (lower) edge of the road.

Since caster is a directional control angle, it is commonly used to offset the effect of road crown. The left front wheel may be set with slightly more positive caster than the right. This counteracts the forces caused by the road crown and the car will travel straight ahead.

Note! Always refer to the service manual for exact caster specifications. They vary with vehicle design.

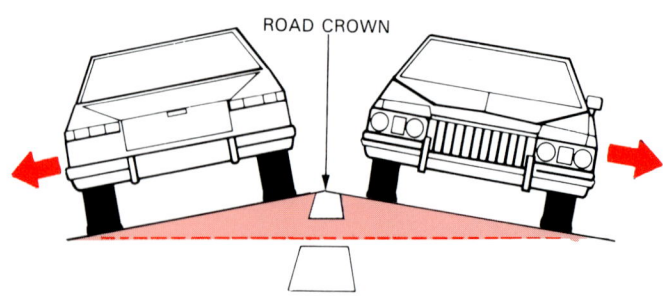

Fig. 32-4. Most roads are crowned in centre to help water flow off surface. Unequal caster can be used to offset road crown and keep car from trying to steer off road.

CAMBER

Camber is the inward or outward tilt of the wheel and tyre assembly when viewed from the front of the

490

car. It controls whether the tyre tread touches the road surface evenly. This is pictured in Fig. 32-5. Camber is a tyre-wearing angle measured in degrees.

The purposes of camber are:

1. To prevent tyre wear on the outer or inner tread.
2. To load the larger inner wheel bearing.
3. To aid steering by placing vehicle weight on the inner end of the spindle.

With positive camber, the tops of the wheels tilt outward when viewed from the front, Fig. 32-5A.

With negative camber, the tops of the wheels tilt inward when viewed from the front. Refer to Fig. 32-5B.

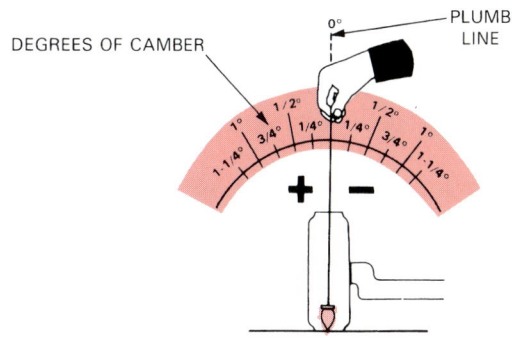

Fig. 32-6. Camber is measured in degrees as shown.

TOE

Toe is determined by the difference in distance between the front and rear of the left and right-hand wheels. Look at Fig. 32-7. Measured in millimeters, toe controls whether the wheels roll in the same direction. Toe is very critical to TYRE WEAR. If the wheels do NOT have the correct toe setting, the tyres will scuff or skid sideways.

Toe-in is produced when the front of the wheels are closer then the rear. As shown in Fig. 32-7A, toe-in causes the wheels to point inward at the front.

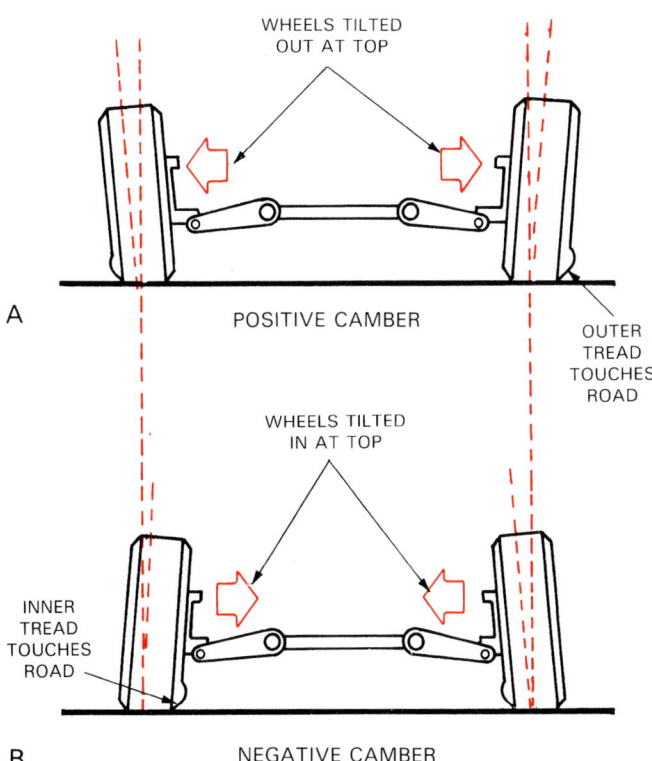

Fig. 32-5. Camber is determined by inward or outward tilt of wheels when viewed from front of car. Positive camber has wheels tilted out at top. Negative camber has wheels tilted out at bottom.

Negative and positive camber are measured from the true vertical (plumb line). As in Fig. 32-6, if the wheel is parallel with the plumb line, camber is zero.

Camber settings

Most vehicle manufacturers suggest a slight positive camber setting (about 1/4 to 1/2 deg.). Suspension wear and above-normal curb weight caused by several passengers or extra luggage tends to increase negative camber. Positive camber counteracts this.

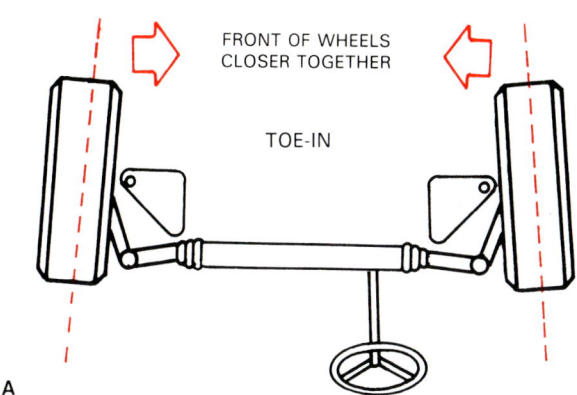

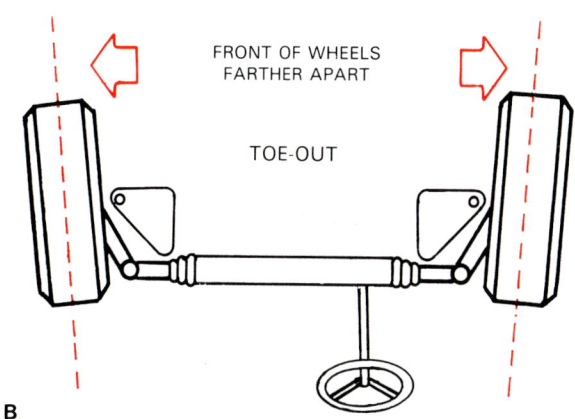

Fig. 32-7. Toe is inward or outward angle of wheels. Toe-in is produced when front of wheels are closer together than the rear. Toe-out results when front of wheels are farther apart than the rear.

Toe-out results when the front of the wheels are farther apart than the rear. See Fig. 32-7B. Toe-out causes the front of the wheels to point away from each other.

Toe settings

Rear-wheel drive cars are usually set to have TOE-IN at the front wheels. This is because the front wheels tend to toe-out while driving. Toe-in is needed to compensate for the action of tyre rolling resistance, play in the steering system, and suspension system action.

As the tyres roll over the road, they are pushed rearward. This turns the tyres outward at the front, causing toe-out. By adjusting the wheels for a slight toe-in (approximately 2 to 6 mm), the wheels and tyres roll straight ahead when driving.

Front-wheel drive cars require different adjustment for toe. Since the front wheels propel the car, they are pushed forward by engine torque. This makes the wheels toe in or point inward while driving.

To compensate for this action, front-wheel drive vehicles normally have front wheels adjusted for a slight toe-out (approximately 2 mm). Theoretically, this will give the front end a zero toe setting when the car moves down the road.

STEERING AXIS INCLINATION

Steering axis inclination is the angle formed by the inward tilt of the ball joints, king pin, or MacPherson strut tube. It is the difference in the angle between the centreline of these suspension parts and true vertical. Refer to Fig. 32-8.

Steering axis inclination is NOT a tyre wearing angle. Like caster, it aids directional stability by helping the steering wheel to return to the straight-ahead position.

Steering axis inclination is NOT adjustable. It is designed into the suspension system by the vehicle manufacturer. If the angle is NOT correct, you must usually replace parts to correct the problem.

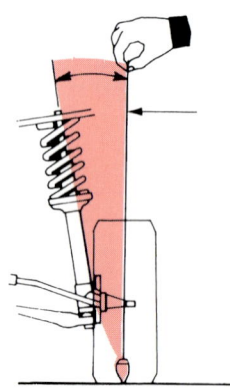

Fig. 32-8. Steering axis inclination is determined by plumb or vertical line and projected line through upper ball joint, king pin, or MacPherson strut tube.

TOE-OUT ON TURNS (TURNING RADIUS ANGLE)

Toe-out on turns, also termed turning radius angle, is the amount the front wheels toe-out when turning corners. As the car goes around a turn, the inside tyre must travel in a smaller radius circle than the outside tyre. The steering system is designed to turn the inside wheel sharper than the outside wheel.

Fig. 32-9 illustrates toe-out on turns. Note how

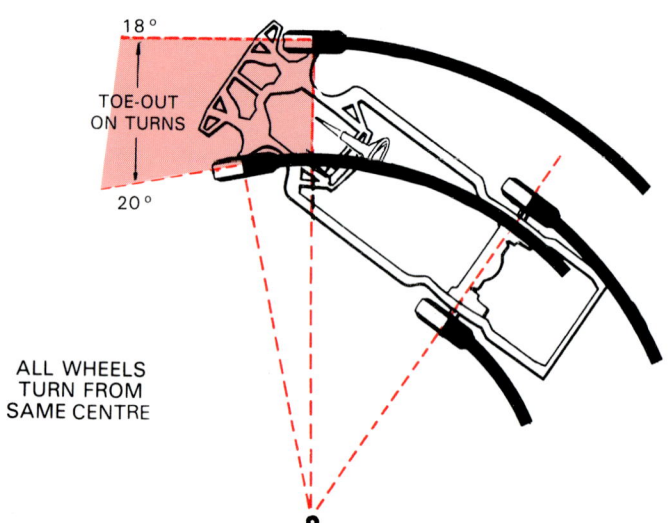

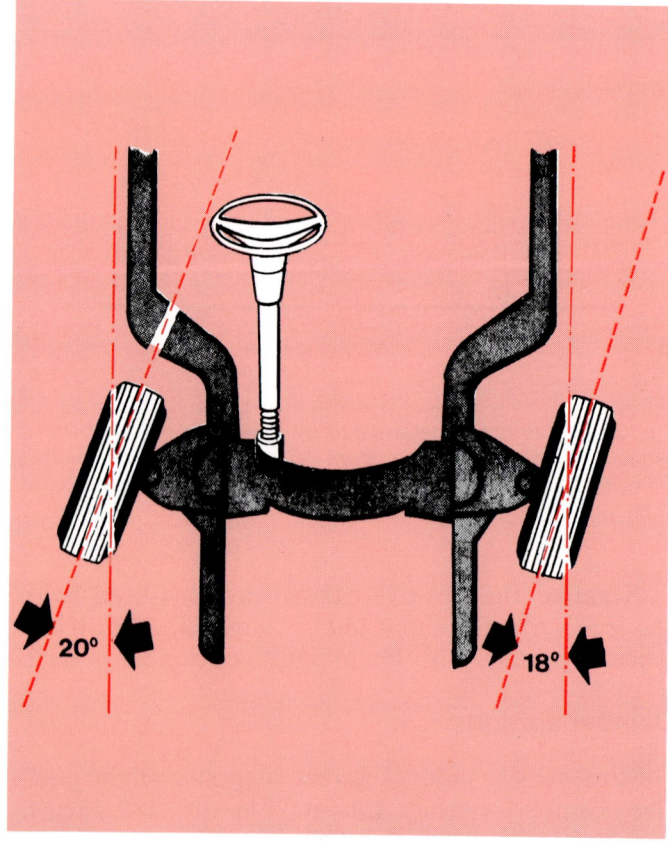

Fig. 32-9. When rounding corners, inside tyre must turn more sharply. Angles built into steering system produces proper toe-out on turns. (Hunter)

each front wheel turns a different number of degrees. This eliminates tyre scrubbing and squeal by keeping the tyres rolling in the right direction on corners.

Toe-out on turns is NOT an adjustable angle. It is controlled by the built-in angle of the steering arms. If incorrect, it indicates bent or damaged steering parts.

TRACKING

Tracking refers to the position or direction of the two front wheels in relation to the two rear wheels. With proper tracking, the rear tyres follow in the tracks of the front tyres, with the car moving straight ahead. Look at Fig. 32-10.

With improper tracking, the rear tyres do NOT follow the tracks of the front tyres. This causes the car body or frame to actually shift partially sideways when moving down the road. Poor tracking will increase tyre wear, lower fuel economy, and upset handling.

ADJUSTING WHEEL ALIGNMENT

Caster, camber, and toe are the three commonly adjustable wheel alignment angles.

Before studying wheel alignment equipment, you should have a basic understanding of how alignment angles are changed. Then you can relate this knowledge to the use of specific alignment equipment.

The basic sequence for wheel alignment is:
1. Inspect and correct tyre, steering, and suspension problems.
2. Adjust caster.
3. Adjust camber and recheck caster.
4. Adjust toe.
5. Check toe-out on turns (needed if there is damage).
6. Check caster, camber, and toe on rear wheels (if needed).
7. Check tracking (if needed).

Caster adjustment methods

Caster is adjusted by moving the control arm so that the ball joint moves to the front or rear of the car. Depending on suspension system type, a control arm can be moved by adding or removing SHIMS, adjusting the STRUT ROD, turning an ECCENTRIC BOLT, or by shifting the control arm shaft bolts in SLOTTED HOLES. See Fig. 32-11.

If the upper control arm ball joint is moved forward, it increases negative caster, Fig. 32-12. If the upper control arm is adjusted to move its ball joint rearward, it would increase positive caster. The opposite is true for the lower control arm, Fig. 32-12.

Figs. 32-11 and 32-12 show various means for caster adjustment.

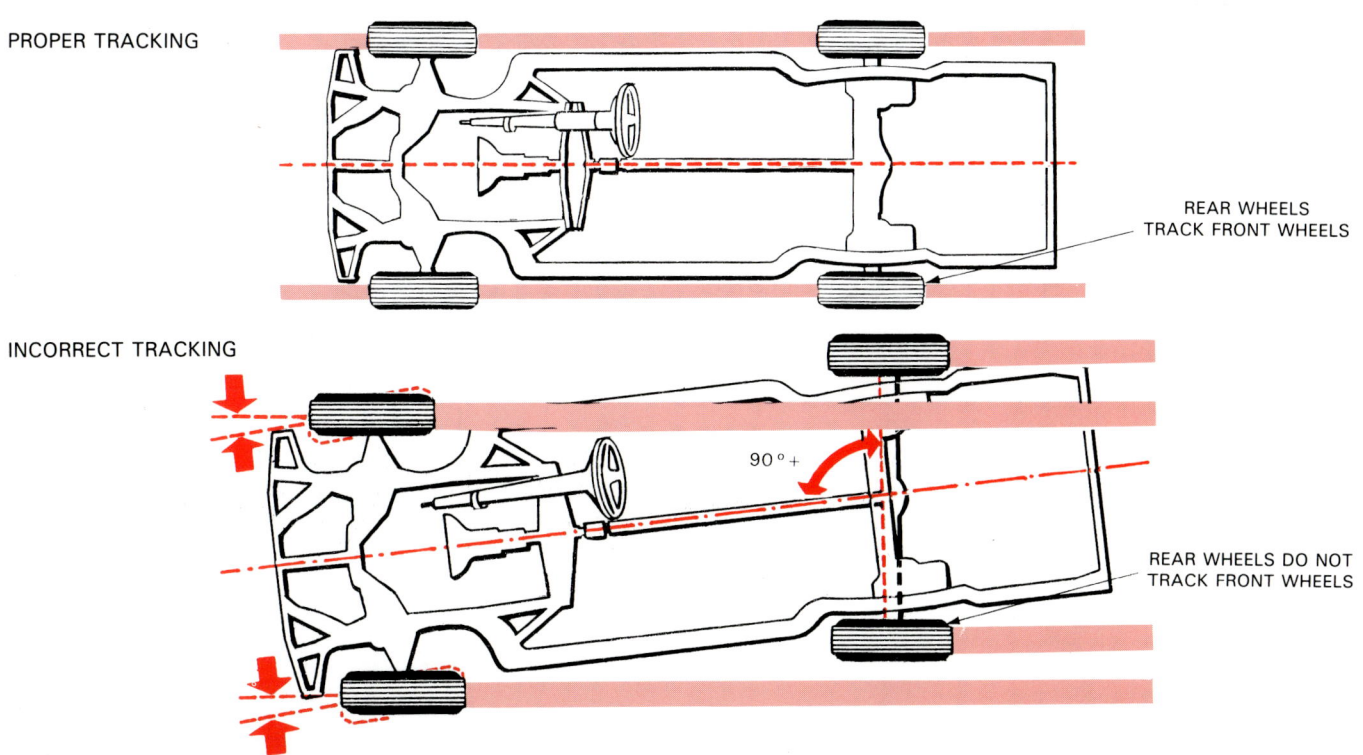

PROPER TRACKING

REAR WHEELS TRACK FRONT WHEELS

INCORRECT TRACKING

90° +

REAR WHEELS DO NOT TRACK FRONT WHEELS

Fig. 32-10. Proper tracking causes the rear wheels to follow directly behind front wheels. (Hunter)

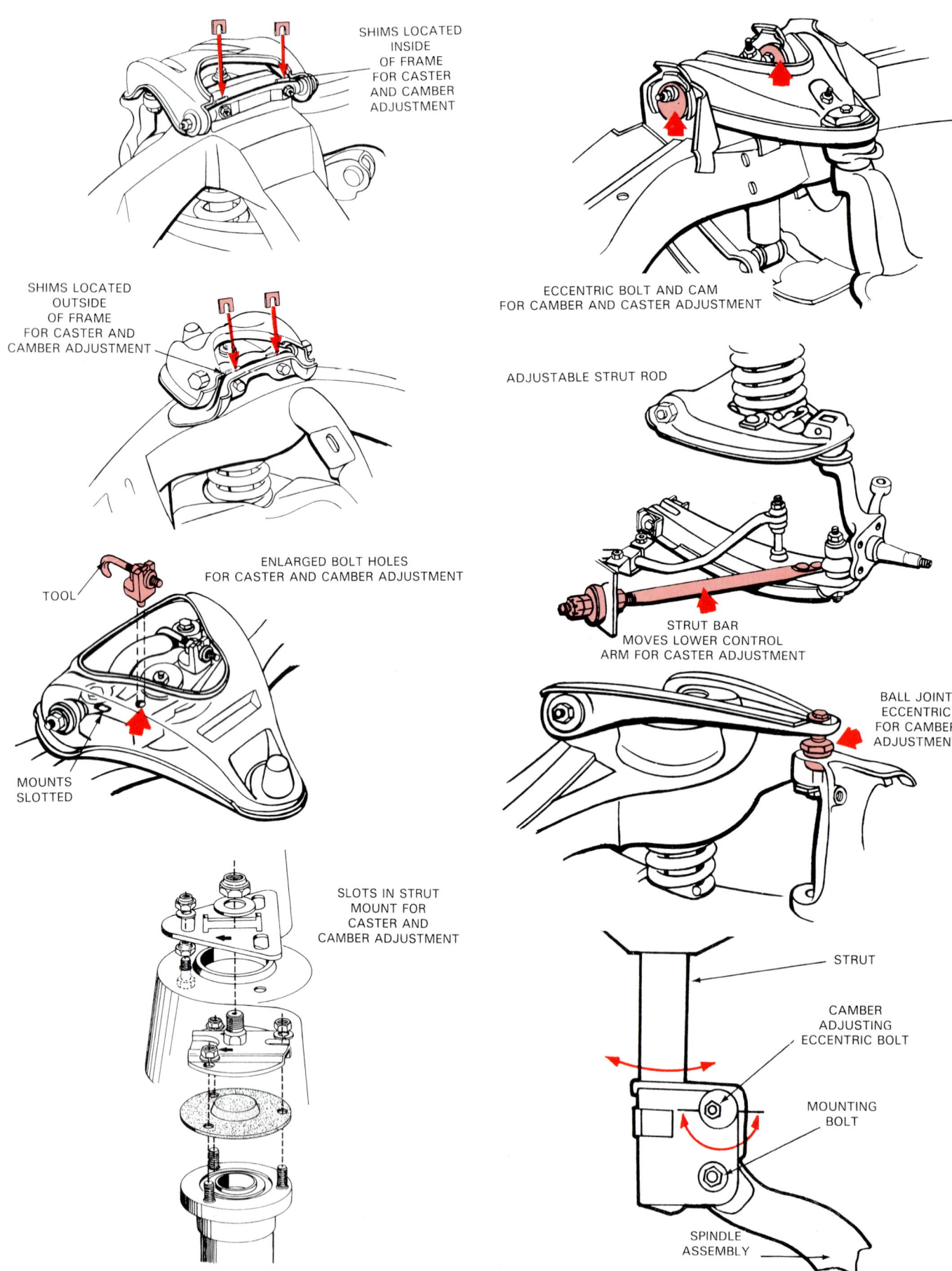

Fig. 32-11. Study various methods used to change caster and camber settings.

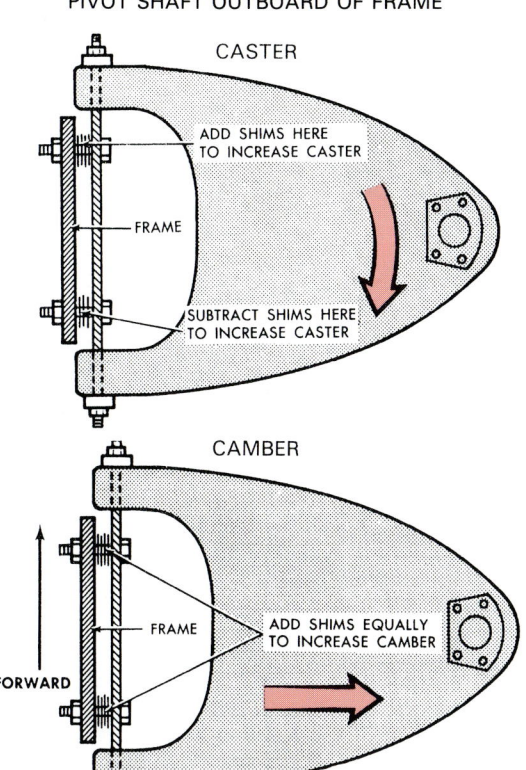

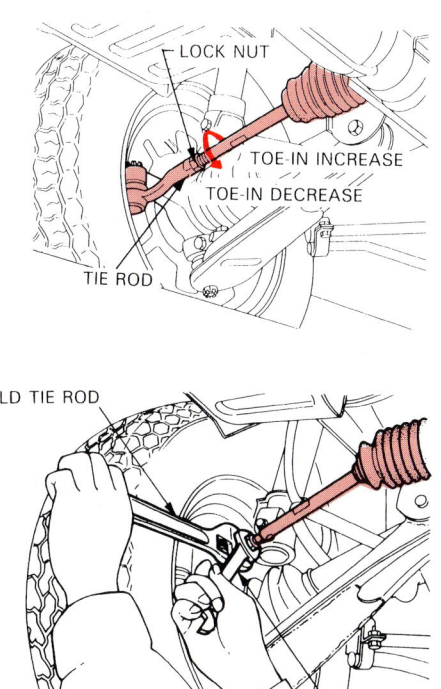

Fig. 32-12. Caster is adjusted by moving either upper or lower control arm to front or rear of car.

Fig. 32-13. Toe is adjusted by lengthening or shortening tie rods. (Subaru)

Camber adjustment methods

Camber is usually adjusted right after setting caster. Camber is changed by moving the control arm in or out so that the ball joint does NOT move forward or rearward. Again, refer to Fig. 32-12. SHIMS or SLOTS in the control arm mount and ECCENTRIC BOLTS are the most common methods for adjustment.

Some MacPherson strut suspensions do not have provisions for caster and camber adjustments. However, other strut type suspension systems have a camber adjustment at the connection between the steering knuckle and strut. See Fig. 32-11.

The top of the steering knuckle and bottom of the strut can be pivoted in or out. The upper bolt on the steering knuckle may have an eccentric that moves the knuckle when turned.

Toe adjustment

Toe is adjusted by lengthening or shortening the TIE RODS. On most rack and pinion steering systems, the tie rod is threaded into the outer ball socket, Fig. 32-13. Linkage type steering systems normally have a sleeve threaded on a two-piece tie rod, Fig. 32-14.

When the steering arms point to the rear of the car, lengthen each tie rod to increase toe-in and shorten them to increase toe out. The opposite is true when the steering knuckle arms are at the front.

Centring a steering wheel

To keep the steering wheel spokes centred shorten or lengthen each tie rod the same amount. Changing one more than the other will rotate the steering wheel spokes. When the car is travelling straight ahead, the

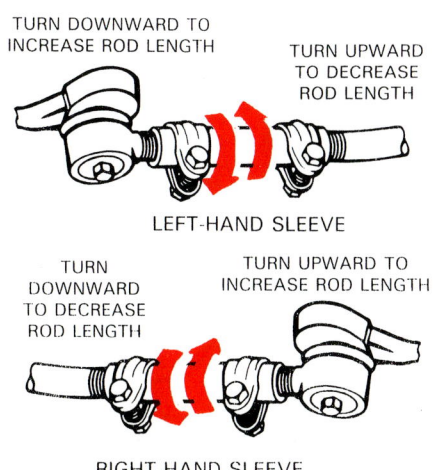

Fig. 32-14. This linkage type steering system uses a sleeve to lengthen or shorten tie rod. Note adjustment method. (Ford)

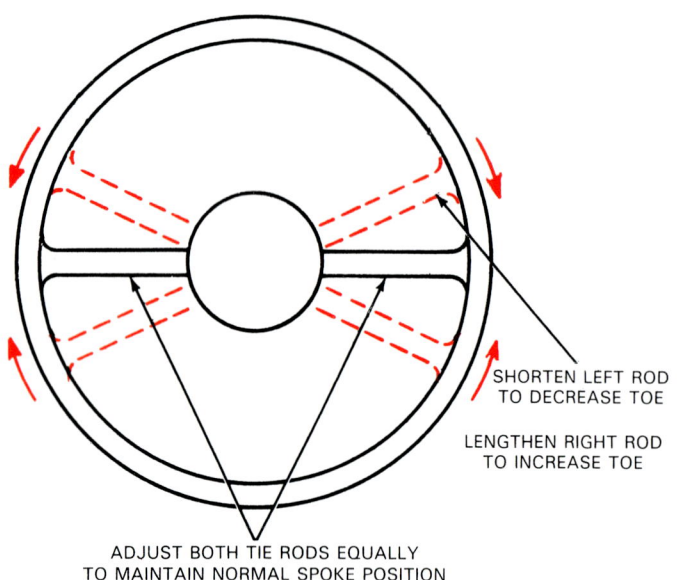

Fig. 32-15. When adjusting toe, steering wheel must be kept in centre position. Study how turning each tie rod end affects position of steering wheel spokes. (Ford)

wheel spokes should be positioned correctly. Fig. 32-15 shows a service manual illustration for centring a steering wheel.

Adjusting rear wheel alignment

Depending upon vehicle make and model, a car may or may NOT have provisions for adjusting rear wheel alignment. If the wheels fail to track properly, it may point to frame, unibody, or rear suspension damage. The car might have been in an accident that shifted the rear wheels out of place. Worn suspension system bushings can also upset tracking, Fig. 32-16. It shows how shims can be used to align the

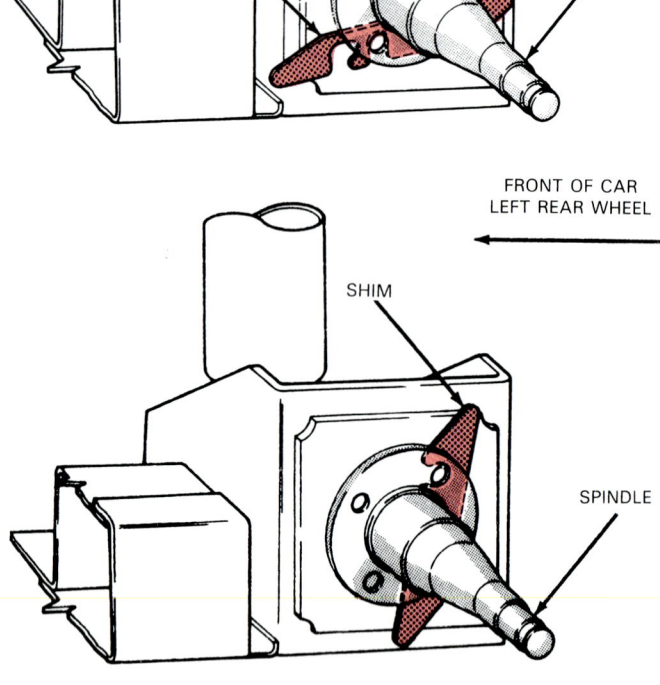

Fig. 32-16. Note how shim can be used to adjust alignment angles on rear axle of this front-wheel drive car. Shim can be placed at bottom, top, front, or rear to change any alignment angle.

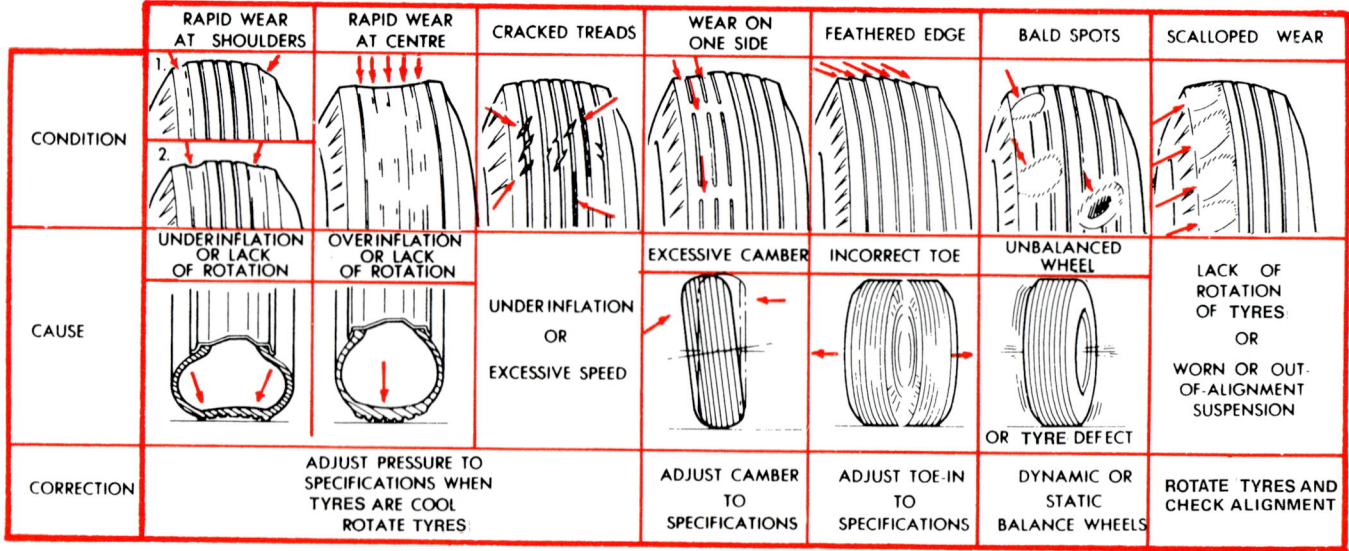

Fig. 32-17. Tyre wear patterns should be read to help determine which steering or suspension parts are worn and to help with alignment checks.

rear wheels of one type of front-wheel drive car. A shim of the correct thickness can be added to adjust camber and toe.

Other methods are sometimes used for aligning the rear wheels of a car. They normally use the principles already covered for aligning the front wheels.

PREALIGNMENT INSPECTION

Now that you understand alignment principles and basic adjustment methods, you are ready to learn how to inspect a car before measuring and setting

wheel alignment. Inspection of the tyres, wheels, suspension and steering system is essential. If any front end part is worn, you must adjust or replace that part before wheel alignment.

Reading tyre wear

Reading tyres is done by inspecting tyre tread wear and diagnosing the cause for any abnormal wear. Fig. 32-17 shows a chart for reading tyre wear patterns. Note the incorrect camber and toe show up as specific tread wear patterns.

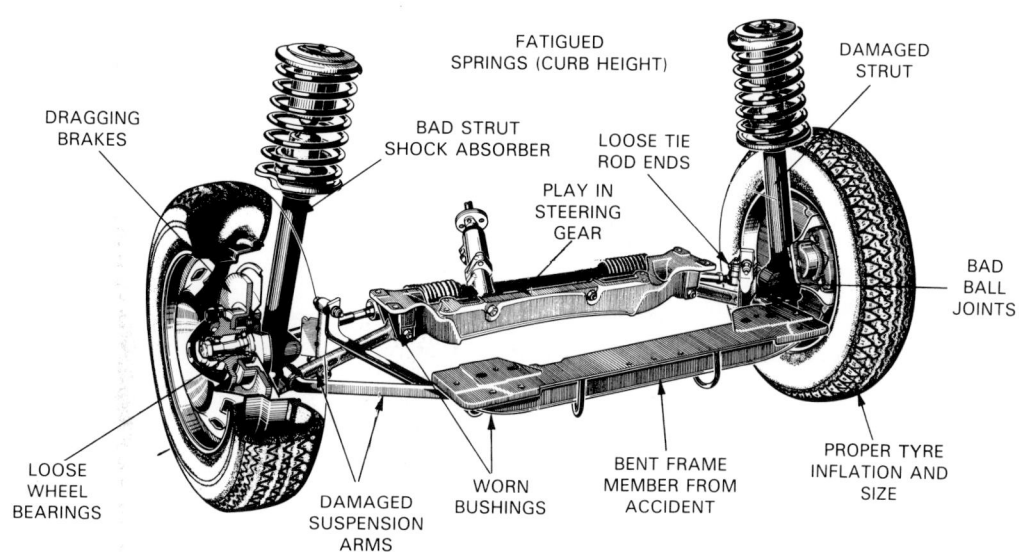

STEERING LINKAGE CHECKS			
Tie Rod End	OK: L__ R__	Comments:
Tie Rods and Inner Ball/Stud Sockets	OK: L__ R__	
Steering Arms	OK: L__ R__	
Tie Rod Adjusting Sleeves	OK: L__ R__	Comments:
Relay Rod	OK: ___	
Pitman Arm	OK: ___	
Idler Arm and Bracket	OK: ___	
Steering Shock Absorber and Bushings	OK: ___	
Steering Gear Mountings	OK: ___	

ALIGNMENT INSPECTION REPORT FORM

Name_____ Date_____, 19___
Address _____ Phone: Bus._____ Home_____
Make _____ Yr. and Model_____ License_____ Odometer_____

TYRE AND WHEEL CHECKS

			Comments:
Condition- Inspection	OK: LF__ RF__ LR__ RR__	
Pressure	OK: LF__ RF__ LR__ RR__	
Wheel Bearings- Adjustment	OK: LF__ RF__ LR__ RR__	
Roughness	OK: LF__ RF__ LR__ RR__	
Runout-Lateral	OK: LF__ RF__ LR__ RR__	
Radial	OK: LF__ RF__ LR__ RR__	
Wheel Balance	OK: LF__ RF__ LR__ RR__	
Shock Absorbers-Operational	OK: LF__ RF__ LR__ RR__	
Leakage and Bushings	OK: LF__ RF__ LR__ RR__	
Riding Height	OK: LF__ RF__ LR__ RR__	

Fig. 32-18. Study types of problems that can affect front wheel alignment. All of these must be corrected first.

Incorrect camber produces wear on one side of the tyre tread, Fig. 32-17. Too much negative camber would wear the INSIDE of the tyre tread. Too much positive camber would wear the OUTER tread only. Correct camber will wear the FULL tread area evenly.

Incorrect toe will cause a feathered edge to form on the tyre tread. A feather edge is a tyre wear pattern with one side of each tread rib worn sharp and raised and the other side worn rounded or recessed. See Fig. 32-17.

With too much toe-in the sharp feathered edge points inward. With too much toe-out, the sharp edge on the tread ribs point away from the centre of the car.

Tyre wear patterns can also indicate incorrect wheel balance, incorrect tyre inflation pressure, tyre construction defects, and tyre damage.

Common front end problems

Before adjusting alignment, always check the car for problems that could affect wheel alignment. You should check for:
1. Loose wheel bearings.
2. Wheel and tyre runout.
3. Worn tyres or tyres of different sizes and types.

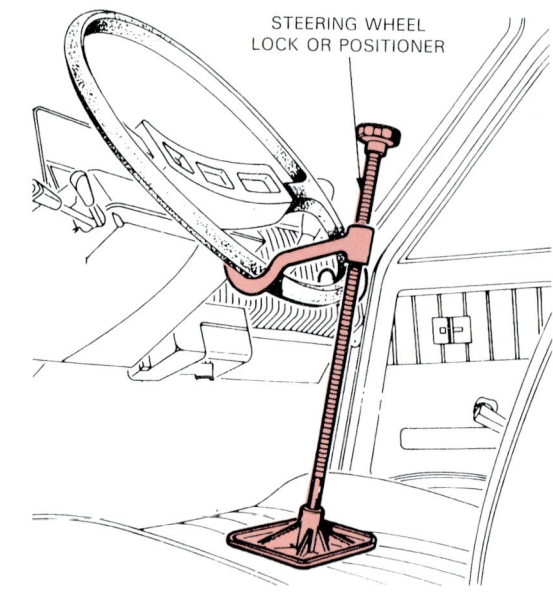

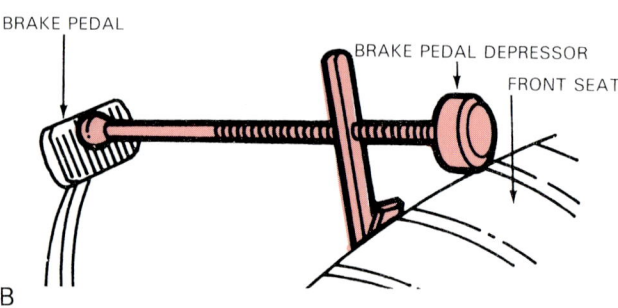

Fig. 32-20. A — Steering wheel lock will hold front wheels straight ahead. B — Brake pedal lock will keep car from rolling.

4. Incorrect tyre inflation.
5. Worn steering components.
6. Worn suspension components.
7. Incorrect kerb height and weight.

Fig. 32-18 shows several components that frequently cause problems during alignment.

WHEEL ALIGNMENT TOOLS AND EQUIPMENT

Various special tools and equipment are needed to adjust wheel alignment. Fig. 32-19 shows several special tools. Fig. 32-20 shows two other commonly used tools. Note the name and basic function of each.

The most basic equipment for wheel alignment are the turning radius gauge, caster-camber gauge, and the trammel. These are the least complicated of all alignment equipment and illustrate the fundamentals for wheel alignment easily.

Later in this chapter, the wheel alignment and geometry will be covered using the Wheel Alignment Bay facilities.

This will involve the use of dedicated measuring instruments and equipment.

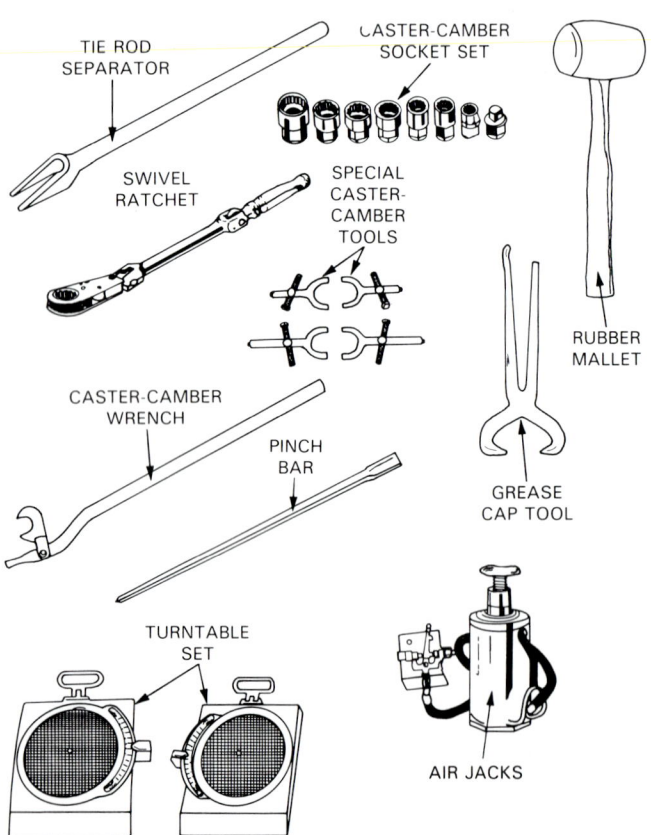

Fig. 32-19. These are common front end alignment tools. (Snap-On)

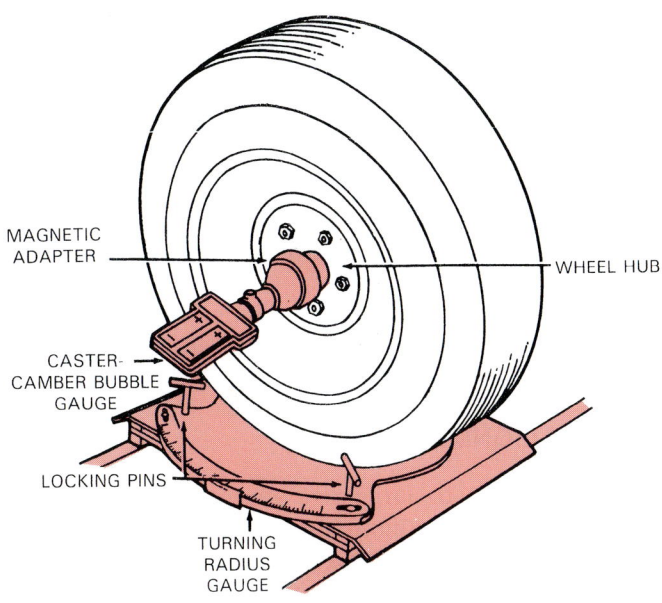

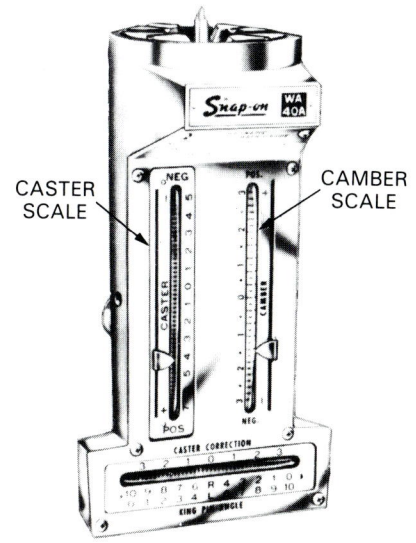

A

Fig. 32-21. Turning radius gauge will measure number of degrees wheels are turned. Also note caster-camber gauge mounted on hub.

TURNING RADIUS GAUGES

Turning radius gauges measure how many degrees the front wheels are turned right or left. Look at Fig. 32-21. They are commonly used when measuring caster, camber, and toe-out on turns.

Turning radius gauges may be portable units. However, they are quite often mounted in the Wheel Alignment Bay equipment as integral units.

The front wheels of the car are centred on the turning radius gauges. Then, when the locking pins are pulled out, the gauge and tyre turn together. The pointer on the gauge will indicate how many degrees the wheels have been turned.

Checking toe-out on turns

To check toe-out on turns, centre the front tyres of the car on the turning radius gauges. Turn one of the front wheels until the gauge reads 20 deg. Then, read the number of degrees showing on the other gauge. Check toe-out on turns on both the right and left sides. If not within specs, check for bent or damaged parts.

CASTER-CAMBER GAUGE

A caster-camber gauge is used with the turning radius gauge to measure caster and camber in degrees. The gauge either fits on the wheel hub magnetically, Fig. 32-21, or may fasten on the wheel rim, Fig. 32-22. Normally, caster and camber are adjusted together since one affects the other.

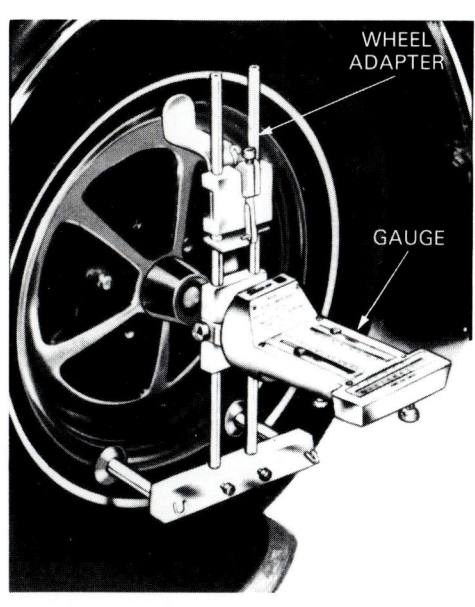

B

Fig. 32-22. Caster-camber gauge. A — Gauge has bubbles that read tilt of hub in degrees. Note scales for caster and camber. B — Gauge may mount on wheel with adapter or may have magnet that sticks to hub. (Snap-On)

Measuring caster

To measure caster with a bubble type caster-camber gauge, turn one of the front wheels inward until the radius gauge reads 20 deg. Turn the adjustment knob on the caster-camber gauge until the bubble is centred on zero. Then, turn the wheel out 20 deg.

The degree marking next to the bubble will equal the caster of that front wheel. Compare your reading to specifications and adjust as needed. Repeat this operation on the other side of the car.

Measuring camber

To measure camber with a bubble type caster-camber gauge, turn the front wheels straight ahead (radius gauge on zero). The car must be on a perfectly level surface (wheel alignment bay floor).

Read the number of degrees next to the bubble on the camber scale of the gauge. It will show camber for that wheel. If outside specifications, adjust camber.

If shims are used, add or remove the same amount of shims from the front and rear of the control arm. This will keep the caster set correctly. Double-check caster, especially when an excessive amount of camber adjustment is needed.

TRAMMEL

A trammel is used to compare the distance between the front and rear of a car's tyres for toe adjustment. Look at Fig. 32-23.

A trammel is a metal rod or shaft with two pointers. The pointers slide on the gauge so that they can be set to the distance between the tyres. The trammel gauge will indicate toe-out or toe-in in millimeters.

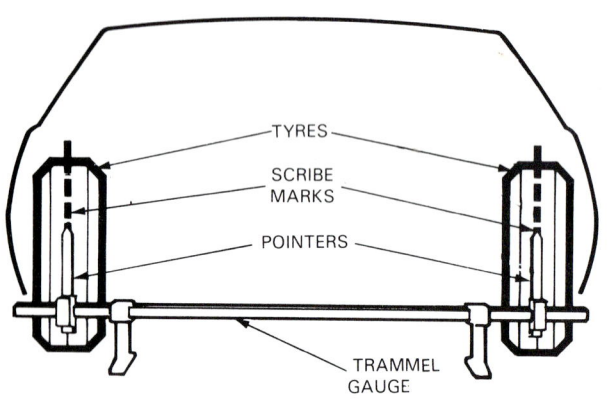

Fig. 32-23. Trammel provides simple method of adjusting toe. Lines are scribed on chalked tyre tread. Then trammel is used to measure distance between lines at front and rear of tyres. Difference equals toe.

Measuring toe

To measure toe with a trammel, raise the wheels and rub a chalk line all the way around the centre rib on each tyre. Then, using a scribing tool, rotate each tyre and scribe a fine line on the chalk line. This will give you a very thin reference line for measuring the distance between the tyres. Lower the car back on the radius gauges.

First, position the trammel gauge at the back of the tyres. Move the pointers until they line up with the lines you scribed on the tyres. Then, without bumping the gauge, position the gauge at the front of the tyres.

The difference in the distance between the lines on the front and rear of the tyres shows toe.

For example, if the lines on the front of the tyres are closer together than on the rear, the wheels are toed-in. If the lines are the same distance apart at the front and rear, toe is zero.

Using service manual instructions, adjust the tie rods until the trammel reads within specifications.

WHEEL ALIGNMENT BAY

Most medium to large size garages have a wheel alignment bay. The alignment bay consists of ramps, turning radius gauges, and one of several kinds of equipment for measuring alignment angles. Refer to Fig. 32-24. The ramps are adjusted perfectly level so that all equipment readings are accurate.

Fig. 32-24. Wheel Alignment Bay contains all equipment needed to set alignment angles. Note turning radius gauges, ramps, and other equipment on wall board.

To use the alignment equipment bay, the car is driven up on the ramps. The front tyres must be carefully centred on the turning radius gauges, Fig. 32-25. Once the vehicle is positioned, the rear wheels are blocked to keep the car from accidentally rolling off the turning radius gauges.

CAUTION! Use extreme care when positioning a car in the alignment bay. Ask a friend to guide you up the ramps and onto the turning radius gauges. Block the rear wheels!

Since there are so many types of alignment equipment designs, always follow the operating

Fig. 32-25. This modern alignment machine is electronic and reads caster and camber on meter face.

instructions provided by the manufacturer. Remember that alignment principles are the same. Apply your knowledge of wheel alignment to the specific type of equipment.

Fig. 32-25 pictures a computerised alignment machine. Instead of a bubble type caster-camber gauge, it uses electronic sensing devices to measure the tilt of the steering knuckle and wheel. Note how the instrument mounts on the wheel.

Look at Fig. 32-26. It shows modern alignment equipment being used to measure toe. A light beam shines across from the front wheels. When the light beam shines directly into the sensor on the opposite

wheel, toe is zero. This is an easy-to-use type machine since you can adjust toe without crawling out from under the car.

Fig. 32-27 shows another type of alignment machine. It projects a light beam (line) onto a large scale mounted next to the wall at the end of the rack. Caster, camber, toe, and other alignment angles are read off the wall mounted scale.

A computerised wheel alignment machine is being used to measure tracking in Fig. 32-28. Note how

Fig. 32-27. Mechanic is using alignment equipment to reflect light up to scale on board for toe adjustment. (Hunter)

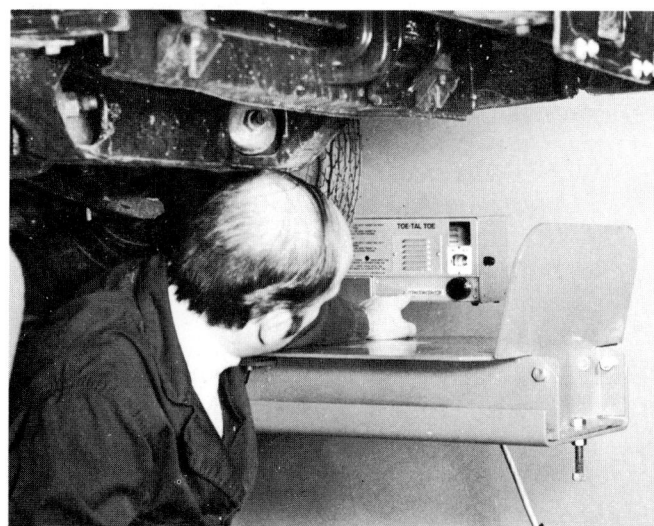

Fig. 32-26. This alignment machine uses light beams shining between wheels to set toe.

Fig. 32-28. State of the art electronic alignment equipment is being used to check track of front and rear wheels. (Hunter)

both the alignment of the front and rear wheels is being compared. A light beam projects back to mirrors mounted on the rear wheels. With proper tracking, the light beam will reflect off the mirrors and back into the sensor on the front wheels.

WARNING! Always remember to use the operating manual provided with the specific alignment equipment. Procedures vary and the slightest mistake could upset proper wheel alignment.

KNOW THESE TERMS

Wheel alignment, Caster, Camber, Toe, Steering axis inclination, Toe-out on turns, Tracking, Reading tyres, Incorrect camber, Incorrect toe, Feathered edge, Turning radius gauge, Caster-camber bubble gauge, Trammel, Wheel alignment bay.

REVIEW QUESTIONS

1. Define the term "alignment".
2. What is the main purpose of wheel alignment?
3. _____ is the forward or rearward tilt of the steering knuckle or steering support when viewed from the side of the car.
4. List the three basic functions of caster.
5. Explain the difference between positive and negative caster.
6. Which of the following pertains to caster?
   a. Measured in millimetres.
   b. A directional control angle.
   c. Can be used to offset road crown.
   d. All of the above are correct.
   e. All of the above are incorrect.
7. _____ is the inward or outward tilt of the wheel and tyre assembly when viewed from the front of the car.
8. List three functions of camber.
9. Explain the difference between positive and negative camber.
10. Some vehicle manufacturer's suggest a slight negative camber setting. True or False?
11. _____ is determined by the distance between the front and rear of the left and right-hand wheels.
12. Explain the difference between toe-in and toe-out.
13. Rear-wheel drive cars commonly use toe-in and front-wheel drive cars commonly use toe-out. True or False?
14. Define the term "tracking".
15. List the six basic steps for wheel alignment.
16. How do you change caster?
17. How do you change camber?
18. How do you adjust toe?
19. _____ _____ is done by inspecting tread wear and diagnosing the cause of abnormal wear.
20. Name seven possible problems you should check before attempting to align a car's front end.
21. Turning radius gauges measure how many degrees the front wheels are turned right or left. True or False?
22. Summarise the use of a caster-camber gauge.
23. A _____ is used to compare the distance between the front and rear of a car's tyres for toe adjustment.
24. A customer complains of excess tyre wear. When checked, the tyres show a feather edge wear pattern. Mechanic A says that the toe is improperly adjusted.
    Mechanic B says that a thorough front end alignment is needed.
    Who is correct?
    a. Mechanic A.
    b. Mechanic B.
    c. Both Mechanic A and B.
    d. Neither Mechanic A nor B.
25. What is a wheel alignment bay?

Chapter 33

TYRE, WHEEL, HUB, WHEEL BEARING FUNDAMENTALS

After studying this chapter, you will be able to:
☐ Identify the parts of a tyre and wheel.
☐ Describe different methods of tyre construction.
☐ Explain tyre and wheel sizes.
☐ Describe tyre ratings.
☐ Identify the parts of driving and non-driving hub and wheel bearing assemblies.

This chapter introduces the various tyre designs used on modern cars. It explains how tyres and wheels are constructed to give safe and dependable service. The chapter also covers hub and wheel bearing construction for both rear-wheel and front-wheel drive vehicles.

TYRES

Motor car tyres perform two basic functions: they act as a soft CUSHION between the road and the metal wheel; tyres must also provide adequate TRACTION (friction) with the road surface.

Tyres must transmit driving, braking, and cornering forces to the road in good weather, when raining, and in snow. At the same time, they should resist punctures and wear.

Parts of a tyre

Although there are several tyre designs, the basic parts of a tyre are the same. See Fig. 33-1. Refer to this illustration as each part is introduced.

1. TYRE BEADS (two rings made of steel wires encased in rubber that hold tyre sidewalls snugly against wheel rim).
2. BODY PLIES (rubberised fabric and cords wrapped around beads; they form the carcass or body of the tyre).

3. TREAD (outer surface of tyre that contacts road).
4. SIDEWALL (outer part of tyre extending from bead to tread; it contains information about tyre).
5. BELTS (sometimes used to strengthen plies and stiffen tread; they lie between tread and inner plies).
6. LINER (thin layer of rubber bonded to inside of plies; provides leakproof membrane for modern tubeless tyre).

Pneumatic tyres

Car tyres are pneumatic which means they are

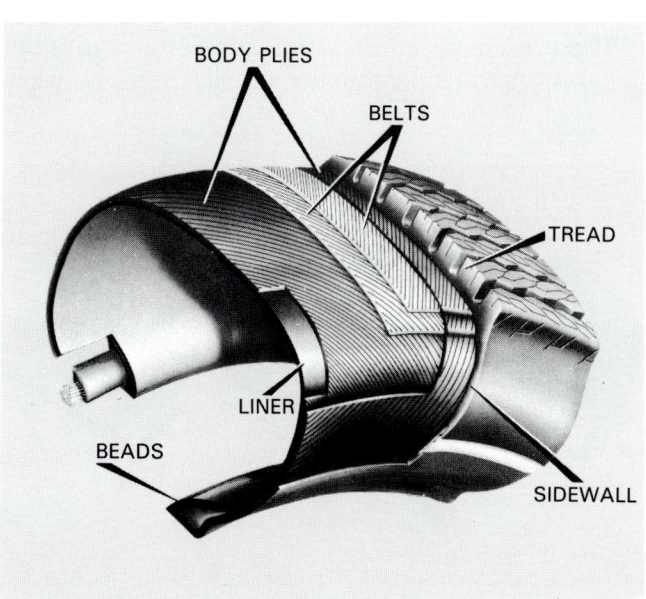

Fig. 33-1. Study basic parts of a tyre. (Uniroyal)

503

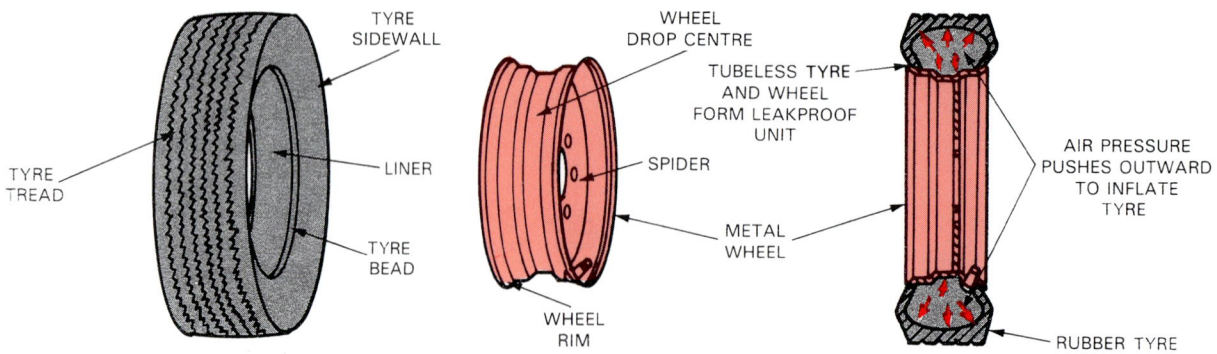

Fig. 33-2. Tyre fits over wheel. With tubeless tyre, tyre and wheel form leakproof unit. Air pressure pushes outward on inside of tyre for inflation.

filled with air. Pictured in Fig. 33-2, internal air pressure pushes out on the inside of the tyre to support the weight of the car.

Tubeless tyres

Today's cars use tubeless tyres that do NOT have a separate inner tube. The tyre and wheel form an airtight unit.

Older cars used inner tubes (soft, thin, leakproof rubber liner) that fit inside the tyre and wheel assemblies.

Rolling resistance

Tyre rolling resistance is a measurement of the amount of friction produced as the tyre operates on the road surface. A high rolling resistance would increase fuel consumption and wear. Typically, tyre rolling resistance is reduced by higher inflation pressure, proper tyre design, and a lighter car.

TYRE CONSTRUCTION

There are many construction and design variations in tyres. A different number of plies may be used.

The plies run at different angles. Also, different materials may be utilised.

Three types of tyres are found on late model motor cars: bias ply tyre, belted bias tyre, and radial tyre.

Bias ply tyre

A bias ply tyre has plies running at an angle from bead to bead. See Fig. 33-3A. The cord angle is also reversed from ply to ply. The tread is bonded directly to the top ply.

A bias ply tyre is one of the oldest designs and it does NOT use belts. The position of the cords in a bias ply tyre allows the body of the tyre to flex easily. This tends to improve cushioning action. A bias ply tyre provides a very smooth ride on rough roads.

One disadvantage is that the weakness of the plies and tread reduce traction at high speeds and increase rolling resistance.

Belted bias tyre

A belted bias tyre is a bias tyre with belts added to increase tread stiffness. Look at Fig. 33-3B. The plies and belts normally run at different angles. The belts

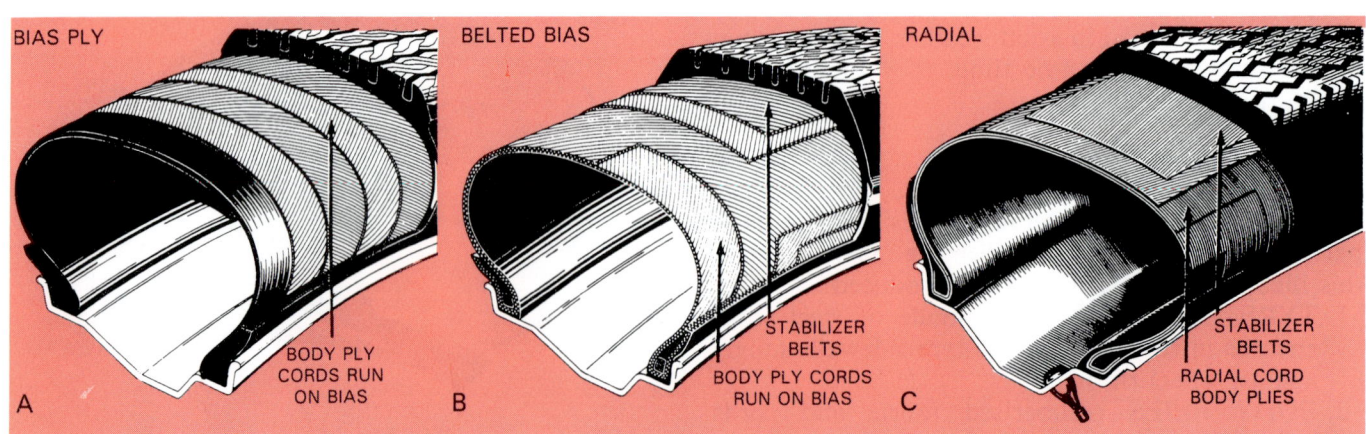

Fig. 33-3. Three tyre types: bias ply, belted bias, and radial. (Firestone)

do NOT run around to the sidewalls but only lie under the tread area.

Usually, two stabiliser belts and two or more plies are used to increase tyre performance.

A belted bias tyre provides a smooth ride, good traction, and offers some improvement in rolling resistance over a bias ply tyre.

Radial ply tyre

A radial ply tyre has plies running straight across from bead to bead with stabiliser belts directly beneath the tread. This is illustrated in Fig. 33-3C.

A radial tyre has a very flexible sidewall, but a stiff tread. The belts can be made of steel, flexten, fibreglass, or other materials.

Radial tyres have a very stable footprint (shape and amount of tread touching road surface). This improves safety, cornering, braking, and wear.

One possible disadvantage of a radial tyre is that it may produce a harder ride at low speeds. The stiff tread area does NOT give or flex as much on rough roads.

TYRE MARKINGS

Tyre markings on the sidewalls of a tyre give information about tyre size, load carrying ability, inflation pressure, number of plies, identification numbers, and manufacturer. It is important that you understand these tyre markings. Refer to Fig. 33-4.

DOT NUMBER
PLY RATING
STYLE
NAME OF TYRE
RADIAL
MANUFACTURER'S NAME
SIZE
LOAD AND INFLATION
TYPE
BRAND

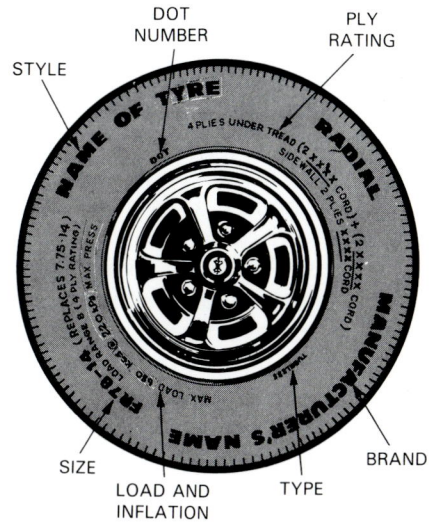

Fig. 33-4. Sidewall will give information about tyre. Study what is given on this tyre.

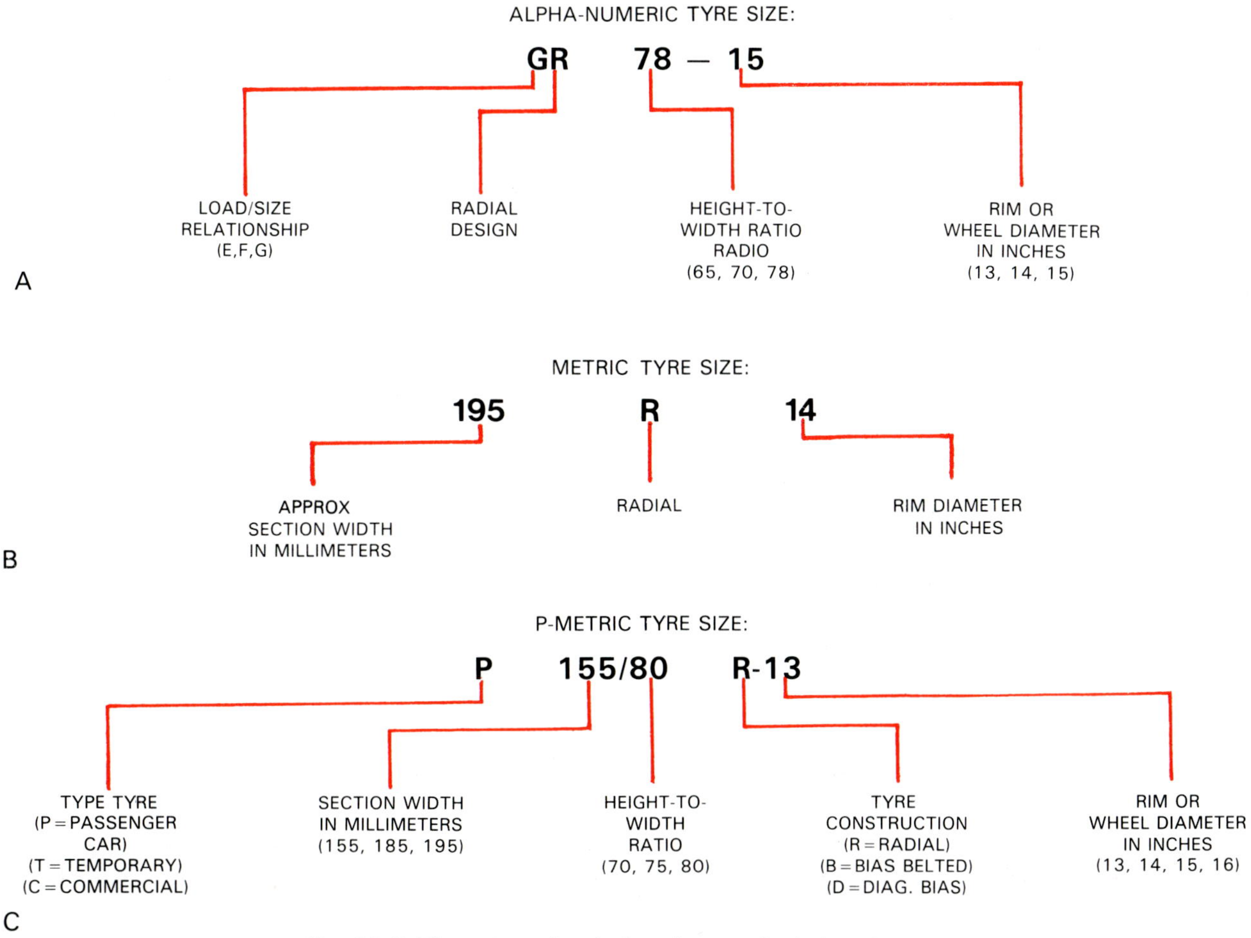

ALPHA-NUMERIC TYRE SIZE:
GR 78 — 15
LOAD/SIZE RELATIONSHIP (E,F,G)
RADIAL DESIGN
HEIGHT-TO-WIDTH RATIO RADIO (65, 70, 78)
RIM OR WHEEL DIAMETER IN INCHES (13, 14, 15)
A
METRIC TYRE SIZE:
195 R 14
APPROX SECTION WIDTH IN MILLIMETERS
RADIAL
RIM DIAMETER IN INCHES
B
P-METRIC TYRE SIZE:
P 155/80 R-13
TYPE TYRE (P = PASSENGER CAR) (T = TEMPORARY) (C = COMMERCIAL)
SECTION WIDTH IN MILLIMETERS (155, 185, 195)
HEIGHT-TO-WIDTH RATIO (70, 75, 80)
TYRE CONSTRUCTION (R = RADIAL) (B = BIAS BELTED) (D = DIAG. BIAS)
RIM OR WHEEL DIAMETER IN INCHES (13, 14, 15, 16)
C

Fig. 33-5. Three tyre size designation numbering systems.

Tyre size

Tyre size is given on the sidewall as a letter-number sequence. There are three common size designations: alpha-numeric (conventional measuring system), metric (the older metric measuring system) and P-metric (the new international standard metric system).

The alpha-numeric (alphabetical-numerical) tyre size rating system uses letters and numbers to denote tyre size in inches and load-carrying capacity in pounds. An example is given in Fig. 33-5A.

The first letter, G, indicates load and size relationship. The higher the letter, the larger the size and load-carrying ability. G is smaller than H, for example. An R in the designation means radial. The first number, 78, is the height-to-width ratio. The last number, 15, is the rim diameter in inches.

The metric system tyre uses a three digit number (195 for instance) to indicate the approximate section width in millimetres and '14' for the rim diameter in inches. An R in the designation means radial. See Fig. 33-5B.

Another variation of metric designation includes the aspect ratio, ie. 195/70HR14.

The P-metric tyre is the newest tyre identification system using metric values and international standards. Look at Fig. 33-5C.

The letter P indicates passenger car tyre. The first number, 155, gives section width in millimetres. The second number, 80, is the height-to-width ratio. The R indicates radial (B means bias belted, D means bias ply). The last number, 13, shows the rim diameter in inches (not metric values).

Fig. 33-6 gives the points of measure for a tyre. Study each carefully.

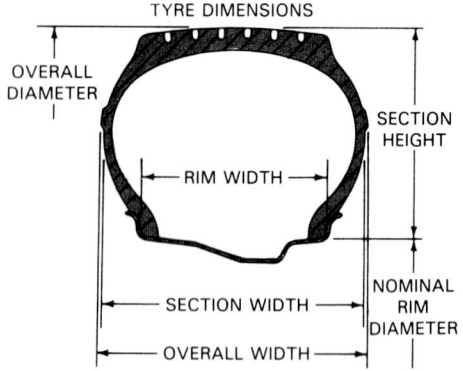

Fig. 33-6. Points of measurement on a tyre. These dimensions are important when ordering new tyres or wheels. (BF Goodrich)

Aspect ratio

The aspect ratio or height-to-width ratio in the tyre size designation is the most difficult value to understand. Fig. 33-7 illustrates different aspect ratios.

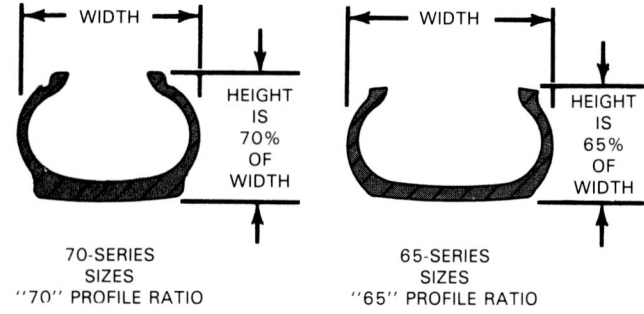

Fig. 33-7. Aspect ratio indicates height and width ratio of tyre. Note how 65 series is wider and shorter than 70 series.

Note that, as the number becomes smaller, the tyre becomes more squat (wider and shorter). The aspect ratio is the comparison of the tyre's height (bead-to-tread) and width (sidewall-to-sidewall).

A 70-series tyre, for example, has a profile ratio of 70; the height of the tyre is 70 percent of the width. A 60 series tyre would be "short" and "fat". A 78 tyre would be "narrower" and "taller".

Maximum load rating

The maximum load rating of a tyre indicates the amount of WEIGHT the tyre can carry at recommended inflation pressure. The maximum load value, 690 kg (1520 lbs), for example, is printed on the sidewall.

Most alpha numeric size tyres are load range B. They are restricted to the load specified at 220 kPa (32 psi). Where a greater load carrying ability is needed, load range C or D tyres are used.

Maximum inflation pressure

The maximum inflation pressure, printed on the tyre sidewall, is the highest air pressure that should be installed in the tyre. Many tyres have a maximum recommended inflation pressure of 220 kPa (32 psi). However, higher load range tyres can hold higher pressures and carry more weight.

Tread plies

The tyre sidewall also states the number of plies and ply rating. The tyre may be a 2-ply, 2-ply with 4-ply rating (plies are made stronger than normal), or 4-ply for example. A greater number of plies or higher ply rating generally increases load-carrying ability.

DOT number

DOT stands for DEPARTMENT OF TRANSPORT. When you see "DOT" on the tyre sidewall, the tyre has passed prescribed safety tests.

Following the letters DOT is the dot number that identifies the particular tyre (manufacturer, plant location, type tyre construction, and date of manufacture). The DOT number is stamped into the tyre sidewall.

Speed Category Symbol

The speed category symbol which is normally a letter in the tyre size marking indicates the maximum speed for which the tyre is rated by the manufacturer.

Speed categories symbols have now been brought into line with international standards and symbols now have the following meanings:

Symbol		Max. Speed
L	=	120
M	=	130
N	=	140
P	=	150
Q	=	160
R	=	170
S	=	180
T	=	190
U	=	200
H	=	210
V	=	210 and over

Note: Speed Categories on tyres are based strictly on tyres being inflated to correct pressures, correctly fitted and in sound condition.

SPECIAL TYRES AND TYRE FEATURES

You should be familiar with several types of special tyres and tyre features.

Wear bars

Wear bars are used to indicate a critical amount of tread wear. Look at Fig. 33-8. When too much tread has worn off, solid bars of rubber will show up across the tread. This tells the customer and mechanic that tyre replacement is needed.

Compact spare tyre

Some imported new cars use a compact spare tyre to save space in the boot. One is shown in Fig. 33-9. Some types of compact tyres are very small in diameter. Others are NOT inflated when in storage. A small bottle of compressed air may be used to inflate the tyre when needed.

WARNING! Most compact spare tyres are designed only for TEMPORARY USE. Refer to Manufacturer's specifications on inflation pressure, maximum driving speed, and distance it can be driven.

Fig. 33-8. When wear bars show up, tyre is worn enough to be unsafe. It should be renewed. (Goodyear)

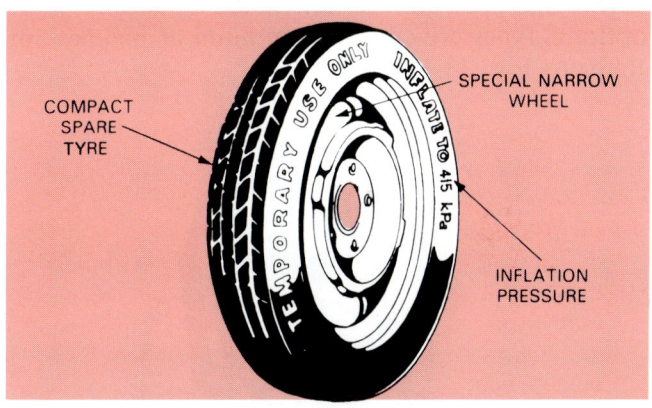

Fig. 33-9. Compact spare tyre is for temporary use only. This one should be inflated to 415 kPa. It requires a special narrow wheel.

Self-sealing tyres

Some tyres are self-sealing (seal small punctures) because of a coating of sealing compound applied to the tyre liner. Refer to Fig. 33-10. If a nail punctures the tyre, air pressure pushes the soft compound into the hole to stop air leakage.

Retreads

Retreads, or remoulds, are used tyres that have had a new tread vulcanised (applied using heat and pressure) to the old carcass or body.

A car tyre which has been retreaded or remoulded by a recognised retreading organisation is regarded as being safe for normal motoring.

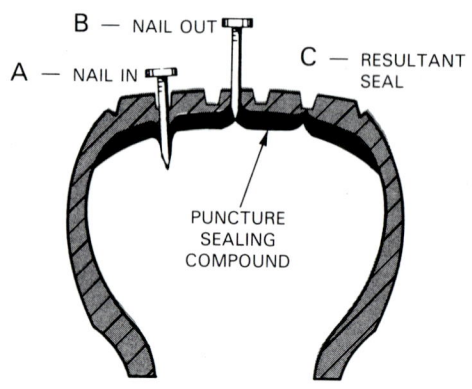

Fig. 33-10. Puncture sealing tyre action. A — Nail punctures hole in tyre. B — Nail is pulled out. C — Sealing compound flows in and clogs hole in tyre to prevent flat.

Retreaded or remoulded tyres are NOT recommended for sustained high speed motoring. Maximum advisable speeds for quality retreads is 120 km/h for diagonal ply and 135 km/h for radial ply.

WHEELS

Wheels must be designed to support the tyre while withstanding loads from acceleration, braking, and cornering. Most wheels are made of steel. A few optional types are cast in aluminium or magnesium. Refer to Fig. 33-11.

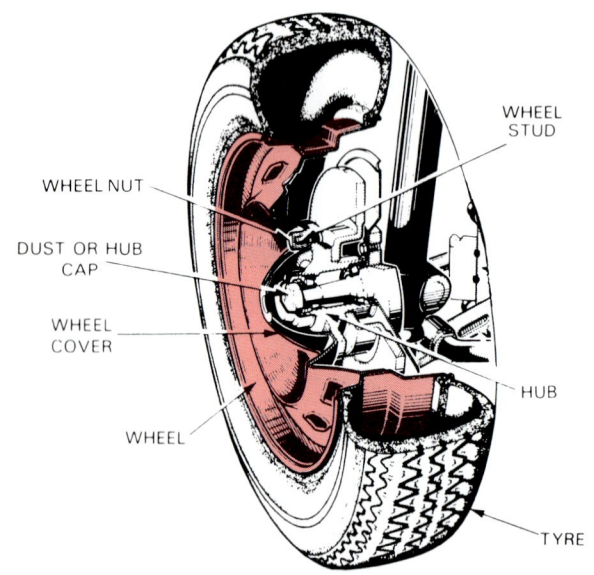

Fig. 33-11. Cutaway view shows many components relating to wheel and tyre assembly. (Peugeot)

"Mag wheels" or "mags" is a nickname for aluminium or magnesium wheels. They do not need wheel covers as do conventional steel wheels. One is pictured in Fig. 33-12.

A drop centre wheel is commonly used on passenger cars because it allows for easy installation

Fig. 33-12. Aluminium or magnesium wheel is often called "mag". It does not need a wheel cover.

and removal of the tyre. See Fig. 33-13. Since the centre of the wheel is smaller in diameter (dropped), the tyre bead can fall into the recess. Then, the other side of the tyre bead can be forced over the rim for removal.

A standard wheel consists of the rim (outer lip that contacts tyre bead) and the spider (centre section that bolts to car hub). Normally, the spider is welded to the rim.

Refer to Fig. 33-13. Fig. 33-14 illustrates the various dimensions of a wheel. Compare the two illustrations closely.

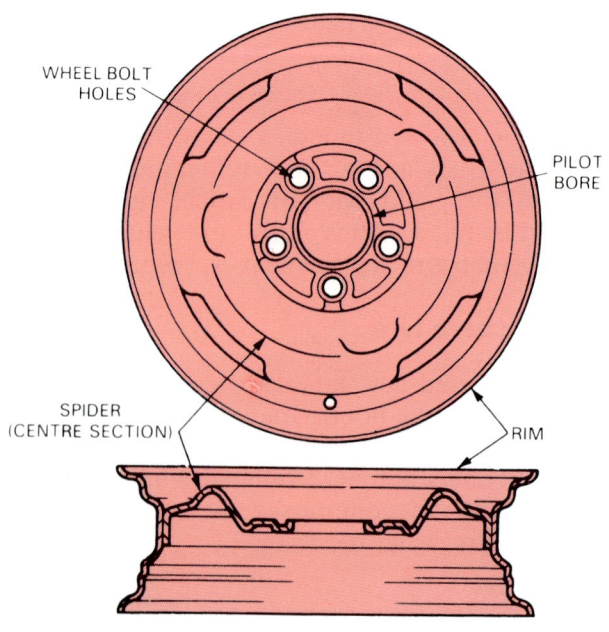

Fig. 33-13. Two views show parts of a conventional drop centre wheel. Study part names.

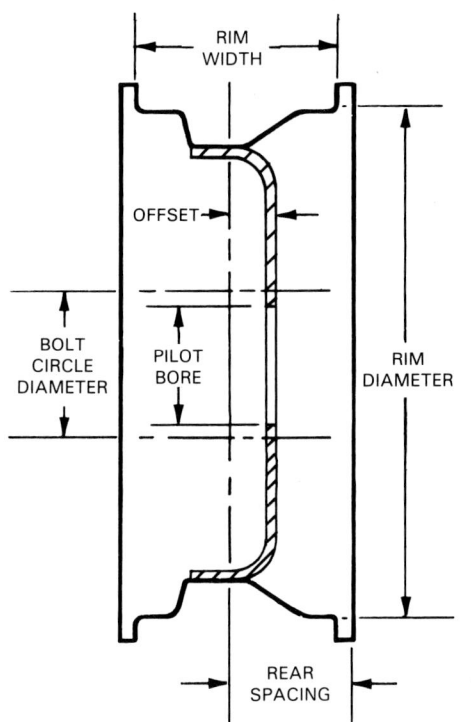

Fig. 33-14. Study basic points of measurement for a wheel. (Goodyear)

Safety rim

A safety rim has small ridges that hold the tyre beads on the wheel during a tyre blow-out (instant rupture and air loss) or flat (slow leak reduces inflation pressure). Look at Fig. 33-15. Small raised lips around the rim keep the tyre beads from sliding into the drop centre section. This improves safety by keeping the tyre from coming off the wheel.

VALVE STEM AND CORE

A valve stem snaps into a hole in the wheel rim of the tubeless tyre assembly to allow inflation and deflation, Fig. 33-15.

The stem is made of rubber. A threaded metal end is formed in the end of the stem.

The valve core is an air valve inside the valve stem. It is a spring-loaded valve, Fig. 33-15.

The valve core allows air to be added to inflate the tyre. However, when the tyre inflator (tool for filling tyre with air) is removed, the spring pushes the valve closed to prevent air leakage.

To remove air from the tyre, push the centre of the valve core inward. The valve will open and air will blow out of the tyre for deflation.

A valve cap screws over the valve stem to keep dirt and moisture out of the valve stem, Fig. 33-15.

WHEEL NUTS, STUDS, AND BOLTS

Wheel nuts hold the wheel and tyre assembly on the car. They screw onto special studs. The inner face of the wheel nut is tapered to help centre the wheel. Refer to Figs. 33-11 and 33-12.

Wheel studs are the special studs that accept the wheel nuts. The studs are pressed through the back of the hub or axle flange. See Fig. 33-16.

Normally, the wheel nuts and studs have right-hand threads (turn clockwise to tighten). When left-hand threads are used, the nut or stud will be marked with an "L". Metric threads will be identified with an "M" or the word "Metric".

A few cars use wheel bolts instead of wheel nuts. The bolts screw into threaded holes in the hub or axle flange.

Wheel weights

Wheel weights are small lead weights attached to the wheel rim to balance the wheel-tyre assembly and prevent vibration. The weights are used to balance out a heavy area of the wheel and tyre.

WHEEL BEARING AND HUB ASSEMBLY

Wheel bearings allow the wheel and tyre to turn freely around the spindle, in the steering knuckle, or

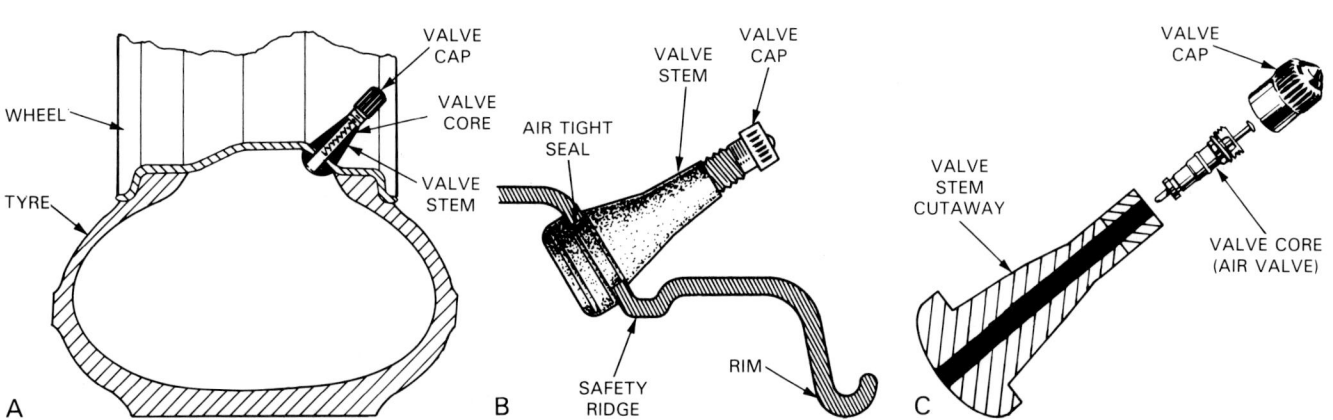

Fig. 33-15. A — Valve stem snaps into hole in wheel. B — Press fit between stem and wheel forms airtight seal. C — Valve core is air valve screwed into valve stem. Valve cap screws over end of stem. (Toyota)

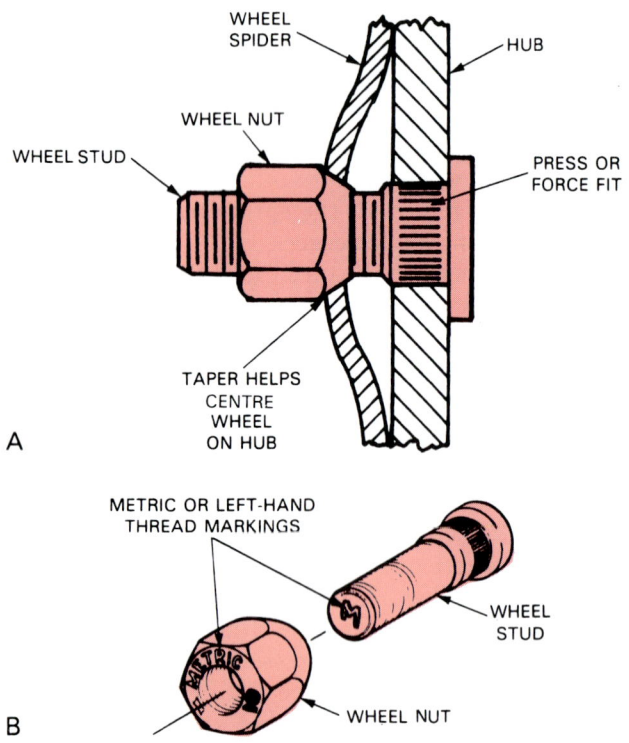

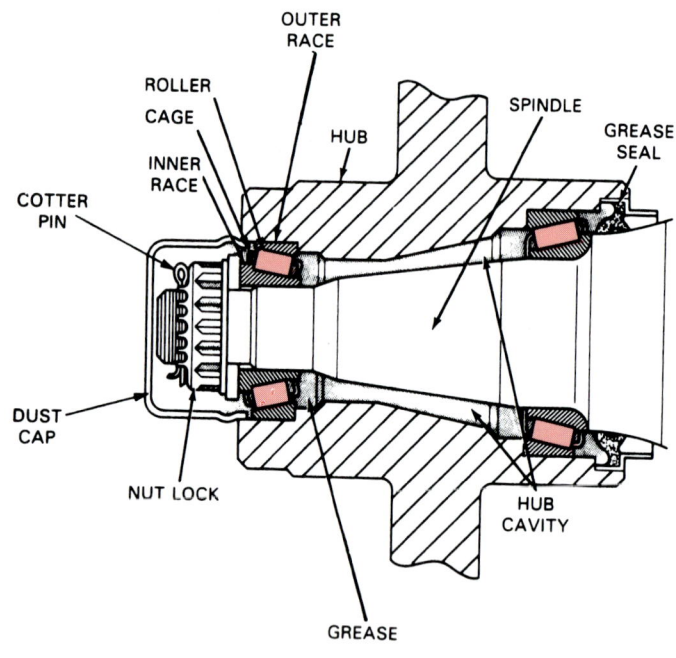

Fig. 33-16. A — Wheel nut screws onto wheel stud to secure wheel to hub. Tapered end of nut must contact and centre the wheel. Stud presses into hub. B — If metric or if threads are left-hand, markings will normally be given on nut or stud.

Fig. 33-18. Typical freewheeling or non-driving wheel bearing assembly for front or rear of car. Two tapered roller bearings allow hub and wheel to revolve around stationary spindle. Grease partially fills hub to lubricate bearings. Inner seal prevents loss of grease. Nut on end of spindle allows adjustment of bearing pre-load.

in the bearing support. Most wheel bearings are either tapered roller, or ball bearing types, Fig. 33-17.

The wheel bearings are lubricated with heavy, high-temperature grease. This allows the elements (rollers or balls) operate with very little friction and wear.

The basic parts of a wheel bearing are:
1. OUTER RACE (cup or cone pressed into hub, steering knuckle, or bearing support).
2. BALLS OR ROLLERS (antifriction elements that fit between inner and outer races).
3. INNER RACE (cup or cone that rests on spindle or drive axle shaft).

There are two basic wheel bearing and hub

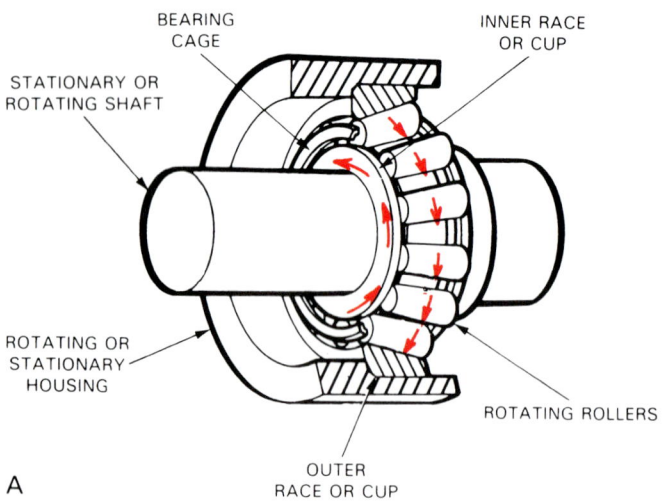

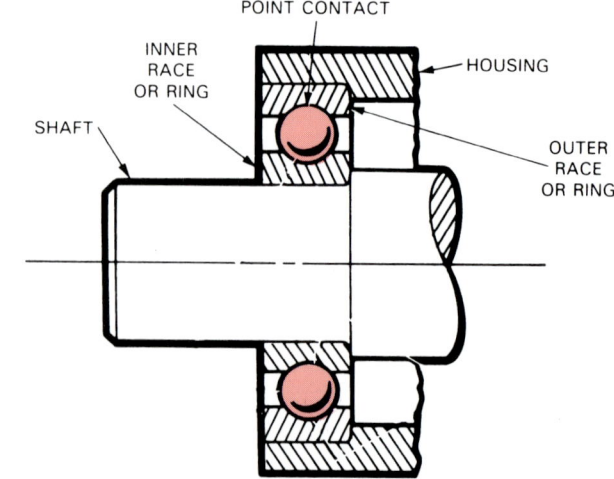

Fig. 33-17. Two basic wheel bearing configurations. A — Tapered roller bearing has cylindrical shaped rollers operating between the inner and outer races. If shaft is stationary, bearing will allow outer housing or hub to turn. If outer bearing mount is stationary, shaft or axle can turn bearing. B — Ball bearings are also used as wheel bearings, especially on front-wheel drive, front bearings or driving hub and bearing assemblies. Balls allow parts to rotate with a minimum amount of friction and wear.

designs: those for the non-driving wheels and those for driving wheels. For example, the front wheels on a rear-wheel drive car would be non-driving. However, the front wheels on a front-wheel drive car are the driving wheels (hubs transfer power to wheels and tyres).

Wheel bearing and hub assembly (non-driving wheels)

A wheel bearing and hub assembly for a car's NON-DRIVING WHEELS would, typically, include the following, Fig. 33-18.

1. SPINDLE (stationary shaft extending outwards from the steering knuckle or suspension system).
2. WHEEL BEARINGS (usually tapered roller bearings mounted on spindle and in wheel hub).
3. HUB (outer housing that holds brake disc or drum, front wheel, grease, and wheel bearings).
4. GREASE SEAL (seal that prevents loss of lubricant from inner end of spindle and hub).
5. SAFETY WASHER (flat washer that keeps outer wheel bearing from rubbing on and possibly turning adjusting nut).
6. SPINDLE ADJUSTING NUT (nut threaded on end of spindle for adjusting wheel bearing).
7. NUT LOCK (thin, slotted nut that fits over main spindle nut).
8. COTTER PIN (soft metal pin that fits through hole in spindle, adjusting nut, and nut lock to keep adjusting nut from turning in service).
9. DUST CAP (metal cap that fits over outer end of hub to keep grease in and road dirt out of bearings).

Since this wheel bearing and hub assembly does NOT transfer driving power, the spindle is stationary. It simply extends outward and provides a mounting place for the wheel bearings, hub, and wheel. With the car moving, the wheel and hub spin on the bearings and spindle. The hub simply freewheels. Note in Fig. 33-18 how the hub is partially filled with grease to lubricate the bearings.

Fig. 33-19 shows a disassembled view of a non-

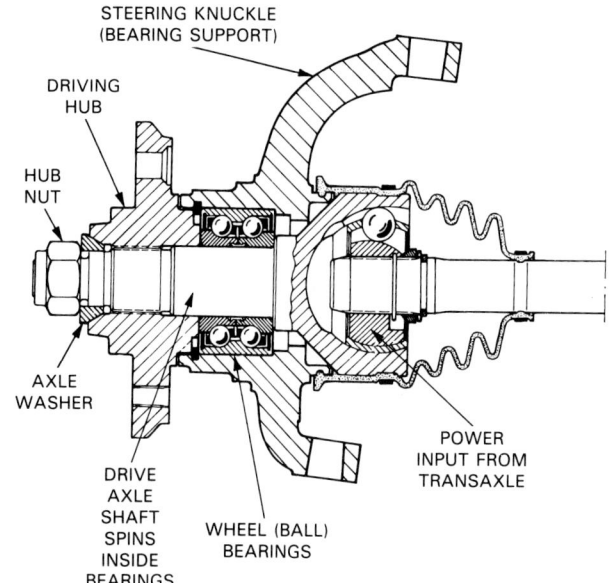

Fig. 33-20. Driving hub and wheel bearing assembly has bearings mounted in stationary steering knuckle or bearing support. Drive axle shaft fits through centre of bearings. Hub is splined to axle shaft. Ball bearings are lubricated by thick, high temperature grease.

driving front bearing and hub assembly. Compare these parts to the ones in Fig. 33-18.

Wheel bearing and hub assembly (driving wheels)

Fig. 33-20 illustrates the basic parts of a typical driving hub and wheel bearing assembly:

1. OUTER DRIVE AXLE (stub axle shaft extending through bearings and splined to hub).
2. BALL OR ROLLER BEARINGS (antifriction elements that allow drive axle to turn in steering knuckle or bearing support).
3. STEERING KNUCKLE or BEARING SUPPORT (steering or suspension component that holds wheel bearings, axle stub shaft, and hub).
4. DRIVE HUB (mounting place for wheel, transfers driving power from stub axle to wheel).

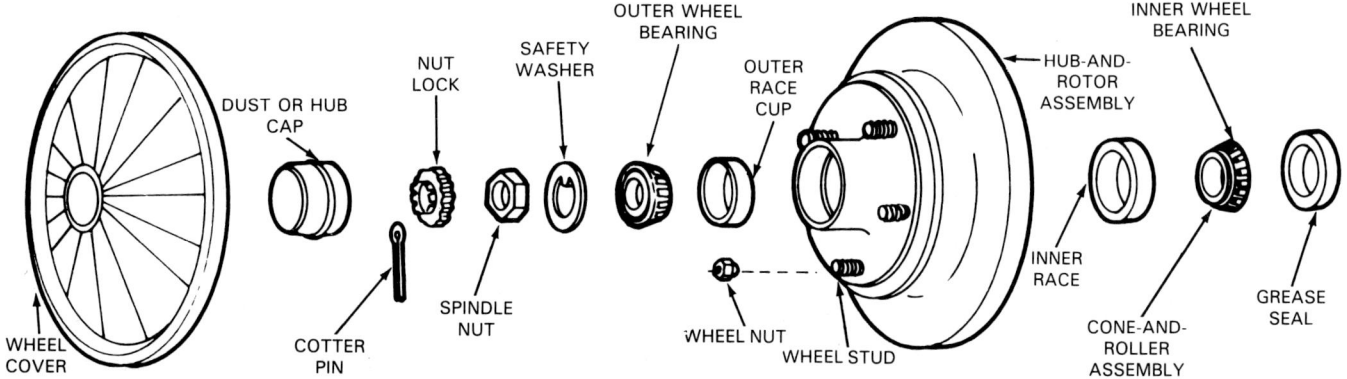

Fig. 33-19. Disassembled view of non-driving wheel bearing and hub assembly. Note names and relationship between parts. This type assembly can be used on front of rear-wheel drive car or rear of front-wheel drive car.

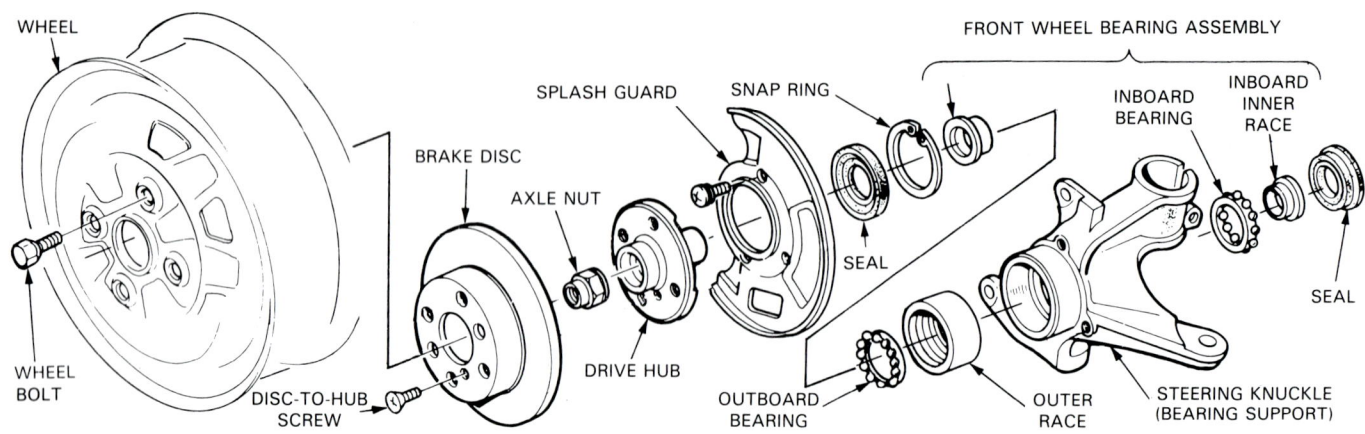

Fig. 33-21. Dismantled view of driving hub and wheel bearing assembly. Study names of parts. This type assembly is commonly used on front of front-wheel drive car. However, it can also be found on rear engine, rear-wheel drive sports cars. (Honda)

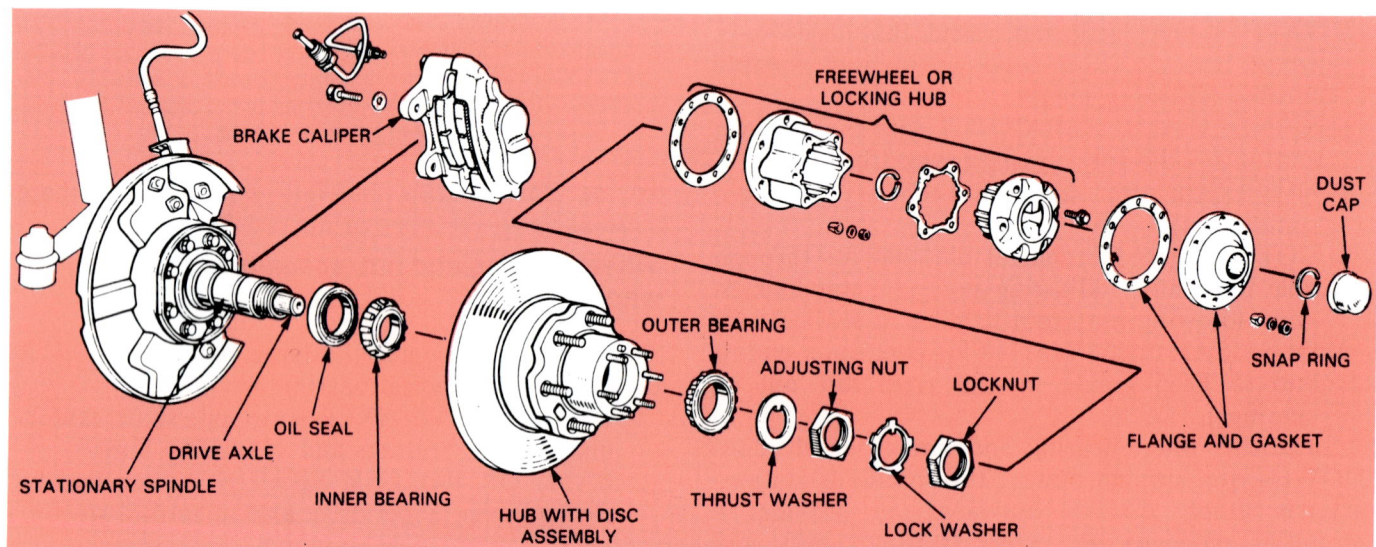

Fig. 33-22. Front hub and wheel bearing assembly for a four-wheel drive vehicle. Compare this unit to ones shown earlier. Note how drive axle sticks through stationary spindle. Freewheel or adjustable hub allows drive axle to be connected and disconnected from hub and wheel assembly for two and four-wheel drive. (Toyota)

5. AXLE WASHER (special washer that fits between hub and locknut).
6. HUB or AXLE LOCKNUT (nut that screws on end of drive axle stub shaft to secure hub and other parts of assembly).
7. GREASE SEAL (prevents lubricant loss between inside of axle and knuckle or bearing support).

As you can see in Fig. 33-20, a wheel bearing and hub assembly for a driving axle is very different from a non-driving unit. Instead of a stationary spindle, the axle shaft spins inside a stationary support.

Fig. 33-21 shows a dismantled view of a driving hub and bearing assembly. Compare them to Fig. 33-20.

Other hub and wheel bearing assemblies

A four-wheel drive hub and wheel bearing assembly is given in Fig. 33-22. Note that it has a driving axle extending through a stationary spindle. A special freewheel or locking hub transfers power from the axle to the hub-disc assembly.

Fig. 33-23 shows a rear wheel bearing assembly for a front-wheel drive vehicle. It is almost identical to a front wheel bearing for a rear-wheel drive car.

Modern cars use a wide variation of hub and wheel bearing assemblies. This is due to the increased use of front-wheel drive cars and imported cars. When you want more information on a specific car, always

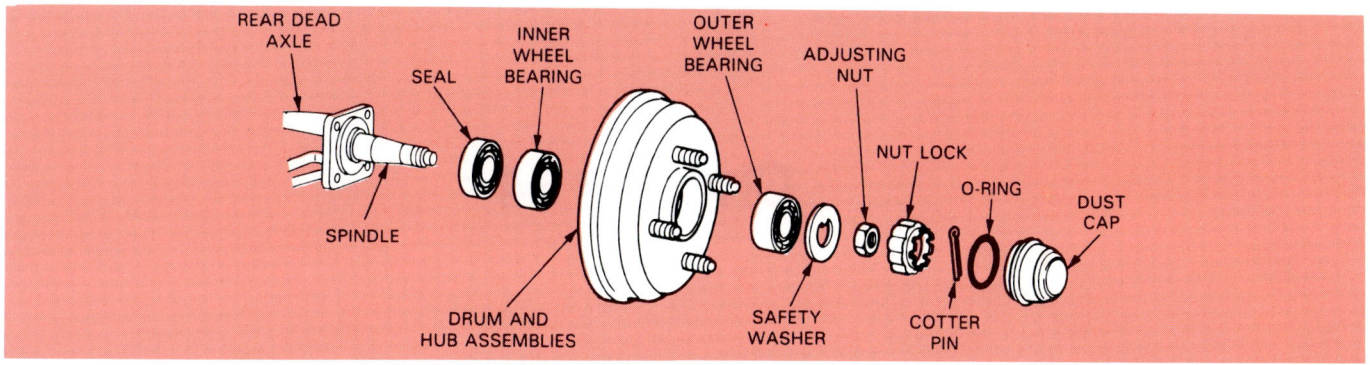

Fig. 33-23. This is a drum, hub, and wheel bearing assembly for a rear dead axle. Dead axle is rear axle on front-wheel drive car. (Subaru)

refer to the vehicle service manual. It will explain and illustrate the hub and wheel bearing assembly clearly. Most designs, however, will be similar.

KNOW THESE TERMS

Tyre bead, Tyre ply, Tread, Sidewall, Belts, Liner, Pneumatic, Tubeless, Rolling resistance, Bias ply tyre, Belted bias tyre, Radial tyre, Tyre markings, Aspect ratio, Load rating, Inflation pressure, DOT number, Wear bar, Compact spare, Self-sealing tyre, Retreads, Safety rim, Valve stem, Valve core, Wheel nut, Wheel stud, Wheel weight, Wheel bearing, Driving hub, Non-driving hub, Spindle, Hub, Grease seal, Safety washer, Spindle adjusting nut, Nut lock, Cotter pin, Dust cap.

REVIEW QUESTIONS

1. What are the two basic functions of a tyre?
2. List and explain the six major parts of a tyre.
3. Car tyres are _____ which means that they are filled with air.
4. Tyre _____ _____ is a measurement of the amount of friction produced as the tyre operates on the road surface.
5. This is NOT a type of tyre commonly used on modern passenger cars.
   a. Radial.
   b. Bias ply.
   c. Lateral ply.
   d. Belted bias.
6. What information is commonly given on the tyre sidewall.
7. A typical tyre inflation pressure would be 150 kPa. True or False?
8. How does a self-sealing tyre work?
9. A _____ _____ has small ridges that hold the tyre on the wheel during a "blow-out".
10. Explain why a valve core is needed.
11. _____ _____ are attached to the rim to balance the wheel-tyre assembly and prevent vibration.
12. Name and describe the basic parts of a wheel bearing.
13. List and explain the nine basic parts of a non-driving hub assembly.
14. List and explain the seven basic parts of a driving hub assembly.
15. A driving hub and a non-driving hub are almost identical. True or False?

Chapter 34

BRAKE SYSTEM FUNDAMENTALS

After studying this chapter, you will be able to:

☐ Explain the hydraulic and mechanical principles of a brake system.
☐ Identify the major parts of an automotive brake system.
☐ Define the basic function of the major parts of a brake system.
☐ Compare drum and disc type brakes.
☐ Describe the operation of a parking brake.
☐ Explain the operation of power brakes.
☐ Summarise the operation of anti-skid systems.

Automotive brakes provide a means of using friction to either slow down, stop, or hold the wheels of the car. When a car is moving down the highway, it has a tremendous amount of stored energy in the form of inertia (tendency to keep moving). To stop the car, the brakes convert kinetic (moving) energy into heat.

As you will learn, a modern motor car uses numerous devices to improve braking ability. For example, dual hydraulic brake systems, hydraulic valves to equalise braking pressure, and computer controlled anti-skid systems are found on today's cars.

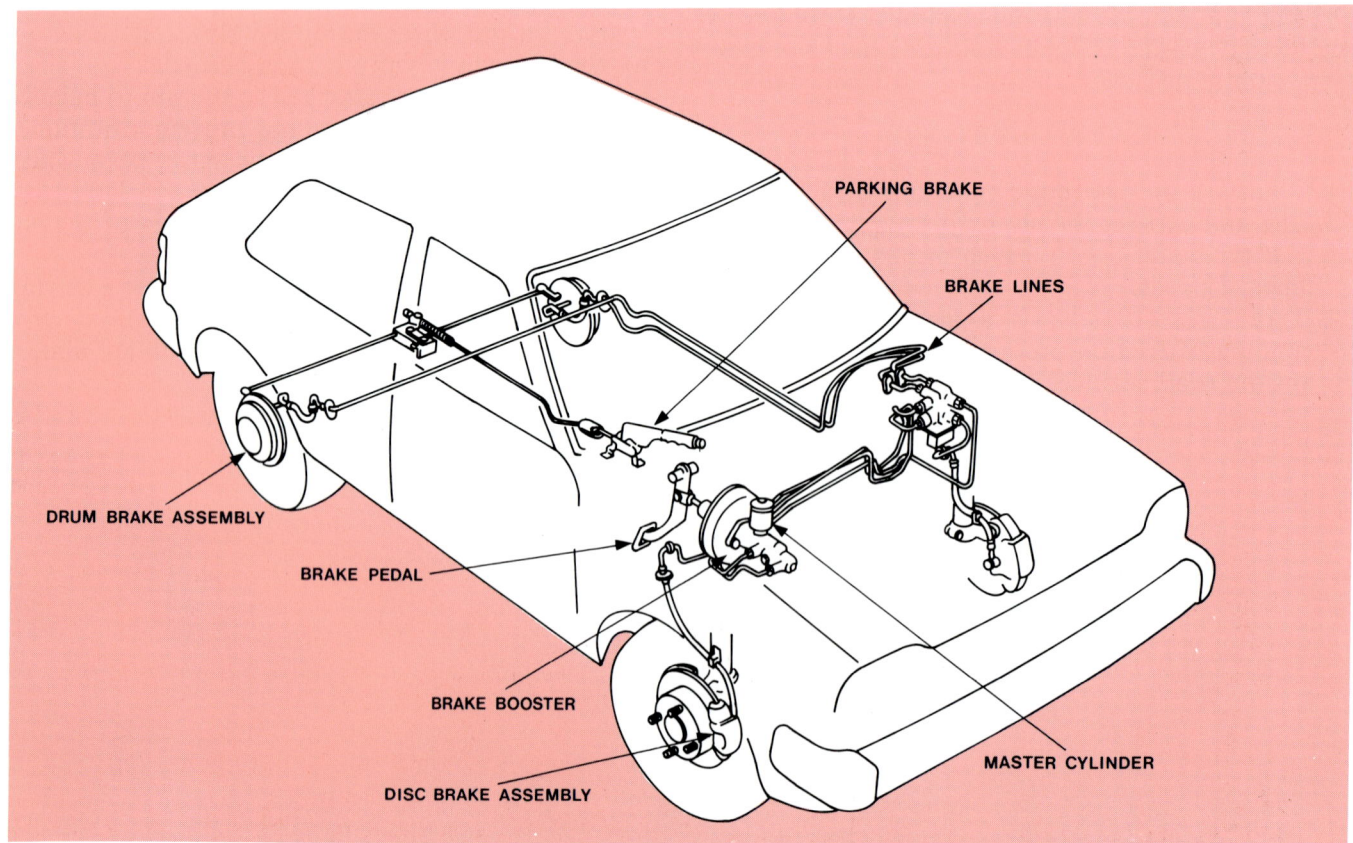

Fig. 34-1. These are the basic parts of an automotive brake system. Study them! (Honda)

BASIC BRAKE SYSTEM

Before detailing the construction and operation of each part, you should have a basic understanding of a brake system. Look at Fig. 34-1. It shows the major components of a simple system. Study the location of the parts as they are introduced:

1. BRAKE PEDAL ASSEMBLY (foot lever for operating master cylinder and power booster).
2. MASTER CYLINDER (hydraulic piston type pump that develops pressure for brake system).
3. BRAKE BOOSTER (vacuum operated device for assisting brake pedal application).
4. BRAKE LINES (metal tubing and rubber hose for transmitting pressure to wheel brake assemblies).
5. WHEEL BRAKE ASSEMBLIES (devices that use system pressure to produce friction for slowing or stopping wheel rotation).
6. PARKING BRAKE (mechanical system for applying rear wheel brake assemblies).

When the driver pushes on the brake pedal, lever action pushes a rod into the brake booster and master cylinder. This produces hydraulic pressure in the master cylinder. Fluid flows through the brake lines to the wheel brake assemblies. The brake assemblies use this pressure to cause friction for braking.

An emergency or parking brake system uses cables or rods to mechanically apply the rear brakes. This provides a system for holding the wheels on hills or during complete hydraulic brake system failure.

Drum and disc type brakes

There are two common types of brake assemblies used on modern cars: disc and drum brakes. Refer to Fig. 34-2.

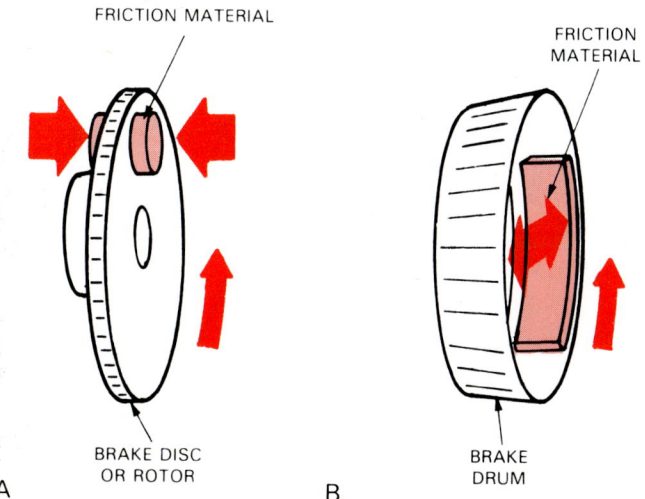

FRICTION MATERIAL

FRICTION MATERIAL

BRAKE DISC OR ROTOR

BRAKE DRUM

A

B

Fig. 34-2. Compare disc and drum brakes. A — Disc brakes are commonly used on front of car and sometimes on rear. B — Drum brakes are normally used on rear of cars.

Disc brakes are frequently used on the two front wheels of a car. Drum brakes are commonly used on the rear wheels. However, disc or drum brakes may be used on all four wheels.

A disc brake assembly basically includes:
1. CALIPER (holds wheel cylinder piston and brake pads).
2. CALIPER CYLINDER (machined hole in caliper, piston fits into the cylinder).
3. BRAKE PADS (friction members pushed against rotor by action of the master cylinder, wheel cylinder, and piston).
4. ROTOR (metal disc that uses friction from brake pads to stop or slow wheel rotation).

A drum brake assembly basically includes:
1. WHEEL CYLINDER ASSEMBLY (hydraulic piston forced outward by fluid pressure).
2. BRAKE SHOES (friction units pushed against the rotating brake drum by action of the wheel cylinder assembly).
3. BRAKE DRUM (rubs against brake shoes to stop wheel rotation and vehicle movement).

BRAKE SYSTEM HYDRAULICS

A hydraulic system is basically a system that uses a liquid to transmit motion or pressure from one part to another. Modern brake systems are hydraulic. They use a confined brake fluid to transfer brake pedal pressure and motion to each of the wheel brake assemblies.

Several principles apply to the operation of a hydraulic system. These include:
1. Liquids in a confined area will NOT compress. However, air in a confined area does compress.
2. When pressure is applied to a closed system, pressure is exerted equally in all directions.
3. A hydraulic system can be used to increase or decrease force or motion.

Hydraulic system action

Suppose two cylinders of equal diameter are placed side by side with a tube connecting them. The system is filled with liquid. Pistons are in each of the cylinders. If you push down on one of the pistons, the other piston will move an equal distance and with equal force.

Since the liquid will not compress and the cylinders are the same diameter, the same amount of liquid is moved from one cylinder to the other.

When pistons of different sizes are used, motion and force can be increased or decreased.

Suppose a small piston acts on a larger piston. The larger piston will move with more force. However, it will move a shorter distance.

When a large diameter piston acts on a smaller piston, the opposite is true. The smaller piston slides farther in its cylinder, but with less force.

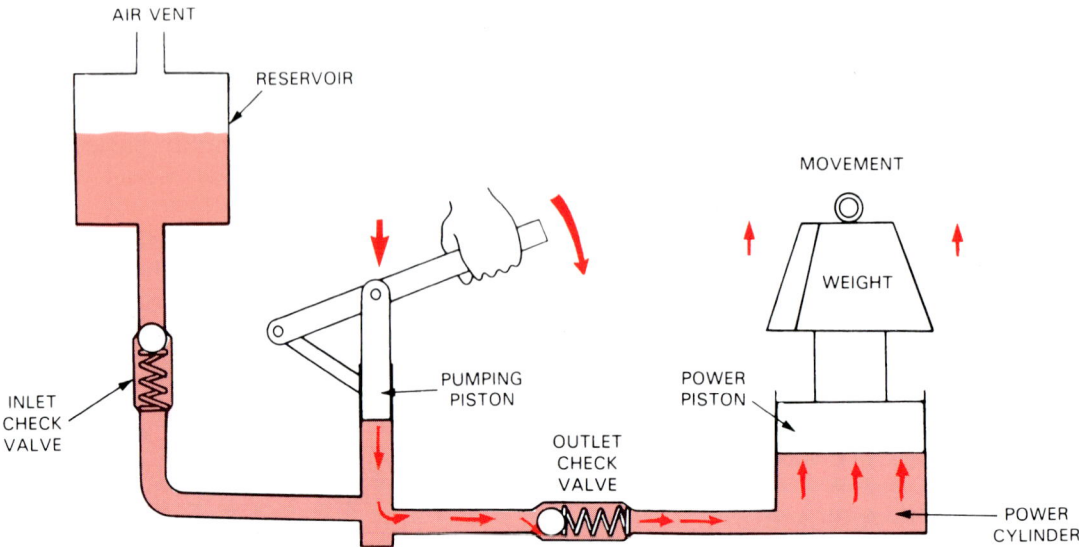

Fig. 34-3. Drawing of hydraulic jack demonstrates how small piston, acting on large piston, can increase force tremendously. Study how check valves only allow oil flow in one direction.

A simple hydraulic jack demonstrates the principles just discussed. Look at Fig. 34-3.

Applying these hydraulic principles to a brake system, you can see how stopping force is transmitted from the master cylinder to each wheel brake assembly.

The master cylinder, Fig. 34-4, acts as the pumping piston that supplies system pressure. The wheel cylinders act as the power piston to move the friction linings into contact with the rotating drums or discs.

BRAKE PEDAL ASSEMBLY

The brake pedal assembly acts as a lever arm to increase the force applied to the master cylinder

piston. A manual master cylinder bolts directly to the engine firewall. The brake pedal assembly bolts under the dash, Fig. 34-5.

The pedal swings on a pivot bolt in the pedal support bracket. A push rod connects the brake pedal to the master cylinder piston.

MASTER CYLINDER

A master cylinder is a foot-operated pump that forces fluid into the brake lines and wheel cylinders. Refer to Fig. 34-6. A master cylinder can have four basic functions:

1. It develops pressure, causing the wheel cylinder pistons to move towards the rotors or drums.

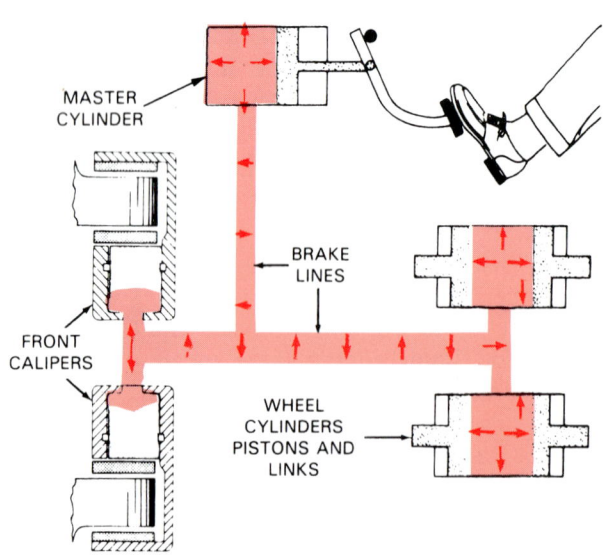

Fig. 34-4. Like basic hydraulic jack, master cylinder acts as pumping piston to move pistons at wheel brake assemblies.

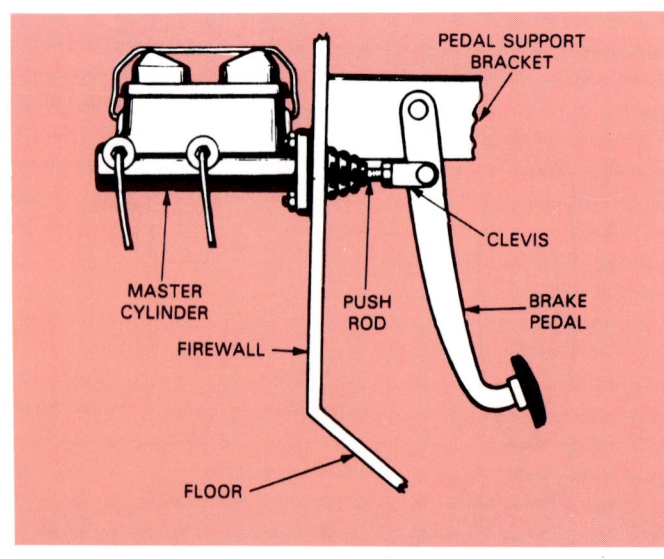

Fig. 34-5. Brake pedal assembly bolts under dash. Push rod transfers pedal movement into master cylinder and operates piston in master cylinder. (Bendix)

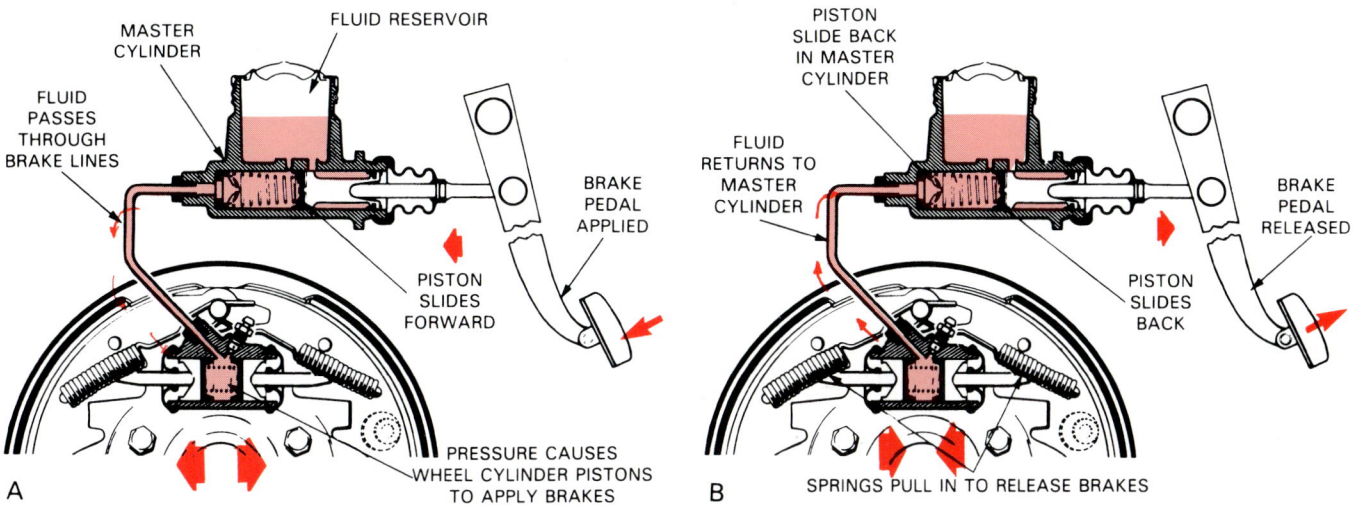

Fig. 34-6. Older, single piston master cylinder demonstrates action in system. A — Application of pedal moves push rod into piston. Master cylinder piston pressurises fluid in cylinder and line. Pressure pushes wheel cylinder pistons apart and brakes are applied. B — Release of brake pedal allows retracting springs to pull brake shoes away from brake drum. Fluid flows back through line and into master cylinder. (Ford)

2. After all of the shoes or pads produce sufficient friction, the master cylinder helps equalise the pressure required for braking.
3. It keeps the system full of fluid as the brake linings wear.
4. It can maintain a slight pressure to keep contaminants (air and water) from entering the system.

Master cylinder components

In its simplest form, a master cylinder consists of a housing, reservoir, piston, rubber cup, return spring, and a rubber boot. Look at Fig. 34-7.

A cylinder is machined in the housing of the master cylinder. The spring, cup, and metal piston slide in this cylinder. Two ports are drilled between the reservoir and cylinder.

The cup and piston in the master cylinder are used to pressurise the brake system. When they are pushed forward, they trap the fluid, building up pressure.

The master cylinder intake port or vent allows fluid to enter the rear of the cylinder as the piston slides forward. Refer to Figs. 34-8A and 34-8B. Fluid flows out of the reservoir through the intake port, and into the area behind the piston and cup.

Then, when the brake pedal is released, the spring forces the piston and cup back in the cylinder. If needed, the rubber cup flexes forward allowing fluid to enter the area in front of the piston and cup. See

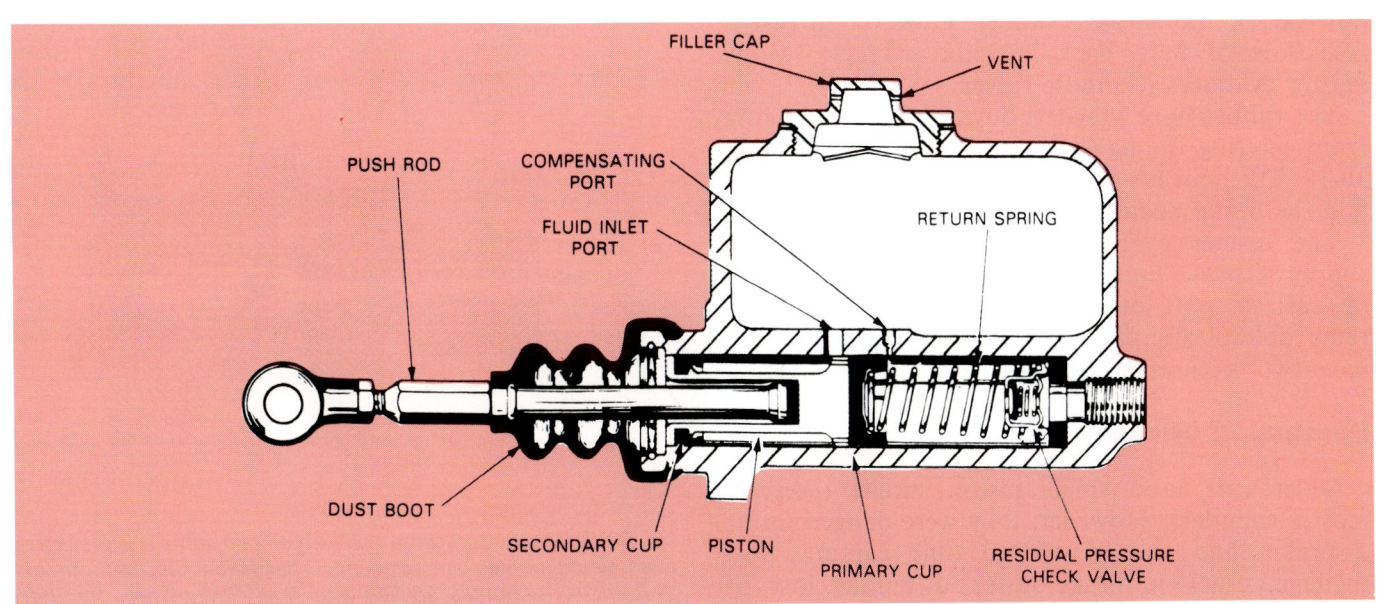

Fig. 34-7. Study the basic parts of a master cylinder. (Bendix)

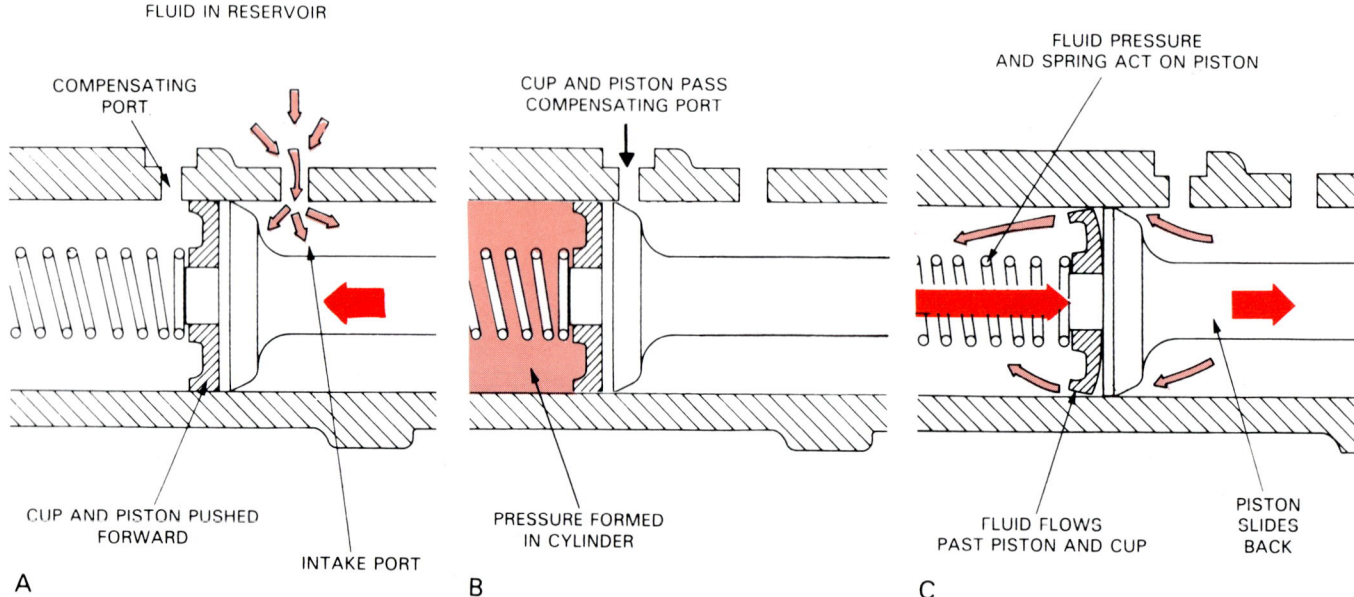

Fig. 34-8. Piston and cup action inside master cylinder. A — Piston slides forward. Fluid flows into area behind piston. Excess fluid flows into reservoir through compensating port. B — Piston and cup move past compensating port and pressure forms in front section of cylinder to apply brakes. C — when brake pedal is released, cup flexes forward so that fluid can flow to front of piston for release of brakes.

Fig. 34-8C. Usually, small holes are drilled in the edge of the piston so that fluid can flow past the cup.

The compensating port releases extra pressure when the piston returns to the released position. Fluid can flow back into the reservoir through the compensating port. The action of the intake port and the compensating port keeps the system full of fluid.

Residual pressure valves maintain residual fluid pressure of approximately 70 kPa to help keep contaminants out of the system.

Fig. 34-9 gives a cutaway view of a typical residual valve. One is in each outlet fitting to the brake lines.

Note how the residual pressure valve allows fluid flow out of the master cylinder. However, it resists free flow of fluid back into the cylinder. Many master cylinders use these valves.

The rubber boot prevents dust, dirt, and moisture from entering the back of the master cylinder, Fig. 34-10. The boot fits over the master cylinder housing and the brake pedal push rod.

The master cylinder reservoir stores an extra supply of brake fluid, Fig. 34-10. The reservoir may be cast as part of the housing or it may be a removable plastic part. Today's reservoirs normally have two sections or compartments.

Dual master cylinder

Older cars used single piston, single reservoir master cylinders. However, they were dangerous. If a brake fluid leak developed (line rupture, seal damage, crack in brake hose), a sudden loss of braking ability could occur. Modern cars use a dual master cylinder for added safety, Fig. 34-10.

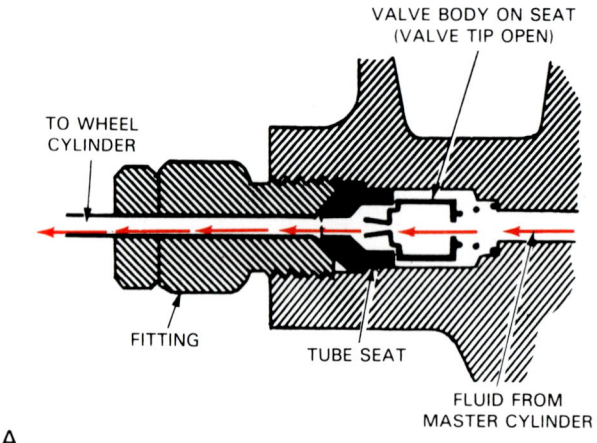

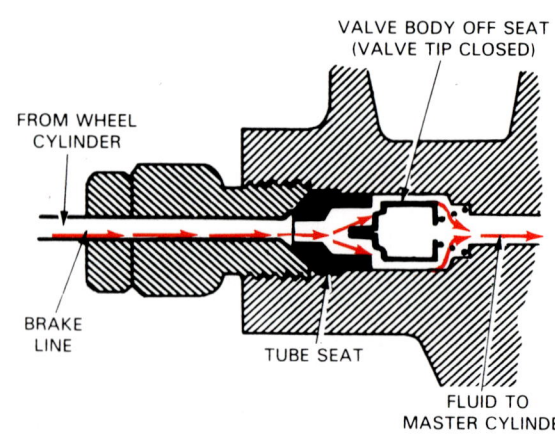

Fig. 34-9. Residual pressure valves maintain some line pressure even when brakes are released. This helps keep air out of system. A — Brakes applied and fluid flows freely through valve. B — After brake release, valve closes to restrict return of fluid to master cylinder. (Ford)

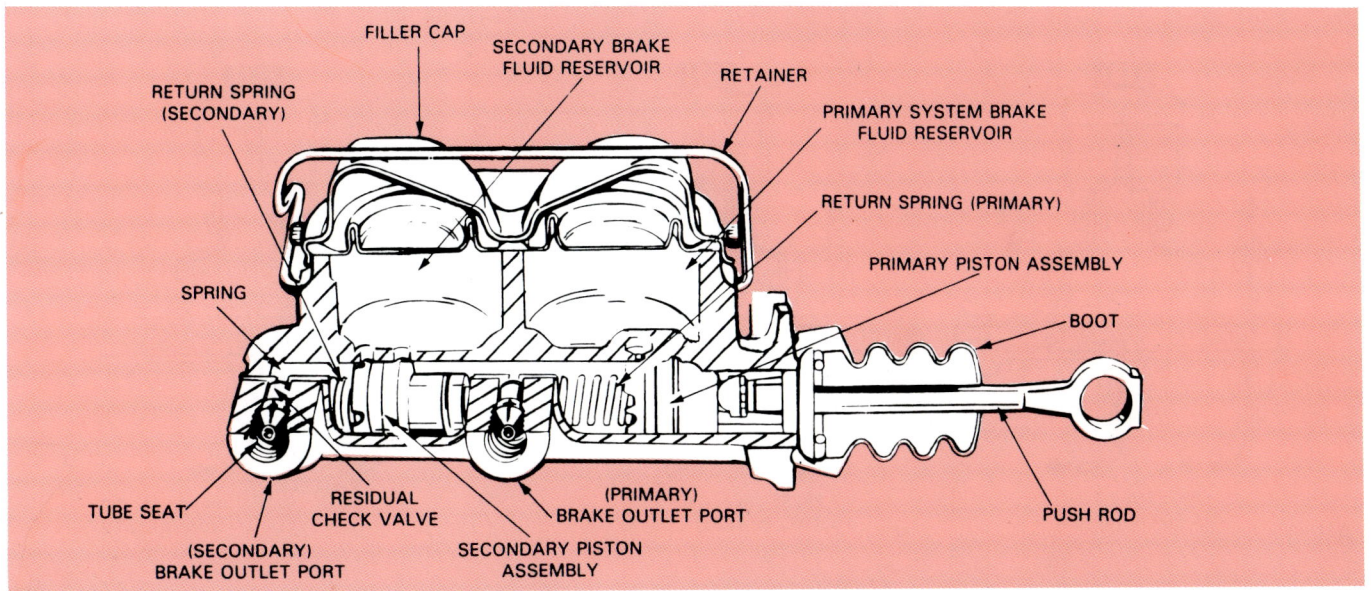

Fig. 34-10. Dual master cylinders are now used because they will still provide braking action when a major hydraulic leak develops. With single-piston master cylinder, a leak could cause sudden and complete loss of brakes. (Delco)

The dual master cylinder, also called a tandem master cylinder, has two separate hydraulic pistons and two fluid reservoirs. See Fig. 34-11. One piston normally operates two of the wheel cylinders. The other piston operates the two other wheel cylinders. Then, if there is a system leak, the other master cylinder piston can still provide braking action on two wheels.

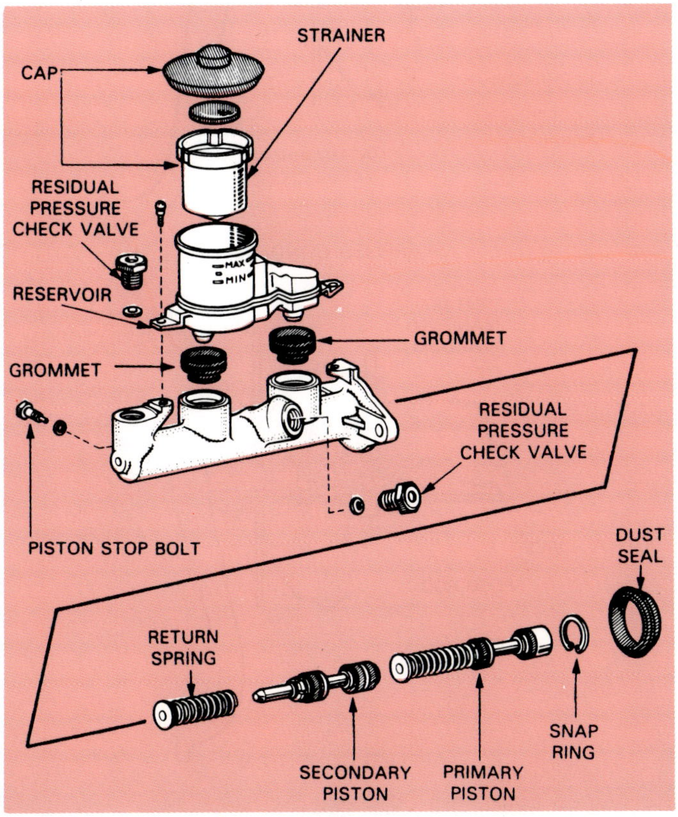

In the dual master cylinder, the rear piston assembly is called the primary piston. The front piston is termed the secondary piston.

Dual master cylinder operation

The action of the pistons, cups, and ports in the dual master cylinder is similar to a single piston unit.

Look at Fig. 34-12A. When both systems are intact (no fluid leaks), both pistons produce and supply pressure to all four of the wheel cylinders.

If there is a pressure loss in the primary section of the brake system (rear section of master cylinder), the primary piston slides forward and pushes on the secondary piston. As shown in Fig. 34-12B, this forces the secondary piston forward mechanically, building pressure in two of the wheel brake assemblies.

When a brake line, wheel cylinder, or other component leaks in the secondary circuit (parts fed by secondary piston), the secondary piston slides completely forward in the cylinder. See Fig. 34-12C. Then, the rear, primary piston provides hydraulic pressure for the other two brake assemblies.

It is very unlikely that both systems should fail at the same time.

POWER BRAKES

Power brakes use engine vacuum or a vacuum pump to assist brake pedal application. The booster (assisting mechanism) is located between the brake pedal linkage and the master cylinder. When the driver presses the brake pedal, the brake booster helps push on the piston.

Fig. 34-11. Note major parts of modern dual master cylinder. (Toyota)

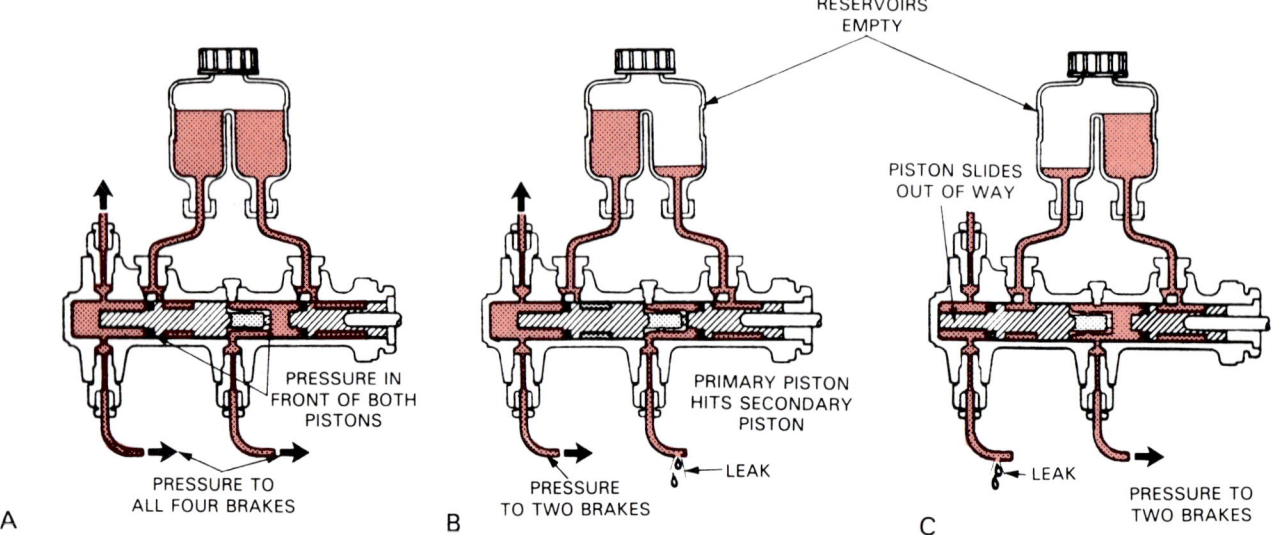

Fig. 34-12. Operation of dual master cylinder. A — No problem in brake system. Both pistons produce pressure for all four brake assemblies. B — Rear brake circuit leaking. Primary piston pushes on secondary piston and two brake assemblies still work to stop car. C — With front brake line leak, secondary piston slides forward in cylinder. Primary piston then operates normally to apply two brake assemblies. (Delco)

Power brake vacuum boosters

A power brake vacuum booster uses engine vacuum (or vacuum pump action on diesel engine) to apply the hydraulic brake system. The principles of a vacuum booster are shown in Fig. 34-13.

A vacuum booster basically consists of a round housing that encloses a diaphragm or a piston. When vacuum is applied to one side of the booster, the piston or diaphragm moves towards the low vacuum area. This movement is used to help force the piston into the master cylinder.

Vacuum booster types

There are two general types of vacuum brake boosters: atmospheric suspended type and vacuum suspended type.

An atmospheric suspended brake booster has normal air pressure on both sides of the diaphragm or piston when the brake pedal is released. As the brakes are applied, a vacuum is formed in one side of the booster. Atmospheric pressure then pushes on and moves the piston or diaphragm.

A vacuum suspended brake booster has vacuum on both sides of the piston or diaphragm when the brake pedal is released. Pushing down on the brake pedal releases vacuum on one side of the booster. The difference in pressure pushes the piston or diaphragm for braking action.

Figs. 34-14 and 34-15 show views of two vacuum brake boosters.

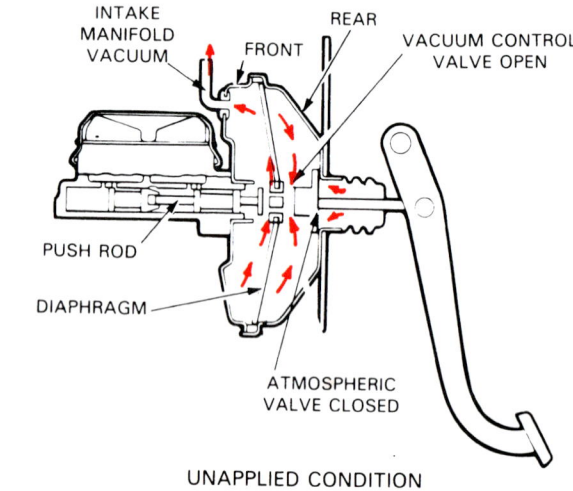

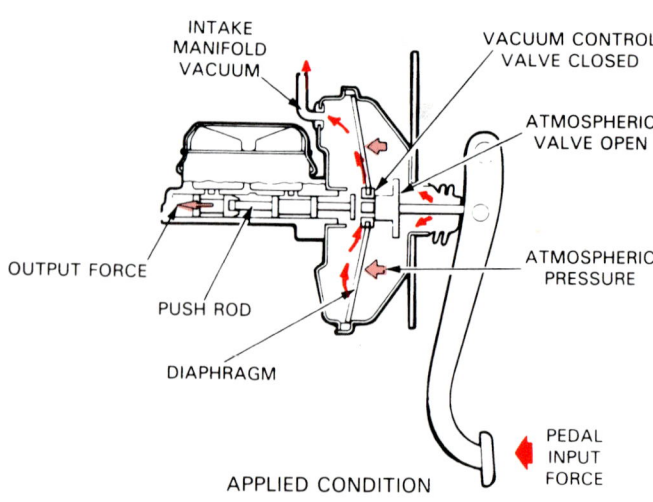

Fig. 34-13. Vacuum booster helps apply brake pedal. It uses engine vacuum or vacuum pump to act on diaphragm which pushes on push rod.

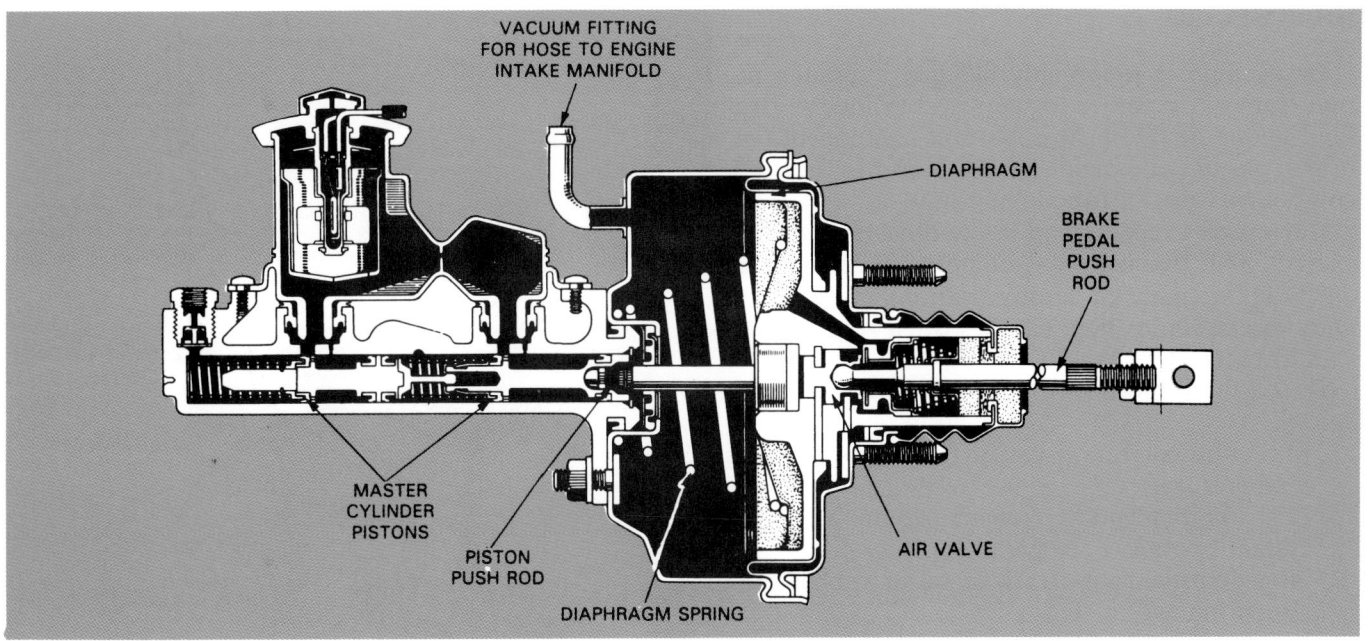

Fig. 34-14. Study internal parts of brake booster and master cylinder. (Toyota)

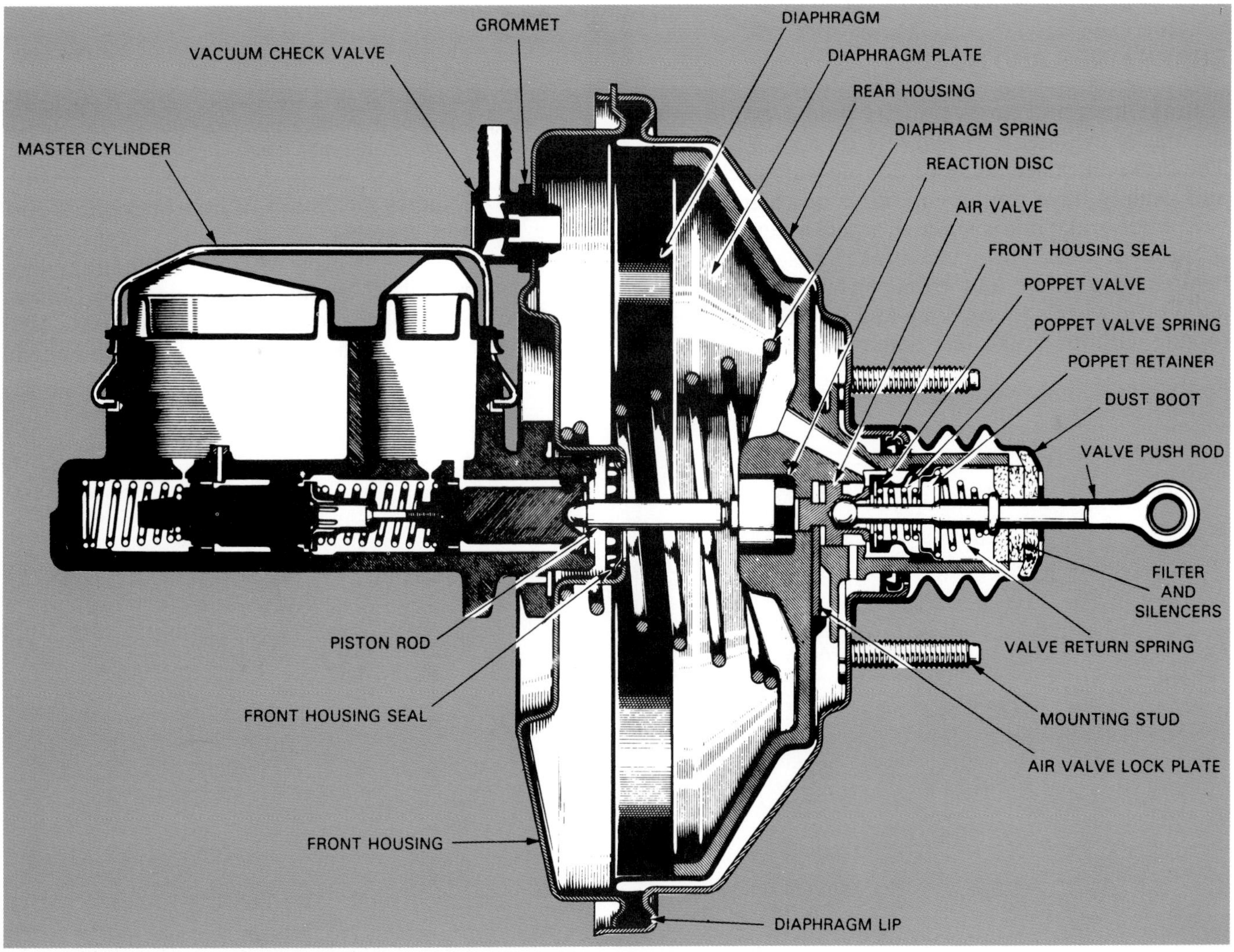

VACUUM FITTING
FOR HOSE TO ENGINE
INTAKE MANIFOLD

DIAPHRAGM

BRAKE
PEDAL
PUSH
ROD

MASTER
CYLINDER
PISTONS

PISTON
PUSH ROD

DIAPHRAGM SPRING

AIR VALVE

VACUUM CHECK VALVE

GROMMET

DIAPHRAGM

DIAPHRAGM PLATE

REAR HOUSING

DIAPHRAGM SPRING

REACTION DISC

AIR VALVE

FRONT HOUSING SEAL

POPPET VALVE

POPPET VALVE SPRING

POPPET RETAINER

DUST BOOT

VALVE PUSH ROD

MASTER CYLINDER

PISTON ROD

FRONT HOUSING SEAL

FRONT HOUSING

FILTER
AND
SILENCERS

VALVE RETURN SPRING

MOUNTING STUD

AIR VALVE LOCK PLATE

DIAPHRAGM LIP

Fig. 34-15. Another type of vacuum brake booster. (Bendix)

BRAKE FLUID

Brake fluid is a specially blended hydraulic fluid that transfers pressure to the wheel cylinders. Brake fluid is one of the most important components of a brake system because it ties all of the other components into a functioning unit.

Car makers recommend brake fluid that meets or exceeds SAE (Society of Automotive Engineers) and DOT (Department of Transportation) specifications. Only brake fluid that satisfies their requirements should be used.

Brake fluid must have the following characteristics:

1. Maintain correct viscosity (free flowing at all temperatures).
2. High boiling point (remain liquid at highest system operating temperature).
3. Noncorrosive (not attack metal or rubber brake system parts).
4. Water tolerance (absorb moisture that collects in system).
5. Lubricate (reduce wear of pistons and cups).
6. Low freezing point (not freeze in cold weather).

BRAKE LINES AND HOSES

Brake lines and hoses transfer fluid pressure from the master cylinder to the wheel cylinders, Fig. 34-16. The brake lines are made of double-wall steel tubing and usually have double-lap flares on their ends.

Rubber brake hoses are used where a flexing action is needed. For instance, brake hoses are used between the frame and front wheel cylinders. This allows the wheels to move up and down or from side to side without brake line damage.

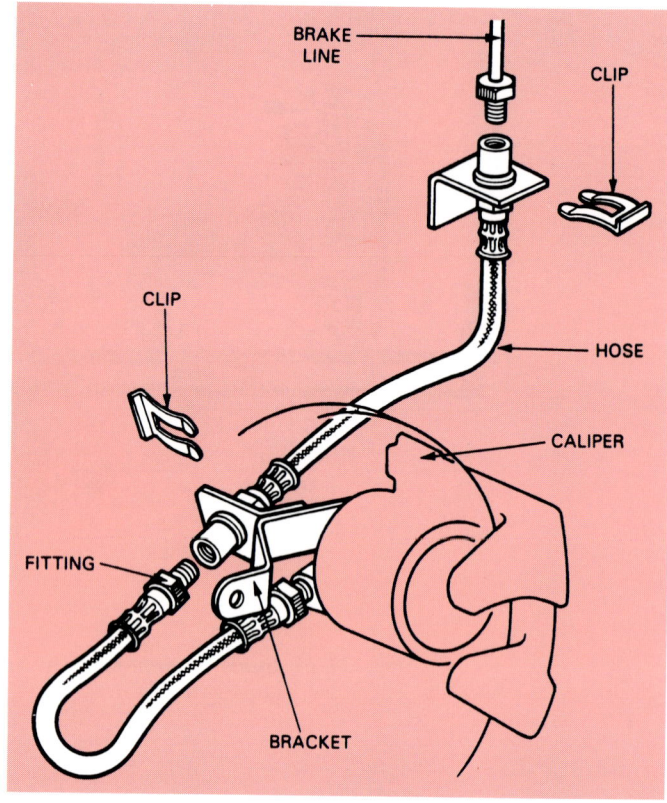

Fig. 34-17. Brackets and clips are used to secure brake hoses and lines to frame or unibody. Lines must not vibrate or they can fatigue and break. (Toyota)

Fig. 34-17 shows the details of how brake lines and brake hoses fit together.

A junction block is used where a single brake line must feed two wheel cylinders. It is simply a hollow fitting with one inlet and two or more outlets.

A longitudinally (front to rear) split brake system

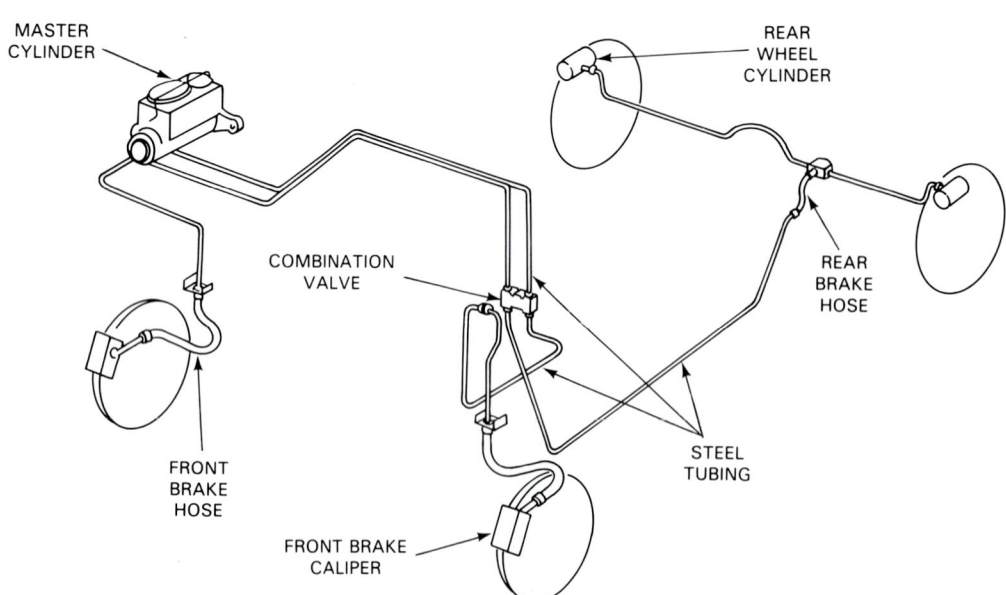

Fig. 34-16. Note routing and location of metal brake lines and flexible hoses. Hoses are needed for suspension movement without line breakage. (Jeep)

LONGITUDINALLY SPLIT

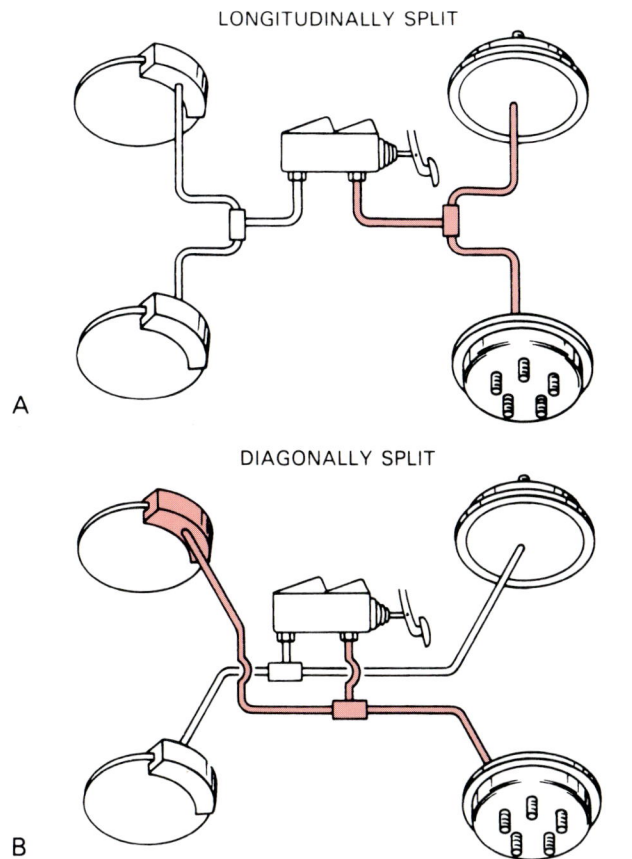

A

DIAGONALLY SPLIT

B

Fig. 34-18. Study how each master cylinder piston operates different wheel brake assemblies with longitudinally and diagonally split system.

has one master cylinder piston operating the front wheel brake assemblies and the other operates the rear brakes. This is shown in Fig. 34-18A.

A diagonally (corner to corner) split brake system has each master cylinder piston operating a brake assembly on opposite corners of the car, Fig. 34-18B.

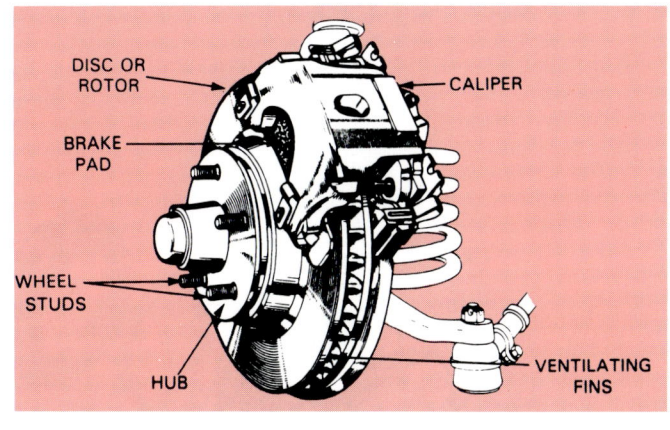

Fig. 34-19. Note externally visible parts of disc brake unit. (Bendix)

DISC BRAKES

Disc brakes are basically like the brakes on a ten-speed bicycle. The friction elements are shaped like pads and are squeezed inward to clamp against a rotating disc or wheel.

A disc brake assembly consists of a caliper, brake pads, rotor, and related hardware (bolts, clips, springs). See Fig. 34-19.

Fig. 34-20 shows sectioned views of typical disc brake assemblies. Note how the caliper pistons move inward to clamp the brake pads against the rotor. The single piston type caliper is much more common.

Brake caliper

The brake caliper assembly includes the caliper housing, piston, piston seal, dust boot, brake pads or shoes, special hardware (clips, springs), and a bleeder screw. These parts are pictured in Fig. 34-21.

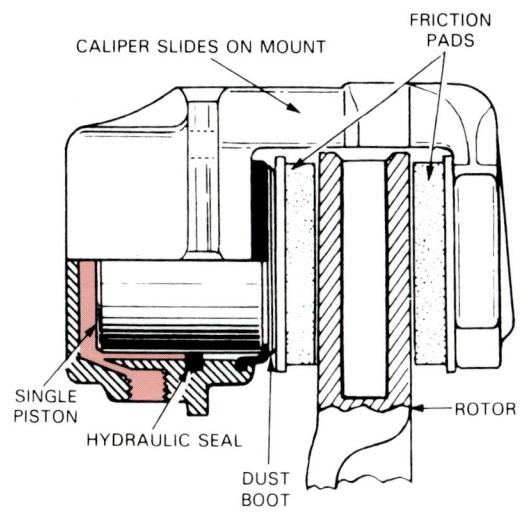

A

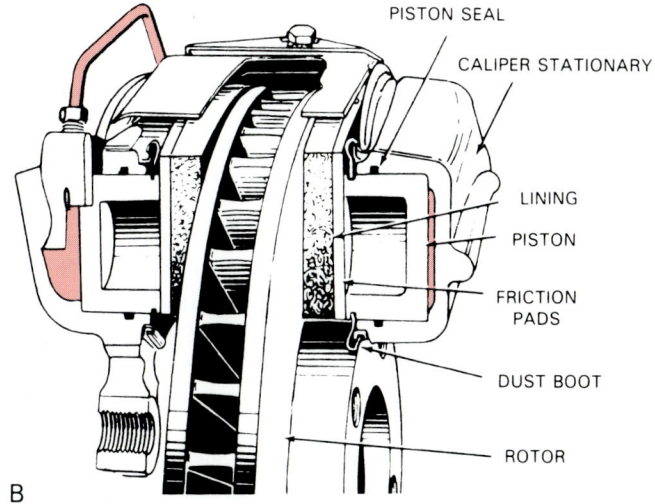

B

Fig. 34-20. Caliper piston pushes brake pad into revolving disc to slow or stop car. A — Single piston caliper is very common. Caliper floats or slides on mount so both pads contact disc. B — Fixed caliper uses pistons on both sides of disc. Caliper is mounted stationary.

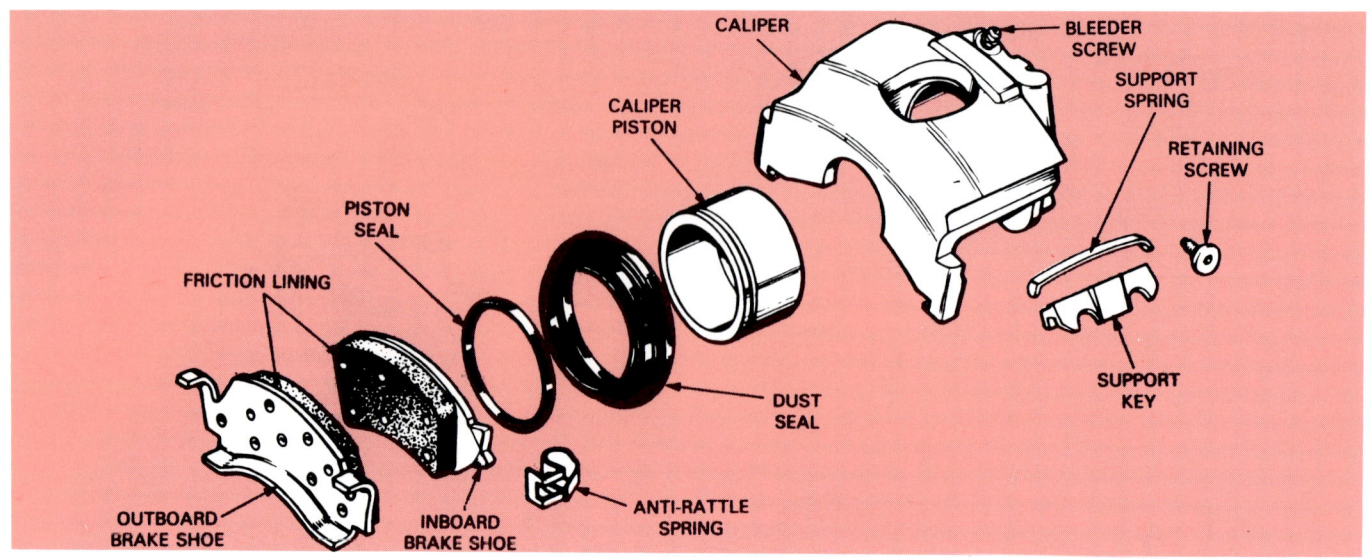

Fig. 34-21. Disassembled view of caliper shows parts. Note how they fit together. (Jeep)

Fig. 34-22 shows how a brake caliper piston operates. When the brake pedal is applied, brake fluid flows into the caliper cylinder. The piston is then pushed outward by fluid pressure to force the brake pads against the rotor.

The piston seal in the caliper prevents pressure leakage between the piston and cylinder. The piston seal also helps pull the piston back into the cylinder when the brakes are NOT applied. The elastic action of the seal acts as a spring to retract the piston.

The piston boot keeps road dirt and water off the caliper piston and wall of the cylinder. Pictured in Fig. 34-22, the boot and seal usually fit into grooves cut in the caliper cylinder and piston.

A bleeder screw allows air to be removed from the hydraulic brake system. It is threaded into the top or side of the caliper housing. Fig. 34-21. When loosened, system pressure can be used to force fluid and air out of the bleeder screw.

Disc brake pads

Disc brake pads consist of steel shoes to which linings are riveted. Brake pads are shown in Fig. 34-21.

Brake pad linings are normally made of either asbestos (asbestos fibre filled) or semi-metallic (metal particle filled) friction material. Many new cars, especially those with front-wheel drive, use semi-metallic linings on the front. Semi-metallic linings can withstand higher operating temperatures without losing their frictional properties.

Anti-rattle clips are frequently used to keep the brake pads from vibrating and rattling, Fig. 34-23A. The clip snaps onto the brake pad to produce a force fit in the caliper. Sometimes, an anti-rattle spring is used instead of a clip.

A pad wear sensor is a metal tab on the brake pad that informs driver of worn brake pad linings. The wear sensor tab will emit a loud squeal or squeak when it scrapes against the brake disc. The sensor only touches the disc when the brake lining has worn too thin.

Brake disc or rotor

The brake disc, also called brake rotor, uses friction from the brake pads to slow or stop wheel rotation. The brake disc is normally made of cast iron. It may be an integral part of the wheel hub. However, many front-wheel drive cars have the disc and the hub as separate units. Refer to Figs. 34-19 and 34-23A.

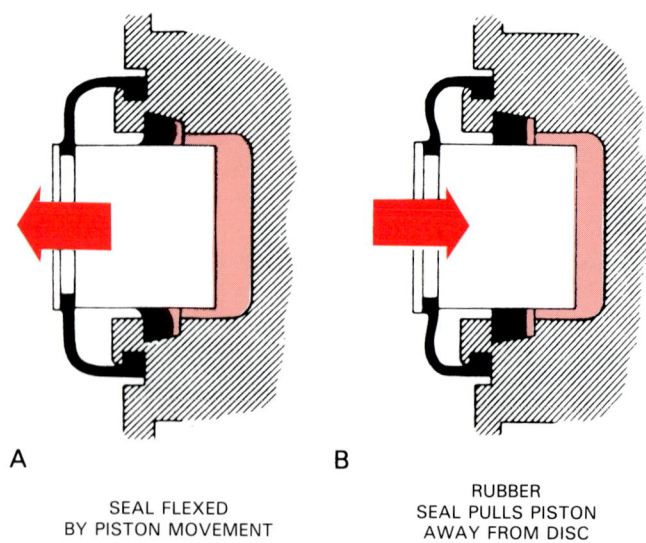

Fig. 34-22. Operation of caliper piston. A — Brakes applied and piston is pushed partially out of cylinder. B — Stretched piston seal pulls piston back after brake release. This keeps pads from rubbing on disc.

The brake disc may be solid or ventilated rib type. The ventilated rib disc is hollow which allows cooling air to circulate inside the disc.

Disc brake types

Disc brakes can be classified as floating, sliding, and fixed caliper types. Floating and sliding calipers are common. The fixed caliper was used on some older passenger cars.

The floating caliper disc brake is mounted on two bolts supported by rubber bushings. The one-piston caliper is free to shift or float in the rubber bushings.

The sliding caliper type disc brake is mounted in slots machined in the caliper adapter. The one-piston caliper is free to slide sideways in the slots or grooves as the linings wear. Look at Fig. 34-23A.

The fixed caliper type disc brake normally uses more than one piston and cylinder. The caliper is bolted directly to the steering knuckle. It is NOT free to move in relation to the disc. Pistons on both sides of the disc push brake pads in from both directions, Fig. 34-20B.

Floating and sliding calipers are used to avoid vibration problems. With a fixed caliper, severe vibrations can occur with a slight runout (wobble) of the disc.

DRUM BRAKES

Drum brakes use many of the same principles already covered for disc brakes. However, drum brakes have a large drum that surrounds the brake shoes and hydraulic wheel cylinder.

A drum brake assembly consists of a backing plate, wheel cylinder, brake shoes and linings, retracting springs, hold-down springs, brake drum, and automatic adjusting mechanism, Fig. 34-23B.

Backing plate

The brake backing plate holds the shoes, springs, wheel cylinder, and other parts inside the brake drum. It also helps keep road dirt and water off the brakes. The backing plate bolts to the axle housing or spindle support, Fig. 34-23B.

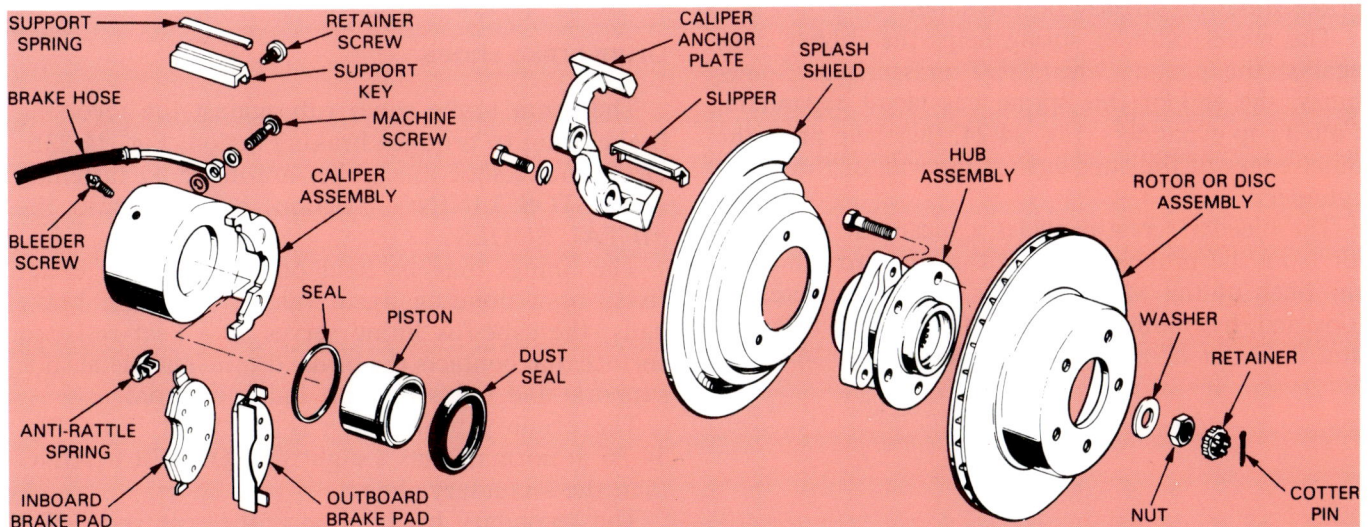

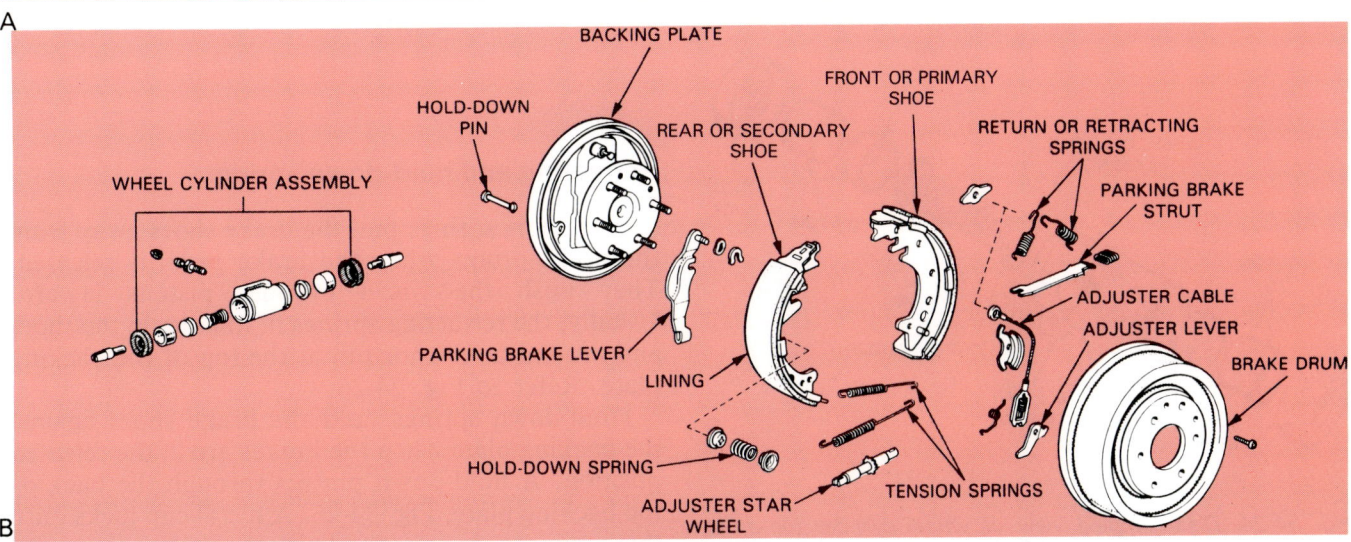

Fig. 34-23. Compare complete disc and drum brake assemblies. A — Disc brake. B — Drum brake. (Jeep and Toyota)

Wheel cylinder assembly

The wheel cylinder assemblies use master cylinder pressure to force the brake shoes out against the brake drums. They bolt to the top of the backing plates.

A wheel cylinder consists of a cylinder or housing, expander spring, rubber cups, pistons, dust boots, and a bleeder screw. See Fig. 34-24.

The wheel cylinder housing forms the enclosure for the other parts of the assembly. It has a precision hole or cylinder in it for the pistons, cups, and spring.

The wheel cylinder boots keep road dirt and water out of the cylinder. They snap into grooves on the outside of the housing, Fig. 34-24.

The wheel cylinder cups are special rubber seals that keep fluid from leaking past the pistons. They fit in the cylinder and against the pistons, as in Fig. 34-25.

The wheel cylinder pistons are metal or plastic plungers that transfer force out of the wheel cylinder assembly. They act on push rods connected to the brake shoes or directly on the shoes.

The wheel cylinder spring helps the rubber cups against the pistons when NOT pressurised. Sometimes, the end of this spring has metal expanders. Called cup expanders, Fig. 34-24, they help press the outer edges of the cups against the wall of the wheel cylinder.

The bleeder screw provides a means of removing air from the brake system. It threads into a hole in the back of the wheel cylinder. When the screw is loosened, hydraulic pressure can be used to force air and fluid out of the system. Refer to Fig. 34-25.

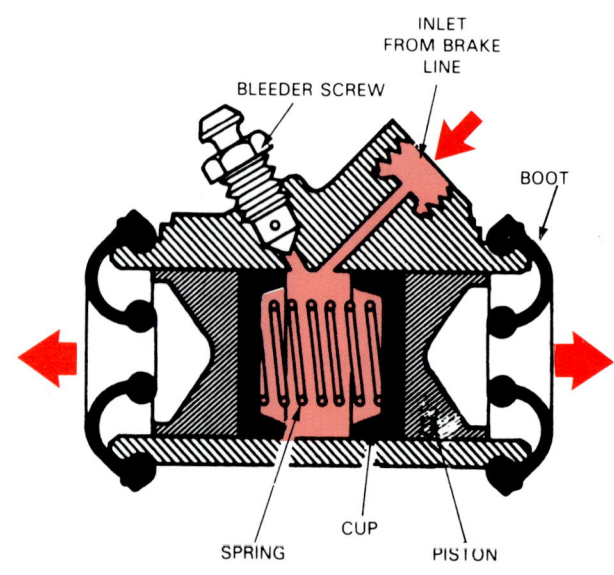

Fig. 34-25. Cutaway view of drum brake type wheel cylinder shows fluid passage into cylinder and bleeder screw. Pressure inside pushes cups and pistons outward to force linings into drum.

Drum brake shoes

The drum brake shoes rub against the revolving brake drum to produce braking action, Fig. 34-23B. Drum brake shoe assemblies are made by fastening ASBESTOS LINING (friction) material onto the METAL SHOES.

The linings may be held on the shoes by either rivets or a bonding agent (glue). Like disc brake pads, the asbestos lining serves as a heat resistant surface that contacts the brake drum. The metal shoe supports and holds the soft lining material.

The primary brake shoe is the front shoe, Fig. 34-26. It normally has a slightly SHORTER LINING than the secondary shoe.

The secondary brake shoe is the rear shoe, Fig. 34-26. It has the LARGEST LINING surface area.

Retracting and hold-down springs

Retracting springs pull the brake shoes away from the brake drums when the brake pedal is released. They push the wheel cylinder pistons inward. Usually, the retracting springs fit in holes in the shoes and around an anchor pin at the top of the backing plate. Refer to Fig. 34-26.

Hold-down springs hold the brake shoes against the backing plate when the brakes are in the released position. A hold-down pin fits through the back of the backing plate, Fig. 34-26. A metal cup locks onto these pins to secure the hold-down springs to the shoes.

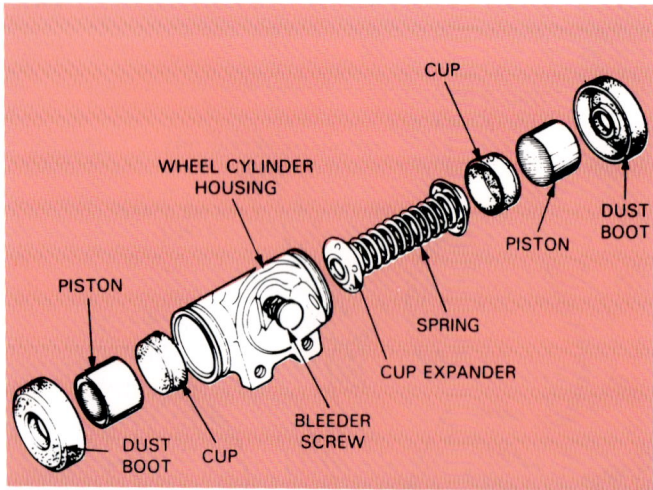

Fig. 34-24. Disassembled view of wheel cylinder for drum brake. Cups prevent fluid leakage out of cylinder. Boots keep debris out of cylinder. (Renault)

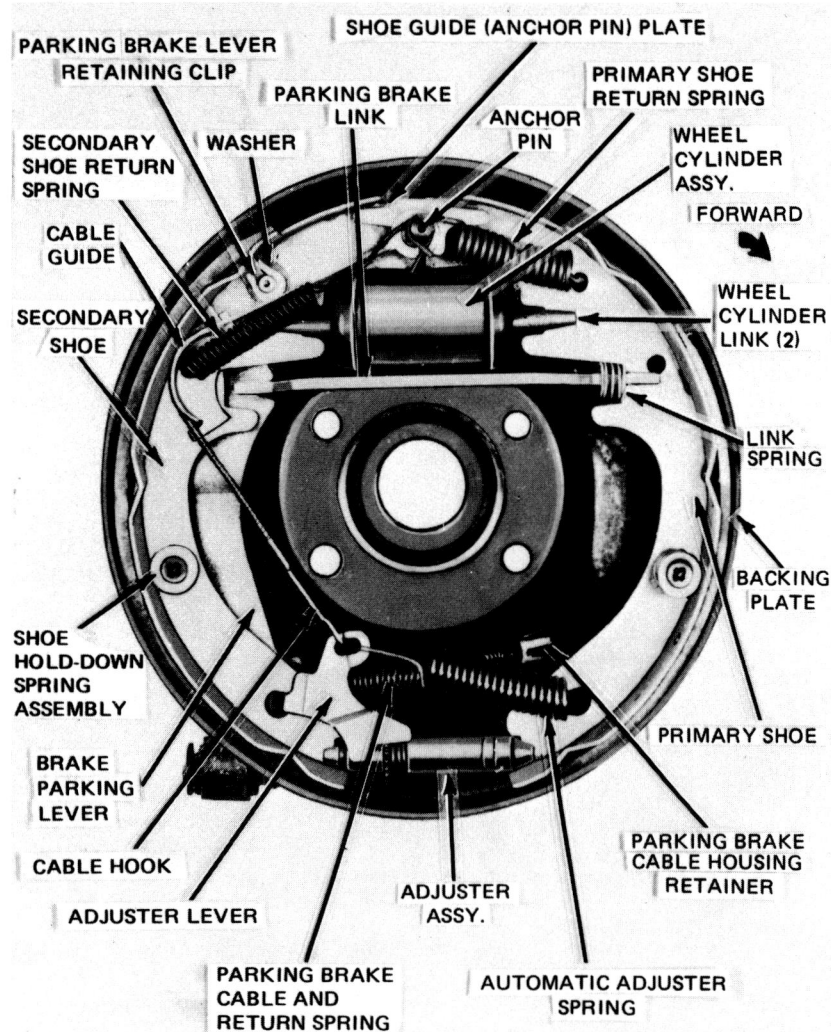

Fig. 34-26. Study parts of common brake assembly mounted on backing plate. (Ford)

As shown in Fig. 34-27, other springs are used on the automatic adjusting mechanism. Brake springs are high quality, capable of withstanding the high temperatures encountered inside the brake drum.

Brake shoe adjusters

Brake shoe adjusters maintain correct drum-to-lining clearance as the brake linings wear. Look at Fig. 34-27.

Many cars use a star wheel (screw) type brake shoe adjusting mechanism. This type includes a star wheel (adjusting screw assembly), adjuster lever, adjuster spring, and either an adjuster cable, lever arm, or link (rods). See 34-27A through D.

Automatic brake shoe adjusters normally function when the brakes are applied with the car moving in reverse. If there is too much lining clearance, the brake shoes move outward and rotate with the drum enough to operate the adjusting lever. This lengthens

the star wheel assembly. The linings are moved closer to the brake drum, maintaining the correct lining-to-drum clearance.

Fig. 34-27E and 34-27F picture brake assemblies that use modern latch type adjusters.

Brake drums

Brake drums provide a rubbing surface for the brake shoe linings, Fig. 34-28. The drum usually fits over the wheel nut studs. A large hole in the middle of the drum centres the drum on the front hub or rear axle flange. The wheel and drum turn together as a unit.

Brake shoe energisation

When the brake shoes are forced against the rotating drum, they are pulled away from their pivot point by friction. This movement, called self-

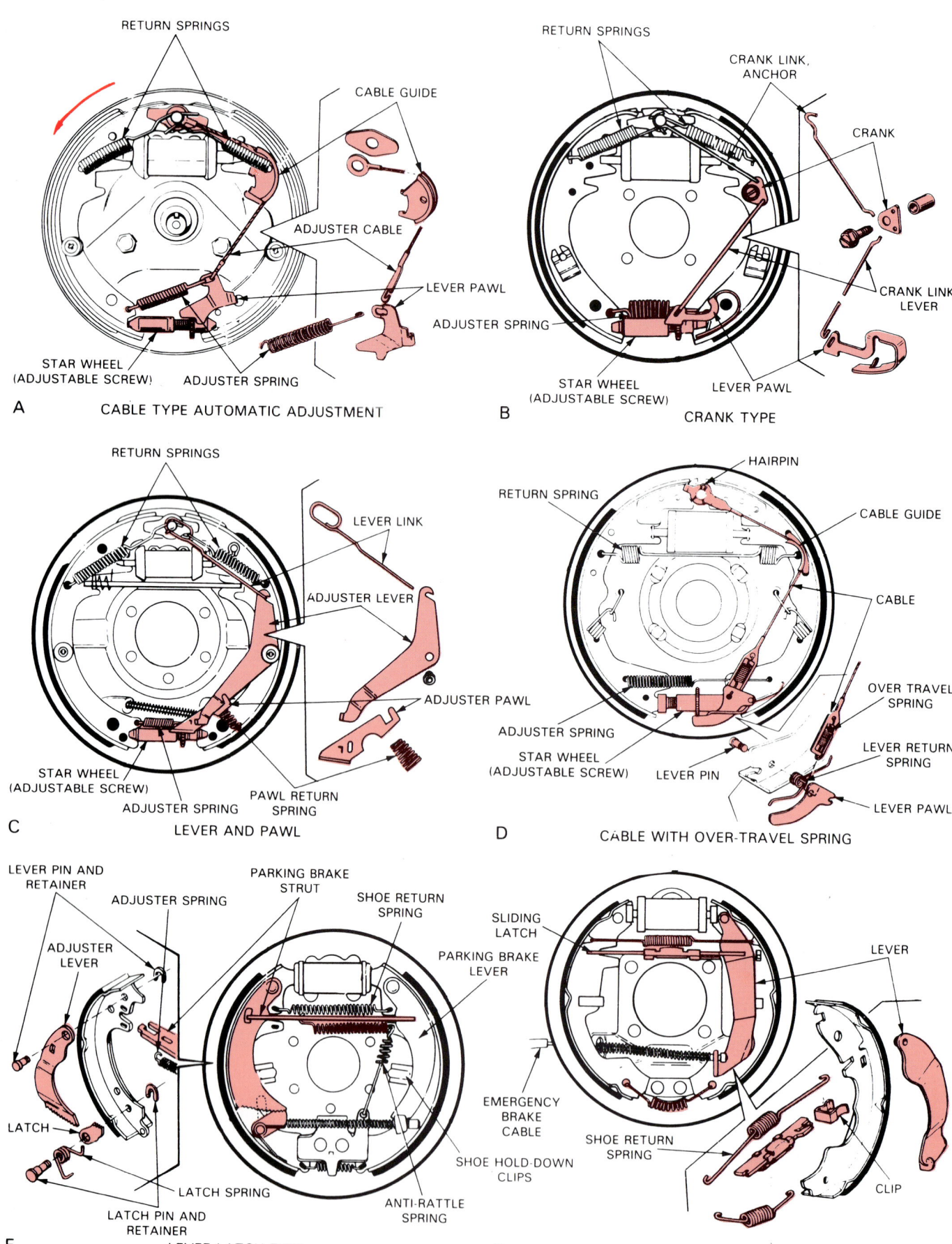

Fig. 34-27. Major variation in drum brake design is in automatic adjusting mechanism. A — Cable type. B — Crank type. C — Lever and pawl type. D — Cable with over-travel spring type. E — Lever-latch type. F — Ratcheting type. (Ford)

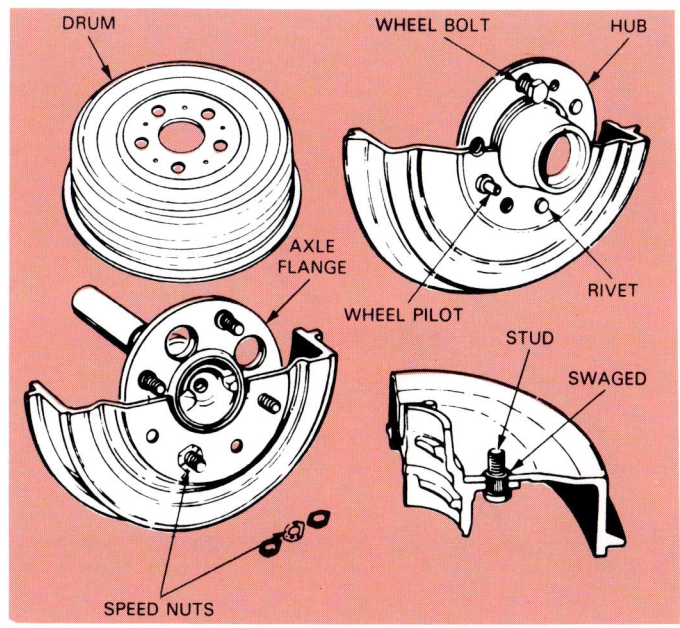

Fig. 34-28. Brake drum provides friction surface for brake shoe linings. Note construction of drum. (Ford)

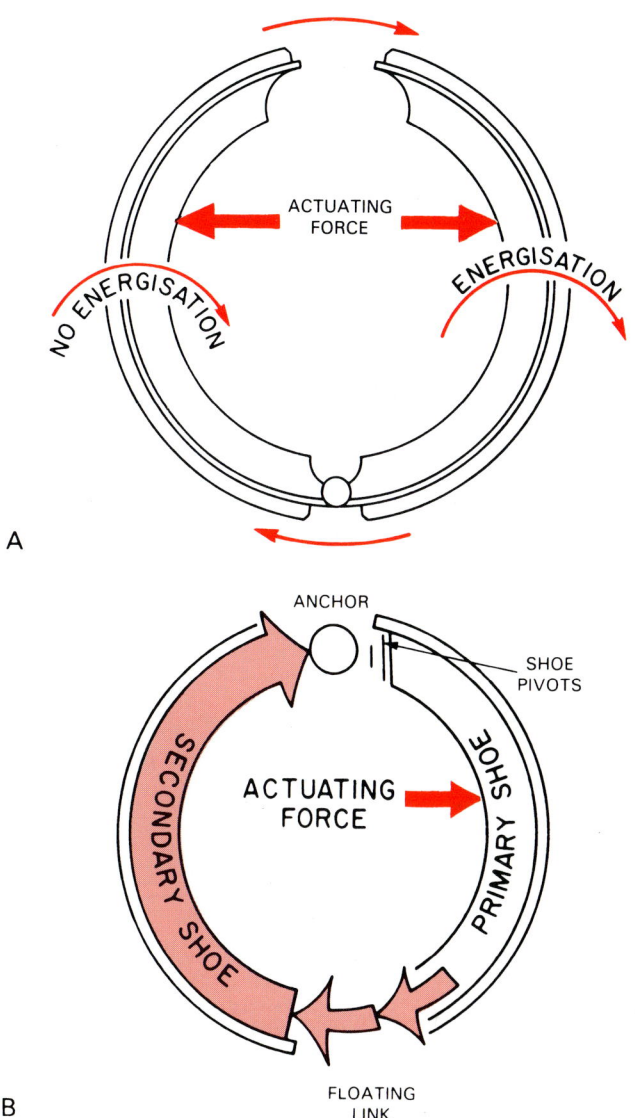

Fig. 34-29. A — Self-energising brakes use friction to force one brake shoe tighter against drum. B — Servo action results when both shoes are free to swing into drum and primary shoe helps apply secondary shoe.

energising action, draws the shoes tighter against the drum. Fig. 34-29.

With most drum brake designs, shoe energisation is supplemented by servo action. Servo action results when the primary (front) shoe helps apply the secondary (rear) shoe. Look at Fig. 34-29B.

The backing plate anchor pin holds the secondary shoe during brake application. However, the primary shoe is free to float out and push against the end of the secondary shoe through the star wheel assembly. This action presses the secondary shoe into the drum with extra force.

Less wheel cylinder hydraulic pressure is needed to apply the brakes because of servo action.

Fig. 34-29A shows a non-servo type brake assembly. Fig. 34-29B shows the more common servo or floating type brakes.

PARKING BRAKES

Parking brakes, provide a mechanical means (cable and levers) of applying the brakes. Fig. 34-30 pictures one type.

When the parking brake hand lever is activated, it pulls a steel cable that runs through a housing. The movement of the cable pulls on a lever inside the drum or disc brake assembly. The lever action forces the brake linings against the rear drums or discs to resist vehicle movement. See Fig. 34-31 which shows a typical dash mounted parking brake lever and brake shoe arrangement.

When disc brakes are used on the rear, a thrust screw and lever can be added to the brake caliper. Then, when the emergency brake is applied, the cable pulls on the caliper lever. The caliper lever turns the large thrust screw which pushes on the caliper piston and applies the brake pads. See Fig. 34-32.

BRAKING RATIO

Braking ratio refers to the comparison of front wheel to rear wheel braking effort. When a car stops, its weight tends to transfer onto the front wheels. The front tyres are pressed against the road with greater force. The rear tyres lose some of their grip on the road. As a result, the front wheels do more of the braking than the rear.

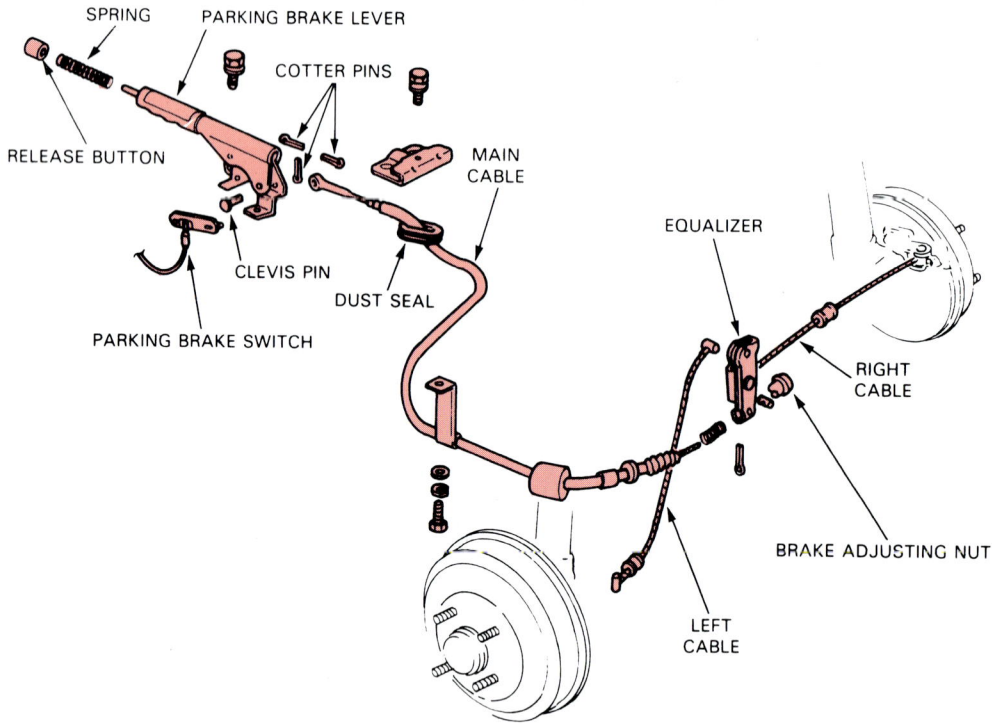

Fig. 34-30. Study parts of typical floor mounted parking brake mechanism. (Toyota)

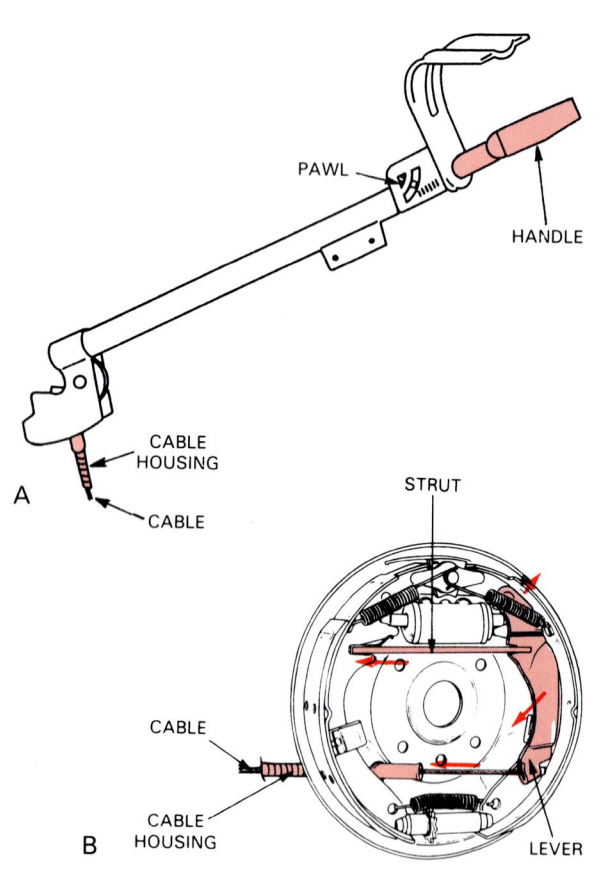

Fig. 34-31. Parts of a typical dash mounted parking brake system. A — Dash mounted lever assembly. B — Cable activates lever in brake assembly that pries out against drum. (Toyota and Ford)

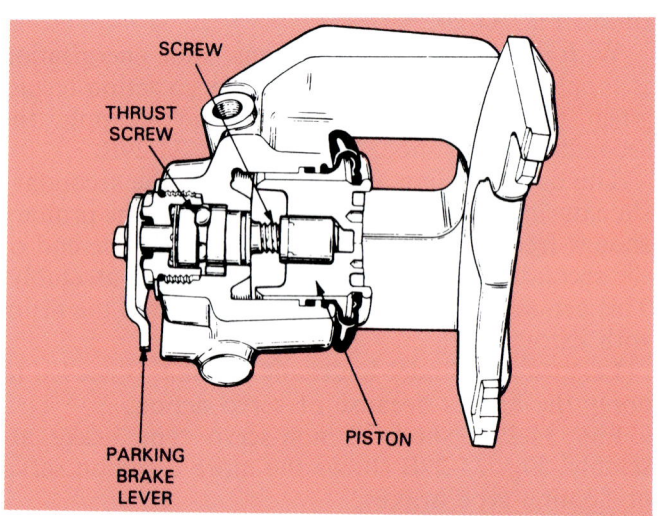

Fig. 34-32. Cross sectional view of caliper that has parking brake mechanism. Cable pulls on and rotates lever. Lever turns screw that pushes piston outward to apply brake. (Bendix)

For this reason, many cars have disc brakes on the front and drum brakes on the rear. Disc brakes are capable of producing more stopping effort than drum brakes. If drum brakes are used on both the front and rear wheels, the front shoe linings and drums normally have a larger surface area.

Typically, front wheel brakes handle 60 to 70 percent of the braking power. Rear wheels handle 30 to 40 percent of the braking. Front-wheel drive cars, having even more weight on the front wheels, can have an even higher braking ratio at the front wheels.

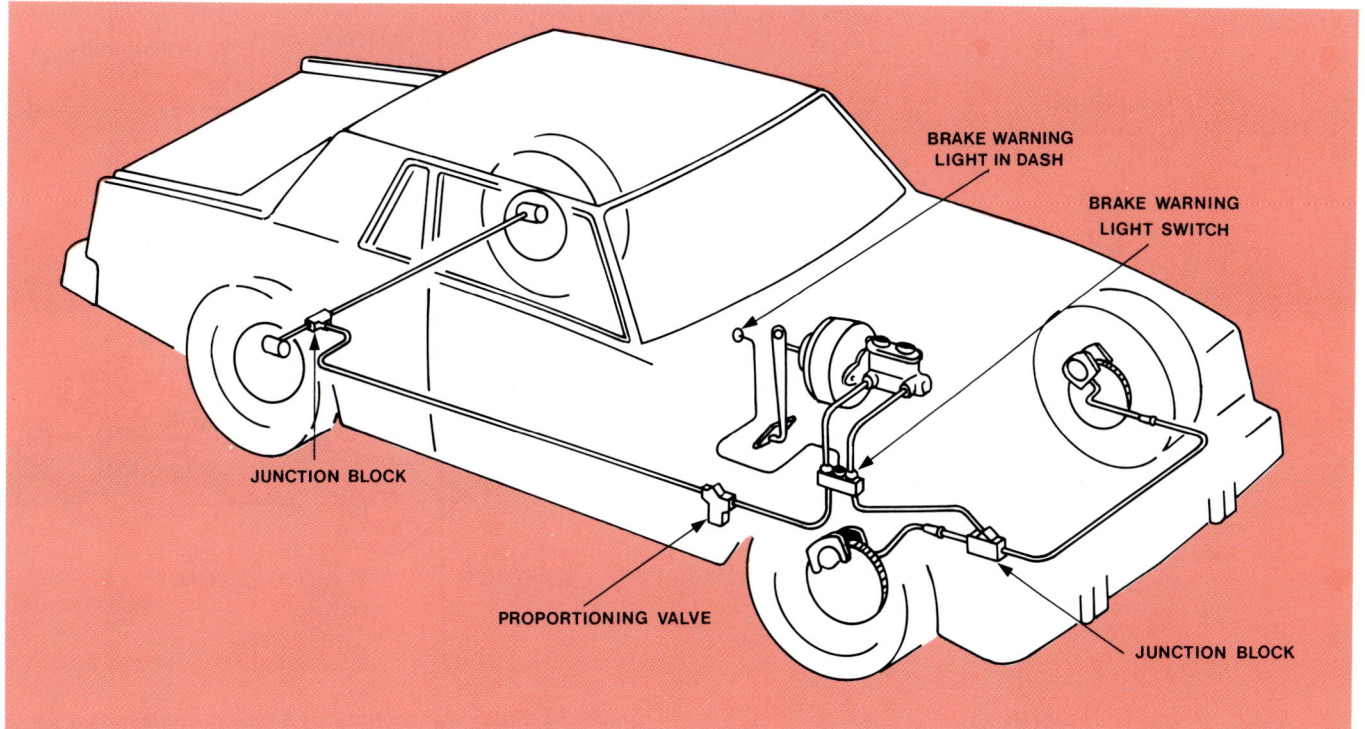

Fig. 34-33. Note location of various valves used in brake system.

BRAKE SYSTEM SWITCHES

There are two types of switches commonly used in a brake system: the stop light switch and the brake warning light switch.

Stop light switch

The stop light switch is a spring-loaded electrical switch that operates the rear stop lights of the car. Most modern cars use a mechanical switch on the brake pedal mechanism. The switch is normally open. When the brake pedal is pressed, it closes the switch and turns on the brake lights.

Hydraulically operated stop light switches are used on some older cars. Brake system pressure pushed on a switch diaphragm and closed the switch to operate the brake lights.

Brake warning light switch (Pressure differential valve)

The brake warning light switch, also called a pressure differential valve, warns the driver of a pressure loss on one side of the dual brake system. Look at Fig. 34-33.

If a leak develops in either the primary or secondary brake system, unequal pressure acts on each side of the warning light switch piston. This pushes the piston to one side, grounding the indicator, Fig. 34-34.

BRAKE SYSTEM CONTROL VALVES

Many brake systems use control valves to regulate the pressure going to each wheel cylinder. The two types of valves are the proportioning valve, and the combination valve. Fig. 34-33 shows the general locations of these valves.

Proportioning valve

A proportioning valve is used to equalise braking action with front disc and rear drum brakes. It is commonly located in the brake line to the rear drum brakes. Look at Fig. 34-33.

The function of the proportioning valve is to limit pressure at the rear drum brake when high pressure is needed to apply the front disc brakes. Thus the proportioning valve prevents rear wheel lockup and skid during heavy brake applications.

Combination valve

A combination valve serves as two valves in one. It can function as a:

1. Proportioning valve.
2. Brake warning light switch.

Many late model cars use a combination valve. Fig. 34-35 shows a cutaway view of a combination valve. Study the two sections.

With some master cylinders, the proportioning and warning lamp valves are mounted inside the master cylinder housing. This design uses the same operating principles, Fig. 34-36.

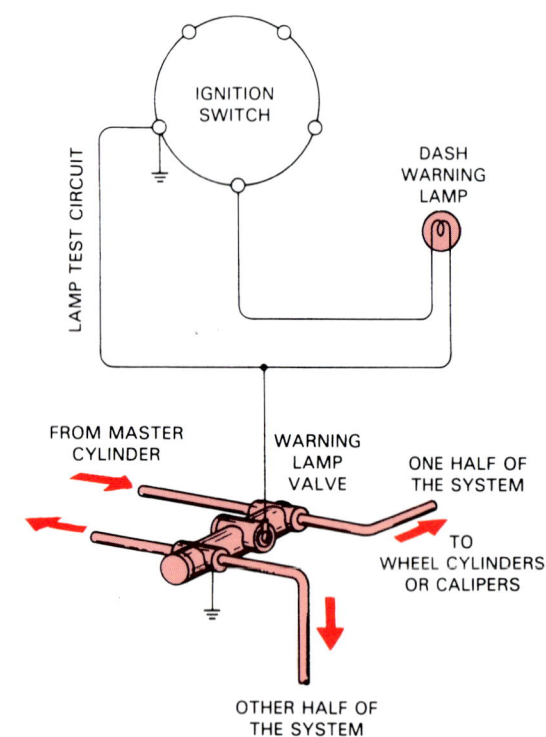

A

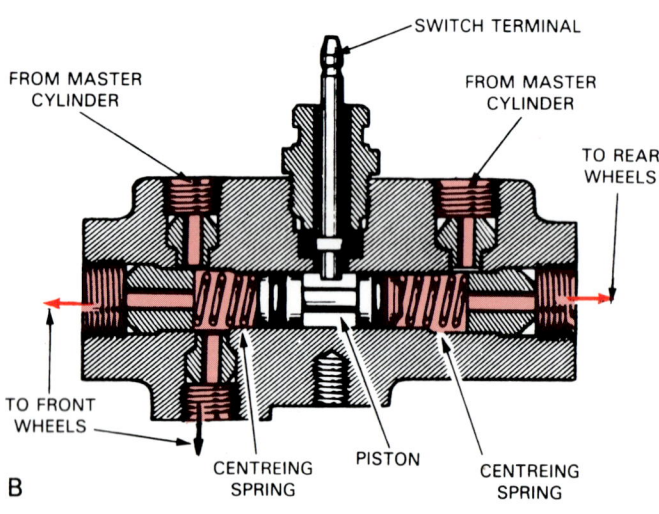

B

Fig. 34-34. Brake warning light switch is activated by difference in pressures in primary and secondary systems. Pressure difference pushes small piston in valve to close warning lamp circuit. A — Brake warning lamp circuit. B — Cutaway view of brake warning lamp switch. (Bendix)

SKID CONTROL BRAKE SYSTEM

A skid control brake system, also termed anti-lock brake system, normally uses wheel rpm sensors, hydraulic valves, and the on-board computer to prevent or limit tyre lockup. One system is given in Fig. 34-37.

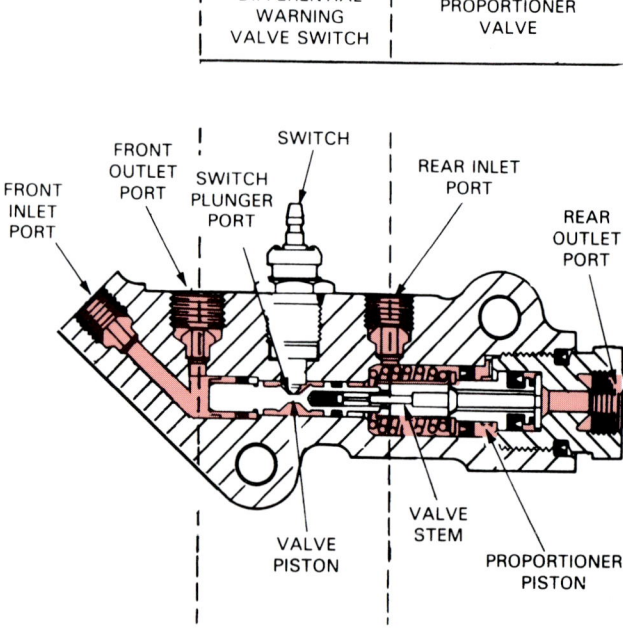

Fig. 34-35. Study the sections of a typical combination valve.

An electrical sensor is mounted at each wheel to measure wheel and tyre rpm. The sensors send alternating or pulsing current signals to the computer. If one wheel slows, the sensor signals reduced frequency and the computer activates the hydraulic valves to reduce pressure to that wheel's brake assembly. This prevents that tyre from skidding.

If a car's tyre were to lock up and slide, the car would NOT stop efficiently. A car stops the fastest when the tyres are almost ready to skid. The skid control system can detect when the wheel speed drops rapidly (ready to skid). The control unit can then send control pulses to the actuator. The actuator then cycles the brakes ON and OFF very quickly, for a controlled stop.

Since exact skid control systems vary, refer to a service manual for more details of system operation. Most systems, however, use the principles just discussed.

KNOW THESE TERMS

Brake pedal assembly, Master cylinder, Brake booster, Brake lines, Wheel brake assemblies, Parking brake, Disc brakes, Drum brakes, Caliper, Brake pads, Rotor, Wheel cylinder, Brake shoes, Hydraulic system Dual master cylinder, Primary and secondary pistons, Vacuum booster, Diagonally split, Bleeder screw, Anti-rattle clips, Pad wear sensor, Floating caliper, Fixed caliper, Backing plate, Asbestos lining, Primary and secondary shoes, Retracting and hold-down springs, Star wheel, Servo action, Braking ratio, Brake warning light, Proportioning valve, Combination Valve, Anti-skid system.

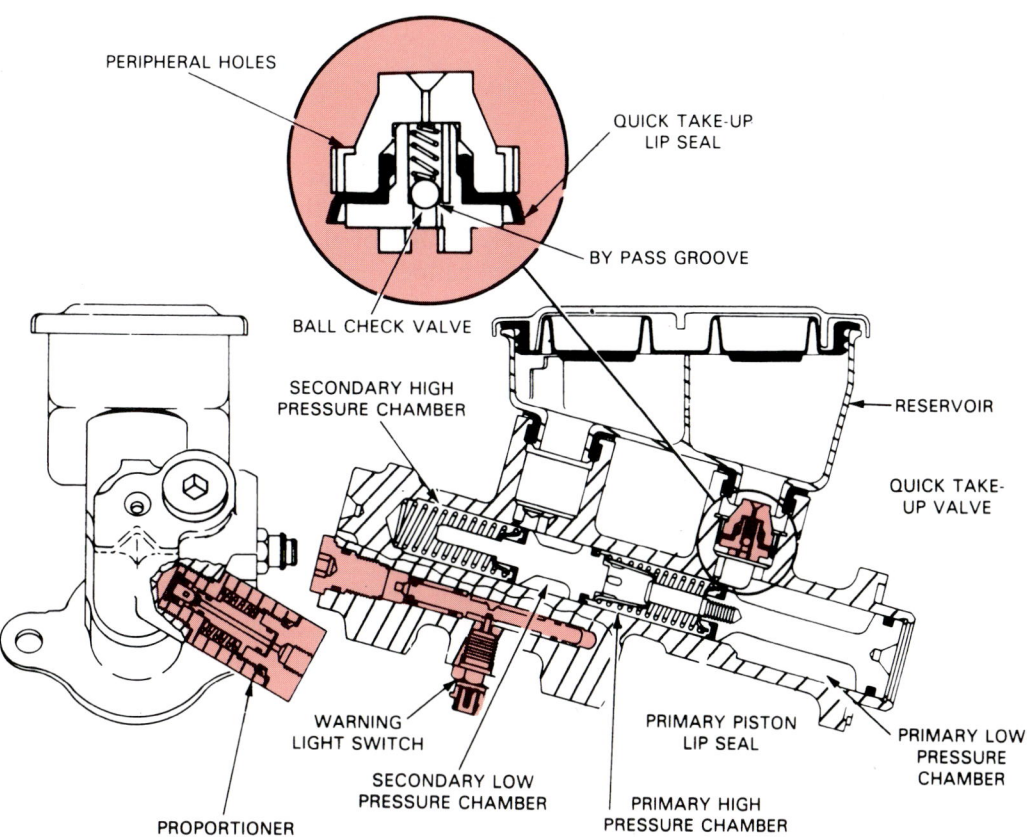

Fig. 34-36. This master cylinder has proportioning and warning lamp valves mounted internally.

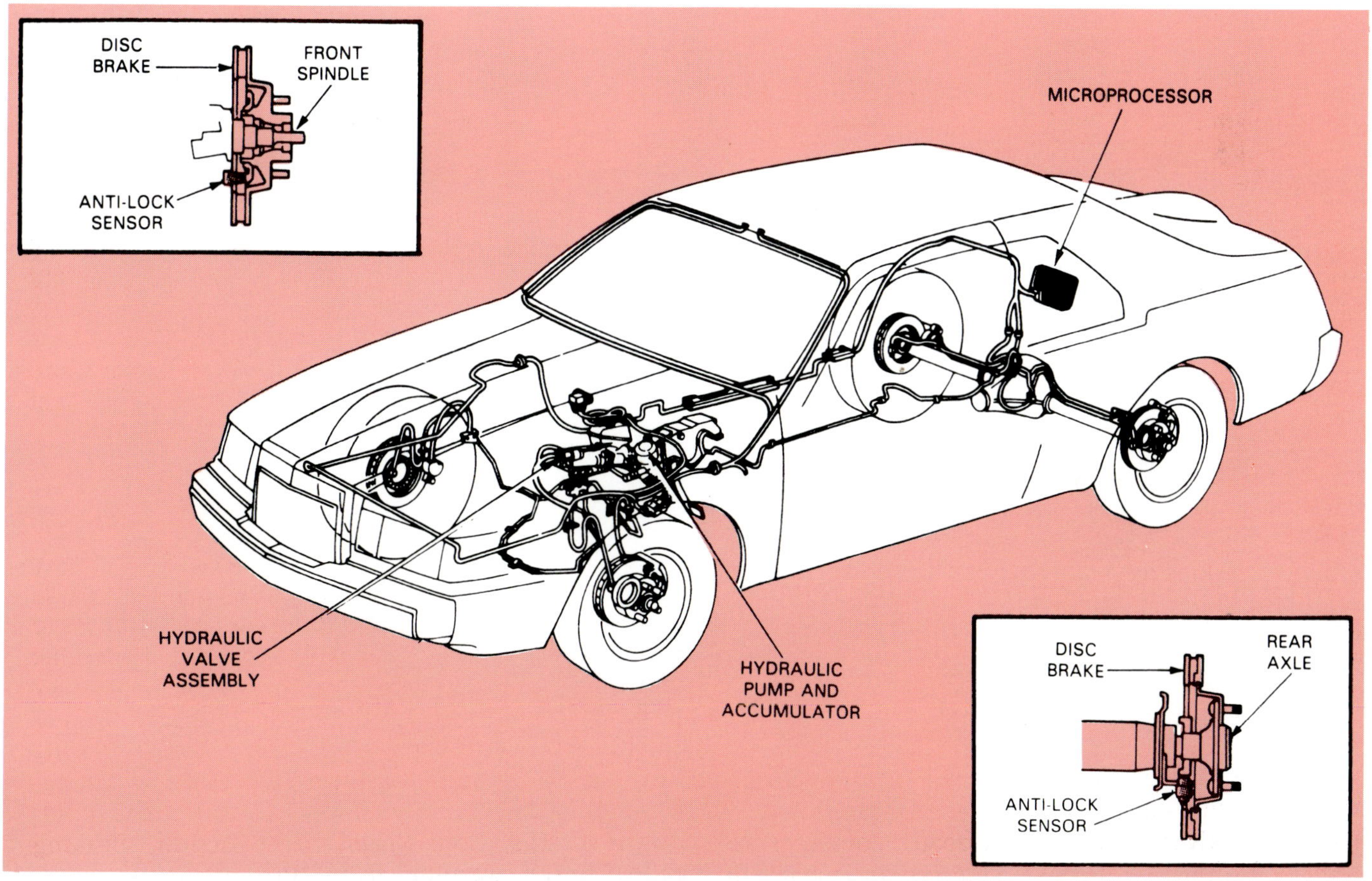

Fig. 34-37. Note typical parts of skid control system. Computer uses sensor data to prevent wheel lockup.

REVIEW QUESTIONS

1. List and explain the six major parts of a brake system.
2. Describe the four major parts of a disc brake assembly.
3. Which of the following is NOT part of a drum brake assembly?
   a. Wheel cylinder.
   b. Booster.
   c. Shoes.
   d. Drum.
4. A _____ system uses a liquid to transmit motion or pressure from one part to another.
5. What arc four functions of a master cylinder?
6. Why is a dual master cylinder used?
7. A power brake _____ _____ uses pedal pressure and engine vacuum to assist brake application.
8. There are two general types of vacuum brake boosters: atmospheric suspended type and vacuum suspended type. True or False?
9. In a diagonally split brake system, each master cylinder cup and piston assembly operates a brake assembly on opposite sides and corners of the car. True or False?

10. What causes a brake caliper piston to retract away from the rotor after brake application?
11. Which of the following is NOT part of a brake caliper?

   a. Piston seal.
   b. Bleeder screw.
   c. Piston boot.
   d. All of the above are correct.

12. Why are floating and sliding calipers more common than fixed calipers?
13. The _____ _____ provides a means of removing air from the wheel cylinder after repairs.
14. Explain the difference between a primary and secondary brake shoe.
15. The brake _____ _____ _____, also called _____ _____ _____, warns the driver of a pressure loss on one side of a dual brake system.
16. How does a proportioning valve equalise braking action?
17. Describe a combination valve.
18. Summarise the operation of a typical skid control system.

Chapter 35

IGNITION SYSTEM FUNDAMENTALS

After studying this chapter, you will be able to:
□ Explain the operating principles of an automotive ignition system.
□ Compare contact point, electronic, and computer-controlled ignition systems.
□ Describe the function of major ignition system components.
□ Explain vacuum, centrifugal, and electronic ignition timing advance.
□ Sketch the primary and secondary sections of an ignition system.
□ Compare ignition coil, spark plug, and distributor design variations.

The car's ignition system produces the high voltage needed to ignite the fuel charges in the cylinders of a petrol engine. The system must create an electric arc across the gaps at the spark plugs. These events must be timed so they happen exactly as each piston nears the top of its compression stroke. The heat of each spark starts combustion and produces the engine's power strokes.

In recent years, different types of ignition systems have been developed to improve engine performance, fuel economy, and dependability, Fig. 35-1. This chapter compares the older contact point type with more modern electronic and computer-coil (distributorless) ignitions. This should give you a sound knowledge of all automotive ignition systems.

FUNCTIONS OF AN IGNITION SYSTEM

An automotive ignition system has several functions:
1. Provide a method of turning a spark ignition or petrol engine ON and OFF.
2. Be capable of operating on various supply voltages (battery or alternator voltage).
3. Produce a high voltage arc at the spark plug electrodes to start combustion.
4. Distribute high voltage pulses to each spark plug in the correct sequence.

5. Time the spark so that it occurs as the piston nears TDC on the compression stroke.
6. Vary spark timing with engine speed, load, and other conditions.

Various ignition system parts and designs are used to achieve these functions.

BASIC IGNITION SYSTEM

An ignition system must change low battery voltage into very high voltage and then send the high voltage to the spark plugs. The parts needed to do this are shown in Fig. 35-2.
1. BATTERY (provides power for system).
2. IGNITION SWITCH (allows driver to turn ignition and engine on and off).
3. IGNITION COIL (changes battery voltage into 30,000 volts or more).
4. SWITCHING DEVICE (contact points or electronic circuit that operates ignition coil).
5. SPARK PLUG (air gap in combustion chamber for electric arc).
6. IGNITION SYSTEM WIRES (conductors that connect components).

With the ignition switch ON, current flows to the parts of the ignition system. When the switching device is closed (conducting current), current flows through and energises the ignition coil.

When the piston is nearing TDC on the compression stroke, the switching device opens. This causes high voltage to shoot out of the ignition coil and to the spark plug.

The electric arc at the plug ignites the fuel mixture. The mixture begins to burn, forming pressure in the cylinder for the engine's power stroke.

When the ignition key is turned to OFF, the battery-to-coil circuit is broken. Without current to the ignition coil, sparks are NOT produced at the spark plugs and the engine stops running.

An actual ignition system is much more complex than the one just discussed. Cars have multiple cylinder engines and the timing of the sparks must vary with operating conditions.

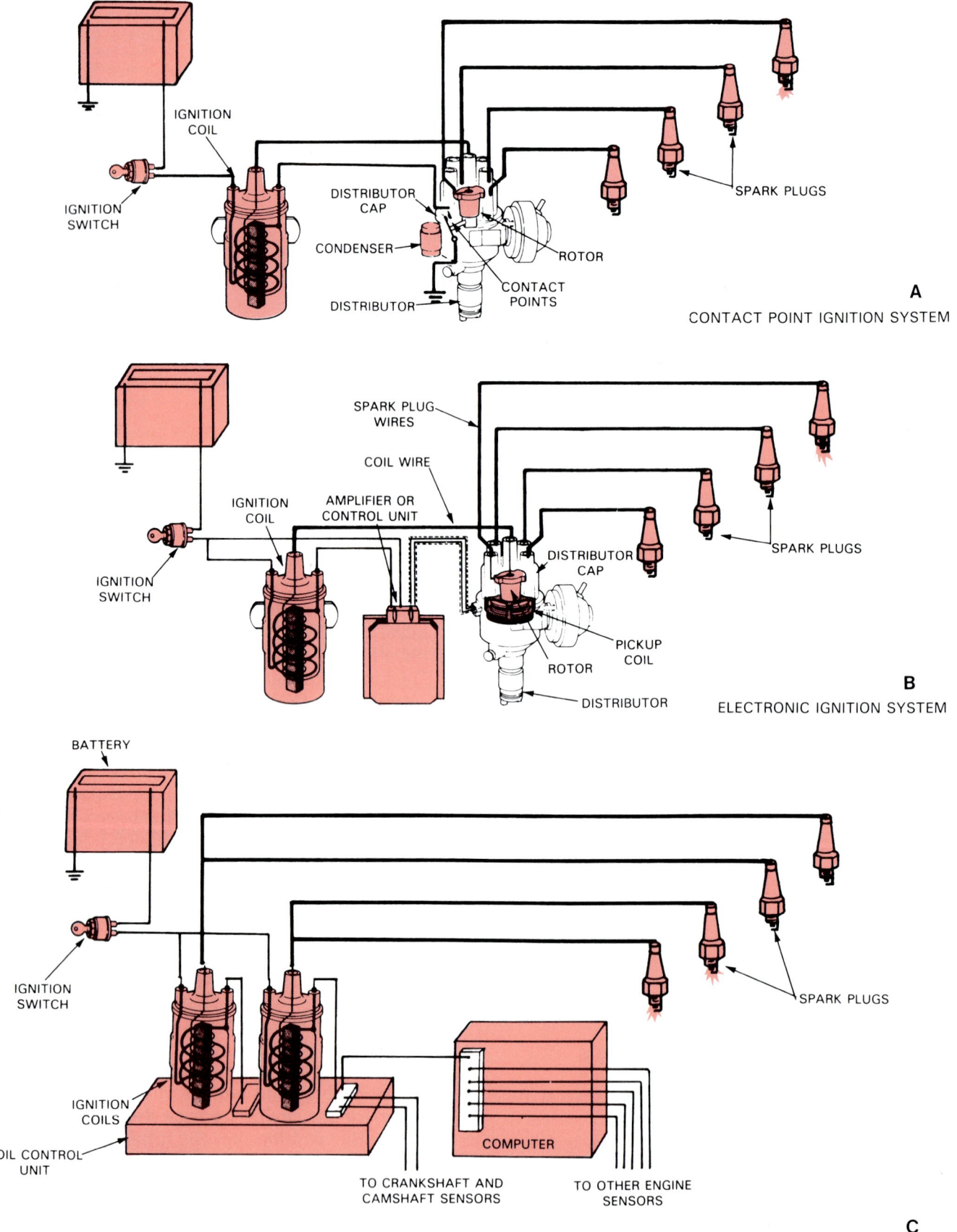

Fig. 35-1. There are three basic types of automotive ignition systems. A — Older contact point. B — Modern electronic type with distributor. C — Latest computer-coil ignition does not use a distributor. (Saab)

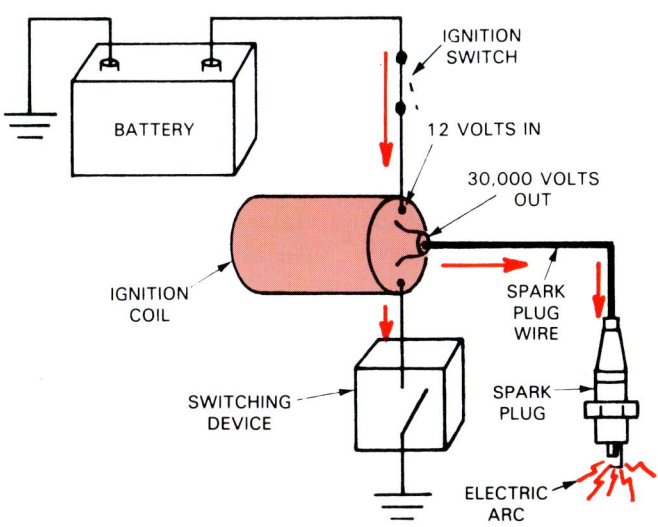

Fig. 35-2. Basic ignition system for one cylinder engine. Battery voltage is stepped up to about 30,000 volts by the coil before it is sent to the spark plug. Switching device times voltage to coil. It can be a set of mechanical breaker points or an electronic switching circuit.

IGNITION SYSTEM SUPPLY VOLTAGE

The ignition system supply voltage is fed to the ignition system by the battery or alternator. The battery provides electricity when starting the engine. After the engine is running, the alternator supplies a slightly higher voltage to the battery and ignition system.

Ignition switch

The ignition switch is a key-operated switch in the driver's compartment. A hot wire (voltage supply wire) connects the switch to the battery. Other terminals on the switch are connected to the ignition system, starter solenoid, and other electrical devices.

Bypass and resistance circuits

An ignition system bypass circuit is sometimes used to supply direct battery voltage to the ignition system during starter motor operation.

When the engine is being started, the ignition switch is in the start or fully clockwise position. Shown in Fig. 35-3A, this connects the battery to the starter motor and to the ignition system. The starter motor rotates the engine until the engine begins to run.

The starter motor draws high current and causes battery voltage to drop below 12.6 volts. The bypass circuit assures that there is still enough voltage and current for ignition operation and easy engine starting.

A resistor circuit may be used in the ignition system to limit supply voltage to the ignition during alternator operation. Look at Fig. 35-3B.

After the engine starts, the ignition key switch is released. A spring inside the switch causes it to return to the RUN POSITION.

To protect the ignition from damage, a resistor circuit is sometimes placed between the switch and ignition coil to limit current flow.

Either a special resistance wire (wire having internal resistance) or a ballast resistor (heat sensitive resistor that can regulate voltage to ignition coil) is used in the resistance circuit. This circuit assures that a relatively steady voltage of about 9.5 to 10.5 volts is applied to the ignition system.

Note! Many electronic ignition systems do not use bypass or resistance circuits.

PRIMARY AND SECONDARY CIRCUITS

The two main sections of an ignition system are the primary and secondary circuits.

The primary circuit of the ignition system includes all of the components and wires operating on low

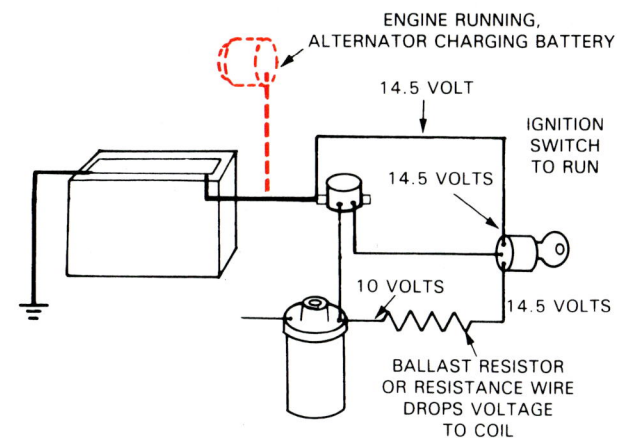

Fig. 35-3. Some ignition systems use resistance and bypass circuits to feed current to the ignition coil. A — When cranking, bypass feeds direct battery voltage to coil. B — After starting, resistance circuit feeds controlled voltage to coil.

voltage (battery or alternator voltage). See Fig. 35-4A.

The secondary circuit of the ignition system is the high voltage (30,000 volt or more) section. It consists of the wires and parts between the coil output and the spark plug ground, Fig. 35-4B.

The primary circuit of the ignition system uses conventional wire, similar to the wire used in the other electrical systems of the car. The secondary wiring however, must have much THICKER INSU-LATION to prevent leakage (arcing) of the high voltage.

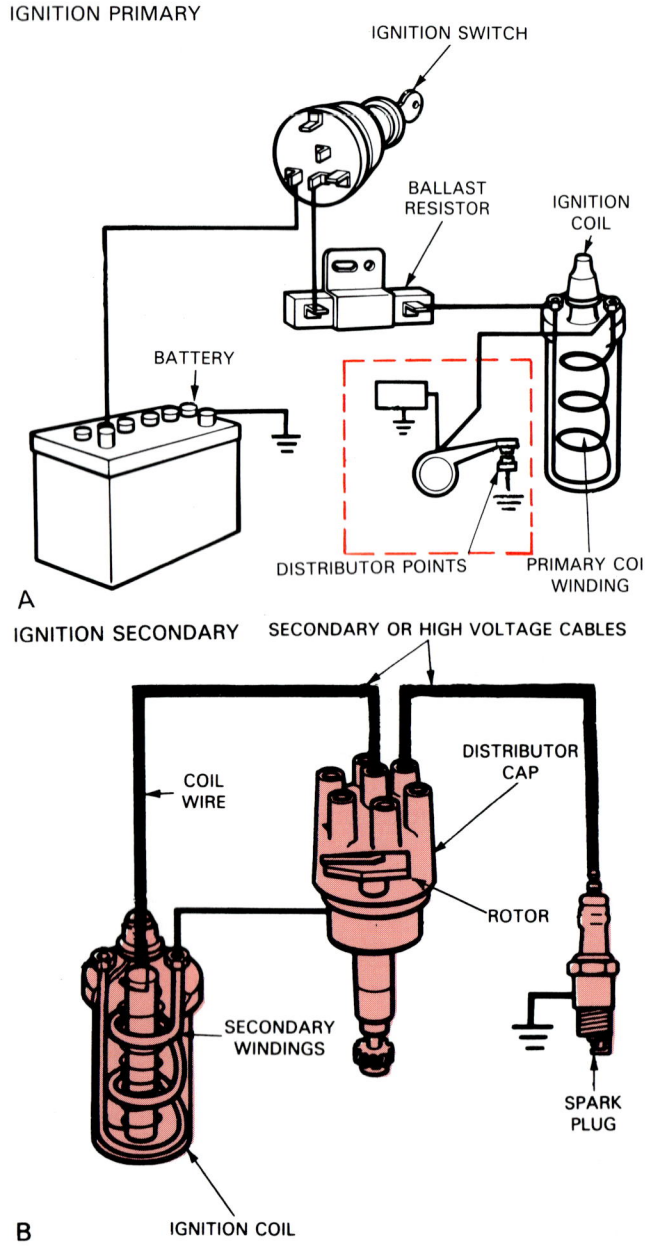

Fig. 35-4. Two major sections of an ignition system. A — Primary circuit includes all parts working on battery voltage. B — Secondary circuit consists of parts carrying high coil output voltage.

IGNITION COIL

An ignition coil produces the high voltage (30,000 volts or more) needed to make current jump the gap at the spark plugs. It is a pulse type transformer capable of producing a short burst of high voltage for starting combustion.

As in Fig. 35-4B, coil output voltage usually passes through the coil wire, distributor, plug wire, and spark plug before starting the burning process in the engine.

Ignition coil construction

Shown in Fig. 35-5, the ignition coil consists of two sets of windings (insulated wire wrapped in circular pattern). The coil has two primary terminals (low voltage connections), an iron core (long piece of iron inside windings), and a high voltage terminal (output or coil wire connection).

The primary windings of the coil are several hundred turns of heavy wire, wrapped around or near the secondary windings.

The secondary windings are several thousand turns of very fine wire located inside or near the primary windings.

Both windings are wrapped around an iron core and are housed inside the coil case.

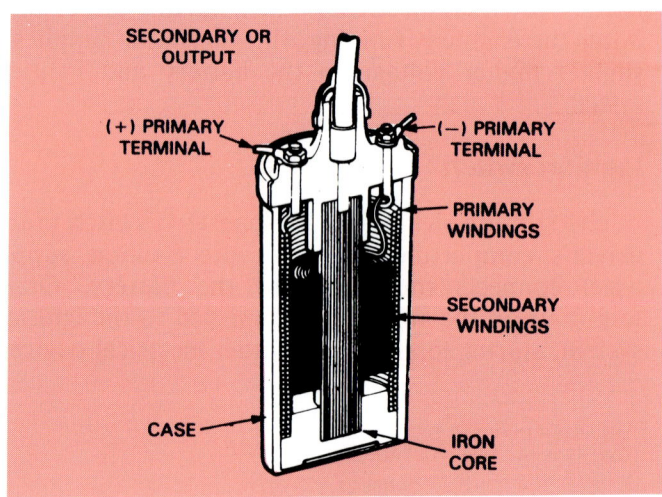

Fig. 35-5. Cutaway of ignition coil shows basic parts. Primary windings surround secondary windings. Iron core is mounted in centre of windings.

Ignition coil operation

When battery current flows through the ignition coil primary windings, a strong magnetic field is produced. Look at Fig. 35-6A. The action of the iron core helps concentrate and strengthen the field.

When the current flowing through the coil is broken, the magnetic field COLLAPSES across the secondary windings. See Fig. 35-6B.

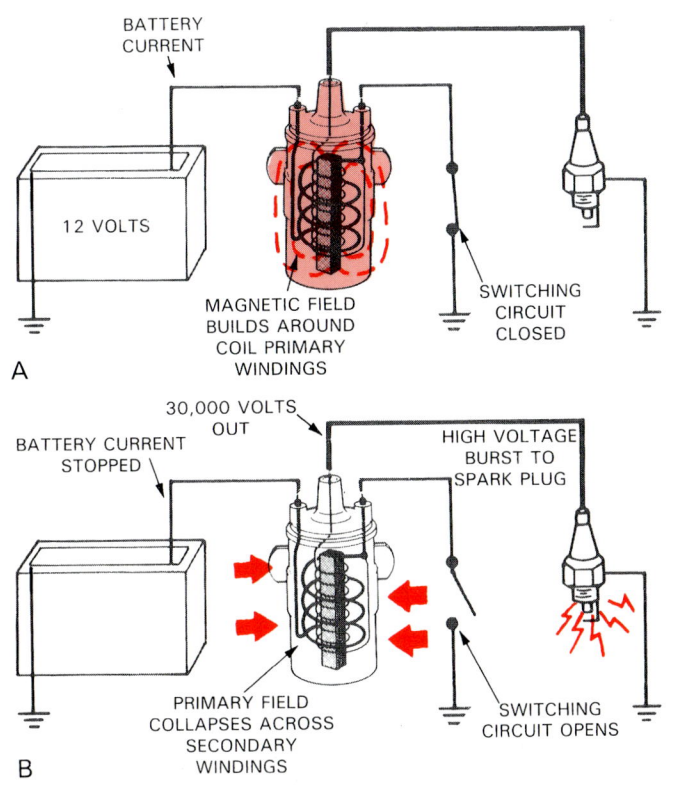

Fig. 35-6. Ignition coil operation. A — With switching device (points or electronic circuit) closed, current flows through ignition coil primary windings. Strong magnetic field builds in coil. B — When switching device opens, current flow stops and magnetic field collapses across secondary windings. This induces high voltage in secondary windings of coil. The spark plug fires. (Saab)

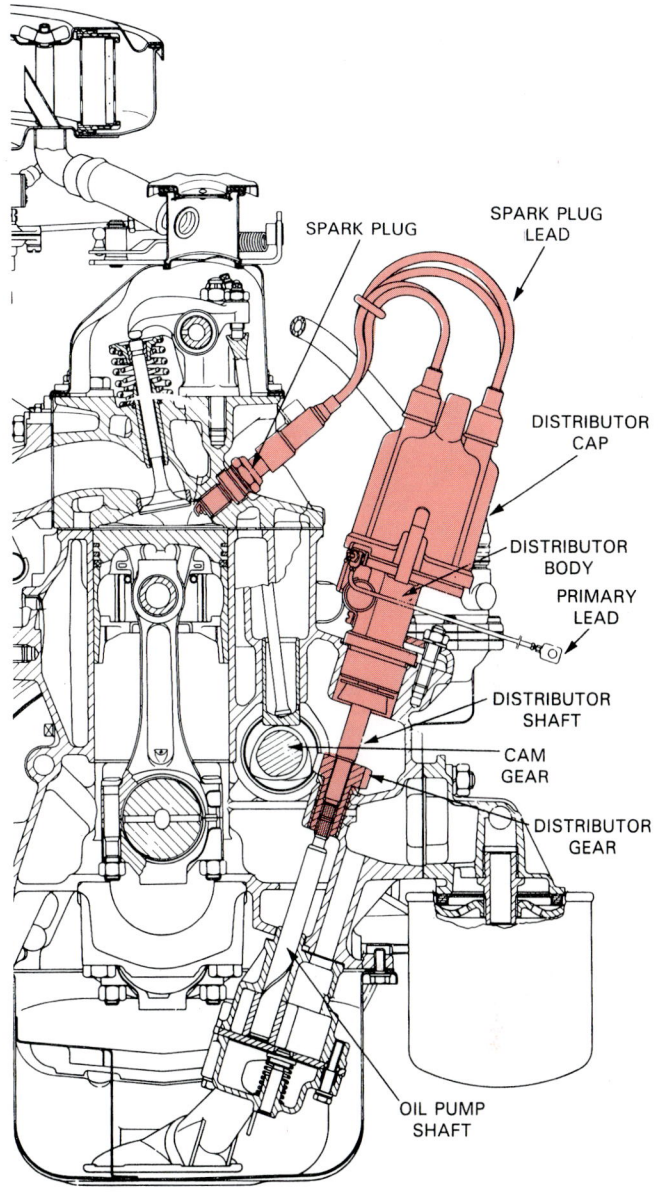

Fig. 35-7. Ignition distributor is usually driven by engine camshaft. Small gear on cam drives gear on distributor at one-half engine rpm. Main purpose of distributor is to feed coil voltage to spark plugs. (Fiat)

Since the secondary windings have more turns than the primary, 30,000 volts is induced into the secondary windings. High voltage shoots out of the top terminal and to a spark plug.

There are two common methods used to break current flow and fire the coil: mechanical breaker points or an electronic switching current.

IGNITION DISTRIBUTORS

Typically, an ignition distributor, Fig. 35-7, has several functions:

1. It actuates the ON/OFF cycles of current flow through the ignition coil primary windings.
2. It distributes the coil's high voltage pulses to each spark plug wire.
3. It must cause the spark to occur at each plug earlier on the compression stroke as engine speed increases and vice versa.
4. It changes spark timing with changes in engine load. As more load is placed on the engine, the spark timing must occur later in the compression stroke to prevent spark knock (abnormal combustion).
5. Sometimes, the bottom of the distributor shaft powers the engine oil pump.
6. Some distributors (unitised distributors) house the ignition coil and electronic switching circuit. Refer to Fig. 35-8.

Distributor types

An ignition distributor can be a contact point (mechanical) or pickup coil (used with electronic switching circuit) type. A contact point distributor is commonly used on older cars. The pickup coil type distributor is used on many modern automobiles.

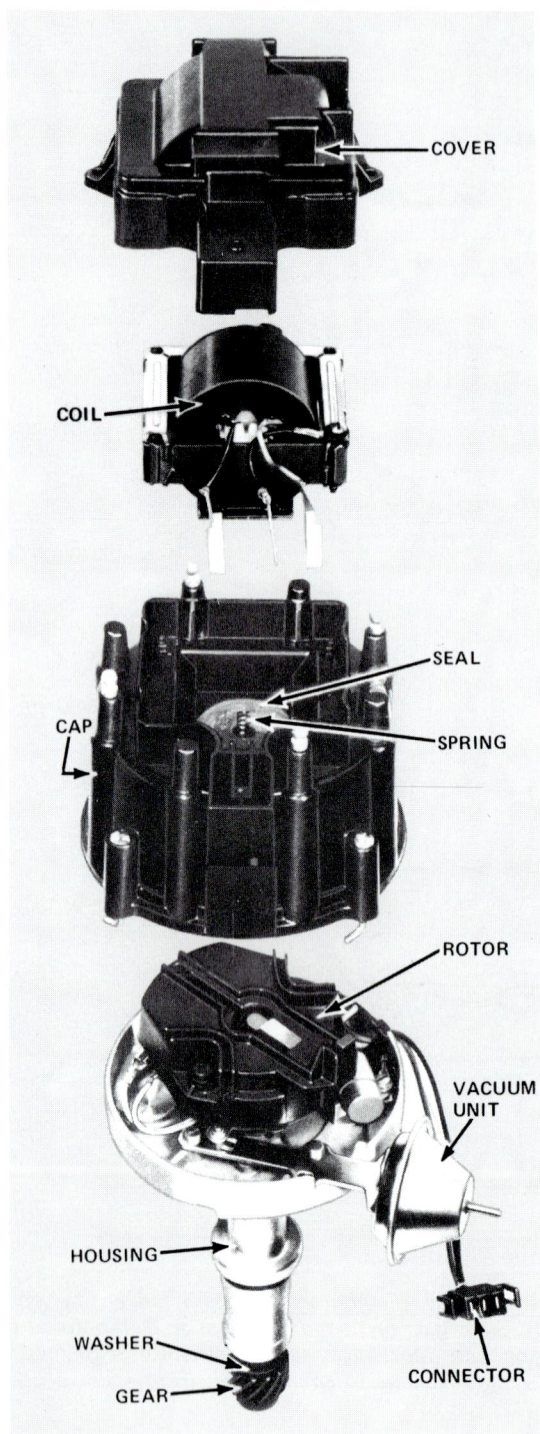

Fig. 35-8. Modern unitised distributor has ignition coil and amplifier (electronic switching circuit) mounted inside. Note part names and locations.

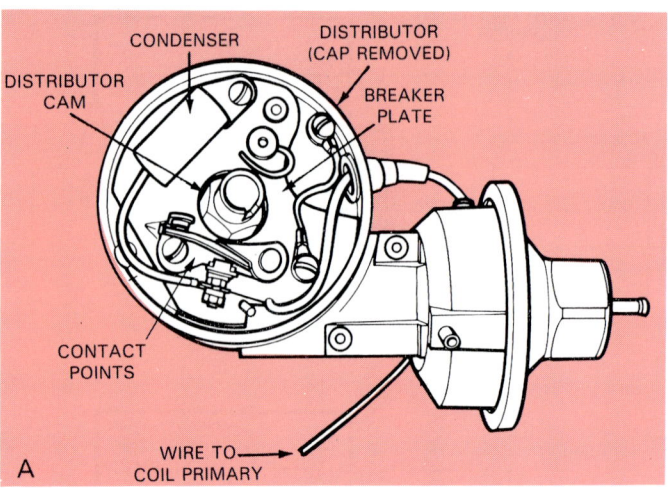

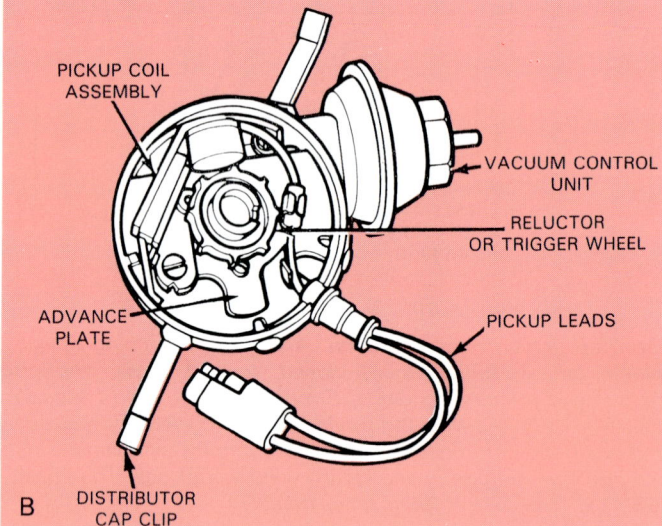

Fig. 35-9. Compare distributors. Trigger wheel and pickup coil replace points in modern electronic ignition. A — Contact point distributor. B — Pickup coil distributor for electronic ignition.

Note the basic differences between the two in Fig. 35-9.

A contact point distributor uses mechanical breaker points to interrupt the flow of primary current through the ignition coil. See Fig. 35-10.

A pickup coil distributor has a trigger wheel and a pickup coil instead of contact points. Wires from the pickup coil are connected to an ECU (electronic control unit). Refer to Fig. 35-9 and 35-10. The trigger wheel pickup coil, and ECU perform the same function as contact points.

CONTACT POINT IGNITION SYSTEM

Before going on to study today's electronic ignition systems, you should have a basic understanding of contact point systems. The operation of each is similar in many ways. Fig. 35-10 compares the two.

The distributor for a contact point ignition consists of the following:

The distributor cam is the lobed part on the distributor shaft that opens the contact points. The cam turns with the shaft at one-half engine speed. One lobe is normally provided for each spark plug. See Fig. 35-9.

The contact points, also called breaker points, act like a spring-loaded electrical switch in the distributor. Small screws hold the contact points on the distributor advance plate. A rubbing block, of fibre

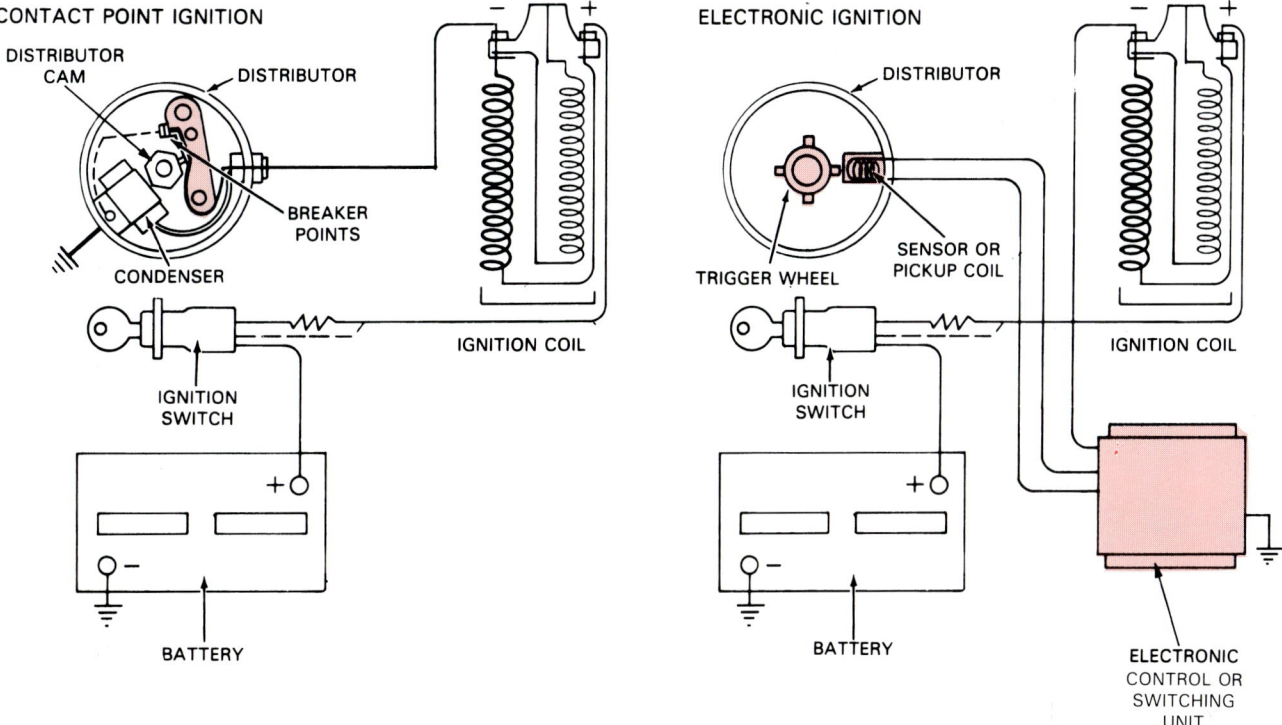

Fig. 35-10. Compare a contact point and an electronic ignition system. Note that pickup coil and control unit replace contact points in modern system.

material, rides on the distributor cam. Wires from the condenser and ignition coil primary connect to the points.

The condenser or capacitor prevents the contact points from arcing and burning. It also provides a storage place for electricity as the points open. This electricity is fed back into the primary when the points reclose. Refer to Fig. 35-10.

Contact point ignition system operation

With the engine running, the distributor shaft and distributor cam rotate. This causes the cam to open and close the points.

Since the points are wired to the primary windings of the ignition coil, the points make and break the ignition coil primary circuit. When the points are closed, a magnetic field builds in the coil. When the points open, the field collapses and voltage is sent to one of the spark plugs.

With the distributor rotating at one-half engine rpm and with one cam lobe per engine cylinder, each spark plug fires once during a complete revolution of the distributor cam.

Point dwell (cam angle)

Dwell or cam angle is the amount of time, given in degrees of distributor rotation, that the points remain closed between each opening. Look at Fig. 35-11.

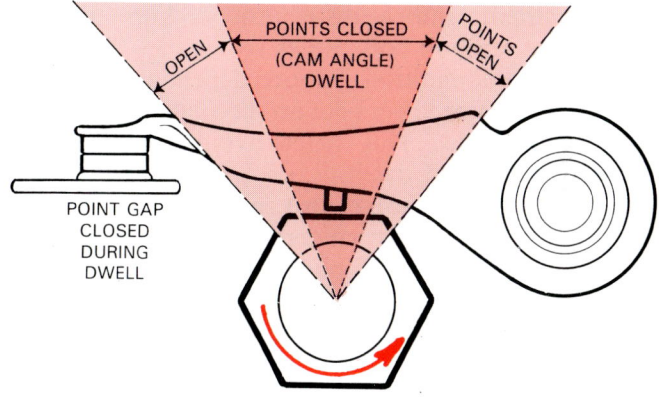

Fig. 35-11. Dwell is time points remain closed in degrees of distributor rotation. Points gap is distance between two points in fully open position. Dwell affects point gap and vice versa.

A dwell period is needed to assure that the coil has enough time to build up a strong magnetic field.

Without enough point dwell, a weak spark would be produced. With too much dwell, the point gap (distance between fully open points) would be too narrow. Point arcing and burning could result.

ELECTRONIC IGNITION SYSTEM

An electronic ignition system, also called a solid state or transistor ignition system, uses an electronic

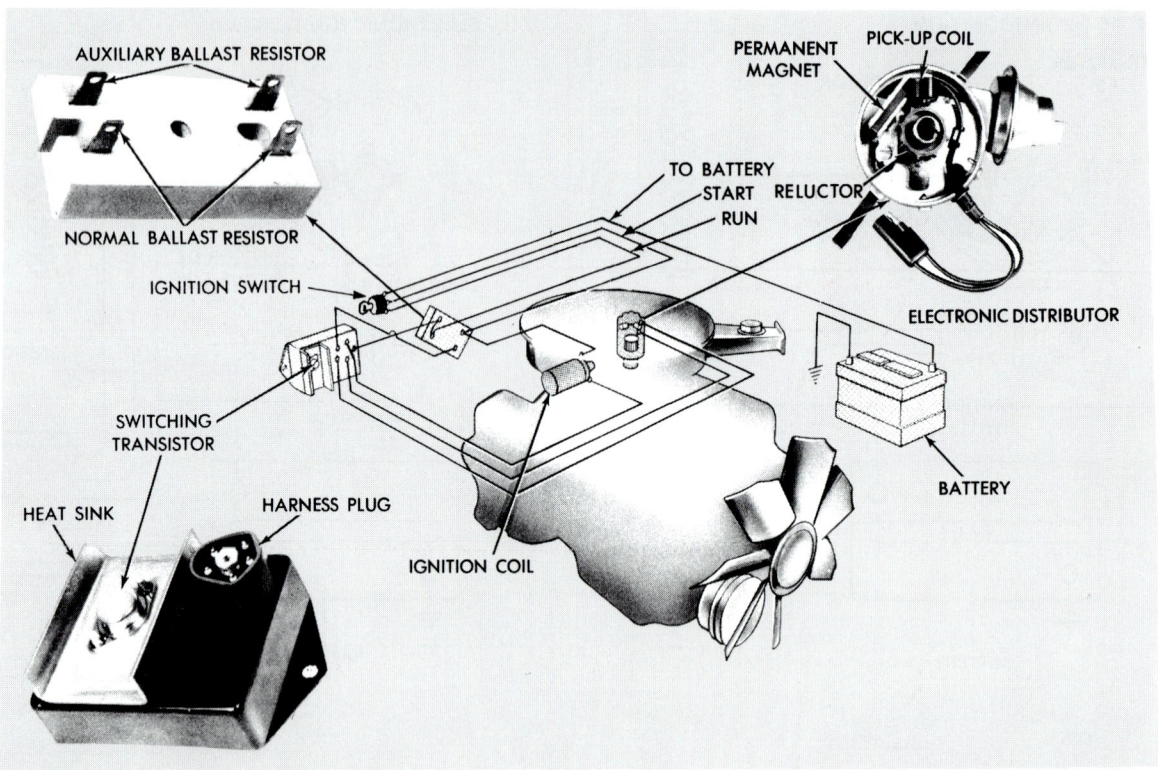

Fig. 35-12. Amplifier or control module contains switching circuit or transistor that operates ignition coil. Note relationship of parts.

control circuit and a distributor pickup coil to operate the ignition coil. Refer to Fig. 35-12.

An electronic ignition is more dependable than a contact point type. There are no mechanical breakers to wear or burn. This helps avoid trouble with ignition timing and dwell.

An electronic ignition is also capable of producing much higher secondary voltages. This is an advantage because wider spark plug gaps and higher voltages are needed to ignite lean air-fuel mixtures. Lean mixtures are now used for reduced exhaust emissions and fuel consumption.

Trigger wheel

The trigger wheel, also called reluctor or pole piece, is fastened to the upper end of the distributor shaft. See Fig. 35-12. The trigger wheel replaces the distributor cam used in a contact point distributor. One tooth is normally provided on the wheel for each engine cylinder.

Pickup coil

The pickup coil, also termed sensor assembly or sensor coil, produces tiny voltage pulses for the ignition system's electronic control unit (module or amplifier). Look at Fig. 35-13. The sensor assembly is a small set of windings forming a coil.

As a trigger wheel tooth passes the pickup coil, it strengthens the magnetic field around the coil. This

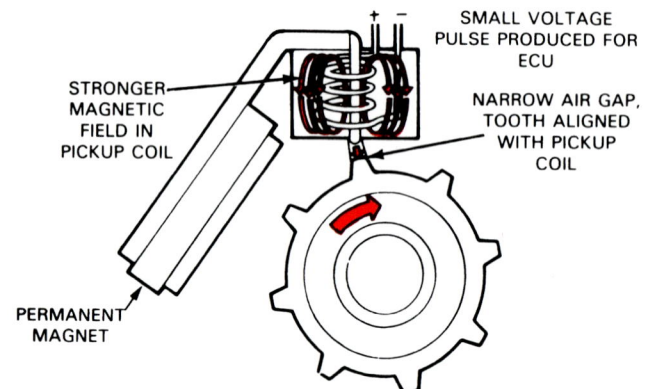

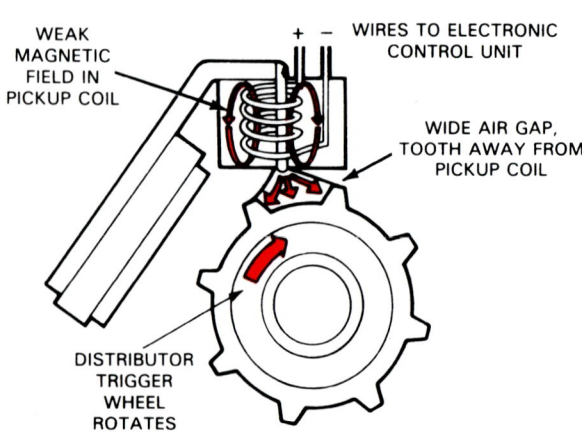

Fig. 35-13. Distributor magnetic pickup coil operation.

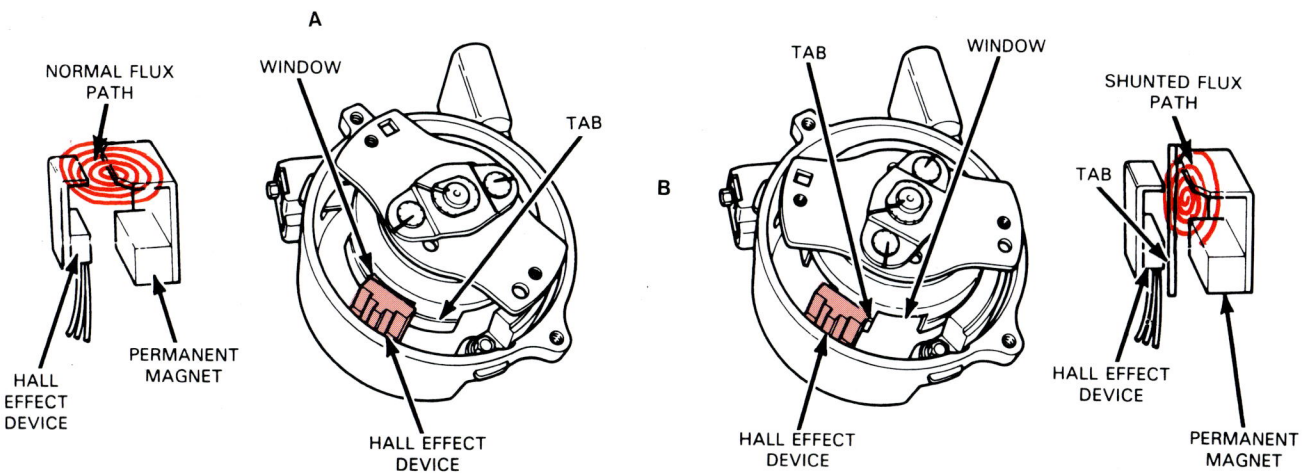

Fig. 35-14. Hall effect pickup chip is similar to magnetic pickup. A — Trigger wheel window (opening) allows strong magnetic field to develop around pickup. B — As trigger wheel rotates, tab or tooth moves between pickup and permanent magnet. This decreases pickup field strength and voltage. (Ford)

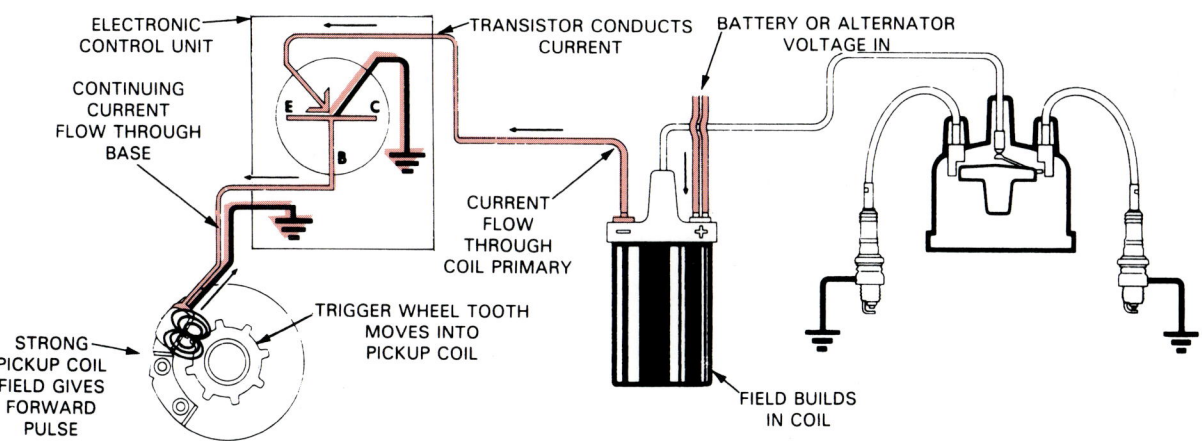

A — As trigger wheel tooth aligns with pickup coil, current flow through base of transistor turns transistor on. Current flows through ignition coil primary and through emitter-collector of transistor. Strong field builds in ignition coil primary windings.

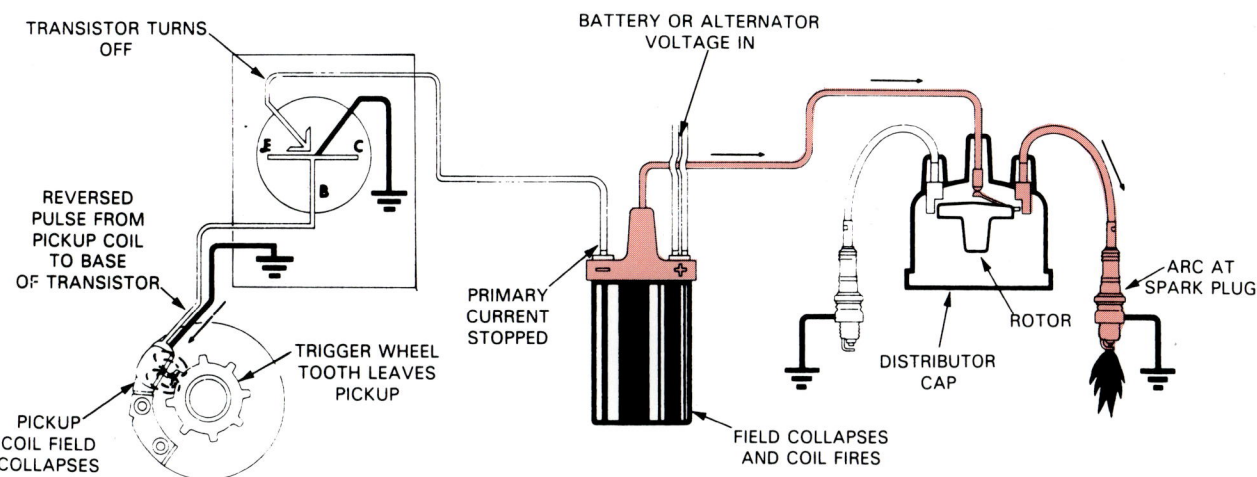

B — Just as trigger wheel passes pickup coil, current pulse flows out of pickup and to base of transistor. Electrical pulse is opposite the polarity of emitter-base voltage. This turns transistor off. Without current flow through emitter and collector, field collapses in ignition coil and 30,000 volts are induced into coil secondary windings. Spark plug fires.

Fig. 35-15. Pickup coil and amplifier switching action.

causes a change in the current flow through the coil. As a result, an electrical pulse (voltage or current change) is sent to the electronic control unit (module or amplifier) as the trigger wheel teeth pass the pickup unit.

Hall effect pickup

Fig. 35-14 shows the action of similar Hall effect pickup. This pickup is a solid state chip or module. A constant amount of current is sent through the device. A permanent magnet is located next to the Hall effect chip.

When the trigger wheel passes between the permanent magnet and the Hall effect chip, the magnetic field is blocked from the chip, decreasing output voltage (sensor or switch OFF). When the trigger wheel tooth moves out from between the magnet and chip, magnetic field action on the chip increases its voltage output (sensor or switch ON). This ON/OFF action operates the ECU.

Ignition switch ECU

The ignition system electronic control unit, amplifier or control module, is an "electronic switch" that turns the ignition coil primary current on and off. See Figs. 35-15A and 35-15B. The ECU does the same thing as contact points.

An ignition ECU is a network of transistors, resistors, capacitors, and other electronic components. A typical module and related wiring is shown in Fig. 35-16. The circuit is sealed in a plastic or metal housing.

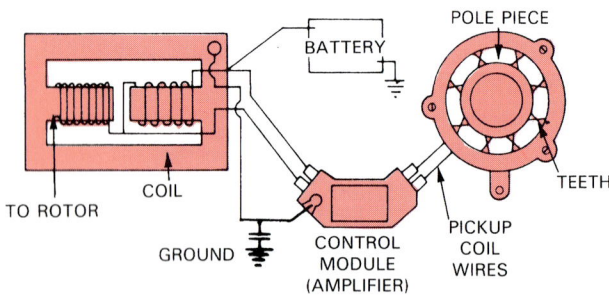

Fig. 35-16. Note electrical connections to typical amplifier. Amplifier contains electronic circuit with several transistors.

The ECU can be located:
1. On the side of the distributor, Fig. 35-17.
2. Inside the distributor, Fig. 35-17.
3. In the engine compartment, Fig. 35-17.
4. Under the car dash.

Electronic ignition system operation

With the engine running, the trigger wheel rotates

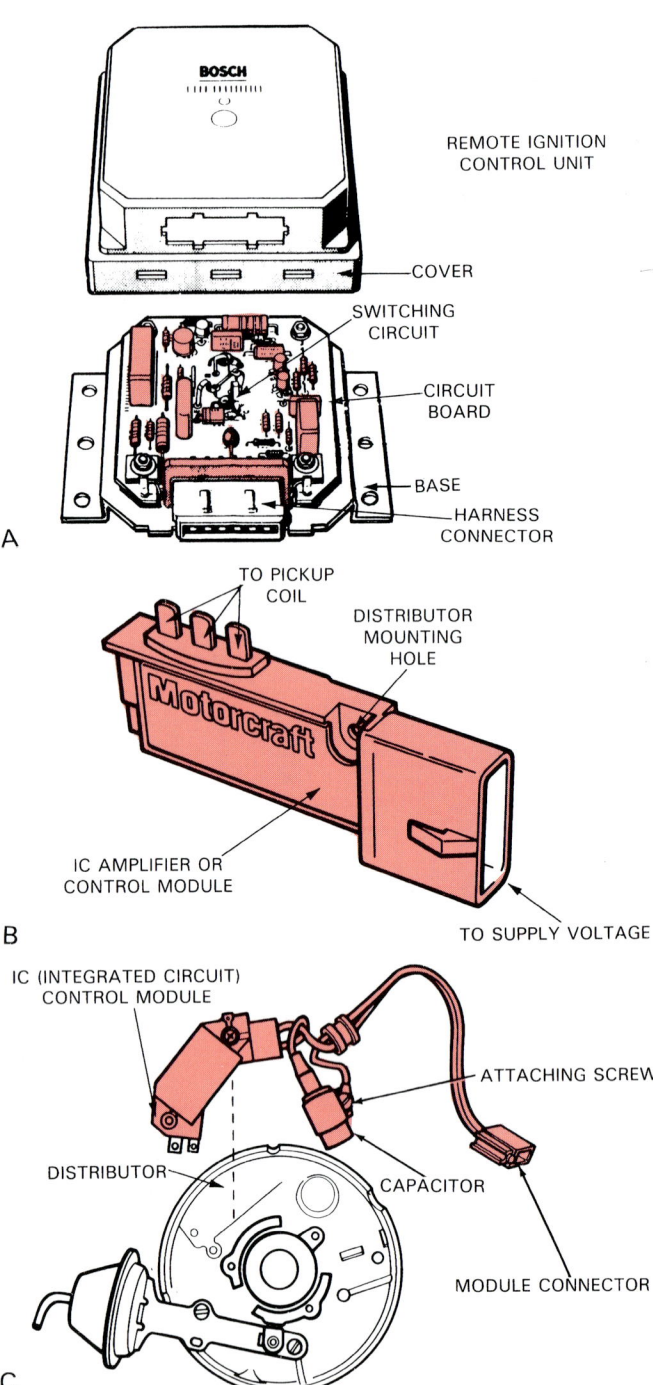

Fig. 35-17. Ignition control unit variations. A — Remote mounted ignition amplifier. B — Amplifier mounted on outside of distributor body. C — Amplifier mounted inside distributor. (Bosch, Ford)

inside the distributor. As the teeth pass the pickup, a change in the magnetic field causes a change in output voltage or current. This results in engine rpm electrical signals entering the ECU, Fig. 35-16.

The ECU increases these tiny pulses into ON/OFF current cycles for the ignition coil. When the ECU is ON, current flows through the primary windings of the ignition coil, developing a magnetic field. Then, when the trigger wheel and pickup turn the ECU OFF, the ignition coil field collapses and fires a spark plug.

ECU dwell time (number of degrees circuit conducts current to ignition coil) is designed into the ECU's electronic circuit. It is not adjustable.

SECONDARY (HIGH VOLTAGE) COMPONENTS

The secondary components of the ignition system operates on the high voltage from the ignition coil. Most of these parts are the same in both contact point and electronic ignition systems.

Since electronic ignitions produce much higher voltages, their parts are designed to withstand higher voltages without electrical leakage (electricity sparking through insulation or from one component to another).

The coil lead carries high voltage from the high voltage (high tension) terminal of the ignition coil to the centre terminal of the distributor cap, Fig. 35-18. It is constructed like a very short spark plug lead.

With a unitised distributor or distributorless ignition, a coil lead is NOT needed.

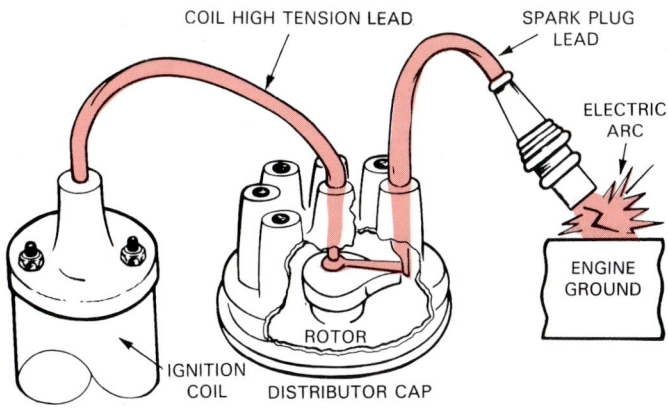

Fig. 35-18. With a distributor, ignition coil output is fed through coil lead to distributor cap centre terminal. Rotating rotor then feeds high voltage to each spark plug lead.

Distributor cap and rotor

The distributor cap is an insulating, plastic component that fits over the top of the distributor housing. Its centre terminal transfers voltage from the coil lead to the rotor, Fig. 35-18.

The distributor cap also has outer or side terminals that send electric arcs to the spark plug leads. Metal terminals are moulded into the plastic cap to make the electrical connections.

The distributor rotor transfers voltage from the coil lead (distributor cap centre terminal) to the spark plug leads (distributor cap outer terminals). Look at Fig. 35-18 again.

The rotor is mounted on top of the distributor shaft. It is a rotating electrical switch that feeds voltage to each spark plug lead.

A metal terminal on the rotor touches the distributor cap centre terminal. The outer end of the rotor terminal ALMOST touches the outer cap terminals. Fig. 35-18.

Voltage is high enough so that it can jump the air space between the rotor and cap. About 3000 volts is used as the spark jumps this rotor-to-cap gap.

Spark plug leads

Spark plug leads carry coil voltage from the distributor cap side terminals to each spark plug. See Fig. 35-18. In more modern computer-coil (distributorless) ignitions, the spark plug leads carry coil voltage directly to the plugs.

Spark plug lead boots protect the metal connectors from corrosion, oil, and moisture. Boots usually fit over both ends of the secondary leads.

Solid wire spark plug leads are used on racing engines and very old automobiles. The wire conductor is simply a stranded metal wire. Solid wires are no longer used because they cause radio interference (noise or static in speakers).

Resistance spark plug leads are now used because they contain internal resistance that prevents radio noise. They use carbon-impregnated strands of rayon braid. Look at Fig. 35-19.

Also called radio suppression leads, they have about 33,000 ohms per metre resistance. This avoids high voltage induced popping or cracking in the radio speakers.

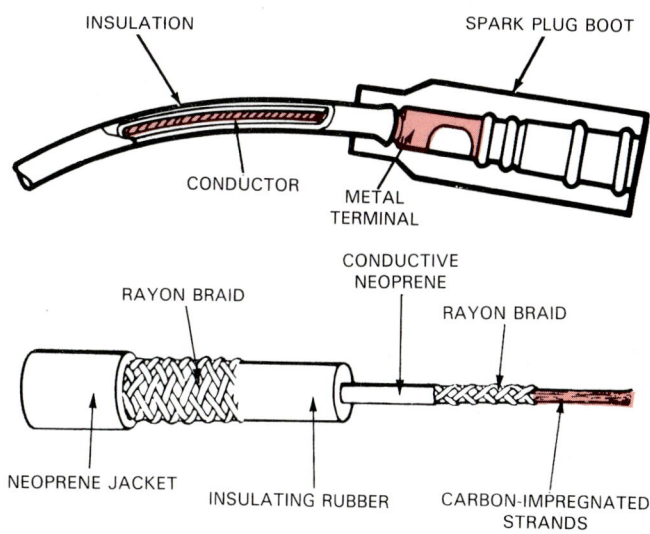

Fig. 35-19. Secondary leads have very thick insulation. Modern types contain carbon-impregnated strands that provide resistance to prevent radio interference. (Champion Spark Plugs)

SPARK PLUGS

The spark plugs use ignition coil high voltage to ignite the fuel mixture. Somewhere between 4000 and 10,000 volts are needed to make current jump the gap at the plug electrodes. This is much lower than the coil's output potential.

Fig. 35-20 shows the basic parts of a spark plug. The centre terminal conducts electricity into the combustion chamber.

The grounded side electrode causes the electricity to jump the gap and return to the battery through frame ground.

The ceramic insulator keeps the high voltage at the plug lead from shorting to ground before producing a spark in the engine cylinder.

The steel shell supports the other parts of the plug and has threads for screwing the plug into the engine cylinder head.

Spark plug reach

Spark plug reach is the distance between the end of the plug threads and the seat or sealing surface on the plug. Refer to Fig. 35-20. Plug reach determines how far the plug extends through the cylinder head.

If spark plug reach is too long, the plug electrode may be struck by the piston at TDC. If reach is too short, the plug electrodes may not extend far enough into the chamber and combustion efficiency may be reduced.

Resistor and non-resistor spark plugs

A resistor spark plug, like a resistor plug lead, has internal resistance (around 10,000 ohms) designed to

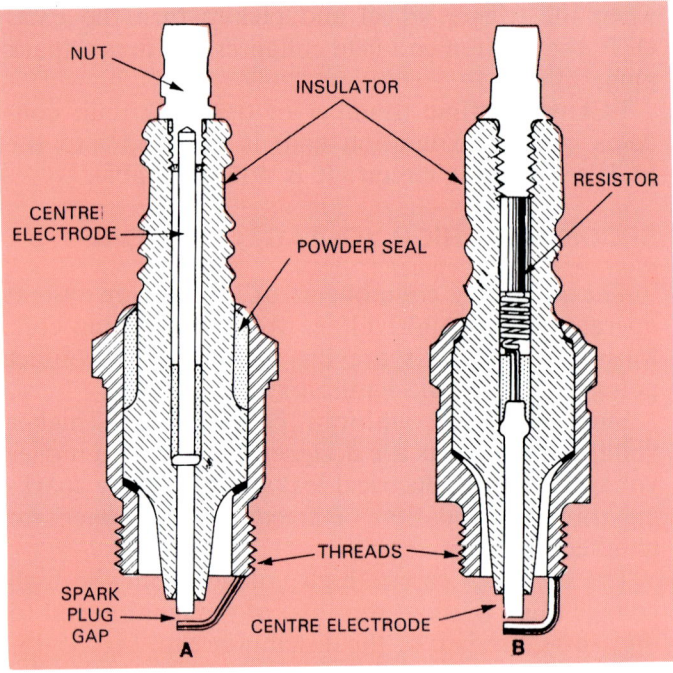

Fig. 35-21. Cutaway shows internal parts of plug. A — Non-resistor plug has solid metal centre electrode. B — Resistor plug has small resistor between two-piece centre electrode. Plug gap is space between side and centre electrodes.

reduce static in radios and televisions. Most new cars require resistor plugs.

Fig. 35-21 illustrates resistor, non-resistor, and other spark plug types.

A non-resistor spark plug has a solid metal rod forming the centre electrode. This type is NOT commonly used, except for racing or off-road applications.

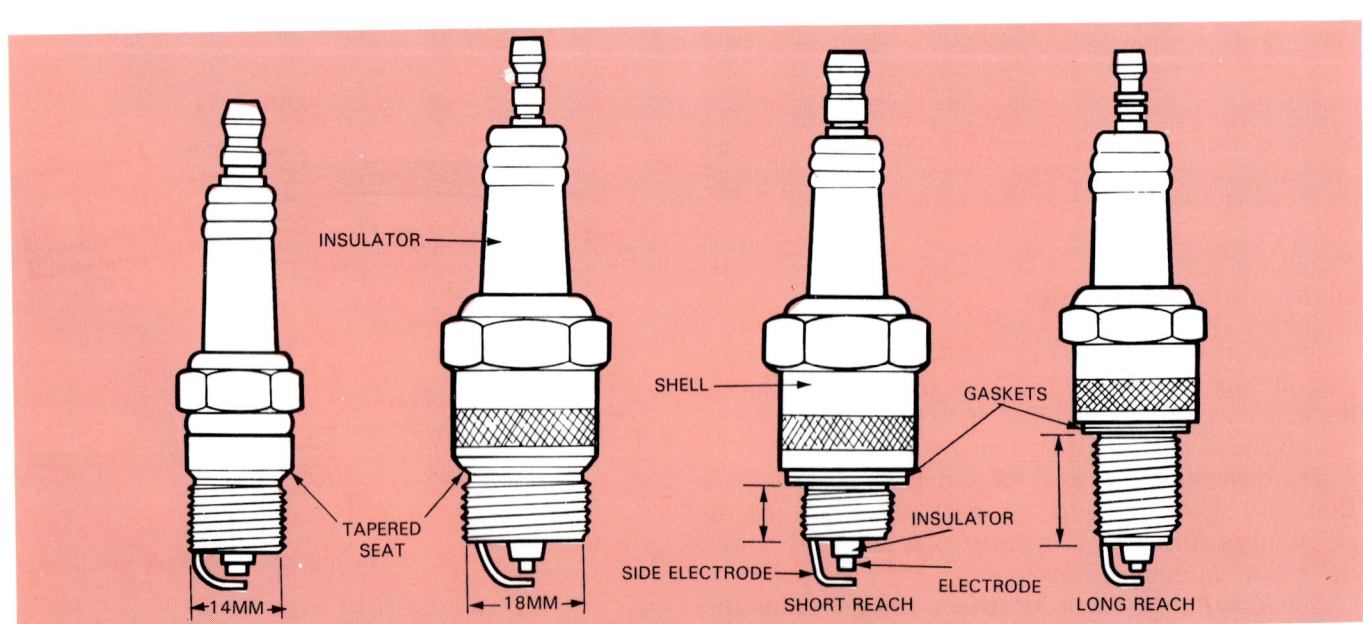

Fig. 35-20. Note spark plug variations. Small 14 mm plug is commonly used in today's engines. Larger 18 mm are for older engines. Reach is length of plug threads.

Spark plug gap

Spark plug gap is the distance between the centre and side electrodes, Fig. 35-21. Normal gap specs range from 0.5 mm to 1.0 mm.

Smaller spark plug gaps are used on older cars equipped with contact point ignition systems. Larger spark plug gaps are now used with modern electronic ignition systems.

Spark plug heat range

Spark plug heat range is a rating of the operating temperature of the spark plug tip. Plug heat range is basically determined by the length and diameter of the insulator tip and ability of the plug to transfer heat into the cooling system. Refer to Fig. 35-22.

A hot spark plug has a long insulator tip and will tend to burn off deposits. This provides a self-cleaning action.

A cold spark plug has a shorter insulator tip; its tip operates at a cooler temperature. A cold plug is used in engines operated at high speeds. The cooler tip will help prevent tip overheating and pre-ignition.

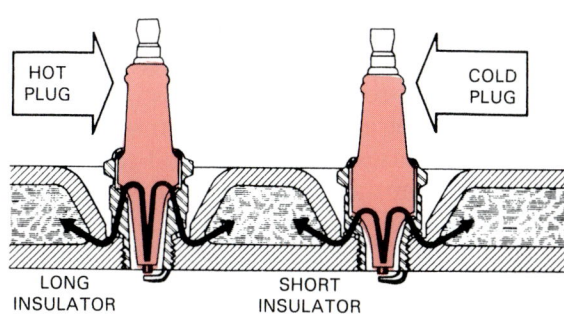

Fig. 35-22. Hot plug has long insulator that prevents heat transfer into water jackets. It will burn off oil deposits. Cold plug has shorter insulator.

Heat range ratings

Auto manufacturers normally recommend a specific spark plug heat range for their engines. The heat range will normally be coded and given as a number on the plug insulator.

Generally, the larger the number on the plug, the hotter the spark plug tip will operate. For instance, a 52 plug would be hotter than a 42 or 32.

The only time you should deviate from plug heat range specs is when abnormal engine or driving conditions are encountered. For example, a hotter plug may be installed in an old, worn out, oil-burning engine. The hotter plug will help burn off oil deposits and prevent oil fouling of the plug.

IGNITION TIMING

Ignition timing, also called spark timing, refers to how early or late the spark plugs fire in relation to the position of the engine pistons. Ignition timing must vary with engine speed, load, and temperature.

Timing advance occurs when the spark plugs fire sooner on the engine's compression strokes. The timing is set several degrees before TDC. More timing advance is needed at higher engine speeds to give combustion enough time to develop pressure on the power stroke.

Timing retard occurs when the spark plugs fire later on the compression strokes. It is the opposite of timing advance. Spark retard is needed at lower engine speeds and under high load conditions. Timing retard prevents the fuel from burning too much on the compression stroke, causing a spark knock or ping (abnormal combustion).

There are three basic methods used to control ignition system spark timing:

1. DISTRIBUTOR CENTRIFUGAL ADVANCE (controlled by engine speed), Fig. 35-23.

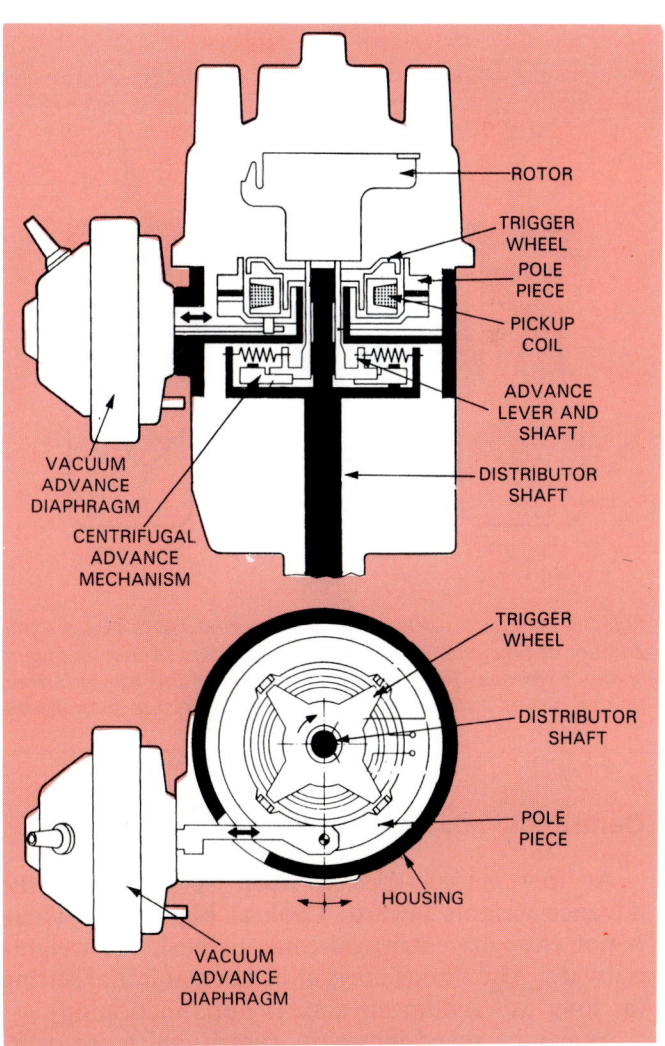

Fig. 35-23. Study parts of vacuum and centrifugal advance mechanisms. (Bosch)

2. DISTRIBUTOR VACUUM ADVANCE (controlled by engine intake manifold vacuum and engine load), Fig. 35-23.

3. ELECTRONIC (COMPUTER) ADVANCE (controlled by various engine sensors: engine rpm, temperature, intake manifold vacuum, throttle position, etc.).

DISTRIBUTOR CENTRIFUGAL ADVANCE

The distributor centrifugal advance makes the ignition coil and spark plugs fire sooner as engine speed increases. See Fig. 35-23. It uses spring-loaded weights, centrifugal force, and lever action to rotate the distributor cam, or trigger wheel on the distributor shaft. By rotating the cam or trigger wheel against distributor shaft rotation, spark timing is advanced.

Fig. 35-24 illustrates how ignition timing must be advanced with engine speed. It helps maintain correct ignition timing for maximum ENGINE POWER.

A distributor centrifugal advance mechanism basically consists of two advance weights, two springs, and an advance lever.

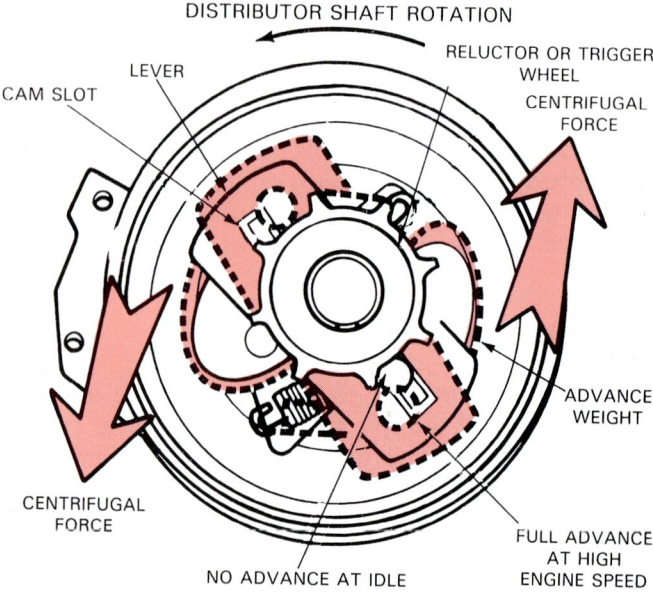

Fig. 35-25. Springs hold weights in at low engine speeds, producing no centrifugal advance. When engine speed increases, weights swing outward. Weights push on and rotate cam or trigger wheel lever. This advances ignition timing.

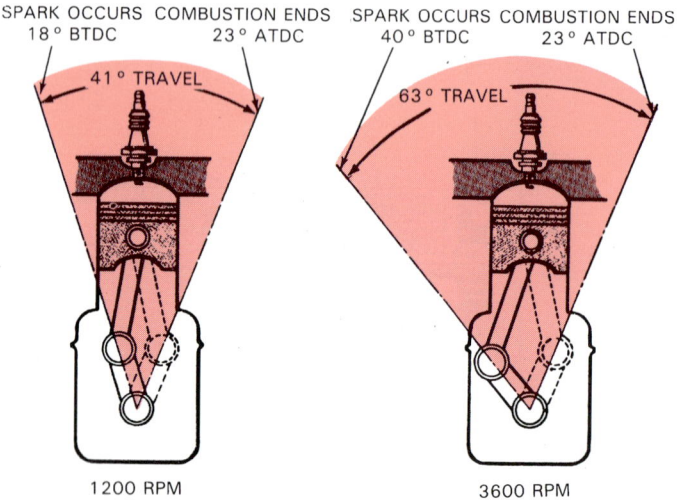

Fig. 35-24. Since each combustion period takes about same amount of time, spark must start combustion sooner as engine speed increases. This will assure that all of the fuel is burned on the power stroke and that sufficient pressure acts on the piston. (Sun)

Centrifugal advance operation

At low engine speeds, small springs hold the advance weights inward. Look at Fig. 35-25. There is not enough centrifugal force to push the weights outward. The timing stays at its normal initial setting (as long as vacuum advance is not functioning).

As engine speed increases, centrifugal force overcomes spring tension. The weights are thrown outward. The edge of the weights acts on the cam or

trigger wheel lever. The lever is rotated on the distributor shaft. This also rotates the distributor cam or trigger wheel.

Since the cam or trigger wheel is turned against distributor shaft rotation, the points open sooner, or the trigger wheel and pickup coil turn off the ECU sooner. This causes the ignition coil to fire with the engine pistons not as far up in their cylinders on the compression stroke.

As engine speed increases more, the weights fly out more and timing is advanced by a greater amount. At a preset engine rpm, the lever strikes a stop and centrifugal advance reaches maximum.

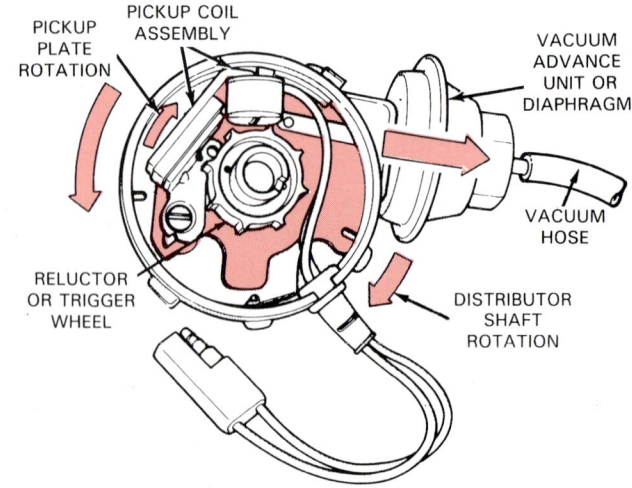

Fig. 35-26. Vacuum advance uses vacuum advance diaphragm to rotate pickup coil or contract points against direction of distributor shaft rotation.

DISTRIBUTOR VACUUM ADVANCE

The distributor vacuum advance provides additional spark advance when engine load is low at part (medium) throttle positions. Refer to Fig. 35-26. It is a method of matching ignition timing with engine load.

The vacuum advance mechanism increases FUEL ECONOMY because it helps maintain ideal spark advance at all times.

A distributor vacuum advance mechanism consists of a vacuum diaphragm, link, movable distributor plate, and a vacuum supply hose. These parts are shown in Fig. 35-27.

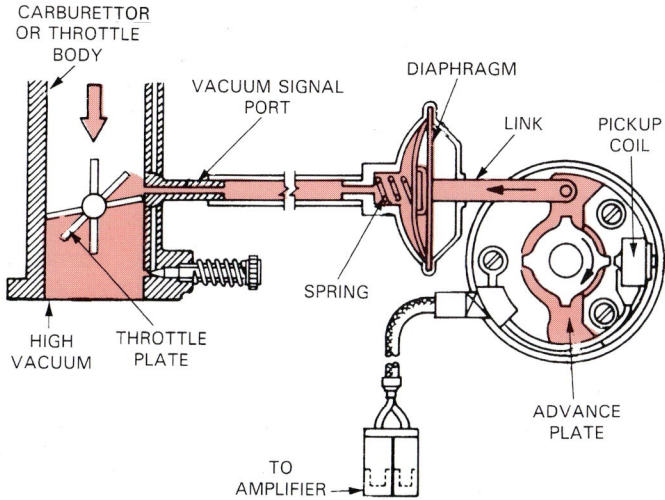

Fig. 35-27. In most units, distributor vacuum advance diaphragm is connected to ported vacuum source on carburettor or throttle body. When throttle plate swings open, vacuum is applied to diaphragm. Diaphragm flexes toward vacuum and pulls on advance plate. Pickup coil or points are rotated against distributor shaft rotation for timing advance. (Fiat)

Vacuum advance operation

At idle, the vacuum port to the distributor advance is covered. Look at Fig. 35-27. Vacuum (suction) is NOT applied to the vacuum diaphragm. Spark timing is NOT advanced.

At part throttle, the throttle valve uncovers the vacuum port and the port is exposed to engine vacuum. This causes the distributor diaphragm to be pulled toward the vacuum. The distributor plate (points or pickup coil) is rotated against distributor shaft rotation and spark timing is advanced.

During acceleration and full throttle, engine vacuum drops. Thus, vacuum is NOT applied to the distributor diaphragm and the vacuum advance does NOT operate. See Fig. 35-27.

DUAL-DIAPHRAGM DISTRIBUTOR

Used on some distributors, the dual-diaphragm vacuum advance mechanism contains two separate vacuum chambers: an advance chamber and a retard chamber. See Fig. 35-28.

Sometimes, a vacuum control switch is used in the distributor vacuum line to alter vacuum diaphragm action.

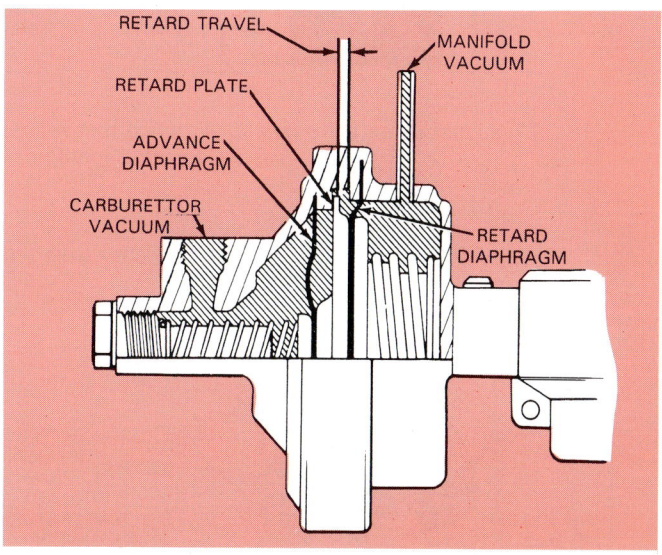

Fig. 35-28. Dual-diaphragm distributor diaphragm has two vacuum chambers. One provides vacuum advance, the other retard. It provides more positive control of ignition timing than single chamber diaphragm.

Vacuum delay valve

A vacuum delay valve restricts the flow of air to slow down the vacuum action on a vacuum device. Fig. 35-29 shows one in the line to the distributor vacuum advance diaphragm.

Note how the delay valve has a small orifice (opening) for vacuum. It also has a check valve that only allows flow in one direction.

The vacuum delay valve keeps the vacuum advance from working too quickly, preventing possible knock or ping. The check valve allows free release of vacuum from the diaphragm when returning to the retard position.

ELECTRONIC SPARK ADVANCE

An electronic spark advance system uses engine sensors, and a computer to control ignition timing. A distributor may be used but it does NOT contain centrifugal or vacuum advance mechanisms. Refer to Fig. 35-30 for an example.

The engine sensors check various engine operating conditions and send electrical data to the computer. The computer can then change ignition timing for maximum engine efficiency.

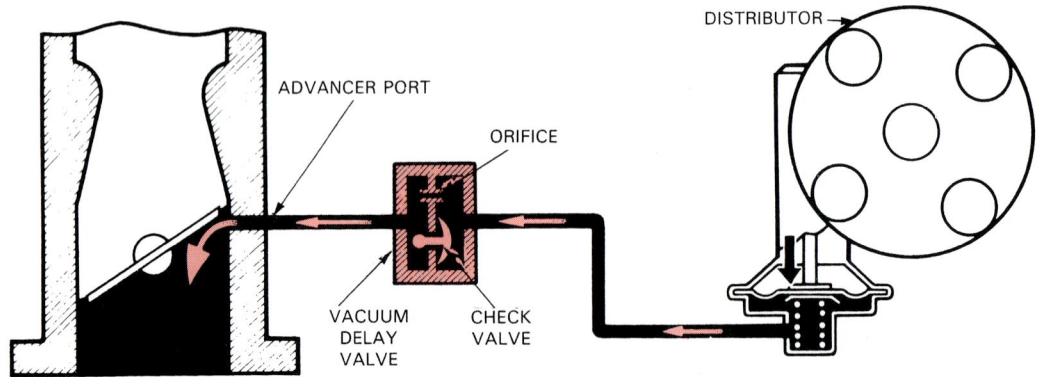

Fig. 35-29. Vacuum delay valve has small orifice that restricts access of vacuum to diaphragm. Check valve allows free flow of air out of diaphragm for quick timing retard. (Toyota)

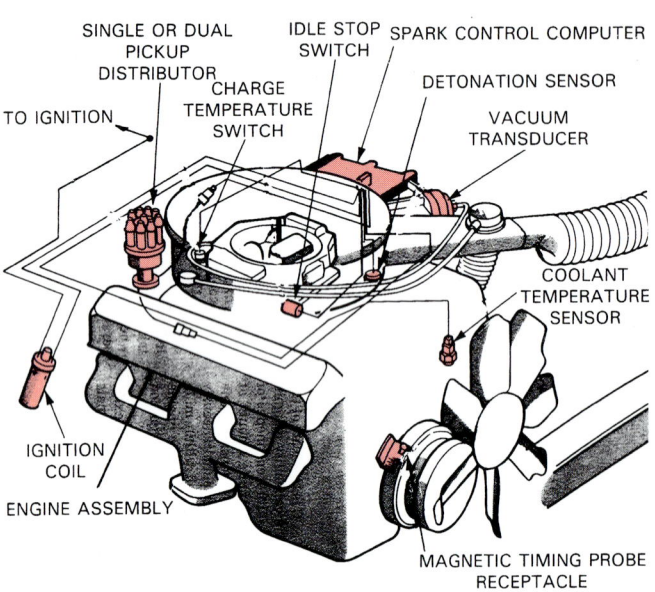

Fig. 35-30. This engine has electronic spark timing system. Computer and engine sensors replace centrifugal and vacuum advance mechanisms in distributor. Note part names and locations.

Ignition system engine sensors typically include:
1. ENGINE SPEED SENSOR (reports engine rpm to computer).
2. CRANKSHAFT POSITION SENSOR (reports piston position).
3. INTAKE VACUUM SENSOR (measures engine vacuum, an indicator of load).
4. INLET AIR TEMPERATURE SENSOR (check temperature of air entering engine).
5. ENGINE COOLANT TEMPERATURE SENSOR (measures operating temperature of engine).
6. DETONATION SENSOR (allows computer to retard timing when engine pings or knocks).
7. THROTTLE POSITION SWITCH (notes position of throttle).

The spark control computer receives input signals (different current or voltage levels) from these sensors. It is programmed (preset) to adjust ignition timing to meet different engine conditions. The computer may be mounted on the air cleaner, fender inner panel, under the car dash, or under a seat.

Electronic spark advance operation

For an example of electronic spark advance, imagine a car travelling down the highway at 80 km/h. The speed sensor would detect moderate engine rpm. The throttle position sensor would detect part throttle. The air inlet and coolant temperature sensors would report normal operating temperatures. The intake manifold pressure sensor would send high vacuum signals to the computer.

The computer could then calculate that the engine would need maximum spark advance. The timing would occur several degrees before TDC on the compression stroke. This would assure that the engine attained high fuel economy on the highway.

If the driver began to pass a car, engine intake manifold vacuum would drop to a very low level. The vacuum sensor signal would be fed to the computer. The throttle position sensor would detect WOT (wide open throttle). Other sensor outputs would stay about the same. The computer could then retard ignition timing to prevent spark knock or ping.

Since computer systems vary, refer to a service manual for more information. The manual will detail the operation of the specific system.

CRANKSHAFT TRIGGERED IGNITION

A crankshaft triggered ignition system places the pickup coil and trigger wheel (pulse ring) unit on the front of the engine. These parts are NOT located inside the distributor. Fig. 35-31 shows a simplified illustration of a crankshaft triggered ignition.

A pulse ring is mounted on the crankshaft damper to provide engine speed information to the pickup

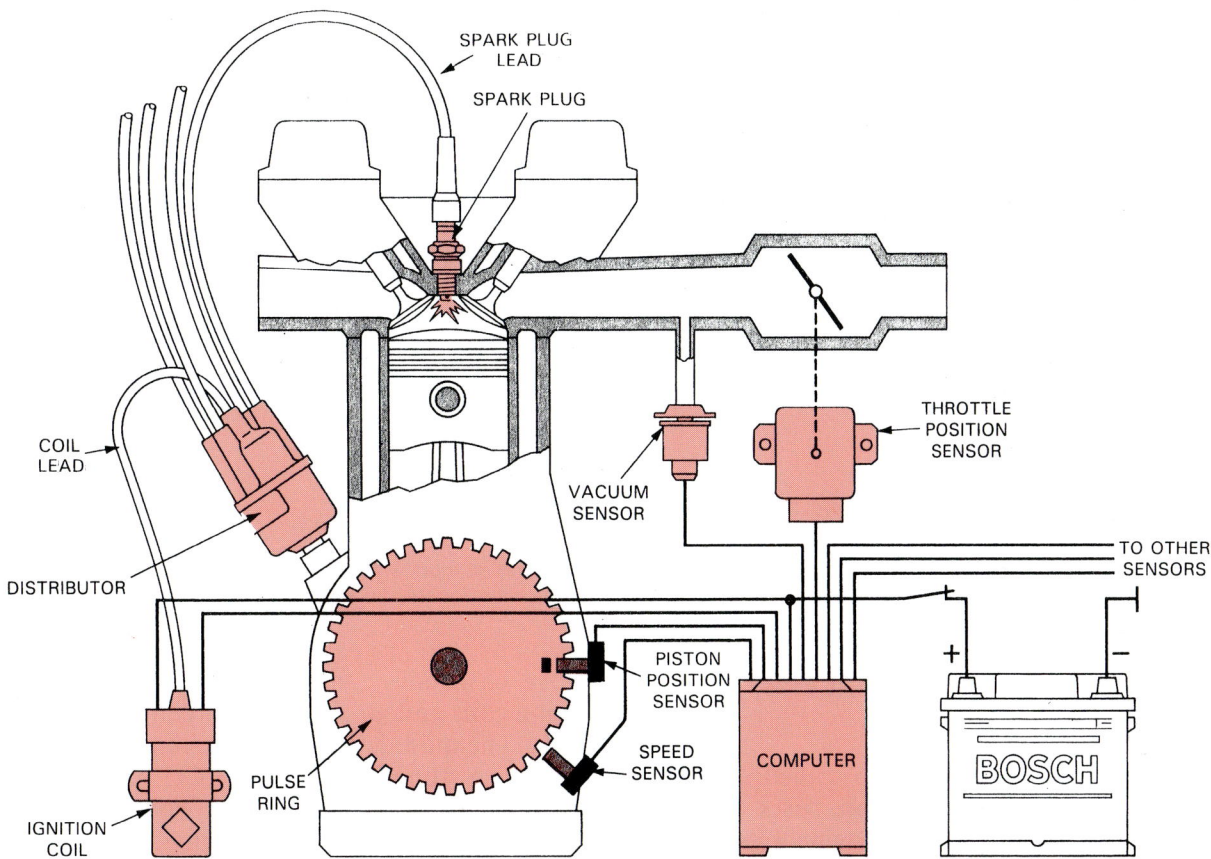

Fig. 35-31. *Simplified crankshaft-triggered ignition system places pickup coil or coils next to engine crankshaft damper. Teeth on damper act as trigger wheel to send electrical pulses to computer. Computer can then operate ignition coil and control spark advance or retard. Study parts and wiring. (Bosch)*

unit. It performs the same function as the trigger wheel in a distributor for an electronic ignition. The teeth on the pulse ring correspond to the number of engine cylinders, Fig. 35-32.

The crankshaft position sensor is mounted next to the crank pulse ring and sends electrical pulses to the system computer. It does the same thing as a distributor pickup. See Fig. 35-32.

Other sensors are commonly used to also feed data to the computer. Look at Fig. 35-33.

The distributor for a crankshaft triggered ignition is simply used to transfer high voltage to each spark plug lead.

Crankshaft triggered ignition operation

The operation of a crank triggered ignition is similar to the other electronic systems already covered. Refer to Fig. 35-31.

A crank triggered ignition can maintain more precise ignition timing than a system with a distributor-mounted pickup coil. There is no backlash or play in the distributor drive gear, timing chain, or gears to upset ignition timing. Crank and piston position is "read" right off the crankshaft.

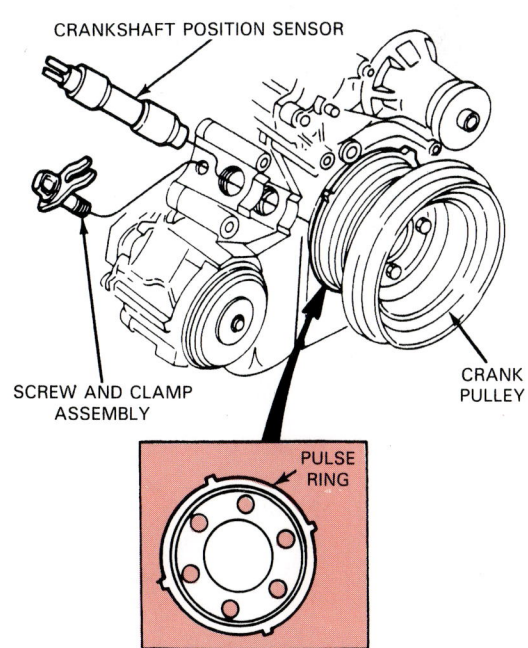

Fig. 35-32. *Pulse ring for this crankshaft-triggered ignition mounts behind crank pulley. Crank position sensor fits in hole in front cover. Wires from sensor connect to on-board computer. Teeth on pulse ring change magnetic field around sensor to produce electric pulse. (Ford)*

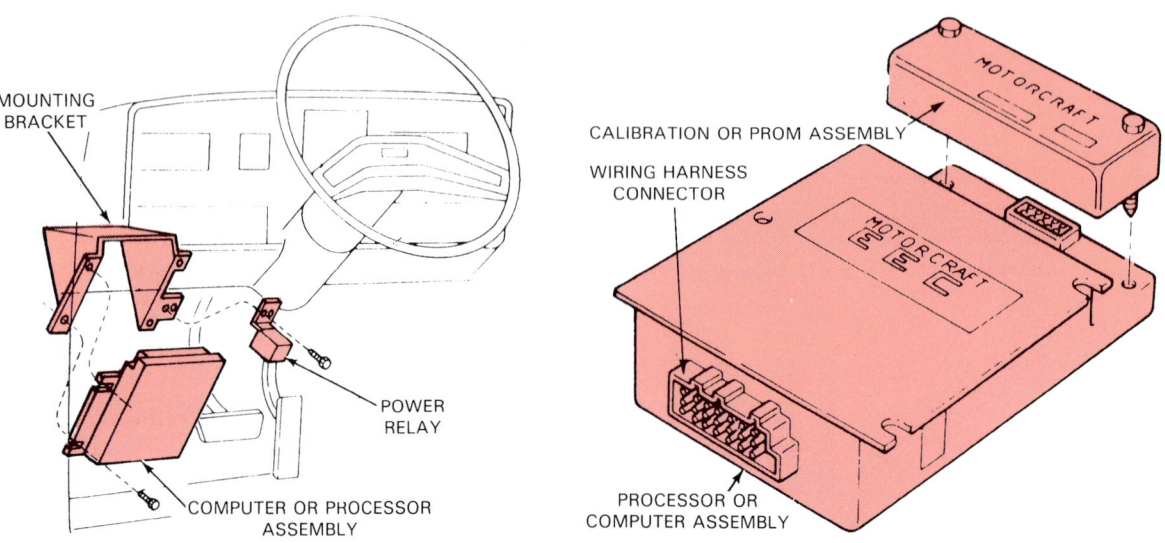

Fig. 35-33. Computer may control ignition system operation as well as fuel ignition, emission control systems, and other critical functions. It is usually located in safe place under car dash. It may also be located in engine compartment or on air cleaner. (Ford)

COMPUTER-COIL (DISTRIBUTORLESS) IGNITION

A computer-coil ignition, also called a distributorless (no distributor) ignition, uses multiple ignition coils, a coil control unit, engine sensors, and a computer to operate the spark plugs. A distributor is NOT needed. See Fig. 35-34.

An electronic coil module consists of several ignition coils and an electronic circuit for operating the coils. The module's electronic circuit performs about the same function as the ECU in an electronic ignition. It is more complex, however, because it must analyse data from engine sensors and the system computer.

A four-cylinder engine would need an electronic

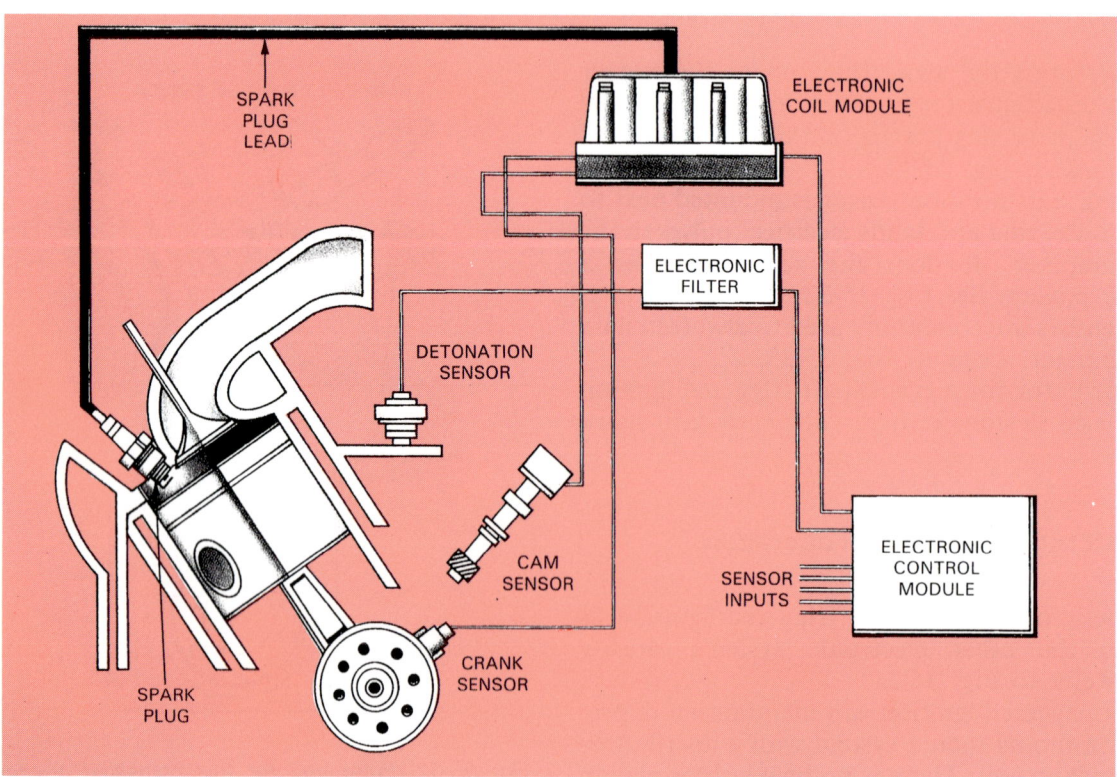

Fig. 35-34. Newest ignition system design does not use a distributor. The on-board computer and an electronic coil control module operate multiple ignition coils. Crank sensor and cam sensor send electrical signals to help control the time when coils fire.

coil module with two ignition coils. A six-cylinder engine would need a module with three ignition coils.

The coils are wired so that they fire TWO SPARK PLUGS at once. One spark plug is on the power stroke. The other is on the exhaust stroke, and has no effect on engine operation.

A cam sensor is usually installed in place of the ignition distributor. It sends electrical pulses to the coil module giving data on camshaft and valve position.

The crank sensor, as discussed, feeds pulses to the module which show engine speed and piston position.

A detonation sensor may be used to allow the system to retard timing if the engine begins to ping or knock.

Distributorless ignition operation

Fig. 35-35 illustrates how a distributorless ignition system works. The on-board computer monitors engine operating conditions and controls ignition timing. Some sensor data is also fed to the electronic coil module.

When the computer and sensors send correct electrical pulses to the coil module, the module fires one of the ignition coils.

Since each coil secondary output is connected to two spark plugs, both spark plugs fire. One produces the power stroke. The other spark plug arc does nothing because that cylinder is on the exhaust stroke. Burned gases are simply being pushed out of the cylinder.

When the next pulse ring tooth aligns with the crank sensor, the next ignition coil fires. Another two spark plugs arc for one more power stroke. This process is repeated over and over as the engine runs.

Advantages of a distributorless ignition

A distributorless ignition system has several possible advantages over other ignition types. Some of these include:

1. No rotor nor distributor cap to burn, crack, or fail.
2. Computer controlled advance. No mechanical weights to stick or wear. No vacuum advance diaphragm to rupture and leak.
3. Play in timing chain and distributor drive gear eliminated as a problem that could upset ignition timing. The crank sensor is not affected by slack in timing chain or gears.
4. More dependable because there are fewer moving parts to wear and malfunction.
5. Requires less maintenance. Ignition timing is usually NOT adjustable.

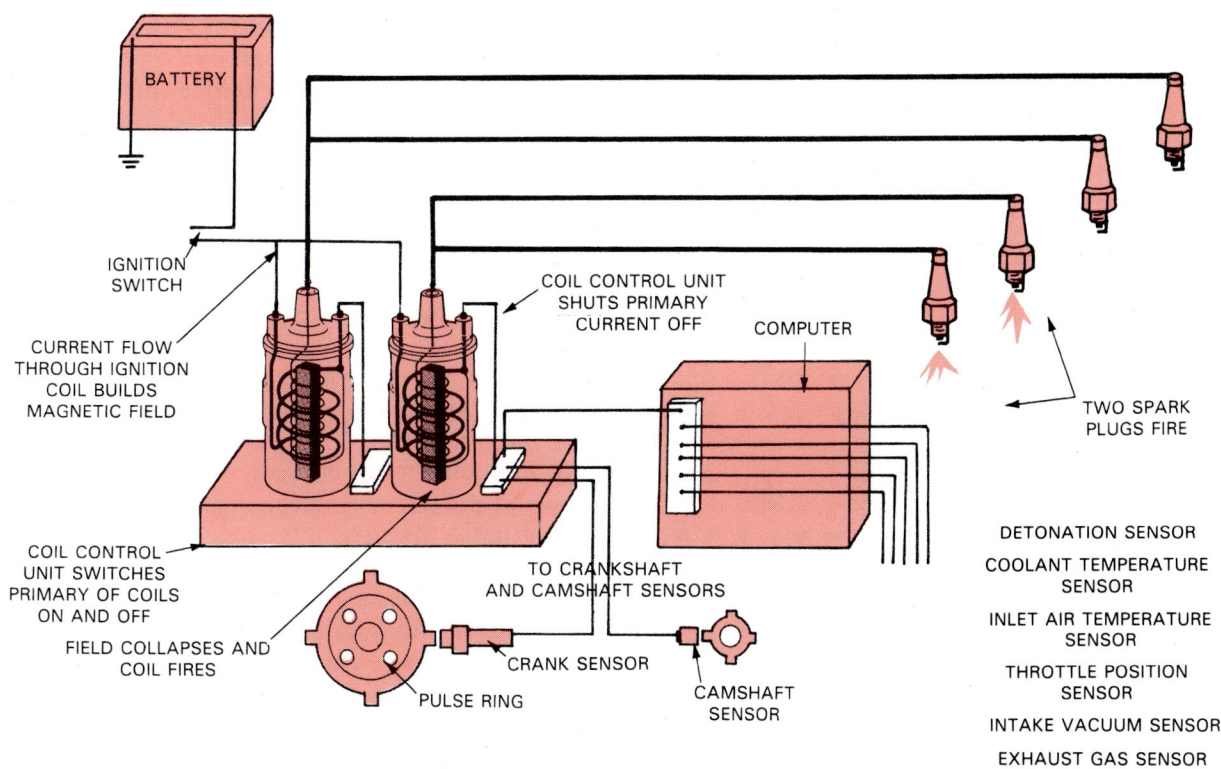

Fig. 35-35. Simplified illustration shows operation of ignition system using no distributor. Input to coil module includes signals from crank sensor, cam sensor, and computer. With correct input, coil module fires one of the ignition coils and its two spark plugs. One plug produces power stroke. Other plug sparks as burned exhaust leaves cylinder. Two coils would operate ignition for a four-cylinder engine. Three coils would be needed for a six-cylinder engine.

ENGINE FIRING ORDER

Engine firing order refers to the sequence in which the spark plugs fire to cause combustion in each cylinder. A four-cylinder engine may have one of two firing orders: 1-3-4-2 or 1-2-4-3. See Fig. 35-36. The cylinders are numbered 1-2-3-4 starting at the front of the engine. In this way, you can tell which cylinders will fire in sequence. Firing orders and cylinder numbers for V-6 and V-8 engines vary.

The engine firing order is sometimes cast into the top of the intake manifold. When not on the manifold, the firing order can be found in a service manual.

The engine firing order is commonly used when installing spark plug leads and when doing other tune-up tasks.

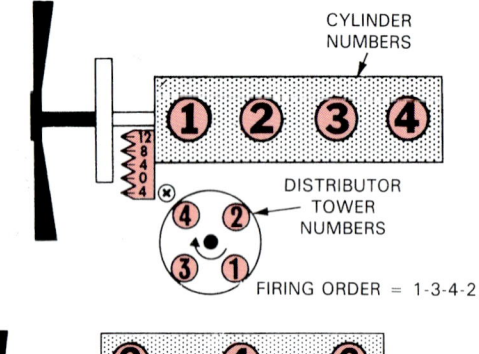

FIRING ORDER = 1-3-4-2

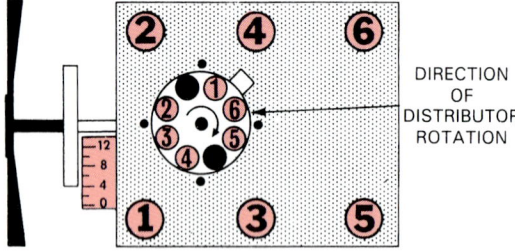

FIRING ORDER = 1-6-5-4-3-2

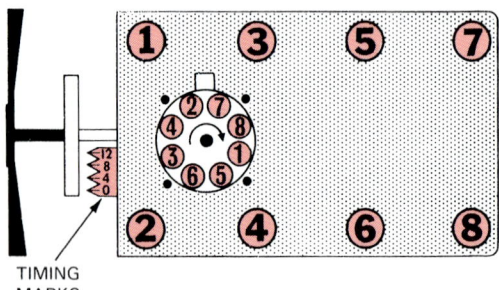

FIRING ORDER = 1-5-6-3-4-2-7-8

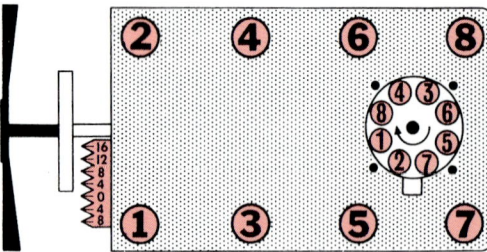

FIRING ORDER = 1-8-4-3-6-5-7-2

Fig. 35-36. Firing order is sequence that spark plugs fire in each cylinder. Firing order information is used when installing plug leads, installing distributor, setting ignition timing, and other operations.

KNOW THESE TERMS

Primary circuit, Secondary circuit, Ignition coil, Ignition distributor, Contact points, Condenser, Dwell, Electronic ignition system, Trigger wheel, Pickup coil, Hall effect, ECU, Coil wire, Distributor cap, Rotor, Resistance plug lead, Spark plug, Plug reach, Plug heat range, Plug gap, Hot plug, Cold plug, Ignition timing, Timing advance, Timing retard, Centrifugal advance, Vacuum advance, Electronic advance, Vacuum delay valve, Crankshaft triggered ignition, Pulse ring, Crank position sensor, Computer-coil ignition, Electronic coil module, Engine firing order.

REVIEW QUESTIONS

1. What are the five basic functions of an ignition system?
2. List and explain the six major parts of an ignition system.
3. An ignition system _____ circuit is used to supply direct battery voltage to the system during starter motor operation.
4. Define the terms "ignition system primary" and "ignition system secondary".
5. An ignition coil is capable of producing this voltage output.
   a. 10 kV.
   b. 20 kV.
   c. 30 kV.
   d. None of the above.
6. The primary windings of an ignition coil are several hundred turns of heavy wire. True or False?
7. When the current flowing through the ignition coil is broken, the magnetic field _____ and induces high _____ into the secondary.
8. Explain the differences between a contact point and pickup coil (electronic) distributor.
9. What is dwell?
10. An electronic ignition system uses an _____ circuit and a distributor _____ _____ to operate the ignition coil.
11. The trigger wheel or reluctor causes a tiny voltage pulse in the pickup coil to operate the ECU and ignition coil. True or False?
12. How does the pickup coil produce signals for the electronic control unit?
13. The ignition system _____ or _____ _____ _____ is the "electronic switch" that turns the ignition coil primary on and off.

14. The ignition system ECU is NOT normally located in the distributor. True or False?
15. Define the term "ECU dwell time".
16. What is electrical leakage?
17. The _____ _____ is an insulating plastic component that fits over the distributor housing and sends voltage to each spark plug lead.
18. Explain the function of the distributor rotor.
19. Why do spark plug leads need internal resistance?
20. It normally takes about _____ to _____ volts to operate a spark plug.
21. Spark plug gap is the distance between the centre electrode and side electrode. True or False?
22. A colder spark plug might be beneficial in an older engine that burns some oil. True or False and Why?
23. Describe the difference between timing advance and timing retard.
24. List and explain the three methods of controlling ignition timing.

25. The distributor centrifugal advance depends upon engine _____ and the vacuum advance depends upon intake manifold pressure (vacuum), an indicator of engine _____.
26. Electronic spark advance uses engine _____ and a _____ to control ignition timing.
27. How does a crankshaft triggered electronic ignition system work?
28. Which of the following does NOT relate to a computer-coil or distributorless ignition system?
    a. No rotor.
    b. No centrifugal or vacuum advance.
    c. Two ignition coils per cylinder.
    d. No coil lead.
    e. No spark plug leads.
29. With a computer-coil or distributorless ignition, two spark plugs fire at once. True or False?
30. List five possible advantages of a distributorless ignition.
31. Engine _____ _____ refers to the sequence in which the spark plugs operate to cause combustion in each cylinder.

Chapter 36

WIRE AND WIRING

After studying this chapter, you will be able to:
- ☐ Identify different types of automotive wiring.
- ☐ Select the correct type of wiring for the job.
- ☐ Make basic wiring repairs.
- ☐ Read wiring diagrams.
- ☐ Perform basic circuit tests.

New wiring, properly installed, is relatively trouble free. As the car ages, the wires tend to deteriorate from exposure to heat, oil, petrol, fumes, acid, and vibration. Vehicles damaged by collision or fire often require extensive rewiring. The auto mechanic should become familiar with types of wire, sizes, insulation, connections, and installation procedures.

PRIMARY WIRE

The primary wiring of a modern motor car handles 12 volts, battery voltage. It has sufficient insulation to prevent current loss at this voltage. All wiring circuits in the car, with the exception of the ignition high tension circuit, use primary wire. NEVER USE PRIMARY WIRE FOR SPARK PLUG LEADS.

SECONDARY WIRE

Secondary wire is used in the ignition system high tension circuit — coil to distributor, distributor to plugs. It has a heavy layer of insulation to protect against excessive corona (loss of electrons to the surrounding air) which could impart sufficient

current into an adjacent wire to cause it to fire a plug. This action is known as cross-firing. Even with good insulation it is important to arrange spark plug leads so that leads to cylinders that fire consecutively are separated. Fig. 36-1 shows the relative difference in the amount of insulation on primary and secondary wires.

STRANDING MATERIAL

Soft copper is widely used for wire stranding. It is an excellent conductor, bends easily, and solders readily. Aluminium also is used to some extent. Copper, stainless steel, carbon impregnated thread, and elastomer type conductors are used for secondary wire stranding. The carbon impregnated thread and elastomer type (Duoprene G, for example) impart a controlled resistance (about 16000 ohms per 1 metre) in the secondary circuit to reduce radio interference. WHEN WORKING ON THE IGNITION SYSTEM, HANDLE RESISTANCE TYPE HIGH TENSION WIRES CAREFULLY. SHARP BENDING AND JERKING CAN SEPARATE THE CONDUCTOR, THUS RUINING THE WIRE. WHEN REMOVING OR INSTALLING SUCH LEADS, GRIP THE INSULATION BOOT — NOT THE WIRE!

Resistance type wires may be identified by such letters as IRS and TVRS.

WARNING: ELECTRONIC IGNITION SYSTEMS PRODUCE EXTREMELY HIGH VOLTAGES IN BOTH THE PRIMARY AND SECONDARY CIRCUITS WHICH ARE CAPABLE OF GIVING A FATAL SHOCK. EXTREME CARE MUST BE TAKEN AT ALL TIMES WHEN WORKING NEAR THESE SYSTEMS.

INSULATION

Plastic of various kinds is used for automotive wire insulation. Rubber is sometimes used. Plastic is highly resistant to heat, cold, fumes, and aging. It

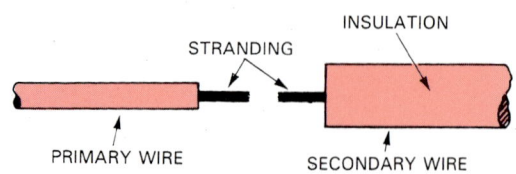

Fig. 36-1. More insulation is required on secondary wires.

strips (peels off) easily and offers excellent dielectic (nonconducting) properties. Silicone secondary wire insulation is very heat resistant.

TERMINAL TYPES

Wire end terminals (connecting devices) are offered in various shapes and sizes. In general, primary terminals may be classified as spade, lug, flag, roll, slide, blade, ring, and bullet types. They may either be solderable or solderless. They are generally made of tin plated copper. See Fig. 36-2.

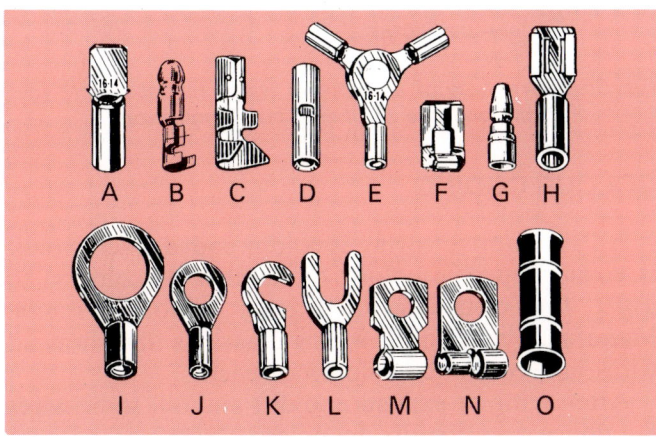

Fig. 36-2. Common primary wire terminal types. A — Male slide. B — Bullet or snap-in. C — Female snap-on. D — Butt connector (must be crimped). E — Three-way connector. F — Female slide. G — Bullet. H — Female slide. I — Lug. J — Ring. K — Hook. L — Spade. M — Roll. N — Flag. O — Female bullet connector. (Belden)

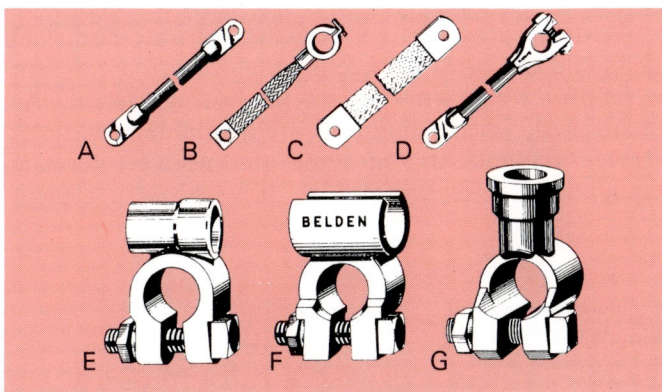

Fig. 36-3. Typical battery cables and terminals. A — Solenoid-to-starter cable. B — Battery ground cable. C — Engine ground strap. D — Battery-to-solenoid cable. E — Closed barrel terminal. F — Open-split barrel terminal. G — Closed barrel terminal. Note that ground cables have no insulation and are of a woven construction. Regular insulated battery cable is also used for ground cables.

BATTERY CABLE TERMINALS

New battery cables (with factory installed terminals) are generally used to replace a used cable with a corroded, useless terminal. However, it is occasionally desirable to replace only the terminal. A number of different types are available, Fig. 36-3.

Terminals on battery cables should be SOLDERED ON. This will ensure a good connection with very little voltage drop (lowering of line voltage due to loose, dirty, or corroded connections). It will also protect against the entry of battery acid and fumes. Soldering will be covered later in this chapter.

TERMINAL BLOCKS

The terminal block is used to supply current to several circuits from one feeder source. The hot wire (wire connected to source of electricity) is attached to one terminal. This terminal is connected to all others by a bus bar (metal plate), Fig. 36-4.

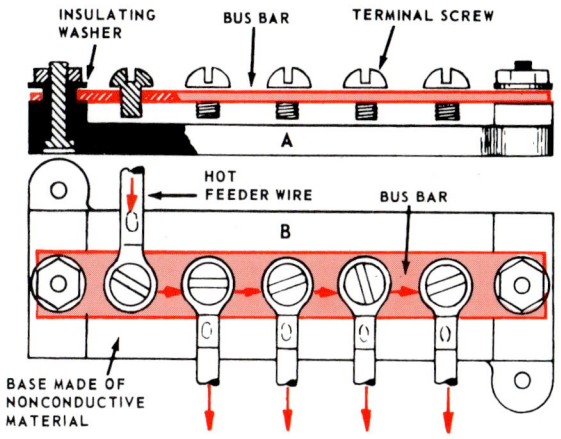

Fig. 36-4. One type of terminal block. Notice how one hot wire is attached to bus bar, thus supplying current to other leads.

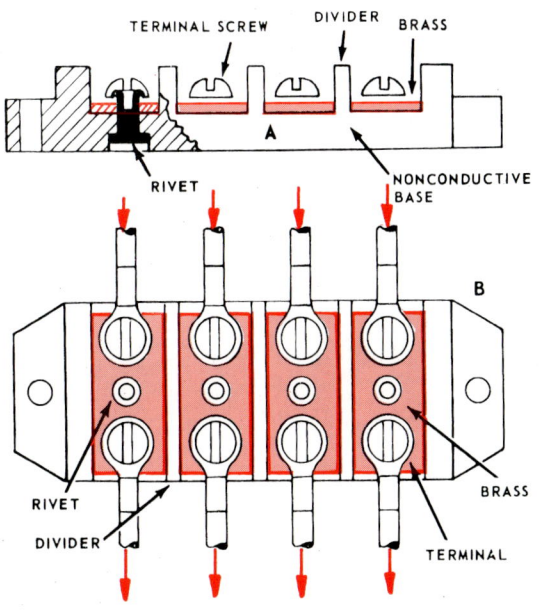

Fig. 36-5. Screw type junction block.

JUNCTION BLOCK

The junction block serves as a common connection point for a number of wires. It may be of the terminal screw or the plug-in type. Unlike the terminal block, the junction block merely connects one wire to a corresponding wire on the other side. There is no common bus bar, Fig. 36-5.

FUSE BLOCK

The fuse block is similar to the junction block except that a fuse is inserted between the connecting points. This protects each circuit against electrical overloads, and groups a number of fuses in one location. A simple fuse block is shown in Fig. 36-6. Fig. 36-7 illustrates a fuse block utilising the compact miniature fuse.

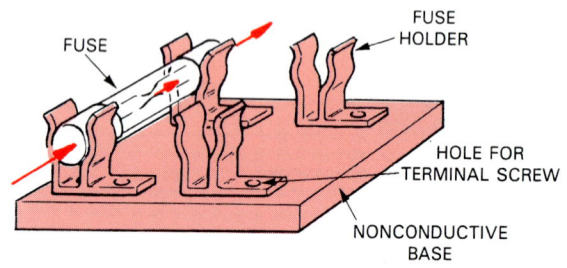

Fig. 36-6. Fuse block. Fuse blocks often contain a number of fuses.

WIRING HARNESS

In an automobile, various sections of wiring are made up in units with common wires (located in same areas) either pulled through a loom (soft woven insulation tube) or taped or tied together. This speeds installation, makes a neat package, and provides proper securing with a greatly reduced number of clamps or clips. Fig. 36-8 shows a typical wiring harness.

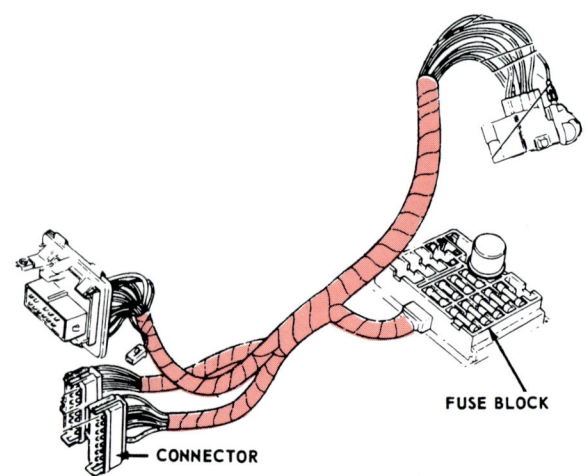

Fig. 36-8. One type of automotive wiring harness. Note fuse block and use of plug-in type connectors.

COLOUR CODING

All automotive wiring is colour coded (each circuit is given a specific colour or number of colours) to assist the mechanic in tracing various circuits. Manufacturers publish wiring diagrams that show all wires and colour or colours of each.

After aging or exposure to dirt and oil, some wires are difficult to identify by colour. In this case, trace the wire back to where it enters the harness. Then, cut away a small portion of the harness covering. This will expose a clean portion of the wire so the colour may be determined.

WIRING DIAGRAMS

A wiring diagram is a drawing showing electrical units and the wires connecting them. Such a diagram is helpful when working on the wiring system. As mentioned, wiring diagrams are available in various service manuals and in some automotive reference type books. Use them! Fig. 36-9 shows a typical

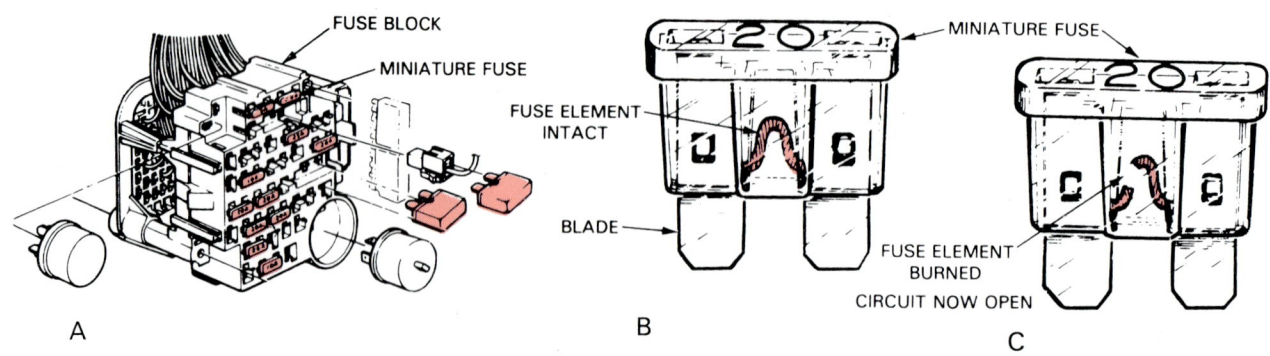

Fig. 36-7. A — Fuse block incorporating a number of miniaturised fuses. B — "Good" miniaturised fuse. Note that element is sound. Current flows from one blade, through element, and out other blade. C — Fuse is "Blown". Element is burned in half, thus opening circuit.

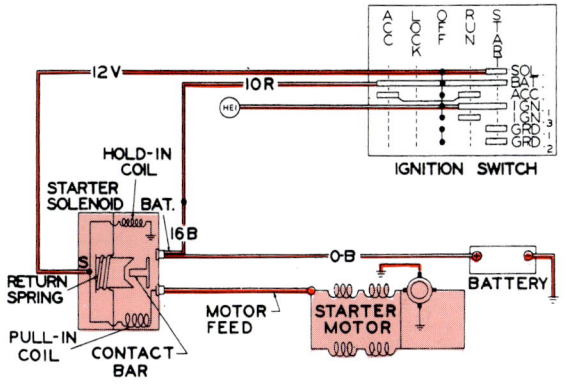

Fig. 36-9. A typical wiring diagram for a starter system.

wiring diagram for a specific unit. The modern auto electrical system is becoming more complicated each year. Many manufacturers break down the various circuits into separate diagrams, Fig. 36-9. They usually also provide an overall diagram showing the entire electrical system, Fig. 36-10.

ELECTRICAL WIRING SYMBOLS

There is a wide variation in the use of automotive electrical symbols. Some companies use their own drawings for some units and standard symbols for others. The unit's basic internal circuit is sometimes shown. In other diagrams, symbols are used for all

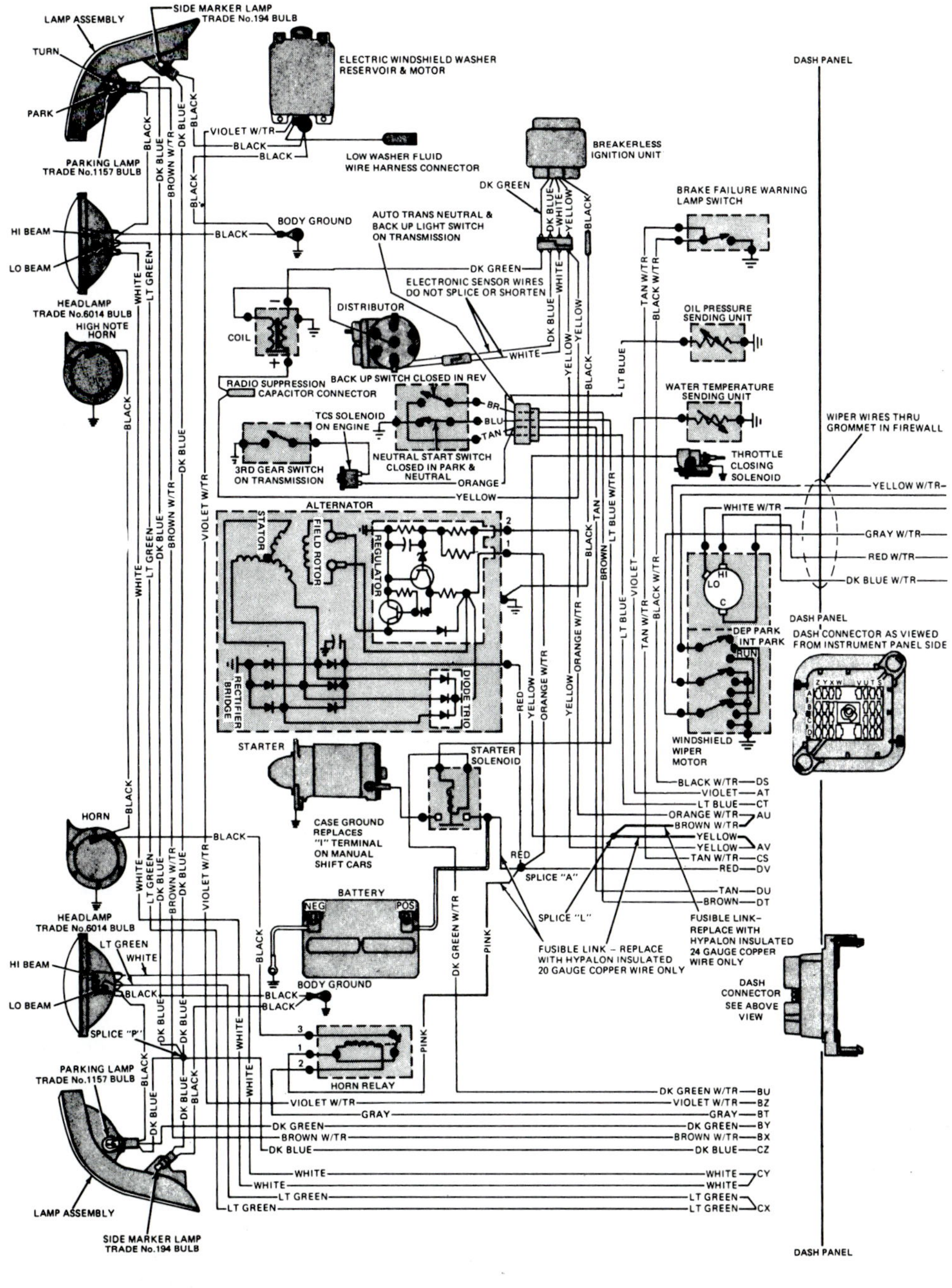

Fig. 36-10. Overall wiring diagram for the front half of car. Note use of symbols and colour coding. (Jeep)

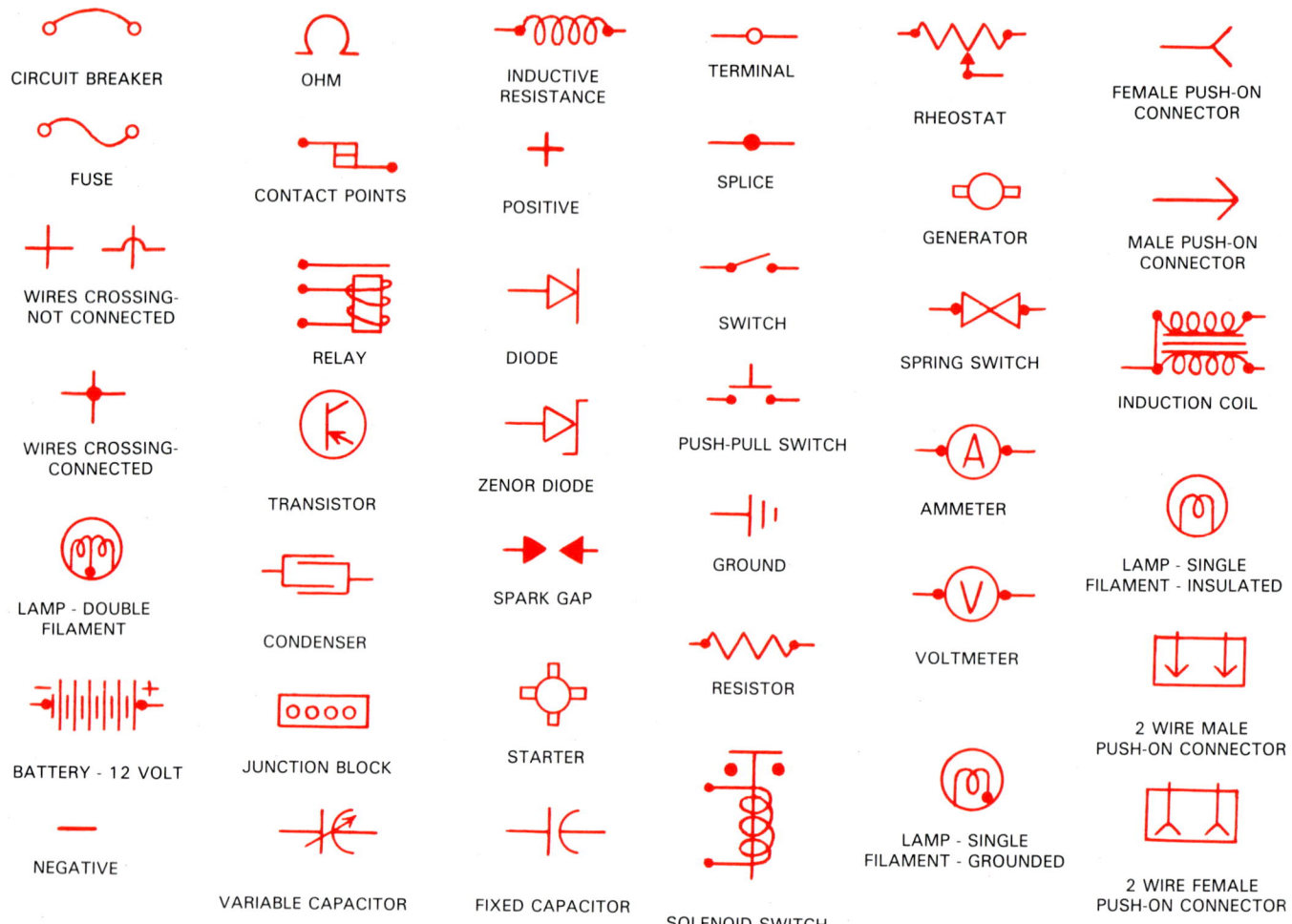

Fig. 36-11. Electrical symbols commonly used in automotive wiring diagrams.

units. Fig. 36-11 illustrates a number of typical symbols widely used in automotive electrical diagrams.

SELECTING CORRECT WIRE SIZE

Wire voltage, electrical load, and wire length are the three important factors in determining correct wire size.

Keep in mind that as wire length INCREASES resistance (with resultant voltage drop) INCREASES. Resistance causes the conductor to heat. Excessive resistance can heat it to the point where the insulation will melt and burn.

As wire size INCREASES, resistance DECREASES. A simple rule then would be: to prevent high resistance and voltage drop wire size must be increased as length is increased. It is obvious then that with a given voltage and load a wire 5 metres long must be of a larger size than a wire 1 metre long.

The electrical load imposed on a wire is merely the sum of the individual loads of each unit serviced by that wire. Common automotive system voltage is now 12 volts.

Automotive cable manufacturers furnish comprehensive product and technical information charts. These charts include all the relevent information a mechanic would need in selecting a particular size cable for a particular job. They can also be used as a means of identifying the size of a particular cable on a vehicle if it needs to be replaced as the chart shows the diameter of the individual strands of wire and the number of strands that go to make up the conductor of any particular size cable.

To use the charts shown, Fig. 36-12 and Fig. 36-13, first compute the total electrical load the wire will be subjected to. Be certain to figure the load of all units concerned. If the load will fluctuate use the peak load figures.

Note: To calculate the ampere load in lamp circuits add the wattages of all lamps in the circuit and divide by the voltage.

When the load has been calculated use the first chart, Fig. 36-12, to select the appropriate size cable.

AUTOMOTIVE CABLE CHARTS

RECOMMENDED CURRENT RATINGS						
Nominal conductor area mm²	Number & diameter of wires No./mm	Equiv. SAE wire gauge	Approximate overall diameter mm	Continuous Duty Amps		
				Single wire	Multiwire	Battery cable
0.85	11/0.32	18	2.5	5	3	
1.25	16/0.32	16	2.9	10	6	
2.00	26/0.32	14	3.1	15	10	
3.00	41/0.32	12	3.8	20	15	
5.00	65/0.32	10	4.6	25	18	
8	94/0.32	8	5.5			45
14	182/0.32	6	7.3			70
19	247/0.32	4	9.0			90
25	323/0.32	3	10.3			100
36.5	455/0.32	2	11.7			120
40	520/0.32	1	12.7			130
50	627/0.32	0	13.2			150
62	779/0.32	00	14.6			180

Fig. 36-12. Chart showing recommended current rating for the different size automotive wiring.

NOMINAL VOLTAGE DROP DATA				
Number & diameter of wires No./mm	Voltage drop per Amp/ metre volts	Load Amps	Residual voltage at equipment terminals on 12 volt system	
			15 metres volts	30 metres volts
11/0.32	0.0208	3	11.1	10.1
16/0.32	0.0143	6	10.7	9.4
26/0.32	0.0088	15	10.0	8.0
41/0.32	0.0056	20	10.3	8.6
65/0.32	0.0035	25	10.7	9.4
94/0.32	0.0024	45	10.4	8.8
182/0.32	0.0014	70	10.5	9.1
247/0.32	0.0009	90	10.8	9.6
323/0.32	0.0007	100	11.0	9.9
455/0.32	0.0005	120	11.1	10.2

Fig. 36-13. Chart for working out voltage drop for a particular size wire.

Example: Computed load is 6 amps. If we look at the first chart we see that the first cable listed on the chart is 0.85 mm² which is rated at 5 amps which is unsuitable. If we go to the second cable on the chart, which is 1.25 mm², it is rated at 10 amps and is suitable.

Now that we have selected the appropriate size wire we can go to the second chart, Fig. 36-13, and work out what the voltage drop will be for the length of wire we are going to use. As can be seen the chart shows what the voltage drop per amp per metre would be for a particular size of wire. If the voltage drop is excessive then a heavier wire will need to be selected.

Using a larger wire than necessary will cause no particular harm unless the wire being replaced MUST produce a specific resistance in the circuit.

Fig. 36-15. Crimping tool.

SELECTING PROPER TERMINALS

After selecting a correct size cable a proper size and type of terminal must be selected. The terminal must be suitable for the unit connecting post or prongs. It must have sufficient current carrying capacity and should be heavy enough to prevent breakage through normal wire flexing and vibration. Fig. 36-14 shows some common errors in terminal selection.

Arrange terminals so they have clearance from metal parts that could ground or short them out. On

CRIMPING TERMINALS

A crimping tool is shown in Fig. 36-15. It will cut and strip the wire as well as form a proper crimp.

The first step is to strip the insulation back for a distance equal to the length of the terminal barrel. The wire is then pushed into the barrel. While being held in, the crimping tool is placed over the spot to be crimped. Be sure to use the proper crimping edge. The handles are squeezed together and the terminal barrel firmly crimped to the wire. Follow the tool manufacturer's instructions. Use the correct barrel

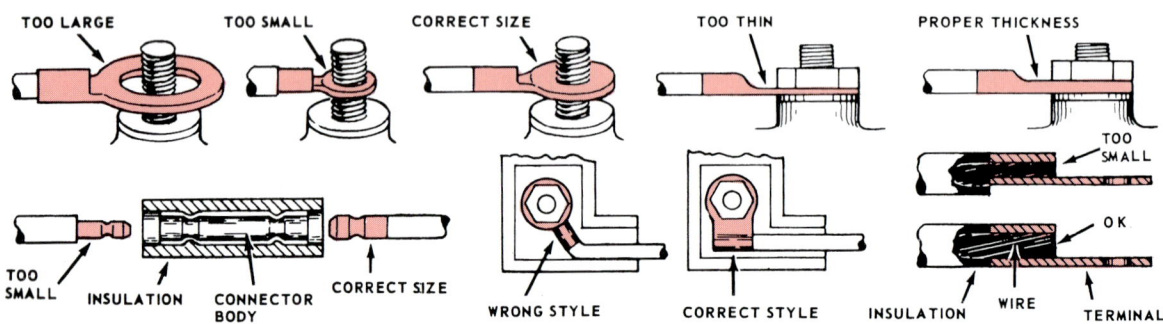

Fig. 36-14. Some common errors in terminal selection.

critical applications or where heavy vibration is present, use a terminal such as the ring type that completely encircles the post. If it loosens, the wire will not fall off.

ATTACHING TERMINALS

Terminals may be either soldered or crimped in place. Crimping is fast and forms a good connection. Soldering, if properly done, forms an excellent connection and, in some cases, may be desired. It is possible to both solder and crimp a connection. Solder forms an electrical path and is not depended on for strength.

Aluminium wire requires crimped terminals.

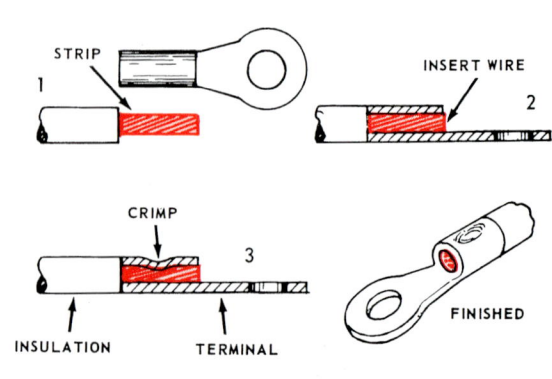

Fig. 36-16. Crimping a terminal.

size for the wire used. NEVER CRIMP A WIRE WITH THE CUTTING EDGE OF A PAIR OF PLIERS. This would crimp the barrel but weaken it, Fig. 36-16.

SOLDERING TERMINALS

Terminals do not have to be especially made for soldering but the lip-type terminal tang lends itself to soldering better than the closed or open barrel tang, Fig. 36-17.

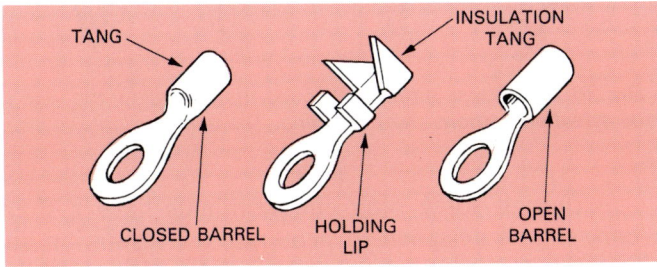

Fig. 36-17. Terminal tangs.

To solder the lip type, strip the wire back as shown in A, Fig. 36-18. Insert the wire as shown in B. Crimp the wire holding lips, one after the other, tightly over the wire. Then, carefully fold the insulation tang around the insulated portion of the wire, as in C and D.

Using ROSIN CORE (NOT ACID CORE) wire solder, place a drop of solder on the holding lips. Hold the iron in contact with the lips until it flows into the lips and wire. Do not hold the iron in contact with the terminal any longer than necessary. This tends to melt the insulation.

When soldering the open barrel type, strip as for crimping. Tin the exposed wire end (coat with a thin layer of solder). Insert in the barrel. While holding the exposed end upright, heat the socket with the iron. While heating, keep wire solder against socket end. When the solder melts, flow it into the barrel. Make certain a sufficient amount enters. Hold the

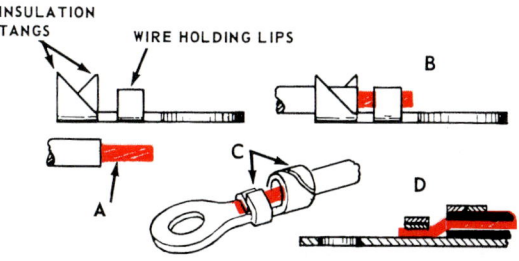

Fig. 36-18. Soldering lip type terminal.

iron in place for a few seconds longer to allow the solder to bond to both the barrel and wire. The barrel may also be crimped if so desired. Crimp before soldering! See Fig. 36-19.

The closed barrel type should be heated and a small amount of solder flowed into the hole. While keeping the barrel hot, press the tinned wire into the hole. Hold the iron in place for several seconds to ensure bonding.

When an insulator boot is to cover the terminal tang or when attaching slide type terminals that will be snapped back into a housing, always slide the boot, housing, etc., on the wire before soldering.

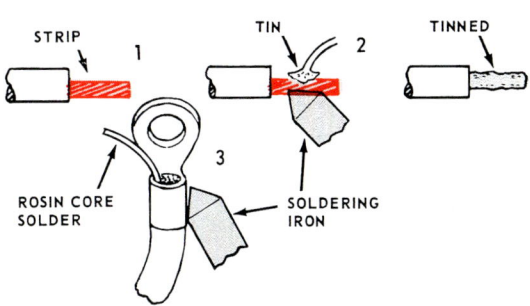

Fig. 36-19. Soldering barrel type terminal tang.

SOLDERING BATTERY TERMINALS

The common practice is to replace the entire battery cable when the terminals are no longer fit for use. However, if it becomes necessary to install a new terminal, use the following procedure.

Cut the cable back far enough to remove the corroded section. Peel the insulation (ground cables often have none) back equal to the depth of the terminal barrel. Place the terminal in a vice, open barrel end up.

Using an acetylene torch (low heat, flame rich in acetylene), heat the stripped cable end. Using rosin core wire solder, flow solder freely into the wire until all strands have been tinned. It may help to rub on a little rosin type soldering paste to assist with tinning.

Place a dab of soldering paste in the terminal barrel. Heat with the torch (keep flame on outside of terminal). When hot, flow solder into the barrel until about one quarter full. While retaining the heat with the torch, force the tinned cable down into the socket. When it slips in the full depth, solder will flow up and over the lip of the barrel. Hold the heat, moving the flame around the terminal outside, for a few seconds longer to allow the heavy cable to heat up and bond firmly to the barrel. Remove heat and hold cable steady until solder sets. Cool under a cold tap. Dry terminal and cable insulation. Then, apply

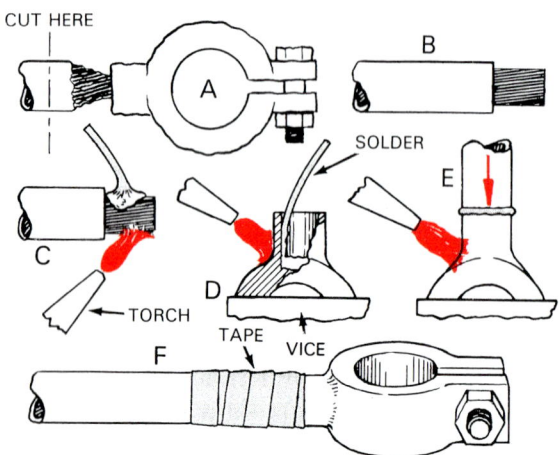

Fig. 36-20. Soldering a battery terminal. A — Cut off corroded section. B — Strip. C — Tin. D — Tin barrel and add solder. E — Insert cable. F — Tape.

plastic tape as shown in Fig. 36-20. For open barrel terminals, tin both cable and inside of barrel heavily. While heating, slide together as above. Do not try to solder battery terminals with a soldering iron — it will not produce sufficient heat.

ATTACHING SPARK PLUG WIRE TERMINALS

Fig. 36-21 shows various spark plug wire terminals. The boots protect against moisture and dirt that can cause flashover (spark jumping to ground along

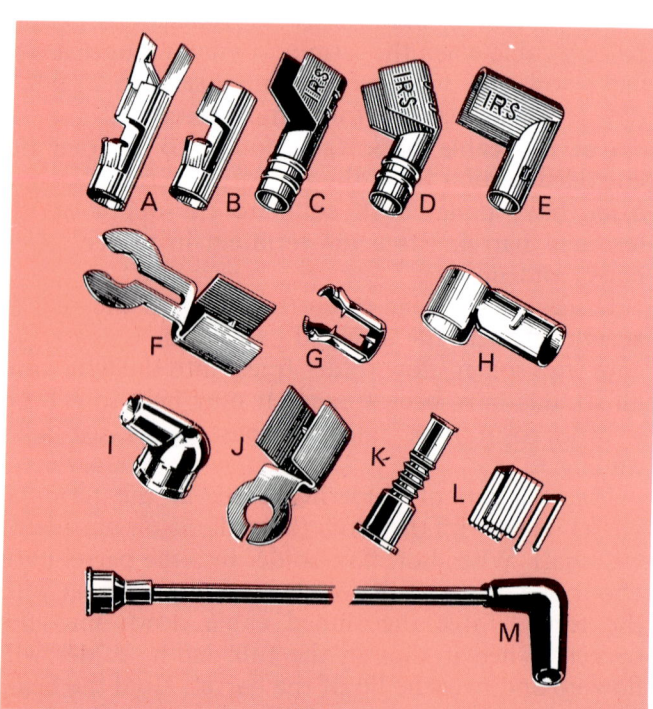

Fig. 36-21. Spark plug wire terminals and boots. A, B, G — Distributor end terminals. C, D, E, F, G, H, J, — Spark plug end terminals. I — Right angle distributor end boot. K — Flexible plug end boot. L — Staples for use with resistance type wire. M — Replacement plug wire with boots bonded to wire.

outside of plug porcelain top). Ready-made sets often bond the boots to both the terminal and wire for added protection against flashover.

When selecting plug end terminals, choose a shape that will snap on the plug without bending the wire sharply. Fig. 36-22. The same applies to distributor terminals.

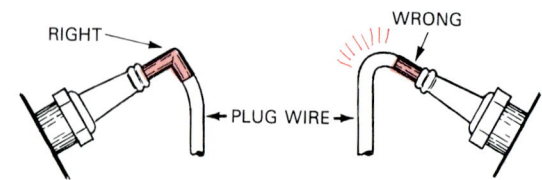

Fig. 36-22. Choose a terminal shape that will allow the wire to be attached without sharp bending. The wire on right will soon fail.

Although some plug end terminals have a sharp barb that is designed to penetrate the insulation and contact the wire (as well as providing holding power), it is good practice to strip the insulation enough to allow the wire to be bent around and laid against the outside of the insulation. This ensures a good electrical contact. See A, Fig. 36-23. Some distributor end terminals, such as in B, Fig. 36-23, have the barbs both at the sides and end. Wire stripping is not necessary if the barb is carefully inserted into the wire end. When attaching terminals to resistance type plug wires, always use staples. The staple is pushed into the wire, thus ensuring a large contact area with the special conductor, C, Fig. 36-23.

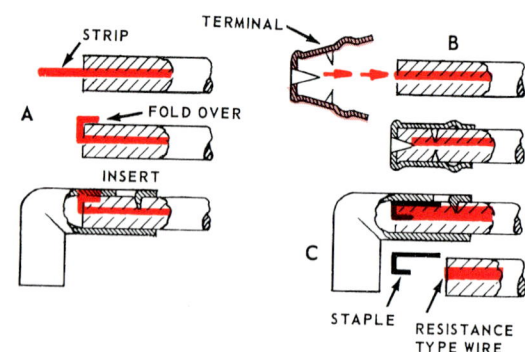

Fig. 36-23. Attaching secondary wire terminals. A — Attaching a plug end terminal to a regular (non-resistance) spark plug wire. B — Attaching a distributor end terminal. C — Using a staple when attaching a terminal to resistance type wire.

JOINING WIRE ENDS

In addition to the terminal, fuse, and junction block, wires may be connected by soldering, crimped butt connectors, and slide or bullet type connectors. If the wire ends are being joined permanently, soldering or butt connectors work very well, Fig. 36-24.

The slide and the bullet type connectors are used where the wires must be separated at some future time. The appropriate slide or bullet terminals are crimped or soldered to the wires. They are then snapped into the connector body and the two halves plugged together, Fig. 36-25.

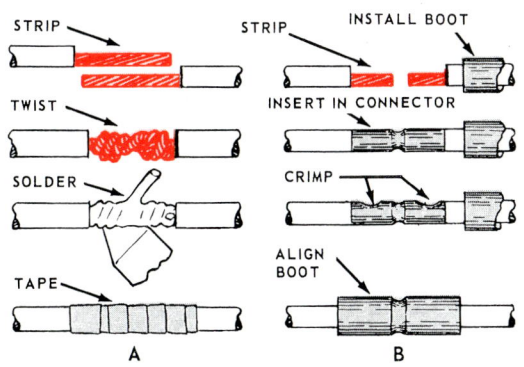

Fig. 36-24. A — Joining wires by soldering. B — Using a crimp type butt connector.

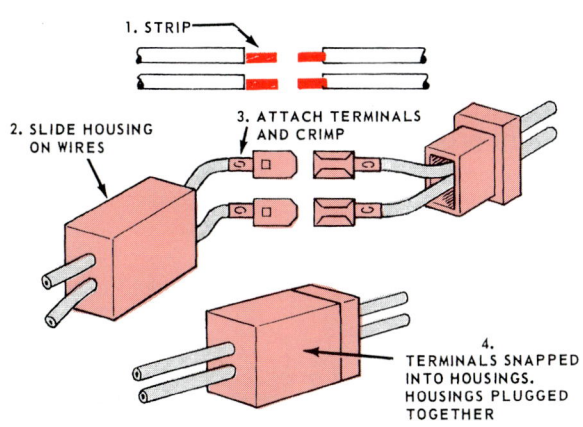

Fig. 36-25. Joining wires by using the slide type quick-connect.

INSTALLING WIRE

Install the wire. Make certain terminals and posts are clean. Connect terminals and tighten securely. Lock washers should be used on screw and post connections. Slip insulator boots, where used, over exposed terminal tang. If of the slide or bullet types, push together tightly and check to see that the connection is secure.

Keep all wiring away from the exhaust system, oily areas, and moving parts. Secure in place with mounting clips or clamps. Fasten in enough spots to prevent excessive vibration and chafing. Where the wire must pass through a hole in sheet metal, install a rubber grommet, Fig. 36-26. When a wire must pass from the fender well or splash shield to the engine, leave enough slack to allow the engine to rock on the mounts without pulling the wire tight.

When installing spark plug leads, avoid sharp bends. If the wires pass through a metal conduit (tube), the conduit should be securely grounded. Install or remove the plug wires by grasping the insulation boots and not the wire proper. Make sure the terminals snap tightly on the plugs and that the distributor ends are all the way in the housing towers. Follow the manufacturer's instructions in arranging the plug wires. If two leads are together going to cylinders that fire consecutively (one after the other) there is a danger of cross firing — especially as the wires age.

If a number of primary wires travel in a common path, pull them through a loom (woven fibre conduit) or tape them together, Fig. 36-26.

FUSE WHEN NEEDED

When adding accessory units such as spotlights and heaters and no provision was made for them in

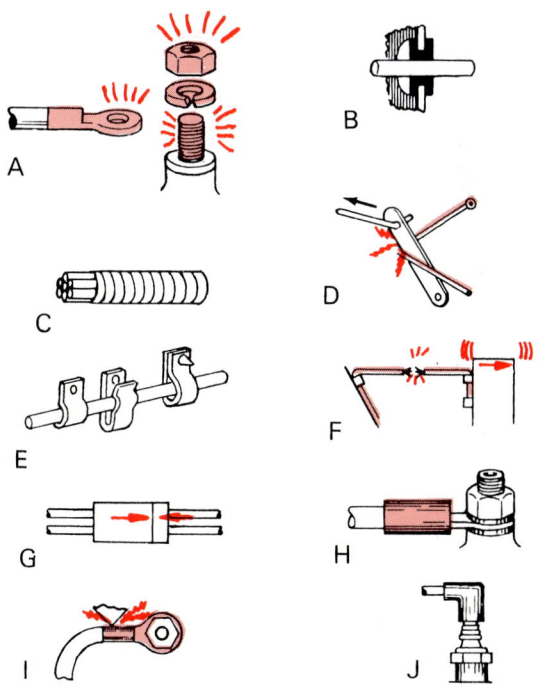

Fig. 36-26. Wiring installation hints. A — Connections must be CLEAN and BRIGHT. B — Use grommets to protect wire passing through thick metal. C — Tape common wires together. D — Avoid moving parts when locating wires. E — Support with suitable clamps. F — Allow some slack when wire runs to a unit that moves. G — Connectors must be pushed together tightly. H — Use boots on terminal tangs and select terminals heavy enough for job. I — Tighten terminals in a position AWAY FROM metal — use boots also. J — Handle resistance plug wires by grasping the boots.

the original wiring, be certain to place a fuse in the circuit. Fuse as closely as possible to the electrical source. This will reduce the possibility of a short between the fuse and source. A small fuse block may be used or the popular in-line fuse can be installed. Be sure to inform the owner as to the location of the new fuse, Fig. 36-27.

NEVER TAP (CONNECT) INTO THE HEAD-LIGHT CIRCUIT TO POWER AN ACCESSORY. THIS COULD OVERLOAD THE HEADLIGHT CIRCUIT BREAKER AND CAUSE TROUBLE. If it is desired to have the unit inoperative when the ignition key is OFF, the hot wire must be connected to the key switch circuit.

PRINTED CIRCUIT

Instead of using a maze of wires to connect numerous small and often complex components, printed circuits may be employed.

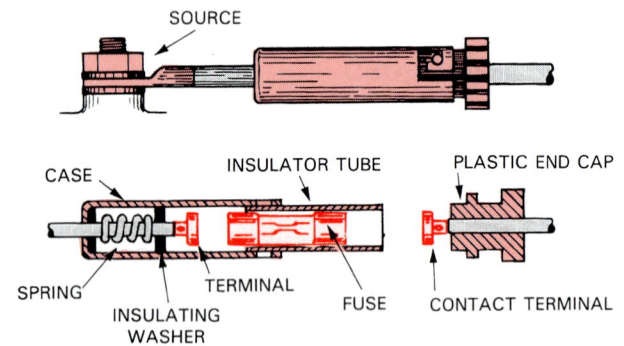

Fig. 36-27. Typical in-line fuse. Fuse as close to the source as practical.

The printed circuit uses a nonconducting panel upon which the electrical units are affixed. The units are then connected by either thin conducter strips cemented to the panel or by a special electrically conductive material "printed" on in the desired circuit patterns. This permits a great number of individual circuits in a very small area.

The modern auto is making increasingly greater use of the printed circuit in such areas as electronic emission control computers, audio accessories, and dash instrumentation. Fig. 36-28 illustrates the use of a printed circuit in a dash instrument cluster.

CHECKING WIRING

Many problems throughout the car can be traced to faulty wiring. Loose or corroded terminals, frayed and bare spots, oil soaked, broken wires, and cracked and porous insulation are the most frequent causes.

When troubleshooting a problem, check the wires, fuses, and connections carefully. Remember that wires can separate with no break in the insulation (especially resistance type secondary wire). A terminal may be tight and still be corroded. A fuse link may burn out at one end instead of in the centre where it will be visible.

CHECKING FOR CONTINUITY

A small test light (battery operated) may be used to test wires for internal breaks. See Fig. 36-29. The test point prods can be pushed through the insulation

Pin Terminals
1. Fasten belts
2. Oil pressure warning
3. Fuel gauge
4. Cluster ignition feed
5. Temperature gauge
6. Emission maintenance
7. Not used
8. Not used
9. Clock
10. Right turn
11. Not used
12. Not used
13. Left turn
14. High beam
15. High beam ground
16. Low washer fluid
17. Ground
18. Alternator warning
19. Brake

Lamps
A. Fasten belt
B. Oil pressure warning
C. Alternator warning
D. Brake
E. Right turn
F. Not used
G. Not used
H. Left turn
J. Not used
K. Emission maintenance
L. Low washer fluid
M. High beam

Other
S1 Temperature gauge sender terminal
A2 Fuel gauge ignition feed terminal
A1 Temperature gauge ignition feed terminal
G2 Fuel gauge ground terminal
G1 Temperature gauge ground terminal
P Clock feed
S2 Fuel gauge sender terminal
Q Clock ground

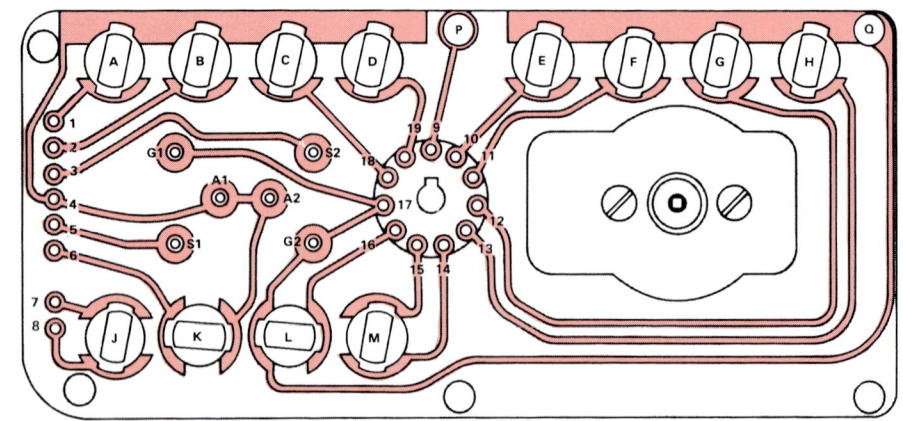

Fig. 36-28. This instrument cluster uses a printed circuit. (Jeep)

if desired (not on plug wires). Hold one prod against one end of the wire and place the other prod against the other end. If the test lamp burns, the wire is continuous. This simple test light is also handy for checking fuses, shorted field windings, and for tracing wires where there are no colour codes. Fig. 36-30 illustrates several checks.

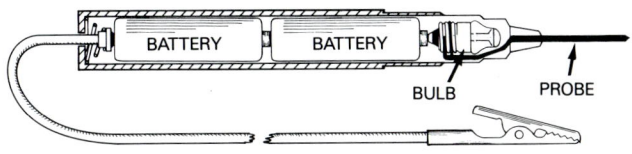

Fig. 36-29. Test light with its own power source. This type of test light is very handy for checking most electrical components for continuity.

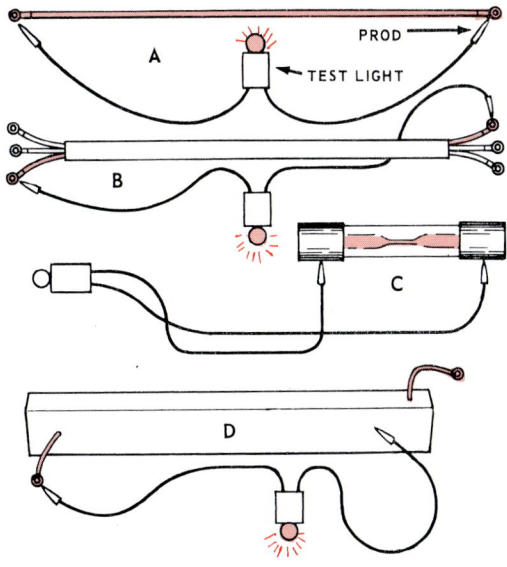

Fig. 36-30. Some wiring checks using a simple test light. A — Prods on ends of wire. Lamp lights indicating wire is continuous. B — Prod held on the end of one wire and the other prod touched to various wire ends. When lamp lights, proper wire end is identified. C — Checking a fuse. Prods in place, lamp does not light. This indicates a faulty fuse. In this case, fuse will be burnt out at end instead of usual narrow centre section. D — One prod touched to a wire end and the other prod to ground. If lamp lights, wire is shorted out.

OTHER CHECKS

Wires and connections must occasionally be checked for resistance, voltage drop, short or near-short circuits. These checks are made with precision instruments: ohmmeter, voltmeter, or ammeter. This will be discussed in the chapters where these tests pertain.

SUMMARY

Primary wire (copper stranding, relatively thin insulation) is used for circuits handling battery voltage. Secondary wire (stainless steel, carbon impregnated thread, and elastomer stranding with very heavy insulation) is used on the ignition high tension circuit. Plastic is widely used for insulation.

All automotive wire uses a stranded (not solid) wire conductor.

Spade, lug, flag, roll, slide, ring, and bullet terminal types are used.

Terminal blocks allow one feeder wire to service a number of other wires. These can be of the screw, bullet, or slide type.

Junction blocks provide a central connecting point for a number of wires.

Fuse blocks give protection against circuit overloads. A wiring harness contains a number of wires either taped together or pulled through a loom. This keeps common wires neatly arranged and facilitates installation.

Automotive electrical systems are colour coded. Use an accurate wiring diagram for troubleshooting or replacing wires.

Line voltage, wire length and electrical load must be taken into consideration when choosing wire size. A wire current rating chart will assist in making the right selection. Remember that undersize wires increase resistance, reduce unit efficiency, and can overheat or burn. On two-wire circuits (one wire for ground) count the length of both wires.

Be certain that terminals are of the correct style and size. They may be soldered or crimped to the wire. Battery replacement terminals should be soldered. When crimping, use a suitable crimping tool. If soldering, use rosin core wire solder. Always slide insulation boots and housings on the wire before attaching the terminal.

Use staples when installing terminals on resistance type secondary leads. Handle secondary resistance wire carefully.

Wire ends may be joined by soldering, using butt connectors, or by attaching bullet or slide connectors.

When installing wires, keep away from heat, oily areas, and moving parts. Terminals must be clean and tight. Use clips to prevent chafing and excessive vibration.

When adding accessories, fuse the circuit as close to the source as possible. Do not tap into the headlight circuit for an accessory.

Clean, tight connections, with proper size wire and good insulation, are imperative. When troubleshooting, always check connections and insulation. Replace cracked, spongy, or frayed wires.

Many wiring checks can be made with a simple test light.

Printed circuits find some application on the auto.

KNOW THESE TERMS

Primary wire, Secondary wire, Fuse block, Wiring harness, Colour coding, Circuit breaker, Filament, Amperes, Watts, Crimped terminal, Rosin core, Acid core, Mini fuse, Printed circuit, Continuity, Voltage drop, Short circuit, Conductor, Fusible link, Diagnostic plug-in, Insulation, Resistance wire, Wiring diagram, Rubber grommet, In-line fuse, Test lamp, Ohmmeter, Voltmeter, Feeder wire.

REVIEW QUESTIONS

1. Primary wire makes excellent spark plug leads. True or False?
2. The most commonly used insulation material is _____.
3. Resistance type spark plug wires are used to provide a hotter spark. True or False?
4. Name three materials used for secondary wire stranding.
5. Stranding for primary wire is made of _____.
6. Resistor spark plug cables are easily damaged by sharp bends and jerking. True or False?
7. All primary automotive wire uses a stranded conductor. True or False?
8. Name five common primary terminal types.
9. Replacement battery cable terminals should be _____.
10. One feeder wire can service several others through the use of _____ block.
11. A number of wires can be connected together in a common location by using a _____ block.
12. The _____ protects a circuit from an overload.
13. A number of common wires, taped together, with leads leaving at various spots, is referred to as a wiring _____ _____.
14. Automotive wiring is _____ coded.
15. What is a wiring diagram?
16. As wire length increases resistance increases. True or False?
17. An undersize wire will increase _____ and will _____.
18. Common automotive system voltage is now _____ volts.
19. As long as a terminal fits the stud or post, it is OK to use. True or False?
20. _____ terminals to the wire is more widely used than _____.
21. Use _____ when attaching terminals to resistance type secondary wire.
22. Copper or stainless steel secondary wire should have a small portion of the insulation stripped and the wire bent up and around the outside of the insulation. True or False?
23. If, when joining wire ends, it is desirable to be able to disconnect them at a future date, a _____ type connector would be a good choice.
24. As long as a connection is tight, it will be a good conductor. True or False?
25. Grommets are used to protect wire passing through thin sheet metal. True or False?
26. When plug leads pass through a metal conduit, the conduit should be _____.
27. Wires should be held by _____ to prevent chafing and vibration.
28. Spark plug wires can _____ if wires are too close together when they serve cylinders that fire consecutively.
29. As long as the insulation is alright, a wire can be considered OK. True or False?
30. A frayed wire can cause a _____-circuit.
31. A corroded connection will increase _____ to electrical flow.
32. The electrical symbols in the left hand column are all numbered. Write down these numbers, one beneath the other. The right hand column lists the items these symbols stand for. Each item has a letter. Match the items to the symbols by placing the letter of the item you have chosen beside the number of the matching symbol.

Wire and Wiring

1. _____

2. _____

3. _____

4. _____

5. _____

6. _____

7. _____

8. _____

9. _____

10. _____

11. _____

12. _____

13. _____

14. _____

15. _____

A. Resistor.

B. Circuit Breaker.

C. Wires Crossing - Not Connected.

D. Fuse.

E. Diode.

F. Wires Crossing - Connected.

G. Positive.

H. Terminal.

I. Switch.

J. Rheostat.

K. Transistor.

L. Battery.

M. Negative.

N. Condenser.

O. Ground.

Chapter 37

BASIC ELECTRICITY AND ELECTRONICS

After studying this chapter, you will be able to:
□ Explain the principles of electricity.
□ Describe the action of basic electric circuits.
□ Compare voltage, current, and resistance.
□ Describe the principles of magnetism and magnetic fields.
□ Identify basic electric and electronic terms and components.
□ Explain different kinds of automotive wiring.
□ Perform fundamental electrical tests.

Almost every system in a modern auto uses some type of electric or electronic component (part). Electronic ignition systems, electronic fuel injection, computerised engine systems, and other advanced systems require mechanics skilled in electrical repairs. Even specialised mechanics need some background in electricity and electronics in order to repair today's cars.

This chapter covers the most important and basic aspects of automotive electricity and electronics.

ELECTRICITY

Everything is made of atoms, Fig. 37-1. You are made of atoms. This book is made of atoms, so is your chair, air, everything.

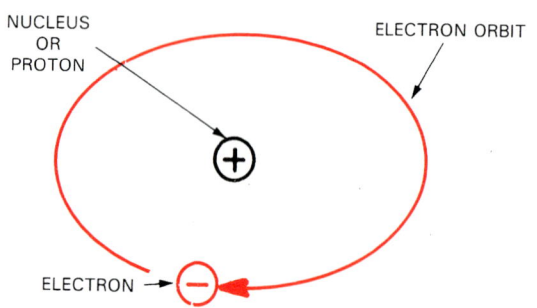

NUCLEUS
OR
PROTON

ELECTRON ORBIT

ELECTRON

Fig. 37-1. Electricity is the flow of free electrons. Conductors contain free electrons, insulators do not. (Ford)

An atom consists of small particles called protons, neutrons, and electrons. Negatively charged electrons circle around the positively charged protons. The makeup of atoms varies in different substances.

Electricity is the movement of electrons from atom to atom. Some substances have atoms that allow electrical flow; others do not.

Conductors (wires, electrical components, and other metal objects) have atoms that allow the flow of electricity. They are substances that contain FREE ELECTRONS (extra electrons not locked to protons in atoms).

Insulators (plastic, rubber, ceramics) do NOT contain free electrons. They resist the flow of electricity. The outside of wire conductors are usually covered with plastic or rubber insulating material.

Simple circuit

A simple circuit consists of:
1. POWER SOURCE (battery, alternator, or generator) which supplies electricity for the circuit.
2. LOAD (electrical device that uses electricity).
3. CONDUCTORS (wires or metal car parts that carry current between power source and load).

Look at Fig. 37-2. The power source feeds electricity to the conductors and load. The conductors carry the electricity out to the load and back to the source. The load changes the electricity into another form of energy (light, heat, or movement).

Current, voltage, and resistance

The three basic elements of electricity are: CURRENT (amps), VOLTAGE (volts), and RESISTANCE (ohms). See Fig. 37-3.

Current (abbreviated I or A) is the FLOW of electrons through a conductor. Just as water flows through a garden hose, electrons flow through a wire in a circuit.

As an example, when current flows through a light bulb, the electrons rub against the atoms in the bulb

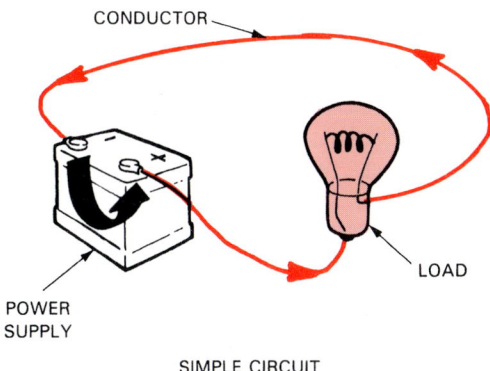

Fig. 37-2. Simple electric circuit consists of power source, load, and a conductor. Electrons will flow through circuit. (British Leyland)

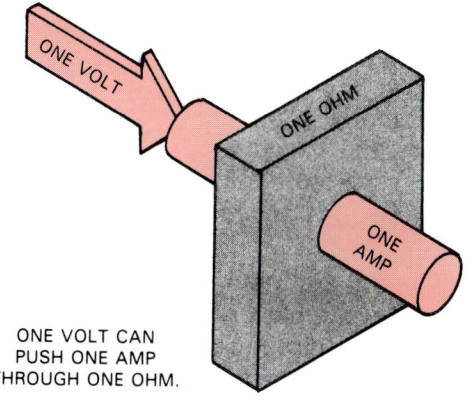

Fig. 37-3. Voltage is pressure or pushing force. Amps are flow of electrons. Ohms oppose current flow.

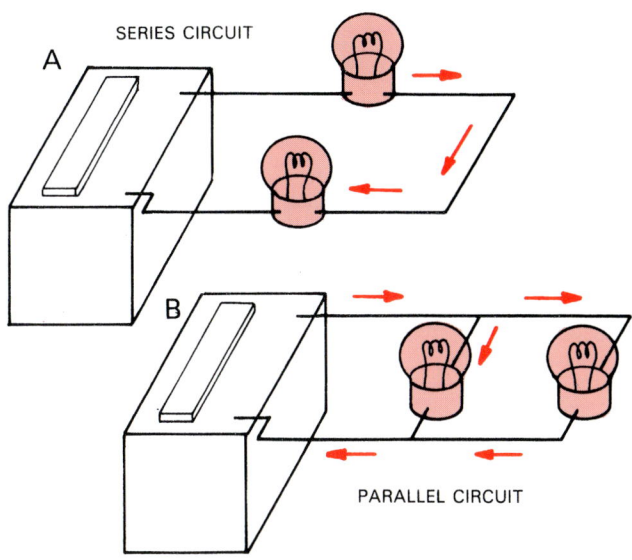

Fig. 37-4. Basic circuit types. A — Series circuit only has one path for current. B — Parallel circuit has separate path for each load.

filament (resistance wire inside bulb). This produces an "electrical friction". The friction heats the filament, making it glow "red hot".

Voltage (abbreviated V or E) is the force or ELECTRICAL PRESSURE that causes current flow. Similarly, water pressure causes water to squirt out the end of the garden hose. An increase in voltage (pressure) causes an increase in current. A decrease in voltage causes a decrease in current. Cars normally use a 12V electrical system.

Resistance (abbreviated R or Ω) is the OPPOSITION to current flow. Resistance is needed to control the flow of current in a circuit. Just as the on/off valve on a garden hose can be opened or closed to control water flow, circuit resistance can be increased or decreased to control the flow of electricity. High resistance reduces current. Low resistance increases current.

Types of circuits

A series circuit, Fig. 37-4A, has more than one load (light bulb or other component) connected in a single electrical path. For example, inexpensive Christmas tree lights can be wired in series. With only one electrical path, if one bulb burns out, all the bulbs stop glowing. The circuit path is broken (opened) and current stops.

A parallel circuit, Fig. 37-4B, has more than one electrical path or leg. Christmas tree lights wired in parallel are not prone to complete failure. One bulb can burn out without affecting the others. The other bulbs have their own leg or path to receive current.

A series-parallel circuit contains both a series circuit and a parallel circuit.

A one-wire circuit uses the car frame or body as a return wire to the power source. An insulated wire connects the battery positive terminal to the load. Then, current returns to the battery's negative terminal through the frame or metal body parts.

The term frame ground refers to the path the current takes when it returns to the negative terminal of the battery. A typical circuit with its symbol is shown in Fig. 37-5. Study it!

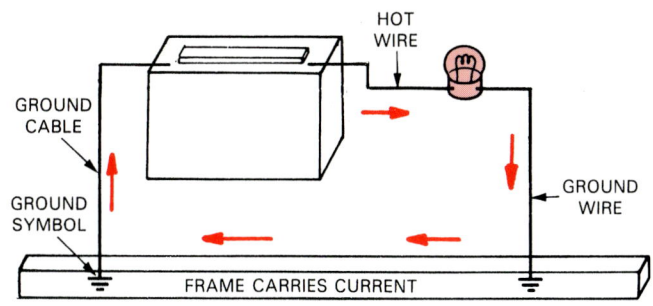

Fig. 37-5. Automotive wiring commonly uses a frame ground. Metal parts of car carry current back to negative of battery. This reduces amount of wires needed.

Ohm's Law

Ohm's Law is a simple formula for calculating an unknown electrical value (volts, amps, or ohms) when two values are given. This is illustrated in Fig. 37-6.

If you know, for example, that a circuit has 12 volts applied and a current flow of 6 amps, Ohm's Law can be used to find circuit resistance. Simply enter the known values into the correct formula.

Resistance = voltage divided by current

$$R = \frac{12 \text{ Volts}}{6 \text{ amps}}$$

$$R = 2 \text{ ohms}$$

This same operation can be used with the other two forms of Ohm's Law to find either volts or amps.

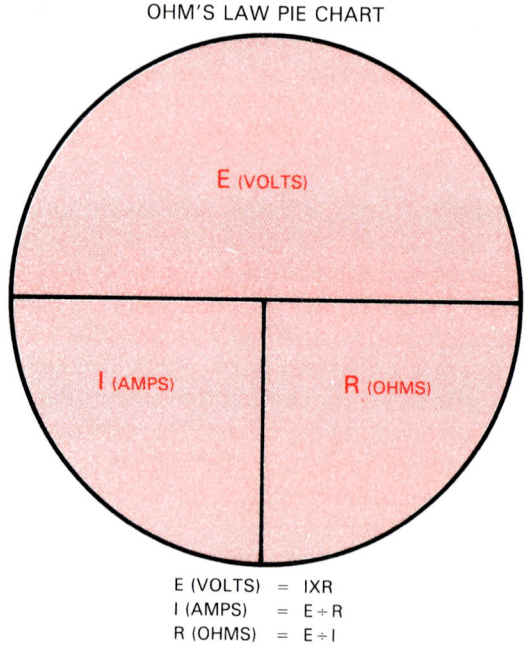

OHM'S LAW PIE CHART

E (VOLTS)

I (AMPS)   R (OHMS)

E (VOLTS) = IXR
I (AMPS) = E ÷ R
R (OHMS) = E ÷ I

Fig. 37-6. Ohm's Law pie chart. Use your finger to cover one of the letters in chart. This will show you Ohm's Law formula.

Magnetic field

You are probably familiar with magnetism from using a simple permanent magnet, Fig. 37-7A. It produces an invisible magnetic field (lines of force) that will attract ferrous (metal containing iron) objects.

A magnetic field can also be created using electricity. A long piece of wire can be wound into a coil. The ends of the wire can be connected to a battery or other source. Then, when current passes through the wire, a magnetic field is produced.

To make the field or lines of force STRONGER, a soft iron bar or core can be inserted into the centre of the coil. The iron core will become magnetised, making an electromagnet (electric magnet).

Magnetism can also create electricity. If a magnetic field is passed over a wire, an electric current is INDUCED or generated in the wire. Look at Fig. 37-7B. The wire cutting the lines of force causes a tiny amount of electricity to flow through the wire. This action is called induction (current generated in wire by a magnetic field).

Many automotive components use the characteristics of magnetism and a magnetic field. Electronic fuel injection, electric motors, relays, ignition systems, and on-board computers are just a few examples.

Electrical terms and components

There are several electrical terms and components which auto mechanics must know. The most important ones are discussed here.

A switch allows an electric circuit to be turned on or off manually (by hand). When the switch is CLOSED (on), the circuit is complete (fully connected) and will operate. When the switch is OPEN (off), the circuit is broken (disconnected) and does not function. See Fig. 37-8.

A short circuit or "short" is caused when a defective live wire or component touches ground. It

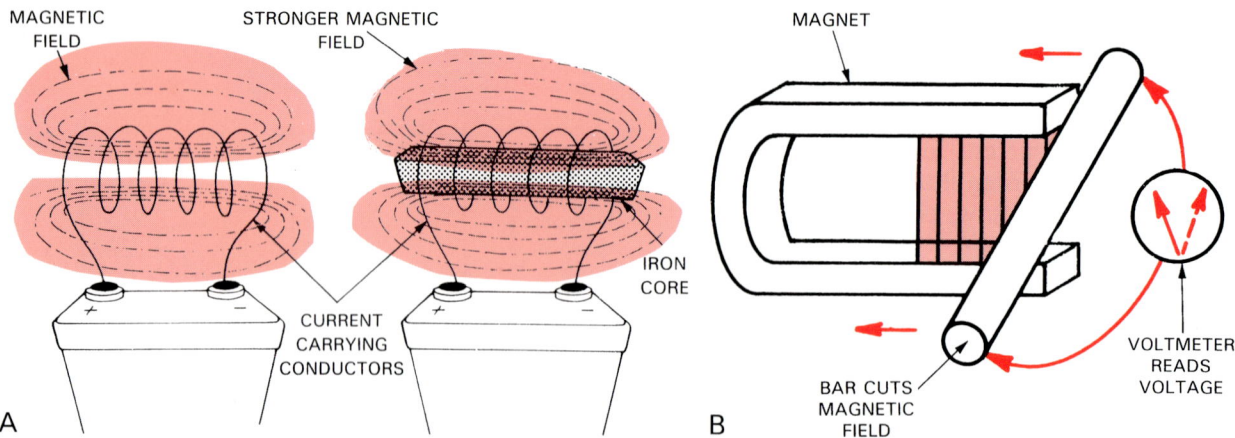

MAGNETIC FIELD   STRONGER MAGNETIC FIELD   MAGNET

IRON CORE

CURRENT CARRYING CONDUCTORS

A

BAR CUTS MAGNETIC FIELD   VOLTMETER READS VOLTAGE

B

Fig. 37-7. A — When current flows through a wire, a magnetic field forms around wire. Wire can be wound into coil to strengthen field. An iron core will strengthen field even more. B — A magnetic field can be used to produce electricity. When iron bar is moved through magnet's field, current is induced in bar and wire.

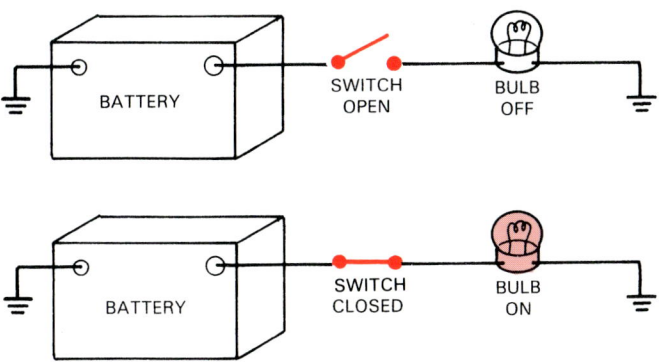

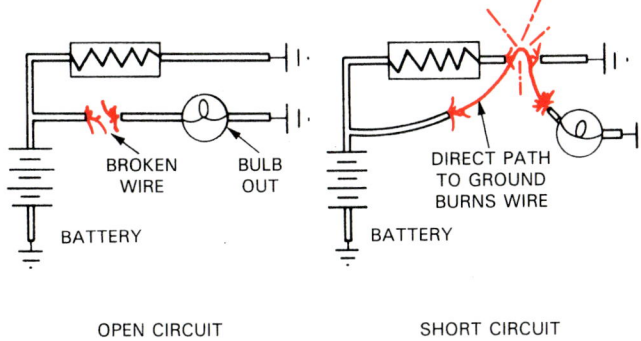

Fig. 37-8. Switch is used to break and complete circuit.

Fig. 37-9. Left. Open circuit has break in wire or electric component. Current stops flowing through circuit. Right. Short circuit has wire touching ground. High amount of current flows through short.

causes excess current flow. See Fig. 37-9. If a short to ground exists between the battery and load UNLIMITED CURRENT FLOW can cause an "electrical fire" (melting and burning the wire insulation).

A fuse protects a circuit against damage caused by a short circuit. The link in the fuse, Fig. 37-10, will melt and burn in half to stop excess current and further circuit damage. A fuse box, Fig. 37-11, is often located under the car's dashboard. It contains fuses for the various circuits.

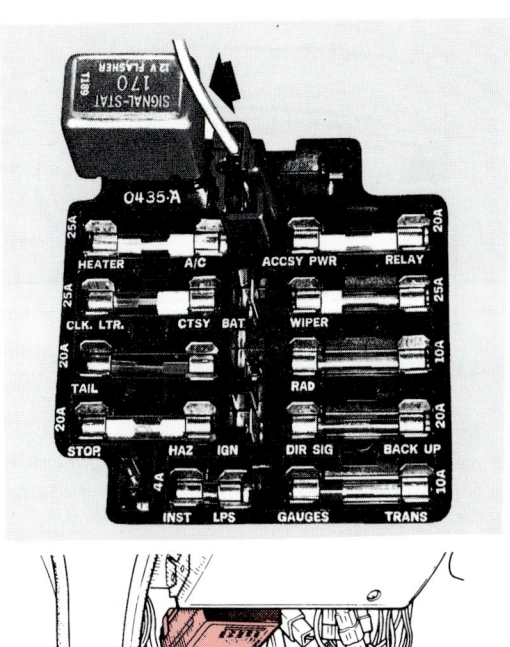

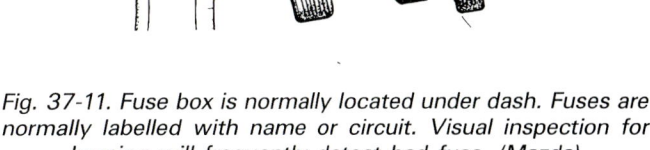

Fig. 37-11. Fuse box is normally located under dash. Fuses are normally labelled with name or circuit. Visual inspection for burning will frequently detect bad fuse. (Mazda)

A circuit breaker performs the same function as a fuse. It disconnects the power source from the circuit when current becomes too high. See Fig. 37-12. Normally, a circuit breaker will automatically reset itself when current returns to normal levels.

A relay is an electrically operated switch. It allows a small dash switch to control another circuit by remote control (control comes from a distant point in circuit). It also allows very small wires to be used

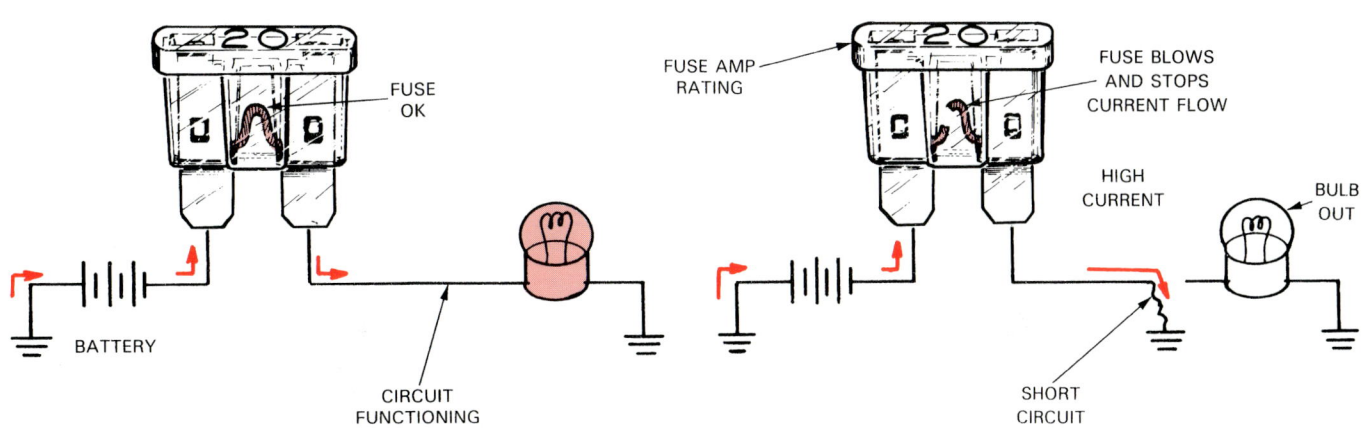

Fig. 37-10. Fuse protects against damage that would be caused by short circuit. High current heats, melts, and opens conductor in fuse. This stops current flow in circuit.

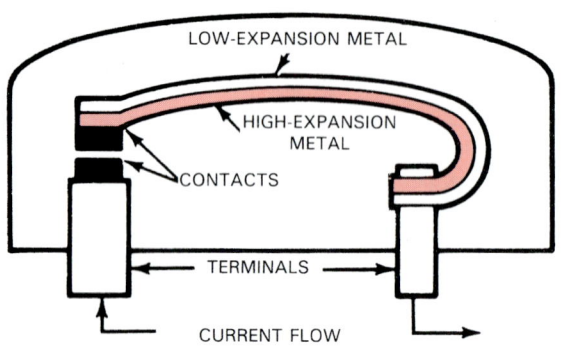

Fig. 37-12. Circuit breaker performs same function as fuse. High current heats bimetal, causing it to deform and open points. This stops current in circuit. When current drops to normal level, breaker cools and closes circuit. (Ford)

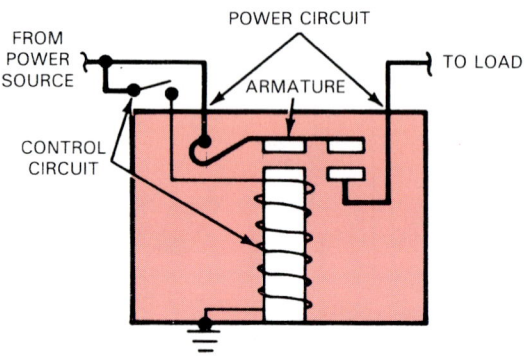

Fig. 37-13. Relay is remote control switch. Small current from dash switch can operate high current circuit elsewhere in car. When current enters control circuit, coil magnetic field pulls points closed. This completes main circuit to load. (Ford)

behind the dash while large wires are needed in the relay operated circuit. Look at Fig. 37-13.

AUTOMOTIVE ELECTRONICS

As you have just learned, some electrical components (relays and circuit breakers, for instance) use moving, mechanical contacts. These contact points can wear, burn, pit, and are relatively slow. In electronic systems, the components are solid state and do NOT have moving parts.

A semiconductor is a special substance capable of acting as both a conductor and an insulator. This characteristic enables electronic, semiconductor devices to control current without mechanical points.

Diode

A diode is an "electronic check valve" that will only allow current to flow in one direction. See Fig. 37-14A.

When forward bias (current entering in right direction), a diode acts as a CONDUCTOR and allows flow.

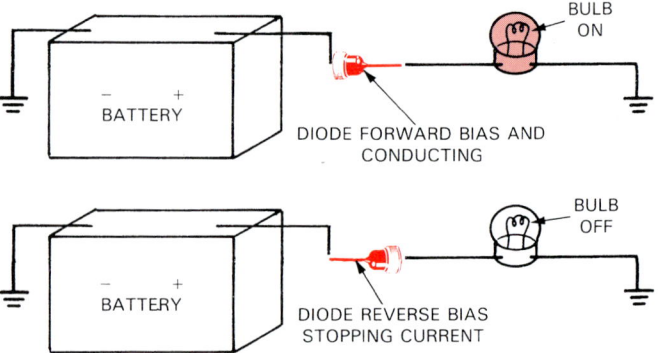

Fig. 37-14A. Diode only allows current flow in one direction. Diodes are used in wide range of electric and electronic circuits.

When reverse bias (current trys to enter wrong way), the diode changes into an INSULATOR. It stops current from passing through the circuit.

Look at Fig. 37-14B which shows a sectional view of an alternator. You will notice diodes set in the heat sink. It is usual for automotive alternators to have two heat sinks, each installed with three diodes.

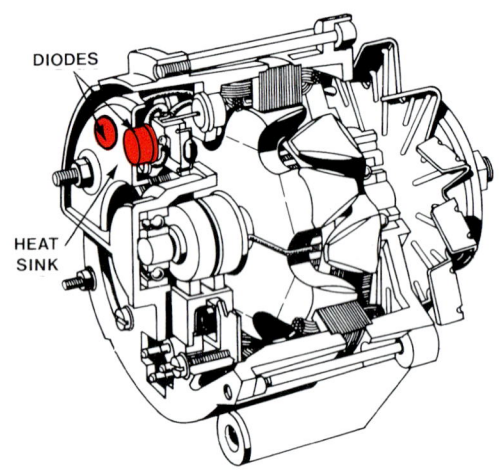

Fig. 37-14B. Sectional view of an alternator, showing diodes installed to heat sink.

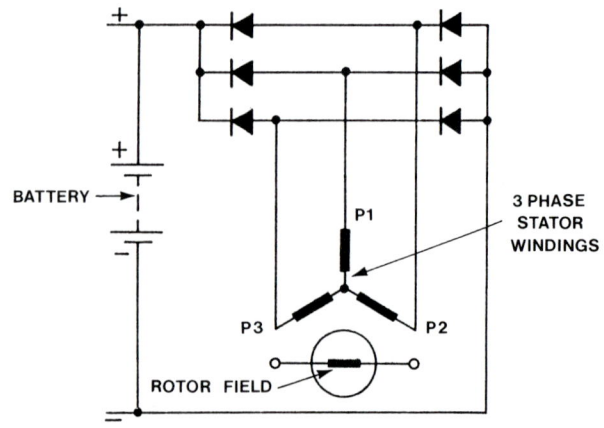

Fig. 37-14C. Line sketch of an alternator, showing two diodes connected into each of the three phases (P1, P2 and P3).

In this application, the diodes are needed to rectify the output of the alternator which is three phase alternating current (AC), to direct current (DC) which is then suitable for charging the battery. For full wave rectification to DC, it requires two diodes in each phase, look at Fig. 37-14C which is a line sketch representing the alternator. It shows how the diodes are connected to the three phase stator for rectification of AC output to DC.

Transistor

A transistor performs the same basic function as a relay: it acts as a remote control switch. It is much more efficient than a relay, however. A transistor can sometimes turn on and off faster than 2000 times a second. It does this without using moving parts which can wear and deteriorate.

Look at Fig. 37-15. A transistor amplifies (increases) a small control or base current. The small base current energises the semiconductor material, changing it from an insulator to a conductor. This allows the much larger circuit current to pass through the transistor.

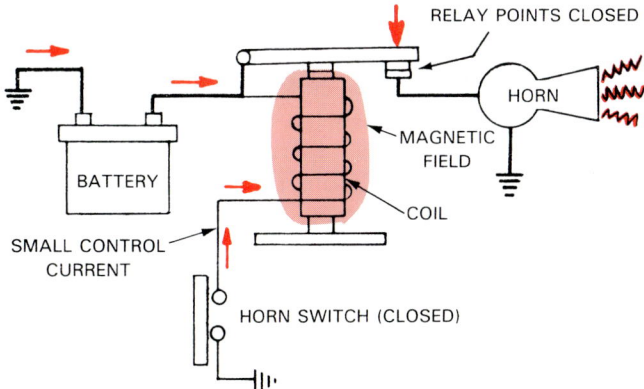

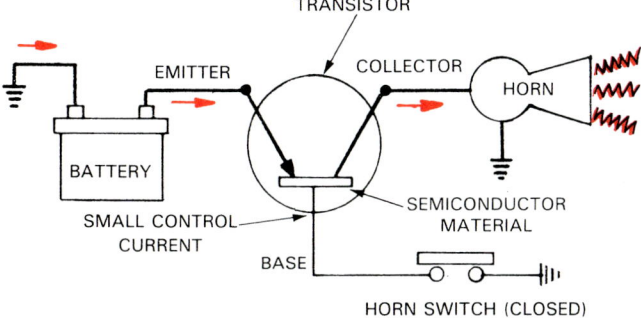

Fig. 37-15. Basically, relay and transistor perform same function. They allow small control current to operate larger current to load. A — When horn button is pressed, small current enters relay coil. Coil field attracts point arm. Then battery current can reach and operate horn through relay. B — When horn button is pressed, small base current enters transistor. This changes semiconductor material in transistor from insulator to conductor. Then, current can flow through transistor and to horn.

Other electronic devices

A condenser or capacitor is a device used to absorb unwanted electrical pulses (voltage fluctuations) in a circuit. They are used in various types of electrical and electronic circuits.

A capacitor is often connected into the supply wires going to a car radio. The capacitor absorbs any NOISE (electrical voltage pulses from alternator or ignition system) which could be heard in the radio speakers as a buzzing noise.

An integrated circuit, abbreviated IC, contains almost microscopic diodes, transistors, resistors, and capacitors in a wafer-like chip (small plastic housing with metal terminals). See Fig. 37-16. Integrated circuits are used in very complex electronic circuits.

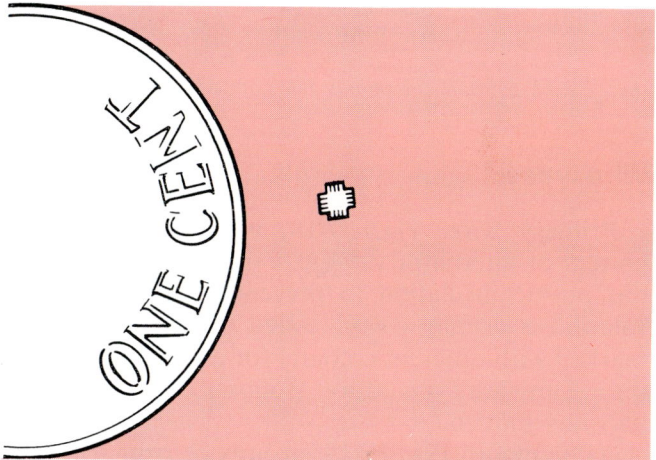

Fig. 37-16. Integrated circuit is tiny chip containing microscopic components: transistors, diodes, resistors, conductors, ICs are used in modern electronic circuits.

Printed circuits do NOT use conventional, round wires; they use flat conductor strips mounted on an insulating board. This is pictured in Fig. 37-17. Printed circuits are normally used instead of wires on the back of the instrument panel. This eliminates the need for a bundle of wires going to the indicators, gauges, and instrument bulbs.

An amplifier is an electronic circuit designed to use a very small current to control a very large current. Its function is much the same as a transistor. However, higher output currents are possible.

A good example of an amplifier is an ignition system control unit (amplifier), introduced in Chapter 35. It uses small electrical pulses from the distributor to produce strong on/off cycles to operate the ignition coil.

AUTOMOTIVE WIRING

An automobile uses various types of wiring in its many electrical systems. It is important that you learn the different types, how they are used, and how to repair them.

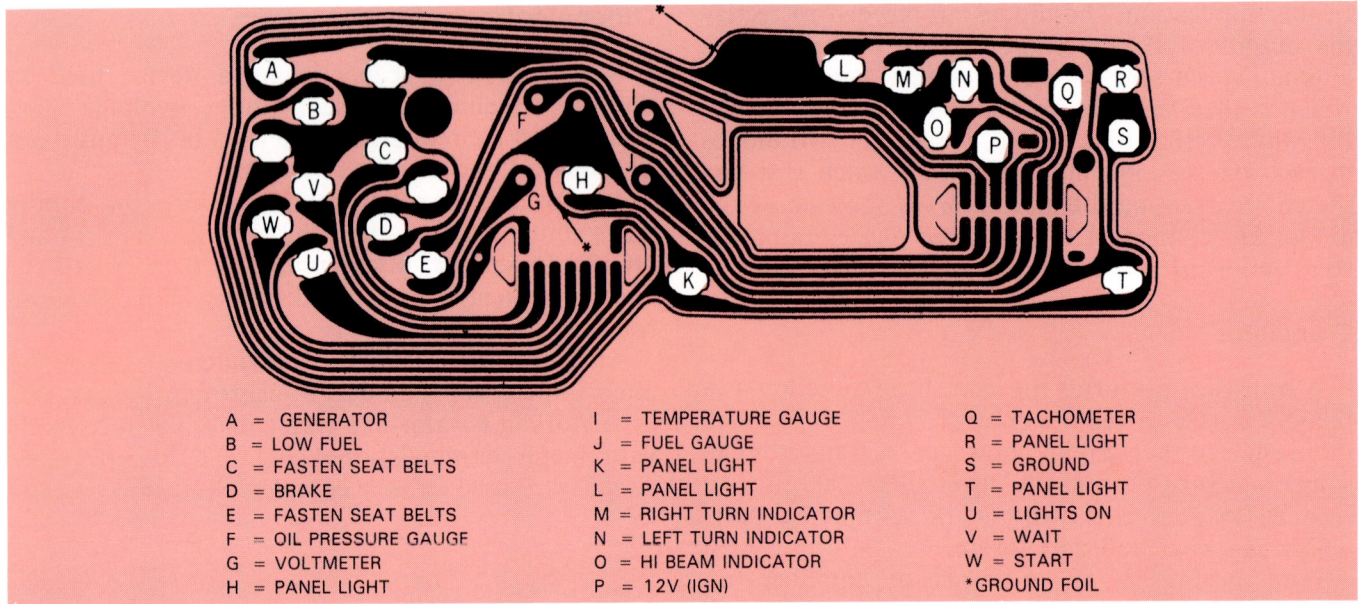

A = GENERATOR		I = TEMPERATURE GAUGE		Q = TACHOMETER	
B = LOW FUEL		J = FUEL GAUGE		R = PANEL LIGHT	
C = FASTEN SEAT BELTS		K = PANEL LIGHT		S = GROUND	
D = BRAKE		L = PANEL LIGHT		T = PANEL LIGHT	
E = FASTEN SEAT BELTS		M = RIGHT TURN INDICATOR		U = LIGHTS ON	
F = OIL PRESSURE GAUGE		N = LEFT TURN INDICATOR		V = WAIT	
G = VOLTMETER		O = HI BEAM INDICATOR		W = START	
H = PANEL LIGHT		P = 12V (IGN)		*GROUND FOIL	

Fig. 37-17. Printed circuit has flat conductor strips mounted on insulating board. Rear of this dash instrument panel is good example.

Wire types

Primary wire, Fig. 37-18, is small and carries battery or alternator voltage. Primary wire normally has plastic insulation to prevent shorting. The insulation is usually colour coded (different wires are marked with different colours) for easy troubleshooting. This lets you trace (follow) wires that are partially hidden.

As shown in Fig. 37-19, groups of primary wires are often enclosed in a wiring harness. A wiring harness is a plastic or tape covering that helps protect and organise the wires.

Wire size is determined by the area of the wire's metal conductor in square millimetres (mm²). The conductor part of automotive wiring is made up of a number of wire strands twisted together. See the automotive cable charts in Chapter 36 which shows the common metric wire sizes and the number and diameter of strands that go to make up each size wire.

When replacing a section of wire, always use wire of equal size. If a smaller wire is used the circuit could malfunction (not work) due to high resistance. Undersize wire could heat up and melt its protective insulation. An electrical fire could result.

Secondary wire, also called high tension, cable, spark plug wire or coil wire, is only used in a car's ignition system. It has extra thick insulation for carrying high voltage from the ignition coil to the spark plugs. The conductor, however, is designed for very small currents.

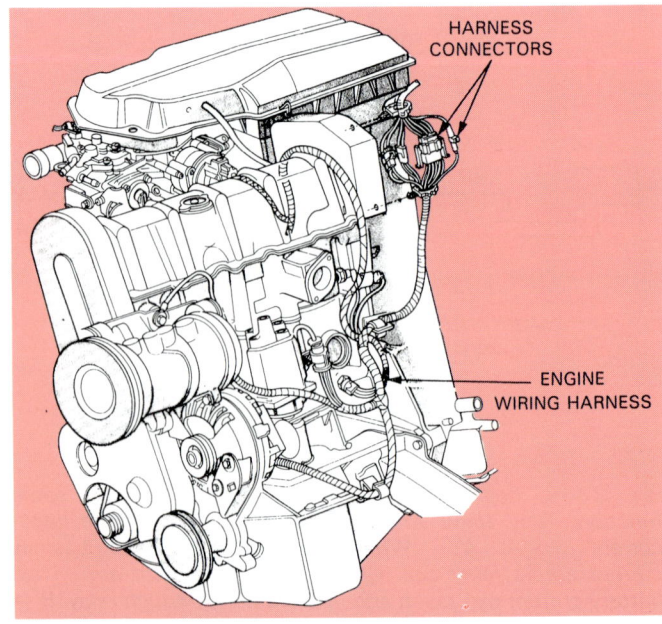

CODE	
B	BLACK
Br	BROWN
G	GREEN
Gy	GRAY
L	BLUE
Lb	LIGHT BLUE
Lg	LIGHT GREEN
O	ORANGE
R	RED
W	WHITE
Y	YELLOW

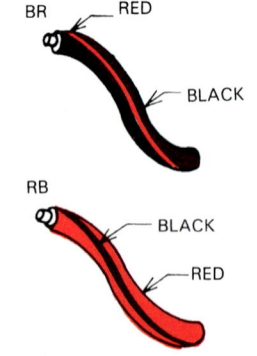

Fig. 37-18. Primary wires are colour coded with different colours. This lets you trace wire through car.

Fig. 37-19. Wiring harness is usually plastic covering around a group of primary wires. It organises and protects wires. Also note harness connectors.

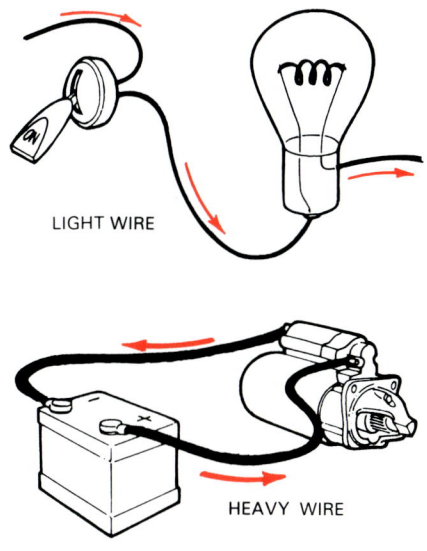

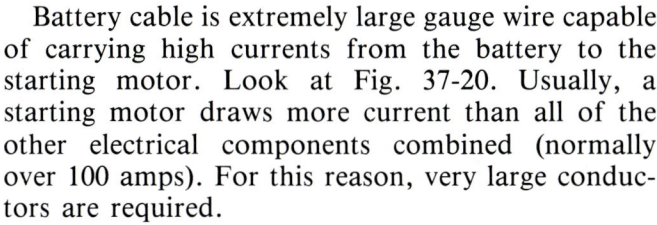

Fig. 37-20. Wire size (conductor area) is matched to current draw. Light wire will only handle small current. Heavy wire is needed for high current draws, starter motors for instance. (British Leyland)

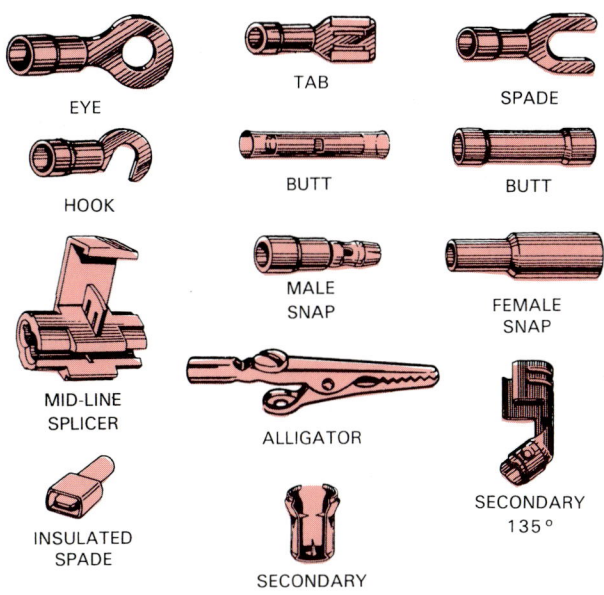

Fig. 37-21. Note various types of wire terminals and connectors. (Belden)

Battery cable is extremely large gauge wire capable of carrying high currents from the battery to the starting motor. Look at Fig. 37-20. Usually, a starting motor draws more current than all of the other electrical components combined (normally over 100 amps). For this reason, very large conductors are required.

Ground wires or ground straps connect electrical components to the chassis or ground of the car. Since they connect circuits or parts to ground, insulation is NOT needed.

Wiring repairs

Numerous methods can be used to repair automotive wiring. The most important, general methods will be introduced.

Crimp connectors and terminals can be used to quickly repair automotive wiring. See Fig. 37-21. Terminals allow a wire to be connected to an electrical component. Connectors or splicers allow a wire to be connected to another wire.

Crimping pliers are used to deform the connector or terminal around the wire. Fig. 37-22 shows a mechanic installing a crimp terminal.

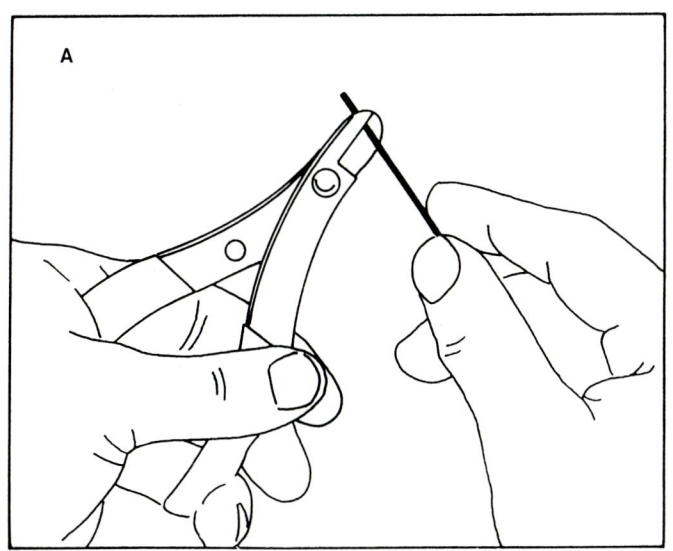

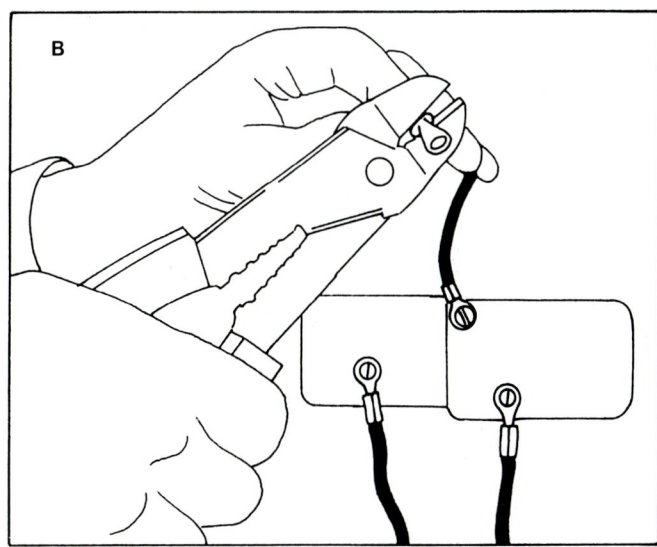

Fig. 37-22. Installing crimp type connectors and terminals. A — Strip off a short section of insulation. B — Use right size crimping jaw to form terminal or connector around wire. Tug on wire lightly to check connection.

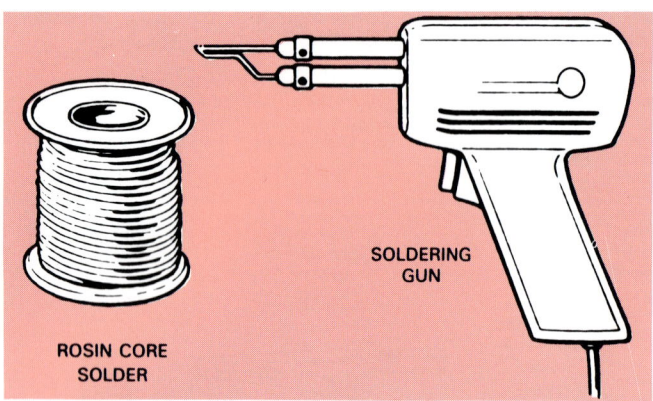

Fig. 37-23. Rosin core solder and soldering gun will make permanent connections between wires and components.

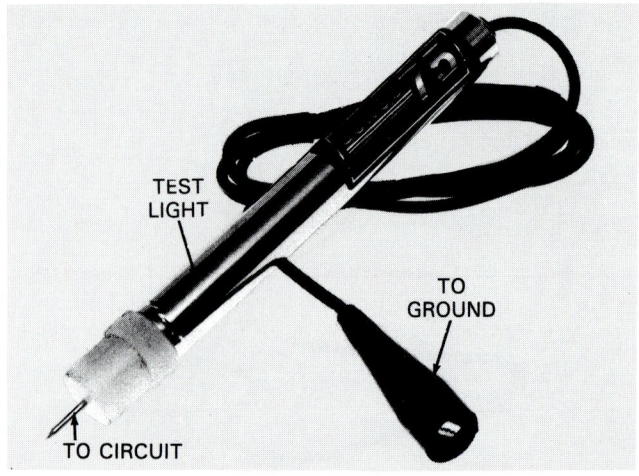

Fig. 37-25. Test light will quickly check for power in circuit. Connect alligator clip to ground and touch tip to circuit. Light will glow if there is power in circuit.

A soldering gun or iron can also be used to permanently fasten wires to terminals or to other wires, Fig. 37-23. The soldering gun produces enough heat to melt solder (lead and tin alloy). The soldering gun is touched to the wire and other component to preheat them. Then, the solder is touched to the joint and melts. When cooled, the solder makes a solid connection between the electrical components.

Rosin core solder should be used on all electrical repairs. It is usually purchased in a roll form for easy use and handling.

Acid core solder can cause corrosion of electrical components. It is recommended for non-electrical repairs (radiator and heater core repairs, for example).

BASIC ELECTRICAL TESTS

Various electrical tests and testing devices are used by an auto mechanic. You should have a general understanding of these tools and how to use them.

A jumper wire is handy for testing switches, relays, solenoids, wires, and other components. The "jumper" can be substituted for the components, as shown in Fig. 37-24. If the circuit begins to function with the jumper in place, then the component being bypassed is defective.

A test light is a fast method of checking a circuit for power or voltage. It has an alligator clip that connects to ground, Fig. 37-25. Then, the pointed tip can be touched to the circuit to check the power. If there is voltage, the light will glow. If it does not glow, there is an open circuit or break between the power source and the test point. See Fig. 37-26.

A self-powered test light is similar to a flashlight with a lead attached. It contains batteries and is used to check for circuit continuity (whether circuit is complete). To use this type test light, the normal source of power (car battery or feed wire) must be disconnected. If the light glows, the circuit or part has continuity (low ohms). If it does NOT glow, there is an open circuit or break (high ohms) between the two test points.

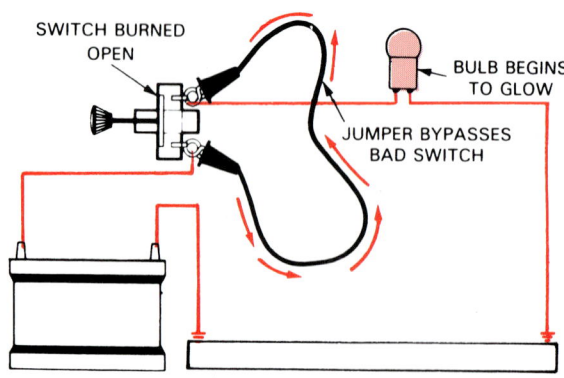

Fig. 37-24. Jumper wire is handy for bypassing electrical components. It can also be used to supply power to section of circuit. (Ford)

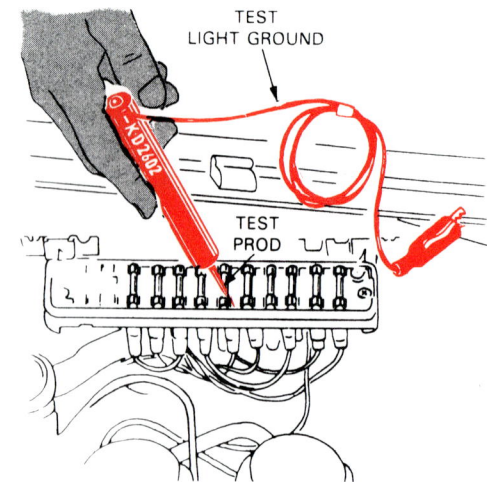

Fig. 37-26. Example of test light use is testing for a bad fuse. Blown fuse will show up when test light only glows on one side of fuse. Good fuse will cause light to glow on both sides of fuse.

Voltmeter, ammeter, ohmmeter

A voltmeter is used to measure the amount of voltage (volts) in a circuit, Fig. 37-27A. It is normally connected across or in parallel to the circuit. The voltmeter reading can be compared to specifications to determine whether an electrical problem exists.

An ammeter measures the amount of current (amps) in a circuit. Fig. 37-27B. Conventional types must be connected in SERIES with the circuit. All of the current in the circuit must pass through the ammeter.

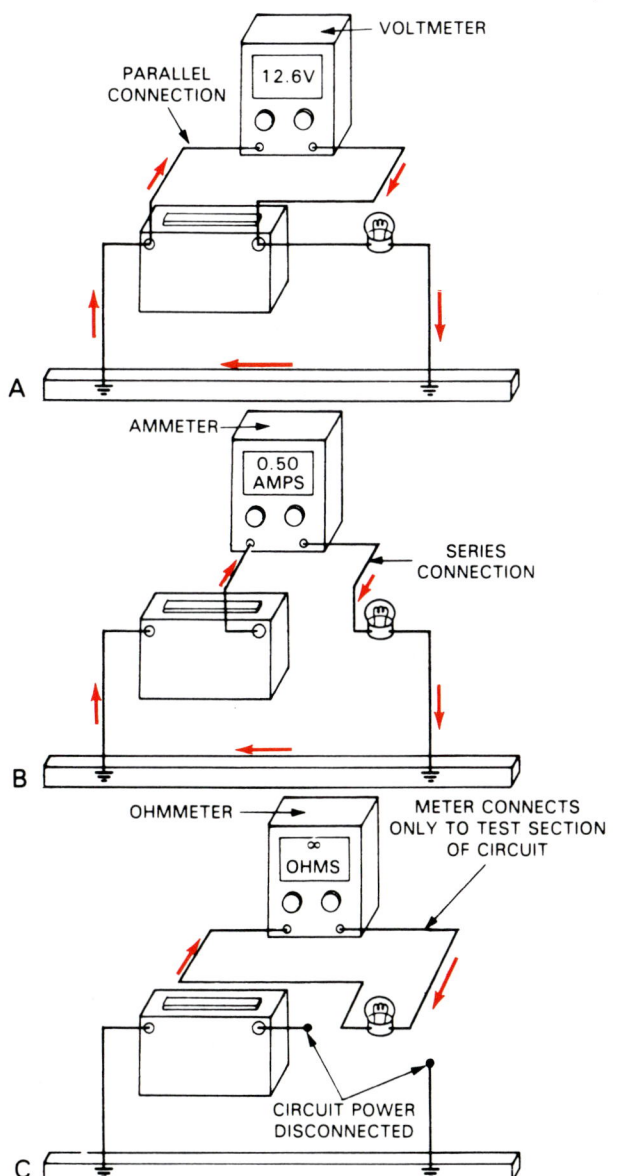

Fig. 37-27. Three basic meter connections. A — Voltmeter connects in parallel. It measures amount of electrical pressure or potential in circuit. B — Ammeter connects in series with circuit. Current flows through meter and circuit. C — Ohmmeter is connected to circuit with power disconnected. Voltage can damage some meters.

A modern inductive or clip-on ammeter is simply slipped over the outside of the wire insulation. It uses the magnetic field around the outside of the wire to determine the amount of current in the wire. An inductive ammeter is very fast and easy to use.

An ohmmeter will measure the amount of resistance (ohms) in a circuit or component. To prevent damage, an ohmmeter must NEVER be connected to a source of voltage. The wire or part being tested must be disconnected from the car's battery.

As in Fig. 37-27C, the ohmmeter is connected across the wire or component being tested. Then, the ohmmeter reading can be compared to specifications. If too high or low, the part is defective.

A multimeter, also called a VOM, is an ohmmeter, ammeter, and voltmeter combined into one case. As pictured in Fig. 37-28, a function knob (control knob) can be turned to select the type measurement to be made (volts, amps, or ohms). It must be connected to the circuit as described for each individual meter.

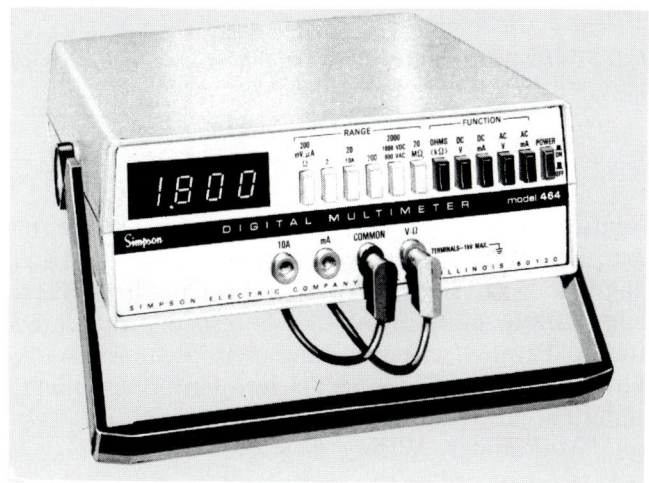

Fig. 37-28. Multimeter or VOM is voltmeter, ammeter, and ohmmeter combined. This is a digital meter because it has number display, not an indicating needle.

Wiring diagrams

A road map shows various cities are connected by roads and highways. Similarly, a wiring diagram shows how electrical components are connected by wires. Look at Fig. 37-29. It serves as an "electrical map" which helps the mechanic with difficult electrical repairs.

Wiring diagrams use symbols to represent the electrical components in a circuit. The lines on the diagram represent the wires. In this way, you can trace each wire and see how it connects to each component.

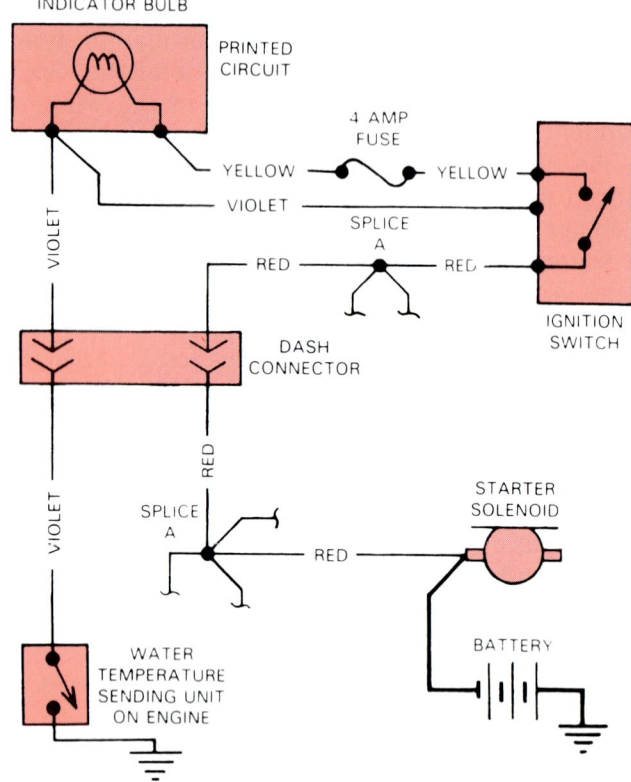

Fig. 37-29. Note how wiring diagram uses symbols to represent parts of electrical circuit.

KNOW THESE TERMS

Conductor, Insulator, Simple circuit, Current, Voltage, Resistance, One wire circuit, Ohm's Law, Magnetic field, Short circuits, Fuse, Circuit breaker, Relay, Semiconductor, Diode, Transistor, Integrated circuit, Printed circuit, Amplifier, Primary wire, Wiring harness, Secondary wire, Crimping pliers, Soldering gun, Rosin core solder, Jumper wire, Test light, Multimeter, Wiring diagram.

REVIEW QUESTIONS

1. A car, this book, and people are made of atoms. True or False?
2. What is electricity?
3. Explain the difference between a conductor and an insulator.

4. Which of the following is NOT part of a simple circuit?

   a. Electric motor.
   b. Load.
   c. Power source.
   d. Conductors.

5. List and explain the three basic elements of electricity.
6. A _____ circuit has more than one load connected in a single electrical path.
7. A _____ circuit has more than one electrical path or leg.
8. What is a one-wire circuit?
9. Using Ohm's Law, find the resistance in a circuit with 12 volts and three amps.
10. A magnetic field can be used to produce electricity and electricity produces a magnetic field. True or False?
11. Define the term "short circuit".
12. Explain the functions of fuses and circuit breakers.
13. A _____ is an electrical, not electronic, device that allows a small current to control a larger current.
14. Explain the difference between an electric component and an electronic component.
15. Which of the following ARE electronic components.

    a. Diode.
    b. Transistor.
    c. Circuit breaker.
    d. IC.

16. An _____ is an electronic circuit that uses a very small current to control a very large current.
17. Secondary wire is used to carry battery voltage and has thick insulation. True or False?
18. Why are wires colour coded?
19. Which of the following should NOT be used for electrical repairs?

    a. Acid core solder.
    b. Rosin core solder.
    c. Crimp connectors.
    d. Soldering gun.

20. Explain the use of a test light, voltmeter, ohmmeter, ammeter, and wiring diagrams.

+	POSITIVE		CONNECTOR
−	NEGATIVE		MALE CONNECTOR
	GROUND		FEMALE CONNECTOR
	FUSE		MULTIPLE CONNECTOR
	CIRCUIT BREAKER		DENOTES WIRE CONTINUES ELSEWHERE
	CAPACITOR		SPLICE
Ω	OHMS		SPLICE IDENTIFICATION
	RESISTOR		OPTIONAL WIRING WITH / WIRING WITHOUT
	VARIABLE RESISTOR		THERMAL ELEMENT (BI-METAL STRIP)
	SERIES RESISTOR		"Y" WINDINGS
	COIL		DIGITAL READOUT
	STEP UP COIL		SINGLE FILAMENT LAMP

	OPEN CONTACT		DUAL FILAMENT LAMP
	CLOSED CONTACT		L.E.D.-LIGHT EMITTING DIODE
	CLOSED SWITCH		THERMISTOR
	OPEN SWITCH		GAUGE
	CLOSED GANGED SWITCH		TIMER
	OPEN GANGED SWITCH		MOTOR
	TWO POLE SINGLE THROW SWITCH		ARMATURE AND BRUSHES
	PRESSURE SWITCH		DENOTES WIRE GOES THROUGH GROMMET
	SOLENOID SWITCH		DENOTES WIRE GOES THROUGH 40 WAY DISCONNECT
	MERCURY SWITCH		DENOTES WIRE GOES THROUGH 25 WAY STEERING COLUMN CONNECTOR
	DIODE OR RECTIFIER		DENOTES WIRE GOES THROUGH 25 WAY INSTRUMENT PANEL CONNECTOR
	BY-DIRECTIONAL ZENER DIODE		

Study wiring diagram symbols carefully. They are important!

Chapter 38

AUTOMOTIVE BATTERIES

After studying this chapter, you will be able to:
- ☐ Explain the operating principles of a lead-acid battery.
- ☐ Describe the basic parts of an automotive battery.
- ☐ Compare conventional and maintenance-free batteries.
- ☐ Explain how temperature and other factors affect battery performance.

Earlier textbook chapters briefly introduced battery operation and electrical fundamentals. This chapter will build upon this knowledge by discussing automotive batteries in more detail.

This chapter prepares you for the next chapter on battery service.

BATTERY PRINCIPLES

An automotive battery is an electro-chemical device for producing and storing electricity. A cutaway view of an auto battery is shown in Fig. 38-1. A battery produces DC (direct current) electricity that flows in only one direction.

When discharging (current flowing out of battery), the battery changes chemical energy into electrical energy. In this way, it releases stored energy.

During charging (current flowing into battery from charging system), electrical energy is converted into chemical energy. The battery can then store energy until needed.

Basic battery cell

A simple battery cell consists of a negative plate, positive plate, container, and electrolyte (battery acid). Look at Fig. 38-2.

The battery plates are made of lead oxide. These act as dissimilar (unlike) metals. The container is usually plastic to resist corrosion. The electrolyte is a mixture of sulphuric acid and water.

If a load (current-using device) is connected to our simple battery cell, current will flow through the load. As in Fig. 38-2, if the load is a light bulb, the bulb will glow because of electron movement.

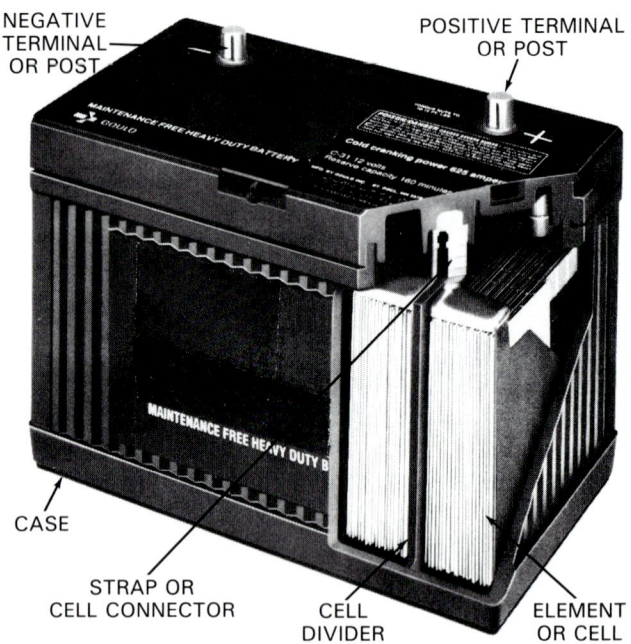

Fig. 38-1. Before learning how a battery works, study in detail its basic parts. Note part names and locations.

Battery cell action

Fig. 38-3 shows the basic chemical-electrical action inside a battery cell. When being charged, the alternator causes free electrons (negative charges) to be deposited on the negative (-) plate. This causes the plates to have a difference in potential (electrical pressure or voltage).

When a load is connected across the terminals, there is a current (flow of electrons) to equalise the difference in charges on the plates. The excess electrons move from the negative to the positively charged plate.

FUNCTIONS OF A BATTERY

A car battery has several important functions. It must:

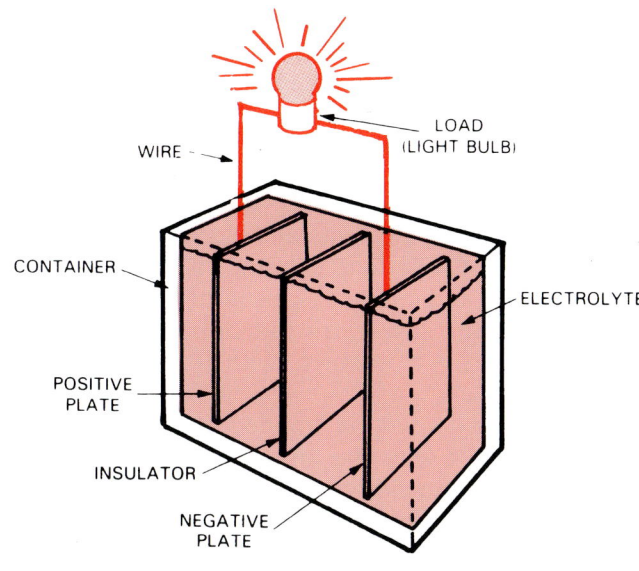

Fig. 38-2. Simple lead-acid battery cell. Positive plate and negative plate are kept apart by separator. Electrolyte causes chemical reaction between plates, producing current flow through circuit. One cell like this would produce 2.1 volts.

1. Operate the starter motor, ignition system, electronic fuel injection system, and other electrical devices for the engine during engine cranking and starting.
2. Supply ALL of the electrical power for the car whenever the engine is NOT running.
3. Help the charging system provide electricity when current demands are above the output limit of the charging system.

4. Act as a capacitor (voltage stabiliser) that smooths current flow through the car's electrical systems.
5. Store energy (electricity) for extended periods.

To illustrate these functions, imagine the following sequence of events: You are sitting in your car with the radio ON, but the engine is NOT running. The battery is supplying the electricity to operate the radio and any indicator lights. It is slowly discharging.

When you start the engine, the battery provides a tremendous amount of current. This energy operates starter motor and essential engine systems. This, too, drains current out of the battery.

As soon as the engine starts, the charging system takes over. It then recharges the battery and feeds current to the electrical units in the car.

If the load becomes too much for the charging system (engine idling slowly, all accessories ON, for example), the battery may also feed current into the electrical system.

BATTERY CONSTRUCTION

An automobile battery is built to withstand severe vibration, cold weather, engine heat, corrosive chemicals, high current discharge, and prolonged periods without use. To properly test and service batteries, you must understand battery construction.

Battery element

A battery element is made up of positive plates, negative plates, straps and separators. The element

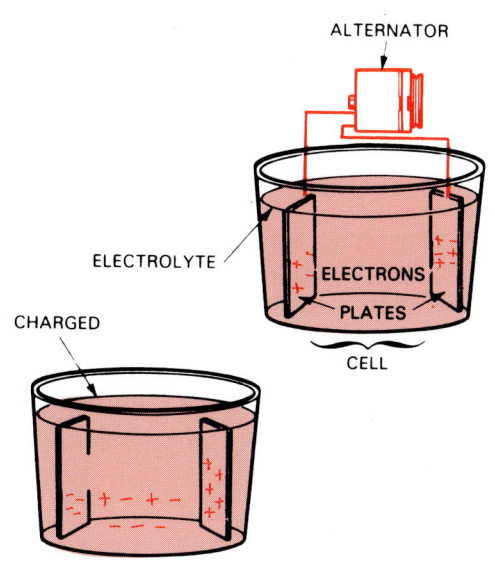

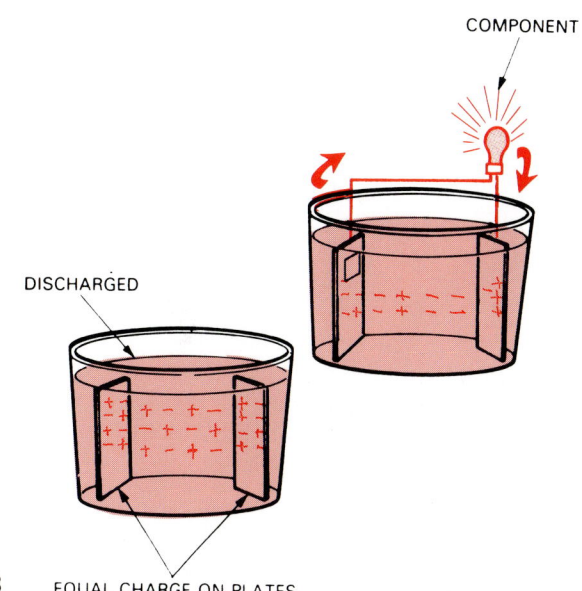

Fig. 38-3. Basic battery action. A — During charging cycle, alternator forces current through cell in reverse direction. This deposits negative charge on one plate and postive charge on other. B — During discharge cycle, current flow through circuit, allows charges to equalise. Movement of electrons produces current flow.

fits into a cell compartment in the battery case. Refer to Fig. 38-4.

The battery plates are made of a GRID (stiff mesh framework) coated with porous LEAD. Shown in Fig. 38-4, several battery plates are needed in each cell to provide enough battery power.

A lead strap connects several negative plates to form a negative plate group. Look at Fig. 38-4. Another lead strap connects the positive plates to form the positive plate group.

The chemically active material in the negative plates is sponge (porous) lead, Fig. 38-4. The active material on the positive plates is lead peroxide. Calcium or antimony is normally added to the lead to increase battery performance and to decrease gassing (acid fumes forming during chemical reaction).

Since the lead on the plates is porous, like a sponge, the battery acid easily penetrates into the lead. This aids the chemical reaction and the production of electricity.

Lead battery straps or connectors run along the upper portion of the case to connect the plates. The battery terminals (posts) are constructed as part of one end of each strap.

Separators fit between the battery plates to keep them from touching and shorting against each other.

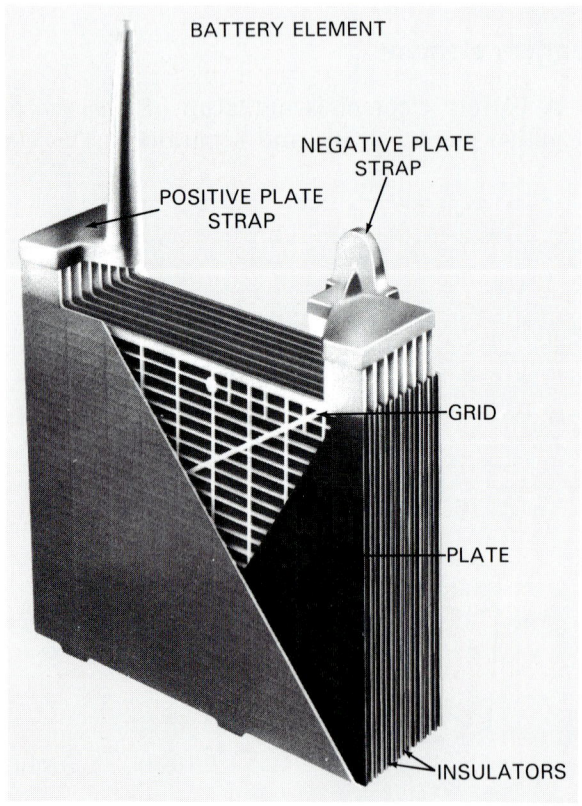

Fig. 38-4. Battery element is made up of positive plate group, negative plate group, separators, and straps. Most auto batteries have six elements. The elements fit into the battery case.

The separators are made of insulating material. They have openings which allow free circulation of the electrolyte around the battery plates.

Battery case, cover, caps

The battery case, usually made of high quality plastic, holds the elements and electrolyte, Fig. 38-5. The case must withstand extreme vibration, temperature change, and corrosive action of the battery acid. Dividers in the case form individual containers for each element. A container, with its element, is one cell.

The battery cover is bonded to the top of the battery case. It seals the top of the case. There is an opening above each battery cell for battery caps or a cell cover. Refer to Fig. 38-5.

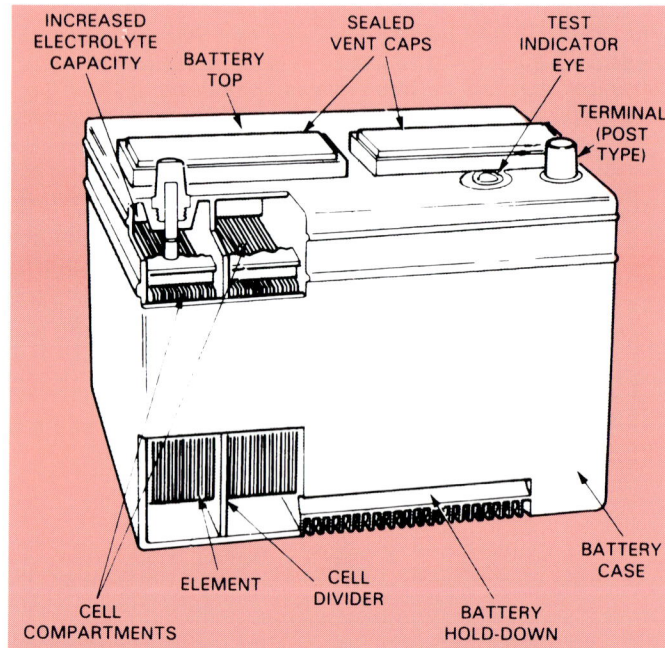

Fig. 38-5. Battery case holds elements and electrolyte. Note part names.

Battery caps snap into the holes in the battery cover. They keep electrolyte from splashing out of the cover. Battery caps also serve as spark arrestors (keep sparks or flames from igniting gases inside battery). Maintenance-free batteries have a large cover that is not removed during normal service.

DANGER: Hydrogen gas can collect at the top of batteries. If this gas is exposed to a flame or spark, it can explode.

Discussed in the next chapter, there are several safety precautions to follow when servicing batteries.

Electrolyte (battery acid)

Electrolyte, often called battery acid, is a mixture of sulphuric acid and distilled water, Fig. 38-6.

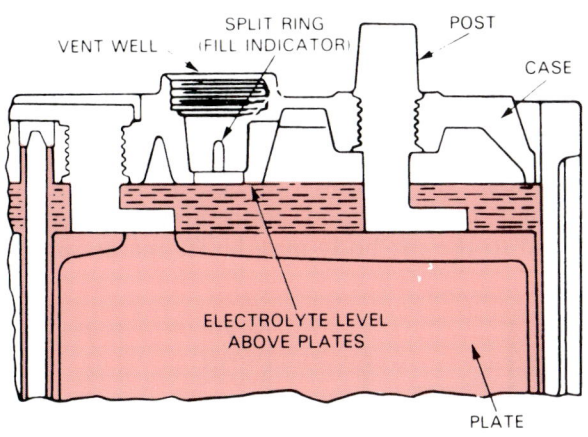

Fig. 38-6. Electrolyte or battery acid covers plates. Acid should just touch split ring in top of case. Vent allows gases to leave case.

Battery acid is poured into each cell until plates are covered.

DANGER! Avoid having electrolyte come in contact with your skin or eyes. The sulphuric acid in the electrolyte can cause serious skin burns or even blindness.

Battery charge indicator

A battery charge indicator, also called an eye or test indicator, shows the general charge condition of the battery. One is pictured in Fig. 38-7.

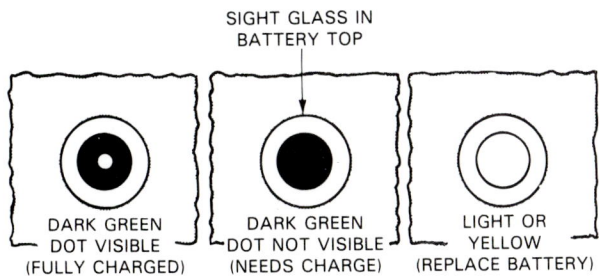

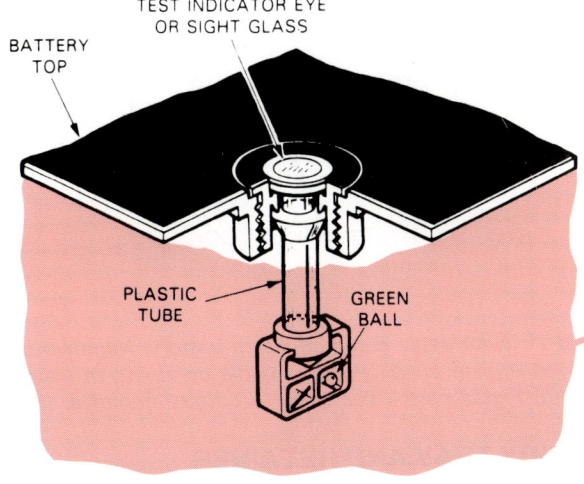

Fig. 38-7. Charge indicator provides easy way of checking battery condition.

The charge indicator changes colour with changes in battery charge. For example, the indicator may be green with the battery fully charged. It may turn black when discharged or yellow when the battery needs replacement.

Battery terminals

Battery terminals provide a means of connecting the battery plates to the car's electrical system. Either post terminals or blade terminals can be used, as shown in Fig. 38-8.

Battery posts are round or flat metal terminals projecting through the top of the battery cover. They serve as connection points for battery cable ends.

Round posts are slightly different in size. The positive post is the larger one, it may be marked with red paint and a (+) symbol. The negative post is smaller and may be marked with black paint. It normally has a negative (–) symbol on, or near it.

Flat posts are usually the same size, with a hole through them to accept the retaining nuts and bolts for the battery cables. The posts, or the battery case will normally be marked with paint and/or symbols, similar to round post batteries to indicate terminal polarity.

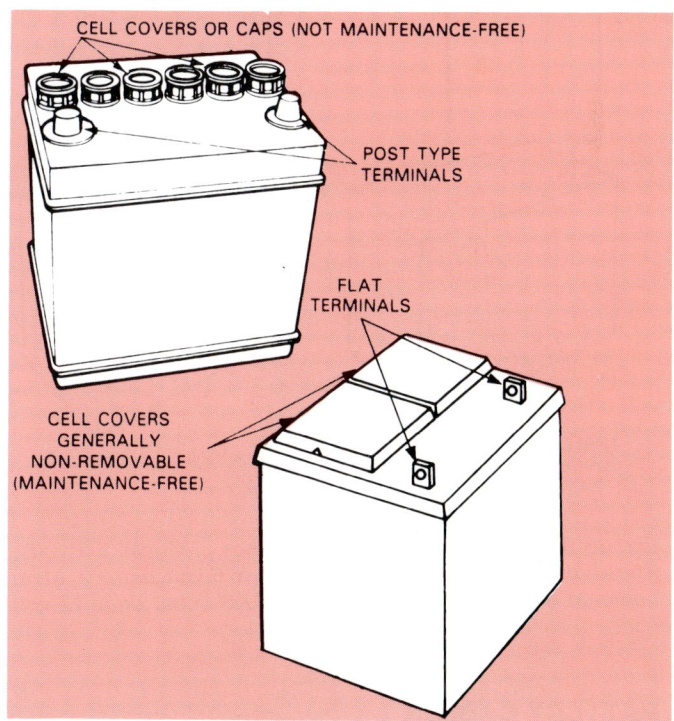

Fig. 38-8. Study differences between batteries. Maintenance-free battery does not have vent caps. Post or blade terminals may be used on conventional or maintenance-free batteries.

Battery voltage

Battery open circuit (no load) cell voltage is 2.1 volts, often rounded off to 2.0 volts. Since the cells in a car battery are connected in series, battery voltage depends upon the number of cells. Refer to Fig. 38-9.

A 12-volt battery has 6 cells that produce an open circuit voltage of 12.6 volts. Modern autos use a 12 V battery and 12 V electrical system.

A 6-volt battery only has 3 cells with an open circuit voltage of 6.3 V. Fig. 38-9. Older cars are designed to use 6 V batteries.

Some cars with diesel engines use TWO 12 V batteries connected in parallel. When two batteries are in parallel, their output voltage stays the same. Current output increases. Dual batteries may be needed to crank and start a compression ignition, diesel engine.

When two batteries are connected in series, their output voltage DOUBLES. Keep this in mind when working with batteries. Two 12 V batteries in series produce 24 volts that could damage electrical devices.

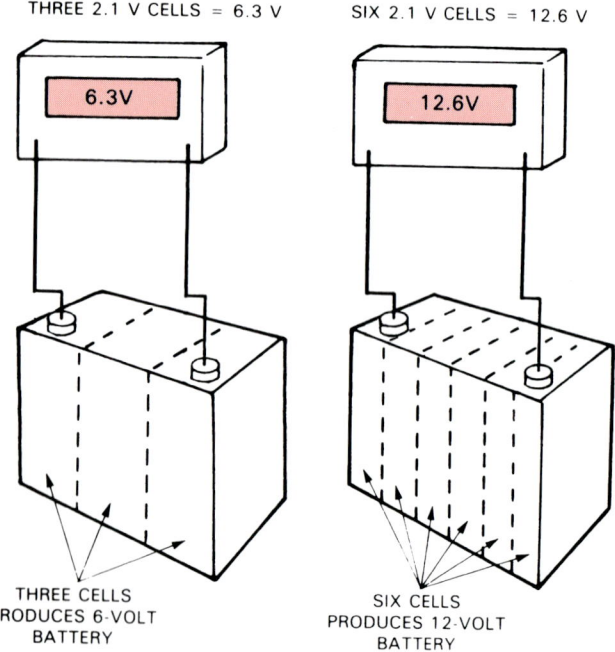

Fig. 38-9. Three cells connected in series produces 6.3 volts. The three cells are rated at or called a 6-volt battery. More common unit of six cells produces 12.6 volts, or a 12-volt battery.

BATTERY CABLES

Battery cables are large wires that connect the battery terminals to the electrical system of the car.

The positive cable from the battery is normally red and fastens to the starter solenoid (introduced in Chapter 1). The negative battery cable from the battery is usually black and connects to earth on the engine block.

Various types of battery cables are pictured in Fig. 38-10. Note that earth cables do not always use insulation.

Sometimes, the negative battery cable will have a body earth wire which ensures that the vehicle body is earthed. One is shown in Fig. 38-11. If this wire does not make a good connection, a component earthed to the car body may NOT operate properly.

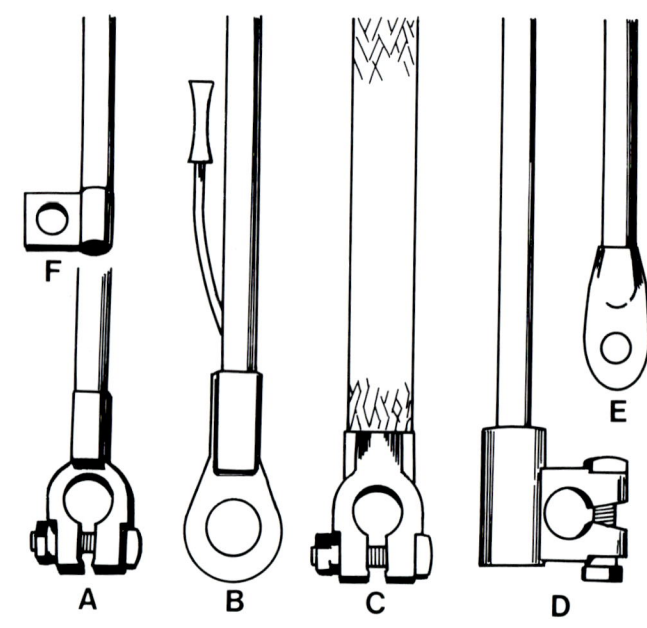

Fig. 38-10. Battery cable types: A — End clamp post type battery cable. B — Battery cable, with auxiliary cable for earth, or accessory connections. C — End clamp, braided earth cable. D — Side clamp type battery cable. E — Spade terminal type battery cable. F — Flat post type battery cable. Note the large conductor size, for carrying large amounts of current to the starter motor.

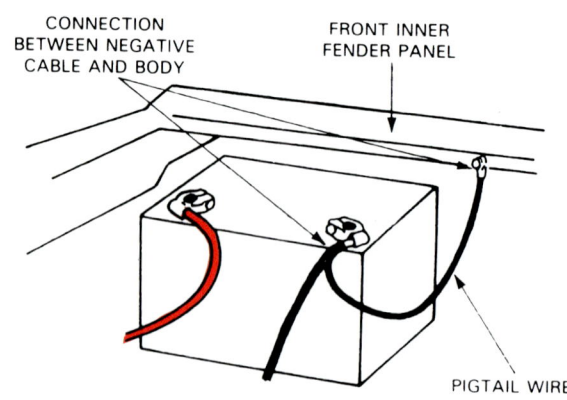

Fig. 38-11. Note cable connections to battery. Negative cable earths on engine block. Positive cable connects to electrical system. Pigtail earths car body to battery negative. (Sun)

BATTERY TRAY AND RETAINER

A battery tray and the retaining clamp hold the battery securely in place. They keep the battery from

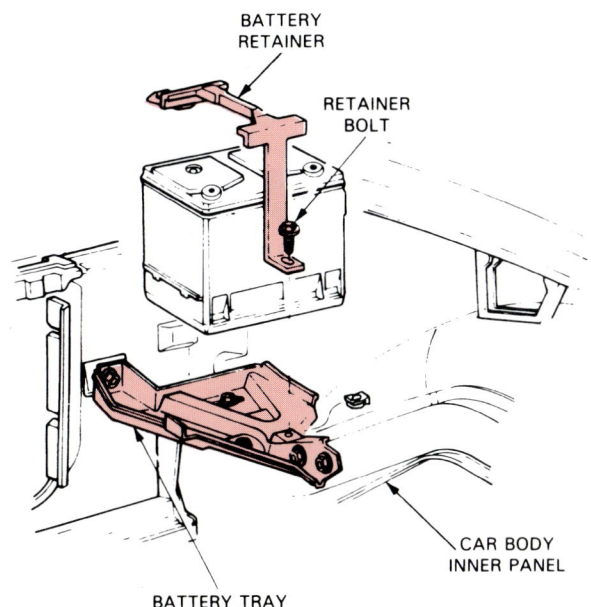

Fig. 38-12. Battery tray and retainer hold battery securely in place. Tray usually mounts on inner body panel.

bouncing around during vehicle movement. Look at Fig. 38-12. It is important that the tray and retaining clamp are in good condition and tight to prevent battery damage.

WET AND DRY CHARGED BATTERIES

There is no difference in the materials used in wet and dry charged batteries. The difference is in how the batteries are prepared for service.

With a wet charged battery, the battery is filled with electrolyte and charged at the factory. The battery is then tested and placed in stock, ready for service.

A dry charged battery contains fully charged elements but does not contain electrolyte. It leaves the factory in a dry state. Before use, the battery is filled with electrolyte. A dry charged battery is commonly used because it has a much longer shelf life than a wet charged battery.

Activation of a dry charged battery is covered in the next chapter.

MAINTENANCE-FREE BATTERY

A maintenance-free battery is easily identified because if does NOT use removable filler caps. Since calcium is used to make the battery plates, water does not have to be added to the electrolyte periodically.

The calcium in the plates reduces the production of battery gases. As a result, battery gas does not carry as much of the chemicals out of the battery. This increases battery service life and decreases service requirements.

BATTERY RATINGS

Battery ratings are set according to national test standards for battery performance. They let the mechanic and consumer compare the cranking power of one battery to another.

Two methods of rating lead-acid storage batteries are common. They were developed by the Society of Automotive Engineers (SAE) and the Battery Council International (BCI). These ratings are the SAE cranking current rating and reserve capacity rating.

SAE cranking current rating

Internationally recognised SAE Cranking Performance Test, (also known as the Cold Cranking Amps):

The discharge load in Amperes which a new fully charged battery at $-18°C$ can deliver for 30 seconds and maintain a voltage of 1.2 volts per cell, or higher.

This rating indicates the battery's ability to crank a specific engine (based on starter current draw) at a specified temperature.

For example, one auto manufacturer recommends a battery with 280 cold cranking amps for a small 4-cylinder engine but a 350 cold cranking amp battery for a larger 6-cylinder engine. A more powerful battery is needed to handle the heavier starter current draw of the larger engine.

Reserve capacity rating (minutes)

The time in minutes that a new fully charged battery will supply a constant load of 25 amps without the voltage falling below 10.5 volts for a 12 volt battery and 5.25 volts for a 6 volt battery.

BATTERY TEMPERATURE AND EFFICIENCY

As battery temperature drops, battery power is reduced. At low temperatures the chemical action inside the battery is slowed down. It will not produce as much current as when warm. This affects the ability of a battery to start an engine in extremely cold weather.

Also, when an engine is cold, the engine oil is very thick. This increases the amount of current needed to crank the engine with the starting motor.

Fig. 38-13 shows a chart comparing battery efficiency and required starting power. Note that at $-18°C$, a battery may only have 40 percent of its normal cranking power. In addition, starter current draw will be up approximately 200 percent. The engine could be very difficult to start on a cold morning. The battery, starter, and electrical connections must be in almost perfect condition.

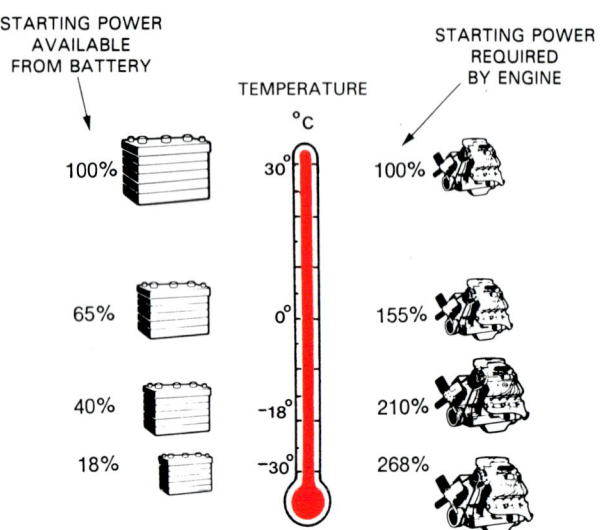

STARTING POWER AVAILABLE FROM BATTERY

TEMPERATURE °C

STARTING POWER REQUIRED BY ENGINE

Fig. 38-13. Study how temperature affects battery power and starter current draw. This is why engines crank slowly in very cold weather. (Champion Spark Plugs)

KNOW THESE TERMS

Discharging, Charging, Cell, Element, Plate, Separator, Strap, Case, Caps, Electrolyte, Charge indicator, Battery voltage, Terminals, Battery cable, Wet charged, Dry charged, Maintenance-free battery, Cranking current rating (Cold cranking amps), Reserve capacity rating.

REVIEW QUESTIONS

1. An _____ _____ is an electro-chemical device for producing and storing electricity.
2. Define the terms battery "discharging" and "charging".
3. Which of the following is NOT part of a basic battery cell?
   a. Positive plate.
   b. Negative plate.
   c. Electrolyte.
   d. All of the above are correct.
   e. All of the above are incorrect.
4. List five functions of a car battery.
5. Explain the function of spark arresters in an automotive battery.
6. _____ gas can collect around the top of batteries. If this is exposed to a flame or spark, it can _____!
7. Electrolyte, also called battery acid, is a mixture of sulphuric acid and distilled water. True or False?
8. What is the purpose of a charge indicator or eye?
9. Describe the difference between battery posts and flat or blade terminals.
10. A 12-volt battery has _____ cells that produce an open circuit voltage of _____ volts.
11. Most modern car batteries are 6-volts. True or False?
12. The battery positive cable normally connects to the _____ _____ and the negative cable connects to _____ on the engine _____.
13. Explain the difference between a wet and a dry charged battery.
14. A _____ battery is easily identified because it usually does NOT have removable filler caps or covers.
15. Which of the following is NOT a conventional battery rating?
   a. Hot cranking amps.
   b. Reserve capacity rating.
   c. Cranking current rating.
   d. Cold cranking amps.

Chapter 39

BATTERY TESTING AND SERVICE

After studying this chapter, you will be able to:
☐ Visually inspect a battery for obvious problems.
☐ Perform common battery tests.
☐ Clean a battery case and terminals.
☐ Charge a battery.
☐ Jump start a car using a second battery.
☐ Replace a defective battery.

A "dead battery" (discharged battery) is a very common problem. The engine will usually fail to crank and start. Even though the lights and horn may work, there is not enough "juice" (current) in the battery to operate the starting motor.

Since this is a common trouble, it is important for you to know how to inspect, test, and service vehicle batteries. This chapter covers the most common tasks relating to battery service.

BATTERY MAINTENANCE

If a battery is not maintained properly, its service life will be reduced. Battery maintenance should be done periodically — during tune-ups, grease jobs, or anytime symptoms indicate battery problems.

Battery maintenance typically includes:
1. Checking electrolyte level or indicator eye.
2. Cleaning battery terminal connections.
3. Cleaning battery top.
4. Checking battery hold-down and tray.
5. Inspecting for physical damage to case and terminals.

Inspecting battery condition

Inspect the battery anytime the engine bonnet is opened. Check for the types of problems shown in Fig. 39-1. Look for a dirt build up on the battery top. Look for case damage, loose or corroded

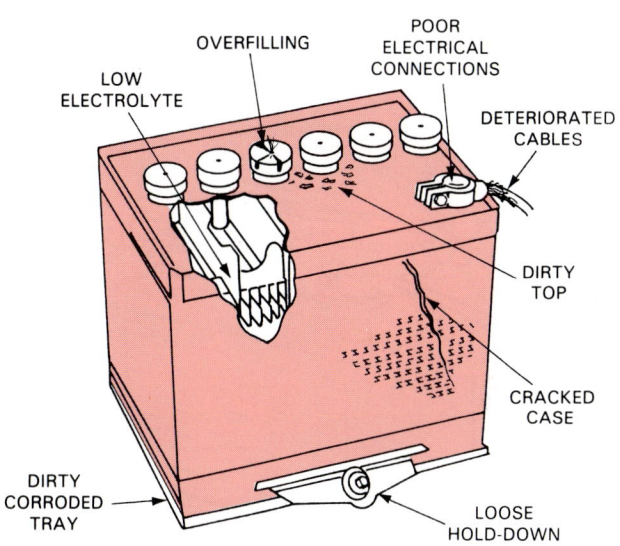

Fig. 39-1. Visually inspect batteries for these kinds of problems. If any are found, correct them.

connections, or any other trouble that could upset battery operation. If any problems are found, correct them before they get worse.

DANGER! Wear eye protection when working around batteries. Batteries contain acid that could cause blindness. Even the film build up on a battery can contain acid.

Battery leakage test

A battery leakage test will find out if current is discharging across the top of the battery case. A dirty battery can run down (discharge) when not in use. This can shorten battery life and cause starting problems.

To do a battery leakage test, set a voltmeter on a low setting. Touch the acid resistant probes on the battery as shown in Fig. 39-2. If the meter registers voltage, current is leaking out of the battery cells. You need to clean the battery top.

Cleaning battery case

If the top of the battery is dirty, wash it down with baking soda and warm water. See Fig. 39-3. This will

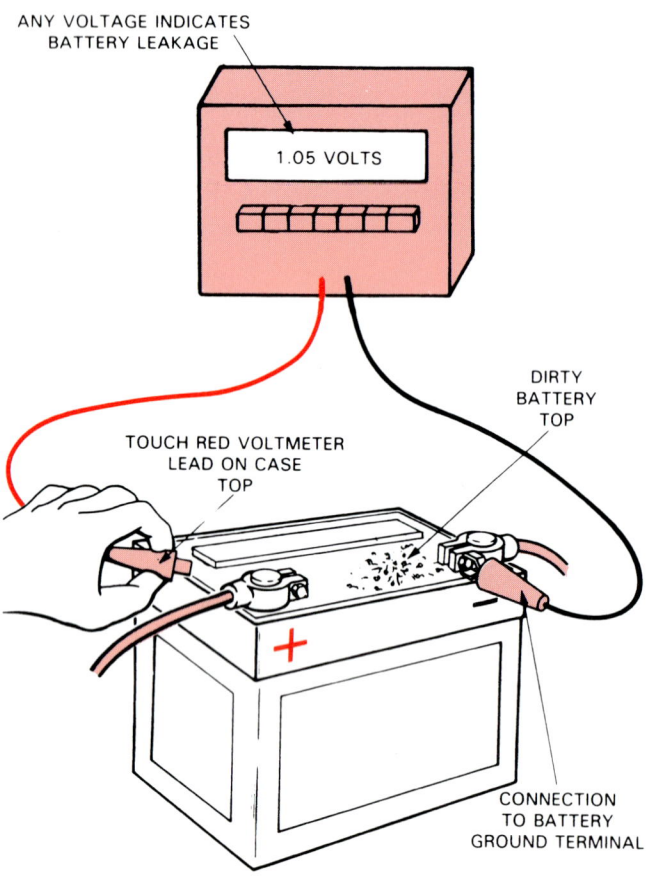

Fig. 39-2. Leak test will quickly show electrical leakage across top of battery. If voltmeter registers, clean battery.

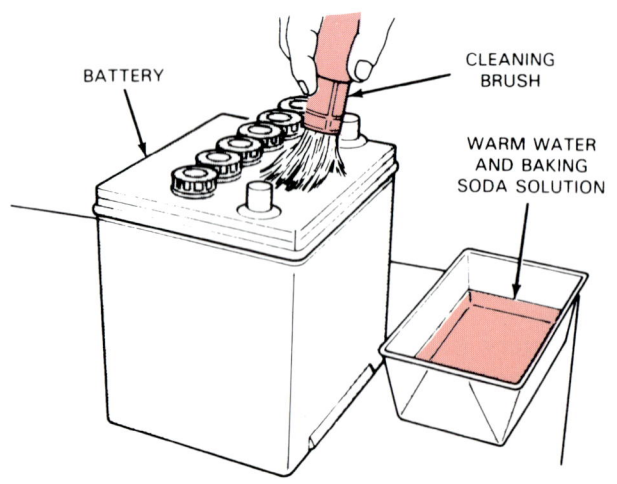

Fig. 39-3. Clean battery with baking soda-water solution and brush. Keep dirt out of filler openings.

neutralise and remove the acid-dirt mixture. If not a maintenance-free battery, be careful not to let debris enter the filler openings.

Battery terminal test

A battery terminal test quickly checks for a poor electrical connection between the battery cables and terminals. A voltmeter is used to measure voltage drop across the cables and terminals, as in Fig. 39-4.

Connect the negative meter lead to the cable end. Touch the positive meter lead on the battery terminal. Disable the ignition or injection system so that the engine will not start. Then, crank the engine while watching the voltmeter reading.

If the voltmeter shows more than 0.5 volt, there is a high resistance at the cable connection. This would tell you to clean the battery connections. A clean, good electrical connection would have less than 0.5 volt drop.

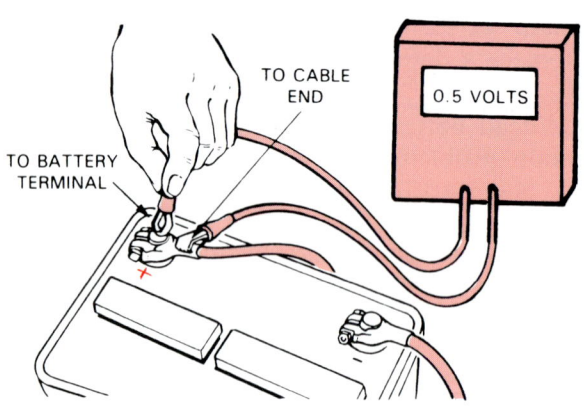

Fig. 39-4. To quickly find out if battery terminal needs cleaning, measure voltage drop across cable-to-terminal connection. Crank engine with ignition disconnected. A reading of over 0.5 volt would require terminal and cable end cleaning.

Cleaning battery terminals

To clean the terminals, remove the battery cables. There are several types of cable fasteners, as pictured in Fig. 39-5. Use a six-point wrench if the bolt or nut is extremely tight. Use pliers only on a spring type cable end or when the fastener head is badly corroded and rounded off. Be careful not to damage the terminal post with excessive side force.

To clean post type terminals, use a cleaning tool like the one in Fig. 39-6. Use the female end to clean the post. Use the male end on the terminal. Twist the tool to remove the oxidised outer surface on the connections.

Do NOT use a knife or scraper to clean battery

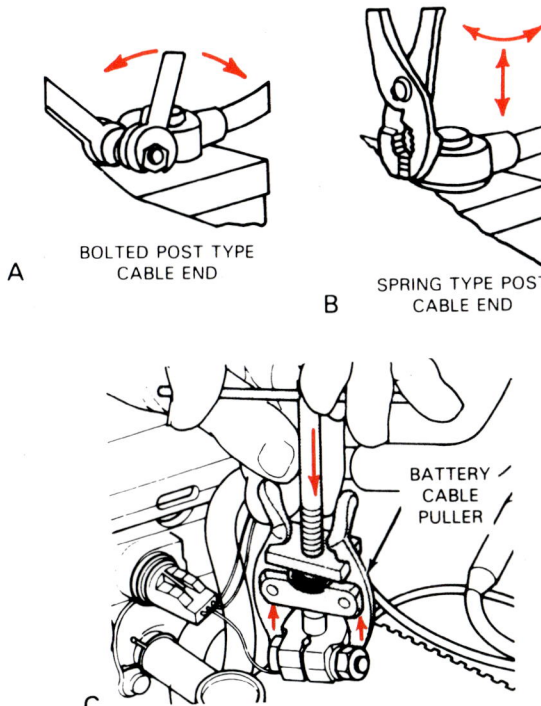

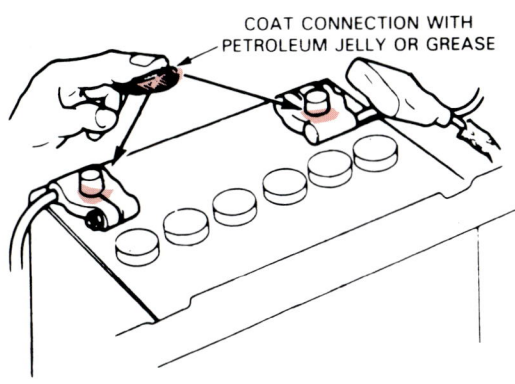

Fig. 39-7. Before reconnecting battery cables, coat connection with petroleum jelly or white grease. This will help prevent corrosion from battery gases. Do not overtighten cable fasteners or damage may result. Most terminals are made of very soft lead. (Honda)

Fig. 39-5. Note methods of removing battery cable from battery terminal. Be careful not to damage post or blade terminal. A — Use two wrenches if cable bolt begins to turn in cable end. A six-point wrench may be needed if fastener head is partially rounded. Use pliers only on badly damaged fasteners. B — Pliers are used to open spring-type battery cable ends. C — A battery cable puller may be needed if cable is stuck on post. This will prevent loosening of post in case.

will keep acid fumes off the connections and keep them from corroding again. Tighten the fasteners just enough to secure the connection. Overtightening can strip the cable bolt threads.

Checking battery electrolyte level

Unlike an older style battery, a maintenance-free battery does NOT need periodic electrolyte service under normal conditions. It is designed to operate for long periods without loss of electrolyte. Older batteries with removable vent caps, however, must have their electrolyte level checked.

DANGER! The invisible HYDROGEN GAS produced by the chemical reaction in a battery is very FLAMMABLE. Keep all sparks and flames away from the top of a battery. Batteries can EXPLODE if the gas is ignited!

terminals. This removes too much metal and can ruin the terminal connection.

When connecting the cables, coat the terminals with petroleum jelly or white grease, Fig. 39-7. This

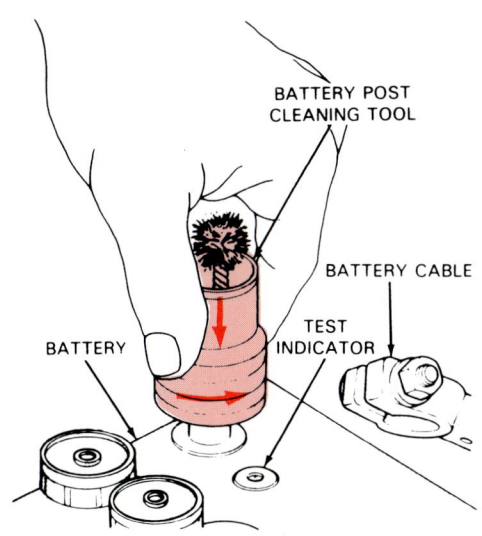

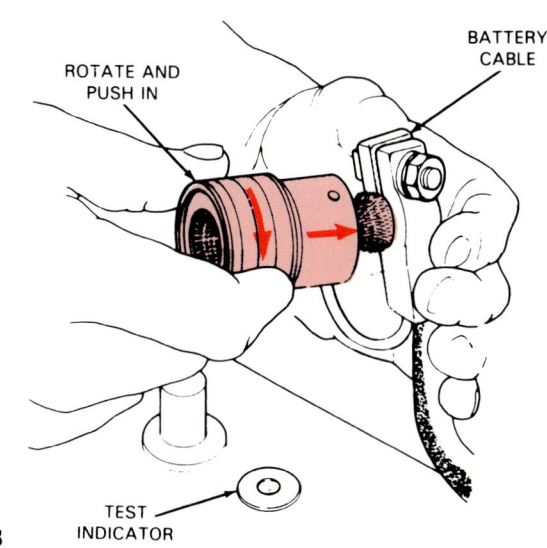

Fig. 39-6. Cleaning battery posts and cable ends. A — Use female end of cleaner on post. Rotate tool on post. B — Use male end of cleaner on cable end. Turn it until all corrosion is gone.

Many old style batteries must have their vent caps removed when checking the electrolyte. The electrolyte should just cover the top of the battery plates and separators. Most batteries have a fill ring (electrolyte level indicator) inside the filler cap opening. The electrolyte should be even with the fill ring.

If the electrolyte is low, fill the cells to the correct level with DISTILLED WATER (purified water). Distilled water should be used because it does not contain many of the impurities found in tap water. Water taken directly out of a tap can contain chemicals that reduce battery life. The chemicals can contaminate the electrolyte and collect in the bottom of the battery case. If enough contaminants collect in the battery, the cell plates can SHORT OUT, ruining the battery.

Battery overcharging

If water must be added to the battery at frequent intervals, the charging system may be overcharging the battery. A faulty charging system can force excessive current into the battery. Battery gassing can then remove water from the battery.

CHECKING BATTERY CHARGE

When you measure battery charge, you check the condition of the battery electrolyte and battery plates. For example, if lights are left ON without the engine running, the battery will run down (discharge). Current flow out of the battery will steadily reduce available battery power. There are several ways to measure battery charge.

Some modern batteries use a charge indicator eye that shows battery charge. You simply look at the eye in the battery cover to determine battery charge. This was covered in the previous chapter.

Hydrometer check

A hydrometer measures the specific gravity (weight or density) of a liquid. A battery hydrometer measures the specific gravity or the state of the charge for battery electrolyte. Look at Fig. 39-8.

Water has a specific gravity standard of ONE (1.000). Fully charged electrolyte has a specific gravity of between 1.265 and 1.299. The larger number denotes that electrolyte is more dense or heavier than water.

As a battery becomes discharged, its electrolyte has a larger percentage of water. Thus, a discharged battery's electrolyte will have a lower specific gravity number than a fully charged battery. This rise and drop in specific gravity can be used to check the charge in a battery.

There are several types of hand held hydrometers.

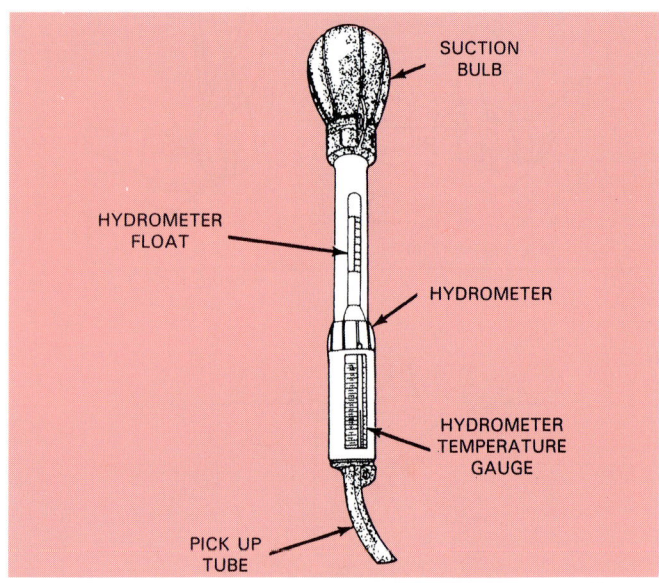

Fig. 39-8. Note basic parts of battery hydrometer. It can be used to check the state of the charge in batteries with vent caps.

Three of these are the float type, ball type, and the needle type.

To use a float type hydrometer, squeeze the hydrometer bulb. Immerse the end of the hydrometer in the electrolyte. Then release the bulb, Fig. 39-9. This will fill the hydrometer with electrolyte.

Shown in Fig. 39-9, compare the numbers on the hydrometer float with the top of the electrolyte. Hold the hydrometer even with your line of sight. Wear safety glasses and do NOT drip acid on anything.

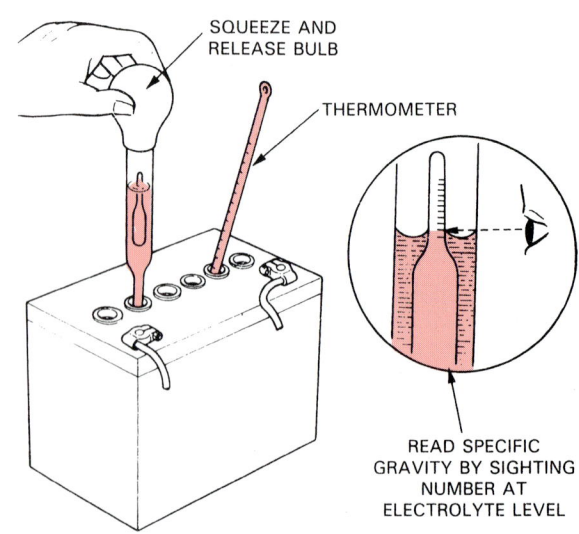

Fig. 39-9. To check battery charge, draw electrolyte into hydrometer by squeezing and releasing bulb. Read specific gravity on float at top level of electrolyte. Temperature of electrolyte will affect reading. (Mazda)

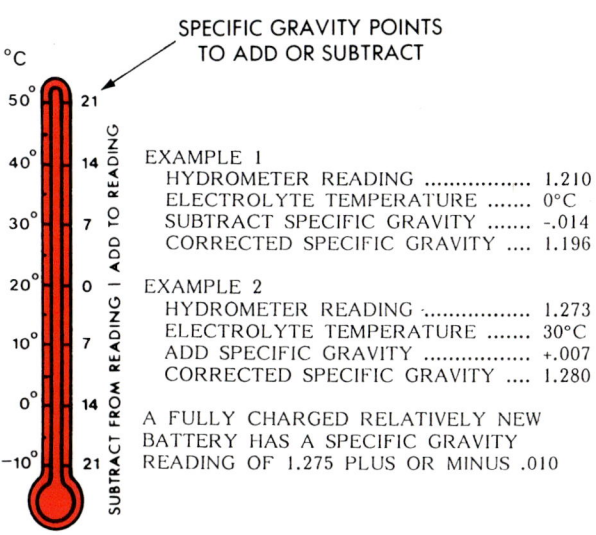

EXAMPLE 1
HYDROMETER READING 1.210
ELECTROLYTE TEMPERATURE .... 0°C
SUBTRACT SPECIFIC GRAVITY .... -.014
CORRECTED SPECIFIC GRAVITY 1.196

EXAMPLE 2
HYDROMETER READING 1.273
ELECTROLYTE TEMPERATURE .... 30°C
ADD SPECIFIC GRAVITY .... +.007
CORRECTED SPECIFIC GRAVITY 1.280

A FULLY CHARGED RELATIVELY NEW
BATTERY HAS A SPECIFIC GRAVITY
READING OF 1.275 PLUS OR MINUS .010

Fig. 39-10. Study examples of using a hydrometer correction chart.

Most float type hydrometers are NOT temperature correcting. However, the better models will have a built-in thermometer and a conversion chart, Fig. 39-10. This will let you compensate for battery temperature.

The ball type battery hydrometer is gaining popularity becauuse you do not have to use a temperature correction chart. The balls change temperature when submerged in the electrolyte. This allows for any temperature offset.

To use the ball type hydrometer, draw electrolyte into the hydrometer with the rubber bulb. Then note the number of balls floating in the electrolyte. Instructions on the hydrometer will tell you whether the battery is fully charged or discharged.

A needle type hydrometer uses the same principle as the ball type. When battery acid is sucked into the hydrometer, the electrolyte causes the plastic needle to register specific gravity.

Hydrometer readings

A fully charged battery should have a hydrometer reading of at least 1.265 or higher. If below 1.265, the battery needs recharging or it may be defective. Look at Fig. 39-11.

A discharged battery could be caused by:

1. Defective battery.
2. Charging system problem (loose alternator belt, for example).
3. Starting system problem.
4. Poor cable connections.
5. Engine performance problem requiring excessive cranking time.

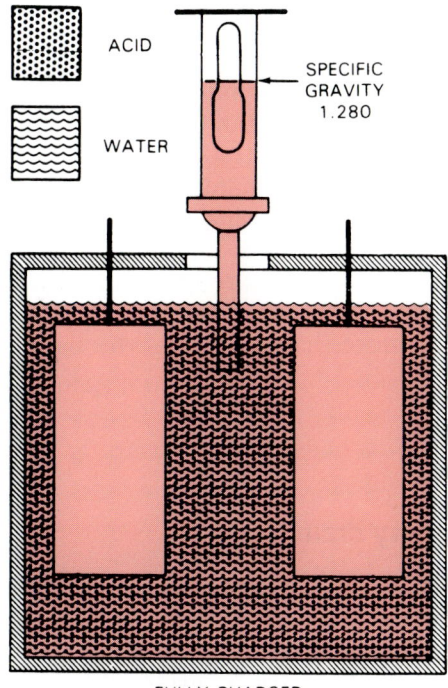

FULLY CHARGED

**ACID IN WATER GIVES ELECTROLYTE
SPECIFIC GRAVITY OF 1.280**

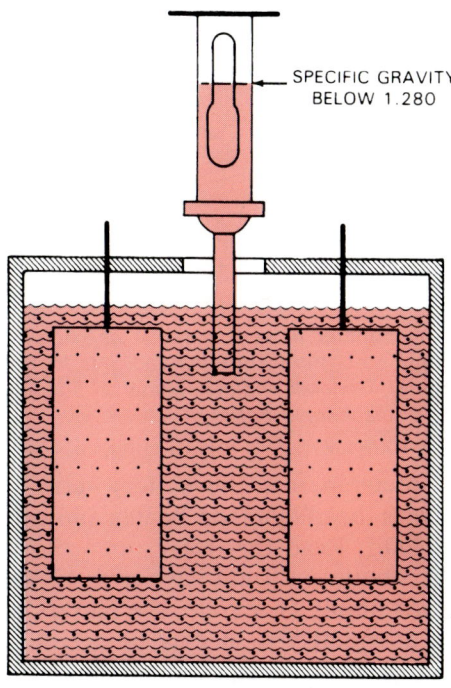

GOING DOWN

**AS BATTERY DISCHARGES, ACID BEGINS TO
LODGE IN PLATES. SPECIFIC GRAVITY DROPS.**

Fig. 39-11. A fully charged battery will have a hydrometer reading of 1.280. If below 1.270 specific gravity, battery needs recharging. (Sun)

6. Electrical problem drawing current out of battery with ignition key OFF.

A defective battery can be found with a hydrometer by checking the electrolyte in every cell. If the specific gravity in any cells VARIES EXCESSIVELY from other cells (25 to 50 points), the battery is usually ruined. The cells with the low readings may be shorted.

When all of the cells have an equal gravity, even if all of them are low, the battery can usually be regenerated by recharging.

With maintenance-free batteries, the hydrometer is not commonly used. A voltmeter and ammeter or load tester, covered later in the chapter, is needed to quickly determine battery condition.

Battery voltage test

A battery voltage test is done by measuring total battery voltage with an accurate voltmeter or special tester. It will determine general state of charge and battery condition quickly. Look at Fig. 39-12.

BATTERY STATE OF CHARGE SPECIFIC GRAVITY AND VOLTAGE					
LOAD ON BATTERY	1.265 FULL CHARGE	1.250 95% CHARGE	1.230 ¾ CHARGE	1.200 ½ CHARGE	1.175 ¼ CHARGE
NO LOAD	12.7 VOLTS	12.6 VOLTS	12.5 VOLTS	12.4 VOLTS	12.2 VOLTS
5 AMPERES	12.5 VOLTS	12.4 VOLTS	12.3 VOLTS	12.1 VOLTS	11.8 VOLTS
15 AMPERES	12.3 VOLTS	12.2 VOLTS	12.0 VOLTS	11.7 VOLTS	11.3 VOLTS
25 AMPERES	12.1 VOLTS	11.9 VOLTS	11.6 VOLTS	11.2 VOLTS	10.7 VOLTS

THIS IS THE RANGE IN WHICH MOST VEHICLE BATTERIES NORMALLY OPERATE IN CUSTOMER SERVICE.

AT 1.180 AND BELOW, STARTING WILL BE UNRELIABLE AND FUNCTION OF OTHER CIRCUITS MAY BE ERRATIC.

Fig. 39-13. Chart shows how battery voltage depends on specific gravity or hydrometer readings. Voltage test is needed on maintenance-free batteries that do not have removable filler caps.

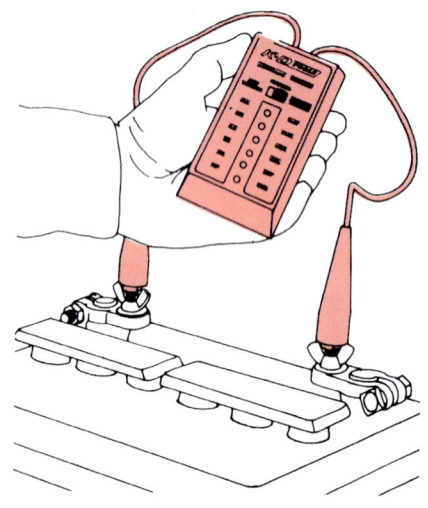

Fig. 39-12. Digital voltmeter or special tester as shown will check general charge on battery. Turn on headlights for a light load. Then read meter. Generally, voltage below 11.5 indicates discharged battery.

Connect the meter across the battery terminals. Turn on the car's headlights or heater blower to provide a light load. Read the meter.

A well charged battery should have over 12 volts. If the meter reads about 11.5 volts, the battery may not be charged adequately or it may be defective.

A battery voltage test is used on maintenance-free batteries. These batteries do not have filler caps that can be easily removed for testing with a hydrometer.

Other tests are needed to find the actual problem when a battery fails a voltage test. The chart in Fig. 39-13 compares specific gravity (hydrometer readings) and battery voltage. Note the relationship.

Cell voltage test

A cell voltage test will let you know if the battery is defective or just discharged. Just like a hydrometer cell test, if the reading on one or more cells is lower (voltage less) than the others, the battery must be replaced.

To do a cell voltage test, insert the special cadmium (acid resistant metal) tips of a low voltage reading voltmeter into each cell. This is pictured in Fig. 39-14. Start at one end of the battery. Work your way down, testing each cell carefully.

Note! Some manufacturers recommend battery fast charging during this test. Refer to a service manual for details.

If cell voltages are low, but equal, recharging will usually restore the battery. If cell voltage readings vary more than 0.2 volts, the battery is FAILING.

DANGER! Make sure you do not drip battery acid on the car or your skin when using a hydrometer or cell voltage tester. The acid will eat the car's paint or burn your skin.

Battery drain test

A battery drain test will check for an abnormal current draw with the ignition key off. When a battery goes dead when not being used, you may need to check for a current drain. It is possible that there is a short or other problem constantly discharging the battery.

A battery can be discharged if an electrical accessory remains ON when the ignition switch is turned OFF. For example, a short in a switch could cause a glove box light to always stay on. This could

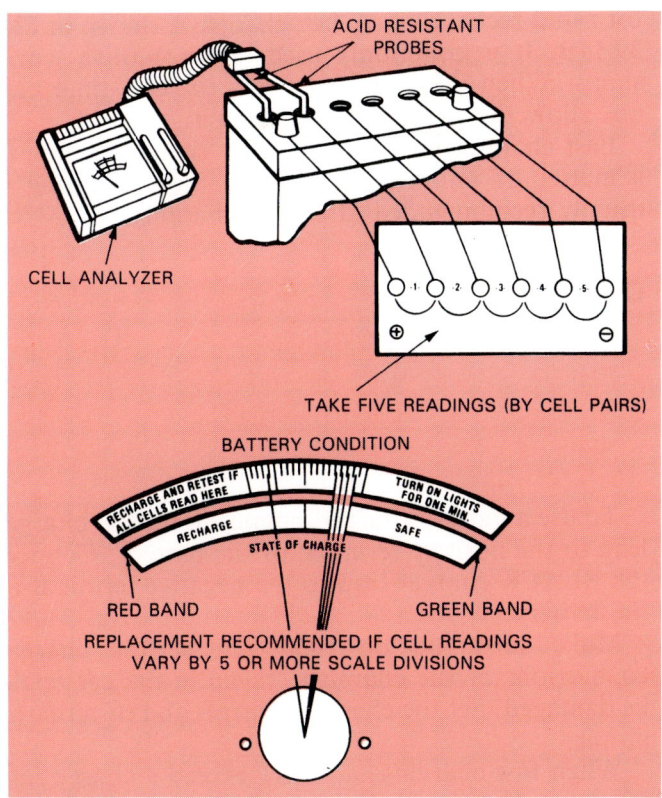

Fig. 39-14. To check battery condition, measure voltage in each cell as shown. Unequal readings indicate a faulty battery.

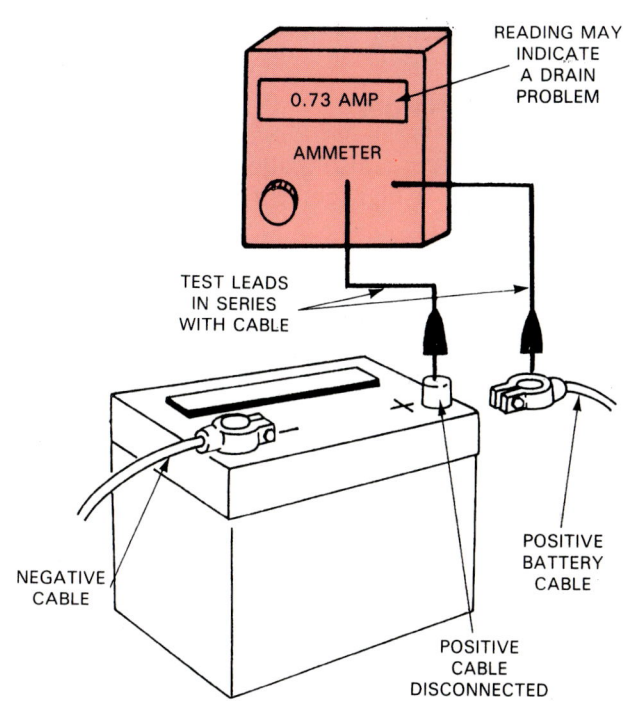

Fig. 39-15. If battery runs down after sitting unused in car, perform a battery drain test. Connect ammeter in series with positive cable. If current is flowing out of battery with everthing turned off, an electrical problem is discharging battery.

slowly drain the battery and cause a no-crank problem.

To perform a battery current drain test, make the ammeter connections shown in Fig. 39-15. Temporarily remove the fuse for the dash clock. Close the doors and boot lid. Then read the ammeter. If everything is OFF (good condition), the ammeter should read zero. However, an ammeter reading would point to a drain and a problem.

To help pinpoint a drain, remove fuses one at a time. When the ammeter reads zero, the problem is in the circuit on that fuse.

BATTERY CHARGERS

When tests show that the battery is discharged, a battery charger may be used to re-energise the battery. The battery charger will force current back into the battery to restore the charge on the battery plates and in the electrolyte. It contains a step-down transformer and rectifier that changes mains wall outlet voltage (around 240 volts a.c.), to slightly above battery voltage (14 to 15 volts d.c.). Refer to Fig. 39-16. It shows a battery charger.

There are two basic types of battery chargers: The slow charger and the fast charger.

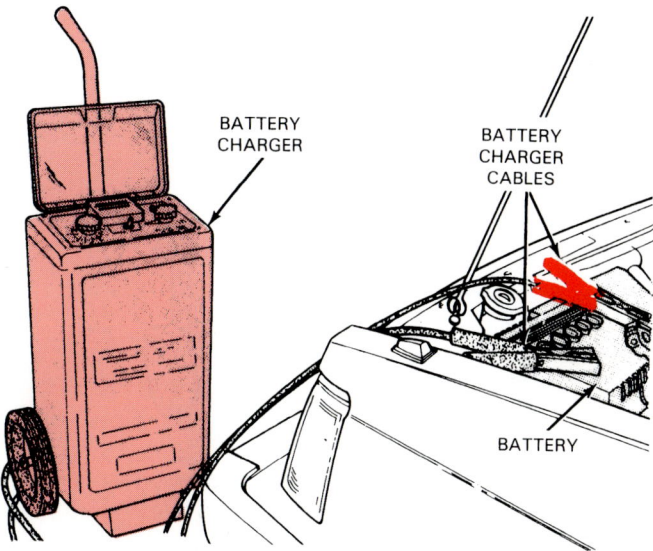

Fig. 39-16. Battery charger forces current through battery to restore charge on plates.

Slow (trickle) charger

A slow charger, also called a trickle charger, feeds a small amount of current into the battery. Charging time is longer (about 12 hours at 10 amps). However, the chemical action inside the battery is improved.

The active materials are plated back on the battery plates better. When time allows, use a slow charger. Look at Fig. 39-17A.

Fast (quick) charger

A fast charger, also called a quick or boost charger, forces a high current flow into the battery for rapid recharging. A fast charger is shown in Fig. 39-17B. It is commonly used in automotive workshops. When the customer needs the car, time may not allow the use of a slow charger.

Fast charging will usually allow engine starting in a matter of minutes. If possible, slow charging is usually recommended after fast charging.

Charging a battery

DANGER! Before connecting a battery charger to a battery, make sure the charger is turned OFF. Also, check that the work area is well ventilated. If a spark ignites any battery gas, the battery could EXPLODE. Wear eye protection!

To use a battery charger, connect the RED charger lead to the positive terminal of the battery. Connect the BLACK charger lead to the negative terminal of the battery.

Make sure you do NOT reverse the charger connections or the charging system in the car could be damaged. Set the charger controls and turn on the power.

When fast charging, do NOT exceed a charge rate of about 35 amps. Also, battery temperature must NOT exceed around 52°C. Exceeding either could damage the battery.

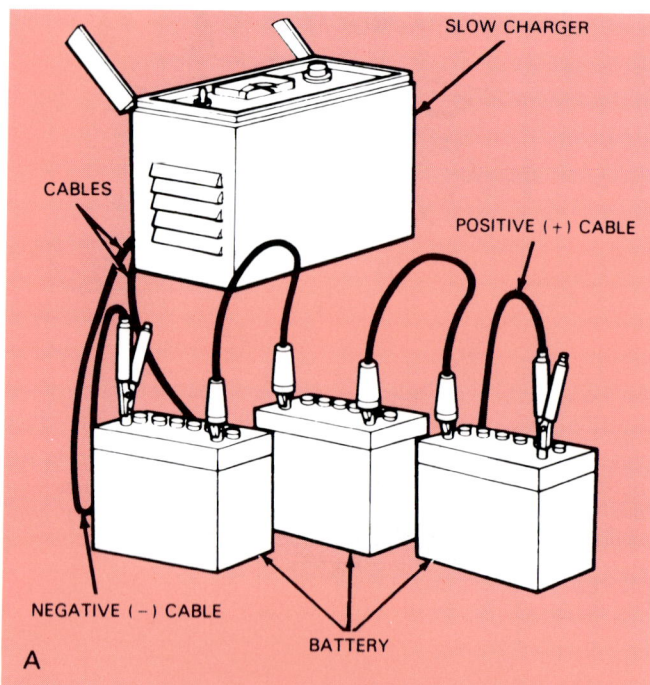

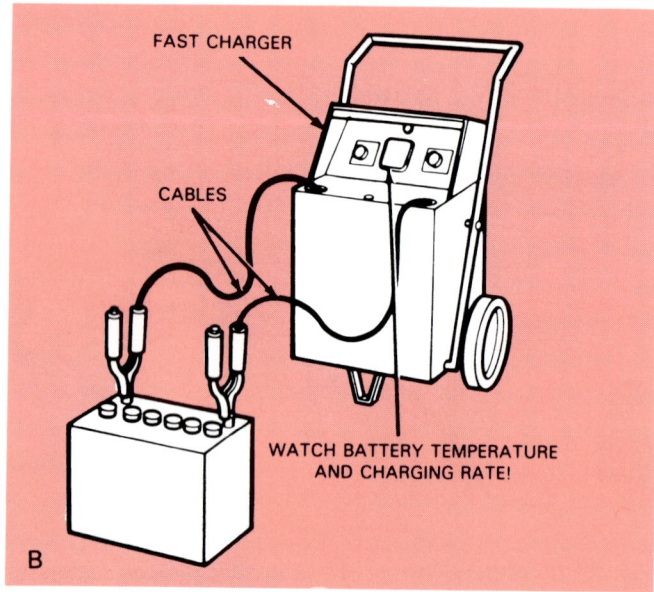

Fig. 39-17. A — Slow charger only forces small amount of current through battery. Since slow charging requires several hours, several batteries may be connected to get more done. B — Fast charging is ok in an emergency. Slow charging should follow fast charging to restore battery properly. Do not allow battery temperature to go above about 52°C or battery damage may occur.

JUMP STARTING

In emergency situations, it may be necessary to jump start the car by connecting another battery to the discharged battery. Look at Fig. 39-18. The two batteries are connected POSITIVE TO POSITIVE and NEGATIVE TO NEGATIVE.

Connect the red jumper cable to the positive terminal of both batteries. Then, connect the black jumper cable to any ground on both vehicles. See Fig. 39-19.

WARNING! Do not short jumper cables together or connect them backwards. This could cause serious damage to the charging and computer system where installed.

BATTERY LOAD TEST

A battery load test, also termed a battery capacity test, is one of the BEST methods of checking battery condition. It tests the battery under full current load.

The hydrometer and voltage tests were general indicators of battery condition. The battery load test, however, actually measures the current output and performance of the battery. It is one of the most common and informative battery tests used in modern automotive workshops. Refer to Fig. 39-20.

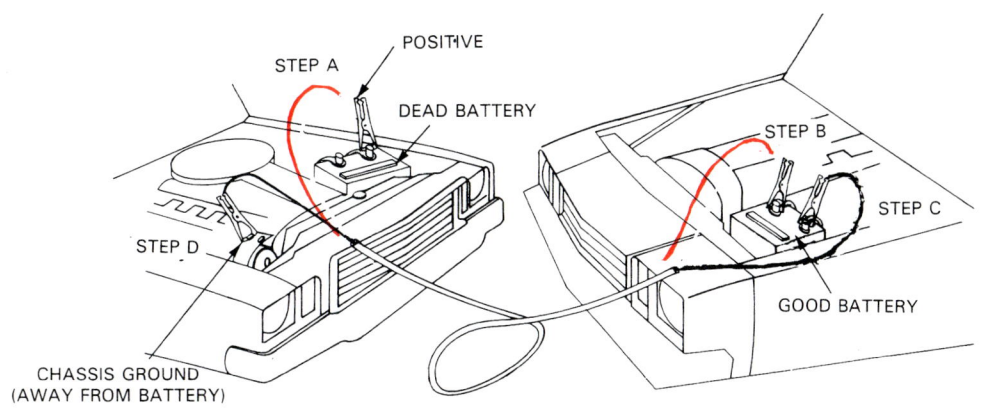

Fig. 39-18. Jumper cables can be used to start car with dead battery. A — Connect red jumper to positive terminal of dead battery. B — Connect other end of red jumper to positive terminal of good battery. C — Connect black jumper to negative terminal of good battery. D — Connect other end of black jumper to good ground, away from dead battery. This will keep any sparks away from battery gases. Run engine and turn on headlights in car with good battery while starting.

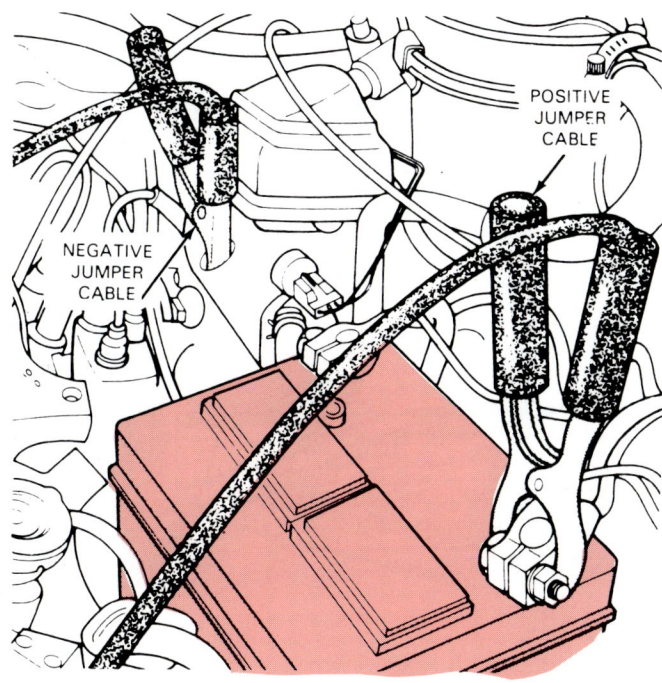

Fig. 39-19. Close up of jumper connections shows connection of negative or black jumper to chassis ground. A spark near dead battery could make battery gases explode.

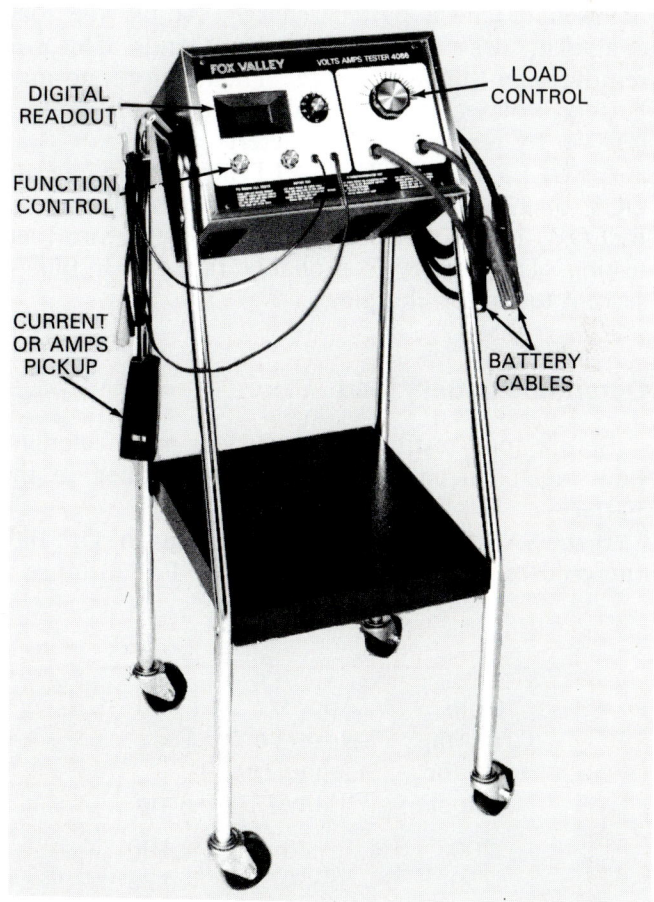

Fig. 39-20. Battery load tester is most accurate method of determining battery condition. It is a commonly used testing device that measures actual battery performance.

Connecting load tester

Connect the load tester to the battery terminals. If the tester is an inductive type (clip-on ammeter lead senses field around outside of cables), use the connections shown in Fig. 39-21. If the tester is NOT inductive, you must connect the ammeter in series.

Control settings and exact procedures vary. Follow the directions provided with the testing equipment.

Double-check battery charge

Before load testing, make sure the battery is adequately charged. Use an hydrometer, or digital voltmeter covered earlier. The load tester, itself, can

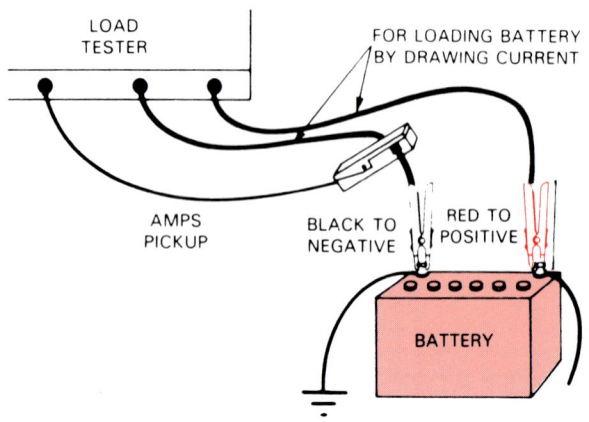

Fig. 39-21. Modern load testers are connected as shown. Clamp large cables to battery. Clip inductive amps pickup around negative tester cable. Large cables load battery by drawing current through tester. Inductive pickup operates ammeter in tester.

be used to check battery charge. Adjust the load control to draw 50 amps for 10 seconds. This will remove any surface charge. Then, check no-load battery voltage, also called open circuit voltage (OCV), by reading the voltmeter.

A FULLY CHARGED BATTERY should have an OCV of 12.4 volts or higher. If battery voltage is BELOW 12.4 volts, charge the battery before load testing. The battery is probably faulty if it fails a second test after charging.

Determine battery load

Before load testing a battery you must calculate how much current draw should be applied to the battery.

If the SAE Cold Cranking Amps is given, DIVIDE the cold cranking rating BY TWO. For instance a battery with 400 cold cranking amps rating should be loaded to 200 amps (400 ÷ 2 = 200).

A load chart will normally be provided with the load testing equipment. See Fig. 39-22.

Loading the battery

After checking battery charge and finding the amp load value, you are ready to test battery output. Double-check that the tester is connected properly. Then, turn the load control knob until the ammeter reads the correct load for your battery, Fig. 39-22.

Hold the load for 15 seconds. Then, read the VOLTMETER while the load is applied. Then, turn the load control completely OFF so the battery will not be discharged.

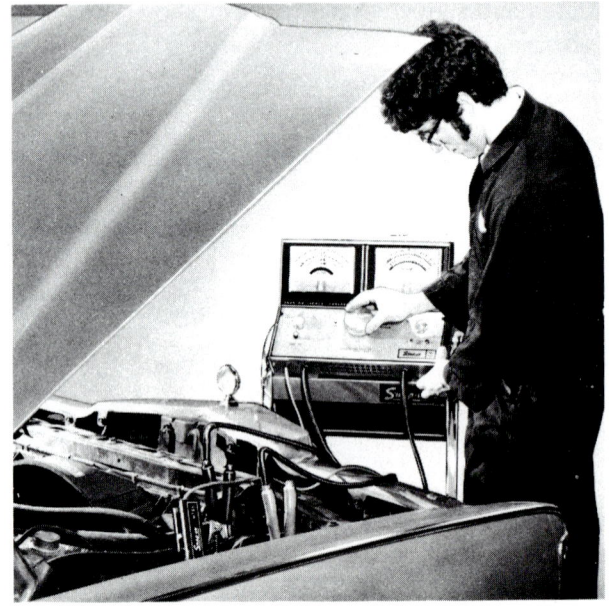

Cold Cranking Current (Amps)	Approx. Reserve Capacity (Mins)	Load Test (Amps)
200	40	100
250	50	125
300	60	150
350	72	175
400	85	200
450	105	225
500	125	250
550	160	275

Fig. 39-22. Chart shows different battery ratings and calculated current values for load testing.

APPROXIMATE ELECTROLYTE TEMPERATURE	MINIMUM ACCEPTABLE VOLTAGE UNDER LOAD FOR GOOD BATTERY
16 °C	9.5
10 °C	9.4
4 °C	9.3
−1 °C	9.1
−7 °C	8.9
−12 °C	8.7
−18 °C	8.5

Fig. 39-23. To load test battery, turn load control knob until ammeter reads calculated test current (Fig. 39-22). Hold load or current for 15 seconds and read voltmeter. If reading is below voltages in chart for specific temperature, battery is probably faulty. (Snap-On Tools)

Load test results

If the voltmeter reads 9.5 Volts or MORE at room temperature, the battery is good. This voltage is based on a battery temperature above 20°C.

A cold battery may show a lower voltage. You will need a temperature compensation chart, like the one in Fig. 39-23. It allows for any reduced battery performance caused by a low temperature.

If the voltmeter reads below 9.5 volts at room temperature, battery performance is POOR. This would show that the battery is not producing enough current to properly run the starter motor. Before replacing the battery, however, a quick charge test should be completed.

Quick (3 minute) charge test

A quick charge test, also termed 3 minute charge test, will determine if the battery is sulphated (plates ruined). If the battery load test results are poor, fast charge the battery. Charge for 3 minutes at 30 to 40 amps. Test the voltage while charging, as shown in Fig. 39-24. If the voltage goes ABOVE 15.5 volts, the battery plates are sulphated and ruined. A new battery should be installed in the car.

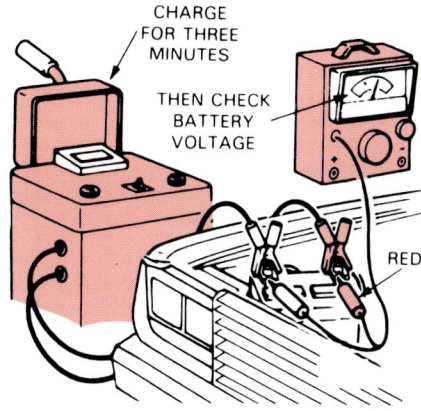

Fig. 39-24. Three minute charge test will double-check load test. Charge battery at about 40 amps while measuring battery voltage for three minutes. If voltage increases above 15.5 volts, replace battery.

Other battery-related problems

If the battery passes all of its tests but the battery does not perform properly (starter motor does not crank for example), the following are a few problems to check out:

1. Defective charging system.
2. Battery drain (light or other accessory ON).
3. Loose allternator belt.
4. Corroded, loose, or defective battery cables.
5. Defective starting system.

ACTIVATING DRY CHARGED BATTERY

A new, dry charged battery must be activated (prepared for service) before installing. Put on safety glasses and rubber gloves. Remove the cell caps or covers. Using a plastic funnel, not a metal funnel, pour electrolyte into each cell. Pour in enough electrolyte to just cover the plates and separators.

Replace the caps. Charge the battery as recommended by the manufacturer. After charging, recheck the electrolyte level. Install the battery.

REMOVING AND REPLACING BATTERY

To remove a battery, first disconnect the battery cables. Then loosen the battery retaining clamp. Using a battery strap or lifting tool, Fig. 39-25, lift the battery carefully out of the car.

DANGER! Always wear safety glasses when carrying a battery. If you were to drop a battery, acid could squirt out of the vent caps or broken case and into your face and eyes.

To install a battery, gently place the battery into its clean tray or box. Check that the battery fits properly. The box must not cut through and rupture the plastic case. Bolt on the retaining clamp and install the cables.

The replacement battery should have a starting power rating EQUAL to factory recommendations. If an undersize battery (lower starting power rating) is installed, battery performance and service life will be reduced.

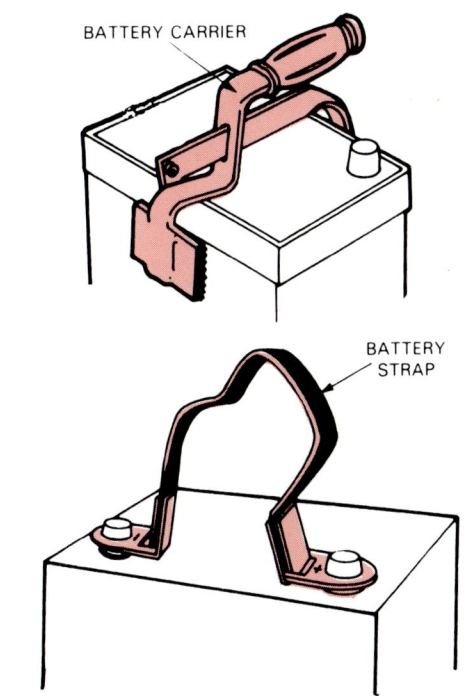

Fig. 39-25. Should you lose your grip and drop a battery, acid could splash out causing eye or skin injury. Always use a battery strap or carrier for safety.

KNOW THESE TERMS

"Dead battery", Battery leakage test, Battery terminal test, Fill ring, Distilled water, Battery charge condition, Hydrometer, Specific gravity, Battery voltage test, Cell voltage test, Battery drain test, Battery charger, Slow charger, Fast charger, Jump starting, Battery load test, Quick charge test, Battery activation.

REVIEW QUESTIONS

1. What five tasks does battery maintenance typically include?
2. A _____ _____ _____ will find out if current is discharging across the top of the battery case.
3. If a voltmeter shows over 0.7 volt drop across the battery post-to-cable connection, what should be done?
4. A knife or scraper is a good method of cleaning corroded battery terminals. True or False?
5. When you measure _____ _____, you check the condition of the battery electrolyte and plates.
6. List six reasons for a discharged battery.
7. A battery _____ _____ is done by measuring total battery voltage with an accurate voltmeter. A _____ _____ battery should have over 12 volts.
8. A customer complains that her battery goes dead repeatedly when the car is not driven for an extended period. A new battery has been installed and tested by another shop.

Mechanic A says that starter motor current draw should be measured. A shorted starting motor could be draining the battery.

Mechanic B says that a battery drain test should be done. An electrical short could be discharging the battery, even with the ignition key off.

Who is correct?

a. Mechanic A.
b. Mechanic B.
c. Both mechanic A and B are correct.
d. Neither mechanic A nor B are correct.

9. When using a battery charger, connect the red lead to positive and the black lead to negative. True or False?
10. Explain how excessive fast charging can ruin a battery.
11. How do you connect jumper cables safely?
12. A _____ _____ _____, also termed a _____ _____ _____, is one of the best methods of checking battery condition.
13. What is an inductive type ammeter lead?
14. Explain how to do a battery load test.
15. If a 12-volt battery shows below _____ volts during a load test at 20°C, the battery is faulty.

a. 9.9 volts.
b. 9.5 volts.
c. 9.7 volts.
d. 10.0 volts.

INDEX

AUTOMOTIVE LUBRICANTS AND FLUIDS

PRODUCT	COMPOSITION	PROPERTIES	APPLICATION
ENGINE OIL	VARIOUS VISCOSITIES OF MULTI GRADE AND MONO GRADE SUCH AS SAE. 15W-30, 15W-40, 20W-40, 20W-50, 10W, 20, 30, 40, 50 PLUS ADDITIVES FOR: ANTI-WEAR	PROLONGS ENGINE LIFE (FRICTION MODIFIED) IMPROVES FUEL CONSUMPTION	ENGINE CRANKCASES SOME GEARBOXES AND TRANS-AXLES
	ANTI-RUST	CONTROLS ENGINE CORROSION	
	ANTI-SLUDGING (DETERGENTS)	CONTROLS PRECIPITATION OF THE OIL	
	ANTI-OXIDATION	REDUCES THE RATE OF OXYGEN ABSORBTION AT HIGH TEMPERATURES	
	ANTI-FOAMING	REDUCES FROTHING AND BUBBLING OF THE OIL	
	ANTI-DEPOSIT	HOLDS DEPOSITS IN SUSPENSION	
GEAR OIL	NON- E.P. TYPE: VISCOSITY NORMALLY SAE. 80 TO 90 PLUS VISCOSITY STABILISER ADDITIVE	NON-EXTREME PRESSURE LUBRICANT OVERCOMES COLD SHIFT BAULK, AND HOT RATTLE	4 AND 5 SPEED MANUAL TRANSMISSIONS, TRANSAXLES, TRANSFER CASES AND LIGHT DUTY FINAL DRIVES
	E.P. TYPE: VISCOSITY NORMALLY SAE. 80 TO 140	EXTREME PRESSURE LUBRICANT	FINAL DRIVES WITH HYPOID GEARS, AND MANUAL STEERING BOXES
	L.S. TYPE: VISCOSITY NORMALLY SAE. 80 TO 140 PLUS FRICTION MODIFIER ADDITIVE	SPECIAL LUBRICANT TO PREVENT CLUTCH CHATTER	LIMITED SLIP FINAL DRIVE DIFFERENTIALS
AUTOMATIC TRANSMISSION FLUID	LOW VISCOSITY, FORMULATED TO WITHSTAND HIGH TEMPERATURES, PLUS ADDITIVES TO REDUCE FOAMING AND TO REDUCE OXIDATION	CHOICE OF FLUID GRADE IS CRITICAL (THERE ARE EMULSIFYING AND NON-EMULSIFYING GRADES). SERVES AS A HYDRAULIC FLUID	AUTOMATIC TRANSMISSIONS AND POWER STEERING SYSTEMS, ALSO SOME MANUAL TRANSMISSIONS
GREASES	LITHIUM SOAP BASED, PLUS MOLYBDENUM DISULPHIDE	HIGH LOAD CARRYING CAPACITY	STEERING BALL JOINTS, UNIVERSAL JOINTS (DRIVESHAFTS), CONSTANT VELOCITY JOINTS (DRIVESHAFTS), SUSPENSION BALL JOINTS
	LITHIUM SOAP BASED, PLUS ZINC OXIDE	WATER RESISTANT, WITH ANTI-SCUFF QUALITIES	DRUM BRAKE, AND BODY HARDWARE COMPONENTS
	NON-MELT, WITH RUST AND OXIDATION INHIBITORS	SPECIFIED FOR ANTI-FRICTION BEARING TEMPERATURES UP TO 240°C	WHEEL BEARINGS (ESPECIALLY WHERE DISC BRAKES ARE FITTED)
ANTI-FREEZE (WITH INHIBITOR)	ETHYLENE GLYCOL, PLUS INHIBITOR ADDITIVE	PROTECTS COOLING SYSTEM AGAINST FREEZING AND CORROSION	ENGINE COOLING SYSTEM
BRAKE FLUID	MANUFACTURED TO MEET DOT. 4, REQUIREMENTS FOR BRAKING SYSTEMS	IT IS FORMULATED TO REDUCE THE EFFECT OF WATER ABSORBTION. IT ALSO HAS A RELATIVELY HIGH VAPOUR LOCK TEMPERATURE	HYDRAULIC BRAKES AND CLUTCHES
STICK LUBRICANT	WAX BASED GREASE	WATER RESISTANT QUALITES	DOOR LOCKS AND CATCHES
PENETRATING OIL	VERY LOW VISCOSITY MACHINE OIL, WITH RELEASING AGENT ADDITIVES	POSSESSES EXCELLENT PENETRATING QUALITIES	FREEING OF SEIZED, AND RUSTED COMPONENTS, LUBRICATION OF LEAF SPRINGS